A TEXTBOOK OF PHARMACEUTICAL MATHEMATICS

(Remedial Mathematics, Vol. I)

A TEXTBOOK OF PHARMACEUTICAL MATHEMATICS

(Remedial Mathematics, Vol. I)

(For B.Pharmacy First Year Students)

[Strictly according to the latest syllabii of U.P.T.U., P.T.U., P.U., K.U.K., M.D.U., G.J.U., C.D.U., G.N.D.U., B.P.T.U., J.N.T.U., R.G.P.V., and other Indian Technical Universities]

By

N.P. BALI
Formerly, Principal
S.B. College, Gurgaon
Haryana

P.N. GUPTA
B.Sc. (Hons.), M.Sc.(Delhi)
M.Phil (KU), Pre.Ph.D (IIT Delhi)
Head, Deptt. of Mathematics
Indira Gandhi National College, Ladwa
Kurukshetra University
Haryana

Dr. C.P. GANDHI
(M.Sc. U.G.C.-NET, Ph.D)
Associate Professor in Mathematics,
Rayat & Bahra Institute of Engineering & Bio-technology, Kharar
Chandigarh

UNIVERSITY SCIENCE PRESS
(An Imprint of Laxmi Publications Pvt. Ltd.)
An ISO 9001:2015 Company

BENGALURU • CHENNAI • GUWAHATI • HYDERABAD
KOCHI • KOLKATA • LUCKNOW • RANCHI
NEW DELHI

Printed and bound in India

First Edition: 2007, *Reprint:* 2008, 2009, *Second Edition:* 2014 *Edition* : 2025

ISBN **:** 978-81-31807-79-8

Branches

Branch	Phone
Bengaluru	080-26 75 69 30
Chennai	044-24 34 47 26
Guwahati	0361-254 36 69
Hyderabad	040-27 55 53 83
Kochi	0484-405 13 03
Kolkata	033-40 04 77 79
Lucknow	0522-430 36 13
Ranchi	0651-224 24 64

PUBLISHED IN INDIA BY

UNIVERSITY SCIENCE PRESS

(An Imprint of Laxmi Publications Pvt. Ltd.)

An ISO 9001:2015 Company

113, GOLDEN HOUSE, GURUDWARA ROAD, DARYAGANJ,
NEW DELHI - 110002, INDIA

Corporate Office : B-33, Sector-5, Noida,
Gautambuddha Nagar (U.P.) - 201 301, INDIA

Telephone : +91-120-44 46 900

www.laxmipublications.com

info@laxmipublications.com

CONTENTS

* Chapters 15A and 15B are included only in the current syllabus of P.U., Chandigarh ; M.D.U., Rohtak ; G.J.U., Hisar and K.U., Kurukshetra.

OTHER BOOKS BY THE SAME AUTHORS

1. A Textbook of Pharmaceutical Mathematics, Vol-I

   By N.P. Bali, P.N. Gupta & Dr. C.P. Gandhi

   Contents. Algebra, Measures of Central Tendency, Trigonometry, Analytical Plane Geometry, Differential Calculus, Integral Calculus, Evaluation of Σn, Σn^2, Σn^3, Partial Fractions.

2. A Textbook of Pharmaceutical Mathematics, Vol-II

   By N.P. Bali, P.N. Gupta & Dr. C.P. Gandhi

   Contents. Differential Equations, Laplace Transform, Diagrammatical and graphical representation of data, Measures of Central Tendency and Dispersion, Moments, Skewness and Kurtosis, Correlation and Regression. Curve Fitting, Probability and Bayes' theorem Binomial, Poisson and Normal distributions. Sampling and Testing of Hypothesis. Statistical Inference, ANOVA. Central Limit theorem.

PREFACE TO THE SECOND EDITION

"A Textbook of Pharmaceutical Mathematics" has been revised by keeping in mind the various suggestions and recommendation invited by the readers. Some more technical examples from various university papers have been solved in this edition.

The subject matter has been made more comprehensive and lucid.

—Authors

PREFACE TO THE FIRST EDITION

"A Textbook of Pharmaceutical Mathematics" has been written strictly according to the latest syllabus of **"Remedial Mathematics" for B.Pharmacy, Ist Year students of the following technical universities of India.**

* U.P.T.U., Lucknow
* G.B.T.U., Lucknow
* K.U.K.; M.D.U. ; G.J.U. ; C.D.U., Haryana
* Punjab University, Chandigarh
* P.T.U., Jalandhar
* G.N.D.U., Amritsar
* Punjabi University, Patiala
* W.B.U., Kolkata
* B.P.T.U., Orissa
* J.N.T.U., Hyderabad
* R.G.P.V., Bhopal

The above cited book has been prepared by keeping in mind the students' capability of solving the mathematical problems. The subject-matter has been presented in a comprehensive and lucid way.

Suggestions and recommendations for the improvement of the book, are invited.

We hope that this book will prove to be an asset to the students.

—Authors

Syllabus

B.PHARMACY [REMEDIAL MATHEMATICS]

(For U.P.T.U., G.B.T.U., P.T.U., B.P.T.U., J.N.T.U., R.G.P.V. and Other Indian Technical Universities)

1. **Algebra :** Equations reducible to quadratics. Simultaneous equations (linear and quadratic). Determinants, properties of simultaneous equations by Cramer's rule. Matrices, definition of special kinds of matrices, arithmetic operations on matrices, inverse of a matrix, solution of simultaneous equations by matrices. Pharmaceutical applications of determinants and matrices. Evaluation of Σn, Σn^2 and Σn^3. Mensuration and its pharmaceutical applications.
2. **Measures of Central Value :** Objectives and pre-requisites of an ideal measure, mean, mode and median.
3. **Trigonometry :** Measurement of angle, T-ratios, addition, subtraction and transformation formulae, T-ratios of multiple, submultiple, allied and certain angles. Applications of logarithms in pharmaceutical computations.
4. **Analytical Plane Geometry :** Cartesian co-ordinates, distance between two points, area of triangle, locus of a point, straight line, slope and intercept form, double-intercept form, slope-point and two point form, general equation of first degree.
5. **Differential Calculus :** Limits and functions, differential coefficient, differentiation of standard functions including function of a function (Chain rule). Differentiation of implicit functions, logarithmic differentiation, parametric differentiation, successive differentiation.
6. **Integral Calculus :** Integration as inverse of differentiation, indefinite integrals of standard forms, integration by parts, substitution and partial fractions, formal evaluation of definite integrals.

Syllabus

B.PHARMACY [REMEDIAL MATHEMATICS]

(K.U.K., M.D.U., G.J.U., C.D.U., P.U. and Other Indian Technical Universities)

Section A

1. **Algebra :** Revision of equations reducible to quadratics and simultaneous equations (linear and quadratic) up to two variables only.

   Determinants and their six important properties, solution of simultaneous equations by Cramer's rule. Matrices. Definition of special matrices (like unit, singular , diagonal matrices etc.), arithmetic operation on matrices, transpose, adjoint, reciprocal and inverse of a matrix, solution of simultaneous equations using matrices. Partial fractions and resolution of linear and quadratic (non-repeated) partial functions. Evaluation of Σn, Σn^2 and Σn^3. Pharmaceutical applications.
2. **Trigonometry :** Revision on angle measurement and T-ratios. Addition, subtraction and transformation formulae. T-ratio of multiple, sub-multiple and allied angles. Solution of simple trigonometric identities based on above concepts. Pharmaceutical application of logarithms.
3. **Analytical Plane Geometry :** Cartesian co-ordinates, distance between two points, area of triangle, locus of point, straight line, slope and intercept form, general equation of first degree.

Section B

4. **Differential Calculus :** Limits and functions, differential coefficient, differentiation of standard functions, including function of a function (Chain rule). Differentiation of implicit functions, logarithmic differentiation, parametric differentiation, elements of successive differentiation.
5. **Integral Calculus :** Integration as inverse of differentiation, indefinite integrals of standard forms, integration by parts, by partial fractions and substitution. Formal evaluation of definite integrals.
6. **Differential Equations :** Definition and formation of ordinary differential equations, equations of first order and first degree, variable separable, homogeneous equations, linear equations (Leibnitz form) and differential equations reducible to these types. Linear differential equations of order greater than one with constant coefficients , complementary function and particular integrals of e^x, x^m, $\sin(ax + b)$ or $\cos(ax + b)$ types of functions, solution of simple simultaneous linear differential equations. Pharmaceutical transforms.
7. **Laplace Transforms :** Definition, properties of linearity and shifting, transforms of elementary functions (without proof) and inverse Laplace transforms not involving Euler's theorem, transforms of derivatives, solutions of ordinary and simultaneous differential equations.

Section C

8. **Pharmaceutical Statistics :** Concept, mathematical computations (wherever applicable and pharmaceutical applications (wherever possible) on :
9. Significant digits and rounding-off numbers, collection of primary and secondary data through experiments of surveys sampling and complete enumeration survey, merits and limitations of various random and non-random sampling methods, data organisation including frequency distributions and tabulation, diagramatic representations of data organization including frequency distributions and tabulation, diagrammatic representations of data, simple, multiple, sub-divided and floating bar diagrams, pie diagrams, 2-D and 3-D pictographic representation, graphs of frequency distributions.
10. Measures of central tendency, ideal characteristics, mean, median, mode, G.M., H.M. and weighted arithmetic mean from discrete and continuous frequency distribution, quartiles, deciles and percentiles, measures of dispersion, range, quartile deviation, mean deviation, standard deviation, calculation of standard deviation from discrete and continuous frequency distributions, standard error of means, coefficient of variation.
11. Probability and events. Bayes' theorem, probability theorems, probability distribution, elements of Binomial, Poisson and Normal distribution, Normal distribution curve and properties, calculation of areas under Normal curve and standard Normal curve (Z statistic), confidence limits, deviations from normality, kurtosis and skewness, elements of Central Limit theorem.
12. Linear correlation and regression analysis, scatter plots, methods of least squares. Pearsonian coefficients of correlation and determination, definitions of amount of explained variance, standard error of estimate and significance of regression (F).
13. Statistical inference, type I and II errors, *t*-test (paired and unpaired)

Note. (*i*) Each question will carry 16 marks. Candidates will be required to attempt at least one question from each of the three Sections and remaining two from any Section(*s*).

A Textbook of Pharmaceutical Mathematics has two volumes, namely, Vol. I and Vol. II. The first volume covers unit-1 to unit-5 and its second volume covers unit-6 to unit-13.

0

List of Important Formulae

1. Some useful constants

$e = 2.7183$

$\pi = 3.1416$

$\sqrt{2} = 1.414$

$\sqrt{3} = 1.732$

$\sqrt{10} = 3.162$

$1 \text{ radian} = 57°17'\,44.8''$

$1° = 0.0174 \text{ radian}$

$\log_e \pi = 1.447$

$\log_e 2 = 0.6931$

$\log_e 3 = 1.0986$

$\log_e 10 = 2.3026$

$\log_{10} e = 0.4343$

2. Some conversion factors

1 ft. = 30.48 cm = 0.3048 m

1 inch = 2.54 cm

1 cm = 0.3937 inch

1 m = 100 cm = 3.2804 ft.

1 ft^2 = 0.0929 m^2

1 acre = 4840 yd^2 = 4046.77 m^2

1 H.P. = 0.7457 kw

F° = C° × 1.8 + 32

1 ft^3 = 0.0283 m^3

1 m^3 = 35.32 ft^3

1 gallon = 3785.43 cm^3

1 Joule = 10^7 ergs

3. Some useful indeterminant forms

(*i*) $\dfrac{\text{Zero}}{\text{Non-zero}} = 0$ *e.g.*, $\dfrac{0}{8} = 0$

(*ii*) $\dfrac{\text{Non-zero}}{\text{zero}} = \infty$ *e.g.*, $\dfrac{8}{0} = \infty$

(*iii*) $a^{\infty} = \begin{cases} \infty, & a > 1 \\ 1, & a = 1 \end{cases}$ *e.g.*, $e^{\infty} = \infty$

$1^{\infty} = 1$

(*iv*) $\log \infty = \infty$

$\log 0 = -\infty$

(*v*) $e^{-\infty} = \dfrac{1}{e^{\infty}} = \dfrac{1}{\infty} = 0$

L' Hospital Rule. If $\underset{x \to a}{\text{Lt}} \dfrac{f(x)}{g(x)} = \dfrac{0}{0}$ or $\dfrac{\infty}{\infty}$ form,

then $$\underset{x \to a}{\text{Lt}} \frac{f(x)}{g(x)} = \underset{x \to a}{\text{Lt}} \frac{f'(x)}{g'(x)}$$

4. Nature of roots

(*a*) The roots of the equation $ax^2 + bx + c = 0$, $0 \neq a, b, c \in R$, are given by

$$x = \frac{-b \pm \sqrt{b^2 - 4ac}}{2a}.$$

(*i*) If $b^2 - 4ac > 0$, then the roots are real and distinct.

(*ii*) If $b^2 - 4ac = 0$, the roots are real and equal.

(*iii*) If $b^2 - 4ac < 0$, the roots are non-real and distinct.

(*b*) The quadratic equation whose roots are α, β, is given by $x^2 - Sx + P = 0$, where S = sum of roots, P = product of roots.

5. Sequences and series

(*i*) *n^{th} term of an A.P. (Arithmetical Progression)* T_n is given by

$$T_n = a + (n - 1)\, d, \quad \text{where} \quad a = \text{first term},\ d = \text{common difference}$$

(*ii*) *Sum of n terms of an A.P.* is given by

$$S_n = \frac{n}{2}\,(2a + (n - 1)\, d)$$

or

$$S_n = \frac{n}{2}\,(a + a + (n - 1)\, d)$$

$$= \frac{n}{2}(a + l), \quad \text{where} \quad l = \text{last term} = a + (n - 1)\, d.$$

(*iii*) *n^{th} term of G.P. (Geometrical Progression)* is given by

$$T_n = ar^{n-1}, \qquad \text{where} \quad a = \text{first term},\ r = \text{common ratio}$$

(*iv*) *Sum of n terms of G.P.* is given by

$$S_n = \frac{a(1-r^n)}{1-r} \quad \text{or} \quad \frac{a(r^n - 1)}{r - 1}, \quad r \neq 1$$

(*v*) *Sum of Infinite G.P.* is given by

$$S_\infty = \frac{a}{1-r}, \quad \text{where} \quad a = \text{first term},\ r = \text{common ratio}.$$

(*vi*) $1 + 2 + 3 + ... + n = \Sigma n = \dfrac{n(n+1)}{2}$

(*vii*) $1^2 + 2^2 + 3^2 + ... + n^2 = \Sigma n^2 = \dfrac{n(n+1)(2n+1)}{6}$

(*viii*) $1^3 + 2^3 + 3^3 + ... + n^3 = \Sigma n^3 = \left(\dfrac{n(n+1)}{2}\right)^2$

6. Binomial theorem

(*i*) *If n is a + ve integer*, then for all values of a and x,

$$(a + x)^n = {}^nC_0\, a^n + {}^nC_1\, a^{n-1} x + {}^nC_2\, a^{n-2} x^2 + {}^nC_3\, a^{n-3} x^3 + ... + {}^nC_n\, x^n$$

(*ii*) *If n is – ve integer or fraction*, then

$$(1 + x)^n = 1 + nx + \frac{n(n-1)}{2!}x^2 + \frac{n(n-1)(n-2)}{3!}x^3 + ... , \text{ provided } |x| < 1$$

(*iii*) $(1+x)^{-1} = 1 - x + x^2 - x^3 + x^4 - \ldots\ldots \infty$

(*iv*) $(1-x)^{-1} = 1 + x + x^2 + x^3 + x^4 + \ldots\ldots \infty$

(*v*) $(1+x)^{-2} = 1 - 2x + 3x^2 - 4x^3 + \ldots\ldots \infty$

(*vi*) $(1-x)^{-2} = 1 + 2x + 3x^2 + 4x^3 + \ldots\ldots \infty$

7. Permutations and Combinations formulae

(*i*) ${}^nP_r = \dfrac{n!}{(n-r)!}$, (*ii*) ${}^nC_r = \dfrac{n!}{r!(n-r)!}$

(*iii*) ${}^nC_0 = 1$, (*iv*) ${}^nC_n = 1$

(*v*) ${}^nC_1 = n$, (*vi*) ${}^nC_2 = \dfrac{n(n-1)}{2!}$

(*vii*) ${}^nC_3 = \dfrac{n(n-1)(n-2)}{3!}$, and so on

(*viii*) ${}^nC_r = {}^nC_{n-r}$ *e.g.*, ${}^8C_7 = {}^8C_1 = 8$

8. Identities

(*i*) $(a+b+c)^2 = a^2 + b^2 + c^2 + 2(ab + bc + ca)$

(*ii*) $a^3 + b^3 + c^3 - 3abc = (a+b+c)(a^2 + b^2 + c^2 - ab - bc - ca)$

(*iii*) If $a + b + c = 0$, then $a^3 + b^3 + c^3 - 3abc = 0$

(*iv*) $(a+b)^3 = a^3 + b^3 + 3ab(a+b)$

(*v*) $(a-b)^3 = a^3 - b^3 - 3ab(a-b)$

(*vi*) $a^3 + b^3 = (a+b)(a^2 - ab + b^2)$

(*vii*) $a^3 - b^3 = (a-b)(a^2 + ab + b^2)$

(*viii*) $a^2 - b^2 = (a+b)(a-b)$

(*ix*) $a^2 + b^2 = (a+ib)(a-ib)$

9. Logarithms

(*i*) *Natural logarithm* : $\log_e x$, generally denoted as $\log x$

(*ii*) Common logarithm : $\log_{10} x$ (*iii*) $\log mn = \log m + \log n$

(*iv*) $\log \dfrac{m}{n} = \log m - \log n$ (*v*) $\log m^n = n \log m$

(*vi*) $\log_a b = \dfrac{\log a}{\log b}$ (*vii*) $\log_{10} e = 0.43429$

(*viii*) $\log \pi = 0.49715$ (*ix*) $\log_e 10 = 2.30259$

(*x*) $\log_a 1 = 0$ (*xi*) $\log_a 0 = -\infty$, $a > 1$

10. Complex numbers

If $z = x + iy$; $x, y \in R$, then

(*i*) $\bar{z} = x - iy$

(*ii*) $|z| = \sqrt{(\text{Re}(z))^2 + (\text{Im}(z))^2}$

(*iii*) $z\bar{z} = |z|^2$

(*iv*) Polar form of z is given by

$$z = r(\cos\theta + i\sin\theta), \quad \ldots(1) \quad \text{where} \quad \theta = \arg z \quad \text{or} \quad \text{amp } z$$

If the value of θ satisfies $-\pi < \theta \le \pi$,

then $\theta = \text{Amp}(z)$ or $\text{Arg}(z)$ and the form (1) is called the standard form of z.

11. Mensuration

(*i*) Area of triangle $= \sqrt{s(s-a)(s-b)(s-c)}$

where $s = \dfrac{a+b+c}{2}$ and a, b, c are the sides of the triangle.

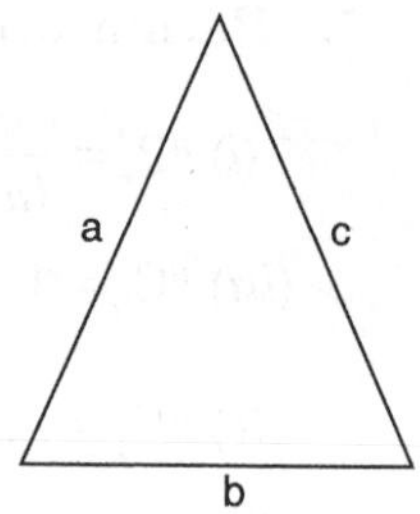

(*ii*) Area of an ellipse $\dfrac{x^2}{a^2} + \dfrac{y^2}{b^2} = 1$ is given by

$$\text{Area} = \pi ab$$

(*iii*) Area of a circle $x^2 + y^2 = r^2$ is given by

$$\text{Area} = \pi r^2$$

(*iv*) Volume of the sphere $x^2 + y^2 + z^2 = R^2$ is given by

$$V = \frac{4}{3}\pi R^3$$

and surface area of this sphere is given by

$$S = 4\pi R^2$$

(*v*) Volume of right circular cone $= \dfrac{1}{3}\pi r^2 h$

Curved surface area of the right circular cone

$$= \pi r\sqrt{r^2 + h^2}.$$

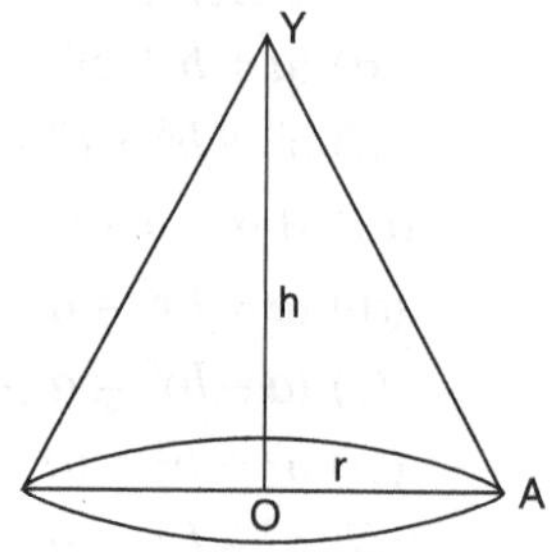

Right circular cone

(*vi*) Volume of right circular cylinder $= \pi r^2 h$

Curved surface area of the right circular cylinder $= 2\pi rh$.

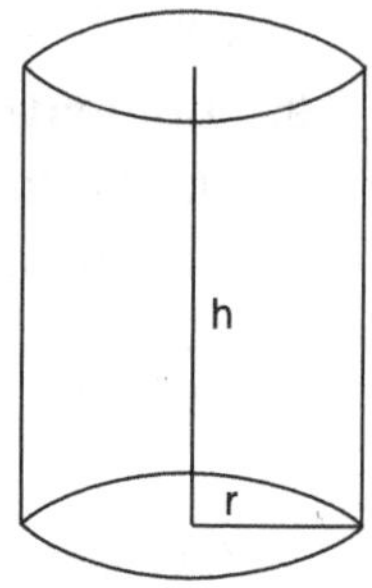

12. Trigonometry

(*i*)

$\cos 0° = 1$	$\sin 90° = 1$
$\cos 30° = \dfrac{\sqrt{3}}{2}$	$\sin 60° = \dfrac{\sqrt{3}}{2}$
$\cos 45° = \dfrac{1}{\sqrt{2}}$	$\sin 45° = \dfrac{1}{\sqrt{2}}$
$\cos 60° = \dfrac{1}{2}$	$\sin 30° = \dfrac{1}{2}$
$\cos 90° = 0$	$\sin 0° = 0$

(*ii*)

$$\sin(n\pi + \theta) = (-1)^n \sin\theta, \quad n \in Z$$

$$\cos(n\pi + \theta) = (-1)^n \cos\theta, \; n \in Z$$

$$\cos n\pi = (-1)^n, \; n \in Z$$

$$\sin n\pi = 0, \quad n \in N$$

(*iii*) If $\sin\theta = 0$, then $\theta = n\pi,\ n \in Z$
If $\cos\theta = 0$, then $\theta = (2n + 1)\pi/2,\ n \in Z$
If $\tan\theta = 0$, then $\theta = n\pi,\ n \in Z$

(*iv*) $\sin(-\theta) = -\sin\theta$
$\cos(-\theta) = \cos\theta$
$\tan(-\theta) = -\tan\theta$

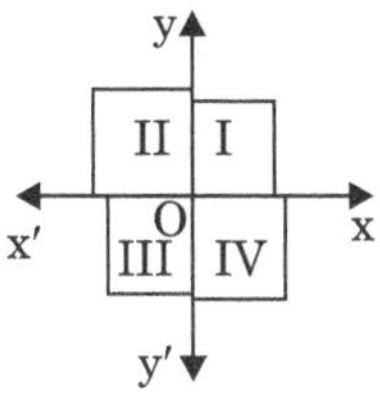

(*v*) *Quadrants.* There are four quadrants.
In Ist quadrant, the values of θ satisfies $0 \le \theta \le 90°$.
In 2nd quadrant, the values of θ satisfies $90° \le \theta \le 180°$
In 3rd quadrant, the values of θ satisfies $180° \le \theta \le 270°$
In 4th quadrant, the values of θ satisfies $270° \le \theta \le 360°$

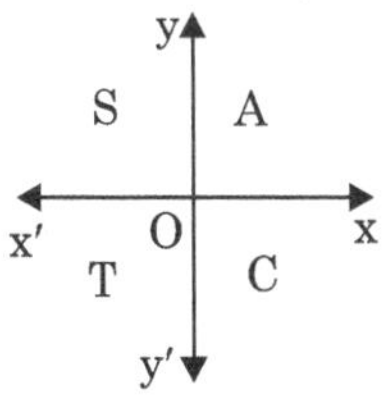

(*vi*) *Signs of trigonometrical functions in each quadrant.* This can be remembered by the following statement "AFTER SCHOOL TO COLLEGE". It means :
In first quadrant, All are +ve.
In 2nd quadrant, sine and cosec are +ve and all others are –ve.
In 3rd quadrant, tan and cot are +ve and all others are –ve.
In 4th quadrant, cos and sec are +ve and all others are –ve.

Note. If we wish to find the values of any of sin, cosine, tan, cot, sec, cosec, etc. at any angles greater than 90°. Then we adopt the following procedure.

Step I. Split the required angles into two parts such that one part should be out of 90°, 270°, 180° or 360° and the 2nd part should be between 0 and 90°.

Step II. Find the quadrant in which the required angle lies and put the sign + ve or – ve.

Step III. When the first angle is 90° or 270°, then
(*a*) $\sin\theta \leftrightarrow \cos\theta$ (*i.e.*, sin θ will change to cos θ and cos θ to sin θ).
(*b*) $\tan\theta \leftrightarrow \cot\theta$
(*c*) $\sec\theta \leftrightarrow \text{cosec}\,\theta$.

Step IV. When the first angle is 180° or 360°, then no change.

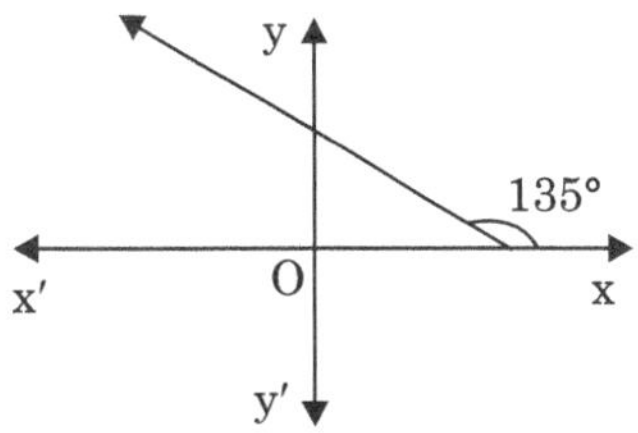

Example 1. *Find the value of sin 135°.*

Sol. $\sin 135° = \sin(90° + 45°)$

$$= +\cos 45° = \frac{1}{\sqrt{2}}.$$

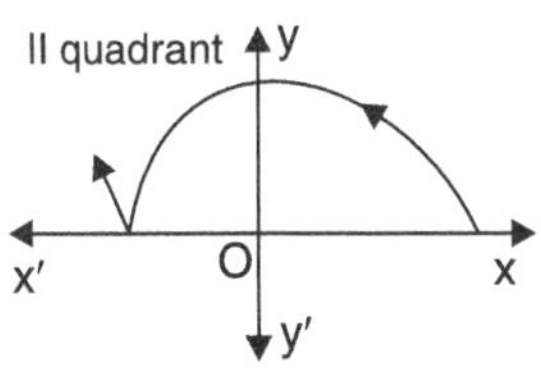

Explanation. Write 135° = 90° + 45°. Now 90° + 45° lies in 2nd quadrant and in 2nd quadrant, sin θ is +ve. Also because of 90°, sin θ will change to cos θ.

Also $135° = 180° - 45°$

$\therefore$ $\sin(135°) = \sin(180° - 45°)$

$$= \sin 45° = \frac{1}{\sqrt{2}}$$

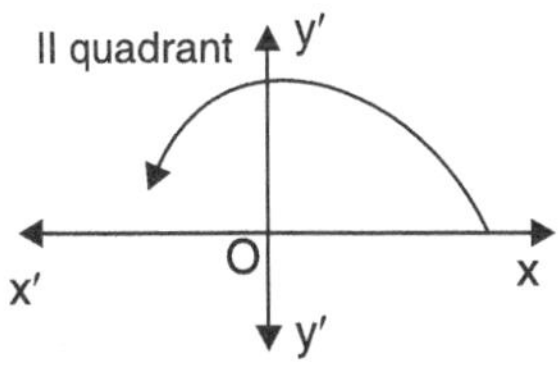

Explanation. 180° – 45° lies in 2nd quadrant and in 2nd quadrant, sin θ is always +ve. Also because of 180°, sin θ will not change.

Example 2. *Find the value of tan 135° and cos 135°.*

Sol. $\tan 135° = \tan(90 + 45°)$
$= -\cot 45° = -1$

Explanation. As 90° + 45° lies in 2nd quadrant and in 2nd quadrant, tan is always –ve. Also because of 90°, tan will change to cot.

Now $$\cos 135° = \cos(90° + 45°)$$
$$= -\sin 45 = -\frac{1}{\sqrt{2}}.$$

Example 3. *Find the values of tan 210°, sin 210°, cosec 210°, sec 210°.*

Sol.
$$\tan 210° = \tan(180° + 30°)$$
$$= +\tan 30°$$
$$= \frac{\sin 30°}{\cos 30°} = \frac{1}{2}\cdot\frac{2}{\sqrt{3}}$$
$$= \frac{1}{\sqrt{3}}$$

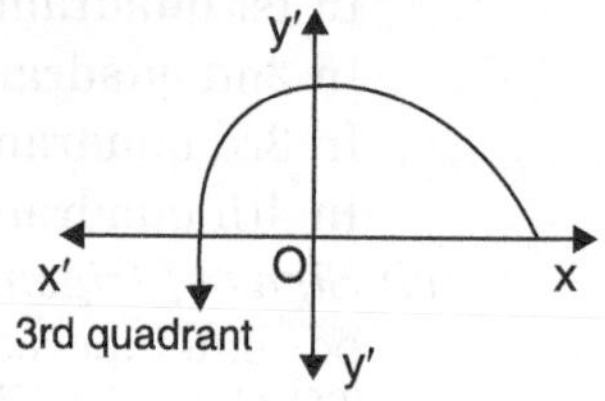

Also, $$\sin 210° = \sin(180° + 30°)$$
$$= -\sin 30°$$
$$= -\frac{1}{2}$$

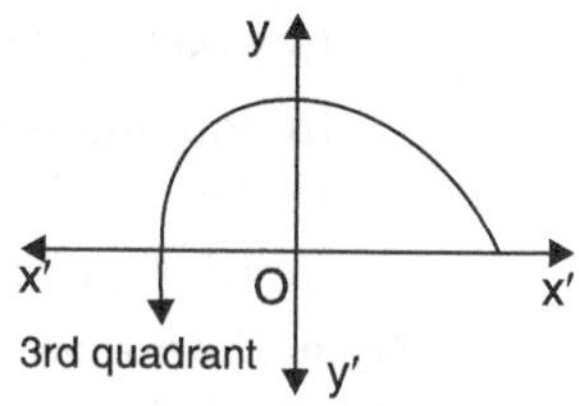

$$\text{cosec } 210° = \text{cosec}(270° - 60°)$$
$$= -\sec 60°$$
$$= -\frac{1}{\cos 60°}$$
$$= -2.$$

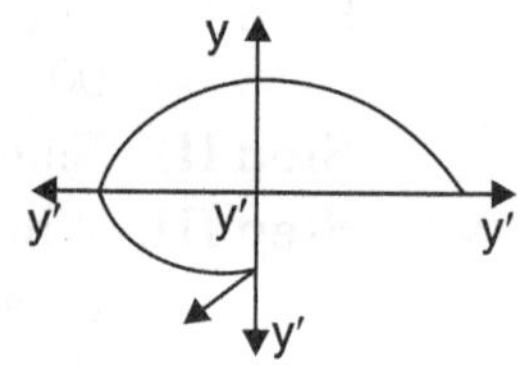

$$\sec 210° = \sec(180° + 30°)$$
$$= -\sec 30°$$
$$= -\frac{1}{\cos 30°}$$
$$= \frac{-2}{\sqrt{3}}.$$

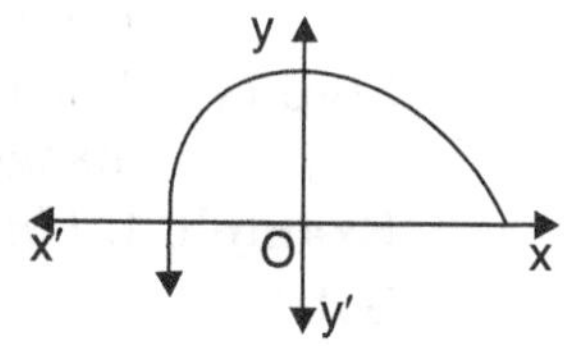

Example 4. *Find tan 420°, cot 420°, sec 420°, cosec 420°.*

Sol.
$$\tan 420° = \tan(360° + 60°) = +\tan 60°$$
$$= \frac{\sin 60°}{\cos 60°} \quad | \text{ tan is + ve in Ist quad.}$$
$$= \frac{\sqrt{3}}{2}\cdot\frac{2}{1} = \sqrt{3}$$

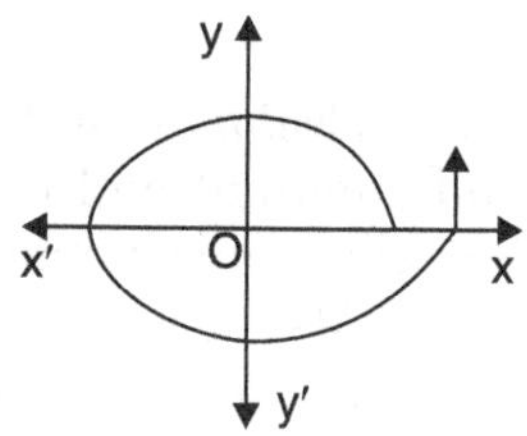

Similarly, we can find cot 420°, sec 420°, cosec 420°.

(vii)
$$\sin(A + B) = \sin A \cos B + \cos A \sin B$$
$$\sin(A - B) = \sin A \cos B - \cos A \sin B$$
$$\cos(A + B) = \cos A \cos B - \sin A \sin B$$
$$\cos(A - B) = \cos A \cos B + \sin A \sin B$$
$$\tan(A + B) = \frac{\tan A + \tan B}{1 - \tan A \tan B}$$

$$\tan (A - B) = \frac{\tan A - \tan B}{1 + \tan A \tan B}$$

$$\tan (\pi/4 + \theta) = \frac{1 + \tan \theta}{1 - \tan \theta}$$

$$\tan (\pi/4 - \theta) = \frac{1 - \tan \theta}{1 + \tan \theta}$$

(*viii*) **C—D Formulae** $S + S = 2SC$

$$S - S = 2CS$$

$$C + C = 2CC$$

$$C - C = -2SS$$

i.e.,

$$\sin C + \sin D = 2 \sin \frac{C + D}{2} \cos \frac{C - D}{2}$$

$$\sin C - \sin D = 2 \cos \frac{C + D}{2} \sin \frac{C - D}{2}$$

$$\cos C + \cos D = 2 \cos \frac{C + D}{2} \cos \frac{C - D}{2}$$

$$\cos C - \cos D = -2 \sin \frac{C + D}{2} \sin \frac{C - D}{2}$$

or

$$= 2 \sin \frac{C + D}{2} \sin \frac{D - C}{2}$$

A—B Formulae

$$2 \sin A \cos B = \sin (A + B) + \sin (A - B)$$

$$2 \cos A \sin B = \sin (A + B) - \sin (A - B)$$

$$2 \cos A \cos B = \cos (A + B) + \cos (A - B)$$

$$2 \sin A \sin B = \cos (A - B) - \cos (A + B)$$

(*ix*) $\sin^2 \theta + \cos^2 \theta = 1$

$$\sec^2 \theta - \tan^2 \theta = 1$$

$$\operatorname{cosec}^2 \theta - \cot^2 \theta = 1$$

$$\sin 2\theta = 2 \sin \theta \cos \theta = \frac{2 \tan \theta}{1 + \tan^2 \theta}$$

$$\cos 2\theta = \cos^2 \theta - \sin^2 \theta$$

$$= 2 \cos^2 \theta - 1 = 1 - 2 \sin^2 \theta = \frac{1 - \tan^2 \theta}{1 + \tan^2 \theta}$$

$$\tan 2\theta = \frac{2 \tan \theta}{1 - \tan^2 \theta}$$

$$\sin 3\theta = 3 \sin \theta - 4 \sin^3 \theta$$

$$\cos 3\theta = 4 \cos^3 \theta - 3 \cos \theta$$

$$\tan 3\theta = \frac{3 \tan \theta - \tan^3 \theta}{1 - 3 \tan^2 \theta}$$

13. Hyperbolic functions

(*i*) $\sinh x = \dfrac{e^x - e^{-x}}{2}$

(*ii*) $\cosh x = \dfrac{e^x + e^{-x}}{2}$

(*iii*) $\dfrac{d}{dx}(\sinh x) = \dfrac{d}{dx}\left(\dfrac{e^x - e^{-x}}{2}\right) = \dfrac{e^x + e^{-x}}{2} = \cosh x$

(*iv*) $\dfrac{d}{dx}(\cosh x) = \sinh x$

(*v*) $\cosh^2 x - \sinh^2 x = 1.$

(*vi*) $\sinh^{-1} z = \log (z + \sqrt{1+z^2})$

(*vii*) $\cosh^{-1} z = \log (z + \sqrt{z^2 - 1})$

(*viii*) $\tanh^{-1} z = \dfrac{1}{2} \log \left(\dfrac{1+z}{1-z}\right)$

14. Limits

(*i*) $\underset{\theta \to 0}{\text{Lt}} \dfrac{\sin \theta}{\theta} = 1$

(*ii*) $\underset{\theta \to 0}{\text{Lt}} \dfrac{\tan \theta}{\theta} = 1$

(*iii*) $\underset{x \to \infty}{\text{Lt}} \left(1+\dfrac{1}{x}\right)^x = e$

(*iv*) $\underset{x \to \infty}{\text{Lt}} \left(1+\dfrac{a}{x}\right)^x = e^a$

15. Differential calculus

(*i*) $\dfrac{d}{dx}(x^n) = nx^{n-1}$

(*ii*) $\dfrac{d}{dx}\left(\dfrac{1}{x}\right) = -\dfrac{1}{x^2}$

(*iii*) $\dfrac{d}{dx}(a^x) = a^x \log_e a$

(*iv*) $\dfrac{d}{dx}(e^x) = e^x \log_e e = e^x$

(*v*) $\dfrac{d}{dx}(\log_a x) = \dfrac{1}{x} \log_a e$

(*vi*) $\dfrac{d}{dx}(\log x) = \dfrac{1}{x}$

(*vii*) $\dfrac{d}{dx}(\sin x) = \cos x$

(*viii*) $\dfrac{d}{dx}(\cos x) = -\sin x$

(*ix*) $\dfrac{d}{dx}(\tan x) = \sec^2 x$

(*x*) $\dfrac{d}{dx}(\cot x) = -\text{cosec}^2 x$

(*xi*) $\dfrac{d}{dx}(\sec x) = \sec x \tan x$

(*xii*) $\dfrac{d}{dx}(\text{cosec } x) = -\text{cosec } x \cot x$

(*xiii*) $\dfrac{d}{dx}(\sin^{-1} x) = \dfrac{1}{\sqrt{1-x^2}}$

(*xiv*) $\dfrac{d}{dx}(\cos^{-1} x) = \dfrac{-1}{\sqrt{1-x^2}}$

(*xv*) $\dfrac{d}{dx}(\tan^{-1} x) = \dfrac{1}{1+x^2}$

(*xvi*) $\dfrac{d}{dx}(\cot^{-1} x) = \dfrac{-1}{1+x^2}$

(*xvii*) $\dfrac{d}{dx}(\sec^{-1} x) = \dfrac{1}{x\sqrt{x^2-1}}$

(*xviii*) $\dfrac{d}{dx}(\text{cosec}^{-1} x) = \dfrac{-1}{x\sqrt{x^2-1}}$

16. Product Rule

(i) $\frac{d}{dx}(\text{I . II}) = \text{I} . \frac{d}{dx}(\text{II}) + \text{II} . \frac{d}{dx}(\text{I})$

(ii) $\frac{d}{dx}(u . v . w) = \frac{du}{dx} . v . w + u . \frac{dv}{dx} . w + u . v. \frac{dw}{dx}$

(iii) $\frac{d}{dx}\left(\frac{u}{v}\right) = \frac{v . \frac{du}{dx} - u . \frac{dv}{dx}}{v^2}$

STANDARD RESULTS OF INTEGRATION

The reader has already learnt integration in Class XII. Here we remind the reader of the following standard results upto the product Rule of Integration.

1. $\int x^n \, dx = \frac{x^{n+1}}{n+1} \; (n \neq -1).$

2. $\int 1 \, dx = x.$

3. $\int \frac{1}{x} \, dx = \log x.$

4. $\int e^x \, dx = e^x.$

5. $\int a^x \, dx = \frac{a^x}{\log a}.$

6. $\int \sin x \, dx = -\cos x.$

7. $\int \cos x \, dx = \sin x.$

8. $\int \tan x \, dx = -\log \cos x$ or $\log \sec x.$

9. $\int \cot x \, dx = \log |\sin x|.$

10. $\int \sec x \, dx = \log |\sec x + \tan x|$ or $\log \left|\tan\left(\frac{\pi}{4} + \frac{x}{2}\right)\right|.$

11. $\int \operatorname{cosec} x \, dx = \log |\operatorname{cosec} x - \cot x|$ or $\log \left|\tan \frac{x}{2}\right|.$

12. $\int \sec^2 x \, dx = \tan x.$

13. $\int \operatorname{cosec}^2 x \, dx = -\cot x.$

14. $\int \sec x \tan x \, dx = \sec x.$

15. $\int \operatorname{cosec} x \cot x \, dx = -\operatorname{cosec} x.$

16. If $\int f(x) \, dx = \phi(x)$, then $\int f(ax+b) \, dx = \frac{\phi(ax+b)}{a}.$

The above result can be stated in words as :

If the integral of a function of x is known, then if x is changed to $ax + b$, where a, b are constants, **the integral is of the same form** (x being changed to $ax + b$) **but is divided by the co-efficient of x.**

17. $\int \sinh x\,dx = \cosh x.$

18. $\int \cosh x\,dx = \sinh x.$

19. $\int \text{sech}^2 x\,dx = \tanh x.$

20. $\int \text{cosech}^2 x\,dx = -\coth x.$

21. $\int \text{sech}\,x \tanh x\,dx = \text{sech}\,x.$

22. $\int \text{cosech}\,x \coth x\,dx = -\text{cosech}\,x.$

23. $\int \dfrac{f'(x)}{f(x)}\,dx = \log|f(x)|.$

24. $\int (f(x))^n f'(x)\,dx = \dfrac{(f(x))^{n+1}}{n+1}.$ $(n \neq -1)$

25. $\int \dfrac{1}{\sqrt{a^2 - x^2}}\,dx = \sin^{-1}\dfrac{x}{a}.$

26. $\int \dfrac{1}{x^2 + a^2}\,dx = \dfrac{1}{a}\tan^{-1}\dfrac{x}{a}$

27. $\int \dfrac{1}{x\sqrt{x^2 - a^2}}\,dx = \dfrac{1}{a}\sec^{-1}\dfrac{x}{a}.$

28. $\int \dfrac{1}{\sqrt{x^2 + a^2}}\,dx = \sinh^{-1}\dfrac{x}{a}$ or $\log\left|x + \sqrt{x^2 + a^2}\right|.$

29. $\int \dfrac{1}{\sqrt{x^2 - a^2}}\,dx = \cosh^{-1}\dfrac{x}{a}$ or $\log\left|x + \sqrt{x^2 - a^2}\right|.$

30. Product Rule of Integration or Rule of Integration by parts.

The integral of the product of two functions

$$= (\text{Ist function}) \times (\text{Integral of II}) - \int \left[\frac{d}{dx}(\text{I}) \times \text{Integral of II}\right] dx.$$

Rule for choice of First Function in Product Rule

The order of preference for taking **first** function is being given below :

1. Logarithmic functions such as $\log x$, $\log(1 + x)$ etc., should be taken first.
2. Inverse circular functions as $\sin^{-1} x$, $\cos^{-1} x$ etc., should be taken first.
3. Polynomial functions as x, x^2, x^3, ... etc., should be taken first.

31. $\int \sqrt{a^2 - x^2}\,dx = \dfrac{x}{2}\sqrt{a^2 - x^2} + \dfrac{a^2}{2}\sin^{-1}\dfrac{x}{a}.$

32. $\int \sqrt{x^2 + a^2}\,dx = \dfrac{x}{2}\sqrt{x^2 + a^2} + \dfrac{a^2}{2}\sinh^{-1}\dfrac{x}{a}.$

33. $\int \sqrt{x^2 - a^2}\,dx = \dfrac{x}{2}\sqrt{x^2 - a^2} - \dfrac{a^2}{2}\cosh^{-1}\dfrac{x}{a}.$

34. $\int e^{ax}\sin bx\,dx = \dfrac{e^{ax}}{a^2 + b^2}(a \sin bx - b \cos bx)$ or

$$= \frac{e^{ax}}{\sqrt{a^2 + b^2}}\sin\left(bx - \tan^{-1}\frac{b}{a}\right).$$

35. $\int e^{ax}\cos bx\,dx = \dfrac{e^{ax}}{a^2 + b^2}(a \cos bx + b \sin bx)$ or

$$= \frac{e^{ax}}{\sqrt{a^2 + b^2}}\cos\left(bx - \tan^{-1}\frac{b}{a}\right).$$

36. $\displaystyle\int \frac{dx}{a^2 - x^2} = \frac{1}{2a} \log \left| \frac{a + x}{a - x} \right|.$

37. $\displaystyle\int \frac{dx}{x^2 - a^2} = \frac{1}{2a} \log \left| \frac{x - a}{x + a} \right|.$

Remark. We are omitting the arbitrary constant in all the above results but it is always understood to be there.

38. Definite integral $\displaystyle\int_a^b f(x)\, dx$ = (value of antiderivative of $f(x)$ at $x = b$) – (value of antiderivative of $f(x)$ at $x = a$).

Note. No arbitrary constant is added in definite integrals as that cancels.

39. Rule to integrate $\displaystyle\int_a^b$ f(x) dx by method of substitution.

If in the integrand, we put $x = \phi(t)$, then $dx = \phi'(t)\, dt$

When $x = a$, find the value of t, say $t = \alpha$.

When $x = b$, find the value of t, say $t = \beta$.

$\therefore \displaystyle\int_a^b f(x)\, dx$ becomes $\displaystyle\int_\alpha^\beta f(\phi(t))\, \phi'(t)\, dt$.

40. $\displaystyle\int_a^b f(x)\, dx = \int_a^b f(z)\, dz.$

41. $\displaystyle\int_a^b f(x)\, dx = -\int_b^a f(x)\, dx.$

42. $\displaystyle\int_a^b f(x)\, dx = \int_a^c f(x)\, dx + \int_c^b f(x)\, dx.$

43. $\displaystyle\int_0^a f(x)\, dx = \int_0^a f(a - x)\, dx.$

44. $\displaystyle\int_0^{2a} f(x)\, dx = \begin{cases} 2\displaystyle\int_0^a f(x)\, dx & \text{if } f(2a - x) = f(x) \\ 0 & \text{if } f(2a - x) = -f(x). \end{cases}$

45. (*a*) If $f(x)$ is an even function of x, *i.e.*, $f(-x) = f(x)$, then

$$\int_{-a}^a f(x)\, dx = 2\int_0^a f(x)\, dx.$$

(*b*) If $f(x)$ is an odd function of x, *i.e.*, $f(-x) = -f(x)$, then

$$\int_{-a}^a f(x)\, dx = 0.$$

46. A short-cut method of integration by parts

$$\int \underset{\text{I}}{u} \, . \, \underset{\text{II}}{v}\, dx = u \, . \, v^{(1)} - u' \, . \, v^{(2)} + u'' \, . \, v^{(3)} - u''' \, . \, v^{(4)} + \ldots\ldots$$

where $v^{(1)}$ = Integration of v and u' = derivative of u

$v^{(2)}$ = Integration of $v^{(1)}$ u'' = derivative of u'

$v^{(3)}$ = Integration of $v^{(2)}$ u''' = derivative of u'', and so on.

$v^{(4)}$ = Integration of $v^{(3)}$ and so on

47. Definite Integrals

(*i*) $$\int_0^{\pi/2} \sin^n x\, dx = \frac{(n-1) \times \text{go on diminishing by } 2}{n \times \text{go on diminishing by } 2} \times \frac{\pi}{2}$$

only if n is a +ve even integer $\left(\text{otherwise no } \frac{\pi}{2}\right)$

(*ii*) $$\int_0^{\pi/2} \cos^n x\, dx = \frac{(n-1) \times \text{go on diminishing by } 2}{n \times \text{go on diminishing by } 2} \times \frac{\pi}{2}$$

only if n is a +ve even integer $\left(\text{otherwise no } \frac{\pi}{2}\right)$

(*iii*) $$\int_0^{\pi/2} \sin^p x \cos^q x\, dx$$

$$= \frac{[(p-1) \times \text{go on diminishing by } 2] \times [(q-1) \times \text{go on diminishing by } 2]}{(p+q) \times \text{go on diminishing by } 2} \times \frac{\pi}{2}$$

only if both p and q are +ve even integers $\left(\text{otherwise no } \frac{\pi}{2}\right)$.

1

Equations Reducible to Quadratic Equations

1.1. INTRODUCTION

In the present chapter, we shall study the methods of finding real and complex roots of the polynomial equations of the type $ax^2 + bx + c = 0$ where a, b, c are arbitrary complex numbers.

1.2. IDENTITY AND EQUATION

An **identity** is a statement of equality between two expressions which is true for all values of the variable involved.

For example, $$(x-1)^2 + 4(x-1) + 10 - x^2 - 2x - 7 \equiv 0$$
is an identity, because the above statement is true for all values of x.

An **equation** is a statement of equality between two expressions which is not true for all values of the variable involved.

For example,

(*i*) $\sin x = 0$ is true for $x = 0, \pm\pi, \pm 2\pi, \ldots\ldots$ (*ii*) $x^2 - 5x + 6 = 0$ is true for $x = 2, 3$.

A polynomial equated to zero is called a **polynomial equation.** The **degree** of a polynomial equation is same as the *degree* of the corresponding polynomial.

For example, $2x^2 - 7x + 6 = 0$ is polynomial equation of degree 2.

A polynomial equation of degree 2 is called a **quadratic equation.**

1.3. ROOT OF AN EQUATION

A value of the variable for which an equation is satisfied is called a **root** of the equation, under consideration.

For example, 1 is not a root of $4x^2 - 16x + 15 = 0$, because $4(1)^2 - 16(1) + 15 = 4 - 16 + 15 = 3 \neq 0$.

1.4. FACTORIZATION METHOD OF SOLVING A QUADRATIC EQUATION

The principle underlying this method is that if $xy = 0$ then either $x = 0$ or $y = 0$.

Let $ax^2 + bx + c = 0$, $a \neq 0$ be the given quadratic equation.

If $b = 0$, then $ax^2 + c = 0$ *i.e.*, $x^2 = -c/a$ or $x = \pm\sqrt{-c/a}$. So, let us assume that $b \neq 0$. The **method of factorization** is applicable only if we can write b as the sum of two numbers whose product is ac. The value of b is changed in the given equation and the factorization is carried.

For example, consider the quadratic equation
$$x^2 - x - 6 = 0. \qquad \ldots(1)$$

Here $b = -1$. We write $b = -1 = (-3) + 2$, because $(-3)(2) = -6 = (1)(-6)$.

$\therefore$ (1) $\Rightarrow x^2 - 3x + 2x - 6 = 0 \Rightarrow x(x-3) + 2(x-3) = 0$

$\Rightarrow (x-3)(x+2) = 0 \Rightarrow x = 3, -2.$

ILLUSTRATIVE EXAMPLES

Example 1. *Solve the equation :* $5x^2 + 15 = 0$.

Sol. We have $5x^2 + 15 = 0$.

$\Rightarrow 5x^2 = -15 \Rightarrow x^2 = -3 \Rightarrow x = \pm\sqrt{-3} = \pm i\sqrt{3}$

$\therefore$ The roots are $-\mathbf{i\sqrt{3}}$ and $\mathbf{i\sqrt{3}}$.

Example 2. *Solve the equation :* $\dfrac{x+2}{x+3} = \dfrac{x+4}{2x+3}$.

Sol. We have $\dfrac{x+2}{x+3} = \dfrac{x+4}{2x+3}$.

$\Rightarrow (x+2)(2x+3) = (x+3)(x+4)$

$\Rightarrow 2x^2 + 3x + 4x + 6 = x^2 + 4x + 3x + 12 \Rightarrow x^2 = 6 \Rightarrow x = \pm\sqrt{6}.$

$\therefore$ The roots are $-\mathbf{\sqrt{6}}$ and $\mathbf{\sqrt{6}}$.

Example 3. *Solve the equation :* $4x^2 - 16x + 15 = 0$.

Sol. We have $4x^2 - 16x + 15 = 0$. ...(1)

We write $-16 = (-6) + (-10)$, because $(-6)(-10) = 60 = (4)(15)$.

$\therefore$ (1) $\Rightarrow 4x^2 - 6x - 10x + 15 = 0 \Rightarrow 2x(2x-3) - 5(2x-3) = 0$

$\Rightarrow (2x-3)(2x-5) = 0 \Rightarrow x = \dfrac{3}{2}, \dfrac{5}{2}.$

$\therefore$ The roots are **3/2** and **5/2.**

Example 4. *Solve the equation :* $\dfrac{1}{p+q+x} = \dfrac{1}{p} + \dfrac{1}{q} + \dfrac{1}{x}$.

(P.T.U. B.Pharma, May 2007)

Sol. We have $\dfrac{1}{p+q+x} = \dfrac{1}{p} + \dfrac{1}{q} + \dfrac{1}{x}$.

$\Rightarrow \dfrac{1}{p+q+x} - \dfrac{1}{x} = \dfrac{1}{p} + \dfrac{1}{q}$ **(Note this step)**

$\Rightarrow \dfrac{x-p-q-x}{(p+q+x)x} = \dfrac{q+p}{pq} \Rightarrow \dfrac{-(p+q)}{(p+q+x)x} = \dfrac{p+q}{pq}$

$\Rightarrow \dfrac{-1}{(p+q+x)x} = \dfrac{1}{pq} \Rightarrow -pq = (p+q+x)x$

$\Rightarrow x^2 + px + qx + pq = 0 \Rightarrow x(x+p) + q(x+p) = 0$

$\Rightarrow (x+p)(x+q) = 0 \Rightarrow x+p = 0$ or $x+q = 0$

$x + p = 0$ implies $x = -p$ and $x + q = 0$ implies $x = -q$.

$\therefore$ The roots are $-\mathbf{p}$ and $-\mathbf{q}$.

Example 5. *Solve the equation :* $\dfrac{a}{ax-1}+\dfrac{b}{bx-1}=a+b.$

Sol. The given equation is $\dfrac{a}{ax-1}+\dfrac{b}{bx-1}=a+b.$

$$\Rightarrow \quad \left(\frac{a}{ax-1}-b\right)+\left(\frac{b}{bx-1}-a\right)=0 \quad \Rightarrow \quad \frac{a-abx+b}{ax-1}+\frac{b-abx+a}{bx-1}=0$$

$$\Rightarrow \quad (a+b-abx)\left[\frac{1}{ax-1}+\frac{1}{bx-1}\right]=0$$

$\therefore$ Either $\quad a+b-abx=0 \quad$ ...(1) $\quad$ or $\quad \dfrac{1}{ax-1}+\dfrac{1}{bx-1}=0 \quad$ ...(2)

$$(1) \Rightarrow \quad a+b=abx \quad \Rightarrow \quad x=\frac{a+b}{ab}.$$

$$(2) \Rightarrow \quad \frac{(bx-1)+(ax-1)}{(ax-1)(bx-1)}=0 \quad \Rightarrow \quad bx-1+ax-1=0$$

$$\Rightarrow \quad (a+b)x=2 \quad \Rightarrow \quad x=\frac{2}{a+b}.$$

$\therefore$ The roots are $\mathbf{\dfrac{a+b}{ab}}$ and $\mathbf{\dfrac{2}{a+b}}$.

Remark. The above method of factorization has very limited utility, because it is not always possible to express b in $ax^2+bx+c=0$ as the sum of numbers l and m so as to satisfy the conditions $l+m=b$ and $lm=ac$.

WORKING RULES FOR SOLVING PROBLEMS

Rule I. *If the equation to solve is* $ax^2+c=0$*, then write* $x^2=-c/a$ *and then* $x=\pm\sqrt{-c/a}$.

The roots are $-\sqrt{-c/a}$ *and* $\sqrt{-c/a}$.

Rule II. *If the equation to solve is* $ax^2+bx+c=0$*, then find two numbers* l *and* m *such that* $l+m=b$ *and* $lm=ac$*. Put* $b=l+m$ *in the equation and factorize the L.H.S.*

EXERCISE 1.1

Solve the following equations by the method of factorization :

1. $7x^2+49=0$ **2.** $2x^2+1=0$

3. $2z^2-10=z$ **4.** $2x^2+3ix+2=0$ **5.** $x^2-7ix-12=0$

6. $\dfrac{1}{3-2x}+\dfrac{1}{5+2x}=\dfrac{1}{2}$ **7.** $abx^2+(b^2-ac)x-bc=0$ **8.** $a(x^2+4)=x(a^2+4)$

9. $abx^2 - (a + b)x + 1 = 0$ 10. $\frac{x-p}{q} + \frac{x-q}{p} = \frac{q}{x-p} + \frac{p}{x-q}$.

Answers

1. $\pm\sqrt{7}i$ 2. $\pm\sqrt{2}i/2$ 3. $-2, 5/2$ 4. $i/2, -2i$
5. $3i, 4i$ 6. $-1/2, -1/2$ 7. $c/b, -b/a$ 8. $a, 4/a$
9. $\frac{1}{a}, \frac{1}{b}$ 10. $0, p+q, \frac{p^2+q^2}{p+q}$.

Hints

2. $2x^2 + 1 = 2x^2 - i^2 = (\sqrt{2}x - i)(\sqrt{2}x + i)$
4. $3i = 4i + (-i)$ and $(4i)(-i) = 4$ which is also equal to 2×2.
10. We have $\left(\frac{x-p}{q} - \frac{p}{x-q}\right) + \left(\frac{x-q}{p} - \frac{q}{x-p}\right) = 0 \Rightarrow \frac{x^2 - px - qx}{qx - q^2} + \frac{x^2 - qx - px}{px - p^2} = 0$

$\Rightarrow \quad x(x - p - q)\,[px - p^2 + qx - q^2] = 0 \Rightarrow x[x - (p + q)][(p + q)x - (p^2 + q^2)] = 0.$

1.5. FORMULA METHOD OF SOLVING A QUADRATIC EQUATION

The 'formula method' of solving a quadratic equation is used when the 'factorization method' is not easily applicable.

Let $\quad ax^2 + bx + c = 0, \quad a \neq 0 \quad$ ...(1)

be the given quadratic equation where a, b, c are *complex numbers.*

$$(1) \Rightarrow ax^2 + bx = -c \quad \Rightarrow x^2 + \frac{b}{a}x = -\frac{c}{a} \quad (\because a \neq 0)$$

$$\Rightarrow x^2 + 2\left(\frac{b}{2a}\right)x = -\frac{c}{a} \quad \Rightarrow x^2 + 2\left(\frac{b}{2a}\right)x + \left(\frac{b}{2a}\right)^2 = \frac{b^2}{4a^2} - \frac{c}{a}$$

$$\Rightarrow \left(x + \frac{b}{2a}\right)^2 = \frac{b^2 - 4ac}{4a^2} \quad \Rightarrow x + \frac{b}{2a} = \pm\sqrt{\frac{b^2 - 4ac}{4a^2}}$$

$$\Rightarrow x = -\frac{b}{2a} \pm \frac{\sqrt{b^2 - 4ac}}{2a} \quad \Rightarrow \mathbf{x = \frac{-b \pm \sqrt{b^2 - 4ac}}{2a}}.$$

These are the required roots of the given equation.

Theorem I. Let $ax^2 + bx + c = 0$, $a \neq 0$ be a quadratic equation where a, b, c be any complex numbers. Prove that α is a root of $ax^2 + bx + c = 0$ if and only if $x - \alpha$ is a factor of $ax^2 + bx + c$.

Proof. Let α be a root of $ax^2 + bx + c = 0$.

$\therefore \quad a\alpha^2 + b\alpha + c = 0.$

Now, $\quad ax^2 + bx + c = (ax^2 + bx + c) - (a\alpha^2 + b\alpha + c)$

$\qquad = (ax^2 - a\alpha^2) + (bx - b\alpha) = (x - \alpha)\,[a(x + \alpha) + b]$

$\therefore \quad x - \alpha$ is a factor of $ax^2 + bx + c$.

Conversely, let $x - \alpha$ be a factor of $ax^2 + bx + c$.

$\therefore \qquad ax^2 + bx + c = (x - \alpha) \cdot p(x) \qquad \ldots(1)$

where $p(x)$ is a polynomial of degree one.

Putting $x = \alpha$ in (1), we get

$$a\alpha^2 + b\alpha + c = (\alpha - \alpha) \cdot p(\alpha) = 0.p(\alpha) = 0$$

$\therefore$ α is a root of $ax^2 + bx + c = 0$.

Theorem II. Prove that a quadratic equation has exactly two roots.

Proof. Let $ax^2 + bx + c = 0$, $a \neq 0$ be a quadratic equation where a, b, c are any complex numbers.

We know that $\dfrac{-b \pm \sqrt{b^2 - 4ac}}{2a}$ are the roots of the given equation.

Let $\alpha = \dfrac{-b + \sqrt{b^2 - 4ac}}{2a}$ and $\beta = \dfrac{-b - \sqrt{b^2 - 4ac}}{2a}$.

$\therefore$ $x - \alpha$ and $x - \beta$ are factors of $ax^2 + bx + c$.

Let $\qquad ax^2 + bx + c = \lambda(x - \alpha)(x - \beta)$.

Comparing, coefficients of x^2, we get $\lambda = a$.

$\therefore \qquad ax^2 + bx + c = a(x - \alpha)(x - \beta) \qquad \ldots(1)$

If possible, let $x = \gamma$ be any other root of $ax^2 + bx + c = 0$ other than α and β.

Putting $x = \gamma$ in (1), we get

$$a\gamma^2 + b\gamma + c = a(\gamma - \alpha)(\gamma - \beta)$$

$\Rightarrow \qquad 0 = a(\gamma - \alpha)(\gamma - \beta) \qquad \ldots(2)$

Now $\qquad a \neq 0, \quad \gamma - \alpha \neq 0, \quad \gamma - \beta \neq 0$

$\therefore$ Equation (2) is impossible $\qquad \therefore$ Our supposition is wrong.

$\therefore$ **A quadratic equation cannot have more than two roots.**

ILLUSTRATIVE EXAMPLES

Example 1. *Solve the equation :* $x^2 - 4x + 7 = 0$.

Sol. We have $x^2 - 4x + 7 = 0$. Here $a = 1$, $b = -4$, $c = 7$

Now $\qquad x = \dfrac{-b \pm \sqrt{b^2 - 4ac}}{2a}$

$\therefore \qquad x = \dfrac{-(-4) \pm \sqrt{(-4)^2 - 4(1)(7)}}{2(1)} = \dfrac{4 \pm 2\sqrt{3}i}{2} = \mathbf{2 \pm \sqrt{3}i}.$

Example 2. *Solve the equation :* $\sqrt{3}y^2 + 10y - 8\sqrt{3} = 0$.

Sol. We have $\qquad \sqrt{3}\, y^2 + 10y - 8\sqrt{3} = 0$. $\quad$ Here $a = \sqrt{3}$, $b = 10$, $c = -8\sqrt{3}$.

Now $\quad y = \dfrac{-b \pm \sqrt{b^2 - 4ac}}{2a}$

$$\therefore \quad y = \frac{-10 \pm \sqrt{(10)^2 - 4(\sqrt{3})(-8\sqrt{3})}}{2\sqrt{3}} = \frac{-10 \pm \sqrt{100+96}}{2\sqrt{3}} = \frac{-10 \pm 14}{2\sqrt{3}} = \mathbf{\frac{2}{\sqrt{3}}, -\frac{12}{\sqrt{3}}}.$$

Example 3. *Determine the value of m for which the equations $3x^2 + 4mx + 2 = 0$ and $2x^2 + 3x - 2 = 0$ may have a common root.*

Sol. We have $3x^2 + 4mx + 2 = 0$ and $2x^2 + 3x - 2 = 0$.

Let α be the common root of given equations.

$$\therefore \quad 3\alpha^2 + 4m\alpha + 2 = 0 \quad \text{...(1)}$$

and

$$2\alpha^2 + 3\alpha - 2 = 0 \quad \text{...(2)}$$

Solving (1) and (2), we get $\dfrac{\alpha^2}{-8m-6} = \dfrac{\alpha}{4+6} = \dfrac{1}{9-8m}$

$$\Rightarrow \quad \frac{\alpha^2}{\alpha} = -\frac{8m+6}{10} \text{ and } \frac{\alpha}{1} = \frac{10}{9-8m} \Rightarrow \alpha = -\frac{4m+3}{5} \text{ and } \alpha = \frac{10}{9-8m}$$

$$\therefore \quad -\frac{4m+3}{5} = \frac{10}{9-8m} \text{ or } (4m+3)(9-8m) + 50 = 0$$

$$\Rightarrow \quad 32m^2 - 12m - 77 = 0 \Rightarrow m = \mathbf{-\frac{11}{8}, \frac{7}{4}}.$$

Example 4. *Solve the equation :* $\dfrac{x-1}{x-2} + \dfrac{x-3}{x-4} = \dfrac{x-5}{x-6} + \dfrac{x-7}{x-8}$.

Sol. We have $\dfrac{x-1}{x-2} + \dfrac{x-3}{x-4} = \dfrac{x-5}{x-6} + \dfrac{x-7}{x-8}$.

$$\Rightarrow \quad \frac{(x-2)+1}{x-2} + \frac{(x-4)+1}{x-4} = \frac{(x-6)+1}{x-6} + \frac{(x-8)+1}{x-8} \quad \textbf{(Note this step)}$$

$$\Rightarrow \quad \left(1 + \frac{1}{x-2}\right) + \left(1 + \frac{1}{x-4}\right) = \left(1 + \frac{1}{x-6}\right) + \left(1 + \frac{1}{x-8}\right)$$

$$\Rightarrow \quad \frac{1}{x-2} + \frac{1}{x-4} = \frac{1}{x-6} + \frac{1}{x-8} \Rightarrow \frac{1}{x-2} - \frac{1}{x-6} = \frac{1}{x-8} - \frac{1}{x-4}$$

(Note this step)

$$\Rightarrow \quad \frac{x-6-x+2}{(x-2)(x-6)} = \frac{x-4-x+8}{(x-8)(x-4)} \Rightarrow \frac{-4}{x^2-6x-2x+12} = \frac{4}{x^2-4x-8x+32}$$

$$\Rightarrow \quad \frac{-1}{x^2-8x+12} = \frac{1}{x^2-12x+32}$$

$$\Rightarrow \quad -x^2 + 12x - 32 = x^2 - 8x + 12 \Rightarrow x^2 - 10x + 22 = 0$$

$$\therefore \quad x = \frac{-(-10) \pm \sqrt{(-10)^2 - 4(1)(22)}}{2(1)} = \frac{10 \pm \sqrt{100-88}}{2} = \frac{10 \pm 2\sqrt{3}}{2} = \mathbf{5 \pm \sqrt{3}}.$$

WORKING RULES FOR SOLVING PROBLEMS

Step I. *Simplify the given equation and express it in the form* $ax^2 + bx + c = 0$.

Step II. *Identify the values of a, b and c.*

Step III. *Use the formula :* $x = \dfrac{-b \pm \sqrt{b^2 - 4ac}}{2a}$ *and simplify it.*

Step IV. *The values of x are the roots of the given equation.*

EXERCISE 1.2

Solve the following equations by the formula method :

1. $x^2 - 9x + 20 = 0$ **2.** $x^2 + 5x + 4 = 0$ (*P.T.U. B. Pharma, Dec. 2006*)

3. $x^2 + x + 1 = 0$ **4.** $x^2 + 2x + 2 = 0$.

5. $3x^2 - 7x + 5 = 0$ **6.** $9x^2 + 10x + 3 = 0$

7. $21x^2 - 29x + 11 = 0$ **8.** $x^2 + 4ix - 4 = 0$

9. $x^2 - ix + 6 = 0$.

10. $\dfrac{x}{x+1} + \dfrac{x+1}{x+2} + \dfrac{x+2}{x+3} = 3$ **11.** $\dfrac{x-a}{x-b} + \dfrac{x-b}{x-a} = \dfrac{a}{b} + \dfrac{b}{a}$

12. $\dfrac{1}{x-a} + \dfrac{1}{x-b} = \dfrac{1}{a} + \dfrac{1}{b}$.

13. (*i*) Find the value of : $\sqrt{20 + \sqrt{20 + \sqrt{20 + \ldots\ldots \infty}}}$

(*ii*) $\sqrt{6 + \sqrt{6 + \sqrt{6 + \sqrt{6} \ldots\ldots \infty}}}$ (*P.T.U. B.Pharma, Dec. 2004, May 2002*)

14. Find the value of : $\sqrt{8 - \sqrt{8 - \sqrt{8 - \ldots\ldots}}}$

15. Find the value of : $2 + \cfrac{1}{2 + \cfrac{1}{2 + \ldots\ldots \infty}}$

16. Find the value of : $2 - \cfrac{1}{2 - \cfrac{1}{2 - \ldots\ldots \infty}}$.

17. Find the condition that the quadratic equations $x^2 + ax + b = 0$ and $x^2 + bx + a = 0$ may have a common root.

Answers

1. 4, 5 **2.** $-4, -1$ **3.** $-\dfrac{1}{2} \pm \dfrac{\sqrt{3}}{2} i$ **4.** $-1 \pm i$

5. $\frac{7}{6} \pm \frac{\sqrt{11}}{6}i$	**6.** $-\frac{5}{9} \pm \frac{\sqrt{2}}{9}i$	**7.** $\frac{29}{42} \pm \frac{\sqrt{83}}{42}i$	**8.** $-2i, -2i$
9. $-2i, 3i$	**10.** $\frac{-6 \pm \sqrt{3}}{3}$	**11.** $0, a+b$	**12.** $a+b, \frac{2ab}{a+b}$
13. (*i*) 5	(*ii*) 3	**14.** $\frac{\sqrt{33}-1}{2}$	**15.** $1+\sqrt{2}$
16. 1	**17.** $a+b+1=0.$		

Hints

10. Given equation is $\left(\frac{x}{x+1} - 1\right) + \left(\frac{x+1}{x+2} - 1\right) + \left(\frac{x+2}{x+3} - 1\right) = 0.$

13. (*i*) Let $x = \sqrt{20 + \sqrt{20 + \sqrt{20 + \ldots\ldots \infty}}}$. $\therefore$ $x = \sqrt{20 + x}$ or $x^2 - x - 20 = 0.$

1.6. EQUATIONS REDUCIBLE TO QUADRATIC FORM

In this section, we shall solve equations which are not quadratic, but could be reduced to quadratic form by certain substitutions.

Type I. Equation of the form $aX^{2n} + bX^n + c = 0$, where X is some function of x.

WORKING RULES FOR SOLVING $aX^{2n} + bX^n + c = 0$, WHERE X IS SOME FUNCTION OF X

Step I. *Put $X^n = y$ and get the quadratic equation $ay^2 + by + c = 0$.*

Step II. *Solve this equation and get two values of y.*

Step III. *Find the values of x by putting $y = X^n$.*

ILLUSTRATIVE EXAMPLES

Example 1. *Solve the equations :*

$$x^{2/3} + x^{1/3} - 2 = 0.$$

Sol. We have $x^{2/3} + x^{1/3} - 2 = 0.$

$\Rightarrow$ $(x^{1/3})^2 + x^{1/3} - 2 = 0$ ...(1)

Let $y = x^{1/3}$ $\therefore$ (1) $\Rightarrow$ $y^2 + y - 2 = 0$

$\Rightarrow$ $y = \frac{-1 \pm \sqrt{1+8}}{2} = \frac{-1 \pm 3}{2} = 1, -2$

$\therefore$ Either $y = 1$	or $y = -2$
$\therefore$ $x^{1/3} = 1$	$\therefore$ $x^{1/3} = -2$
$\Rightarrow$ $x = (1)^3 = 1$	$\Rightarrow$ $x = (-2)^3 = -8$

$\therefore$ The roots are 1, – 8.

Type II. Equation of the form $a(p^x)^2 + b(p^x) + c = 0$.

WORKING RULES FOR SOLVING $a(p^x)^2 + b(p^x) + c = 0$

Step I. *Put $p^x = y$ and get the quadratic equation $ay^2 + by + c = 0$.*
Step II. *Solve this equation and get two values of y.*
Step III. *Find the values of x by putting $y = p^x$.*

Example 2. *Solve the equations :*

(*i*) $5^{2x} - 5^{x+3} + 125 = 5^x$ (*ii*) $2^{2x+2} - 6^x - 2(3^{2x+2}) = 0$.

Sol. (*i*) We have $5^{2x} - 5^{x+3} + 125 = 5^x$.

$\Rightarrow \quad (5^x)^2 - 5^x . 5^3 + 125 - 5^x = 0 \quad \Rightarrow \quad (5^x)^2 - 5^x(5^3 + 1) + 125 = 0$

$\Rightarrow \quad (5^x)^2 - 126.5^x + 125 = 0$ ...(1)

Let $y = 5^x$ $\therefore$ (1) $\Rightarrow$ $y^2 - 126y + 125 = 0$

$$\therefore \quad y = \frac{126 \pm \sqrt{(-126)^2 - 4(1)(125)}}{2(1)} = 1, 125$$

$\therefore$ Either $y = 1$	or $y = 125$
$\Rightarrow \quad 5^x = 1 \Rightarrow 5^x = 5^0 \Rightarrow x = 0$	$\Rightarrow \quad 5^x = 125 \Rightarrow 5^x = 5^3 \Rightarrow x = 3$

$\therefore$ The roots are **0, 3.**

(*ii*) We have $2^{2x+2} - 6^x - 2(3^{2x+2}) = 0$.

$\Rightarrow \quad 2^{2x} . 2^2 - 2^x . 3^x - 2 . 3^{2x} . 3^2 = 0 \quad \Rightarrow \quad 4 . 2^{2x} - 2^x . 3^x - 18 . 3^{2x} = 0$

Dividing by 3^{2x}, we get $\frac{4 . 2^{2x}}{3^{2x}} - \frac{2^x . 3^x}{3^{2x}} - \frac{18 . 3^{2x}}{3^{2x}} = 0$

$$\Rightarrow \quad 4\left(\frac{2}{3}\right)^{2x} - \left(\frac{2}{3}\right)^x - 18 = 0 \qquad ...(1)$$

Let $y = \left(\frac{2}{3}\right)^x$ $\therefore$ (1) $\Rightarrow$ $4y^2 - y - 18 = 0$

$$\therefore \quad y = \frac{1 \pm \sqrt{1 + 4(4)(18)}}{8} = \frac{1 \pm 17}{8} = -2, \frac{9}{4}$$

$\therefore$ Either $y = -2$	or $y = 9/4$.
$\Rightarrow \quad \left(\frac{2}{3}\right)^x = -2$	$\Rightarrow \quad \left(\frac{2}{3}\right)^x = \frac{9}{4} \Rightarrow \left(\frac{2}{3}\right)^x = \left(\frac{2}{3}\right)^{-2}$
This is impossible, because $(2/3)^x$ is non-negative for each value of x.	$\Rightarrow \quad x = -2$

$\therefore$ The root is **– 2.**

Type III. Equation of the form $aX + \frac{b}{X} + c = 0$, where X is some function of x.

WORKING RULES FOR SOLVING $aX + \frac{b}{X} + c = 0$, WHERE X IS SOME FUNCTION OF x

Step I. *Put $X = y$ and get the equation $ay + \frac{b}{y} + c = 0$.*

Step II. *Multiply both sides by y and get the quadratic equation $ay^2 + cy + b = 0$.*

Step III. *Solve this equation and get two values of y.*

Step IV. *Find the values of x by putting $y = X$.*

Example 3. *Solve the equations :*

(*i*) $8\sqrt{\frac{x}{x+3}} - \sqrt{\frac{x+3}{x}} = 2$ (K.U.K. 2006 ; G.J.U.B.Pharma 2002)

(*ii*) $\sqrt{\frac{x}{x-1}} + \sqrt{\frac{x-1}{x}} = 2\frac{1}{6}$ (P.T.U. B.Pharma, June 2003)

(*iii*) $(15 + 4\sqrt{14})^{x^2-2x} + (15 - 4\sqrt{14})^{x^2-2x} = 30$.

Sol. (*i*) We have $8\sqrt{\frac{x}{x+3}} - \sqrt{\frac{x+3}{x}} = 2$. ...(1)

Let $\mathbf{y = \sqrt{\frac{x}{x+3}}}$ $\therefore$ (1) $\Rightarrow$ $8y - \frac{1}{y} = 2 \Rightarrow 8y^2 - 1 = 2y \Rightarrow 8y^2 - 2y - 1 = 0$

$\therefore \quad y = \frac{2 \pm \sqrt{4+32}}{16} = \frac{2 \pm 6}{16} = \frac{1}{2}, -\frac{1}{4}$

$\therefore$ Either $y = 1/2$	or $y = -1/4$.
$\Rightarrow \sqrt{\frac{x}{x+3}} = \frac{1}{2}$	$\Rightarrow \sqrt{\frac{x}{x+3}} = -\frac{1}{4}$.
$\Rightarrow \frac{x}{x+3} = \frac{1}{4} \Rightarrow 4x = x + 3 \Rightarrow x = \frac{3}{3} = 1.$	This is impossible, because L.H.S. is non-negative.

$\therefore$ The root is **1**.

(*ii*) Let $\sqrt{\frac{x}{x-1}} = t$, then $\sqrt{\frac{x-1}{x}} = \frac{1}{\sqrt{\frac{x}{x-1}}} = \frac{1}{t}$

$\therefore$ The given equation reduces to

$$t + \frac{1}{t} = \frac{13}{6}$$

$$\Rightarrow \quad \frac{t^2+1}{t} = \frac{13}{6}$$

$\Rightarrow \quad 6(t^2+1) = 13t$

$\Rightarrow \quad 6t^2 - 13t + 6 = 0$

$$\Rightarrow \quad t = \frac{-(-13) \pm \sqrt{(-13)^2 - 4.6.6}}{2(6)} = \frac{13 \pm \sqrt{169-144}}{12}$$

$$= \frac{13 \pm \sqrt{25}}{12} = \frac{13 \pm 5}{12} = \frac{18}{12}, \frac{8}{12} = \frac{3}{2}, \frac{2}{3}$$

When $t = \frac{3}{2}$	$t = \frac{2}{3}$
$\sqrt{\frac{x}{x-1}} = \frac{3}{2}$ Squaring	$\sqrt{\frac{x}{x-1}} = \frac{2}{3}$ Squaring
$\frac{x}{x-1} = \frac{9}{4}$	$\frac{x}{x-1} = \frac{4}{9}$
$\Rightarrow \quad 4x = 9(x-1)$	$\Rightarrow \quad 9x = 4(x-1)$
$\Rightarrow \quad 4x - 9x = -9$	$\Rightarrow \quad 9x - 4x = -4$
$\Rightarrow \quad -5x = -9$	$\Rightarrow \quad 5x = -4$
$\Rightarrow \quad x = \frac{9}{5}$	$\Rightarrow \quad x = -4/5.$

Hence the required solution is given by $x = -\frac{4}{5}$

| As $x = \frac{9}{5}$ does not satisfy the given equation

(*iii*) We have $\quad (15+4\sqrt{14})^{x^2-2x} + (15-4\sqrt{14})^{x^2-2x} = 30 \quad$ ...(1)

Now $\quad (15+4\sqrt{14})(15-4\sqrt{14}) = 225 - 224 = 1. \quad \therefore \quad 15 - 4\sqrt{14} = \frac{1}{15+4\sqrt{14}}$

$$\therefore \quad (1) \Rightarrow \quad (15+4\sqrt{14})^{x^2-2x} + \frac{1}{(15+4\sqrt{14})^{x^2-2x}} = 30 \quad ...(2)$$

Let $\mathbf{y = (15+4\sqrt{14})^{x^2-2x}}$ $\quad \therefore \quad (2) \Rightarrow y + \frac{1}{y} = 30 \Rightarrow y^2 + 1 = 30y \Rightarrow y^2 - 30y + 1 = 0$

$$\therefore \quad y = \frac{30 \pm \sqrt{900-4}}{2} = \frac{30 \pm \sqrt{896}}{2} = 15 \pm 4\sqrt{14}$$

$\therefore$ Either $\quad y = 15 + 4\sqrt{14}$	or $\quad y = 15 - 4\sqrt{14}$
$\Rightarrow \quad (15+4\sqrt{14})^{x^2-2x} = 15 + 4\sqrt{14}$	$(15+4\sqrt{14})^{x^2-2x} = 15 - 4\sqrt{14}$
$\Rightarrow \quad x^2 - 2x = 1 \quad \Rightarrow \quad x^2 - 2x - 1 = 0$	$\Rightarrow \quad (15+4\sqrt{14})^{x^2-2x} = (15+4\sqrt{14})^{-1}$
	$\Rightarrow \quad x^2 - 2x = -1 \quad \Rightarrow \quad x^2 - 2x + 1 = 0$
$\therefore \quad x = \frac{2 \pm \sqrt{4+4}}{2} = 1 \pm \sqrt{2}$	$\therefore \quad (x-1)^2 = 0$ or $x = 1, 1.$

$\therefore$ The roots are $\mathbf{1 \pm \sqrt{2}, 1, 1.}$

Type IV. Equation of the form $\lambda(x + a)(x + b)(x + c)(x + d) = k$.

WORKING RULES FOR SOLVING $\lambda(x + a)(x + b)(x + c)(x + d) = k$

Step I. *Express the given equation in the form $\lambda(x + a)(x + b)(x + c)(x + d) = k$, if already not so. Note that the coefficient of x in each factor should be '1'.*

Step II. *Group a, b, c, d into two pairs having equal sums. Let $a + b = c + d$.*

Step III. *Write the equation in the form : $\lambda[(x + a)(x + b)][(x + c)(x + d)] = k$. Multiply the factors and get $\lambda[x^2 + (a + b)x + ab][x^2 + (c + d)x + cd] = k$ i.e., $\lambda[x^2 + (a + b)x + ab][x^2 + (a + b)x + cd] = k$, because $a + b = c + d$.*

Step IV. *Put $x^2 + (a + b)x = y$ and get the equation $\lambda(y + ab)(y + cd) = k$. Simplify and get the quadratic equation $\lambda y^2 + \lambda(ab + cd)y + \lambda abcd - k = 0$.*

Step V. *Solve the equation and get two values of y.*

Step VI. *Find the values of x by putting $y = x^2 + (a + b)x$.*

Example 4. *Solve the equations :*

(*i*) $x(x + 1)^2 (x + 2) = 72$ (*ii*) $(2x + 7)(x^2 - 9)(2x - 5) = 91.$

Sol. (*i*) We have $x(x + 1)^2 (x + 2) = 72.$

$\Rightarrow \quad (x + 0)(x + 1)(x + 1)(x + 2) = 72 \quad ...(1)$

Now $0 + 2 = 1 + 1$ (each = 2)

$\therefore$ (1) implies $[(x + 0)(x + 2)][(x + 1)(x + 1)] = 72$

$\Rightarrow \quad (x^2 + 2x)(x^2 + 2x + 1) = 72 \quad ...(2)$

Let $y = x^2 + 2x$ $\therefore$ (2) $\Rightarrow y(y + 1) = 72 \Rightarrow y^2 + y - 72 = 0$

$$\Rightarrow \quad y = \frac{-1 \pm \sqrt{1 + 288}}{2} = \frac{-1 \pm 17}{2} = 8, -9$$

$\therefore$ Either $y = 8$

$\Rightarrow x^2 + 2x = 8 \Rightarrow x^2 + 2x - 8 = 0$

$$\Rightarrow x = \frac{-2 \pm \sqrt{4 + 32}}{2} = \frac{-2 \pm 6}{2} = 2, -4$$

or $y = -9$

$\Rightarrow x^2 + 2x = -9 \Rightarrow x^2 + 2x + 9 = 0$

$$\Rightarrow x = \frac{-2 \pm \sqrt{4 - 36}}{2} = \frac{-2 \pm 4\sqrt{2}i}{2} = -1 \pm 2\sqrt{2}i.$$

$\therefore$ The roots are **$2, -4, -1 \pm 2\sqrt{2}i$.**

(*ii*) We have $(2x + 7)(x + 3)(x - 3)(2x - 5) = 91. \quad ...(1)$

$$\Rightarrow \quad 2\left(x + \frac{7}{2}\right)(x + 3)(x - 3)2\left(x - \frac{5}{2}\right) = 91$$

Now $\frac{7}{2} + (-3) = 3 + \left(-\frac{5}{2}\right)$ $\left(\text{each} = \frac{1}{2}\right)$

$\therefore$ We take I, III factors and II, IV factors together.

$\therefore$ (1) $\Rightarrow [(2x + 7)(x - 3)][(x + 3)(2x - 5)] = 91$

$\Rightarrow \quad (2x^2 + x - 21)(2x^2 + x - 15) = 91 \quad ...(2)$

Let $y = 2x^2 + x$ $\therefore$ (2) $\Rightarrow (y - 21)(y - 15) = 91$

$\Rightarrow \quad y^2 - 15y - 21y + 315 - 91 = 0 \qquad \Rightarrow \quad y^2 - 36y + 224 = 0$

$\Rightarrow \quad (y - 8)(y - 28) = 0 \qquad \Rightarrow \quad y = 8, 28$

$\therefore$ Either $y = 8$	or $y = 28$
$\Rightarrow 2x^2 + x = 8 \Rightarrow 2x^2 + x - 8 = 0$	$\Rightarrow 2x^2 + x = 28 \Rightarrow 2x^2 + x - 28 = 0$
$\Rightarrow x = \dfrac{-1 \pm \sqrt{1+64}}{4} = \dfrac{-1 \pm \sqrt{65}}{4}$	$\Rightarrow x = \dfrac{-1 \pm \sqrt{1+224}}{4} = \dfrac{-1 \pm 15}{4} = \dfrac{7}{2}, -4$

$\therefore$ The roots are $\mathbf{\dfrac{-1 \pm \sqrt{65}}{4}, \dfrac{7}{2}, -4}$.

Type V. Equation of the form $\sqrt{ax+b} \pm \sqrt{cx+d} = k$ or $\sqrt{ex+f}$.

WORKING RULES FOR SOLVING $\sqrt{ax+b} \pm \sqrt{cx+d} = k$ or $\sqrt{ex+f}$

Step I. *Square both sides of the equation.*
Step II. *Transpose the terms, so that the expression under radical sign is on one side.*
Step III. *Square both sides again and solve it and get the values of x.*
Step IV. *Test all values of x so obtained and reject those values which do not satisfy the given equation.*

Example 5. *Solve the equations :*

(*i*) $\sqrt{1-5x} + \sqrt{1-3x} = 2$ (*ii*) $\sqrt{x+5} + \sqrt{x+21} = \sqrt{6x+40}$.

Sol. (*i*) We have $\sqrt{1-5x} + \sqrt{1-3x} = 2$. ...(1)

Squaring, we get $(1 - 5x) + (1 - 3x) + 2\sqrt{1-5x}\,\sqrt{1-3x} = 4$

$\Rightarrow \quad 2\sqrt{(1-5x)(1-3x)} = 2 + 8x \quad \Rightarrow \quad \sqrt{(1-5x)(1-3x)} = 1 + 4x$

Squaring again, we get $1 - 5x - 3x + 15x^2 = 1 + 16x^2 + 8x$

$\Rightarrow \quad x^2 + 16x = 0 \quad \Rightarrow \quad x(x + 16) = 0$

$\therefore \quad x = 0, -16$

$x = 0$ is a root of (1) if $\sqrt{1-5(0)} + \sqrt{1-3(0)} = 2$

or if $\sqrt{1} + \sqrt{1} = 2$ if $2 = 2$, which is true. $\therefore$ $x = 0$ is a root.

$x = -16$ is a root of (1) if $\sqrt{1-5(-16)} + \sqrt{1-3(-16)} = 2$

or if $\sqrt{81} + \sqrt{49} = 2$ if $9 + 7 = 2$ if $16 = 2$, which is not true.

$\therefore$ $x = -16$ is an extraneous root. $\therefore$ The only root is **0**.

(*ii*) We have $\sqrt{x+5} + \sqrt{x+21} = \sqrt{6x+40}$. ...(1)

Squaring, we get $(x + 5) + (x + 21) + 2\sqrt{x+5}\,\sqrt{x+21} = 6x + 40$

$\Rightarrow \quad 2\sqrt{(x+5)(x+21)} = 4x + 14 \quad \Rightarrow \quad \sqrt{(x+5)(x+21)} = 2x + 7$

Squaring again, we get $x^2 + 21x + 5x + 105 = 4x^2 + 49 + 28x$

$$\Rightarrow \quad 3x^2 + 2x - 56 = 0$$

$$\therefore \quad x = \frac{-2 \pm \sqrt{4+672}}{6} = \frac{-2 \pm 26}{6} = 4, -\frac{14}{3}.$$

$x = 4$ is a root of (1) if $\sqrt{4+5} + \sqrt{4+21} = \sqrt{6(4)+40}$.

or if $3 + 5 = 8$, which is true. $\therefore$ $x = 4$ is a root of (1).

$x = -\dfrac{14}{3}$ is a root of (1) if $\sqrt{-\dfrac{14}{3}+5} + \sqrt{-\dfrac{14}{3}+21} = \sqrt{6\left(-\dfrac{14}{3}\right)+40}$

or if $\sqrt{\dfrac{1}{3}} + \sqrt{\dfrac{49}{3}} = \sqrt{12}$ if $\dfrac{8}{\sqrt{3}} = 2\sqrt{3}$ if $8 = 6$, which is not true.

$\therefore$ $x = -14/3$ is an extraneous root. $\therefore$ The only root is **4**.

Example 6. *Solve the equation :* $\sqrt{x^2 - 5x + 6} - \sqrt{x^2 - 9} = \sqrt{2x^2 - 11x + 15}$.

Sol. We have $\sqrt{x^2 - 5x + 6} - \sqrt{x^2 - 9} = \sqrt{2x^2 - 11x + 15}$. ...(1)

Now $x^2 - 5x + 6 = x^2 - 3x - 2x + 6 = x(x-3) - 2(x-3) = (x-3)(x-2),$

$x^2 - 9 = (x-3)(x+3)$

and $2x^2 - 11x + 15 = 2x^2 - 6x - 5x + 15 = 2x(x-3) - 5(x-3) = (x-3)(2x-5).$

$\therefore$ (1) $\Rightarrow \sqrt{(x-3)(x-2)} - \sqrt{(x-3)(x+3)} = \sqrt{(x-3)(2x-5)}$

$$\Rightarrow \quad \sqrt{x-3}\,[\sqrt{x-2} - \sqrt{x+3} - \sqrt{2x-5}] = 0$$

$\therefore$ Either $\sqrt{x-3} = 0$ ...(2)

or $\sqrt{x-2} - \sqrt{x+3} - \sqrt{2x-5} = 0$ ...(3)

(2) $\Rightarrow$ $x - 3 = 0 \Rightarrow x = 3$

(3) $\Rightarrow$ $\sqrt{x-2} - \sqrt{x+3} = \sqrt{2x-5}$

Squaring, we get $(x-2) + (x+3) - 2\sqrt{x-2}\sqrt{x+3} = 2x - 5$

$$\Rightarrow \quad -2\sqrt{(x-2)(x+3)} = -6 \quad \Rightarrow \quad \sqrt{(x-2)(x+3)} = 3$$

Squaring again, we get $x^2 + 3x - 2x - 6 = 9$

$$\Rightarrow \quad x^2 + x - 15 = 0 \quad \Rightarrow \quad x = \frac{-1 \pm \sqrt{61}}{2}$$

It may be checked that $x = \dfrac{-1 \pm \sqrt{61}}{2}$ does not satisfy (3) and hence the given equation.

$\therefore$ The only root is **3**.

Type VI. Equation of the form $p(ax^2 + bx + c) + q\sqrt{ax^2 + bx + c} = r$.

WORKING RULES FOR SOLVING $p(ax^2 + bx + c) + q\sqrt{ax^2 + bx + c} = r$.

Step I. *Put $\sqrt{ax^2 + bx + c} = y$ and get the quadratic equation $py^2 + qy - r = 0$.*

Step II. *Solve this equation and get two values of y. If any or both values of y are negative, then we reject those values, because $y = \sqrt{ax^2 + bx + c}$ is always non-negative.*

Step III. *Find the values of x by using $y = \sqrt{ax^2 + bx + c}$.*

Example 7. *Solve the equations :*

(*i*) $8 + 9\sqrt{(3x - 1)(x - 2)} = 3x^2 - 7x$ (*ii*) $\sqrt{x^2 - 2x + 9} - \dfrac{x^2}{2} = 3 - x$.

Sol. (*i*) We have $8 + 9\sqrt{(3x - 1)(x - 2)} = 3x^2 - 7x$.

$\Rightarrow \quad 9\sqrt{3x^2 - 6x - x + 2} = 3x^2 - 7x - 8$

$\Rightarrow \quad 9\sqrt{3x^2 - 7x + 2} = (3x^2 - 7x + 2) - 10$ ...(1)

Let $\mathbf{y = \sqrt{3x^2 - 7x + 2}}$ $\therefore$ (1) $\Rightarrow$ $9y = y^2 - 10$ $\Rightarrow$ $y^2 - 9y - 10 = 0$

$\Rightarrow \quad y = \dfrac{9 \pm \sqrt{81 + 40}}{2} = \dfrac{9 \pm 11}{2} = -1, 10$

$\therefore$ Either $y = -1$	or $y = 10$.
$\Rightarrow \sqrt{3x^2 - 7x + 2} = -1$	$\Rightarrow \sqrt{3x^2 - 7x + 2} = 10$
This is impossible because L.H.S. is	$\Rightarrow 3x^2 - 7x + 2 = 100 \Rightarrow 3x^2 - 7x - 98 = 0$
non-negative.	$\therefore x = \dfrac{7 \pm \sqrt{49 + 1176}}{6} = 7, -\dfrac{14}{3}$.

$\therefore$ The roots are **7, – 14/3.**

(*ii*) We have $\sqrt{x^2 - 2x + 9} - \dfrac{x^2}{2} = 3 - x$. $\therefore$ $2\sqrt{x^2 - 2x + 9} - x^2 = 6 - 2x$

$\Rightarrow \quad 2\sqrt{x^2 - 2x + 9} = (x^2 - 2x + 9) - 3$ ...(1)

Let $\mathbf{y = \sqrt{x^2 - 2x + 9}}$ $\therefore$ (1) $\Rightarrow$ $2y = y^2 - 3$ $\Rightarrow$ $y^2 - 2y - 3 = 0$

$\Rightarrow \quad y = \dfrac{2 \pm \sqrt{4 + 12}}{2} = \dfrac{2 \pm 4}{2} = -1, 3.$

$\therefore$ Either $y = -1$	or $y = 3$
$\Rightarrow \sqrt{x^2 - 2x + 9} = -1$	$\Rightarrow \sqrt{x^2 - 2x + 9} = 3 \Rightarrow x^2 - 2x + 9 = 9$
This is impossible, because L.H.S. is	$\Rightarrow x^2 - 2x = 0$
non-negative.	$\Rightarrow x = 0, 2.$

$\therefore$ The roots are **0, 2.**

Type VII. Equation of the form $\sqrt{ax^2 + bx + c} \pm \sqrt{dx^2 + ex + f} = k$ or $gx + h$.

WORKING RULES FOR SOLVING $\sqrt{ax^2 + bx + c} \pm \sqrt{dx^2 + ex + f} = k$ (or $gx + h$).

Step I. *Let the given equation be* $\sqrt{ax^2 + bx + c} - \sqrt{dx^2 + ex + f} = k$.

Step II. *Put* $\sqrt{ax^2 + bx + c} = A$ *and* $\sqrt{dx^2 + ex + f} = B$ *and get the equation*

$$A - B = k \quad ...(1)$$

Step III. *Simplify* $A^2 - B^2$ *and let it be* $p(x)$.

$$\therefore \quad A^2 - B^2 = p(x) \quad ...(2)$$

where p(x) is either a quadratic polynomial or a linear polynomial or a constant.

Step IV. *Divide (2) by (1) and get* $\dfrac{A^2 - B^2}{A - B} = \dfrac{p(x)}{k}$ *i.e.,* $A + B = \dfrac{p(x)}{k}$ ...(3)

Step V. *Solve (1) and (3) and get the value of A (or of B).*

Step VI. *Find the value of x by putting* $A = \sqrt{ax^2 + bx + c}$. *The value of B will also give the same values of x.*

Other forms of the given equation are also solved by following the same method.

Example 8. *Solve the equations :*

$$\sqrt{5x^2 - 6x + 8} - \sqrt{5x^2 - 6x - 7} = 1.$$

Sol. We have $\sqrt{5x^2 - 6x + 8} - \sqrt{5x^2 - 6x - 7} = 1$ ...(1)

Let $A = \sqrt{5x^2 - 6x + 8}$ and $B = \sqrt{5x^2 - 6x - 7}$

$\therefore$ (1) $\Rightarrow$ $A - B = 1$ ...(2)

Now, $A^2 - B^2 = (5x^2 - 6x + 8) - (5x^2 - 6x - 7) = 15$

$\therefore$ $A^2 - B^2 = 15$ ...(3)

Dividing (3) by (2), we get $\dfrac{A^2 - B^2}{A - B} = \dfrac{15}{1}$.

$\Rightarrow$ $A + B = 15$ ...(4)

(1) + (4) $\Rightarrow$ $2A = 16$ *i.e.,* $A = 8$ $\therefore$ $\sqrt{5x^2 - 6x + 8} = 8$

or $5x^2 - 6x + 8 = 64$ *i.e.,* $5x^2 - 6x - 56 = 0$

$$\therefore \quad x = \frac{6 \pm \sqrt{36 + 1120}}{10} = \frac{6 \pm 34}{10} = 4, -\frac{14}{5}.$$

$\therefore$ The roots are **4, – 14/5.**

EXERCISE 1.3

Solve the following equations :

1. (*i*) $x^4 - 8x^2 - 9 = 0$ (*ii*) $(x^2 - 5x)^2 - 30(x^2 - 5x) - 216 = 0$

(*iii*) $\left(x - \dfrac{x}{x+1}\right)^2 + 2x\left(\dfrac{x}{x+1}\right) = 3$ (*iv*) $28x^{-2} + 17x^{-1} - 3 = 0$

(*v*) $x + \sqrt{x} = \dfrac{6}{25}$ (*P.T.U. B.Pharma, May 2006*)

2. (*i*) $2^{x+1} + 4^x = 8$ (*P.T.U. B.Pharma, June 2003*)

(*ii*) $7^{1+x} + 7^{1-x} = 50$ (*iii*) $5^{x+1} + 5^{2-x} = 5^3 + 1$

(*iv*) $2^{x^2} \div 4^x = 8 \div 1$ (*P.T.U. B.Pharma, Dec. 2005*)

(*v*) $5^{1+x} + 5^{1-x} = 26$ (*P.T.U. B.Pharma, May 2005*)

3. (*i*) $\dfrac{2x+3}{x+1} + 6\left(\dfrac{x+1}{2x+3}\right) = 7$ (*ii*) $\dfrac{7x^2+1}{x^2-1} - 4\left(\dfrac{x^2-1}{7x^2+1}\right) + 3 = 0$

4. (*i*) $(x-1)(x+5)(x+8)(x+2) = 880$ (*ii*) $(x-1)(x+1)(2x+3)(2x-1) = 3$

(*iii*) $(x+1)(x+2)(x+3)(x+4) = 120$ (*iv*) $(x^2 + x - 6)(x^2 - 3x - 4) = 24$

5. (*i*) $\sqrt{3x+1} - \sqrt{x-1} = 2$ (*ii*) $\sqrt{x+2} + \sqrt{x+7} = \sqrt{6x+13}$

(*P.T.U. B.Pharma May. 2006*)

(*iii*) $\sqrt{2x+1} - \sqrt{x+2} = \sqrt{5x+3}$ (*P.T.U. B.Pharma, Dec. 2001*)

(*iv*) $\sqrt{x^2-16} - (x-4) = \sqrt{x^2-5x+4}$ (*v*) $\sqrt{x^2+4x-21} + \sqrt{x^2-x-6} = \sqrt{6x^2-5x-39}$

6. (*i*) $3x^2 + 15x - 2 = 2\sqrt{x^2+5x+1}$ (*ii*) $12 + 9\sqrt{(x-1)(3x+2)} = 3x^2 - x$

(*iii*) $x(x+3) + 3\sqrt{2x^2+6x+5} = 25$ (*iv*) $x^2 - 4x - 12\sqrt{x^2-4x+19} + 51 = 0$

(*P.T.U. B.Pharma, Dec. 2002*)

7. (*i*) $\sqrt{x^2+3x+32} + \sqrt{x^2+3x+5} = 9$ (*ii*) $\sqrt{x^2-3x+36} - \sqrt{x^2-3x+9} = 3.$

Answers

1. (*i*) $\pm 3, \pm i$ (*ii*) $-4, 2, 3, 9$ (*iii*) $\dfrac{1 \pm \sqrt{5}}{2}, \dfrac{-3 \pm \sqrt{3}i}{2}$

(*iv*) $7, -\dfrac{4}{3}$ (*v*) $\dfrac{1}{25}$

2. (*i*) 1 (*ii*) ± 1 (*iii*) $-1, 2$

(*iv*) $-1, 3$ (*v*) $-1, 1$

3. (*i*) $-3/4, -2$ (*ii*) $\pm \sqrt{33}/11, \pm \sqrt{3}i/3$

4. (*i*) $-10, 3, \dfrac{-7 \pm \sqrt{79}i}{2}$ (*ii*) $0, -\dfrac{1}{2}, \dfrac{-1 \pm \sqrt{33}}{4}$ (*iii*) $-6, 1, \dfrac{-5 \pm \sqrt{39}i}{2}$

(*iv*) $0, 1, \dfrac{1 \pm \sqrt{57}}{2}$ **5.** (*i*) 1, 5 (*ii*) 2 (*iii*) $\dfrac{-5 \pm \sqrt{17}}{2}$

(*iv*) 4, 5 (*v*) 2, 3 **6.** (*i*) – 16/3, 1/3 (*ii*) $-\frac{17}{3}$

(*iii*) – 5, 2 (*iv*) $x = 9, -5, 3, 1$ **7.** (*i*) – 4, 1 (*ii*) 0, 3.

Hint

1. (*iii*) Given equation is $\left(\frac{x^2+x-x}{x+1}\right)^2 + \frac{2x^2}{x+1} = 3$ or $\left(\frac{x^2}{x+1}\right)^2 + 2\left(\frac{x^2}{x+1}\right) = 3.$

1.7. SIMULTANEOUS EQUATIONS IN TWO VARIABLES

The different methods of solving simultaneous equations are illustrated below :

ILLUSTRATIVE EXAMPLES

Example 1. *Solve the equations :*

(*i*) $x^2 + y^2 = 185$, $x + y = 19$

(*ii*) $\sqrt{\frac{x}{y}} + \sqrt{\frac{y}{x}} = \frac{10}{3}$, $x + y = 10$ (G.J.U. B.Pharma 2003)

(*iii*) $x^3 + y^3 = 91$, $x + y = 7$

(*iv*) $x^2 - 3xy + 2y^2 = 6$, $x^2 - xy + y^2 = 21.$

Sol. (*i*) We have $x^2 + y^2 = 185$ ...(1) and $x + y = 19$ ...(2)

(2) implies $y = 19 - x.$

Putting this value of y in (1), we get $x^2 + (19 - x)^2 = 185$ *i.e.,* $x^2 + 361 + x^2 - 38x = 185$

$\Rightarrow \quad 2x^2 - 38x + 176 = 0 \quad \Rightarrow \quad x^2 - 19x + 88 = 0$

$\Rightarrow \quad (x - 8)(x - 11) = 0 \quad \Rightarrow \quad x = 8, 11$

∴ Either $x = 8$	or $y = 11.$
∴ $y = 19 - x = 19 - 8 = 11$	∴ $y = 19 - x = 19 - 11 = 8$

∴ The solution is $x = $ **8,** $y = $ **11 ;** $x = $ **11,** $y = $ **8.**

(*ii*) We have $\sqrt{\frac{x}{y}} + \sqrt{\frac{y}{x}} = \frac{10}{3}$ ...(1) and $x + y = 10$ ...(2)

(1) $\Rightarrow \quad \frac{\sqrt{x}}{\sqrt{y}} + \frac{\sqrt{y}}{\sqrt{x}} = \frac{10}{3} \quad \Rightarrow \quad \frac{x+y}{\sqrt{xy}} = \frac{10}{3} \quad \Rightarrow \quad \frac{10}{\sqrt{xy}} = \frac{10}{3}$ [using (2)]

$\Rightarrow \quad \sqrt{xy} = 3 \quad \Rightarrow \quad xy = 9$

$\Rightarrow \quad x(10 - x) = 9$ [using (2)]

$\Rightarrow \quad x^2 - 10x + 9 = 0 \quad \Rightarrow \quad x = 1, 9$

∴ Either $x = 1$	or $x = 9.$
∴ $y = 10 - x = 10 - 1 = 9$	∴ $y = 10 - x = 10 - 9 = 1$

∴ The solution is $x = $ **1,** $y = $ **9 ;** $x = $ **9,** $y = $ **1.**

(*iii*) We have $x^3 + y^3 = 91$ ...(1) and $x + y = 7$ ...(2)

(2) $\Rightarrow \quad y = 7 - x$

$\therefore$ (1) implies $x^3 + (7 - x)^3 = 91 \Rightarrow x^3 + (343 - x^3 - 21x(7 - x)) = 91$

$\Rightarrow 343 - 147x + 21x^2 = 91 \Rightarrow 21x^2 - 147x + 252 = 0$

$\Rightarrow x^2 - 7x + 12 = 0 \Rightarrow x = 3, 4$

$\therefore$ Either $x = 3$	or $x = 4$.
$\therefore$ $y = 7 - x = 7 - 3 = 4$	$\therefore$ $y = 7 - x = 7 - 4 = 3$

$\therefore$ The solution is $\mathbf{x = 3, y = 4\ ;\ x = 4, y = 3.}$

(*iv*) We have $x^2 - 3xy + 2y^2 = 6$ ...(1) and $x^2 - xy + y^2 = 21$ ...(2)

We shall obtain a homogeneous equation by dividing (1) and (2) and cross-multiplying.

Dividing (1) by (2), we get

$$\frac{x^2 - 3xy + 2y^2}{x^2 - xy + y^2} = \frac{6}{21} \quad \text{i.e.,} \quad \frac{x^2 - 3xy + 2y^2}{x^2 - xy + y^2} = \frac{2}{7}$$

$$\Rightarrow 7x^2 - 21xy + 14y^2 = 2x^2 - 2xy + 2y^2 \Rightarrow 5x^2 - 19xy + 12y^2 = 0$$

Dividing by x^2, we get $5 - 19\frac{y}{x} + 12\left(\frac{y}{x}\right)^2 = 0 \Rightarrow 12\left(\frac{y}{x}\right)^2 - 19\left(\frac{y}{x}\right) + 5 = 0$

$$\therefore \frac{y}{x} = \frac{19 \pm \sqrt{361 - 240}}{24} = \frac{19 \pm 11}{24} = \frac{30}{24}, \frac{8}{24} = \frac{5}{4}, \frac{1}{3}$$

Case I. $\mathbf{\frac{y}{x} = \frac{5}{4}}$. $\therefore$ $y = \frac{5}{4}x$

Putting this value in (1), we get $x^2 - 3x\left(\frac{5}{4}x\right) + 2\left(\frac{5}{4}x\right)^2 = 6$

$$\Rightarrow x^2 - \frac{15}{4}x^2 + \frac{25}{8}x^2 = 6 \Rightarrow \frac{3x^2}{8} = 6 \Rightarrow x^2 = 16 \Rightarrow x = \pm 4$$

$x = 4$ implies $y = \frac{5}{4}(4) = 5$ and $x = -4$ implies $y = \frac{5}{4}(-4) = -5$.

Case II. $\mathbf{\frac{y}{x} = \frac{1}{3}}$. $\therefore$ $y = \frac{x}{3}$.

Putting this value in (1), we get $x^2 - 3x\left(\frac{x}{3}\right) + 2\left(\frac{x}{3}\right)^2 = 6$

$$\Rightarrow x^2 - x^2 + \frac{2}{9}x^2 = 6 \Rightarrow \frac{2}{9}x^2 = 6 \Rightarrow x^2 = 27 \Rightarrow x = \pm 3\sqrt{3}$$

$x = 3\sqrt{3}$ implies $y = \frac{1}{3}(3\sqrt{3}) = \sqrt{3}$ and $x = -3\sqrt{3}$ implies $y = \frac{1}{3}(-3\sqrt{3}) = -\sqrt{3}$

$$\therefore \left.\begin{matrix}\mathbf{x = 4}\\ \mathbf{y = 5}\end{matrix}\right], \left.\begin{matrix}\mathbf{x = -4}\\ \mathbf{y = -5}\end{matrix}\right], \left.\begin{matrix}\mathbf{x = 3\sqrt{3}}\\ \mathbf{y = \sqrt{3}}\end{matrix}\right], \left.\begin{matrix}\mathbf{x = -3\sqrt{3}}\\ \mathbf{y = -\sqrt{3}}\end{matrix}\right]$$

EXERCISE 1.4

Solve the following simultaneous equations :

1. (*i*) $x + 2y = 1, x^2 + y^2 = 10$ (*ii*) $x + y = 20, xy = 64$

   (*iii*) $x + \frac{4}{y} = 1, y + \frac{4}{x} = 25$ (*iv*) $\frac{1}{x} + \frac{1}{y} = \frac{5}{6}, 3x + 4y = 18$

2. (*i*) $\frac{1}{x} + \frac{1}{y} = \frac{3}{4}, \frac{1}{x^2} + \frac{1}{y^2} = \frac{5}{16}$ (*ii*) $x^2 + xy = 15, x^2 - y^2 = 5$

   (*iii*) $x + y = 10, \sqrt{\frac{x}{y}} + \sqrt{\frac{y}{x}} = \frac{5}{2}$ (*iv*) $4x - 3y = 1, 12xy + 13x^2 = 25$

   (*G.J.U. B. Pharma 2003*)

3. $x + y = a + b, \frac{a}{x} + \frac{b}{y} = 2$

4. $(x + y)^2 - 2(x + y) = 15, xy = 6.$

Answers

1. (*i*) $x = 3, y = -1 ; x = -13/5, y = 9/5$ (*ii*) $x = 16, y = 4 ; x = 4, y = 16$

   (*iii*) $x = 4/5, y = 20 ; x = 1/5, y = 5$ (*iv*) $x = 2, y = 3 ; x = 18/5, y = 9/5.$

2. (*i*) $x = 2, y = 4 ; x = 4, y = 2$ (*ii*) $x = 3, y = 2 ; x = -3, y = -2$

   (*iii*) $x = 8, y = 2 ; x = 2, y = 8$ (*iv*) $x = 1, y = 1 ; x = -\frac{25}{29}, y = -\frac{43}{29}.$

3. $x = a, y = b ; x = \frac{a+b}{2}, y = \frac{a+b}{2}$

4. $x = 3, y = 2 ; x = 2, y = 3, y = 3 ; x = \frac{-3 + i\sqrt{15}}{2}, y = \frac{-3 - i\sqrt{15}}{2} ; x = \frac{-3 - i\sqrt{15}}{2}, y = \frac{-3 + i\sqrt{15}}{2}$

Hints

1. (*iii*) Find the value of x from the first equation and the value of $4/x$ from the second equation. Multiply these values of x and $4/x$.
2. (*i*) Let $A = 1/x$ and $B = 1/y$. (*ii*) Divide the equations.
4. The first equation being quadratic in $x + y$, implies $x + y = 5, -3$.
   In the first case, consider $x + y = 5, xy = 6$ and in the second case, consider $x + y = -3, xy = 6$.

1.8. PRACTICAL PROBLEMS

In this section, we shall learn the method of solving practical problems by using quadratic equation.

ILLUSTRATIVE EXAMPLES

Example 1. *The sum of the squares of two positive numbers is 233 and one of the numbers is 3 less than twice the other number. Find the numbers.* (P.T.U. B.Pharma, Dec. 2002)

Sol. Let one number be x. $\therefore$ The other number is $2x - 3$.

But the given condition, $x^2 + (2x - 3)^2 = 233$.

$\Rightarrow \quad x^2 + 4x^2 + 9 - 12x = 233 \quad \Rightarrow \quad 5x^2 - 12x - 224 = 0$

$$\Rightarrow \quad x = \frac{12 \pm \sqrt{(-12)^2 - 4(5)(-224)}}{2(5)} = \frac{12 \pm \sqrt{144 + 4480}}{10} = \frac{12 \pm 68}{10} = \frac{80}{10}, -\frac{56}{10} = 8, -5.6$$

$\therefore$ Either $x = 8$	or $x = -5.6$.
$\therefore$ The numbers are 8, $2(8) - 3 = 13$.	This is impossible.

$\therefore$ The numbers are **8** and **13.**

Example 2. *A two digit number is four times the sum and three times the product of its digits. Find the number.*

Sol. Let digit at the ten's place $= x$ and digit at the unit's place $= y$.

$\therefore$ By the given conditions,

$10x + y = 4(x + y) \quad \ldots(1) \quad$ and $\quad 10x + y = 3xy \quad \ldots(2)$

(1) $\Rightarrow \quad 10x + y = 4x + 4y \quad \Rightarrow \quad 6x = 3y \quad \Rightarrow \quad y = 2x.$

Putting the value of y in (2), we get $10x + 2x = 3x(2x)$.

$\Rightarrow \quad 3x^2 - 6x = 0 \quad \Rightarrow \quad 3x(x - 2) = 0 \quad \Rightarrow \quad x = 0, 2$

$\therefore$ Either $x = 0$	or $x = 2$.
This is impossible because the number is	$\Rightarrow$ $y = 2x = 2(2) = 4$
two digited.	$\therefore$ The number is $10x + y = 10(2) + 4 = 24$

$\therefore$ The number is **24.**

Example 3. *The area of a rectangle is 255 sq. m. If its length be diminished by 1 m. and breadth increased by 1 m., it becomes a square. Find its dimensions.*

Sol. Let length $= x$ mt. and breadth $= y$ m.

Area = 255 sq. m.

$\therefore \quad xy = 255 \qquad \ldots(1)$

New length $= (x - 1)$ mt. and new breadth $= (y + 1)$ m.

Now, the rectangle has become a square.

$\therefore \quad x - 1 = y + 1 \qquad \ldots(2)$

(2) $\Rightarrow \quad x = y + 2.$

Putting this value of x in (1), we get $(y + 2)y = 255$.

$\Rightarrow \quad y^2 + 2y - 255 = 0 \quad \Rightarrow \quad y = -17, 15$

$\therefore \quad y = 15 \qquad (\because \; y$ is positive)

$\therefore \quad x = y + 2 = 15 + 2 = 17$

$\therefore$ Length = **17 m.** and breadth = **15 m.**

Example 4. *A factory kept increasing its output by the same percentage every year. Find the percentage, if it is known that the output has doubled in the last two years.*

Sol. Let output two years ago $= x$ units and annual increase $= r\%$

$\therefore$ Output in the last year $= x\left(\frac{100 + r}{100}\right)$ units.

and output in the present year $= x\left(\frac{100 + r}{100}\right)^2$ units.

$\therefore$ By the given condition, $x\left(\frac{100+r}{100}\right)^2 = 2x.$

$$\Rightarrow \quad \left(\frac{100+r}{100}\right)^2 = 2 \Rightarrow (100+r)^2 = 20{,}000 \Rightarrow r^2 + 200r - 10{,}000 = 0$$

$$\Rightarrow \quad r = \frac{-200 \pm \sqrt{40000+40000}}{2} = \frac{-200 \pm 200\sqrt{2}}{2} = 100(\sqrt{2}-1),\ 100(-1-\sqrt{2}).$$

$\therefore \quad r = 100(\sqrt{2}-1) = 100(1.4142-1) = \mathbf{41.42\%}.$ $(\because\ r > 0)$

Example 5. *A group of students decided to buy a tape-recorder from ₹170 to 195. But at the last moment, two students backed out of the decision so that the remaining students had to pay 1 rupee more than they had planned. What was the price of the tape-recorder, if the students paid equal shares.*

Sol. Let cost of tape recorder = ₹ x and number of students = n

$\therefore \quad 170 \le x \le 195$ ...(1)

Contribution per student = ₹ $\frac{x}{n}$

In the second case, number of students = $n - 2$

New contribution per head = $\frac{x}{n-2}$

By the given condition, $\frac{x}{n-2} = \frac{x}{n} + 1$

$$\Rightarrow \quad \frac{x}{n-2} - \frac{x}{n} = 1 \Rightarrow x\left[\frac{1}{n-2} - \frac{1}{n}\right] = 1 \Rightarrow x = \frac{n^2 - 2n}{2}$$

$\therefore \quad 170 \le \frac{n^2-2n}{2} \le 195$ [using (1)]

$\Rightarrow \quad 340 \le n^2 - 2n \le 390$ ...(2)

$\Rightarrow \quad 341 \le n^2 - 2n + 1 \le 391 \Rightarrow 341 \le (n-1)^2 \le 391$

$\Rightarrow \quad \sqrt{341} \le n - 1 \le \sqrt{391}$ $(\because\ n-1$ is non-negative)

$\Rightarrow \quad 18.47 \le n - 1 \le 19.77 \Rightarrow 19.47 \le n \le 20.77$

$\therefore \quad n = 20$, since it is a natural number.

$\therefore \quad x = \frac{n^2-2n}{2} = \frac{(20)^2 - 2(20)}{2} = 180.$ $\therefore$ Cost of tape-recorder = **₹ 180.**

Example 6. *Students of B.Pharma in an engineering college, are made to stand in rows. If 4 students are extra in a row, there would be 2 rows less. If 4 students are less in a row there would be 4 more rows. Find the number of students of B.Pharma.*

Sol. Take x, the number of rows and y, the number of students.

Then, number of students in a row = $\frac{y}{x}$

As given, $\frac{y}{x} + 4 = \frac{y}{x-2}$ ...(1) [From Ist condition] and $$\frac{y}{x} - 4 = \frac{y}{x+4}$$ ...(2) [From 2nd condition]

Now from (1), $\frac{y}{x} - \frac{y}{x-2} = -4$

$$\Rightarrow \quad y\left(\frac{1}{x} - \frac{1}{x-2}\right) = -4$$

$$= y\left(\frac{x-2-x}{x(x-2)}\right) = -4 \Rightarrow \frac{-2y}{x(x-2)} = -4$$

$$\Rightarrow \quad y = 2x(x-2) \qquad ...(3)$$

Also from (2), $y\left(\frac{1}{x} - \frac{1}{x+4}\right) = 4$

$$\Rightarrow \quad \left(\frac{x+4-x}{x(x+4)}\right) = 4 \Rightarrow \frac{4y}{x(x+4)} = 4$$

$$\Rightarrow \quad y = +\,x(x+4) \qquad ...(4)$$

Equating the value of y, from (3) and (4), we have

$$2x(x-2) = x(x+4) = 0$$

$$\Rightarrow \quad x(2x - 4 - x - 4) = 0$$

$$\Rightarrow \quad x(x-8) = 0$$

$$\Rightarrow \quad x = 8 \qquad | \; x \neq 0$$

From (4), $y = 8(12) = 96$

$\therefore$ Required number of students = **96.**

Example 7. *₹ 6500 were divided equally among a certain number of villagers. Had there been 15 more villagers, each would have got ₹ 30 less. Find the original number of villagers.*

Sol. Let x = The required number of villagers,

From first condition, each villager would get = ₹$\frac{6500}{x}$

Also from second condition,

$$(x+15)\left(\frac{6500}{x} - 30\right) = 6500$$

$$\Rightarrow \quad 6500 - 30x + \frac{97500}{x} - 450 = 6500$$

$$\Rightarrow \quad -30x + \frac{97500}{x} = 450$$

$$\Rightarrow \quad -30x^2 + 97500 = 450x$$

$$\Rightarrow \quad 30x^2 + 450x - 97500 = 0$$

$$\Rightarrow \quad x^2 + 15x - 3250 = 0$$

Here $a = 1$, $b = 15$, $c = -3250$

$$\therefore \quad x = \frac{-15 \pm \sqrt{225 + 13000}}{2} = \frac{-15 \pm 115}{2}$$

$$= \frac{100}{2}, \frac{-130}{2} = 50, -65$$

But $x \neq -65$ $\therefore$ $\mathbf{x = 50}$ *i.e.*, the original number of villagers is **50.**

Example 8. *Suman purchased a number of books, each at the same Price, for a total of ₹750. He sold each book at ₹42 with the total received money. She could by 10 more books than previous. What is the number of books did she buy ?*

Sol. Let x = The total number of books,

then Price of each book $= \frac{750}{x}$

Profit per book $= 42 - \frac{750}{x}$

As given, $\frac{750}{x} \times 100 = \left(42 - \frac{750}{x}\right) \times x$

$$\Rightarrow \quad \frac{7500}{x} = 42x - 750 \Rightarrow 7500 = 42x^2 - 750x$$

$$\Rightarrow \quad 7x^2 - 125x - 1250 = 0$$

Here $a = 7, b = -125, c = -1250$

$$\therefore \quad x = \frac{125 \pm \sqrt{15625 + 35000}}{14}$$

$$= \frac{125 \pm \sqrt{51625}}{14} = \frac{125 \pm 225}{14} = 25, -\frac{50}{7}$$

But $x \neq \frac{-50}{7}$ (no. of books)

$\therefore$ The required number of books purchased by Suman = **25.**

EXERCISE 1.5

1. Divide 27 into two parts whose product is 72.
2. Two positive numbers are such that their sum is 54 and product is 629. Find the numbers.
3. The sum of a number and its reciprocal is 41/20. Find the number.
4. A number exceeds its positive square root by 12. Find the number.
5. The sum of squares of two consecutive odd numbers is 394. Find the integers.
6. The perimeter of a rectangle is 54 metres and its area is 180 sq. metres. Find its length and breadth.
7. Find the side of a square room such that if its length and width are each increased by 4 mt., the area is increased by 120 sq. mt.
8. A piece of cloth costs ₹ 35. If the piece were 4 m longer and each metre costs Re. 1 less, the cost would remain the same. How long is the piece ?
9. A number of points are marked on a plane and are connected pairwise by a line segment. If the total number of line segments is 10, how many points are marked on the plane ?

10. A swimming pool is fitted with three pipes with uniform flow. The first two pipes operating simultaneously, fill the pool in the same time during which the pool is filled by the third pipe alone. The second pipe fills the pool five hour faster than the first pipe and four hours slower than the third pipe. Find the time required by each pipe to fill the pool individually.

11. A pilot covers a distance of 800 km. He could have saved 40 minutes by increasing the speed of the air bus by 40 km/hr. Find the speed of the air bus.

12. A picture 20 cm × 30 cm is placed in a frame with a border of uniform width. If the area of the border is equal to the area of the picture, find the width of the border.

13. The area of a rectangular plot is 1500 sq. metre. If the length of the plot be 10 metre less and breadth 10 metre more, the plot would be a square. Find the length and breadth of the plot.

14. Four gallons are drawn from a tank full of liquid. It is then filled with water then 6 gallons of the mixture are drawn and the tank is again filled with water. If the ratio of the liquid left in the tank to the quantity of water is to be 1 : 2, how much does the tank hold ?

Answers

1. 3, 24 **2.** 17, 37 **3.** 4/5 or 5/4 **4.** 16

5. (– 15 and – 13) or (13 and 15) **6.** 15 m, 12 m **7.** 13 m

8. 10 m **9.** 5 **10.** 15 hours, 10 hours, 6 hours

11. 200 km/hr **12.** 5 cm **13.** length = 50 m, breadth = 30 m

14. 12 gallons.

Hints

3. Let the number be x. $\therefore$ $x + \dfrac{1}{x} = \dfrac{41}{20}$ *i.e.*, $20x^2 - 41x + 20 = 0$.

4. Let the number be x. $\therefore$ $x = \sqrt{x} + 12$ *i.e.*, $(x - 12)^2 = x$.

5. Let one odd integer be x. $\therefore$ Next odd integer $= x + 2$. $\therefore$ $x^2 + (x + 2)^2 = 394$.

7. Let side of square be x mt. $\therefore$ $(x + 4)^2 = x^2 + 120$.

8. Let length $= x$ mt. $\therefore$ Rate per meter = ₹ $\dfrac{35}{x}$. By the given condition, $(x + 4)\left(\dfrac{35}{x} - 1\right) = 35$.

9. Let number of points be n.

$\therefore$ Number of line segments = Number of combinations of points taking 2 at a time $= {}^nC_2$.

10. Let V be the volume of the pool. Let the second pipe fills the pool in x hours.

$\therefore$ Ist and IIIrd pipes takes $x + 5$ and $x - 4$ hours to fill the pool respectively.

Also, $\dfrac{V}{x+5} + \dfrac{V}{x} = \dfrac{V}{x-4}$ $\Rightarrow$ $\dfrac{1}{x+5} + \dfrac{1}{x} = \dfrac{1}{x-4}$.

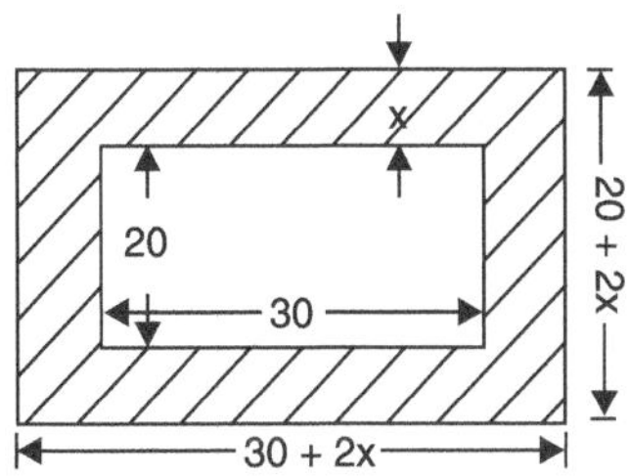

11. Let the speed be x km/hr

$\therefore$ $\dfrac{800}{x} - \dfrac{800}{x+40} = \dfrac{40}{60}$.

12. Let the width be x cm

$\therefore$ $(30 + 2x)(20 + 2x) - (30 \times 20) = 30 \times 20$.

13. If x be the length of the plot, then breadth $= \dfrac{1500}{x}$. As given, $x - 10 = \dfrac{1500}{x} + 10$

14. Let the tank holds x gallons.

After first trial, liquid left = $(x - 4)$ gallons and the ratio = $(x - 4) : 4$

After second trial,

$$\text{liquid let } t = (x - 4) - \frac{x - 4}{x} . 6$$

$$= \frac{x^2 - 10x + 24}{x} \text{ gallons}$$

$$\text{and water} \quad = 4 - \frac{4}{x} . 6 + 6 = \frac{10x - 24}{x}$$

As per given $\dfrac{\dfrac{x^2 - 10x + 24}{x}}{\dfrac{10x - 24}{x}} = \dfrac{1}{2}$, find x.

2

Matrices and Determinants

2.1. INTRODUCTION

The use of matrices and determinants help a lot in mathematical investigations and has become an integral part of mathematics.

2.2. DEFINITION OF A MATRIX

A **matrix** is defined as a rectangular array (arrangement) of numbers or functions and subject to certain rules of operations.

If mn numbers or functions are arranged in the form of a rectangular array A, having m rows and n columns, then A is called a $m \times n$ matrix. A $m \times n$ matrix is generally written as

$$A = \begin{bmatrix} a_{11} & a_{12} \ldots\ldots a_{1j} \ldots\ldots\ldots a_{1n} \\ a_{21} & a_{22} \ldots\ldots a_{2j} \ldots\ldots\ldots a_{2n} \\ \ldots\ldots\ldots\ldots\ldots\ldots\ldots\ldots\ldots \\ \ldots\ldots\ldots\ldots\ldots\ldots\ldots\ldots\ldots \\ a_{i1} & a_{i2} \ldots\ldots a_{ij} \ldots\ldots\ldots a_{in} \\ \ldots\ldots\ldots\ldots\ldots\ldots\ldots\ldots\ldots \\ \ldots\ldots\ldots\ldots\ldots\ldots\ldots\ldots\ldots \\ a_{m1} & a_{m2} \ldots\ldots a_{mj} \ldots\ldots a_{mn} \end{bmatrix}$$

The numbers or functions a_{ij} are called the **elements** of the matrix A. The elements of A are having double subscript notation. The first subscript indicates the row and the second subscript indicates the column in which the elements stands. Thus, the elements in the first row have 1 as first subscript and all the elements in the second column have 2 as the second subscript.

A $m \times n$ matrix is also called as a matrix of **order m × n** or of **order m by n.**

In the short form, the above matrix A is also written as $[a_{ij}]_{m \times n}$.

ILLUSTRATIVE EXAMPLES

Example 1. *If a matrix has 24 elements, what are the possible dimensions (order) it can have ? What, if it has 13 elements ?*

Sol. Let the matrix $A = [a_{ij}]_{m \times n}$ has 24 elements. $\therefore \quad mn = 24$

$\therefore$ The possible values of (m, n) are

(1, 24), (2, 12), (3, 8), (4, 6), (6, 4), (8, 3), (12, 2), (24, 1).

$\therefore$ The possible orders matrix containing 24 elements are

1 × 24, 2 × 12, 3 × 8, 4 × 6, 6 × 4, 8 × 3, 12 × 2 or 24 × 1.

Let $B = [b_{ij}]_{m \times n}$ contains 13 elements. $\therefore \quad m \times n = 13$

$\therefore$ The possible orders of B are **1 × 13** or **13 × 1.**

Example 2. *Construct a 3 × 4 matrix $A = [a_{ij}]$, whose element a_{ij} is given by $\frac{(i+j)^2}{2}$.*

Sol. We have $a_{ij} = \frac{(i+j)^2}{2}$, $1 \le i \le 3$, $1 \le j \le 4$.

$$\therefore \quad a_{11} = \frac{(1+1)^2}{2} = 2, \quad a_{12} = \frac{(1+2)^2}{2} = \frac{9}{2}, \quad a_{13} = \frac{(1+3)^2}{2} = 8, \quad a_{14} = \frac{(1+4)^2}{2} = \frac{25}{2},$$

$$a_{21} = \frac{(2+1)^2}{2} = \frac{9}{2}, \quad a_{22} = \frac{(2+2)^2}{2} = 8, \quad a_{23} = \frac{(2+3)^2}{2} = \frac{25}{2}, \quad a_{24} = \frac{(2+4)^2}{2} = 18,$$

$$a_{31} = \frac{(3+1)^2}{2} = 8, \quad a_{32} = \frac{(3+2)^2}{2} = \frac{25}{2}, \quad a_{33} = \frac{(3+3)^2}{2} = 18, \quad a_{34} = \frac{(3+4)^2}{2} = \frac{49}{2}.$$

$$\therefore \quad A = \begin{bmatrix} a_{11} & a_{12} & a_{13} & a_{14} \\ a_{21} & a_{22} & a_{23} & a_{24} \\ a_{31} & a_{32} & a_{33} & a_{34} \end{bmatrix} = \begin{bmatrix} 2 & 9/2 & 8 & 25/2 \\ 9/2 & 8 & 25/2 & 18 \\ 8 & 25/2 & 18 & 49/2 \end{bmatrix}.$$

2.3. TYPES OF MATRICES

(P.T.U. B.Pharma Dec. 2006)

(i) **Square Matrix.** A matrix is called a **square matrix** if the number of rows and columns in the matrix are equal.

A square matrix having n rows (or columns) is called a square matrix of order n or an n-square matrix.

In the square matrix $[a_{ij}]_{n \times n}$, the line containing elements $a_{11}, a_{22}, \ldots\ldots, a_{nn}$ is called the **principal diagonal** of $[a_{ij}]$. The elements $a_{11}, a_{22}, \ldots\ldots, a_{nn}$ are called the **diagonal elements** of square matrix $[a_{ij}]_{n \times n}$.

Example : $\begin{bmatrix} 1 & 4 & -3 \\ 4 & 6 & 7 \\ 2 & 8 & 8 \end{bmatrix}$ is a square matrix, because number of rows and columns are equal. But $\begin{bmatrix} 1 & 4 \\ 4 & 6 \\ 2 & 8 \end{bmatrix}$ is not a square matrix, because number of rows (3) and columns (2) are not equal.

(ii) **Row Matrix.** A matrix is called a **row matrix** if it contains only one row. [1 5 4 3] is a 1 × 4 row matrix. [4] is also a 1 × 1 row matrix.

(iii) **Column Matrix.** A matrix is called a **column matrix** if it contains only one column.

For example, $\begin{bmatrix} 1 \\ 4 \\ 2 \end{bmatrix}$ is a 3 × 1 column matrix.

(iv) **Zero (or Null) Matrix.** A matrix is called a **zero matrix** if all its elements are zero. For example, [0 0 0] is a zero matrix. A zero matrix is denoted by 'O'.

(v) **Diagonal Matrix.** A square matrix is called a **diagonal matrix** if all its non-diagonal elements are zero.

For example, $\begin{bmatrix} 4 & 0 & 0 \\ 0 & -2 & 0 \\ 0 & 0 & 6 \end{bmatrix}$ is a 3 × 3 diagonal matrix.

(vi) **Scalar Matrix.** A square matrix is called a **scalar matrix** if all its non-diagonal elements are zero and all diagonal elements equal.

For example, $\begin{bmatrix} 4 & 0 & 0 \\ 0 & 4 & 0 \\ 0 & 0 & 4 \end{bmatrix}$ is a scalar matrix.

(vii) **Unit Matrix.** A square matrix is called a **unit matrix** if all its non-diagonal elements are zero and all diagonal elements unity.

For example, $\begin{bmatrix} 1 & 0 \\ 0 & 1 \end{bmatrix}$ is a unit matrix of order 2. This matrix is denoted by I_2.

Similarly, $I_3 = \begin{bmatrix} 1 & 0 & 0 \\ 0 & 1 & 0 \\ 0 & 0 & 1 \end{bmatrix}$ is a unit matrix of order 3. Also I_3 is a scalar matrix.

A unit matrix is also known as an **identity matrix.**

Remark. Every identity matrix and null square matrix is a scalar matrix and every scalar matrix is a diagonal matrix.

(viii) **Triangular Matrices.** There are two types of triangular matrices :

(a) A square matrix is called an **upper triangular matrix** if all its elements below the principal diagonal are zero.

Thus, $[a_{ij}]_{m \times n}$ is an upper triangular matrix if *(i)* $m = n$ and *(ii)* $a_{ij} = 0$ for $i > j$.

For example, $\begin{bmatrix} 4 & 3 & 7 \\ 0 & 8 & 11 \\ 0 & 0 & -5 \end{bmatrix}$ is an upper triangular matrix.

(b) A square matrix is called a **lower triangular matrix** if all its elements above the principal diagonal are zero.

Thus, $[a_{ij}]_{m \times n}$ is a lower triangular matrix if *(i)* $m = n$ and *(ii)* $a_{ij} = 0$ for $i < j$.

For example, $\begin{bmatrix} -2 & 0 & 0 \\ 7 & 3 & 0 \\ 9 & -6 & 8 \end{bmatrix}$ is a lower triangular matrix.

(ix) **Comparable Matrices.** Two matrices are called **comparable** if the matrices are of same order.

Thus, the matrices $[a_{ij}]_{m \times n}$ and $[b_{ij}]_{p \times q}$ are comparable if *(i)* $m = p$ and *(ii)* $n = q$.

For example, the matrices $\begin{bmatrix} 1 & 4 & 5 \\ 6 & 9 & 7 \end{bmatrix}$ and $\begin{bmatrix} 3 & 6 & 8 \\ -2 & 5 & 7 \end{bmatrix}$ are comparable.

(x) **Equal Matrices.** Two matrices are called **equal** if the matrices are of same order and their corresponding elements are equal.

Thus, the matrices $[a_{ij}]_{m \times n}$ and $[b_{ij}]_{p \times q}$ are equal if

(i) $m = p$ *(ii)* $n = q$ and *(iii)* $a_{ij} = b_{ij}$ $\forall$* i, j.

For example, the matrices $\begin{bmatrix} a & b & 4 \\ c & x & 6 \end{bmatrix}$ and $\begin{bmatrix} 0 & 3 & y \\ 6 & 7 & z \end{bmatrix}$ are equal if $a = 0$, $b = 3$ $y = 4$, $c = 6$, $x = 7$, $z = 6$.

Remark. Equal matrices are always comparable matrices but comparable matrices may not be equal matrices.

---

*$\forall$ means 'for all'.

Example 3. *Find the values of x, y, z if :*

$$\begin{bmatrix} 2x+y & x-y \\ x-z & x+y+z \end{bmatrix} = \begin{bmatrix} 10 & -1 \\ 2 & 8 \end{bmatrix}$$

Sol. We have $\begin{bmatrix} 2x+y & x-y \\ x-z & x+y+z \end{bmatrix} = \begin{bmatrix} 10 & -1 \\ 2 & 8 \end{bmatrix}$

By equality of matrices :

$2x + y = 10$ ...(1) $\quad x - y = -1$ ...(2)

$x - z = 2$ ...(3) $\quad x + y + z = 8$ ...(4)

(1) + (2) $\Rightarrow$ $3x = 9$ $\Rightarrow$ $x = 3$

$\therefore$ (2) $\Rightarrow$ $3 - y = -1$. $\therefore$ $y = 3 + 1 = 4$

(3) $\Rightarrow$ $3 - z = 2$. $\therefore$ $z = 3 - 2 = 1$.

$\therefore$ $\mathbf{x = 3, y = 4, z = 1.}$

(*xi*) **Negative of a Matrix.** The **negative** of a matrix $A = [a_{ij}]$ is the matrix $[-a_{ij}]$ and is denoted by $-A$.

For example, if $A = \begin{bmatrix} 4 & 3 & 7 \\ 6 & 8 & -4 \end{bmatrix}$, then $-A = \begin{bmatrix} -4 & -3 & -7 \\ -6 & -8 & 4 \end{bmatrix}$.

EXERCISE 2.1

1. If a matrix has 18 elements, what are the possible orders it can have ? What, if it has 11 elements ? (*P.T.U. B.Pharma, Dec. 2005*)
2. If $A = [a_{ij}]$ be a matrix of order $m \times n$, then write A in the expanded form :
   (*i*) $m = 1, n = 2$ (*ii*) $m = 3, n = 3$.
3. Construct a 2×2 matrix $A = [a_{ij}]$ whose element a_{ij} is given by :
   (*i*) $a_{ij} = \dfrac{(2i+j)^2}{2}$ (*ii*) $a_{ij} = \dfrac{|2i - 3j|}{2}$
4. Construct a $m \times n$ matrix $A = [a_{ij}]$ whose element a_{ij} is given by
   (*i*) $a_{ij} = \dfrac{3i - j}{2}$, $m = 2, n = 3$ (*ii*) $a_{ij} = \dfrac{2i + 3j}{2}$, $m = 2, n = 3$.
   (*P.T.U. B.Pharma, May 2005*)
   (*iii*) $a_{ij} = \dfrac{(i - 2j)^2}{2}$, $m = 2, n = 3$ (*iv*) $a_{ij} = \dfrac{(i + 2j)^2}{2}$, $m = 3, n = 2$
5. If $\begin{bmatrix} a+b & 2 \\ 5 & ab \end{bmatrix} = \begin{bmatrix} 6 & 2 \\ 5 & 8 \end{bmatrix}$, find the values of a and b.
6. Find the values of a, b, c, d from the following matrix equations :
   (*i*) $\begin{bmatrix} 2a+5 & b+7 \\ 3c-8 & 3d+9 \end{bmatrix} = \begin{bmatrix} 1 & 2 \\ 1 & 3 \end{bmatrix}$ (*ii*) $\begin{bmatrix} 2a+b & a-2b \\ 5c-d & 4c+3d \end{bmatrix} = \begin{bmatrix} 4 & -3 \\ 11 & 24 \end{bmatrix}$

7. Find the values of x, y, z, w from the matrix equations :

(i) $\begin{bmatrix} x & 3x-y \\ 2x+z & 3y-w \end{bmatrix} = \begin{bmatrix} 3 & 2 \\ 4 & 7 \end{bmatrix}$ (ii) $\begin{bmatrix} 2x-3y & z-w & 3 \\ 1 & x+4y & 3z+4w \end{bmatrix} = \begin{bmatrix} 1 & -2 & 3 \\ 1 & 6 & 29 \end{bmatrix}$

Answers

1. $1 \times 18, 2 \times 9, 3 \times 6, 6 \times 3, 9 \times 2, 18 \times 1 ; 1 \times 11, 11 \times 1$

2. (i) $[a_{11} \quad a_{12}]$ (ii) $\begin{bmatrix} a_{11} & a_{12} & a_{13} \\ a_{21} & a_{22} & a_{23} \\ a_{31} & a_{32} & a_{33} \end{bmatrix}$

3. (i) $\begin{bmatrix} 9/2 & 8 \\ 25/2 & 18 \end{bmatrix}$ (ii) $\begin{bmatrix} 1/2 & 2 \\ 1/2 & 1 \end{bmatrix}$

4. (i) $\begin{bmatrix} 1 & 1/2 & 0 \\ 5/2 & 2 & 3/2 \end{bmatrix}$ (ii) $\begin{bmatrix} 5/2 & 8/3 & 11/2 \\ 7/2 & 5 & 13/2 \end{bmatrix}$ (iii) $\begin{bmatrix} 1/2 & 9/2 & 25/2 \\ 0 & 2 & 8 \end{bmatrix}$

(iv) $\begin{bmatrix} 9/2 & 25/2 \\ 8 & 18 \\ 25/2 & 49/2 \end{bmatrix}$

5. $a = 2, b = 4$ or $a = 4, b = 2$

6. (i) $a = -2, b = -5, c = 3, d = -2$ (ii) $a = 1, b = 2, c = 3, d = 4$

7. (i) $x = 3, y = 7, z = -2, w = 14$ (ii) $x = 2, y = 1, z = 3, w = 5$.

Hint

2. (i) $m = 1$ implies that there is only one row in A and $n = 2$ implies that there are two columns in A.
$\therefore$ A $= [a_{11} \quad a_{12}]$.

2.4. ADDITION OF MATRICES

If A and B are two matrices of the same order, then their **sum** A + B is defined and is obtained by adding the corresponding elements of A and B.

In other words, if $A = [a_{ij}]_{m \times n}$ and $B = [b_{ij}]_{m \times n}$ are matrices of same order, then A + B is defined and is given by

$$A + B = [c_{ij}]_{m \times n}, \quad \text{where } c_{ij} = a_{ij} + b_{ij}.$$

For example, if $A = \begin{bmatrix} 2 & 3 & 6 \\ 4 & 0 & 9 \end{bmatrix}$ and $B = \begin{bmatrix} 0 & 6 & 8 \\ 7 & 0 & 8 \end{bmatrix}$

then $A + B = \begin{bmatrix} 2+0 & 3+6 & 6+8 \\ 4+7 & 0+0 & 9+8 \end{bmatrix} = \begin{bmatrix} 2 & 9 & 14 \\ 11 & 0 & 17 \end{bmatrix}$.

If $C = \begin{bmatrix} 1 & 4 \\ 2 & 3 \end{bmatrix}$, then neither A + C nor B + C is defined.

Remark. If two matrices are not of the same order, their sum is not defined.

ILLUSTRATIVE EXAMPLES

Example 1. *Evaluate A + B, if defined, for the given matrices A and B :*

(i) $A = \begin{bmatrix} 1 & 4 & 3 \\ 6 & 8 & 9 \end{bmatrix}, B = \begin{bmatrix} 6 & 8 & 3 \\ 3 & -3 & 16 \end{bmatrix}$ (ii) $A = \begin{bmatrix} 3 & 4 \\ 6 & 8 \end{bmatrix}, B = \begin{bmatrix} 3 \\ 4 \end{bmatrix}$

(iii) $A = \begin{bmatrix} \sin^2\theta & -\cos^2\theta \\ 1 & \sec^2\theta \end{bmatrix}, B = \begin{bmatrix} \cos^2\theta & -\sin^2\theta \\ 0 & -\tan^2\theta \end{bmatrix}$.

Sol. (*i*) $A = \begin{bmatrix} 1 & 4 & 3 \\ 6 & 8 & 9 \end{bmatrix}$, $B = \begin{bmatrix} 6 & 8 & 3 \\ 3 & -3 & 16 \end{bmatrix}$. A and B are of same order.

$\therefore$ A + B is defined.

$$A + B = \begin{bmatrix} 1+6 & 4+8 & 3+3 \\ 6+3 & 8-3 & 9+16 \end{bmatrix} = \begin{bmatrix} \mathbf{7} & \mathbf{12} & \mathbf{6} \\ \mathbf{9} & \mathbf{5} & \mathbf{25} \end{bmatrix}.$$

(*ii*) $A = \begin{bmatrix} 3 & 4 \\ 6 & 8 \end{bmatrix}$, $B = \begin{bmatrix} 3 \\ 4 \end{bmatrix}$. A and B are of different orders. $\therefore$ A + B is not defined.

(*iii*) $A = \begin{bmatrix} \sin^2\theta & -\cos^2\theta \\ 1 & \sec^2\theta \end{bmatrix}$, $B = \begin{bmatrix} \cos^2\theta & -\sin^2\theta \\ 0 & -\tan^2\theta \end{bmatrix}$. A and B are of same order.

$\therefore$ A + B is defined.

$$A + B = \begin{bmatrix} \sin^2\theta + \cos^2\theta & -\cos^2\theta - \sin^2\theta \\ 1+0 & \sec^2\theta - \tan^2\theta \end{bmatrix} = \begin{bmatrix} \mathbf{1} & \mathbf{-1} \\ \mathbf{1} & \mathbf{1} \end{bmatrix}.$$

$\because \sin^2\theta + \cos^2\theta = 1$
$\sec^2\theta - \tan^2\theta = 1$

2.5. DIFFERENCE OF MATRICES

If A and B are two matrices of the same order, then their **difference** A – B is defined and is obtained by subtracting the elements of B from the corresponding elements of A.

For example, if $A = \begin{bmatrix} 4 & 3 \\ 6 & 8 \\ 9 & 15 \end{bmatrix}$, $B = \begin{bmatrix} 3 & 8 \\ 11 & 7 \\ 6 & 8 \end{bmatrix}$,

then $$A - B = \begin{bmatrix} 4-3 & 3-8 \\ 6-11 & 8-7 \\ 9-6 & 15-8 \end{bmatrix} = \begin{bmatrix} 1 & -5 \\ -5 & 1 \\ 3 & 7 \end{bmatrix}.$$

Remark. By definition A – B is same as the matrix A + (– B).

2.6. PROPERTIES OF ADDITION

(*i*) **Commutative law of addition.** If A and B are matrices of same order then

A + B = B + A.

(*ii*) **Associative law of addition.** If A, B and C are matrices of same order then

A + (B + C) = (A + B) + C.

(*iii*) **Existence of additive identity.** If A is any matrix then

A + O = A = O + A.

Here 'O' is a zero matrix of same order as that of A.

$\therefore$ The zero matrix is called as *additive identity* for the matrices.

(*iv*) **Existence of additive inverse.** If A is any matrix, then

A + (– A) = O = (– A) + A.

The matrix '– A' is called as the *additive inverse* of the matrix A.

2.7. CANCELLATION LAWS IN ADDITION

Theorem. *If A, B, C are matrices of same order then*

(*i*) $A + B = A + C \Rightarrow B = C$ *(Left cancellation law)*

(*ii*) $B + A = C + A \Rightarrow B = C$ *(Right cancellation law)*

Proof. (*i*) We have $A + B = A + C$

$\Rightarrow \quad (-A) + (A + B) = (-A) + (A + C)$

$\Rightarrow \quad (-A + A) + B = (-A + A) + C$ (By associative law of addition)

$\Rightarrow \quad O + B = O + C$ ($\because$ '– A' is the additive inverse of A)

$\Rightarrow \quad B = C.$ ($\because$ O is the additive identity)

Proof of part (*ii*) is left for the reader as an exercise.

Example 2. *If* $A = \begin{bmatrix} 4 & 7 \\ 6 & 5 \end{bmatrix}$, $B = \begin{bmatrix} 3 & 8 \\ 7 & 2 \end{bmatrix}$, $C = \begin{bmatrix} 8 & 11 \\ 6 & 1 \end{bmatrix}$, *then show that :*

(*i*) $A + B = B + A$ (*ii*) $(A + B) + C = A + (B + C)$

(*iii*) $A + O = O + A = A$ (*iv*) $A + (-A) = (-A) + A = O.$

Sol. (*i*) $A + B = \begin{bmatrix} 4 & 7 \\ 6 & 5 \end{bmatrix} + \begin{bmatrix} 3 & 8 \\ 7 & 2 \end{bmatrix} = \begin{bmatrix} 4+3 & 7+8 \\ 6+7 & 5+2 \end{bmatrix} = \begin{bmatrix} 7 & 15 \\ 13 & 7 \end{bmatrix}$

$$B + A = \begin{bmatrix} 3 & 8 \\ 7 & 2 \end{bmatrix} + \begin{bmatrix} 4 & 7 \\ 6 & 5 \end{bmatrix} = \begin{bmatrix} 3+4 & 8+7 \\ 7+6 & 2+5 \end{bmatrix} = \begin{bmatrix} 7 & 15 \\ 13 & 7 \end{bmatrix}$$

$\therefore \quad$ **A + B = B + A.**

(*ii*) We have $A + B = \begin{bmatrix} 7 & 15 \\ 13 & 7 \end{bmatrix}$.

$$\therefore \quad (A + B) + C = \begin{bmatrix} 7 & 15 \\ 13 & 7 \end{bmatrix} + \begin{bmatrix} 8 & 11 \\ 6 & 1 \end{bmatrix} = \begin{bmatrix} 7+8 & 15+11 \\ 13+6 & 7+1 \end{bmatrix} = \begin{bmatrix} 15 & 26 \\ 19 & 8 \end{bmatrix}$$

$$B + C = \begin{bmatrix} 3 & 8 \\ 7 & 2 \end{bmatrix} + \begin{bmatrix} 8 & 11 \\ 6 & 1 \end{bmatrix} = \begin{bmatrix} 3+8 & 8+11 \\ 7+6 & 2+1 \end{bmatrix} = \begin{bmatrix} 11 & 19 \\ 13 & 3 \end{bmatrix}$$

$$\therefore \quad A + (B + C) = \begin{bmatrix} 4 & 7 \\ 6 & 5 \end{bmatrix} + \begin{bmatrix} 11 & 19 \\ 13 & 3 \end{bmatrix} = \begin{bmatrix} 4+11 & 7+19 \\ 6+13 & 5+3 \end{bmatrix} = \begin{bmatrix} 15 & 26 \\ 19 & 8 \end{bmatrix}$$

$\therefore \quad$ **(A + B) + C = A + (B + C).**

(*iii*) Since A is of order 2 × 2, we take $O = \begin{bmatrix} 0 & 0 \\ 0 & 0 \end{bmatrix}$ so that A + O and O + A may be defined.

$$A + O = \begin{bmatrix} 4 & 7 \\ 6 & 5 \end{bmatrix} + \begin{bmatrix} 0 & 0 \\ 0 & 0 \end{bmatrix} = \begin{bmatrix} 4+0 & 7+0 \\ 6+0 & 5+0 \end{bmatrix} = \begin{bmatrix} 4 & 7 \\ 6 & 5 \end{bmatrix} = A$$

$$O + A = \begin{bmatrix} 0 & 0 \\ 0 & 0 \end{bmatrix} + \begin{bmatrix} 4 & 7 \\ 6 & 5 \end{bmatrix} = \begin{bmatrix} 0+4 & 0+7 \\ 0+6 & 0+5 \end{bmatrix} = \begin{bmatrix} 4 & 7 \\ 6 & 5 \end{bmatrix} = A$$

$\therefore \quad$ **A + O = O + A = A.**

(*iv*) $A = \begin{bmatrix} 4 & 7 \\ 6 & 5 \end{bmatrix} \Rightarrow -A = \begin{bmatrix} -4 & -7 \\ -6 & -5 \end{bmatrix}$

$$\therefore \quad A + (-A) = \begin{bmatrix} 4 & 7 \\ 6 & 5 \end{bmatrix} + \begin{bmatrix} -4 & -7 \\ -6 & -5 \end{bmatrix} = \begin{bmatrix} 4+(-4) & 7+(-7) \\ 6+(-6) & 5+(-5) \end{bmatrix} = \begin{bmatrix} 0 & 0 \\ 0 & 0 \end{bmatrix} = O$$

$$(-A) + A = \begin{bmatrix} -4 & -7 \\ -6 & -5 \end{bmatrix} + \begin{bmatrix} 4 & 7 \\ 6 & 5 \end{bmatrix} = \begin{bmatrix} -4+4 & -7+7 \\ -6+6 & -5+5 \end{bmatrix} = \begin{bmatrix} 0 & 0 \\ 0 & 0 \end{bmatrix} = O$$

$\therefore \quad$ **A + (– A) = (– A) + A = O.**

Example 3. *If* $A = \begin{bmatrix} 1 & 4 \\ 3 & 2 \\ 2 & 5 \end{bmatrix}$ *and* $B = \begin{bmatrix} -1 & 2 \\ 0 & 5 \\ 3 & 1 \end{bmatrix}$, *then find the matrix X for which*

$$A + B - X = O.$$

Sol. We have $\quad A + B - X = O \quad \Rightarrow \quad A + B = O + X$

$\Rightarrow \quad A + B = X \quad \Rightarrow \quad X = A + B$

$$\Rightarrow \quad X = \begin{bmatrix} 1 & 4 \\ 3 & 2 \\ 2 & 5 \end{bmatrix} + \begin{bmatrix} -1 & 2 \\ 0 & 5 \\ 3 & 1 \end{bmatrix} = \begin{bmatrix} \mathbf{0} & \mathbf{6} \\ \mathbf{3} & \mathbf{7} \\ \mathbf{5} & \mathbf{6} \end{bmatrix}.$$

EXERCISE 2.2

1. Find the sum of the following matrices :

   (*i*) $\begin{bmatrix} 2 & 6 & 5 \\ 3 & 9 & 7 \end{bmatrix}$ and $\begin{bmatrix} 3 & 11 & 8 \\ 7 & 15 & 3 \end{bmatrix}$ (*ii*) $\begin{bmatrix} -1 & 4 & -6 \\ 8 & 5 & 14 \\ 2 & 8 & 5 \end{bmatrix}$ and $\begin{bmatrix} 12 & 7 & 6 \\ 8 & 0 & 5 \\ 3 & 2 & 4 \end{bmatrix}$

2. If $A = \begin{bmatrix} 2 & 3 & 6 \\ 8 & 0 & 4 \end{bmatrix}$ and $B = \begin{bmatrix} 7 & 6 & 8 \\ -4 & -3 & 2 \end{bmatrix}$, then find A + B and A – B.

3. Find A + B if defined, for the following matrices :

   (*i*) $A = [4 \quad 3]$, $B = [-3 \quad 6]$ (*ii*) $A = \begin{bmatrix} 1 \\ 6 \\ 8 \end{bmatrix}$, $B = [3 \quad 6 \quad 18]$

   (*iii*) $A = \begin{bmatrix} 1 & 6 & 8 \\ 5 & 7 & 10 \\ 10 & 12 & 9 \end{bmatrix}$, $B = \begin{bmatrix} -12 & 5 & 9 \\ 17 & 6 & 5 \\ 4 & 3 & 17 \end{bmatrix}$

   (*iv*) $A = \begin{bmatrix} 7 & -11 & 7 \\ 6 & 9 & 8 \end{bmatrix}$, $B = \begin{bmatrix} 1 & 6 \\ 8 & 7 \end{bmatrix}$.

4. Solve for X : $\begin{bmatrix} 2 & -3 \\ 4 & -2 \end{bmatrix} - X = \begin{bmatrix} -3 & 4 \\ 5 & -1 \end{bmatrix}$.

5. Verify commutative law of addition for the matrices :

   $A = \begin{bmatrix} 3 & 4 & 6 & 8 \\ 1 & 2 & 6 & 0 \\ 3 & 11 & 2 & 7 \end{bmatrix}$, $B = \begin{bmatrix} 2 & 0 & 1 & 5 \\ 6 & 8 & 7 & 12 \\ 8 & 9 & 11 & 12 \end{bmatrix}$.

6. If $A = \begin{bmatrix} 2 & 3 & 7 \\ 6 & 5 & 9 \\ 11 & -18 & 7 \end{bmatrix}$, then show that A + (– A) = O = (– A) + A.

7. (*i*) If $A = \begin{bmatrix} 1 & -3 & 2 \\ 2 & 0 & 2 \end{bmatrix}$ and $B = \begin{bmatrix} 2 & -1 & -1 \\ 1 & 0 & -1 \end{bmatrix}$, then find the matrix C such that A + B + C is a zero matrix.

(*ii*) If $A = \begin{bmatrix} 1 & 3 \\ 2 & 1 \\ 3 & -1 \end{bmatrix}$ and $B = \begin{bmatrix} 2 & 1 \\ 1 & 2 \\ 1 & 1 \end{bmatrix}$, then find the matrix C such that A + B + C is a zero matrix.

8. (*i*) Find the values of x and y from the following equation :

$$\begin{bmatrix} x-y & 2 & -2 \\ 4 & x & 6 \end{bmatrix} + \begin{bmatrix} 3 & -2 & 2 \\ 1 & 0 & -1 \end{bmatrix} = \begin{bmatrix} 6 & 0 & 0 \\ 5 & 2x+y & 5 \end{bmatrix}$$

(*ii*) Find the values of x, y and z from the following equation :

$$\begin{bmatrix} x+y+z \\ x+y \\ y+z \end{bmatrix} + \begin{bmatrix} 1 \\ 3 \\ 7 \end{bmatrix} = \begin{bmatrix} 8 \\ 8 \\ 10 \end{bmatrix}.$$

Answers

1. (*i*) $\begin{bmatrix} 5 & 17 & 13 \\ 10 & 24 & 10 \end{bmatrix}$ (*ii*) $\begin{bmatrix} 11 & 11 & 0 \\ 16 & 5 & 19 \\ 5 & 10 & 9 \end{bmatrix}$

2. $\begin{bmatrix} 9 & 9 & 14 \\ 4 & -3 & 6 \end{bmatrix}, \begin{bmatrix} -5 & -3 & -2 \\ 12 & 3 & 2 \end{bmatrix}$ **3.** (*i*) $[1 \quad 9]$

(*ii*) Not defined (*iii*) $\begin{bmatrix} -11 & 11 & 17 \\ 22 & 13 & 15 \\ 14 & 15 & 26 \end{bmatrix}$ (*iv*) Not defined.

4. $\begin{bmatrix} 5 & -7 \\ -1 & -1 \end{bmatrix}$ **7.** (*i*) $\begin{bmatrix} -3 & 4 & -1 \\ -3 & 0 & -1 \end{bmatrix}$ (*ii*) $\begin{bmatrix} -3 & -4 \\ -3 & -3 \\ -4 & 0 \end{bmatrix}$.

8. (*i*) $x = \frac{3}{2}, \ y = -\frac{3}{2}$ (*ii*) $x = 4, \quad y = 1, \quad z = 2.$

2.8. SCALAR MULTIPLICATION OF A MATRIX

If A is a matrix of order $m \times n$ and k be a scalar (any complex number), then the **scalar multiple** kA of A by k is defined to be a $m \times n$ matrix whose elements are obtained by multiplying the corresponding elements of A by k.

In other words, if $A = [a_{ij}]_{m \times n}$

then $$kA = [c_{ij}]_{m \times n}, \text{ where } c_{ij} = ka_{ij}.$$

For example, if $A = \begin{bmatrix} 4 & 7 & 3 \\ 2 & 6 & 8 \end{bmatrix}$ and $k = 7$, then

$$kA = 7\begin{bmatrix} 4 & 7 & 3 \\ 2 & 6 & 8 \end{bmatrix} = \begin{bmatrix} 7(4) & 7(7) & 7(3) \\ 7(2) & 7(6) & 7(8) \end{bmatrix} = \begin{bmatrix} 28 & 49 & 21 \\ 14 & 42 & 56 \end{bmatrix}.$$

If $A_1, A_2, \ldots\ldots, A_k$ be all $m \times n$ matrices, then for numbers $\lambda_1, \lambda_2, \ldots\ldots, \lambda_k$, the matrix

$$\lambda_1 A_1 + \lambda_2 A_2 + \ldots\ldots + \lambda_k A_k$$

is called a **linear combination** of the matrices $A_1, A_2, \ldots\ldots, A_k$.

2.9. PROPERTIES OF SCALAR MULTIPLICATION

If A and B are matrices of the same order and k, l are numbers, then

(*i*) $k(A + B) = kA + kB$

(*ii*) $(k + l)A = kA + lA$

(*iii*) $k(lA) = (kl)A$

(*iv*) $1A = A$.

ILLUSTRATIVE EXAMPLES

Example 1. *If* $A = \begin{bmatrix} 3 & 8 & 11 \\ 6 & -3 & 8 \end{bmatrix}$ *and* $B = \begin{bmatrix} 1 & -6 & 15 \\ 3 & 8 & 17 \end{bmatrix}$, *find 7A + 5B.*

Sol.

$$7A + 5B = 7\begin{bmatrix} 3 & 8 & 11 \\ 6 & -3 & 8 \end{bmatrix} + 5\begin{bmatrix} 1 & -6 & 15 \\ 3 & 8 & 17 \end{bmatrix}$$

$$= \begin{bmatrix} 7(3) & 7(8) & 7(11) \\ 7(6) & 7(-3) & 7(8) \end{bmatrix} + \begin{bmatrix} 5(1) & 5(-6) & 5(15) \\ 5(3) & 5(8) & 5(17) \end{bmatrix}$$

$$= \begin{bmatrix} 21 & 56 & 77 \\ 42 & -21 & 56 \end{bmatrix} + \begin{bmatrix} 5 & -30 & 75 \\ 15 & 40 & 85 \end{bmatrix}$$

$$= \begin{bmatrix} 21+5 & 56-30 & 77+75 \\ 42+15 & -21+40 & 56+85 \end{bmatrix} = \begin{bmatrix} \mathbf{26} & \mathbf{26} & \mathbf{152} \\ \mathbf{57} & \mathbf{19} & \mathbf{141} \end{bmatrix}.$$

Example 2. *If* $A = \begin{bmatrix} 2 & -2 \\ 4 & 2 \\ -5 & 1 \end{bmatrix}$, $B = \begin{bmatrix} 8 & 0 \\ 4 & -2 \\ 3 & 6 \end{bmatrix}$, *find the matrix X, such that*

2A + 3X = 5B.

Sol. We have 2A + 3X = 5B.

$$\Rightarrow \quad 2\begin{bmatrix} 2 & -2 \\ 4 & 2 \\ -5 & 1 \end{bmatrix} + 3X = 5\begin{bmatrix} 8 & 0 \\ 4 & -2 \\ 3 & 6 \end{bmatrix} \quad \Rightarrow \quad \begin{bmatrix} 4 & -4 \\ 8 & 4 \\ -10 & 2 \end{bmatrix} + 3X = \begin{bmatrix} 40 & 0 \\ 20 & -10 \\ 15 & 30 \end{bmatrix}$$

$$\Rightarrow \quad 3X = \begin{bmatrix} 40 & 0 \\ 20 & -10 \\ 15 & 30 \end{bmatrix} - \begin{bmatrix} 4 & -4 \\ 8 & 4 \\ -10 & 2 \end{bmatrix} = \begin{bmatrix} 36 & 4 \\ 12 & -14 \\ 25 & 28 \end{bmatrix}$$

$$\Rightarrow \quad X = \frac{1}{3}\begin{bmatrix} 36 & 4 \\ 12 & -14 \\ 25 & 28 \end{bmatrix} = \begin{bmatrix} 12 & 4/3 \\ 4 & -14/3 \\ 25/3 & 28/3 \end{bmatrix}.$$

Example 3. *Find X and Y if* $2X + Y = \begin{bmatrix} 4 & 4 & 7 \\ 7 & 3 & 4 \end{bmatrix}$ *and* $X - 2Y = \begin{bmatrix} -3 & 2 & 1 \\ 1 & -1 & 2 \end{bmatrix}$.

Sol. We have

$$2X + Y = \begin{bmatrix} 4 & 4 & 7 \\ 7 & 3 & 4 \end{bmatrix} \quad ...(1)$$

and

$$X - 2Y = \begin{bmatrix} -3 & 2 & 1 \\ 1 & -1 & 2 \end{bmatrix} \quad ...(2)$$

$$(1) \times 2 \Rightarrow \quad 4X + 2Y = 2\begin{bmatrix} 4 & 4 & 7 \\ 7 & 3 & 4 \end{bmatrix} = \begin{bmatrix} 8 & 8 & 14 \\ 14 & 6 & 8 \end{bmatrix} \quad ...(3)$$

(2) + (3) $\Rightarrow$ $$5X = \begin{bmatrix} -3 & 2 & 1 \\ 1 & -1 & 2 \end{bmatrix} + \begin{bmatrix} 8 & 8 & 14 \\ 14 & 6 & 8 \end{bmatrix} = \begin{bmatrix} 5 & 10 & 15 \\ 15 & 5 & 10 \end{bmatrix}$$

$\therefore$ $$X = \frac{1}{5}\begin{bmatrix} 5 & 10 & 15 \\ 15 & 5 & 10 \end{bmatrix} = \begin{bmatrix} \mathbf{1} & \mathbf{2} & \mathbf{3} \\ \mathbf{3} & \mathbf{1} & \mathbf{2} \end{bmatrix}$$

Putting the value of X in (1), we get

$$Y = \begin{bmatrix} 4 & 4 & 7 \\ 7 & 3 & 4 \end{bmatrix} - 2X = \begin{bmatrix} 4 & 4 & 7 \\ 7 & 3 & 4 \end{bmatrix} - 2\begin{bmatrix} 1 & 2 & 3 \\ 3 & 1 & 2 \end{bmatrix}$$

$$= \begin{bmatrix} 4 & 4 & 7 \\ 7 & 3 & 4 \end{bmatrix} - \begin{bmatrix} 2 & 4 & 6 \\ 6 & 2 & 4 \end{bmatrix} = \begin{bmatrix} \mathbf{2} & \mathbf{0} & \mathbf{1} \\ \mathbf{1} & \mathbf{1} & \mathbf{0} \end{bmatrix}.$$

EXERCISE 2.3

1. If $A = \begin{bmatrix} 2 & 4 & 3 \\ 1 & 0 & 6 \\ 0 & -2 & -3 \end{bmatrix}$ and $B = \begin{bmatrix} 0 & 2 & -3 \\ 1 & 3 & -4 \\ 0 & 7 & 1 \end{bmatrix}$, find 7A – 4B.

2. Solve the equation : $2\begin{bmatrix} x & z \\ y & t \end{bmatrix} + 3\begin{bmatrix} 1 & -1 \\ 0 & 2 \end{bmatrix} = 3\begin{bmatrix} 3 & 5 \\ 4 & 6 \end{bmatrix}$.

3. Find X and Y if $X + Y = \begin{bmatrix} 5 & 2 \\ 0 & 9 \end{bmatrix}$ and $X - Y = \begin{bmatrix} 3 & 6 \\ 0 & -1 \end{bmatrix}$.

4. Find the values of x and y from the equation :

$$2\begin{bmatrix} x & 5 \\ 7 & y-3 \end{bmatrix} + \begin{bmatrix} 3 & 4 \\ 1 & 2 \end{bmatrix} = \begin{bmatrix} 7 & 14 \\ 15 & 14 \end{bmatrix}.$$

5. If $2X + 3Y = \begin{bmatrix} 2 & 3 \\ 4 & 0 \end{bmatrix}$ and $3X + 2Y = \begin{bmatrix} -2 & 2 \\ 1 & -5 \end{bmatrix}$, find X and Y.

6. (*i*) Find a matrix X such that 2A + B + X = O, where $A = \begin{bmatrix} -1 & 2 \\ 3 & 4 \end{bmatrix}$, $B = \begin{bmatrix} 3 & -2 \\ 1 & 5 \end{bmatrix}$.

(*ii*) Given $A = \begin{bmatrix} 1 & 2 & -3 \\ 5 & 0 & 2 \\ 1 & -1 & 1 \end{bmatrix}$ and $B = \begin{bmatrix} 3 & -1 & 2 \\ 4 & 2 & 5 \\ 2 & 0 & 3 \end{bmatrix}$, find a matrix C such that A + 2C = B.

Answers

1. $\begin{bmatrix} 14 & 20 & 33 \\ 3 & -12 & 58 \\ 0 & -42 & -25 \end{bmatrix}$

2. $x = 3, y = 6, z = 9, t = 6$

3. $X = \begin{bmatrix} 4 & 4 \\ 0 & 4 \end{bmatrix}, Y = \begin{bmatrix} 1 & -2 \\ 0 & 5 \end{bmatrix}$

4. $x = 2, y = 9$

5. $X = \begin{bmatrix} -2 & 0 \\ -1 & -3 \end{bmatrix}, Y = \begin{bmatrix} 2 & 1 \\ 2 & 2 \end{bmatrix}$

6. (*i*) $X = \begin{bmatrix} -1 & -2 \\ -7 & -13 \end{bmatrix}$ (*ii*) $C = \begin{bmatrix} 1 & -3/2 & 5/2 \\ -1/2 & 1 & 3/2 \\ 1/2 & 1/2 & 1 \end{bmatrix}$.

2.10. MULTIPLICATION OF MATRICES

Let A and B matrices of orders $m \times n$ and $n \times p$ respectively. The **product** AB of matrices A and B is defined only if the number of columns of A is equal to the number of rows of B. The product AB will be a $m \times p$ matrix.

For example, let $A = \begin{bmatrix} a_{11} & a_{12} \\ a_{21} & a_{22} \end{bmatrix}$ and $B = \begin{bmatrix} b_{11} & b_{12} & b_{13} \\ b_{21} & b_{22} & b_{23} \end{bmatrix}$.

Here the product AB is defined, because the number of columns (= 2) of A equals the number of rows (= 2) of B. In this case,

$$AB = \begin{bmatrix} a_{11} & a_{12} \\ \xrightarrow{\quad\quad} & \\ a_{21} & a_{22} \end{bmatrix} \begin{bmatrix} a_{11} \big\downarrow & b_{12} & b_{13} \\ b_{21} & b_{22} & b_{23} \end{bmatrix}.$$

$$= \begin{bmatrix} a_{11}b_{11} + a_{12}b_{21} & a_{11}b_{12} + a_{12}b_{22} & a_{11}b_{13} + a_{12}b_{23} \\ a_{21}b_{11} + a_{22}b_{21} & a_{21}b_{12} + a_{22}b_{22} & a_{21}b_{13} + a_{22}b_{23} \end{bmatrix}.$$

WORKING RULES TO FIND THE PRODUCT AB

Step I. *Identify the number of columns of A and the number of rows of B. If these are not equal, then AB is not defined. In case, these are equal, then AB is defined.*

Step II. *The order of AB is m × n, where m and n are numbers of rows of A and columns of B respectively.*

Step III. *Draw a horizontal arrow below the first row of A and a vertical arrow on the right of the first column of B. Verify that the number of elements above the horizontal arrow and on the left of the vertical arrow are equal. These arrows will also indicate that only rows of A and only columns of B are to be used for finding the elements of the product AB.*

Step IV. *To find the (i, j)th element of AB, multiply the elements of the ith row of A with the corresponding elements of the jth column of B and then add these products.*

ILLUSTRATIVE EXAMPLES

Example 1. *If* $A = \begin{bmatrix} 4 & 7 \\ 6 & 2 \end{bmatrix}$ *and* $B = \begin{bmatrix} 1 & 8 \\ 3 & 7 \end{bmatrix}$, *find AB.*

Sol. The product AB is defined because number of columns of A and number of rows of B are equal (each = 2).

$$AB = \begin{bmatrix} 4 \;\; 7 \\ \xrightarrow{\quad} \\ 6 \;\; 2 \end{bmatrix} \begin{bmatrix} 1 & \big\downarrow & 8 \\ 3 & & 7 \end{bmatrix} = \begin{bmatrix} 4(1) + 7(3) & 4(8) + 7(7) \\ 6(1) + 2(3) & 6(8) + 2(7) \end{bmatrix} = \begin{bmatrix} \mathbf{25} & \mathbf{81} \\ \mathbf{12} & \mathbf{62} \end{bmatrix}.$$

Example 2. *Find the matrix AB, if defined, when :*

(*i*) $A = \begin{bmatrix} 1 \\ 2 \end{bmatrix}$, $B = \begin{bmatrix} 2 \\ 4 \end{bmatrix}$ (*ii*) $A = \begin{bmatrix} 1 \\ 2 \end{bmatrix}$, $B = [2 \quad 4]$

(*iii*) $A = [2 \quad 4]$, $B = [1 \quad 2]$ (*iv*) $A = \begin{bmatrix} 1 & 1 \\ 0 & 5 \end{bmatrix}$, $B = \begin{bmatrix} 6 & 7 \\ 9 & 8 \end{bmatrix}$

(v) $A = \begin{bmatrix} 1 & 5 & 7 \\ 0 & 6 & 8 \end{bmatrix}$, $B = \begin{bmatrix} 7 & 8 \\ 9 & 8 \\ 6 & 5 \end{bmatrix}$ (vi) $A = \begin{bmatrix} 7 & 8 \\ 9 & 8 \\ 6 & 5 \end{bmatrix}$, $B = \begin{bmatrix} 1 & 5 & 7 \\ 0 & 6 & 8 \end{bmatrix}$.

Sol. (*i*) We have $A = \begin{bmatrix} 1 \\ 2 \end{bmatrix}$, $B = \begin{bmatrix} 2 \\ 4 \end{bmatrix}$.

No. of columns of A = 1 and no. of rows of B = 2. These are not equal.

∴ AB is not defined.

(*ii*) We have $A = \begin{bmatrix} 1 \\ 2 \end{bmatrix}$, $B = [2 \quad 4]$.

No. of columns of A = 1 and no. of rows of B = 1. These are equal.

∴ AB is defined and

$$AB = \begin{bmatrix} \overset{1}{\longrightarrow} \\ 2 \end{bmatrix} [2\downarrow \quad 4] = \begin{bmatrix} 1(2) & 1(4) \\ 2(2) & 2(4) \end{bmatrix} - \begin{bmatrix} \mathbf{2} & \mathbf{4} \\ \mathbf{4} & \mathbf{8} \end{bmatrix}.$$

(*iii*) We have $A = [2 \quad 4]$, $B = [1 \quad 2]$.

No. of columns of A = 2 and no. of rows of B = 1. These are not equal.

∴ AB is not defined.

(*iv*) We have $A = \begin{bmatrix} 1 & 1 \\ 0 & 5 \end{bmatrix}$, $B = \begin{bmatrix} 6 & 7 \\ 9 & 8 \end{bmatrix}$.

No. of columns of A = 2 and no. of rows of B = 2. These are equal.

∴ AB is defined and

$$AB = \begin{bmatrix} 1 \quad 1 \\ \longrightarrow \\ 0 \quad 5 \end{bmatrix} \begin{bmatrix} 6 & \downarrow & 7 \\ 9 & \downarrow & 8 \end{bmatrix} = \begin{bmatrix} 1(6)+1(9) & 1(7)+1(8) \\ 0(6)+5(9) & 0(7)+5(8) \end{bmatrix} = \begin{bmatrix} \mathbf{15} & \mathbf{15} \\ \mathbf{45} & \mathbf{40} \end{bmatrix}.$$

(*v*) We have $A = \begin{bmatrix} 1 & 5 & 7 \\ 0 & 6 & 8 \end{bmatrix}$, $B = \begin{bmatrix} 7 & 8 \\ 9 & 8 \\ 6 & 5 \end{bmatrix}$.

The matrices A and B are of orders 2 × 3 and 3 × 2 respectively, so the product AB is defined and will be of order 2 × 2.

$$AB = \begin{bmatrix} 1 \quad 5 \quad 7 \\ \longrightarrow \\ 0 \quad 6 \quad 8 \end{bmatrix} \begin{bmatrix} 7 & & 8 \\ 9 & \downarrow & 8 \\ 6 & & 5 \end{bmatrix}$$

$$= \begin{bmatrix} 1(7)+5(9)+7(6) & 1(8)+5(8)+7(5) \\ 0(7)+6(9)+8(6) & 0(8)+6(8)+8(5) \end{bmatrix} = \begin{bmatrix} \mathbf{94} & \mathbf{83} \\ \mathbf{102} & \mathbf{88} \end{bmatrix}$$

(*vi*) We have $A = \begin{bmatrix} 7 & 8 \\ 9 & 8 \\ 6 & 5 \end{bmatrix}$, $B = \begin{bmatrix} 1 & 5 & 7 \\ 0 & 6 & 8 \end{bmatrix}$

The matrices A and B are of orders 3 × 2 and 2 × 3 respectively, so the product AB is defined and will be of order 3 × 3.

$$AB = \begin{bmatrix} 7 \quad 8 \\ \longrightarrow \\ 9 \quad 8 \\ 6 \quad 5 \end{bmatrix} \begin{bmatrix} 1 & \downarrow & 5 & 7 \\ 0 & \downarrow & 6 & 8 \end{bmatrix}$$

$$= \begin{bmatrix} 7(1)+8(0) & 7(5)+8(6) & 7(7)+8(8) \\ 9(1)+8(0) & 9(5)+8(6) & 9(7)+8(8) \\ 6(1)+5(0) & 6(5)+5(6) & 6(7)+5(8) \end{bmatrix} = \begin{bmatrix} \mathbf{7} & \mathbf{83} & \mathbf{113} \\ \mathbf{9} & \mathbf{93} & \mathbf{127} \\ \mathbf{6} & \mathbf{60} & \mathbf{82} \end{bmatrix}.$$

2.11. POSITIVE INTEGRAL POWERS OF MATRICES

Let A be a $m \times n$ matrix. By definition of matrix multiplication, the product AA is defined only when $m = n$.

$\therefore$ AA is defined, if A is a square matrix.

The product AA is denoted by A^2.

If A is a n-square matrix, than A^2 is also a n-square matrix.

$\therefore$ AA^2 and A^2 A are both defined and we also have

$$AA^2 = A(AA) = (AA)A = A^2A$$

$$\therefore \qquad AA^2 = A^2\, A = AAA$$

We write AAA as A^3.

Similarly, we have $A^n = AAA \ldots\ldots n$ times, $n \in \mathbf{N}$.

A square matrix A is said to be an **idempotent matrix** if $A^2 = A$.

A square matrix A is said to be an **involutory matrix** if $A^2 = I$.

2.12. MATRIX POLYNOMICAL

Let A be a square matrix of order n. An expression of the form $a_0 A^m + a_1 A^{m-1} + \ldots\ldots + a_m I_n$, where $a_0, a_1, \ldots\ldots, a_m$ are scalars, is called a **matrix polynomical.**

If $f(x) = \alpha_0 x^m + \alpha_1 x^{m-1} + \ldots + \alpha_m$, then for a square matrix A of order n, we define $f(A) = \alpha_0 A^m + \alpha_1 A^{m-1} + \ldots + \alpha_m I_n$.

Example 3. *If* $A = \begin{bmatrix} 2 & -1 \\ 3 & 2 \end{bmatrix}$, $B = \begin{bmatrix} 0 & 4 \\ -1 & 7 \end{bmatrix}$, *find* $3A^2 - 2B$.

Sol.

$$A^2 = \begin{bmatrix} 2 & -1 \\ 3 & 2 \end{bmatrix} \begin{bmatrix} 2 & -1 \\ 3 & 2 \end{bmatrix} = \begin{bmatrix} 4-3 & -2-2 \\ 6+6 & -3+4 \end{bmatrix} = \begin{bmatrix} 1 & -4 \\ 12 & 1 \end{bmatrix}$$

$$\therefore \quad 3A^2 - 2B = 3\begin{bmatrix} 1 & -4 \\ 12 & 1 \end{bmatrix} - 2\begin{bmatrix} 0 & 4 \\ -1 & 7 \end{bmatrix} = \begin{bmatrix} 3 & -12 \\ 36 & 3 \end{bmatrix} - \begin{bmatrix} 0 & 8 \\ -2 & 14 \end{bmatrix} = \begin{bmatrix} \mathbf{3} & \mathbf{-20} \\ \mathbf{38} & \mathbf{-11} \end{bmatrix}.$$

Example 4. (*i*) *If A and B are two matrices such that A + B and AB are both defined, show that A and B are square matrices of same order.*

(*ii*) *If A and B are two matrices such that AB and BA are both defined and order of A be* $m \times n$, *show that order of B is* $n \times m$.

Sol. (*i*) Let orders of A and B be $m \times n$ and $p \times q$ respectively. Since A + B is defined, we have $m = p$ and $n = q$.

$\therefore$ Order of B is also $m \times n$.

Since AB is defined, we have : no. of columns of A = no. of rows of B

$\therefore$ $n = m$ *i.e.,* A and B are both square matrices of orders $n \times n$ each.

(*ii*) Let order of B be $p \times q$.

Since AB is defined, we have : no. of columns of A = no. of rows of B

$$\therefore \qquad n = p$$

Also, BA is defined. $\therefore$ No. of columns of B = no. of rows of A

$\therefore$ $q = m$.

$\therefore$ The orders of B is $n \times m$.

Example 5. *If* $A = \begin{bmatrix} 1 & -1 \\ 2 & 3 \end{bmatrix}$, $B = \begin{bmatrix} 2 & 1 \\ 1 & 0 \end{bmatrix}$, *verify that* $(A+B)^2 \neq A^2 + 2AB + B^2$.

Sol. $$A + B = \begin{bmatrix} 1 & -1 \\ 2 & 3 \end{bmatrix} + \begin{bmatrix} 2 & 1 \\ 1 & 0 \end{bmatrix} = \begin{bmatrix} 3 & 0 \\ 3 & 3 \end{bmatrix}$$

$$\therefore \quad (A+B)^2 = \begin{bmatrix} 3 & 0 \\ 3 & 3 \end{bmatrix} \begin{bmatrix} 3 & 0 \\ 3 & 3 \end{bmatrix} = \begin{bmatrix} 9+0 & 0+0 \\ 9+9 & 0+9 \end{bmatrix} = \begin{bmatrix} 9 & 0 \\ 18 & 9 \end{bmatrix} \quad \ldots(1)$$

$$A^2 = \begin{bmatrix} 1 & -1 \\ 2 & 3 \end{bmatrix} \begin{bmatrix} 1 & -1 \\ 2 & 3 \end{bmatrix} = \begin{bmatrix} 1-2 & -1-3 \\ 2+6 & -2+9 \end{bmatrix} = \begin{bmatrix} -1 & -4 \\ 8 & 7 \end{bmatrix}$$

$$AB = \begin{bmatrix} 1 & -1 \\ 2 & 3 \end{bmatrix} \begin{bmatrix} 2 & 1 \\ 1 & 0 \end{bmatrix} = \begin{bmatrix} 2-1 & 1-0 \\ 4+3 & 2+0 \end{bmatrix} = \begin{bmatrix} 1 & 1 \\ 7 & 2 \end{bmatrix}$$

$$B^2 = \begin{bmatrix} 2 & 1 \\ 1 & 0 \end{bmatrix} \begin{bmatrix} 2 & 1 \\ 1 & 0 \end{bmatrix} = \begin{bmatrix} 4+1 & 2+0 \\ 2+0 & 1+0 \end{bmatrix} = \begin{bmatrix} 5 & 2 \\ 2 & 1 \end{bmatrix}$$

$$\therefore \quad A^2 + 2AB + B^2 = \begin{bmatrix} -1 & -4 \\ 8 & 7 \end{bmatrix} + 2\begin{bmatrix} 1 & 1 \\ 7 & 2 \end{bmatrix} + \begin{bmatrix} 5 & 2 \\ 2 & 1 \end{bmatrix}$$

$$= \begin{bmatrix} -1+2+5 & -4+2+2 \\ 8+14+2 & 7+4+1 \end{bmatrix} = \begin{bmatrix} 6 & 0 \\ 24 & 12 \end{bmatrix} \quad \ldots(2)$$

$\therefore$ From (1) and (2), $\mathbf{(A+B)^2 \neq A^2 + 2AB + B^2}$.

Example 6. *If* $A = \begin{bmatrix} 2 & 1 \\ 3 & 4 \end{bmatrix}$ *and* $B = \begin{bmatrix} 1 & 3 \\ 7 & 5 \end{bmatrix}$, *show that* $AB \neq BA$.

Sol. No. of columns in A = no. of rows in B(each = 2)

$\therefore$ AB is defined and

$$AB = \begin{bmatrix} 2 & 1 \\ 3 & 4 \end{bmatrix} \begin{bmatrix} 1 & 3 \\ 7 & 5 \end{bmatrix} = \begin{bmatrix} 2(1)+1(7) & 2(3)+1(5) \\ 3(1)+4(7) & 3(3)+4(5) \end{bmatrix} = \begin{bmatrix} 9 & 11 \\ 31 & 29 \end{bmatrix} \quad \ldots(1)$$

Also, no. of columns in B = no. of rows in A (each = 2)

$\therefore$ BA is defined and

$$BA = \begin{bmatrix} 1 & 3 \\ 7 & 5 \end{bmatrix} \begin{bmatrix} 2 & 1 \\ 3 & 4 \end{bmatrix} = \begin{bmatrix} 1(2)+3(3) & 1(1)+3(4) \\ 7(2)+5(3) & 7(1)+5(4) \end{bmatrix} = \begin{bmatrix} 11 & 13 \\ 29 & 27 \end{bmatrix} \quad \ldots(2)$$

$\therefore$ From (1) and (2), $\mathbf{AB \neq BA}$.

2.13. PROPERTIES OF MATRIX MULTIPLICATION

(*i*) **Matrix multiplication is not commutative in general.** For matrices A and B, we need not have AB = BA.

For example, if :

(*a*) A and B are 2×3 and 3×4 matrices, then AB is a 2×4 matrix whereas BA is not defined.

(*b*) A and B are 2×3 and 3×2 matrices, then AB is a 2×2 matrix and BA is a 3×3 matrix.

(*c*) A and B are 2×2 matrices, then both AB and BA are 2×2 matrices. Even in this case, we may not have AB = BA.

(*ii*) **Associative law of multiplication.** If A, B and C be matrices of the type $m \times n$, $n \times p$ and $p \times q$ respectively, then (AB)C = A(BC).

(*iii*) **Matrix multiplication is distributive with respect to addition**

(*a*) If A, B and C are matrices of the type $m \times n$, $n \times p$ and $n \times p$ respectively, then

$$A(B + C) = AB + AC.$$

(*b*) If A, B and C are matrices of the type $m \times n$, $m \times n$ and $n \times p$ respectively, then

$$(A + B)C = AC + BC.$$

(*iv*) **The product of non-zero matrices may be a zero matrix.** For matrices A and B, it may be possible that AB = O and neither A nor B is a zero matrix.

For example, let $A = \begin{bmatrix} 2 & 1 \\ 0 & 0 \end{bmatrix}$ and $B = \begin{bmatrix} 1 & 2 \\ -2 & -4 \end{bmatrix}$. Here $A \neq O$, $B \neq O$

and $$AB = \begin{bmatrix} 2 & 1 \\ 0 & 0 \end{bmatrix} \begin{bmatrix} 1 & 2 \\ -2 & -4 \end{bmatrix} = \begin{bmatrix} 2(1)+1(-2) & 2(2)+1(-4) \\ 0(1)+0(-2) & 0(2)+0(-4) \end{bmatrix} = \begin{bmatrix} 0 & 0 \\ 0 & 0 \end{bmatrix} = O.$$

Example 7. *If* $A = [1 \;\; 3 \;\; 4]$, $B = \begin{bmatrix} 2 & 1 \\ 3 & 7 \\ 6 & 8 \end{bmatrix}$, $C = \begin{bmatrix} 3 & 4 \\ 5 & 6 \\ 1 & 0 \end{bmatrix}$, *verify that*

$A(B + C) = AB + AC.$

Sol. $$B + C = \begin{bmatrix} 2 & 1 \\ 3 & 7 \\ 6 & 8 \end{bmatrix} + \begin{bmatrix} 3 & 4 \\ 5 & 6 \\ 1 & 0 \end{bmatrix} = \begin{bmatrix} 5 & 5 \\ 8 & 13 \\ 7 & 8 \end{bmatrix}$$

$$\therefore \quad \text{L.H.S.} = A(B + C) = [1 \;\; 3 \;\; 4] \begin{bmatrix} 5 & 5 \\ 8 & 13 \\ 7 & 8 \end{bmatrix} \qquad ...(1)$$

$$= [1(5) + 3(8) + 4(7) \quad 1(5) + 3(13) + 4(8)]$$
$$= [5 + 24 + 28 \quad 5 + 39 + 32] = [57 \quad 76]$$

$$AB = [1 \;\; 3 \;\; 4] \begin{bmatrix} 2 & 1 \\ 3 & 7 \\ 6 & 8 \end{bmatrix} = [1(2) + 3(3) + 4(6) \quad 1(1) + 3(7) + 4(8)]$$

$$= [2 + 9 + 24 \quad 1 + 21 + 32] = [35 \quad 54] \qquad ...(2)$$

$$AC = \begin{bmatrix} 1 & 3 & 4 \\ \multicolumn{3}{c}{\longrightarrow} \end{bmatrix} \begin{bmatrix} 3 & 4 \\ 5 \ \downarrow & 6 \\ 1 & 0 \end{bmatrix} = [1(3) + 3(5) + 4(1) \quad 1(4) + 3(6) + 4(0)]$$

$$= [3 + 15 + 4 \quad 4 + 18 + 0] = [22 \quad 22]$$

$\therefore$ R.H.S. = AB + AC = [35 54] + [22 22] = [57 76].

$\therefore$ From (1), (2) and (3), **A(B + C) = AB + AC.**

Example 8. *If* $A = \begin{bmatrix} 3 & -2 \\ 4 & -2 \end{bmatrix}$ *and* $I = \begin{bmatrix} 1 & 0 \\ 0 & 1 \end{bmatrix}$, *find k such that* $A^2 = kA - 2I$.

Sol. We have $A = \begin{bmatrix} 3 & -2 \\ 4 & -2 \end{bmatrix}$.

$$\therefore \quad A^2 = AA = \begin{bmatrix} 3 & -2 \\ \multicolumn{2}{c}{\longrightarrow} \\ 4 & -2 \end{bmatrix} \begin{bmatrix} 3 \ \downarrow & -2 \\ 4 & -2 \end{bmatrix} = \begin{bmatrix} 9-8 & -6+4 \\ 12-8 & -8+4 \end{bmatrix} = \begin{bmatrix} 1 & -2 \\ 4 & -4 \end{bmatrix}$$

$$kA - 2I = k\begin{bmatrix} 3 & -2 \\ 4 & -2 \end{bmatrix} - 2\begin{bmatrix} 1 & 0 \\ 0 & 1 \end{bmatrix} = \begin{bmatrix} 3k & -2k \\ 4k & -2k \end{bmatrix} - \begin{bmatrix} 2 & 0 \\ 0 & 2 \end{bmatrix} = \begin{bmatrix} 3k-2 & -2k \\ 4k & -2k-2 \end{bmatrix}$$

We have $A^2 = kA - 2I$.

$$\therefore \quad \begin{bmatrix} 1 & -2 \\ 4 & -4 \end{bmatrix} = \begin{bmatrix} 3k-2 & -2k \\ 4k & -2k-2 \end{bmatrix}$$

$$\Rightarrow \quad 3k - 2 = 1 \quad \ldots(1) \qquad -2k = -2 \quad \ldots(2)$$

$$4k = 4 \quad \ldots(3) \qquad -2k - 2 = -4 \quad \ldots(4)$$

(1) $\Rightarrow k = 1$, (2) $\Rightarrow k = 1$, (3) $\Rightarrow k = 1$, (4) $\Rightarrow k = 1$.

$\therefore$ $k = \mathbf{1}$.

Example 9. *If* $A = \begin{bmatrix} \cos\theta & \sin\theta \\ -\sin\theta & \cos\theta \end{bmatrix}$, *then prove that* $A^n = \begin{bmatrix} \cos n\theta & \sin n\theta \\ -\sin n\theta & \cos n\theta \end{bmatrix}$, $n \in \mathbf{N}$.

Sol. We shall prove the result by using P.M.I.*

We have $A = \begin{bmatrix} \cos\theta & \sin\theta \\ -\sin\theta & \cos\theta \end{bmatrix}$. $\therefore \quad A^1 = \begin{bmatrix} \cos 1.\theta & \sin 1.\theta \\ -\sin 1.\theta & \cos 1.\theta \end{bmatrix}$

$\therefore$ The result is true for $n = 1$. Let the result be true for $n = k$.

$$\therefore \quad A^k = \begin{bmatrix} \cos k\theta & \sin k\theta \\ -\sin k\theta & \cos k\theta \end{bmatrix}$$

Let $\quad n = k + 1$.

$$\text{Now} \quad A^{k+1} = A.A^k = \begin{bmatrix} \cos\theta & \sin\theta \\ -\sin\theta & \cos\theta \end{bmatrix} \begin{bmatrix} \cos k\theta & \sin k\theta \\ -\sin k\theta & \cos k\theta \end{bmatrix}$$

$$= \begin{bmatrix} \cos\theta\cos k\theta - \sin\theta\sin k\theta & \cos\theta\sin k\theta + \sin\theta\cos k\theta \\ -\sin\theta\cos k\theta - \cos\theta\sin k\theta & -\sin\theta\sin k\theta + \cos\theta\cos k\theta \end{bmatrix}$$

$$= \begin{bmatrix} \cos(\theta + k\theta) & \sin(k\theta + \theta) \\ -\sin(\theta + k\theta) & \cos(\theta + k\theta) \end{bmatrix} = \begin{bmatrix} \cos(k+1)\theta & \sin(k+1)\theta \\ -\sin(k+1)\theta & \cos(k+1)\theta \end{bmatrix}$$

| cos (A + B) = cos A cos B − sin A sin B
| sin (A + B) = sin A cos B + cos A sin B

$\therefore$ The result is true for $n = k + 1$.

---

* **Principle of Mathematical Induction (P.M.I).** This principle states that if a statement is true for $n = 1$ and the truth of the statement for any arbitarary but fixed $k \in \mathbf{N}$ implies the truth of the statement for the next natural number $k + 1$, then the statement is true for all natural numbers.

$\therefore$ By P.M.I., we have $A^n = \begin{bmatrix} \cos n\theta & \sin n\theta \\ -\sin n\theta & \cos n\theta \end{bmatrix}$, $n \in \mathbf{N}$.

Example 10. *If* $f(x) = \begin{pmatrix} \cos x & -\sin x & 0 \\ \sin x & \cos x & 0 \\ 0 & 0 & 1 \end{pmatrix}$, *show that* $f(x)\, f(y) = f(x + y)$.

(P.T.U. B. Pharma, May 2003)

Sol. Given $$f(x) = \begin{pmatrix} \cos x & -\sin x & 0 \\ \sin x & \cos x & 0 \\ 0 & 0 & 1 \end{pmatrix}$$

$\therefore$ $$f(y) = \begin{pmatrix} \cos y & -\sin y & 0 \\ \sin y & \cos y & 0 \\ 0 & 0 & 1 \end{pmatrix}$$

Now $$f(x)\, f(y) = \begin{pmatrix} \cos x & -\sin x & 0 \\ \sin x & \cos x & 0 \\ 0 & 0 & 1 \end{pmatrix} \begin{pmatrix} \cos y & -\sin y & 0 \\ \sin y & \cos y & 0 \\ 0 & 0 & 1 \end{pmatrix}$$

$$= \begin{pmatrix} \cos x \cos y - \sin x \sin y & -\cos x \sin y - \sin x \cos y & 0 \\ \sin x \cos y + \cos x \sin y & -\sin x \sin y + \cos x \cos y & 0 \\ 0 & 0 & 1 \end{pmatrix}$$

$$= \begin{pmatrix} \cos (x + y) & -\sin (x + y) & 0 \\ \sin (x + y) & \cos (x + y) & 0 \\ 0 & 0 & 1 \end{pmatrix} = f(x + y).$$ Hence Proved.

| $\cos (A + B) = \cos A \cos B - \sin A \sin B$
| $\sin(A + B) = \sin A \cos B + \cos A \sin B$

EXERCISE 2.4

1. Find the order of A × B if the orders of A and B are respectively :
(*i*) 2 × 2 and 2 × 3 (*ii*) 4 × 1 and 1 × 3 (*iii*) 4 × 4 and 4 × 1.

2. Compute AB if $A = \begin{bmatrix} 1 & 2 & 3 \\ 4 & 5 & 6 \end{bmatrix}$ and $B = \begin{bmatrix} 2 & 5 & 3 \\ 3 & 6 & 4 \\ 4 & 7 & 5 \end{bmatrix}$.

3. If $A = \begin{bmatrix} 2 & -3 & -5 \\ -1 & 4 & 5 \\ 1 & -3 & -4 \end{bmatrix}$ and $B = \begin{bmatrix} 2 & -2 & -4 \\ -1 & 3 & 4 \\ 1 & -2 & -3 \end{bmatrix}$, show that : AB = A and BA = B.

4. Find AB, if defined, for the following matrices :

(*i*) $A = [1 \quad 4]$, $B = \begin{bmatrix} 3 \\ 4 \\ 5 \end{bmatrix}$ (*ii*) $A = \begin{bmatrix} 4 & 1 \\ 3 & 5 \end{bmatrix}$ $B = \begin{bmatrix} 5 \\ 9 \end{bmatrix}$

(*iii*) $A = \begin{bmatrix} 1 & 5 & 7 \\ 0 & 6 & 9 \end{bmatrix}$, $B = \begin{bmatrix} 3 & 0 \\ 6 & 8 \\ 7 & 0 \end{bmatrix}$ (*iv*) $A = [1 \quad 2]$, $B = [4 \quad 7]$.

5. If $A = \begin{bmatrix} \cos \alpha & \sin \alpha \\ -\sin \alpha & \cos \alpha \end{bmatrix}$, show that $A^2 = \begin{bmatrix} \cos 2\alpha & \sin 2\alpha \\ -\sin 2\alpha & \cos 2\alpha \end{bmatrix}$.

6. (*i*) Given that $A = \begin{bmatrix} 2 & -3 & -5 \\ -1 & 4 & 5 \\ 1 & -3 & -4 \end{bmatrix}$ and $B = \begin{bmatrix} -1 & 3 & 5 \\ 1 & -3 & -5 \\ -1 & 3 & 5 \end{bmatrix}$, show that AB = O.

(*ii*) If $A = \begin{pmatrix} 1 & 2 & -1 \\ 2 & 3 & -1 \\ 3 & 4 & -1 \end{pmatrix}, B = \begin{pmatrix} 1 & -3 & -5 \\ -1 & 3 & -5 \\ -1 & 3 & -5 \end{pmatrix}$, Calculate AB

7. If $A = \begin{bmatrix} \cos\theta & \sin\theta \\ \sin\theta & \cos\theta \end{bmatrix}$, $B = \begin{bmatrix} \cos\phi & \sin\phi \\ \sin\phi & \cos\phi \end{bmatrix}$, show that AB = BA.

8. If $\begin{bmatrix} x & y \\ 3y & x \end{bmatrix}\begin{bmatrix} 1 \\ 2 \end{bmatrix} = \begin{bmatrix} 3 \\ 5 \end{bmatrix}$, find x and y.

9. If $A = \begin{bmatrix} 2 & -1 & 1 \\ 0 & 1 & 2 \\ 1 & 0 & 1 \end{bmatrix}$, find A^2 and A^3.

10. Find the value of x for which the matrix product

$$\begin{bmatrix} 2 & 0 & 7 \\ 0 & 1 & 0 \\ 1 & -2 & 1 \end{bmatrix}\begin{bmatrix} -x & 14x & 7x \\ 0 & 1 & 0 \\ x & -4x & -2x \end{bmatrix}$$

equals an identity matrix.

11. Show that the product AB of the matrices

$A = \begin{bmatrix} \cos^2\theta & \cos\theta\sin\theta \\ \cos\theta\sin\theta & \sin^2\theta \end{bmatrix}$ and $B = \begin{bmatrix} \cos^2\phi & \cos\phi\sin\phi \\ \cos\phi\sin\phi & \sin^2\phi \end{bmatrix}$ is a null matrix, where θ and φ differ by an odd multiple of $\frac{\pi}{2}$.

12. Give an example of two matrices A and B such that AB = O, where neither A = O nor B = O.

13. If $A = \begin{bmatrix} 3 & 1 \\ -1 & 2 \end{bmatrix}$, show that $A^2 - 5A + 7I = O$. *(P.T.U. B. Pharma, May 2006)*

14. If $A = \begin{bmatrix} 2 & 3 \\ -1 & 2 \end{bmatrix}$, find $f(A)$ where $f(x) = x^3 + 3x^2 - 4x$.

15. If $A = \begin{bmatrix} 3 & 2 & 0 \\ 1 & 4 & 0 \\ 0 & 0 & 5 \end{bmatrix}$, show that $A^2 - 7A + 10I_3 = O$.

16. For what value of x :

(*i*) $[1 \quad 1 \quad x]\begin{bmatrix} 1 & 0 & 2 \\ 0 & 2 & 1 \\ 2 & 1 & 0 \end{bmatrix}\begin{bmatrix} 1 \\ 1 \\ 1 \end{bmatrix} = O$?

(*ii*) $[x \quad 4 \quad 1]\begin{bmatrix} 2 & 1 & 0 \\ 1 & 0 & 2 \\ 0 & 2 & 4 \end{bmatrix}\begin{bmatrix} x \\ 4 \\ 1 \end{bmatrix} = O$?

17. If $A = \begin{bmatrix} 1 & -3 & 2 \\ 2 & 1 & -3 \\ 4 & -3 & -1 \end{bmatrix}$, $B = \begin{bmatrix} 1 & 4 & 1 & 0 \\ 2 & 1 & 1 & 1 \\ 1 & -2 & 1 & 2 \end{bmatrix}$ and $C = \begin{bmatrix} 2 & 1 & -1 & -2 \\ 3 & -2 & -1 & -1 \\ 2 & -5 & -1 & 0 \end{bmatrix}$,

show that AB = AC. Thus, AB = AC does not necessarily imply B = C.

18. If $A = \begin{bmatrix} 3 & -4 \\ 1 & -1 \end{bmatrix}$, prove that $A^n = \begin{bmatrix} 1+2n & -4n \\ n & 1-2n \end{bmatrix}$, $n \in \mathbf{N}$.

19. If $A = \begin{bmatrix} \lambda & 1 \\ 0 & \lambda \end{bmatrix}$, prove that $A^n = \begin{bmatrix} \lambda^n & n\lambda^{n-1} \\ 0 & \lambda^n \end{bmatrix}$, $n \in \mathbf{N}$.

Answers

1. (*i*) 2 × 3 (*ii*) 4 × 3 (*iii*) 4 × 1

2. $\begin{bmatrix} 20 & 38 & 26 \\ 47 & 92 & 62 \end{bmatrix}$

4. (*i*) Not defined (*ii*) $\begin{bmatrix} 29 \\ 60 \end{bmatrix}$ (*iii*) $\begin{bmatrix} 82 & 40 \\ 99 & 48 \end{bmatrix}$ (*iv*) Not defined

6. (*ii*) $\begin{pmatrix} 0 & 0 & -10 \\ 0 & 0 & -20 \\ 0 & 0 & -30 \end{pmatrix}$ **8.** $x = 1, y = 1$ **9.** $\begin{bmatrix} 5 & -3 & 1 \\ 2 & 1 & 4 \\ 3 & -1 & 2 \end{bmatrix}, \begin{bmatrix} 11 & -8 & 0 \\ 8 & -1 & 8 \\ 8 & -4 & 3 \end{bmatrix}$

10. $x = 1/5$ **12.** $A = \begin{bmatrix} 2 & 0 \\ 0 & 0 \end{bmatrix}, B = \begin{bmatrix} 0 & 0 \\ 0 & 3 \end{bmatrix}$ **14.** $\begin{bmatrix} -15 & 51 \\ -17 & -15 \end{bmatrix}$

16. (*i*) -2 (*ii*) $-2 \pm \sqrt{6}i$

Hint

11. Use : $\cos(\theta - \phi) = 0$.

2.14. TRANSPOSE OF A MATRIX

(*P.T.U. B.Pharma, May 2007*)

Let A be a matrix of order $m \times n$. The $n \times m$ matrix obtained from A by interchanging its rows and columns is called the **transpose** of A and is denoted by A′ or by A^T.

ILLUSTRATIVE EXAMPLES

Example 1. *Find the transpose of the following matrices :*

(*i*) $[1 \quad 2 \quad 6]$ (*ii*) $\begin{bmatrix} 5 \\ 9 \\ 0 \end{bmatrix}$ (*iii*) $\begin{bmatrix} 2 & 3 \\ 0 & 8 \end{bmatrix}$ (*iv*) $\begin{bmatrix} 5 & 8 & -7 & 6 \\ 2 & 3 & -1 & 0 \end{bmatrix}$.

Sol. (*i*) Let $A = [1 \quad 2 \quad 6]$ $\therefore A' = [1 \quad 2 \quad 6]' = \begin{bmatrix} \mathbf{1} \\ \mathbf{2} \\ \mathbf{6} \end{bmatrix}$.

(*ii*) Let $A = \begin{bmatrix} 5 \\ 9 \\ 0 \end{bmatrix}$ $\therefore A' = \begin{bmatrix} 5 \\ 9 \\ 0 \end{bmatrix}' = [\mathbf{5} \quad \mathbf{9} \quad \mathbf{0}]$.

(*iii*) Let $A = \begin{bmatrix} 2 & 3 \\ 0 & 8 \end{bmatrix}$ $\therefore A' = \begin{bmatrix} 2 & 3 \\ 0 & 8 \end{bmatrix}' = \begin{bmatrix} \mathbf{2} & \mathbf{0} \\ \mathbf{3} & \mathbf{8} \end{bmatrix}$.

(*iv*) Let $A = \begin{bmatrix} 5 & 8 & -7 & 6 \\ 2 & 3 & -1 & 0 \end{bmatrix}$. $\therefore A' = \begin{bmatrix} 5 & 8 & -7 & 6 \\ 2 & 3 & -1 & 0 \end{bmatrix}' = \begin{bmatrix} \mathbf{5} & \mathbf{2} \\ \mathbf{8} & \mathbf{3} \\ \mathbf{-7} & \mathbf{-1} \\ \mathbf{6} & \mathbf{0} \end{bmatrix}$.

2.15. PROPERTIES OF TRANSPOSE

(*i*) (A′)′ = A, where A is any matrix.

(*ii*) (A + B)′ = A′ + B′, where A and B are matrices of the same order.

(*iii*) (kA)′ = kA′, where A is any matrix and k is any number.

(*iv*) (AB)′ = B′A′, where A and B are matrices for which AB is defined.

Remark. Property (*iv*) is known as the *reversal law for the transpose of the product.*

Example 2. *If* $A = \begin{bmatrix} 2 & 1 \\ 3 & 5 \end{bmatrix}$ *and* $B = \begin{bmatrix} 9 & 0 \\ 6 & 8 \end{bmatrix}$, *then verify that :*

(*i*) $(A')' = A$ (*ii*) $(A + B)' = A' + B'$ (*iii*) $(kA)' = kA'$ (*iv*) $(AB)' = B'A'$.

Sol. (*i*) We have $A = \begin{bmatrix} 2 & 1 \\ 3 & 5 \end{bmatrix}$.

$$\therefore \quad A' = \begin{bmatrix} 2 & 1 \\ 3 & 5 \end{bmatrix}' = \begin{bmatrix} 2 & 3 \\ 1 & 5 \end{bmatrix} \text{ and } (A')' = \begin{bmatrix} 2 & 3 \\ 1 & 5 \end{bmatrix}' = \begin{bmatrix} 2 & 1 \\ 3 & 5 \end{bmatrix} = A$$

$\therefore$ **(A′)′ = A.**

$$(ii)\ A + B = \begin{bmatrix} 2 & 1 \\ 3 & 5 \end{bmatrix} + \begin{bmatrix} 9 & 0 \\ 6 & 8 \end{bmatrix} = \begin{bmatrix} 11 & 1 \\ 9 & 13 \end{bmatrix}$$

$$\therefore \quad \text{L.H.S.} = (A + B)' = \begin{bmatrix} 11 & 1 \\ 9 & 13 \end{bmatrix}' = \begin{bmatrix} 11 & 9 \\ 1 & 13 \end{bmatrix}$$

$$\text{R.H.S.} = A' + B' = \begin{bmatrix} 2 & 1 \\ 3 & 5 \end{bmatrix}' + \begin{bmatrix} 9 & 0 \\ 6 & 8 \end{bmatrix}' = \begin{bmatrix} 2 & 3 \\ 1 & 5 \end{bmatrix} + \begin{bmatrix} 9 & 6 \\ 0 & 8 \end{bmatrix} = \begin{bmatrix} 11 & 9 \\ 1 & 13 \end{bmatrix}$$

$\therefore$ **(A + B)′ = A′ + B′.**

$$(iii) \quad kA = k\begin{bmatrix} 2 & 1 \\ 3 & 5 \end{bmatrix} = \begin{bmatrix} 2k & k \\ 3k & 5k \end{bmatrix}$$

$$\therefore \quad \text{L.H.S.} = (kA)' = \begin{bmatrix} 2k & k \\ 3k & 5k \end{bmatrix}' = \begin{bmatrix} 2k & 3k \\ k & 5k \end{bmatrix}$$

$$\text{R.H.S.} = kA' = k\begin{bmatrix} 2 & 1 \\ 3 & 5 \end{bmatrix}' = k\begin{bmatrix} 2 & 3 \\ 1 & 5 \end{bmatrix} = \begin{bmatrix} 2k & 3k \\ k & 5k \end{bmatrix}.$$

$\therefore$ **(kA)′ = kA′.**

$$(iv) \quad AB = \begin{bmatrix} 2 & 1 \\ 3 & 5 \end{bmatrix} \begin{bmatrix} 9 & 0 \\ 6 & 8 \end{bmatrix} = \begin{bmatrix} 18+6 & 0+8 \\ 27+30 & 0+40 \end{bmatrix} = \begin{bmatrix} 24 & 8 \\ 57 & 40 \end{bmatrix}$$

$$\therefore \quad \text{L.H.S.} = (AB)' = \begin{bmatrix} 24 & 8 \\ 57 & 40 \end{bmatrix}' = \begin{bmatrix} 24 & 57 \\ 8 & 40 \end{bmatrix}.$$

$$\text{R.H.S.} = B'A' = \begin{bmatrix} 9 & 0 \\ 6 & 8 \end{bmatrix}' \begin{bmatrix} 2 & 1 \\ 3 & 5 \end{bmatrix}' = \begin{bmatrix} 9 & 6 \\ 0 & 8 \end{bmatrix} \begin{bmatrix} 2 & 3 \\ 1 & 5 \end{bmatrix}$$

$$= \begin{bmatrix} 18+6 & 27+30 \\ 0+8 & 0+40 \end{bmatrix} = \begin{bmatrix} 24 & 57 \\ 8 & 40 \end{bmatrix}.$$

$\therefore$ **(AB)′ = B′A′.**

Example 3. *Verify that (AB)′ = B′A′, where*

(i) $A = \begin{bmatrix} 3 \\ 1 \\ -2 \end{bmatrix}, B = [1 \quad -5 \quad 7]$ (ii) $A = \begin{bmatrix} 2 & 3 \\ 4 & 1 \end{bmatrix}, B = \begin{bmatrix} 1 & 0 & -1 \\ 2 & 1 & 3 \end{bmatrix}$.

Sol. (i) We have $A = \begin{bmatrix} 3 \\ 1 \\ -2 \end{bmatrix}$ and $B = [1 \quad -5 \quad 7]$

$$\therefore \quad AB = \begin{bmatrix} 3 \\ \to \\ 1 \\ -2 \end{bmatrix} [1 \downarrow \quad -5 \quad 7] = \begin{bmatrix} 3 & -15 & 21 \\ 1 & -5 & 7 \\ -2 & 10 & -14 \end{bmatrix}$$

$$\therefore \quad \text{L.H.S.} = (AB)' = \begin{bmatrix} 3 & -15 & 21 \\ 1 & -5 & 7 \\ -2 & 10 & -14 \end{bmatrix}' = \begin{bmatrix} 3 & 1 & -2 \\ -15 & -5 & 10 \\ 21 & 7 & -14 \end{bmatrix}$$

$$\text{R.H.S.} = B'A' = [1 \quad -5 \quad 7]' \begin{bmatrix} 3 \\ 1 \\ -2 \end{bmatrix}' = \begin{bmatrix} 1 \\ \to \\ -5 \\ 7 \end{bmatrix} [3 \downarrow \quad 1 \quad -2] = \begin{bmatrix} 3 & 1 & -2 \\ -15 & -5 & 10 \\ 21 & 7 & -14 \end{bmatrix}.$$

$\therefore$ **(AB)′ = B′A′.**

(ii) We have $A = \begin{bmatrix} 2 & 3 \\ 4 & 1 \end{bmatrix}$ and $B = \begin{bmatrix} 1 & 0 & -1 \\ 2 & 1 & 3 \end{bmatrix}$

$$\therefore \quad AB = \begin{bmatrix} 2 & 3 \\ \multicolumn{2}{c}{\longrightarrow} \\ 4 & 1 \end{bmatrix} \begin{bmatrix} 1 \downarrow & 0 & -1 \\ 2 \downarrow & 1 & 3 \end{bmatrix} = \begin{bmatrix} 2+6 & 0+3 & -2+9 \\ 4+2 & 0+1 & -4+3 \end{bmatrix} = \begin{bmatrix} 8 & 3 & 7 \\ 6 & 1 & -1 \end{bmatrix}$$

$$\therefore \quad \text{L.H.S.} = (AB)' = \begin{bmatrix} 8 & 3 & 7 \\ 6 & 1 & -1 \end{bmatrix}' = \begin{bmatrix} 8 & 6 \\ 3 & 1 \\ 7 & -1 \end{bmatrix}$$

$$\text{R.H.S.} = B'A' = \begin{bmatrix} 1 & 0 & -1 \\ 2 & 1 & 3 \end{bmatrix}' \begin{bmatrix} 2 & 3 \\ 4 & 1 \end{bmatrix}' = \begin{bmatrix} 1 & 2 \\ \multicolumn{2}{c}{\longrightarrow} \\ 0 & 1 \\ -1 & 3 \end{bmatrix} \begin{bmatrix} 2 \downarrow & 4 \\ 3 \downarrow & 1 \end{bmatrix}$$

$$= \begin{bmatrix} 2+6 & 4+2 \\ 0+3 & 0+1 \\ -2+9 & -4+3 \end{bmatrix} = \begin{bmatrix} 8 & 6 \\ 3 & 1 \\ 7 & -1 \end{bmatrix}$$

$\therefore$ **(AB)′ = B′A′.**

2.16. SYMMETRIC MATRIX

(P.T.U. B. Pharma, May 2006)

A square matrix $A = [a_{ij}]$ is said to be a **symmetric matrix** if its (*i*, *j*)th element is the same as its (*j*, *i*)th element *i.e.*, $a_{ij} = a_{ji} \ \forall \ i, j$.

Theorem. *A necessary and sufficient condition for the matrix A to be symmetric is that A′ = A.*

Proof. A is a symmetric matrix

$\Leftrightarrow$ A is a square matrix and (*i*, *j*)th element of A = (*j*, *i*)th element of A for all *i*, *j*.

$\Leftrightarrow$ A is a square matrix and (*i*, *j*)th element of A = (*i*, *j*)th element of A′ for all *i*, *j*.

$\Leftrightarrow$ A = A′ *i.e.*, A′ = A.

2.17. SKEW-SYMMETRIC MATRIX

(P.T.U. B. Pharma, May 2006, May 2007)

A square matrix A = $[a_{ij}]$ is said to be a **skew-symmetric matrix** if its (i, j)th element is the negative of its (j, i)th element *i.e.*, $a_{ij} = -a_{ji} \ \forall \ i, j$.

For example, $A = \begin{bmatrix} 0 & 9 & -6 \\ -9 & 0 & 8 \\ 6 & -8 & 0 \end{bmatrix}$ is a skew-symmetric matrix, because

$$a_{12} = -a_{21} = 9, \quad a_{13} = -a_{31} = -6, \quad a_{23} = -a_{32} = 8,$$
$$a_{11} = -a_{11} = 0, \quad a_{22} = -a_{22} = 0, \quad a_{33} = -a_{33} = 0.$$

In a skew-symmetric matrix, we have $a_{ij} = -a_{ji} \ \forall \ i, j$.

∴ In particular, $a_{ii} = -a_{ii} \ \forall \ i$.

⇒ $2a_{ii} = 0 \ \forall \ i$ *i.e.*, $a_{ii} = 0 \ \forall \ i$.

∴ **All the elements on the principal diagonal must be zero.**

The following matrices are skew-symmetric matrices :

$$\begin{bmatrix} 0 & 5 \\ -5 & 0 \end{bmatrix}, \begin{bmatrix} 0 & 6 & -5 \\ -6 & 0 & 17 \\ 5 & -17 & 0 \end{bmatrix}, \begin{bmatrix} 0 & 2 & -3 & 7 \\ -2 & 0 & 9 & -8 \\ 3 & -9 & 0 & 5 \\ -7 & 8 & -5 & 0 \end{bmatrix}.$$

Theorem I. *A necessary and sufficient condition for a matrix A to be skew-symmetric is that A′ = – A.*

Proof. A is a skew-symmetric matrix

⇔ A is a square matrix and (i, j)th element of A = – (j, i)th element of A for all i, j.

⇔ A is a square matrix and (i, j)th element of A = – (i, j)th element of A′ for all i, j.

⇔ A = – A′ *i.e.*, A′ = – A.

Example 4. *If* $A = \begin{bmatrix} 2 & 4 \\ 5 & 6 \end{bmatrix}$, *show that :*

(*i*) *A + A′ is symmetric.* (*ii*) *A – A′ is skew-symmetric.*

Sol. We have $A = \begin{bmatrix} 2 & 4 \\ 5 & 6 \end{bmatrix}$. ∴ $A' = \begin{bmatrix} 2 & 4 \\ 5 & 6 \end{bmatrix}' = \begin{bmatrix} 2 & 5 \\ 4 & 6 \end{bmatrix}$

(*i*) $$A + A' = \begin{bmatrix} 2 & 4 \\ 5 & 6 \end{bmatrix} + \begin{bmatrix} 2 & 5 \\ 4 & 6 \end{bmatrix} = \begin{bmatrix} 4 & 9 \\ 9 & 12 \end{bmatrix}.$$

Now $(A + A')' = \begin{bmatrix} 4 & 9 \\ 9 & 12 \end{bmatrix}' = \begin{bmatrix} 4 & 9 \\ 9 & 12 \end{bmatrix}$. ∴ A + A′ is symmetric.

(*ii*) $$A - A' = \begin{bmatrix} 2 & 4 \\ 5 & 6 \end{bmatrix} - \begin{bmatrix} 2 & 5 \\ 4 & 6 \end{bmatrix} = \begin{bmatrix} 0 & -1 \\ 1 & 0 \end{bmatrix}$$

Now $(A - A')' = \begin{bmatrix} 0 & -1 \\ 1 & 0 \end{bmatrix}' = \begin{bmatrix} 0 & 1 \\ -1 & 0 \end{bmatrix} = -\begin{bmatrix} 0 & -1 \\ 1 & 0 \end{bmatrix} = -(A - A)'$

∴ A – A′ is skew-symmetric.

Theorem II. *Prove that every square matrix is uniquely expressible as the sum of a symmetric matrix and a skew-symmetric matrix.*

Proof. Let A be a square matrix.

∴ $$A = \frac{A + A}{2} = \frac{A + A' + A - A'}{2} = \frac{A + A'}{2} + \frac{A - A'}{2}$$

Now $\left(\dfrac{A+A'}{2}\right)' = \dfrac{(A+A')'}{2} = \dfrac{A'+(A')'}{2} = \dfrac{A'+A}{2} = \dfrac{A+A'}{2}$

| $(KA)' = KA'$
$A'' = A$

and $\left(\dfrac{A-A'}{2}\right)' = \dfrac{(A-A')'}{2} = \dfrac{A'-(A')'}{2} = \dfrac{A'-A}{2} = -\dfrac{A-A'}{2}.$

$\therefore$ $\dfrac{A+A'}{2}$ is symmetric and $\dfrac{A-A'}{2}$ is skew-symmetric.

$\therefore$ A is expressible as the sum of symmetric and skew-symmetric matrices.

Uniqueness. Let A = P + Q, where P is symmetric and Q is skew-symmetric.

$\therefore$ $A' = (P+Q)' = P' + Q' = P + (-Q) = P - Q.$ | $Q' = -R$

Solving, we get $P = \dfrac{A+A'}{2}$ and $Q = \dfrac{A-A'}{2}$

$\therefore$ Representation of A as the sum of $\dfrac{A+A'}{2}$ and $\dfrac{A-A'}{2}$ is unique.

Example 5. *Express* $\begin{bmatrix} 4 & 3 & 7 \\ 6 & 5 & -8 \\ 1 & 2 & 6 \end{bmatrix}$ *as the sum of a symmetric matrix and a skew-symmetric matrix.*

Sol. Let $A = \begin{bmatrix} 4 & 3 & 7 \\ 6 & 5 & -8 \\ 1 & 2 & 6 \end{bmatrix}$. $\therefore$ $A' = \begin{bmatrix} 4 & 3 & 7 \\ 6 & 5 & -8 \\ 1 & 2 & 6 \end{bmatrix}' = \begin{bmatrix} 4 & 6 & 1 \\ 3 & 5 & 2 \\ 7 & -8 & 6 \end{bmatrix}$

$\therefore$ $$A + A' = \begin{bmatrix} 4 & 3 & 7 \\ 6 & 5 & -8 \\ 1 & 2 & 6 \end{bmatrix} + \begin{bmatrix} 4 & 6 & 1 \\ 3 & 5 & 2 \\ 7 & -8 & 6 \end{bmatrix} = \begin{bmatrix} 8 & 9 & 8 \\ 9 & 10 & -6 \\ 8 & -6 & 12 \end{bmatrix}$$

and $$A - A' = \begin{bmatrix} 4 & 3 & 7 \\ 6 & 5 & -8 \\ 1 & 2 & 6 \end{bmatrix} - \begin{bmatrix} 4 & 6 & 1 \\ 3 & 5 & 2 \\ 7 & -8 & 6 \end{bmatrix} = \begin{bmatrix} 0 & -3 & 6 \\ 3 & 0 & -10 \\ -6 & 10 & 0 \end{bmatrix}$$

Now $\dfrac{1}{2}(A+A') = \dfrac{1}{2}\begin{bmatrix} 8 & 9 & 8 \\ 9 & 10 & -6 \\ 8 & -6 & 12 \end{bmatrix} = \begin{bmatrix} 4 & 9/2 & 4 \\ 9/2 & 5 & -3 \\ 4 & -3 & 6 \end{bmatrix}$, which is symmetric,

and $\dfrac{1}{2}(A-A') = \dfrac{1}{2}\begin{bmatrix} 0 & -3 & 6 \\ 3 & 0 & -10 \\ -6 & 10 & 0 \end{bmatrix} = \begin{bmatrix} 0 & -3/2 & 3 \\ 3/2 & 0 & -5 \\ -3 & 5 & 0 \end{bmatrix}$, which is skew-symmetric.

$\therefore$ $$A = \frac{1}{2}(A+A') + \frac{1}{2}(A-A') = \begin{bmatrix} \mathbf{4} & \mathbf{9/2} & \mathbf{4} \\ \mathbf{9/2} & \mathbf{5} & \mathbf{-3} \\ \mathbf{4} & \mathbf{-3} & \mathbf{6} \end{bmatrix} + \begin{bmatrix} \mathbf{0} & \mathbf{-3/2} & \mathbf{3} \\ \mathbf{3/2} & \mathbf{0} & \mathbf{-5} \\ \mathbf{-3} & \mathbf{5} & \mathbf{0} \end{bmatrix}.$$

EXERCISE 2.5

1. Find the transpose of the following matrices :

(i) $\begin{bmatrix} 4 & 3 \\ 6 & 9 \end{bmatrix}$ (ii) $[1 \quad 6 \quad 8 \quad 9]$

2. Verify that $(A')' = A$ where

(i) $A = [3 \quad 4 \quad 7]$ (ii) $A = \begin{bmatrix} 2 & 6 & 8 \\ 5 & 3 & 8 \\ 11 & 9 & 7 \end{bmatrix}$

3. Show that the following matrices are symmetric matrices :

(i) $\begin{bmatrix} 2 & 3 \\ 3 & 11 \end{bmatrix}$ (ii) $\begin{bmatrix} -3 & 5 & -6 \\ 5 & 11 & 15 \\ -6 & 15 & 12 \end{bmatrix}$.

4. Show that the following matrices are skew-symmetric :

(i) $\begin{bmatrix} 0 & -3 \\ 3 & 0 \end{bmatrix}$ (ii) $\begin{bmatrix} 0 & -7 & -15 \\ 7 & 0 & 6 \\ 15 & -6 & 0 \end{bmatrix}$.

5. (i) If $A = \begin{bmatrix} 2 & 1 \\ -3 & -1 \end{bmatrix}$, show that $A + A'$ is a symmetric matrix.

(ii) If $A = \begin{bmatrix} 3 & -4 \\ 7 & 8 \end{bmatrix}$, show that $A - A'$ is a skew-symmetric matrix.

6. For matrix $A = \begin{bmatrix} -3 & 6 & 0 \\ 4 & -5 & 8 \\ 0 & -7 & -2 \end{bmatrix}$, find $\frac{1}{2}(A - A')$.

7. Verify $(A + B)' = A' + B'$ for the matrices $A = \begin{bmatrix} -7 & -8 & 6 \\ 8 & 5 & 9 \\ 4 & 6 & 7 \end{bmatrix}$ and $B = \begin{bmatrix} 8 & 9 & 11 \\ 1 & 2 & 3 \\ 6 & 8 & 2 \end{bmatrix}$.

8. (i) If $A = \begin{bmatrix} \cos\alpha & \sin\alpha \\ -\sin\alpha & \cos\alpha \end{bmatrix}$, show that $A'A = I = AA'$. *(P.T.U. B.Pharma, Dec. 2006)*

(ii) If $A = \frac{1}{3}\begin{bmatrix} 1 & 2 & 2 \\ 2 & 1 & -2 \\ -2 & 2 & -1 \end{bmatrix}$, show that $AA' = I = A'A$.

9. If $A = \begin{bmatrix} 1 & 2 & 2 \\ 2 & 1 & -2 \\ a & 2 & b \end{bmatrix}$ is a matrix such that $AA' = 9I_3$, then find the values of a and b.

10. Verify that $(AB)' = B'A'$ where

(i) $A = \begin{bmatrix} -1 \\ 2 \\ 3 \end{bmatrix}$, $B = [-2 \quad -1 \quad -4]$ (ii) $A = \begin{bmatrix} 2 & -1 \\ 4 & 2 \end{bmatrix}$, $B = \begin{bmatrix} 4 & 3 \\ -2 & 1 \end{bmatrix}$.

(iii) $A = \begin{bmatrix} 0 & -1 & 5 \\ 6 & 3 & -4 \end{bmatrix}$, $B = \begin{bmatrix} 4 & -3 \\ 1 & 0 \\ 7 & -6 \end{bmatrix}$ (iv) $A = \begin{bmatrix} 1 & -4 \\ 0 & 5 \\ 6 & 7 \end{bmatrix}$, $B = \begin{bmatrix} 2 & 3 & -1 \\ 1 & 0 & -7 \end{bmatrix}$.

11. If A is any square matrix, then show that :

(i) $\frac{1}{2}(A + A')$ is symmetric (ii) $\frac{1}{2}(A - A')$ is skew-symmetric.

12. Show that the elements on the main diagonal of a skew-symmetric matrix are all zero.

13. Show that a matrix which is both symmetric and skew-symmetric, must be a null matrix.

14. Express the following matrices as the sum of a symmetric matrix and a skew-symmetric matrix :

(i) $\begin{bmatrix} 3 & -4 \\ 1 & -1 \end{bmatrix}$ (ii) $\begin{bmatrix} p & q \\ r & s \end{bmatrix}$ (iii) $\begin{bmatrix} 7 & 1 & 0 \\ 1 & 5 & 6 \\ 0 & 8 & 4 \end{bmatrix}$

(iv) $\begin{bmatrix} 14 & 17 & 18 \\ 19 & 6 & -7 \\ 1 & 2 & 5 \end{bmatrix}$ (v) $\begin{bmatrix} 2 & -1 \\ 4 & 2 \end{bmatrix}$ *(P.T.U. B.Pharma, May 2005)*

15. (i) If A is any square matrix, show that AA′ and A′A are symmetric matrices.

(ii) If A is any symmetric matrix, show that A^n, $n \in \mathbf{N}$ is also symmetric.

(iii) If A and B are symmetric matrices of same order, then show that AB + BA is symmetric.

(iv) If A and B are symmetric matrices of same order, then show that AB – BA is skew-symmetric.

Answers

1. (i) $\begin{bmatrix} 4 & 6 \\ 3 & 9 \end{bmatrix}$ (ii) $\begin{bmatrix} 1 \\ 6 \\ 8 \\ 9 \end{bmatrix}$ 6. $\begin{bmatrix} 0 & 1 & 0 \\ -1 & 0 & 15/2 \\ 0 & -15/2 & 0 \end{bmatrix}$

9. $a = -2$, $b = -1$ 14. (i) $\begin{bmatrix} 3 & -3/2 \\ -3/2 & 1 \end{bmatrix} + \begin{bmatrix} 0 & -5/2 \\ 5/2 & 0 \end{bmatrix}$

(ii) $\frac{1}{2}\begin{bmatrix} 2p & q+r \\ q+r & 2s \end{bmatrix} + \frac{1}{2}\begin{bmatrix} 0 & q-r \\ r-q & 0 \end{bmatrix}$ (iii) $\begin{bmatrix} 7 & 1 & 0 \\ 1 & 5 & 7 \\ 0 & 7 & 4 \end{bmatrix} + \begin{bmatrix} 0 & 0 & 0 \\ 0 & 0 & -1 \\ 0 & 1 & 0 \end{bmatrix}$

(iv) $\begin{bmatrix} 14 & 18 & 19/2 \\ 18 & 6 & -5/2 \\ 19/2 & -5/2 & 5 \end{bmatrix} + \begin{bmatrix} 0 & -1 & 17/2 \\ 1 & 0 & -9/2 \\ -17/2 & 9/2 & 0 \end{bmatrix}$ (v) $A = \begin{bmatrix} 2 & 3/2 \\ 3/2 & 2 \end{bmatrix} + \begin{bmatrix} 0 & -\frac{5}{2} \\ \frac{5}{2} & 0 \end{bmatrix}$.

Hints

12. Let $A = [a_{ij}]_{n \times n}$ be a skew-symmetric matrix.

$\therefore$ By definition, $a_{ij} = -a_{ji}$ $\forall$ i, j. $\therefore$ In particular $a_{ii} = -a_{ii}$ $\forall$ i.

$\therefore$ $2a_{ii} = 0$ *i.e.*, $a_{ii} = 0$ $\forall$ i. $\therefore$ $a_{11} = 0, a_{22} = 0, \ldots\ldots, a_{nn} = 0$.

15. (ii) The result is trivially true for $n = 1$. Let A^k be symmetric.

Now $(A^{k+1})' = (AA^k)' = (A^k)'A' = A^kA = A^{k+1}$.

(iv) $(AB - BA)' = (AB)' - (BA)' = B'A' - A'B' = BA - AB = -(AB - BA)$.

DETERMINANTS

2.18. INTRODUCTION

In this chapter, we shall associate a number (or function) with each square matrix. This quantity is called the *determinant* of the matrix, under consideration. A matrix which is not square, does not possess determinant.

Let $A = [a_{ij}] = \begin{bmatrix} a_{11} & a_{12} \ldots a_{1n} \\ a_{21} & a_{22} \ldots a_{2n} \\ \ldots\ldots & \ldots\ldots \\ \ldots\ldots & \ldots\ldots \\ a_{n1} & a_{n2} \ldots a_{nn} \end{bmatrix}$ be a square matrix of order n.

The determinant of A is denoted by any of the following :

$$\det A,\ |A|,\ |a_{ij}|,\ \begin{vmatrix} a_{11} & a_{12} \ldots\ldots a_{1n} \\ a_{21} & a_{22} \ldots\ldots a_{2n} \\ \ldots\ldots & \ldots\ldots \\ \ldots\ldots & \ldots\ldots \\ a_{n1} & a_{n2} \ldots\ldots a_{nn} \end{vmatrix}$$

For example, if $A = \begin{bmatrix} 4 & 3 & 6 \\ 8 & 7 & 5 \\ -1 & 4 & 6 \end{bmatrix}$, then $|A| = \begin{vmatrix} 4 & 3 & 6 \\ 8 & 7 & 5 \\ -1 & 4 & 6 \end{vmatrix}$.

The determinant of a square matrix of order n is called a *determinant of order n*.

If the elements of square matrix A are numbers, then $|A|$ is a number and in case, the elements of A are functions, then $|A|$ is a function.

2.19. DETERMINANT OF SQUARE MATRIX OF ORDER 1

Let $A = [a_{11}]$ be a square matrix of order 1. The quantity 'a_{11}' is defined as the value of the determinant of A.

$\therefore \quad |A| = |a_{11}| = a_{11}$.

For example, if $A = [-5]$, then $|A| = \det[-5] = -5$.

2.20. DETERMINANT OF SQUARE MATRIX OF ORDER 2

Let $A = \begin{bmatrix} a_{11} & a_{12} \\ a_{21} & a_{22} \end{bmatrix}$ be a square matrix of order 2. The quantity $a_{11}a_{22} - a_{21}a_{12}$ is defined as the determinant of A.

$$\therefore \quad |A| = \begin{vmatrix} a_{11} & a_{12} \\ a_{21} & a_{22} \end{vmatrix} \qquad \text{(Replacing square brackets by vertical lines)}$$

$$= a_{11}a_{22} - a_{21}a_{12}$$

ILLUSTRATIVE EXAMPLES

Example 1. *Find* $|A|$ *if A is given by :*

(*i*) $\begin{bmatrix} 2 & -3 \\ 4 & 9 \end{bmatrix}$ (*ii*) $\begin{bmatrix} \cos\alpha & -\sin\alpha \\ \sin\alpha & \cos\alpha \end{bmatrix}$ (*iii*) $\begin{bmatrix} 1 & \omega \\ \omega & -\omega \end{bmatrix}$ (*iv*) $\begin{bmatrix} 2x+3 & x+2 \\ 2x+1 & x+1 \end{bmatrix}$.

Sol. (*i*) $|A| = \begin{vmatrix} 2 & -3 \\ 4 & 9 \end{vmatrix} = 18 - (-12) = \mathbf{30}$.

(*ii*) $|A| = \begin{vmatrix} \cos\alpha & -\sin\alpha \\ \sin\alpha & \cos\alpha \end{vmatrix} = \cos^2\alpha - (-\sin^2\alpha) = \mathbf{1}.$

(*iii*) $|A| = \begin{vmatrix} 1 & \omega \\ \omega & -\omega \end{vmatrix} = -\omega - \omega^2 = -(\omega + \omega)^2 = -(-1) = \mathbf{1}. \quad (\because \quad 1 + \omega + \omega^2 = 0)$

(*iv*) $|A| = \begin{vmatrix} 2x+3 & x+2 \\ 2x+1 & x+1 \end{vmatrix} = (2x + 3)(x + 1) - (2x + 1)(x + 2)$

$= (2x^2 + 2x + 3x + 3) - (2x^2 + 4x + x + 2) = 2x^2 + 5x + 3 - 2x^2 - 5x - 2 = \mathbf{1}.$

Example 2. *Solve for x* : $\begin{vmatrix} x & 3 \\ 5 & 2x \end{vmatrix} = \begin{vmatrix} 5 & -4 \\ 5 & 3 \end{vmatrix}$.

Sol. We have $\begin{vmatrix} x & 3 \\ 5 & 2x \end{vmatrix} = \begin{vmatrix} 5 & -4 \\ 5 & 3 \end{vmatrix}$.

$\Rightarrow \quad 2x^2 - 15 = 15 + 20 \Rightarrow 2x^2 = 50 \Rightarrow x^2 = 25 \Rightarrow x = \pm \mathbf{5}.$

EXERCISE 2.6

1. Evaluate the following determinants :

(*i*) $\begin{vmatrix} 4 & 3 \\ 6 & 9 \end{vmatrix}$ (*ii*) $\begin{vmatrix} 5 & 0 \\ 6 & 7 \end{vmatrix}$ (*iii*) $\begin{vmatrix} -18 & 9 \\ 27 & 15 \end{vmatrix}$ (*iv*) $\begin{vmatrix} 35 & 16 \\ 19 & -31 \end{vmatrix}$

2. Evaluate the following determinants :

(*i*) $\begin{vmatrix} \cos x & -\sin x \\ \sin x & \cos x \end{vmatrix}$ (*ii*) $\begin{vmatrix} \sin 10^\circ & -\cos 10^\circ \\ \sin 80^\circ & \cos 80^\circ \end{vmatrix}$ (*iii*) $\begin{vmatrix} \log_a b & 1 \\ 1 & \log_b a \end{vmatrix}$

(*iv*) $\begin{vmatrix} ad+bc & bd-ac \\ ac-bd & ad+bc \end{vmatrix}$ (*v*) $\begin{vmatrix} a+ib & c+id \\ -c+id & a-ib \end{vmatrix}$ (*vi*) $\begin{vmatrix} \log_3 5/2 & \log_4 3 \\ \log_3 8 & \log_4 9 \end{vmatrix}$

3. Solve for x : (*i*) $\begin{vmatrix} 2x & 5 \\ 1 & 3 \end{vmatrix} = 7$ (*ii*) $\begin{vmatrix} 4 & x \\ -3 & 5 \end{vmatrix} = 8.$

Answers

1. (*i*) 18 (*ii*) 35 (*iii*) – 513 (*iv*) – 1389
2. (*i*) 1 (*ii*) 1 (*iii*) 0 (*iv*) $(a^2 + b^2)(c^2 + d^2)$
(*v*) $a^2 + b^2 + c^2 + d^2$ (*vi*) 15/2 **3.** (*i*) 2 (*ii*) – 4

Hint

2. (*vi*) Given det. $= (\log_3 2^9)(\log_4 3^2) - (\log_3 2^3)(\log_4 3)$

$= (18 - 3) \log_3 2 \log_4 3 = 15 \log_3 2 (\log_2 3 \log_4 2) = 15 \times 1 \times \frac{1}{2} \log_4 4.$

2.21. MINOR AND COFACTORS

The value of determinant of square matrix of order 3 or more is defined in terms of *cofactors* of its elements.

Let $A = [a_{ij}]$ be a square matrix of order n.

$\therefore \qquad | A | = | a_{ij} |.$

The determinant obtained by removing its ith row and jth column in $| A |$ is denoted by M_{ij} and is called the **minor** of the element a_{ij}. The minor M_{ij} of the element a_{ij} is called a minor of the matrix A and also of the determinant $| A |$.

Thus, we see that, the determinant obtained from a determinant by omitting the row and the column in which a particular element lies is the minor of that element.

For example, if $| A | = \begin{vmatrix} 4 & 3 & 8 \\ 6 & 7 & 5 \\ 3 & 1 & 2 \end{vmatrix}$, then

$$M_{12} = \text{minor of } a_{12} \ (= 3) = \begin{vmatrix} 6 & 5 \\ 3 & 2 \end{vmatrix} = 12 - 15 = -3.$$

$$M_{31} = \text{minor of } a_{31} \ (= 3) = \begin{vmatrix} 3 & 8 \\ 7 & 5 \end{vmatrix} = 15 - 56 = -41 \text{ etc.}$$

If M_{ij} is the minor of a_{ij}, then $(-1)^{i+j} M_{ij}$ is called the **cofactor** of a_{ij} and is denoted by A_{ij}.

$\therefore$ **Cofactor of $a_{ij} = A_{ij} = (-1)^{i+j} M_{ij}$.**

In the above example, $\quad A_{12} = (-1)^{1+2} M_{12} = (-1)(-3) = 3$

$\qquad\qquad A_{31} = (-1)^{3+1} M_{31} = (1)(-41) = -41.$

Remark. $(-1)^{i+j}$ is 1 or -1 according as $i + j$ is even or odd.

$\therefore$ A_{ij} and M_{ij} coincides if $i + j$ is even. And if $i + j$ is odd then we have $A_{ij} = -M_{ij}$.

ILLUSTRATIVE EXAMPLES

Example 1. *Find all the minors and cofactors of the elements in* $\begin{vmatrix} 4 & 3 & 1 \\ 1 & 3 & 2 \\ 2 & 1 & 5 \end{vmatrix}$.

Sol. Let $\quad a_{11} = 4, \qquad a_{12} = 3, \qquad a_{13} = 1,$

$\qquad a_{21} = 1, \qquad a_{22} = 3, \qquad a_{23} = 2,$

$\qquad a_{31} = 2, \qquad a_{32} = 1, \qquad a_{33} = 5.$

$$M_{11} = \text{minor of } a_{11} \ (= 4) = \begin{vmatrix} 3 & 2 \\ 1 & 5 \end{vmatrix} = 15 - 2 = \mathbf{13}$$

$$M_{12} = \text{minor of } a_{12} \ (= 3) = \begin{vmatrix} 1 & 2 \\ 2 & 5 \end{vmatrix} = 5 - 4 = \mathbf{1}$$

$$M_{13} = \text{minor of } a_{13} \ (= 1) = \begin{vmatrix} 1 & 3 \\ 2 & 1 \end{vmatrix} = 1 - 6 = \mathbf{-5}$$

$$M_{21} = \text{minor of } a_{21} \ (= 1) = \begin{vmatrix} 3 & 1 \\ 1 & 5 \end{vmatrix} = 15 - 1 = \mathbf{14}$$

$$M_{22} = \text{minor of } a_{22} \ (= 3) = \begin{vmatrix} 4 & 1 \\ 2 & 5 \end{vmatrix} = 20 - 2 = \mathbf{18}$$

$$M_{23} = \text{minor of } a_{23} \ (= 2) = \begin{vmatrix} 4 & 3 \\ 2 & 1 \end{vmatrix} = 4 - 6 = \mathbf{-2}$$

$$M_{31} = \text{minor of } a_{31} \ (= 2) = \begin{vmatrix} 3 & 1 \\ 3 & 2 \end{vmatrix} = 6 - 3 = \mathbf{3}$$

$$M_{32} = \text{minor of } a_{32} (= 1) = \begin{vmatrix} 4 & 1 \\ 1 & 2 \end{vmatrix} = 8 - 1 = \mathbf{7}$$

$$M_{33} = \text{minor of } a_{33} (= 5) = \begin{vmatrix} 4 & 3 \\ 1 & 3 \end{vmatrix} = 12 - 3 = \mathbf{9}.$$

The following are the cofactors :

$$A_{11} = \text{cofactor of } a_{11} (= 4) = (-1)^{1+1} M_{11} = 1 \times 13 = \mathbf{13}$$
$$A_{12} = \text{cofactor of } a_{12} (= 3) = (-1)^{1+2} M_{12} = -1 \times 1 = \mathbf{-1}$$
$$A_{13} = \text{cofactor of } a_{13} (= 1) = (-1)^{1+3} M_{13} = 1 \times (-5) = \mathbf{-5}$$
$$A_{21} = \text{cofactor of } a_{21} (= 1) = (-1)^{2+1} M_{21} = -1 \times 14 = \mathbf{-14}$$
$$A_{22} = \text{cofactor of } a_{22} (= 3) = (-1)^{2+2} M_{22} = 1 \times 18 = \mathbf{18}$$
$$A_{23} = \text{cofactor of } a_{23} (= 2) = (-1)^{2+3} M_{23} = -1 \times (-2) = \mathbf{2}$$
$$A_{31} = \text{cofactor of } a_{31} (= 2) = (-1)^{3+1} M_{31} = 1 \times 3 = \mathbf{3}$$
$$A_{32} = \text{cofactor of } a_{32} (= 1) = (-1)^{3+2} M_{32} = -1 \times 7 = \mathbf{-7}$$
$$A_{33} = \text{cofactor of } a_{33} (= 5) = (-1)^{3+3} M_{33} = 1 \times 9 = \mathbf{9}.$$

Remark. If $i + j$ is *even*, then $A_{ij} = (-1)^{i+j} M_{ij} = 1 \times M_{ij} = M_{ij}$ *i.e.*, cofactor and minor of a_{ij} are equal. If $i + j$ is *odd*, then $A_{ij} = (-1)^{i+j} M_{ij} = (-1) M_{ij} = -M_{ij}$ *i.e.*, cofactor and minor differ by minus sign.

2.22. DETERMINANT OF SQUARE MATRIX OF ORDER 3

The value of the determinant of a square matrix of order n (≥ 2) is defined as the sum of the products of the elements of any row (respectively column) with their corresponding cofactors.

Let $A = \begin{bmatrix} a_{11} & a_{12} & a_{13} \\ a_{21} & a_{22} & a_{23} \\ a_{31} & a_{32} & a_{33} \end{bmatrix}$ be a square matrix of order 3.

Using the first row, we have

$$|A| = a_{11} A_{11} + a_{12} A_{12} + a_{13} A_{13}.$$

$$\therefore \quad |A| = a_{11} (-1)^{1+1} \begin{vmatrix} a_{22} & a_{23} \\ a_{32} & a_{33} \end{vmatrix} + a_{12} (-1)^{1+2} \begin{vmatrix} a_{21} & a_{23} \\ a_{31} & a_{33} \end{vmatrix} + a_{13} (-1)^{1+3} \begin{vmatrix} a_{21} & a_{22} \\ a_{31} & a_{32} \end{vmatrix}$$

$$= a_{11} (a_{22} a_{33} - a_{32} a_{23}) - a_{12} (a_{21} a_{33} - a_{31} a_{23}) + a_{13} (a_{21} a_{32} - a_{31} a_{22}).$$

We say that $|A|$ has been expanded along R_1. If $|A|$ is expanded along any other row (or column), same result would be obtained.

Example 2. *Write the cofactors of elements of the second row of the following determinants and hence evaluate them :*

(*i*) $\begin{vmatrix} 1 & 2 & 3 \\ -4 & 3 & 6 \\ 2 & -7 & 9 \end{vmatrix}$ (*ii*) $\begin{vmatrix} 1 & a & bc \\ 1 & b & ca \\ 1 & c & ab \end{vmatrix}$

Sol. (*i*) Let $\Delta = \begin{vmatrix} 1 & 2 & 3 \\ -4 & 3 & 6 \\ 2 & -7 & 9 \end{vmatrix}$

The cofactors of elements of second row are A_{21}, A_{22}, A_{23}.

$$A_{21} = \text{cofactor of } -4 = (-1)^{2+1}\begin{vmatrix} 2 & 3 \\ -7 & 9 \end{vmatrix} = -(18+21) = \mathbf{-39}$$

$$A_{22} = \text{cofactor of } 3 = (-1)^{2+2}\begin{vmatrix} 1 & 3 \\ 2 & 9 \end{vmatrix} = 9-6 = \mathbf{3}$$

$$A_{23} = \text{cofactor of } 6 = (-1)^{2+3}\begin{vmatrix} 1 & 2 \\ 2 & -7 \end{vmatrix} = -(-7-4) = \mathbf{11.}$$

$\therefore$ $\Delta = -4A_{21} + 3A_{22} + 6A_{23}$ (Expanding along second row)

$= -4(-39) + 3(3) + 6(11) = 156 + 9 + 66 = \mathbf{231.}$

(*ii*) Let $\Delta = \begin{vmatrix} 1 & a & bc \\ 1 & b & ca \\ 1 & c & ab \end{vmatrix}$.

$$A_{21} = (-1)^{2+1}\begin{vmatrix} a & bc \\ c & ab \end{vmatrix} = \mathbf{-(a^2b - bc^2),} \qquad A_{22} = (-1)^{2+2}\begin{vmatrix} 1 & bc \\ 1 & ab \end{vmatrix} = \mathbf{ab - bc,}$$

$$A_{23} = (-1)^{2+3}\begin{vmatrix} 1 & a \\ 1 & c \end{vmatrix} = -(c-a) = \mathbf{a - c.}$$

$\therefore$ $\Delta = 1.A_{21} + bA_{22} + caA_{23} = -(a^2b - bc^2) + b(ab - bc) + ca(a - c)$

$= \mathbf{bc^2 - a^2b + ab^2 - b^2c + a^2c - ac^2.}$

Example 3. *Find the value of* :

(*i*) $\begin{vmatrix} 1 & 2 & 3 \\ 4 & 5 & 6 \\ 7 & 8 & 9 \end{vmatrix}$ (*ii*) $\begin{vmatrix} a & h & g \\ h & b & f \\ g & f & c \end{vmatrix}$ (P.T.U. B. Pharma, May 2006)

(*iii*) $\begin{vmatrix} 0 & 1 & \sec\theta \\ \tan\theta & -\sec\theta & \tan\theta \\ 1 & 0 & 1 \end{vmatrix}$.

Sol. (*i*) Let $\Delta = \begin{vmatrix} 1 & 2 & 3 \\ 4 & 5 & 6 \\ 7 & 8 & 9 \end{vmatrix}$. We expand Δ along R_1.

$\therefore$ $\Delta = 1.A_{11} + 2.A_{12} + 3.A_{13}$

$$= 1(-1)^{1+1}\begin{vmatrix} 5 & 6 \\ 8 & 9 \end{vmatrix} + 2(-1)^{1+2}\begin{vmatrix} 4 & 6 \\ 7 & 9 \end{vmatrix} + 3(-1)^{1+3}\begin{vmatrix} 4 & 5 \\ 7 & 8 \end{vmatrix}$$

$= 1(45 - 48) - 2(36 - 42) + 3(32 - 35) = -3 + 12 - 9 = \mathbf{0.}$

(*ii*) Let $\Delta = \begin{vmatrix} a & h & g \\ h & b & f \\ g & f & c \end{vmatrix}$. We expand Δ along R_1.

$\therefore$ $\Delta = a.A_{11} + h.A_{12} + g.A_{13}$

$$= a(-1)^2\begin{vmatrix} b & f \\ f & c \end{vmatrix} + h(-1)^3\begin{vmatrix} h & f \\ g & c \end{vmatrix} + g(-1)^4\begin{vmatrix} h & b \\ g & f \end{vmatrix}$$

$= a(bc - f^2) - h(hc - fg) + g(hf - bg)$
$= abc - af^2 - ch^2 + fgh + fgh - bg^2 = \mathbf{abc + 2fgh - af^2 - bg^2 - ch^2}.$

(*iii*) Let $\Delta = \begin{vmatrix} 0 & 1 & \sec\theta \\ \tan\theta & -\sec\theta & \tan\theta \\ 1 & 0 & 1 \end{vmatrix}$. We expand Δ along R_1.

$\therefore \quad \Delta = 0.A_{11} + 1.A_{12} + \sec\theta.A_{13}$

$= 0 + 1(-1)^3 \begin{vmatrix} \tan\theta & \tan\theta \\ 1 & 1 \end{vmatrix} + \sec\theta(-1)^4 \begin{vmatrix} \tan\theta & -\sec\theta \\ 1 & 0 \end{vmatrix}$

$= -(\tan\theta - \tan\theta) + \sec\theta(0 + \sec\theta) = 0 + \sec^2\theta = \mathbf{\sec^2\theta}.$

Example 4. *Given that* $x = -9$ *is a root of* $\begin{vmatrix} x & 3 & 7 \\ 2 & x & 2 \\ 7 & 6 & x \end{vmatrix} = 0$. *Find the other roots.*

Sol. We have $\begin{vmatrix} x & 3 & 7 \\ 2 & x & 2 \\ 7 & 6 & x \end{vmatrix} = 0$ ...(1)

$\Rightarrow \quad x A_{11} + 3A_{12} + 7A_{13} = 0$ (Expanding along R_1)

$\Rightarrow \quad x\begin{vmatrix} x & 2 \\ 6 & x \end{vmatrix} - 3\begin{vmatrix} 2 & 2 \\ 7 & x \end{vmatrix} + 7\begin{vmatrix} 2 & x \\ 7 & 6 \end{vmatrix} = 0$

$\Rightarrow \quad x(x^2 - 12) - 3(2x - 14) + 7(12 - 7x) = 0$

$\Rightarrow \quad x^3 - 12x - 6x + 42 + 84 - 49x = 0 \Rightarrow x^3 - 67x + 126 = 0$ ...(2)

$x = -9$ is a root of (2). $\quad | \; (-9)^3 - 67(-9) + 126 = -729 + 603 + 126 = 0$

$\therefore \quad x + 9$ is a factor of $x^3 - 67x + 126$.

$\therefore \quad (2) \Rightarrow (x + 9)(x^2 - 9x + 14) = 0$

Solving $x^2 - 9x + 14 = 0$, we get $x = \dfrac{9 \pm \sqrt{81 - 56}}{2} = \dfrac{9 \pm 5}{2} = 2, 7.$

$\therefore$ The other roots of (1) are **2 and 7.**

Remark 1. The value of a determinant is not changed if it is expanded along any row or column.

Remark 2. When no reference of the corresponding matrix is needed, we may denote a determinant by Δ.

Remark 3. It is advisable to expand a determinant along that row (or column) which contains maximum number of zeros.

Remark 4. The determinant of a square zero matrix is zero.

Important Observations :

(*i*) **If each element of a particular row (or column) of a determinant is zero, then the value of the determinant is zero.**

For example, let $\Delta = \begin{vmatrix} 4 & 6 & 0 \\ -11 & 5 & 0 \\ 18 & -27 & 0 \end{vmatrix}$

C_3 of Δ is of zeros. Expanding Δ along C_3, we get

$$\Delta = 0.\begin{vmatrix} -11 & 5 \\ 18 & -27 \end{vmatrix} - 0.\begin{vmatrix} 4 & 6 \\ 18 & -27 \end{vmatrix} + 0.\begin{vmatrix} 4 & 6 \\ -11 & 5 \end{vmatrix} = 0 - 0 + 0 = 0.$$

***(ii)* If each element above (or below) of the principal diagonal of a determinant is zero, then the value of the determinant is equal to the product of its diagonal elements.**

For example, let $\Delta = \begin{vmatrix} 7 & 0 & 0 \\ -18 & 5 & 0 \\ 29 & 31 & 2 \end{vmatrix}$

In Δ, each element above the principal diagonal is zero. Expanding Δ along R_1, we get

$$\Delta = 7\begin{vmatrix} 5 & 0 \\ 31 & 2 \end{vmatrix} - 0 + 0 = 7(10 - 0) = 70.$$

This is same as the product of 7, 5 and 2.

***(iii)* The value of determinant of a skew-symmetric matrix of odd order is always zero.**

For example, let $A = \begin{bmatrix} 0 & a & b \\ -a & 0 & c \\ -b & -c & 0 \end{bmatrix}$ be a skew-symmetric matrix of odd order 3.

$$\therefore \qquad |A| = \begin{vmatrix} 0 & a & b \\ -a & 0 & c \\ -b & -c & 0 \end{vmatrix}.$$

Expanding along R_1, we get $\quad |A| = 0 - a\begin{vmatrix} -a & c \\ -b & 0 \end{vmatrix} + b\begin{vmatrix} -a & 0 \\ -b & -c \end{vmatrix}$

$$= -a(0 + bc) + b(ac - 0) = -abc + bac = 0.$$

***(iv)* The sum of the products of the elements of any row (respectively column) of a determinant with the cofactors of the corresponding elements of any other row (respectively column) is zero.**

Let $\Delta = \begin{vmatrix} a_{11} & a_{12} & a_{13} \\ a_{21} & a_{22} & a_{23} \\ a_{31} & a_{32} & a_{33} \end{vmatrix}$ be a determinant of order 3.

Sum of the products of elements of R_1 of Δ with the cofactors of the corresponding elements of R_2

$$= a_{11}A_{21} + a_{12}A_{22} + a_{13}A_{23}$$

$$= a_{11} \cdot (-1)^{2+1}\begin{vmatrix} a_{12} & a_{13} \\ a_{32} & a_{33} \end{vmatrix} + a_{12} \cdot (-1)^{2+2}\begin{vmatrix} a_{11} & a_{13} \\ a_{31} & a_{33} \end{vmatrix} + a_{13} \cdot (-1)^{2+3}\begin{vmatrix} a_{11} & a_{12} \\ a_{31} & a_{32} \end{vmatrix}$$

$$= -a_{11}(a_{12}a_{33} - a_{32}a_{13}) + a_{12}(a_{11}a_{33} - a_{31}a_{13}) - a_{13}(a_{11}a_{32} - a_{31}a_{12})$$

$$= -a_{11}a_{12}a_{33} + a_{11}a_{32}a_{13} + a_{12}a_{11}a_{33} - a_{12}a_{31}a_{13} - a_{13}a_{11}a_{32} + a_{13}a_{31}a_{12}$$

$$= 0.$$

Similarly, other sum of products will come out to be zero.

2.23. SARRUS RULE

Let $A = \begin{bmatrix} a_{11} & a_{12} & a_{13} \\ a_{21} & a_{22} & a_{23} \\ a_{31} & a_{32} & a_{33} \end{bmatrix}$ be a square matrix of order 3. The value of $|A|$ can also be found by using **Sarrus rule** as explained below.

The first column and the second column of | A | are written after the third column :

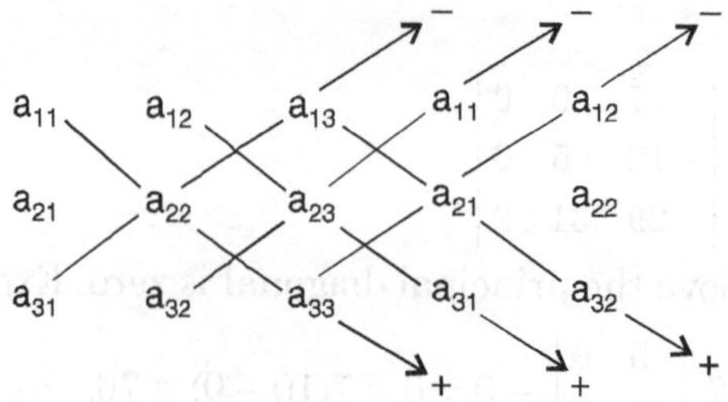

In this method, the products of elements along the diagonals are used. There are 6 such products. By Sarrus rule

$$| A | = \begin{pmatrix} \text{Sum of products of elements} \\ \text{along diagonals } \textbf{descending} \\ \textbf{to the right} \end{pmatrix} - \begin{pmatrix} \text{Sum of products of elements} \\ \text{along diagonals } \textbf{ascending} \\ \textbf{to the right} \end{pmatrix}$$

$$\therefore \quad | A | = (a_{11}a_{22}a_{33} + a_{12}a_{23}a_{31} + a_{13}a_{21}a_{32}) - (a_{31}a_{22}a_{13} + a_{32}a_{23}a_{11} + a_{33}a_{21}a_{12}).$$

For example, if $A = \begin{bmatrix} 2 & 1 & 3 \\ 5 & 3 & 7 \\ 1 & 2 & 6 \end{bmatrix}$, then

$$| A | = \begin{vmatrix} 2 & 1 & 3 \\ 5 & 3 & 7 \\ 1 & 2 & 6 \end{vmatrix}$$

$$= 2(3)(6) + 1(7)(1) + 3(5)(2) - 1(3)(3) - 2(7)(2) - 6(5)(1)$$

$$= 36 + 7 + 30 - 9 - 28 - 30 = 6.$$

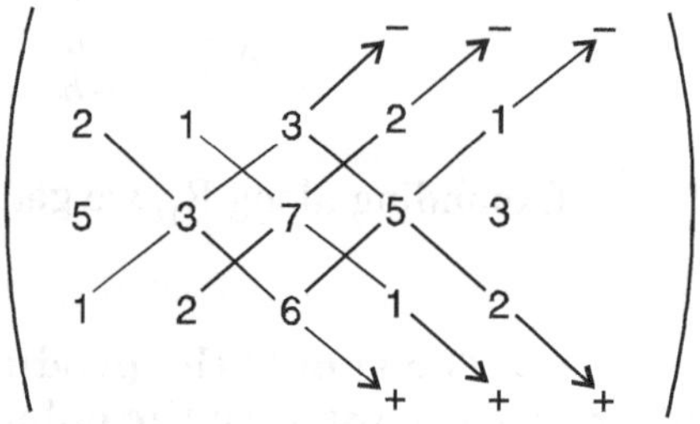

2.24. SINGULAR AND NON-SINGULAR MATRICES

A square matrix is called a **singular matrix** if its determinant is zero, otherwise it is called a **non-singular matrix.**

Example 5. *Which of the following matrices are singular :*

(*i*) $\begin{bmatrix} 4 & 3 \\ 6 & 9 \end{bmatrix}$ (*ii*) $\begin{bmatrix} 14 & 7 \\ 4 & 2 \end{bmatrix}$ (*iii*) $\begin{bmatrix} 1 & 4 & 3 \\ 6 & 8 & -5 \\ 2 & 8 & 6 \end{bmatrix}$.

Sol. (*i*) Let $A = \begin{bmatrix} 4 & 3 \\ 6 & 9 \end{bmatrix}$. $\therefore \quad | A | = \begin{vmatrix} 4 & 3 \\ 6 & 9 \end{vmatrix} = 36 - 18 = 18 \neq 0.$

$\therefore$ A is not singular *i.e.*, A is **non-singular.**

(*ii*) Let $A = \begin{bmatrix} 14 & 17 \\ 4 & 2 \end{bmatrix}$. $\therefore \quad | A | = \begin{vmatrix} 14 & 7 \\ 4 & 2 \end{vmatrix} = 28 - 28 = 0.$

$\therefore$ A is **singular.**

(*iii*) Let $A = \begin{bmatrix} 1 & 4 & 3 \\ 6 & 8 & -5 \\ 2 & 8 & 6 \end{bmatrix}$. $\therefore \quad | A | = \begin{vmatrix} 1 & 4 & 3 \\ 6 & 8 & -5 \\ 2 & 8 & 6 \end{vmatrix}$

Expanding along R_1, we get

$$| A | = 1\begin{vmatrix} 8 & -5 \\ 8 & 6 \end{vmatrix} - 4\begin{vmatrix} 6 & -5 \\ 2 & 6 \end{vmatrix} + 3\begin{vmatrix} 6 & 8 \\ 2 & 8 \end{vmatrix}$$

$$= (48 + 40) - 4(36 + 10) + 3(48 - 16) = 88 - 184 + 96 = 0.$$

$\therefore$ A is **singular.**

EXERCISE 2.7

1. Write the cofactors of the second column of the following determinants and hence evaluate :

(i) $\begin{vmatrix} 5 & 2 & 1 \\ 3 & 0 & 2 \\ 8 & 1 & 3 \end{vmatrix}$ (ii) $\begin{vmatrix} 18 & 40 & 58 \\ 16 & 36 & 52 \\ 12 & 28 & 40 \end{vmatrix}$.

2. Find the values of the determinants of the identity matrices I_2 and I_3.

3. Evaluate the following determinants :

(i) $\begin{vmatrix} 2 & 0 & 0 \\ 5 & 9 & 3 \\ 1 & 6 & 7 \end{vmatrix}$ (ii) $\begin{vmatrix} 6 & 1 & -3 \\ 1 & 3 & -2 \\ 2 & 1 & 4 \end{vmatrix}$

(iii) $\begin{vmatrix} 85 & 25 & 35 \\ 88 & 26 & 36 \\ 91 & 27 & 37 \end{vmatrix}$. *(P.T.U. B. Pharma, Dec. 2005)*

4. Show that the following matrices are singular :

(i) $\begin{bmatrix} 4 & 3 \\ 8 & 6 \end{bmatrix}$ (ii) $\begin{bmatrix} -31 & 160 & 161 \\ 0 & 0 & 0 \\ -15 & 79 & 12 \end{bmatrix}$.

5. Find the value of λ, so that the following matrices may be singular :

(i) $\begin{bmatrix} 7 & 1 \\ 2 & \lambda \end{bmatrix}$ (ii) $\begin{bmatrix} \lambda+1 & -3 & 4 \\ -5 & \lambda+2 & 2 \\ 4 & 1 & \lambda-6 \end{bmatrix}$.

6. Evaluate each of the following determinant by (i) sarrus method (ii) expanding along R_1, (iii) expanding along C_3. Verify that the same answer is obtained :

(i) $\begin{vmatrix} 9 & 6 & 1 \\ 3 & 5 & 2 \\ 7 & 3 & 4 \end{vmatrix}$ (ii) $\begin{vmatrix} 5 & 6 & 7 \\ 11 & -8 & 9 \\ 1 & 3 & 7 \end{vmatrix}$

Answers

1. (i) $A_{12} = 7, A_{22} = 7, A_{32} = -7$; 7 (ii) $A_{12} = -16, A_{22} = 24, A_{32} = -8$; 0
2. 1, 1 3. (i) 90 (ii) 91 (iii) 0
5. (i) 2/7 (ii) 0, $\dfrac{3 \pm \sqrt{205}}{2}$ 6. (i) 112 (ii) – 536.

Hint

5. (ii) $\begin{vmatrix} \lambda+1 & -3 & 4 \\ -5 & \lambda+2 & 2 \\ 4 & 1 & \lambda-6 \end{vmatrix}$

$$= (\lambda + 1)(-1)^2\begin{vmatrix} \lambda+2 & 2 \\ 1 & \lambda-6 \end{vmatrix} + (-3)(-1)^3\begin{vmatrix} -5 & 2 \\ 4 & \lambda-6 \end{vmatrix} + 4(-1)^4\begin{vmatrix} -5 & \lambda+2 \\ 4 & 1 \end{vmatrix}$$

$$= (\lambda + 1)(\lambda^2 - 4\lambda - 14) + 3(-5\lambda + 22) + 4(-4\lambda - 13)$$

$$= \lambda(\lambda^2 - 3\lambda - 49).$$

2.25. PROPERTIES OF DETERMINANTS

In this section, we shall learn some of the properties of determinants. So far as proofs are concerned, we shall restrict ourselves to the determinants of order 3.

Property 1. *If the rows and columns of a determinant are interchanged, the value of the determinant remains unchanged.*

Proof. Let $\Delta = \begin{vmatrix} a_1 & b_1 & c_1 \\ a_2 & b_2 & c_2 \\ a_3 & b_3 & c_3 \end{vmatrix}$ be any determinant of order 3.

$$\therefore \quad \Delta = a_1 \begin{vmatrix} b_2 & c_2 \\ b_3 & c_3 \end{vmatrix} - b_1 \begin{vmatrix} a_2 & c_2 \\ a_3 & c_3 \end{vmatrix} + c_1 \begin{vmatrix} a_2 & b_2 \\ a_3 & b_3 \end{vmatrix} \quad \text{(Expanding along I}^{\text{st}}\text{ row)}$$

$$= a_1(b_2c_3 - b_3c_2) - b_1(a_2c_3 - a_3c_2) + c_1(a_2b_3 - a_3b_2).$$

Let $\Delta' = \begin{vmatrix} a_1 & a_2 & a_3 \\ b_1 & b_2 & b_3 \\ c_1 & c_2 & c_3 \end{vmatrix}$

be the determinant obtained from Δ by interchanging rows and columns.

$$\therefore \quad \Delta' = a_1A_{11} + b_1A_{21} + c_1A_{31} \quad \text{(Expanding along I}^{\text{st}}\text{ column)}$$

$$= a_1(-1)^2 \begin{vmatrix} b_2 & b_3 \\ c_2 & c_3 \end{vmatrix} + b_1(-1)^3 \begin{vmatrix} a_2 & a_3 \\ c_2 & c_3 \end{vmatrix} + c_1(-1)^4 \begin{vmatrix} a_2 & a_3 \\ b_2 & b_3 \end{vmatrix}$$

$$= a_1(b_2c_3 - b_3c_2) - b_1(a_2c_3 - a_3c_2) + c_1(a_2b_3 - a_3b_2) = \Delta.$$

Hence the result holds.

Illustration. Let $\Delta = \begin{vmatrix} 1 & 2 \\ 4 & 3 \end{vmatrix}$. $\quad \therefore \quad \Delta = 1(3) - 4(2) = 3 - 8 = -5$

Let $\Delta' = \begin{vmatrix} 1 & 4 \\ 2 & 3 \end{vmatrix}$. $\quad \therefore \quad \Delta' = 1(3) - 2(4) = 3 - 8 = -5$

$\therefore \quad \Delta' = \Delta.$

Remark. If A is a square matrix, then A′ is obtained by interchanging rows and columns of A.

$\therefore$ The above property asserts that $|A'| = |A|$.

Property 2. *If two rows (or columns) of a determinant are interchanged, the value of the determinant is multiplied by – 1.*

Proof. Let $\Delta = \begin{vmatrix} a_1 & b_1 & c_1 \\ a_2 & b_2 & c_2 \\ a_3 & b_3 & c_3 \end{vmatrix}$ be any determinant of order 3.

Expanding Δ along the Ist row, we get

$$\Delta = a_1 \begin{vmatrix} b_2 & c_2 \\ b_3 & c_3 \end{vmatrix} - b_1 \begin{vmatrix} a_2 & c_2 \\ a_3 & c_3 \end{vmatrix} + c_1 \begin{vmatrix} a_2 & b_2 \\ a_3 & b_3 \end{vmatrix}$$

$$= a_1(b_2c_3 - b_3c_2) - b_1(a_2c_3 - a_3c_2) + c_1(a_2b_3 - a_3b_2).$$

Let Δ' be the determinant obtained from Δ by interchanging any two rows (or columns). To be specific, let first and third columns be interchanged.

$$\therefore \qquad \Delta' = \begin{vmatrix} c_1 & b_1 & a_1 \\ c_2 & b_2 & a_2 \\ c_3 & b_3 & a_3 \end{vmatrix}$$

Expanding Δ' along the first row, we get

$$\Delta' = c_1 \begin{vmatrix} b_2 & a_2 \\ b_3 & a_3 \end{vmatrix} - b_1 \begin{vmatrix} c_2 & a_2 \\ c_3 & a_3 \end{vmatrix} + a_1 \begin{vmatrix} c_2 & b_2 \\ c_3 & b_3 \end{vmatrix}$$

$$= c_1(b_2a_3 - b_3a_2) - b_1(c_2a_3 - c_3a_2) + a_1(b_3c_2 - b_2c_3)$$

$$= -[c_1(a_2b_3 - a_3b_2) - b_1(a_2c_3 - a_3c_2) + a_1(b_2c_3 - b_3c_2)] = -\Delta.$$

Hence the result holds.

Illustration. Let $\Delta = \begin{vmatrix} 1 & 4 \\ 3 & 2 \end{vmatrix}$. $\quad \therefore \quad \Delta = 1(2) - 3(4) = 2 - 12 = -10$

Let $\quad \Delta' = \begin{vmatrix} 3 & 2 \\ 1 & 4 \end{vmatrix}$ be obtained from Δ by interchanging first and second row.

$$\therefore \qquad \Delta' = 3(4) - 1(2) = 12 - 2 = 10 = -(-10) = -\Delta.$$

Remark. If Δ' is obtained from Δ by interchanging ith and jth row, then this operation is denoted by $R_i \leftrightarrow R_j$. Similarly, if ith and jth columns are interchanged, then we shall denote that operation by $C_i \leftrightarrow C_j$.

Property 3. *If two rows (or columns) of a determinant are identical, then the value of the determinant is zero.*

Proof. Let $\Delta = \begin{vmatrix} a_1 & b_1 & c_1 \\ a_2 & b_2 & c_2 \\ a_2 & b_2 & c_2 \end{vmatrix}$ be any determinant of order 3 in which second and third rows are identical.

Expanding Δ along the Ist row, we get

$$\Delta = a_1 \begin{vmatrix} b_2 & c_2 \\ b_2 & c_2 \end{vmatrix} - b_1 \begin{vmatrix} a_2 & c_2 \\ a_2 & c_2 \end{vmatrix} + c_1 \begin{vmatrix} a_2 & b_2 \\ a_2 & b_2 \end{vmatrix}$$

$$= a_1(b_2c_2 - b_2c_2) - b_1(a_2c_2 - a_2c_2) + c_1(a_2b_2 - a_2b_2) = a_1(0) - b_1(0) + c_1(0) = 0.$$

Hence the result holds.

Illustration. Let

$$\Delta = \begin{vmatrix} 1 & 2 & 1 \\ 7 & 17 & 7 \\ 2 & 10 & 2 \end{vmatrix} \qquad (\text{In } \Delta,\ C_1 = C_3)$$

$$\therefore \qquad \Delta = 1 \begin{vmatrix} 17 & 7 \\ 10 & 2 \end{vmatrix} - 2 \begin{vmatrix} 7 & 7 \\ 2 & 2 \end{vmatrix} + 1 \begin{vmatrix} 7 & 17 \\ 2 & 10 \end{vmatrix} \qquad (\text{Expanding along the first row})$$

$$= 1(34 - 70) - 2(14 - 14) + 1(70 - 34) = -36 - 0 + 36 = 0.$$

Property 4. *If the elements of a row (or column) of a determinant are multiplied by a scalar, then the value of the new determinant is equal to same scalar times the value of the original determinant.*

Proof. Let $\Delta = \begin{vmatrix} a_1 & b_1 & c_1 \\ a_2 & b_2 & c_2 \\ a_3 & b_3 & c_3 \end{vmatrix}$ be any determinant of order 3.

Expanding Δ along the first row, we get

$$\Delta = a_1 \begin{vmatrix} b_2 & c_2 \\ b_3 & c_3 \end{vmatrix} - b_1 \begin{vmatrix} a_2 & c_2 \\ a_3 & c_3 \end{vmatrix} + c_1 \begin{vmatrix} a_2 & b_2 \\ a_3 & b_3 \end{vmatrix}$$

$$= a_1(b_2c_3 - b_3c_2) - b_1(a_2c_3 - a_3c_2) + c_1(a_2b_3 - a_3b_2).$$

Let Δ′ be the determinant obtained from Δ by multiplying the elements of a row (or column) by some scalar. To be specific, let the elements of third column be multiplied by k.

$$\therefore \quad \Delta' = \begin{vmatrix} a_1 & b_1 & kc_1 \\ a_2 & b_2 & kc_2 \\ a_3 & b_3 & kc_3 \end{vmatrix}$$

Expanding Δ′ along the first row, we get

$$\Delta' = a_1 \begin{vmatrix} b_2 & kc_2 \\ b_3 & kc_3 \end{vmatrix} - b_1 \begin{vmatrix} a_2 & kc_2 \\ a_3 & kc_3 \end{vmatrix} + kc_1 \begin{vmatrix} a_2 & b_2 \\ a_3 & b_3 \end{vmatrix}$$

$$= a_1(kb_2c_3 - kb_3c_2) - b_1(ka_2c_3 - ka_3c_2) + kc_1(a_2b_3 - a_3b_2)$$

$$= k[a_1(b_2c_3 - b_3c_2) - b_1(a_2c_3 - a_3c_2) + c_1(a_2b_3 - a_3b_2)] = k\Delta.$$

Hence the result holds.

Illustration. Let $\Delta = \begin{vmatrix} 1 & 5 & 7 \\ 2 & 6 & 8 \\ 9 & 3 & 7 \end{vmatrix}$.

To find Δ, we write $\begin{vmatrix} 1 & 5 & 7 \\ 2 & 6 & 8 \\ 9 & 3 & 7 \end{vmatrix} \begin{matrix} 1 & 5 \\ 2 & 6 \\ 9 & 3 \end{matrix}$

$$\therefore \quad \Delta = 1.6.7 + 5.8.9 + 7.2.3 - 9.6.7 - 3.8.1 - 7.2.5$$

$$= 42 + 360 + 42 - 378 - 24 - 70 = -28.$$

Let $\Delta' = \begin{vmatrix} 1 & 5 & 7 \\ 2 & 6 & 8 \\ 5\times 9 & 5\times 3 & 5\times 7 \end{vmatrix}$

To find Δ′, we write $\begin{vmatrix} 1 & 5 & 7 \\ 2 & 6 & 8 \\ 45 & 15 & 35 \end{vmatrix} \begin{matrix} 1 & 5 \\ 2 & 6 \\ 45 & 15 \end{matrix}$

$$\therefore \quad \Delta' = 1.6.35 + 5.8.45 + 7.2.15 - 45.6.7 - 15.8.1 - 35.2.5$$

$$= 210 + 1800 + 210 - 1890 - 120 - 350 = -140 = 5(-28) = 5\Delta.$$

Remark 1. If Δ′ is obtained from Δ by multiplying ith row by the scalar k, then this operation is denoted by $R_i \to kR_i$. Similarly argument also work for columns.

Remark 2. Let A be a square matrix of order n. The matrix kA is obtained from A by multiplying its each element by k.

∴ The elements of $|kA|$ are k times the corresponding element of $|A|$. If we take k common from each of the n rows, we get $|kA| = (k . k . \ldots n \text{ times}) |A|$ *i.e.*, $\mathbf{|kA| = k^n |A|}$.

Property 5. *If each element of any row (or column) of a determinant is the sum of two numbers, then the determinant is expressible as the sum of two determinants of the same order.*

Proof. Let $\Delta = \begin{vmatrix} a_1 + \alpha_1 & b_1 + \beta_1 & c_1 + \gamma_1 \\ a_2 & b_2 & c_2 \\ a_3 & b_3 & c_3 \end{vmatrix}$

Expanding Δ along the first row, we get

$$\Delta = (a_1 + \alpha_1)\begin{vmatrix} b_2 & c_2 \\ b_3 & c_3 \end{vmatrix} - (b_1 + \beta_1)\begin{vmatrix} a_2 & c_2 \\ a_3 & c_3 \end{vmatrix} + (c_1 + \gamma_1)\begin{vmatrix} a_2 & b_2 \\ a_3 & b_3 \end{vmatrix}$$

$$= \left[a_1\begin{vmatrix} b_2 & c_2 \\ b_3 & c_3 \end{vmatrix} - b_1\begin{vmatrix} a_2 & c_2 \\ a_3 & c_3 \end{vmatrix} + c_1\begin{vmatrix} a_2 & b_2 \\ a_3 & b_3 \end{vmatrix}\right]$$

$$+ \left[\alpha_1\begin{vmatrix} b_2 & c_2 \\ b_3 & c_3 \end{vmatrix} - \beta_1\begin{vmatrix} a_2 & c_2 \\ a_3 & c_3 \end{vmatrix} + \gamma_1\begin{vmatrix} a_2 & b_2 \\ a_3 & b_3 \end{vmatrix}\right]$$

$$= \begin{vmatrix} a_1 & b_1 & c_1 \\ a_2 & b_2 & c_2 \\ a_3 & b_3 & c_3 \end{vmatrix} + \begin{vmatrix} \alpha_1 & \beta_1 & \gamma_1 \\ a_2 & b_2 & c_2 \\ a_3 & b_3 & c_3 \end{vmatrix}.$$ Hence the result holds.

Property 6. *If to each elements of a row (or column) of a determinant be added equi-multiples of the corresponding elements of one or more rows (or columns), then the value of the determinant is not changed.*

Proof. Let $\Delta = \begin{vmatrix} a_1 & b_1 & c_1 \\ a_2 & b_2 & c_2 \\ a_3 & b_3 & c_3 \end{vmatrix}$.

Let Δ′ be the determinant obtained from Δ by adding p times the elements of the second column and q times the elements of the third column to the corresponding elements of the first column.

$$\therefore \quad \Delta' = \begin{vmatrix} a_1 + pb_1 + qc_1 & b_1 & c_1 \\ a_2 + pb_2 + qc_2 & b_2 & c_2 \\ a_3 + pb_3 + qc_3 & b_3 & c_3 \end{vmatrix}$$

Expanding Δ′ along the first column, we get

$$\Delta' = (a_1 + pb_1 + qc_1)\begin{vmatrix} b_2 & c_2 \\ b_3 & c_3 \end{vmatrix} - (a_2 + pb_2 + qc_2)\begin{vmatrix} b_1 & c_1 \\ b_3 & c_3 \end{vmatrix} + (a_3 + pb_3 + qc_3)\begin{vmatrix} b_1 & c_1 \\ b_2 & c_2 \end{vmatrix}$$

$$= \left[a_1\begin{vmatrix} b_2 & c_2 \\ b_3 & c_3 \end{vmatrix} - a_2\begin{vmatrix} b_1 & c_1 \\ b_3 & c_3 \end{vmatrix} + a_3\begin{vmatrix} b_1 & c_1 \\ b_2 & c_2 \end{vmatrix}\right]$$

$$+ p[b_1(b_2c_3 - b_3c_2) - b_2(b_1c_3 - b_3c_1) + b_3(b_1c_2 - b_2c_1)]$$

$$+ q[c_1(b_2c_3 - b_3c_2) - c_2(b_1c_3 - b_3c_1) + c_3(b_1c_2 - b_2c_1)]$$

$$= \begin{vmatrix} a_1 & a_2 & a_3 \\ b_1 & b_2 & b_3 \\ c_1 & c_2 & c_3 \end{vmatrix} + p(b_1b_2c_3 - b_1b_3c_2 - b_2b_1c_3 + b_2b_3c_1 + b_3b_1c_2 - b_3b_2c_1)$$

$$+ q(c_1b_2c_3 - c_1b_3c_2 - c_2b_1c_3 + c_2b_3c_1 + c_3b_1c_2 - c_3b_2c_1)$$

$= \Delta + p(0) + q(0) = \Delta$. Hence the result holds.

Remark. The operation performed in the above property is denoted by $C_1 \to C_1 + pC_2 + qC_3$. For example, if three times the elements of first row are added to the corresponding elements of the second row, then this will be denoted by $R_2 \to R_2 + 3R_1$.

Illustration. Let $\Delta = \begin{vmatrix} 1 & 5 & 7 \\ 6 & 7 & 2 \\ 1 & 2 & 3 \end{vmatrix}$.

$$\therefore \quad \Delta = 1\begin{vmatrix} 7 & 2 \\ 2 & 3 \end{vmatrix} - 5\begin{vmatrix} 6 & 2 \\ 1 & 3 \end{vmatrix} + 7\begin{vmatrix} 6 & 7 \\ 1 & 2 \end{vmatrix}$$

$$= (21 - 4) - 5(18 - 2) + 7(12 - 7) = 17 - 80 + 35 = -28.$$

Let Δ' be obtained by operating $R_2 \to R_2 - 3R_1 + 4R_3$

$$\therefore \quad \Delta' = \begin{vmatrix} 1 & 5 & 7 \\ 6-3(1)+4(1) & 7-3(5)+4(2) & 2-3(7)+4(3) \\ 1 & 2 & 3 \end{vmatrix}$$

$$= \begin{vmatrix} 1 & 5 & 7 \\ 7 & 0 & -7 \\ 1 & 2 & 3 \end{vmatrix} = 1\begin{vmatrix} 0 & -7 \\ 2 & 3 \end{vmatrix} - 5\begin{vmatrix} 7 & -7 \\ 1 & 3 \end{vmatrix} + 7\begin{vmatrix} 7 & 0 \\ 1 & 2 \end{vmatrix}$$

$$= (0+14) - 5(21+7) + 7(14-0) = 14 - 140 + 98 = -28.$$

$$\therefore \quad \Delta' = \Delta.$$

Property 7. *The determinant of product of square matrices of same order is equal to the product of the determinants of matrices.*

In other words, if A and B are square matrices of same order, then

$$|AB| = |A|.|B|.$$

The proof of this property of determinants is beyond the scope of this book.

Illustration :

$$\text{Let } A = \begin{bmatrix} 1 & 2 \\ 5 & 7 \end{bmatrix} \text{ and } B = \begin{bmatrix} 2 & 7 \\ 8 & 9 \end{bmatrix}.$$

$$AB = \begin{bmatrix} 1 & 2 \\ \longrightarrow & \\ 5 & 7 \end{bmatrix} \begin{bmatrix} 2\downarrow & 7 \\ 8\downarrow & 9 \end{bmatrix} = \begin{bmatrix} 1(2)+2(8) & 1(7)+2(9) \\ 5(2)+7(8) & 5(7)+7(9) \end{bmatrix} = \begin{bmatrix} 18 & 25 \\ 66 & 98 \end{bmatrix}.$$

$$\therefore \quad |AB| = \begin{vmatrix} 18 & 25 \\ 66 & 98 \end{vmatrix} = 18(98) - 66(25) = 114.$$

$$\text{Also, } |A|.|B| = \begin{vmatrix} 1 & 2 \\ 5 & 7 \end{vmatrix} . \begin{vmatrix} 2 & 7 \\ 8 & 9 \end{vmatrix} = (7-10)(18-56) = (-3)(-38) = 114.$$

$$\therefore \quad |AB| = |A|.|B|.$$

Remark. $|A|.|B| = |AB|$ implies that the product of two determinants of same order is equal to the determinant of the product of the corresponding matrices.

For example :

$$\begin{vmatrix} a & b \\ c & d \end{vmatrix} \begin{vmatrix} \alpha & \beta \\ \gamma & \delta \end{vmatrix} = \text{det. of} \begin{bmatrix} a & b \\ \longrightarrow & \\ c & d \end{bmatrix} \begin{bmatrix} \alpha\downarrow & \beta \\ \gamma\downarrow & \delta \end{bmatrix} = \begin{vmatrix} a\alpha + b\gamma & a\beta + b\delta \\ c\alpha + d\gamma & c\beta + d\delta \end{vmatrix}.$$

$\therefore$ **Determinants of same order can be multiplied like square matrices of same order.**

WORKING RULES FOR EVALUATING DETERMINANTS

Rule I. *To find the value of a determinant of order 2, use the formula :*

$$\begin{vmatrix} a & b \\ c & d \end{vmatrix} = ad - bc.$$

Rule II. *To find the value of a determinant of order 3, 'Sarrus method' may be used.*

Rule III. *In order to find the value of a determinant, properties of determinant should be used to make zeros at maximum number of places in any particular row (or column) and then expand the determinant along that row (or column).*

Rule IV. *If all elements of a row are equal, then to create zeros, subtract any column from the remaining columns.*

Rule V. *If all elements of a column are equal, then to create zeros, subtract any row from the remaining rows.*

ILLUSTRATIVE EXAMPLES

Example 1. *Without expanding, show that the value of each determinant is zero :*

(*i*) $\begin{vmatrix} 9 & 9 & 12 \\ 1 & -3 & -4 \\ 1 & 9 & 12 \end{vmatrix}$ (*ii*) $\begin{vmatrix} 1 & 3 & 5 \\ 2 & 6 & 10 \\ 31 & 11 & 38 \end{vmatrix}$

(*iii*) $\begin{vmatrix} 42 & 1 & 6 \\ 28 & 7 & 4 \\ 14 & 3 & 2 \end{vmatrix}$.

Sol. (*i*) Let Δ be the given determinant. Taking 3 common from C_2 and 4 common from C_3, we get

$$\Delta = 3 \times 4 \begin{vmatrix} 9 & 3 & 3 \\ 1 & -1 & -1 \\ 1 & 3 & 3 \end{vmatrix} = 12 \times 0 = \mathbf{0}. \qquad (\because \ C_2 \text{ and } C_3 \text{ are same})$$

(*ii*) Let Δ be the given determinant. Operating $R_2 \to R_2 - 2R_1$, we get

$$\Delta = \begin{vmatrix} 1 & 3 & 5 \\ 2-2 & 6-6 & 10-10 \\ 31 & 11 & 38 \end{vmatrix} = \begin{vmatrix} 1 & 3 & 5 \\ 0 & 0 & 0 \\ 31 & 11 & 38 \end{vmatrix} = \mathbf{0}. \quad (\because \ R_2 \text{ consists of all zeros})$$

(*iii*) Let Δ be the given determinant. Operating $C_1 \to C_1 - 7C_3$, we get

$$\Delta = \begin{vmatrix} 42-42 & 1 & 6 \\ 28-28 & 7 & 4 \\ 14-14 & 3 & 2 \end{vmatrix} = \begin{vmatrix} 0 & 1 & 6 \\ 0 & 7 & 4 \\ 0 & 3 & 2 \end{vmatrix} = \mathbf{0}. \qquad (\because \ C_1 \text{ consists of all zeros})$$

Example 2. *Evaluate the following determinants :*

(*i*) $\begin{vmatrix} 1 & a & bc \\ 1 & b & ca \\ 1 & c & ab \end{vmatrix}$ (*ii*) $\begin{vmatrix} 1 & x & x^3 \\ 1 & y & y^3 \\ 1 & z & z^3 \end{vmatrix}$.

Sol. (*i*) Let $\Delta = \begin{vmatrix} 1 & a & bc \\ 1 & b & ca \\ 1 & c & ab \end{vmatrix}$. Operating $R_2 \to R_2 - R_1$, $R_3 \to R_3 - R_1$, we get

$$\Delta = \begin{vmatrix} 1 & a & bc \\ 0 & b-a & ca-bc \\ 0 & c-a & ab-bc \end{vmatrix}$$

$\therefore$ $\Delta = \begin{vmatrix} 1 & a & bc \\ 0 & b-a & -c(b-a) \\ 0 & c-a & -b(c-a) \end{vmatrix}$. Taking $b - a$, $c - a$ common from R_2, R_3 respectively, we get

$$\Delta = (b-a)(c-a) \begin{vmatrix} 1 & a & bc \\ 0 & 1 & -c \\ 0 & 1 & -b \end{vmatrix}$$

$$\therefore \quad \Delta = (b-a)(c-a) \left[1 . \begin{vmatrix} 1 & -c \\ 1 & -b \end{vmatrix} - 0 + 0 \right] \qquad (\text{Expanding along } C_1)$$

$$= (b-a)(c-a)(-b+c) = \mathbf{(b-a)(c-a)(c-b)}.$$

(*ii*) Let $\Delta = \begin{vmatrix} 1 & x & x^3 \\ 1 & y & y^3 \\ 1 & z & z^3 \end{vmatrix}$. Operating $R_2 \to R_2 - R_1$, $R_3 \to R_3 - R_1$, we get

$$\Delta = \begin{vmatrix} 1 & x & x^3 \\ 0 & y-x & y^3-x^3 \\ 0 & z-x & z^3-x^3 \end{vmatrix}$$. Taking out $y-x$ from R_2 and $z-x$ from R_3, we get

$$\Delta = (y-x)(z-x)\begin{vmatrix} 1 & x & x^3 \\ 0 & 1 & y^2+yx+x^2 \\ 0 & 1 & z^2+zx+x^2 \end{vmatrix}$$. Operating $R_3 \to R_3 - R_2$, we get

$$\Delta = (y-x)(z-x)\begin{vmatrix} 1 & x & x^3 \\ 0 & 1 & y^2+yx+x^2 \\ 0 & 0 & z^2+zx-y^2-yx \end{vmatrix}$$

$$= (y-x)(z-x)\begin{vmatrix} 1 & x & x^3 \\ 0 & 1 & y^2+yx+x^2 \\ 0 & 0 & (z-y)(z+y+x) \end{vmatrix} \quad \left(\begin{aligned} \because\ & z^2+zx-y^2-yx \\ &= (z-y)(z+y)+x(z-y) \\ &= (z-y)(x+y+z) \end{aligned}\right)$$

$$= (y-x)(z-x)\left[0-0+(z-y)(z+y+x)\begin{vmatrix} 1 & x \\ 0 & 1 \end{vmatrix}\right]$$ (Expanding along R_3)

$$= (y-x)(z-x)(z-y)(x+y+z)(1) = \mathbf{(x-y)(y-z)(z-x)(x+y+z).}$$

Example 3. *Show that :*

(*i*) $$\begin{vmatrix} a & b-c & c+b \\ a+c & b & c-a \\ a-b & b+a & c \end{vmatrix} = (a^2+b^2+c^2)(a+b+c)$$

(*ii*) $$\begin{vmatrix} (b+c)^2 & a^2 & bc \\ (c+a)^2 & b^2 & ca \\ (a+b)^2 & c^2 & ab \end{vmatrix} = (a^2+b^2+c^2)(a+b+c)(b-c)(c-a)(a-b)$$

(*iii*) $$\begin{vmatrix} 2ab & a^2 & b^2 \\ a^2 & b^2 & 2ab \\ b^2 & 2ab & a^2 \end{vmatrix} = -(a^3+b^3)^2.$$

Sol. (*i*) Let $$\Delta = \begin{vmatrix} a & b-c & c+b \\ a+c & b & c-a \\ a-b & b+a & c \end{vmatrix}$$. Operating $C_1 \to aC_1$, we get

$$\Delta = \frac{1}{a}\begin{vmatrix} a^2 & b-c & c+b \\ a^2+ac & b & c-a \\ a^2-ab & b+a & c \end{vmatrix}$$. Operating $C_1 \to C_1 + bC_2 + cC_3$, we get

$$\Delta = \frac{1}{a}\begin{vmatrix} a^2+b^2+c^2 & b-c & c+b \\ a^2+b^2+c^2 & b & c-a \\ a^2+b^2+c^2 & b+a & c \end{vmatrix}$$

Taking $a^2 + b^2 + c^2$ common from C_1, we get

$$\Delta = \frac{a^2+b^2+c^2}{a}\begin{vmatrix} 1 & b-c & c+b \\ 1 & b & c-a \\ 1 & b+a & c \end{vmatrix}$$. Operating $R_2 \to R_2 - R_1$ and $R_3 \to R_3 - R_1$, we get

$$\Delta = \frac{a^2+b^2+c^2}{a}\begin{vmatrix} 1 & b-c & c+b \\ 0 & c & -a-b \\ 0 & a+c & -b \end{vmatrix}$$. Expanding along C_1, we get

$$\Delta = \frac{a^2+b^2+c^2}{a}.1\begin{vmatrix} c & -a-b \\ a+c & -b \end{vmatrix} = \frac{a^2+b^2+c^2}{a}(-bc+a^2+ab+ac+bc)$$

$$= \frac{a^2+b^2+c^2}{a}.a(a+b+c) = \mathbf{(a^2+b^2+c^2)(a+b+c)}.$$

(*ii*) Let $\Delta = \begin{vmatrix} (b+c)^2 & a^2 & bc \\ (c+a)^2 & b^2 & ca \\ (a+b)^2 & c^2 & ab \end{vmatrix}$. Operating $C_1 \to C_1 - 2C_3$, we get

| $(b+c)^2 - 2bc = b^2+c^2+2bc-2bc = b^2+c^2$

$$\Delta = \begin{vmatrix} b^2+c^2 & a^2 & bc \\ c^2+a^2 & b^2 & ca \\ a^2+b^2 & c^2 & ab \end{vmatrix}$$. Operating $C_1 \to C_1 + C_2$, we get

$$\Delta = \begin{vmatrix} b^2+c^2+a^2 & a^2 & bc \\ c^2+a^2+b^2 & b^2 & ca \\ a^2+b^2+c^2 & c^2 & ab \end{vmatrix}$$. Taking $a^2+b^2+c^2$ common from C_1, we get

$$\Delta = (a^2+b^2+c^2)\begin{vmatrix} 1 & a^2 & bc \\ 1 & b^2 & ca \\ 1 & c^2 & ab \end{vmatrix}$$. Operating $R_2 \to R_2 - R_1$ and $R_3 \to R_3 - R_1$, we get

$$= (a^2+b^2+c^2)\begin{vmatrix} 1 & a^2 & bc \\ 0 & b^2-a^2 & ca-bc \\ 0 & c^2-a^2 & ab-bc \end{vmatrix}.$$

Taking $b-a$, $c-a$ common from R_2, R_3 respectively, we get

$$\Delta = (a^2+b^2+c^2)(b-a)(c-a)\begin{vmatrix} 1 & a^2 & bc \\ 0 & b+a & -c \\ 0 & c+a & -b \end{vmatrix}$$. Operating $R_3 \to R_3 - R_2$, we get

$$\Delta = (a^2+b^2+c^2)(b-a)(c-a)\begin{vmatrix} 1 & a^2 & bc \\ 0 & b+a & -c \\ 0 & c-b & c-b \end{vmatrix},$$ Taking $(c-b)$ out from R_3,

$$= (a^2+b^2+c^2)(b-a)(c-a)(c-b)\begin{vmatrix} 1 & a^2 & bc \\ 0 & b+a & -c \\ 0 & 1 & 1 \end{vmatrix}$$

Expanding along C_1, we get $\Delta = \mathbf{(a^2+b^2+c^2)(b-a)(c-a)(c-b)(a+b+c)}$.

(*iii*) Let $\Delta = \begin{vmatrix} 2ab & a^2 & b^2 \\ a^2 & b^2 & 2ab \\ b^2 & 2ab & a^2 \end{vmatrix}$. Operating $R_1 \to R_1 + R_2 + R_3$, we get

$$\Delta = \begin{vmatrix} (a+b)^2 & (a+b)^2 & (a+b)^2 \\ a^2 & b^2 & 2ab \\ b^2 & 2ab & a^2 \end{vmatrix}$$ | $(a+b)^2 = a^2+b^2+2ab$

$$\Rightarrow \quad \Delta = (a+b)^2 \begin{vmatrix} 1 & 1 & 1 \\ a^2 & b^2 & 2ab \\ b^2 & 2ab & a^2 \end{vmatrix}. \text{ Operating } C_2 \to C_2 - C_1, C_3 \to C_3 - C_1, \text{ we get}$$

$$\Delta = (a+b)^2 \begin{vmatrix} 1 & 0 & 0 \\ a^2 & b^2 - a^2 & 2ab - a^2 \\ b^2 & 2ab - b^2 & a^2 - b^2 \end{vmatrix}. \text{ Expanding along } R_1, \text{ we get}$$

$$\Delta = (a+b)^2 \left[1.\begin{vmatrix} b^2 - a^2 & 2ab - a^2 \\ 2ab - b^2 & a^2 - b^2 \end{vmatrix} - 0 + 0 \right]$$

$$= (a+b)^2 \; [-(a^2 - b^2)^2 - (2ab - a^2)(2ab - b^2)]$$

$$= -(a+b)^2 \, [a^4 + b^4 - 2a^2b^2 + 4a^2b^2 - 2a^3b - 2ab^3 + a^2b^2]$$

$$= -(a+b)^2 \, [a^4 + b^4 + a^2b^2 - 2a^3b - 2ab^3 + 2a^2b^2]$$

$$= -(a+b)^2 (a^2 - ab + b^2)^2 = -\mathbf{(a^3 + b^3)^2}. \qquad | \; a^3 + b^3 = (a+b)(a^2 - ab + b^2)$$

Example 4. *Show that* $\begin{vmatrix} a+b & b+c & c+a \\ b+c & c+a & a+b \\ c+a & a+b & b+c \end{vmatrix} = 2\begin{vmatrix} a & b & c \\ b & c & a \\ c & a & b \end{vmatrix}$.

Sol. Let $\Delta = \begin{vmatrix} a+b & b+c & c+a \\ b+c & c+a & a+b \\ c+a & a+b & b+c \end{vmatrix}$. Applying $C_1 \to C_1 + C_2 + C_3$, we get

$$\Delta = \begin{vmatrix} 2(a+b+c) & b+c & c+a \\ 2(a+b+c) & c+a & a+b \\ 2(a+b+c) & a+b & b+c \end{vmatrix}. \text{ Taking 2 common from } C_1, \text{ we get}$$

$$\Delta = 2\begin{vmatrix} a+b+c & b+c & c+a \\ a+b+c & c+a & a+b \\ a+b+c & a+b & b+c \end{vmatrix}. \text{ Applying } C_2 \to C_2 - C_1, C_3 \to C_3 - C_1, \text{ we get}$$

$$\Delta = 2\begin{vmatrix} a+b+c & -a & -b \\ a+b+c & -b & -c \\ a+b+c & -c & -a \end{vmatrix} = 2(-1)(-1)\begin{vmatrix} a+b+c & a & b \\ a+b+c & b & c \\ a+b+c & c & a \end{vmatrix}$$

Applying $C_1 \to C_1 - C_2 - C_3$, we get

$$\Delta = 2\begin{vmatrix} c & a & b \\ a & b & c \\ b & c & a \end{vmatrix} = -2\begin{vmatrix} a & c & b \\ b & a & c \\ c & b & a \end{vmatrix} \qquad (C_1 \leftrightarrow C_2)$$

$$= 2\begin{vmatrix} a & b & c \\ b & c & a \\ c & a & b \end{vmatrix} \qquad (C_2 \leftrightarrow C_3)$$

∴ The result holds.

Example 5. *Evaluate the following determinants :*

(*i*) $\begin{vmatrix} b+c & a & a \\ b & c+a & b \\ c & c & a+b \end{vmatrix}$ (*ii*) $\begin{vmatrix} x+\lambda & x & x \\ x & x+\lambda & x \\ x & x & x+\lambda \end{vmatrix}$

(*iii*) $\begin{vmatrix} a & a+b & a+b+c \\ 2a & 3a+2b & 4a+3b+2c \\ 3a & 6a+3b & 10a+6b+3c \end{vmatrix}$.

Sol. (*i*) Let $\Delta = \begin{vmatrix} b+c & a & a \\ b & c+a & b \\ c & c & a+b \end{vmatrix}$. Expanding Δ along R_1, we get

$$\Delta = (b+c)\begin{vmatrix} c+a & b \\ c & a+b \end{vmatrix} - a\begin{vmatrix} b & b \\ c & a+b \end{vmatrix} + a\begin{vmatrix} b & c+a \\ c & c \end{vmatrix}$$

$$= (b+c)(ca+cb+a^2+ab-bc) - a(ba+b^2-bc) + a(bc-c^2-ca)$$

$$= bca + b^2c + a^2b + ab^2 - b^2c + c^2a + c^2b + a^2c + abc - bc^2 - a^2b - ab^2 + abc + abc - ac^2 - a^2c$$

$= \mathbf{4abc}$.

(*ii*) $$\begin{vmatrix} x+\lambda & x & x \\ x & x+\lambda & x \\ x & x & x+\lambda \end{vmatrix} = \begin{vmatrix} x+\lambda+x+x & x & x \\ x+x+\lambda+x & x+\lambda & x \\ x+x+x+\lambda & x & x+\lambda \end{vmatrix} \qquad (C_1 \to C_1 + C_2 + C_3)$$

$$= (3x+\lambda)\begin{vmatrix} 1 & x & x \\ 1 & x+\lambda & x \\ 1 & x & x+\lambda \end{vmatrix} \qquad \text{(Taking } 3x+\lambda \text{ common from } C_1)$$

$$= (3x+\lambda)\begin{vmatrix} 1 & x & x \\ 0 & \lambda & 0 \\ 0 & 0 & \lambda \end{vmatrix} \qquad (R_2 \to R_2 - R_1, R_3 \to R_3 - R_1)$$

$$= (3x+\lambda) \,.\, 1 \begin{vmatrix} \lambda & 0 \\ 0 & \lambda \end{vmatrix} \qquad \text{(Expanding along } C_1)$$

$= \boldsymbol{\lambda^2(3x+\lambda)}$.

Remark. In evaluating determinants, the properties of determinants are used to make zeros at maximum number of places in any particular row (or column).

(*iii*) $$\begin{vmatrix} a & a+b & a+b+c \\ 2a & 3a+2b & 4a+3b+2c \\ 3a & 6a+3b & 10a+6b+3c \end{vmatrix} = \begin{vmatrix} a & a+b & a+b+c \\ 0 & a & 2a+b \\ 0 & 3a & 7a+3b \end{vmatrix}$$

(Operating $R_2 \to R_2 - 2R_1$, $R_3 \to R_3 - 3R_1$)

$$= \begin{vmatrix} a & a+b & a+b+c \\ 0 & a & 2a+b \\ 0 & 0 & a \end{vmatrix} \qquad \text{(Operating } R_3 \to R_3 - 3R_2)$$

$$= a\begin{vmatrix} a & 2a+b \\ 0 & a \end{vmatrix} = a(a^2 - 0) = \mathbf{a^3}. \qquad \text{(Expanding along } C_1)$$

Example 6. *Show that :*

(*i*) $$\begin{vmatrix} -a^2 & ab & ac \\ ba & -b^2 & bc \\ ac & bc & -c^2 \end{vmatrix} = 4a^2b^2c^2$$

(*ii*) $$\begin{vmatrix} b^2+c^2 & ab & ac \\ ba & c^2+a^2 & bc \\ ca & cb & a^2+b^2 \end{vmatrix} = 4a^2b^2c^2$$

(*iii*) $$\begin{vmatrix} (b+c)^2 & a^2 & a^2 \\ b^2 & (c+a)^2 & b^2 \\ c^2 & c^2 & (a+b)^2 \end{vmatrix} = 2abc(a+b+c)^3.$$

Sol. (*i*) Let $\Delta = \begin{vmatrix} -a^2 & ab & ac \\ ba & -b^2 & bc \\ ac & bc & -c^2 \end{vmatrix}$

Taking a, b, c common from R_1, R_2, R_3 respectively, we get

$$\therefore \quad \Delta = a.b.c \begin{vmatrix} -a & b & c \\ a & -b & c \\ a & b & -c \end{vmatrix} = (abc)(abc) \begin{vmatrix} -1 & 1 & 1 \\ 1 & -1 & 1 \\ 1 & 1 & -1 \end{vmatrix}$$

(Taking a, b, c common from C_1, C_2, C_3 respectively)

$$= a^2b^2c^2 \begin{vmatrix} -1 & 0 & 0 \\ 1 & 0 & 2 \\ 1 & 2 & 0 \end{vmatrix} \qquad (C_2 \to C_2 + C_1, C_3 \to C_3 + C_1)$$

$$= a^2b^2c^2 \left[-1 \begin{vmatrix} 0 & 2 \\ 2 & 0 \end{vmatrix} - 0 + 0 \right] = a^2b^2c^2 \times 4 = \mathbf{4a^2b^2c^2}. \text{ (Expanding along } R_1)$$

(*ii*) Let $\Delta = \begin{vmatrix} b^2 + c^2 & ab & ac \\ ba & c^2 + a^2 & bc \\ ca & cb & a^2 + b^2 \end{vmatrix}$.

Multiplying R_1, R_2, R_3 by a, b, c respectively and dividing Δ by abc, we get

$$\Delta = \frac{1}{abc} \begin{vmatrix} a(b^2 + c^2) & a^2b & a^2c \\ b^2a & b(c^2 + a^2) & b^2c \\ c^2a & c^2b & c(a^2 + b^2) \end{vmatrix}$$

Taking a, b, c common from C_1, C_2, C_3 respectively, we get

$$\Delta = \frac{abc}{abc} \begin{vmatrix} b^2 + c^2 & a^2 & a^2 \\ b^2 & c^2 + a^2 & b^2 \\ c^2 & c^2 & a^2 + b^2 \end{vmatrix}. \text{ Operating } R_1 \to R_1 + R_2 + R_3, \text{ we get}$$

$$\Delta = \begin{vmatrix} 2(b^2 + c^2) & 2(c^2 + a^2) & 2(a^2 + b^2) \\ b^2 & c^2 + a^2 & b^2 \\ c^2 & c^2 & a^2 + b^2 \end{vmatrix}. \text{ Taking 2 common from } R_1, \text{ we get}$$

$$\Delta = 2 \begin{vmatrix} b^2 + c^2 & c^2 + a^2 & a^2 + b^2 \\ b^2 & c^2 + a^2 & b^2 \\ c^2 & c^2 & a^2 + b^2 \end{vmatrix}. \text{ Operating } R_2 \to R_2 - R_1, R_3 \to R_3 - R_1, \text{ we get}$$

$$\Delta = 2 \begin{vmatrix} b^2 + c^2 & c^2 + a^2 & a^2 + b^2 \\ -c^2 & 0 & -a^2 \\ -b^2 & -a^2 & 0 \end{vmatrix}. \text{ Operating } R_1 \to R_1 + R_2 + R_3, \text{ we get}$$

$$\Delta = 2 \begin{vmatrix} 0 & c^2 & b^2 \\ -c^2 & 0 & -a^2 \\ -b^2 & -a^2 & 0 \end{vmatrix}$$

$= 2[0 - c^2(0 - a^2b^2) + b^2(a^2c^2 - 0)]$ (Expanding along R_1)

$= 2\,(a^2b^2c^2 + a^2b^2c^2) = \mathbf{4a^2b^2c^2}$.

(*iii*) Let $\Delta = \begin{vmatrix} (b+c)^2 & a^2 & a^2 \\ b^2 & (c+a)^2 & b^2 \\ c^2 & c^2 & (a+b)^2 \end{vmatrix}$. Operating $C_1 \to C_1 - C_3$ and $C_2 \to C_2 - C_3$, we get

$$\Delta = \begin{vmatrix} (b+c)^2 - a^2 & 0 & a^2 \\ 0 & (c+a)^2 - b^2 & b^2 \\ c^2 - (a+b)^2 & c^2 - (a+b)^2 & (a+b)^2 \end{vmatrix}$$

$$= \begin{vmatrix} (b+c+a)(b+c-a) & 0 & a^2 \\ 0 & (c+a+b)(c+a-b) & b^2 \\ (c+a+b)(c-a-b) & (c+a+b)(c-a-b) & (a+b)^2 \end{vmatrix}$$

Taking $(a + b + c)$ common from C_1 and C_2 each, we get

$$\Delta = (a+b+c)^2 \begin{vmatrix} b+c-a & 0 & a^2 \\ 0 & c+a-b & b^2 \\ c-a-b & c-a-b & (a+b)^2 \end{vmatrix}$$

Operating $R_3 \to R_3 - R_1 - R_2$, we get

$$\Delta = (a+b+c)^2 \begin{vmatrix} b+c-a & 0 & a^2 \\ 0 & c+a-b & b^2 \\ -2b & -2a & 2ab \end{vmatrix}$$

Operating $C_1 \to C_1 + \frac{1}{a} C_3$, $C_2 \to C_2 + \frac{1}{b} C_3$, we get

$$\Delta = (a+b+c)^2 \begin{vmatrix} b+c & a^2/b & a^2 \\ b^2/a & c+a & b^2 \\ 0 & 0 & 2ab \end{vmatrix}$$

Expanding along R_3, we get

$$\Delta = (a+b+c)^2 \,.\, 2ab \begin{vmatrix} b+c & a^2/b \\ b^2/a & c+a \end{vmatrix} = (a+b+c)^2\ 2ab((b+c)(c+a) - ab))$$

$$= (a+b+c)^2\ 2ab(bc + c^2 + ca) = \mathbf{2abc\ (a + b + c)^3}.$$

Example 7. (*i*) *If x, y, z are all distinct and if* $\begin{vmatrix} x & x^2 & 1+x^3 \\ y & y^2 & 1+y^3 \\ z & z^2 & 1+z^3 \end{vmatrix} = 0$, *show that* $xyz = -1$.

(*ii*) *If x, y, z are all distinct and if* $\begin{vmatrix} x & x^3 & x^4-1 \\ y & y^3 & y^4-1 \\ z & z^3 & z^4-1 \end{vmatrix} = 0$, *show that*

$$xyz(yz + zx + xy) = x + y + z.$$

Sol. (*i*) Let $\Delta = \begin{vmatrix} x & x^2 & 1+x^3 \\ y & y^2 & 1+y^3 \\ z & z^2 & 1+z^3 \end{vmatrix}$.

$\therefore \qquad \Delta = 0 \qquad$ ...(1)

Operating $R_2 \to R_2 - R_1$ and $R_3 \to R_3 - R_1$, we get

$$\Delta = \begin{vmatrix} x & x^2 & 1+x^3 \\ y-x & y^2-x^2 & y^3-x^3 \\ z-x & z^2-x^2 & z^3-x^3 \end{vmatrix}$$

Taking out $y - x$ from R_2 and $z - x$ from R_3, we get

$$\Delta = (y-x)(z-x)\begin{vmatrix} x & x^2 & 1+x^3 \\ 1 & y+x & y^2+yx+x^2 \\ 1 & z+x & z^2+zx+x^2 \end{vmatrix}.$$ Operating $R_3 \to R_3 - R_2$, we get

$$\Delta = (y-x)(z-x)\begin{vmatrix} x & x^2 & 1+x^3 \\ 1 & y+x & y^2+yx+x^2 \\ 0 & z-y & (z-y)(x+y+z) \end{vmatrix}$$

$[\because\ z^2 + zx + x^2 - y^2 - yx - x^2 = z^2 - y^2 + x(z-y)$
$= (z-y)(z+y) + x(z-y) = (z-y)(x+y+z)]$

Taking out $z - y$ from R_3, we get

$$\Delta = (z-y)(y-x)(z-x)\begin{vmatrix} x & x^2 & 1+x^3 \\ 1 & y+x & y^2+yx+x^2 \\ 0 & 1 & x+y+z \end{vmatrix}$$

Expanding along R_1, we get

$$\Delta = (z-y)(y-x)(z-x)\begin{bmatrix} x[(y+x)(x+y+z)-(y^2+yx+x^2)] \\ -x^2[x+y+z-0]+(1+x^3)[1-0] \end{bmatrix}$$

$\therefore$ $\Delta = (z-y)(y-x)(z-x)\,[(x+y+z)(xy+x^2-x^2) - xy^2 - yx^2 - x^3 + 1 + x^3]$

$\therefore$ $\Delta = (z-y)(y-x)(z-x)(xyz+1)$

$\therefore$ (1) $\Rightarrow (z-y)(y-x)(z-x)(xyz+1) = 0$

$\Rightarrow$ $xyz + 1 = 0$ $\qquad (\because\ x \neq y \neq z)$

$\therefore$ $\mathbf{xyz = -1.}$

(*ii*) Let $\Delta = \begin{vmatrix} x & x^3 & x^4-1 \\ y & y^3 & y^4-1 \\ z & z^3 & z^4-1 \end{vmatrix}.$

$\therefore$ $\Delta = 0$ ...(1)

Operating $R_2 \to R_2 - R_1$, $R_3 \to R_3 - R_1$, we get

$$\Delta = \begin{vmatrix} x & x^3 & x^4-1 \\ y-x & y^3-x^3 & y^4-x^4 \\ z-x & z^3-x^3 & z^4-x^4 \end{vmatrix}.$$

Taking out $y - x$ from R_2 and $z - x$ from R_3, we get

$$\Delta = (y-x)(z-x)\begin{vmatrix} x & x^3 & x^4-1 \\ 1 & y^2+yx+x^2 & (y+x)(y^2+x^2) \\ 1 & z^2+zx+x^2 & (z+x)(z^2+x^2) \end{vmatrix}$$

$$= (y-x)(z-x)\begin{vmatrix} x & x^3 & x^4-1 \\ 1 & y^2+yx+x^2 & x^3+x^2y+xy^2+y^3 \\ 1 & z^2+zx+x^2 & x^3+z^2x+zx^2+z^3 \end{vmatrix}$$

Operating $R_3 \to R_3 - R_2$, we get

$$\Delta = (y-x)(z-x)\begin{vmatrix} x & x^3 & x^4-1 \\ 1 & x^2+xy+y^2 & x^3+x^2y+xy^2+y^3 \\ 0 & (z-y)(x+y+z) & (z-y)(x^2+y^2+z^2+xy+yz+zx) \end{vmatrix}$$

Taking out $z - y$ from R_3, we get

$$\Delta = (z-y)(y-x)(z-x)\begin{vmatrix} x & x^3 & x^4-1 \\ 1 & x^2+xy+y^2 & x^3+x^2y+xy^2+y^3 \\ 0 & x+y+z & x^2+y^2+z^2+xy+yz+zx \end{vmatrix}$$

Expanding along R_3, we get

$$\begin{aligned}\Delta &= (z-y)(y-x)(z-x)\,[0-(x+y+z)(x^4+x^3y+x^2y^2+xy^3-x^4+1) \\ &\qquad + (x^2+y^2+z^2+xy+yz+zx)(x^3+x^2y+xy^2-x^3)] \\ &- (z-y)(y-x)(z-x)\,[-(x+y+z)(x^3y+x^2y^2+xy^3+1) \\ &\qquad + (x^2+y^2+z^2+xy+yz+zx)(x^2y+xy^2)] \\ &= (z-y)(y-x)(z-x)\,[-(x+y+z)(x^3y+x^2y^2+xy^3) \\ &\qquad -x-y-z+(x^2+y^2+z^2+xy+yz+zx)\,x^2y \\ &\qquad + (x^2+y^2+z^2+xy+yz+zx)\,xy^2] \\ &= (z-y)(y-x)(z-x)\,[-x^4y-x^3y^2-x^2y^3-x^3y^2-x^2y^3 \\ &\qquad -xy^4-x^3yz-x^2y^2z-xy^3z-x-y-z+x^4y+x^2y^3+x^2yz^2 \\ &\qquad + x^3y^2+x^2y^2z+x^3yz+x^3y^2+xy^4+xy^2z^2+x^2y^3 \\ &\qquad + xy^3z+x^2y^2z] \\ &= (z-y)(y-x)(z-x)\,[-x-y-z+x^2yz^2+x^2y^2z+xy^2z^2] \\ &= (z-y)(y-x)(z-x)\,[-x-y-z+xyz(xz+xy+yz)]\end{aligned}$$

$\therefore$ (1) $\Rightarrow$ $(z-y)(y-x)(z-x)\left[-x-y-z+xyz(yz+zx+xy)\right] = 0$

$\Rightarrow$ $\left[-x-y-z+xyz(yz+zx+xy)\right] = 0 \quad (\because \; x \neq y \neq z)$

$\Rightarrow$ $\mathbf{xyz(yz+zx+xy) = x+y+z.}$

Example 8. *Solve the equation :* $\begin{vmatrix} 3x-8 & 3 & 3 \\ 3 & 3x-8 & 3 \\ 3 & 3 & 3x-8 \end{vmatrix} = 0.$

Sol. The given equation is $\begin{vmatrix} 3x-8 & 3 & 3 \\ 3 & 3x-8 & 3 \\ 3 & 3 & 3x-8 \end{vmatrix} = 0.$

$\Rightarrow \begin{vmatrix} 3x-2 & 3x-2 & 3x-2 \\ 3 & 3x-8 & 3 \\ 3 & 3 & 3x-8 \end{vmatrix} = 0$ (By applying $R_1 \to R_1 + R_2 + R_3$)

$\Rightarrow (3x-2)\begin{vmatrix} 1 & 1 & 1 \\ 3 & 3x-8 & 3 \\ 3 & 3 & 3x-8 \end{vmatrix} = 0$

$\Rightarrow (3x-2)\begin{vmatrix} 0 & 0 & 1 \\ 0 & 3x-11 & 3 \\ 11-3x & 11-3x & 3x-8 \end{vmatrix} = 0$ (By applying $C_1 \to C_1 - C_3$, $C_2 \to C_2 - C_3$)

$\Rightarrow (3x-2)\,1\,.\begin{vmatrix} 0 & 3x-11 \\ 11-3x & 11-3x \end{vmatrix} = 0$ (Expanding along R_1)

$\Rightarrow (3x-2)(0+(3x-11)^2) = 0 \quad \Rightarrow \quad (3x-2)(3x-11)^2 = 0$

$$\mathbf{x = \frac{2}{3}, \frac{11}{3}, \frac{11}{3}}.$$

EXERCISE 2.8

1. Without expanding, show that each of the following determinants is equal to zero :

(i) $\begin{vmatrix} 3 & 8 & 6 \\ 6 & 15 & 12 \\ 7 & 17 & 14 \end{vmatrix}$ (ii) $\begin{vmatrix} 1 & 10 & 11 \\ 1 & 12 & 13 \\ 1 & 9 & 10 \end{vmatrix}$ (iii) $\begin{vmatrix} 43 & 1 & 6 \\ 35 & 7 & 4 \\ 17 & 3 & 2 \end{vmatrix}$

(iv) $\begin{vmatrix} x+1 & x+2 & x+a \\ x+2 & x+3 & x+b \\ x+3 & x+4 & x+c \end{vmatrix} = 0$, where a, b, c are in A.P. (*P.T.U. B.Pharma, Dec. 2005*)

2. Without expanding, show that each of the following determinants is equal to zero :

(i) $\begin{vmatrix} 1 & a & abc \\ 1 & b & abc \\ 1 & c & abc \end{vmatrix}$ (ii) $\begin{vmatrix} 1 & a & b+c \\ 1 & b & c+a \\ 1 & c & a+b \end{vmatrix}$ (iii) $\begin{vmatrix} x+y & y+z & z+x \\ z & x & y \\ 1 & 1 & 1 \end{vmatrix}$

(*P.T.U. B.Pharma, May 2007 ; K.U.K. 2006*)

(iv) $\begin{vmatrix} 1 & bc & a(b+c) \\ 1 & ca & b(c+a) \\ 1 & ab & c(a+b) \end{vmatrix}$ (v) $\begin{vmatrix} 1/a & a^2 & bc \\ 1/b & b^2 & ca \\ 1/c & c^2 & ab \end{vmatrix}$ (vi) $\begin{vmatrix} b^2c^2 & bc & b+c \\ c^2a^2 & ca & c+a \\ a^2b^2 & ab & a+b \end{vmatrix}$.

3. Without expanding, show that :

(i) $(a+b+c)$ is factor of $\begin{vmatrix} a & b & c \\ b & c & a \\ c & a & b \end{vmatrix}$ (ii) $(a-b)$ is a factor of $\begin{vmatrix} 1 & a & a^2 \\ 1 & b & b^2 \\ 1 & c & c^2 \end{vmatrix}$

(iii) $\left(\frac{1}{a}+\frac{1}{b}+\frac{1}{c}+1\right)$ is a factor of $\begin{vmatrix} 1+a & 1 & 1 \\ 1 & 1+b & 1 \\ 1 & 1 & 1+c \end{vmatrix}$.

4. Show that :

(i) $\begin{vmatrix} a & b & c \\ b & c & a \\ c & a & b \end{vmatrix} = (a+b+c)(ab+bc+ca-a^2-b^2-c^2)$

(ii) $\begin{vmatrix} x & a & a \\ a & x & a \\ a & a & x \end{vmatrix} = (x-a)^2(x+2a)$ (iii) $\begin{vmatrix} a^2+2a & 2a+1 & 1 \\ 2a+1 & a+2 & 1 \\ 3 & 3 & 1 \end{vmatrix} = (a-1)^3$

(iv) $\begin{vmatrix} x+a & b & c \\ a & x+b & c \\ a & b & x+c \end{vmatrix} = x^2(x+a+b+c)$ (v) $\begin{vmatrix} x+4 & 2x & 2x \\ 2x & x+4 & 2x \\ 2x & 2x & x+4 \end{vmatrix} = (5x+4)(x-4)^2$

(vi) $\begin{vmatrix} y+z & z & y \\ z & z+x & x \\ y & x & x+y \end{vmatrix} = 4xyz$ (vii) $\begin{vmatrix} x & y & x+y \\ y & x+y & x \\ x+y & x & y \end{vmatrix} = -2(x^3+y^3)$.

5. Show that :

(i) $\begin{vmatrix} b+c & a-b & a \\ c+a & b-c & b \\ a+b & c-a & c \end{vmatrix} = 3abc-a^3-b^3-c^3$ (ii) $\begin{vmatrix} a-b & b-c & c-a \\ b+c & c+a & a+b \\ a & b & c \end{vmatrix} = a^3+b^3+c^3-3abc$.

6. Show that $\begin{vmatrix} 1+a & 1 & 1 \\ 1 & 1+b & 1 \\ 1 & 1 & 1+c \end{vmatrix} = abc\left(\frac{1}{a}+\frac{1}{b}+\frac{1}{c}+1\right)$.

7. Show that :

(i) $\begin{vmatrix} 1 & 1 & 1 \\ \alpha & \beta & \gamma \\ \beta\gamma & \gamma\alpha & \alpha\beta \end{vmatrix} = (\beta-\gamma)(\gamma-\alpha)(\alpha-\beta)$ (ii) $\begin{vmatrix} 1 & a & a^2 \\ 1 & b & b^2 \\ 1 & c & c^2 \end{vmatrix} = (b-c)(c-a)(a-b)$

(iii) $\begin{vmatrix} x & y & z \\ x^2 & y^2 & z^2 \\ x^3 & y^3 & z^3 \end{vmatrix} = xyz(x-y)(y-z)(z-x)$

(iv) $\begin{vmatrix} \alpha & \alpha^2 & \beta+\gamma \\ \beta & \beta^2 & \gamma+\alpha \\ \gamma & \gamma^2 & \alpha+\beta \end{vmatrix} = (\beta-\gamma)(\gamma-\alpha)(\alpha-\beta)(\alpha+\beta+\gamma)$

(v) $\begin{vmatrix} 1 & a & a^2-bc \\ 1 & b & b^2-ca \\ 1 & c & c^2-ab \end{vmatrix} = 0$ (vi) $\begin{vmatrix} 1 & x+y & x^2+y^2 \\ 1 & y+z & y^2+z^2 \\ 1 & z+x & z^2+x^2 \end{vmatrix} = (x-y)(y-z)(z-x)$.

8. Show that :

$$\begin{vmatrix} 3a & -a+b & -a+c \\ -b+a & 3b & -b+c \\ -c+a & -c+b & 3c \end{vmatrix} = 3(a+b+c)(ab+bc+ca).$$

9. Evaluate the following determinants :

(i) $\begin{vmatrix} 1 & a & a^3 \\ 1 & b & b^3 \\ 1 & c & c^3 \end{vmatrix}$ (ii) $\begin{vmatrix} x & y & z \\ x^2 & y^2 & z^2 \\ yz & zx & xy \end{vmatrix}$

(iii) $\begin{vmatrix} a & a^2 & bc \\ b & b^2 & ca \\ c & c^2 & ab \end{vmatrix}$ (iv) $\begin{vmatrix} y+z & x & y \\ z+x & z & x \\ x+y & y & z \end{vmatrix}$.

10. Show that :

(i) $\begin{vmatrix} a^2 & bc & ac+c^2 \\ a^2+ab & b^2 & ac \\ ab & b^2+bc & c^2 \end{vmatrix} = 4a^2b^2c^2$ (ii) $\begin{vmatrix} ab & -b^2 & bc \\ ca & bc & -c^2 \\ -a^2 & ab & ca \end{vmatrix} = 4a^2b^2c^2$

(iii) $\begin{vmatrix} a+b+c & -c & -b \\ -c & a+b+c & -a \\ -b & -a & a+b+c \end{vmatrix} = 2(a+b)(b+c)(c+a)$.

(iv) $\begin{vmatrix} a^2+1 & ab & ac \\ ab & b^2+1 & bc \\ ac & bc & c^2+1 \end{vmatrix} = \begin{vmatrix} a^2+1 & b^2 & c^2 \\ a^2 & b^2+1 & c^2 \\ a^2 & b^2 & c^2+1 \end{vmatrix} = 1+a^2+b^2+c^2$

11. Without expanding, show that $\begin{vmatrix} a & b & c \\ x & y & z \\ p & q & r \end{vmatrix} = \begin{vmatrix} x & y & z \\ p & q & r \\ a & b & c \end{vmatrix} = \begin{vmatrix} y & b & q \\ x & a & p \\ z & c & r \end{vmatrix}$.

12. Show that :

(i) $\begin{vmatrix} b+c & c+a & a+b \\ q+r & r+p & p+q \\ y+z & z+x & x+y \end{vmatrix} = 2 \begin{vmatrix} a & b & c \\ p & q & r \\ x & y & z \end{vmatrix}$

(ii) $\begin{vmatrix} x+y & y+z & z+x \\ y+z & z+x & x+y \\ z+x & x+y & y+z \end{vmatrix} = 2 \begin{vmatrix} x & y & z \\ y & z & x \\ z & x & y \end{vmatrix}$

(iii) $\begin{vmatrix} b+c & c+a & a+b \\ c+a & a+b & b+c \\ a+b & b+c & c+a \end{vmatrix} = 2\,(a+b+c)(ab+bc+ca-a^2-b^2-c^2).$

13. Show that :

(i) $\begin{vmatrix} a-b-c & 2a & 2a \\ 2b & b-c-a & 2b \\ 2c & 2c & c-a-b \end{vmatrix} = (a+b+c)^3$

(ii) $\begin{vmatrix} a+b+2c & a & b \\ c & b+c+2a & b \\ c & a & c+a+2b \end{vmatrix} = 2(a+b+c)^3.$

14. Show that :

(i) $\begin{vmatrix} 1+a^2-b^2 & 2ab & -2b \\ 2ab & 1-a^2+b^2 & 2a \\ 2b & -2a & 1-a^2-b^2 \end{vmatrix} = (1+a^2+b^2)^3$

(ii) $\begin{vmatrix} 1 & a & a^2 \\ a^2 & 1 & a \\ a & a^2 & 1 \end{vmatrix} = (a^3-1)^2$ (iii) $\begin{vmatrix} b^2-ab & b-c & bc-ac \\ ab-a^2 & a-b & b^2-ab \\ bc-ac & c-a & ab-a^2 \end{vmatrix} = 0.$

15. Solve the equations :

(i) $\begin{vmatrix} x-1 & 2x-3 & 1 \\ x-3 & 2x-1 & 2 \\ x-5 & 2x-5 & 3 \end{vmatrix} = 0$ (ii) $\begin{vmatrix} x+2 & 2x+3 & 3x+4 \\ 2x+3 & 3x+4 & 4x+5 \\ 3x+5 & 5x+8 & 10x+17 \end{vmatrix} = 0$

(iii) $\begin{vmatrix} x^2 & 0 & 3 \\ x & 1 & -4 \\ 1 & 2 & 0 \end{vmatrix} = 11$ (iv) $\begin{vmatrix} a+x & a-x & a-x \\ a-x & a+x & a-x \\ a-x & a-x & a+x \end{vmatrix} = 0.$

Answers

9. (i) $(b-c)(c-a)(a-b)(a+b+c)$ (ii) $(y-z)(z-x)(x-y)(xy+yz+zx)$

(iii) $(a-b)(b-c)(c-a)(ab+bc+ca)$ (iv) $(x+y+z)(x-z)^2$

15. (i) -1 (ii) $-1, -1, -2$ (iii) $-7/4, 1$ (iv) $0, 3a$.

Hints

1. (i) Apply $C_3 \to C_3 - 2C_1$. (ii) Apply $C_3 \to C_3 - C_2$. (iii) Apply $C_1 \to C_1 - 7C_3$.

2. (ii) Apply $C_3 \to C_3 + C_2$. (iv) Apply $C_3 \to C_3 + C_2$.

(v) Apply $R_1 \to aR_1$, $R_2 \to bR_2$, $R_3 \to cR_3$ and then take abc common from C_3.

(vi) Let Δ be the given determinant.

Multiplying R_1, R_2, R_3 by a, b, c respectively.

$$\therefore \quad \Delta = \frac{1}{abc}\begin{vmatrix} ab^2c^2 & abc & a(b+c) \\ bc^2a^2 & abc & b(c+a) \\ ca^2b^2 & abc & c(a+b) \end{vmatrix} = \frac{(abc)^2}{abc}\begin{vmatrix} bc & 1 & ab+ac \\ ca & 1 & bc+ba \\ ab & 1 & ca+cb \end{vmatrix}$$

Now use $C_3 \to C_3 + C_1$.

3. (*ii*) Apply $R_1 \to R_1 - R_2$ and take $a - b$ common.

$$(iii) \begin{vmatrix} 1+a & 1 & 1 \\ 1 & 1+b & 1 \\ 1 & 1 & 1+c \end{vmatrix} = abc\begin{vmatrix} \frac{1}{a}+1 & \frac{1}{a} & \frac{1}{a} \\ \frac{1}{b} & \frac{1}{b}+1 & \frac{1}{b} \\ \frac{1}{c} & \frac{1}{c} & \frac{1}{c}+1 \end{vmatrix}$$

$$= abc\begin{vmatrix} \frac{1}{a}+1+\frac{1}{b}+\frac{1}{c} & \frac{1}{a}+\frac{1}{b}+1+\frac{1}{c} & \frac{1}{a}+\frac{1}{b}+\frac{1}{c}+1 \\ \frac{1}{b} & \frac{1}{b}+1 & \frac{1}{b} \\ \frac{1}{c} & \frac{1}{c} & \frac{1}{c}+1 \end{vmatrix}.$$

[Using $R_1 \to R_1 + R_2 + R_3$]

4. (*iii*) Apply $R_1 \to R_1 - R_3$, $R_2 \to R_2 - R_3$ and then expand along C_3.

(*vi*) Apply $R_1 \to R_1 + R_2 + R_3$, $R_1 \to \frac{1}{2}R_1$, $R_1 \to R_1 - R_2$, $C_3 \to C_3 - C_1$.

(*vii*) Apply $C_1 \to C_1 + C_2 + C_3$.

5. (*i*) Apply $C_1 \to C_1 + C_3$.

7. (*iii*) Take x, y, z common from C_1, C_2, C_3 respectively.

(*vi*) Apply $R_2 \to R_2 - R_1$, $R_3 \to R_3 - R_1$. (*vii*) Apply $R_2 \to R_2 - R_1$, $R_3 \to R_3 - R_1$.

8. Apply $C_1 \to C_1 + C_2 + C_3$.

9. (*ii*) Apply $C_2 \to C_2 - C_1$, $C_3 \to C_3 - C_1$. (*iv*) Apply $R_1 \to R_1 + R_2 + R_3$.

10. (*i*) Take a, b, c common from C_1, C_2, C_3 respectively. In the next step, apply $C_1 \to C_1 - C_2 - C_3$.

(*ii*) Take a, b, c common from C_1, C_2, C_3 respectively. In the next step, take a, b, c common from R_1, R_2, R_3 respectively.

(*iii*) Apply $R_2 \to R_2 + R_1$, $R_3 \to R_3 + R_1$.

12. (*i*) Apply $C_1 \to C_1 + C_2 + C_3$, $C_1 \to \frac{1}{2}C_1$, $C_2 \to C_2 - C_1$, $C_3 \to C_3 - C_1$ in this given order.

(*iii*) Apply $C_1 \to C_1 + C_2 + C_3$, $R_2 \to R_2 - R_1$, $R_3 \to R_3 - R_1$.

13. (*i*) Apply $R_1 \to R_1 + R_2 + R_3$ and then take $a + b + c$ common from R_1.

14. (*i*) Apply $C_1 \to C_1 - bC_3$, $C_2 \to C_2 + aC_3$ (*ii*) Apply $C_1 \to C_1 + C_2 + C_3$.

(*iii*) Given determinant

$$= \begin{vmatrix} b(b-a) & b-c & c(b-a) \\ a(b-a) & a-b & b(b-a) \\ c(b-a) & c-a & a(b-a) \end{vmatrix}.$$

Take $b - a$ common from C_1 and C_3.

15. (*i*) Apply $R_2 \to R_2 - 2R_1$, $R_3 \to R_3 - 3R_1$.

(*ii*) Apply $R_2 \to R_2 - R_1$.

(*iv*) Apply $C_1 \to C_1 + C_2 + C_3$.

APPLICATIONS OF DETERMINANTS

2.26. AREA OF A TRIANGLE

We know that if $A(x_1, y_1)$, $B(x_2, y_2)$, $C(x_3, y_3)$ are the vertices of a triangle ABC, then the area of the triangle is given by

$$\frac{1}{2}[(x_1y_2 - x_2y_1) + (x_2y_3 - x_3y_2) + (x_3y_1 - x_1y_3)].$$

Let Δ denote the area of the triangle ABC.

$$\therefore \quad \Delta = \frac{1}{2}[(x_1y_2 - x_2y_1) + (x_2y_3 - x_3y_2) + (x_3y_1 - x_1y_3)]$$

$$= \frac{1}{2}[(x_1y_2 - x_1y_3) + (x_2y_3 - x_2y_1) + (x_3y_1 - x_3y_2)]$$

$$= \frac{1}{2}[x_1(y_2 - y_3) - x_2(y_1 - y_3) + x_3(y_1 - y_2)]$$

$$\therefore \quad \Delta = \frac{1}{2}\begin{vmatrix} x_1 & y_1 & 1 \\ x_2 & y_2 & 1 \\ x_3 & y_3 & 1 \end{vmatrix}.$$

Remark 1. Since the area is always positive, we shall take the absolute value of the above determinant. If the points A, B, C are in the anti-clockwise direction, then the value of Δ will come out to be positive itself.

Remark 2. The points $A(x_1, y_1)$, $B(x_2, y_2)$ and $C(x_3, y_3)$ are collinear if and only if area of the Δ ABC is zero.

$\therefore$ The points A, B and C are collinear if and only if $\frac{1}{2}\begin{vmatrix} x_1 & y_1 & 1 \\ x_2 & y_2 & 1 \\ x_3 & y_3 & 1 \end{vmatrix} = 0$ *i.e.*, $\begin{vmatrix} x_1 & y_1 & 1 \\ x_2 & y_2 & 1 \\ x_3 & y_3 & 1 \end{vmatrix} = 0.$

Remark 3. The point $P(x, y)$ lies on the line joining $A(x_1, y_1)$ and $B(x_2, y_2)$ if and only if area of the ΔPAB is zero.

$\therefore$ The equation of the line passing through the point $A(x_1, y_1)$ and $B(x_2, y_2)$ is

$$\frac{1}{2}\begin{vmatrix} x & y & 1 \\ x_1 & y_1 & 1 \\ x_2 & y_2 & 1 \end{vmatrix} = 0 \text{ i.e., } \begin{vmatrix} x & y & 1 \\ x_1 & y_1 & 1 \\ x_2 & y_2 & 1 \end{vmatrix} = 0.$$

ILLUSTRATIVE EXAMPLES

Example 1. *Using determinants, find the area of the triangle with vertices (– 2, 4), (2, – 6) and (5, 4). Are the given points collinear ?*

Sol. Area of triangle $= \frac{1}{2}\begin{vmatrix} -2 & 4 & 1 \\ 2 & -6 & 1 \\ 5 & 4 & 1 \end{vmatrix}$

$$= \frac{1}{2}\left[-2\begin{vmatrix} -6 & 1 \\ 4 & 1 \end{vmatrix} - 4\begin{vmatrix} 2 & 1 \\ 5 & 1 \end{vmatrix} + 1\begin{vmatrix} 2 & -6 \\ 5 & 4 \end{vmatrix}\right]$$

$$= \frac{1}{2}[-2(-10) - 4(-3) + (38)] = \frac{1}{2}(70) = \mathbf{35\ sq.\ units.}$$

Since area is non-zero, the points are **non-collinear.**

Example 2. *Show that the following points are collinear : (b, c + a), (c, a + b), (a, b + c).*

Sol. Area of triangle $= \frac{1}{2}\begin{vmatrix} b & c+a & 1 \\ c & a+b & 1 \\ a & b+c & 1 \end{vmatrix} = \frac{1}{2}\begin{vmatrix} b & b+c+a & 1 \\ c & c+a+b & 1 \\ a & a+b+c & 1 \end{vmatrix}$

(Applying $C_2 \to C_2 + C_1$)

$$= \frac{a+b+c}{2}\begin{vmatrix} b & 1 & 1 \\ c & 1 & 1 \\ a & 1 & 1 \end{vmatrix} = \frac{a+b+c}{2}(0) = 0.$$

∴ The given points are **collinear.**

Example 3. *Find the value of x so that the points (3, – 2), (x, 2) and (8, 8), lie on a line.*

Sol. Let the given points be A(3, – 2), B(x, 2) and C(8, 8).

Since the points are collinear, so the area of triangle ABC must be zero.

$$\Rightarrow \quad \frac{1}{2}\begin{vmatrix} 3 & -2 & 1 \\ x & 2 & 1 \\ 8 & 8 & 1 \end{vmatrix} = 0 \quad \Rightarrow \quad \frac{1}{2}\begin{vmatrix} 3 & -2 & 1 \\ x-3 & 4 & 0 \\ 5 & 10 & 0 \end{vmatrix} = 0$$

(Applying $R_2 \to R_2 - R_1$, $R_3 \to R_3 - R_1$)

$$\therefore \quad 1 \cdot \begin{vmatrix} x-3 & 4 \\ 5 & 10 \end{vmatrix} = 0 \qquad \text{(Expanding along } C_3\text{)}$$

$\therefore \quad 10x - 30 - 20 = 0 \quad \Rightarrow \quad 10x = 50 \quad \Rightarrow \quad$ $\boldsymbol{x = 5}$.

Example 4. *Find the value of x, if the area of the triangle with vertices (x, 4), (2, – 6) and (5, 4) be 35 cm².*

Sol. The vertices of the triangle are (x, 4), (2, – 6) and (5, 4).

$$\therefore \quad \text{Absolute value of } \frac{1}{2}\begin{vmatrix} x & 4 & 1 \\ 2 & -6 & 1 \\ 5 & 4 & 1 \end{vmatrix} = 35$$

$$\Rightarrow \quad \frac{1}{2}\begin{vmatrix} x & 4 & 1 \\ 2 & -6 & 1 \\ 5 & 4 & 1 \end{vmatrix} = \pm\, 35$$

$$\Rightarrow \quad x\begin{vmatrix} -6 & 1 \\ 4 & 1 \end{vmatrix} - 4\begin{vmatrix} 2 & 1 \\ 5 & 1 \end{vmatrix} + 1 \cdot \begin{vmatrix} 2 & -6 \\ 5 & 4 \end{vmatrix} = \pm\, 70$$

$$\Rightarrow \quad x(-6-4) - 4(2-5) + (8+30) = \pm\, 70$$

$$\Rightarrow \quad -10x + 12 + 38 = \pm\, 70$$

$$\Rightarrow \quad -10x = \pm\, 70 - 50$$

$\therefore \quad -10x = 70 - 50$	or $\quad -10x = -70 - 50$
$\Rightarrow \quad x = \dfrac{20}{-10} = -2$	$\Rightarrow \quad x = \dfrac{-120}{-10} = 12$

∴ $\boldsymbol{x = -2}$ or **12.**

EXERCISE 2.9

1. Find the area of the triangles whose vertices are :
   (*i*) (– 3, 5), (3, – 6), (7, 2) (*P.T.U. Dec. 2005*) (*ii*) (2, – 7), (1, 3), (10, 8)
   (*iii*) (1, 1), (0, – 1), (1, 2) (P.T.U. B. Pharma May 2006)
2. Using determinants, find the area of triangles with vertices given below. Are the given points collinear ?
   (*i*) (1, 4), (2, 3), (– 5, – 3) (*ii*) (1, – 1), (2, 4), (– 3, 5).
3. (*i*) Points $(a, 0)$, $(0, b)$, $(1, 1)$ are collinear, show that $\frac{1}{a} + \frac{1}{b} = 1$.
   (*ii*) Points $(a, 0)$, $(0, b)$ and $(3a, y)$ are collinear. Find the value of y, $(a \neq 0)$.
   (*iii*) For what value of k do the points $(-1, 4)$, $(-3, 8)$ and $(-k + 1, 3k)$ lie on a straight line ?
   (*iv*) Points $(a, 0)$, $(0, b)$ and (x, y) are collinear. Show that $\frac{x}{a} + \frac{y}{b} = 1$.
4. If the points (a_1, b_1), (a_2, b_2) and $(a_1 + a_2, b_1 + b_2)$ are collinear, show that $a_1b_2 = a_2b_1$.

Answers

1. (*i*) 46 square units (*ii*) 47.5 square units (*iii*) $\frac{1}{2}$ square units
2. (*i*) 6.5 square units, No (*ii*) 13 square units, No
3. (*ii*) $-2b$ (*iii*) 0.

2.27. SOLUTION OF SYSTEMS OF LINEAR EQUATIONS BY CRAMER'S RULE

Consider the system of equations :

$$x + y = 7$$
$$x - y = 1.$$

By simple algebraic manipulation, one can see that there is only *one* solution to this system namely, $x = 4$, $y = 3$.

The system of equations :

$$x + y = 7$$
$$2x + 2y = 14$$

has *infinitely many* solutions, namely $x = k$, $y = 7 - k$, where k is any number.

The system of equations :

$$x + y = 7$$
$$2x + 2y = 15$$

has *no* solution.

Thus, we see that a system of linear equations may have either *one solution* or *infinitely many solutions* or *no solution at all.*

If a system of equations has at least one solution, then the system is called **consistent,** otherwise the system is called **inconsistent.**

In the above examples, the first two systems are *consistent* and the third system is *inconsistent.*

In this section, we shall use determinants in solving *systems of linear equations.*

Let us consider the system

$$a_1x + b_1y + c_1z = d_1$$
$$a_2x + b_2y + c_2z = d_2$$
$$a_3x + b_3y + c_3z = d_3$$

of three linear equations in three variables x, y, z.

Define $\Delta = \begin{vmatrix} a_1 & b_1 & c_1 \\ a_2 & b_2 & c_2 \\ a_3 & b_3 & c_3 \end{vmatrix}$

Let Δ_i $(i = 1, 2, 3)$ be the determinant obtained from Δ by replacing its ith column by the column containing d_1, d_2, d_3.

$$\therefore \quad \Delta_1 = \begin{vmatrix} d_1 & b_1 & c_1 \\ d_2 & b_2 & c_2 \\ d_3 & b_3 & c_3 \end{vmatrix}, \Delta_2 = \begin{vmatrix} a_1 & d_1 & c_1 \\ a_2 & d_2 & c_2 \\ a_3 & d_3 & c_3 \end{vmatrix} \text{ and } \Delta_3 = \begin{vmatrix} a_1 & b_1 & d_1 \\ a_2 & b_2 & d_2 \\ a_3 & b_3 & d_3 \end{vmatrix}$$

Putting the values of d_1, d_2, d_3 in Δ_1, we get

$$\Delta_1 = \begin{vmatrix} a_1x + b_1y + c_1z & b_1 & c_1 \\ a_2x + b_2y + c_2z & b_2 & c_2 \\ a_3x + b_3y + c_3z & b_3 & c_3 \end{vmatrix}$$

$$= x\begin{vmatrix} a_1 & b_1 & c_1 \\ a_2 & b_2 & c_2 \\ a_3 & b_3 & c_3 \end{vmatrix} + y\begin{vmatrix} b_1 & b_1 & c_1 \\ b_2 & b_2 & c_2 \\ b_3 & b_3 & c_3 \end{vmatrix} + z\begin{vmatrix} c_1 & b_1 & c_1 \\ c_2 & b_2 & c_2 \\ c_3 & b_3 & c_3 \end{vmatrix}$$

$$= x\Delta + y(0) + z(0) = x\Delta.$$

$$\therefore \quad \Delta_1 = x\Delta.$$

Similarly, we have $\Delta_2 = y\Delta$ and $\Delta_3 = z\Delta$.

Case I. $\Delta \neq 0$

In this case, $x = \dfrac{\Delta_1}{\Delta}, \quad y = \dfrac{\Delta_2}{\Delta}, \quad z = \dfrac{\Delta_3}{\Delta}.$

$\therefore$ The system is **consistent** and has **unique** solution.

Case II. $\Delta = 0$

In this case, $\Delta_1 = x(0), \quad \Delta_2 = y(0), \quad \Delta_3 = z(0).$

If $\Delta_1, \Delta_2, \Delta_3$ are *not* all zero, then at least one of the equations :

$$\Delta_1 = x(0), \quad \Delta_2 = y(0) \quad \Delta_3 = z(0)$$

will not be possible for any value of the variable, because right side of each equation is zero. In this sub-case, the system is **inconsistent** and has *no* solution.

If $\Delta_1, \Delta_2, \Delta_3$ are all zero, then $0 = x(0), 0 = y(0), 0 = z(0)$.

Here, the values of x, y, z assume indeterminant forms. In this sub-case, the system may or may not be consistent, because the system will have either **infinitely many** solutions or *no* solution.

For example, consider the system of equations :

$$x + y + z = 2$$
$$x + y + z = 3$$
$$x + y + z = 4.$$

For this system $\Delta = 0, \Delta_1 = 0, \Delta_2 = 0, \Delta_3 = 0$ and the system is having no solution.

The given findings can be shown diagrammatically as follows :

Cramer's rule

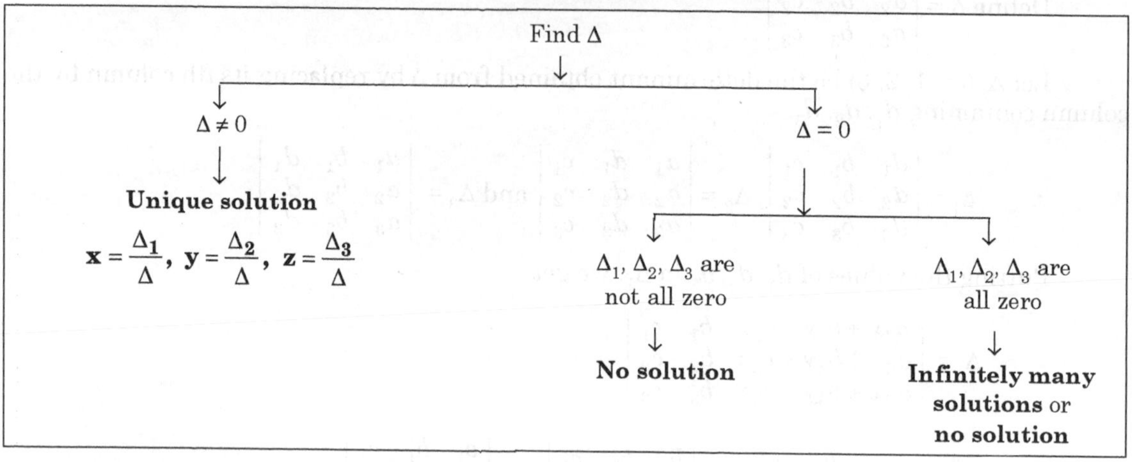

Remark 1. If d_1, d_2, d_3 are all zero, then the system is called **homogeneous,** otherwise, it is called **non-homogeneous.**

In case of a homogeneous system, each of Δ_1, Δ_2, Δ_3 will contain a column of zeros.

$\therefore \qquad \Delta_1 = \Delta_2 = \Delta_3 = 0.$

If $\Delta \neq 0$, then $\quad x = \dfrac{\Delta_1}{\Delta} = \dfrac{0}{\Delta} = 0, \quad y = \dfrac{\Delta_2}{\Delta} = \dfrac{0}{\Delta} = 0, \quad z = \dfrac{\Delta_3}{\Delta} = \dfrac{0}{\Delta} = 0$

and if $\Delta = 0$, then the system will have infinitely many solutions*.

$\therefore$ A homogeneous system of linear equations is always consistent and $x = 0$, $y = 0$, $z = 0$ will always be solution of a homogeneous system of equations in the variables x, y, z.

The solution $x = 0, y = 0, z = 0$ is called the **trivial solution** of the system of equations in x, y and z.

Remark 2. The above method of solving three equations in three variables is general and is applicable to systems containing n (≥ 2) linear equations in n variables.

The above method of solving linear equations is called **Cramer's rule** after the Swiss mathematician **Gabriel Cramer** (1704–1752).

WORKING RULES FOR SOLVING EQUATIONS BY CRAMER'S RULE

Step I. *Express the given system in the standard form :*

$$a_1x + b_1y + c_1z = d_1$$
$$a_2x + b_2y + c_2z = d_2$$
$$a_3x + b_3y + c_3z = d_3.$$

Find the value of the determinant $\Delta = \begin{vmatrix} a_1 & b_1 & c_1 \\ a_2 & b_2 & c_2 \\ a_3 & b_3 & c_3 \end{vmatrix}$.

Step II. *If $\Delta \neq 0$, then the given system will have unique solution given by $x = \dfrac{\Delta_1}{\Delta}$, $y = \dfrac{\Delta_2}{\Delta}$, $z = \dfrac{\Delta_3}{\Delta}$, where Δ_1, Δ_2, Δ_3 are the determinants obtained from Δ by replacing respectively Ist column, IInd column, IIIrd column by the column containing elements d_1, d_2, d_3.*

*__Why this step.__ The possibility of 'no solution' is 'ruled' out because $x = 0$, $y = 0$, $z = 0$ is always a solution of any homogeneous system of equations in the variables x, y, z.

Step III. *If $\Delta = 0$, then put x (or y or z) equal to k and find the solution of any two of the given equations. Substitute these values of x, y and z in terms of k in the third equation.*

Step IV. *If the third equation is satisfied by this solution, then these values represent the solution of the given system. These are infinitely many solutions.*

If the third equation is not satisfied then declare that the given system has no solution.

ILLUSTRATIVE EXAMPLES

Example 1. *Solve the linear equations $2x + 3y = 5$, $3x - 2y = 1$ with the help of determinants.*

Sol. The given system is
$$2x + 3y = 5$$
$$3x - 2y = 1.$$

Let
$$\Delta = \begin{vmatrix} 2 & 3 \\ 3 & -2 \end{vmatrix} = 2(-2) - 3(3) = -4 - 9 = -13 \neq 0.$$

$\therefore$ The system has unique solution.

$$\Delta_1 = \begin{vmatrix} 5 & 3 \\ 1 & -2 \end{vmatrix} = 5(-2) - 3(1) = -10 - 3 = -13$$

$$\Delta_2 = \begin{vmatrix} 2 & 5 \\ 3 & 1 \end{vmatrix} = 2(1) - 3(5) = 2 - 15 = -13.$$

$\therefore$ By **Cramer's rule,** $x = \dfrac{\Delta_1}{\Delta} = \dfrac{-13}{-13} = 1,\quad y = \dfrac{\Delta_2}{\Delta} = \dfrac{-13}{-13} = 1$

$\therefore$ The solution is $\mathbf{x = 1, y = 1}$.

Example 2. (*i*) *Solve the linear equations, if consistent, $x + 3y = 2$, $2x + 6y = 7$ with the help of determinants.*

(*ii*) *Solve, using Cramer's rule*
$$x - 3y + z = 2$$
$$3x + y + z = 6$$
$$5x + y + 3z = 3$$

Sol. (*i*) The given system is
$$x + 3y = 2 \qquad \ldots(1)$$
$$2x + 6y = 7. \qquad \ldots(2)$$

Let
$$\Delta = \begin{vmatrix} 1 & 3 \\ 2 & 6 \end{vmatrix} = 6 - 6 = 0.$$

Let $y = k$. $\therefore$ (1) implies $x = 2 - 3k$.

Putting the values of x and y in (2), we get
$$2(2 - 3k) + 6k = 7 \text{ i.e., } 4 = 7, \text{ which is not true.}$$

$\therefore$ The given system is **inconsistent.**

(*ii*) Here $\Delta = \begin{vmatrix} 1 & -3 & 1 \\ 3 & 1 & 1 \\ 5 & 1 & 3 \end{vmatrix} = 1(3 - 1) + 3(9 - 5) + 1(3 - 5) = 2 + 12 - 2 = 12 \neq 0$

$\therefore$ The system has a unique solution.

$$\Delta_1 = \begin{vmatrix} 2 & -3 & 1 \\ 6 & 1 & 1 \\ 3 & 1 & 3 \end{vmatrix} = 2(3 - 1) + 3(18 - 3) + 1(6 - 3) = 4 + 45 + 3 = 52$$

$$\Delta_2 = \begin{vmatrix} 1 & 2 & 1 \\ 3 & 6 & 1 \\ 5 & 3 & 3 \end{vmatrix} = 1(18-3) - 2(9-5) + 1(9-30) = 15 - 8 - 21 = -14$$

$$\Delta_3 = \begin{vmatrix} 1 & -3 & 2 \\ 3 & 1 & 6 \\ 5 & 1 & 3 \end{vmatrix} = 1(3-6) + 3(9-30) + 2(3-5) = -3 - 63 - 4 = -70$$

∴ By Cramer's rule

$$x = \frac{\Delta_1}{\Delta} = \frac{52}{12} = \frac{13}{3}$$

$$y = \frac{\Delta_2}{\Delta} = -\frac{14}{12} = -\frac{7}{6}$$

$$z = \frac{\Delta_3}{\Delta} = -\frac{70}{12} = -\frac{35}{6}.$$

Example 3. *Solve the linear equations $2x + 7y = 9$, $4x + 14y = 18$, with the help of determinants.*

Sol. The given system is $2x + 7y = 9$ ...(1)

$4x + 14y = 18.$ ...(2)

Let $\Delta = \begin{vmatrix} 2 & 7 \\ 4 & 14 \end{vmatrix} = 28 - 28 = 0.$

Let $x = k$. ∴ (1) implies $2k + 7y = 9$ *i.e.*, $y = \frac{9-2k}{7}$.

Putting the values of x and y in (2), we get $4(k) + 14\left(\frac{9-2k}{7}\right) = 18$

⇒ $4k + 18 - 4k = 18$ ⇒ $18 = 18$, which is true.

∴ The given system has infinitely many solutions given by $x = \mathbf{k}$, $y = \mathbf{\frac{9-2k}{7}}$, where **k** is any number.

Example 4. *Using determinants, solve the following system of equations :*

$x + y + 2z = 0$, $x - 2y + 2z = 0$, $x - z = 0$.

Sol. The given equations are

$$x + y + 2z = 0$$
$$x - 2y + 2z = 0$$
$$x + 0.y - z = 0.$$

∴ $$\Delta = \begin{vmatrix} 1 & 1 & 2 \\ 1 & -2 & 2 \\ 1 & 0 & -1 \end{vmatrix} = 1(2-0) - 1(-1-2) + 2(0+2) = 2 + 3 + 4 = 9 \neq 0.$$

∴ The system has unique solution.

$$\Delta_1 = \begin{vmatrix} 0 & 1 & 2 \\ 0 & -2 & 2 \\ 0 & 0 & -1 \end{vmatrix} = 0$$

Similarly, $\Delta_2 = 0, \Delta_3 = 0$.

$\therefore$ By **Cramer's rule,** $x = \dfrac{\Delta_1}{\Delta} = \dfrac{0}{9} = 0,$

$$y = \frac{\Delta_2}{\Delta} = \frac{0}{9} = 0, z = \frac{\Delta_3}{\Delta} = \frac{0}{9} = 0.$$

$\therefore$ $\quad x = \mathbf{0}, y = \mathbf{0}, z = \mathbf{0}.$

Example 5. *Using determinants, solve the following system of equations :*

$$2x - 3y - z = 0, x + 3y - 2z = 0, x - 3y = 0.$$

Sol. The given equations are $2x - 3y - z = 0$ ...(1)

$$x + 3y - 2z = 0 \quad ...(2)$$

$$x - 3y + 0.z = 0. \quad ...(3)$$

$$\therefore \quad \Delta = \begin{vmatrix} 2 & -3 & -1 \\ 1 & 3 & -2 \\ 1 & -3 & 0 \end{vmatrix} = 2(0 - 6) + 3(0 + 2) - 1(-3 - 3) = -12 + 6 + 6 = 0.$$

Putting $z = k$ in (1) and (2), we get

$$2x - 3y = k \quad ...(4)$$

$$x + 3y = 2k. \quad ...(5)$$

In fact, any two equations can be used. To solve (4) and (5),

Here $\quad \Delta = \begin{vmatrix} 2 & -3 \\ 1 & 3 \end{vmatrix} = 6 + 3 = 9 \neq 0 \qquad \Delta_1 = \begin{vmatrix} k & -3 \\ 2k & 3 \end{vmatrix} = 3k + 6k = 9k$

$$\Delta_2 = \begin{vmatrix} 2 & k \\ 1 & 2k \end{vmatrix} = 4k - k = 3k$$

$$\therefore \quad x = \frac{\Delta_1}{\Delta} = \frac{9k}{9} = k, \quad y = \frac{\Delta_2}{\Delta} = \frac{3k}{9} = \frac{k}{3} \text{ (Cramer's rule)}$$

Putting $x = k,\ y = \dfrac{k}{3}, z = k$ in (3), we get $k - 3\left(\dfrac{k}{3}\right) = 0$ *i.e.*, $0 = 0$, which is true.

$\therefore$ The given system has infinitely many solutions given by

$$x = \mathbf{k}, y = \frac{\mathbf{k}}{\mathbf{3}}, \ z = \mathbf{k}, \text{ where } \mathbf{k} \text{ is any number.}$$

Example 6. *Find the value of λ for which the homogeneous system of equations :*

$$2x + 3y - 2z = 0, 2x - y + 3z = 0, 7x + \lambda y - z = 0$$

has non-trivial solution. Find the solution.

Sol. The given equations are

$$2x + 3y - 2z = 0 \quad ...(1)$$

$$2x - y + 3z = 0 \quad ...(2)$$

$$7x + \lambda y - z = 0 \quad ...(3)$$

$$\therefore \quad \Delta = \begin{vmatrix} 2 & 3 & -2 \\ 2 & -1 & 3 \\ 7 & \lambda & -1 \end{vmatrix} = 2(1 - 3\lambda) - 3(-2 - 21) - 2(2\lambda + 7) = -10\lambda + 57.$$

Let the given system has non-trivial solution.

$\therefore \quad \Delta = 0. \quad \therefore \quad -10\lambda + 57 = 0 \quad \text{or} \quad \lambda = \mathbf{57/10.}$

Let $z = k$.

$\therefore \quad (1) \Rightarrow \quad 2x + 3y = 2k$ ...(3)

$(2) \Rightarrow \quad 2x - y = -3k$ ...(4)

To solve (3) and (4),

Here $$\Delta = \begin{vmatrix} 2 & 3 \\ 2 & -1 \end{vmatrix} = -2 - 6 = -8 \neq 0$$

$$\Delta_1 = \begin{vmatrix} 2k & 3 \\ -3k & -1 \end{vmatrix} = -2k + 9k = 7k$$

$$\Delta_2 = \begin{vmatrix} 2 & 2k \\ 2 & -3k \end{vmatrix} = -6k - 4k = -10k$$

$$\therefore \quad x = \frac{\Delta_1}{\Delta} = \frac{7k}{-8}, \quad y = \frac{\Delta_2}{\Delta} = \frac{-10k}{-8} = \frac{5}{4}k$$

Putting $x = -\frac{7}{8}k,\ y = \frac{5}{4}k,\ z = k$ in (3), we get

$$7\left(-\frac{7}{8}k\right) + \frac{57}{10}\left(\frac{5}{4}k\right) - k = 0 \quad \text{or} \quad 0 = 0, \text{ which is true.}$$

$\therefore$ The given system has infinitely many solutions given by

$$x = -\frac{\mathbf{7}}{\mathbf{8}}\mathbf{k},\ y = \frac{\mathbf{5}}{\mathbf{4}}\mathbf{k},\ z = \mathbf{k},$$

where **k** is any number.

Example 7. *By using determinants, prove that there is no solution for the equations :*

$$x + 4y - 2z = 3,\ 3x + y + 5z = 7,\ 2x + 3y + z = 5.$$

Sol. The given equations are
$$x + 4y - 2z = 3$$
$$3x + y + 5z = 7$$
$$2x + 3y + z = 5.$$

$$\therefore \quad \Delta = \begin{vmatrix} 1 & 4 & -2 \\ 3 & 1 & 5 \\ 2 & 3 & 1 \end{vmatrix} = 1(1 - 15) - 4(3 - 10) - 2(9 - 2) = -14 + 28 - 14 = 0.$$

$\therefore$ The system has either infinitely many solutions or no solution.

Putting $z = k$ in the first two equations, we get

$x + 4y = 3 + 2k$ ...(1)

$3x + y = 7 - 5k.$ ...(2)

To solve (1) and (2),

$$\therefore \quad \Delta = \begin{vmatrix} 1 & 4 \\ 3 & 1 \end{vmatrix} = 1 - 12 = -11 \neq 0$$

$$\Delta_1 = \begin{vmatrix} 3 + 2k & 4 \\ 7 - 5k & 1 \end{vmatrix} = 3 + 2k - 28 + 20k = 22k - 25$$

$$\Delta_2 = \begin{vmatrix} 1 & 3 + 2k \\ 3 & 7 - 5k \end{vmatrix} = 7 - 5k - 9 - 6k = -11k - 2.$$

$$\therefore \quad x = \frac{\Delta_1}{\Delta} = \frac{22k - 25}{-11}, \quad y = \frac{\Delta_2}{\Delta} = \frac{-11k - 2}{-11}. \text{ (Cramer's rule)}$$

Putting the values of x, y, z in the last equation, we get

$$2\left(\frac{22k-25}{-11}\right)+3\left(\frac{-11k-2}{-11}\right)+k=5.$$

$\Rightarrow$ $44k - 50 - 33k - 6 - 11k = -55 \Rightarrow -56 = -55$, which is not true.

$\therefore$ The given system is **inconsistent.**

Remark. In order to show that the given system of equations is inconsistent, it is also sufficient to show that $\Delta = 0$ and at least one of $\Delta_1, \Delta_2, \Delta_3$ is non-zero.

EXERCISE 2.10

1. Find whether the following systems of equations have unique solution or not :

   (*i*) $x + 2y = 4$, $3x + 4y = 7$ (*ii*) $x + 2y = 4$, $2x + 4y = 8$ (*iii*) $5x + 9y = 3$, $x - y = 7$.

2. Solve the following equations using Cramer's rule :

   (*i*) $2x - y = 17$, $3x + 5y = 6$ (*ii*) $x + 2y = 1$, $3x + y = 4$ (*iii*) $8x - 3y = 14$, $2x + 9y = -3$.

   (*iv*) $9x + 5y = 10$, $3y - 2x = 8$ (*v*) $5x + 7y = -2$, $4x + 6y = -3$ (*vi*) $2x - y = -2$, $3x + 4y + 8 = 0$

   (*vii*) $3x + ay = 4$, $2x + ay = 2, a \neq 0$ (*viii*) $3x + 4y = 1$, $6x + 8y = 2$ (*ix*) $5x - y = 7$, $15x - 3y = 21$.

3. Find whether the following systems have unique solution or not :

   (*i*) $x + y + 3z = 0$, $x + 3y - 3z = 0$, $-2x - 4y - 4z = 0$ (*ii*) $4x + 7y - z = 0$, $7x - 8y + 11z = 0$, $8x + 14y - 2z = 0$ (*iii*) $x + 2y - 2z = 5$, $-x + 3y = 2$, $-2y + z = 7$.

4. Solve the following equations using Cramer's rule :

   (*i*) $3x - 4y + 5z = -6$, $x + y - 2z = -1$, $2x + 3y + z = 5$ (*ii*) $2x + 3y + 4z = 8$, $3x + y - z = -2$, $4x - y - 5z = -9$ (*P.T.U. B. Pharma, May 2005*) (*iii*) $5x - y + 4z = 0$, $2x + 3y + 5z = 0$, $5x - 2y + 6z = 0$

   (*iv*) $3x - 4y + 2z = -1$, $2x + 3y + 5z = 7$, $x + z = 2$ (*v*) $3x + 4y + 2z = 8$, $2y - 3z = 3$, $x - 2y + 6z = -2$ (*vi*) $x + y - 2z = 0$, $3x + 2y - z = 0$, $2x + y + z = 0$

   (*vii*) $2x - y + 3z = 0$, $x + 2y - z = 0$, $5y - 5z = 0$ (*viii*) $x - y + z = 3$, $2x + y - z = 2$, $-x - 2y + 2z = 1$ (*ix*) $x - y + 3z = 6$, $x + 3y - 3z = -4$, $5x + 3y + 3z = 10$.

5. Find the value of λ so that the system of equations $x - y + 3z = 0$, $x + \lambda y - 3z = 0$ and $5x - 3y + 3z = 0$ has a non-trivial solution. For this value of λ, also find the non-trivial solutions.

6. If the system of equations $x = ay + z$, $y = z + ax$, $z = x + y$ has a non-trivial solution, find the value of a.

7. For what value of k, do the following system of equations possesses a non-trivial solution over the set of rational numbers :

$$x + ky + 3z = 0$$

$$3x + ky - 2z = 0$$
$$2x + 3y - 4z = 0.$$

For that value of k, find the solution of the system.

Answers

1. (*i*) Yes (*ii*) No (*iii*) Yes
2. (*i*) $x = 7, y = -3$ (*ii*) $x = 7/5, y = -1/5$ (*iii*) $x = 3/2, y = -2/3$
   (*iv*) $x = -10/37, y = 92/37$ (*v*) $x = 9/2, y = -7/2$ (*vi*) $x = -16/11, y = -10/11$
   (*vii*) $x = 2, y = -2/a$ (*viii*) $x = k, y = \dfrac{1-3k}{4}$, where k is any number
   (*ix*) $x = k, y = 5k - 7$, where k is any number.
3. (*i*) Yes (*ii*) No (*iii*) Yes
4. (*i*) $x = -1, y = 2, z = 1$ (*ii*) $x = 1, y = -2, z = 3$ (*iii*) $x = 0, y = 0, z = 0$
   (*iv*) $x = 3, y = 2, z = -1$ (*v*) $x = -2, y = 3, z = 1$
   (*vi*) $x = -3k, y = 5k, z = k$, where k is any number.
   (*vii*) $x = -k, y = k, z = k$, where k is any number.
   (*viii*) $x = \dfrac{5}{3}, y = k - \dfrac{4}{3}, z = k$, where k is any number.
   (*ix*) $x = 1 - k, y = k, z = \dfrac{2k+5}{3}$, where k is any number.
5. $\lambda = 3, x = -3k/2, y = 3k/2, z = k$, where k is any number. **6.** -1
7. $k = 33/2, x = 5\lambda/2, y = -\lambda/3, z = \lambda$, where λ is any rational number.

Hints

5. Given homogeneous system will have non-trivial solution if $\begin{vmatrix} 1 & -1 & 3 \\ 1 & \lambda & -3 \\ 5 & 3 & 3 \end{vmatrix} = 0.$

6. $\begin{vmatrix} 1 & -a & -1 \\ a & -1 & 1 \\ 1 & 1 & -1 \end{vmatrix} = 0 \Rightarrow 1(0) + a(-a-1) - 1(a+1) = 0.$

ADJOINT AND INVERSE OF MATRICES

2.28. INTRODUCTION

In the present chapter, we shall learn the method of finding inverse of matrices (non-singular). This knowledge will help us to solve the system of linear equations.

2.29. ADJOINT OF A MATRIX

(P.T.U. B. Pharma, May 2007)

The concept of adjoint is defined only for square matrices. If A is a square matrix, then the **adjoint** of A is defined as the transpose of the matrix obtained by replacing the elements of A by their corresponding cofactors in | A |. The adjoint of square matrix A is denoted by '*adj A*'.

Thus, if $A = \begin{bmatrix} a_{11} & a_{12} & \cdots & \cdots & a_{1n} \\ a_{21} & a_{22} & \cdots & \cdots & a_{2n} \\ \cdots & \cdots & \cdots & \cdots & \cdots \\ \cdots & \cdots & \cdots & \cdots & \cdots \\ a_{n1} & a_{n2} & \cdots & \cdots & a_{nn} \end{bmatrix}$

then
$$\text{adj } A = \begin{bmatrix} A_{11} & A_{12} & \cdots & \cdots & A_{1n} \\ A_{21} & A_{22} & \cdots & \cdots & A_{2n} \\ \cdots & \cdots & \cdots & \cdots & \cdots \\ \cdots & \cdots & \cdots & \cdots & \cdots \\ A_{n1} & A_{n2} & \cdots & \cdots & A_{nn} \end{bmatrix}' = \begin{bmatrix} A_{11} & A_{21} & \cdots & \cdots & A_{n1} \\ A_{12} & A_{22} & \cdots & \cdots & A_{n2} \\ \cdots & \cdots & \cdots & \cdots & \cdots \\ \cdots & \cdots & \cdots & \cdots & \cdots \\ A_{1n} & A_{2n} & \cdots & \cdots & A_{nn} \end{bmatrix},$$

where A_{ij} is the cofactor of a_{ij} in $|A|$, $1 \le i, j \le n$.

In particular, if $A = \begin{bmatrix} a_{11} & a_{12} & a_{13} \\ a_{21} & a_{22} & a_{23} \\ a_{31} & a_{32} & a_{33} \end{bmatrix}$, then

$$\text{adj } A = \begin{bmatrix} A_{11} & A_{12} & A_{13} \\ A_{21} & A_{22} & A_{23} \\ A_{31} & A_{32} & A_{33} \end{bmatrix}' = \begin{bmatrix} A_{11} & A_{21} & A_{31} \\ A_{12} & A_{22} & A_{32} \\ A_{13} & A_{23} & A_{33} \end{bmatrix},$$

where A_{ij} is the cofactor of a_{ij} in $|A|$, $1 \le i, j \le 3$.

ILLUSTRATIVE EXAMPLES

Example 1. *Find the adjoint of the matrix* $A = \begin{bmatrix} 2 & 1 \\ 4 & -1 \end{bmatrix}$.

Sol. $A = \begin{bmatrix} 2 & 1 \\ 4 & -1 \end{bmatrix}$. We find all cofactors :

$$A_{11} = \text{cofactor of } a_{11} \, (= 2) = (-1)^{1+1} \, |-1| = -1$$
$$A_{12} = \text{cofactor of } a_{12} \, (= 1) = (-1)^{1+2} \, |4| = -4$$
$$A_{21} = \text{cofactor of } a_{21} \, (= 4) = (-1)^{2+1} \, |1| = -1$$
$$A_{22} = \text{cofactor of } a_{22} \, (= -1) = (-1)^{2+2} \, |2| = 2.$$

$$\therefore \quad \text{adj } A = \begin{bmatrix} A_{11} & A_{12} \\ A_{21} & A_{22} \end{bmatrix}' = \begin{bmatrix} -1 & -4 \\ -1 & 2 \end{bmatrix}' = \begin{bmatrix} \mathbf{-1} & \mathbf{-1} \\ \mathbf{-4} & \mathbf{2} \end{bmatrix}.$$

Example 2. *Find adjoint of A, where* $A = \begin{bmatrix} 1 & -1 & 2 \\ 2 & 3 & 5 \\ -2 & 0 & 1 \end{bmatrix}$.

Sol. $A = \begin{bmatrix} 1 & -1 & 2 \\ 2 & 3 & 5 \\ -2 & 0 & 1 \end{bmatrix}$.

We know that adj A is the transpose of the matrix obtained by replacing the elements of A by their corresponding cofactors.

$$A_{11} = \text{cofactor of } a_{11} \, (= 1) = (-1)^{1+1} \begin{vmatrix} 3 & 5 \\ 0 & 1 \end{vmatrix} = 3 - 0 = 3$$

$$A_{12} = \text{cofactor of } a_{12}\ (= -1) = (-1)^{1+2}\begin{vmatrix} 2 & 5 \\ -2 & 1 \end{vmatrix} = -(2+10) = -12$$

$$A_{13} = \text{cofactor of } a_{13}\ (= 2) = (-1)^{1+3}\begin{vmatrix} 2 & 3 \\ -2 & 0 \end{vmatrix} = 0+6 = 6$$

$$A_{21} = \text{cofactor of } a_{21}\ (= 2) = (-1)^{2+1}\begin{vmatrix} -1 & 2 \\ 0 & 1 \end{vmatrix} = -(-1-0) = 1$$

$$A_{22} = \text{cofactor of } a_{22}\ (= 3) = (-1)^{2+2}\begin{vmatrix} 1 & 2 \\ -2 & 1 \end{vmatrix} = 1+4 = 5$$

$$A_{23} = \text{cofactor of } a_{23}\ (= 5) = (-1)^{2+3}\begin{vmatrix} 1 & -1 \\ -2 & 0 \end{vmatrix} = -(0-2) = 2$$

$$A_{31} = \text{cofactor of } a_{31}\ (= -2) = (-1)^{3+1}\begin{vmatrix} -1 & 2 \\ 3 & 5 \end{vmatrix} = -5-6 = -11$$

$$A_{32} = \text{cofactor of } a_{32}\ (= 0) = (-1)^{3+2}\begin{vmatrix} 1 & 2 \\ 2 & 5 \end{vmatrix} = -(5-4) = -1$$

$$A_{33} = \text{cofactor of } a_{33}\ (= 1) = (-1)^{3+3}\begin{vmatrix} 1 & -1 \\ 2 & 3 \end{vmatrix} = 3+2 = 5.$$

$$\therefore \quad \text{adj } A = \begin{bmatrix} 3 & -12 & 6 \\ 1 & 5 & 2 \\ -11 & -1 & 5 \end{bmatrix}' = \begin{bmatrix} \mathbf{3} & \mathbf{1} & \mathbf{-11} \\ \mathbf{-12} & \mathbf{5} & \mathbf{-1} \\ \mathbf{6} & \mathbf{2} & \mathbf{5} \end{bmatrix}.$$

Theorem I. *If A is a square matrix of order n, then prove that*

$$A(\text{adj } A) = (\text{adj } A)\, A = |A|\, I_n,$$

where I_n is the identity matrix of order n.

Proof. Let $A = \begin{bmatrix} a_{11} & a_{12} & \cdots\cdots & a_{1n} \\ a_{21} & a_{22} & \cdots\cdots & a_{2n} \\ \cdots & \cdots & \cdots & \cdots \\ \cdots & \cdots & \cdots & \cdots \\ a_{n1} & a_{n2} & \cdots\cdots & a_{nn} \end{bmatrix}$ be a square matrix of order n.

$\therefore$ In short, we write $A = [a_{ij}]_{n \times n}$.

Let adj $A = [b_{ij}]_{n \times n}$.

$\therefore$ $b_{ij} = A_{ji}$ = cofactor of a_{ji} in $|A|$.

The matrices A and adj A are both square matrices of order n, therefore, A(adj A) and (adj A)A are both defined and are square matrices of order n.

For $1 \le i, j \le n$, the $(i, j)^{\text{th}}$ element of A(adj A)

$$= \sum_{k=1}^{n} a_{ik}\, b_{kj} \qquad \text{(By def. of product of matrices)}$$

$$= \sum_{k=1}^{n} a_{ik}\, A_{jk}. \qquad (\because \ b_{kj} = A_{jk})$$

Case I. i = j

In this case, $\sum_{k=1}^{n} a_{ik}\, A_{jk} = \sum_{k=1}^{n} a_{ik}\, A_{ik}$

$= a_{i1} A_{i1} + a_{i2} A_{i2} + \ldots\ldots + a_{in} A_{in}$

= Value of | A | expressed in terms of elements of ith row

= | A |.

Case II. i ≠ j

In this case, $\sum_{k=1}^{n} a_{ik}\, A_{jk} = a_{i1} A_{j1} + a_{i2} A_{j2} + \ldots\ldots + a_{in} A_{jn}$

= Value of determinant of the matrix obtained from A by replacing j^{th} row by i^{th} row. The determinant of this new matrix vanishes, because of identical i^{th} and j^{th} rows.

= 0.

$$\therefore \quad (i, j)^{th} \text{ element of } A(\text{adj } A) = \begin{cases} |A| & \text{if} \quad i = j \\ 0 & \text{if} \quad i \neq j \end{cases}$$

∴ The elements of A(adj A) along the principal diagonal are each | A | and the non-diagonal elements are 0 each.

$$\therefore \quad A(\text{adj } A) = \begin{bmatrix} |A| & 0 & \cdots\cdots & 0 \\ 0 & |A| & \cdots\cdots & 0 \\ \cdots & \cdots & \cdots & \cdots \\ \cdots & \cdots & \cdots & \cdots \\ 0 & 0 & \cdots\cdots & |A| \end{bmatrix} = |A| \begin{bmatrix} 1 & 0 & \cdots\cdots & 0 \\ 0 & 1 & \cdots\cdots & 0 \\ \cdots & \cdots & \cdots & \cdots \\ \cdots & \cdots & \cdots & \cdots \\ 0 & 0 & \cdots\cdots & 1 \end{bmatrix} = |A|\, I_n.$$

∴ $A(\text{adj } A) = |A|\, I_n$. Similarly, $(\text{adj } A)A = |A|\, I_n$.

∴ $\mathbf{A(adj\ A) = (adj\ A)A = |\ A\ |\ I_n}$.

Corollary 1. *If A is a non-singular square matrix of order n, then*

$$|\ adj\ A\ | = |\ A\ |^{n-1}.$$

Proof. We have $A\,(\text{adj } A) = |A|\, I_n$

$\Rightarrow \quad |A \text{ adj } A| = ||A|\, I_n|$

$\Rightarrow \quad |A|\; |\text{adj } A| = |A|^n\, |I_n|$*

$\Rightarrow \quad |A|\; |\text{adj } A| = |A|^n$

$\Rightarrow \quad \mathbf{|\ adj\ A\ | = |\ A\ |^{n-1}}$. $\qquad (\because \ |A| \neq 0)$

Remark. If A is non-singular, then adj. A is also non-singular, because $|A| \neq 0$ implies $|A|^{n-1} \neq 0$.

Corollary 2. *If A is a non-singular square matrix of order n, then*

$$adj\ (adj\ A) = |\ A\ |^{n-2} A.$$

Proof. For any non-singular square matrix B of orders n, we have

$$B(\text{adj } B) = |B|\, I_n \qquad \ldots(1)$$

Since A is non-singular, the square matrix adj A is also non-singular and $|\text{adj } A| = |A|^{n-1}$.

***Why this step.** We have | AB | = | A |. | B | for square matrices of same order. Also, if A is a square matrix of order n, then $|kA| = k^n\, |A|$.

Putting B = adj A in (1), we get

$$(\text{adj } A)(\text{adj } (\text{adj } A)) = |\text{ adj } A\ | I_n$$
$$\Rightarrow \quad (\text{adj } A)(\text{adj } (\text{adj } A)) = |A|^{n-1} I_n$$

Pre-multiplying by A, we get

$$A[(\text{adj } A)(\text{adj } (\text{adj } A))] = A(|A|^{n-1} I_n)$$
$$\Rightarrow \quad (A (\text{adj } A))(\text{adj } (\text{adj } A)) = |A|^{n-1} (A I_n)$$
$$\Rightarrow \quad (|A| I_n)(\text{adj } (\text{adj } A)) = |A|^{n-1} A$$
$$\Rightarrow \quad |A| (I_n (\text{adj } (\text{adj } A))) = |A|^{n-1} A$$
$$\Rightarrow \quad \textbf{adj (adj A)} = |A|^{n-2} A. \qquad (\because |A| \neq 0)$$

Theorem II. *If A is a square matrix, then prove that adj A′ = (adj A)′.*

Proof. Let A be a square matrix of order *n*.

∴ adj A′ and (adj A)′ are square matrices of order *n*.

(i, j)th element of (adj A)′ = $(j, i)^{th}$ element of the adj A.
= cofactor of $(i, j)^{th}$ element of A
= cofactor of $(j, i)^{th}$ element of A′
= $(i, j)^{th}$ element of adj A′.

∴ (adj A)′ = adj A′ *i.e.*, **adj A′ = (adj A)′.**

Corollary. *If A is a symmetric matrix, then prove that adj A is also symmetric.*

Proof. (adj A)′ = adj A′ (By above theorem)
= adj A (∵ A is symmetric)

∴ (adj A)′ = adj A *i.e.*, **adj A is symmetric.**

Example 3. *If* $A = \begin{bmatrix} 1 & 2 \\ 3 & 4 \end{bmatrix}$, *then verify that A(adj A) = (adj A)A = | A |* I_2.

Sol. We have $A = \begin{bmatrix} 1 & 2 \\ 3 & 4 \end{bmatrix}$.

Here A_{11} = cofactor of $a_{11}(= 1) = (-1)^{1+1} | 4 | = 4$
A_{12} = cofactor of $a_{12}(= 2) = (-1)^{1+2} | 3 | = -3$
A_{21} = cofactor of $a_{21}(= 3) = (-1)^{2+1} | 2 | = -2$
A_{22} = cofactor of $a_{22}(= 4) = (-1)^{2+2} | 1 | = 1.$

$$\therefore \quad \text{adj } A = \begin{bmatrix} 4 & -3 \\ -2 & 1 \end{bmatrix}' = \begin{bmatrix} 4 & -2 \\ -3 & 1 \end{bmatrix}$$

$$\therefore \quad A(\text{adj } A) = \begin{bmatrix} 1 & 2 \\ 3 & 4 \end{bmatrix}\begin{bmatrix} 4 & -2 \\ -3 & 1 \end{bmatrix} = \begin{bmatrix} 4-6 & -2+2 \\ 12-12 & -6+4 \end{bmatrix} = \begin{bmatrix} -2 & 0 \\ 0 & -2 \end{bmatrix}.$$

$$(\text{adj } A)A = \begin{bmatrix} 4 & -2 \\ -3 & 1 \end{bmatrix}\begin{bmatrix} 1 & 2 \\ 3 & 4 \end{bmatrix} = \begin{bmatrix} 4-6 & 8-8 \\ -3+3 & -6+4 \end{bmatrix} = \begin{bmatrix} -2 & 0 \\ 0 & -2 \end{bmatrix}$$

$$|A| I_2 = \begin{vmatrix} 1 & 2 \\ 3 & 4 \end{vmatrix} \cdot \begin{bmatrix} 1 & 0 \\ 0 & 1 \end{bmatrix} = (4-6) \cdot \begin{bmatrix} 1 & 0 \\ 0 & 1 \end{bmatrix} = -2\begin{bmatrix} 1 & 0 \\ 0 & 1 \end{bmatrix} = \begin{bmatrix} -2 & 0 \\ 0 & -2 \end{bmatrix}$$

∴ **A(adj A) = (adj A)A = | A |** $\mathbf{I_2}$**.**

Example 4. *If* $A = \begin{bmatrix} 2 & 1 & 3 \\ 3 & 1 & 2 \\ 1 & 2 & 3 \end{bmatrix}$, *verify that* $A(adj\ A) = (adj\ A)A = |\ A\ |\ I_3$.

Sol. We have $A = \begin{bmatrix} 2 & 1 & 3 \\ 3 & 1 & 2 \\ 1 & 2 & 3 \end{bmatrix}$.

$$A_{11} = \text{cofactor of } a_{11}(= 2) = (-1)^{1+1} \begin{vmatrix} 1 & 2 \\ 2 & 3 \end{vmatrix} = 3 - 4 = -1$$

$$A_{12} = \text{cofactor of } a_{12}(= 1) = (-1)^{1+2} \begin{vmatrix} 3 & 2 \\ 1 & 3 \end{vmatrix} = -(9 - 2) = -7$$

$$A_{13} = \text{cofactor of } a_{13}(= 3) = (-1)^{1+3} \begin{vmatrix} 3 & 1 \\ 1 & 2 \end{vmatrix} = 6 - 1 = 5$$

$$A_{21} = \text{cofactor of } a_{21}(= 3) = (-1)^{2+1} \begin{vmatrix} 1 & 3 \\ 2 & 3 \end{vmatrix} = -(3 - 6) = 3$$

$$A_{22} = \text{cofactor of } a_{22}(= 1) = (-1)^{2+2} \begin{vmatrix} 2 & 3 \\ 1 & 3 \end{vmatrix} = 6 - 3 = 3$$

$$A_{23} = \text{cofactor of } a_{23}(= 2) = (-1)^{2+3} \begin{vmatrix} 2 & 1 \\ 1 & 2 \end{vmatrix} = -(4 - 1) = -3$$

$$A_{31} = \text{cofactor of } a_{31}(= 1) = (-1)^{3+1} \begin{vmatrix} 1 & 3 \\ 1 & 2 \end{vmatrix} = 2 - 3 = -1$$

$$A_{32} = \text{cofactor of } a_{32}(= 2) = (-1)^{3+2} \begin{vmatrix} 2 & 3 \\ 3 & 2 \end{vmatrix} = -(4 - 9) = 5$$

$$A_{33} = \text{cofactor of } a_{33}(= 3) = (-1)^{3+3} \begin{vmatrix} 2 & 1 \\ 3 & 1 \end{vmatrix} = 2 - 3 = -1.$$

$$\therefore \quad \text{adj } A = \begin{bmatrix} -1 & -7 & 5 \\ 3 & 3 & -3 \\ -1 & 5 & -1 \end{bmatrix}' = \begin{bmatrix} -1 & 3 & -1 \\ -7 & 3 & 5 \\ 5 & -3 & -1 \end{bmatrix}.$$

$$\therefore \quad A(\text{adj } A) = \begin{bmatrix} 2 & 1 & 3 \\ 3 & 1 & 2 \\ 1 & 2 & 3 \end{bmatrix} \begin{bmatrix} -1 & 3 & -1 \\ -7 & 3 & 5 \\ 5 & -3 & -1 \end{bmatrix}$$

$$= \begin{bmatrix} -2 - 7 + 15 & 6 + 3 - 9 & -2 + 5 - 3 \\ -3 - 7 + 10 & 9 + 3 - 6 & -3 + 5 - 2 \\ -1 - 14 + 15 & 3 + 6 - 9 & -1 + 10 - 3 \end{bmatrix} = \begin{bmatrix} 6 & 0 & 0 \\ 0 & 6 & 0 \\ 0 & 0 & 6 \end{bmatrix}.$$

$$(\text{adj } A)A = \begin{bmatrix} -1 & 3 & -1 \\ -7 & 3 & 5 \\ 5 & -3 & -1 \end{bmatrix} \begin{bmatrix} 2 & 1 & 3 \\ 3 & 1 & 2 \\ 1 & 2 & 3 \end{bmatrix}$$

$$= \begin{bmatrix} -2 + 9 - 1 & -1 + 3 - 2 & -3 + 6 - 3 \\ -14 + 9 + 5 & -7 + 3 + 10 & -21 + 6 + 15 \\ 10 - 9 - 1 & 5 - 3 - 2 & 15 - 6 - 3 \end{bmatrix} = \begin{bmatrix} 6 & 0 & 0 \\ 0 & 6 & 0 \\ 0 & 0 & 6 \end{bmatrix}.$$

$$| A | = \begin{vmatrix} 2 & 1 & 3 \\ 3 & 1 & 2 \\ 1 & 2 & 3 \end{vmatrix} = 2(3-4) - 1(9-2) + 3(6-1) = -2 - 7 + 15 = 6.$$

$$\therefore \quad | A | I_3 = 6 \begin{bmatrix} 1 & 0 & 0 \\ 0 & 1 & 0 \\ 0 & 0 & 1 \end{bmatrix} = \begin{bmatrix} 6 & 0 & 0 \\ 0 & 6 & 0 \\ 0 & 0 & 6 \end{bmatrix}.$$

$\therefore$ **A(adj A) = (adj A)A = | A | I_3.**

EXERCISE 2.11

1. Find the adjoint of the following square matrices :

   (i) $\begin{bmatrix} 7 & 4 \\ 3 & -6 \end{bmatrix}$ (ii) $\begin{bmatrix} a & b \\ c & d \end{bmatrix}$.

2. Find the adjoint of the following square matrices :

   (i) $\begin{bmatrix} x+1 & x-1 \\ x^2 & 1-x \end{bmatrix}$ (ii) $\begin{bmatrix} 2i & -1+i \\ -i & 3+i \end{bmatrix}$.

3. Compute the adjoint of the matrix $A = \begin{bmatrix} 3 & 4 \\ 5 & 7 \end{bmatrix}$ and verify that (adj A) A = | A | I.

4. If $A = \begin{bmatrix} 5 & -2 \\ 3 & -2 \end{bmatrix}$, verify that A (adj A) = (adj A) A = | A | I_2.

5. If $A = \begin{bmatrix} 1 & 2 \\ 3 & 5 \end{bmatrix}$ and $B = \begin{bmatrix} 2 & 0 \\ 1 & 5 \end{bmatrix}$, then verify that adj (AB) = (adj B)(adj A).

6. If $A = \begin{bmatrix} -1 & -2 & -2 \\ 2 & 1 & -2 \\ 2 & -2 & 1 \end{bmatrix}$, show that adj A = 3A′.

7. For the matrix $A = \begin{bmatrix} 1 & -1 & 1 \\ 2 & 3 & 0 \\ 18 & 2 & 10 \end{bmatrix}$, show that A(adj A) = O. *(P.T.U. B.Pharma, Dec. 2004)*

8. For the matrix $A = \begin{bmatrix} \cos\alpha & -\sin\alpha & 0 \\ \sin\alpha & \cos\alpha & 0 \\ 0 & 0 & 1 \end{bmatrix}$, verify that A(adj A) = | A | I.

9. Find the adjoint of the matrix $\begin{bmatrix} 1 & 2 & 3 \\ 2 & 3 & 2 \\ 3 & 3 & 4 \end{bmatrix}$. Also verify that

   A(adj A) = (adj A)A = | A | I, where A is the given matrix.

10. If A is a square matrix of order 3 × 3, then prove that A(adj A) = | A | I_3.

Answers

1. (i) $\begin{bmatrix} -6 & -4 \\ -3 & 7 \end{bmatrix}$ (ii) $\begin{bmatrix} d & -b \\ -c & a \end{bmatrix}$

2. (*i*) $\begin{bmatrix} 1-x & 1-x \\ -x^2 & 1+x \end{bmatrix}$ (*ii*) $\begin{bmatrix} 3+i & 1-i \\ i & 2i \end{bmatrix}$

3. $\begin{bmatrix} 7 & -4 \\ -5 & 3 \end{bmatrix}$ 9. $\begin{bmatrix} 6 & 1 & -5 \\ -2 & -5 & 4 \\ -3 & 3 & -1 \end{bmatrix}$.

2.30. INVERSE OF A MATRIX

Let A be a square matrix of order n. If there exists a square matrix B of order n such that $AB = BA = I_n$, then B is called the **inverse** of A.

It is clear from the definition that if B is inverse of A, then A is inverse of B. The concept of inverse of matrix is defined only for square matrices.

If inverse of A exists, then A is called an **invertible matrix.**

Remark. The inverse of identity matrix I is I itself because I . I = I = I . I.

Theorem I. *The inverse of a square matrix, if it exists, is unique.*

Proof. Let A be a square matrix of order n such that inverse of A exists.

Let B and C be any two inverses of A.

$\therefore$ By definition, $AB = BA = I_n$ and $AC = CA = I_n$.

We have $B = BI_n = B(AC) = (BA)C = I_nC = C$.

$\therefore$ $B = C$.

$\therefore$ Any two inverses of A are equal matrices.

$\therefore$ The inverse of A is unique.

Remark. The inverse of an invertible matrix A is denoted by A^{-1}.

Theorem II. *A square matrix is invertible if and only if it is non-singular.*

Proof. Necessity. Let the square matrix A be invertible.

$\therefore$ There exists a square matrix B such that AB = BA = I.

$\Rightarrow$ $|AB| = |BA| = |I|$

$\Rightarrow$ $|A|.|B| = |B|.|A| = 1$

($\because$ Determinant of the product of square matrices of same order is the product of their determinants.)

$\therefore$ The number $|A|$ is non-zero, for otherwise $|A|.|B|$ will become zero.

$\therefore$ The matrix A is non-singular.

Sufficiency. Let A be non-singular. Since A is a square matrix, we have

$$A(\text{adj } A) = (\text{adj } A)A = |A|\, I$$

$$\Rightarrow \frac{1}{|A|}[A(\text{adj } A)] = \frac{1}{|A|}[(\text{adj } A)A] = \frac{1}{|A|}. \,|A|\, I$$

(Since, $|A| \neq 0$, we can divide by $|A|$)

$$\Rightarrow A\left(\frac{\text{adj } A}{|A|}\right) = \left(\frac{\text{adj } A}{|A|}\right)A = I.$$

$\therefore$ The matrix A is invertible and $A^{-1} = \dfrac{\text{adj } A}{|A|}$.

$\therefore$ The result holds.

Theorem III. *If A is an invertible matrix, then prove that $(A')^{-1} = (A^{-1})'$.*

Proof. Since A is an invertible matrix, A^{-1} exits and we have

$$AA^{-1} = A^{-1}A = I$$

$$\Rightarrow \quad (AA^{-1})' = (A^{-1}A)' = I'$$

$$\Rightarrow \quad (A^{-1})'A' = A'(A^{-1})' = I \qquad (\because (AB)' = B'A' \text{ and } I' = I)$$

$\therefore$ By definition, $(A^{-1})'$ is the inverse of A'.

$$\therefore \quad \mathbf{(A')^{-1} = (A^{-1})'}.$$

Theorem IV. *If A and B are invertible matrices of order n, then show that AB is also invertible and $(AB)^{-1} = B^{-1}A^{-1}$.*

Proof. The matrices A and B are invertible.

$\therefore$ A^{-1} and B^{-1} exists and $AA^{-1} = A^{-1}A = I_n$, $BB^{-1} = B^{-1}B = I_n$.

A and B are square matrices of order *n*, therefore AB is defined.

Also $|AB| = |A|.|B| \neq 0$, because A and B are invertible and so $|A| \neq 0$, $|B| \neq 0$.

$\therefore$ AB is invertible, *i.e.*, $(AB)^{-1}$ exists.

Now $\quad (AB)(B^{-1}A^{-1}) = A(BB^{-1})A^{-1} = A(I_n)A^{-1} = AA^{-1} = I_n$

and $\quad (B^{-1}A^{-1})(AB) = B^{-1}(A^{-1}A)B = B^{-1}(I_n)B = B^{-1}B = I_n$.

$\therefore$ We have $\quad (AB)(B^{-1}A^{-1}) = (B^{-1}A^{-1})(AB) = I_n$.

$\therefore$ By definition, $\quad \mathbf{(AB)^{-1} = B^{-1}A^{-1}}$.

Remark. $(AB)^{-1} = B^{-1}A^{-1}$ is called the **reversal law** of inverse of product of matrices.

WORKING RULES FOR FINDING THE INVERSE OF SQUARE MATRIX A

Step I. *Find the value of* $|A|$.

Step II. *If* $|A| = 0$*, then A cannot have its inverse.*

Step III. *If* $|A| \neq 0$*, then A has its inverse and is given by* $\dfrac{adj\ A}{|A|}$.

Step IV. *Find cofactors of all elements of A and compute adj A'. Find* A^{-1} *by multiplying 'adj A' by '*$\dfrac{1}{|A|}$*'.*

ILLUSTRATIVE EXAMPLES

Example 1. *Find the inverse of* $\begin{bmatrix} 3 & 6 \\ 7 & 2 \end{bmatrix}$*. Also verify your answer.*

Sol. Let $A = \begin{bmatrix} 3 & 6 \\ 7 & 2 \end{bmatrix}$.

$\therefore \quad |A| = \begin{vmatrix} 3 & 6 \\ 7 & 2 \end{vmatrix} = 6 - 42 = -36 \neq 0. \qquad \therefore$ A is non-singular.

$\therefore$ A^{-1} exists and $A^{-1} = \dfrac{\text{adj A}}{|A|}$.

We find adj A :

$A_{11} = (-1)^{1+1}\ |2| = 2, \qquad A_{12} = (-1)^{1+2}\ |7| = -7$

$A_{21} = (-1)^{2+1}\ |6| = -6, \qquad A_{22} = (-1)^{2+2}\ |3| = 3.$

$$\therefore \quad \text{adj A} = \begin{bmatrix} 2 & -7 \\ -6 & 3 \end{bmatrix}' = \begin{bmatrix} 2 & -6 \\ -7 & 3 \end{bmatrix}$$

$$\therefore \quad A^{-1} = \frac{\text{adj } A}{|A|} = \frac{1}{-36}\begin{bmatrix} 2 & -6 \\ -7 & 3 \end{bmatrix} = \begin{bmatrix} \mathbf{-1/18} & \mathbf{1/16} \\ \mathbf{7/36} & \mathbf{-1/12} \end{bmatrix}.$$

Verification :

$$AA^{-1} = \begin{bmatrix} 3 & 6 \\ 7 & 2 \end{bmatrix} \times \frac{1}{-36}\begin{bmatrix} 2 & -6 \\ -7 & 3 \end{bmatrix} = -\frac{1}{36}\begin{bmatrix} 6-42 & -18+18 \\ 14-14 & -42+6 \end{bmatrix}$$

$$= -\frac{1}{36}\begin{bmatrix} -36 & 0 \\ 0 & -36 \end{bmatrix} = \frac{-36}{-36}\begin{bmatrix} 1 & 0 \\ 0 & 1 \end{bmatrix} = I_2.$$

Similarly, $A^{-1}A = I_2$.

Example 2. *If* $A = \begin{bmatrix} 3 & -5 \\ -4 & 2 \end{bmatrix}$, *verify that* $A^2 - 5A - 14I = O$ *and hence find* A^{-1}.

Sol. $A^2 - 5A - 14I = \begin{bmatrix} 3 & -5 \\ -4 & 2 \end{bmatrix}\begin{bmatrix} 3 & -5 \\ -4 & 2 \end{bmatrix} - 5\begin{bmatrix} 3 & -5 \\ -4 & 2 \end{bmatrix} - 14\begin{bmatrix} 1 & 0 \\ 0 & 1 \end{bmatrix}$

$$= \begin{bmatrix} 9+20 & -15-10 \\ -12-8 & 20+4 \end{bmatrix} - \begin{bmatrix} 15 & -25 \\ -20 & 10 \end{bmatrix} - \begin{bmatrix} 14 & 0 \\ 0 & 14 \end{bmatrix}$$

$$= \begin{bmatrix} 29-15-14 & -25+25-0 \\ -20+20-0 & 24-10-14 \end{bmatrix} = \begin{bmatrix} 0 & 0 \\ 0 & 0 \end{bmatrix} = O$$

$\therefore \quad A^2 - 5A - 14I = O$ ...(1)

(1) $\Rightarrow \quad A^2 - 5A = 14I \quad \Rightarrow \quad A(A - 5I) = 14I \quad \Rightarrow \quad A\left(\frac{A-5I}{14}\right) = I$

(1) $\Rightarrow \quad A^2 - 5A = 14I \quad \Rightarrow \quad (A - 5I)A = 14I \quad \Rightarrow \quad \left(\frac{A-5I}{14}\right)A = I$

$$\therefore \quad A\left(\frac{A-5I}{14}\right) = \left(\frac{A-5I}{14}\right)A = I$$

$\therefore$ By definition $A^{-1} = \dfrac{A-5I}{14}$.

Hence $\quad A^{-1} = \frac{1}{14}(A - 5I) = \frac{1}{14}\left(\begin{bmatrix} 3 & -5 \\ -4 & 2 \end{bmatrix} - 5\begin{bmatrix} 1 & 0 \\ 0 & 1 \end{bmatrix}\right) = \mathbf{\frac{1}{14}\begin{bmatrix} -2 & -5 \\ -4 & -3 \end{bmatrix}}.$

Example 3. *Find the inverse of the matrix* $\begin{bmatrix} 1 & -1 & 2 \\ 0 & 2 & -3 \\ 3 & -2 & 4 \end{bmatrix}$.

Sol. Let $\quad A = \begin{bmatrix} 1 & -1 & 2 \\ 0 & 2 & -3 \\ 3 & -2 & 4 \end{bmatrix}$

$$\therefore \quad |A| = \begin{vmatrix} 1 & -1 & 2 \\ 0 & 2 & -3 \\ 3 & -2 & 4 \end{vmatrix} = 1(8-6) + 1(0+9) + 2(0-6) = 2 + 9 - 12 = -1 \neq 0.$$

$\therefore$ A is invertible and $A^{-1} = \dfrac{\text{adj } A}{|A|}$.

We find adj A :

$$A_{11} = (-1)^2 \begin{vmatrix} 2 & -3 \\ -2 & 4 \end{vmatrix} = 8 - 6 = 2$$

$$A_{12} = (-1)^3 \begin{vmatrix} 0 & -3 \\ 3 & 4 \end{vmatrix} = -(0 + 9) = -9$$

$$A_{13} = (-1)^4 \begin{vmatrix} 0 & 2 \\ 3 & -2 \end{vmatrix} = 0 - 6 = -6$$

$$A_{21} = (-1)^3 \begin{vmatrix} -1 & 2 \\ -2 & 4 \end{vmatrix} = -(-4 + 4) = 0$$

$$A_{22} = (-1)^4 \begin{vmatrix} 1 & 2 \\ 3 & 4 \end{vmatrix} = 4 - 6 = -2$$

$$A_{23} = (-1)^5 \begin{vmatrix} 1 & -1 \\ 3 & -2 \end{vmatrix} = -(-2 + 3) = -1$$

$$A_{31} = (-1)^4 \begin{vmatrix} -1 & 2 \\ 2 & -3 \end{vmatrix} = 3 - 4 = -1$$

$$A_{32} = (-1)^5 \begin{vmatrix} 1 & 2 \\ 0 & -3 \end{vmatrix} = -(-3 - 0) = 3$$

$$A_{33} = (-1)^6 \begin{vmatrix} 1 & -1 \\ 0 & 2 \end{vmatrix} = 2 + 0 = 2.$$

$$\therefore \quad \text{adj } A = \begin{bmatrix} 2 & -9 & -6 \\ 0 & -2 & -1 \\ -1 & 3 & 2 \end{bmatrix}' = \begin{bmatrix} 2 & 0 & -1 \\ -9 & -2 & 3 \\ -6 & -1 & 2 \end{bmatrix}$$

$$\therefore \quad A^{-1} = \frac{\text{adj } A}{|A|} = \frac{1}{-1} \begin{bmatrix} 2 & 0 & -1 \\ -9 & -2 & 3 \\ -6 & -1 & 2 \end{bmatrix} = \begin{bmatrix} \mathbf{-2} & \mathbf{0} & \mathbf{1} \\ \mathbf{9} & \mathbf{2} & \mathbf{-3} \\ \mathbf{6} & \mathbf{1} & \mathbf{-2} \end{bmatrix}.$$

Verification :

$$AA^{-1} = \begin{bmatrix} 1 & -1 & 2 \\ 0 & 2 & -3 \\ 3 & -2 & 4 \end{bmatrix} \begin{bmatrix} -2 & 0 & 1 \\ 9 & 2 & -3 \\ 6 & 1 & -2 \end{bmatrix}$$

$$= \begin{bmatrix} -2-9+12 & 0-2+2 & 1+3-4 \\ 0+18-18 & 0+4-3 & 0-6+6 \\ -6-18+24 & 0-4+4 & 3+6-8 \end{bmatrix} = \begin{bmatrix} 1 & 0 & 0 \\ 0 & 1 & 0 \\ 0 & 0 & 1 \end{bmatrix} = I.$$

Similarly, it can be checked that $A^{-1} A = I$.

Example 4. *Find the inverse of* $\begin{bmatrix} 1 & 2 & 4 \\ 3 & -19 & 7 \\ 2 & 4 & 8 \end{bmatrix}$, *if exists.*

Sol.

$$A = \begin{bmatrix} 1 & 2 & 4 \\ 3 & -19 & 7 \\ 2 & 4 & 8 \end{bmatrix}$$

$$\therefore \quad |A| = \begin{vmatrix} 1 & 2 & 4 \\ 3 & -19 & 7 \\ 2 & 4 & 8 \end{vmatrix} = 1(-152 - 28) - 2(24 - 14) + 4(12 + 38) = -180 - 20 + 200 = 0.$$

$\therefore$ A is singular. $\therefore$ The inverse of A does not exist.

Example 5. *If* $A = \begin{bmatrix} 1 & 2 & 2 \\ 2 & 1 & 2 \\ 2 & 2 & 1 \end{bmatrix}$, *find* A^{-1} *and hence prove that* $A^2 - 4A - 5I = O$.

Sol. We have $A = \begin{bmatrix} 1 & 2 & 2 \\ 2 & 1 & 2 \\ 2 & 2 & 1 \end{bmatrix}$.

$$\therefore \quad |A| = \begin{vmatrix} 1 & 2 & 2 \\ 2 & 1 & 2 \\ 2 & 2 & 1 \end{vmatrix} = 1(1-4) - 2(2-4) + 2(4-2) = -3 + 4 + 4 = 5 \neq 0$$

(Expanding along R_1)

$\therefore$ A is invertible and $A^{-1} = \dfrac{\text{adj } A}{|A|}$.

Now, $A_{11} = \begin{vmatrix} 1 & 2 \\ 2 & 1 \end{vmatrix} = -3$, $A_{12} = -\begin{vmatrix} 2 & 2 \\ 2 & 1 \end{vmatrix} = 2$, $A_{13} = \begin{vmatrix} 2 & 1 \\ 2 & 2 \end{vmatrix} = 2$

$A_{21} = -\begin{vmatrix} 2 & 2 \\ 2 & 1 \end{vmatrix} = 2$, $A_{22} = \begin{vmatrix} 1 & 2 \\ 2 & 1 \end{vmatrix} = -3$, $A_{23} = -\begin{vmatrix} 1 & 2 \\ 2 & 2 \end{vmatrix} = 2$

$A_{31} = \begin{vmatrix} 2 & 2 \\ 1 & 2 \end{vmatrix} = 2$, $A_{32} = -\begin{vmatrix} 1 & 2 \\ 2 & 2 \end{vmatrix} = 2$, $A_{33} = \begin{vmatrix} 1 & 2 \\ 2 & 1 \end{vmatrix} = -3$

$$\therefore \quad \text{adj } A = \begin{bmatrix} -3 & 2 & 2 \\ 2 & -3 & 2 \\ 2 & 2 & -3 \end{bmatrix}' = \begin{bmatrix} -3 & 2 & 2 \\ 2 & -3 & 2 \\ 2 & 2 & -3 \end{bmatrix}$$

$$\therefore \quad A^{-1} = \frac{\text{adj } A}{|A|} = \mathbf{\frac{1}{5}\begin{bmatrix} -3 & 2 & 2 \\ 2 & -3 & 2 \\ 2 & 2 & -3 \end{bmatrix}}.$$

Now, $A^2 - 4A - 5I = AA - 4AI - 5I = A(A - 4I) - 5I = 5\left[\frac{1}{5}A(A - 4I) - I\right]$

$$\therefore \quad A^2 - 4A - 5I = 5\left[A\left(\frac{A - 4I}{5}\right) - I\right] \qquad \text{...(1)}$$

Here, $\dfrac{A - 4I}{5} = \dfrac{1}{5}\left(\begin{bmatrix} 1 & 2 & 2 \\ 2 & 1 & 2 \\ 2 & 2 & 1 \end{bmatrix} - 4\begin{bmatrix} 1 & 0 & 0 \\ 0 & 1 & 0 \\ 0 & 0 & 1 \end{bmatrix}\right) = \dfrac{1}{5}\begin{bmatrix} -3 & 2 & 2 \\ 2 & -3 & 2 \\ 2 & 2 & -3 \end{bmatrix} = A^{-1}$

$\therefore$ (1) $\Rightarrow$ $A^2 - 4A - 5I = 5[AA^{-1} - I] = 5[I - I] = 5O = O$

$\therefore$ $\mathbf{A^2 - 4A - 5I = O.}$

EXERCISE 2.12

1. For what value of k, the matrix $\begin{bmatrix} k & 2 \\ 3 & 4 \end{bmatrix}$ has no inverse ?

2. Find the inverse of the following matrices :

   (i) $\begin{bmatrix} 2 & 5 \\ -3 & 1 \end{bmatrix}$ (ii) $\begin{bmatrix} 2 & -3 \\ 4 & 5 \end{bmatrix}$.

3. If $A = \begin{bmatrix} 3 & 2 \\ 7 & 5 \end{bmatrix}$ and $B = \begin{bmatrix} 6 & 7 \\ 8 & 9 \end{bmatrix}$, find $(AB)^{-1}$.

4. Verify that $(AB)^{-1} = B^{-1}A^{-1}$, where :

   (i) $A = \begin{bmatrix} 2 & -1 \\ 4 & 2 \end{bmatrix}$, $B = \begin{bmatrix} 4 & 3 \\ -2 & 1 \end{bmatrix}$ (ii) $A = \begin{bmatrix} 2 & 1 \\ 5 & 3 \end{bmatrix}$, $B = \begin{bmatrix} 4 & 5 \\ 3 & 4 \end{bmatrix}$.

5. (i) If $A = \begin{bmatrix} 3 & 2 \\ 1 & 1 \end{bmatrix}$, show that $A^2 - 4A + I = O$. Hence find A^{-1}.

   (ii) If $A = \begin{bmatrix} 3 & 2 \\ 2 & 1 \end{bmatrix}$, show that $A^2 - 4A - I_2 = O$. Hence find A^{-1}. *(P.T.U. B. Pharma, May 2005)*

6. Find the inverse of the following matrices :

   (i) $\begin{bmatrix} 1 & 3 & 3 \\ 1 & 4 & 3 \\ 1 & 3 & 4 \end{bmatrix}$ (ii) $\begin{bmatrix} 2 & -3 & 3 \\ 2 & 2 & 3 \\ 3 & -2 & 2 \end{bmatrix}$

   (P.T.U. B. Pharma, May 2006)

   (iii) $\begin{bmatrix} 1 & 1 & 3 \\ 1 & 3 & -3 \\ -2 & -4 & -4 \end{bmatrix}$. *(P.T.U. B. Pharma, Dec. 2006)*

7. (a) If $A = \begin{bmatrix} 1 & -1 & 1 \\ 2 & -1 & 0 \\ 1 & 0 & 0 \end{bmatrix}$, find A^{-1} and show that $A^{-1} = A^2$.

   (b) If $A = \frac{1}{9}\begin{pmatrix} -8 & 1 & 4 \\ 4 & 4 & 7 \\ 1 & -8 & 4 \end{pmatrix}$, prove that $A^{-1} = A^T$. A^T being the transpose of A.

   | **Hint.** We show $AA^T = I$ |

8. If $A = \begin{bmatrix} 1 & 2 & 2 \\ 2 & 1 & 2 \\ 2 & 2 & 1 \end{bmatrix}$, show that $A^2 - 4A - 5I = O$ and hence find A^{-1}.

Answers

1. $\frac{3}{2}$ **2.** (i) $\begin{bmatrix} 1/17 & -5/17 \\ 3/17 & 2/17 \end{bmatrix}$ (ii) $\begin{bmatrix} 5/22 & 3/22 \\ -2/11 & 1/11 \end{bmatrix}$ **3.** $\begin{bmatrix} -47 & 39/2 \\ 41 & -17 \end{bmatrix}$

5. (i) $\begin{bmatrix} 1 & -2 \\ -1 & 3 \end{bmatrix}$ (ii) $\begin{bmatrix} -1 & 2 \\ 2 & -3 \end{bmatrix}$

6. (i) $\begin{bmatrix} 7 & -3 & -3 \\ -1 & 1 & 0 \\ -1 & 0 & 1 \end{bmatrix}$ (ii) $\frac{1}{5}\begin{bmatrix} -2 & 0 & 3 \\ -1 & 1 & 0 \\ 2 & 1 & -2 \end{bmatrix}$ (iii) $\begin{bmatrix} 3 & 1 & \frac{3}{2} \\ \frac{-5}{4} & \frac{-1}{4} & \frac{-3}{4} \\ \frac{1}{4} & \frac{1}{4} & \frac{1}{4} \end{bmatrix}$

7. (a) $\begin{bmatrix} 0 & 0 & 1 \\ 0 & -1 & 2 \\ 1 & -1 & 1 \end{bmatrix}$ 8. $\frac{1}{5}\begin{bmatrix} -3 & 2 & 2 \\ 2 & -3 & 2 \\ 2 & 2 & -3 \end{bmatrix}$.

2.31. SOLUTION OF SYSTEM OF LINEAR EQUATIONS BY MATRIX METHOD

In this section, we shall use inverse of matrices in solving systems of linear equations.

Let us consider the system

$$a_1x + b_1y + c_1z = d_1$$
$$a_2x + b_2y + c_2z = d_2$$
$$a_3x + b_3y + c_3z = d_3$$

of three linear equations in three variables x, y, z.

This system can be expressed in the form of matrices as

$$\begin{bmatrix} a_1x + b_1y + c_1z \\ a_2x + b_2y + c_2z \\ a_3x + b_3y + c_3z \end{bmatrix} = \begin{bmatrix} d_1 \\ d_2 \\ d_3 \end{bmatrix}$$

$$\Rightarrow \quad \begin{bmatrix} a_1 & b_1 & c_1 \\ a_2 & b_2 & c_2 \\ a_3 & b_3 & c_3 \end{bmatrix}\begin{bmatrix} x \\ y \\ z \end{bmatrix} = \begin{bmatrix} d_1 \\ d_2 \\ d_3 \end{bmatrix} \quad \Rightarrow \quad AX = B$$

where $A = \begin{bmatrix} a_1 & b_1 & c_1 \\ a_2 & b_2 & c_2 \\ a_3 & b_3 & c_3 \end{bmatrix}$, $X = \begin{bmatrix} x \\ y \\ z \end{bmatrix}$, $B = \begin{bmatrix} d_1 \\ d_2 \\ d_3 \end{bmatrix}$.

$|A|$ may or may not be zero.

Case I. $|A| \neq 0$.

In this case, A^{-1} exists.

$\therefore$ $AX = B$ implies $A^{-1}(AX) = A^{-1}B$

$\Rightarrow (A^{-1}A)X = A^{-1}B \quad \Rightarrow \quad IX = A^{-1}B \quad \Rightarrow \quad X = A^{-1}B.$

$\therefore$ The system is **consistent** and has **unique** solution given by $\mathbf{X = A^{-1}B}$.

Case II. $|A| = 0$.

In this case, we calculate the matrix (adj A) B.

If (adj A)B $\neq$ O, then the system is **inconsistent** and has **no** solution.

If (adj A)B = O, then the system is either **consistent** or **inconsistent** because the system will have either **infinitely many** solutions or **no** solution.

For example, consider the system of equations :

$$x + y + z = 5$$
$$x + y + z = 6$$
$$x + y + z = 7.$$

For this system (adj A)B = O and the system is having no solution.

The given findings can be shown diagrammatically as below :

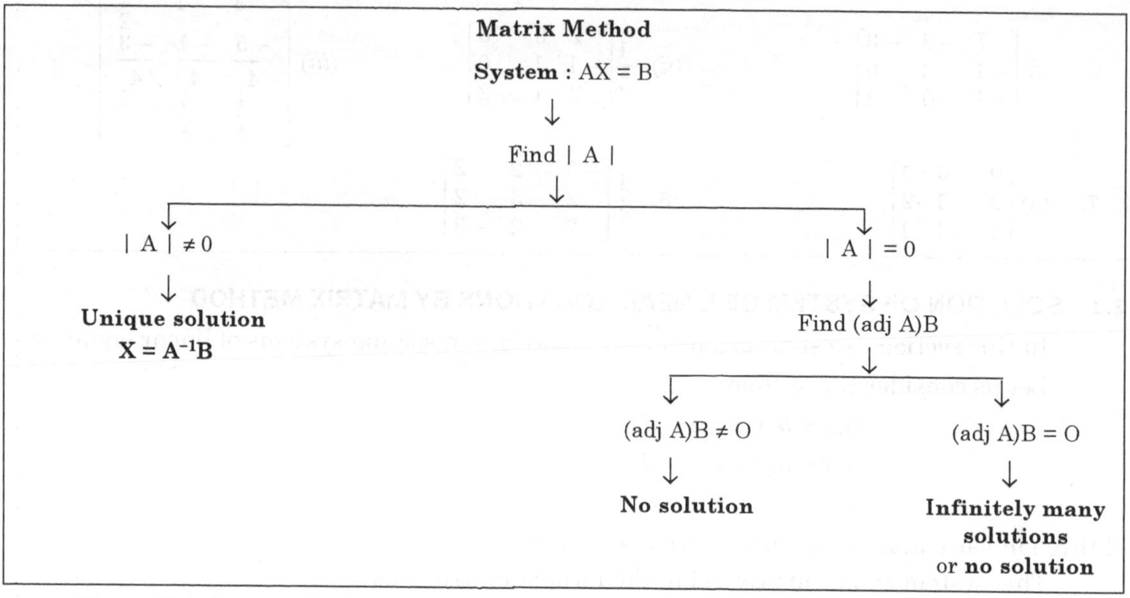

Remark 1. In case of a *homogeneous system,* we have B = O.

If | A | ≠ 0, then

$$X = A^{-1}B = A^{-1}O = O \quad i.e., \quad x = 0 \; y = 0, \quad z = 0$$

and if | A | = 0, then the system will have infinitely many solutions.

∴ A homogeneous system of linear equations is always consistent and $x = 0$, $y = 0$, $z = 0$ will always be solution of a homogeneous system of equations in the variables x, y, z.

Remark 2. The above method of solving three equations in three variables is general and is applicable to systems containing n (≥ 2) linear equations in n variables.

WORKING RULES FOR SOLVING EQUATIONS BY THE MATRIX METHOD

Step I. *Express the given system in the standard form :*

$$a_1x + b_1y + c_1z = d_1$$
$$a_2x + b_2y + c_2z = d_2$$
$$a_3x + b_3y + c_3z = d_3.$$

Write $A = \begin{bmatrix} a_1 & b_1 & c_1 \\ a_2 & b_2 & c_2 \\ a_3 & b_3 & c_3 \end{bmatrix}, X = \begin{bmatrix} x \\ y \\ z \end{bmatrix}, B = \begin{bmatrix} d_1 \\ d_2 \\ d_3 \end{bmatrix}$

Find the value of the determinant | A |.

Step II. *If | A | ≠ 0, then the given system will have unique solution given by* $X = A^{-1}B$.

Step III. *If | A | = 0, then put x (or y or z) equal to k and find the solution of any two of the given equations. Substitute these values of x, y and z in terms of k in the third equation.*

Step IV. *If the third equation is satisfied by this solution, then these values represent the solution of the given system. These are infinitely many solutions.*

If the third equation is not satisfied then declare that the given system has no solution.

ILLUSTRATIVE EXAMPLES

Example 1. *Solve the equations :*

$$x + 2y = 4, \quad 2x + 5y = 9.$$

Sol. The given equations are $x + 2y = 4$

$2x + 5y = 9$

Let $A = \begin{bmatrix} 1 & 2 \\ 2 & 5 \end{bmatrix}, X = \begin{bmatrix} x \\ y \end{bmatrix}, B = \begin{bmatrix} 4 \\ 9 \end{bmatrix}$. $\therefore$ The system is AX = B.

$$|A| = \begin{bmatrix} 1 & 2 \\ 2 & 5 \end{bmatrix} = 5 - 4 = 1 \neq 0.$$

$\therefore$ The system has unique solution, $X = A^{-1}B$.

For $A = \begin{bmatrix} 1 & 2 \\ 2 & 5 \end{bmatrix}$, we have $A_{11} = 5, \quad A_{12} = -2, \quad A_{21} = -2, \quad A_{22} = 1$

$$\therefore \quad \text{adj } A = \begin{bmatrix} 5 & -2 \\ -2 & 1 \end{bmatrix}' = \begin{bmatrix} 5 & -2 \\ -2 & 1 \end{bmatrix}$$

$$\therefore \quad A^{-1} = \frac{\text{adj } A}{|A|} = \frac{1}{1}\begin{bmatrix} 5 & -2 \\ -2 & 1 \end{bmatrix} = \begin{bmatrix} 5 & -2 \\ -2 & 1 \end{bmatrix}$$

The solution is $X = A^{-1}B$.

$$\therefore \quad \begin{bmatrix} x \\ y \end{bmatrix} = \begin{bmatrix} 5 & -2 \\ -2 & 1 \end{bmatrix}\begin{bmatrix} 4 \\ 9 \end{bmatrix} = \begin{bmatrix} 20 - 18 \\ -8 + 9 \end{bmatrix} = \begin{bmatrix} 2 \\ 1 \end{bmatrix}$$

$\therefore$ $\mathbf{x = 2, y = 1.}$

Example 2. *Use matrix method to solve the equations :*

$$6x + 4y = 2, \quad 9x + 6y = 3.$$

Sol. The given equations are $6x + 4y = 2$ ...(1)

$9x + 6y = 3.$ ...(2)

Let $A = \begin{bmatrix} 6 & 4 \\ 9 & 6 \end{bmatrix}, X = \begin{bmatrix} x \\ y \end{bmatrix}, B = \begin{bmatrix} 2 \\ 3 \end{bmatrix}$. $\therefore$ The given system is AX = B

Now $|A| = \begin{vmatrix} 6 & 4 \\ 9 & 6 \end{vmatrix} = 36 - 36 = 0.$

$\therefore$ The system has either infinitely many solutions or no solution.

Let $x = k$. $\therefore$ (1) $\Rightarrow$ $6k + 4y = 2$ *i.e.*, $y = \dfrac{1 - 3k}{2}$.

Putting the values of x and y in (2), we get

$$9k + 6\left(\frac{1 - 3k}{2}\right) = 3.$$

$\Rightarrow$ $9k + 3 - 9k = 3$ $\Rightarrow$ $3 = 3$, which is true.

$\therefore$ The given system has infinitely many solutions given by

$$x = \mathbf{k}, y = \frac{\mathbf{1 - 3k}}{\mathbf{2}}, \text{ where } \mathbf{k} \text{ is any number.}$$

Example 3. *Solve the equations :*

$$\frac{2}{x}+\frac{3}{y}+\frac{10}{z}=4,\ \frac{4}{x}-\frac{6}{y}+\frac{5}{z}=1,\ \frac{6}{x}+\frac{9}{y}-\frac{20}{z}=2.$$

Sol. The given equations are

$$2\left(\frac{1}{x}\right)+3\left(\frac{1}{y}\right)+10\left(\frac{1}{z}\right)=4$$

$$4\left(\frac{1}{x}\right)-6\left(\frac{1}{y}\right)+5\left(\frac{1}{z}\right)=1$$

$$6\left(\frac{1}{x}\right)+9\left(\frac{1}{y}\right)-20\left(\frac{1}{z}\right)=2$$

Let $A=\begin{bmatrix}2 & 3 & 10\\ 4 & -6 & 5\\ 6 & 9 & -20\end{bmatrix},\ X=\begin{bmatrix}1/x\\ 1/y\\ 1/z\end{bmatrix},\ B=\begin{bmatrix}4\\ 1\\ 2\end{bmatrix}\quad \therefore\ AX=B$

$$|A|=\begin{vmatrix}2 & 3 & 10\\ 4 & -6 & 5\\ 6 & 9 & -20\end{vmatrix}=2(120-45)-3(-80-30)+10(36+36)$$
$$=150+330+720=1200\neq 0.$$

$\therefore$ The system has unique solution $X=A^{-1}B$.

To find adj for A, we have

$A_{11}=120-45=75,\quad A_{12}=-(-80-30)=110,\quad A_{13}=36+36=72,$

$A_{21}=-(-60-90)=150,\quad A_{22}=-40-60=-100,\quad A_{23}=-(18-18)=0,$

$A_{31}=15+60=75,\quad A_{32}=-(10-40)=30,\quad A_{33}=-12-12=-24.$

$$\text{adj } A=\begin{bmatrix}75 & 110 & 72\\ 150 & -100 & 0\\ 75 & 30 & -24\end{bmatrix}'=\begin{bmatrix}75 & 150 & 75\\ 110 & -100 & 30\\ 72 & 0 & -24\end{bmatrix}$$

$$\therefore\quad A^{-1}=\frac{\text{adj } A}{|A|}=\frac{1}{1200}\begin{bmatrix}75 & 150 & 75\\ 110 & -100 & 30\\ 72 & 0 & -24\end{bmatrix}$$

The solution is $X=A^{-1}B$.

$$\therefore\quad \begin{bmatrix}1/x\\ 1/y\\ 1/z\end{bmatrix}=\frac{1}{1200}\begin{bmatrix}75 & 150 & 75\\ 110 & -100 & 30\\ 72 & 0 & -24\end{bmatrix}\begin{bmatrix}4\\ 1\\ 2\end{bmatrix}$$

$$=\frac{1}{1200}\begin{bmatrix}300+150+150\\ 440-100+60\\ 288+0-48\end{bmatrix}=\frac{1}{1200}\begin{bmatrix}600\\ 400\\ 240\end{bmatrix}=\begin{bmatrix}1/2\\ 1/3\\ 1/5\end{bmatrix}$$

$$\therefore\quad \frac{1}{x}=\frac{1}{2},\ \frac{1}{y}=\frac{1}{3},\ \frac{1}{z}=\frac{1}{5}.\qquad \therefore\ \boldsymbol{x=2,\ y=3,\ z=5.}$$

Example 4. *Solve the equations :*

$$2x-y+3z=5,\ 3x+2y-z=7,\ 4x+5y-5z=9.$$

Sol. The given equations are $2x - y + 3z = 5$ ...(1)

$3x + 2y - z = 7$ ...(2)

$4x + 5y - 5z = 9$ ...(3)

Let $A = \begin{bmatrix} 2 & -1 & 3 \\ 3 & 2 & -1 \\ 4 & 5 & -5 \end{bmatrix}$, $X = \begin{bmatrix} x \\ y \\ z \end{bmatrix}$, $B = \begin{bmatrix} 5 \\ 7 \\ 9 \end{bmatrix}$

$\therefore$ The given system is AX = B.

$$|A| = \begin{vmatrix} 2 & -1 & 3 \\ 3 & 2 & -1 \\ 4 & 5 & -5 \end{vmatrix} = 2(-10+5) + 1(-15+4) + 3(15-8) = -10 - 11 + 21 = 0$$

$\therefore$ The system has either infinitely many solutions or no solution.

Let $z = k$.

$\therefore$ (1) $\Rightarrow$ $2x - y = 5 - 3k$ ...(4)

(2) $\Rightarrow$ $3x + 2y = 7 + k$ ...(5)

This represents a system of two equations in two variables x, y. We apply matrix method to solve (4) and (5).

Let $A = \begin{bmatrix} 2 & -1 \\ 3 & 2 \end{bmatrix}$, $X = \begin{bmatrix} x \\ y \end{bmatrix}$, $B = \begin{bmatrix} 5-3k \\ 7+k \end{bmatrix}$. $\therefore$ AX = B

Now $|A| = \begin{vmatrix} 2 & -1 \\ 3 & 2 \end{vmatrix} = 4 + 3 = 7 \neq 0$

$\therefore$ The system has unique solution, $X = A^{-1}B$.

For $A = \begin{bmatrix} 2 & -1 \\ 3 & 2 \end{bmatrix}$, we have $A_{11} = 2$, $A_{12} = -3$, $A_{21} = 1$, $A_{22} = 2$.

$$\therefore \quad \text{adj } A = \begin{bmatrix} 2 & -3 \\ 1 & 2 \end{bmatrix}' = \begin{bmatrix} 2 & 1 \\ -3 & 2 \end{bmatrix}$$

$$\therefore \quad A^{-1} = \frac{\text{adj } A}{|A|} = \frac{1}{7}\begin{bmatrix} 2 & 1 \\ -3 & 2 \end{bmatrix}$$

The solution is $X = A^{-1}B$.

$$\therefore \quad \begin{bmatrix} x \\ y \end{bmatrix} = \frac{1}{7}\begin{bmatrix} 2 & 1 \\ -3 & 2 \end{bmatrix}\begin{bmatrix} 5-3k \\ 7+k \end{bmatrix} = \frac{1}{7}\begin{bmatrix} -5k+17 \\ 11k-1 \end{bmatrix} = \begin{bmatrix} \frac{-5k+17}{7} \\ \frac{11k-1}{7} \end{bmatrix}$$

$$\therefore \quad x = \frac{-5k+17}{7}, \quad y = \frac{11k-1}{7}$$

Putting the values of x, y and z in (3), we get $4\left(\frac{-5k+17}{7}\right) + 5\left(\frac{11k-1}{7}\right) - 5k = 9$.

$\Rightarrow$ $-20k + 68 + 55k - 5 - 35k = 63$

$\Rightarrow$ $63 = 63$, which is true.

$\therefore$ The given system has infinitely many solutions given by

$$x = \frac{-5k+17}{7},\ y = \frac{11k-1}{7},\ z = k, \text{ where } k \text{ is any number.}$$

Example 5. *Find* A^{-1} *if* $A = \begin{bmatrix} 1 & 2 & -3 \\ 2 & 3 & 2 \\ 3 & -3 & -4 \end{bmatrix}$.

Hence solve the equations :

$$x + 2y - 3z = -4,\ 2x + 3y + 2z = 2,\ 3x - 3y - 4z = 11.$$

Sol. $A = \begin{bmatrix} 1 & 2 & -3 \\ 2 & 3 & 2 \\ 3 & -3 & -4 \end{bmatrix}$

$$\therefore\ |A| = \begin{vmatrix} 1 & 2 & -3 \\ 2 & 3 & 2 \\ 3 & -3 & -4 \end{vmatrix} = 1(-12+6) - 2(-8-6) - 3(-6-9) = -6 + 28 + 45 = 67 \neq 0.$$

$\therefore$ A^{-1} exists and $A^{-1} = \dfrac{\text{adj } A}{|A|}$.

We find adj A :

$$A_{11} = -6, \quad A_{12} = 14, \quad A_{13} = -15,$$
$$A_{21} = 17, \quad A_{22} = 5, \quad A_{23} = 9,$$
$$A_{31} = 13, \quad A_{32} = -8, \quad A_{33} = -1.$$

$$\therefore\ \text{adj } A = \begin{bmatrix} -6 & 14 & -15 \\ 17 & 5 & 9 \\ 13 & -8 & -1 \end{bmatrix}' = \begin{bmatrix} -6 & 17 & 13 \\ 14 & 5 & -8 \\ -15 & 9 & -1 \end{bmatrix}$$

$$\therefore\ A^{-1} = \frac{\text{adj } A}{|A|} = \frac{1}{67}\begin{bmatrix} -6 & 17 & 13 \\ 14 & 5 & -8 \\ -15 & 9 & -1 \end{bmatrix}.$$

The equations are

$$x + 2y - 3z = -4$$
$$2x + 3y + 2z = 2$$
$$3x - 3y - 4z = 11.$$

$$\Rightarrow\ AX = B, \quad \text{where } X = \begin{bmatrix} x \\ y \\ z \end{bmatrix},\ B = \begin{bmatrix} -4 \\ 2 \\ 11 \end{bmatrix}$$

$|A| = 67 \neq 0$.

$\therefore$ The system AX = B has unique solution : $X = A^{-1}B$.

$$\therefore\ \begin{bmatrix} x \\ y \\ z \end{bmatrix} = \frac{1}{67}\begin{bmatrix} -6 & 17 & 13 \\ 14 & 5 & -8 \\ -15 & 9 & -1 \end{bmatrix}\begin{bmatrix} -4 \\ 2 \\ 11 \end{bmatrix} = \frac{1}{67}\begin{bmatrix} 24 + 34 + 143 \\ -56 + 10 - 88 \\ 60 + 18 - 11 \end{bmatrix} = \frac{1}{67}\begin{bmatrix} 201 \\ -134 \\ 67 \end{bmatrix} = \begin{bmatrix} 3 \\ -2 \\ 1 \end{bmatrix}$$

$\therefore$ $x = 3,\ y = -2,\ z = 1.$

EXERCISE 2.13

1. Find whether the following systems of equations have unique solution or not :

(i) $x + 2y = 4$, $3x + 4y = 7$ (ii) $x + 2y = 4$, $2x + 4y = 8$ (iii) $5x + 9y = 3$, $x - y = 7$

2. Solve the following equations using matrix method :

(i) $5x + 2y = 3$, $3x + 2y = 5$ (ii) $3x + 4y = -1$, $2x + 5y = 4$ (iii) $3x + 4y = 5$, $x - y = -3$

(iv) $x + 5y = 3$, $2x + 10y = 6$ (v) $7x - y = 8$, $14x - 2y = 16$ (vi) $x - y = 1$, $3x - 3y = 3$.

3. Find whether the following systems have unique solution or not :

(i) $x + y + 3z = 0$, $x + 3y - 3z = 0$, $-2x - 4y - 4z = 0$ (ii) $4x + 7y - z = 0$, $7x - 8y + 11z = 0$, $8x + 14y - 2z = 0$ (iii) $x + 2y - 2z = 5$, $-x + 3y = 2$, $-2y + z = 7$

4. Solve the following equations using matrix method :

(i) $x + 3y = 1$, $3x - z = -8$, $2x - y + z = -3$ (ii) $x + 2y - 3z = 6$, $3x + 2y - 2z = 3$, $2x - y + z = 2$ (iii) $x - y = 3$, $2x + 3y + 4z = 17$, $y + 2z = 1$

(iv) $x + y + 2z = 4$, $2x - y + 3z = 9$, $3x - y - z = 2$ (v) $u - 2v + w = 1$, $2u + v + w = 1$, $u + v - 2w = -2$ (vi) $3x - y + 2z = 3$, $2x - 3y - z = -3$, $x + 2y + z = 4$.

(*P.T.U. B.Pharma, Dec. 2006*) (*P.T.U. B.Pharma, May 2006*)

5. Solve the following equations using matrix method :

(i) $5x + 3y + z = 16$, $2x + y + 3z = 19$, $x + 2y + 4z = 25$ (ii) $x - y + z = 4$, $2x + y - 3z = 0$, $x + y + z = 2$ (iii) $x - y + z = 2$, $2x - y = 0$, $2y - z = 1$

(*G.J.U. B.Pharma, 2004 ; P.T.U. Dec. 2005*)

(iv) $x + y + z = 0$, $2x - y + z = 0$, $x - 2y + 3z = 0$ (v) $3x - 2y + 3z = 0$, $2x + y - z = 0$, $4x - 3y + 2z = 0$ (vi) $6x - 12y + 25z = 0$, $4x + 15y - 20z = 0$, $2x + 18y + 15z = 0$.

6. Solve the following equations using matrix method :

(i) $x + y + 4z = 0$, $4x + y + 14z = 0$, $2x - y + 6z = 0$ (ii) $x - y - z = 0$, $2x + y + 3z = 0$, $x + 2y + 4z = 0$ (iii) $x - y - 3z = 7$, $4x + 3y + 7z = 11$, $6x + y + z = 25$

(iv) $3x + y - z = 1$, $x - y + 2z = -4$, $19x + 5y - 4z = 1$ (v) $x + y + z = 6$, $x + 2y + 3z = 14$, $x + 4y + 7z = 30$ (vi) $3x - y + 2z = 3$, $x - 2y - z = 1$, $2x + y + 3z = 5$.

7. Solve the following system of equations by matrix method, where $x \neq 0, y \neq 0, z \neq 0$:

$$\frac{2}{x} - \frac{3}{y} + \frac{3}{z} = 10, \frac{1}{x} + \frac{1}{y} + \frac{1}{z} = 10, \quad \frac{3}{x} - \frac{1}{y} + \frac{2}{z} = 13.$$

8. Find whether the following equations are consistent and if consistent, find the solution.

$$3x - y + 2z = 3$$
$$2x + y + 3z = 5$$
$$x - 2y - z = 1.$$

9. (i) Find A^{-1}, where $A = \begin{bmatrix} 2 & -1 & 3 \\ 1 & 3 & 2 \\ 3 & -4 & -1 \end{bmatrix}$. Hence solve the equations :

$$2x - y + 3z = 13, \quad x + 3y + 2z = 1, \quad 3x - 4y - z = 8.$$

(*ii*) Find A^{-1}, where $A = \begin{bmatrix} 1 & 1 & 1 \\ 1 & 2 & -3 \\ 2 & -1 & 3 \end{bmatrix}$. Hence solve the equations :

$$x + y + 2z = 0, \quad x + 2y - z = 9, \quad x - 3y + 3z = -14.$$

Answers

1. (*i*) Yes (*ii*) No (*iii*) Yes
2. (*i*) $x = -1, y = 4$ (*ii*) $x = -3, y = 2$ (*iii*) $x = -1, y = 2$
   (*iv*) $x = 3 - 5k, y = k$, where k is any number
   (*v*) $x = k, y = 7k - 8$, where k is any number
   (*vi*) $x = k, y = k - 1$, where k is any number.
3. (*i*) Yes (*ii*) No (*iii*) Yes
4. (*i*) $x = -2, y = 1, z = 2$ (*ii*) $x = 1, y = -5, z = -5$ (*iii*) $x = 6, y = 3, z = -1$
   (*iv*) $x = 1, y = -1, z = 2$ (*v*) $u = 0, v = 0, w = 1$ (*vi*) $x = 1, y = 2, z = -1$
5. (*i*) $x = 1, y = 2, z = 5$ (*ii*) $x = 2, y = -1, z = 1$
   (*iii*) $x = 1, y = 2, z = 3$ (*iv*) $x = 0, y = 0, z = 0$ (*v*) $x = 0, y = 0, z = 0$
   (*vi*) $x = 0, y = 0, z = 0$
6. (*i*) $x = -\frac{10}{3}k, y = -\frac{2}{3}k, z = k$, where k is any number.
   (*ii*) $x = -\frac{2}{3}k, y = -\frac{5}{3}k, z = k$, where k is any number.
   (*iii*) $x = \frac{2k + 32}{7}, y = \frac{-19k - 17}{7}, z = k$, where k is any number (*iv*) No solution
   (*v*) $x = k - 2, y = 8 - 2k, z = k$, where k is any number (*vi*) No solution.
7. $x = \frac{1}{2}, y = \frac{1}{3}, z = \frac{1}{5}$ **8.** Inconsistent
9. (*i*) $A^{-1} = \frac{-1}{36}\begin{bmatrix} 5 & -13 & -11 \\ 7 & -11 & -1 \\ -13 & 5 & 7 \end{bmatrix}$; $x = 1, y = -2, z = 3$
   (*ii*) $A^{-1} = \frac{1}{11}\begin{bmatrix} -3 & 4 & 5 \\ 9 & -1 & -4 \\ 5 & -3 & -1 \end{bmatrix}$; $x = 1, y = 3, z = -2$.

Hint

9. (*ii*) Use $A'X = B \Rightarrow X = (A')^{-1}B \Rightarrow X = (A^{-1})'B$.

3

Measures of Central Tendency

3.1. INTRODUCTION

In the present chapter, we shall discuss some important measures of central tendency and their merits and demerits. These measures play an important role in analysing the statistical data.

3.2. AVERAGES OR MEASURES OF CENTRAL TENDENCY OR MEASURE OF LOCATION

An average or a central value of a data or of a statistical series is the value of the variate which describes the characteristic of the entire data or associated frequency distribution. In other words, an average of a distribution is the value of the variable which is representative of the entire distribution. Since an average represents the entire data, therefore, its values lie somewhere in between two extremes *i.e.,* the largest and the smallest items. For this reason, an average is called a measure of location or a measure of central tendency.

There are five measures of central tendency :

(*a*) A. M. or mean (*b*) Median (*c*) Mode (*d*) Geometric mean
(*e*) Harmonic mean.

3.3. REQUISITES FOR AN IDEAL MEASURE

According to Prof. Yule, the following are the characteristic to be satisfied for an ideal measure for central tendency.

(*a*) It should be rigidly defined.

(*b*) It should be easily calculated.

(*c*) It should be based on all observations.

(*d*) It should be suitable for further mathematical treatment. It means, if we are given the averages and sizes of a number of series, we should be able to calculate the averages of the composite series obtained on combining the given series.

(*e*) It should be affected as little as possible by fluctuations of sampling. In addition to above, we may add the following which is not due to Prof. Yule.

(*f*) It should not be affected much by extreme values.

3.4. ARITHMETIC MEAN

In the case of Individual observations (when frequency is not given)

(*a*) Direct Method. If $x_1, x_2, \ldots\ldots, x_n$ are the given observations, then A.M. $\bar{x}$ is given by

$$\bar{x} = \frac{x_1 + x_2 + \ldots\ldots + x_n}{n} = \sum_{i=1}^{n} x_i$$

(*b*) Short Cut Method (Shift of origin). Shifting the origin to an *arbitrary point a,* the formula

$$\bar{x} = \frac{1}{n}\Sigma x \quad \text{becomes} \quad \bar{x} - a = \frac{1}{n}\Sigma(x - a)$$

$$\bar{x} = a + \frac{1}{n}\Sigma d_x, \text{ where } \quad d_x = x - a$$

Thus,
$$\bar{x} = a + \frac{1}{n}\Sigma d_x,$$

where a = arbitrary number, called *Assumed Mean*

$$\Sigma d_x = \Sigma(x - a) = (x_1 - a) + (x_2 - a) + \ldots\ldots + (x_n - a)$$

= sum of the deviations of the variate x from a

n = number of observations.

In the case of Discrete Series :

(a) Direct Method. If the frequency distribution is

$x : x_1, x_2, \ldots\ldots, x_n.$

$f : f_1, f_2, \ldots\ldots, f_n.$ then

$$\bar{x} = \frac{f_1x_1 + f_2x_2 + \ldots\ldots + f_nx_n}{f_1 + f_2 + \ldots\ldots + f_n} = \frac{\Sigma fx}{N}, \quad \text{where } N = f_1 + f_2 + \ldots\ldots + f_n = \Sigma f$$

(b) Short Cut Method (Shift of origin). Shifting the origin to an arbitrary point 'a', the formula

$$\bar{x} = \frac{1}{N}\Sigma fx \quad \text{becomes} \quad \bar{x} - a = \frac{1}{N}\Sigma f(x - a)$$

$$\bar{x} = a + \frac{1}{N}\Sigma fd_x, \quad \text{where} \quad d_x = x - a$$

Thus,
$$\bar{x} = a + \frac{1}{N}\Sigma fd_x, \quad \text{where} \quad a = \text{assumed mean}$$

$$\Sigma fd_x = \Sigma f(x - a)$$

$$= f_1(x_1 - a) + f_2(x_2 - a) + \ldots\ldots + f_n(x_n - a)$$

= sum of the products of f and the deviation of the corresponding variate x from a.

$$N = f_1 + f_2 + \ldots\ldots + f_n = \Sigma f.$$

Note. If the frequencies are given in terms of class intervals, the mid-values of class intervals are considered as x and then the above formulae are applied.

In the case of Continuous Series having equal class intervals, say of width h, we use a different formula (shift of origin and change of scale; Step Deviation Method).

Let
$$u = \frac{x - a}{h}, \text{ then } x = a + hu$$

$\therefore$
$$\Sigma fx = \Sigma f(a + hu) = a\,\Sigma f + h\Sigma fu$$

Dividing both sides by $N = \Sigma f$, we get

$$\frac{\Sigma fx}{N} = a + \frac{h\Sigma fu}{N} \quad \text{or} \quad \bar{x} = a + h\frac{\Sigma fu}{N}, \text{ where } u = \frac{x - a}{h}.$$

ILLUSTRATIVE EXAMPLES

Example 1. *Find the mean from the following data :*

Marks	*No. of students*	*Marks*	*No. of students*
Below 10	*5*	*Below 60*	*60*
Below 20	*9*	*Below 70*	*70*
Below 30	*17*	*Below 80*	*78*
Below 40	*29*	*Below 90*	*83*
Below 50	*45*	*Below 100*	*85*

Sol. The frequency distribution table can be written as :

Marks	*Mid-values (x)*	*f*	*x* – 55	$u = \frac{x-55}{10}$	*fu*
0—10	5	5	– 50	– 5	– 25
10—20	15	4	– 40	– 4	– 16
20—30	25	8	– 30	– 3	– 24
30—40	35	12	– 20	– 2	– 24
40—50	45	16	– 10	– 1	– 16
50—60	55	15	0	0	0
60—70	65	10	10	1	10
70—80	75	8	20	2	16
80—90	85	5	30	3	15
90—100	95	2	40	4	8
		N = Σ*f* = 85			Σ*fu* = – 56

$$\text{Mean } \bar{x} = a + h\frac{\Sigma fu}{N} = 55 + 10 \times \left(\frac{-56}{86}\right) \qquad [\text{Here } a = 55,\ h = 10]$$

$$= 55 - \frac{112}{17} = 48.41 \text{ marks.}$$

Example 2. *Find the mean wage from the following data :*

Wages	*800*	*820*	*860*	*900*	*920*	*980*	*1000*
No. of workers	*7*	*14*	*19*	*25*	*20*	*10*	*5*

Sol. Applying step deviation method, here h = common magnitude of the class interval is not known.

So, divide the range by the number of classes. The nearest integer to the value obtained gives the magnitude of the class interval.

Here range = 1000 – 800 = 200

No. of classes = 7

$$\therefore \qquad h = \frac{200}{7} = 28.57 \equiv 29$$

Also the deviations (d_i) should be divisible by h. So we take $h = 20$ (The nearest value to 29). Take A = 900.

We consider the following table :

Wages x_i	*Frequency* f_i	$d_i = x_i - A$	$u_i = \frac{x_i - A}{h}$	$f_i u_i$
800	7	– 100	– 5	– 35
820	14	– 80	–4	– 56
860	19	– 40	– 2	– 38
900	25	0	0	0
920	20	20	1	20
980	10	80	4	40
1000	5	100	5	25
	N = 100			Total = – 44

Required mean $$\bar{x} = A + h\frac{\sum_{i=1}^{n} f_i u_i}{N} = 900 + 20\left(-\frac{44}{100}\right)$$

$$= 900 - \frac{880}{100} = \mathbf{891.2.}$$

Example 3. *Find the missing frequency in the following frequency distribution if the mean of the distribution is 1.46.*

No. of accidents	*0*	*1*	*2*	*3*	*4*	*5*	*Total*
Frequency	*46*	–	–	*25*	*10*	*5*	*200*

Sol. Let f_1, f_2 be the required frequencies.

Since sum of all frequencies = 200 (given)

$\Rightarrow$ $46 + f_1 + f_2 + 25 + 10 + 5 = 200$

$\Rightarrow$ $f_1 + f_2 = 200 - 86 = 114$ ...(1)

Also, mean of the distribution = 1.46

$$\Rightarrow \frac{\sum_{i=1}^{n} f_i x_i}{\sum_{i=1}^{n} f_i} = 1.46$$

$$\Rightarrow \frac{0 \times 46 + 1 \times f_1 + 2 \times f_2 + 3 \times 25 + 4 \times 10 + 5 \times 5}{200} = 1.46$$

$\Rightarrow$ $f_1 + 2f_2 + 75 + 40 + 25 = 292$

$\Rightarrow$ $f_1 + 2f_2 = 292 - 140 = 152$ ...(2)

Solving (1) and (2), $f_1 = \mathbf{76}$, $f_2 = \mathbf{38}$.

Example 4. *The mean of 200 items was 50. Later on it was discovered that two items were misread as 92 and 8 instead of 192 and 88. Find out the correct mean.*

Sol. Given incorrected value of $\bar{x} = 50$, $n = 200$

Using $\bar{x} = \dfrac{\Sigma x}{n}$, we have

Incorrected value of $\Sigma x = \bar{x} \times n = 50 \times 200$
$= 10000$

Corrected value of Σx = Incorrected value of Σx – (Incorrected items) + (Corrected items)
$= 10000 - (92 + 8) + (192 + 88)$
$= 10000 - 100 + 280 = 10180$

$\therefore$ Corrected mean $= \dfrac{\text{Corrected } \Sigma x}{n} = \dfrac{10180}{200} = \mathbf{50.9.}$

3.5. PROPERTIES OF ARITHMETIC MEAN

Property I. *The algebraic sum of the deviations of all the variates from their arithmetic mean is zero.*

Proof. Let d_x be the deviation of the variate x from the mean $\bar{x}$, then $d_x = x - \bar{x}$

$$\therefore \quad \Sigma f d_x = \Sigma f (x - \bar{x}) = \Sigma fx - \bar{x}\,\Sigma f$$

$$= N\bar{x} - N\bar{x} = 0 \qquad \left|\ \because \bar{x} = \frac{\Sigma fx}{N}, \text{ where } N = \Sigma f.\right.$$

Property II. *The sum of the squares of the deviations of a set of values is minimum when taken about mean.*

Proof. Let the frequency distribution be x_i/f_i, $i = 1, 2, \ldots\ldots, n$. Let z be the sum of the squares of the deviations of the given values from an arbitrary point a (say).

$$\Rightarrow \text{ Let } \quad z = \sum_{i=1}^{n} f (x - a)^2.$$

We have to show that z is minimum when $a = \bar{x}$.

z will be minimum when $\dfrac{dz}{da} = 0$ and $\dfrac{d^2 z}{da^2} > 0$

$$\text{Now,} \quad \frac{dz}{da} = \sum_{i=1}^{n} 2f(x-a).(-1) = -2\sum_{i=1}^{n} f(x-a)$$

$$\therefore \quad \frac{dz}{da} = 0 \quad \Rightarrow \quad -2\Sigma f(x-a) = 0$$

$$\Rightarrow \quad \Sigma fx - a\Sigma f = 0$$

$$\Rightarrow \quad N\bar{x} - aN = 0 \qquad \left[\because \bar{x} = \frac{\Sigma fx}{N}, \Sigma f = N\right]$$

$$\Rightarrow \quad \bar{x} - a = 0 \qquad (\because \; N = \Sigma f \neq 0)$$

$$\Rightarrow \quad a = \bar{x}$$

$$\text{Also,} \quad \frac{d^2 z}{da^2} = -2\sum_{i=1}^{n} f(-1) = 2\Sigma f = 2N > 0$$

Hence z is minimum when $a = \bar{x}$.

Property III. (*Mean of the composite series*)

If $\bar{x}_i$ *(i = 1, 2, ..., k) be the arithmetic means of k distributions with respective frequencies* n_i *(i = 1, 2, ..., k), then mean* $\bar{x}$ *of the whole distribution obtained by combining the k distributions is given by*

$$\bar{x} = \frac{n_1\bar{x}_1 + n_2\bar{x}_2 + \ldots + n_k\bar{x}_k}{n_1 + n_2 + \ldots + n_k} = \frac{\sum_{i=1}^{k} n_i\bar{x}_i}{\sum_{i=1}^{k} n_i}$$

Proof. Let $x_{11}, x_{12}, x_{13}, \ldots\ldots, x_{1n_1}$ be the variables of the first distribution, $x_{21}, x_{22}, \ldots\ldots, x_{2n_2}$ be the variables of the second distribution, and so on. Then by definition

$$\left.\begin{aligned} \bar{x}_1 &= \frac{1}{n_1}(x_{11} + x_{12} + \ldots\ldots + x_{1n_1}) \\ \bar{x}_2 &= \frac{1}{n_2}(x_{21} + x_{22} + \ldots\ldots + x_{2n_2}) \\ &\ldots\ldots\ldots\ldots\ldots\ldots\ldots\ldots\ldots\ldots \\ \bar{x}_k &= \frac{1}{n_k}(x_{k1} + x_{k2} + \ldots\ldots + x_{kn_k}) \end{aligned}\right\} \quad \ldots(A)$$

The mean $\bar{x}$ of the whole distribution of size $(n_1 + n_2 + \ldots\ldots + n_k)$ is given by

$$\bar{x} = \frac{(x_{11} + x_{12} + \ldots\ldots + x_{1n_1}) + (x_{21} + x_{22} + \ldots\ldots + x_{2n_2}) + \ldots\ldots + (x_{k1} + x_{k2} + \ldots\ldots + x_{kn_k})}{n_1 + n_2 + \ldots\ldots + n_k}$$

$$= \frac{n_1\bar{x}_1 + n_2\bar{x}_2 + \ldots\ldots + n_k\bar{x}_k}{n_1 + n_2 + \ldots\ldots + n_k} = \frac{\sum_{i=1}^{k} n_i\bar{x}_i}{\sum_{i=1}^{k} n_i}.$$

Example 5. *The mean annual salary paid to all employees of a company was ₹ 5000. The mean annual salaries paid to male and female employees were ₹ 5200 and ₹ 4200 respectively. Determine the percentage of males and females employed by the company.*

Sol. Let p_1 and p_2 represent percentage of males and females respectively.

Then $p_1 + p_2 = 100$ ...(1)

Mean annual salary of all employees ($\bar{x}$) = ₹ 5000

Mean annual salary of all males ($\bar{x}_1$) = ₹ 5200

Mean annual salary of all females ($\bar{x}_2$) = ₹ 4200

Using $\bar{x} = \dfrac{p_1\bar{x}_1 + p_2\bar{x}_2}{p_1 + p_2}$, we get $5000 = \dfrac{5200p_1 + 4200p_2}{100}$

or $52p_1 + 42p_2 = 5000$ or $26p_1 + 21p_2 = 2500$ or $26p_1 + 21(100 - p_1) = 2500$ [Using (1)]

or $5p_1 = 2500 - 2100 = 400$ $\therefore$ $p_1 = 80$ and $p_2 = 100 - 80 = 20$

Hence the percentage of males and females is 80 and 20 respectively.

3.6. MERITS OF A.M

The arithmetic mean (A.M.) satisfies all the properties for an ideal measure.

3.7. DEMERITS OF A.M

(*i*) It cannot be determined graphically or by inspection.

(*ii*) It cannot be determined if a single observation is missing.

(*iii*) A.M. is affected very much by extreme values. In case of extreme values A.M. gives a disorted picture of the distribution and hence no longer remains the representative of the distribution.

(*iv*) In extremely asymmetrical distribution. A.M is not a suitable measure of location.

(*v*) A.M. cannot be used if we are dealing with qualitative characteristic such as intelligence, honesty, beauty, etc.

(*vi*) A.M. cannot be calculated if the extreme class is open *i.e.,* if below 10 or above 90 is given in a frequency distribution.

3.8. WEIGHTED MEAN

Let w_1, w_2, ... w_n, are the weights attached to the observations x_1, x_2 ...... x_n, then

weighted A.M. $= \frac{\Sigma w_i x_i}{\Sigma w_i}$.

Weighted mean gives the result equal to the simple mean if the weights assigned to each of the variate values are equal. It results in higher value than the simple mean if the smaller weights are given to smaller observations and larger weights to larger observations.

If the weights attached to larger observations are small and those attached to smaller observations are large, then the weighted mean results in smaller value than the simple mean.

Example 6. *Find the weighted mean of first n natural number, the weights being the corresponding number.*

Sol. Given

x_i	1	2	3	......	n
w_i	1	2	3	......	n

Required weighted mean $= \frac{\Sigma w_i x_i}{\Sigma w_i} = \frac{1^2 + 2^2 + 3^2 + ... + n^2}{1 + 2 + 3 + ... + n}$

$$= \frac{\Sigma n^2}{\Sigma n} = \frac{\frac{n(n+1)(2n+1)}{6}}{\frac{n(n+1)}{2}} = \frac{2n+1}{3}$$

3.9. PROPERTIES OF WEIGHTED MEAN

(*i*) Weighted mean is useful in Problems relating to

(*a*) Construction of index numbers

(*b*) Standardised birth and death rates

(*ii*) The simple A.M. is equal to the weighted mean if the weights are equal.

(*iii*) The simple A.M. is less than the weighted mean if greater weights are assigned to greater values and smaller weights are assigned to small values.

(*iv*) The simple A.M. is greater than the weighted mean if the smaller weights are attached to greater values and greater weights to smaller values.

Example 7. *A train runs 25 miles at a speed of 30 m/h. Another 50 miles at a speed of 40 m/h, then due to repair of tracks, for 6 minutes at a speed of 10 m/h and finally covers the* *remaining distance of 24 miles at a speed of 24 m/h. What is the average speed in m/h ?*

Sol. Take the time as weights, we consider the following table :

Speed (m/h) x_i	*time* w_i	$w_i x_i$
30	50	1500
40	75	3000
10	6	60
24	60	1440
	191	6000

(As the first train, covers 30 miles in 60 minute

$$\Rightarrow \quad 1 \text{ mile} = \frac{60}{30}$$

$$\Rightarrow \quad 25 \text{ mile} = \frac{60}{30} \times 25 = 50 \text{ minute})$$

$$\text{Average speed} = \frac{\Sigma w_i x_i}{\Sigma w_i} = \frac{6000}{191} = \mathbf{31.47\ m/h.}$$

MEDIAN

3.10. MEDIAN

(P.T.U. B.Pharma, Dec. 2006)

The **median** of a statistical series is defined as the size of the middle most item (or the A.M. of two middle most items), provided the items are in order of magnitude.

For example, the median for the series 4, 6, 10, 12, 18 is 10 and for the series 4, 6, 10, 12, 18, 22, the value of median is $\frac{10+12}{2} = 11$.

It can be observed that 50% items in the series would have value less than or equal to median and 50% items would be with values greater than or equal to the value of the median.

For an individual series, the median is given by :

$$\textbf{median} = \textbf{size of } \frac{\mathbf{n+1}}{\mathbf{2}} \textbf{th item,}$$

where $x_1, x_2, ..., x_n$ are the values of the variable under consideration. The values $x_1, x_2, ..., x_n$ are supposed to have been arranged in order of magnitude.

If $\frac{n+1}{2}$ comes out to be in decimal, then we take median as the A.M. of size of $\left(\frac{n}{2}\right)^{th}$ and $\left(\frac{n}{2}+1\right)^{th}$ items.

WORKING RULES FOR FINDING MEDIAN FOR AN INDIVIDUAL SERIES

Step I. *Arrange the given items in order of magnitude.*

Step II. *Find the total number 'n' of items.*

Step III. *Write : median = size of* $\frac{n+1}{2}$ *th item.*

Step IV. (*i*) *If* $\frac{n+1}{2}$ *is a whole number, then* $\frac{n+1}{2}$ *th item gives the value of median.*

(*ii*) *If* $\frac{n+1}{2}$ *is in fraction, then the A.M. of* $\frac{n}{2}$ *th and* $\left(\frac{n}{2}+1\right)$ *th items gives the value of median.*

For a frequency distribution, in which frequency (f) of different values (x) of the variable are given, we have

$$\textbf{median} = \textbf{size of } \frac{\mathbf{N+1}}{\mathbf{2}}\textbf{th item.}$$

The values of the variable are supposed to have been arranged in order of magnitude.

WORKING RULES FOR FINDING MEDIAN FOR A FREQUENCY DISTRIBUTION

Step I. *Arrange the values of the variable in order of magnitude and find the cumulative frequencies (c.f.).*

Step II. *Find the total 'N' of all frequencies and check that it is equal to the last c.f.*

Step III. *Write : median = size of* $\frac{N+1}{2}$ *th item.*

Step IV. (*a*) *If* $\frac{N+1}{2}$ *is a whole number, then* $\frac{N+1}{2}$ *th item gives the value of median. For this, look at the cumulative frequency column and find that total which is either equal to* $\frac{N+1}{2}$ *or the next higher than* $\frac{N+1}{2}$ *and determine the value of the variable corresponding to this. That gives the value of median.*

(*b*) *If* $\frac{N+1}{2}$ *is in fraction, then the A.M. of* $\frac{N}{2}$ *th and* $\left(\frac{N}{2}+1\right)$ *th items gives the value of median.*

In case, the values of the variable are given in the form of classes, we shall assume that items in the classes are uniformly distributed in the corresponding classes. We define :

$$\textbf{median} = \textbf{size of } \frac{\mathbf{N}}{\mathbf{2}}\textbf{th item.}$$

Here we shall get the class in which *N*/2th item is present. This is called the **median class.** To ascertain the value of median in the median class, the following formula is used.

$$\textbf{median} = \mathbf{L} + \left(\frac{\mathbf{N/2 - C}}{\mathbf{f}}\right)\mathbf{h},$$

where L = lower limit of the median class

C = cumulative frequency of the class preceding the median class

f = simple frequency of the median class

h = width of the median class.

The following points must be taken care of while calculating median :

1. The values of the variable must be in order of magnitude. In case of classes of values of the variable, the classes must be strictly in *ascending order* of magnitude.

2. If the classes are in *inclusive form,* then the actual limits of the *median class* are to be taken for finding L and *h*.

3. The classes may not be of equal width.

4. In case of open end classes, it is advisable to find average by using median.

WORKING RULES FOR FINDING MEDIAN FOR A FREQUENCY DISTRIBUTION WITH CLASS INTERVALS

Step I. *Arrange the classes in the ascending order of magnitude. The classes must be in 'exclusive form'. The widths of classes may not be equal. Find the cumulative frequencies (c.f.).*

Step II. *Find the total 'N' of all frequencies and check that it is equal to the last c.f.*

Step III. *Write : median = size of* $\frac{N}{2}$ *th item.*

Step IV. *Look at the cumulative frequency column and find that total which is either equal to* $\frac{N}{2}$ *or the next higher than* $\frac{N}{2}$ *and determine the class corresponding to this. That gives the 'median class'.*

Step V. *Write : median* $= L + \left(\frac{N/2 - C}{f}\right) h.$

Put the values of L, N/2, C, f, h and calculate the value of median.

ILLUSTRATIVE EXAMPLES

Example 1. *Find the median for the series : 4, 6, 9, 4, 2, 8, 10.*

Sol. The values of the variable arranged in ascending order are x : 2, 4, 4, 6, 8, 9, 10.

Here $n = 7$. $\therefore \quad \frac{n+1}{2} = \frac{7+1}{2} = 4$

$\therefore$ Median = size of 4th item = **6.**

Example 2. *Find the median for the series : 205, 170, 131, 174, 153, 142, 147, 157, 196, 148.*

Sol. The values of the variable arranged in ascending order are x : 131, 142, 147, 148, 153, 157, 170, 174, 196, 205.

Here $n = 10$. $\therefore \quad \frac{n+1}{2} = \frac{10+1}{2} = 5.5$

$\therefore$ Median = size of 5.5th item

$$= \frac{\text{size of 5th item} + \text{size of 6th item}}{2}$$

$$= \frac{153 + 157}{2} = \mathbf{155.}$$

Example 3. *The following are the marks obtained by a batch of 10 students in a certain class test in Statistics and Accountancy :*

Roll No.	*1*	*2*	*3*	*4*	*5*	*6*	*7*	*8*	*9*	*10*
Marks in Statistics	*63*	*64*	*62*	*32*	*30*	*60*	*47*	*46*	*35*	*28*
Marks in Accountancy	*68*	*65*	*35*	*42*	*26*	*85*	*44*	*80*	*33*	*72*

In which subject is the level of knowledge of students higher ?

Sol. In this problem, median would be the most suitable average.

Marks in Statistics arranged in ascending order are : 28, 30, 32, 35, 46, 47, 60, 62, 63, 64.

Here $n = 10$. $\therefore \quad \dfrac{n+1}{2} = \dfrac{10+1}{2} = 5.5$

$\therefore$ Median = size of 5.5th item

$$= \frac{\text{size of 5th item} + \text{size of 6th item}}{2} = \frac{46+47}{2} = \mathbf{46.5\ marks.}$$

Marks in Accountancy arranged in ascending order are :
26, 33, 35, 42, 44, 65, 68, 72, 80, 85.

Here, $n = 10$. $\therefore \quad \dfrac{n+1}{2} = \dfrac{10+1}{2} = 5.5$

$\therefore$ Median = size of 5.5th item

$$= \frac{\text{size of 5th item} + \text{size of 6th item}}{2} = \frac{44+65}{2} = \mathbf{54.5\ marks.}$$

$\therefore$ Level of knowledge of students is higher in Accountancy.

Example 4. *Find the median for the following frequency distribution :*

x	*0*	*1*	*2*	*3*	*4*	*5*	*6*
f	*5*	*9*	*10*	*12*	*6*	*4*	*2*

Sol. **Calculation of Median**

x	*f*	*c.f.*
0	5	5
1	9	14
2	10	24
3	12	36
4	6	42
5	4	46
6	2	48 = N
	N = 48	

$$\frac{N+1}{2} = \frac{48+1}{2} = 24.5$$

$\therefore$ Median = size of 24.5th item

$$= \frac{2+3}{2} = \mathbf{2.5.}$$

Example 5. *The number of students absenting from schools were recorded for 100 days as below :*

No. of students who absented	*5*	*6*	*7*	*8*	*9*	*10*	*11*	*12*	*13*
No. of days	*12*	*14*	*16*	*25*	*9*	*8*	*7*	*5*	*4*

Find the median number of students absenting per day.

Sol. **Calculation of Median**

No. of students who absented x	*No. of days* f	*c.f.*
5	12	12
6	14	26
7	16	42
8	25	67
9	9	76
10	8	84
11	7	91
12	5	96
13	4	100
	N = 100	

$$\frac{N+1}{2} = \frac{100+1}{2} = 50.5$$

∴ Median $= \text{size of } 50.5\text{th item}$

$$= \frac{8+8}{2} = \mathbf{8\ students.}$$

Example 6. *Find the median for the following wage distribution for a certain factory :*

Monthly wage (₹)	*50–80*	*80–100*	*100–110*	*110–120*	*120–130*	*130–150*	*150–180*	*180–200*
No. of workers	*30*	*127*	*140*	*240*	*176*	*135*	*20*	*3*

Sol. **Calculation of Median**

Monthly wage (₹)	*No. of workers* f	*c.f.*
50–80	30	30
80–100	127	157
100–110	140	C = 297
L = 110–120	f = 240	537
120–130	176	713
130–150	135	848
150–180	20	868
180–200	3	871 = N
	N = 871	

$$\frac{N}{2} = \frac{871}{2} = 435.5$$

∴ Median = size of 435.5th item

∴ Median class is 110–120.

$$\therefore \text{Median} = L + \left(\frac{N/2 - C}{f}\right)h$$

$$= 110 + \left(\frac{435.5 - 297}{240}\right)10$$

$$= 110 + 5.77 = ₹\ \mathbf{115.77.}$$

Example 7. *Find the median score of the scores on Reading Comprehension Test (Age 12 $^+$) of 1000 students which are given below :*

Score (out of 75)	*Frequency*	*Score (out of 75)*	*Frequency*
0–5	*6*	*25–30*	*250*
5–10	*12*	*30–35*	*185*
10–15	*50*	*35–40*	*110*
15–20	*120*	*40–45*	*32*
20–25	*225*	*45–50*	*10*

Sol. **Calculation of Median**

Score (out of 75)	*f*	*c.f.*
0–5	6	6
5–10	12	18
10–15	50	68
15–20	120	188
20–25	225	C = 413
L = 25–30	*f* = 250	663
30–35	185	848
35–40	110	958
40–45	32	990
45–50	10	1000 = N
	N = 1000	

$$\frac{N}{2} = \frac{1000}{2} = 500$$

∴ Median = size of 500th item

∴ Median class is 25–30.

$$\therefore \quad \text{Median} = L + \left(\frac{N/2 - C}{f}\right)h$$

$$= 25 + \left(\frac{500 - 413}{250}\right)5$$

$= 25 + 1.74 =$ **26.74 marks.**

Example 8. *Find the median for the following data :*

Weekly wage (in ₹)	*60–69*	*70–79*	*80–89*	*90–99*	*100–109*	*110–119*
No. of workers	*5*	*15*	*20*	*30*	*20*	*8*

Sol. Here the classes are given in the 'inclusive form'. To find median, we shall have to convert, these class in the 'exclusive form'.

$70 - 69 = 1$ and $\frac{1}{2}$ of '1' $= 0.5$.

Also, $60 - 0.5 = 59.5$ and $69 + 0.5 = 69.5$.

∴ Actual limits of the class 60–69 are 59.5 and 69.5.

∴ 60–69 in the exclusive form is 59.5–69.5. Similarly, we write all classes in the 'exclusive form'.

Calculation of Median

Weekly wage (in ₹)	*No. of workers f*	*c.f.*
59.5–69.5	5	5
69.5–79.5	15	20
79.5–89.5	20	C = 40
L = 89.5–99.5	*f* = 30	70
99.5–109.5	20	90
109.5–119.5	8	98 = N
	N = 98	

$$\frac{N}{2} = \frac{98}{2} = 49$$

∴ Median = size of 49th item

∴ Median class is 89.5–99.5.

$$\therefore \quad \text{Median} = L + \left(\frac{N/2 - C}{f}\right)h$$

$$= 89.5 + \left(\frac{49 - 40}{30}\right)10$$

$= 89.5 + 3 =$ **₹ 92.5.**

Example 9. *The following table gives the distribution of marks secured by some students in a certain examination :*

Marks	*0–20*	*21–30*	*31–40*	*41–50*	*51–60*	*61–70*	*71–80*
No. of students	*42*	*38*	*120*	*84*	*48*	*36*	*31*

Find the median marks.

Sol. Calculation of Median

Marks	*No. of students* *f*	*c.f.*
0–20	42	42
21–30	38	C = 80
31–40	*f* = 120	200
41–50	84	284
51–60	48	332
61–70	36	368
71–80	31	399 = N
	N = 399	

$$\frac{N}{2} = \frac{399}{2} = 199.5$$

∴ Median = size of 199.5th item

∴ Median class is 31–40.

Median class in the 'exclusive form' is 30.5–40.5.

$$\therefore \text{ Median} = L + \left(\frac{N/2 - C}{f}\right)h$$

$$= 30.5 + \left(\frac{199.5 - 80}{120}\right)10$$

$$= 30.5 + 9.96 = \textbf{40.46 Marks.}$$

Example 10. *Find the median for the following frequency distribution :*

Height (above) (in inches)	*60*	*62*	*64*	*66*	*68*	*70*	*72*	*74*
No. of students	*793*	*712*	*687*	*530*	*465*	*387*	*142*	*0*

Sol. Calculation of Median

Height (in inches)	*f*	*c.f.*
60–62	793 – 712 = 81	81
62–64	712 – 687 = 25	81 + 25 = 106
64–66	687 – 530 = 157	106 + 157 = 263
66–68	530 – 465 = 65	*c* = 263 + 65 = 328
68–70	*f* = 465 – 387 = 78	328 + 78 = 406
70–72	387 – 142 = 245	406 + 245 = 651
72–74	142 – 0 = 142 = N	651 + 142 = 793
	N = 793	

$$\frac{N}{2} = \frac{793}{2} = 396.5$$

∴ Median = size of 396.5^{th} item

∴ Median class is 68–70.

$$\therefore \text{ Median} = L + \left(\frac{N/2 - C}{f}\right)h$$

$$= 68 + \left(\frac{396.5 - 328}{78}\right)2$$

$$= 68 + 1.76 = \textbf{69.76 inches.}$$

Example 11. *You are given the following incomplete frequency distribution. It is known that the total frequency is 1000 and that the median is 413.11. Estimate the missing frequencies.*

Value	*Frequency*	*Value*	*Frequency*
300–325	*5*	*400–425*	*326*
325–350	*17*	*425–450*	*?*
350–375	*80*	*450–475*	*88*
375–400	*?*	*475–500*	*9*

Sol. Let the missing frequencies of the classes 375–400 and 425–450 be a and b respectively. Median is given to be 413.11.

Value	Frequency f	c.f.
300–325	5	5
325–350	17	22
350–375	80	102
375–400	a	$102 + a = C$
400–425	$326 = f$	$428 + a$
425–450	b	$428 + a + b$
450–475	88	$516 + a + b$
475–500	9	$525 + a + b = 1000$
	N = 1000	

∴ Median class is 400–425.

Now, median $= L + \left(\dfrac{N/2 - C}{f}\right)h$

Here, $L = 400$, $\dfrac{N}{2} = 500$, $f = 326$, $C = 102 + a$, $h = 25$.

$$\therefore \quad 413.11 = 400 + \left(\frac{500 - (102 + a)}{326}\right)25$$

$$\therefore \quad (13.11)326 = (500 - 102 - a)\ 25$$

or $4273.86 = (398 - a)\ 25$ or $398 - a = 170.95$

or $a = 227.05 = 227$ (nearly)

Also, $525 + a + b = 1000$ ∴ $b = 1000 - 525 - 227 = 248$

∴ The missing frequencies are **227** and **248.**

3.11. MATHEMATICAL PROPERTY OF MEDIAN

In a statistical data, the sum of numerical values of deviation of items from median is always least, *i.e.*, $\sum_{i=1}^{n} f_i\,|x_i - \text{Median}|$ is always least, where f_i is the frequency of x_i $(1 \le i \le n)$.

3.12. MERITS OF MEDIAN

1. It is rigidly defined.
2. It is well defined.
3. It is not affected by the presence of extreme items.
4. It is best suited in the presence of open end classes.
5. It can also be located graphically.

3.13. DEMERITS OF MEDIAN

1. It is not based on all the observations *e.g.*, the median of 10, 25, 50, 60 and 65, is 50. We can replace the observation 10 and 25 by any two values which are smaller than 50% without affecting the value of the median. This property is called – **median is insensitive.** Due to this property, median is also called a positional average.
2. As compared to mean, it is affected much by fluctuation of sampling.
3. It is not capable of further algebric treatment.
4. It can only be calculated when the data is in order of magnitude.

Uses. Median is the only average to be used while dealing with qualitative data, *i.e.,* to find average intelligence, or average honesty etc.

EXERCISE 3.1

1. Find the median for the series : 4, 6, 7, 11, 18.
2. (*i*) Find the median for the series : 3, 6, 6, 9, 10, 12.
   (*ii*) The marks obtained by 12 students are : 25, 20, 32, 23, 40, 27, 30, 25, 20, 10, 15, 41. Find the median.
3. (*i*) Mean of a finite number of observations is 5. If each observation is increased by 2. Find the new mean.
   (*ii*) A contractor employs three types of workers, male, female and children. To a male worker, he pays ₹ 40 per day, to a female worker ₹ 32 per day and to a child worker ₹ 21 per day. What is the average wage per day paid by the contractor ?
   (*iii*) Find the mean of 50 observations when it is given that the mean of 32 of them is 28 and the mean of the remaining 18 observations is 30.
   (*iv*) The mean of 35 observations was found to be 88.60. But later on it came to knowledge that 97 was misread as 79. Find the correct mean.
   (*v*) A factory employs 100 workers of whom 80 work in the first shift and 20 work in the second shift. The average wage of all the workers is ₹ 50. If the average wage of 60 workers in the first shift is ₹ 52, find the average wage of the remaining 40 workers of the second shift. (*P.T.U. Dec.2005*)
4. Find the median of the following observations : 46, 64, 87, 41, 58, 77, 35, 90, 55, 92, 33. If 92 is replaced by 99 and 41 by 43 in the above data, find the new median.
5. Find the value of median for the following data :

x	1	2	3	4	5	6	7	8	9
f	5	6	7	2	2	1	2	1	1

6. The number of students absent in a school was recorded every day for 147 days and the raw data was presented in the form of frequency distribution given below :

No. of students absent	5	6	7	8	9	10	11	12	13	15	18	20
No. of days	1	5	11	14	16	13	10	70	4	1	1	1

Find the median and describe what information it conveys ?

7. (*i*) In a survey of 304 families, the following frequency distribution of the number of children was obtained :

No. of children	0	1	2	3	4	5
No. of families	43	55	60	64	48	34

Find the median for this distribution.

   (*ii*) The marks obtained out of 50 in a class of 102 students in a test were as given in the frequency table. Find the median.

Marks	20	22	23	24	26	31	38	43
No. of students	8	15	28	27	20	2	1	1

8. Calculate the median for the following data :

Class	*Frequency*	*Class*	*Frequency*	*Class*	*Frequency*
10–20	6	40–50	2	70–80	9
20–30	8	50–60	4	80–90	4
30–40	10	60–70	7	90–100	6

9. Find the value of median for the following series :

Class	1–10	11–20	21–30	31–40	41–50
No. of items	10	21	51	45	26

10. Find the value of median for the following data :

Weight (in kg)	60–64	65–74	75–79	80–84	85–94	95–99	100–104
No. of items	10	39	35	20	26	9	6

11. Calculate median and arithmetic average for the following data :

Class interval	10–20	10–30	10–40	10–50	10–60	10–70	10–80	10–90
Frequency	4	6	56	97	124	137	146	150

12. Calculate median for the following data :

Mid-value	115	125	135	145	155	165	175	185	195
Frequency	6	25	48	72	116	60	38	22	3

13. Find the median marks :

Marks less than	10	20	30	40	50	60
No. of students	4	10	30	40	47	50

14. An incomplete frequency distribution is given below :

Variable	10–20	20–30	30–40	40–50	50–60	60–70	70–80
Frequency	12	30	?	65	?	25	18

It is given that mean value is 46 and the total number of items is 229. You are required to find the missing frequencies.

15. The median for the following data is 525.

Class interval	0–100	100–200	200–300	300–400	400–500	500–600	600–700	700–800	800–900	900–1000
Frequency	2	5	x	12	17	20	y	9	7	4

Find the values of x and y if the total frequency is 100.

Answers

1. 7 **2.** (*i*) 7.5 (*ii*) 25 **3.** (*i*) 7
(*ii*) ₹ 31 (*iii*) 28.72 (*iv*) 89.11 (*v*) ₹ 42
4. 58, 58 **5.** 3
6. 12 students ; For about half the number of days, 12 or less students were absent.
7. (*i*) 2 children (*ii*) 26 **8.** 55 **9.** 29.42
10. 77.86 kg **11.** 44.63, 47 **12.** 153.79 **13.** 27.5 marks
14. 32, 47 **15.** 9, 15.

MODE

3.14. DEFINITION

The **mode** of a statistical series is defined as that value of the variable around which the values of the variable tend to be most heavily concentrated. It can also be defined as that value of the variable whose own frequency is dominating and at the same time, the frequencies of its neighbouring items are also dominating. Thus, we see that mode is that value of the variable around which the items of the series cluster densely. Let us consider the data regarding the sale of readymade shirts.

Size (in inches)	30	32	34	36	38	40	42
No. of shirts sold	5	22	24	38	16	8	2

Here we see that the frequency of 36 is highest and the frequencies of its neighbouring items (34, 38) are also dominating. Here the most fashionable, modal size is 36 inches. Technically, we shall say that the mode of the distribution is 36 inches.

In case of mode, we are to deal with the frequencies of values of the items, thus if we are to find the value of mode for an individual series, we will have to see the repetition of different items, *i.e.*, we would be in a way expressing it in the form of frequency distribution. Thus, we start our discussion for evaluating mode for frequency distributions. There are two methods of finding mode of a frequency distribution.

3.15. MODE BY INSPECTION

Sometimes the frequencies in a frequency distribution are so distributed that we would be able to find the value of mode just, by inspection. For example, let us consider the frequency distribution :

x	4	5	6	7	8	9	10	11	12
f	1	2	1	5	12	4	2	2	1

Here we can say, at once, that mode is 8.

3.16. MODE BY GROUPING

Let us consider the distribution :

x	4	5	6	7	8	9	10	11	12
f	4	5	7	14	8	15	2	2	1

Here the frequency of 9 is more than the frequency of 7, whereas the frequencies of neighbouring items of 7 are more than that for 9. In such a case, we would not be able to judge the value of mode just by inspecting the data. In case there is even slight doubt as to which is the value of mode, we go for this method. In this method, two tables are drawn. These tables are called 'Grouping Table' and 'Analysis Table'. In the grouping table, six columns are drawn. The column of frequencies is taken as the column I. In the column II, the sum of two frequencies are taken at a time. In the column III, we exclude the first frequency and take the sum of two frequencies at a time. In the column IV, we take the sum of three frequencies at a time. In the column V, we exclude the first frequency and take the sum of frequencies, taking three at a time. In the last column, we exclude the first two frequencies and take the sum of three frequencies at a time. The next step is to mark the maximum sums in each of the six columns.

In the analysis table, six rows are drawn corresponding to each column in the grouping table. In this table, columns are made for those values of the variable whose frequencies accounts for giving maximum totals in the columns of the grouping table. In this table, marks are given to the values of the variable as often as their frequencies are added to make the total maximum in the columns of the grouping table. The value of the variable which get the maximum marks is declared to be the mode of the distribution.

In case, the values of the variable are given in the form of classes, we shall assume that the items in the classes are uniformly distributed in the corresponding classes. Here we shall get a 'class' either by the *method of inspection* or the *method of grouping*. This class is called the **modal class.** To ascertain the value of mode in the modal class, the following formula is used :

$$\textbf{mode} = \mathbf{L} + \left(\frac{\Delta_1}{\Delta_1 + \Delta_2}\right)\mathbf{h},$$

where L = lower limit of the modal class

Δ_1 = difference of frequencies of modal class and pre-modal class

Δ_2 = difference of frequencies of modal class and post-modal class

h = width of the modal class.

The following points must be taken care of while calculating mode :

1. The values (or classes of values) of the variable must be in ascending order of magnitude.

2. If the classes are in inclusive form, then the actual limits of the modal class are to be taken for finding L and h.

3. The classes must be of equal width.

It may be noted that while analysing the analysis table, we may find two or more values (or classes of values) of the variable getting equal marks. In such a case, the grouping method fails. Such a distribution is called a **multi-modal distribution.**

3.17. EMPIRICAL MODE

In case of a multi-modal distribution, we find the value of mode by using the relation :

$$\textbf{mode = 3 median – 2 A.M.}$$

This mode is called **empirical mode** in the sense that this relation can not be established algebraically. But it is generally observed that in distributions, the value of mode is approximately equal to 3 median – 2 A.M. That is why, this mode is called *empirical mode*.

WORKING RULES FOR FINDING MODE

Step I. *If mode is not evident by the 'method of inspection', then the 'method of grouping' should be used.*

Step II. *In case, the values of variable are given in terms of classes of equal width, then **step I**, will give the 'modal class'.*

Step III. *To find value of the mode, use the formula :* $\text{mode} = L + \left(\frac{\Delta_1}{\Delta_1 + \Delta_2}\right) h.$

Step IV. *In case, the distribution is multimodal, then find the value of mode by using the formula : 'Mode = 3 Median – 2 A.M'.*

ILLUSTRATIVE EXAMPLES

Example 1. *Find the mode for the following individual series :*

x	*5*	*7*	*3*	*5*	*2*	*1*	*5*	*8*	*5*

Sol. **Calculation of mode**

x	*Tally bars*	*f*
5	||||	4
7	|	1
3	|	1
2	|	1
1	|	1
8	|	1

x (*in ascending order*)	*f*
1	1
2	1
3	1
5	4
7	1
8	1

By inspection, mode = **5.**

Example 2. (*i*) *Find the mode for the following data :*

25, 16, 19, 48, 19, 20, 34, 15, 19, 20, 21, 24, 19, 16, 22, 16, 18, 20, 16, 19.

(*ii*) *If one of the 19's is changed to 16 in the above data, find the new mode.*

Sol. (*i*) **Calculation of mode**

x	*Tally bars*	*f*
25	|	1
16	||||	4
19	~~||||~~	5
48	|	1
20	|||	3
34	|	1
15	|	1
21	|	1
24	|	1
22	|	1
18	|	1

x (*in ascending order*)	*f*
15	1
16	4
18	1
19	5
20	3
21	1
22	1
24	1
25	1
34	1
48	1

By inspection, mode = **19.** This is so because the frequency of 19 is maximum and the frequencies of the neighbouring items are also dominating.

(*ii*) Here one of the 19's is changed to 16.

∴ The new frequency distribution is as follows :

x	15	16	18	19	20	21	22	24	25	34	48
f	1	4 + 1 = 5	1	5 – 1 = 4	3	1	1	1	1	1	1

Here the frequency of 16 is maximum but the frequencies of neighbouring items of 4 are more dominating. In this case, use apply the method of grouping to find the value of mode.

Grouping Table

x	*I* f	*II*	*III*	*IV*	*V*	*VI*
15	1	**6**		7		
16	**5**		6		10	
18	1	5				**8**
19	4		**7**	**8**		
20	3	4			5	
21	1		2			3
22	1	2		3		
24	1		2		3	
25	1	2				3
34	1		2			
48	1					

Analysis Table

Column	15	16	19	20	21	18
I	1					
II	1	1				
III			1	1		
IV			1	1	1	
V		1	1			1
VI			1	1		1
	2	2	4	3	1	2

∴ Mode = **19.**

Example 3. *Find the mode for the following distribution :*

Profit ('000 ₹)	*28*	*29*	*30*	*31*	*32*	*33*
No. of firms	*4*	*7*	*10*	*6*	*2*	*1*

Sol. **Calculation of Mode**

Profit ('000 ₹) x	*No. of firms* f
28	4
29	7
30	10
31	6
32	2
33	1

By inspection, mode is **₹ 30,000.** This is so because the frequency of 30,000 is very high as compared with the frequencies of other values of x. Moreover, the frequencies of the neighbouring items are also dominating.

Example 4. *Find the value of mode for the following distribution :*

x	4	5	6	7	8	9	10	11	12	13
f	2	5	8	9	12	14	14	15	11	13

Sol. For the given distribution, we cannot judge the value of mode, just by inspection. In this case, we shall apply the method of grouping. The values of the variable are already in order of magnitude.

Sol. Method of Grouping

Size (x)	Frequency					
	I	II	III	IV	V	VI
4	2	7		15		
5	5		13		22	
6	8	17				29
7	9		21	35		
8	12	26			**40**	
9	14		**28**			**43**
10	14	**29**		**40**		
11	15		26		39	
12	11	24				
13	13					

Explanation :

In column I, *original* frequencies are written.

In column II, frequencies of column I are combined *two by two.*

In column III, *leave the first frequency of column I* and combine the others *two by two.*

In column IV, frequencies of column I are combined *three by three.*

In column V, *leave the first frequency of column I* and combine the others *three by three.*

In column VI, *leave the first two frequencies in column I* and combine the others *three by three.*

In all these columns, the maximum frequency is written in bold black type.

Note. All operations are done on column I.

Now we frame another table in which against every maximum item of columns I to VI, we write down the corresponding size or sizes. The size (x) which occurs maximum number of times is the mode.

Columns	*Size of item having max. frequency*
I	11
II	10, 11
III	9, 10
IV	10, 11, 12
V	8, 9, 10
VI	9, 10, 11

Since the item 10 occurs maximum number of times (*i.e.*, 5 times), hence the mode is 10.

Example 5. *Calculate the mode from the following data :*

Class	*20–24*	*25–29*	*30–34*	*35–39*	*40–44*	*45–49*	*50–54*	*55–59*
F	*3*	*5*	*10*	*20*	*12*	*6*	*3*	*1*

(P.T.U. B. Pharma, Dec. 2005)

Sol. Here the greatest frequency 20 lies in the class 35–39. Hence modal class is 35–39. But the actual limits of this class are 34.5–39.5.

i.e., $l = 34.5 - 39.5,\ f_m = 20,$

$f_1 = 10,\ f_2 = 12,\ h = 4$

$$\therefore \quad \text{Mode} = l + \frac{f_m - f_1}{2f_m - f_1 - f_2} \times h$$

$$= 34.5 + \frac{20 - 10}{40 - 10 - 12} \times 4$$

$$= 34.5 + \frac{10}{18} \times 4 = 34.5 + \frac{20}{9} = \mathbf{36.7.}$$

Example 6. *Find the mode of the following frequency distribution :*

x_i :	*1*	*2*	*3*	*4*	*5*	*6*	*7*	*8*	*9*	*10*	*11*	*12*
f_i :	*3*	*8*	*15*	*23*	*35*	*40*	*32*	*28*	*20*	*45*	*14*	*6*

Sol. Here we see that the distribution is not regular. As the frequencies are increasing upto 40 and then decreasing. The frequency 45 does not seem to be consistent. So we can't say, maximum frequency is 45 and mode is 10. We apply method of grouping.

In column I, are the original frequencies column II is obtained by combining the frequencies two by two, *i.e.*, $3 + 8 = 11$, $15 + 23 = 38$, etc.

In column III, we leave the first frequency of column I, then combine the frequencies two by two, *i.e.*, $8 + 15 = 23$, $23 + 35 = 58$, etc.

In column IV, we combine the frequencies three by three, *i.e.*, $3 + 8 + 15 = 26$ etc.

In column V, we leave the first frequency of column I, then combine the frquencies three by three, *i.e.*, $8 + 15 + 23 = 46$, etc.

In column VI, we leave the first two frequencies, then combine the freq. three by three, *i.e.*, $15 + 23 + 35 = 73$, etc. and so on.

The maximum frequency in each column is marked 0.

x_i	f_i	*Grouping table*				
	I	*II*	*III*	*IV*	*V*	*VI*
1	3					
		11				
2	8					
			23			
3	15					
		38		26		
4	23					
			58		46	
5	35					
		75		**98**		73
6	40					
			72		**107**	
7	32					
		60		80		**100**
8	28					
			48		93	
9	20					
		65		65		79
10	45					
			59			
11	14					
		20				
12	6					

To find mode, consider the following Analysis table :

Analysis Table

Column no.	*Max. Freq.*	*Value of x giving max. freq.*
I	45	10
II	75	5, 6
III	72	6, 7
IV	98	4, 5, 6
V	107	5, 6, 7
VI	100	6, 7, 8

The value 6 is repeated. $\therefore$ mode = **6.**

Example 7. *Find the mode for the following frequency distribution :*

Class	*10–20*	*20–30*	*30–40*	*40–50*	*50–60*	*60–70*
f	*10*	*15*	*30*	*12*	*8*	*6*

Sol. By 'inspection', modal class is 30–40.

Now, $$\text{mode} = L + \left(\frac{\Delta_1}{\Delta_1 + \Delta_2}\right) h$$

Here L = 30, $\Delta_1 = 30 - 15 = 15$, $\Delta_2 = 30 - 12 = 18$, $h = 10$

$$\therefore \quad \text{Mode} = 30 + \left(\frac{15}{15 + 18}\right) 10 = 30 + 4.545 = \mathbf{34.545.}$$

Example 8. *Find the mode for the following frequency distribution :*

Class	*0–5*	*5–10*	*10–15*	*15–20*	*20–25*	*25–30*	*30–35*	*35–40*	*40–45*	*45–50*
f	*6*	*9*	*4*	*2*	*10*	*8*	*7*	*5*	*1*	*3*

Sol. We find the 'modal class' by using the 'method of grouping'.

Grouping Table

Class	*f* *I*	*II*	*III*	*IV*	*V*	*VI*
0–5	6	15		19		
5–10	9		13		15	
10–15	4	6				16
15–20	2		12	**20**		
20–25	**10**	**18**			**25**	
25–30	8		**15**			**20**
30–35	7	12		13		
35–40	5		6		9	
40–45	1	4				
45–50	3					

Analysis Table

Column	*20–25*	*25–30*	*30–35*	*15–20*	*35–40*
I	1				
II	1	1			
III		1	1		
IV	1	1		1	
V	1	1	1		
VI		1	1		1
Total	4	5	3	1	1

Since the total is maximum for the class 25–30, the modal class is 25–30.

Now $$\text{mode} = L + \left(\frac{\Delta_1}{\Delta_1 + \Delta_2}\right) h$$

Here L = 25, $\Delta_1 = 8 - 10 = -2$, $\Delta_2 = 8 - 7 = 1$, $h = 5$.

$\therefore$ $$\text{Mode} = 25 + \left(\frac{-2}{-2+1}\right)5 = 25 + 10 = \mathbf{+35.}$$

Example 9. *Calculate the value of mode for the following frequency distribution :*

Class	*Frequency*	*Class*	*Frequency*
1–4	*2*	*21–24*	*14*
5–8	*5*	*25–28*	*14*
9–12	*8*	*29–32*	*15*
13–16	*9*	*33–36*	*11*
17–20	*12*	*37–40*	*13*

Sol. In this problem, we shall make use of 'grouping method' to find the modal class.

Grouping Table

Class	*Frequency* *I*	*II*	*III*	*IV*	*V*	*VI*
1–4	2	7		15		
5–8	5		13		22	
9–12	8	17				29
13–16	9		21	35		
17–20	12	26			**40**	
21–24	14		**28**			**43**
25–28	14	**29**		**40**		
29–32	**15**		26		39	
33–36	11	24				
37–40	13					

Analysis Table

Column	*29–32*	*25–28*	*21–24*	*33–36*	*17–20*
I	1				
II	1	1			
III		1	1		
IV	1	1		1	
V		1	1		1
VI	1	1	1		
Total	4	5	3	1	1

$\therefore$ Modal class is 25–28.

Now $$\text{mode} = L + \left(\frac{\Delta_1}{\Delta_1 + \Delta_2}\right) h$$

Here $L = 24.5$, $\Delta_1 = 14 - 14 = 0$, $\Delta_2 = 14 - 15 = -1$, $h = 28.5 - 24.5 = 4$

$$\therefore \qquad \text{Mode} = 24.5 + \left(\frac{0}{0-1}\right)4 = 24.5 - 0 = \mathbf{24.5.}$$

Remark. The modal class 25–28 'expressed in exclusive form' is 24.5–28.5.

Example 10. *If the mode and mean of a moderately asymmetrical series are 16 ″ and 15.6 ″ respectively, what would be its most probable median ?*

Sol. We have mode = 16″ and mean = 15.6″.

The formulae is mode = 3 median – 2 A.M.

$$\therefore \qquad 16 = 3 \text{ median} - 2(15.6)$$

$$\Rightarrow \qquad 3 \text{ median} = 16 + 31.2 = 47.2$$

$$\therefore \qquad \text{median} = \frac{47.2}{3} = \mathbf{15.73''.}$$

3.18. MODE IN CASE OF CLASSES OF UNEQUAL WIDTHS

When the values of the variable are given in the form of classes and the classes are not of equal width, then we would not be able to proceed directly to find the modal class either by the method of inspection or by the method of grouping. In fact, we are to compare the frequencies of different classes in order to observe the concentration of items about some item. If the classes happen to be of unequal width, then we would not be able to compare the frequencies in different classes. To make the comparison meaningful, we will first make classes of equal width by grouping two or more classes or by breaking classes, as per the need.

Example 11. *Calculate mode for the following distribution :*

Marks	*No. of students*	*Marks*	*No. of students*
0–2	*8*	*25–30*	*45*
2–4	*12*	*30–40*	*60*
4–10	*20*	*40–50*	*20*
10–15	*10*	*50–60*	*13*
15–20	*16*	*60–80*	*15*
20–25	*25*	*80–100*	*4*

Sol. In this frequency distribution, the classes are not of equal width. Before ascertaining the modal class, we shall make the widths of classes equal.

Marks	*No. of students*
0–20	8 + 12 + 20 + 10 + 16 = 66
20–40	25 + 45 + 60 = 130
40–60	20 + 13 = 33
60–80	15
80–100	4

By inspection, modal class is 20–40.

$$\text{Now} \qquad \text{mode} = L + \left(\frac{\Delta_1}{\Delta_1 + \Delta_2}\right) h$$

Here $L = 20$, $\Delta_1 = 130 - 66 = 64$, $\Delta_2 = 130 - 33 = 97$, $h = 20$

$$\therefore \quad \text{Mode} = 20 + \left(\frac{64}{64+97}\right)20 = 20 + 7.9503 = \mathbf{27.9503.}$$

Example 12. *Calculate median and mode for the following data :*

Class	*2*	*3*	*4*	*5–7*	*7–10*	*10–15*	*15–20*	*20–25*
Frequency	*1*	*2*	*2*	*3*	*5*	*10*	*8*	*4*

Sol. We make classes as 0–5, 5–10, 10–15, 15–20 and 20–25.

Class	*Frequency* f	*c.f.*
0–5	1 + 2 + 2 = 5	5
5–10	3 + 5 = 8	13
10–15	10	23
15–20	8	31
20–25	4	35 = N
	N = 35	

Calculation of Median

$$\frac{N}{2} = \frac{35}{2} = 17.5$$

$\therefore$ Median = size of 17.5th item

$\therefore$ Median class is 10–15.

$$\therefore \quad \text{Median} = L + \left(\frac{N/2 - C}{f}\right)h = 10 + \left(\frac{17.5 - 13}{10}\right)5 = 10 + 2.25 = \mathbf{12.25.}$$

Calculation of Mode

By inspection, modal class is 10–15.

Now
$$\text{mode} = L + \left(\frac{\Delta_1}{\Delta_1 + \Delta_2}\right)h$$

Here $L = 10,\ \Delta_1 = 10 - 8 = 2,\ \Delta_2 = 10 - 8,\ h = 5.$

$$\therefore \quad \text{Mode} = 10 + \left(\frac{2}{2+2}\right)5 = 10 + 2.5 = \mathbf{12.5.}$$

3.19. LOCATING MODE GRAPHICALLY

In a frequency distribution, the value of mode can also be determined graphically.

Method :

Step I Draw a histogram of the given data.

Step II Draw two lines diagonally in the inside of the modal class bar, starting from each upper corner of the bar to the upper corner of the adjacent bar.

Step III Draw a perpendicular line from the intersection of the two diagonal lines to x-axis, which will give the modal value.

Example 13. *Find the mode of the following frequency distribution, graphically. Also verify the result by direct method.*

Class	*0–100*	*100–200*	*200–300*	*300–400*	*400–500*	*500–600*
f_i	*12*	*18*	*27*	*20*	*17*	*6*

Sol. Here maximum frequency = 27

∴ Modal class : 200–300. Draw the histogram of the given following distribution, which is shown in the figure.

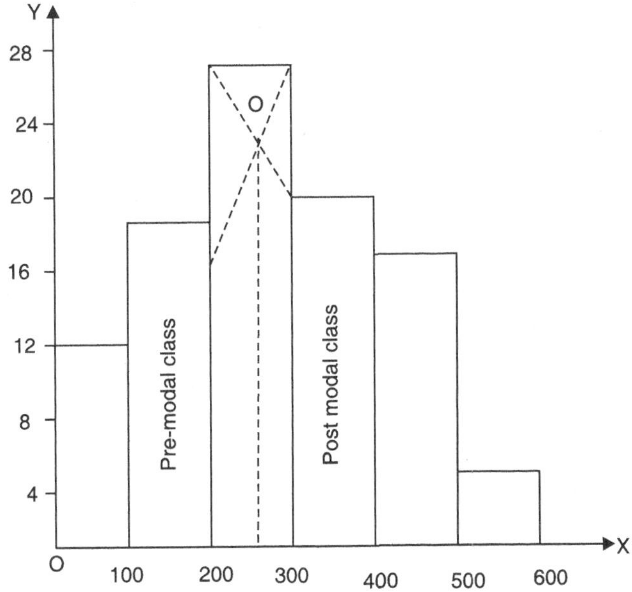

The perpendicular line drawn from 0, meets the x-axis at 256.

∴ mode = 256.

By direct method,

Here modal class is 200–300

$$\therefore \quad l = 200, \quad \Delta_1 = 27 - 18 = 9$$
$$\Delta_2 = 27 - 20 = 7,$$
$$h = 100$$

$$\therefore \quad \text{mode} = l + \left(\frac{\Delta_1}{\Delta_1 + \Delta_2}\right)h = 200 + \frac{9}{9+7} . 100$$

$$= 200 + \frac{9}{16} . 100 = 200 + \frac{225}{4} = 200 + 56.25$$

$$= 256.25 \cong \mathbf{256.}$$

Remarks :

(*i*) Mode can also be determined from a frequency polygon, in which a perpendicular is drawn on the base from the apex of the polygon and the point where it meets the base, gives the modal value.

(*ii*) Graphical method of determining mode can be used only where there is one class containing the highest frequency. If two or more classes have the same highest frequency, mode cannot be determined graphically.

Example 14. *10% of the workers in a firm employing a total of 1000 workers earn less than ₹ 5, 200 earn between ₹ 5 – 9.99, 30% between 10 and 14.99, 250 workers between 15 and 19.99 and the rest 20 and above. What is the median ?*

Sol. Arrange the given data in tabular form, we get

Earnings	*No. of workers*	*c.f.*
Below	$5 \; \frac{10}{100} \times 1000 = 100$	100
5–9.99	200	300
10–14.99	$\frac{30}{100} \times 1000 = 300$	600
15–19.99	250	850
20 above	150	1000

Here $\frac{N}{2} = \frac{1000}{2} = 500$, median class = 9.99 – 14.99

$$l = 9.99, \qquad h = \frac{10 - 9.99}{2} = 0.01$$

$$\therefore \quad \text{Required median} = \frac{l + \frac{N}{2} - C}{f} \times c = 9.99 + \frac{500 - 300}{300} \times 5$$

$$= 9.99 + 3.33 = \mathbf{13.32.}$$

Example 15. *The A.M. mode and median of a group of 75 observations were 27, 34, 29. It was later discovered that one observation was wrongly read as 43, instead of the correct value 53. Find the correct mean, mode and median.*

Sol. Let $x_1, x_2, \ldots\ldots x_n$, be the given observations then as given mean $\bar{x} = 27$

$$\Rightarrow \quad \frac{\sum_{i=1}^{n} x_i}{75} = 27 \Rightarrow \sum_{i=1}^{n} x_i = 75 \times 27 = 2025$$

$$\text{Corrected value of } \sum_{i=1}^{n} x_i = 2025 - \text{incorrected} + \text{corrected}$$

$$= 2025 - 43 + 53 = 2035$$

$$\therefore \quad \text{Corrected mean} = \frac{2035}{75} = 27.13$$

Further, the values of median and mode will not be affected by this error because the median and mode are positional averages. Also the median is 29, which is far away from 43, 53. Similarly the value of mode is 34.

Hence the corrected mode = **34.** Hence the corrected median = **29.**

3.20. MERITS OF MODE

1. It is easy to compute.
2. It is not affected by the extreme items.
3. It can be located graphically.

3.21. DEMERITS OF MODE

1. It is not simple to understand.
2. It is not well defined. There are number of formulae to calculate mode, not necessarily giving the same answer.
3. It is not capable of further algebraic treatment.

Uses. Mode is the only average to be used to find an ideal size, *i.e.,* in business forecasting, in the manufacturing if ready-made garments.

EXERCISE 3.2

1. Find the value of mode for the following series :

10, 12, 17, 12, 10, 12, 16, 11.

2. Find the mode for the following series :

3, 5, 6, 2, 5, 4, 5, 9, 5.

3. Find the value of mode for the following frequency distribution :

x	10	11	12	13	14	15	16	17	18
f	2	4	6	8	10	9	6	2	1

4. Calculate the mode for the following frequency distribution :

x	4	5	6	7	8	9	10	11	12	13
f	2	5	8	9	12	14	14	15	11	13

5. Find the median and mode for the following frequency distribution :

x	4	5	6	7	8	9	10
f	10	11	10	30	28	7	3

(P.T.U. B.Pharma, Dec. 2001)

6. The number of fully ripened apples on 100 plants were counted with the following results :

2	plants	had	0	apples
5	,,	,,	1	,,
7	,,	,,	2	,,
11	,,	,,	3	,,
18	,,	,,	4	,,
24	,,	,,	5	,,
12	,,	,,	6	,,
8	,,	,,	7	,,
6	,,	,,	8	,,
4	,,	,,	9	,,
3	,,	,,	10	,,

(i) How many apples are there ?

(ii) What is the average number of apples per plant ?

(iii) What is the modal number of apples ?

7. Find the mode for the following frequency distribution :

Class	0–4	4–8	8–12	12–16
f	4	8	5	6

8. Find the mode for the following frequency distribution :

Marks	0–5	5–10	10–15	15–20	20–25	25–30	30–35
No. of students	11	20	31	45	30	12	6

9. The following table gives the weight of 50 students of a class. Find the modal weight :

Weight (in kg)	37–41	42–46	47–51	52–56	57–61	62–66	67–71
No. of students	3	7	11	14	7	6	2

10. Calculate the modal value for the following frequency distribution :

Marks	*No. of candidates*	*Marks*	*No. of candidates*
0–9	6	50–59	263
10–19	29	60–69	133
20–29	87	70–79	43
30–39	181	80–89	9
40–49	247	90–99	2

11. Following table shows the marks obtained by 60 students. Calculate (i) median (ii) mode.

Marks	*No. of students*
More than 70%	8
,, 60%	18
,, 50%	40
,, 40%	40
,, 30%	55
,, 20%	60

12. Obtain the median and mode for the following series :

Marks	10–25	25–40	40–55	55–70	70–85	85–100
Frequency	6	20	44	26	3	1

13. Calculate median and arithmetic average for the following data. Also calculate mode with the help of median and A.M.

Variable	100–110	110–120	120–130	130–140	150–160	160–170	170–180
Frequency	4	6	20	32	33	8	2

14. Find out the median and mode for the following :

Wage (in ₹)	5–15	15–25	25–35	35–45	45–55	55–65
No. of employees	4	6	10	5	3	2

15. The following table shows the distribution of 100 families according to their expenditure per week. Number of families corresponding to the groups ₹ 1,000–2,000 and ₹ 3,000–4,000 are missing. The mode is given to be ₹ 2,400. Calculate the missing frequencies :

Expenditure (₹)	0–1,000	1,000–2,000	2,000–3,000	3,000–4,000	4,000–5,000
No. of families	14	?	27	?	15

16. Calculate mode for the following distribution :

Class	0–4	4–6	6–8	8–12	12–18	18–20
Frequency	4	6	8	12	7	2

17. Find the mode for the following data :

Size of items	*Frequency*	*Size of items*	*Frequency*
0–4	5	20–24	14
4–8	7	24–28	6
8–12	9	28–32	3
12–16	17	32–36	1
16–20	15	36–40	0

18. Calculate the median and mode for the following :

Class	*Frequency*	*Class*	*Frequency*
10–20	4	10–60	124
10–30	16	10–70	137
10–40	56	10–80	146
10–50	97	10–90	150

19. The ages of employees in two factories A and B are given below :

Age of employees (in years)	*Number of employees*	
	Factory A	*Factory B*
20–30	5	8
30–40	26	40
40–50	78	58
50–60	104	90
60–70	98	83

Find the modal age of employees in Factory A and Factory B.

Answers

1. 12	**2.** 5	**3.** 15.733	**4.** 10
5. median = 30, mode = 7		**6.** (*i*) 486	(*ii*) 4.86 (*iii*) 5
7. 6.286	**8.** 17.414	**9.** 53	**10.** 47.5488
11. (*i*) 54.54% marks		(*ii*) 56.67% marks	
12. 48.18 marks, 48.57 marks		**13.** 137.03, 140.14, 130.81	
14. ₹ 30, ₹ 29.44		**15.** 23, 21	**16.** 7
17. 18.67		**18.** 44.63, 40.67	**19.** 58.12 years, 58.2 years.

Hint

15. Let the frequencies of the classes 1000–2000 and 3000–4000 be a and b respectively.

$$\therefore \quad 2400 = 2000 + \left(\frac{27-9}{(27-a)+(27-b)}\right) \times 1000 \quad i.e., \quad 400 = \frac{(27-a)100}{54-a-b}$$

Also, $a + b = 44$.

4

Measurement of Angles

4.1. INTRODUCTION

Trigonometry is that branch of mathematics which deals with the measurement of sides and angles of triangles. The study of trigonometry is useful to engineers, scientists, surveyors, astronomers and others. This is also made use of in navigation and seismology.

The starting point of trigonometry is angle and its measurement.

4.2. ANGLES

The figure obtained by rotating a given ray about its end point is called an **angle.** The original position of the ray is called the **initial side** of the angle, whereas the final position of the ray is called the **terminal side** of the angle (Fig. 4.1).

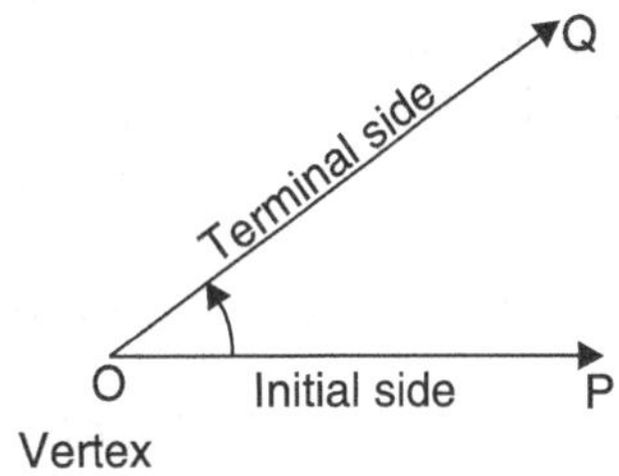

Fig. 4.1

An angle is called **positive** if the direction of rotation of ray from the initial side to the terminal side is *anti-clockwise* (Fig. 4.2).

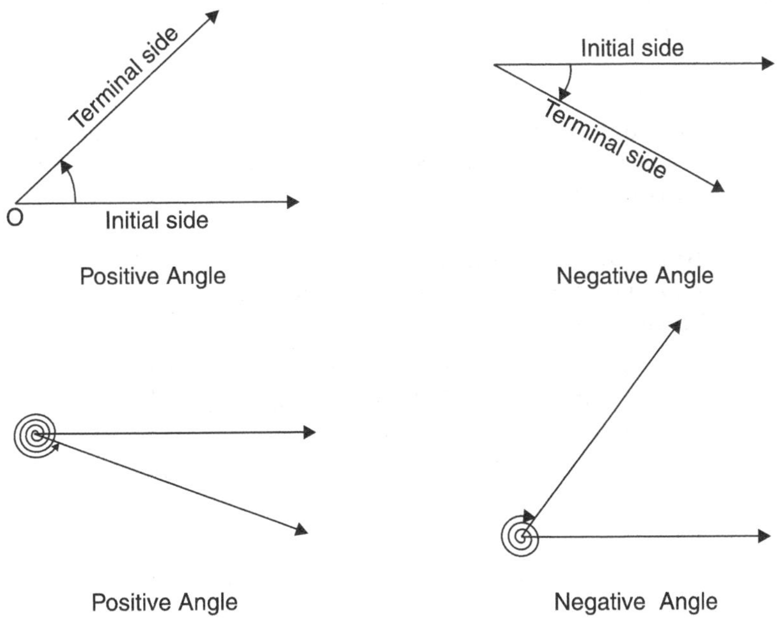

Fig. 4.2 **Fig. 4.3**

An angle is called **negative** if the direction of rotation of ray from the initial side to the terminal side is *clockwise* (Fig. 4.3).

There is neither lower limit nor upper limit for the magnitude of an angle.

4.3. QUADRANTS

Let $X'OX$ and $Y'OY$ be two perpendicular lines intersecting at O. $X'OX$ and $Y'OY$ are respectively called the x-axis and y-axis.

The lines $X'OX$ and $Y'OY$ divide the plane into four parts, **called quadrants** (Fig. 4.4).

(*i*) XOY is called **first quadrant.**

(*ii*) $X'OY$ is called **second quadrant.**

(*iii*) $X'OY'$ is called **third quadrant.**

(*iv*) XOY' is called **fourth quadrant.**

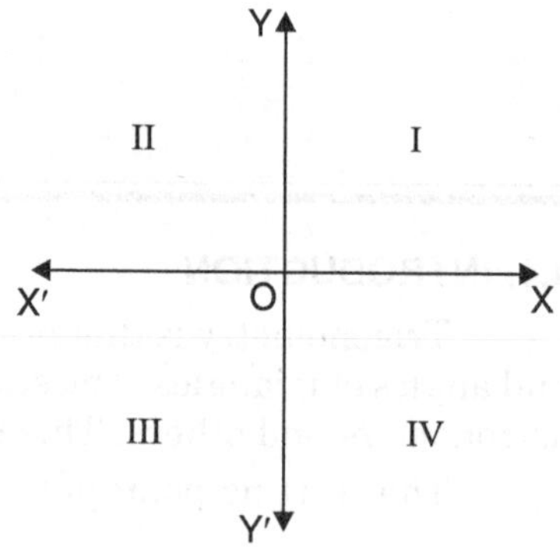

Fig. 4.4

If the vertex of an angle is at O and initial side along X-axis, then the angle is said to be in **standard position** and in that particular quadrant in which the terminal side of the angle lies. For example, the angles 40°, 220° lies in the I and III quadrants respectively.

If the terminal side of an angle in standard form lies along either axis, then the angle is called a **quadrantal angle.**

4.4. SYSTEMS OF MEASURING ANGLES

There are two systems of measuring angles. These are :

(*i*) Sexagesimal system (or English system)

(*ii*) Circular system.

4.5. SEXAGESIMAL SYSTEM

In this system, the unit of measuring an angle is a *degree.* If a right angle is divided in 90 equal parts, then each part is called a **degree.** One degree is divided in 60 equal parts, each part is called a **minute.** One minute is further divided in 60 equal parts, each part is called a **second.**

∴ **1 right angle = 90 degrees (written as 90°)**

1 degree = 60 minutes (written as 60′)

1 minute = 60 seconds (written as 60″)

ILLUSTRATIVE EXAMPLES

Example 1. *Find (i) number of minutes (ii) number of seconds in 4.5 degrees.*

Sol. (*i*) 4.5 degrees = 4.5 × 60 minutes (∵ 1° = 60′)

= **270 minutes.**

(*ii*) 4.5 degrees = 270 minutes = 270 × 60 seconds (∵ 1′ = 60″)

= **16,200 seconds.**

Example 2. *Find the quadrant in which the following angles lie :*

(*i*) *315°* (*ii*) *870°* (*iii*) *– 45°* (*iv*) *– 1520°.*

Sol. (*i*) Angle 315° is +ve, so the terminal side revolves in the anti-clockwise direction.

Since 315° = 270° + 45°, the terminal side lies in the IV quadrant (Fig. 4.5).

∴ 315° is in **IV quadrant.**

(*ii*) Angle 870° is + ve, so the terminal side revolves in the anti-clockwise direction. Since 870° = 2 × 360° + 150°, the terminal side makes two complete revolution and traces further 150°. The terminal side lies in the II quadrant (Fig. 4.5).

∴ 870° is in **II quadrant.**

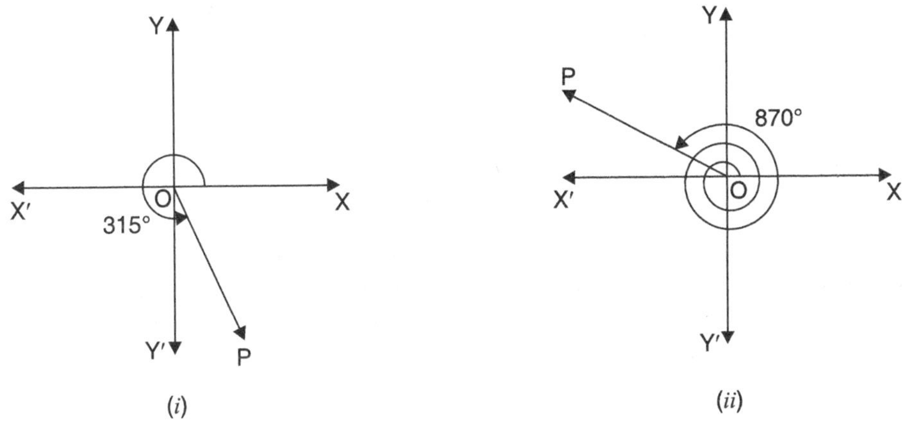

Fig. 4.5

(*iii*) Angle – 45° is –ve, so the terminal side revolves in the clockwise direction. The terminal side lies in IV quadrant (Fig. 4.6).

∴ – 45° is in **IV quadrant.**

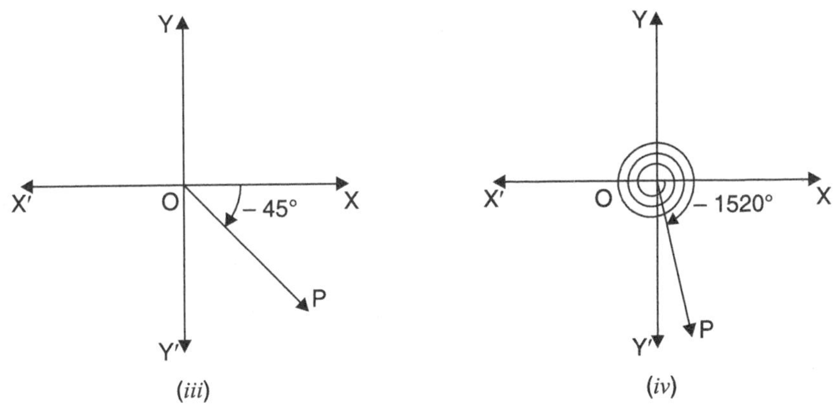

Fig. 4.6

(*iv*) Since – 1520° is – ve, the terminal side revolves in the clockwise direction.

Since – 1520° = – (4 × 360° + 80°) = 4(– 360°) + (– 80°), the terminal side makes four complete revolutions and traces further – 80°. The terminal side lies in IV quadrant.

∴ – 1520° is in **IV quadrant.**

4.6. CIRCULAR SYSTEM

Draw any circle. Let O be its centre and r its radius. Let arc AB be equal to the radius. Produce AO to meet the circle at C (Fig. 4.7).

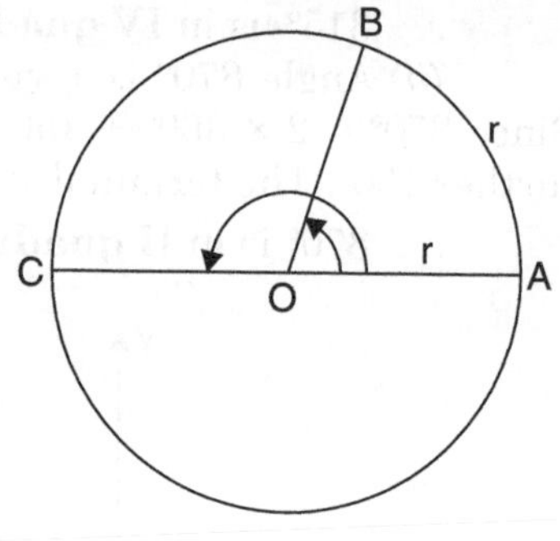

Fig. 4.7

Since angles at the centre of a circle are proportional to the arcs on which they stand, we have

$$\frac{\angle AOB}{\angle AOC} = \frac{\text{arc } AB}{\text{arc } AC}$$

$$\therefore \quad \frac{\angle AOB}{180^\circ} = \frac{r}{\frac{1}{2}(2\pi r)}$$

or $$\angle AOB = \left(\frac{180}{\pi}\right)^\circ, \text{ a constant angle.}$$

$\therefore$ $\angle AOB$ is independent of the radius of the circle. This angle is defined as one *radian*. Thus, one **radian** is the angle subtended at the centre of a circle by a positive arc equal in length to the radius of the circle. The arc is regarded positive if it is measured in anti-clock-wise direction. We have also seen that the measure of a radian is independent of the circle used.

In the circular system of measuring angle, the unit of measurement is a *radian, i.e.,* the circular measure of an angle is the number of radians it contains.

In the above circle, we have $\angle AOB = \left(\frac{180}{\pi}\right)^\circ$

$$\therefore \quad 1 \text{ radian} = \left(\frac{180}{\pi}\right)^\circ \qquad (\because \angle AOB = 1 \text{ radian})$$

$$\therefore \quad \boldsymbol{\pi} \textbf{ radians} = \mathbf{180^\circ}$$

Radian measures of some commonly used angles are given below :

Angle in Degrees	0°	30°	45°	60°	90°	120°	135°	150°
Angle in Radians	0	$\frac{\pi}{6}$	$\frac{\pi}{4}$	$\frac{\pi}{3}$	$\frac{\pi}{2}$	$\frac{2\pi}{3}$	$\frac{3\pi}{4}$	$\frac{5\pi}{6}$

180°	210°	225°	240°	270°	300°	315°	330°	360°
π	$\frac{7\pi}{6}$	$\frac{5\pi}{4}$	$\frac{4\pi}{3}$	$\frac{3\pi}{2}$	$\frac{5\pi}{3}$	$\frac{7\pi}{4}$	$\frac{11\pi}{6}$	2π

Remark 1. π radians $= 180^\circ \Rightarrow 1 \text{ radian} = \frac{180^\circ}{\pi} = \left(\frac{180}{22/7}\right)^\circ = \frac{630^\circ}{11} = 57^\circ + \frac{3^\circ}{11} = 57^\circ + \left(\frac{3}{11} \times 60\right)'$

$$= 57^\circ + 16' + \left(\frac{4}{11}\right)' = 57^\circ + 16' + \left(\frac{4}{11} \times 60\right)'' = 57^\circ + 16' + 22'' \text{ (approx.)}$$

$$\therefore \quad \textbf{1 radian} = \mathbf{57^\circ\ 16'\ 22''} \textbf{ (approx.)}$$

Remark 2. π radians = 180° $\Rightarrow$ $1° = \frac{\pi}{180}$ radians $= \frac{22}{7 \times 180}$ radians = 0.01746 radian (approx.)

$\therefore$ **1° = 0.01746 radian (approx.)**

Remark 3. When the angle is measured in the circular measure, then the word 'radians' with the angle is generally omitted. The angle θ radians is also written as θ^c. For example, the angle $\frac{\pi}{6}$ radians can also be written as $\left(\frac{\pi}{6}\right)^c$ or simply as $\frac{\pi}{6}$.

Example 3. *(i) Express the angles 60°, 75°, 115° in the circular system.*

(ii) Express the angles $\frac{4\pi}{9}$ radians, $\frac{7\pi}{12}$ radians, $\frac{1}{3}$ radians in the sexagesimal system.

Sol. (*i*) We have 180° = π radians.

$$\therefore \quad 60° = \left(\frac{\pi}{180} \times 60\right) \text{ radians} = \mathbf{\frac{\pi}{3} \text{ radians}}$$

$$75° = \left(\frac{\pi}{180} \times 75\right) \text{ radians} = \mathbf{\frac{5\pi}{12} \text{ radians}}$$

$$\text{and} \quad 115° = \left(\frac{\pi}{180} \times 115\right) \text{ radians} = \mathbf{\frac{23\pi}{36} \text{ radians}}$$

(*ii*) We have π radians = 180°.

$$\therefore \quad \frac{4\pi}{9} \text{ radians} = \left(\frac{180}{\pi} \times \frac{4\pi}{9}\right)^\circ = \mathbf{80°}$$

$$\frac{7\pi}{12} \text{ radians} = \left(\frac{180}{\pi} \times \frac{7\pi}{12}\right)^\circ = \mathbf{105°}$$

$$\text{and} \quad \frac{1}{3} \text{ radians} = \left(\frac{180}{\pi} \times \frac{1}{3}\right)^\circ = \left(\frac{60}{22} \times 7\right)^\circ = \left(\frac{210}{11}\right)^\circ = 19° + \left(\frac{1}{11}\right)^\circ$$

$$= 19° + \left(\frac{1}{11} \times 60\right)'$$

$$= 19° + 5' + \left(\frac{5}{11}\right)' = 19° + 5' + \left(\frac{5}{11} \times 60\right)''$$

$$= 19° + 5' + \left(27\frac{3}{11}\right)'' = \mathbf{19°\ 5'\ \left(27\frac{3}{11}\right)''}.$$

Example 4. *Express 104° 36′ in the circular system.*

Sol. $104°\ 36' = 104° + \left(\frac{36}{60}\right)^\circ = \left(\frac{180}{11}\right)^\circ$

Now 180° = π radians

$$\therefore \quad 104^\circ\, 36' = \left(\frac{\pi}{180}\cdot\frac{523}{5}\right) \text{ radians} = \mathbf{\frac{523}{900}\,\pi \text{ radians.}}$$

Example 5. *Express* $\frac{3\pi}{11}$ *radians in the sexagesimal system.*

Sol. We have π radians = 180°. $\therefore$ 1 radian = $\left(\frac{180}{\pi}\right)^\circ$.

$$\therefore \quad \frac{3\pi}{11} \text{ radians} = \left(\frac{3\pi}{11}\times\frac{180}{\pi}\right)^\circ = \left(\frac{540}{11}\right)^\circ = \left(49\frac{1}{11}\right)^\circ$$

$$= 49^\circ + \left(\frac{1}{11}\right)^\circ = 49^\circ + \left(\frac{1}{11}\times 60\right)'$$

$$= 49^\circ + \left(5\frac{5}{11}\right)' = 49^\circ + 5' + \left(\frac{5}{11}\right)'$$

$$= 49^\circ + 5' + \left(\frac{5}{11}\times 60\right)'' = 49^\circ + 5' + \left(27\frac{3}{11}\right)'' = \mathbf{49^\circ\, 5'\left(27\frac{3}{11}\right)''}.$$

Example 6. *The angles of a triangle are in the ratio 2 : 3 : 4. Express the angles in circular measure as well as in degrees.*

Sol. Let the angles of the triangle be $(2x)^\circ$, $(3x)^\circ$, $(4x)^\circ$.

$\therefore$ $(2x)^\circ + (3x)^\circ + (4x)^\circ = 180^\circ \Rightarrow 9x = 180 \Rightarrow x = 20.$

$\therefore$ The angles are **40°, 60°, 80°.**

Now 180° = π radians implies

$$40^\circ = \left(\frac{\pi}{180}.40\right)^c = \left(\frac{2\pi}{9}\right)^c, \quad 60^\circ = \left(\frac{\pi}{180}.60\right)^c = \left(\frac{\pi}{3}\right)^c$$

and
$$80^\circ = \left(\frac{\pi}{180}.80\right)^c = \left(\frac{4\pi}{9}\right)^c$$

$\therefore$ Angles in the sexagesimal system are : **60° , 60°, 80°.**

$\therefore$ Angles in the circular system are : $\mathbf{\left(\frac{2\pi}{9}\right)^c, \left(\frac{\pi}{3}\right)^c, \left(\frac{4\pi}{9}\right)^c}$.

Example 7. *Find in radians the angle of a regular octagon.*

Sol. No. of sides = 8

$\therefore$ No. of exterior angles = 8

Sum of all exterior angles = 360°

$\therefore$ Each exterior angle $= \left(\frac{360}{8}\right)^\circ = 45^\circ$

$\therefore$ Each interior angle $= 180^\circ - 45^\circ = 135^\circ = \left(135\times\frac{\pi}{180}\right)$ radians

$$= \mathbf{\frac{3\pi}{4} \text{ radians.}}$$

WORKING RULES FOR SOLVING PROBLEMS

Rule I. *1 right angle = 90°, 1° = 60′ and 1′ = 60″.*

Rule II. *180° = π radians.*

Rule III. (*i*) *Radian measure* $= \frac{\pi}{180} \times$ *Degree measure.*

(*ii*) *Degree measure* $= \frac{180}{\pi} \times$ *Radian measure.*

Rule IV. *The angle between two consecutive digits in a clock is 30 (= 360/12) degrees.*

Rule V. (*i*) *The hour-hand subtend an angle of 30 (= 360/12) degrees in one hour.*

(*ii*) *The minute-hand subtend an angle of 6 (= 360/60) degrees in one minute.*

Rule VI. *In a regular polygon :*

(*i*) *All the exterior angles are equal* (*ii*) *All the interior angles are equal*

(*iii*) *All the sides are equal* (*iv*) *Sum of all exterior angle = 360°*

(*v*) *Each exterior angle* $= \left(\frac{360}{\text{No. of sides}}\right)^\circ$

(*vi*) *Each interior angle = 180° – exterior angle.*

Rule VII. *To find the angle of a regular polygon means to find the interior angle of the regular polygon.*

EXERCISE 4.1

1. Find the quadrant in which the following angles lie :

   (*i*) 60° (*ii*) 240° (*iii*) 120° (*iv*) 330°.

2. Write the following angles in circular measure :

   (*i*) 75° (*ii*) 240° (*iii*) 530° (*iv*) – 37° 30′.

3. Write the following angles in sexagesimal measure :

   (*i*) $\left(\frac{\pi}{6}\right)^c$ (*ii*) $\left(\frac{3\pi}{5}\right)^c$ (*iii*) $\left(\frac{2\pi}{9}\right)^c$ (*iv*) $\left(\frac{\pi}{5}\right)^c$.

4. Sketch the following angles :

   (*i*) 650° (*ii*) – 565°.

   Also find the quadrants in which these angles lie.

5. Find the quadrants in which the following angles lie :

   (*i*) $\left(\frac{11\pi}{6}\right)^c$ (*ii*) $\left(\frac{17\pi}{3}\right)^c$ (*iii*) $\left(-\frac{16\pi}{3}\right)^c$ (*iv*) $\left(-\frac{5\pi}{6}\right)^c$.

   Also find the quadrants in which these angles lie.

6. Express the following angles in sexagesimal system :

   (*i*) 3/4 radian (*ii*) – 4 radians.

7. In a right angled triangle, the difference between the two acute angles is π/9 in circular measure. Express the acute angles in degrees.

8. Find the magnitude in radians and in degrees of the interior angle of a regular :
(*i*) pentagon (*ii*) polygon of 20 sides
(*iii*) Polygon of 40 sides.

9. The number of sides of two regular polygons are as 5 : 4 and the difference between their angles is 9° ; find the number of sides in the polygons.

Answers

1. (*i*) First (*ii*) Third (*iii*) Second (*iv*) Fourth
2. (*i*) $\frac{5\pi}{12}$ radians (*ii*) $\frac{4\pi}{3}$ radians (*iii*) $\frac{53\pi}{18}$ radians (*iv*) $-\frac{5\pi}{24}$ radians
3. (*i*) 30° (*ii*) 108° (*iii*) 40° (*iv*) 36°
4. (*i*) IV (*ii*) IV (*iii*) II (*iv*) III
5. (*i*) IV (*ii*) IV (*iii*) II (*iv*) III
6. (*i*) 42° 57′ 17″ (*ii*) – 229° 5′ 27″ 7. 55°, 35°
8. (*i*) 108°, $\left(\frac{3\pi}{5}\right)^c$ (*ii*) 162°, $\left(\frac{9\pi}{10}\right)^c$ (*iii*) 171°, $\frac{171}{180}\pi$ radians
9. 10, 8.

Hints

7. If the angles are $\alpha°$, 90°, $(90-\alpha)°$, then $\alpha° - (90-\alpha)° = \left(\frac{\pi}{9} \times \frac{180}{\pi}\right)^\circ$, provided $\alpha > 90 - \alpha$.

9. If the numbers of sides be $5k$ and $4k$, then we have

$$\left[180° - \left(\frac{360}{5k}\right)^\circ\right] - \left[180° - \left(\frac{360}{4k}\right)^\circ\right] = 9°.$$

4.7. ARC, RADIUS AND ANGLE RELATION

Theorem. *If in a circle of radius r, an arc of length l subtends an angle θ radians, then* $\theta = \frac{l}{r}$.

Proof. Let O be the centre of the circle of radius r.

Let $AB = l$ and $AC = r$, where units of l and r are same.

Since, angles at the centre of a circle are proportional to the arcs on which they stand, we have

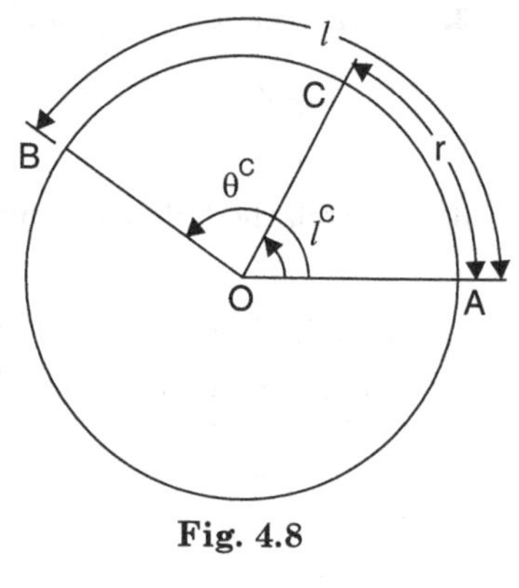

Fig. 4.8

$$\frac{\angle AOB}{\angle AOC} = \frac{\text{arc } AB}{\text{arc } AC}$$

$$\Rightarrow \quad \frac{\theta \text{ radian}}{1 \text{ radian}} = \frac{l}{r}$$

$$\Rightarrow \quad \theta = \frac{l}{r}$$

Remark. While using the formula $\theta = l/r$, the angle θ must be expressed in radians, l and r must be in the same unit of length.

ILLUSTRATIVE EXAMPLES

Example 1. *Find the length of an arc of a circle of radius 10 cm which subtends an angle of 45° at the centre.*

Sol. Let $\angle AOB = 45°$, $OA = 10$ cm, arc $AB = l$ cm.

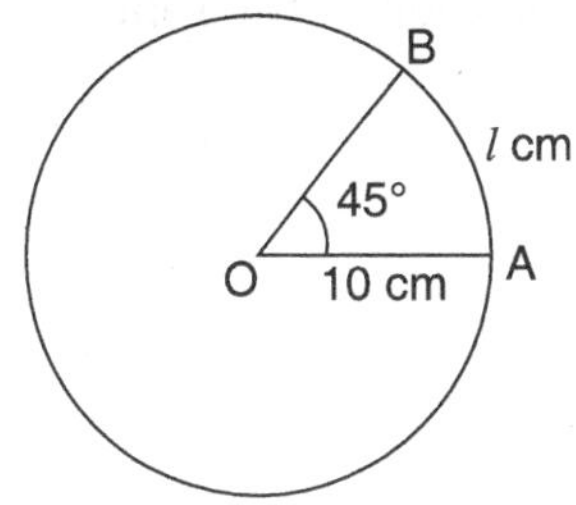

We have $45° = \frac{\pi}{180} \times 45 \text{ radian} = \frac{\pi}{4}$ radian

Using $\frac{l}{r} = \theta$, we have $\frac{l}{10} = \frac{\pi}{4}$

$\therefore \quad l = \frac{10\pi}{4} = \frac{5\pi}{2}$

$\therefore$ Length of arc $= \mathbf{\frac{5\pi}{2}}$ **cm.**

Example 2. *A railway train is travelling on a curve of 1500 mt. radius at the rate of 66 kph. Through what angle will it turn in 20 seconds ?*

Sol. Speed of train = 66 kph

$\therefore$ Distance travelled in 20 seconds

$$= \frac{66 \times 1000 \times 20}{60 \times 60} = \frac{1100}{3} \text{ mt.}$$

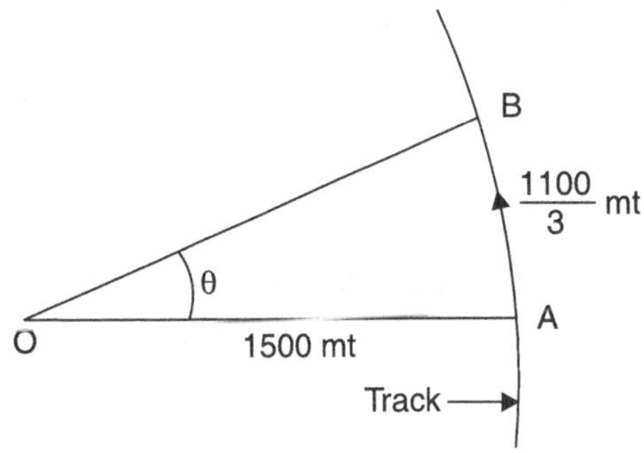

Here $l = AB = 1100/3$, $r = OA = 1500$.

Let angle $AOB = \theta$ radians

$$\therefore \quad \theta = \frac{l}{r} = \frac{1100/3}{1500} = \frac{11}{45}$$

$\therefore$ Angle $= \frac{11}{45}$ radian $= \left(\frac{11}{45} \times \frac{180}{\pi}\right)^\circ = \mathbf{14°.}$

Example 3. *In a circle of diameter 40 cm, the length of a chord is 20 cm. Find the length of the minor arc of the chord.*

Sol. Let O be the centre of the circle of diameter 40 cm.

Let BC be the chord of length 20 cm.

$\therefore$ ΔOBC is equilateral.

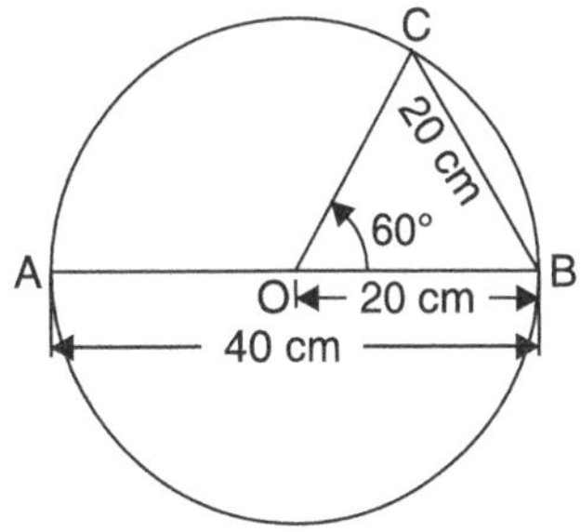

$\therefore \quad \angle BOC = 60° = \left(60 \times \frac{\pi}{180}\right)$ radians $= \frac{\pi}{3}$ radians.

$\therefore \quad \frac{l}{r} = \theta$ implies $\frac{\text{Arc } BC}{20} = \frac{\pi}{3}$.

$\therefore \quad \text{Arc } BC = \frac{20}{3} \times \frac{22}{7} = \mathbf{20\frac{20}{21}}$ **cm.**

Example 4. *A horse is tethered to a stake by a rope 30 m long. If the horse moves along the circumference of a circle always keeping the rope tight, find how far it will have gone when the rope has traced an angle of 105°.*

Sol. Let O denotes the stake and OA, the initial position of the rope. Let OB be the final position of the rope.

$$\therefore \quad \angle AOB = 105° = \left(105 \times \frac{\pi}{180}\right) \text{ radians} = \frac{7\pi}{12} \text{ radians.}$$

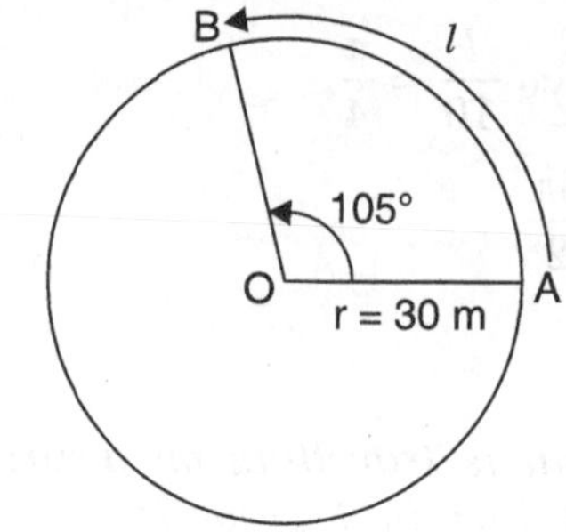

Let l metre be the distance covered by the horse.

$$\therefore \quad \frac{l}{r} = \theta \text{ implies } \frac{l}{30} = \frac{7\pi}{12}$$

$$\therefore \quad l = \frac{7}{12} \times \frac{22}{7} \times 30 \text{ m} = \mathbf{55 \ m.}$$

Example 5. *Find the distance from the eye at which a coin must be placed so as just to hide the full moon. The diameter of coin being 2 cm and the diameter of the moon makes an angle of 31′ at the eye of the observer.*

Sol. The coin will completely hide the moon if the lines OP and OQ touch the coin at A and B. Since angle $AOB = 31'$ is very small, the diameter AB of the coin can be considered as an arc of circle with centre at O and radius OA.

Let $\quad l = \text{arc } AB = 2 \text{ cm}, r = OA = OB.$

$$\angle AOB = 31' = \left(\frac{31}{60}\right)^{\circ} = \left(\frac{31}{60} \times \frac{\pi}{180}\right) \text{ radian} = \frac{31\pi}{10800} \text{ radian}$$

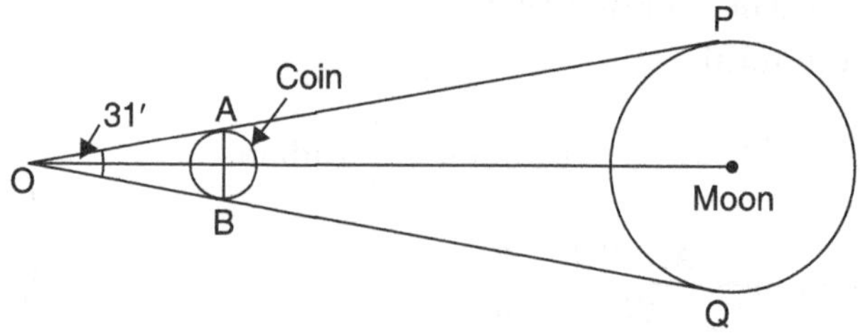

Let $\quad \theta = \frac{31\pi}{10800}. \quad \therefore \quad \frac{l}{r} = \theta \text{ implies } r = \frac{l}{\theta} = \frac{2}{\frac{31\pi}{10800}} = 221.79$

$\therefore$ Distance of the coin from the eye = **221.79 cm.**

EXERCISE 4.2

1. A wire 121 cm long is bent so as to lie along the arc of a circle of 180 cm radius. Find in degrees the angle subtended at the centre of the arc.
2. Find in degrees the angle subtended at the centre of a circle of diameter 200 cm by an arc of length 22 cm.
3. Find the angle in radian through which a pendulum swings if its length is 75 cm and the tip describes an arc of length :

   (*i*) 10 cm (*ii*) 15 cm (*iii*) 21 cm.
4. A railroad curve is to be laid out on a circle. What radius should be used if the track is to change direction by 25° in a distance of 40 metres ?
5. A circular wire 6 cm radius is cut and bent so as to lie along the circumference of a hoop whose radius is 96 cm. Find in radians the angle which is subtended at the centre of the hoop.
6. The radius of a circle is 5 cm. A chord of this circle is equal to the radius. Find the length of the minor arc of the chord.
7. A train is travelling at the rate of 10 km/hr on a circular curve of half a kilometre radius. Through what angle in degrees has it turned in a minute ?
8. Find the ratio of the radii of two circles at the centres of which two equal arcs subtend angles of 60 degrees and 75 degrees.
9. A wheel makes 360 revolutions in one minute. Through how many radians does it turn in one second ?
10. The minute hand of a clock is 1.5 cm long. How far does the tip of the hand move in 50 minutes ?

Answers

1. 38° 30′ **2.** 12° 36′ **3.** (*i*) $\frac{2}{15}$ radian (*ii*) $\frac{1}{5}$ radian

(*iii*) $\frac{7}{25}$ radian **4.** 91.64 mt. **5.** $\frac{\pi}{8}$ radian **6.** 5.24 cm

7. $\left(19\frac{1}{11}\right)^\circ$ **8.** 5 : 4 **9.** 12π **10.** 7.86 cm.

Hints

8. If the radii are r_1 and r_2, then $\frac{60\pi}{180} r_1 = \frac{75\pi}{180} r_2$.

9. Required angle $= \left(\frac{360}{60} \times 360\right)^\circ = \left(6 \times 360 \times \frac{\pi}{180}\right)$ radians.

5

Trigonometric Functions

5.1. INTRODUCTION

In the present chapter, we shall define trigonometric functions of an angle. We shall learn the methods of finding values of trigonometric functions for some specific angles.

5.2. DEFINITION OF TRIGONOMETRIC FUNCTIONS

Let θ be any real number. Construct the angle whose measure is θ radians, with vertex at the origin of a rectangular coordinate system and initial side along the positive x-axis. Let $P(x, y)$ be any point, distinct from the origin, on the terminal side OA of the angle. Let $OP = r$. Some of the possible positions of the terminal side of the angle θ radians are as given in the following figures (Fig. 5.1).

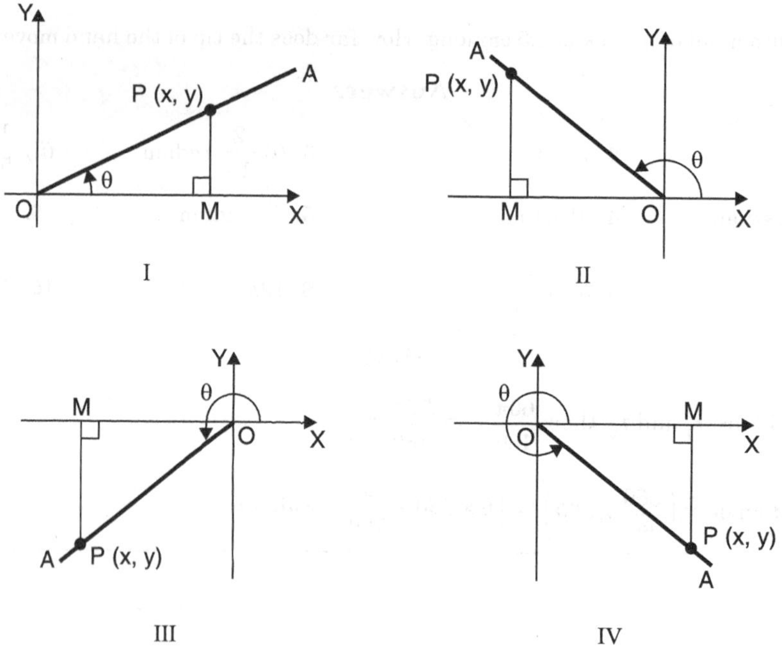

Fig. 5.1

(*i*) $\frac{y}{r}$ is called the **sine of** θ (written as **sin** θ).

(*ii*) $\frac{x}{r}$ is called the **cosine of** θ (written as **cos** θ).

(*iii*) $\frac{y}{x}$ is called the **tangent of** θ (written as **tan** θ), provided θ is not an odd multiple of $\frac{\pi}{2}$.

(*iv*) $\frac{x}{y}$ is called the **cotangent of** θ (written as **cot** θ), provided θ is not an even multiple of $\frac{\pi}{2}$.

(*v*) $\frac{r}{x}$ is called the **secant of** θ (written as **sec** θ), provided θ is not an odd multiple of $\frac{\pi}{2}$.

(*vi*) $\frac{r}{y}$ is called the **cosecant of** θ (written as **cosec** θ), provided θ is not an even multiple of $\frac{\pi}{2}$.

Remark 1. sin θ is read as the '*sine of angle* θ' and it should never be interpreted as the product of 'sin' and 'θ'.

Remark 2. (*i*) tan θ and sec θ are defined only when the terminal side of the angle θ is not coincident with the *y*-axis.

(*ii*) cot θ and cosec θ are defined only when the terminal side of the angle θ is not coincident with the *x*-axis.

Remark 3. The trigonometric functions are all functions of θ, because for each angle θ, there exist *exactly one* value of *t*-ratios, provided that particular *t*-ratio is defined for that angle θ.

Remark 4. If the terminal side of the angle θ (Fig. 5.1) is in the I quadrant, then

$$\sin\theta = \frac{y}{r} = \frac{MP}{OP}, \quad \cos\theta = \frac{x}{r} = \frac{OM}{OP}, \quad \tan\theta = \frac{y}{x} = \frac{MP}{OM},$$

$$\cot\theta = \frac{x}{y} = \frac{OM}{MP}, \quad \sec\theta = \frac{r}{x} = \frac{OP}{OM}, \quad \text{cosec}\,\theta = \frac{r}{y} = \frac{OP}{MP}.$$

Notation $(\cos\theta)^2$ is written as $\cos^2\theta$ (read "cos square θ")

$(\tan\theta)^3$ is written as $\tan^3\theta$ (read "tan cube θ")

$(\text{cosec}\,\theta)^n$ is written as $\text{cosec}^n\,\theta$ (read "cosec *n*th power θ"), *n* being a positive integer etc.

Caution. $(\cos\theta)^2$ should not be written as $\cos\theta^2$ or as $\cos^2\theta^2$.

5.3. TRIGONOMETRICAL IDENTITIES

A trigonometrical equation is called a **trigonometrical identity** if it is true for all angles for which the trigonometrical functions involved are defined.

We prove some fundamental trigonometrical identities.

Let θ be any real number. Construct the angle whose measure is θ radians, with vertex at the origin of a rectangular coordinate system and initial side along the positive *x*-axis. Let *P*(*x*, *y*) be any point, distinct from the origin, on the terminal side *OA* of the angle (Fig. 5.2). Let *OP* = *r*.

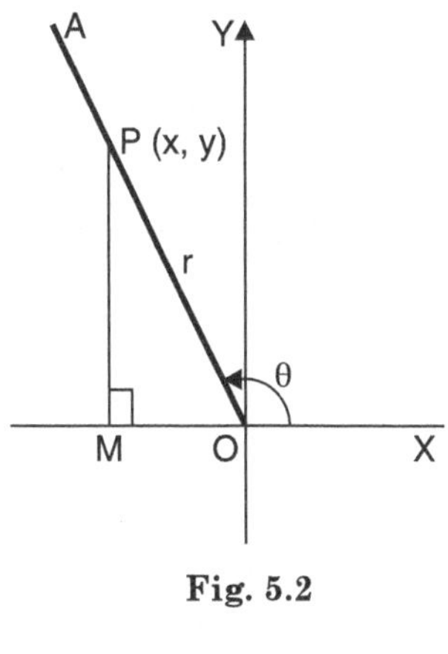

Fig. 5.2

(*i*) $\sin\theta\,\text{cosec}\,\theta = \frac{y}{r} \times \frac{r}{y} = 1$

(*ii*) $\cos\theta\sec\theta = \frac{x}{r} \times \frac{r}{x} = 1$

(*iii*) $\tan\theta\cot\theta = \frac{y}{x} \times \frac{x}{y} = 1$

(iv) $\dfrac{\sin\theta}{\cos\theta} = \dfrac{y/r}{x/r} = \dfrac{y}{x} = \tan\theta$

(v) $\dfrac{\cos\theta}{\sin\theta} = \dfrac{x/r}{y/r} = \dfrac{x}{y} = \cot\theta$

(vi) $\sin^2\theta + \cos^2\theta = (\sin\theta)^2 + (\cos\theta)^2 = \left(\dfrac{y}{r}\right)^2 + \left(\dfrac{x}{r}\right)^2 = \dfrac{y^2+x^2}{r^2} = \dfrac{r^2}{r^2} = 1$

(Using Pythagoras result)

(vii) $1 + \tan^2\theta = 1 + (\tan\theta)^2 = 1 + \left(\dfrac{y}{x}\right)^2 = \dfrac{x^2+y^2}{x^2} = \dfrac{r^2}{x^2} = \left(\dfrac{r}{x}\right)^2 = (\sec\theta)^2 = \sec^2\theta$

(viii) $1 + \cot^2\theta = 1 + (\cot\theta)^2 = 1 + \left(\dfrac{x}{y}\right)^2 = \dfrac{y^2+x^2}{y^2} = \dfrac{r^2}{y^2} = \left(\dfrac{r}{y}\right)^2 = (\text{cosec}\,\theta)^2 = \text{cosec}^2\,\theta.$

Thus, we have the following fundamental identities :

(i) $\mathbf{\sin\theta\ \text{cosec}\,\theta = 1}$ (ii) $\mathbf{\cos\theta\sec\theta = 1}$

(iii) $\mathbf{\tan\theta\cot\theta = 1}$ (iv) $\mathbf{\tan\theta = \dfrac{\sin\theta}{\cos\theta}}$

(v) $\mathbf{\cot\theta = \dfrac{\cos\theta}{\sin\theta}}$ (vi) $\mathbf{\sin^2\theta + \cos^2\theta = 1}$

(vii) $\mathbf{1 + \tan^2\theta = \sec^2\theta}$ (viii) $\mathbf{1 + \cot^2\theta = \text{cosec}^2\,\theta}$.

Remark 1. The values of tan θ, cot θ, sec θ and cosec θ in terms of sin θ and cos θ are as follows :

$$\tan\theta = \frac{\sin\theta}{\cos\theta},\quad \cot\theta = \frac{\cos\theta}{\sin\theta},\quad \sec\theta = \frac{1}{\cos\theta},\quad \text{cosec}\,\theta = \frac{1}{\sin\theta}.$$

Remark 2. (vi) implies $\sin^2\theta = 1 - \cos^2\theta$ and $\cos^2\theta = 1 - \sin^2\theta$.

(vii) implies $\tan^2\theta = \sec^2\theta - 1$ and $\sec^2\theta - \tan^2\theta = 1$.

(viii) implies $\cot^2\theta = \text{cosec}^2\,\theta - 1$ and $\text{cosec}^2\,\theta - \cot^2\theta = 1$.

WORKING RULES FOR SOLVING PROBLEMS

Rule I. *Trigonometrical identities are proved by using following methods :*

(i) *If one side of the identity is a constant or a very simple expression, then the other side is simplified and is shown equal to this side.*

(ii) *If each side of the identity is neither constant nor a very simple expression, then both sides are simplified separately and are equated.*

(iii) *If each side of the identity is neither constant nor a very simple expression, then the identity can also be proved by using the 'if' method.*

Rule II. *If no formula is directly applicable in solving an expression, then all t-ratios occurring in the expression are changed in terms of 'sine' and 'cosine' and the simplification is carried.*

ILLUSTRATIVE EXAMPLES

Example 1. *Show that :*

(i) $\dfrac{1-\cos\theta}{\sin\theta} = \dfrac{\sin\theta}{1+\cos\theta}$ (ii) $\tan\theta + \cot\theta = \sec\theta\ \text{cosec}\,\theta = 2\ \text{cosec}\ 2\theta.$

(P.T.U. B.Pharma Dec. 2006)

Sol. (*i*) L.H.S. $= \dfrac{1-\cos\theta}{\sin\theta} = \dfrac{1-\cos\theta}{\sin\theta} \times \dfrac{1+\cos\theta}{1+\cos\theta} = \dfrac{1-\cos^2\theta}{\sin\theta\,(1+\cos\theta)}$

$$= \frac{\sin^2\theta}{\sin\theta\,(1+\cos\theta)} = \frac{\sin\theta}{1+\cos\theta} = \text{R.H.S.}$$

(*ii*) L.H.S. $= \tan\theta + \cot\theta = \dfrac{\sin\theta}{\cos\theta} + \dfrac{\cos\theta}{\sin\theta} = \dfrac{\sin^2\theta+\cos^2\theta}{\cos\theta\sin\theta}$

$$= \frac{1}{\cos\theta\sin\theta} = \sec\theta\,\text{cosec}\,\theta = \frac{2}{2\sin\theta\cos\theta} = \frac{2}{\sin 2\theta} = 2\,\text{cosec}\,2\theta.$$

Example 2. *Prove that :* $\sin^2 A\cos^2 B + \cos^2 A\sin^2 B + \cos^2 A\cos^2 B + \sin^2 A\sin^2 B = 1.$

Sol. L.H.S. $= \sin^2 A\cos^2 B + \cos^2 A\sin^2 B + \cos^2 A\cos^2 B + \sin^2 A\sin^2 B$

$= (\sin^2 A\cos^2 B + \cos^2 A\cos^2 B) + (\cos^2 A\sin^2 B + \sin^2 A\sin^2 B)$

$= (\sin^2 A + \cos^2 A)\cos^2 B + (\cos^2 A + \sin^2 A)\sin^2 B = (1)\cos^2 B + (1)\sin^2 B$

$= \cos^2 B + \sin^2 B = 1 = \text{R.H.S.}$

Example 3. *Prove that :* $\dfrac{\cot^2\theta}{(1-\text{cosec}\,\theta)^2} = \dfrac{1+\sin\theta}{1-\sin\theta}.$

Sol. L.H.S. $= \dfrac{\cot^2\theta}{(1-\text{cosec}\,\theta)^2} = \dfrac{\text{cosec}^2\theta - 1}{(1-\text{cosec}\,\theta)^2} = \dfrac{(\text{cosec}\,\theta - 1)(\text{cosec}\,\theta + 1)}{(1-\text{cosec}\,\theta)^2}$

$$= -\frac{\text{cosec}\,\theta + 1}{1-\text{cosec}\,\theta} = -\frac{\dfrac{1}{\sin\theta}+1}{1-\dfrac{1}{\sin\theta}} = -\frac{1+\sin\theta}{\sin\theta - 1} = \frac{1+\sin\theta}{1-\sin\theta} = \text{R.H.S.}$$

Example 4. *Prove that :* $\sqrt{\dfrac{1+\cos A}{1-\cos A}} = \text{cosec}\,A + \cot A.$ (P.T.U. B.Pharma Dec. 2006)

Sol. L.H.S. $= \sqrt{\dfrac{1+\cos A}{1-\cos A}} = \sqrt{\dfrac{1+\cos A}{1-\cos A}\times\dfrac{1+\cos A}{1+\cos A}} = \sqrt{\dfrac{(1+\cos A)^2}{1-\cos^2 A}}$

$$= \frac{1+\cos A}{\sin A} = \frac{1}{\sin A} + \frac{\cos A}{\sin A} = \text{cosec}\,A + \cot A = \text{R.H.S.} \quad | \ \sin^2 A = 1-\cos^2 A$$

Example 5. *Prove that :* $\dfrac{\tan\theta + \sec\theta - 1}{\tan\theta - \sec\theta + 1} = \dfrac{1+\sin\theta}{\cos\theta}.$

Sol. L.H.S. $= \dfrac{\tan\theta + \sec\theta - 1}{\tan\theta - \sec\theta + 1} = \dfrac{\tan\theta + \sec\theta - (\sec^2\theta - \tan^2\theta)}{\tan\theta - \sec\theta + 1}$ **(Note this step)**

$$= \frac{\tan\theta + \sec\theta - (\sec\theta + \tan\theta)(\sec\theta - \tan\theta)}{\tan\theta - \sec\theta + 1} \quad | \; a^2 - b^2 = (a+b)(a-b)$$

$$= \frac{(\tan\theta + \sec\theta)\{1 - (\sec\theta - \tan\theta)\}}{\tan\theta - \sec\theta + 1} = \tan\theta + \sec\theta$$

$$= \frac{\sin\theta}{\cos\theta} + \frac{1}{\cos\theta} = \frac{\sin\theta + 1}{\cos\theta} = \text{R.H.S.}$$

Example 6. *Prove that : $2\sec^2\theta - 2\operatorname{cosec}^2\theta - \sec^4\theta + \operatorname{cosec}^4\theta = \cot^4\theta - \tan^4\theta$.*

Sol. L.H.S. $= 2\sec^2\theta - 2\operatorname{cosec}^2\theta - \sec^4\theta + \operatorname{cosec}^4\theta$

$= 2(1 + \tan^2\theta) - 2(1 + \cot^2\theta) - (1 + \tan^2\theta)^2 + (1 + \cot^2\theta)^2$

$$\left| \begin{array}{l} \sec^2\theta - \tan^2\theta = 1, \\ \operatorname{cosec}^2\theta - \cot^2\theta = 1 \end{array} \right.$$

$= 2 + 2\tan^2\theta - 2 - 2\cot^2\theta - 1 - \tan^4\theta - 2\tan^2\theta + 1 + \cot^4\theta + 2\cot^2\theta$

$= \cot^4\theta - \tan^4\theta = \text{R.H.S.}$

Example 7. *Prove that : $(\sec A - \operatorname{cosec} A)(1 + \tan A + \cot A) = \tan A \sec A - \cot A \operatorname{cosec} A$.*

Sol. L.H.S. $= (\sec A - \operatorname{cosec} A)(1 + \tan A + \cot A) = \left(\frac{1}{\cos A} - \frac{1}{\sin A}\right)\left(1 + \frac{\sin A}{\cos A} + \frac{\cos A}{\sin A}\right)$

$$= \left(\frac{\sin A - \cos A}{\cos A \sin A}\right)\left(\frac{\cos A \sin A + \sin^2 A + \cos^2 A}{\cos A \sin A}\right) = \frac{\sin^3 A - \cos^3 A}{\sin^2 A \cos^2 A}$$

$$| \; a^3 - b^3 = (a-b)(a^2 + ab + b^2)$$

R.H.S. $= \tan A \sec A - \cot A \operatorname{cosec} A$

$$= \frac{\sin A}{\cos A} \times \frac{1}{\cos A} - \frac{\cos A}{\sin A} \times \frac{1}{\sin A} = \frac{\sin^3 A - \cos^3 A}{\sin^2 A \cos^2 A}$$

$\therefore$ L.H.S. = R.H.S.

Example 8. *If $\cos\theta + \sin\theta = \sqrt{2}\cos\theta$, show that $\cos\theta - \sin\theta = \sqrt{2}\sin\theta$.*

Sol. We have $\cos\theta + \sin\theta = \sqrt{2}\cos\theta$,

$\therefore \quad \sin\theta = \sqrt{2}\cos\theta - \cos\theta = (\sqrt{2} - 1)\cos\theta$

$\Rightarrow \quad \cos\theta = \frac{1}{\sqrt{2} - 1}\sin\theta = \frac{1}{\sqrt{2} - 1} \times \frac{\sqrt{2} + 1}{\sqrt{2} + 1}\sin\theta$

$\Rightarrow \quad \cos\theta = \frac{\sqrt{2} + 1}{2 - 1}\sin\theta \Rightarrow \cos\theta = \sqrt{2}\sin\theta + \sin\theta$

$\Rightarrow \quad$ **$\cos\theta - \sin\theta = \sqrt{2}\sin\theta$.**

Example 9. *If $\sec^2 A = 2 + 2\tan A$, prove that $\tan A = 1 \pm \sqrt{2}$*

(P.T.U. B.Pharma Dec. 2005)

Sol. Given $\quad \sec^2 A = 2 + 2\tan A$

or $\quad 1 + \tan^2 A = 2 + 2\tan A$

$\Rightarrow \tan^2 A - 2\tan A - 1 = 0$

$\Rightarrow \quad t^2 - 2t - 1 = 0$, where $t = \tan A$

On solving, $t = \dfrac{2 \pm \sqrt{4+4}}{2} = \dfrac{2 \pm \sqrt{8}}{2} = \dfrac{2 \pm 2\sqrt{2}}{2} = 1 \pm \sqrt{2}$

$\Rightarrow$ $\quad \mathbf{\tan A = 1 \pm \sqrt{2}}$.

EXERCISE 5.1

Prove the following identities :

1. $\cos^4 A - \sin^4 A = 1 - 2\sin^2 A$
2. $\sec^4 \theta - \tan^4 \theta = 1 + 2\tan^2 \theta$
3. $\sec^4 \theta - \sec^2 \theta = \tan^4 \theta + \tan^2 \theta$
4. $\sin^4 \theta + \cos^4 \theta = 1 - 2\sin^2 \theta \cos^2 \theta$
5. $\sin^6 A + \cos^6 A = 1 - 3\sin^2 A \cos^2 A$
6. $\sec^6 A = \tan^6 A + 3\tan^2 A \sec^2 A + 1$
7. $\text{cosec}^6 A - \cot^6 A = 1 + 3\cot^2 A + 3\cot^4 A$
8. $\dfrac{1+\cos\alpha}{1-\cos\alpha} = \dfrac{\tan^2\alpha}{(\sec\alpha - 1)^2}$
9. (*i*) $\sqrt{\dfrac{1-\cos\theta}{1+\cos\theta}} = \text{cosec}\,\theta - \cot\theta$ (*ii*) $\sqrt{\dfrac{1+\sin\theta}{1-\sin\theta}} = \sec\theta + \tan\theta$

   (*iii*) $\sqrt{\dfrac{1-\sin\theta}{1+\sin\theta}} = \sec\theta - \tan\theta$
10. $\sin^8\theta - \cos^8\theta = (\sin^2\theta - \cos^2\theta)(1 - 2\sin^2\theta\cos^2\theta)$
11. $\dfrac{\sin\theta + 1 - \cos\theta}{\sin\theta - 1 + \cos\theta} = \dfrac{1+\sin\theta}{\cos\theta}$
12. (*i*) $\dfrac{\tan\theta - \cot\theta}{\sin\theta\cos\theta} = \sec^2\theta - \text{cosec}^2\theta$ (*ii*) $(\sin\theta + \cos\theta)(\tan\theta + \cot\theta) = \sec\theta + \text{cosec}\,\theta$
13. $\dfrac{\sin\theta}{\text{cosec}\,\theta + \cot\theta} - 2 = \dfrac{\sin\theta}{\cot\theta - \text{cosec}\,\theta}$
14. $\dfrac{\cot\theta + \text{cosec}\,\theta - 1}{\cot\theta - \text{cosec}\,\theta + 1} = \text{cosec}\,\theta + \cot\theta$
15. $\dfrac{\text{cosec}\,A}{\text{cosec}\,A - 1} + \dfrac{\text{cosec}\,A}{\text{cosec}\,A + 1} = 2\sec^2 A$
16. $(1 - \tan\theta)^2 + (1 - \cot\theta)^2 = (\sec\theta - \text{cosec}\,\theta)^2$
17. (*i*) $\sec\theta(1 - \sin\theta)(\sec\theta + \tan\theta) = 1$ (*ii*) $\sin\theta(1 + \tan\theta) + \cos\theta(1 + \cot\theta) = \sec\theta + \text{cosec}\,\theta$
18. (*i*) $\dfrac{\sin\theta}{1-\cos\theta} + \dfrac{\tan\theta}{1+\cos\theta} = \sec\theta\,\text{cosec}\,\theta + \cot\theta$

    (*ii*) $(\sec\theta - \cos\theta)(\text{cosec}\,\theta - \sin\theta) = \dfrac{1}{\tan\theta + \cot\theta}$
19. (*i*) $(\text{cosec}\,\theta + \cot\theta)^2 = \dfrac{1+\cos\theta}{1-\cos\theta} = \dfrac{\sec\theta + 1}{\sec\theta - 1}$ (*ii*) $\dfrac{\cos x}{1-\sin x} + \dfrac{1-\sin x}{\cos x} = 2\sec x.$

Hints

2. Put $\sec^2\theta = 1 + \tan^2\theta$ in the L.H.S.
5. Recall : $(a+b)^3 = a^3 + b^3 + 3ab(a+b)$.
10. $a^8 - b^8 = (a^4 - b^4)(a^4 + b^4) = (a^2 - b^2)(a^2 + b^2)((a^2 + b^2)^2 - 2a^2b^2)$.
11. Crossmultiply.
13. Given identity holds if $\dfrac{\sin\theta}{\text{cosec}\,\theta + \cot\theta} - \dfrac{\sin\theta}{\cot\theta - \text{cosec}\,\theta} = 2.$

5.4. ELIMINATION

In this section, we shall learn the method of eliminating 'θ' from two given equations involving trigonometrical functions of 'θ'. By using given equations involving 'θ' and trigonometrical identities, we shall obtain an equation not involving 'θ'.

Remark. If $ax + by + c = 0$

$a'x + b'y + c' = 0,$

then we have $$\frac{x}{bc' - cb'} = \frac{y}{ca' - ac'} = \frac{1}{ab' - ba'}$$

i.e., $$x = \frac{\mathbf{bc' - cb'}}{\mathbf{ab' - ba'}} \text{ and } y = \frac{\mathbf{ca' - ac'}}{\mathbf{ab' - ba'}}.$$

ILLUSTRATIVE EXAMPLES

Example 1. *Eliminate θ from the equations :*

(i) $x = h + a \cos \theta$, $y = k + b \sin \theta$ *(ii)* $x = a \sec^3 \theta$, $y = b \cot^3 \theta$.

Sol. *(i)* We have $x = h + a \cos \theta$ ...(1) $y = k + b \sin \theta$ ...(2)

(1) ⇒ $\cos \theta = \dfrac{x-h}{a}$ (2) ⇒ $\sin \theta = \dfrac{y-k}{b}$

Now $\sin^2 \theta + \cos^2 \theta = 1$

$$\therefore \quad \left(\frac{\mathbf{x - h}}{\mathbf{a}}\right)^2 + \left(\frac{\mathbf{y - k}}{\mathbf{b}}\right)^2 = \mathbf{1},$$ is the required result.

(ii) We have $x = a \sec^3 \theta$ ...(1) $y = b \cot^3 \theta$ ...(2)

$$(1) \Rightarrow \quad \sec^3 \theta = \frac{x}{a} \Rightarrow (\sec^3 \theta)^{2/3} = \left(\frac{x}{a}\right)^{2/3} \Rightarrow \sec^2 \theta = \left(\frac{x}{a}\right)^{2/3}$$

$$(2) \Rightarrow \quad \cot^3 \theta = \frac{y}{b} \Rightarrow \tan^3 \theta = \frac{b}{y} \Rightarrow \tan^2 \theta = \left(\frac{b}{y}\right)^{2/3}$$

Now $\sec^2 \theta - \tan^2 \theta = 1.$

$$\therefore \quad \left(\frac{\mathbf{x}}{\mathbf{a}}\right)^{\mathbf{2/3}} - \left(\frac{\mathbf{b}}{\mathbf{y}}\right)^{\mathbf{2/3}} = \mathbf{1},$$ is the required result.

Example 2. *Eliminate θ from the equations :* $a \sec \theta - c \tan \theta = b$, $b \sec \theta + d \tan \theta = c$.

Sol. We have $a \sec \theta - c \tan \theta - b = 0$

and $b \sec \theta + d \tan \theta - c = 0.$

Solving for sec θ and tan θ, we get $$\frac{\sec \theta}{c^2 + bd} = \frac{\tan \theta}{-b^2 + ac} = \frac{1}{ad + bc}$$

$$\therefore \quad \sec \theta = \frac{c^2 + bd}{ad + bc}, \quad \tan \theta = \frac{ac - b^2}{ad + bc}$$

We know that $\sec^2 \theta = 1 + \tan^2 \theta$ $\therefore$ $\left(\dfrac{c^2 + bd}{ad + bc}\right)^2 = 1 + \left(\dfrac{ac - b^2}{ad + bc}\right)^2$

$$\Rightarrow \quad \mathbf{(c^2 + bd)^2 = (ad + bc)^2 + (ac - b^2)^2},$$ is the required result.

Example 3. *If* $\frac{\sin A}{\sin B} = m$ *and* $\frac{\cos A}{\cos B} = n$, *show that*

$$\tan A = \pm \frac{m}{n}\sqrt{\frac{1-n^2}{m^2-1}} \quad \text{and} \quad \tan B = \pm\sqrt{\frac{1-n^2}{m^2-1}}.$$

Sol. We have $\frac{\sin A}{\sin B} = m$ ...(1) $\qquad \frac{\cos A}{\cos B} = n$ ...(2)

To find tan A, we eliminate B using (1) and (2).

(1) $\Rightarrow \quad \sin B = \frac{\sin A}{m}$ $\qquad$ (2) $\Rightarrow \quad \cos B = \frac{\cos A}{n}$

We know that $\sin^2 B + \cos^2 B = 1$.

$$\therefore \quad \frac{\sin^2 A}{m^2} + \frac{\cos^2 A}{n^2} = 1 \qquad ...(3)$$

Dividing both sides by $\cos^2 A$, we get $\frac{\tan^2 A}{m^2} + \frac{1}{n^2} = \sec^2 A$.

$$\Rightarrow \quad \frac{1}{m^2}\tan^2 A + \frac{1}{n^2} = 1 + \tan^2 A \Rightarrow \left(\frac{1}{m^2} - 1\right)\tan^2 A = 1 - \frac{1}{n^2}$$

$$\Rightarrow \quad \frac{1-m^2}{m^2}\tan^2 A = \frac{n^2-1}{n^2}$$

$$\Rightarrow \quad \tan^2 A = \frac{m^2}{n^2}\cdot\frac{n^2-1}{1-m^2} = \frac{m^2}{n^2}\cdot\frac{1-n^2}{m^2-1} \qquad \therefore \quad \mathbf{\tan A = \pm \frac{m}{n}\sqrt{\frac{1-n^2}{m^2-1}}}.$$

(1) and (2) implies $\frac{\sin A}{\cos A} = \frac{m \sin B}{n \cos B}$

$$\Rightarrow \quad \tan A = \frac{m}{n}\tan B \Rightarrow \pm\frac{m}{n}\sqrt{\frac{1-n^2}{m^2-1}} = \frac{m}{n}\tan B$$

$$\therefore \quad \mathbf{\tan B = \pm\sqrt{\frac{1-n^2}{m^2-1}}}.$$

Example 4. *If* $\tan A + \sin A = m$, $\tan A - \sin A = n$. *Then prove that* $m^2 - n^2 = \pm 4\sqrt{mn}$

(P.T.U. B. Pharma June 2003)

Sol. Given $\quad \tan A + \sin A = m$ ...(1)

$\qquad\qquad \tan A - \sin A = n$ ...(2)

Adding (1) and (2), we get

$$2 \tan A = m + n$$

or $$\tan A = \frac{m+n}{2}$$

$$\Rightarrow \quad \cot A = \frac{2}{m+n} \qquad ...(3)$$

Subtracting (1) and (2), we get

$$2 \sin A = m - n$$

$$\Rightarrow \quad \sin A = \frac{m-n}{2}$$

$$\Rightarrow \quad \text{cosec } A = \frac{2}{m-n} \qquad ...(4)$$

Using $\text{cosec}^2 A - \cot^2 A = 1$, we have

$$\left(\frac{2}{m-n}\right)^2 - \left(\frac{2}{m+n}\right)^2 = 1$$

$$\Rightarrow \quad \frac{4}{(m-n)^2} - \frac{4}{(m+n)^2} = 1$$

$$\Rightarrow \quad 4\left(\frac{(m+n)^2-(m-n)^2}{(m-n)^2\,(m+n)^2}\right) = 1$$

$$\Rightarrow \quad 4\left(\frac{m^2+n^2+2mn-m^2-n^2+2mn}{(m^2-n^2)^2}\right) = 1$$

$$\Rightarrow \quad 16mn = (m^2-n^2)^2$$

$$\Rightarrow \quad m^2-n^2 = \pm 4\sqrt{mn}.$$

EXERCISE 5.2

Eliminate θ from the following equations :

1. $x = \tan\theta,\ y = \sec\theta$
2. $x = a\cos\theta,\ y = b\sin\theta$
3. $x = a\cos^3\theta,\ y = b\sin^3\theta$
4. $x = a\,\text{cosec}^3\theta,\ y = b\cot^2\theta$

Eliminate θ from the following equations (Q. No. 5—8):

5. $\sin\theta + \cos\theta = a,\ \sin\theta - \cos\theta = b$
6. $x\cos\theta - y\sin\theta = a,\ x\sin\theta + y\cos\theta = b$
7. $l\tan\theta + m\sec\theta = n,\ l'\tan\theta + m'\sec\theta = n'$
8. $a\cot\theta + b\,\text{cosec}\,\theta = x^2,\ b\cot\theta + a\,\text{cosec}\,\theta = y^2$
9. If $x = r\sin\theta\cos\phi,\ y = r\sin\theta\sin\phi,\ z = r\cos\theta$, show that $x^2+y^2+z^2 = r^2$.
10. If $m = \sin\theta + \cos\theta,\ n = \sec\theta + \text{cosec}\,\theta$, show that $n(m+1)(m-1) = 2m$.

Answers

1. $y^2 - x^2 = 1$
2. $\left(\frac{x}{a}\right)^2 + \left(\frac{y}{b}\right)^2 = 1$
3. $\left(\frac{x}{a}\right)^{2/3} + \left(\frac{y}{b}\right)^{2/3} = 1$
4. $\left(\frac{x}{a}\right)^{2/3} - \frac{y}{b} = 1$
5. $a^2 + b^2 = 2$
6. $a^2 + b^2 = x^2 + y^2$
7. $(ln' - l'n)^2 - (m'n - mn')^2 = (lm' - l'm)^2$
8. $x^4 - y^4 = b^2 - a^2$.

Hint

6. Square and add.

5.5. SIGNS OF TRIGONOMETRIC FUNCTIONS

Let θ be any real number. Construct the angle whose measure is θ radians with vertex at the origin of a rectangular coordinate system and initial side along the positive x-axis. Let $P(x, y)$ be any point, distinct from the origin, on the terminal side of the angle. Let $OP = r$. Some of the possible positions of the terminal side of the angle θ radians are as given in the figure. 'r' is always taken positive (Fig. 5.3).

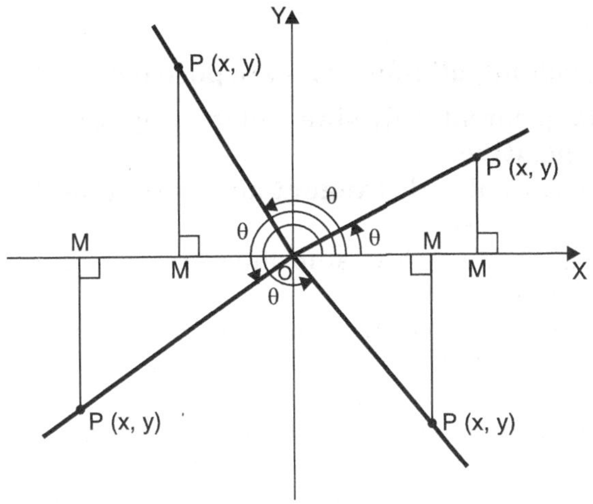

Fig. 5.3

First quadrant. Here x, y and r are all positive. The ratio of any two of these is positive.

$\therefore$ **All t-functions are positive.**

Second quadrant. Here x is negative, y and r are positive.

$$\sin\theta = \frac{y}{r} = \frac{+\text{ve}}{+\text{ve}} = +\text{ve},\ \cos\theta = \frac{x}{r} = \frac{-\text{ve}}{+\text{ve}} = -\text{ve},\ \tan\theta = \frac{y}{x} = \frac{+\text{ve}}{-\text{ve}} = -\text{ve},$$

$$\cot\theta = \frac{x}{y} = \frac{-\text{ve}}{+\text{ve}} = -\text{ve},\ \sec\theta = \frac{r}{x} = \frac{+\text{ve}}{-\text{ve}} = -\text{ve},\ \text{cosec}\,\theta = \frac{r}{y} = \frac{+\text{ve}}{+\text{ve}} = +\text{ve}.$$

$\therefore$ **In the second quadrant, sin θ and cosec θ are positive and all other t-functions are negative.**

Third quadrant. Here x and y are negative, r is positive.

$$\sin\theta = \frac{y}{r} = \frac{-\text{ve}}{+\text{ve}} = -\text{ve},\ \cos\theta = \frac{x}{r} = \frac{-\text{ve}}{+\text{ve}} = -\text{ve},\ \tan\theta = \frac{y}{x} = \frac{-\text{ve}}{-\text{ve}} = +\text{ve},$$

$$\cot\theta = \frac{x}{y} = \frac{-\text{ve}}{-\text{ve}} = +\text{ve},\ \sec\theta = \frac{r}{x} = \frac{+\text{ve}}{-\text{ve}} = -\text{ve},\ \text{cosec}\,\theta = \frac{r}{y} = \frac{+\text{ve}}{-\text{ve}} = -\text{ve}.$$

$\therefore$ **In the third quadrant, tan θ and cot θ are positive and all other t-functions are negative.**

Fourth quadrant. Here x and r are positive, y is negative.

$$\sin\theta = \frac{y}{r} = \frac{-\text{ve}}{+\text{ve}} = -\text{ve},\ \cos\theta = \frac{x}{r} = \frac{+\text{ve}}{+\text{ve}} = +\text{ve},\ \tan\theta = \frac{y}{x} = \frac{-\text{ve}}{+\text{ve}} = -\text{ve},$$

$$\cot\theta = \frac{x}{y} = \frac{+\text{ve}}{-\text{ve}} = -\text{ve},\ \sec\theta = \frac{r}{x} = \frac{+\text{ve}}{+\text{ve}} = +\text{ve},\ \text{cosec}\,\theta = \frac{r}{y} = \frac{+\text{ve}}{-\text{ve}} = -\text{ve}.$$

$\therefore$ **In the fourth quadrant, cos θ and sec θ are positive and all other t-functions are negative.**

Thus, we see that :

(*i*) In the **first** quadrant, **all** *t*-functions are **positive.**

(*ii*) In the **second** quadrant, only **sine** and its reciprocal **cosecant** are **positive.**

(*iii*) In the **third** quadrant, only **tangent** and its reciprocal **cotangent** are **positive.**

(*iv*) In the **fourth** quadrant, only **cosine** and its reciprocal **secant** are **positive.**

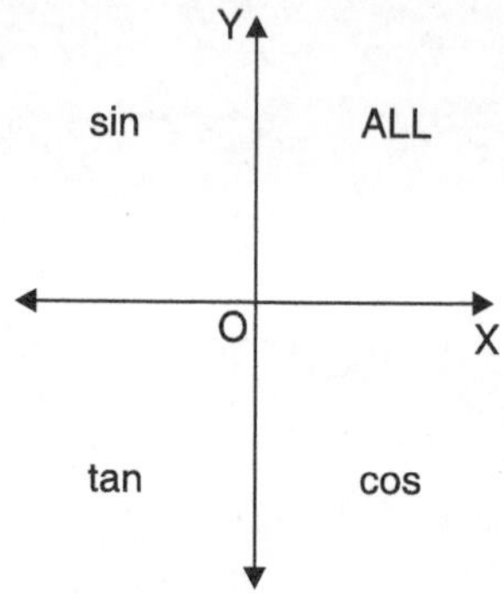

Remark 1. The figure given on the right shows which trigonometrical functions (with their reciprocals) are positive in various quadrants.

ALL-SIN-TAN-COS can be easily recalled by remembering that the first letters *A, S, T* and *C* are also the first letters of the phrase 'After school to college'.

Remark 2. When the angle *A* is expressed in degrees then a *t*-ratio of *A* represents the corresponding *t*-ratio of the radian measure of the angle *A*. For example, sine function of 30° represents the sine function of π/6 and we write sin 30° = sin π/6.

ILLUSTRATIVE EXAMPLES

Example 1. *Which of the trigonometrical ratios are positive for :*

(*i*) *240°* (*ii*) *– 200°* (*iii*) $\left(\frac{10}{3}\pi\right)^c$.

Sol. (*i*) 240° is in the third quadrant. In the third quadrant, only tan θ and cot θ are positive.

∴ **tan 240°** and **cot 240°** are positive.

(*ii*) – 200° lies in the second quadrant.

[∵ All angles between – 180° and – 270° are in the second quadrant]

∴ **sin (– 200°)** and **cosec (– 200°)** are positive.

(*iii*) $\frac{10\pi}{3}$ radians $= \frac{10\pi}{3} \times \frac{180}{\pi}$ degrees = 600° (∵ π radians = 180°)

Dividing 600° by 360°, we get 600° = 360°(1) + 240°.

∴ The revolving line made one complete revolution in the positive direction and has further traced an angle of 240°.

∴ The revolving line is in the third quadrant.

∴ Angle of 600° $\left(=\frac{10}{3}\pi\right)$ lies in the third quadrant.

∴ **tan** $\frac{\mathbf{10\pi}}{\mathbf{3}}$ and **cot** $\frac{\mathbf{10\pi}}{\mathbf{3}}$ are positive.

Example 2. *Find the quadrant in which cos θ is negative and tan θ is positive.*

Sol. cos θ is positive in I and IV quadrants.

∴ cos θ is negative in II and III quadrants.

Also, tan θ is positive in I and III quadrants.

∴ In the **III quadrant,** cos θ is negative and tan θ is positive.

EXERCISE 5.3

1. Which of the following are positive :
(*i*) sin 240° (*ii*) cos 330° (*iii*) tan 315° (*iv*) sec 315°.
2. Which of the following are negative :
(*i*) cos 120° (*ii*) cot 210° (*iii*) sec 240° (*iv*) cosec 250°.
3. Which of the following are positive :
(*i*) cos $(-\pi/3)^c$ (*ii*) tan $(7\pi/6)^c$ (*iii*) sec $(2\pi/3)^c$ (*iv*) cosec $(5\pi/6)^c$.
4. Which of the trigonometric functions are positive for :
(*i*) 70° (*ii*) 125° (*iii*) – 40°.
5. Which of the trigonometric functions are positive for :
(*i*) $(7\pi/4)^c$ (*ii*) $(-\pi/7)^c$ (*iii*) $(3\pi/4)^c$.
6. If $\sin\theta = -1/\sqrt{2}$ and $\tan\theta = 1$, find the quadrant in which θ lies.
7. In which quadrant must θ lie and why, when :
(*i*) $\cos\theta > 0$ and $\tan\theta < 0$ (*ii*) $\tan\theta < 0$ and $\operatorname{cosec}\theta > 0$.

Answers

1. (*ii*), (*iv*) **2.** (*i*), (*iii*), (*iv*) **3.** (*i*), (*ii*), (*iv*) **4.** (*i*) All

(*ii*) sin 125°, cosec 125° (*iii*) cos (– 40°), sec (– 40°) **5.** (*i*) $\cos\frac{7\pi}{4}$, $\sec\frac{7\pi}{4}$

(*ii*) $\cos\left(-\frac{\pi}{7}\right)$, $\sec\left(-\frac{\pi}{7}\right)$ (*iii*) $\sin\frac{3\pi}{4}$, $\operatorname{cosec}\frac{3\pi}{4}$ **6.** Third

7. (*i*) Fourth (*ii*) Second.

5.6. DOMAIN AND RANGE OF TRIGONOMETRIC FUNCTION*

Let θ be any real number. Construct the angle whose measure is θ radians, with vertex at the origin of a rectangular coordinate system and initial line along the positive *x*-axis. Let *P*(*x*, *y*) be any point, distinct from the origin, on the terminal side *OA* of the angle (Fig. 5.4).

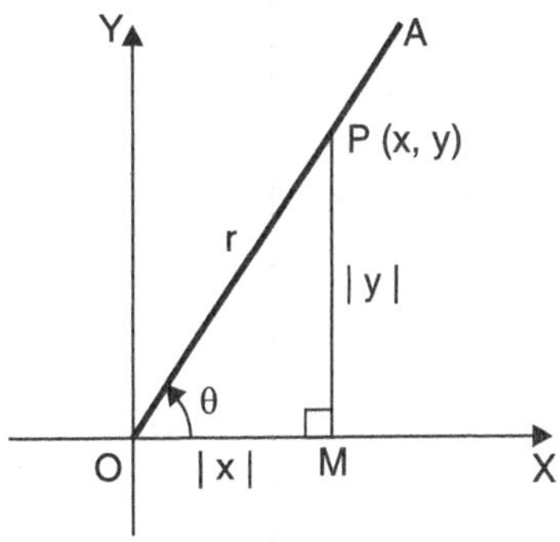

Fig. 5.4

Let $OP = r$.

(*i*) $$\sin\theta = \frac{y}{r}$$

For any θ, $-r \le y \le r$, $\therefore\ -1 \le \frac{y}{r} \le 1$. $\therefore\ -1 \le \sin\theta \le 1$

∴ Domain of sin θ = **R**, Range of sin θ = [– 1, 1].

(*ii*) $$\cos\theta = \frac{x}{r}$$

*If *f* is a function from *A* to *B*, then *A* is called the **domain** of '*f*' and the subset $\{f(x) : x \in A\}$ of *B* is called the **range** of the function '*f*'.

For any θ, $-r \le x \le r$. $\therefore\ -1 \le \frac{x}{r} \le 1$. $\therefore\ -1 \le \cos\theta \le 1$

$\therefore$ Domain of cos θ = **R**, Range of cos θ = [– 1, 1].

(*iii*) tan θ is defined for any real number θ which is not an **odd** multiple of π/2. For any such value of θ, the ratio *y*/*x* can be any real number.

$\therefore$ Domain of tan θ = $\mathbf{R} - \{(2n+1)\frac{\pi}{2} : n \in \mathbf{Z}\}$, Range of tan θ = **R.**

(*iv*) cot θ is defined for any real number θ which is not an **even** multiple of π/2. For any such value of θ, the ratio *x*/*y* can be any real number.

$\therefore$ Domain of cot θ = $\mathbf{R} - \{n\pi : n \in \mathbf{Z}\}$, Range of cot θ = **R.**

(*v*) sec θ is defined for any real number θ which is not an **odd** multiple of π/2. For any such value of θ, we have $|x/r| \le 1$.

$$\Rightarrow \qquad \frac{1}{|x/r|} \ge 1, \left|\frac{r}{x}\right| \ge 1, \ |\sec\theta| \ge 1.$$

$\therefore$ Domain of sec θ = $\mathbf{R} - \{(2n+1)\frac{\pi}{2}, n \in \mathbf{Z}\}$, Range of sec θ = $\mathbf{R} - (-1, 1)$.

(*vi*) cosec θ is defined for any real number θ which is not an **even** multiple of π/2. For any such value of θ, we have $|y/r| \le 1$.

$$\Rightarrow \qquad \frac{1}{|y/r|} \ge 1, \left|\frac{r}{y}\right| \ge 1, \ |\text{cosec}\,\theta| \ge 1.$$

$\therefore$ Domain of cosec θ = $\mathbf{R} - \{n\pi, n \in \mathbf{Z}\}$, Range of cosec θ = $\mathbf{R} - (-1, 1)$.

Remark 1. The maximum (*i.e.,* greatest) value of sin θ and cos θ is 1 and the minimum (*i.e.,* least) value is – 1.

$\therefore$ (*i*) $|\sin\theta| \le 1$ *i.e.,* $-1 \le \sin\theta \le 1$ *i.e.,* $\sin^2\theta \le 1$.

(*ii*) $|\cos\theta| \le 1$ *i.e.,* $-1 \le \cos\theta \le 1$ *i.e.,* $\cos^2\theta \le 1$.

Remark 2. tan θ and cot θ can take any real number value.

Remark 3. sec θ and cosec θ cannot take value in (– 1, 1).

$\therefore$ (*i*) $|\sec\theta| \ge 1$ *i.e.,* $\sec^2\theta \ge 1$.

(*ii*) $|\text{cosec}\,\theta| \ge 1$ *i.e.,* $\text{cosec}^2\theta \ge 1$.

ILLUSTRATIVE EXAMPLES

Example 1. *Show that* $\sec^2\theta + \text{cosec}^2\theta \ge 4$.

Sol. $\sec^2\theta + \text{cosec}^2\theta = (1 + \tan^2\theta) + (1 + \cot^2\theta)$

$$= 2 + (\tan^2\theta + \cot^2\theta) = 2 + (\tan^2\theta + \cot^2\theta - 2) + 2$$

$$= 4 + (\tan\theta - \cot\theta)^2 \ge 4. \qquad [\because\ (\tan\theta - \cot\theta)^2 \ge 0]$$

Example 2. *Prove that* cos θ + sec θ *can never be equal to* 3/2.

Sol. If possible, let $\cos\theta + \sec\theta = \frac{3}{2}$.

$$\therefore \qquad \cos\theta + \frac{1}{\cos\theta} = \frac{3}{2}. \quad \therefore\ 2\cos^2\theta - 3\cos\theta + 2 = 0.$$

$\therefore \qquad \cos\theta = \dfrac{3 \pm \sqrt{9-16}}{4} = \dfrac{3 \pm i\sqrt{7}}{4}$, a complex number.

This is impossible, because $\cos\theta$ is a real number.

$\therefore$ $\cos\theta + \sec\theta$ can never be equal to 3/2.

Example 3. *Is the equation $2\sin^2\theta - \cos\theta + 4 = 0$ possible ?*

Sol. Let $2\sin^2\theta - \cos\theta + 4 = 0$.

$\therefore \quad 2(1-\cos^2\theta) - \cos\theta + 4 = 0. \quad \therefore \quad 2\cos^2\theta + \cos\theta - 6 = 0.$

$\therefore \qquad \cos\theta = \dfrac{-1 \pm \sqrt{1+48}}{4} = \dfrac{-1 \pm 7}{4} = 2, -\dfrac{3}{2}.$

Now $|\cos\theta| \le 1$.

$\therefore \quad \cos\theta = 2 \Rightarrow |\cos\theta| = |2| = 2 \ge 1$, which is impossible

and $\qquad \cos\theta = -\dfrac{3}{2} \Rightarrow |\cos\theta| = \left|-\dfrac{3}{2}\right| = \dfrac{3}{2} \ge 1$, which is also impossible.

$\therefore$ The given equation is not possible.

Example 4. *If x be real, show that $\cos\theta$ cannot be equal to $x + \dfrac{1}{x}$.*

Sol. If possible, let $\cos\theta = x + \dfrac{1}{x}$.

$\Rightarrow \qquad \cos\theta = \dfrac{x^2+1}{x} \Rightarrow x^2 - x\cos\theta + 1 = 0 \Rightarrow x = \dfrac{\cos\theta \pm \sqrt{\cos^2\theta - 4}}{2}.$

Since x is real, we have $\cos^2\theta - 4 \ge 0$ *i.e.*, $\cos^2\theta \ge 4$, which is impossible, because $\cos^2\theta \le 1$.

$\therefore \quad \cos\theta$ cannot be equal to $x + \dfrac{1}{x}$.

Example 5. *If x and y are non-zero real numbers, show that $\sin^2\theta = \dfrac{x^2+y^2}{2xy}$ is possible only when $x = y$.*

Sol. Let $\qquad \sin^2\theta = \dfrac{x^2+y^2}{2xy} \qquad \ldots(1)$

We know that $\sin^2\theta \le 1$. $\therefore \quad \dfrac{x^2+y^2}{2xy} \le 1 \qquad \ldots(2)$

Multiplying by +ve quantity $2xy$, we get $x^2 + y^2 \le 2xy \qquad \left((1) \Rightarrow 2xy = \dfrac{x^2+y^2}{\sin^2\theta} > 0\right)$

$\Rightarrow \quad x^2 + y^2 - 2xy \le 0 \Rightarrow (x-y)^2 \le 0 \Rightarrow (x-y)^2 = 0^* \Rightarrow x - y = 0 \Rightarrow x = y.$

$\therefore \quad \sin^2\theta = \dfrac{x^2+y^2}{2xy}$ is possible only when $x = y$.

***Why this step.** $(x-y)^2 \le 0$ implies $(x-y)^2 = 0$, because square of a real number is not negative.

Example 6. *If x and y be real, show that the equation* $\sec^2 \theta = \frac{4xy}{(x+y)^2}$ *is possible only when* $x = y$.

Sol. We know that $\sec \theta$ is numerically ≥ 1 $\therefore$ $\sec^2 \theta \geq 1$

$$\Rightarrow \quad \frac{4xy}{(x+y)^2} \geq 1 \qquad \left(\because \sec^2 \theta = \frac{4xy}{(x+y)^2}\right)$$

$$\Rightarrow \quad 4xy \geq (x+y)^2 \qquad [\because (x+y)^2 \text{ is +ve}]$$

$$\Rightarrow \quad (x+y)^2 - 4xy \leq 0 \Rightarrow (x-y)^2 \leq 0$$

$$\Rightarrow \quad (x-y)^2 = 0 \qquad [\because \text{Square of a real number cannot be –ve.}]$$

$$\Rightarrow \quad x - y = 0 \Rightarrow x = y.$$

$\therefore$ The equation $\sec^2 \theta = \frac{4xy}{(x+y)^2}$ is possible only when $x = y$.

EXERCISE 5.4

1. Which of the following are possible :
   (*i*) $\sin \theta = 1/4$ (*ii*) $\cos \theta = -2$ (*iii*) $\tan \theta = -15$
   (*iv*) $\cot \theta = 1/3$ (*v*) $\sec \theta = 1/3$ (*vi*) $\text{cosec } \theta = 2$.
2. (*i*) Show that $\text{cosec}^2 \theta + \sin^2 \theta$ can never be less than 2.
   (*ii*) Show that $\tan^2 \theta + \cot^2 \theta$ can never be less than 2.
3. (*i*) Prove that $\sin \theta + \text{cosec } \theta$ can never be equal to 3/2.
   (*ii*) Prove that $\tan \theta + \cot \theta$ can never be equal to 3/2.
4. Is the equation $2 \cos^2 \theta + \cos \theta - 6 = 0$ possible ?
5. Show that no value of $\sec \theta$ can satisfy the equation $6 \sec^2 \theta - 5 \sec \theta + 1 = 0$.
6. Is the equation $\sin \theta = a + \frac{1}{a}$ possible for real values of a ?
7. If x be real, show that $\sin \theta$ cannot be equal to $x + \frac{1}{x}$.
8. Find the value of $\cos \theta$ for which the equation : $2 \cos \theta = x + \frac{1}{x}$ is possible, x being real.
9. Prove that $(a+b)^2 = 4ab \sin^2 \theta$ is possible only when a and b are equal. a and b are non-zero real numbers.
10. Show that for non-zero real numbers p and q, the equation : $\cos^2 \theta = \frac{p^2 + q^2}{2pq}$ is possible only when $p = q$.
11. Is the statement : $\sec \theta = \frac{a^2 + b^2}{2ab}$ possible ?

Answers

1. (*i*), (*iii*), (*iv*), (*vi*) **4.** No **6.** No **8.** $\pm$ 1 **11.** Yes.

Hints

2. (*i*) $\text{cosec}^2\theta + \sin^2\theta = (\text{cosec}\,\theta - \sin\theta)^2 + 2 \geq 2$.

8. $2\cos\theta = x + \dfrac{1}{x} \Rightarrow x^2 - (2\cos\theta)x + 1 = 0 \quad \therefore \quad x = \cos\theta \pm \sqrt{\cos^2\theta - 1}$

Since x is real, we have $\cos^2\theta - 1 \geq 0$. $\therefore$ $\cos^2\theta = 1$, because $\cos^2\theta \leq 1$.

11. Square both sides and use $\sec^2\theta \geq 1$.

5.7. EXPRESSION OF ANY T-RATIO IN TERMS OF A GIVEN T-RATIO

Any trigonometric ratio can be expressed in terms of any other trigonometric ratio by using the following identities :

(*i*) $\sin\theta\,\text{cosec}\,\theta = 1$ (*ii*) $\cos\theta\sec\theta = 1$ (*iii*) $\tan\theta\cot\theta = 1$

(*iv*) $\tan\theta = \dfrac{\sin\theta}{\cos\theta}$ (*v*) $\cot\theta = \dfrac{\cos\theta}{\sin\theta}$ (*vi*) $\sin^2\theta + \cos^2\theta = 1$

(*vii*) $1 + \tan^2\theta = \sec^2\theta$ (*viii*) $1 + \cot^2\theta = \text{cosec}^2\theta$.

ILLUSTRATIVE EXAMPLES

Example 1. *If* $4\sin\theta = 3$, *find the value of* $\dfrac{\sec\theta + 3\tan\theta}{2\sec\theta - 7\tan\theta}$.

Sol. We have $4\sin\theta = 3$. $\Rightarrow$ $\sin\theta = 3/4$.

$$\frac{\sec\theta + 3\tan\theta}{2\sec\theta - 7\tan\theta} = \frac{\dfrac{1}{\cos\theta} + 3\dfrac{\sin\theta}{\cos\theta}}{\dfrac{2}{\cos\theta} - 7\dfrac{\sin\theta}{\cos\theta}} = \frac{1 + 3\sin\theta}{2 - 7\sin\theta} = \frac{1 + 3\left(\dfrac{3}{4}\right)}{2 - 7\left(\dfrac{3}{4}\right)} = \frac{\dfrac{13}{4}}{-\dfrac{13}{4}} = -\mathbf{1}.$$

Example 2. *If* $\cos A = \dfrac{21}{29}$ *and A lies in the fourth quadrant, find* $\sin A$ *and* $\tan A$.

Sol. We have $\cos A = 21/29$

$\therefore$ $$\sin^2 A = 1 - \cos^2 A = 1 - \left(\frac{21}{29}\right)^2 = 1 - \frac{441}{841} = \frac{400}{841} \quad \therefore \quad \sin A = \pm\frac{20}{29}$$

$\Rightarrow$ $$\sin A = -\mathbf{\frac{20}{29}} \qquad (\because \ \sin A \text{ is } -\text{ve in IV quadrant})$$

Also $$\tan A = \frac{\sin A}{\cos A} = \frac{-20/29}{21/29} = -\mathbf{\frac{20}{21}}.$$

Example 3. *If* $\sin\theta\sec\theta = -1$ *and* θ *lies in the second quadrant, find the values of* $\sin\theta$ *and* $\sec\theta$.

Sol. We have $\sin\theta\sec\theta = -1$ $\quad\therefore\quad$ $\sin\theta \,.\, \dfrac{1}{\cos\theta} = -1$ *i.e.,* $\tan\theta = -1$.

Now $$\sec^2\theta = 1 + \tan^2\theta = 1 + (-1)^2 = 1 + 1 = 2$$

$\therefore$ $$\sec\theta = -\sqrt{2} \qquad (\because \ \theta \text{ is in II quadrant})$$

Given equation implies $\sin\theta = \dfrac{-1}{\sec\theta} = \dfrac{-1}{-\sqrt{2}} = \dfrac{1}{\sqrt{2}}$.

$\therefore$ $$\sin\theta = \mathbf{1/\sqrt{2}} \text{ and } \sec\theta = -\mathbf{\sqrt{2}}.$$

Example 4. *If sin θ = 2 cos θ, find the values of other t-ratios of θ.*

Sol. We have $\sin\theta = 2\cos\theta$. $\therefore \dfrac{\sin\theta}{\cos\theta} = 2$ *i.e.,* $\tan\theta = 2$.

(*i*) $$\cot\theta = \frac{1}{\tan\theta} = \mathbf{\frac{1}{2}}.$$

(*ii*) We have $1 + \tan^2\theta = \sec^2\theta$.

$\therefore$ $\sec^2\theta = 1 + (2)^2 = 5$ *i.e.,* $\mathbf{\sec\theta = \pm\sqrt{5}}$.

(*iii*) $$\cos\theta = \frac{1}{\sec\theta} = \frac{1}{\pm\sqrt{5}} = \pm\mathbf{\frac{1}{\sqrt{5}}}.$$

(*iv*) $$\sin\theta = \sin\theta \times \frac{\cos\theta}{\cos\theta} = \tan\theta\cos\theta = 2\left(\pm\frac{1}{\sqrt{5}}\right) = \pm\mathbf{\frac{2}{\sqrt{5}}}.$$ **(Note this step)**

(*v*) $$\text{cosec}\,\theta = \frac{1}{\sin\theta} = \frac{1}{\pm 2/\sqrt{5}} = \pm\mathbf{\frac{\sqrt{5}}{2}}.$$

Example 5. *Find the values of other t-functions of θ if θ lies in II quadrant and cot θ = – 12/5.*

Sol. We have $\cot\theta = -12/5$.

$$\therefore \quad \tan\theta = \frac{1}{\cot\theta} = \frac{1}{-12/5} = -\frac{5}{12}$$

$$\text{cosec}^2\theta = 1 + \cot^2\theta = 1 + \left(-\frac{12}{5}\right)^2 = 1 + \frac{144}{25} = \frac{169}{25}. \quad \therefore \quad \text{cosec}\,\theta = \pm\frac{13}{5}$$

Since cosec θ is +ve in II quadrant, we have $\text{cosec}\,\theta = 13/5$.

$$\sin\theta = \frac{1}{\text{cosec}\,\theta} = \frac{1}{13/5} = \frac{5}{13}$$

$$\cos\theta = \cos\theta \times \frac{\sin\theta}{\sin\theta} = \cot\theta\sin\theta = \left(-\frac{12}{5}\right)\left(\frac{5}{13}\right) = -\frac{12}{13}$$ **(Note this step)**

$$\sec\theta = \frac{1}{\cos\theta} = \frac{1}{-12/13} = -\frac{13}{12}$$

$$\therefore \quad \sin\theta = \mathbf{\frac{5}{13}},\ \cos\theta = -\mathbf{\frac{12}{13}},\ \tan\theta = -\mathbf{\frac{5}{12}},\ \sec\theta = -\mathbf{\frac{13}{12}},\ \text{cosec}\,\theta = \mathbf{\frac{13}{5}}.$$

Example 6. *If* $\sin\theta = \dfrac{21}{29}$, *show that* $\sec\theta + \tan\theta = 2\dfrac{1}{2}$, *θ lying between 0 and* $\dfrac{\pi}{2}$.

Sol. We have $\sin\theta = \dfrac{21}{29}$.

$$\therefore \quad \cos^2\theta = 1 - \sin^2\theta = 1 - \left(\frac{21}{29}\right)^2 = 1 - \frac{441}{841} = \frac{400}{841} \quad \therefore \quad \cos\theta = \pm\sqrt{\frac{400}{841}} = \pm\frac{20}{29}$$

$\therefore$ $\cos\theta = \dfrac{20}{29}$, because cos θ is +ve in I quadrant.

$$\therefore \quad \sec\theta + \tan\theta = \frac{1}{\cos\theta} + \frac{\sin\theta}{\cos\theta} = \frac{1+\sin\theta}{\cos\theta} = \frac{1+(21/29)}{20/29} = \frac{29+21}{20} = \frac{50}{20} = \frac{5}{2} = \mathbf{2\frac{1}{2}}.$$

EXERCISE 5.5

1. If $\sin\theta = 1/3$ and θ lies in the II quadrant, find the value of $\cos\theta$.
2. If $\cos\theta = 21/29$ and θ lies in the I quadrant, find the value of $\sin\theta$.
3. If $\tan\theta = 5$ and θ lies in the III quadrant, find the value of $\sec\theta$.
4. If $\sin\theta = -\dfrac{3}{5}$ and $\sec\theta = \dfrac{5}{4}$, find the value of $\cot\theta$.
5. If $5\cos\theta = 3$, find the value of $\dfrac{\operatorname{cosec}\theta - \cot\theta}{\operatorname{cosec}\theta + \cot\theta}$.
6. If $4\sin^2\theta = 1$, find the value of $\dfrac{2 + 3\cos^2\theta}{1 - 2\cos^2\theta}$.
7. If θ lies in the III quadrant and $\cos\theta = -1/2$, find the values of othor trigonometric functions.
8. If θ lies in the II quadrant and $\sin\theta = 3/5$, find the values of other trigonometric functions.
9. If θ lies in the III quadrant and $\tan\theta = 4/3$, find the values of other trigonometric functions.
10. If $\sin\theta + \cos\theta = 0$ and θ lies in the fourth quadrant, find $\sin\theta$ and $\cos\theta$.
11. If $\tan\theta + \sec\theta = 4$, find the values of $\tan\theta$ and $\sec\theta$.
12. The sine of an angle is to its cosine as 8 : 15. If θ lies in the III quadrant, find the values of $\sin\theta$ and $\cos\theta$.

Answers

1. $-\dfrac{2\sqrt{2}}{3}$ **2.** $\dfrac{20}{29}$ **3.** $-\sqrt{26}$ **4.** $-4/3$

5. $1/4$ **6.** $-17/2$

7. $\sin\theta = -\sqrt{3}/2$, $\tan\theta = \sqrt{3}$, $\cot\theta = 1/\sqrt{3}$, $\sec\theta = -2$, $\operatorname{cosec}\theta = -2\sqrt{3}$

8. $\cos\theta = -4/5$, $\tan\theta = -3/4$, $\cot\theta = -4/3$, $\sec\theta = -5/4$, $\operatorname{cosec}\theta = 5/3$

9. $\sin\theta = -4/5$, $\cos\theta = -3/5$, $\cot\theta = 3/4$, $\sec\theta = -5/3$, $\operatorname{cosec}\theta = -5/4$

10. $\sin\theta = -1/\sqrt{2}$, $\cos\theta = 1/\sqrt{2}$ **11.** 15/8, 17/8

12. $-8/17$, $-15/17$.

Hints

10. $\sin\theta + \cos\theta = 0 \;\Rightarrow\; \sin\theta = -\cos\theta \;\Rightarrow\; \tan\theta = -1.$

11. Use : $1 = (\sec\theta - \tan\theta)(\sec\theta + \tan\theta) = (\sec\theta - \tan\theta) \,.\, 4.$

12. Let $\sin\theta = 8k$ and $\cos\theta = 15k$ and use $\sin^2\theta + \cos^2\theta = 1$.

5.8. VALUES OF TRIGONOMETRIC RATIOS FOR θ = 0, $\pi/6$, $\pi/4$, $\pi/3$, $\pi/2$

(*i*) $\theta = 0°$. In this case, the terminal side of angle θ (= 0°) coincides with the initial side OX. Let $P(x, y)$ (Fig. 5.5) be *any* point on the terminal side OX. Let $OP = r > 0$.

$$\therefore \qquad x = r, \quad y = 0.$$

∴ We have $\sin 0° = \frac{y}{r} = \frac{0}{r} = 0$; $\cos 0° = \frac{x}{r} = \frac{r}{r} = 1$

$\tan 0° = \frac{y}{x} = \frac{0}{r} = 0$; **cot 0° is not defined**

$\sec 0° = \frac{r}{x} = \frac{r}{r} = 1$; **cosec 0° is not defined.**

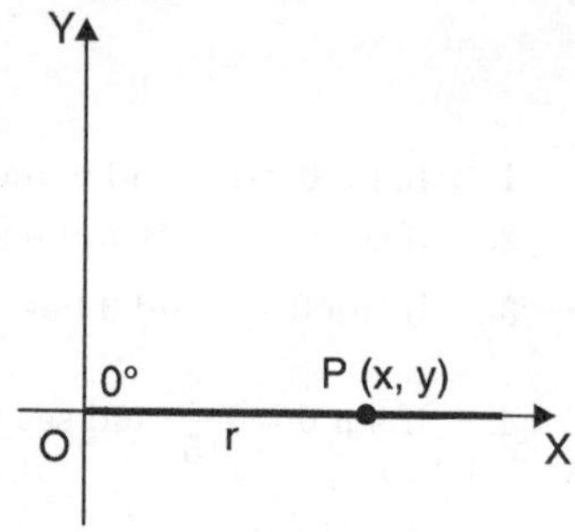

Fig. 5.5

(*ii*) $\theta = \frac{\pi}{6}$ (= 30°). Let angle *XOA* be $\frac{\pi}{6}$.

Let *P*(*x*, *y*) (Fig. 5.6) be *any* point on the terminal side *OA*. Let $OP = r > 0$.

Produce *PM* to *P′* such that $MP' = MP$.

Δs *OMP* and *OMP′* are congruent.

∴ $\angle P'OM = \pi/6$ and so $\Delta OPP'$ is equilateral.

∴ $OP = PP' = PM + MP' = PM + PM = 2PM$

∴ $PM = \frac{OP}{2} = \frac{r}{2}$.

Now $OM^2 = OP^2 - MP^2 = r^2 - \frac{r^2}{4} = \frac{3r^2}{4}$

∴ $OM = \frac{\sqrt{3}}{2} r.$

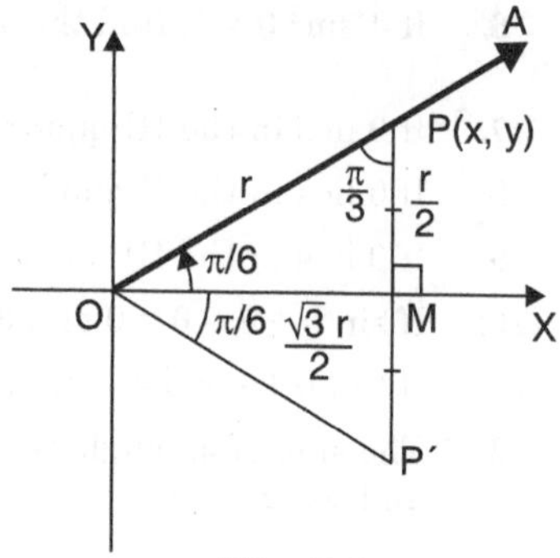

Fig. 5.6

∴ $\sin \frac{\pi}{6} = \frac{y}{r} = \frac{MP}{OP} = \frac{r/2}{r} = \frac{1}{2}$ $\cos \frac{\pi}{6} = \frac{OM}{OP} = \frac{\frac{\sqrt{3}}{2} r}{r} = \frac{\sqrt{3}}{2}$

$\tan \frac{\pi}{6} = \frac{MP}{OM} = \frac{r/2}{\frac{\sqrt{3}}{2} r} = \frac{1}{\sqrt{3}}$ $\cot \frac{\pi}{6} = \frac{OM}{MP} = \frac{\frac{\sqrt{3}}{2}}{r/2} = \sqrt{3}$

$\sec \frac{\pi}{6} = \frac{OP}{OM} = \frac{r}{\frac{\sqrt{3}}{2} r} = \frac{2}{\sqrt{3}}$ $\operatorname{cosec} \frac{\pi}{6} = \frac{OP}{MP} = \frac{r}{\frac{r}{2}} = 2.$

(*iii*) $\theta = \frac{\pi}{4}$ (= 45°). Let angle *XOA* be $\frac{\pi}{4}$.

Let *P*(*x*, *y*) (Fig. 5.7) be any point on the terminal side *OA*. Let $OP = r > 0$.

Triangle *OMP* is isosceles. Also $OM^2 + MP^2 = OP^2$.

∴ $2OM^2 = OP^2$ or $OM = \sqrt{\frac{r^2}{2}} = \frac{r}{\sqrt{2}}$

∴ *MP* is also $\frac{r}{\sqrt{2}}$.

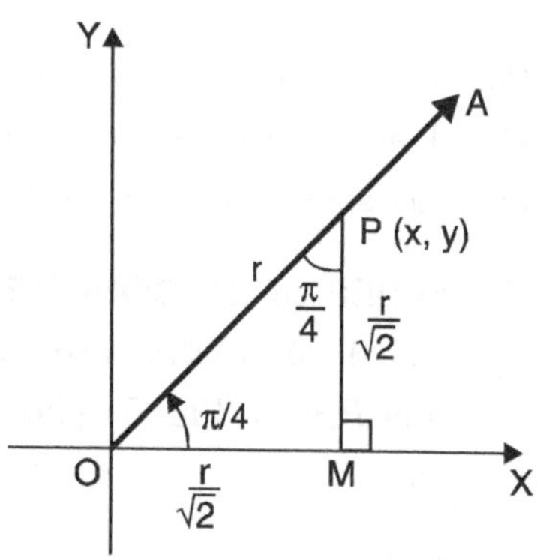

Fig. 5.7

$$\therefore \qquad \sin\frac{\pi}{4} = \frac{MP}{OP} = \frac{r/\sqrt{2}}{r} = \frac{1}{\sqrt{2}}$$

$$\cos\frac{\pi}{4} = \frac{OM}{OP} = \frac{r/\sqrt{2}}{r} = \frac{1}{\sqrt{2}}$$

It is easy to see that $\tan\frac{\pi}{4} = 1$, $\cot\frac{\pi}{4} = 1$, $\sec\frac{\pi}{4} = \sqrt{2}$, $\operatorname{cosec}\frac{\pi}{4} = \sqrt{2}$.

(*iv*) $\theta = \frac{\pi}{3}$ (= **60°**). Let angle XOA be $\frac{\pi}{3}$

Let $P(x, y)$ (Fig. 5.8) be any point on the terminal side OA. Let $OP = r > 0$. Let M' be on OX such that $MM' = OM$.

Δs OMP and $MM'P$ are congruent.

$\therefore \angle MM'P = \frac{\pi}{3}$ and so $\Delta OM'P$ is equilateral.

$$\therefore \qquad OP = OM' = OM + MM' = OM + OM = 2OM$$

$$\therefore \qquad OM = \frac{1}{2} OP = \frac{r}{2}.$$

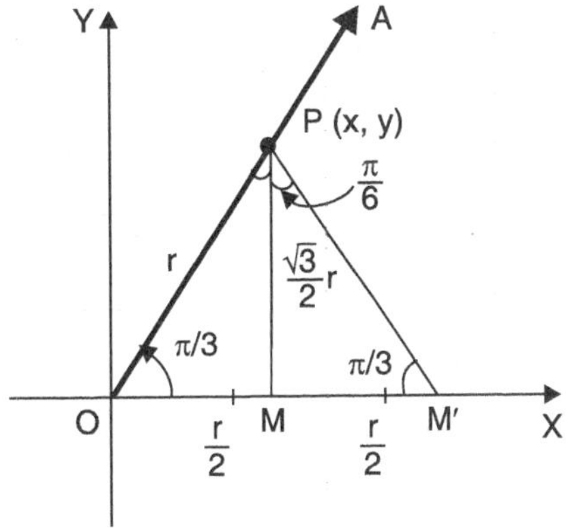

Fig. 5.8

Also $MP^2 = OP^2 - OM^2 = r^2 - \frac{r^2}{4} = \frac{3r^2}{4}$.

$$\therefore \qquad MP = \frac{\sqrt{3}}{2} r.$$

$$\therefore \qquad \sin\frac{\pi}{3} = \frac{MP}{OP} = \frac{\frac{\sqrt{3}}{2} r}{r} = \frac{\sqrt{3}}{2} \qquad\qquad \cos\frac{\pi}{3} = \frac{OM}{OP} = \frac{\frac{r}{2}}{r} = \frac{1}{2}$$

It is easy to see that $\tan\frac{\pi}{3} = \sqrt{3}$, $\cot\frac{\pi}{3} = \frac{1}{\sqrt{3}}$, $\sec\frac{\pi}{3} = 2$, $\operatorname{cosec}\frac{\pi}{3} = \frac{2}{\sqrt{3}}$.

(*v*) $\theta = \frac{\pi}{2}$ (= **90°**). Let angle XOA be $\frac{\pi}{2}$. In this case, the terminal side coincide with OY. Let $P(x, y)$ (Fig. 5.9) be any point on the terminal side OY.

Let $\qquad OP = r > 0.$

$\therefore \qquad x = 0, y = r.$

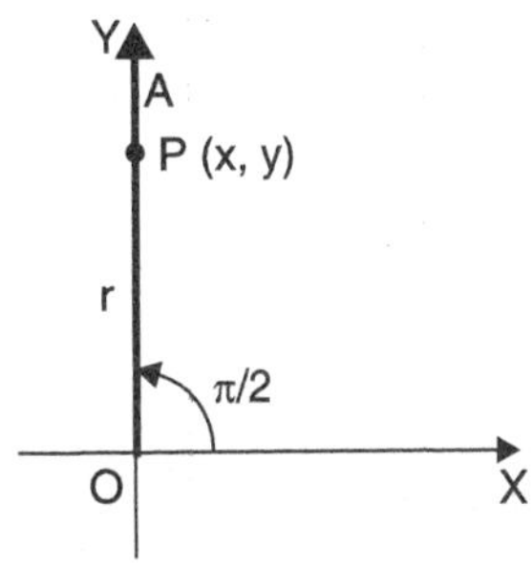

Fig. 5.9

$\therefore$ We have $\sin\frac{\pi}{2} = \frac{y}{r} = \frac{r}{r} = 1 \qquad \cos\frac{\pi}{2} = \frac{x}{r} = \frac{0}{r} = 0$

$\tan\frac{\pi}{2}$ **is not defined** $\qquad \cot\frac{\pi}{2} = \frac{x}{y} = \frac{0}{r} = 0$

$\sec\frac{\pi}{2}$ **is not defined** $\qquad \operatorname{cosec}\frac{\pi}{2} = \frac{r}{y} = \frac{r}{r} = 1.$

Remark. The values of t-ratios for $\theta = 0, \frac{\pi}{6}, \frac{\pi}{4}, \frac{\pi}{3}, \frac{\pi}{2}$ can be easily recalled, by using the following table :

	0 (= 0°)	$\frac{\pi}{6}$ (= 30°)	$\frac{\pi}{4}$ (= 45°)	$\frac{\pi}{3}$ (= 60°)	$\frac{\pi}{2}$ (= 90°)
sin	$\frac{\sqrt{0}}{2} = \mathbf{0}$	$\frac{\sqrt{1}}{2} = \mathbf{\frac{1}{2}}$	$\frac{\sqrt{2}}{2} = \mathbf{\frac{1}{\sqrt{2}}}$	$\mathbf{\frac{\sqrt{3}}{2}}$	$\frac{\sqrt{4}}{2} = \mathbf{1}$
cos	$\frac{\sqrt{4}}{2} = \mathbf{1}$	$\mathbf{\frac{\sqrt{3}}{2}}$	$\frac{\sqrt{2}}{2} = \mathbf{\frac{1}{\sqrt{2}}}$	$\frac{\sqrt{1}}{2} = \mathbf{\frac{1}{2}}$	$\frac{\sqrt{0}}{2} = \mathbf{0}$

The values of the remaining t-ratios can be found out by using the identities :

$$\tan\theta = \frac{\sin\theta}{\cos\theta}, \quad \cot\theta = \frac{\cos\theta}{\sin\theta}, \quad \sec\theta = \frac{1}{\cos\theta}, \quad \text{cosec}\,\theta = \frac{1}{\sin\theta}.$$

ILLUSTRATIVE EXAMPLES

Example 1. *Using tan 60° = $\sqrt{3}$, find the value of sec 60°.*

Sol. We know that $\sec^2 A = 1 + \tan^2 A$.

$$\therefore \quad \sec^2 60° = 1 + \tan^2 60° = 1 + (\sqrt{3})^2 = 1 + 3 = 4.$$

$$\therefore \quad \sec 60° = \pm 2$$

Since 60° lies in I quadrant and secant is +ve in I quadrant.

$$\therefore \quad \sec 60° = \mathbf{2}.$$

Example 2. *Show that $\sec^2 30° + \text{cosec}^2 45° + \cot^2 60° + \sin^2 90° = \frac{14}{3}$.*

Sol. We have that $\sec 30° = \frac{2}{\sqrt{3}}$, $\text{cosec}\, 45° = \sqrt{2}$, $\cot 60° = \frac{1}{\sqrt{3}}$, $\sin 90° = 1$.

$\therefore \quad \sec^2 30° + \text{cosec}^2 45° + \cot^2 60° + \sin^2 90°$

$$= \left(\frac{2}{\sqrt{3}}\right)^2 + (\sqrt{2})^2 + \left(\frac{1}{\sqrt{3}}\right)^2 + (1)^2 = \frac{4}{3} + 2 + \frac{1}{3} + 1 = \mathbf{\frac{14}{3}}.$$

Example 3. *Find the value of θ (0° < θ < 90°), from the equation : $2\cos^2\theta + 5\sin\theta - 4 = 0$.*

Sol. $2\cos^2\theta + 5\sin\theta - 4 = 0 \Rightarrow 2(1 - \sin^2\theta) + 5\sin\theta - 4 = 0$

$$\Rightarrow 2\sin^2\theta - 5\sin\theta + 2 = 0 \quad \therefore \quad \sin\theta = \frac{5 \pm \sqrt{25 - 16}}{4} = \frac{5 \pm 3}{4} = 2, \frac{1}{2}.$$

$\sin\theta = 2$ is impossible, because $-1 \le \sin\theta \le 1$.

$\sin\theta = \frac{1}{2}$ implies $\theta = \mathbf{30°}$.

Example 4. *Given angle C of a triangle ABC to be obtuse, find all angles, when $\sin(A + B) = \frac{\sqrt{3}}{2}$ and $\cos(A - B) = \frac{1}{\sqrt{2}}$, where $A > B$.*

Sol. $\sin(A + B) = \frac{\sqrt{3}}{2}$ ...(1) $\quad \cos(A - B) = \frac{1}{\sqrt{2}}$ ...(2)

Since C is obtuse, angle $A + B$ is acute.

(1) $\Rightarrow$ $A + B = 60°$ $(\because \sin 60° = \sqrt{3}/2)$

(2) $\Rightarrow$ $A - B = 45°$ $(\because \cos 45° = 1/\sqrt{2})$

Adding, we get $2A = 105°$. $\therefore A = \dfrac{105°}{2} = 52°\ 30'$

Subtracting, we get $2B = 15°$. $\therefore B = \dfrac{15°}{2} = 7°\ 30'$

Also $A + B + C = 180°$ $\therefore C = 180° - (A + B) = 180° - 60° = 120°$.

$\therefore$ The angles are **52° 30′, 7° 30′, 120°.**

EXERCISE 5.6

1. Find the value of : cosec 30° cot 45° $\tan^2 60°$. **2.** Find the value of : $\sin^2 45° + \tan^2 45°$.

3. If $A = 30°$, show that $\sin 2A \neq 2 \sin A$. **4.** Using $\sin 30° = 1/2$, find the value of cos 30°.

5. If $A = 30°$, verify that :

(*i*) $\sin 2A = 2 \sin A \cos A$ (*ii*) $\cos 2A = \cos^2 A - \sin^2 A$ (*iii*) $\cos 2A = 1 - 2 \sin^2 A$

(*iv*) $\cos 2A = 2 \cos^2 A - 1$ (*v*) $\sin 2A = \dfrac{2 \tan A}{1 + \tan^2 A}$

(*vi*) $\cos 2A = \dfrac{1 - \tan^2 A}{1 + \tan^2 A}$ (*vii*) $\tan 2A = \dfrac{2 \tan A}{1 - \tan^2 A}$.

6. If $A = 30°$, verify that :

(*i*) $\sin 3A = 3 \sin A - 4 \sin^3 A$

(*ii*) $\cos 3A = 4 \cos^3 A - 3 \cos A$.

7. Find the value of θ $(0° \le \theta \le 90°)$ from the equations :

(*i*) $2 \cos^2 \theta - 5 \cos \theta + 2 = 0$ (*ii*) $2 \cos^2 \theta = 3 \sin \theta$

(*iii*) $3 \sec^4 \theta - 10 \sec^2 \theta + 8 = 0$ (*iv*) $3 \tan \theta + \cot \theta = 5 \operatorname{cosec} \theta$.

8. Show that : *(P.T.U. B. Pharma, May 2004)*

(*i*) $\cot^2 60° + \sin^2 45° + \sin^2 30° + \cos^2 90° = \dfrac{13}{12}$

(*ii*) $\dfrac{4}{3} \cot^2 30° + 3 \sin^2 60° - 2 \operatorname{cosec}^2 60° - \dfrac{3}{4} \tan^2 30° = \dfrac{10}{3}$.

(*iii*) $\sin^2 30° + \cos^2 60° + \tan^2 45° + \sec^2 60° - \operatorname{cosec}^2 30° = \dfrac{3}{2}$.

(*iv*) $\tan^2 60° + 4 \cos^2 45° + 3 \sec^2 30° + 5 \cos^2 90° = 9$.

9. Find the value of :

(*i*) $\dfrac{\sin 30°}{\cos 45°} + \dfrac{\cot 45°}{\sec 60°} - \dfrac{\sin 60°}{\sin 45°} - \dfrac{\cos 30°}{\sin 90°}$ (*ii*) $\operatorname{cosec} \dfrac{\pi}{4} \operatorname{cosec} \dfrac{\pi}{3} \left(\operatorname{cosec} \dfrac{\pi}{6} - \operatorname{cosec} \dfrac{\pi}{2}\right)$.

10. (*i*) If $A - B$ and $A + B$ are positive acute angles and $\sin (A - B) = \frac{1}{2}$ and $\cos (A + B) = \frac{1}{2}$, find A and B.

(*ii*) If $A - B$ and $A + B$ are positive acute angles and $\tan (A + B) = \sqrt{3}$ and $\cot (A - B) = 1$, find A and B.

(*iii*) If $A + B + C = \pi$, show that $\cos^2 \frac{A}{2} + \cos^2 \frac{B}{2} + \cos^2 \frac{C}{2} = 2\left(1 + \sin \frac{A}{2} \sin \frac{B}{2} \sin \frac{C}{2}\right)$

Answers

1. 6 **2.** 3/2 **4.** $\sqrt{3}/2$

7. (*i*) 60° (*ii*) 30° (*iii*) 30° or 45° (*iv*) 60°

9. (*i*) $\frac{(\sqrt{2}+1)(1-\sqrt{3})}{2}$ (*ii*) $\frac{2\sqrt{6}}{3}$

10. (*i*) $A = 45°, B = 15°$ (*ii*) $A = 52° 30', B = 7° 30'$.

Hint

7. (*iii*) We have $3 \sec^4 \theta - 10 \sec^2 \theta + 8 = 0$. $\therefore$ $\sec^2 \theta = \frac{10 \pm \sqrt{100 - 96}}{6} = \frac{10 \pm 2}{6} = 2, \frac{4}{3}$

$$\sec^2 \theta = 2 \Rightarrow \sec \theta = \sqrt{2} \Rightarrow \theta = 45°, \quad \sec^2 \theta = \frac{4}{3} \Rightarrow \sec \theta = \frac{2}{\sqrt{3}} \Rightarrow \theta = 30°.$$

(*iv*) We have $3 \tan \theta + \cot \theta = 5 \operatorname{cosec} \theta$. $\therefore$ $\frac{3 \sin \theta}{\cos \theta} + \frac{\cos \theta}{\sin \theta} = \frac{5}{\sin \theta}$

$$\Rightarrow \quad 3 \sin^2 \theta + \cos^2 \theta = 5 \cos \theta \quad \Rightarrow \quad 3 - 3 \cos^2 \theta + \cos^2 \theta - 5 \cos \theta = 0$$

$$\Rightarrow \quad 2 \cos^2 \theta + 5 \cos \theta - 3 = 0 \quad \Rightarrow \quad \cos \theta = \frac{-5 \pm \sqrt{25 + 24}}{4} = \frac{-5 \pm 7}{4} = -3, \frac{1}{2}.$$

$\cos \theta = -3$ is impossible, $\cos \theta = 1/2$ implies $\theta = 60°$.

5.9. T-RATIOS OF ALLIED ANGLES

(*P.T.U. B. Pharma, May 2007*)

An angle is said to be **allied to another angle** if either :

(*i*) their sum is zero

or (*ii*) their sum or difference is a multiple of $\pi/2$.

For example : $-\theta$ is allied to angle θ $[\because \ -\theta + \theta = 0]$

$\frac{\pi}{2} + \theta$ is allied to angle θ $\left[\because \ \left(\frac{\pi}{2} + \theta\right) - \theta = \frac{\pi}{2}\right]$

$180° - \theta$ is allied to angle θ $[\because \ (180° - \theta) + \theta = 180° = 2 \times 90°]$

Now, we prove some results to find the *t*-ratios of angles allied to θ, in terms of *t*-ratios of θ.

Theorem. I. *If θ is a positive acute angle, then prove that*

$$\sin(-\theta) = -\sin\theta, \quad \cos(-\theta) = \cos\theta, \quad \tan(-\theta) = -\tan\theta,$$
$$\cot(-\theta) = -\cot\theta, \quad \sec(-\theta) = \sec\theta, \quad \operatorname{cosec}(-\theta) = -\operatorname{cosec}\theta.$$

Proof. Let θ $(0 < \theta < \pi/2)$ be any real number. Construct the angles whose measures are θ radians and $(-\theta)$ radians with vertices at the origin of a rectangular coordinate system and initial sides along the positive x-axis.

Let $P(x, y)$ and $P'(x', y')$ be any points on the terminal sides of the angle θ radians and $(-\theta)$ radians respectively and equidistant from the origin (Fig. 5.10).

Let $OP = r$ and $OP' = r'$.

$\therefore$ $r' = r$.

Δs $OP'M'$ and OPM are congruent.

$\therefore$ $OM' = OM$ and $M'P' = MP$.

$\therefore$ $x' = x$ and $y' = -y$.

Now,
$$\sin(-\theta) = \frac{y'}{r'} = \frac{-y}{r} = -\sin\theta$$
$$\cos(-\theta) = \frac{x'}{r'} = \frac{x}{r} = \cos\theta$$
$$\tan(-\theta) = \frac{y'}{x'} = \frac{-y}{x} = -\tan\theta$$
$$\cot(-\theta) = \frac{x'}{y'} = \frac{x}{-y} = -\cot\theta$$
$$\sec(-\theta) = \frac{r'}{x'} = \frac{r}{x} = \sec\theta$$
$$\text{cosec}(-\theta) = \frac{r'}{y'} = \frac{r}{-y} = -\text{cosec}\,\theta$$

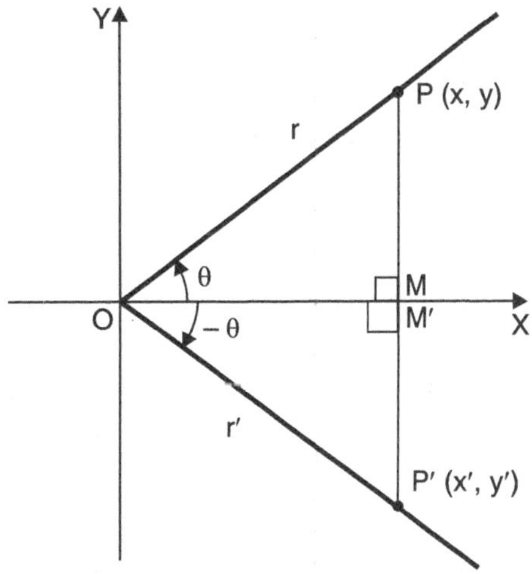

Fig. 5.10

Theorem. II. If θ is a positive acute angle, then prove that

$$\sin\left(\frac{\pi}{2}-\theta\right) = \cos\theta,\quad \cos\left(\frac{\pi}{2}-\theta\right) = \sin\theta,\quad \tan\left(\frac{\pi}{2}-\theta\right) = \cot\theta,$$
$$\cot\left(\frac{\pi}{2}-\theta\right) = \tan\theta,\quad \sec\left(\frac{\pi}{2}-\theta\right) = \text{cosec}\,\theta,\quad \text{cosec}\left(\frac{\pi}{2}-\theta\right) = \sec\theta.$$

Proof. Let θ $(0 < \theta < \pi/2)$ be any real number. Construct the angles whose measures are θ radians and $\left(\frac{\pi}{2}-\theta\right)$ radians with vertices at the origin of a rectangular coordinate system and initial side along the positive x-axis.

Let $P(x, y)$ and $P'(x', y')$ be any points on the terminal sides of angles θ radians and $\left(\frac{\pi}{2}-\theta\right)$ radians respectively and equidistant from the origin (Fig. 5.11).

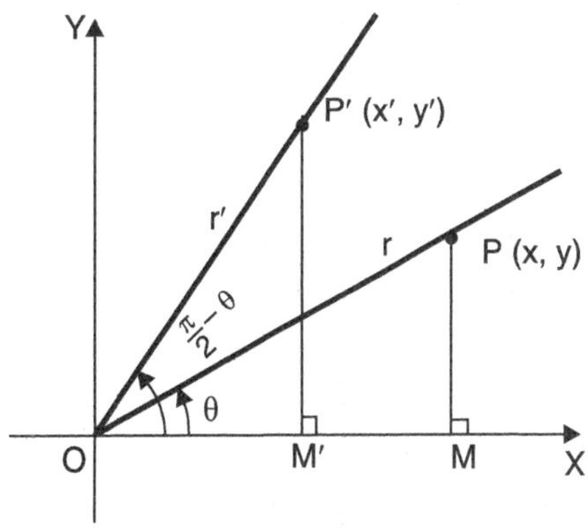

Fig. 5.11

Let $OP = r$ and $OP' = r'$.

$\therefore$ $r' = r$

As $OM'P'$ and PMO are congruent.

$\therefore \quad OM' = MP$ and $M'P' = OM$

$\therefore \quad x' = OM' = MP = y$ and $y' = M'P' = OM = x$.

Now, $\quad \sin\left(\frac{\pi}{2}-\theta\right) = \frac{y'}{r'} = \frac{x}{r} = \cos\theta; \qquad \cos\left(\frac{\pi}{2}-\theta\right) = \frac{x'}{r'} = \frac{y}{r} = \sin\theta$

$$\tan\left(\frac{\pi}{2}-\theta\right) = \frac{y'}{x'} = \frac{x}{y} = \cot\theta; \qquad \cot\left(\frac{\pi}{2}-\theta\right) = \frac{x'}{y'} = \frac{y}{x} = \tan\theta$$

$$\sec\left(\frac{\pi}{2}-\theta\right) = \frac{r'}{x'} = \frac{r}{y} = \text{cosec}\,\theta; \qquad \text{cosec}\left(\frac{\pi}{2}-\theta\right) = \frac{r'}{y'} = \frac{r}{x} = \sec\theta.$$

Theorem. III. *If θ is a positive acute angle, then prove that*

$$\sin\left(\frac{\pi}{2}+\theta\right) = \cos\theta, \quad \cos\left(\frac{\pi}{2}+\theta\right) = -\sin\theta, \quad \tan\left(\frac{\pi}{2}+\theta\right) = -\cot\theta,$$

$$\cot\left(\frac{\pi}{2}+\theta\right) = -\tan\theta, \quad \sec\left(\frac{\pi}{2}+\theta\right) = -\text{cosec}\,\theta, \quad \text{cosec}\left(\frac{\pi}{2}+\theta\right) = \sec\theta.$$

Proof. In this case, $OM' = MP$ and $M'P' = OM$.

$\therefore \quad x' = -OM' = -MP = -y$ and $y' = M'P' = OM = x$ (Fig. 5.12).

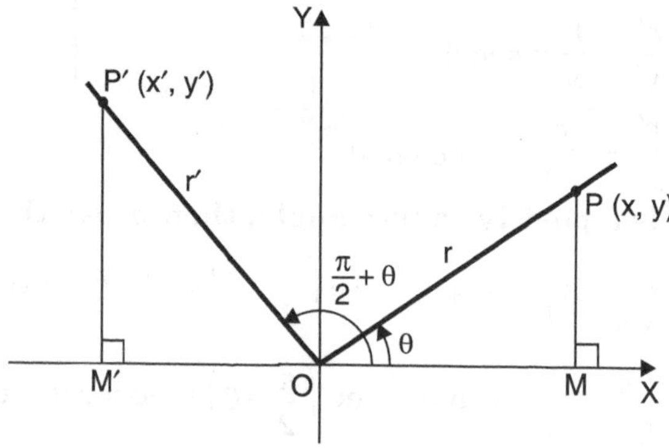

Fig. 5.12

$\therefore \quad \sin\left(\frac{\pi}{2}+\theta\right) = \frac{y'}{r'} = \frac{x}{r} = \cos\theta; \qquad \cos\left(\frac{\pi}{2}+\theta\right) = \frac{x'}{r'} = \frac{-y}{r} = -\sin\theta$

$$\tan\left(\frac{\pi}{2}+\theta\right) = \frac{y'}{x'} = \frac{x}{-y} = -\cot\theta; \qquad \cot\left(\frac{\pi}{2}+\theta\right) = \frac{x'}{y'} = \frac{-y}{x} = -\tan\theta$$

$$\sec\left(\frac{\pi}{2}+\theta\right) = \frac{r'}{x'} = \frac{r}{-y} = -\text{cosec}\,\theta; \quad \text{cosec}\left(\frac{\pi}{2}+\theta\right) = \frac{r'}{y'} = \frac{r}{x} = \sec\theta.$$

Theorem. IV. *If θ is a positive acute angle, then prove that*

$$\sin(\pi-\theta) = \sin\theta, \quad \cos(\pi-\theta) = -\cos\theta, \quad \tan(\pi-\theta) = -\tan\theta$$

$$\cot(\pi-\theta) = -\cot\theta, \quad \sec(\pi-\theta) = -\sec\theta, \quad \text{cosec}(\pi-\theta) = \text{cosec}\,\theta.$$

Proof. In this case $OM' = OM$ and $M'P' = MP$.

$\therefore \quad x' = -OM' = -OM = -x$

and $\quad y' = M'P' = MP = y$. (Fig. 5.13)

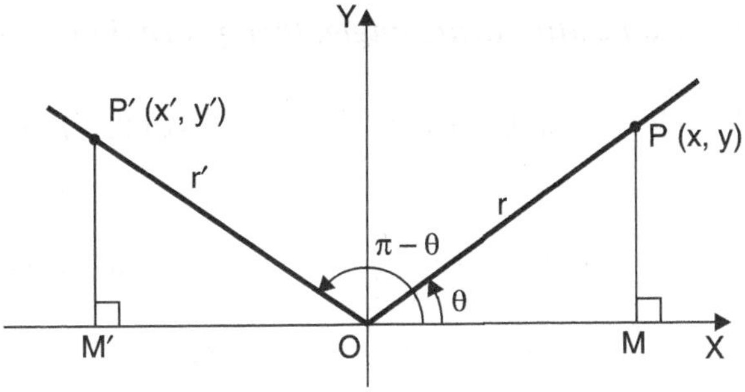

Fig. 5.13

$$\therefore \quad \sin(\pi-\theta)=\frac{y'}{r'}=\frac{y}{r}=\sin\theta; \qquad \cos(\pi-\theta)=\frac{x'}{r'}=\frac{-x}{r}=-\cos\theta$$

$$\tan(\pi-\theta)=\frac{y'}{x'}=\frac{y}{-x}=-\tan\theta; \qquad \cot(\pi-\theta)=\frac{x'}{y'}=\frac{-x}{y}=-\cot\theta$$

$$\sec(\pi-\theta)=\frac{r'}{x'}=\frac{r}{-x}=-\sec\theta; \qquad \text{cosec}(\pi-\theta)=\frac{r'}{y'}=\frac{r}{y}=\text{cosec}\,\theta.$$

Theorem. V. *If θ is a positive acute angle, then prove that*

$$\sin(\pi+\theta)=-\sin\theta,\quad \cos(\pi+\theta)=-\cos\theta,\quad \tan(\pi+\theta)=\tan\theta$$
$$\cot(\pi+\theta)=\cot\theta,\quad \sec(\pi+\theta)=-\sec\theta,\quad \text{cosec}(\pi+\theta)=-\text{cosec}\,\theta.$$

Proof. In this case $OM' = OM$ and $M'P' = MP$.

$\therefore$ $x' = -OM' = -OM = -x'$ and $y' = -M'P' = -MP = -y$ (Fig. 5.14).

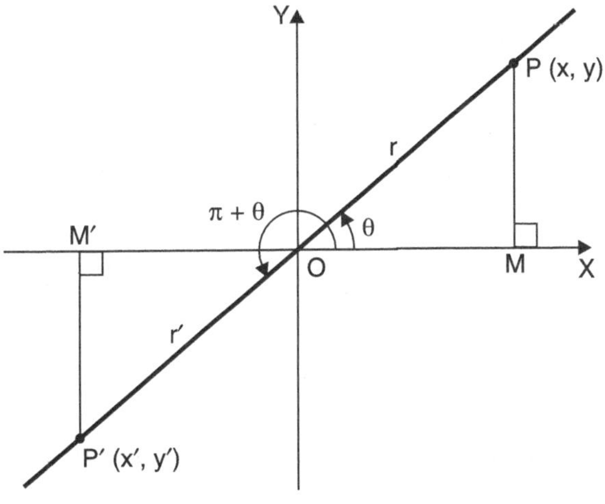

Fig. 5.14

$$\therefore \quad \sin(\pi+\theta)=\frac{y'}{r'}=\frac{-y}{r}=-\sin\theta; \qquad \cos(\pi+\theta)=\frac{x'}{r'}=\frac{-x}{r}=-\cos\theta$$

$$\tan(\pi+\theta)=\frac{y'}{x'}=\frac{-y}{-x}=\tan\theta; \qquad \cot(\pi+\theta)=\frac{x'}{y'}=\frac{-x}{-y}=\cot\theta$$

$$\sec(\pi+\theta)=\frac{r'}{x'}=\frac{r}{-x}=-\sec\theta; \qquad \text{cosec}(\pi+\theta)=\frac{r'}{y'}=\frac{r}{-y}=-\text{cosec}\,\theta.$$

Theorem. VI. *If θ is a positive acute angle, then prove that*

$$\sin\left(\frac{3\pi}{2}-\theta\right) = -\cos\theta, \quad \cos\left(\frac{3\pi}{2}-\theta\right) = -\sin\theta, \quad \tan\left(\frac{3\pi}{2}-\theta\right) = \cot\theta,$$

$$\cot\left(\frac{3\pi}{2}-\theta\right) = \tan\theta, \quad \sec\left(\frac{3\pi}{2}-\theta\right) = -\operatorname{cosec}\theta, \quad \operatorname{cosec}\left(\frac{3\pi}{2}-\theta\right) = -\sec\theta.$$

Proof. In this case, $OM' = MP$ and $M'P' = OM$.

$\therefore$ $x' = -OM' = -MP = -y$ and $y' = -M'P' = -OM = -x$ (Fig. 5.15).

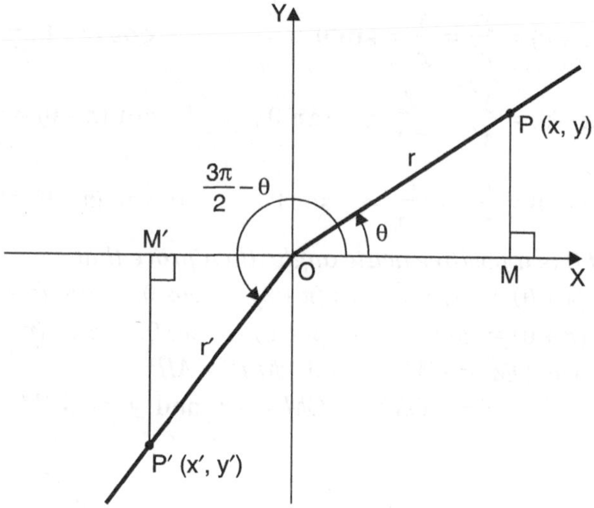

Fig. 5.15

$$\therefore \quad \sin\left(\frac{3\pi}{2}-\theta\right) = \frac{y'}{r'} = \frac{-x}{r} = -\cos\theta; \qquad \cos\left(\frac{3\pi}{2}-\theta\right) = \frac{x'}{r'} = \frac{-y}{r} = -\sin\theta$$

$$\tan\left(\frac{3\pi}{2}-\theta\right) = \frac{y'}{x'} = \frac{-x}{-y} = \cot\theta; \qquad \cot\left(\frac{3\pi}{2}-\theta\right) = \frac{x'}{y'} = \frac{-y}{-x} = \tan\theta$$

$$\sec\left(\frac{3\pi}{2}-\theta\right) = \frac{r'}{x'} = \frac{r}{-y} = -\operatorname{cosec}\theta; \qquad \operatorname{cosec}\left(\frac{3\pi}{2}-\theta\right) = \frac{r'}{y'} = \frac{r}{-x} = -\sec\theta.$$

Theorem. VII. *If θ is a positive acute angle, then prove that*

$$\sin\left(\frac{3\pi}{2}+\theta\right) = -\cos\theta, \quad \cos\left(\frac{3\pi}{2}+\theta\right) = \sin\theta, \quad \tan\left(\frac{3\pi}{2}+\theta\right) = -\cot\theta,$$

$$\cot\left(\frac{3\pi}{2}+\theta\right) = -\tan\theta, \quad \sec\left(\frac{3\pi}{2}+\theta\right) = \operatorname{cosec}\theta, \quad \operatorname{cosec}\left(\frac{3\pi}{2}+\theta\right) = -\sec\theta.$$

Proof. In this case, $OM' = MP$ and $M'P' = OM$.

$\therefore$ $x' = OM' = MP = y$ and $y' = -M'P' = -OM = -x$ (Fig. 5.16).

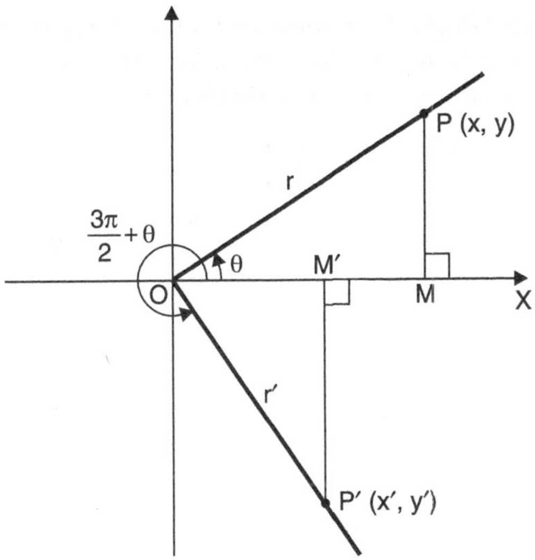

Fig. 5.16

$$\therefore \quad \sin\left(\frac{3\pi}{2}+\theta\right)=\frac{y'}{r'}=\frac{-x}{r}=-\cos\theta\,; \qquad \cos\left(\frac{3\pi}{2}+\theta\right)=\frac{x'}{r'}=\frac{y}{r}=\sin\theta$$

$$\tan\left(\frac{3\pi}{2}+\theta\right)=\frac{y'}{x'}=\frac{-x}{y}=-\cot\theta\,; \qquad \cot\left(\frac{3\pi}{2}+\theta\right)=\frac{x'}{y'}=\frac{y}{-x}=-\tan\theta$$

$$\sec\left(\frac{3\pi}{2}+\theta\right)=\frac{r'}{x'}=\frac{r}{y}=\text{cosec}\,\theta\,; \qquad \text{cosec}\left(\frac{3\pi}{2}+\theta\right)=\frac{r'}{y'}=\frac{r}{-x}=-\sec\theta.$$

Theorem. VIII. *If θ is a positive acute angle, then prove that*

$$\sin(2\pi-\theta)=-\sin\theta,\quad \cos(2\pi-\theta)=\cos\theta,\quad \tan(2\pi-\theta)=-\tan\theta$$
$$\cot(2\pi-\theta)=-\cot\theta,\quad \sec(2\pi-\theta)=\sec\theta,\quad \text{cosec}(2\pi-\theta)=-\text{cosec}\,\theta.$$

Proof. $\sin(2\pi-\theta)=\sin(-\theta)$ [∵ The position of terminal side of an angle is not changed by one or more complete revolutions]

$=-\sin\theta$ [By Theorem I]

Similarly, $\cos(2\pi-\theta)=\cos(-\theta)=\cos\theta$

$\tan(2\pi-\theta)=\tan(-\theta)=-\tan\theta$ $\qquad$ $\cot(2\pi-\theta)=\cot(-\theta)=-\cot\theta$

$\sec(2\pi-\theta)=\sec(-\theta)=\sec\theta$ $\qquad$ $\text{cosec}(2\pi-\theta)=\text{cosec}(-\theta)=-\text{cosec}\,\theta.$

WORKING RULES TO EXPRESS T-RATIOS OF ALLIED ANGLES OF θ IN TERMS OF T-RATIOS OF θ

Step I. *If the angle is negative, use the t-ratios of '– θ' to make the angle positive. For example, we write* $\sin(-240°)=-\sin 240°$.

Step II. *If the angle is greater than 2π (i.e., 360°), subtract the greatest possible multiple of 2π from it, remembering that the t-ratios of* $(n(2\pi)+\theta)$ *are exactly the same as those of θ.*

Step III. *Consider θ lying in the first quadrant (even if it actually does not lie) and find the quadrant in which the allied angle lie. Then determine the sign of the given t-ratio of the allied angle by the rule shown in (Fig. 5.17) below :*

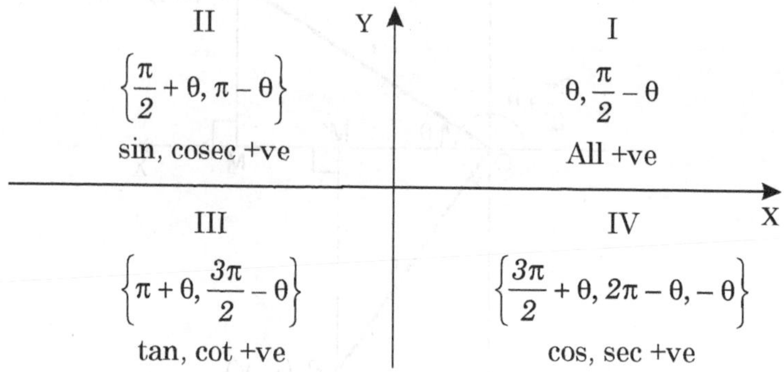

Fig. 5.17

Step IV. (*i*) *Now, if allied angle is* $-\theta, \pi-\theta, \pi+\theta, 2\pi-\theta$; *t-ratio remains the same i.e., sine remains sine, cosine remains cosines and so on.*

(*ii*) *If allied angle is* $\frac{\pi}{2}-\theta, \frac{\pi}{2}+\theta, \frac{3\pi}{2}-\theta, \frac{3\pi}{2}+\theta$; *the t-ratio is changed as follows :*

sine changes to cosine and vice-versa ; tangent changes to cotangent and vice-versa ; secant changes to cosecant and vice-versa.

ILLUSTRATIVE EXAMPLES

Example 1. *Find the values of :*

(*i*) *sin* $(90° + \theta)$ (*ii*) *cos* $(180° - \theta)$ (*iii*) *tan* $(270° - \theta)$ (*iv*) *sec* $(180° + \theta)$.

Sol. (*i*) $90° + \theta$ involves 90°, the *t*-ratio *sine* is changed to *cosine.* Also, assuming θ to be in I quadrant, $90° + \theta$ lies in II quadrant and in this quadrant *sine* is +ve.

∴ $\mathbf{\sin (90° + \theta) = \cos \theta}$.

(*ii*) $180° - \theta$ involves 180°, the *t*-ratio *cosine* remains same. Also, assuming θ to be in I quadrant, $180° - \theta$ lies in II quadrant and in this quadrant, *cosine* is –ve.

∴ $\mathbf{\cos (180° - \theta) = -\cos \theta}$.

(*iii*) $270° - \theta$ involves 270°, the *t*-ratio *tangent* is changed to *cotangent.* Also, assuming θ to be in I quadrant, $270° - \theta$ lies in III quadrant and in this quadrant *tangent* is +ve.

∴ $\mathbf{\tan (270° - \theta) = \cot \theta}$.

(*iv*) $180° + \theta$ involves 180°, the *t*-ratio *secant* remains same. Also, assuming θ to be in I quadrant, $180° + \theta$ lies in III quadrant and in this quadrant *secant* is –ve.

∴ $\mathbf{\sec (180° + \theta) = -\sec \theta}$.

Example 2. *Find the values of :*

(*i*) *sin 225°* (*ii*) *cos (– 405°)* (*iii*) *tan (– 240°)* (*iv*) *cosec 840°.*

Sol. (*i*) $225° = 180° + 45°$ involves 180°, the *t*-ratio is not changed. Also, 225° lies in III quadrant and in this quadrant *sine* is –ve.

∴ $\sin 225° = \sin (180° + 45°) = -\sin 45° = \mathbf{-1/\sqrt{2}}$.

Alternative method

225° = 270° – 45° involves 270°, the *t*-ratio *sine* is changed to *cosine.* Also, 225° lies in III quadrant and in this quadrant *sine* is –ve.

$$\therefore \quad \sin 225° = \sin (270° - 45°) = -\cos 45° = -\mathbf{1/\sqrt{2}}.$$

(*ii*) $\cos(-405°) = \cos 405°$ $\quad (\because \cos(-\theta) = \cos\theta)$

$= \cos(360° + 45°) = \cos 45° = \mathbf{1/\sqrt{2}}.$ $\quad (\because \cos(360° + \theta) = \cos\theta)$

(*iii*) $\tan(-240°) = -\tan 240°$ $\quad (\because \tan(-\theta) = -\tan\theta)$

240° = 180° + 60° involves 180°, the *t*-ratio is not changed. Also 240° lies in III quadrant and in this quadrant, *tangent* is +ve.

$$\tan 240° = \tan(180° + 60°) = \tan 60° = \sqrt{3}$$

$$\therefore \quad \tan(-240°) = -\tan 240° = -\mathbf{\sqrt{3}}.$$

(*iv*) $\text{cosec } 840° = \text{cosec }(2 \times 360° + 120°) = \text{cosec } 120°.$

120° = 180° – 60° involves 180°, the *t*-ratio is not changed. Also 120° lies in II quadrant and in this quadrant, *cosecant* is +ve.

$$\therefore \quad \cos 120° = \cos(180° - 60°) = \text{cosec } 60° = \frac{2}{\sqrt{3}}$$

$$\therefore \quad \text{cosec } 840° = \text{cosec } 120° = \mathbf{2/\sqrt{3}}.$$

Example 3. *Evaluate* :

(*i*) *tan 120°* (*ii*) *cos 1050°* (*iii*) *cosec (– 585°)* (*iv*) *cot 495°.*

Sol. (*i*) $\tan 120° = \tan(90° + 30°) = -\cot 30° - -\mathbf{\sqrt{3}}.$

Alternatively, we can write

$$\tan 120° = \tan(180° - 60°) = -\tan 60° = -\mathbf{\sqrt{3}}.$$

(*ii*) $\cos 1050° = \cos(2(360°) + 330°) = \cos 330° = \cos(360° - 30°) = \cos 30° = \mathbf{\dfrac{\sqrt{3}}{2}}.$

(*iii*) $\text{cosec }(-585°) = -\text{cosec } 585°$ $\quad [\because \text{cosec }(-\theta) = -\text{cosec }\theta, \text{ for all } \theta]$

$= -\text{cosec }(360° + 225°) = -\text{cosec } 225°$

$= -\text{cosec }(180° + 45°) = -(-\text{cosec } 45°) = \mathbf{\sqrt{2}}.$

$[\because \text{cosec }(180° + \theta) = -\text{cosec }\theta]$

(*iv*) $\cot 495° = \cot(360° + 135°) = \cot 135°$

$= \cot(180° - 45°) = -\cot 45° = -\mathbf{1}.$ $\quad [\because \cot(180° - \theta) = -\cot\theta]$

Example 4. *Show that* : $\left(3\cos\dfrac{\pi}{3}\sec\dfrac{\pi}{3} - 4\sin\dfrac{5\pi}{6}\tan\dfrac{\pi}{4}\right)\cos 2\pi = 1.$

Sol. $\cos\dfrac{\pi}{3} = \dfrac{1}{2}, \quad \sec\dfrac{\pi}{3} = 2, \quad \sin\dfrac{5\pi}{6} = \sin\left(\pi - \dfrac{\pi}{6}\right) = \sin\dfrac{\pi}{6} = \dfrac{1}{2},$

$$\tan\frac{\pi}{4} = 1, \cos 2\pi = \cos(2\pi + 0) = \cos 0 = 1.$$

$\therefore$ $\quad$ L.H.S. $= \left(3\cos\frac{\pi}{3}\sec\frac{\pi}{3} - 4\sin\frac{5\pi}{6}\tan\frac{\pi}{4}\right)\cos 2\pi = \left[3\left(\frac{1}{2}\right)(2) - 4\left(\frac{1}{2}\right)(1)\right](1)$

$= 3 - 2 = 1 =$ R.H.S.

Example 5. *Show that : tan 315° cot (– 405°) + cot 495° tan (– 585°) = 2.*

Sol. $\tan 315° = \tan(360° - 45°) = -\tan 45° = -1$

$\cot(-405°) = -\cot 405° = -\cot(360° + 45°) = -\cot 45° = -1$

$\cot 495° = \cot(360° + 135°) = \cot 135° = \cot(180° - 45°) = -\cot 45° = -1$

$\tan(-585°) = -\tan 585° = -\tan(360° + 225°) = -\tan 225° = -\tan(180° + 45°)$

$= -\tan 45° = -1.$

$\therefore$ $\quad$ L.H.S. $= \tan 315° \cot(-405°) + \cot 495° \tan(-585°)$

$= (-1)(-1) + (-1)(-1) = 1 + 1 = 2 =$ R.H.S.

Example 6. *Simplify* $\dfrac{\tan(90° + \theta)\sin(180° + \theta)\sec(270° + \theta)}{\cos(270° - \theta)\,\text{cosec}(180° - \theta)\cot(360° - \theta)}$.

Sol. $\tan(90° + \theta) = -\cot\theta$; $\quad \sin(180° + \theta) = -\sin\theta$

$\sec(270° + \theta) = \text{cosec}\,\theta$; $\quad \cos(270° - \theta) = -\sin\theta$

$\text{cosec}(180° - \theta) = \text{cosec}\,\theta$; $\quad \cot(360° - \theta) = -\cot\theta$

$\therefore$ Given expression $= \dfrac{(-\cot\theta)(-\sin\theta)(\text{cosec}\,\theta)}{(-\sin\theta)(\text{cosec}\,\theta)(-\cot\theta)} = \mathbf{1}.$

Example 7. *Find the value of x from the equation tan (90° + A) sin A + cosec (90° + A) + x cot (90° + A) = 0.*

Sol. Given $\tan(90° + A)\sin A + \text{cosec}(90° + A) + x\cot(90° + A) = 0$

$\Rightarrow \quad -\cot A \sin A + \sec A - x\tan A = 0$

$\Rightarrow \quad -\dfrac{\cos A}{\sin A}\sin A + \dfrac{1}{\cos A} - x\tan A = 0$

$\Rightarrow \quad \dfrac{1}{\cos A} - \cos A = x\tan A$

$\Rightarrow \quad \dfrac{1 - \cos^2 A}{\cos A} = x\tan A$

$\Rightarrow \quad \dfrac{\sin^2 A}{\cos A} = x\dfrac{\sin A}{\cos A}$

$\Rightarrow \quad x = \mathbf{\sin A}.$

Example 8. *Show that :* $\sin^2\dfrac{\pi}{18} + \sin^2\dfrac{\pi}{9} + \sin^2\dfrac{7\pi}{18} + \sin^2\dfrac{4\pi}{9} = 2.$

Sol. L.H.S. $= \sin^2\dfrac{\pi}{18} + \sin^2\dfrac{2\pi}{18} + \sin^2\dfrac{7\pi}{18} + \sin^2\dfrac{8\pi}{18}$

$= \sin^2\dfrac{\pi}{18} + \sin^2\dfrac{2\pi}{18} + \sin^2\left(\dfrac{9\pi - 2\pi}{18}\right) + \sin^2\left(\dfrac{9\pi - \pi}{18}\right)$

$= \sin^2\dfrac{\pi}{18} + \sin^2\dfrac{2\pi}{18} + \left(\sin\left(\dfrac{\pi}{2} - \dfrac{\pi}{9}\right)\right)^2 + \left(\sin\left(\dfrac{\pi}{2} - \dfrac{\pi}{18}\right)\right)^2$

$$= \sin^2 \frac{\pi}{18} + \sin^2 \frac{\pi}{9} + \cos^2 \frac{\pi}{9} + \cos^2 \frac{\pi}{18}$$

$$= \left(\sin^2 \frac{\pi}{18} + \cos^2 \frac{\pi}{18}\right) + \left(\sin^2 \frac{\pi}{9} + \cos^2 \frac{\pi}{9}\right)$$

$$= 1 + 1 = 2 = \text{R.H.S.}$$

EXERCISE 5.7

Evaluate :

1. (*i*) sin 120° (*ii*) cos 150° (*iii*) tan 240°
(*iv*) cot (– 30°) (*v*) sec (– 60°) (*vi*) cosec (– 90°)

2. (*i*) sin 300° (*ii*) tan 270° (*iii*) $\cos \frac{5\pi}{6}$
(*iv*) cos 750° (*v*) tan (– 480°) (*vi*) cosec (– 1410°)

3. (*i*) $\cos \frac{3\pi}{4}$ (*ii*) $\sin \left(-\frac{\pi}{4}\right)$ (*iii*) $\sec \frac{25\pi}{6}$
(*iv*) $\tan \frac{13\pi}{3}$ (*v*) $\cot \left(-\frac{15\pi}{4}\right)$ (*vi*) $\operatorname{cosec} \frac{11\pi}{4}$.

4. Show that :
$\cos \theta + \sin (270° + \theta) - \sin (270° - \theta) + \cos (180° + \theta) = 0$.

5. Show that :

$$\cos \left(\frac{3\pi}{2} + \theta\right) \cos (2\pi + \theta) \left[\cot \left(\frac{3\pi}{2} - \theta\right) + \cot (2\pi + \theta)\right] = 1.$$

6. Show that in any triangle ABC :

(*i*) $\sin (A + B) = \sin C$ (*ii*) $\cos \frac{A+B}{2} = \sin \frac{C}{2}$ (*iii*) $\tan \frac{A+B}{2} = \cot \frac{C}{2}$.

7. In any quadrilateral $ABCD$, show that :
(*i*) $\sin (A + B) + \sin (C + D) = 0$ (*ii*) $\cos (A + B) = \cos (C + D)$.

8. If $ABCD$ be a cyclical quadrilateral, show that $\cos A + \cos B + \cos C + \cos D = 0$.

9. Show that : $\cos 24° + \cos 55° + \cos 125° + \cos 204° + \cos 300° = \frac{1}{2}$.

10. Solve for x, $\sec (90° + A) + x \sin A \tan (90° + A) = \cos (90° + A)$.

11. Show that : (*i*) $\cos 105° + \cos 15° = \sin 75° - \sin 15°$ (*ii*) $\sin^2 54° - \sin^2 72° = \sin^2 18° - \sin^2 36°$.

12. Find all the t-ratios of 870°.

13. Show that :
(*i*) $\sin 150° \cos 120° + \cos 330° \sin 660° = -1$
(*ii*) $\sin -420° \cos 390° + \cos (-660°) \sin (330°) = -1$
(*iii*) $\sin 600° \tan (-690°) + \sec 840° \cot (-945°) = 3/2$

(*iv*) $\sin(-690°)\cos(-300°) + \cos(-750°)\sin(-240°) = 1$.

(*v*) $\sin(-420°)\cos 390° + \cos(-660°)\sin(330°) = -1$.

(*P.T.U. B.Pharma, Dec. 2004*)

14. Show that :

(*i*) $\dfrac{\cos(90° + \theta)\sec(-\theta)\tan(180° - \theta)}{\sec(360° - \theta)\sin(180° + \theta)\cot(90° - \theta)} = -1$

(*ii*) $\dfrac{\tan(90° - \theta)\sec(180° - \theta)\sin(-\theta)}{\sin(180° + \theta)\cot(360° - \theta)\operatorname{cosec}(90° + \theta)} = 1$ (*P.T.U. B.Pharma, Dec. 2004*)

(*iii*) $\dfrac{\sin(-\theta)\tan(90° + \theta)\sin(180° + \theta)\sec(270° + \theta)}{\sin(360° - \theta)\cos(270° - \theta)\operatorname{cosec}(180° - \theta)\cot(360° - \theta)} = 1.$

15. (*i*) Find all angles between 0° and 360° which satisfy the equation $\sin\theta = -1/2$.

(*ii*) Find all positive angles less than 2π which satisfy the equation $\cos^2 x = 1/4$.

16. Show that :

(*i*) $\cos^2\dfrac{\pi}{8} + \cos^2\dfrac{3\pi}{8} + \cos^2\dfrac{5\pi}{8} + \cos^2\dfrac{7\pi}{8} = 2$

(*ii*) $\sin^2\dfrac{\pi}{8} + \sin^2\dfrac{3\pi}{8} + \sin^2\dfrac{5\pi}{8} + \sin^2\dfrac{7\pi}{8} = 2.$

Answers

1. (*i*) $\sqrt{3}/2$ (*ii*) $-\sqrt{3}/2$ (*iii*) $\sqrt{3}$ (*iv*) $-\sqrt{3}$
(*v*) 2 (*vi*) -1

2. (*i*) $-\sqrt{3}/2$ (*ii*) Not defined (*iii*) $-\sqrt{3}/2$ (*iv*) $\sqrt{3}/2$ (*v*) $\sqrt{3}$ (*vi*) 2

3. (*i*) $-1/\sqrt{2}$ (*ii*) $-1/\sqrt{2}$ (*iii*) $2/\sqrt{3}$ (*iv*) $\sqrt{3}$
(*v*) 1 (*vi*) $\sqrt{2}$

10. $-\cot A$

12. $\sin 870° = 1/2$, $\cos 870° = -\sqrt{3}/2$, $\tan 870° = -1/\sqrt{3}$, $\cot 870° = -\sqrt{3}$, $\sec 870° = -2/\sqrt{3}$, $\operatorname{cosec} 870° = 2$

15. (*i*) 210°, 330° (*ii*) $\pi/3, 2\pi/3, 4\pi/3, 5\pi/3$.

Hints

6. (*i*) $\sin(A + B) = \sin(180° - C) = \sin C$.

7. (*i*) $\sin(A + B) = \sin[360° - (C + D)] = -\sin(C + D)$.

8. We have $A + C = 180°$, $B + D = 180°$

$\therefore$ $\cos A = \cos(180° - C) = -\cos C$ and $\cos B = \cos(180° - D) = -\cos D$.

16. (*i*) $\cos^2\dfrac{7\pi}{8} = \left(\cos\dfrac{4\pi + 3\pi}{8}\right)^2 = \left(\cos\left(\dfrac{\pi}{2} + \dfrac{3\pi}{8}\right)\right)^2 = \left(-\sin\dfrac{3\pi}{8}\right)^2 = \sin^2\dfrac{3\pi}{8}$.

6

Trigonometric Functions of Sum and Difference of Angles

6.1. INTRODUCTION

In this chapter, we shall study the methods of finding the values of trigonometric functions for sum and difference of angles. These formulae would help us to find the values such as sin 15°, cos 105° etc. The formulae for finding the values of trigonometric functions for sum of angles and for difference of angles are called **addition formulae** and **subtraction formulae** respectively.

6.2. ADDITION FORMULAE

We shall prove that

(*i*) $\mathbf{\sin (A + B) = \sin A \cos B + \cos A \sin B}$ (*P.T.U. B.Pharma, May 2007*)

(*ii*) $\mathbf{\cos (A + B) = \cos A \cos B - \sin A \sin B}$

(*iii*) $\mathbf{\tan (A + B) = \dfrac{\tan A + \tan B}{1 - \tan A \tan B}}$.

Proof. Let a revolving line, starting from its initial position OX, trace out $\angle XOY = A$ and take the position OY.

Let it (the revolving line) revolve *further* from its position OY and trace out $\angle YOZ = B$ and take the position OZ (Fig. 6.1).

$\therefore \quad \angle XOZ = A + B$

Let P be any point on the final position OZ, of the revolving line.

From P, draw $PQ \perp OX$ and $PR \perp OY$

From R, draw $RS \perp OX$ and $RT \perp PQ$.

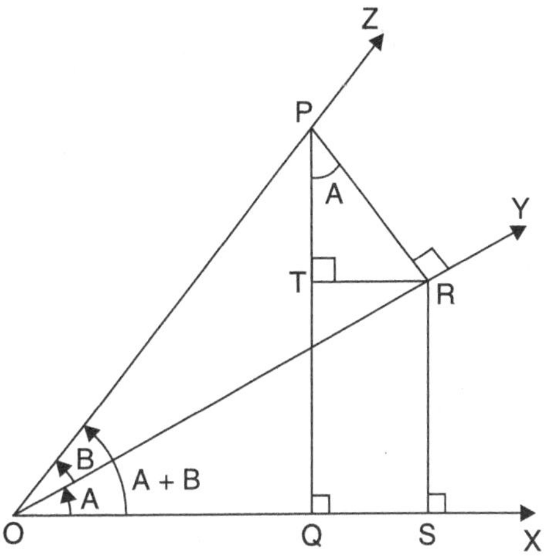

Fig. 6.1

In right angled triangle PRT,

$$\angle TPR = 90° - \angle TRP = \angle PRO - \angle TRP$$
$$= \angle TRO = A$$

(*i*) $$\sin (A + B) = \frac{QP}{OP} = \frac{QT + TP}{OP} = \frac{SR + TP}{OP} = \frac{SR}{OP} + \frac{TP}{OP} = \frac{SR}{OP} \cdot \frac{OR}{OR} + \frac{TP}{OP} \cdot \frac{RP}{RP}$$

(Multiplying and dividing the numerator of each fraction by the hypotenuse of right angled triangle of which it is a side.)

$$= \frac{SR}{OR} \cdot \frac{OR}{OP} + \frac{TP}{RP} \cdot \frac{RP}{OP} = \sin A \cos B + \cos A \sin B$$

$\therefore$ **sin (A + B) = sin A cos B + cos A sin B.**

Caution. sin $(A + B)$ is not equal to sin A + sin B.

(*ii*) $\cos (A + B) = \frac{OQ}{OP} = \frac{OS - QS}{OP} = \frac{OS - TR}{OP} = \frac{OS}{OP} - \frac{TR}{OP}$

$$= \frac{OS}{OP} \cdot \frac{OR}{OR} - \frac{TR}{OP} \cdot \frac{RP}{RP} = \frac{OS}{OR} \cdot \frac{OR}{OP} - \frac{TR}{RP} \cdot \frac{RP}{OP}$$

$$= \cos A \cos B - \sin A \sin B$$

$\therefore$ **cos (A + B) = cos A cos B – sin A sin B.**

Caution. cos $(A + B)$ is not equal to cos A + cos B.

(*iii*) $\tan (A + B) = \frac{QP}{OQ} = \frac{QT + TP}{OS - QS} = \frac{SR + TP}{OS - TR} = \frac{\frac{SR}{OS} + \frac{TP}{OS}}{\frac{OS}{OS} - \frac{TR}{OS}} = \frac{\tan A + \frac{TP}{OS}}{1 - \frac{TR}{TP} \cdot \frac{TP}{OS}}$.

$\left(\text{Multiplying and dividing the numerator of the fraction } \frac{TR}{OS} \text{ by } TP\text{, so as to get tangents} \right)$

As TPR and SOR are similar. $\therefore \quad \frac{TP}{OS} = \frac{PR}{OR} = \tan B$

$\therefore$ $$\mathbf{\tan (A + B) = \frac{\tan A + \tan B}{1 - \tan A \tan B}}.$$

Remark. In the proofs of **addition formulae**, we have assumed that A, B and $A + B$ are positive acute angles. These **addition formulae** are infact true for all possible values of angles A and B.

Corollary 1. *Deduction of formula for tan (A + B) from those of sin (A + B) and cos (A + B).*

Proof. **tan (A + B)** $= \frac{\sin (A + B)}{\cos (A + B)} = \frac{\sin A \cos B + \cos A \sin B}{\cos A \cos B - \sin A \sin B}$

$$= \frac{\frac{\sin A \cos B}{\cos A \cos B} + \frac{\cos A \sin B}{\cos A \cos B}}{\frac{\cos A \cos B}{\cos A \cos B} - \frac{\sin A \sin B}{\cos A \cos B}} = \frac{\frac{\sin A}{\cos A} + \frac{\sin B}{\cos B}}{1 - \frac{\sin A}{\cos A} \cdot \frac{\sin B}{\cos B}} = \mathbf{\frac{\tan A + \tan B}{1 - \tan A \tan B}}.$$

Corollary 2. *$\tan (45° + \theta) = \frac{1 + \tan \theta}{1 - \tan \theta}$.*

Proof. **tan (45° + θ)** $= \frac{\tan 45° + \tan A}{1 - \tan 45° \tan A} = \mathbf{\frac{1 + \tan \theta}{1 - \tan \theta}}.$

6.3. SUBTRACTION FORMULAE

We shall prove that

(*i*) $\sin (A - B) = \sin A \cos B - \cos A \sin B$

(*ii*) $\cos (A - B) = \cos A \cos B + \sin A \sin B$

(*iii*) $\tan (A - B) = \dfrac{\tan A - \tan B}{1 + \tan A \tan B}$.

Proof. (*i*) Let a revolving line, starting from its initial position OX, trace out $\angle XOY = A$ and take the position OY.

Let it (the revolving line) revolve *back* from its position OY and trace out $\angle YOZ = B$ and take the position OZ. (Fig. 6.2)

$\therefore \quad \angle XOZ = A - B.$

Let P be any point on the final position OZ, of the revolving line.

From P, draw $PQ \perp OX$ and $PR \perp OY$.

From R, draw $RS \perp OX$ and $RT \perp QP$, produced.

In right angled triangle PRT,

$$\angle TPR = 90° - \angle PRT = \angle PRY - \angle PRT$$
$$= \angle TRY = A.$$

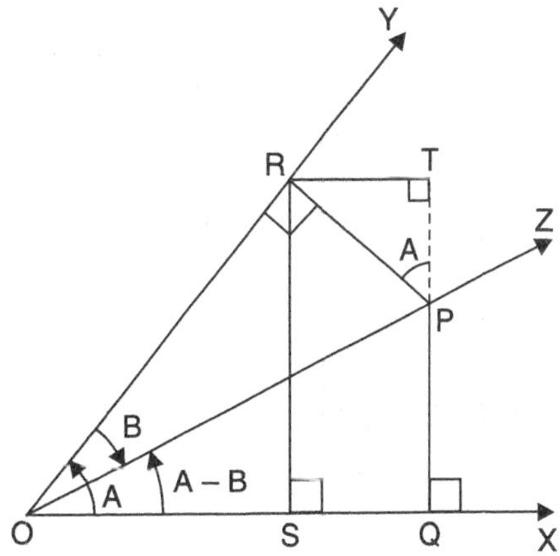

Fig. 6.2

$$\therefore \quad \sin (A - B) = \frac{QP}{OP} = \frac{QT - TP}{OP} = \frac{SR - TP}{OP}$$
$$= \frac{SR}{OP} - \frac{TP}{OP}$$
$$= \frac{SR}{OP} \cdot \frac{OR}{OR} - \frac{TP}{OP} \cdot \frac{RP}{RP}$$
$$= \frac{SR}{OR} \cdot \frac{OR}{OP} - \frac{TP}{RP} \cdot \frac{RP}{OP}$$
$$= \sin A \cos B - \cos A \sin B$$

$\therefore \quad$ **sin (A – B) = sin A cos B – cos A sin B.**

Caution. $\sin (A - B)$ is not equal to $\sin A - \sin B$.

$$(ii) \quad \cos (A - B) = \frac{OQ}{OP} = \frac{OS + QS}{OP} = \frac{OS + TR}{OP} = \frac{OS}{OP} + \frac{TR}{OP}$$
$$= \frac{OS}{OP} \cdot \frac{OR}{OR} + \frac{TR}{OP} \cdot \frac{RP}{RP} = \frac{OS}{OR} \cdot \frac{OR}{OP} + \frac{TR}{RP} \cdot \frac{RP}{OP}$$
$$= \cos A \cos B + \sin A \sin B$$

$\therefore \quad$ **cos (A – B) = cos A cos B + sin A sin B.**

Caution. $\cos (A - B)$ is not equal to $\cos A - \cos B$.

$$(iii) \quad \tan (A - B) = \frac{QP}{OQ} = \frac{QT - TP}{OS + QS} = \frac{SR - TP}{OS + TR} = \frac{\dfrac{SR}{OS} - \dfrac{TP}{OS}}{\dfrac{OS}{OS} + \dfrac{TR}{OS}} = \frac{\tan A - \dfrac{TP}{OS}}{1 + \dfrac{TR}{TP} \cdot \dfrac{TP}{OS}}$$

As TPR and SOR are similar.

$$\therefore \quad \frac{TP}{OS} = \frac{PR}{OR} = \tan B$$

$$\therefore \quad \tan (A - B) = \frac{\tan A - \tan B}{1 + \tan A \tan B}.$$

Remark. In the proofs of **subtraction formulae**, we have assumed that A, B and $A - B$ are positive acute angles. These **subtraction formulae** are infact true for all possible values of angles A and B.

Corollary 1. *Deduction of formula for tan (A – B) from those of sin (A – B) and cos (A – B).*

Proof. $\tan (A - B) = \dfrac{\sin (A - B)}{\cos (A - B)} = \dfrac{\sin A \cos B - \cos A \sin B}{\cos A \cos B + \sin A \sin B}$

$$= \frac{\dfrac{\sin A \cos B}{\cos A \cos B} - \dfrac{\cos A \sin B}{\cos A \cos B}}{\dfrac{\cos A \cos B}{\cos A \cos B} + \dfrac{\sin A \sin B}{\cos A \cos B}} = \frac{\dfrac{\sin A}{\cos A} - \dfrac{\sin B}{\cos B}}{1 + \dfrac{\sin A}{\cos A} \cdot \dfrac{\sin B}{\cos B}} = \frac{\tan A - \tan B}{1 + \tan A \tan B}.$$

Corollary 2. $\tan (45° - A) = \dfrac{1 - \tan A}{1 + \tan A}$.

Proof. $\tan (45° - A) = \dfrac{\tan 45° - \tan A}{1 + \tan 45° \tan A} = \dfrac{1 - \tan A}{1 + \tan A}$.

6.4. THEOREM

Prove that

(*i*) $\sin (A + B) \sin (A - B) = \sin^2 A - \sin^2 B$

(*ii*) $\cos (A + B) \cos (A - B) = \cos^2 A - \sin^2 B$.

Proof. (*i*) $\sin (A + B) \sin (A - B) = (\sin A \cos B + \cos A \sin B) (\sin A \cos B - \cos A \sin B)$

$$= \sin^2 A \cos^2 B - \cos^2 A \sin^2 B = \sin^2 A (1 - \sin^2 B) - (1 - \sin^2 A) \sin^2 B$$

$$= \sin^2 A - \sin^2 A \sin^2 B - \sin^2 B + \sin^2 A \sin^2 B = \sin^2 A - \sin^2 B$$

$\therefore$ **$\sin (A + B) \sin (A - B) = \sin^2 A - \sin^2 B$.**

(*ii*) $\cos (A + B) \cos (A - B) = (\cos A \cos B - \sin A \sin B) (\cos A \cos B + \sin A \sin B)$

$$= \cos^2 A \cos^2 B - \sin^2 A \sin^2 B = \cos^2 A (1 - \sin^2 B) - (1 - \cos^2 A) \sin^2 B$$

$$= \cos^2 A - \cos^2 A \sin^2 B - \sin^2 B + \cos^2 A \sin^2 B = \cos^2 A - \sin^2 B.$$

$\therefore$ **$\cos (A + B) \cos (A - B) = \cos^2 A - \sin^2 B$.**

Corollary 1. $\sin (A + B) \sin (A - B) = \cos^2 B - \cos^2 A$.

Proof. We have $\sin (A + B) \sin (A - B) = \sin^2 A - \sin^2 B$

$$= (1 - \cos^2 A) - (1 - \cos^2 B) = \cos^2 B - \cos^2 A.$$

Corollary 2. $\cos (A + B) \cos (A - B) = \cos^2 B - \sin^2 A$.

Proof. We have $\cos (A + B) \cos (A - B) = \cos^2 A - \sin^2 B$

$$= (1 - \sin^2 A) - (1 - \cos^2 B) = \cos^2 B - \sin^2 A.$$

WORKING RULES FOR SOLVING PROBLEMS

Rule I. (*i*) $\sin (A + B) = \sin A \cos B + \cos A \sin B$

(*ii*) $\cos (A + B) = \cos A \cos B - \sin A \sin B$

(*iii*) $\tan (A + B) = \dfrac{\tan A + \tan B}{1 - \tan A \tan B}$.

Rule II. (*i*) $\sin (A - B) = \sin A \cos B - \cos A \sin B$

(*ii*) $\cos (A - B) = \cos A \cos B + \sin A \sin B$

(*iii*) $\tan (A - B) = \dfrac{\tan A - \tan B}{1 + \tan A \tan B}$.

Rule III. (*i*) $\tan (45° + A) = \dfrac{1 + \tan A}{1 - \tan A}$ (*ii*) $\tan (45° - A) = \dfrac{1 - \tan A}{1 + \tan A}$.

Rule IV. (*i*) $\sin (A + B) \sin (A - B) = \sin^2 A - \sin^2 B$

(*ii*) $\cos (A + B) \cos (A - B) = \cos^2 A - \sin^2 B$.

ILLUSTRATIVE EXAMPLES

Example 1. *Calculate sin 105° and cos 15°.*

Sol. $\sin 105° = \sin (60° + 45°) = \sin 60° \cos 45° + \cos 60° \sin 45°$

$$= \frac{\sqrt{3}}{2} \times \frac{1}{\sqrt{2}} + \frac{1}{2} \times \frac{1}{\sqrt{2}} = \frac{\sqrt{3}+1}{2\sqrt{2}}.$$

$$\cos 15° = \cos (45° - 30°) = \cos 45° \cos 30° + \sin 45° \sin 30°$$

$$= \frac{1}{\sqrt{2}} \times \frac{\sqrt{3}}{2} + \frac{1}{\sqrt{2}} \times \frac{1}{2} = \frac{\sqrt{3}+1}{2\sqrt{2}}.$$

Remark. The values of sin 105° and cos 15° are equal, because $\sin 105° = \sin (90° + 15°) = \cos 15°$.

Example 2. *Show that tan 75° + cot 75° = 4.*

Sol. $$\tan 75° = \tan (45° + 30°) = \frac{\tan 45° + \tan 30°}{1 - \tan 45° \tan 30°} = \frac{1 + \frac{1}{\sqrt{3}}}{1 - 1\left(\frac{1}{\sqrt{3}}\right)} = \frac{\sqrt{3}+1}{\sqrt{3}-1}. \quad ...(1)$$

Also $$\cot 75° = \frac{1}{\tan 75°} = \frac{\sqrt{3}-1}{\sqrt{3}+1} \quad ...(2)$$

$$\text{L.H.S.} = \tan 75° + \cot 75° = \frac{\sqrt{3}+1}{\sqrt{3}-1} + \frac{\sqrt{3}-1}{\sqrt{3}+1} \quad \text{[Using (1) and (2)]}$$

$$= \frac{(\sqrt{3}+1)^2 + (\sqrt{3}-1)^2}{(\sqrt{3}-1)(\sqrt{3}+1)} = \frac{(3+1+2\sqrt{3}) + (3+1-2\sqrt{3})}{3-1} = \frac{8}{2} = 4 = \text{R.H.S.}$$

Example 3. *Show that cot x cot 2x – cot 2x cot 3x – cot 3x cot x = 1.*

Sol. We have $3x = 2x + x$

$\therefore$ $\cot 3x = \cot (2x + x)$

$$\Rightarrow \qquad \cot 3x = \frac{\cot 2x \cot x - 1}{\cot x + \cot 2x}$$

$$\tan(A+B) = \frac{\tan A + \tan B}{1 - \tan A \tan B}$$

$$\Rightarrow \quad \cot(A+B) = \frac{1-\tan A \tan B}{\tan A + \tan B} = \frac{1 - \frac{1}{\cot A}\cdot\frac{1}{\cot B}}{\frac{1}{\cot A} + \frac{1}{\cot B}} = \frac{\cot A \cot B - 1}{\cot B + \cot A}$$

$\Rightarrow \quad \cot 3x \cot x + \cot 3x \cot 2x = \cot 2x \cot x - 1$

$\Rightarrow \quad$ **$\cot x \cot 2x - \cot 2x \cot 3x - \cot 3x \cot x = 1$.**

Example 4. *If $\sin A = \frac{3}{5}$, $0 < A < \frac{\pi}{2}$ and $\cos B = -\frac{12}{13}$, $\pi < B < \frac{3\pi}{2}$, find the values of :*

(i) $\sin(A+B)$ (ii) $\cos(A+B)$ (iii) $\tan(A+B)$
(iv) $\sin(A-B)$ (v) $\cos(A-B)$ (vi) $\tan(A-B)$.

Sol. We have $\sin A = \frac{3}{5}$, $0 < A < \frac{\pi}{2}$ and $\cos B = -\frac{12}{13}$, $\pi < B < \frac{3\pi}{2}$.

$$\therefore \quad \cos A = +\sqrt{1-\sin^2 A} = \sqrt{1-\left(\frac{3}{5}\right)^2} = \frac{4}{5}*$$

and $$\sin B = -\sqrt{1-\cos^2 B} = -\sqrt{1-\left(-\frac{12}{13}\right)^2} = -\frac{5}{13}*.$$

(*i*) $\sin(A+B) = \sin A \cos B + \cos A \sin B = \left(\frac{3}{5}\right)\left(-\frac{12}{13}\right) + \left(\frac{4}{5}\right)\left(-\frac{5}{13}\right) = -\mathbf{\frac{56}{65}}$

(*ii*) $\cos(A+B) = \cos A \cos B - \sin A \sin B = \left(\frac{4}{5}\right)\left(-\frac{12}{13}\right) - \left(\frac{3}{5}\right)\left(-\frac{5}{13}\right) = -\mathbf{\frac{33}{65}}$

(*iii*) $\tan(A+B) = \frac{\sin(A+B)}{\cos(A+B)} = \frac{-56/65}{-33/65} = \mathbf{\frac{56}{33}}$

(*iv*) $\sin(A-B) = \sin A \cos B - \cos A \sin B = \left(\frac{3}{5}\right)\left(-\frac{12}{13}\right) - \left(\frac{4}{5}\right)\left(-\frac{5}{13}\right) = -\mathbf{\frac{16}{65}}$

(*v*) $\cos(A-B) = \cos A \cos B + \sin A \sin B = \left(\frac{4}{5}\right)\left(-\frac{12}{13}\right) + \left(\frac{3}{5}\right)\left(-\frac{5}{13}\right) = -\mathbf{\frac{63}{65}}$

(*vi*) $\tan(A-B) = \frac{\sin(A-B)}{\cos(A-B)} = \frac{-16/65}{-63/65} = \mathbf{\frac{16}{63}}$.

Example 5. *Show that : tan 70° = 2 tan 50° + tan 20°.* **(G.T.U. B. Pharma 2003)**

Sol. $\tan 70° = \tan(50° + 20°)$

$$\therefore \quad \tan 70° = \frac{\tan 50° + \tan 20°}{1 - \tan 50° \tan 20°}$$

$\Rightarrow \quad \tan 70°(1 - \tan 50° \tan 20°) = \tan 50° + \tan 20°$

***Why this step.** cos A is +ve because A is in the first quadrant.
sin B is –ve because B is in the third quadrant.

$\Rightarrow$ $\tan 70° - \tan 70° \tan 50° \tan 20° = \tan 50° + \tan 20°$

$\Rightarrow$ $\tan 70° - \dfrac{1}{\tan 20°} \tan 50° \tan 20° = \tan 50° + \tan 20°$

$$\left[\because \tan 70° = \tan (90° - 20°) = \cot 20° = \frac{1}{\tan 20°}\right]$$

$\Rightarrow$ $\tan 70° - \tan 50° = \tan 50° + \tan 20°$

$\Rightarrow$ $\mathbf{\tan 70° = 2 \tan 50° + \tan 20°}$.

Example 6. *Show that :*

(*i*) $\cos\theta \cos\phi = \cos^2 \dfrac{\theta - \phi}{2} - \sin^2 \dfrac{\theta + \phi}{2}$.

(*ii*) $\cos^2 A + \cos^2 B - 2 \cos A \cos B \cos (A + B) = \sin^2 (A + B)$.

(*iii*) $\dfrac{\tan (45° + \theta) - \tan (45° - \theta)}{\tan (45° + \theta) + \tan (45° - \theta)} = \sin 2\theta$.

Sol. (*i*) We know that $\cos^2 A - \sin^2 B = \cos (A + B) \cos (A - B)$

$\therefore$ R.H.S. $= \cos^2 \dfrac{\theta - \phi}{2} - \sin^2 \dfrac{\theta + \phi}{2} = \cos\left(\dfrac{\theta - \phi}{2} + \dfrac{\theta + \phi}{2}\right) \cos\left(\dfrac{\theta - \phi}{2} - \dfrac{\theta + \phi}{2}\right)$

$= \cos\theta \cos(-\phi) = \cos\theta \cos\phi =$ L.H.S. | $\cos(-\theta) = \cos\theta$

(*ii*) L.H.S. $\cos^2 A + \cos^2 B - 2 \cos A \cos B \cos (A + B)$

$= \cos^2 A + (1 - \sin^2 B) - 2 \cos A \cos B \cos (A + B)$

$= 1 + (\cos^2 A - \sin^2 B) - 2 \cos A \cos B \cos (A + B)$

$= 1 + \cos (A + B) \cos (A - B) - 2 \cos A \cos B \cos (A + B)$

$= 1 + \cos (A + B) [\cos (A - B) - 2 \cos A \cos B]$

$= 1 + \cos (A + B) [\cos A \cos B + \sin A \sin B - 2 \cos A \cos B]$

$= 1 + \cos (A + B) [\sin A \sin B - \cos A \cos B]$

$= 1 + \cos (A + B) [-\cos (A + B)] = 1 - \cos^2 (A + B) = \sin^2 (A + B) =$ R.H.S.

(*iii*) L.H.S. $= \dfrac{\tan (45° + \theta) - \tan (45° - \theta)}{\tan (45° + \theta) + \tan (45° - \theta)} = \dfrac{\dfrac{1 + \tan\theta}{1 - \tan\theta} - \dfrac{1 - \tan\theta}{1 + \tan\theta}}{\dfrac{1 + \tan\theta}{1 - \tan\theta} + \dfrac{1 - \tan\theta}{1 + \tan\theta}}$

$$= \frac{\dfrac{(1 + \tan\theta)^2 - (1 - \tan\theta)^2}{1 - \tan^2\theta}}{\dfrac{(1 + \tan\theta)^2 + (1 - \tan\theta)^2}{1 - \tan^2\theta}} = \frac{(1 + \tan\theta)^2 - (1 - \tan\theta)^2}{(1 + \tan\theta)^2 + (1 + \tan\theta)^2} = \frac{4 \tan\theta}{2 (1 + \tan^2\theta)}$$

$$= \frac{2 \tan\theta}{1 + \tan^2\theta} = \frac{\dfrac{2 \sin\theta}{\cos\theta}}{1 + \dfrac{\sin^2\theta}{\cos^2\theta}} = \frac{2 \sin\theta \cos\theta}{\cos^2\theta + \sin^2\theta} = \frac{\sin\theta \cos\theta + \cos\theta \sin\theta}{1}$$

$= \sin (\theta + \theta) = \sin 2\theta =$ R.H.S.

Example 7. *If* $\theta + \phi = \alpha$ *and* $\tan\theta = k\tan\phi$, *then prove that* $\sin(\theta - \phi) = \frac{k-1}{k+1}\sin\alpha$.

Sol. We have $\tan\theta = k\tan\phi$.

$$\therefore \quad \frac{\tan\theta}{\tan\phi} = \frac{k}{1} \Rightarrow \frac{\tan\theta + \tan\phi}{\tan\theta - \tan\phi} = \frac{k+1}{k-1} \text{ (By Componendo and Dividendo rule)}$$

$$\Rightarrow \quad \frac{\frac{\sin\theta}{\cos\theta} + \frac{\sin\phi}{\cos\phi}}{\frac{\sin\theta}{\cos\theta} - \frac{\sin\phi}{\cos\phi}} = \frac{k+1}{k-1} \Rightarrow \frac{\sin\theta\cos\phi + \cos\theta\sin\phi}{\sin\theta\cos\phi - \cos\theta\sin\phi} = \frac{k+1}{k-1}$$

$$\Rightarrow \quad \frac{\sin(\theta+\phi)}{\sin(\theta-\phi)} = \frac{k+1}{k-1} \Rightarrow \frac{\sin(\theta-\phi)}{\sin\alpha} = \frac{k-1}{k+1} \quad [\because \ \alpha = \theta + \phi]$$

$$\Rightarrow \quad \mathbf{\sin(\theta - \phi) = \frac{k-1}{k+1}\sin\alpha.}$$

Example 8. *Evaluate :* $\sin^2(15° + A) - \sin^2(15° - A)$

Sol. $\sin^2(15° + A) - \sin^2(15° - A)$ | Using $\sin^2 A - \sin^2 B = \sin(A+B)\sin(A-B)$

$$= \sin(15° + A + 15° - A)\sin(15° + A - 15° + A)$$

$$= \sin 30° \sin 2A = \frac{1}{2}\sin 2A = \frac{2\sin A\cos A}{2} = \sin A\cos A.$$

Example 9. *If* $\theta + \phi = \frac{\pi}{4}$, *then* $(1 + \tan\theta)(1 + \tan\phi) = 2$

Sol. Given $\theta + \phi = \frac{\pi}{4}$

$$\Rightarrow \quad \phi = \frac{\pi}{4} - \theta$$

$$\therefore \quad (1+\tan\theta)(1+\tan\phi) = (1+\tan\theta)\left(1+\tan\left(\frac{\pi}{4} - \theta\right)\right)$$

$$= (1+\tan\theta)\left(1 + \frac{\tan\frac{\pi}{4} - \tan\theta}{1 + \tan\frac{\pi}{4}\tan\theta}\right)$$

$$= (1+\tan\theta)\left(1 + \frac{1-\tan\theta}{1+\tan\theta}\right)$$

$$= (1+\tan\theta)\frac{(1+\tan\theta+1-\tan\theta)}{1+\tan\theta} = 2.$$

EXERCISE

Evaluate :

1. (*i*) $\sin 12^\circ \cos 18^\circ + \cos 18^\circ \sin 12^\circ$ (*ii*) $\sin 70^\circ \cos 10^\circ - \cos 70^\circ \sin 10^\circ$

(*iii*) $\cos 40^\circ \cos 20^\circ - \sin 40^\circ \sin 20^\circ$ (*iv*) $\cos 68^\circ \cos 38^\circ + \sin 68^\circ \sin 38^\circ$

(*v*) $\dfrac{\tan 69^\circ + \tan 66^\circ}{1 - \tan 69^\circ \tan 66^\circ}$ (*vi*) $\dfrac{\tan 35^\circ - \tan 5^\circ}{1 + \tan 35^\circ \tan 5^\circ}$

2. (*i*) $\sin 75^\circ$ (*ii*) $\cos 75^\circ$

(*iii*) $\cos 105^\circ$ (*iv*) $\sin 15^\circ$

(*v*) $\tan 75^\circ$ (*vi*) $\tan 13\pi/12$.

Show that : (Q. N. 3–6)

3. (*i*) $\cos\left(\dfrac{\pi}{4} - A\right) \cos\left(\dfrac{\pi}{4} - B\right) - \sin\left(\dfrac{\pi}{4} - A\right) \sin\left(\dfrac{\pi}{4} - B\right) = \sin (A + B)$

(*ii*) $\sin (60^\circ + A) \cos (30^\circ - B) + \cos (60^\circ + A) \sin (30^\circ - B) = \cos (A - B)$

(*iii*) $\sin (n + 1)\theta \sin (n - 1)\theta + \cos (n + 1)\theta \cos (n - 1)\theta = \cos 2\theta$

(*iv*) $\dfrac{\tan (A - B) + \tan B}{1 - \tan (A - B) \tan B} = \tan A$ (*v*) $\dfrac{\cos 36^\circ - \sin 36^\circ}{\cos 36^\circ + \sin 36^\circ} = \tan 9^\circ$

(P.T.U. B. Pharma, Dec. 2005)

4. (*i*) $\sqrt{3} \cos 23^\circ - \sin 23^\circ = 2 \cos 53^\circ$ (*ii*) $\cos 15^\circ - \sin 15^\circ = \dfrac{1}{\sqrt{2}}$

5. (*i*) $\tan 2\theta - \tan \theta = \tan \theta \sec 2\theta$ (*ii*) $\cot 2\alpha - \tan \alpha = \dfrac{\cos 3\alpha}{\cos \alpha \sin 2\alpha}$

6. (*i*) $\cot (A + B) = \dfrac{\cot A \cot B - 1}{\cot B + \cot A}$ (*ii*) $\cot (A - B) = \dfrac{\cot A \cot B + 1}{\cot B - \cot A}$

Show that (Q. No. 7–12) :

7. (*i*) $\sin 105^\circ + \cos 105^\circ = \dfrac{1}{\sqrt{2}}$ (*ii*) $\tan 15^\circ + \cot 15^\circ = 4$

(*iii*) $\cos (36^\circ - A) \cos (36^\circ + A) + \cos (54^\circ + A) \cos (54^\circ - A) = \cos A$ *(P.T.U. B.Pharma May 2005)*

8. $\dfrac{\sin (A - B)}{\cos A \cos B} + \dfrac{\sin (B - C)}{\cos B \cos C} + \dfrac{\sin (C - A)}{\cos C \cos A} = 0$

9. (*i*) $\tan 3\theta - \tan 2\theta - \tan \theta = \tan \theta \tan 2\theta \tan 3\theta$

(*ii*) $\tan 9A - \tan 5A - \tan 4A = \tan 4A \tan 5A \tan 9A$

10. (*i*) $1 + \tan 69^\circ = \tan 69^\circ \tan 66^\circ - \tan 66^\circ$ (*ii*) $\tan 15^\circ + \tan 30^\circ + \tan 15^\circ \tan 30^\circ = 1$

11. (*i*) $\sin (2n + 1)A \sin A = \sin^2 (n + 1)A - \sin^2 nA$

(*ii*) $\cos^2 (45^\circ + x) - \sin^2 (45^\circ - x)$ is independent of x

12. (*i*) $\cos 2\theta \cos 2\phi + \sin^2 (\theta - \phi) - \sin^2 (\theta + \phi) = \cos (2\theta + 2\phi)$

(*ii*) $\dfrac{\tan (A + B)}{\cot (A - B)} = \dfrac{\sin^2 A - \sin^2 B}{\cos^2 A - \sin^2 B}$.

13. If $\sin A = \dfrac{1}{\sqrt{10}}$, $\sin B = \dfrac{1}{\sqrt{5}}$ (A, B and $A + B$ being positive acute angles), show that $A + B = \dfrac{\pi}{4}$.

14. If $\sin \alpha = \dfrac{15}{17}$, $\cos \beta = \dfrac{12}{13}$, find the values of $\sin(\alpha + \beta)$, $\cos(\alpha - \beta)$ and $\tan(\alpha + \beta)$ (α and β are positive acute angles).

Answers

1. (*i*) $\dfrac{1}{2}$ (*ii*) $\dfrac{\sqrt{3}}{2}$ (*iii*) $\dfrac{1}{2}$ (*iv*) $\dfrac{\sqrt{3}}{2}$

(*v*) -1 (*vi*) $\dfrac{1}{\sqrt{3}}$ **2.** (*i*) $\dfrac{\sqrt{3}+1}{2\sqrt{2}}$ (*ii*) $\dfrac{\sqrt{3}-1}{2\sqrt{2}}$

(*iii*) $\dfrac{1-\sqrt{3}}{2\sqrt{2}}$ (*iv*) $\dfrac{\sqrt{3}-1}{2\sqrt{2}}$ (*v*) $2+\sqrt{3}$ (*vi*) $2-\sqrt{3}$

14. $\dfrac{220}{221}, \dfrac{171}{221}, \dfrac{220}{21}$.

Hints

4. (*i*) $53° = 30° + 23°$.

6. (*i*) L.H.S. $= \dfrac{\cos A \cos B - \sin A \sin B}{\sin A \cos B + \cos A \sin B}$. Divide Num. and Den. by $\sin A \sin B$ and simplify.

8. Use $\sin(A - B) = \sin A \cos B - \cos A \sin B$ etc.

9. (*i*) $\tan 3\theta = \tan(2\theta + \theta) = \dfrac{\tan 2\theta + \tan \theta}{1 - \tan 2\theta \tan \theta}$. Now cross-multiply and simplify.

10. (*i*) $-1 = \tan 135° = \tan(69° + 66°)$. Now expand and simplify.

7

Transformation Formulae

7.1. INTRODUCTION

In the present chapter, we shall learn some transformation formulae for writing the product of two trigonometric functions as the sum (or difference) of two trigonometric functions.

7.2. TRANSFORMATION OF PRODUCTS INTO SUM OR DIFFERENCE OF T-FUNCTIONS

In this section, we shall learn the method of expressing the product of *t*-functions as sum or difference of *t*-functions.

Theorem I. If A and B are arbitrary angles, then prove that

(*i*) **2 sin A cos B = sin (A + B) + sin (A – B)**

(*ii*) **2 cos A sin B = sin (A + B) – sin (A – B)**

(*iii*) **2 cos A cos B = cos (A + B) + cos (A – B)**

(*iv*) **2 sin A sin B = cos (A – B) – cos (A + B).**

Proof. (*i*) $\sin (A + B) + \sin (A - B)$

$$= (\sin A \cos B + \cos A \sin B) + (\sin A \cos B - \cos A \sin B) = 2 \sin A \cos B.$$

$\therefore$ **2 sin A cos B = sin (A + B) + sin (A – B)**

(*ii*) $\sin (A + B) - \sin (A - B)$

$$= (\sin A \cos B + \cos A \sin B) - (\sin A \cos B - \cos A \sin B) = 2 \cos A \sin B.$$

$\therefore$ **2 cos A sin B = sin (A + B) – sin (A – B).**

(*iii*) $\cos (A + B) + \cos (A - B)$

$$= (\cos A \cos B - \sin A \sin B) + (\cos A \cos B + \sin A \sin B) = 2 \cos A \cos B.$$

$\therefore$ **2 cos A cos B = cos (A + B) + cos (A – B).**

(*iv*) $\cos (A - B) - \cos (A + B)$

$$= (\cos A \cos B + \sin A \sin B) - (\cos A \cos B - \sin A \sin B) = 2 \sin A \sin B.$$

$\therefore$ **2 sin A sin B = cos (A – B) – cos (A + B).**

The formulae (*i*)—(*iv*) are called **product formulae.**

Caution. In the last formula (*iv*) *i.e.*, for $2 \sin A \sin B$; the R.H.S. is

$$\cos (A - B) - \cos (A + B) \text{ and not } \cos (A + B) - \cos (A - B).$$

Remark. The above formulae are also called "*A*, *B* " formulae

Aid to Memory

1. In the "*A*, *B*" formulae, 2 must be there with the product. If not, we create it : *e.g.*,

$$\sin A \cos B = \frac{1}{2} (2 \sin A \cos B), \quad \frac{1}{5} \cos A \sin B = \frac{1}{10} (2 \cos A \sin B) \text{ etc.}$$

2. (*i*) $2 \sin A \cos B = \sin (sum) + \sin (difference)$

(*ii*) $2 \cos A \sin B = \sin (sum) - \sin (difference)$

(*iii*) $2 \cos A \cos B = \cos (sum) + \cos (difference)$

(*iv*) $2 \sin A \sin B = \cos (difference) - \cos (sum)$.

Here *sum* stands for $A + B$ and *difference* for $A - B$.

ILLUSTRATIVE EXAMPLES

Example 1. *Change the following products into sum or difference of t-ratios :*

(*i*) $2 \sin 3\theta \cos 5\theta$ (*ii*) $\cos 9\theta \sin 2\theta$

(*iii*) $4 \cos 8\theta \cos 3\theta$ (*iv*) $5 \sin \theta \sin 8\theta$.

Sol. (*i*) $2 \sin 3\theta \cos 5\theta = \sin (3\theta + 5\theta) + \sin (3\theta - 5\theta) = \sin 8\theta + \sin (-2\theta)$
$= \mathbf{\sin 8\theta - \sin 2\theta}$.

(*ii*) $\cos 9\theta \sin 2\theta = \frac{1}{2}[2 \cos 9\theta \sin 2\theta] = \frac{1}{2}[\sin (9\theta + 2\theta) - \sin (9\theta - 2\theta)]$

$= \mathbf{\frac{1}{2}[\sin 11\theta - \sin 7\theta]}$.

(*iii*) $4 \cos 8\theta \cos 3\theta = 2 [2 \cos 8\theta \cos 3\theta] = 2 [\cos (8\theta + 3\theta) + \cos (8\theta - 3\theta)]$
$= \mathbf{2 [\cos 11\theta + \cos 5\theta]}$.

(*iv*) $5 \sin \theta \sin 8\theta = \frac{5}{2}[2 \sin 8\theta \sin \theta] = \frac{5}{2}[\cos (8\theta - \theta) - \cos (8\theta + \theta)]$

[Note this step]

$= \mathbf{\frac{5}{2}[\cos 7\theta - \cos 9\theta]}$.

Example 2. *Show that* $\sin (45° + A) \sin (45° - A) = \frac{1}{2} \cos 2A$.

Sol. $\sin (45° + A) \sin (45° - A) = \frac{1}{2}[2 \sin (45° + A) \sin (45° - A)]$

$= \frac{1}{2}\{\cos [(45° + A) - (45° - A)] - \cos [(45° + A) + (45° - A)]\}$

$= \frac{1}{2}[\cos 2A - \cos 90°] = \frac{1}{2}[\cos 2A - 0] = \mathbf{\frac{1}{2} \cos 2A}$.

Example 3. *Show that* $\frac{2 \cos 2A + 1}{2 \cos 2A - 1} = \tan (60° + A) \tan (60° - A)$.

Sol. R.H.S. $= \tan (60° + A) \tan (60° - A)$

$$= \frac{\sin (60° + A) \sin (60° - A)}{\cos (60° + A) \cos (60° - A)} = \frac{2 \sin (60° + A) \sin (60° - A)}{2 \cos (60° + A) \cos (60° - A)}$$

$$= \frac{\cos [(60° + A) - (60° - A)] - \cos [(60° + A) + (60° - A)]}{\cos [(60° + A) + (60° - A)] + \cos [(60° + A) - (60° - A)]} = \frac{\cos 2A - \cos 120°}{\cos 120° + \cos 2A}$$

$$= \frac{\cos 2A - \left(-\frac{1}{2}\right)}{\left(-\frac{1}{2}\right) + \cos 2A} = \frac{2 \cos 2A + 1}{2 \cos 2A - 1} = \text{L.H.S.} \qquad \left| \cos 120° = \frac{-1}{2} \right.$$

Example 4. *Show that :* $\sin 10° \sin 50° \sin 60° \sin 70° = \frac{\sqrt{3}}{16}$.

Sol. L.H.S. $= \sin 10° \sin 50° \sin 60° \sin 70°$

$$= \frac{\sqrt{3}}{2} \sin 10° \sin 50° \sin 70° = \frac{\sqrt{3}}{4}[2 \sin 50° \sin 10°] \sin 70°$$

$$= \frac{\sqrt{3}}{4}\ ((\cos 50° - 10°) - \cos (50° + 10°)) \sin 70°$$

$$= \frac{\sqrt{3}}{4}\ [\cos 40° - \cos 60°] \sin 70° = \frac{\sqrt{3}}{4}\left[\cos 40° - \frac{1}{2}\right] \sin 70°$$

$$= \frac{\sqrt{3}}{4} \cos 40° \sin 70° - \frac{\sqrt{3}}{8} \sin 70° = \frac{\sqrt{3}}{8}\ [2 \sin 70° \cos 40°] - \frac{\sqrt{3}}{8} \sin 70°$$

$$= \frac{\sqrt{3}}{8}[\sin 110° + \sin 30°] - \frac{\sqrt{3}}{8} \sin 70° = \frac{\sqrt{3}}{8} \sin 110° + \frac{\sqrt{3}}{8}\left(\frac{1}{2}\right) - \frac{\sqrt{3}}{8} \sin 70°$$

[2 sin A cos B = sin (A + B) + sin (A – B)]

$$= \frac{\sqrt{3}}{8} \sin (180° - 70°) + \frac{\sqrt{3}}{16} - \frac{\sqrt{3}}{8} \sin 70° = \frac{\sqrt{3}}{8} \sin 70° + \frac{\sqrt{3}}{16} - \frac{\sqrt{3}}{8} \sin 70°$$

$$= \frac{\sqrt{3}}{16} = \text{R.H.S.}$$

Example 5. *Show that* $\cos 20° \cos 30° \cos 40° \cos 80° = \frac{\sqrt{3}}{16}$.

Sol. L.H.S. = cos 20° cos 30° cos 40° cos 80°

$$= \frac{\sqrt{3}}{2} \cos 20° \cos 40° \cos 80° = \frac{\sqrt{3}}{4}\ [2 \cos 40° \cos 20°] \cos 80°$$

$$= \frac{\sqrt{3}}{4}[\cos 60° + \cos 20°] \cos 80° = \frac{\sqrt{3}}{4}\left[\frac{1}{2} + \cos 20°\right] \cos 80°$$

[2 cos A cos B = cos (A + B) + cos (A – B)]

$$= \frac{\sqrt{3}}{8} \cos 80° + \frac{\sqrt{3}}{4} \cos 20° \cos 80° = \frac{\sqrt{3}}{8} \cos 80° + \frac{\sqrt{3}}{8}[2 \cos 80° \cos 20°]$$

$$= \frac{\sqrt{3}}{8} \cos 80° + \frac{\sqrt{3}}{8}[\cos 100° + \cos 60°] = \frac{\sqrt{3}}{8} \cos 80° + \frac{\sqrt{3}}{8} \cos (180° - 80°) + \frac{\sqrt{3}}{8} \times \frac{1}{2}$$

$$= \frac{\sqrt{3}}{8} \cos 80° - \frac{\sqrt{3}}{8} \cos 80° + \frac{\sqrt{3}}{16} = \frac{\sqrt{3}}{16} = \text{R.H.S.}$$

EXERCISE 7.1

1. Change the following products as sum or difference of *t*-ratios :

   (*i*) $2 \sin 7\theta \cos 2\theta$ (*ii*) $2 \cos 5\theta \sin \theta$ (*iii*) $\frac{1}{2} \cos 2\theta \cos \theta$ (*iv*) $\frac{1}{7} \sin 8\theta \sin 2\theta$.

Show that :

2. (*i*) $\sin (45° + A) \sin (45° - A) = \frac{1}{2} \cos 2A$ (*ii*) $\sec \left(\frac{\pi}{4} + \theta\right) \sec \left(\frac{\pi}{4} - \theta\right) = 2 \sec 2\theta$
3. (*i*) $2 \sin (2\theta + \phi) \cos (\theta - 2\phi) = \sin (3\theta - \phi) + \sin (\theta + 3\phi)$

   (*ii*) $\cos (60° + \alpha) \sin (60° - \alpha) = \frac{1}{4} (\sqrt{3} - 2 \sin 2\alpha)$
4. $\cos (A + B) + \sin (A - B) = 2 \sin (45° + A) \cos (45° + B)$
5. $2 \cos \frac{\pi}{13} \cos \frac{9\pi}{13} + \cos \frac{3\pi}{13} + \cos \frac{5\pi}{13} = 0$
6. $\frac{2 \sin (\alpha - \gamma) \cos \gamma - \sin (\alpha - 2\gamma)}{2 \sin (\beta - \gamma) \cos \gamma - \sin (\beta - 2\gamma)} = \frac{\sin \alpha}{\sin \beta}$.

7. (*i*) $\tan (A + 30°) + \cot (A - 30°) = \dfrac{1}{\sin 2A - \sin 60°}$

(*ii*) $\tan (45° + \theta) + \tan (45° - \theta) = 2 \sec 2\theta$

8. (*i*) $\cos^2 A + \cos^2 B - 2 \cos A \cos B \cos (A + B) = \sin^2 (A + B)$

(*ii*) $\sin^2 A + \sin^2 (A - B) - 2 \sin A \cos B \sin (A - B) = \sin^2 B$

9. $4 \cos A \cos (60° - A) \cos (60° + A) = \cos 3A$

10. (*i*) $\sin 10° \sin 30° \sin 50° \sin 70° = \dfrac{1}{16}$ (*ii*) $\sin 20° \sin 40° \sin 60° \sin 80° = \dfrac{3}{16}$

(*iii*) $\cos 20° \cos 40° \cos 60° \cos 80° = \dfrac{1}{16}$ (*iv*) $\cos 10° \cos 30° \cos 50° \cos 70° = \dfrac{3}{16}$

11. $\tan 20° \tan 40° \tan 60° \tan 80° = 3$

12. $\sin A \sin (B - C) + \sin B \sin (C - A) + \sin C \sin (A - B) = 0.$

Answer

1. (*i*) $\sin 9\theta + \sin 5\theta$ (*ii*) $\sin 6\theta - \sin 4\theta$

(*iii*) $\dfrac{1}{4} [\cos 3\theta + \cos \theta]$ (*iv*) $\dfrac{1}{14} [\cos 6\theta - \cos 10\theta]$.

Hint

8. (*i*) Use : $2 \cos A \cos B = \cos (A + B) + \cos (A - B)$.

7.2. TRANSFORMATION OF SUM OR DIFFERENCE INTO PRODUCT OF T-FUNCTIONS

In this section, we shall learn the method of expressing the sum or difference of *t*-functions into product of *t*-functions.

Theorem. *If C and D are arbitrary angles, then prove that*

(*i*) $\sin C + \sin D = 2 \sin \dfrac{C + D}{2} \cos \dfrac{C - D}{2}$

(*ii*) $\sin C - \sin D = 2 \cos \dfrac{C + D}{2} \sin \dfrac{C - D}{2}$

(*iii*) $\cos C + \cos D = 2 \cos \dfrac{C + D}{2} \cos \dfrac{C - D}{2}$

(*iv*) $\cos C - \cos D = 2 \sin \dfrac{C + D}{2} \sin \dfrac{D - C}{2} = -2 \sin \dfrac{C + D}{2} \sin \dfrac{C - D}{2}$.

Proof. Let A and B be angles such that $A + B = C$ and $A - B = D$.

Solving, we get $A = \dfrac{C + D}{2}$ and $B = \dfrac{C - D}{2}$

(*i*) $\sin C + \sin D = \sin (A + B) + \sin (A - B)$

$= (\sin A \cos B + \cos A \sin B) + (\sin A \cos B - \cos A \sin B)$

$= 2 \sin A \cos B = 2 \sin \dfrac{C + D}{2} \cos \dfrac{C - D}{2}$

$\therefore$ $\mathbf{\sin C + \sin D = 2 \sin \dfrac{C + D}{2} \cos \dfrac{C - D}{2}}$.

(*ii*) $\sin C - \sin D = \sin (A + B) - \sin (A - B)$

$= (\sin A \cos B + \cos A \sin B) - (\sin A \cos B - \cos A \sin B)$

$= 2 \cos A \sin B = 2 \cos \dfrac{C + D}{2} \sin \dfrac{C - D}{2}$

$\therefore \quad \mathbf{\sin C - \sin D = 2 \cos \frac{C+D}{2} \sin \frac{C-D}{2}}.$

(*iii*) $\cos C + \cos D = \cos (A + B) + \cos (A - B)$

$$= (\cos A \ \cos B - \sin A \sin B) + (\cos A \cos B + \sin A \sin B)$$

$$= 2 \cos A \cos B = 2 \cos \frac{C+D}{2} \cos \frac{C-D}{2}$$

$\therefore \quad \mathbf{\cos C + \cos D = 2 \cos \frac{C+D}{2} \cos \frac{C-D}{2}}.$

(*iv*) $\cos C - \cos D = \cos (A + B) - \cos (A - B)$

$$= (\cos A \ \cos B - \sin A \sin B) - (\cos A \cos B + \sin A \sin B)$$

$$= -2 \sin A \sin B = -2 \sin \frac{C+D}{2} \sin \frac{C-D}{2}$$

$$= -2 \sin \frac{C+D}{2} \sin \left(-\frac{D-C}{2}\right)$$

$$= -2 \sin \frac{C+D}{2} \times \left(-\sin \frac{D-C}{2}\right) = 2 \sin \frac{C+D}{2} \sin \frac{D-C}{2}$$

$\therefore \quad \mathbf{\cos C - \cos D = 2 \sin \frac{C+D}{2} \sin \frac{D-C}{2}}.$

Caution. The R.H.S. of (*iv*) contains $\sin \frac{D-C}{2}$ and not $\sin \frac{C+D}{2}$.

Remark 1. The above formulae are called '*C, D*' formulae.

Remark 2. In each of the four '*C, D*' formulae, both *t*-ratios on the L.H.S. are same.
If however, we have sin *C* and cos *D*, then either 'sin' is changed into 'cos' or 'cos' into 'sin' as below :

$$\sin C + \cos D = \sin C + \sin \left(\frac{\pi}{2} - D\right) \quad \text{or} \quad \sin C + \cos D = \cos \left(\frac{\pi}{2} - C\right) + \cos D.$$

ILLUSTRATIVE EXAMPLES

Example 1. *Express as products of t-ratios :*

(*i*) *sin 4θ + sin 3θ* (*ii*) *sin 7θ – sin 4θ* (*iii*) *cos 9θ + cos 3θ* (*iv*) *cos 3θ – cos 7θ.*

Sol. (*i*) $\sin 4\theta + \sin 3\theta = 2 \sin \frac{4\theta + 3\theta}{2} \cos \frac{4\theta - 3\theta}{2} = \mathbf{2 \sin \frac{7\theta}{2} \cos \frac{\theta}{2}}$

(*ii*) $\sin 7\theta - \sin 4\theta = 2 \cos \frac{7\theta + 4\theta}{2} \sin \frac{7\theta - 4\theta}{2} = \mathbf{2 \cos \frac{11\theta}{2} \sin \frac{3\theta}{2}}$

(*iii*) $\cos 9\theta + \cos 3\theta = 2 \cos \frac{9\theta + 3\theta}{2} \cos \frac{9\theta - 3\theta}{2} = \mathbf{2 \cos 6\theta \cos 3\theta}$

(*iv*) $\cos 3\theta - \cos 7\theta = 2 \sin \frac{3\theta + 7\theta}{2} \sin \frac{7\theta - 3\theta}{2} = \mathbf{2 \sin 5\theta \sin 2\theta}.$

Example 2. *Show that cos 52° + cos 68° + cos 172° = 0.*

Sol. $\cos 52° + \cos 68° + \cos 172° = (\cos 68° + \cos 52°) + \cos 172°$

$$= 2 \cos \frac{68° + 52°}{2} \cos \frac{68° - 52°}{2} + \cos 172° = 2 \cos 60° \cos 8° + \cos (180° - 8°)$$

$$= 2 \left(\frac{1}{2}\right) \cos 8° - \cos 8° = \cos 8° - \cos 8° = \mathbf{0}.$$

Example 3. *Show that* $\frac{\cos 2B - \cos 2A}{\sin 2B + \sin 2A} = \tan (A - B)$. (G.N.D.U. B.Pharma, 2005)

Sol. L.H.S. $= \dfrac{\cos 2B - \cos 2A}{\sin 2B + \sin 2A} = \dfrac{2 \sin \dfrac{2B+2A}{2} \sin \dfrac{2A-2B}{2}}{2 \sin \dfrac{2B+2A}{2} \cos \dfrac{2B-2A}{2}} = \dfrac{\sin (A-B)}{\cos (B-A)}$

$= \dfrac{\sin (A-B)}{\cos (-(A-B))} = \dfrac{\sin (A-B)}{\cos (A-B)} = \tan (A-B) =$ R.H.S.

Example 4. *If cosec A + sec A = cosec B + sec B, then show that* $\tan A \tan B = \cot \dfrac{A+B}{2}$.

Sol. We have cosec A + sec A = cosec B + sec B.

$\Rightarrow \dfrac{1}{\sin A} + \dfrac{1}{\cos A} = \dfrac{1}{\sin B} + \dfrac{1}{\cos B} \qquad \Rightarrow \dfrac{1}{\sin A} - \dfrac{1}{\sin B} = \dfrac{1}{\cos B} - \dfrac{1}{\cos A}$

$\Rightarrow \dfrac{\sin B - \sin A}{\sin A \sin B} = \dfrac{\cos A - \cos B}{\cos B \cos A} \qquad \Rightarrow \dfrac{\sin B - \sin A}{\cos A - \cos B} = \dfrac{\sin A \sin B}{\cos B \cos A}$

$\Rightarrow \dfrac{2 \cos \dfrac{B+A}{2} \sin \dfrac{B-A}{2}}{2 \sin \dfrac{A+B}{2} \sin \dfrac{B-A}{2}} = \tan A \tan B \qquad \therefore\ \mathbf{\cot \dfrac{A+B}{2} = \tan A \tan B.}$

Example 5. *Prove that* $(\cos \alpha + \cos \beta)^2 + (\sin \alpha + \sin \beta)^2 = 4 \cos^2 \dfrac{\alpha - \beta}{2}$.

Sol. L.H.S. $= (\cos \alpha + \cos \beta)^2 + (\sin \alpha + \sin \beta)^2$

$= \left(2 \cos \dfrac{\alpha+\beta}{2} \cos \dfrac{\alpha-\beta}{2}\right)^2 + \left(2 \sin \dfrac{\alpha+\beta}{2} \cos \dfrac{\alpha-\beta}{2}\right)^2$

$= 4 \cos^2 \dfrac{\alpha-\beta}{2} \left(\cos^2 \dfrac{\alpha+\beta}{2} + \sin^2 \dfrac{\alpha+\beta}{2}\right) = 4 \cos^2 \dfrac{\alpha-\beta}{2} =$ R.H.S.

Example 6. *Show that* $\sin 10° + \sin 20° + \sin 40° + \sin 50° = \sin 70° + \sin 80°$.

Sol. L.H.S. $= \sin 10° + \sin 20° + \sin 40° + \sin 50°$

$= (\sin 50° + \sin 10°) + (\sin 40° + \sin 20°)$ **(Note this step)**

$= 2 \sin 30° \cos 20° + 2 \sin 30° \cos 10°$

$= 2 \sin 30° [\cos 20° + \cos 10°] = 2 \times \dfrac{1}{2} [2 \cos 15° \cos 5°]$

$= 2 \cos 15° \cos 5°$

R.H.S. $= \sin 70° + \sin 80° = \sin 80° + \sin 70°$

$= 2 \sin \dfrac{80° + 70°}{2} \cos \dfrac{80° - 70°}{2}$

$= 2 \sin 75° \cos 5° = 2 \sin (90° - 15°) \cos 5° = 2 \cos 15° \cos 5°$

$\therefore$ L.H.S. = R.H.S.

Example 7. *Show that* $\dfrac{\sin A - \sin 3A + \sin 5A - \sin 7A}{\cos A - \cos 3A - \cos 5A + \cos 7A} = \cot 2A$.

Sol. L.H.S. $= \dfrac{\sin A - \sin 3A + \sin 5A - \sin 7A}{\cos A - \cos 3A - \cos 5A + \cos 7A} = \dfrac{-(\sin 3A - \sin A) - (\sin 7A - \sin 5A)}{(\cos A - \cos 3A) - (\cos 5A - \cos 7A)}$

$= \dfrac{-2 \cos \dfrac{4A}{2} \sin \dfrac{2A}{2} - 2 \cos \dfrac{12A}{2} \sin \dfrac{2A}{2}}{2 \sin \dfrac{4A}{2} \sin \dfrac{2A}{2} - 2 \sin \dfrac{12A}{2} \sin \dfrac{2A}{2}} = \dfrac{-2 \cos 2A \sin A - 2 \cos 6A \sin A}{2 \sin 2A \sin A - 2 \sin 6A \sin A}$

$$= \frac{-2\sin A(\cos 2A + \cos 6A)}{2\sin A(\sin 2A - \sin 6A)} = \frac{-(\cos 2A + \cos 6A)}{\sin 2A - \sin 6A}$$

$$= \frac{\cos 6A + \cos 2A}{\sin 6A - \sin 2A} = \frac{2\cos\frac{8A}{2}\cos\frac{4A}{2}}{2\cos\frac{8A}{2}\sin\frac{4A}{2}} = \frac{\cos 2A}{\sin 2A} = \cot 2A = \text{R.H.S.}$$

Example 8. *Show that* $\frac{\sin 8A\cos A - \cos 3A\sin 6A}{\cos 2A\cos A - \sin 3A\sin 4A} = \tan 2A.$

Sol. L.H.S. $= \frac{\sin 8A\cos A - \cos 3A\sin 6A}{\cos 2A\cos A - \sin 3A\sin 4A} = \frac{2\sin 8A\cos A - 2\sin 6A\cos 3A}{2\cos 2A\cos A - 2\sin 4A\sin 3A}$

$$= \frac{(\sin 9A + \sin 7A) - (\sin 9A + \sin 3A)}{(\cos 3A + \cos A) - (\cos A - \cos 7A)} = \frac{\sin 7A - \sin 3A}{\cos 7A + \cos 3A} = \frac{2\cos\frac{10A}{2}\sin\frac{4A}{2}}{2\cos\frac{10A}{2}\cos\frac{4A}{2}}$$

$$= \frac{\sin 2A}{\cos 2A} = \tan 2A = \text{R.H.S.}$$

Example 9. *Show that*

$$\sin 51^\circ + \cos 81^\circ = \cos 21^\circ$$

Sol. $\sin 51^\circ + \cos 81^\circ = \sin 51^\circ + \cos(90 - 9)^\circ$

$$= \sin 51^\circ + \sin 9^\circ$$

$$= 2\sin\left(\frac{51+9}{2}\right)^\circ \cos\left(\frac{51-9}{2}\right)^\circ = 2\sin 30^\circ\cos 21^\circ$$

$$= 2\cdot\frac{1}{2}\cos 21^\circ = \cos 21^\circ.$$

Example 10. *If* $\cos(\theta + 2\alpha) = m\cos\theta$, *show that*

$$\cot\alpha = \frac{1+m}{1-m}\tan(\theta + \alpha).$$

Sol. Given $\cos(\theta + 2\alpha) = m\cos\theta$

$\Rightarrow$ $\frac{\cos(\theta + 2\alpha)}{\cos\theta} = \frac{m}{1}$. Applying componendo and dividendo,

$$\frac{\cos(\theta + 2\alpha) + \cos\theta}{\cos(\theta + 2\alpha) - \cos\theta} = \frac{m+1}{m-1}$$

$$\Rightarrow \quad \frac{2\cos\frac{\theta + 2\alpha + \theta}{2}\cos\frac{\theta + 2\alpha - \theta}{2}}{-2\sin\frac{\theta + 2\alpha + \theta}{2}\sin\frac{\theta + 2\alpha - \theta}{2}} = \frac{m+1}{m-1}$$

$$\Rightarrow \quad \frac{\cos(\theta + \alpha)\cos\alpha}{-\sin(\theta + \alpha)\sin\alpha} = \frac{m+1}{m-1} \quad \Rightarrow \quad \cot\alpha = \frac{1+m}{1-m}\cdot\tan(\alpha + \theta).$$

Example 11. *If* $A + B + C = 180^\circ$, *show that*

$$\cos^2 A + \cos^2 B + \cos^2 C = 1 - 2\cos A\cos B\cos C.$$

Sol. Given $\quad A + B + C = 180^\circ \quad \Rightarrow \quad C = 180^\circ - (A + B)$

$\therefore \quad \cos^2 A + \cos^2 B + \cos^2 C = \cos^2 A + \cos^2 B + \cos^2(180^\circ - \overline{A + B})$

$$= \cos^2 A + \cos^2 B + \cos^2 (A + B) \qquad | \cos (180° - \theta) = -\cos \theta$$
$$= \cos^2 A + \cos^2 (A + B) + \cos^2 B$$
$$= \cos^2 A - \sin^2(A + B) + 1 + \cos^2 B \qquad | \cos^2 \theta = 1 - \sin^2 \theta$$
$$= \cos (A + A + B) \cos (A - \overline{A + B}) + 1 + \cos^2 B$$
$$| \cos^2 A - \sin^2 B = \cos (A + B) \cos (A - B)$$
$$= \cos (2A + B) \cos (- B) + 1 + \cos^2 B$$
$$= \cos (2A + B) \cos B + \cos^2 B + 1$$
$$= \cos B (\cos (2A + B) + \cos B) + 1$$
$$= \cos B \left(2 \cos \frac{2A + B + B}{2} \cos \frac{2A + B - B}{2}\right) + 1$$
$$\left| \cos C + \cos D = 2 \cos \frac{C + D}{2} \cos \frac{C - D}{2}\right.$$
$$= 2 \cos B \cos (A + B) \cos A + 1 = 2 \cos B \cos (180° - C) \cos A + 1$$
$$= - 2 \cos B \cos C \cos A + 1 = 1 - 2 \cos A \cos B \cos C.$$

EXERCISE 7.2

1. Express the following as product of *t*-ratios :

(*i*) $\sin 9\theta + \sin 5\theta$ (*ii*) $\frac{1}{2}(\sin 8\theta - \sin 2\theta)$ (*iii*) $\cos 4\theta + \cos \theta$ (*iv*) $\frac{1}{2}(\cos 2\theta - \cos 20\theta)$.

2. Prove the following :

(*i*) $\cos\left(\frac{\pi}{4} + x\right) + \cos\left(\frac{\pi}{4} - x\right) = \sqrt{2} \cos x$

(*ii*) $\cos\left(\frac{3\pi}{4} + x\right) - \cos\left(\frac{3\pi}{4} - x\right) = -\sqrt{2} \sin x$

(*iii*) $\sin 105° + \cos 105° = \cos 45°$. (*P.T.U. B.Pharma Dec. 2005*)

Prove the following :

3. (*i*) $\frac{\sin x - \sin y}{\cos x + \cos y} = \tan \frac{x - y}{2}$ (*ii*) $\frac{\sin 5x + \sin 3x}{\cos 5x + \cos 3x} = \tan 4x$

(*iii*) $\frac{\sin x + \sin 3x}{\cos x + \cos 3x} = \tan 2x$ (*iv*) $\frac{\cos 9x - \cos 5x}{\sin 17x - \sin 3x} = -\frac{\sin 2x}{\cos 10x}$

(*v*) $\frac{\cos 11° + \sin 11°}{\cos 11° - \sin 11°} = \tan 56°$ (*P.T.U. B.Pharma May. 2005*)

4. (*i*) $\sin 3x + \sin 2x - \sin x = 4 \sin x \cos \frac{x}{2} \cos \frac{3x}{2}$

(*ii*) $\cos 55° + \cos 65° + \cos 175° = 0$

(*iii*) $\sin 10° + \sin 50° - \sin 70° = 0$ (*P.T.U. B.Pharma, Dec. 2005*)

(*iv*) $\cos 20° + \cos 100° + \cos 140° = 0$ (*P.T.U. B.Pharma, May 2006*)

5. (*i*) $\cos 3A \cos 2A + \sin 4A \sin A = \cos A \cos 2A$

(*ii*) $\sin \frac{11\theta}{4} \sin \frac{\theta}{4} + \sin \frac{7\theta}{4} \sin \frac{3\theta}{4} = \sin 2\theta \sin \theta$

6. (*i*) $(\cos \alpha - \cos \beta)^2 + (\sin \alpha - \sin \beta)^2 = 4 \sin^2 \frac{\alpha - \beta}{2}$

(*ii*) $\cos^2 2x - \cos^2 6x = \sin 4x \sin 8x$

7. (*i*) $\sin \alpha + \sin \left(\alpha + \frac{2\pi}{3}\right) + \sin \left(\alpha + \frac{4\pi}{3}\right) = 0$ (*ii*) $\cos \alpha + \cos \left(\alpha + \frac{2\pi}{3}\right) + \cos \left(\alpha + \frac{4\pi}{3}\right) = 0.$

Prove the following (8—14) :

8. $\cos \alpha + \cos \beta + \cos \gamma + \cos (\alpha + \beta + \gamma) = 4 \cos \frac{\alpha + \beta}{2} \cos \frac{\beta + \gamma}{2} \cos \frac{\gamma + \alpha}{2}$

9. $\frac{\sin A + \sin 3A + \sin 5A + \sin 7A}{\cos A + \cos 3A + \cos 5A + \cos 7A} = \tan 4A$

10. $\frac{(\cos \theta - \cos 3\theta)(\sin 8\theta + \sin 2\theta)}{(\sin 5\theta - \sin \theta)(\cos 4\theta - \cos 6\theta)} = 1$

11. $\frac{\cos \theta + \cos 3\theta - \cos 2\theta}{\sin \theta + \sin 3\theta - \sin 2\theta} = \cot 2\theta$

12. $\frac{\cos 4x + \cos 3x + \cos 2x}{\sin 4x + \sin 3x + \sin 2x} = \cot 3x$

13. $\frac{\sin 11A \sin A + \sin 7A \sin 3A}{\cos 11A \sin A + \cos 7A \sin 3A} = \tan 8A$

14. $\frac{\sin 5x - 2 \sin 3x + \sin x}{\cos 5x - \cos x} = \tan x.$

15. If $\sin x + \sin y = a$ and $\cos x + \cos y = b$, find the values of :

(*i*) $\tan \frac{x + y}{2}$ (*ii*) $\tan \frac{x - y}{2}.$

Answers

1. (*i*) $2 \sin 7\theta \cos 2\theta$ (*ii*) $\cos 5\theta \sin 3\theta$ (*iii*) $2 \cos \frac{5\theta}{2} \cos \frac{3\theta}{2}$ (*iv*) $\sin 11\theta \sin 9\theta$

15. (*i*) $\frac{a}{b}$ (*ii*) $\pm \sqrt{\frac{4 - a^2 - b^2}{a^2 + b^2}}.$

Hints

7. (*i*) L.H.S. $= \sin \alpha + 2 \sin \frac{2\alpha + 2\pi}{2} \cos \left(-\frac{2\pi}{6}\right) = \sin \alpha + 2 (-\sin \alpha) \left(\frac{1}{2}\right) = 0.$

15. Given equations reduce to $2 \sin \frac{x + y}{2} \cos \frac{x - y}{2} = a$ ...(1)

and $2 \cos \frac{x + y}{2} \cos \frac{x - y}{2} = b$ ...(2)

Dividing (1) by (2), we get $\tan \frac{x + y}{2} = \frac{a}{b}.$

Squaring (1) and (2) and adding, we get $4\cos^2 \frac{x - y}{2} = a^2 + b^2$. $\therefore \tan^2 \frac{x - y}{2} = \frac{1}{\cos^2 \frac{x - y}{2}} - 1$ etc.

8

Trigonometric Functions of Multiple and Sub-Multiple Angles

8.1. INTRODUCTION

In this chapter, we shall learn the methods of finding the values of trigonometric functions of multiple angles $2A$, $3A$, etc. and submultiple angles $A/2$, $A/3$, etc. in terms of trigonometric functions of angle A. These formulae would lead us to evaluate the values of trigonometric functions of angles like 18°, 36°, $7\frac{1}{2}^\circ$, $142\frac{1}{2}^\circ$, etc.

8.2. TRIGONOMETRIC FUNCTIONS OF MULTIPLE ANGLES

The angles $2A$, $3A$, ... are called **multiple angles** of A.

Theorem I. *Prove that :*

(*i*) $\sin 2A = 2\sin A \cos A$ (*ii*) $\cos 2A = \begin{cases} \cos^2 A - \sin^2 A \\ 1 - 2\sin^2 A \\ 2\cos^2 A - 1 \end{cases}$

(*iii*) $\tan 2A = \dfrac{2\tan A}{1-\tan^2 A}$.

Proof. (*i*) $\sin 2A = \sin (A + A) = \sin A \cos A + \cos A \sin A = 2 \sin A \cos A$

$\therefore$ $\mathbf{\sin 2A = 2 \sin A \cos A.}$

(*ii*) $\cos 2A = \cos (A + A) = \cos A \cos A - \sin A \sin A = \cos^2 A - \sin^2 A$

$\therefore$ $\mathbf{\cos 2A = \cos^2 A - \sin^2 A.}$ ...(1)

(1) $\Rightarrow$ $\cos 2A = (1 - \sin^2 A) - \sin^2 A = 1 - 2\sin^2 A$

$\therefore$ $\mathbf{\cos 2A = 1 - 2 \sin^2 A.}$

(1) $\Rightarrow$ $\cos 2A = \cos^2 A - (1 - \cos^2 A) = 2\cos^2 A - 1$

$\therefore$ $\mathbf{\cos 2A = 2 \cos^2 A - 1.}$

(*iii*) $$\tan 2A = \tan (A + A) = \frac{\tan A + \tan A}{1 - \tan A \tan A} = \frac{2\tan A}{1-\tan^2 A}$$

$\therefore$ $$\mathbf{\tan 2A = \frac{2\tan A}{1-\tan^2 A}}.$$

Corollary 1. *Prove that :*

(*i*) $1 - \cos 2A = 2\sin^2 A$ (*ii*) $1 + \cos 2A = 2\cos^2 A$.

Proof. (*i*) We have $\cos 2A = 1 - 2\sin^2 A$. $\therefore$ $\mathbf{1 - \cos 2A = 2\sin^2 A}$.

(*ii*) We have $\cos 2A = 2\cos^2 A - 1$ $\therefore$ $\mathbf{1 + \cos 2A = 2\cos^2 A}$.

Aid to memory

(*i*) 1 – cos (*double angle*) = 2 sin² (*angle*) (*ii*) 1 + cos (*double angle*) = 2 cos² (*angle*).

Corollary 2. *Prove that :*

(*i*) $\sin^2 A = \dfrac{1 - \cos 2A}{2}$ (*ii*) $\cos^2 A = \dfrac{1 + \cos 2A}{2}$.

Proof. (*i*) $\dfrac{1-\cos 2A}{2} = \dfrac{1-(1-2\sin^2 A)}{2} = \dfrac{2\sin^2 A}{2} = \sin^2 A$ $\therefore$ $\mathbf{\sin^2 A = \dfrac{1-\cos 2A}{2}}$.

(*ii*) $\dfrac{1+\cos 2A}{2} = \dfrac{1+(2\cos^2 A - 1)}{2} = \dfrac{2\cos^2 A}{2} = \cos^2 A$ $\therefore$ $\mathbf{\cos^2 A = \dfrac{1+\cos 2A}{2}}$.

Aid to memory

(*i*) $\sin^2 (angle) = \dfrac{1-\cos(double\ angle)}{2}$ (*ii*) $\cos^2 (angle) = \dfrac{1+\cos(double\ angle)}{2}$.

Theorem II. *Prove that :*

(*i*) $\sin 2A = \dfrac{2\tan A}{1+\tan^2 A}$ (*ii*) $\cos 2A = \dfrac{1-\tan^2 A}{1+\tan^2 A}$.

Proof. (*i*) $\sin 2A = 2\sin A\cos A = \dfrac{2\sin A\cos A}{1} = \dfrac{2\sin A\cos A}{\cos^2 A + \sin^2 A}$

$$= \frac{\dfrac{2\sin A\cos A}{\cos^2 A}}{\dfrac{\cos^2 A}{\cos^2 A} + \dfrac{\sin^2 A}{\cos^2 A}} = \frac{2\tan A}{1+\tan^2 A} \quad \therefore \quad \mathbf{\sin 2A = \frac{2\tan A}{1+\tan^2 A}}.$$

(*ii*) $\cos 2A = \cos^2 A - \sin^2 A = \dfrac{\cos^2 A - \sin^2 A}{1} = \dfrac{\cos^2 A - \sin^2 A}{\cos^2 A + \sin^2 A}$

$$= \frac{\dfrac{\cos^2 A}{\cos^2 A} - \dfrac{\sin^2 A}{\cos^2 A}}{\dfrac{\cos^2 A}{\cos^2 A} + \dfrac{\sin^2 A}{\cos^2 A}} = \frac{1-\tan^2 A}{1+\tan^2 A} \quad \therefore \quad \mathbf{\cos 2A = \frac{1-\tan^2 A}{1+\tan^2 A}}.$$

Caution. In the denominator of sin 2*A*, we have 1 + tan² *A*, whereas in the denominator of tan 2*A*, there is 1 – tan² *A*.

Theorem III. Prove that :

(*i*) $\sin 3\theta = 3\sin\theta - 4\sin^3\theta$ (*ii*) $\cos 3\theta = 4\cos^3\theta - 3\cos\theta$

(*iii*) $\tan 3\theta = \dfrac{3\tan\theta - \tan^3\theta}{1 - 3\tan^2\theta}$.

Proof. (*i*) $\sin 3\theta = \sin(\theta + 2\theta) = \sin\theta\cos 2\theta + \cos\theta\sin 2\theta$

$= \sin\theta(1 - 2\sin^2\theta) + \cos\theta \,.\, 2\sin\theta\cos\theta = \sin\theta - 2\sin^3\theta + 2\sin\theta\cos^2\theta$

$= \sin\theta - 2\sin^3\theta + 2\sin\theta(1 - \sin^2\theta) = \sin\theta - 2\sin^3\theta + 2\sin\theta - 2\sin^3\theta$

$= 3 \sin \theta - 4 \sin^3 \theta.$

$\therefore \quad \mathbf{\sin 3\theta = 3 \sin \theta - 4 \sin^3 \theta.}$

(*ii*) $\cos 3\theta = \cos (\theta + 2\theta) = \cos \theta \cos 2\theta - \sin \theta \sin 2\theta$

$$= \cos \theta (2 \cos^2 \theta - 1) - \sin \theta \,.\, 2 \sin \theta \cos \theta = 2 \cos^3 \theta - \cos \theta - 2 \sin^2 \theta \cos \theta$$

$$= 2 \cos^3 \theta - \cos \theta - 2 (1 - \cos^2 \theta) \cos \theta = 2 \cos^3 \theta - \cos \theta - 2 \cos \theta + 2 \cos^3 \theta$$

$$= 4 \cos^3 \theta - 3 \cos \theta$$

$\therefore \quad \mathbf{\cos 3\theta = 4 \cos^3 \theta - 3 \cos \theta.}$

(*iii*) $\tan 3\theta = \tan (\theta + 2\theta) = \dfrac{\tan \theta + \tan 2\theta}{1 - \tan \theta \tan 2\theta}$

$$= \frac{\tan \theta + \dfrac{2 \tan \theta}{1 - \tan^2 \theta}}{1 - \tan \theta \dfrac{2 \tan \theta}{1 - \tan^2 \theta}} = \frac{\tan \theta (1 - \tan^2 \theta) + 2 \tan \theta}{1 - \tan^2 \theta - 2 \tan^2 \theta} = \frac{3 \tan \theta - \tan^3 \theta}{1 - 3 \tan^2 \theta}.$$

$\therefore \quad \mathbf{\tan 3\theta = \dfrac{3 \tan \theta - \tan^3 \theta}{1 - 3 \tan^2 \theta}}.$

8.3. TRIGONOMETRIC FUNCTIONS OF SUB-MULTIPLE ANGLES

The angles $A/2$, $A/3$, ... are called **sub-multiple angles** of A.

Theorem. Prove that :

(*i*) $\mathbf{\sin \dfrac{A}{2} = \pm \sqrt{\dfrac{1 - \cos A}{2}}}$ (*ii*) $\mathbf{\cos \dfrac{A}{2} = \pm \sqrt{\dfrac{1 + \cos A}{2}}}$

(*iii*) $\mathbf{\tan \dfrac{A}{2} = \pm \sqrt{\dfrac{1 - \cos A}{1 + \cos A}}}$ (*iv*) $\mathbf{\tan \dfrac{A}{2} = \dfrac{\pm \sqrt{1 + \tan^2 A} - 1}{\tan A}}$

(*v*) $\mathbf{\tan \dfrac{A}{2} = \dfrac{1 - \cos A}{\sin A}}$ (*vi*) $\mathbf{\cot \dfrac{A}{2} = \dfrac{1 + \cos A}{\sin A}}.$

Proof. (*i*) We have $\cos A = \cos 2\left(\dfrac{A}{2}\right) = 1 - 2 \sin^2 \dfrac{A}{2}$

$\Rightarrow \quad 2 \sin^2 \dfrac{A}{2} = 1 - \cos A \quad \Rightarrow \quad \sin^2 \dfrac{A}{2} = \dfrac{1 - \cos A}{2}$

$\therefore \quad \mathbf{\sin \dfrac{A}{2} = \pm \sqrt{\dfrac{1 - \cos A}{2}}}.$

(*ii*) We have $\cos A = \cos 2\left(\dfrac{A}{2}\right) = 2 \cos^2 \dfrac{A}{2} - 1$

$\Rightarrow \quad 2 \cos^2 \dfrac{A}{2} = 1 + \cos A \quad \Rightarrow \quad \cos^2 \dfrac{A}{2} = \dfrac{1 + \cos A}{2}$

$\therefore \quad \mathbf{\cos \dfrac{A}{2} = \pm \sqrt{\dfrac{1 + \cos A}{2}}}.$

(*iii*) We have $\cos A = \cos 2\left(\frac{A}{2}\right) = \dfrac{1-\tan^2\frac{A}{2}}{1+\tan^2\frac{A}{2}}$

$$\Rightarrow \quad \frac{\cos A + 1}{\cos A - 1} = \frac{\left(1-\tan^2\frac{A}{2}\right)+\left(1+\tan^2\frac{A}{2}\right)}{\left(1-\tan^2\frac{A}{2}\right)-\left(1+\tan^2\frac{A}{2}\right)}$$

$$= \frac{2}{-2\tan^2\frac{A}{2}} = -\frac{1}{\tan^2\frac{A}{2}}$$

(Using componendo and dividendo formula)

$$\Rightarrow \quad \tan^2\frac{A}{2} = -\frac{\cos A - 1}{\cos A + 1} = \frac{1-\cos A}{1+\cos A}$$

$$\therefore \quad \mathbf{\tan\frac{A}{2} = \pm\sqrt{\frac{1-\cos A}{1+\cos A}}}.$$

Alternatively, $\tan\frac{A}{2} = \dfrac{\sin\frac{A}{2}}{\cos\frac{A}{2}} = \dfrac{\pm\sqrt{\frac{1-\cos A}{2}}}{\pm\sqrt{\frac{1+\cos A}{2}}} = \pm\sqrt{\dfrac{1-\cos A}{1+\cos A}}$

$$\therefore \quad \mathbf{\tan\frac{A}{2} = \pm\sqrt{\frac{1-\cos A}{1+\cos A}}}.$$

(*iv*) We have $\tan A = \tan 2\left(\frac{A}{2}\right) = \dfrac{2\tan\frac{A}{2}}{1-\tan^2\frac{A}{2}}$.

$$\Rightarrow \quad \tan A\left(1-\tan^2\frac{A}{2}\right) = 2\tan\frac{A}{2} \qquad \Rightarrow \quad (\tan A)\tan^2\frac{A}{2} + 2\tan\frac{A}{2} - \tan A = 0$$

$$\Rightarrow \quad \tan\frac{A}{2} = \frac{-2\pm\sqrt{4+4\tan^2 A}}{2\tan A} = \frac{\pm\sqrt{1+\tan^2 A}-1}{\tan A}$$

$$\therefore \quad \mathbf{\tan\frac{A}{2} = \frac{\pm\sqrt{1+\tan^2 A}-1}{\tan A}}.$$

(*v*) $$\tan\frac{A}{2} = \frac{\sin\frac{A}{2}}{\cos\frac{A}{2}} = \frac{2\sin^2\frac{A}{2}}{2\sin\frac{A}{2}\cos\frac{A}{2}} = \frac{1-\cos A}{\sin A}$$

$\therefore \qquad \mathbf{\tan \frac{A}{2} = \frac{1 - \cos A}{\sin A}}.$

(*vi*) $$\cot \frac{A}{2} = \frac{\cos \frac{A}{2}}{\sin \frac{A}{2}} = \frac{2\cos^2 \frac{A}{2}}{2 \sin \frac{A}{2} \cos \frac{A}{2}} = \frac{1 + \cos A}{\sin A}$$

$\therefore \qquad \mathbf{\cot \frac{A}{2} = \frac{1 + \cos A}{\sin A}}.$

WORKING RULES FOR SOLVING PROBLEMS

Rule I. (*i*) $\sin 2A = 2 \sin A \cos A = \dfrac{2 \tan A}{1 + \tan^2 A}$

(*ii*) $\cos 2A = \cos^2 A - \sin^2 A = 1 - 2 \sin^2 A = 2 \cos^2 A - 1 = \dfrac{1 - \tan^2 A}{1 + \tan^2 A}$

(*iii*) $\tan 2A = \dfrac{2 \tan A}{1 - \tan^2 A}$.

Rule II. (*i*) $\sin^2 A = \dfrac{1 - \cos 2A}{2}$ (*ii*) $\cos^2 A = \dfrac{1 + \cos 2A}{2}$.

Rule III. (*i*) $\sin 3A = 3 \sin A - 4 \sin^3 A$ (*ii*) $\cos 3A = 4 \cos^3 A - 3 \cos A$

(*iii*) $\tan 3A = \dfrac{3 \tan A - \tan^3 A}{1 - 3 \tan^2 A}$

Rule IV. (*i*) $\sin \dfrac{A}{2} = \pm \sqrt{\dfrac{1 - \cos A}{2}}$ (*ii*) $\cos \dfrac{A}{2} = \pm \sqrt{\dfrac{1 + \cos A}{2}}$

(*iii*) $\tan \dfrac{A}{2} = \pm \sqrt{\dfrac{1 - \cos A}{1 + \cos A}}$ (*iv*) $\tan A = \dfrac{\pm\sqrt{1 + \tan^2 A} - 1}{\tan A}$

(*v*) $\tan \dfrac{A}{2} = \dfrac{1 - \cos A}{\sin A}$ (*vi*) $\cot \dfrac{A}{2} = \dfrac{1 + \cos A}{\sin A}$.

ILLUSTRATIVE EXAMPLES

Example 1. *Find sin 2A, cos 2A, tan 2A if :*

(*i*) $\sin A = \dfrac{4}{5}$ (*ii*) $\cos A = \dfrac{5}{13}$ (*iii*) $\tan A = \dfrac{12}{5}$.

Sol. (*i*) We have $\sin A = \dfrac{4}{5}$.

$$\therefore \qquad \cos^2 A = 1 - \sin^2 A = 1 - \left(\frac{4}{5}\right)^2 = \frac{9}{25} \quad i.e., \quad \cos A = \pm \frac{3}{5}$$

$$\therefore \qquad \sin 2A = 2 \sin A \cos A = 2\left(\frac{4}{5}\right)\left(\pm \frac{3}{5}\right) = \pm \mathbf{\frac{24}{25}}$$

$$\cos 2A = 1 - 2\sin^2 A = 1 - 2\left(\frac{4}{5}\right)^2 = 1 - \frac{32}{25} = -\frac{\mathbf{7}}{\mathbf{25}}$$

Alternatively, $\cos 2A = 2\cos^2 A - 1 = 2\left(\pm\frac{3}{5}\right)^2 - 1 = \frac{18}{25} - 1 = -\frac{\mathbf{7}}{\mathbf{25}}.$

We have $\tan A = \dfrac{\sin A}{\cos A} = \dfrac{4/5}{\pm 3/5} = \pm\dfrac{4}{3}.$

$$\therefore \quad \tan 2A = \frac{2\tan A}{1-\tan^2 A} = \frac{2(\pm 4/3)}{1-(\pm 4/3)^2} = \frac{\pm 8/3}{1-(16/9)} = \frac{\pm 8/3}{-7/9} = \mp\frac{\mathbf{24}}{\mathbf{7}}$$

Alternatively, $\tan 2A = \dfrac{\sin 2A}{\cos 2A} = \dfrac{\pm 24/25}{-7/25} = \mp\dfrac{\mathbf{24}}{\mathbf{7}}.$

Remark. The sign '$\mp$' means that if $\tan A = 4/3$ then $\tan 2A = -24/7$ and $\tan 2A = 24/7$ when $\tan A = -4/3$.

(*ii*) We have $\cos A = \dfrac{5}{13}.$

$$\therefore \quad \sin^2 A = 1 - \cos^2 A = 1 - \left(\frac{5}{13}\right)^2 = 1 - \frac{25}{169} = \frac{144}{169} \quad \textit{i.e.,} \quad \sin A = \pm\frac{12}{13}$$

$$\therefore \quad \sin 2A = 2\sin A\cos A = 2\left(\pm\frac{12}{13}\right)\left(\frac{5}{13}\right) = \pm\frac{\mathbf{120}}{\mathbf{169}}$$

$$\cos 2A = 2\cos^2 A - 1 = 2\left(\frac{5}{13}\right)^2 - 1 = \frac{50}{169} - 1 = -\frac{\mathbf{119}}{\mathbf{169}}$$

$$\tan 2A = \frac{\sin 2A}{\cos 2A} = \frac{\pm 120/169}{-119/169} = \mp\frac{\mathbf{120}}{\mathbf{119}}.$$

(*iii*) We have $\tan A = \dfrac{12}{5}.$

$$\therefore \quad \sin 2A = \frac{2\tan A}{1+\tan^2 A} = \frac{2(12/5)}{1+(12/5)^2} = \frac{24/5}{1+(144/25)} = \frac{24/5}{169/25} = \frac{\mathbf{120}}{\mathbf{169}}$$

$$\cos 2A = \frac{1-\tan^2 A}{1+\tan^2 A} = \frac{1-(12/5)^2}{1+(12/5)^2} = \frac{1-(144/25)}{1+(144/25)} = -\frac{\mathbf{119}}{\mathbf{169}}$$

$$\tan 2A = \frac{2\tan A}{1-\tan^2 A} = \frac{2(12/5)}{1-(12/5)^2} = \frac{24/5}{1-(144/25)} = \frac{24/5}{-119/25} = -\frac{\mathbf{120}}{\mathbf{119}}$$

Alternatively, $\tan 2A = \dfrac{\sin 2A}{\cos 2A} = \dfrac{120/169}{-119/169} = -\dfrac{\mathbf{120}}{\mathbf{119}}.$

Example 2. *Prove that* $\sqrt{2+\sqrt{2+2\cos 4\theta}} = 2\cos\theta.$

Sol. L.H.S. $= \sqrt{2+\sqrt{2+2\cos 4\theta}} = \sqrt{2+\sqrt{2+2(2\cos^2 2\theta - 1)}}$

$$= \sqrt{2 + \sqrt{4\cos^2 2\theta}} = \sqrt{2 + 2\cos 2\theta}$$

$$= \sqrt{2 + 2(2\cos^2\theta - 1)} = \sqrt{4\cos^2\theta} = 2\cos\theta = \text{R.H.S.}$$

Example 3. *Prove that* $\cos^2 A + \cos^2 (A + 120°) + \cos^2 (A - 120°) = \frac{3}{2}$.

Sol. L.H.S. $= \cos^2 A + \cos^2 (A + 120°) + \cos^2 (A - 120°)$

$$= \frac{1 + \cos 2A}{2} + \frac{1 + \cos (2A + 240°)}{2} + \frac{1 + \cos (2A - 240°)}{2}$$

$$= \frac{3}{2} + \frac{1}{2}[\cos 2A + \cos (2A + 240°) + \cos (2A - 240°)]$$

$$= \frac{3}{2} + \frac{1}{2}\left(\cos 2A + 2\cos\frac{4A}{2}\cos\frac{480°}{2}\right) = \frac{3}{2} + \frac{1}{2}[\cos 2A + 2\cos 2A \cos (180° + 60°)]$$

$$= \frac{3}{2} + \frac{1}{2}[\cos 2A + 2\cos 2A(-\cos 60°)] = \frac{3}{2} + \frac{1}{2}\left[\cos 2A - 2\cos 2A\left(\frac{1}{2}\right)\right]$$

$$= \frac{3}{2} + \frac{1}{2}(0) = \frac{3}{2} = \text{R.H.S.}$$

Example 4. *Show that* $\frac{\sec 8A - 1}{\sec 4A - 1} = \frac{\tan 8A}{\tan 2A}$.

Sol. L.H.S. $= \dfrac{\sec 8A - 1}{\sec 4A - 1} = \dfrac{\frac{1}{\cos 8A} - 1}{\frac{1}{\cos 4A} - 1} = \dfrac{1 - \cos 8A}{1 - \cos 4A} \cdot \dfrac{\cos 4A}{\cos 8A}$

$$= \frac{2\sin^2 4A}{2\sin^2 2A} \cdot \frac{\cos 4A}{\cos 8A} = \frac{(2\sin 4A\cos 4A)\sin 4A}{2\sin^2 2A\cos 8A}$$

$$= \frac{\sin 8A\,(2\sin 2A\cos 2A)}{2\sin^2 2A\cos 8A} = \frac{\sin 8A\cos 2A}{\sin 2A\cos 8A} = \frac{\sin 8A/\cos 8A}{\sin 2A/\cos 2A} = \frac{\tan 8A}{\tan 2A} = \text{R.H.S.}$$

Example 5. *If* $2\cos\theta = x + \frac{1}{x}$*, prove that* $2\cos 3\theta = x^3 + \frac{1}{x^3}$.

Sol. We have $2\cos\theta = x + \frac{1}{x}$.

$$2\cos 3\theta = 2(4\cos^3\theta - 3\cos\theta) = 8\cos^3\theta - 6\cos\theta = (2\cos\theta)^3 - 3(2\cos\theta)$$

$$= \left(x + \frac{1}{x}\right)^3 - 3\left(x + \frac{1}{x}\right) = x^3 + \frac{1}{x^3} + 3x\frac{1}{x}\left(x + \frac{1}{x}\right) - 3\left(x + \frac{1}{x}\right) = \mathbf{x^3 + \frac{1}{x^3}}.$$

Example 6. *Find the values of* $\sin\frac{x}{2}$, $\cos\frac{x}{2}$ *and* $\tan\frac{x}{2}$ *if*:

(*i*) $\cos x = -\frac{1}{3}$, *x is in quadrant III* (*ii*) $\sin x = \frac{1}{4}$, *x is in quadrant II.*

Sol. (*i*) We have $\cos x = -\dfrac{1}{3}$. ...(1)

(1) $\Rightarrow 1 - 2\sin^2 \dfrac{x}{2} = -\dfrac{1}{3} \qquad \Rightarrow 2\sin^2 \dfrac{x}{2} = \dfrac{4}{3}$

$\Rightarrow \qquad \sin \dfrac{x}{2} = \pm \sqrt{\dfrac{2}{3}} = \pm \dfrac{\sqrt{6}}{3}$

$\therefore \qquad \sin \dfrac{x}{2} = \dfrac{\mathbf{\sqrt{6}}}{\mathbf{3}}. \qquad \left(\because \pi < x < \dfrac{3\pi}{2} \Rightarrow \dfrac{\pi}{2} < \dfrac{x}{2} < \dfrac{3\pi}{4} \Rightarrow \sin \dfrac{x}{2} > 0\right)$

(1) $\Rightarrow 2\cos^2 \dfrac{x}{2} - 1 = -\dfrac{1}{3} \qquad \Rightarrow 2\cos^2 \dfrac{x}{2} = \dfrac{2}{3}$

$\Rightarrow \qquad \cos \dfrac{x}{2} = \pm \dfrac{1}{\sqrt{3}} = \pm \dfrac{\sqrt{3}}{3}$

$\therefore \qquad \cos \dfrac{x}{2} = -\dfrac{\mathbf{\sqrt{3}}}{\mathbf{3}} \qquad \left(\because \pi < x < \dfrac{3\pi}{2} \Rightarrow \dfrac{\pi}{2} < \dfrac{x}{2} < \dfrac{3\pi}{4} \Rightarrow \cos \dfrac{x}{2} < 0\right)$

$$\tan \frac{x}{2} = \frac{\sin \frac{x}{2}}{\cos \frac{x}{2}} = \frac{\frac{\sqrt{6}}{3}}{-\frac{\sqrt{3}}{3}} = -\mathbf{\sqrt{2}}.$$

(*ii*) We have $\sin x = \dfrac{1}{4}$.

$\Rightarrow \qquad \cos^2 x = 1 - \sin^2 x = 1 - \dfrac{1}{16} = \dfrac{15}{16} \quad \Rightarrow \cos x = \pm \dfrac{\sqrt{15}}{4}$

$\therefore \qquad \cos x = -\dfrac{\sqrt{15}}{4} \qquad \left(\because \dfrac{\pi}{2} < x < \pi \Rightarrow \cos x < 0\right)$

$\Rightarrow \qquad 1 - 2\sin^2 \dfrac{x}{2} = -\dfrac{\sqrt{15}}{4} \qquad \Rightarrow 2\sin^2 \dfrac{x}{2} = 1 + \dfrac{\sqrt{15}}{4} = \dfrac{4+\sqrt{15}}{4}$

$\Rightarrow \qquad \sin^2 \dfrac{x}{2} = \dfrac{4+\sqrt{15}}{8}$

$\Rightarrow \qquad \sin \dfrac{x}{2} = \pm \dfrac{\sqrt{4+\sqrt{15}}}{\sqrt{8}} = \pm \sqrt{\dfrac{4+\sqrt{15}}{8}} = \pm \sqrt{\dfrac{8+2\sqrt{15}}{16}} = \pm \dfrac{\sqrt{8+2\sqrt{15}}}{4}$

$\therefore \qquad \sin \dfrac{x}{2} = \dfrac{\mathbf{\sqrt{8+2\sqrt{15}}}}{\mathbf{4}}. \qquad \left(\because \dfrac{\pi}{2} < x < \pi \Rightarrow \dfrac{\pi}{4} < \dfrac{x}{2} < \dfrac{\pi}{2} \Rightarrow \sin \dfrac{x}{2} > 0\right)$

Also, $\qquad \cos^2 \dfrac{x}{2} = 1 - \sin^2 \dfrac{x}{2} = 1 - \dfrac{8+2\sqrt{15}}{16} = \dfrac{8-2\sqrt{15}}{16}$

$\Rightarrow \qquad \cos \dfrac{x}{2} = \dfrac{\mathbf{\sqrt{8-2\sqrt{15}}}}{\mathbf{4}}. \qquad \left(\because \dfrac{\pi}{2} < x < \pi \Rightarrow \dfrac{\pi}{4} < \dfrac{x}{2} < \dfrac{\pi}{2} \Rightarrow \cos \dfrac{x}{2} > 0\right)$

$$\tan\frac{x}{2} = \frac{\sin\frac{x}{2}}{\cos\frac{x}{2}} = \frac{\frac{\sqrt{8+2\sqrt{15}}}{4}}{\frac{\sqrt{8-2\sqrt{15}}}{4}} = \sqrt{\frac{8+2\sqrt{15}}{8-2\sqrt{15}}} = \sqrt{\frac{8+2\sqrt{15}}{8-2\sqrt{15}} \times \frac{8+2\sqrt{15}}{8+2\sqrt{15}}}$$

$$= \sqrt{\frac{(8+2\sqrt{15})^2}{64-60}} = \frac{8+2\sqrt{15}}{2} = \mathbf{4+\sqrt{15}}.$$

Example 7. *Evaluate :*

(*i*) $\sin 7\frac{1^\circ}{2}$ (*ii*) $\tan 11\frac{1^\circ}{4}$.

Sol. (*i*)

$$\sin 7\frac{1^\circ}{2} = +\sqrt{\frac{1-\cos 15^\circ}{2}} = \sqrt{\frac{1-\cos(45^\circ-30^\circ)}{2}} \qquad \left(\because \sin\frac{A}{2} = \pm\sqrt{\frac{1-\cos A}{2}}\right)$$

$$= \sqrt{\frac{1-\left(\frac{1}{\sqrt{2}}.\frac{\sqrt{3}}{2}+\frac{1}{\sqrt{2}}.\frac{1}{2}\right)}{2}} = \sqrt{\frac{2\sqrt{2}-\sqrt{3}-1}{4\sqrt{2}}}$$

$$= \sqrt{\frac{\sqrt{2}\,(2\sqrt{2}-\sqrt{3}-1)}{(4\sqrt{2})\sqrt{2}}} = \frac{\sqrt{\mathbf{4-\sqrt{6}-\sqrt{2}}}}{\mathbf{2\sqrt{2}}}.$$

(*ii*)

$$\tan 11\frac{1^\circ}{4} = \frac{\sin 11\frac{1^\circ}{4}}{\cos 11\frac{1^\circ}{4}} = \frac{2\sin^2 11\frac{1^\circ}{4}}{2\sin 11\frac{1^\circ}{4}\cos 11\frac{1^\circ}{4}} = \frac{1-\cos 22\frac{1^\circ}{2}}{\sin 22\frac{1^\circ}{2}}$$

$$= \frac{1-\sqrt{\frac{1+\cos 45^\circ}{2}}}{\sqrt{\frac{1-\cos 45^\circ}{2}}} = \frac{\sqrt{2}-\sqrt{1+\frac{1}{\sqrt{2}}}}{\sqrt{1-\frac{1}{\sqrt{2}}}} = \frac{\sqrt{2}-\sqrt{\frac{\sqrt{2}+1}{\sqrt{2}}}}{\sqrt{\frac{\sqrt{2}-1}{\sqrt{2}}}}$$

$$= \frac{\sqrt{2\sqrt{2}}-\sqrt{\sqrt{2}+1}}{\sqrt{\sqrt{2}-1}} \times \frac{\sqrt{\sqrt{2}+1}}{\sqrt{\sqrt{2}+1}} = \frac{\sqrt{4+2\sqrt{2}}-\sqrt{2}+1}{\sqrt{2-1}}$$

$$= \mathbf{\sqrt{4+2\sqrt{2}}-\sqrt{2}+1}.$$

Example 8. *Show that :*

(*i*) $\cot 7\frac{1^\circ}{2} = \sqrt{2}+\sqrt{3}+\sqrt{4}+\sqrt{6}$ (*ii*) $\tan 142\frac{1^\circ}{2} = \sqrt{2}-\sqrt{3}+\sqrt{4}-\sqrt{6}$.

Sol. (*i*)

$$\cot 7\frac{1^\circ}{2} = \frac{\cos 7\frac{1^\circ}{2}}{\sin 7\frac{1^\circ}{2}} = \frac{2\cos^2 7\frac{1^\circ}{2}}{2\sin 7\frac{1^\circ}{2}\cos 7\frac{1^\circ}{2}} = \frac{1+\cos 15^\circ}{\sin 15^\circ}$$

$$= \frac{1+\cos(45^\circ-30^\circ)}{\sin(45^\circ-30^\circ)} = \frac{1+\left(\left(\frac{1}{\sqrt{2}}\right)\left(\frac{\sqrt{3}}{2}\right)+\left(\frac{1}{\sqrt{2}}\right)\left(\frac{1}{2}\right)\right)}{\left(\frac{1}{\sqrt{2}}\right)\left(\frac{\sqrt{3}}{2}\right)-\left(\frac{1}{\sqrt{2}}\right)\left(\frac{1}{2}\right)}$$

$$= \frac{2\sqrt{2}+\sqrt{3}+1}{\sqrt{3}-1} = \frac{2\sqrt{2}+\sqrt{3}+1}{\sqrt{3}-1} \times \frac{\sqrt{3}+1}{\sqrt{3}+1}$$

$$= \frac{2\sqrt{6}+2\sqrt{2}+3+\sqrt{3}+\sqrt{3}+1}{3-1} = \frac{2\sqrt{6}+2\sqrt{2}+2\sqrt{3}+4}{2}$$

$$= \sqrt{6}+\sqrt{2}+\sqrt{3}+2 = \boldsymbol{\sqrt{2}+\sqrt{3}+\sqrt{4}+\sqrt{6}}.$$

(ii) $$\tan 142\frac{1^\circ}{2} = \tan\left(\frac{285}{2}\right)^\circ = \tan\left(\frac{360-75}{2}\right)^\circ = \tan\left(180^\circ - 37\frac{1^\circ}{2}\right)$$

$$= -\tan 37\frac{1^\circ}{2} = -\tan\left(\frac{75}{2}\right)^\circ = -\tan\left(\frac{90-15}{2}\right) \times \frac{\cos 7\frac{1^\circ}{2} - \sin 7\frac{1^\circ}{2}}{\cos 7\frac{1^\circ}{2} - \sin 7\frac{1^\circ}{2}}$$

$$= -\tan\left(45^\circ - 7\frac{1^\circ}{2}\right) = -\frac{1-\tan 7\frac{1^\circ}{2}}{1+\tan 7\frac{1^\circ}{2}}$$

$$= -\frac{\cos 7\frac{1^\circ}{2} - \sin 7\frac{1^\circ}{2}}{\cos 7\frac{1^\circ}{2} + \sin 7\frac{1^\circ}{2}} = -\frac{\left(\cos 7\frac{1^\circ}{2} - \sin 7\frac{1^\circ}{2}\right)^2}{\cos^2 7\frac{1^\circ}{2} - \sin^2 7\frac{1^\circ}{2}}$$

$$= \frac{\cos^2 7\frac{1^\circ}{2} + \sin^2 7\frac{1^\circ}{2} - 2\sin 7\frac{1^\circ}{2}\cos 7\frac{1^\circ}{2}}{\cos\left(7\frac{1^\circ}{2} + 7\frac{1^\circ}{2}\right)} = -\frac{1-\sin 15^\circ}{\cos 15^\circ}$$

$$= \frac{1-\sin(45^\circ - 30^\circ)}{\cos(45^\circ - 30^\circ)} = -\frac{1-\left(\left(\frac{1}{\sqrt{2}}\right)\left(\frac{\sqrt{3}}{2}\right) - \left(\frac{1}{\sqrt{2}}\right)\left(\frac{1}{2}\right)\right)}{\left(\frac{1}{\sqrt{2}}\right)\left(\frac{\sqrt{3}}{2}\right) + \left(\frac{1}{\sqrt{2}}\right)\left(\frac{1}{2}\right)}$$

$$= -\frac{2\sqrt{2}-\sqrt{3}+1}{\sqrt{3}+1} = -\frac{2\sqrt{2}-\sqrt{3}+1}{\sqrt{3}+1} \times \frac{\sqrt{3}-1}{\sqrt{3}-1}$$

$$= \frac{-2\sqrt{6}+2\sqrt{2}+3-\sqrt{3}-\sqrt{3}+1}{3-1} = \frac{-2\sqrt{6}-2\sqrt{3}+2\sqrt{2}+4}{2}$$

$$= -\sqrt{6}-\sqrt{3}+\sqrt{2}+2 = \boldsymbol{\sqrt{2}-\sqrt{3}+\sqrt{4}-\sqrt{6}}.$$

Example 9. *If* $\cos\theta = \dfrac{\cos\phi - e}{1 - e\cos\phi}$, *then* $\tan\dfrac{\theta}{2} = \pm\sqrt{\dfrac{1+e}{1-e}}\tan\dfrac{\phi}{2}$.

Sol. Given $$\cos\theta = \frac{\cos\phi - e}{1 - e\cos\phi}$$

Now $$\tan\frac{\theta}{2} = \pm\sqrt{\frac{1-\cos\theta}{1+\cos\theta}} \qquad \text{...(1)}$$

$$= \pm \sqrt{\frac{1 - \dfrac{\cos\phi - e}{1 - e\cos\phi}}{1 + \dfrac{\cos\phi - e}{1 - e\cos\phi}}} = \pm \sqrt{\frac{1 - e\cos\phi - \cos\phi + e}{1 - e\cos\phi + \cos\phi - e}}$$

$$= \pm \sqrt{\frac{1 + e - \cos\phi\,(e+1)}{1 - e - \cos\phi\,(e-1)}} = \pm \sqrt{\frac{(1+e)(1-\cos\phi)}{(1-e)(1+\cos\phi)}}$$

$$= \pm \sqrt{\frac{1+e}{1-e}}\sqrt{\frac{1-\cos\phi}{1+\cos\phi}} = \pm \sqrt{\frac{1+e}{1-e}}\left(\pm \tan\frac{\phi}{2}\right) \qquad \text{(Using (1))}$$

$$= \pm \sqrt{\frac{1+e}{1-e}} \tan\frac{\phi}{2}.$$

8.4. TRIGONOMETRIC FUNCTIONS OF 18° AND 36°

Theorem. *Find the values of trigonometric functions of 18° and 36°.*

Proof. $\theta = 18° \Rightarrow 5\theta = 90° \Rightarrow 2\theta = 90° - 3\theta \Rightarrow \sin 2\theta = \sin(90° - 3\theta) = \cos 3\theta$

$\Rightarrow 2\sin\theta\cos\theta = 4\cos^3\theta - 3\cos\theta \Rightarrow \cos\theta\,(4\cos^2\theta - 3) - \cos\theta\,(2\sin\theta) = 0$

$\Rightarrow \cos\theta\,(4(1-\sin^2\theta) - 3 - 2\sin\theta) = 0 \Rightarrow 4\sin^2\theta + 2\sin\theta - 1 = 0$ [$\because \cos 18° \neq 0$]

$$\Rightarrow \sin\theta = \frac{-1 \pm \sqrt{4+16}}{8} = \frac{-1 \pm \sqrt{5}}{4} \Rightarrow \sin 18° = \frac{\sqrt{5}-1}{4} \qquad (\because \sin 18° \text{ is + ve})$$

$$\cos 18° = \sqrt{1 - \sin^2 18°} = \sqrt{1 - \frac{5+1-2\sqrt{5}}{16}} = \sqrt{\frac{10+2\sqrt{5}}{16}} = \sqrt{\frac{10+2\sqrt{5}}{4}}.$$

Now, the other *t*-functions of 18° can easily be found out.

The identity $\cos 2\theta = 1 - 2\sin^2\theta$ implies

$$\cos 36° = 1 - 2\left(\frac{5+1-2\sqrt{5}}{16}\right) = \frac{\sqrt{5}+1}{4}.$$

Also $$\sin 36° = \sqrt{1-\cos^2 36°} = \sqrt{1 - \frac{5+1+2\sqrt{5}}{16}} = \frac{\sqrt{10-2\sqrt{5}}}{4}.$$

The other *t*-functions of 36° can now be found out easily.

Corollary 1. (*i*) $\mathbf{\sin 72°} = \sin(90° - 18°) = \cos 18° = \dfrac{\sqrt{10+2\sqrt{5}}}{4}$

(*ii*) $\mathbf{\cos 72°} = \cos(90° - 18°) = \sin 18° = \dfrac{\sqrt{5}-1}{4}$

Corollary 2. (*i*) $\mathbf{\sin 54°} = \sin(90° - 36°) = \cos 36° = \dfrac{\sqrt{5}+1}{4}$

(*ii*) $\mathbf{\cos 54°} = \cos(90° - 36°) = \sin 36° = \dfrac{\sqrt{10-2\sqrt{5}}}{4}.$

Example 10. *Prove that* $\sin 6^\circ \sin 42^\circ \sin 66^\circ \sin 78^\circ = \frac{1}{16}$.

Sol. L.H.S. $= \sin 6^\circ \sin 42^\circ \sin 66^\circ \sin 78^\circ$

$$= \frac{1}{4}(2 \sin 66^\circ \sin 6^\circ)(2 \sin 78^\circ \sin 42^\circ)$$

(Note that 66° – 6° = 60° and 78° + 42° = 120°)

$$= \frac{1}{4}(\cos 60^\circ - \cos 72^\circ)(\cos 36^\circ - \cos 120^\circ)$$

$$= \frac{1}{4}\left(\frac{1}{2} - \cos(90^\circ - 18^\circ)\right)(\cos 36^\circ - \cos(180^\circ - 60^\circ))$$

$$= \frac{1}{4}\left(\frac{1}{2} - \sin 18^\circ\right)(\cos 36^\circ - (-\cos 60^\circ)) = \frac{1}{4}\left(\frac{1}{2} - \sin 18^\circ\right)\left(\cos 36^\circ + \frac{1}{2}\right)$$

$$= \frac{1}{4}\left(\frac{1}{2} - \frac{\sqrt{5}-1}{4}\right)\left(\frac{\sqrt{5}+1}{4} + \frac{1}{2}\right) = \frac{1}{4}\left(\frac{3-\sqrt{5}}{4}\right)\left(\frac{3+\sqrt{5}}{4}\right)$$

$$= \frac{1}{4}\left(\frac{9-5}{16}\right) = \frac{1}{16} = \text{R.H.S.}$$

Example 11. *Show that* $\cos 36^\circ \cos 72^\circ \cos 108^\circ \cos 144^\circ = \frac{1}{16}$.

Sol. L.H.S. $= \cos 36^\circ \cos 72^\circ \cos 108^\circ \cos 144^\circ$

$$= \cos 36^\circ \cos(90^\circ - 18^\circ) \cos(90^\circ + 18^\circ) \cos(180^\circ - 36^\circ)$$

$$= \cos 36^\circ \sin 18^\circ(-\sin 18^\circ)(-\cos 36^\circ)$$

$$= \sin^2 18^\circ \cos^2 36^\circ = \left(\frac{\sqrt{5}-1}{4}\right)^2\left(\frac{\sqrt{5}+1}{4}\right)^2 = \left(\frac{5-1}{16}\right)^2 = \left(\frac{1}{4}\right)^2 = \frac{1}{16} = \text{R.H.S.}$$

EXERCISE 8.1

1. Show that $\frac{\sin x}{1+\cos x} = \tan\frac{x}{2}$. *(P.T.U. B. Pharma, Dec. 2006)*
2. Show that $\frac{1-\cos 2A}{1+\cos 2A} = \tan^2 A$.
3. If $\cos A = 4/5$, find the value of $\cos 2A$.
4. If $\tan A = 2/3$, find the value of $\tan 2A$.
5. Show that $\frac{1-\tan^2(45^\circ + A)}{1+\tan^2(45^\circ + A)} = -\sin 2A$.
6. Show that $\cot\frac{A}{2} = \frac{1+\cos A}{\sin A}$
7. Show that $\cos 4A = 1 - 8\sin^2 A \cos^2 A$.
8. If $\cos\theta = \frac{1}{2}\left(a + \frac{1}{a}\right)$, then show that $\cos 2\theta = \frac{1}{2}\left(a^2 + \frac{1}{a^2}\right)$.

9. Find the values of sin 2*A* and cos 2*A*, when sin *A* = 3/5. (*P.T.U. B. Pharma, Dec. 2006*)

10. Find the values of sin 2*A* and cos 2*A*, when cos *A* = 20/29.

11. Show that :

(*i*) $\frac{\sin 3\theta}{\sin \theta} - \frac{\cos 3\theta}{\cos \theta} = 2$ (*ii*) $\frac{2 \sin \theta - \sin 2\theta}{2 \sin \theta + \sin 2\theta} = \tan^2 \frac{\theta}{2}$

(*iii*) $\frac{\cot^2 A - 1}{2 \cot A} = \cot 2A$ (*iv*) $\frac{\cot A - \tan A}{\cot A + \tan A} = \cos 2A.$

12. Find the values of $\sin \frac{x}{2}$, $\cos \frac{x}{2}$ and $\tan \frac{x}{2}$ if $\tan x = \frac{3}{4}$ and $\pi < x < \frac{3\pi}{2}$.

13. If $\tan A = \frac{1}{2}$, $\tan B = \frac{1}{3}$, find the value of tan (2*A* + *B*).

14. If sin θ + cos θ = 1, find the value of sin 2θ.

15. Show that $\frac{\cos A + \sin A}{\cos A - \sin A} - \frac{\cos A - \sin A}{\cos A + \sin A} = 2 \tan 2A.$

16. Show that $\cos 2\theta = \frac{1 - \tan^2 \theta}{1 + \tan^2 \theta}$. Hence find the value of tan 30°, given that $\cos 60° = \frac{1}{2}$.

17. Show that $\frac{\tan^2 \theta}{\tan^2 \theta - 1} + \frac{\text{cosec}^2 \theta}{\sec^2 \theta - \text{cosec}^2 \theta} = -\sec 2\theta.$

18. Show that $\frac{\tan 5\theta + \tan 3\theta}{\tan 5\theta - \tan 3\theta} = 4 \cos 2\theta \cos 4\theta.$

19. Show that tan *A* + cot *A* = 2 cosec 2*A* and deduce that tan 75° + cot 75° = 4.

20. If $\sin \theta = \frac{1}{2}\left(x + \frac{1}{x}\right)$, show that $\sin 3\theta + \frac{1}{2}\left(x^3 + \frac{1}{x^3}\right) = 0.$

21. If $\tan A = \frac{1}{7}$ and $\tan B = \frac{1}{3}$, show that cos 2*A* = sin 4*B*.

22. Find the values of :

(*i*) $\sin 11\frac{1°}{4}$ (*ii*) $\cos 7\frac{1°}{2}$ (*iii*) $\tan 7\frac{1°}{2}$.

23. Show that :

(*i*) $\sin^2 24° - \sin^2 6° = \frac{\sqrt{5} - 1}{8}$ (*ii*) $\sin^2 72° - \sin^2 60° = \frac{\sqrt{5} - 1}{8}.$

24. Show that :

(*i*) cos 6° cos 42° cos 66° cos 78° = 1/16 (*ii*) $\sin 36° \sin 72° \sin 108° \sin 144° = \frac{5}{16}$

(*M.D.U. B. Pharma, 2006*)

(*iii*) $\cos \frac{2\pi}{15} \cos \frac{4\pi}{15} \cos \frac{8\pi}{15} \cos \frac{16\pi}{15} = \frac{1}{64}.$

Answers

3. 7/25 **4.** 12/5 **9.** ± 24/25, 7/25 **10.** ± 840/841, – 41/841

12. $\frac{3\sqrt{10}}{10}, -\frac{\sqrt{10}}{10}, -3$ **13.** 3 **14.** 0 **16.** $\frac{1}{\sqrt{3}}$.

22. (*i*) 0.195 (*ii*) 0.991 (*iii*) 0.132.

Hints

12. $\cos^2 x = \frac{1}{1+\tan^2 x} = \frac{1}{1+(3/4)^2} = \frac{16}{25}$.

13. $\tan(2A+B) = \dfrac{\dfrac{2\tan A}{1-\tan^2 A} + \tan B}{1-\left(\dfrac{2\tan A}{1-\tan^2 A}\right)\tan B}$.

14. Use $(\sin\theta + \cos\theta)^2 = 1 + 2\sin\theta\cos\theta$.

9

Logarithms and Its Pharmaceutical Computations

I. LOGARITHM OF A NUMBER

9.1. INTRODUCTION

The technique of logarithms is used to simplify the algebraic calculations to a great extent. Multiplication, division of large numbers are done very easily by using the methods of logarithms. The *n*th root of a number cannot be found by using simple arithmetical tools for every value of *n*, but the use of logarithms also solve this problem. The technique of logarithms is very essential for anyone, who deal with mathematical manipulations. By using logarithms, results are obtained correct to certain decimal places. The technique of logarithms was invented by **John Napier (1550–1617).**

9.2. DEFINITION OF LOGARITHM

If $a > 0$, $a \neq 1$ and $a^x = y$, then we define the **logarithm** of y to the base a as x and write $\log_a y = x$.

Equivalently, the logarithm of a number to a given base is the index or the power to which the base must be raised in order to obtain the given number.

Illustrations :

(*i*) $\log_4 64$ is equal to 3 because $(4)^3 = 64$.

(*ii*) $\log_3 243$ is equal to 5 because $(3)^5 = 243$

Remark 1. 'log' is the abbreviation of the word 'logarithm'.

Remark 2. If $a > 0$, $a \neq 1$ and $a^x = y$, then y is positive for every value of x.

Also, $a^x = y$ implies $\log_a y = x$.

$\therefore$ **Logarithm is defined only for positive numbers.**

Remark 3. If $a > 0$, $a \neq 1$, $y > 0$, then $\log_a y$ is that number (x) for which $a^x = y$.

Remark 4. Logarithms of zero and negative numbers are not defined.

For example, the expressions $\log_a 0$, $\log_a (-7)$, $\log_a (2 - \sqrt{5})$ are meaningless.

Remark 5. We have $\log_5 625 = 4$, because $(5)^4 = 625$.

$\log_5 625 = 4$ is called the *logarithmic* form of $(5)^4 = 625$.

$(5)^4 = 625$ is called the *exponential form* of $\log_5 625 = 4$.

9.3. UNIQUENESS OF LOGARITHM OF A NUMBER

Let $a > 0$, $a \neq 1$, $y > 0$. $\therefore$ $\log_a y$ is defined.

If possible, let $\log_a y = x_1$ and $\log_a y = x_2$.

$\therefore \quad a^{x_1} = y$ and $a^{x_2} = y$

$\Rightarrow \quad a^{x_1} = a^{x_2} \Rightarrow x_1 = x_2$. $\quad (\because a \neq 1)$

$\therefore$ **The value of $\log_a y$ is unique.**

9.4. BASIC RESULTS

(*i*) Let $a > 0$, $a \neq 1$. We have $a^0 = 1$.

$\therefore \quad \mathbf{\log_a 1 = 0.}$

$\therefore$ **The logarithm of 1 to any base is '0'.**

(*ii*) Let $a > 0$, $a \neq 1$. We have $a^1 = a$.

$\therefore \quad \mathbf{\log_a a = 1.}$

$\therefore$ **The logarithm of a to itself as base is 1.**

(*iii*) Let $a > 0$, $a \neq 1$ and $a^x = y$.

$\therefore \quad \log_a y = x$.

$\Rightarrow \quad a^x = y \Rightarrow \mathbf{a^{(\log_a y)} = y} \quad (\because x = \log_a y)$

and $\quad \log_a y = x \Rightarrow \mathbf{\log_a (a^x) = x.}$

ILLUSTRATIVE EXAMPLES

Example 1. *Write the following in logrithmic form :*

(*i*) $10^4 = 10000$ (*ii*) $(0.5)^2 = 0.25$.

Sol. (*i*) The logarithmic form of $10^4 = 10000$ is $\mathbf{\log_{10} 10000 = 4.}$

(*ii*) The logarithmic form of $(0.5)^2 = 0.25$ is $\mathbf{\log_{0.5} 0.25 = 2.}$

Example 2. *Write the following in exponential form :*

(*i*) $\log_7 343 = 3$ (*ii*) $\log_9 6561 = 4$.

Sol. (*i*) The exponential form of $\log_7 343 = 3$ is $\mathbf{7^3 = 343.}$

(*ii*) The exponential form of $\log_9 6561 = 4$ is $\mathbf{9^4 = 6561.}$

Example 3. *Find the values of :*

(*i*) $\log_8 128$ (*ii*) $\log_{0.5} 256$ (*iii*) $\log_{19} 6859$.

Sol. (*i*) Let $\log_8 128 = x$. $\therefore$ $(8)^x = 128$

$\Rightarrow \quad (2^3)^x = 2^7 \Rightarrow 2^{3x} = 2^7 \Rightarrow 3x = 7 \Rightarrow x = \frac{7}{3}$

$\therefore \quad \log_8 128 = \mathbf{\frac{7}{3}}$.

(*ii*) Let $\log_{0.5} 256 = x$ $\therefore$ $(0.5)^x = 256$

$\Rightarrow \quad \left(\frac{1}{2}\right)^x = (2)^8 \Rightarrow 2^{-x} = (2)^8 \Rightarrow -x = 8 \Rightarrow x = -8$

$\therefore \quad \log_{0.5} 256 = \mathbf{-8.}$

(*iii*) Let $\log_{19} 6859 = x$. $\therefore$ $(19)^x = 6859$

$\Rightarrow \quad (19)^x = (19)^3 \Rightarrow x = 3$

$\therefore \quad \log_{19} 6859 = \mathbf{3.}$

Example 4. *Solve for x :*

(i) $\log_x = 0.001 = -3$ *(ii)* $\log_{0.4} 0.0256 = x$.

Sol. *(i)* We have $\log_x 0.001 = -3$. $\therefore$ $x^{-3} = 0.001$

$$\Rightarrow \quad \frac{1}{x^3} = \frac{1}{1000} \Rightarrow x^3 = (10)^3 \Rightarrow x = \mathbf{10}.$$

(ii) We have $\log_{0.4} 0.0256 = x$. $\therefore$ $(0.4)^x = 0.0256$

$$\Rightarrow \quad \left(\frac{4}{10}\right)^x = \frac{256}{10000} \Rightarrow \left(\frac{4}{10}\right)^x = \left(\frac{4}{10}\right)^4 \Rightarrow x = \mathbf{4}.$$

Example 5. *Find the value of x if* $\log_2 (x^2 - 1) = \log_2 8$.

Sol. We have $\log_2 (x^2 - 1) = \log_2 8$.

$$\Rightarrow \quad x^2 - 1 = 8 \Rightarrow x^2 = 9 \Rightarrow x = \pm \mathbf{3}.$$

WORKING RULES FOR SOLVING PROBLEMS

Rule I. *If* $a > 0$, $a \neq 1$ *and* $a^x = y$, *then* $\log_a y = x$.

Rule II. $\log_a x_1 = \log_a x_2$ *implies* $x_1 = x_2$.

Rule III. *(i)* $\log_a 1 = 0$ *(ii)* $\log_a a = 1$.

EXERCISE 9.1

1. Show that $(27)^{-4/3} = \dfrac{1}{81}$ and express it in the logarithmic form.
2. Find the values of the following :

   (i) $\mathrm{Log}_b b$ *(ii)* $\log_n 1$ *(iii)* $\log_2 \sqrt{32}$ *(iv)* $\log_7 \sqrt[3]{7}$.
3. Find the values of the following :

   (i) $\log_9 243$ *(ii)* $\log_{10} (0.001)$ *(iii)* $\log_{\sqrt{2}} 16$ *(iv)* $\log_{81} 243$

   (v) $\log_{3\sqrt{2}} 324$ *(vi)* $\log_{0.2} 0.008$ *(vii)* $\log_5 \sqrt[3]{5}$ *(viii)* $\log_r r^4$.
4. Find the value of x from the following equations :

   (i) $\log_{16} x = \dfrac{5}{4}$ *(ii)* $\log_2 x = -3$ *(iii)* $\log_x 81 = 4$ *(iv)* $\log_2 \left(\dfrac{1}{16}\right) = x$

   (v) $\log_x 32 = \dfrac{5}{4}$ *(vi)* $\log_{1/9} 27\sqrt{3} = x$ *(vii)* $\log_{0.2} x = 3$ *(viii)* $\log_{0.4} x = 4$.
5. If $\log_{10} (a^2 - 4a + 5) = 0$, then show that $a = 2$.
6. Show that $\log_2 \{\log_2 (\log_2 16)\} = 1$.

Answers

1. $\log_{27}\left(\dfrac{1}{81}\right) = -\dfrac{4}{3}$ **2.** *(i)* 1 *(ii)* 0 *(iii)* $\dfrac{5}{2}$ *(iv)* $\dfrac{1}{3}$

3. (*i*) $\frac{5}{2}$ (*ii*) -3 (*iii*) 8 (*iv*) $\frac{5}{4}$ (*v*) 4 (*vi*) 3

(*vii*) $\frac{1}{3}$ (*viii*) 4 **4.** (*i*) 32 (*ii*) $\frac{1}{8}$ (*iii*) 3 (*iv*) -4

(*v*) 16 (*vi*) $-\frac{7}{4}$ (*vii*) 0.008 (*viii*) 0.0256.

Hints

2. (*i*) $\log_b b = x \Rightarrow b^x = b \Rightarrow b^x = b^1 \Rightarrow x = 1$

(*ii*) $\log_n 1 = x \Rightarrow n^x = 1 \Rightarrow n^x = n^0 \Rightarrow x = 0$

(*iv*) $\log_7 \sqrt[3]{7} = x \Rightarrow 7^x = \sqrt[3]{7} \Rightarrow x = \frac{1}{3}$

3. (*i*) $\log_9 243 = x \Rightarrow 9^x = 243 \Rightarrow 3^{2x} = 3^5$

(*v*) $\log 324 = x \Rightarrow (\sqrt[3]{2})^x = 324 \Rightarrow (\sqrt[3]{2})^x = 3^4.2^2 = (\sqrt[3]{2})^4$

(*vi*) $\log_{0.2} 0.008 = x \Rightarrow \left(\frac{2}{10}\right)^x = \frac{8}{1000} \Rightarrow \left(\frac{2}{10}\right)^x = \left(\frac{2}{10}\right)^3$

4. (*ii*) $\log_2 x = -3 \Rightarrow 2^{-3} = x$

(*v*) $\log_x 32 = \frac{5}{4} \Rightarrow x^{5/4} = 32 = 2^5 \Rightarrow x^{5/4} = (2^4)^{5/4}$

6. $\log \{\log (\log_2 16)\} = \log_2 \{\log_2 4\} = \log_2 2.$

II. LAWS OF LOGARITHMS

9.5. PRODUCT FORMULA

To prove that $\log_a mn = \log_a m + \log_a n$.

Proof. Let $\log_a m = x$ and $\log_a n = y$.

$\therefore \quad a^x = m$ and $a^y = n$.

$\Rightarrow \quad mn = a^x . a^y \Rightarrow mn = a^{x+y}$

$\therefore \quad \log_a mn = x + y$

$\therefore \quad \mathbf{\log_a mn = \log_a m + \log_a n.}$

Thus, the logarithm of the product of two numbers to any base is equal to the sum of the logarithms of the numbers to the same base.

Caution. $\log_a (m + n) \neq \log_a m + \log_a n$

and $\log_a (m - n) \neq \log_a m - \log_a n$

$\therefore$ The prefix $\log_a$ is not a multiplier.

In fact there is no formula for expressing $\log_a (m \pm n)$ in terms of $\log_a m$ and $\log_a n$.

Generalization. The above product formula is applicable to the product of any finite number of factors.

i.e., $\log_a (lmn \ldots\ldots) = \log_a l + \log_a m + \log_a n + \ldots\ldots$

9.6. QUOTIENT FORMULA

To prove that $\log_a \frac{m}{n} = \log_a m - \log_a n$.

Proof. Let $\log_a m = x$ and $\log_a n = y$.

$\therefore \quad a^x = m$ and $a^y = n$.

$\Rightarrow \quad \frac{m}{n} = \frac{a^x}{a^y} \Rightarrow \frac{m}{n} = a^{x-y}$

$\therefore \quad \log_a \frac{m}{n} = x - y$

$\therefore \quad \mathbf{\log_a \frac{m}{n} = \log_a m - \log_a n.}$

Thus, the logarithm of the quotient of two numbers to any base is equal to the difference of the logarithms of the numerator and the denominator to the same base.

9.7. POWER FORMULA

To prove that $\log_a m^n = n \log_a m$. *(P.T.U. B.Pharma, Dec. 2001)*

Proof. Let $\log_a m = x$.

$\therefore \quad a^x = m$

$\Rightarrow \quad m^n = (a^x)^n \Rightarrow m^n = a^{nx}$

$\therefore \quad \log_a m^n = nx$

$\therefore \quad \mathbf{\log_a m^n = n \log_a m.}$

Thus, the logarithm of a number raised to a power is equal to the index of the power times the logarithm of the number to the same base.

9.8. BASE CHANGING FORMULA

To prove that $\log_a m = \log_b m \times \log_a b$.

Proof. Let $\log_b m = x$ and $\log_a b = y$.

$\therefore \quad b^x = m$ and $a^y = b$

Eliminating b, we get $(a^y)^x = m$.

$\Rightarrow \quad m = a^{xy}$

$\therefore \quad \log_a m = xy$

$\therefore \quad \mathbf{\log_a m = \log_b m \times \log_a b.}$

Remark. The base changing formula can also be put as :

$$\log_b m = \frac{\log_a m}{\log_a b}.$$

Corollary. We have $\log_a m = \log_b m \times \log_a b$.

In particular let $m = a$.

$\therefore \quad \log_a a = \log_b a \times \log_a b$

$\Rightarrow \quad \mathbf{\log_b a \times \log_a b = 1.}$ $\quad (\because \log_a a = 1)$

ILLUSTRATIVE EXAMPLES

Example 1. *Express* $\log_a \dfrac{x^3 \sqrt[5]{y^2}}{\sqrt{z^3}}$ *in terms of* $\log_a x$, $\log_a y$, $\log_a z$.

Sol.
$$\log_a \frac{x^3 \sqrt[5]{y^2}}{\sqrt{z^3}} = \log_a \frac{x^3 y^{2/5}}{z^{3/2}}$$
$$= \log_a (x^3 y^{2/5}) - \log_a z^{3/2} \qquad \textit{(By Quotient Formula)}$$
$$= \log_a x^3 + \log_a y^{2/5} - \log_a z^{3/2} \qquad \textit{(By Product Formula)}$$
$$= \mathbf{3 \log_a x + \frac{2}{5} \log_a y - \frac{3}{2} \log_a z.} \qquad \textit{(By Power Formula)}$$

Example 2. *Find the values of:*

(*i*) $\log_2 \left(\dfrac{\sqrt[3]{4}}{(4)^2 \sqrt{8}} \right)$ (*ii*) $\log_{11} \left(\dfrac{121\sqrt{14641}}{\sqrt[3]{1331}} \right)$.

Sol. (*i*)
$$\log_2 \frac{\sqrt[3]{4}}{(4)^2 \sqrt{8}} = \log_2 \sqrt[3]{4} - \log_2 \left((4)^2 \sqrt{8}\right)$$
$$= \log_2 (4)^{1/3} - [\log_2 (4)^2 + \log_2 \sqrt{8}\,] = \log_2 (2^2)^{1/3} - \log_2 (2^2)^2 - \log_2 (2^3)^{1/2}$$
$$= \log_2 2^{2/3} - \log_2 2^4 - \log_2 2^{3/2} = \frac{2}{3} \log_2 2 - 4 \log_2 2 - \frac{3}{2} \log_2 2$$
$$= \frac{2}{3}(1) - 4(1) - \frac{3}{2}(1) = \frac{2}{3} - 4 - \frac{3}{2} = \mathbf{-\frac{29}{6}}.$$

(*ii*)
$$\log_{11} \left(\frac{121\sqrt{14641}}{\sqrt[3]{1331}} \right) = \log_{11} \left(121\sqrt{14641}\right) - \log_{11} \sqrt[3]{1331}$$
$$= \log_{11} 121 + \log_{11} \sqrt{14641} - \log_{11} \sqrt[3]{1331}$$
$$= \log_{11} (11)^2 + \log_{11} ((11)^4)^{1/2} - \log_{11} ((11)^3)^{1/3}$$
$$= 2 \log_{11} 11 + 2 \log_{11} 11 - \log_{11} 11$$
$$= 2(1) + 2(1) - 1 = \mathbf{3}.$$

Example 3. *Given* $\log_2 16 = 4$. *Find* $\log_{16} 2$.

Sol. Using $\log_b m = \dfrac{\log_a m}{\log_a b}$, we have
$$\log_{16} 2 = \frac{\log_2 2}{\log_2 16} = \mathbf{\frac{1}{4}}.$$

Example 4. *Show that* $3 \log 4 - 2 \log 6 + \log (18)^{3/2} = \log (96\sqrt{2})$.

Sol.
$$\text{L.H.S.} = 3 \log 4 - 2 \log 6 + \log (18)^{3/2} = \log (4)^3 - \log (6)^2 + \log (18)^{3/2}$$
$$= (\log (4)^3 + \log (18)^{3/2}) - \log (6)^2 = \log (4)^3 (18)^{3/2} - \log (6)^2$$
$$= \log \frac{(4)^3 (18)^{3/2}}{(6)^2} = \log \frac{64 \times 18 \times \sqrt{18}}{36}$$
$$= \log (32 \times 3 \times \sqrt{2}) = \mathbf{\log (96\sqrt{2})}.$$

Example 5. *Prove that* $7 \log \frac{16}{15} + 5 \log \frac{25}{24} + 3 \log \frac{81}{80} = \log 2.$

Sol. L.H.S. $= 7 \log \frac{16}{15} + 5 \log \frac{25}{24} + 3 \log \frac{81}{80}$

$= 7[\log 16 - \log 15] + 5 [\log 25 - \log 24] + 3 [\log 81 - \log 80]$

$= 7[\log 2^4 - \log (3 \times 5)] + 5 [\log 5^2 - \log (2^3 . 3)] + 3 [\log 3^4 - \log (2^4 . 5)]$

$= 7[4 \log 2 - (\log 3 + \log 5)] + 5 [2 \log 5 - (\log 2^3 + \log 3)] + 3[4 \log 3 - (\log 2^4 + \log 5)]$

$= 7[4 \log 2 - \log 3 - \log 5] + 5 [2 \log 5 - (3 \log 2 + \log 3)] + 3[4 \log 3 - (4 \log 2 + \log 5)]$

$= 28 \log 2 - 7 \log 3 - 7 \log 5 + 10 \log 5 - 15 \log 2 - 5 \log 3 + 12 \log 3 - 12 \log 2 - 3 \log 5$

$= \log 2 =$ R.H.S.

Remark. If base is not mentioned in a problem, it is the same base throughout (whatever it may be).

Example 6. *Solve the equation :*

$$\log (x + 4) - \log 7 = 3 \log 2 - \log (x + 5).$$

Sol. We have $\log (x + 4) - \log 7 = 3 \log 2 - \log (x + 5).$

$\Rightarrow \quad \log \frac{x+4}{7} = \log (2)^3 - \log (x + 5)$

$\Rightarrow \quad \log \frac{x+4}{7} = \log \frac{8}{x+5} \quad \Rightarrow \quad \frac{x+4}{7} = \frac{8}{x+5}$

$\Rightarrow \quad (x + 4) (x + 5) = 56 \quad \Rightarrow \quad x^2 + 9x - 36 = 0$

$\therefore \quad x = -12, 3$

$x = -12 \quad \Rightarrow \quad \log (x + 4) = \log (-12 + 4) = \log (-8)$, which is not defined.

$\therefore \quad x = -12$ is rejected.

$\therefore \quad$ **$x = 3$.**

Example 7. *If a, b, c, d are any four positive numbers different from '1', prove that* $\log_b a \times \log_c b \times \log_d c = \log_d a.$

Sol. L.H.S. $= \log_b a \times \log_c b \times \log_d c$

$= (\log_b a \times \log_c b) \log_d c$

$= \log_c a \times \log_d c$ (Using base changing formula)

$= \log_d a =$ R.H.S.

Example 8. *If* $x = \log_{2a} a$, $y = \log_{3a} 2a$, $z = \log_{4a} 3a$, *show that* $xyz + 1 = 2yz.$

Sol. $xyz + 1 = (\log_{2a} a) (\log_{3a} 2a) (\log_{4a} 3a) + 1$

$= (\log_{2a} a \times \log_{3a} 2a) \log_{4a} 3a + 1 = \log_{3a} a \times \log_{4a} 3a + 1$

$= \log_{4a} a + \log_{4a} 4a$ $\quad (\because \log_{4a} 4a = 1)$

$= \log_{4a} (a \times 4a) = \log_{4a} 4a^2.$

$2yz = 2 \log_{3a} 2a \times \log_{4a} 3a = 2 \log_{4a} 2a$

$= \log_{4a} (2a)^2 = \log_{4a} 4a^2$

$\therefore \quad$ **$xyz + 1 = 2yz$.**

WORKING RULES FOR SOLVING PROBLEMS

Rule I. $log_a mn = log_a m + log_a n.$

Rule II. $log_a \frac{m}{n} = log_a m - log_a n.$

Rule III. $log_a m^n = n \, log_a m.$

Rule IV. $log_a m = log_b m \times log_a b$ *or equivalently* $log_b m = \frac{log_a m}{log_a b}.$

Rule V. $log_b a \times log_a b = 1.$

EXERCISE 9.2

1. Express in terms of log x, log y and log z :

(*i*) $\log \frac{x^3}{y^2 z^4}$ (*ii*) $\log \frac{\sqrt[n]{x^m \, y^r}}{\sqrt[3]{z}}$.

2. Find the values of :

(*i*) $\log_7 \sqrt[3]{343}$ (*ii*) $\log_3 27\sqrt{729}$

(*iii*) $\log_2 \frac{\sqrt{64}}{\sqrt{8}}$ (*iv*) $\log_5 \frac{\sqrt[4]{25}}{625}$.

3. It is known that log 2 = 0.3010. Find the values of log 4 and log 8.

4. (*i*) Show that $\log \frac{a^2}{bc} + \log \frac{b^2}{ca} + \log \frac{c^2}{ab} = 0.$

(*ii*) Show that $\log_a (1 + 2 + 3) = \log_a 1 + \log_a 2 + \log_a 3.$

(*iii*) Show that $\log \frac{35}{78} = \log 7 + \log 5 - \log 2 - \log 3 - \log 13.$

5. (*i*) Show that $\frac{1}{2} \log_{10} 25 - 2 \log_{10} 3 + \log_{10} 18 = 1.$

(*ii*) Show that $\log_{10} 2 + 16 \log_{10} \frac{16}{15} + 12 \log_{10} \frac{25}{24} + 7 \log_{10} \frac{81}{80} = 1.$

6. Solve the following equations :

(*i*) $\log (4x - 3) = \log (x + 1) + \log 3$ (*ii*) $\log x = \log 3 + 2 \log 2 - \frac{3}{4} \log 16$

(*iii*) $\log 2 + \log (x + 2) - \log (3x - 5) = \log 3$ (*iv*) $\log x = 2 \log 3 + \frac{1}{2} \log 25 - \log 7.$

7. If $\log \frac{x+y}{2} = \frac{1}{2} (\log x + \log y)$, show that $x = y$.

8. Show that $x^{\log y - \log z} \cdot y^{\log z - \log x} \cdot z^{\log x - \log y} = 1.$

9. If a, b, c are three consecutive natural numbers, then show that $\log (1 + ac) = 2 \log b.$

10. Show that :

(*i*) $\log_b a \times \log_c b \times \log_a c = 1$

(*ii*) $\log_2 3 \times \log_3 4 \times \log_4 5 \times \log_5 6 \times \log_6 7 \times \log_7 8 = 3$.

Answers

1. (*i*) $3 \log x - 2 \log y - 4 \log z$ (*ii*) $\frac{m}{n} \log x + \frac{r}{n} \log y - \frac{1}{3} \log z$

2. (*i*) 1 (*ii*) 6 (*iii*) $\frac{3}{2}$ (*iv*) $-\frac{7}{2}$

3. 0.6020, 0.9030 **6.** (*i*) 6 (*ii*) $\frac{3}{2}$ (*iii*) $\frac{19}{7}$

(*iv*) $\frac{45}{7}$.

Hints

3. Use $\log 4 = \log (2)^2 = 2 \log 2$.

4. (*ii*) $\log_a (1 + 2 + 3) = \log_a 6 = \log_a (1 . 2 . 3)$.

(*iii*) R.H.S. $= (\log 7 + \log 5) - (\log 2 + \log 3 + \log 13)$

$= \log (7 \times 5) - \log (2 \times 3 \times 13) = \log 35 - \log 78$.

5. (*iii*) L.H.S. $= \log_{10} 2 + 16 (\log_{10} 2^4 - \log_{10} (3 \times 5)) + 12 (\log_{10} 5^2 - \log_{10} (2^3 \times 3))$

$+ 7 (\log_{10} 3^4 - \log_{10} (2^4 \times 5))$

7. We have $\log \frac{x+y}{2} = \log \sqrt{xy} \Rightarrow \frac{x+y}{2} = \sqrt{xy} \Rightarrow (x+y)^2 = 4xy \Rightarrow (x-y)^2 = 0$.

8. $x^{\log y - \log z} . y^{\log z - \log x} . z^{\log x - \log y} = 1$.

if $\log [x^{\log y - \log z} . y^{\log z - \log x} . z^{\log x - \log y}] = \log 1$.

i.e., if $(\log y - \log z) \log x + (\log z - \log x) \log y + (\log x - \log y) \log z = 0$

i.e., if $\log y \log x - \log z \log x + \log z \log y - \log x \log y + \log x \log z - \log y \log z = 0$.

9. We have $b = a + 1$ and $c = a + 2$.

III. COMMON LOGARITHMS

9.9. DEFINITION OF COMMON LOGARITHM

The logarithms to the base '10' are called **common logarithms.**

In all practical calculations, common logarithms are used. In the rest of the present chapter, we shall be discussing and using only the common logarithms and thus if no base is mentioned with logarithm of a number, it would be taken to be as '10'.

ILLUSTRATIVE EXAMPLES

Example 1. *Given log 2 = 0.30103, calculate log 5.*

Sol. log 5 means $\log_{10} 5$.

$$\therefore \quad \log 5 = \log_{10} 5 = \log_{10} \frac{10}{2} = \log_{10} 10 - \log_{10} 2$$

$$= 1 - \log_{10} 2 \qquad [\because \log_{10} 10 = 1]$$

$$= 1 - 0.30103 = \mathbf{0.69897}.$$

Example 2. *Given log 2 = 0.3010, log 3 = 0.4771, find log 360.*

Sol. We have $360 = 36 \times 10 = 4 \times 9 \times 10 = 2^2 \times 3^2 \times 10.$

$$\therefore \quad \log 360 = \log (2^2 \times 3^2 \times 10) = \log (2)^2 + \log (3)^2 + \log 10$$
$$= 2 \log 2 + 2 \log 3 + \log 10 = 2\,(0.3010) + 2(0.4771) + 1$$
$$= 0.6020 + 0.9542 + 1 = \mathbf{2.5562.}$$

Example 3. *Given log 2 = 0.3010, log 3 = 0.4771; find log $(0.405)^{1/2}$.*

Sol.
$$\log (0.405)^{1/2} = \frac{1}{2} \log \frac{405}{1000} = \frac{1}{2} \log \frac{81}{200}$$
$$= \frac{1}{2} \log \frac{3^4}{2 \times 10^2} = \frac{1}{2} \; [\log 3^4 - \log (2 \times 10^2)]$$
$$= \frac{1}{2} \; [4 \log 3 - (\log 2 + \log 10^2)] = \frac{1}{2} \; [4 \log 3 - \log 2 - 2 \log 10]$$
$$= \frac{1}{2} \; [4(0.4771) - (0.3010) - 2 \times 1] = \frac{1}{2} \; [1.9084 - 2.3010]$$
$$= \frac{1}{2} \; (-0.3926) = \mathbf{-0.1963.}$$

EXERCISE 9.3

1. Given log 2 = 0.3010, find :

   (*i*) $\log_{10} 64$ (*ii*) $\log \dfrac{1000}{256}$ (*iii*) log 0.0005 (*iv*) $\log_8 25$.

2. Given log 2 = 0.30103, log 3 = 0.47712, find log 162.
3. (*i*) Given $\log_{10} 2 = 0.3010$, $\log_{10} 3 = 0.4771$; find $\log_5 216$.

   (*ii*) Given log 2 = 0.30103, log 3 = 0.47712, find log 360.
4. Given log 2 = 0.30103 and log 7 = 0.845098, find :

   (*i*) $\log \left(\dfrac{4}{343}\right)^{1/2}$ (*ii*) $\log \left(\dfrac{4}{0.49}\right)^{1/2}$.

5. Given $\log_{10} 2 = 0.30103$, $\log_{10} 3 = 0.47712$, $\log_{10} 7 = 0.845098$; find :

   (*i*) $\log_{10} 162$ (*ii*) $\log_{10} (2056)^{1/5}$ (*iii*) $\log \dfrac{2^6 \times 5^3}{7^2}$.

6. Given log 4 = 0.60206, log 3 = 0.47712, find $\log \sqrt[8]{0.00018}$.
7. Given log 2 = 0.3010300 and log 3 = 0.4771213, then find $\log \left(3\dfrac{3}{8}\right)^{1/2}$.
8. Given log 2 = 0.3010 and log 3 = 0.4771 ; solve the equation $2^x.3^{2x} = 100$.

Answers

1. (*i*) 1.806 (*ii*) 0.5920 (*iii*) – 3.3010 (*iv*) 1.55 **2.** 2.20951
3. (*i*) 3.3395 (*ii*) 2.5563 **4.** (*i*) – 0.66617 (*ii*) 0.455932
5. (*i*) 2.20951 (*ii*) 0.66269 (*iii*) 2.212894 **6.** – 0.46809 **7.** 0.26413695
8. 1.593.

Hints

1. (*iii*) $\log 0.0005 = \log \frac{5}{10000} = \log \frac{1}{2000} = \log 1 - \log (2 \times 1000) = 0 - (\log 2 + 3)$.

(*iv*) $\log_8 25 = \frac{\log_{10} 25}{\log_{10} 8} = \frac{\log_{10} 100 - 2\log_{10} 2}{3 \log_{10} 2}$.

6. $\log \sqrt[8]{0.00018} = \frac{1}{8} \log \frac{18}{100000} = \frac{1}{8} \log \frac{2 \times 3^2}{10^5} = \frac{1}{8} [\log 2 + 2 \log 3 - 5 \log 10]$

Also $\log 2 = \log \sqrt{4} = \frac{1}{2} \log 4 = \frac{1}{2} (0.60206) = 0.30103$.

8. $2^x . 3^{2x} = 100 \Rightarrow \log 2^x + \log 3^{2x} = \log 100$

$\Rightarrow x (\log 2 + 2 \log 3) = 2 \Rightarrow x = \frac{2}{\log 2 + 2 \log 3}$.

9.10. TWO PARTS OF COMMON LOGARITHM OF A NUMBER

The common logarithm of a number (positive of course) consists of two parts : characteristic and mantissa.

9.11. CHARACTERISTIC

The integral part of the logarithm of a number, after expressing the decimal part as non-negative (if already not so), is called the **characteristic** of the logarithm of the number under considerations.

The characteristic of the logarithm of a number is either positive or zero or negative.

Illustrations. (*i*) If $\log x = 4.3729$, then characteristic of $\log x = 4$.

(*ii*) If $\log y = -2.8719$, then we write

$$\log y = -2 - 0.8719 = (-2 - 1) + (1 - 0.8719) = \bar{3}.1281$$

and say that the characteristic of $\log y = -3$.

Note. $\bar{3}.1281$ means that the integral part 3 is negative and the decimal part .1281 is positive.

9.12. METHOD OF FINDING CHARACTERISTIC

Case I. Number is equal or greater than 1.

We have $10^0 = 1$, $10^1 = 10$, $10^2 = 100$, $10^3 = 1000$ etc.

$\therefore$ $\log 1 = 0$, $\log 10 = 1$, $\log 100 = 2$, $\log 1000 = 3$ etc.

Since $\log 1 = 0$, the characteristic of $\log 1 = 0$.

The logarithm of any number between 1 and 10 (*i.e.,* a number of one digit in its integral part) lies between 0 and 1 ; hence its characteristic is zero.

The logarithm of any number between 10 and 100 (*i.e.*, a number of two digits in its integral part) lies between 1 and 2 ; hence its characteristic is 1.

Similarly, 2 is the characteristic of the logarithm of any number between 100 and 1000 (*i.e.*, a number of three digits in its integral part).

In general, $(n-1)$ is the characteristic of the logarithm of any number of n digits in its integral part.

∴ **The characteristic of the logarithm of '1' is zero and of any number greater than unity is one less than the number of digits in its integral part.**

Illustrations :

(*i*) Characteristic of log 5.39 = 1 – 1 = 0.

(*ii*) Characteristic of log 936.1 = 3 – 1 = 2.

(*iii*) Characteristic of log 236987 = 6 – 1 = 5.

Case II. Number is positive and less than one.

We have $10^0 = 1,\ 10^{-1} = 0.1,\ 10^{-2} = 0.01,\ 10^{-3} = 0.001$ etc.

∴ log 1 = 0, log 0.1 = – 1, log 0.01 = – 2, log 0.001 = – 3 etc.

Thus the logarithm of any number between 0.1 and 1 lies between – 1 and 0 and is therefore, equal to – 1 + some proper fraction.

∴ Characteristic = – 1.

The logarithm of any number between 0.01 and 0.1 lies between – 2 and – 1 and is therefore, equal to – 2 + some proper fraction.

∴ Characteristic = – 2.

Similarly, – 3 is the characteristic of the logarithm of any number between 0.001 and 0.01 (*i.e.*, a number in which there are two zeros immediately after the decimal point).

In general $-(n+1)$ is the characteristic of the logarithm of a number in which there are n zeros immediately after the decimal point.

∴ **The characteristic of the logarithm of any positive number less than 1, is negative and is numerically one more than the number of zeros immediately after the decimal point.**

Illustrations :

(*i*) Characteristic of log 0.236 = – (0 + 1) = – 1.

(*ii*) Characteristic of log 0.07 = – (1 + 1) = – 2.

(*iii*) Characteristic of log 0.0000693 = – (4 + 1) = – 5.

Remark. A negative integer $-n$ is also denoted by $\bar{n}$ and is read as 'n bar'.

Caution. $\bar{3}.637$ means – 3 + 0.637 whereas – 3.637 means – 3 – 0.637.

ILLUSTRATIVE EXAMPLES

Example 1. *Find the characteristics of the following logarithms :*

(*i*) *log 37.3* (*ii*) *log 8.0067*

(*iii*) *log 23679* (*iv*) *log 87.00967.*

Sol. (*i*) 37.3 is greater than 1 and has 2 digits in its integral part.

∴ Characteristic of log 37.3 = 2 – 1 = **1.**

(*ii*) 8.0067 is greater than 1 and has 1 digit in its integral part.

∴ Characteristic of log 8.0067 = 1 – 1 = **0.**

(*iii*) 23679 is greater than 1 and has 5 digits in its integral part.

∴ Characteristic of log 23679 = 5 – 1 = **4.**

(*iv*) 87.00967 is greater than 1 and has 2 digits in its integral part.

∴ Characteristic of log 87.00967 = 2 – 1 = **1.**

Example 2. *Find the characteristic of the following logarithms :*

(*i*) *log 0.237* (*ii*) *log 0.00087*

(*iii*) *log 0.09378* (*iv*) *log 0.000005.*

Sol. (*i*) 0.237 is a positive number less than 1 and has no zero immediately after its decimal point.

∴ Characteristic of log 0.237 = – (0 + 1) = **– 1.**

(*ii*) 0.00087 is a positive number less than 1 and has 3 zeros immediately after its decimal point.

∴ Characteristic of log 0.00087 = – (3 + 1) = **– 4.**

(*iii*) 0.09378 is a positive number less than 1 and has one zero immediately after its decimal point.

∴ Characteristic of log 0.09378 = – (1 + 1) = **– 2.**

(*iv*) 0.000005 is a positive number less than 1 and has 5 zeros immediately after its decimal point.

∴ Characteristic of 0.000005 = – (5 + 1) = **– 6.**

Example 3. *Given log 2 = 0.3010, find the number of digits in* 2^{64}.

Sol. $\log 2^{64} = 64 \log 2 = 64(0.3010) = 19.2640$

∴ Characteristic of $\log 2^{64} = 19$

∴ Number of digits in $2^{64} = 19 + 1 = \mathbf{20}$.

Example 4. *Given log 2 = 0.3010 ; find position of the first significant digit in the value of* $(0.5)^{10}$.

Sol.
$$\log (0.5)^{10} = \log \left(\frac{1}{2}\right)^{10} = \log 2^{-10}$$
$$= -10 \log 2 = -10\,(0.3010) = -3.010$$
$$= -3 - 0.010 = (-3 - 1) + (1 - 0.010) = -4 + 0.990 = \bar{4}.990$$

∴ Characteristic of $\log (0.5)^{10} = \bar{4}$ *i.e.,* – 4.

∴ Number of zeros immediately after the decimal point = 4 – 1 = 3

∴ First significant digit is at **4th place** after decimal.

9.13. STANDARD FORM OF A POSITIVE NUMBER

Let n be a positive number. If n can be expressed as $m \times 10^p$, where $1 \le m < 10$ and $p \in \mathbf{Z}$, then $m \times 10^p$ is called the **standard form** of the positive number n.

Let $m \times 10^p$, $1 \le m < 10$, $p \in \mathbf{Z}$ be the standard form of a positive number n.

$$\therefore \quad \log n = \log (m \times 10^p) = \log m + \log 10^p$$
$$= \log m + p \log 10 = \log m + p\,.$$
$$\therefore \quad \log n = p + \log m \qquad \ldots(1)$$

Now $1 \le m < 10$ implies $0 \le \log m < 1$.

∴ $\log m$ is the decimal part of $\log n$ and is also non-negative.

∴ The integral part of $\log n$ after expressing the decimal part as non-negative is p.

∴ By definition, characteristic of $\log n = p$.

∴ **If $m \times 10^p$, $1 \le m < 10$, $p \in Z$ is the standard form of a positive number, then the characteristic of $\log (m \times 10^p)$ is p.**

Illustrations :

(*i*) $5.39 = 5.39 \times 10^0$ ⇒ Characteristic of log 5.39 = 0

(*ii*) $936.1 = 9.361 \times 10^2$ ⇒ Characteristic of log 936.1 = 2

(*iii*) $236987 = 2.36987 \times 10^5$ ⇒ Characteristic of log 236987 = 5

(*iv*) $0.236 = 2.36 \times 10^{-1}$ ⇒ Characteristic of log 0.235 = – 1

(*v*) $0.07 = 7 \times 10^{-2}$ ⇒ Characteristic of log 0.07 = – 2

(*vi*) $0.0000693 = 6.93 \times 10^{-5}$ ⇒ Characteristic of log 0.0000693 = – 5.

Example 5. *Find the characteristics of the logarithm of the following numbers by expressing them in standard form :*

(*i*) *3862* (*ii*) *4.385*

(*iii*) *0.00023* (*iv*) *0.376.*

Sol. (*i*) $3862 = 3.862 \times 10^3$ $(1 \le 3.862 < 10)$

∴ Characteristic of 3862 = Characteristic of 3.862×10^3 = **3.**

(*ii*) $4.385 = 4.385 \times 10^0$ $(1 \le 4.385 < 10)$

∴ Characteristic of 4.385 = Characteristic of 4.385×10^0 = **0.**

(*iii*) $0.00023 = 2.3 \times 10^{-4}$ $(1 \le 2.3 < 10)$

∴ Characteristic of 0.00023 = Characteristic of 2.3×10^{-4} = **– 4.**

(*iv*) $0.376 = 3.76 \times 10^{-1}$ $(1 \le 3.76 < 10)$

∴ Characteristic of 0.376 = Characteristic 3.76×10^{-1} = **– 1.**

WORKING RULES FOR SOLVING PROBLEMS

Rule I. *Characteristic of log 1 is 0.*

Rule II. *If x > 1, then the characteristic of log x is one less than the number of digits in the integral part of x.*

Rule III. *If 0 < x < 1, then the characteristic of log x is negative and one more than the number of zeros immediately after the decimal point.*

Rule IV. *If $m \times 10^p$, $1 \le m < 10$, $p \in Z$ is the standard form of a positive number, then the characteristic of $\log (m \times 10^p)$ is p.*

EXERCISE 9.4

1. Find the characteristics of the following logarithms :

(*i*) log 2.2 (*ii*) log 343.6 (*iii*) log 935698 (*iv*) log 6.0007.

2. Find the characteristics of the following logarithms :

(*i*) log 0.057 (*ii*) log 0.9 (*iii*) log 0.000368 (*iv*) log 0.000001.

3. Find the characteristics of the logarithm of the following numbers by expressing them in standard form :

(*i*) 38.62 (*ii*) 910 (*iii*) 555.2 (*iv*) 346857.2

(*v*) 0.12 (*vi*) 0.0056 (*vii*) 0.000004 (*viii*) 0.000000106.

4. Given log 3 = 0.4771, find the number of digits in 3^{62}.
5. Given log 2 = 0.30103, log 3 = 0.4771, find the number of digits in $3^{12} \times 2^{18}$.
6. Given log 2 = 0.30103, find the position of the first significant digit in the value of 2^{-37}.
7. Given log 3 = 0.47712, find the position of the first significant digit in the value of 3^{-32}.

Answers

1. (*i*) 0	(*ii*) 2	(*iii*) 5	(*iv*) 0	**2.** (*i*) – 2
(*ii*) – 1	(*iii*) – 4	(*iv*) – 6	**3.** (*i*) 1	(*ii*) 2
(*iii*) 2	(*iv*) 5	(*v*) – 1	(*vi*) – 3	(*vii*) – 6
(*viii*) – 7	**4.** 30	**5.** 12	**6.** 12th	**7.** 16th.

Hints

3. (*i*) $38.62 = \frac{38.62}{10} \times 10 = 3.862 \times 10^1$

∴ Ch. of log 38.62 = 1

(*vi*) $0.0056 = 0.0056 \times 1000 \times \frac{1}{1000} = 5.6 \times 10^{-3}$

∴ Ch. of log 0.0056 = – 3

5. $\log (3^{12} \times 2^{18}) = 12 \log 3 + 18 \log 2 = 12(0.4771) + 18(0.30103) = 11.14374$

∴ Ch. of $\log (3^{12} \times 2^{18}) = 11$

∴ No. of digits in $3^{12} \times 2^{18} = 11 + 1 = 12$.

9.14. MANTISSA

The positive decimal part of the logarithm of a number after expressing the decimal part as non-negative (if already not so), is called the **mantissa** of the logarithm of the number under consideration.

By definition, the mantissa of the logarithm of a number is either positive or zero.

Illustrations. (*i*) If $\log x = 2.376$, then mantissa of $\log x = 0.376$.

(*ii*) If $\log y = -4.1786$, then we write $\log y = (-4-1) + (1-0.1786) = -5 + 0.8214 = \bar{5}.8214$ and say that the mantissa of $\log y = 0.8214$.

9.15. METHOD OF FINDING MANTISSA

The mantissa of logarithm of a number is found by using a table named 'LOGARITHMS' given at the end of the book. For convenience in printing, the decimals have been dropped in the table.

This table gives the value of the mantissa of logarithm of a number correct to four places of decimals and is therefore, called 'four figure table'.

The following is the procedure of using 'Logarithms' table for finding the mantissa of logarithm of a number :

Step I. Remove the decimal point from the given number and make it a number of four digits (if already not so).

Illustrations :

(*i*) For mantissa of log 2.379, we shall consider the number 2379.

(*ii*) For mantissa of log 34.872, we shall consider the number 3487.

(*iii*) For mantissa of log 34.876, we shall consider the number 3488.

(*iv*) For mantissa of log 93, we shall consider the number 9300.

(*v*) For mantissa of log 0.07, we shall consider the number 7000.

(*vi*) For mantissa of log 0.0169834, we shall consider the number 1698.

Step II. In the table of logarithm, run your eye down the extreme left hand column till you arrive at the **first two** digits of the number from the left.

Step III. Trace the **third digit** from the left in the top horizontal row. Mark the number at the intersection of the column and the row under consideration.

Step IV. Trace the **fourth digit** from the left in the column of mean differences. Run your eye vertically down this digit and horizontally along the row in which the number in step III lies. Find the number situated at the intersection of the column and the row. Add this number to the number obtained in step III.

If the fourth digit is zero, then there is no need of consulting mean differences.

Step V. Prefix a decimal point of this number. This gives the required mantissa.

ILLUSTRATIVE EXAMPLES

Example 1. *Find log 143.69.*

Sol. Characteristic. 143.69 is greater than 1 and has 3 digits in its integral part.

∴ Characteristic of 143.69 = 3 – 1 = 2.

Mantissa. Removing the decimal point from 143.69, we get 14369. The four digit number to be used is 1437 (∵ 9 ≥ 5). The first two digits from the left form the number **14**, the third digit is **3** and the fourth digit is **7.**

	0→	3	→9	Mean Differences		
				1→	7	→9
10 ↓						
14		1553			22	
↓ 99						

In the table of 'Logarithms', we run our eye down the extreme left hand column, till we reach the number 14.

At the intersection of the horizontal row beginning with 14 and column headed by 3, we find the number 1553.

At the intersection of the horizontal row beginning with 14, the small mean difference column on the right headed by 7, we find the number 22.

Adding 22 to 1553, we get 1575.

Prefixing the decimal point, the required mantissa is 0.1575.

∴ Combining characteristic and mantissa, we get

log 143.69 = **2.1575.**

Example 2. *Find log 0.0071683.*

Sol. Characteristic. 0.0071683 is a positive number less than 1 and has 2 zeros immediately after its decimal point.

∴ Characteristic of log 0.0071683 = – (2 + 1) = – 3.

Mantissa. Removing the decimal point from 0.0071683, we get 71683. The four digit number to be used is 7168 (∵ 3 < 5). The first-two digits from the left form the number **71,** the third digit is **6** and the fourth digit is 8.

In the table of 'logarithms', we run our eye down the extreme left hand column, till we reach the number 71.

At the intersection of the horizontal row beginning with 71 and column headed by 6, we find the number 8549.

At the intersection of the horizontal row beginning with 71 and the small mean difference column on the right headed by 8, we find the number 5.

Adding 5 to 8549, we get 8554.

Prefixing the decimal point, the required mantissa is 0.8554.

∴ Combining characteristic and mantissa, we get

$$\log 0.0071683 = \mathbf{\bar{3}.8554}.$$

Caution. It is wrong to write log 0.0071683 as – 3.8554.

Example 3. *Find log 873920.*

Sol. Characteristic. 873920 is greater than 1 and has 6 digits in its integral part.

∴ Characteristic of log 873920 = 6 – 1 = 5.

Mantissa. The four digit number to be used is 8739 (∵ 2 < 5).

	0→	3	→9	Mean Differences	
				1→	9
10 ↓					
87		9410			4
↓ 99					

Adding 4 to 9410, we get 9414.

∴ Mantissa of log 873920 = 0.9414.

∴ log 873920 = **5.9414.**

Example 4. *Evaluate* $\log_{43} 57$.

Sol. $$\log_{43} 57 = \frac{\log_{10} 57}{\log_{10} 43} \qquad \textit{(By base changing formula)}$$

$$= \frac{1.7559}{1.6335} = \mathbf{1.074.} \qquad \textit{(From log table)}$$

Example 5. *Solve the equation :*

$$2^x \,.\, 3^{2x+1} = 7^{4x+3}.$$

Sol. Given equation is $2^x \,.\, 3^{2x+1} = 7^{4x+3}$.

Taking logarithms of both sides, we get

$$\log (2^x \,.\, 3^{2x+1}) = \log 7^{4x+3}.$$

$$\Rightarrow \quad \log 2^x + \log 3^{2x+1} = \log 7^{4x+3}.$$

$$\Rightarrow \quad x \log 2 + (2x + 1) \log 3 = (4x + 3) \log 7$$

$$\Rightarrow \quad x [\log 2 + 2 \log 3 - 4 \log 7] = 3 \log 7 - \log 3$$

$$\therefore \quad x = \frac{3 \log 7 - \log 3}{\log 2 + 2 \log 3 - 4 \log 7} = \frac{3(0.8451) - 0.4771}{0.3010 + 2(0.4771) - 4(0.8451)} \quad \text{(using log table)}$$

$$= \frac{2.5353 - 0.4771}{0.3010 + 0.9542 - 3.3804} = \frac{2.0582}{1.2552 - 3.3804} = \frac{2.0582}{-2.1242} = -0.9687$$

$$\therefore \quad x = \mathbf{-0.9687.}$$

EXERCISE 9.5

Find the values of the following by using 'Logarithms' table (Q. 1 to 12) :

1. log 25	**2.** log 10000	**3.** log 893
4. log 7650	**5.** log 12.678	**6.** log 23687.2
7. log 0.768	**8.** log 0.0003	**9.** log 0.005698
10. log 0.00007	**11.** $\log_{33} 44$	**12.** $\log_{67} 85$.

13. Solve the equation : $3^x \times 7^{2x+1} = (0.11)^{x+5}$.

14. Find the number of digits in the integral part of $(58.13)^{20} \div (46.15)^{15}$.

Answers

1. 1.3979	**2.** 4.0000	**3.** 2.9509	**4.** 3.8837	**5.** 1.1031	**6.** 4.3746
7. $\bar{1}.8854$	**8.** $\bar{4}.4771$	**9.** $\bar{3}.7557$	**10.** $\bar{5}.8451$	**11.** 1.0824	**12.** 1.057
13. – 1.8 (nearly)		**14.** 11.			

9.16. ANTILOGARITHMS

(P.T.U. B. Pharma, May 2007)

The number whose logarithm is x is called the **antilogarithm** of x and is written as **antilog x** or just as **AL x.**

Thus, if $\log y = x$, then we have antilog $x = y$.

Remark. $\log y = x \Rightarrow y = \text{antilog } x$ and $y = 10^x$.

$\therefore$ **Antilog x = 10^x.**

9.17. METHOD OF FINDING ANTILOGARITHM

The antilogarithm of a number is found by using a table named 'ANTILOGARITHMS' given at the end of the book. This table gives the value of the antilogarithm of a number correct to four places of decimal and is therefore called 'four figure table'.

For finding antilogarithm, we take into consideration only the mantissa (*i.e.*, the positive decimal part). The characteristic (*i.e.*, the integral part) is used only to determine the number of digits in the integral part or the number of zeros immediately after the decimal point.

The method of consulting the table of antilogarithms is the same as that of the table of logarithms discussed above.

ILLUSTRATIVE EXAMPLES

Example 1. *Find the number whose logarithm is 2.1575.*

Sol. Let $\log x = 2.1575.$

$\therefore$ $x = \text{Antilog } 2.1575$

$\therefore$ Characteristic of $\log x = 2$

$\therefore$ x contains $3(= 2 + 1)$ digits in its integral part.

Mantissa of $\log x = .1575$.

	0→	7	→ 9	Mean Differences		
				1→	5	→ 9
.00 ↓						
.15		1435			2	
↓ .99						

In the table of antilogarithm, in the row starting with 15 and under the column headed by 7, the number is 1435. In the same row and under the small column on the right (called the mean difference column) headed by 5, the number is 2.

Adding 2 to 1435, we get 1437. Since x contain 3 digits in its integral part, we put decimal after 3 digits and get x *i.e.*, antilog 2.1575 = **143.7.**

Example 2. *Find the value of antilog* $\bar{3}.8554$.

Sol. Let $x = \text{antilog } \bar{3}.8554.$

$\therefore$ $\log x = \bar{3}.8554.$

$\therefore$ Characteristic of $\log x = -3$

$\therefore$ x contains $2\ (= 3 - 1)$ zeros immediately after the decimal point.

Mantissa of $\log x = .8554$.

In the table of antilogarithms, in the row starting with .85 and under the column headed by 5, the number is 7161. In the same row and under the small column on the right headed by 4, the number is 7.

Adding 7 to 7161, we get 7168.

Since x contains 2 zeros immediately after the decimal point, we get

$$x = \text{antilog } \bar{3}.8554 = \mathbf{0.007168.}$$

Example 3. *Find antilog 5.9414.*

Sol. Mantissa = .9414.

	0→	1	→9	Mean Differences		
				1→	4	→9
.00 ↓						
.94		8730			8	
↓ .99						

Adding 8 to 8730, we get 8738. Since, characteristic is 5, the integral part of antilog 5.9414 must have 6(= 5 + 1) digits.

∴ Antilog 5.9414 = **873800.**

EXERCISE 9.6

Find the values of the following by using 'Antilogarithms' table :

1. Antilog 1.2400 **2.** Antilog 2.5690 **3.** Antilog 0.6987

4. Antilog 1.67 **5.** Antilog 6.972 **6.** Antilog 8

7. Antilog $\bar{1}.2$ **8.** Antilog $\bar{3}.653$ **9.** Antilog $\bar{4}.1236$

10. Antilog $\bar{2}.5999$ **11.** Antilog (– 0.7214) **12.** Antilog (– 2.5514).

Answers

1. 17.38 **2.** 370.7 **3.** 4.997 **4.** 46.77

5. 9376000 **6.** 100000000 **7.** 0.1585 **8.** 0.004498

9. 0.0001329 **10.** 0.03980 **11.** 0.1900 **12.** 0.002809.

Hints

11. $-0.7214 = -1 + 1 - 0.7214 = \bar{1}.2786$. ∴ Antilog (– 0.7214) = Antilog $\bar{1}.2786$

12. $-2.5514 = -2 - .5514 = -3 + 1 - .5514 = \bar{3}.4486$. ∴ Antilog (– 2.5514) = Antilog $\bar{3}.4486$.

9.18. APPLICATIONS OF LOGARITHMS TO PROBLEMS OF COMPOUND INTEREST, GROWTH AND DECAY (DEPRECIATION)

The technique of logarithms is used to simplify the algebraic calculation to a great extent. The *n*th root of a number can also be found out very easily by using logarithms.

ILLUSTRATIVE EXAMPLES

Example 1. *Find the seventh root of 0.00001427.*

Sol. Let $x = (0.00001427)^{1/7}$.

Taking logarithms of both sides, we get

$$\log x = \log (0.00001427)^{1/7} = \frac{1}{7}\log (0.00001427)$$

$$= \frac{1}{7}(\bar{5}.1545) = \frac{1}{7}(-5 + 0.1545)$$

$$= \frac{1}{7}(-5 - 2 + 2 + 0.1545) \quad \text{(Note this step)}$$

$$= \frac{1}{7}(-7 + 2.1545) = -1 + \frac{1}{7}(2.1545) = -1 + 0.3078 = \bar{1}.3078$$

$\therefore$ $x = \text{antilog}\ (\bar{1}.3078) = \mathbf{0.2032.}$

Example 2. *Find x, if*

$$\log x = 2 \log 3 + \frac{1}{3} \log 5 - \log 7.$$

Sol. We have $\log x = 2 \log 3 + \frac{1}{3} \log 5 - \log 7.$

$\therefore$ $\log x = 2(0.4771) + \frac{1}{3}(0.6990) - 0.8451 = 0.3421$

$\therefore$ $x = \text{antilog}\ 0.3421 = \mathbf{2.199.}$

Example 3. *Evaluate* : $\log \dfrac{\left(\sqrt{\sqrt{625}+11}\right)\left(\sqrt{64}\right)}{\sqrt{\sqrt[5]{3125}+\sqrt[3]{343}}}$.

Sol. $$\log \frac{\left(\sqrt{\sqrt{625}+11}\right)\left(\sqrt{64}\right)}{\sqrt{\sqrt[5]{3125}+\sqrt[3]{343}}} = \log \frac{\left(\sqrt{25+11}\right)(8)}{\sqrt{5+7}}$$

$$= \log \frac{6 \times 8}{2\sqrt{3}} = \log 8\sqrt{3}$$

$$= \log 8 + \frac{1}{2}\log 3 = 0.9031 + \frac{1}{2}(0.4771) = \mathbf{1.14165.}$$

Example 4. *Evaluate* : $\dfrac{(3.142)^3 \times (0.078)^{1/3}}{(0.005)^{1/4}}$.

Sol. Let $$x = \frac{(3.142)^3 \times (0.078)^{1/3}}{(0.005)^{1/4}}$$

Taking logarithms of both sides, we get

$$\log x = \log \left[\frac{(3.142)^3 \times (0.078)^{1/3}}{(0.005)^{1/4}}\right]$$

$$= \log [(3.142)^3 \times (0.078)^{1/3}] - \log (0.005)^{1/4}$$

$$= 3 \log (3.142) + \frac{1}{3} \log \ (0.078) - \frac{1}{4} \log (0.005)$$

$$= 3(0.4972) + \frac{1}{4} (\bar{2}.8921) - \frac{1}{4} (\bar{3}.6990)$$

$$= 1.4916 + \frac{1}{3} [\bar{3} + 1.8921] - \frac{1}{4} (\bar{4} + 1.\ 6990) \quad \text{(Note this step)}$$

$$= 1.4916 + \bar{1} + 0.6307 - \bar{1} - 0.4248$$

$\therefore$ $\log x = 2.1223 - 0.4248 = 1.6975$

$\therefore$ $x = \text{antilog } (1.6975) =$ **49.83.**

9.19. FORMULAE RELATING COMPOUND INTEREST, GROWTH & DECAY (DEPRECIATION)

(*i*) If P = principal, n = no. of periods, r = rate of interest per period, A = amount after n periods, then

$$\mathbf{A = P\left(1+\frac{r}{100}\right)^{n}.}$$

(*ii*) If P_0 = population at beginning , n = no. of periods , r = rate of growth of population per period, P_n = population at the end of n periods, then

$$\mathbf{P_n = P_0\left(1+\frac{r}{100}\right)^{n}.}$$

(*iii*) If P_0 = population at beginning, n = no. of periods, r = rate of decay of population per period, P_n = population at the end of n periods, then

$$\mathbf{P_n = P_0\left(1-\frac{r}{100}\right)^{n}.}$$

(*iv*) If V_0 = value of a machine at beginning, n = no. of periods, r = rate of depreciation of machine per period, V_n = value of machine at the end of n periods, then

$$\mathbf{V_n = V_0\left(1-\frac{r}{100}\right)^{n}.}$$

Example 5. *An investment of ₹ 25,000 earns interest at 9 per cent, compounded annually. What will be the value of the investment at the end of 5 years ?*

Sol. Investment = ₹ 25,000, rate of interest = 9% p.a., time = 5 years.

$\therefore$ $P = 25{,}000,\ r = 9,\ n = 5.$

Using $A = P\left(1+\frac{r}{100}\right)^{n}$, we get $A = 25{,}000\left(1+\frac{9}{100}\right)^{5}.$

$\therefore$ $A = 25{,}000 \times (1.09)^5$

$\therefore$ $\log A = \log [25{,}000 \times (1.09)^5] = \log 25{,}000 + 5 \log 1.09$

$= 4.3979 + 5(0.0374) = 4.5849$

$\therefore$ $A = \text{antilog } 4.5849 = 38{,}450$

$\therefore$ Value of investment after 5 years = **₹ 38,450.**

Example 6. *If ₹ 1,750 are invested at 9 per cent interest per year for 10 years. Find :*

(i) the interest compounded annually,

(ii) the interest compounded half yearly and,

(iii) the difference between interests obtained in (a) and (b).

Sol. (*i*) **Interest compounded annually**

Principal = ₹ 1,750, rate of interest = 9% p.a., time = 10 years.

$\therefore \quad P = 1{,}750,\ r = 9,\ n = 10$

Using $\quad A = P\left(1+\frac{r}{100}\right)^n$, we get $A = 1{,}750\left(1+\frac{9}{100}\right)^{10}$.

$\therefore \quad A = 1{,}750 \times (1.09)^{10}$

$\therefore \quad \log A = \log\,[1{,}750 \times (1.09)^{10}] = \log 1{,}750 + 10 \log 1.09$

$= 3.2430 + 10\,(0.0374) = 3.617$

$\therefore \quad A = \text{antilog } 3.617 = 4{,}140$

∴ Interest = ₹ 4,140 – ₹ 1,750 = **₹ 2,390.**

(*ii*) **Interest compounded half yearly**

Principal = ₹ 1,750, rate of interest = 4.5% $\left(=\frac{9}{2}\%\right)$ per half year, time = 20 (= 10 × 2) half years.

$\therefore \quad P = 1{,}750,\quad r = 4.5,\quad n = 20$

Using $\quad A = P\left(1+\frac{r}{100}\right)^n$, we get $A = 1{,}750\left(1+\frac{4.5}{100}\right)^{20}$.

$\therefore \quad A = 1{,}750 \times (1.045)^{20}$

$\therefore \quad \log A = \log\,[1{,}750 \times (1.045)^{20}] = \log 1{,}750 + 20 \log 1.045$

$= 3.\,2430 + 20\,(0.0191) = 3.625$

$\therefore \quad A = \text{antilog } 3.625 = 4{,}217$

∴ Interest = ₹ 4,217 – ₹ 1,750 = **₹ 2,467.**

(*iii*) Difference of interests = ₹ 2,467 – ₹ 2,390 = **₹ 77.**

Problems Related to Growth/Decay

Example 7. *The bacteria in a culture grows by 7% in the first hour, decreases by 6% in the second hour, and again increases by 5% in the third hour. If at the end of third hour, the number of bacteria is 11270000, find the original number of bacteria in the sample.*

Sol. Let N = the original number of bacteria in the sample.

As given, $N\left(1+\frac{7}{100}\right)\left(1-\frac{6}{100}\right)\left(1+\frac{5}{100}\right) = 11270000$

$\Rightarrow \quad N\cdot\frac{107}{100}\cdot\frac{94}{100}\cdot\frac{105}{100} = 11270000$

⇒ Taking logarithm, (base 10) both sides, we have

$\log N + \log 107 + \log 94 + \log 105 - \log 10^6 = \log 11270000$

$\Rightarrow \quad \log N = \log 11270000 + 6 \log 10 - \log 107 - \log 94 - \log 105$

$= 7.0519 + 6 - 2.0293 - 1.97312 - 2.02118$

$= 13.0519 - 6.0236 = 7.0283$

Taking antilogarithm,

$$N = \text{antilog}\ (7.0283) = 10673331.54$$

$$N \cong \mathbf{106733331.}$$

Example 8. *The population of a city grows at the rate of 2.5% p.a. If the present population is 20,50,000, find its population after 11 years.*

Sol. Present population = 20,50,000, rate of growth = 2.5% p.a., time = 11 years.

$\therefore\ P_0 = 20{,}50{,}000,\ r = 2.5,\ n = 11.$

Using $P_n = P_0\left(1+\dfrac{r}{100}\right)^n$, we get $P_{11} = 20{,}50{,}000\left(1+\dfrac{2.5}{100}\right)^{11}$.

$$\therefore \quad P_{11} = 20{,}50{,}000\ (1.025)^{11}$$
$$= \text{AL}\ (\log 20{,}50{,}000 \times (1.025)^{11}) = \text{AL}\ [\log 20{,}50{,}000 + 11 \log 1.025]$$
$$= \text{AL}\ [6.3118 + 11(0.0107)] = \text{AL}\ (6.4295) = 26{,}88{,}000.$$

$\therefore$ Population after 11 years = **26,88,000.**

Example 9. *India's population according to 1991 census was 8.4×10^7 and it increases each year by 2% of what it had been at the beginning of each year. Find the population in 2011.*

Sol. Present population = 8.4×10^7, rate of growth = 2% p.a., time = 2011 – 1991 = 20 years.

$\therefore\ P_0 = 8.4 \times 10^7,\ r = 2,\ n = 20.$

Using $P_n = P_0\left(1+\dfrac{r}{100}\right)^n$, we get $P_{20} = 8.4 \times 10^7\left(1+\dfrac{2}{100}\right)^{20}$.

$$\therefore \quad P_{20} = 8.4 \times 10^7 \times (1.02)^{20} = \text{AL}\ [\log\ [8.4 \times 10^7 \times (1.02)^{20}]]$$
$$= \text{AL}\ [\log 8.4 + 7\log 10 + 20\log 1.02] = \text{AL}\ [0.9243 + 7(1) + 20\ (0.0086)]$$
$$= \text{AL}\ [8.0963] = 124800000 = 12.48 \times 10^7.$$

$\therefore$ Population in 2011 = $\mathbf{12.48 \times 10^7}$**.**

Problems Related to Depreciation

Example 10. *The price of a motor bike depreciates by 2% of its value at the beginning of each year. Find the sale value of the motor bike after 3 years if its present sale value is ₹ 20000.*

Sol. Let P = The present sale value of the motor bike = ₹ 20000

$$r = -2\%$$
$$n = 3 \text{ years}$$

Let $\quad S$ = The sale value after 3 years.

then,
$$S = P\left(1+\frac{r}{100}\right)^n = 20000\left(1-\frac{2}{100}\right)^3$$
$$= 20000\left(\frac{98}{100}\right)^3$$

Taking logarithm (base 10) both sides

$$\log S = \log 20000 + \log\left(\frac{98}{100}\right)^3$$
$$= \log 20000 + 3\ (\log 98 - \log 100)$$

$= 4.30102 + 3\,(1.99122 - 2)$

$= 4.30102 - 3\,(0.0088)$

$= 4.30102 - 0.0264 = 4.27462$

Taking antilog, both sides

$S = \text{antilog}\,(4.27462)$

$= 18820.0165 \equiv 18820.$

i.e., the sale value of the motor bike 3 years = ₹ **18820.**

Example 11. *A machine is depreciated at the rate of 10% on reducing balance. The original cost was ₹ 10000 and the ultimate scrap value was ₹ 3750. Find the effective life of the machine ?*

Sol. Here r = rate of depreciation = $10\% = \dfrac{10}{100} = 0.1$

V_0 = Value of machine in the beginning = ₹ 10,000.

V_n = Value of machine at the end of n years = ₹ 3750.

n = The effective life of the machine.

Using $$V_n = V_0\left(1 - \frac{r}{100}\right)^n$$

$$3750 = 10000\,(1 - 0.1)^n = 10000\,(0.9)^n$$

Taking logarithm (base 10), both sides, we get

$\log 3750 = \log 10^4 + n \log (0.9)$

$3.57403 = 4 + n(-0.045757)$

$= 4 - (0.045757)n \Rightarrow (0.045757)n = 4 - 3.57403 = 0.42597$

$\Rightarrow \quad n = \dfrac{0.42597}{0.045757}$

$= 9.3093 = \mathbf{9\ years.}$

Example 12. *The cost price of a new car is ₹ 3.7×10^5. The insurance company calculates its price at any subsequent time according to the rule that the price depreciates at the rate of 5% a year during first two years and at the rate of 10% a year thereafter. What will be the price of the car after 5 years ?*

Sol. Cost of new car = ₹ 3.7×10^5, rates of depreciation = 5% p.a. for first 2 years and 10% p.a. thereafter, time = 5 years.

$\therefore \quad V_0 = 3.7 \times 10^5,\ r_1 = 5,\ r_2 = 10,\ n_1 = 2,\ n_2 = 3\ (= 5 - 2).$

$\therefore$ Using $V_n = V_0\left(1 - \dfrac{r_1}{100}\right)^{n_1}\left(1 - \dfrac{r_2}{100}\right)^{n_2}$, we have $V_5 = 3.7 \times 10^5\left(1 - \dfrac{5}{100}\right)^2\left(1 - \dfrac{10}{100}\right)^3.$

$\therefore \quad V_5 = 3.7 \times 10^5\,(0.95)^2\,(0.9)^3$

$= \text{AL}\,[\log\,[3.7 \times 10^5\,(0.95)^2\,(0.9)^3]]$

$= \text{AL}\,[\log 3.7 + 5 \log 10 + 2 \log 0.95 + 3 \log 0.9]$

$= \text{AL}\,[0.5682 + 5(1) + 2(\bar{1}.9777) + 3(\bar{1}.9542)]$

$$= \text{AL } [0.5682 + 5 + 2(-1) + 2(0.9777) + 3(-1) + 3(0.9542)]$$
$$= \text{AL } [5.3862] = 243300 = 2.433 \times 10^5.$$

$\therefore$ Price of car after 5 years = **₹ 2.433 × 10^5.**

Example 13. *The value of a machine depreciates at the rate of 7% p.a. If the present value of the machine is ₹ 70,250, find its value after $4\frac{3}{4}$ years.*

Sol. Present value of machine = ₹ 70,250, rate of depreciation = 7% p.a., time = $4\frac{3}{4}$ years.

$$\therefore \quad V_0 = 70{,}250,\ r_1 = 7,\ r_2 = 7 \times \frac{3}{4} = 5.25,\ n_1 = 4,\ n_2 = 1.$$

Using $V_n = V_0\left(1-\frac{r_1}{100}\right)^{n_1}\left(1-\frac{r_2}{100}\right)^{n_2}$, we have $V_n = 70{,}250\left(1-\frac{7}{100}\right)^4\left(1-\frac{5.25}{100}\right)^1$.

$$\therefore \quad V_n = 70{,}250\,(0.93)^4\,(0.9475)$$
$$= \text{AL } [\log 70{,}250\,(0.93)^4\,(0.9475)]$$
$$= \text{AL } [\log 70{,}250 + 4 \log 0.93 + \log 0.9475]$$
$$= \text{AL } [4.8466 + 4(\bar{1}.9685) + \bar{1}.9765]$$
$$= \text{AL } [4.8466 + 4(-1) + 4(0.9685) - 1 + 0.9765]$$
$$= \text{AL } [4.6971] = 49{,}780.$$

$\therefore$ Value of machine after $4\frac{3}{4}$ years = **₹ 49,780.**

IV. APPLICATIONS OF LOGARITHM TO MENSURATION

(a) Area and Perimeter of a Triangle

(i) Area of a triangle = $\frac{1}{2}$ × Base × Height

(ii) Area of a triangle = $\sqrt{s(s-a)(s-b)(s-c)}$, where a, b, c denote the sides of triangle and $s = \frac{1}{2}(a+b+c)$

(iii) Area of an equilateral triangle of side $a = \frac{\sqrt{3}}{4}a^2$.

Clearly, its perimeter = $3a$ $\left[\text{Base} = a \text{ and Height} = \frac{\sqrt{3}}{2}a\right]$

(iv) Area of an isosceles right angled triangle, where a is the length of each one of the equal sides = $a^2/2$

(b) Quadrilaterals (Areas and Perimeter of Plane Figures)

(i) Area of a rectangle = (length × breadth).

(ii) Diagonal of a rectangle = $\sqrt{(\text{length})^2 + (\text{breadth})^2}$.

(iii) Perimeter of a rectangle = 2(length + breadth).

(iv) Area of a square = a^2, where a = side of the square.

(v) Diagonal of a square = $\sqrt{2}\,a$.

(*vi*) Area of a square = (1/2) (diagonal)^2.
(*vii*) Perimeter of a square = 4 × (side).
(*viii*) Area of parallelogram = (Base × Height).
(*ix*) Perimeter of a parallelogram = 2(sum of adjacent sides).
(*x*) Area of a rhombus = $\frac{1}{2}$ (product of diagonals).
(*xi*) Perimeter of a rhombus = 4 × (side).

Remark : The diagonals of a rhombus bisect each other at right angles.

(*xii*) Area of a Trapezium = $\frac{1}{2}$ × (Sum of parallel sides × perpendicular distance between them).

ILLUSTRATIVE EXAMPLES

Example 1. *A vertical flag-staff stands on a horizontal plane. From a point distant 150.4 m from its foot, the angle of elevation of its top is found to be 32°. Find the height of the flag-staff.*

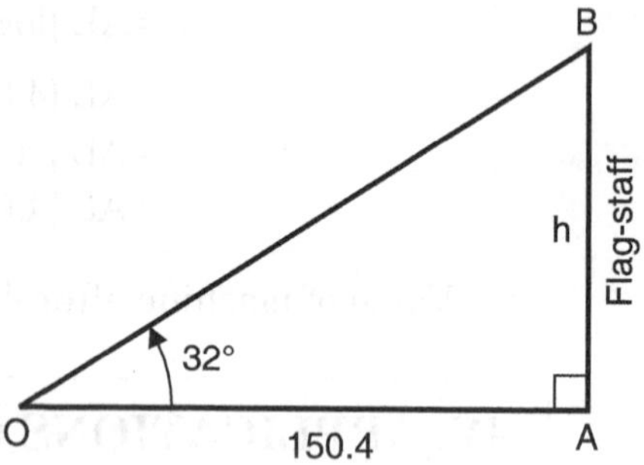

Sol. Let h m be the height of the flag-staff.

In ΔOAB, $\frac{AB}{OA} = \tan 32° \therefore \frac{h}{150.4} = \tan 32°$

$\therefore \quad h = 150.4 \tan 32°$

$\Rightarrow \quad \log h = \log 150.4 + \log \tan 32°$

$= 2.1772 + \bar{1}.7958 = 1.973$

$\therefore \quad h = \text{antilog } 1.973 = 93.97.$

$\therefore$ Height of flag-staff = **93.97 m.**

Example 2. *A town B is 13 km south and 18 km west of a town A. Find the bearing and the distance of B from A.*

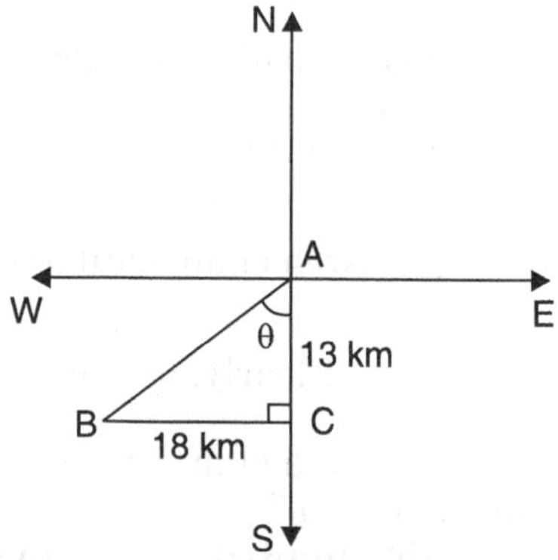

Sol. In ΔABC, we have $AC = 13$ km and $BC = 18$ km.

$\therefore \quad \tan \theta = \frac{BC}{AC} = \frac{18}{13}$

$\therefore \quad \log \tan \theta = \log 18 - \log 13$

$= 1.2553 - 1.1139 = 0.1414$

Also $\log \tan 54°\ 10' = 0.1414$ **(From tables)**

$\therefore \quad \theta = 54°\ 10'$

$\therefore$ The bearing of B from A is **S 54° 10′ W.**

Also $\quad AB^2 = BC^2 + AC^2 = (18)^2 + (13)^2$

$\therefore \quad = 324 + 169 = 493$

$\therefore \quad AB = \sqrt{493} = 22.2$

$\therefore$ The distance of B from A is **22.2 km.**

Example 3. *The sides of a triangular plate are 8 cm, 19 cm, and 15 cm. If its weight be 96 gram, what is the weight of the plate per square cm ?*

Sol. Let area of the triangular plate ABC is

$$A = \sqrt{s(s-a)(s-b)(s-c)}$$

Here $a = 8$ cm, $b = 19$ cm, $c = 15$ cm.

$$s = \frac{a+b+c}{2} = \frac{8+19+15}{2} = \frac{42}{2} = 21$$

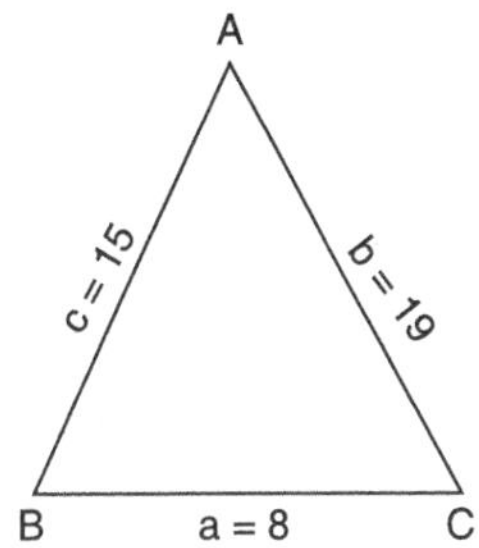

$$A = \sqrt{21(21-8)(21-19)(21-15)}$$

$$= \sqrt{21 \times 13 \times 2 \times 6} = (21 \times 13 \times 2 \times 6)^{1/2}$$

Taking logarithm, of both sides, we get

$$\log A = \frac{1}{2} [\log 21 + \log 13 + \log 2 + \log 6]$$

$$= \frac{1}{2} [1.3222 + 1.1139 + 0.3010 + 0.7782]$$

$$= \frac{1}{2} [3.\ 5153] = 1.7577.$$

$\therefore$ $A = \text{antilog } (1.7577) = 57.24 \text{ cm}^2.$

$\therefore$ Weight of the plate per sq. cm

$$= \frac{96}{57.24} = \mathbf{1.677\ g.}$$

Example 4. *The area of a rectangular field is 2.5 hectares and its sides are in the ratio of 3:2. Find the perimeter of the field.*

Sol. Let the length and breadth of the rectangular field ABCD be $3x$ and $2x$ metres respectively.

Area of the rectangular field = length × breadth = $3x \times 2x = 6x^2$ sq. m. ...(1)

Also, area of the field = 2.5 hectares = 2.5 × 10000 sq. m. ...(2)

From (1), and (2), $6x^2 = 2.5 \times 10000$ or $x^2 = \frac{2.5 \times 10000}{6}$

Taking logarithm on both sides, we get

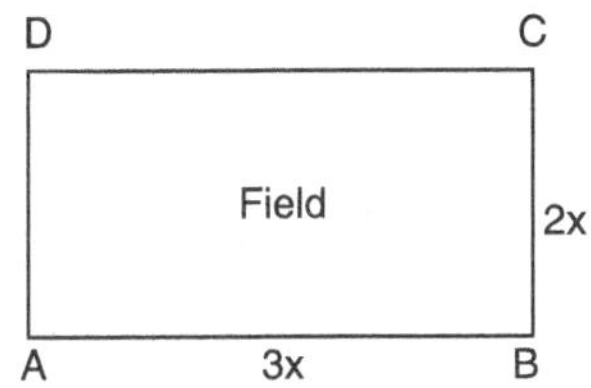

$$\log x^2 = \log \left[\frac{2.5 \times 10000}{6}\right]$$

$$= 2 \log x$$

$$= \log 2.5 + \log 10000 - \log 6$$

$$= 0.3979 + 4.0000 - 0.7782$$

$\Rightarrow$ $2 \log x = 4.3979 - 0.7782 = 3.6197$

or $\log x = \frac{3.6197}{2} = 1.8099.$

$$x = \text{antilog } 1.8099 = 64.55 \text{ (approx.)}$$

$\therefore$ Length of the rectangle = 3 × 64.55 m = 193.65 m

Breadth of the rectangle = 2 × 64.55 m = 129.10 m.

Hence, the perimeter of the rectangular field = $2(l + b)$

= 2(193.65 + 129.10) = 2 × 322.75 = **645.50 metres.**

Example 5. *Find the area of a triangular field the lengths of whose sides are 275 m, 660 m and 715 m. What is the cost of cultivating the field at the rate of ₹ 200 per hectare ?*

Sol. Let $a = 275$ m, $b = 660$ m and $c = 715$ m.

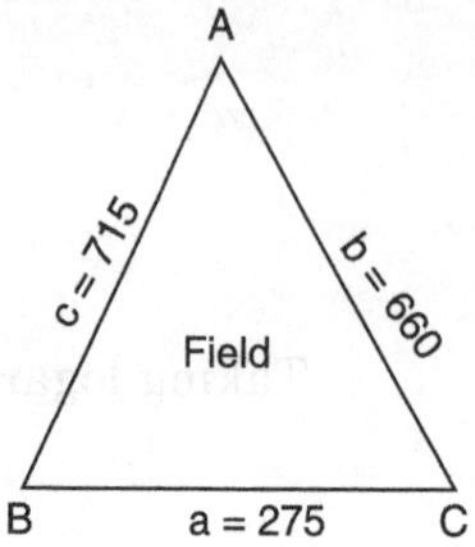

$\therefore \quad s = \frac{1}{2}(a + b + c)$

$= \frac{1}{2}(275 + 660 + 715) = 825.$

$\therefore \quad (s - a) = (825 - 275) = 550,$

$(s - b) = (825 - 660) = 165$

and $(s - c) = (825 - 715) = 110.$

$\therefore$ Area of the triangle is given by :

$$A = \sqrt{s(s-a)(s-b)(s-c)}$$

$$= \sqrt{825 \times 550 \times 165 \times 110}$$

$\therefore \quad \log A = \frac{1}{2}(\log 825 + \log 550 + \log 165 + \log 110)$

$= \frac{1}{2}(2.9165 + 2.7404 + 2.2175 + 2.0414) = \frac{1}{2}(9.9158) = 4.9579.$

$\therefore \quad A = \text{antilog}\ (4.9579) = 90760$

Thus, the area of the field = 90760 m^2.

Now, the cost of cultivating 10000 m^2 = ₹ 200.

$\therefore$ A cost of cultivating the given field = ₹ $\left(\frac{200}{10000} \times 90760\right)$

= **₹ 1815.20.**

Example 6. *Find the length of the diagonal of a square whose area is 24200 m^2.*

Sol. Area of the square *ABCD* = 24200 m^2.

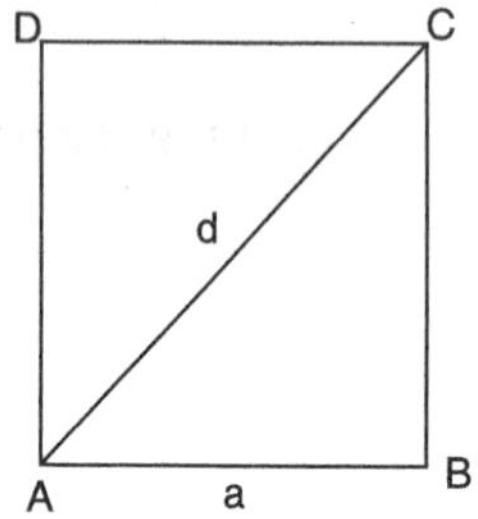

Side of the square (a) = $\sqrt{24200}$ m.

Diagonal of a square: $d = \sqrt{a^2 + a^2}$

$= \sqrt{24200 + 24200} = \sqrt{48400}$

Taking logarithm on both sides, we get

$\log d = \log(\sqrt{48400}) = \log (48400)^{1/2}$

$= \frac{1}{2}\log 48400 = \frac{1}{2} \times 4.6848 = 2.3424.$

$\therefore \quad d = \text{antilog}\ 2.3424 = 220.$

Hence, the length of the diagonal is **220 metres.**

Example 7. *Find the area of a quadrilateral whose sides are 9 m, 40 m, 28 m and 15 m respectively and the angle between first two sides is a right angle.*

Sol. Let ABCD be the given quadrilateral, in which

$\angle B = 90°$. So, $AC^2 = AB^2 + BC^2$.

$AC = \sqrt{AB^2 + BC^2}$

$= \sqrt{(9)^2 + (40)^2} = \sqrt{1681} = 41$ m.

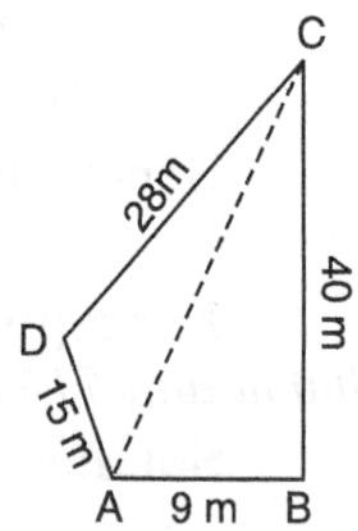

Now, area of Δ ABC $= \left(\frac{1}{2} \times \text{base} \times \text{height}\right)$

$$= \left(\frac{1}{2} \times 9 \times 40\right) \text{m}^2 = 180 \text{ m}^2.$$

In Δ ACD, let $\quad AC = a = 41$ m,

$CD = b = 28$ m and $DA = c = 15$ m.

$\therefore \quad s = \frac{1}{2}(41 + 28 + 15) = 42.$

$\therefore \quad (s - a) = 1, (s - b) = 14$ and $(s - c) = 27.$

$\therefore$ Area of Δ ACD $= \sqrt{42 \times 1 \times 14 \times 27} = 14 \times 3 \times 3 = 126.$

So, area of quadrilateral

ABCD = (area of Δ ABC + area of Δ ACD)

$= (180 + 126)$ m^2 = **306 m^2.**

Example 8. *The lengths of adjacent sides of a parallelogram are 51 m and 37 m. The length of one of the diagonals is 20 m. Find its smaller altitude.*

Sol. From the figure

$$AB = 51 \text{ m}, AD = 37 \text{ m}, BD = 20 \text{ m}$$

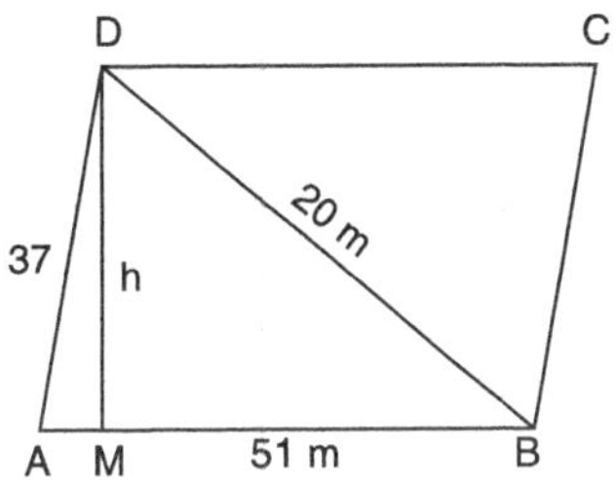

Let A = Area of Δ ABD $= \sqrt{s(s-a)(s-b)(s-c)}$

$$s = \frac{a+b+c}{2} = \frac{51+37+20}{2}$$

$$= \frac{108}{2} = 54.$$

$\therefore \quad A = \sqrt{54(54-51)(54-37)(54-20)}$

$$= \sqrt{54 \times 3 \times 17 \times 34}$$

Taking logarithm,

$$\log A = \frac{1}{2}[\log 54 + \log 3 + \log 17 + \log 34]$$

$$= \frac{1}{2}[1.7324 + 0.4771 + 1.2304 + 1.5315]$$

$$= \frac{1}{2}[4.9714] = 2.4857.$$

$\Rightarrow \quad A = \text{antilog } 2.4857 = 306.$

$\therefore$ Area of the triangle ABD = 306 sq. m.

Again area of the Δ ABD $= \frac{1}{2} \times \text{base} \times \text{altitude}.$

$\therefore \quad 306 = \frac{1}{2} \times 51 \times h$

Taking logarithm on both sides, we get

$$\log 306 = \log\left(\frac{1}{2} \times 51 \times h\right) = \log 51 + \log h - \log 2$$

or $$2.4857 = 1.7076 + \log h - 0.3010 = 1.4066 + \log h.$$

or $$\log h = 2.4857 - 1.4066 = 1.0791.$$

$\therefore$ $$h = \text{antilog } 1.0791 = 11.99.$$

Hence, the altitude of the triangle is **12 m (approx.).**

Example 9. *A parallelogram, the length of whose sides are 62 m and 25 m has one diagonal 68 m long. Find the area of the parallelogram.*

Sol. Let ABCD represent the given parallelogram in which

$$AB = DC = 62 \text{ m},$$
$$BC = AD = 25 \text{ m}$$
and $$AC = 68 \text{ m}.$$

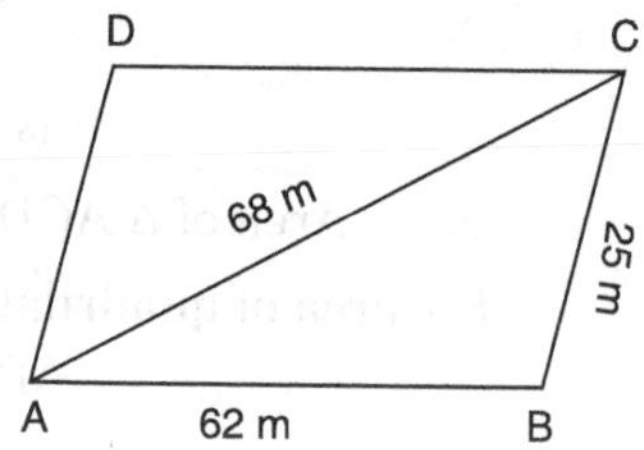

Let A be the area of the parallelogram ABCD.

$\therefore$ $$A = \text{area of } \Delta\, ABC + \text{area of } \Delta\, ACD$$
$$= 2 \text{ area of } \Delta\, ABC \quad (\because \Delta\, ABC = \Delta\, ACD) \quad ...(i)$$

Now, $$s = \frac{62 + 68 + 25}{2} \text{ m} = 77.5 \text{ m}.$$

$$\text{Area of } \Delta\, ABC = \sqrt{s(s-a)(s-b)(s-c)}$$
$$= \sqrt{77.5\,(77.5 - 62)\,(77.5 - 68)\,(77.5 - 25)}$$
$$= \sqrt{77.5 \times 15.5 \times 9.5 \times 52.5} \quad ...(ii)$$

Thus from (*i*) and (*ii*),

$$A = 2\sqrt{77.5 \times 15.5 \times 9.5 \times 52.5}$$

or $$\log A = \log 2 + \frac{1}{2}(\log 77.5 + \log 15.5 + \log 9.5 + \log 52.5)$$
$$= .3010 + \frac{1}{2}(1.8893 + 1.1903 + 0.9777 + 1.7202)$$
$$= .3010 + \frac{1}{2}(5.7775)$$
$$= .3010 + 2.88875 = 3.18975 \simeq 3.1898.$$

$\therefore$ $$A = \text{antilog } 3.1898 = 1548.$$

Thus area of the parallelogram = **1548 sq. m.**

Example 10. *The area of a trapezium is 727 sq. cm. The measure of the parallel sides are 30.6 cm and 14.3 cm. Find the height of the trapezium.*

Sol. Let h be the height of the trapezium, ABCD *i.e.*, h be the distance between the parallel sides, then

$$\text{Area} = \frac{1}{2} \text{ (sum of the parallel sides)} \times \text{(distance between them)}$$

$\therefore$ $$727 = \frac{1}{2}h\,(30.6 + 14.3) = \frac{1}{2}\,h(44.9)$$

or $$h = \frac{2 \times 727}{44.9}$$

$\Rightarrow$ $$\log h = \log\left(\frac{2 \times 727}{44.9}\right)$$

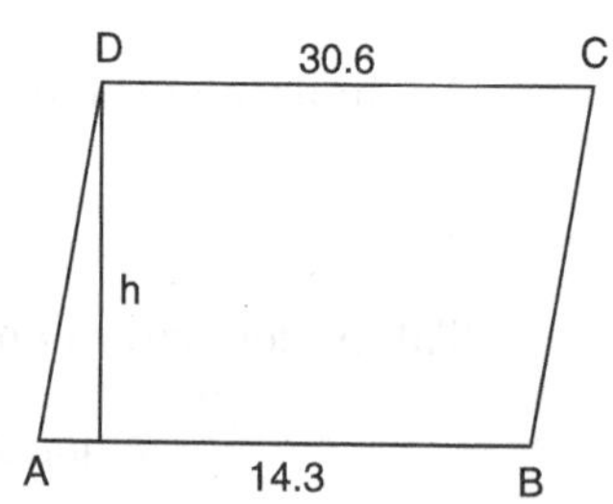

or $$\log h = \log 2 + \log 727 - \log 44.9$$
$$= 0.3010 + 2.8615 - 1.6522 = 1.5103.$$
$$h = \text{antilog}\ (1.5103) = 32.38$$
Hence the height of trapezium = 32.38 cm.

EXERCISE 9.7

1. Find the square root of 0.6254. **2.** Find the fifth root of 8.0125.
3. Find the seventh root of 0.03457.

Using logarithms, compute the following (Q. No. 4–7) :

4. $\dfrac{38.7 \times 0.0021}{0.0189}$ **5.** $(38.56)^{1/4} \times (79.38)^{1/2}$ **6.** $\dfrac{(6.284)(624)^{1/2}}{(0.005)^{1/4}}$

7. $\dfrac{(435)^3 \times \sqrt{0.056}}{(380)^4}$.

8. If an amount of ₹ 572 is invested at 10% per year interest, compounded every year for 5 years, what is the amount realized at the end of 5 years ?
9. Find how many years it will take the sum of ₹ 35,000 to double when invested at 4% p.a. compounded half yearly.
10. A sum of money amounts of ₹ 10,000 in 2 years and ₹ 10,948 in 3 years at compound interest which is being compounded annually. Find the sum and rate percent per annum.
11. At what rate percent compound interest, will ₹ 15,640 yield ₹ 1,950 compound interest in 3 years ?
12. The population of a city grows at the rate of 1.87% p.a. If the present population is 1,20,680, find its population after 8 years.
13. Through depreciation, an automobile loses 20% per year of its values at the beginning of the year. If the original cost of an automobile was ₹ 36,000, determine its value at the end of five years.
14. A machine purchased for ₹ 20,000 depreciates at the rate of 5% per annum, the depreciation being worked out on the value of the machine at the beginning of the year. Find its depreciated value after 7 years.
15. In the 1991 census, the population of Delhi was 9.4×10^7. If the population increases at the rate of 2% every year, what would be the population in 2001 ?
16. The value of a machine depreciates at the rate of 14.3% p.a. If the present value of the machine is ₹ 80,000, find its value after $3\frac{1}{2}$ years.
17. The bacteria in a culture grows by 10% in the first hour, decreases by 10% in the second hour, and again increases by 10% in the third hour. If at the end of third hour, the count of bacteria is 131900000, find the original count of bacteria in the sample.

Answers

1. 0.7909 **2.** 1.517 **3.** 0.6183 **4.** 4.299 **5.** 22.2

6. 590.5 **7.** 0.0009342 **8.** ₹ 921.2 **9.** $17\frac{1}{2}$ years

10. ₹ 8,341, 9.48% p.a. **11.** 4% p.a **12.** 1,39,800 **13.** ₹ 11,800

14. ₹ 13,990 **15.** 1.146×10^8 **16.** ₹ 46,760 **17.** 121121711.

Hints

9. Let the number of years be n.

$$\therefore \quad 70000 = 35000\left(1+\frac{4/2}{100}\right)^{2n} \Rightarrow 2 = (1.02)^{2n} \Rightarrow \log 2 = 2n \log 1.02$$

$$\Rightarrow \quad n = \frac{\log 2}{2\log 1.02} = \frac{0.3010}{2(0.0086)} = \frac{3010}{172}.$$

12. Reqd. population $= 1{,}20{,}680\left(1+\frac{1.87}{100}\right)^8 = 1{,}20{,}680\,(1.0187)^8$

$= \text{AL}\,(\log 1{,}20{,}680 + 8 \log 1.0187) = \text{AL}\,(5.0816 + 8(0.0080))$

$= \text{AL}\ 5.1456 = 1{,}39{,}800.$

10

Analytical Plane Geometry

10.1. INTRODUCTION

In 1637, a French philosopher and mathematician *Rene Descartes* (1596—1665) published his work on geometry in the book named *La Geometric*. He incorporated the use of tools of algebra in studying geometry by establishing 1-1 correspondence between the points in a plane and the ordered pairs of real numbers. He simplified the proofs of geometrical results by introducing the processes of algebra in geometry. The approach of *Descarte* was named '*analytic approach to geometry*'.

10.2. DEFINITION

Coordinate geometry is that branch of mathematics which treats geometry algebraically.

10.3. COORDINATE SYSTEM IN A PLANE AND GRAPH OF COORDINATES OF A POINT

Let $X'OX$ and $Y'OY$ be two perpendicular straight lines intersecting at O. The line $X'OX$ is taken horizontal. The point O is called the **origin.** The horizontal line $X'OX$ and vertical line $Y'OY$ are respectively called the **x-axis** and the **y-axis.** Taking O as the origin, the number scale is made on both axes. The axes divides the plane in four parts, called **quadrants.** The quadrants XOY, $X'OY$, $X'OY'$ and XOY' are respectively called I, II, III and IV quadrants.

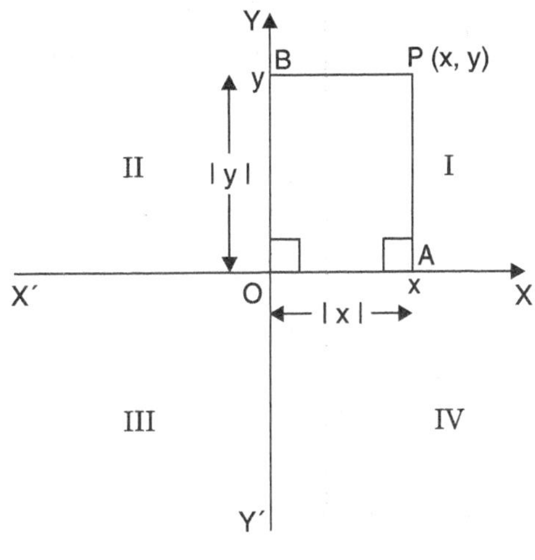

The axes $X'OX$ and $Y'OY$ are called **rectangular coordinate axes** or simply **coordinate axes,** provided there is no involvement of oblique axes in the discussion.

Let P be any point in the plane. Draw $PA \perp X'OX$ and $PB \perp Y'OY$.

Let x and y be the numbers corresponding to the points A and B on the axes $X'OX$ and $Y'OY$ respectively.

$$\therefore \qquad OA = |\,x\,| \quad \text{and} \quad OB = |\,y\,|.$$

Thus, we see that for a point P in the plane, there correspond an ordered pair (x, y) of real numbers.

Conversely, let (x, y) be any ordered pair of real numbers. Let A and B be the points on the axes $X'OX$ and $Y'OY$ corresponding to the real numbers x and y respectively. Let the perpendiculars the A and B meet at P. The point P is unique for a given ordered pair (x, y) of real numbers. Thus, we see that there is 1-1 correspondence between the points in a plane and the ordered pairs of real numbers.

The real number x is called the **x-coordinate** of P or the **abscissa** of P. Similarly, the real number y is called the **y-coordinate** of P or the **ordinate** of P. The real numbers x and y are not reversible in (x, y). The ordered pairs (x, y) and (y, x) represents the same point on the plane if and only if $x = y$.

The coordinates x and y of P are written as (x, y).

The reader would find it interesting to note that if :

(*i*) P is on x-axis, then $y = 0$. (*ii*) P is on y-axis, then $x = 0$.

(*iii*) P is in the I quadrant, then $x > 0, y > 0$.

(*iv*) P is in the II quadrant, then $x < 0, y > 0$.

(*v*) P is in the III quadrant, then $x < 0, y < 0$.

(*vi*) P is in the IV quadrant, then $x > 0, y < 0$.

ILLUSTRATIVE EXAMPLES

Example 1. *If three vertices of a rectangle are (0, 0), (2, 0) and (0, 3), find the coordinates of the fourth vertex.*

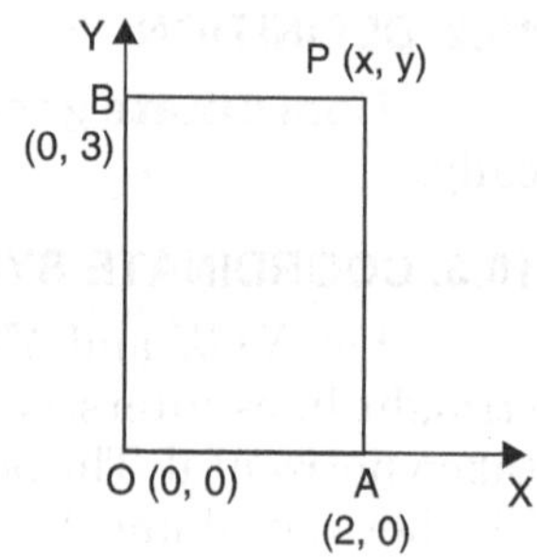

Sol. Let $OAPB$ be the rectangle with vertices $O(0, 0)$, $A(2, 0)$, $P(x, y)$ and $B(0, 3)$.

Now, $x = BP = OA = 2$ and $y = AP = OB = 3$

$\therefore$ Coordinates of fourth vertex = **(2, 3).**

Exmaple 2. *The base of an equilateral triangle with side 10 cm lie along x-axis such that the midpoint of the base is at the origin. Find the vertices of the triangle.*

Sol. Let AB be the base of the equilateral triangle.

$$\therefore \quad OA = OB = \frac{10}{2} \text{ cm} = 5 \text{ cm}.$$

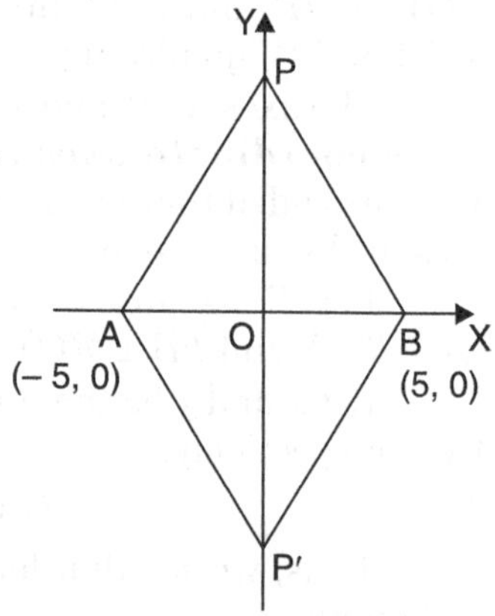

$\therefore$ Coordinates of A and B are $(-5, 0)$ and $(5, 0)$ respectively.

There are two equilateral triangles with base AB.

Triangle ABP. We have $AP = BP = AB = 10$ cm.

By *Pythagoras result,*

$$OP = \sqrt{AP^2 - OA^2} = \sqrt{(10)^2 - (5)^2} = 5\sqrt{3}$$

$\therefore$ Coordinates of P = $\mathbf{(0, 5\sqrt{3})}$.

Triangle ABP′. We have $AP' = BP' = AB = 10$ cm

By *Pythagoras result,*

$$OP' = \sqrt{AP'^2 - OA^2} = \sqrt{(10)^2 - (5)^2} = 5\sqrt{3}$$

$\therefore$ Coordinates of P' = $\mathbf{(0, -5\sqrt{3})}$.

EXERCISE 10.1

1. Plot the following points on a cartesian plane :
   (*i*) (3, 4) (*ii*) (3, – 7) (*iii*) (– 5, – 8)
   (*iv*) (– 6, 2) (*v*) (0, 4) (*vi*) (0, – 6).
2. In which quadrant the following points lie :
   (*i*) (5, 9) (*ii*) (– 6, 8) (*iii*) (15, – 7)
   (*iv*) (– 3, – 4) (*v*) (9, 2) (*vi*) (– 6, 8).
3. Draw the quadrilateral whose vertices are (– 4, 5) , (0, 7), (5, – 5) and (– 4, – 2).
4. If three vertices of a rectangle are (0, 0), (– 4, 0), (0, 5), find the coordinates of the fourth vertex.
5. The base of an equilateral triangle with side 7 cm lie along the y-axis such that the mid-point of the base is at the origin. Find the vertices of the triangle.

Answers

2. (*i*) First (*ii*) second (*iii*) fourth (*iv*) third
   (*v*) first (*vi*) second.
4. (– 4, 5)
5. $\left(0, \frac{7}{2}\right), \left(0, -\frac{7}{2}\right), \left(\frac{7\sqrt{3}}{2}, 0\right)$; $\left(0, \frac{7}{2}\right), \left(0, -\frac{7}{2}\right), \left(-\frac{7\sqrt{3}}{2}, 0\right)$.

10.4. DISTANCE FORMULA

Let $P(x_1, y_1)$ and $Q(x_2, y_2)$ be any two points in the plane. For the sake of exactness, let us assume that the points P and Q are both in the I quadrant. Draw PA and QB perpendicular to x-axis. Draw $PC \perp QB$.

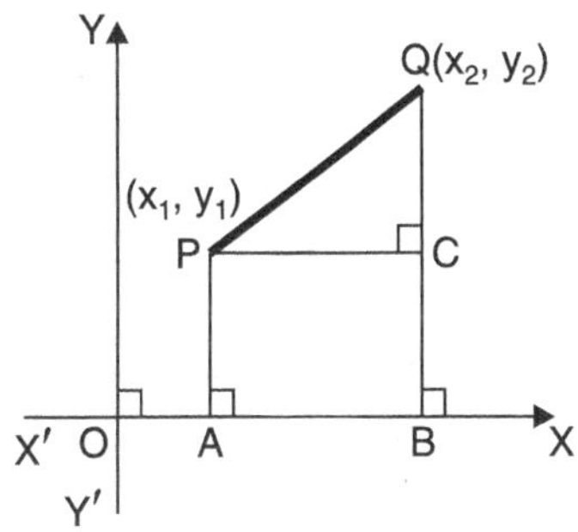

Since, PCQ is a right-angled triangle, therefore by *Pythagorean theorem,*

$$PQ^2 = PC^2 + CQ^2$$

$$\therefore \quad PQ = \sqrt{PC^2 + CQ^2} \quad ...(1)$$

Now, $PC = AB = OB - OA = x_2 - x_1$

and $CQ = BQ - BC = BQ - AP = y_2 - y_1$

$$\therefore \quad (1) \Rightarrow \quad PQ = \sqrt{(x_2 - x_1)^2 + (y_2 - y_1)^2}.$$

$\therefore$ **The distance PQ between the points $P(x_1, y_1)$ and $Q(x_2, y_2)$ is given by**

$$\sqrt{(x_2 - x_1)^2 + (y_2 - y_1)^2}.$$

Note 1. When *three* points are given and it is required to prove that they are **collinear** *i.e.,* they lie on a line, then show that sum of the distances between two point-pairs is equal to the distance between the third point-pair.

Note 2. When *three* points are given and it is required to prove that they form :

(*i*) an **isosceles triangle,** show that two of its sides are equal.

(*ii*) an **equilateral triangle,** show that its all sides are equal.

(*iii*) a **right angle triangle,** show that the sum of the squares of two sides is equal to the square of the third side.

Note 3. When *four* points are given and it is required to prove that they form :

(*i*) a **parallelogram,** show that opposite sides are equal.

(*ii*) a **rectangle,** show that opposite sides are equal and diagonals are also equal.

(*iii*) a **parallelogram but not a rectangle,** show that opposite sides are equal and diagonals are not equal.

(*iv*) a **square,** show that all sides are equal and diagonals are also equal.

(*v*) a **rhombus,** show that all sides are equal and diagonals are not equal.

WORKING RULES FOR SOLVING PROBLEMS

Rule I. *The distance of the points* (x_1, y_1) *and* (x_2, y_2) *is* $\sqrt{(x_2 - x_1)^2 + (y_2 - y_1)^2}$.

Rule II. *The distance of the point* (x_1, y_1) *from the origin is* $\sqrt{x_1^2 + y_1^2}$.

Rule III. *Three points are collinear if the sum of two distances is equal to the third distance.*

Rule IV. *The triangle ABC is a/an :*

(*i*) **equilateral triangle** *if* $AB = BC = CA$.

(*ii*) **isosceles triangle** *if two of its sides are equal.*

(*iii*) **right angled triangle** *if the sum of squares of two sides is equal to the square of the third side.*

Rule V. *The quadrilateral ABCD is a :*

(*i*) **parallelogram** *if* $AB = CD$ *and* $BC = DA$.

(*ii*) **rectangle** *if* $AB = CD$, $BC = DA$ *and* $AC = BD$.

(*iii*) **parallelogram but not a rectangle** *if* $AB = CD$, $BC = DA$ *and* $AC \neq BD$.

(*iv*) **square** *if* $AB = BC = CD = DA$ *and* $AC = BD$.

(*v*) **rhombus** *if* $AB = BC = CD = DA$ *and* $AC \neq BD$.

ILLUSTRATIVE EXAMPLES

Example 1. *Find the distance between the following pair of points :*

(*i*) $A(2, 5)$, $B(-3, 7)$ (*ii*) $A(am_1^2, 2am_1)$, $B(am_2^2, 2am_2)$

(*iii*) $A(a \cos \alpha, a \sin \alpha)$, $B(a \cos \beta, a \sin \beta)$.

Sol. (*i*) $AB = \sqrt{(-3-2)^2 + (7-5)^2} = \sqrt{25+4} = \mathbf{\sqrt{29}}$.

(*ii*) $$AB = \sqrt{(am_2^2 - am_1^2)^2 + (2am_2 - 2am_1)^2} = \sqrt{a^2(m_2^2 - m_1^2)^2 + 4a^2(m_2 - m_1)^2}$$

$$= a\sqrt{(m_2 - m_1)^2\,[(m_2 + m_1)^2 + 4]} = \mathbf{a(m_2 - m_1)\sqrt{(m_2 + m_1)^2 + 4}}.$$

(assuming $m_1 \le m_2$)

(*iii*) $$AB = \sqrt{(a \cos \beta - a \cos \alpha)^2 + (a \sin \beta - a \sin \alpha)^2}$$

$$= \sqrt{(a^2 \cos^2 \beta + a^2 \cos^2 \alpha - 2a^2 \cos \alpha \cos \beta) + (a^2 \sin^2 \beta + a^2 \sin^2 \alpha - 2a^2 \sin \alpha \sin \beta}$$

$$= \sqrt{a^2 (\cos^2 \beta + \sin^2 \beta) + a^2 (\cos^2 \alpha + \sin^2 \alpha) - 2a^2 (\cos \alpha \cos \beta + \sin \alpha \sin \beta)}$$

$$= \sqrt{a^2 + a^2 - 2a^2 \cos(\alpha - \beta)} = \sqrt{2a^2(1 - \cos(\alpha - \beta))}$$

$$= \sqrt{2a^2 . 2\sin^2 \frac{\alpha - \beta}{2}} = \mathbf{2a\left|\sin \frac{\alpha - \beta}{2}\right|}.$$

(assuming $a > 0$)

Example 2. *Find the relation that must exist between x and y so that (x, y) is equidistant from (6, – 1) and (2, 3).*

Sol. Let $P = (x, y)$, $A = (6, -1)$ and $B = (2, 3)$.

By the given condition, $PA = PB$. $\therefore$ $\sqrt{(x-6)^2 + (y+1)^2} = \sqrt{(x-2)^2 + (y-3)^2}$

or $x^2 + 36 - 12x + y^2 + 1 + 2y = x^2 + 4 - 4x + y^2 + 9 - 6y$ or $\mathbf{x - y = 3.}$

Example 3. *Show that the points (5, 1), (1, – 1) and (11, 4) lie on a straight line.*

Sol. Let $A = (5, 1), B = (1, -1)$ and $C = (11, 4)$.

Now,
$$AB = \sqrt{(1-5)^2 + (-1-1)^2} = \sqrt{16+4} = 2\sqrt{5}$$
$$BC = \sqrt{(11-1)^2 + (4-(-1))^2} = \sqrt{100+25} = 5\sqrt{5}$$
$$AC = \sqrt{(11-5)^2 + (4-1)^2} = \sqrt{36+9} = 3\sqrt{5}$$
$$AB + AC = 2\sqrt{5} + 3\sqrt{5} = 5\sqrt{5} = BC. \quad \therefore \quad A, B, C \text{ are on a line.}$$

Example 4. *Show that the quadrilateral with vertices (3, 2), (0, 5), (– 3, 2), (0, – 1), in this order, is a square. Also verify the result by drawing the points on a plane.*

Sol. Let $A(3, 2)$, $B(0, 5)$, $C(-3, 2)$ and $D(0, -1)$ be the given points.

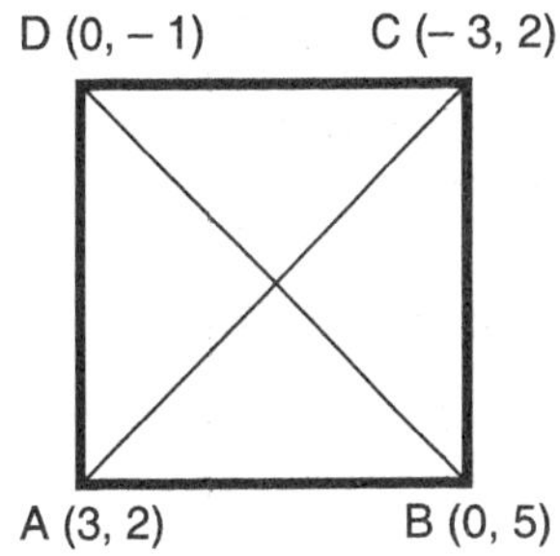

$$AB = \sqrt{(0-3)^2 + (5-2)^2} = \sqrt{9+9} = \sqrt{18} = 3\sqrt{2}.$$
$$BC = \sqrt{(-3-0)^2 + (2-5)^2} = \sqrt{9+9} = \sqrt{18} = 3\sqrt{2}.$$
$$CD = \sqrt{(0+3)^2 + (-1-2)^2} = \sqrt{9+9} = \sqrt{18} = 3\sqrt{2}.$$
$$DA = \sqrt{(3-0)^2 + (2+1)^2} = \sqrt{9+9} = \sqrt{18} = 3\sqrt{2}.$$

$\therefore$ $AB = BC = CD = DA$.

$\therefore$ $ABCD$ is either a rhombus or a square.

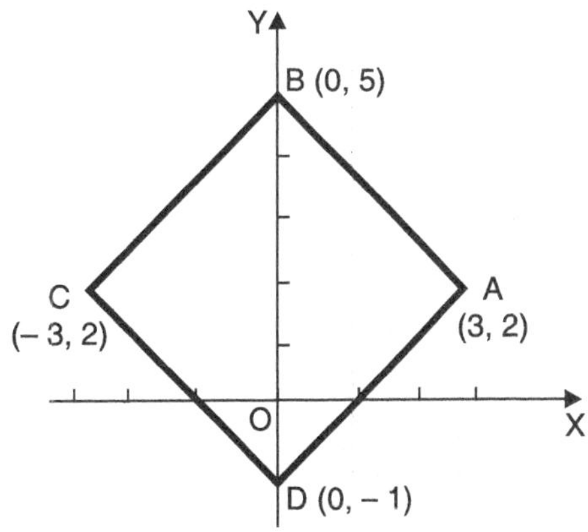

Diagonal $AC = \sqrt{(-3-3)^2 + (2-2)^2}$
$$= \sqrt{36+0} = 6$$

Diagonal $BD = \sqrt{(0-0)^2 + (-1-5)^2}$
$$= \sqrt{0+36} = 6$$

$\therefore$ $AC = BD$.

$\therefore$ All sides are equal and diagonal are also equal.

$\therefore$ $ABCD$ is a square.

This is also verified by drawing the vertices on a plane.

Example 5. *An equilateral triangle has one vertex at (3, 4) and another at (– 2, 3). Find the coordinates of the third vertex.*

Sol. Let $A(3, 4)$ and $B(-2, 3)$ be the two vertices and let the third vertex be $C(x, y)$. The triangle is equilateral.

$\therefore$ $AB = BC = CA$ *i.e.,* $AB^2 = BC^2 = CA^2$

or $(3 + 2)^2 + (4 - 3)^2 = (x + 2)^2 + (y - 3)^2 = (x - 3)^2 + (y - 4)^2$

or $26 = x^2 + y^2 + 4x - 6y + 13 = x^2 + y^2 - 6x - 8y + 25$ ...(1)

From the last two members of (1), we get $4x - 6y + 13 = - 6x - 8y + 25$

or $10x + 2y - 12 = 0$ or $5x + y - 6 = 0$ or $y = 6 - 5x$.

From the first two members of (1), we get $x^2 + y^2 + 4x - 6y = 13$.

Putting $y = 6 - 5x$, we get $x^2 + (6 - 5x)^2 + 4x - 6(6 - 5x) = 13$

or $x^2 + 36 + 25x^2 - 60x + 4x - 36 + 30x = 13$ or $2x^2 - 2x - 1 = 0$

$$x = \frac{2 \pm \sqrt{4+8}}{4} = \frac{2 + 2\sqrt{3}}{4} = \frac{1 \pm \sqrt{3}}{2} = \frac{1+\sqrt{3}}{2} \text{ and } \frac{1-\sqrt{3}}{2}$$

$$x = \frac{1+\sqrt{3}}{2} \text{ implies } y = 6 - 5\left(\frac{1+\sqrt{3}}{2}\right) = \frac{12 - 5 - 5\sqrt{3}}{2} = \frac{7 - 5\sqrt{3}}{2}$$

$$x = \frac{1-\sqrt{3}}{2} \text{ implies } y = 6 - 5\left(\frac{1-\sqrt{3}}{2}\right) = \frac{12 - 5 + 5\sqrt{3}}{2} = \frac{7 + 5\sqrt{3}}{2}$$

$\therefore$ The third vertex is $\left(\frac{1+\sqrt{3}}{2}, \frac{7-5\sqrt{3}}{2}\right)$ or $\left(\frac{1-\sqrt{3}}{2}, \frac{7+5\sqrt{3}}{2}\right)$.

Example 6. *Two opposite vertices of a square are (2, 6) and (0, – 2). Find the coordinates of the other vertices.*

Sol. Let *ABCD* be the square with given vertices *A*(2, 6) and *C*(0, – 2). Let (*x*, *y*) be the coordinates of *B*.

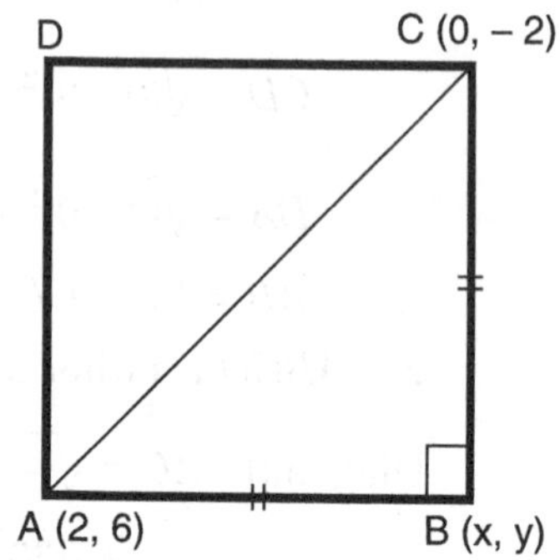

$\therefore$ $AB = BC$ ...(1)

and $AB^2 + BC^2 = AC^2$ ...(2)

(1) $\Rightarrow \sqrt{(x-2)^2 + (y-6)^2} = \sqrt{(x-0)^2 + (y+2)^2}$

$\Rightarrow x^2 + 4 - 4x + y^2 + 36 - 12y = x^2 + y^2 + 4 + 4y$

$\Rightarrow 4x + 16y = 36 \Rightarrow x + 4y = 9$ ...(3)

(2) $\Rightarrow [(x- 2)^2 + (y - 6)^2] + [(x - 0)^2 + (y + 2)^2]$
$= (0 - 2)^2 + (- 2 - 6)^2$

$\Rightarrow x^2 + 4 - 4x + y^2 + 36 - 12y + x^2 + y^2 + 4 + 4y$
$= 4 + 64$

$\Rightarrow 2x^2 + 2y^2 - 4x - 8y = 24$

$\Rightarrow x^2 + y^2 - 2x - 4y = 12$

$\Rightarrow (9 - 4y)^2 + y^2 - 2(9 - 4y) - 4y = 12$ ($\because$ (3) $\Rightarrow x = 9 - 4y$)

$\Rightarrow 81 + 16y^2 - 72y + y^2 - 18 + 8y - 4y - 12 = 0$

$\Rightarrow 17y^2 - 68y + 51 = 0 \Rightarrow y^2 - 4y + 3 = 0 \Rightarrow y = 1, 3$

When $y = 1$, (3) $\Rightarrow x = 9 - 4(1) = 5$

When $y = 3$, (3) $\Rightarrow x = 9 - 4(3) = - 3$

$\therefore$ $(x, y) = (5, 1), (- 3, 3)$. One value of (x, y) correspond to *D*. $\therefore$ The other vertices of the square are **(5, 1)** and **(– 3, 3)**.

EXERCISE 10.2

1. Find the distances between the following pairs of points :
   (*i*) (0, 0) and (4, 5) (*ii*) (5, – 12) and (9, – 9)
   (*iii*) $(x - y, y - x)$ and $(x + y, x + y)$ (*iv*) $(b + c, c + a)$ and $(c + a, a + b)$.
2. Show that each of the following sets of points are the vertices of a right angled triangle :
   (*i*) (4, 4), (3, 5), (– 1, – 1) (*ii*) (6, 2), (3, – 1), (– 2, 4)
3. Show that the following sets of points are collinear :
   (*i*) (– 2, 3), (1, 2), (7, 0) (*ii*) (4, 3), (2, 0), (– 4, – 9)
   Also verify the result by drawing the points on a plane.
4. Show that the points (– 4, – 6), (1, 1.5) and (2, 3) do not form a triangle.
5. Show that the triangles whose vertices given below are equilateral :
   (*i*) $(-1, -1), (1, 1), (-\sqrt{3}, \sqrt{3})$ (*ii*) $(2a, 4a), (2a, 6a), (2a + \sqrt{3}\,a, 5a)$.
6. Show that the triangles whose vertices given below are isosceles :
   (*i*) (7, 8), (– 7, 0), (– 1, – 6) (*ii*) (8, 2), (5, – 3), (0, 0).
   (*P.T.U. B.Pharma, June 2003*)
7. Find the value of x such that $PQ = QR$, where P, Q and R are (6, – 1), (1, 3) and $(x, 8)$ respectively.
8. A line is of length 10 units and one of its ends is (– 2, 3). If the ordinate of the other end is 9, prove that the abscissa of the other end is either 6 or – 10.
9. Which point on x-axis is equidistant from (7, 6) and (– 3, 4) ?
10. Show that the quadrilaterals whose vertices given below are parallelogram :
    (*i*) (– 1, 0), (0, 3), (1, 3), (0, 0) (*ii*) (– 2, – 1), (1, 0), (4, 3), (1, 2).
11. Show that the quadrilaterals whose vertices given below are rectangles :
    (*i*) (0, – 1), (– 2, 3), (6, 7), (8, 3) (*ii*) (3, 2), (11, 8), (8, 12), (0, 6).
12. Show that the quadrilaterals whose vertices given below are rhombuses :
    (*i*) (7, 3), (3, 0), (0, – 4), (4, – 1) (*ii*) (3, – 4), (4, 2), (5, – 4), (4, – 10).
13. Show that the quadrilaterals whose vertices given below are squares :
    (*i*) (0, – 1), (2, 1), (0, 3), (– 2, 1) (*ii*) (2, 6), (5, 1), (0, – 2), (– 3, 3).
14. If the point A form an equilateral triangle with the points $B(0, 0)$ and $C(3, \sqrt{3})$, find the coordinates of the point A.
15. Find the abscissa of the point whose ordinate is 4 and which is at a distance of 5 units from (5, 0).

Answers

1. (*i*) $\sqrt{41}$ (*ii*) 5 (*iii*) $2\sqrt{x^2 + y^2}$
(*iv*) $\sqrt{a^2 + 2b^2 + c^2 - 2ab - 2bc}$ **7.** 5, – 3 **9.** (3, 0)

14. $(0, 2\sqrt{3})$ or $(3, -\sqrt{3})$ **15.** 2 or 8.

Hints

4. It is sufficient to show that the sum of two sides is not greater than the third side.

15. Let the point be with coordinates $(x, 4)$.

10.5. AREA OF A TRIANGLE

Let ABC be a triangle with vertices $A(x_1, y_1)$, $B(x_2, y_2)$ and $C(x_3, y_3)$.

Draw AL, BM, CN perpendiculars on x-axis. Required area ABC

= Area of trap. $ALNC$ + Area of trap. $CNMB$

– Area of trap. $ALMB$

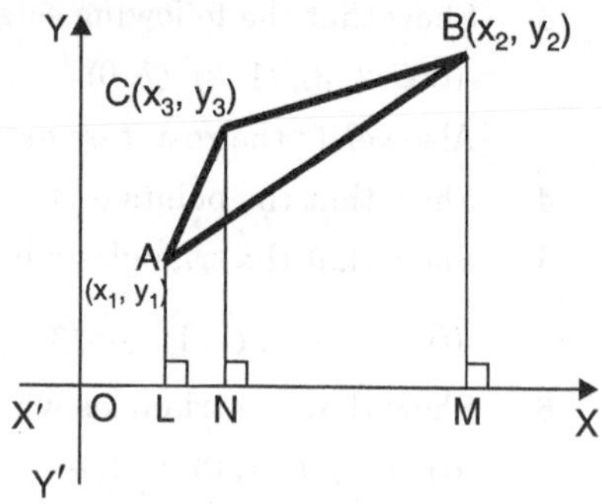

$$= \frac{1}{2}(AL + CN)LN + \frac{1}{2}(CN + BM)NM - \frac{1}{2}(AL + BM)LM$$

Now, $AL = y_1$, $BM = y_2$, $CN = y_3$ and

$$LN = ON - OL = x_3 - x_1$$

$$NM = OM - ON = x_2 - x_3$$

$$LM = OM - OL = x_2 - x_1$$

$$\therefore \quad \text{Area } ABC = \frac{1}{2}(y_1 + y_3)(x_3 - x_1) + \frac{1}{2}(y_3 + y_2)(x_2 - x_3) - \frac{1}{2}(y_1 + y_2)(x_2 - x_1)$$

$$= \frac{1}{2}[x_1(-y_1 - y_3 + y_1 + y_2) + x_2(y_3 + y_2 - y_1 - y_2) + x_3(y_1 + y_3 - y_3 - y_2)]$$

$$= \frac{1}{2}[x_1(y_2 - y_3) + x_3(y_3 - y_1) + x_3(y_1 - y_2)].$$

The area of triangle ΔABC will come out to be a positive quantity only when the vertices A, B, C are taken in anticlock direction.

Thus, in general Δ **ABC** $= \dfrac{1}{2} \mid \mathbf{x_1(y_2 - y_3) + x_2(y_3 - y_1) + x_3(y_1 - y_2)} \mid$.

Remark 1. In terms of a determinant, the area of the triangle with vertices (x_1, y_1), (x_2, y_2) and (x_3, y_3) can also be expressed as :

$$\text{Area of triangle} = \text{Numerical value of } \frac{1}{2}\begin{vmatrix} x_1 & y_1 & 1 \\ x_2 & y_2 & 1 \\ x_3 & y_3 & 1 \end{vmatrix}.$$

Remark 2. The area of a quadrilateral can be found out by dividing the quadrilateral into two triangles.

10.6. CONDITION FOR COLLINEARITY OF THREE POINTS

If the points A, B and C are collinear, then the area of triangle must be *zero*. This also holds *vice-versa*. Thus, to show that given three points are *collinear*, it would be sufficient to show that the area of the triangle with given points as vertices are *zero*.

WORKING RULES FOR SOLVING PROBLEMS

Rule I. *Area of triangle with vertices (x_1, y_1), (x_2, y_2) and (x_3, y_3) is*

$$\frac{1}{2} \mid x_1 (y_2 - y_3) + x_2 (y_3 - y_1) + x_3 (y_1 - y_2) \mid.$$

Rule II. *Three points are collinear if and only if area of triangle formed by these points is zero.*

ILLUSTRATIVE EXAMPLES

Example 1. *Find the area of the triangle whose vertices are (1, 6), (3, 0) and (– 3, – 7).*

Sol. Let $A = (x_1, y_1) = (1, 6)$, $B = (x_2, y_2) = (3, 0)$ and $C = (x_3, y_3) = (-3, -7)$.

$$\therefore \quad \text{Area of } \Delta ABC = \frac{1}{2} \mid x_1(y_2 - y_3) + x_2(y_3 - y_1) + x_3(y_1 - y_2) \mid$$

$$= \frac{1}{2} \mid 1(0 + 7) + 3(-7 - 6) + (-3)(6 - 0) \mid$$

$$= \frac{1}{2} \mid 7 - 39 - 18 \mid = \frac{1}{2} \mid -50 \mid = \frac{1}{2} \times 50 = \textbf{25 sq. units.}$$

Example 2. *Find the area of the quadrilateral whose vertices are (2, 1), (6, 0), (5, – 2) and (– 3,–1).*

Sol. Let the vertices be $A(2, 1)$, $B(6, 0)$, $C(5, -2)$ and $D(-3, -1)$.

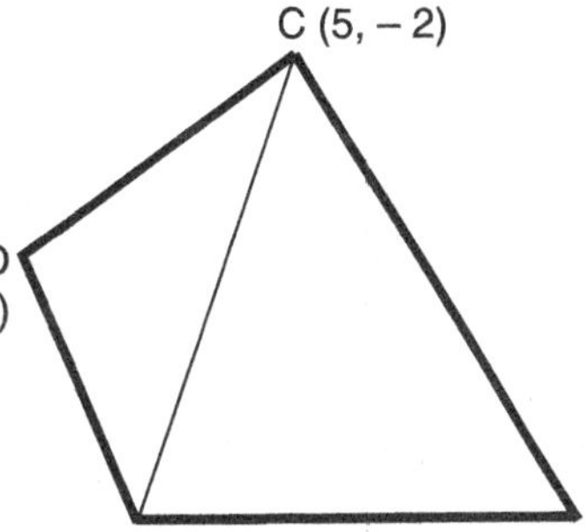

Area of ΔABC

$$= \frac{1}{2} \mid 2(0 + 2) + 6(-2 - 1) + 5(1 - 0) \mid$$

$$= \frac{1}{2} \mid 4 - 18 + 5 \mid = \frac{1}{2} \mid -9 \mid = \frac{1}{2} \times 9$$

= 4.5 sq. units.

Area of ΔACD

$$= \frac{1}{2} \mid 2(-2 + 1) + 5(-1 - 1) + (-3)(1 + 2) \mid$$

$$= \frac{1}{2} \mid -2 - 10 - 9 \mid = \frac{1}{2} \mid -21 \mid = \frac{1}{2} \times 21 = 10.5 \text{ sq. units.}$$

$\therefore$ Area of quadrilateral $ABCD$ = Area of ΔABC + Area of ΔACD

= 4.5 + 10.5 = **15 sq. units.**

Example 3. *Show that the points $A(b, c + a)$, $B(c, a + b)$ and $C(a, b + c)$ are collinear.*

(G.J.U. B.Pharma, 2003)

Sol. The given points are $A(b, c + a)$, $B(c, a + b)$ and $C(a, b + c)$.

$$\text{Area of } \Delta ABC = \frac{1}{2} \mid b(a+b-b-c) + c(b+c-c-a) + a(c+a-a-b) \mid$$

$$= \frac{1}{2} \mid ba - bc + cb - ca + ac - ab \mid = \frac{1}{2} \mid 0 \mid = \frac{1}{2} \times 0 = 0.$$

$\therefore$ Points A, B, C are collinear.

Example 4. *A, B are two points (3, 4) and (5, – 2). Find a point P such that PA = PB and ΔPAB = 10.*

Sol. The given points are $A(3, 4)$ and $B(5, -2)$. Let (x, y) be the coordinates of P.

$\therefore \quad PA = PB \quad \ldots(1) \quad$ and $\quad \Delta PAB = 10. \quad \ldots(2)$

$(1) \Rightarrow \quad PA^2 = PB^2$

$\Rightarrow \quad (x-3)^2 + (y-4)^2 = (x-5)^2 + (y+2)^2$

$\Rightarrow \quad x^2 + 9 - 6x + y^2 + 16 - 8y = x^2 + 25 - 10x + y^2 + 4 + 4y$

$\Rightarrow \quad 4x - 12y = 4 \Rightarrow x = 3y + 1 \quad \ldots(3)$

$(2) \Rightarrow \quad \frac{1}{2} \mid x(4+2) + 3(-2-y) + 5(y-4) \mid = 10$

$\Rightarrow \quad \mid 6x - 6 - 3y + 5y - 20 \mid = 20 \Rightarrow \mid 6x + 2y - 26 \mid = 20$

$\Rightarrow \quad 6x + 2y - 26 = \pm 20 \Rightarrow 3x + y - 13 = \pm 10.$

Case I. 3x + y – 13 = 10

$\Rightarrow \quad 3(3y+1) + y = 23 \Rightarrow 10y = 20 \Rightarrow y = 2 \quad$ [using (3)]

$\therefore$ (3) $\Rightarrow \quad x = 3(2) + 1 = 7$

$\therefore$ Coordinates of P are **(7, 2).**

Case II. 3x + y – 13 = – 10

$\Rightarrow \quad 3(3y+1) + y = 3 \Rightarrow 10y = 0 \Rightarrow y = 0 \quad$ [using (3)]

$\therefore$ (3) $\Rightarrow \quad x = 3(0) + 1 = 1$

$\therefore$ Coordinates of P are **(1, 0).**

Remark. There are two possible positions of the point P: one above AB and one below AB.

Example 5. *The vertices of a quadrilateral are (6, 3), (– 3, 5), (4, – 2) and (x, 3x) and are denoted by A,B,C and D, respectively. Find the values of x so that the area of triangle ABC is double the area of triangle DBC.*

Sol. Given points are : $A(6, 3)$, $B(-3, 5)$, $C(4, -2)$, $D(x, 3x)$.

$$\text{Area of } \Delta ABC = \frac{1}{2} \mid 6(5+2) + (-3)(-2-3) + 4(3-5) \mid = \frac{1}{2} \mid 49 \mid = \frac{49}{2}$$

$$\text{Area of } \Delta DBC = \frac{1}{2} \mid -3(-2-3x) + 4(3x-5) + x(5+2) \mid$$

$$= \frac{1}{2} \mid 28x - 14 \mid = \frac{\mid 14 \mid}{2} \mid 2x - 1 \mid = 7 \mid 2x - 1 \mid.$$

By the given condition, area of ΔABC $2 \times$ area of ΔDBC.

$$\Rightarrow \quad \frac{49}{2} = 2 \times 7 \mid 2x - 1 \mid \Rightarrow \mid 2x - 1 \mid = \frac{7}{4} \Rightarrow 2x - 1 = \pm \frac{7}{4}.$$

$$2x - 1 = \frac{7}{4} \Rightarrow x = \frac{11}{8} \quad \text{and} \quad 2x - 1 = -\frac{7}{4} \Rightarrow x = -\frac{3}{8}$$

$$\therefore \qquad x = \frac{11}{8} \text{ or } -\frac{3}{8}.$$

EXERCISE 10.3

1. Find the area of the triangle whose vertices are :

   (*i*) (3, 4), (2, – 1), (4, – 6) (*ii*) (1, – 1), (– 1, – 1), $(-\sqrt{3}, \sqrt{3})$.

   (*iii*) (– 2, 1), (2, – 3), (4, 4) (*P.T.U. B.Pharma, Dec. 2001*)

   (*iv*) (0, 0), (1, 0), (1, 1). (*P.T.U. B.Pharma, May 2007*)

2. Show that the area of the triangle with vertices (λ, λ – 2), (λ + 3, λ) and (λ + 2, λ + 2) is independent of λ.

3. Show that the following points are collinear :

   (*i*) (1, 4), (3, – 2), (– 3, 16) (*ii*) (– 5, 1), (5, 5), (10, 7)

   (*iii*) (– 3, – 7), (4, 7), (5, 9) (*P.T.U. B.Pharma, Dec. 2004*)

   (*iv*) (3, 4), (8, 5), (13, 6). (*P.T.U. B.Pharma, May 2007*)

4. Find the value of x so that the points (3, – 2), (x, 2), (8, 8) may be collinear. (*P.T.U. B.Pharma, June 2003*)

5. For what value of k do the points (– 1, 4), (– 3, 8) and (– k + 1, 3k) lie on a straight line ?

6. Points (a, 0), (0, b) and (x, y) are collinear, show that $\frac{x}{a} + \frac{y}{b} = 1$.

7. Prove that the points (a, b), (c, d) and ($a – c$, $b – d$) are collinear if $ad = bc$.

8. For what value of x, the area of the triangle formed by the points (5, –1), (x, 4), (6, 3) is 5.5 sq. units ?

9. (*i*) Show that the points (p + 1, 1) (2p + 1, 3) and (2p + 2, 2p) are collinear if p = – 1/2 or 2.

   (*ii*) If the points (a, 0), (0, b) and (1, 1) are collinear, show that $\frac{1}{a} + \frac{1}{b} = 1$.

   (*P.T.U. B.Pharma, Dec. 2005*)

10. If the vertices of a triangle are (1, λ), (4, – 3), (– 9, 7) and its area is 15 sq. units. Find the value of λ. (*P.T.U. B.Pharma, May 2002*)

11. Find the area of the quadrilateral whose vertices, taken in order, are :

    (*i*) (1, 1), (3, 4), (5, – 2) and (4, – 7) (*ii*) (4, 3), (– 5, 6), (– 7, – 2) and (0, – 7).

Answers

1. (*i*) $7\frac{1}{2}$ sq. units (*ii*) $1 + \sqrt{3}$ sq. units (*iii*) 18 sq. units (*iv*) $\frac{1}{2}$ sq. units.

4. 5 **5.** 0 **8.** $\frac{7}{2}$ or 9 **10.** – 3 or $\frac{21}{13}$

11. (*i*) $\frac{41}{2}$ sq. units (*ii*) 84 sq. units.

10.7. SECTION FORMULAE

Definition of internal division. A point R is said to divide the line PQ **internally** in the ratio $m : n$, if R is within PQ and

$$\frac{PR}{RQ} = \frac{m}{n}.$$

In the next theorem, we shall derive a formula to find the coordinates of a point which divide the join of two points internally in a given ratio.

Theorem I. If the point R(x, y) divides the join of points P(x_1, y_1) and Q(x_2, y_2) internally in the ratio m : n, then prove that

$$\mathbf{x = \frac{mx_2 + nx_1}{m + n}, \; y = \frac{my_2 + ny_1}{m + n}}.$$

Proof. The given points are $P(x_1, y_1)$ and $Q(x_2, y_2)$. For the sake of exactness, let us assume that the points P and Q are both in the I quadrant. Let R be such that $PR : RQ = m : n$.

Draw PA, QB and RC ⊥s on the x-axis. Through R, draw $DE \parallel x$-axis to meet AP produced in D and BQ in E.

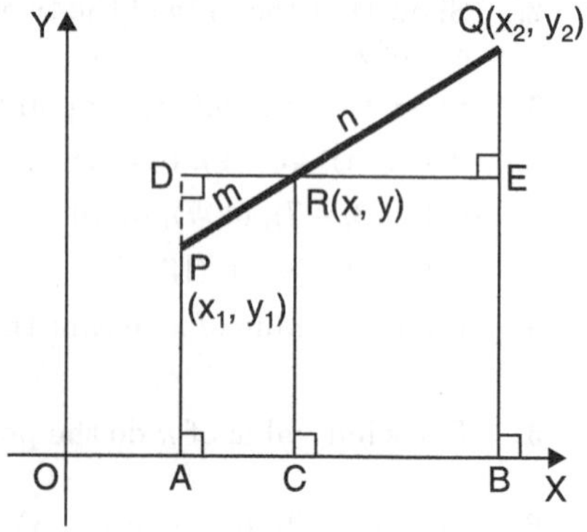

Δs PRD and QRE are similar.

$$\therefore \quad \frac{DR}{RE} = \frac{PD}{EQ} = \frac{PR}{RQ} = \frac{m}{n} \qquad ...(1)$$

$$(1) \Rightarrow \quad \frac{m}{n} = \frac{DR}{RE} = \frac{AC}{CB} = \frac{OC - OA}{OB - OC} = \frac{x - x_1}{x_2 - x}$$

$$\Rightarrow \quad mx_2 - mx = nx - nx_1 \Rightarrow mx_2 + nx_1 = (m + n)\, x$$

$$\Rightarrow \quad x = \frac{mx_2 + nx_1}{m + n}.$$

$$\text{Again (1)} \Rightarrow \quad \frac{m}{n} = \frac{PD}{EQ} = \frac{AD - AP}{BQ - BE} = \frac{y - y_1}{y_2 - y}$$

$$\Rightarrow \quad my_2 - my = ny - ny_1 \Rightarrow my_2 + ny_1 = (m + n)y \Rightarrow y = \frac{my_2 + ny_1}{m + n}$$

$\therefore$ **The coordinates of R are** $\mathbf{\left(\frac{mx_2 + nx_1}{m + n}, \frac{my_2 + ny_1}{m + n}\right)}$.

WORKING RULES TO FIND THE POINT OF INTERNAL DIVISION

Step I. *Multiply m by the x-coordinate of the point remote from m and n by the x-coordinate of the point remote from n as shown by the arrows.*

Step II. *Add the products* mx_2 *and* nx_1.

Step III. *Divide the sum* $mx_2 + nx_1$ *by* $m + n$. *This gives the x-coordinate of the point of internal division.*

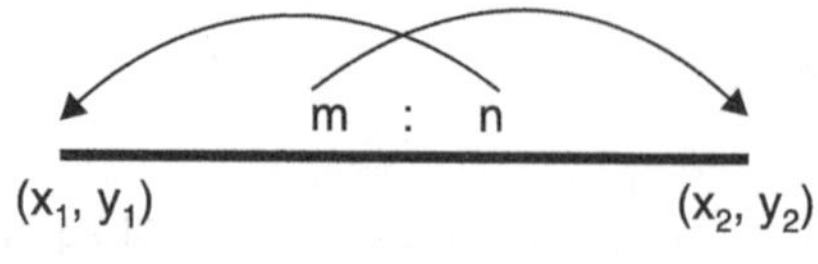

Step IV. *Similarly, find y coordinate.*

Remark 1. The above section formula also holds good even if either point or both points are not in the I quadrant.

Remark 2. The ratio $m : n$ is equivalent to $\frac{m}{n} : \frac{n}{n} (= 1)$ as well as to $\frac{m}{m} (= 1) : \frac{n}{m}$.

Therefore, in a given ratio, 1 can be kept on either side.

Corollary. Mid-point of a line segment. Let $M(x, y)$ be the mid-point of the line joining $P(x_1, y_1)$ and $Q(x_2, y_2)$.

$\therefore \quad PM = MQ$

$\therefore \quad M$ divided PQ in the ratio 1 : 1 internally.

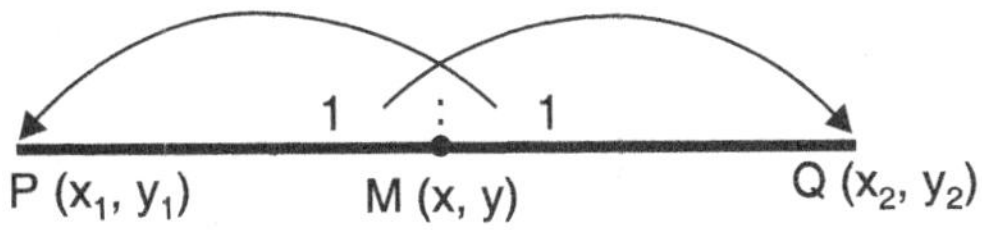

$$\therefore \quad x = \frac{1.x_2 + 1.x_1}{1+1} = \frac{x_1 + x_2}{2}$$

and
$$y = \frac{1.y_2 + 1.y_1}{1+1} = \frac{y_1 + y_2}{2}.$$

$\therefore$ **The mid-point is** $\mathbf{M\left(\frac{x_1 + x_2}{2}, \frac{y_1 + y_2}{2}\right)}$.

ILLUSTRATIVE EXAMPLES

Example 1. *Find the mid-point of the line joining (3, 5) and (– 7, – 3). Also draw the points on a cartesian plane.*

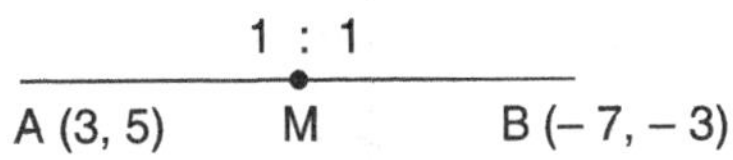

Sol. Let $A = (3, 5)$ and $B = (-7, -3)$.

Let $M(x, y)$ be the mid-point of AB.

$$\therefore \quad x = \frac{3 + (-7)}{2} = \frac{-4}{2} = -2,$$

and
$$y = \frac{5 + (-3)}{2} = \frac{2}{2} = 1.$$

$\therefore$ The mid-point of the given line is **(– 2, 1)**.

In the adjoining figure, the points are shown on a cartesian plane.

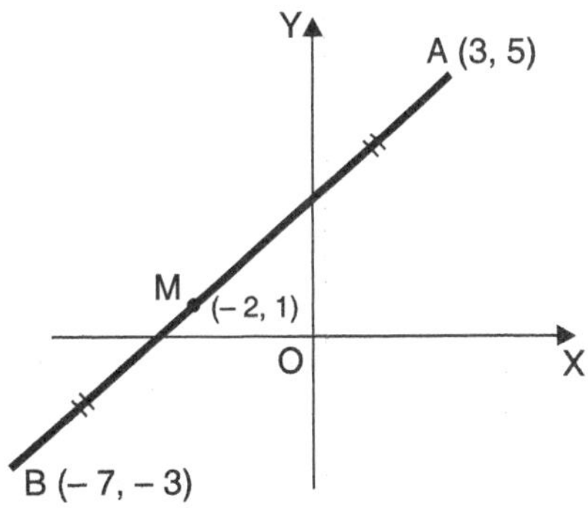

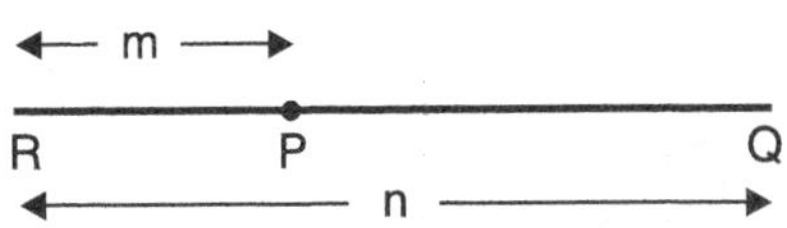

Definition of external division. A point R is said to divide the line PQ **externally** in the ratio $m : n$, where $m \neq n$, if R is outside PQ and

$$\frac{PR}{RQ} = \frac{m}{n}.$$

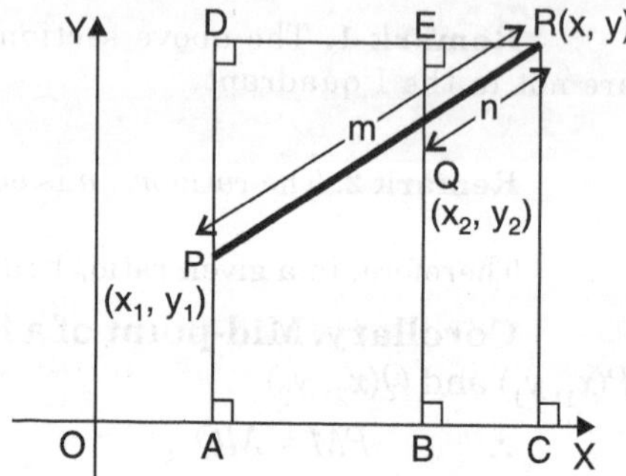

Theorem II. If the point R(x, y) divides the join of points $P(x_1, y_1)$ and $Q(x_2, y_2)$ externally in the ratio m : n, where m ≠ n, then prove that

$$\mathbf{x = \frac{mx_2 - nx_1}{m - n}, \ y = \frac{my_2 - ny_1}{m - n}}.$$

Proof. Let $m > n$. $\therefore$ The point of division R lies on the right of PQ.

In this case, we have similar triangles PRD and QRE.

$$\therefore \quad \frac{DR}{RE} = \frac{PD}{EQ} = \frac{PR}{RQ} = \frac{m}{n} \qquad ...(1)$$

$$(1) \Rightarrow \quad \frac{m}{n} = \frac{DR}{RE} = \frac{AC}{CB} = \frac{OC - OA}{OC - OB} = \frac{x - x_1}{x - x_2}$$

$$\Rightarrow \quad mx - mx_2 = nx - nx_1 \Rightarrow mx_2 - nx_1 = (m - n)x$$

$$\Rightarrow \quad x = \frac{mx_2 - nx_1}{m - n}$$ (Division by $m - n$ is justified, because $m \neq n$)

$$\text{Again (1)} \Rightarrow \quad \frac{m}{n} = \frac{PD}{EQ} = \frac{AD - AP}{BE - BQ} = \frac{y - y_1}{y - y_2}$$

$$\Rightarrow \quad my - my_2 = ny - ny_1$$

$$\Rightarrow \quad my_2 - ny_1 = (m - n)y \Rightarrow y = \frac{my_2 - ny_1}{m - n}.$$

$\therefore$ **The coordinates of R are** $\mathbf{\left(\frac{mx_2 - nx_1}{m - n}, \frac{my_2 - ny_1}{m - n}\right)}$.

Alternative proof. We observe that $Q(x_2, y_2)$ divides PR in the ratio $m - n : n$ internally.

$$\Rightarrow \quad x_2 = \frac{(m - n)x + nx_1}{(m - n) + n} \Rightarrow mx_2 = (m - n)x + nx_1 \Rightarrow x = \frac{mx_2 - nx_1}{m - n}.$$

Also, $$y_2 = \frac{(m - n)y + ny_1}{(m - n) + n} \text{ or } my_2 = (m - n)y + ny_1 \text{ or } y = \frac{my_2 - ny_1}{m - n}$$

$\therefore$ **The coordinates of R are** $\mathbf{\left(\frac{mx_2 - nx_1}{m - n}, \frac{my_2 - ny_1}{m - n}\right)}$.

WORKING RULES TO FIND THE POINT OF EXTERNAL DIVISION

Step I. *Multiply m by the x-coordinate of the point remote from m and n by the x-coordinate of the point remote from n as shown by arrows.*

Step II. *Subtract the product nx_1 from mx_2.*

Step III. *Divide the difference $mx_2 - nx_1$ by $m - n$. This gives the x-coordinate of the point of external division.*

Step IV. *Similarly find y-coordinate.*

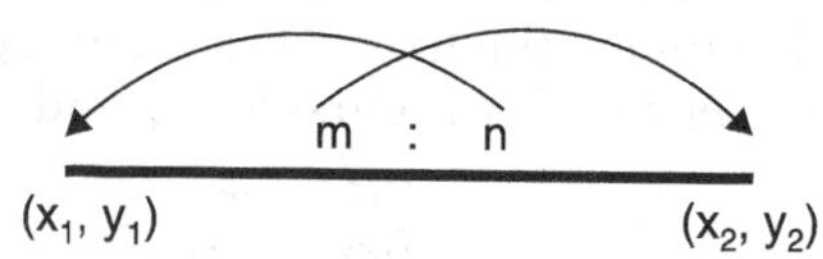

Remark 1. The above section formula also holds good even if either point or both points are not in the I quadrant.

Remark 2. The ratio $m : n$ is equivalent to $\frac{m}{n} : \frac{n}{n}$ (= 1) as well as to $\frac{m}{m}$ (= 1) : $\frac{n}{m}$.

Therefore, in a given ratio, 1 can be kept on either side.

Remark 3. The coordinates $\left(\frac{mx_2 - nx_1}{m-n}, \frac{my_2 - ny_1}{m-n}\right)$ of the point dividing $P(x_1, y_1)$ and $Q(x_2, y_2)$ externally in the ratio $m : n$, where $m \neq n$ can also be expressed as $\left(\frac{mx_2 + (-n)x_1}{m+(-n)}, \frac{my_2 + (-n)y_1}{m+(-n)}\right)$ and this can be thought of as the coordinates of the point dividing PQ internally in the ratio $m : -n$.

Example 2. *Find the coordinates of the points which divides the join of (1, 7) and (– 5, 6) in the ratio (i) 1 : 2 internally (ii) 3 : 2 externally.*

Sol. Let $A = (1, 7)$ and $B = (-5, 6)$.

(*i*) Let $P(x, y)$ divides AB in the ratio 1 : 2 internally.

$$\therefore \quad x = \frac{1\times(-5) + 2\times 1}{1+2} = \frac{-5+2}{3} = -1$$

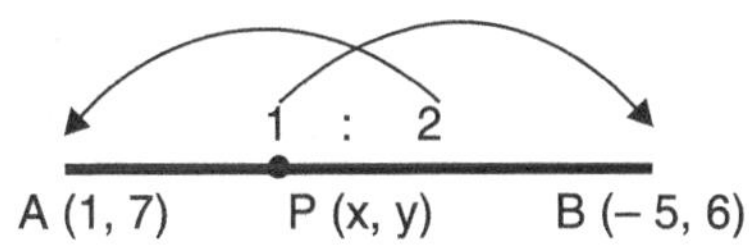

$$\text{and} \quad y = \frac{1\times 6 + 2\times 7}{1+2} = \frac{6+14}{3} = \frac{20}{3}.$$

$\therefore$ The point of division is **(– 1, 20/3).**

(*ii*) Let $Q(x, y)$ divides AB in the ratio 3 : 2 externally. This is equivalent to say that Q divides AB in the ratio 3 : – 2 internally.

$$\therefore \quad x = \frac{3\times(-5) + (-2)\times 1}{3+(-2)} = \frac{-15-2}{1} = -17$$

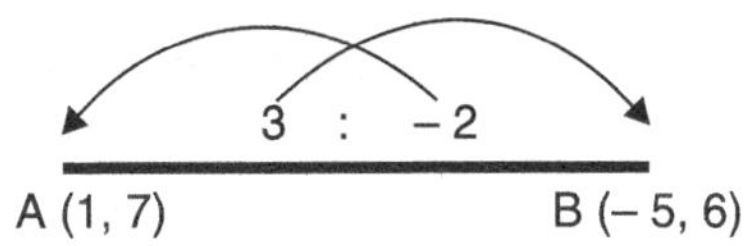

$$\text{and} \quad y = \frac{3\times 6 + (-2)\times 7}{3+(-2)} = \frac{18-14}{1} = 4.$$

$\therefore$ The point of division is **(– 17, 4).**

Example 3. *Find the ratio in which (3, – 9) divides the join of (1, – 3) and (– 3, 9).*

Sol. Let $A = (1, -3)$ and $B = (-3, 9)$. Let $(3, -9)$ divides AB in the ratio $k : 1$ internally.

$$\therefore \quad 3 = \text{abscissa of } (3, -9) = \frac{k(-3) + 1(1)}{k+1}.$$

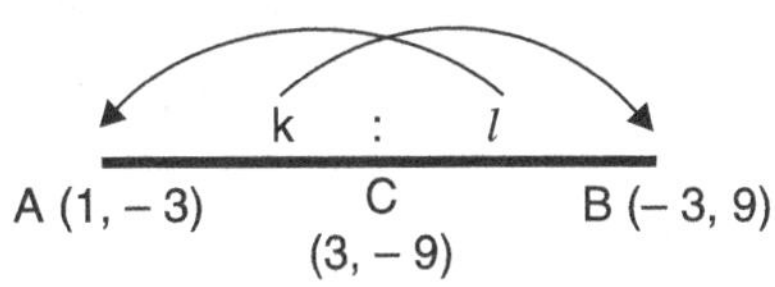

$\Rightarrow \quad 3k + 3 = -3k + 1 \quad \Rightarrow \quad 6k = -2 \quad$ *i.e.*, $\quad k = -1/3$

$\therefore$ The ratio is – 1/3 : 1 internal *i.e.*, **1 : 3 external.**

Remark. Considering ordinates, we get $-9 = \frac{9k-3}{k+1}$. Solving, we get $-9k - 9 = 9k - 3$ or $k = -1/3$. Thus, we get the same value of k. In case, we get different value of k, then it would imply that (3, – 9) is not on the line joining (1, – 3) and (– 3, 9).

Example 4. *Using section formula, show that the points (1,– 1), (2, 1) and (4, 5) are collinear.*

Sol. Let $A = (1, -1)$, $B = (2, 1)$ and $C = (4, 5)$.

Let C divides AB in the ratio $k : 1$ internally.

$$\therefore \quad 4 = \frac{k(2) + 1(1)}{k+1} \quad \ldots(1)$$

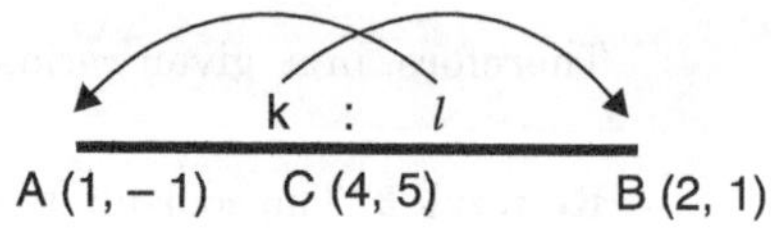

and
$$5 = \frac{k(1) + 1(-1)}{k+1} \quad \ldots(2)$$

(1) $\Rightarrow 4k + 4 = 2k + 1 \Rightarrow k = -3/2$ and (2) $\Rightarrow 5k + 5 = k - 1 \Rightarrow k = -3/2$.

$\therefore$ We get the same value of k. $\therefore$ C is on the line AB *i.e.*, A, B, C are collinear.

Example 5. *In what ratio is the line joining (4, 5) and (1, 2) is divided by the y-axis ? Find also the point of division, and show the points on a coordinate plane.*

(P.T.U. B.Pharma, May 2002)

Sol. Let the given points be $P(4, 5)$ and $Q(1, 2)$. Let PQ be divided by the y-axis at $R(0, y)$ in the ratio $k : 1$ internally.

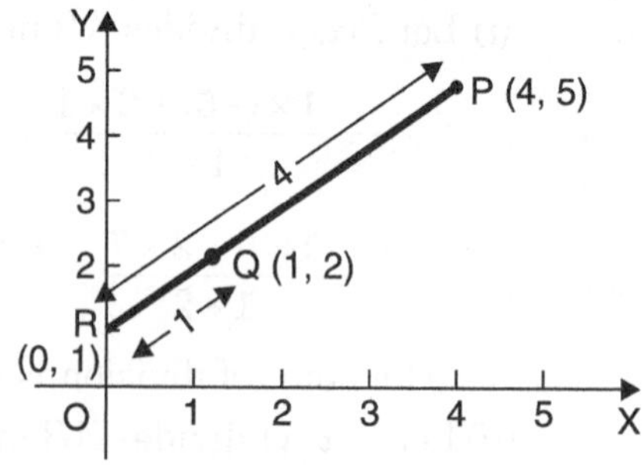

$$\therefore\ 0 = \frac{k \times 1 + 1 \times 4}{k+1} \quad \text{(Considering the abscissa of } R)$$

$\Rightarrow \quad k = -4.$

$\therefore$ The ratio is $-4 : 1$ internal, *i.e.*, $4 : 1$ external.

$$\therefore\ y = \frac{k \times 2 + 1 \times 5}{k+1} = \frac{-4 \times 2 + 5}{-4+1} = \frac{-3}{-3} = 1$$

$\therefore$ The required point is **(0, 1).**

Example 6. (*i*) *In what ratio is the line joining (2, – 3) and (5, 6) is divided by the x-axis (ii) y-axis. Also find the points of division and show the points on a coordinate plane.*

Sol. Let the given points be $P(2, -3)$ and $Q(5, 6)$.

(*i*) Let PQ be divided by the x-axis at $R(x,0)$ is the ratio $k : 1$ internally. Considering the ordinate of R, we have $0 = \dfrac{k \times 6 + 1 \times (-3)}{k+1}$

$\Rightarrow \quad 6k - 3 = 0 \quad$ *i.e.*, $\quad k = 1/2$

$\therefore$ The ratio is $1/2 : 1$ internal *i.e.*, **1 : 2 internal.**

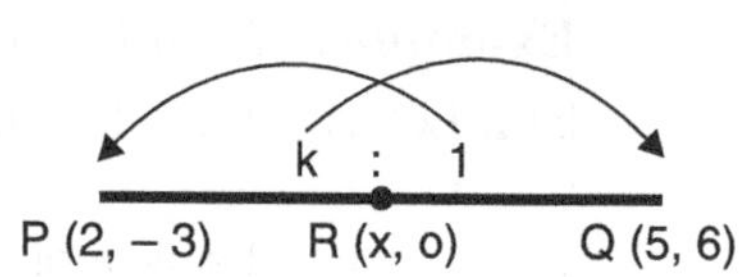

Also,
$$x = \frac{k \times 5 + 1 \times 2}{k+1} = \frac{\frac{1}{2}(5) + 2}{\frac{1}{2} + 1} = \frac{\frac{9}{2}}{\frac{3}{2}} = 3. \quad \therefore$$
The point is **R(3, 0).**

(*ii*) Let PQ be divided by the y-axis at $S(0, y)$ in the ratio $k : 1$ internally. Considering the abscissa of S,

we have
$$0 = \frac{k \times 5 + 1 \times 2}{k+1}$$

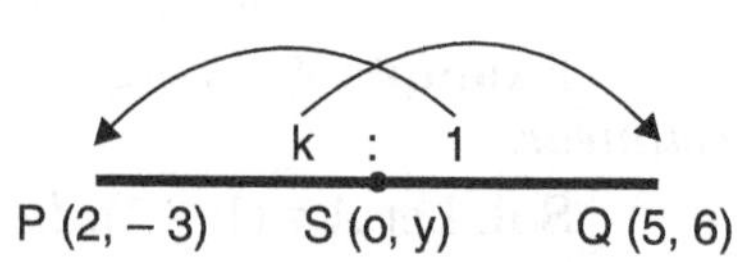

$\Rightarrow \quad 5k + 2 = 0 \quad$ *i.e.*, $\quad k = -2/5.$

∴ The ratio is – 2/5 : 1 internal *i.e.,* **2 : 5 external.**

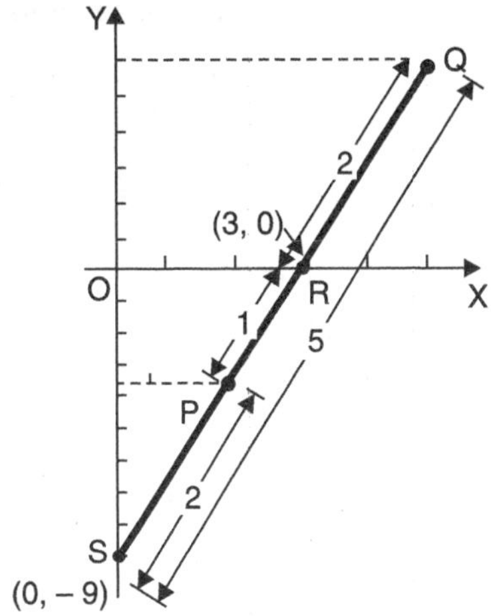

Also,
$$y = \frac{k \times 6 + 1 \times (-3)}{k+1}$$

$$= \frac{\left(-\frac{2}{5}\right) \times 6 - 3}{-\frac{2}{5} + 1}$$

$$= \frac{-\frac{27}{5}}{\frac{3}{5}} = -9.$$

∴ The point is **S (0, – 9).**

Example 7. *Find the ratio and the point in which the line joining (1, 3) and (2, 7) is divided by the line $3x + y - 9 = 0$.*

Sol. Let $A = (1, 3)$ and $B = (2, 7)$. Let AB be divided by the line $3x + y - 9 = 0$ at the point $P(x, y)$ and in the ratio $k : 1$ internally.

∴ By **Section formula,**

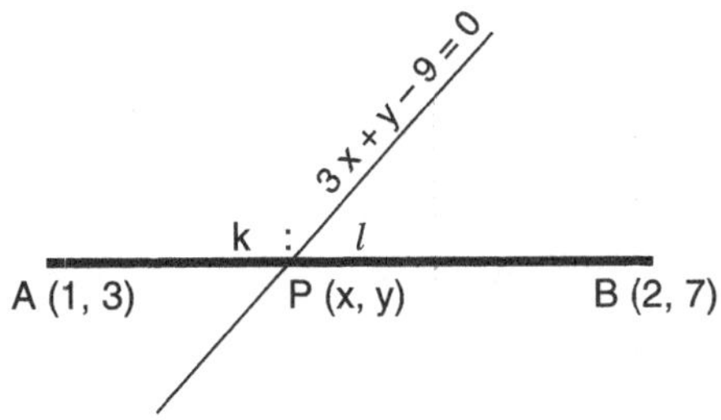

$$x = \frac{k \times 2 + 1 \times 1}{k+1} = \frac{2k+1}{k+1}$$

and
$$y = \frac{k \times 7 + 1 \times 3}{k+1} = \frac{7k+3}{k+1}.$$

∴ Coordinates of P are $\left(\frac{2k+1}{k+1}, \frac{7k+3}{k+1}\right)$

and this point lies on the line $3x + y - 9 = 0$.

$$\therefore \quad 3\left(\frac{2k+1}{k+1}\right) + \left(\frac{7k+3}{k+1}\right) - 9 = 0 \text{ or } 6k + 3 + 7k + 3 - 9k - 9 = 0 \text{ or } 4k - 3 = 0 \text{ or } k = \frac{3}{4}.$$

∴ The line divides AB in the ratio 3/4 : 1 internally *i.e.,* in the ratio **3 : 4 internally.**

Putting the value of k, we get

$$x = \frac{2\left(\frac{3}{4}\right) + 1}{\frac{3}{4} + 1} = \frac{10/4}{7/4} = \frac{10}{7} \quad \text{and} \quad y = \frac{7\left(\frac{3}{4}\right) + 3}{\frac{3}{4} + 1} = \frac{33/4}{7/4} = \frac{33}{7}.$$

∴ The point of division is **(10/7, 33/7).**

Example 8. *Three consecutive vertices of a parallelogram ABCD are A(3, 0), B(5, 2), C(– 2, 6). Find the fourth vertex D.*

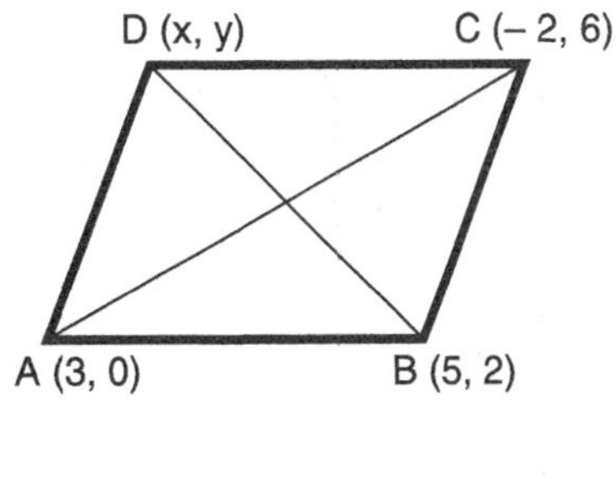

Sol. Let the fourth vertex be $D(x, y)$. Since $ABCD$ is a parallelogram, the diagonals bisect each other.

∴ Mid-point of BD = Mid-point of AC.

$$\therefore \quad \left(\frac{x+5}{2}, \frac{y+2}{2}\right) = \left(\frac{3+(-2)}{2}, \frac{0+6}{2}\right).$$

Equating abscissas and ordinates, we get

$$\frac{x+5}{2} = \frac{3-2}{2} \quad \text{or} \quad x+5=1 \quad \text{or} \quad x=-4.$$

and
$$\frac{y+2}{2} = \frac{0+6}{2} \quad \text{or} \quad y+2=6 \quad \text{or} \quad y=4.$$

∴ The fourth vertex $D(x, y)$ is **(– 4, 4).**

Example 9. *A, B, C are three collinear points where A = (3, 4) and B = (7, 7). If distance between A and C is 10 units, find the coordinates of C.*

Sol. We have $A = (3, 4)$, $B = (7, 7)$. Let (x, y) be the coordinates of C.

$$AB = \sqrt{(7-3)^2 + (7-4)^2} = \sqrt{16+9} = \sqrt{25} = 5.$$

Since A, B, C are collinear and $AC = 10$, the point B must lie between A and C.

Also, $BC = AC - AB = 10 - 5 = 5.$

∴ B is the mid-point of AC.

∴ $$\left(\frac{3+x}{2}, \frac{4+y}{2}\right) = (7, 7) \qquad (\because \text{ Mid-point of } AC = B)$$

∴ $$\frac{3+x}{2} = 7 \quad \text{and} \quad \frac{4+y}{2} = 7.$$ Solving, we get $x = 11$, $y = 10$.

∴ The coordinates of C are **(11, 10).**

10.8. CENTROID OF A TRIANGLE

The point of concurrence of three medians of a triangle is called its **centroid.**

Let $A(x_1, y_1)$, $B(x_2, y_2)$, $C(x_3, y_3)$ be the coordinates of the vertices of a triangle. Let D be the mid-point of BC.

∴ The coordinates of D are $\left(\frac{x_2+x_3}{2}, \frac{y_2+y_3}{2}\right)$.

Let G divides DA in the ratio 1 : 2 internally.

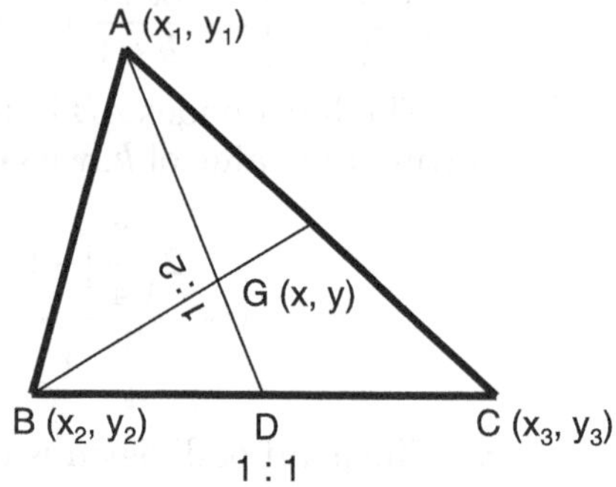

∴ $$x = \frac{1(x_1) + 2\left(\frac{x_2+x_3}{2}\right)}{1+2} = \frac{x_1+x_2+x_3}{3}$$

and
$$y = \frac{1(y_1) + 2\left(\frac{y_2+y_3}{2}\right)}{1+2} = \frac{y_1+y_2+y_3}{3}.$$

∴ $$G = \left(\frac{x_1+x_2+x_3}{3}, \frac{y_1+y_2+y_3}{3}\right).$$

The symmetry of the coordinates of G shows that it also lies on the other two medians through B and C. Thus, all medians meet at G *i.e.*, it is the centroid.

∴ The centroid is $\left(\mathbf{\frac{x_1+x_2+x_3}{3}, \frac{y_1+y_2+y_3}{3}}\right)$.

Example 10. *Find the centroid of the triangle with vertices (1, 7), (– 3, – 4) and (– 6, 4).*

Sol. Let $A = (1, 7)$, $B = (-3, -4)$ and $C = (-6, 4)$ be the vertices of the given triangle. Let $G(x, y)$ be the centroid of the given triangle.

$\therefore$ By using $x = \dfrac{x_1 + x_2 + x_3}{3}$, $y = \dfrac{y_1 + y_2 + y_3}{3}$, we get

$$x = \frac{1 + (-3) + (-6)}{3} = -\frac{8}{3}, \quad y = \frac{7 + (-4) + 4}{3} = \frac{7}{3}$$

$\therefore$ The centroid is **(– 8/3, 7/3).**

10.9. INCENTRE OF A TRIANGLE

The point of concurrence of three internal bisectors of the angles of a triangle is called its **incentre.**

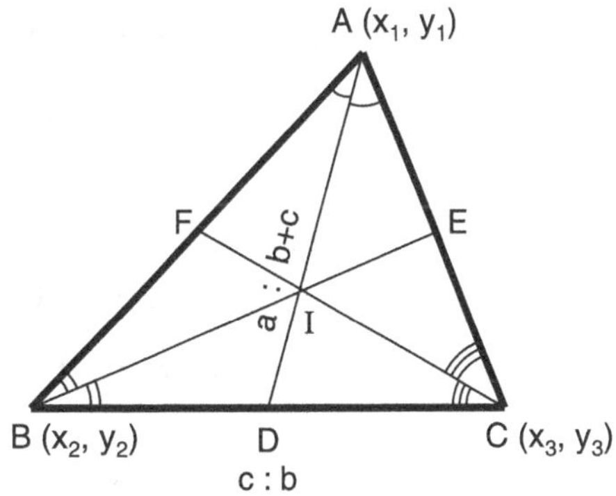

Let $A(x_1, y_1)$, $B(x_2, y_2)$, $C(x_3, y_3)$ be the coordinates of the vertices of a triangle. Let AD, BE and CF be the internal bisectors of the angles A, B, C respectively.

Let the sides BC, CA and AB be denoted by a, b, c respectively.

By geometry, $\dfrac{BD}{DC} = \dfrac{BA}{CA} = \dfrac{c}{b}$ ...(1)

$\therefore$ The coordinates of D are $\left(\dfrac{bx_2 + cx_3}{b + c}, \dfrac{by_2 + cy_3}{b + c}\right)$

$$(1) \Rightarrow \frac{BD}{DC} + 1 = \frac{c}{b} + 1 \Rightarrow \frac{BD + DC}{DC} = \frac{c + b}{b} \Rightarrow \frac{BD}{DC} = \frac{c + b}{b} \Rightarrow DC = \frac{ab}{b + c}.$$

Let AD and CF intersects at I.

Since, CI is the internal bisector of angle C, we have $\dfrac{AI}{ID} = \dfrac{AC}{DC} = \dfrac{b}{ab/(b + c)} = \dfrac{b + c}{a}$.

$\therefore$ I divides AD internally in the ratio $b + c : a$.

$\therefore$ The coordinates of I are $\left(\dfrac{ax_1 + (b + c)\left(\dfrac{bx_2 + cx_3}{b + c}\right)}{a + (b + c)}, \dfrac{ay_1 + (b + c)\left(\dfrac{by_2 + cy_3}{b + c}\right)}{a + (b + c)}\right)$

or $\left(\dfrac{ax_1 + bx_2 + cx_3}{a + b + c}, \dfrac{ay_1 + by_2 + cy_3}{a + b + c}\right)$.

The symmetry of the coordinates of I shows that it also lies on the internal bisector through B. Thus, all internal bisectors meet at I *i.e.*, it is the incentre.

$\therefore$ The incentre is $\left(\dfrac{\mathbf{ax_1 + bx_2 + cx_3}}{\mathbf{a + b + c}}, \dfrac{\mathbf{ay_1 + by_2 + cy_3}}{\mathbf{a + b + c}}\right)$.

Remark. In an equilateral triangle, centroid and incentre are coincident.

Example 11. *Find the incentre of the triangle whose vertices are (– 36, 7), (20, 7) and (0, – 8).*

Sol. Let $A = (-36, 7)$, $B = (20, 7)$ and $C = (0, -8)$.

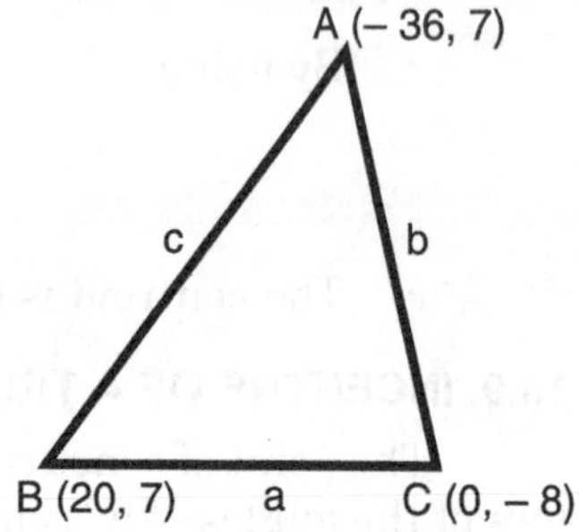

$$\therefore \quad a = BC = \sqrt{(0-20)^2 + (-8-7)^2} = \sqrt{400 + 225} = 25$$

$$b = CA = \sqrt{(-36-0)^2 + (7+8)^2} = \sqrt{1296 + 225} = 39$$

$$c = AB = \sqrt{(20+36)^2 + (7-7)^2} = \sqrt{3136 + 0} = 56.$$

Let I (x, y) be the incentre of the triangle.

$\therefore$ By using $x = \dfrac{ax_1 + bx_2 + cx_3}{a+b+c}$, $y = \dfrac{ay_1 + by_2 + cy_3}{a+b+c}$,

we get $x = \dfrac{25(-36) + 39(20) + 56(0)}{25 + 39 + 56} = \dfrac{-120}{120} = -1$, $y = \dfrac{25(7) + 39(7) + 56(-8)}{25 + 39 + 56} = \dfrac{0}{120} = 0.$

$\therefore$ The incentre is **(– 1, 0).**

EXERCISE 10.4

1. (*i*) Determine the point that bisects the line segment whose end points are (4, – 6) and (12, 15).
   (*ii*) Given A (– 3, 2) and B(5, 4), find the mid-point of AB.
2. Find the centroid of the triangle whose vertices are :
   (*i*) (1, 2), (3, 5), (6, 3) (*ii*) (0, 9), (– 5, 6), (11, – 7).
3. One end of diameter of a circle is (4, 1) and the centre is (3, 3), find the coordinates of the other end of the diameter.
4. Show that the quadrilateral with vertices (1, 4), (– 2, 1), (0, – 1) and (3, 2) is a parallelogram.
5. Prove that the points (1, 1), (4, 4), (4, 8) and (1, 5) forms a parallelogram. Is it a rectangle ?
6. If A(– 1, 3), B(1, – 1) and C(5, 1) are the vertices of a triangle ABC, find the length of the median through A.
7. Find the third vertex of a triangle if two of its vertices are at (– 1, 4) and (5, 2) and the medians meet at (0, – 3).
8. Find the coordinates of the point which divides the join of :
   (*i*) (1, 4) and (2, – 8) in the ratio 2 : 5 internally
   (*ii*) (– 3, 7) and (0, 8) in the ratio 3 : 1 internally.
9. The line joining the points (2, – 2) and (– 2, 4) is trisected. Find the points of trisection.
10. Find the ratio in which the line joining (– 3, – 10) and (3, 5) is divided by the (*i*) x-axis (*ii*) y-axis. Also find the points of division.
11. Find the incentre of the triangle whose vertices are (4, – 2), (– 2, 4) and (5, 5).
12. If the points (– 2, – 1), (1, 0), (x, 3) and (1, y) form a parallelogram, find the values of x and y.
13. Find the fourth vertex of the parallelogram whose three vertices, taken in order, are :
    (*i*) (– 2, – 1), (1, 0), (4, 3) (*ii*) $(a + b, a - b)$, $(2a + b, 2a - b)$, $(a - b, a + b)$.

14. Find the ratio in which the line joining (3, – 1) and (8, 9) is divided by the line $x - y - 2 = 0$. Also find the point of division.

15. Using section formula, show that the following points are collinear :

(*i*) (– 3, – 2), (5, 2), (1, 0) (*ii*) (b, $c + a$), (c, $a + b$), (a, $b + c$).

Answers

1.	(*i*) (8, 9/2)	(*ii*) (1, 3)	**2.** (*i*) (10/3, 10/3)	(*ii*) (2, 8/3)
3.	(2, 5)	**5.** No.	**6.** 5	**7.** (– 4, – 15)
8.	(*i*) (9/7, 4/7)	(*ii*) (– 3/4, 31/4)	**9.** (2/3, 0), (– 2/3, 2)	
10.	(*i*) 2 : 1 internally, (1, 0)		(*ii*) 1 : 1 internally, (0, – 5/2)	
11.	(5/2, 5/2)	**12.** $x = 4$, $y = 2$	**13.** (*i*) (1, 2)	(*ii*) (– b, b)
14.	2 : 3 internally, (5, 3).			

Hints

4. It is enough to show that its diagonals bisects each other.

9. The first point of trisection divides the line in the ratio 1 : 2 internally.

10.10. LOCUS AND ITS EQUATION *(P.T.U. B.Pharma, May 2007)*

Let A and B be any two points. Let P be a point which moves so that its distances from A and B are equal. The point P cannot be at Q, because AQ is not equal to BQ. It may be proved that the point P can be any where on the right bisector of AB. We shall say that this line is the *locus* (path) of the point P, under the above mentioned condition. Let us define the term '*locus*' formally.

When a point moves so as always to satisfy a given condition or conditions, the path it traces out is called its **locus** under these conditions. Technically, a *locus* represents the '*set of all points*' which lies on it.

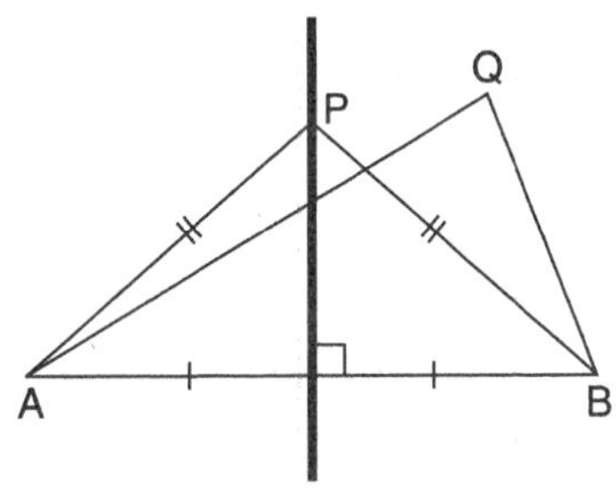

In the above example, the right bisector of AB is the locus under given condition.

If $P(x, y)$ be a general point on the locus, then an equation involving x and y which is satisfied by each point on the locus and such that each point satisfying the equation is on the locus is called the **equation** of the locus.

WORKING RULES FOR FINDING THE EQUATION OF THE LOCUS OF A POINT

Step I. *Take a general point P(x, y) on the locus.*

Step II. *Write down the given geometric conditions, under which the point P moves.*

Step III. *Express the above said conditions in terms of x and y by the help of the formulae and simplify the equation (by squaring both sides if square roots are there and taking L.C.M. to remove the denominators).*

Step IV. *The simplified equation so obtained is the required equation of the locus.*

ILLUSTRATIVE EXAMPLES

Example 1. *Find the equation of the locus of a point which moves so that :*

(*i*) *it is always 2.5 units above x-axis.*

(*ii*) *it is always 4 units from y-axis.*

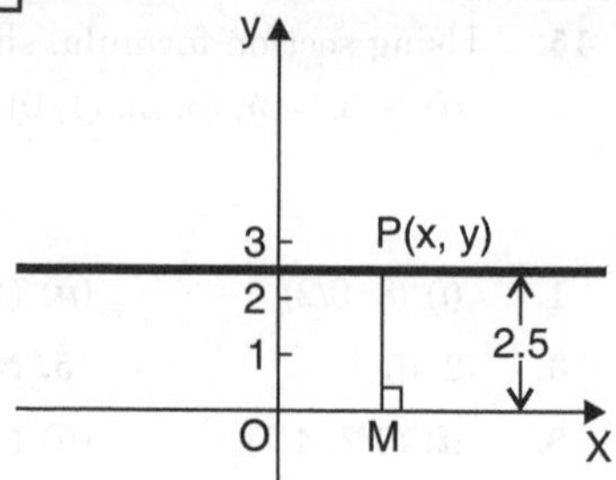

Sol. (*i*) Let $P(x, y)$ be a general point on the locus. Draw $PM \perp x$-axis. By the given condition, $PM = 2.5$.

$\therefore$ $\mathbf{y = 2.5.}$

$\therefore$ This is the required equation of the locus.

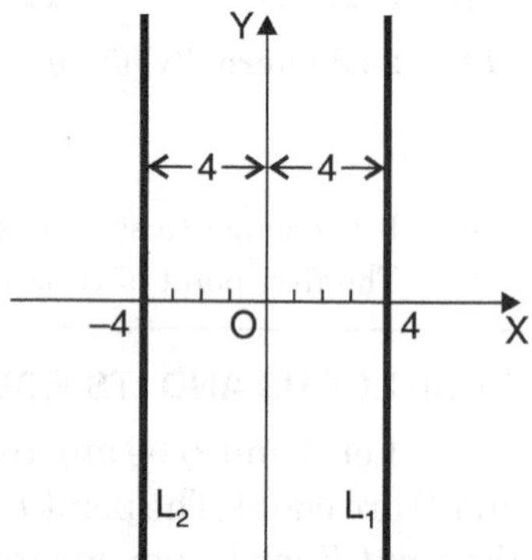

(*ii*) Let $P(x, y)$ be a general point on the locus.

$\therefore$ By the given condition, P is either on L_1, or on L_2 which are lines at distance of 4 units from y-axis. If P is on L_1 then we have $x = 4$ and if P is on L_2, then we have $x = -4$.

$\therefore$ For any point $P(x, y)$, we have $\mathbf{x = \pm 4.}$

This is the required equation of the locus.

Example 2. *Find the equation of the locus of a point which moves so that its ordinate is always greater than its abscissa by a given distance.*

Sol. Let $P(x, y)$ be a general point on the locus.

$\therefore$ By the given condition, ordinate = abscissa + given distance.

$\therefore$ $\mathbf{y = x + k,}$ assuming given distance to be k.

This is the required equation of the locus.

Example 3. *Find the equation of the locus of a point which moves so that its distances from (a, b) and (c, d) are equal.*

Sol. Let the given points be $A(a, b)$ and $B(c, d)$.

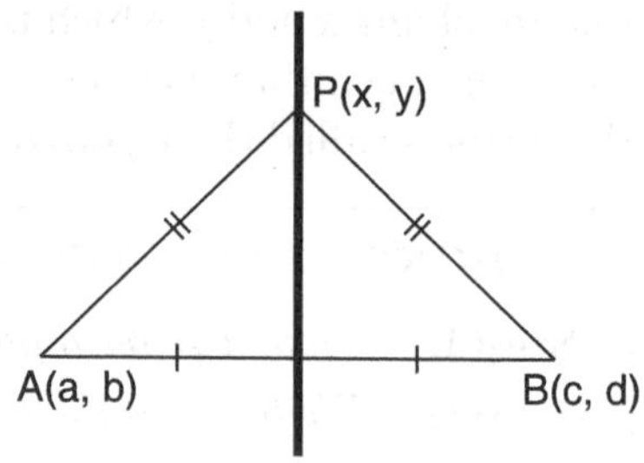

Let $P(x, y)$ be a general point on the locus.

$\therefore$ By the given condition, $PA = PB$.

$$\therefore \quad \sqrt{(x-a)^2 + (y-b)^2} = \sqrt{(x-c)^2 + (y-d)^2}$$

or
$$x^2 + a^2 - 2ax + y^2 + b^2 - 2by = x^2 + c^2 - 2cx + y^2 + d^2 - 2dy$$

or
$$\mathbf{2(c - a)x + 2(d - b)y = (c^2 + d^2) - (a^2 + b^2).}$$

This is the required equation of the locus.

Example 4. *Find the equation of the locus of a point which moves so that its distance from (3, 2) is twice its distance from (1, 1).*

Sol. Let $A(3, 2)$ and $B(1, 1)$ be the given points. Let $P(x, y)$ be a general point on the locus.

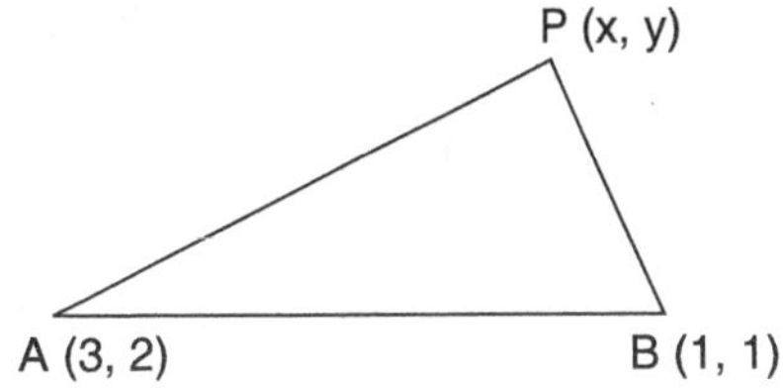

$\therefore$ By the given condition, $PA = 2PB$.

$$\therefore \quad \sqrt{(x-3)^2+(y-2)^2} = 2\sqrt{(x-1)^2+(y-1)^2}$$

$$\Rightarrow \quad x^2+9-6x+y^2+4-4y = 4[x^2+1-2x+y^2+1-2y]$$

$$\Rightarrow \quad \mathbf{3x^2+3y^2-2x-4y-5=0.}$$

This is the required equation of the locus.

Example 5. *Find the equation of the locus of the point P which moves so that $\Delta PAB = 10$, where $A(4, 7)$ and $B(0, -5)$ are fixed points.*

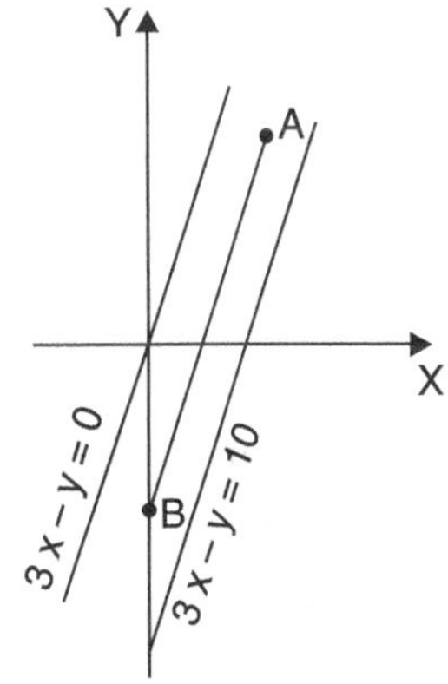

Sol. Let the coordinates of P be (x, y).

$\therefore$ Area of triangle with vertices $P(x, y)$, $A(4, 7)$ and $B(0, -5)$ is 10.

$$\therefore \quad \frac{1}{2} \mid x(7+5)+4(-5-y)+0\,(y-7) \mid = 10$$

$$\Rightarrow \quad \mid 12x-20-4y \mid = 20 \quad \Rightarrow \quad 12x-4y-20 = \pm 20$$

$$\Rightarrow \quad 3x-y-5 = \pm 5 \quad \Rightarrow \quad 3x-y=10 \quad \text{or} \quad 3x-y=0.$$

$\therefore$ The locus of $P(x, y)$ is either $\mathbf{3x - y = 10}$ or $\mathbf{3x - y = 0}$.

Remark. The lines $3x - y = 0$ and $3x - y = 10$ are on opposite sides of AB. If P is any point on either of the lines, the area of ΔPAB will be 10.

Example 6. *Find the equation of the locus of a point which is equidistant from the point (2, 4) and the y-axis.* (P.T.U. B.Pharma, Dec. 2002)

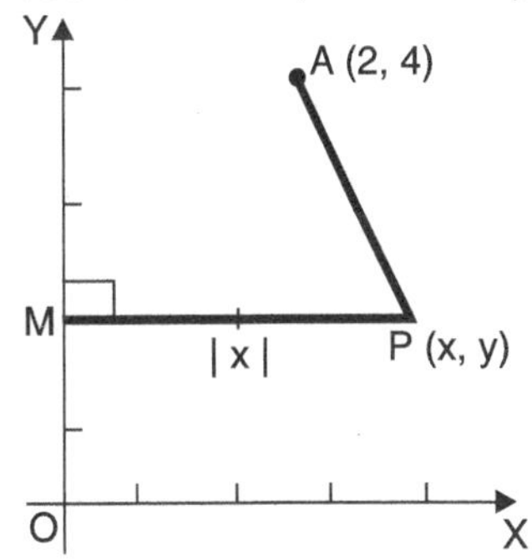

Sol. Let $A(2, 4)$ be the given point. Let $P(x, y)$ be a general point on the locus. Draw $PM \perp y$-axis.

$\therefore$ By the given condition, $PA = PM$.

$$\therefore \quad \sqrt{(x-2)^2+(y-4)^2} = \mid x \mid$$

$$\Rightarrow \quad x^2+4-4x+y^2+16-8y = x^2$$

$$\Rightarrow \quad \mathbf{y^2-4x-8y+20=0.}$$

This is the required equation of the locus.

Example 7. *Show that the equation of the locus of a point such that the difference of their distances from two fixed points (ae, 0) and (– ae, 0) is 2a, is* $\dfrac{x^2}{a^2} - \dfrac{y^2}{a^2(e^2-1)} = 1$.

Sol. Let $P(x, y)$ be a general point on the locus and $A(ae, 0)$ and $B(-ae, 0)$ be the given points.

By the given condition, $\quad \mid PA - PB \mid = 2a.$

$$\Rightarrow \quad PA - PB = \pm 2a$$

$$\Rightarrow \quad \sqrt{(x-ae)^2+y^2} - \sqrt{(x+ae)^2+y^2} = \pm 2a \qquad \ldots(1)$$

Identically, we have $[(x - ae)^2 + y^2] - [(x + ae)^2 + y^2] = -4aex$ ...(2)

$$(2) \div (1) \Rightarrow \sqrt{(x - ae)^2 + y^2} + \sqrt{(x + ae)^2 + y^2} = \mp 2ex \quad ...(3)$$

$$(1) + (3) \Rightarrow 2\sqrt{(x - ae)^2 + y^2} = \pm (2a - 2ex)$$

$$\Rightarrow x^2 + a^2e^2 - 2aex + y^2 = a^2 + e^2x^2 - 2aex \quad [\because (\pm 1)^2 = 1]$$

$$\Rightarrow x^2 - e^2x^2 + y^2 = a^2 - a^2e^2 \Rightarrow (e^2 - 1)x^2 - y^2 = a^2(e^2 - 1)$$

$$\Rightarrow \frac{\mathbf{x^2}}{\mathbf{a^2}} - \frac{\mathbf{y^2}}{\mathbf{a^2(e^2 - 1)}} = \mathbf{1}.$$ This is the required equation of the locus.

EXERCISE 10.5

1. Show that the point (4, 3) lies on the locus of a point whose equation is $x + 3y - 13 = 0$.
2. Show that the points $(a, 0)$, $(-a, 0)$, $(0, b)$, $(0, -b)$ lies on the locus of a point whose equation is $\frac{x^2}{a^2} + \frac{y^2}{b^2} = 1$.
3. Which of the following points lies on the locus of $x^2 + y^2 = 16$:
   (*i*) (4, 1) (*ii*) (4, 0)
   (*iii*) (0, 4) (*iv*) $(2\sqrt{2}, -2\sqrt{2})$.
4. Find the equation of the locus of a point which moves so that :
   (*i*) it is always 2 units above x-axis. (*ii*) it is always 4 units below x-axis.
5. Find the equation of the locus of a point which moves so that :
   (*i*) it is always 6 units on the right of y-axis
   (*ii*) it is always 2 units on the left of y-axis.
6. Find the equation of the locus of a point which moves so that :
   (*i*) the sum of the squares of its distances from the coordinates axes is p^2.
   (*ii*) the sum of the squares of its distances from the points $(-a, 0)$ and $(a, 0)$ is $2k^2$, where k is a constant.
7. Find the equation of the locus of a point which moves so that it is equidistant from the points :
   (*i*) (– 1, – 1) and (4, 2) (*ii*) $(a^2 + b^2, a^2 - b^2)$ and $(a^2 - b^2, a^2 + b^2)$.
8. Find the equation of the locus of a point which moves so that :
   (*i*) its distance from $(ap, 0)$ is p ($p > 0, p \neq 1$) times its distance from $(a/p, 0)$.
   (*ii*) its distances from (1, 2) and (5, 3) are in the ratio 2 : 3.
   (*iii*) Find the equation of the locus of the point which moves such that its distance from the point (– 1, 2) is greater than 3. (*P.T.U. B.Pharma, June 2003*)
9. Find the equation of the locus of a point which moves so that :
   (*i*) the line segment having end points (– 2, 0) and (2, 0) subtend a right angle at this point.
   (*ii*) the join of (– 2, 3) and (2, 3) subtend a right angle at this point.

10. Find the equation of the locus of a point which moves so that :

(*i*) its distance from the axis of x is half its distance from the origin.

(*ii*) its distance from the x-axis is same as its distance from the point (4, 2).

11. Find the equation of the locus of a point which moves so that :

(*i*) the sum of its distances from the points (0, 2) and (0, – 2) is 6.

(*ii*) the sum of its distances from the points (ae, 0) and (– ae, 0) is $2a$.

12. Let A and B be the points (5, 0) and (– 5, 0) respectively. Find the equation of the locus of the point $P(x, y)$ such that $AP - BP = 6$.

13. Find the equation of the locus of the mid-point of :

(*i*) AB where A, B are points on the x-axis and y-axis where the variable line $x \cos \alpha + y \sin \alpha = p$ cuts the axes, p being constant.

(*ii*) OA where A is any point on the locus represented by $y^2 = 4ax$.

Answers

3. (*ii*), (*iii*), (*iv*) **4.** (*i*) $y = 2$ (*ii*) $y = -4$

5. (*i*) $x = 6$ (*ii*) $x = -2$ **6.** (*i*) $x^2 + y^2 = p^2$ (*ii*) $x^2 + y^2 = k^2 - a^2$

7. (*i*) $5x + 3y - 9 = 0$ (*ii*) $x - y = 0$ **8.** (*i*) $x^2 + y^2 = a^2$

(*ii*) $5x^2 + 5y^2 + 22x - 12y - 91 = 0$

(*iii*) the positive half of the circle with centre (– 1, 2) and radius 3.

9. (*i*) $x^2 + y^2 = 4$ (*ii*) $x^2 + y^2 - 6y + 5 = 0$

10. (*i*) $x^2 - 3y^2 = 0$ (*ii*) $x^2 - 8x - 4y + 20 = 0$ **11.** (*i*) $\frac{x^2}{5} + \frac{y^2}{9} = 1$ (*ii*) $\frac{x^2}{a^2} + \frac{y^2}{a^2(1-e^2)} = 1$

12. $\frac{x^2}{9} - \frac{y^2}{16} = 1$ **13.** (*i*) $\frac{1}{x^2} + \frac{1}{y^2} = \frac{4}{p^2}$ (*ii*) $y^2 = 2ax$.

Hints

8. (*iii*) Let P(x, y) and A (– 1, 2) be the given points then, given (AP) > 3. Simplify it.

13. (*i*) For A, put $y = 0$ in $x \cos \alpha + y \sin \alpha = p$. $\therefore$ $A = (p \sec \alpha, 0)$. Similarly, $B = (0, p \operatorname{cosec} \alpha)$.

Let $P(x, y)$ be the mid-point of AB. $\therefore$ $x = \frac{p \sec \alpha}{2}$ and $y = \frac{p \operatorname{cosec} \alpha}{2}$.

$\therefore$ $\cos \alpha = \frac{p}{2x}$ and $\sin \alpha = \frac{p}{2y}$. Now use $\cos^2 \alpha + \sin^2 \alpha = 1$.

10.11. SLOPE OF A NON-VERTICAL LINE

Any line in a coordinate plane may or may not be parallel to x-axis. If a line is not parallel to x-axis, then the angle which is made by the line in the anti-clockwise direction from the x-axis is called the **inclination** of the line. This inclination lies between 0° and 180°. The inclination of a line is generally denoted by θ. The inclination of a line parallel to x-axis is defined to be 0°.

10.12. SLOPE

The concept of slope is defined only for lines which are not parallel to the y-axis. Let AB be a line which is not parallel to y-axis and let θ be the inclination of the line. The **slope** of the line AB is defined as $\tan \theta$.

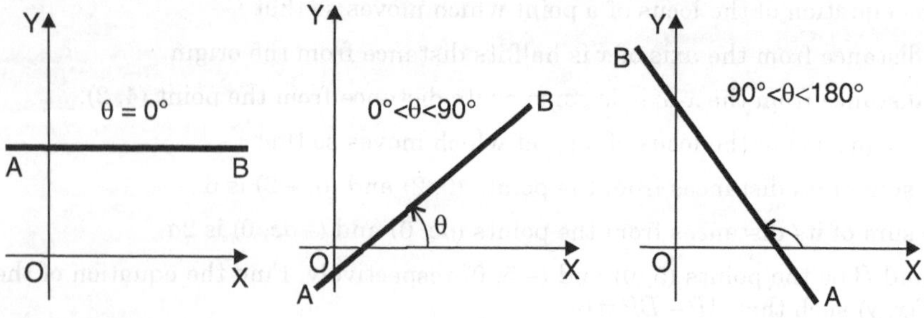

(i) (ii) (iii)

In other words, the slope of a non-vertical line is the tangent of the inclination of the line.

If $\theta = 0°$, then slope of line $= \tan 0° = 0$. [Fig. (*i*)]

If $0° < \theta < 90°$, then slope of line $= \tan \theta > 0$. [Fig. (*ii*)]

If $90° < \theta < 180°$, then slope of line $= \tan \theta < 0$. [Fig. (*iii*)]

The slope of a line is generally, denoted by the letter *m*.

$\therefore$ $\quad$ **m = tan θ.**

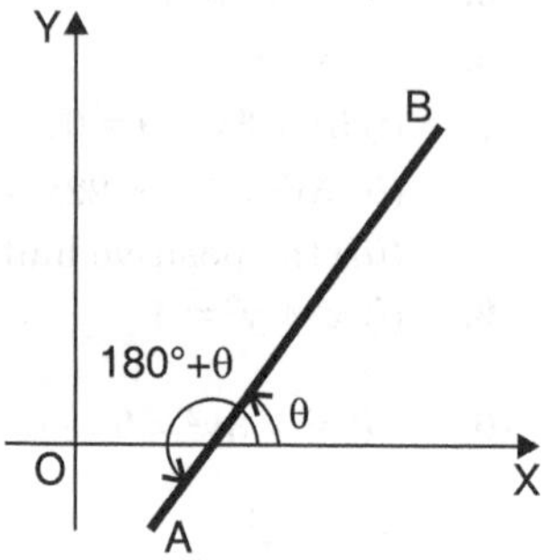

Remark 1. If a line is parallel to *y*-axis, then $\theta = 90°$ and so $\tan \theta = \tan 90°$ is not defined. That is why, we do not define the slope of a vertical line.

Remark 2. Slope of a line segment is independent of its sense, for slope of $\quad BA = \tan (180° + \theta)$

$= \tan \theta =$ slope of *AB*.

ILLUSTRATIVE EXAMPLES

Example 1. *(a) What is the slope of a line whose inclination is :*

(i) 0° $\qquad$ *(ii) 90°*

(iii) $\frac{\pi}{3}$ $\qquad$ (P.T.U. B.Pharma, Dec. 2003)

(iv) 150°.

(b) What is the inclination of a line whose slope is :

(i) 0 $\qquad$ *(ii) 1*

(iii) – 1 $\qquad$ *(iv)* $-1/\sqrt{3}$.

Sol. (*a*) (*i*) Here $\quad \theta = 0°$

$$\text{Slope} = \tan \theta = \tan 0° = \mathbf{0}.$$

(*ii*) Here $\quad \theta = 90°$

$\therefore$ The slope of line is not defined.

(*iii*) Here $\quad \theta = \frac{\pi}{3} = 60°$

$\therefore$ Slope $\quad = \tan \theta = \tan 60° = \mathbf{\sqrt{3}}$.

(*iv*) Here $\theta = 150°$

$\therefore$ Slope $= \tan\theta = \tan 150° = \tan(180° - 30°) = -\tan 30° = -\dfrac{1}{\sqrt{3}}.$

(*b*) (*i*) Here $m = 0$. Let θ be the inclination of the line.

$\therefore$ $\tan\theta = 0 \;\Rightarrow\; \theta = \mathbf{0°}.$

(*ii*) Here $m = 1$. Let θ be the inclination of the line.

$\therefore$ $\tan\theta = 1 \;\Rightarrow\; \theta = \mathbf{45°}$ [$\because$ $0° \le \theta < 180°$]

(*iii*) Here $m = -1$. Let θ be the inclination of the line.

$\therefore$ $\tan\theta = -1 \;\Rightarrow\; \theta = \mathbf{135°}.$

(*iv*) Here $m = -\dfrac{1}{\sqrt{3}}$. Let θ be the inclination of the line.

$\therefore$ $\tan\theta = -\dfrac{1}{\sqrt{3}} \;\Rightarrow\; \theta = \mathbf{150°}.$

Theorem. If a non-vertical line passes through two distinct points (x_1, y_1), (x_2, y_2), then the slope, m, of the line is given by

$$\mathbf{m = \frac{y_2 - y_1}{x_2 - y_1}.}$$

Proof. Let $P(x_1, y_1)$ and $Q(x_2, y_2)$ be the given points on the line. Let θ be the inclination of the line.

Since the line is non-vertical, we have $x_1 \ne x_2$ *i.e.*, $x_2 - x_1 \ne 0$.

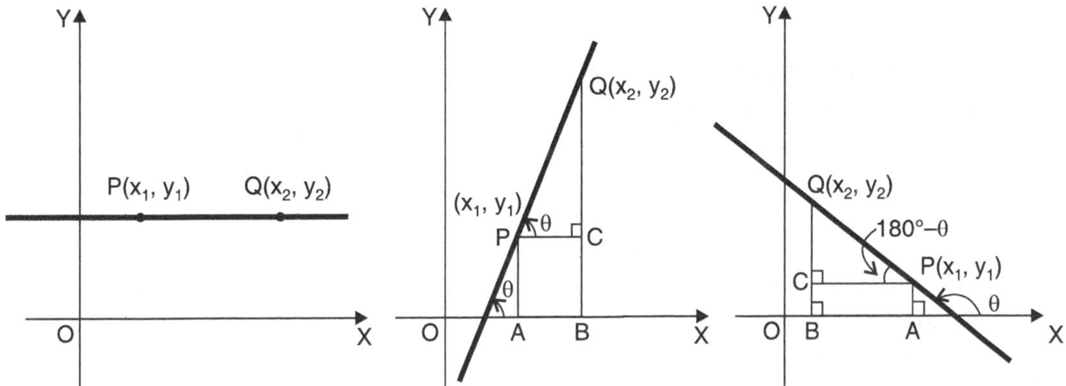

(i) (ii) (iii)

Case I. $\theta = \mathbf{0°}.$

In this case, $m = 0$ [By def.]

Also, $\dfrac{y_2 - y_1}{x_2 - x_1} = \dfrac{0}{x_2 - x_1} = 0$ [$\because$ $y_1 = y_2$]

$\therefore$ $m = \dfrac{y_2 - y_1}{x_2 - x_1}.$

Case II. $\mathbf{0° < \theta < 90°}$. In this case,

$$m = \tan\theta = \frac{QC}{PC} = \frac{QB - BC}{AB} = \frac{QB - AP}{OB - OA} = \frac{y_2 - y_1}{x_2 - x_1}.$$

Case III. **90° < θ < 180°.** In this case,

$$m = \tan\theta = -\tan(180° - \theta) = -\frac{QC}{PC} = -\frac{QB - BC}{AB} = -\frac{QB - AP}{OA - OB} = -\frac{y_2 - y_1}{x_1 - x_2} = \frac{y_2 - y_1}{x_2 - x_1}.$$

∴ **For any non-vertical line,** $\mathbf{m = \frac{y_2 - y_1}{x_2 - x_1}}$**.**

Example 2. *Find the slope of the line passing through the points :*
(i) (1, 6) and (– 4, 2) *(ii) (5, 9) and (2, 9).*

Sol. (*i*) Let $A = (1, 6)$ and $B = (-4, 2)$.

∴ Slope of $AB = \frac{2-6}{-4-1} = \frac{-4}{-5} = \mathbf{\frac{4}{5}}$. $\left(\text{Using slope} = \frac{y_2 - y_1}{x_2 - x_1}\right)$

(*ii*) Let $A = (5, 9), B = (2, 9)$.

∴ Slope of $AB = \frac{9-9}{2-5} = \frac{0}{-3} = \mathbf{0}$.

Example 3. *If the slope of the line joining (2, 5) and (3, λ) is – 2, find the value of λ.*

Sol. Let $A = (2, 5)$ and $B = (3, \lambda)$.

Slope of $AB = -2$.

∴ $-2 = \frac{\lambda - 5}{3-2}$ $\left(\because \ m = \frac{y_2 - y_1}{x_2 - x_1}\right)$

∴ $-2 = \lambda - 5$ *i.e.,* $\lambda = 5 - 2 = \mathbf{3}$.

10.13. PARALLEL AND PERPENDICULAR LINES

In this section, we shall develop relations between slopes of lines which are either parallel or perpendicular to each other.

Theorem I. Two non-vertical lines are parallel if and only if their slopes are equal.

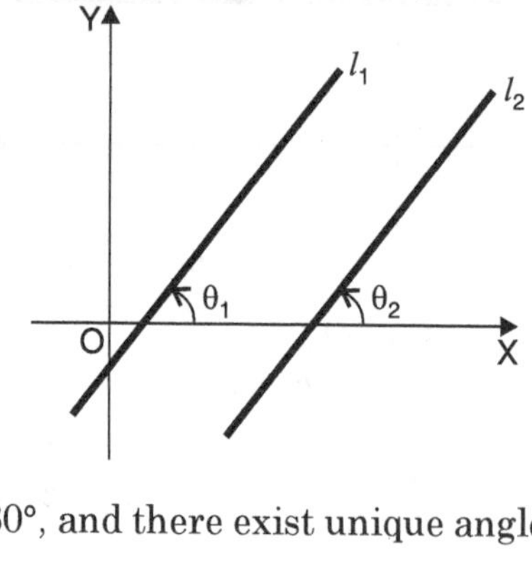

Proof. Let l_1, l_2 be two non-vertical lines with respective inclinations θ_1, θ_2.

Necessity. Let l_1 and l_2 be parallel.

∴ Their inclinations are equal.

∴ $\theta_1 = \theta_2$

∴ $\tan\theta_1 = \tan\theta_2$

∴ Slopes are equal.

Sufficiency. Let the slopes of l_1 and l_2 be equal.

∴ $\tan\theta_1 = \tan\theta_2$

⇒ $\theta_1 = \theta_2$, because θ_1 and θ_2 lies between 0° and 180°, and there exist unique angle θ (0° ≤ θ < 180°) for a given value of tan θ.

∴ The inclinations of the lines are equal.

Theorem II. Two non-vertical lines are perpendicular if and only if the product of their slopes is minus one.

Proof. Let l_1, l_2 be two non-vertical lines with respective inclinations θ_1, θ_2.

Necessity. Let l_1 and l_2 be perpendicular.

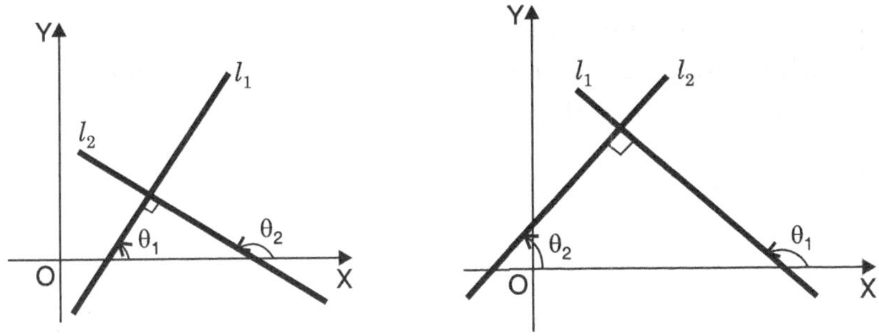

(i) (ii)

$\therefore$ Either $\theta_2 = \theta_1 + 90°$ [Fig. (*i*)] ...(1) or $\theta_1 = \theta_2 + 90°$ [Fig. (*ii*)] ...(2)

(1) $\Rightarrow$ $\tan\theta_2 = \tan(\theta_1 + 90°) = -\cot\theta_1 = -\dfrac{1}{\tan\theta_1}$. $\therefore$ $\tan\theta_1 \tan\theta_2 = -1$.

(2) $\Rightarrow$ $\tan\theta_1 = \tan(\theta_2 + 90°) = -\cot\theta_2 = -\dfrac{1}{\tan\theta_2}$. $\therefore$ $\tan\theta_1 \tan\theta_2 = -1$.

$\therefore$ In either case, product of slopes is -1.

Sufficiency. Let $\tan\theta_1 \tan\theta_2 = -1$.

$\therefore$ $\tan\theta_1 = \dfrac{-1^*}{\tan\theta_2} = -\cot\theta_2 = \tan[-(90° - \theta_2)]$ or $\tan(90° + \theta_2)$

$\Rightarrow$ $\theta_1 = \theta_2 - 90°$ or $\theta_1 = 90° + \theta_2$ $\Rightarrow$ $\theta_2 = \theta_1 + 90°$ or $\theta_1 = 90° + \theta_2$

$\therefore$ The lines are perpendicular.

Remark. Let the slopes of two non-vertical lines be m_1 and m_2, then

(*i*) lines are || iff $m_1 = m_2$ (*ii*) lines are $\perp$ iff $m_1 m_2 = -1$.

WORKING RULES FOR SOLVING PROBLEMS

Rule I. *If m be the slope of a non-vertical line passing through the distinct points* (x_1, y_1) *and* (x_2, y_2) *then* $m = \dfrac{y_2 - y_1}{x_2 - x_1}$.

Rule II. *Two lines are parallel if and only if their slopes are equal.*

Rule III. *Two lines are perpendicular if and only if the product of their slopes is minus one.*

Example 4. *Show that the points (1, 1), (2, 3) and (3, 5) are collinear.*

Sol. Let (1, 1), (2, 3) and (3, 5) be the coordinates of the points *A*, *B* and *C* respectively.

$$\text{Slope of } AB = \frac{3-1}{2-1} = 2 \quad \text{and} \quad \text{Slope of } BC = \frac{5-3}{3-2} = 2$$

$\therefore$ Slope of *AB* = Slope of *BC*

$\therefore$ *AB* and *BC* are parallel. $\therefore$ *A*, *B*, *C* are collinear because *B* is on both lines *AB* and *BC*.

***Why this step.** $\tan\theta_1 \tan\theta_2 = -1$ implies that $\tan\theta_2$ is non-zero.

$\therefore$ Division by $\tan\theta_2$ is justified.

Example 5. *Show that the line joining (2, – 3) and (– 5, 1) is :*

(i) parallel to the line joining (7, – 1) and (0, 3)

(ii) perpendicular to the line joining (4, 5) and (0, – 2).

Also verify this by drawing the line segments on a coordinate plane.

Sol. Let l_1 be the line joining the points (2, – 3) and (– 5, 1).

$\therefore$ Slope of $l_1 = \dfrac{1-(-3)}{-5-2} = -\dfrac{4}{7}$.

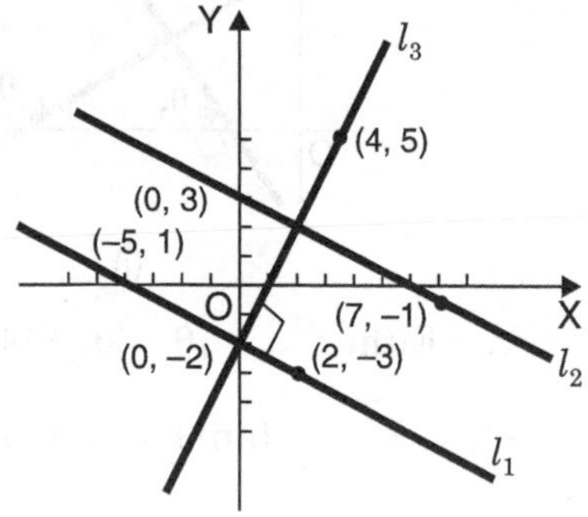

(i) Let l_2 be the line joining the points (7, – 1) and (0, 3).

$\therefore$ Slope of $l_2 = \dfrac{3-(-1)}{0-7} = -\dfrac{4}{7}$.

$\therefore$ Slope of l_1 = slope of l_2. (each = – 4/7)

$\therefore$ The lines l_1 and l_2 are parallel.

(ii) Let l_3 be the line joining the points (4, 5) and (0, – 2).

$\therefore$ Slope of $l_3 = \dfrac{-2-5}{0-4} = \dfrac{-7}{-4} = \dfrac{7}{4}$

$\therefore$ Slope of $l_1 \times$ Slope of $l_3 = -\dfrac{4}{7} \times \dfrac{7}{4} = -1$. $\therefore$ The lines l_1 and l_3 are perpendicular.

EXERCISE 10.6

1. What can be said regarding inclination of a line if its slope is :

   (i) positive (ii) zero

   (iii) negative (iv) not defined ?

2. Find the slope of the line passing through the points :

   (i) (6, 3) and (9, 3) (ii) (0, 9) and (– 3, 0).

3. Find the value of x, if the slope of the line joining (1, 5) and (x, – 7) is 4.

4. Show that the line joining (5, 6) and (2, 3) is parallel to the line through the points (9, – 2) and (6, – 5).

5. Show that the line through the points (0, 0) and (4, – 3) is perpendicular to the line through the points (– 4, – 2) and (– 1, 2). Also verify the result graphically.

6. State whether the lines in each part are parallel or perpendicular or neither :

   (i) through (1, – 12) and (4, 6) ; through (10, 5) and (16, 4).

   (ii) through (– 3, – 4) and (– 1, 0) ; through (6, – 3) and (7, – 1).

   (iii) through (0, 0) and (6, 7) ; through (0, 0) and (7, 6).

   (iv) through (2, – 5) and (– 2, 5) ; through (6, 3) and (1, 1).

   Also verify your result graphically in each case.

7. Show that the points A(2, 3), B(1, 4) and C(– 3, 8) are collinear.

8. Without using Pythagorean theorem, show that (4, 4), (3, 5) and (– 1, – 1) are the vertices of a right-angled triangle. Also verify the result graphically.

9. If the line through $(-4, y)$ and $(1, 3)$ is parallel to the line through $(-2, 5)$ and $(0, 7)$, find the value of y. Also verify your result graphically.

10. $A(5, -3)$, $B(8, 2)$, $C(0, 0)$ are the vertices of a triangle. Show that the median from A is perpendicular to the side BC.

11. A quadrilateral has vertices $(4, 1)$, $(1, 7)$, $(-6, 0)$ and $(-1, -9)$. Show that the mid-points of the sides of this quadrilateral form a parallelogram. *(G.J.U. B.Pharma, 2003)*

12. $A(3, 4)$, $B(-3, 0)$ and $C(7, -4)$ are the vertices of a triangle. Show that the line joining the mid-points of AB and AC is parallel to BC and is half of it.

13. The points $(a, 0)$, $(0, b)$ and (x, y) are collinear. Using slopes, prove that $\frac{x}{a} + \frac{y}{b} = 1$.

Answers

1. (*i*) Inclination is acute (*ii*) Either coincident or parallel to x-axis (*iii*) Inclination is obtuse. (*iv*) Parallel to y-axis.
2. (*i*) 0 (*ii*) 3
3. -2
6. (*i*) perpendicular (*ii*) parallel (*iii*) neither (*iv*) perpendicular
9. -2.

Hint

8. Show that the product of slopes of two of its sides is '-1'.

STRAIGHT LINE AND FAMILY OF STRAIGHT LINES

10.14. INTRODUCTION

In the present chapter, we shall learn the methods of finding the equations of various types of straight lines. The concept of 'slope' would be used quite often in this chapter. We shall conclude this chapter with the study of equation of family of straight lines passing through the intersection of two straight lines.

10.15. DESCRIPTION OF A LINE BY AN EQUATION

An equation is called the **equation of a straight line** if the coordinates of every point on the straight lines satisfies the equation of the straight line and every point whose coordinates satisfies the equation of the straight line is on the straight line.

We shall see that every first degree equation like $ax + by + c = 0$ would be the equation of a certain straight line and conversely, the equation of any straight line would always be of the type $ax + by + c = 0$.

Remark. A straight line is briefly written as a 'line'.

10.16. EQUATION OF A LINE PARALLEL TO X-AXIS

To find the equation of the straight line parallel to x-axis and at a given distance from it.

Let l be a straight line parallel to x-axis and at a distance 'h' from it. Therefore, the line lies above the x-axis if $h > 0$ and if $h < 0$, then the line would lie below the x-axis.

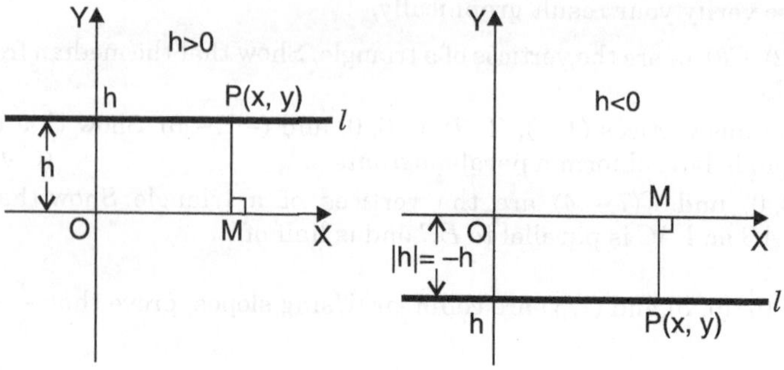

Let $P(x, y)$ be a general point on the line l.

$\therefore \qquad y = h$

For any point on the line, the ordinate of the point is 'h' and conversely every point with ordinate h would be on the line.

$\therefore$ **The equation of the line is y = h.**

Remark. In particular, the equation of x-axis is $y = 0$.

10.17. EQUATION OF A LINE PARALLEL TO Y-AXIS

To find the equation of the straight line parallel to y-axis and at a given distance from it.

Let l be a straight line parallel to y-axis and at a distance 'k' from it. Therefore, the line lies on the right of y-axis if $k > 0$ and if $k < 0$, then the line would lie on the left of y-axis.

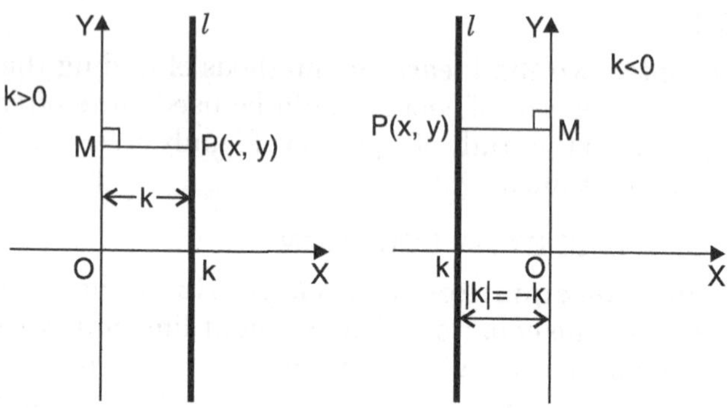

Let $P(x, y)$ be a general point on the line.

$\therefore \qquad \mathbf{x = k.}$

This is the required equation of the line.

Remark. In particular, the equation of y-axis is $x = 0$.

ILLUSTRATIVE EXAMPLES

Example 1. *Find the equation of the straight line which passes through (2, – 3) and is :*

(i) parallel to x-axis. *(ii) parallel to y-axis.*

Sol. (*i*) Let the equation of the required line be $y = k$.

Since, the point (2, – 3) is on the line, so $-3 = k$.

$\therefore$ The required equation is $\mathbf{y = -3}$.

(*ii*) Let the equation of the required line be $x = k$.

Since, the point (2, – 3) is on the line, so $2 = k$.

$\therefore$ The required equation is $\mathbf{x = 2}$.

10.18. POINT-SLOPE FORM

If the slope of a line is given, then the line can take infinitely many positions on a coordinate plane. All these lines will be parallel to each other [(Fig. (*i*)].

Similarly, a line is not fixed just by identifying only one point on a line. There are infinitely many lines passing through a given point, say, $A(x_1, y_1)$ [Fig. (*ii*)]. But there will be one and only one line passing through a given point and having given slope.

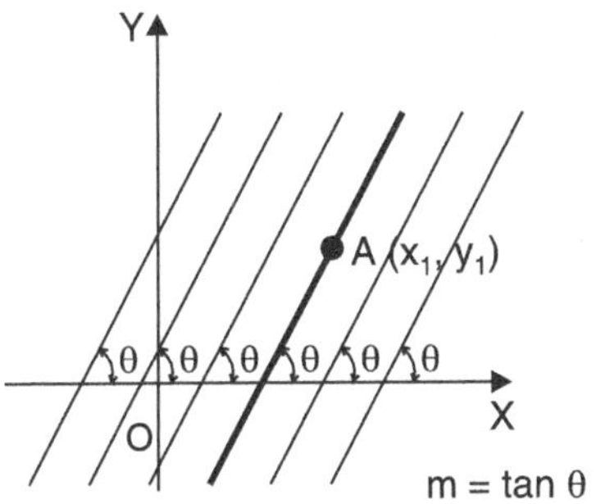

Fig. (*i*)

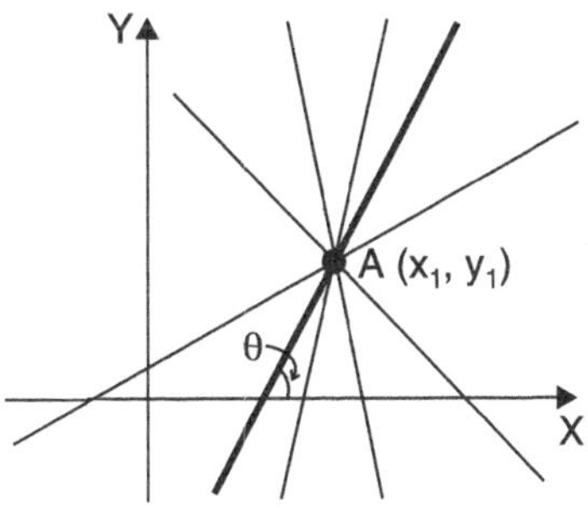

Fig. (*ii*)

Now we shall find the equation of line passing through a given point and having given slope.

To find the equation of the straight line having given one point on the line and its slope.

Let a non-vertical line passes through point $A(x_1, y_1)$ and having slope m.

Let $P(x, y)$ be a general point on the line. Since the line passes through $A(x_1, y_1)$ and $P(x, y)$, so the slope of the line is equal to $\frac{y - y_1}{x - x_1}$, ($x - x_1 \neq 0$, because the point A and P are distinct and the line is non-vertical).

Also, the slope of the line is given to be m.

$$\therefore \qquad \frac{y - y_1}{x - x_1} = m$$

$$\Rightarrow \qquad \mathbf{y - y_1 = m(x - x_1)}.$$

This is the required equation of the line. This represents the equation of a line in **point-slope** form.

Remark. If a vertical line passes through the line point (x_1, y_1), then its equation is given by $x = x_1$.

Example 2. *Find the equation of the straight line passing through (1, 7) and having slope 4.*

Sol. Given point = (1, 7), slope = 4.

$\therefore$ Using $y - y_1 = m(x - x_1)$, the equation of the line is $y - 7 = 4(x - 1)$

or $\qquad y - 7 = 4x - 4 \quad$ or $\quad \mathbf{4x - y + 3 = 0}$.

Example 3. *Find the equation of the line passing through (– 4, – 5) and perpendicular to the line joining (1, 2) and (5, 6).*

Sol. Let the given points be $A(-4, -5)$, $B(1, 2)$ and $C(5, 6)$.

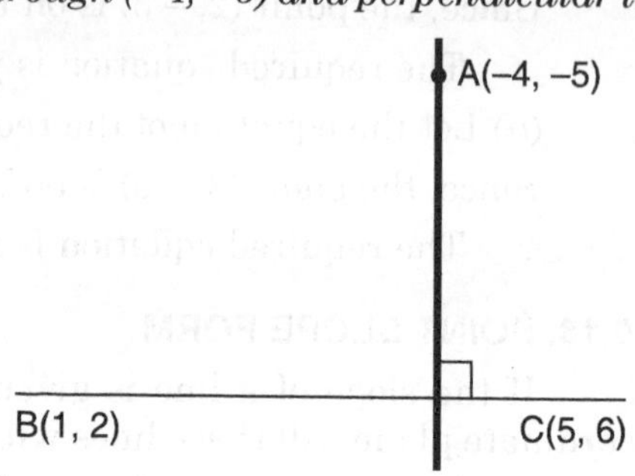

$$\text{Slope of } BC = \frac{6-2}{5-1} = \frac{4}{4} = 1$$

Let m be the slope of the required line. Since the required line is perpendicular to BC, so $m \times 1 = -1$.

$\therefore$ $m = -1$

The required line is also to pass through $A(-4, -5)$.

$\therefore$ The equation of the required line is

$$y - (-5) = -1.(x - (-4)) \qquad [\text{Using } y - y_1 = m(x - x_1)]$$

or $y + 5 = -x - 4$ or $\mathbf{x + y + 9 = 0}$.

Example 4. *Find the equation of the line joining the points (3, – 1) and (2, 3). Also find the equation of the other line which is perpendicular to this line and passing through the point (5, 2).*

Sol. Let the given points be $A(3, -1)$, $B(2, 3)$ and $C(5, 2)$.

Equation of AB

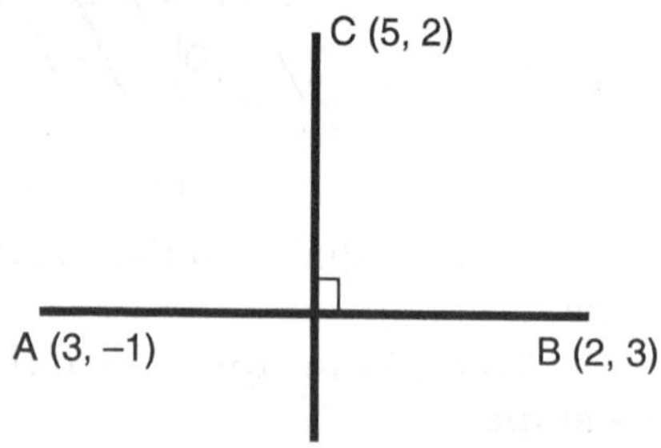

$$\text{Slope of } AB = \frac{3-(-1)}{2-3} = \frac{4}{-1} = -4$$

The equation of AB is $y - y_1 = (\text{Slope of } AB)(x - x_1)$.

$\Rightarrow$ $y - (-1) = -4(x - 3)$ $[\text{Taking } (x_1, y_1) = (3, -1)]$

$\Rightarrow$ $y + 1 = -4x + 12 \Rightarrow \mathbf{4x + y - 11 = 0}$.

Equation of the line ⊥ to AB

Let slope $= m$

$\therefore$ $m \times (-4) = -1$ *i.e.*, $m = 1/4$

The equation of required line is $y - y_1 = m(x - x_1)$.

$$\Rightarrow \quad y - 2 = \frac{1}{4}(x - 5) \qquad [\text{Taking } (x_1, y_1) = (5, 2)]$$

$\Rightarrow$ $4y - 8 = x - 5 \Rightarrow \mathbf{x - 4y + 3 = 0}$.

Example 5. *Show that the perpendicular drawn from the point (4, 1) on the line joining (6, 5) and (2, – 1) divides it in the ratio 8 : 5 internally.*

Sol. Let the given points be $A(4, 1)$, $B(6, 5)$ and $C(2, -1)$. Draw $AM \perp BC$.

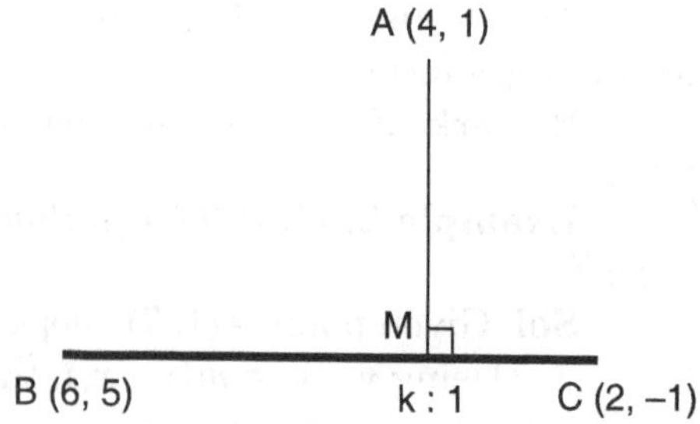

$$\text{Slope of } BC = \frac{-1-5}{2-6} = \frac{-6}{-4} = \frac{3}{2}$$

$$\therefore \text{ Slope of } AM = -\frac{1}{3/2} = -\frac{2}{3}$$

Using $y - y_1 = m(x - x_1)$, the equation of AM is

$$y - 1 = -\frac{2}{3}(x - 4).$$

$\Rightarrow \quad 3y - 3 = -2x + 8 \quad \Rightarrow \quad 2x + 3y - 11 = 0.$

Let M divides BC in the ratio $k : 1$ internally.

$\therefore$ Coordinates of M are $\left(\dfrac{2k+6}{k+1}, \dfrac{-k+5}{k+1}\right)$. M is on AM.

$$\therefore \quad 2\left(\frac{2k+6}{k+1}\right) + 3\left(\frac{-k+5}{k+1}\right) - 11 = 0$$

$$\Rightarrow \quad 4k + 12 - 3k + 15 - 11k - 11 = 0 \Rightarrow -10k + 16 = 0 \Rightarrow k = 8/5$$

$\therefore$ M divides BC in the ratio $8/5 : 1$ internally *i.e.*, **8 : 5 internally.**

10.19. TWO-POINT FORM

To find the equation of the straight line having given two distinct points on the line.

Let a non-vertical line passes through two distinct points $A(x_1, y_1)$ and $B(x_2, y_2)$.

Let $P(x, y)$ be a general point on the line.

Since the line passes through A, B and P, we have

Slope of AB = slope of AP.

$$\Rightarrow \quad \frac{y_2 - y_1}{x_2 - x_1} = \frac{y - y_1}{x - x_1}$$

$$\Rightarrow \quad \mathbf{y - y_1 = \frac{y_2 - y_1}{x_2 - x_1}(x - x_1).}$$

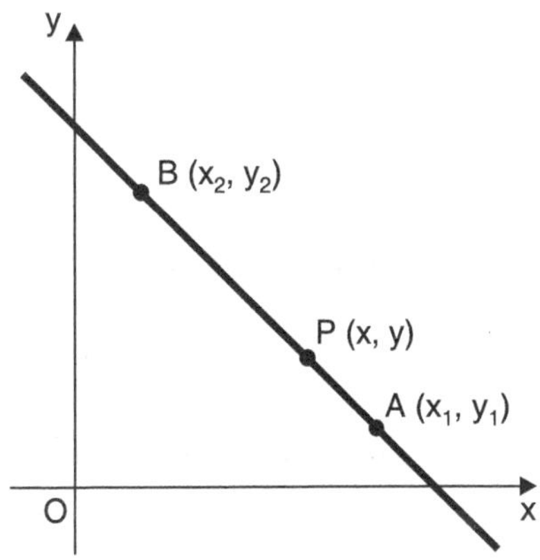

This is the required equation of the line.

Remark. The equation of the line passing through the points (x_1, y_1) and (x_2, y_2) can also be expressed as

$$y - y_2 = \frac{y_2 - y_1}{x_2 - x_1}(x - x_2).$$

The equations $y - y_1 = \dfrac{y_2 - y_1}{x_2 - x_1}(x - x_1)$ and $y - y_2 = \dfrac{y_2 - y_1}{x_2 - x_1}(x - x_2)$ appears to be distinct, but these equations become identical, when expressed in the form $ax + by + c = 0$.

Example 6. *Find the equation of the straight line passing through the points (4, 2) and (– 2, 8).* (P.T.U. B.Pharma, Dec. 2006)

Sol. Let $\quad A = (4, 2) \quad$ and $\quad B = (-2, 8)$.

The equation of the line passing through (x_1, y_1) and (x_2, y_2) is $y - y_1 = \dfrac{y_2 - y_1}{x_2 - x_1}(x - x_1)$.

$\therefore$ The equation of the required line is $y - 2 = \dfrac{8-2}{-2-4}(x - 4)$

(Here $x_1 = 4, y_1 = 2, x_2 = -2, y_2 = 8$)

or $\quad y - 2 = -1 \,.\, (x - 4) \quad$ or $\quad \mathbf{x + y - 6 = 0.}$

Remark. Using $y - y_2 = \dfrac{y_2 - y_1}{x_2 - x_1}(x - x_2)$, we get $\quad y - 8 = \dfrac{8-2}{-2-4}(x - (-2))$.

$\Rightarrow \quad y - 8 = -1 \,.\, (x + 2) \quad \Rightarrow \quad x + y - 6 = 0$. This is the same equation as obtained above.

Example 7. *Find the equations of the sides of the triangle whose vertices are (– 1, 8), (4, – 2) and (– 5, – 3). Also find the equation of the median through (– 1, 8).*

Sol. Let the vertices of the triangle be $A(-1, 8)$, $B(4, -2)$ and $C(-5, -3)$.

Equation of side AB.

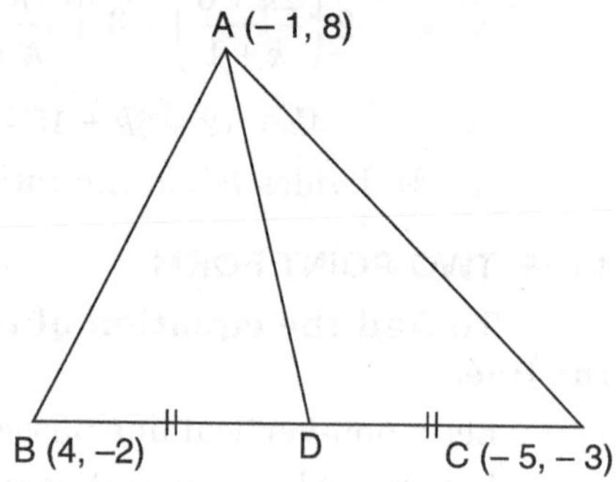

The required equation is $y - 8 = \dfrac{-2-8}{4-(-1)}(x-(-1))$.

$$\left(\text{Using } y - y_1 = \frac{y_2 - y_1}{x_2 - x_1}(x - x_1)\right)$$

$$\Rightarrow \quad y - 8 = \frac{-10}{5}(x+1) \Rightarrow y - 8 = -2x - 2$$

$\Rightarrow$ $\mathbf{2x + y - 6 = 0}$.

Equation of side BC. The required equation is

$$y - (-2) = \frac{-3-(-2)}{-5-4} = (x - 4).$$

$$\Rightarrow \quad y + 2 = \frac{-1}{-9}(x-4) \Rightarrow 9y + 18 = x - 4 \Rightarrow \mathbf{x - 9y - 22 = 0}.$$

Equation of side CA. The required equation is $y - (-3) = \dfrac{8-(-3)}{-1-(-5)} = (x - (-5))$.

$$\Rightarrow \quad y + 3 = \frac{11}{4}(x+5) \Rightarrow 4y + 12 = 11x + 55 \Rightarrow \mathbf{11x - 4y + 43 = 0}.$$

Equation of median through A. Let D be the mid-point of BC.

$$\therefore \quad D = \left(\frac{4+(-5)}{2}, \frac{-2+(-3)}{2}\right) = \left(-\frac{1}{2}, -\frac{5}{2}\right)$$

The equation of AD *i.e.,* the median through A is $y - 8 = \dfrac{-\frac{5}{2} - 8}{-\frac{1}{2} - (-1)}(x - (-1))$.

$$\Rightarrow \quad y - 8 = \frac{-21}{1}(x+1) \Rightarrow y - 8 = -21x - 21 \Rightarrow \mathbf{21x + y + 13 = 0}.$$

Example 8. *Show that the points $(at_1^2, 2at_1)$ $(at_2^2, 2at_2)$ and $(a, 0)$ are collinear if $t_1t_2 = -1$.*

Sol. Given points are $(at_1^2, 2at_1)$, $(at_2^2, 2at_2)$ and $(a, 0)$.

The equation of the straight line passing through $(at_1^2, 2at_1)$ and $(at_2^2, 2at_2)$ is

$$y - 2at_1 = \frac{2at_2 - 2at_1}{at_2^2 - at_1^2}(x - at_1^2) \quad \text{or} \quad y - 2at_1 = \frac{2a(t_2 - t_1)}{a(t_2^2 - t_1^2)}(x - at_1^2)$$

or $$y - 2at_1 = \frac{2}{t_2 + t_1}(x - at_1^2) \quad \text{or} \quad (t_1 + t_2)(y - 2at_1) = 2(x - at_1^2). \qquad ...(1)$$

Given three points are collinear if $(a, 0)$ lies on (1)

or if $(t_1 + t_2)(0 - 2at_1) = 2(a - at_1^2)$ or if $-2at_1^2 - 2at_1t_2 = 2a - 2at_1^2$

or if $-2at_1t_2 = 2a$ or if $t_1t_2 = -1$.

$\therefore$ Given points are collinear if $\mathbf{t_1t_2 = -1}$.

Example 9. *Find the ratio in which the line joining (2, 3) and (4, 1) divides the line joining (1, 2) and (4, 3).*

Sol. Let the given points be $A(1, 2)$, $B(4, 3)$, $C(2, 3)$ and $D(4, 1)$.

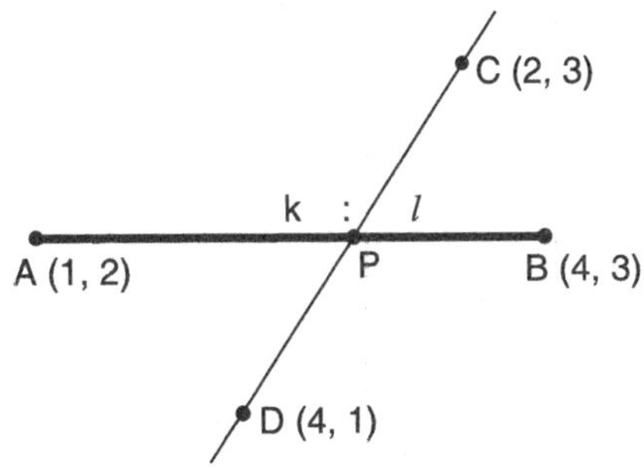

Let CD divides AB at P in the ratio $k : 1$ internally.

$$\therefore \qquad P = \left(\frac{4k+1}{k+1}, \frac{3k+2}{k+1}\right).$$

The equation of CD is $y - 3 = \dfrac{1-3}{4-2}(x-2)$

or $$y - 3 = \frac{-2}{2}(x-2)$$

or $x + y - 5 = 0$. Now P lies on the line CD.

$$\Rightarrow \left(\frac{4k+1}{k+1}\right) + \left(\frac{3k+2}{k+1}\right) - 5 = 0 \Rightarrow 4k + 1 + 3k + 2 - 5k - 5 = 0$$

$\Rightarrow$ $2k - 2 = 0 \Rightarrow k = 1$. $\therefore$ The required ratio is **1 : 1 internal.**

WORKING RULES FOR SOLVING PROBLEMS

Rule I. (*i*) *The equation of the line parallel to x-axis and at a distance 'h' from it is* $\mathbf{y = h}$.

(*ii*) *The equation of the line parallel to y-axis and at a distance 'k' from it is* $\mathbf{x = k}$.

Rule II. *The equation of the line passing through* (x_1, y_1) *and having slope 'm' is* $\mathbf{y - y_1 = m(x - x_1)}$.

Rule III. *The equation of the non-vertical line passing through* (x_1, y_1) *and* (x_2, y_2) *is*

$$\mathbf{y - y_1 = \frac{y_2 - y_1}{x_2 - x_1}(x - x_1)}.$$

EXERCISE 10.7

1. Write the equation of the straight line which is parallel to :

(*i*) x-axis and 4 units above it *(P.T.U. B.Pharma, May 2006)*

(*ii*) x-axis and 3 units below it (*iii*) y-axis and 3 units on the right of it

(*iv*) y-axis and 6 units on the left of it. *(P.T.U. B.Pharma, Dec. 2006)*

2. Find the equation of the straight line passing through $(3, -4)$ and parallel to :

(*i*) x-axis (*ii*) y-axis.

3. Find the equation of the straight line passing through $(2, -7)$ and perpendicular to :

(*i*) x-axis (*ii*) y-axis.

4. Find the equation of the line passing through the mid-point of the line joining (1, 6) and (5, 6) and is parallel to :
   (*i*) x-axis (*ii*) y-axis.
5. Find the equation of a line through the origin which makes an angle of 45° with the positive direction of x-axis.
6. Find the equation of the straight line :
   (*i*) passing through $(\sqrt{2}, 2\sqrt{2})$ and slope 2/3
   (*ii*) passing through (– 1, – 2) and slope 4/7.
7. Find the equation of the straight line passing through (3, – 1) and parallel to the line joining (3, – 2) and (– 6, 5). *(P.T.U. B.Pharma, May 2007)*
8. Find the equation of the right bisector of the line segment joining :
   (*i*) (1, 0) and (2, 3) (*ii*) (1, 1) and (3, 5).
9. The vertices of a triangle are $A(10, 4)$, $B(-4, 9)$ and $C(-2, -1)$. Find the equation of the altitude through A.
10. Find the equation of the straight line passing through the points :
    (*i*) (3, 0) and (0, 4) (*ii*) (1, 2) and (0, 5) (*iii*) $(at_1^2, 2at_1)$ and $(at_2^2, 2at_2)$
    (*iv*) (a, b), $(a + b, a - b)$. *(P.T.U. B.Pharma, May 2006)*
11. Show that the points (1, 4), (3, – 2) and (– 3, 16) are collinear and find the equation of the line passing through these points.
12. Find the equation of the straight line which bisects the join of (5, 4) and (– 7, 0) and also bisects the join of (6, – 5) and (0, – 3).
13. Find the equations of the sides of the triangle whose vertices are (2, 1), (– 2, 3), (4, 5).
14. The mid-points of the sides of a triangle are (2, 2), (2, 3), (4, 6). Find the equations of the sides.
15. $A(-3, -1)$ and $B(2, 2)$ are two adjacent vertices and $M(3, 0)$ is the point of intersection of the diagonals of a parallelogram $ABCD$. Find the equations of the sides of the parallelogram.
16. Find the equations the medians of the triangle whose vertices are (– 1, 6), (– 3, – 9) and (5, – 8).
17. Find the equations of the sides and diagonals of the quadrilateral whose vertices are (– 2, 6), (1, 2), (10, 4) and (7, 8).

Answers

1. (*i*) $y = 4$ (*ii*) $y = -3$ (*iii*) $x = 3$ (*iv*) $x = -6$
2. (*i*) $y = -4$ (*ii*) $x = 3$ **3.** (*i*) $x = 2$ (*ii*) $y = -7$
4. (*i*) $y = 6$ (*ii*) $x = 3$ **5.** $x - y = 0$
6. (*i*) $2x - 3y + 4\sqrt{2} = 0$ (*ii*) $4x - 7y - 10 = 0$ **7.** $7x + 9y - 30 = 0$ **8.** (*i*) $x + 3y - 6 = 0$
   (*ii*) $x + 2y - 8 = 0$ **9.** $x - 5y + 10 = 0$ **10.** (*i*) $4x + 3y - 12 = 0$ (*ii*) $3x + y - 5 = 0$
   (*iii*) $y(t_1 + t_2) = 2x + 2at_1t_2$ (*iv*) $(2b - a)\,x + by = b^2 - a^2 + 2ab$
11. $3x + y - 7 = 0$ **12.** $3x + 2y - 1 = 0$
13. $x + 2y - 4 = 0$, $x - 3y + 11 = 0$, $2x - y - 3 = 0$ **14.** $3x - 2y - 2 = 0$, $2x - y - 1 = 0$, $x = 4$
15. $3x - 5y + 4 = 0$, $x + 7y - 16 = 0$, $3x - 5y - 22 = 0$, $x + 7y + 10 = 0$
16. $29x + 4y + 5 = 0$, $8x - 5y - 21 = 0$, $13x + 14y + 47 = 0$
17. $4x + 3y - 10 = 0$, $2x - 9y + 16 = 0$, $4x + 3y - 52 = 0$, $2x - 9y + 58 = 0$, $x + 6y - 34 = 0$, $x - y + 1 = 0$.

Hint

12. The required line passes through the mid-points of both line segments.

10.20. INTERCEPTS ON AXES

If a line intersect the x-axis at A, then OA (with due regard to sign) is called the **intercept** of the line on x-axis or **x-intercept** of the line.

Similarly, if a line intersect the y-axis at B, then OB (with due regard to sign) is called the **intercept** of the line on y-axis or **y-intercept** of the line.

If a line intersect the axes at A and B, then AB is called the **portion of the line intercepted between the axes.**

A non-vertical line is completely determined if its slope and intercept on the y-axis are given.

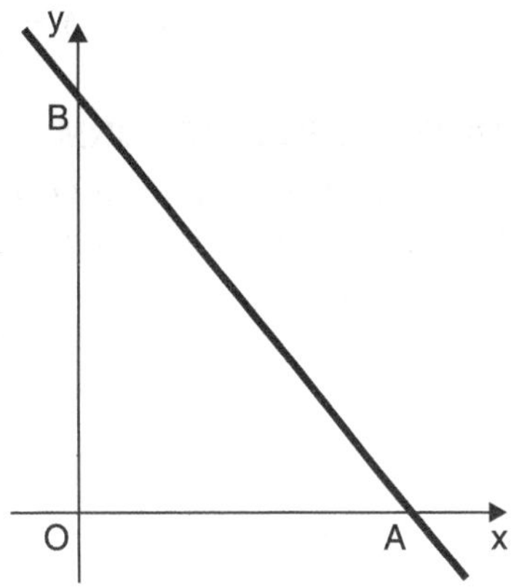

10.21. SLOPE-INTERCEPT FORM

To find the equation of the straight line having given its slope and its intercept on the y-axis. *(P.T.U. B.Pharma, May 2007)*

Let a non-vertical line has slope m and intercept on y-axis equal to c.

Let $P(x, y)$ be a general point on the line. Let the line intercept the y-axis at A.

$\therefore$ The coordinates of A are $(0, c)$.

$\therefore$ Slope of line $= \dfrac{y - c}{x - 0}$ (Assuming, P is not at A.)

Also, the slope of the line is given to be m.

$$\therefore \qquad \frac{y - c}{x} = m$$

$$\Rightarrow \qquad \mathbf{y = mx + c}$$

This is required equation of the line.

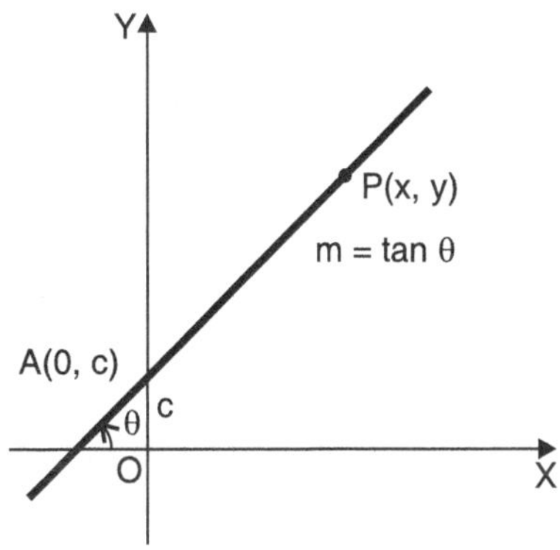

Corollary. If a line passes through origin and has slope m, then its equation is

$$y = mx + 0 \text{ i.e., } \mathbf{y = mx}.$$

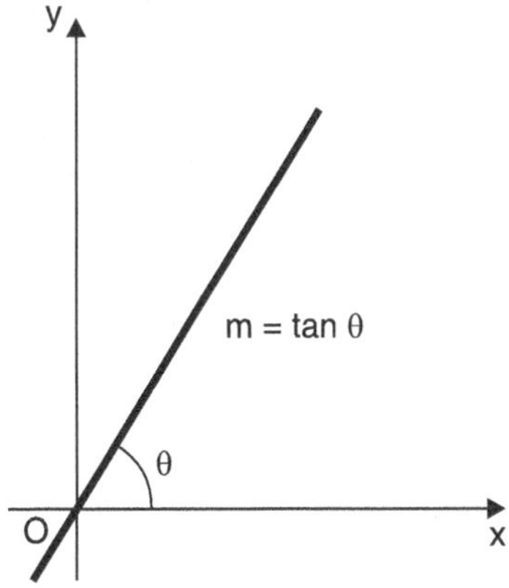

Example *Find the equation of the straight line which makes angle 30° with positive direction of x-axis and cuts intercepts + 5 on the y-axis.*

Sol. We have $m = \tan 30° = 1/\sqrt{3}$ and $c = 5$.

$\therefore$ The equation of the line is $y = mx + c$.

$$\Rightarrow \qquad y = \frac{1}{\sqrt{3}} x + 5 \Rightarrow \mathbf{x - \sqrt{3}y + 5\sqrt{3} = 0.}$$

10.22. INTERCEPT FORM

To find the equation of the straight line having given the intercepts which the line makes on the axes.

Let a line makes intercepts a and b on x-axis and y-axis respectively, where $a \neq 0$.

The line is non-vertical, because b is finite. Let $P(x, y)$ be a general point on the line.

Let the line intersects x-axis and y-axis at A and B respectively.

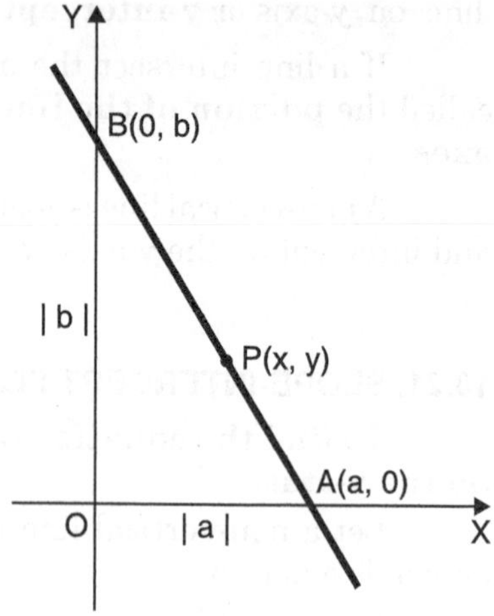

$\therefore \qquad A = (a, 0)$ and $B = (0, b)$.

Since the line passes through A, B and P, we have

Slope of AB = slope of AP

$$\Rightarrow \qquad \frac{b-0}{0-a} = \frac{y-0}{x-a}$$

$$\Rightarrow \qquad bx - ab = -ay \quad \Rightarrow \quad bx + ay = ab$$

$$\Rightarrow \qquad \frac{bx}{ab} + \frac{ay}{ab} = \frac{ab}{ab} \quad \Rightarrow \quad \mathbf{\frac{x}{a} + \frac{y}{b} = 1.}$$

This is the required equation of the line.

ILLUSTRATIVE EXAMPLES

Example 1. *Find the equation of the line which makes intercepts – 4 and 5 on the axes.*

Sol. Here $\qquad a = -4$ and $b = 5$.

The equation of the line in the **intercept form** is $\frac{x}{a} + \frac{y}{b} = 1$.

$$\Rightarrow \qquad \frac{x}{-4} + \frac{y}{5} = 1 \Rightarrow -5x + 4y = 20 \Rightarrow \mathbf{5x - 4y + 20 = 0.}$$

Example 2. *A straight line passes through the point (2, 3) and the portion of the line intercepted between the axes is bisected at this point. Find the equation of the line.*

(P.T.U. B.Pharma, Dec. 2002, May 2002)

Sol. Let the intercepts of the line be a and b. Let the line intersects the x-axis and y-axis at A and B.

$\therefore$ $A(a, 0)$ and $B(0, b)$ are on the line and the mid-point of AB is (2, 3).

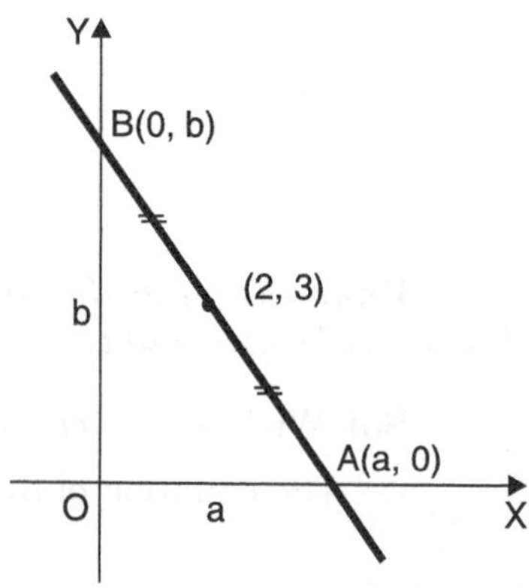

$$\therefore \quad 2 = \frac{a+0}{2} \Rightarrow a = 4 \quad \text{and} \quad 3 = \frac{0+b}{2} \Rightarrow b = 6.$$

$\therefore$ The equation of the line in the **intercept form** is

$$\frac{x}{4} + \frac{y}{6} = 1 \qquad \left(\text{using } \frac{x}{a} + \frac{y}{b} = 1\right)$$

$$\Rightarrow \qquad \mathbf{3x + 2y - 12 = 0.}$$

Example 3. *A line passes through (3, 4) and the sum of its intercepts on the axes is 14. Find the equation of the line.* (P.T.U. B.Pharma, May 2004)

Sol. Let x-intercept of the line $= a$. $\therefore$ y-intercept of the line $= 14 - a$

$\therefore$ The equation of the line in the **intercept form** is $\dfrac{x}{a} + \dfrac{y}{14-a} = 1$ ...(1)

This line passes through (3, 4). $\therefore$ $\dfrac{3}{a} + \dfrac{4}{14-a} = 1$

$\Rightarrow$ $42 - 3a + 4a = 14a - a^2 \Rightarrow a^2 - 13a + 42 = 0 \Rightarrow a = 7, 6.$

Case I. a = 7. In the case, (1) becomes $\dfrac{x}{7} + \dfrac{y}{14-7} = 1$ or $x + y - 7 = 0$.

Case II. a = 6. In this case, (1) becomes $\dfrac{x}{6} + \dfrac{y}{14-6} = 1$ or $4x + 3y - 24 = 0$.

$\therefore$ There are two required lines with equations $\mathbf{x + y - 7 = 0}$ and $\mathbf{4x + 3y - 24 = 0}$.

10.23. NORMAL FORM

To find the equation of the straight line on which the length of the perpendicular from the origin and the angle which this perpendicular makes with the x-axis are given. *(P.T.U. B.Pharma, Dec. 2002)*

Proof. Let l be a non-vertical straight line on which the length of perpendicular from the origin is p and this perpendicular makes an angle α ($\neq 0$) with the positive direction of x-axis.

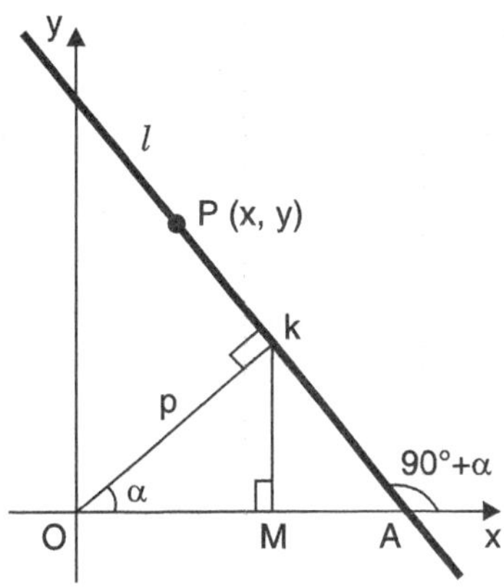

$\angle KAX = 180° - \angle OAK = 180° - (90° - \alpha) = 90° + \alpha.$

$\therefore$ Slope of $l = \tan(90° + \alpha) = -\cot\alpha.$

Also, $OM = OK \cdot \dfrac{OM}{OK} = p\cos\alpha$

and $MK = OK \cdot \dfrac{MK}{OK} = p\sin\alpha.$

$\therefore$ $K = (p\cos\alpha, p\sin\alpha)$

Let $P(x, y)$ be a general point on the line.

$\therefore$ Slope of l = slope of KP

$\Rightarrow$ $-\cot\alpha = \dfrac{y - p\sin\alpha}{x - p\cos\alpha} \Rightarrow -\dfrac{\cos\alpha}{\sin\alpha} = \dfrac{y - p\sin\alpha}{x - p\cos\alpha}$

$\Rightarrow$ $-x\cos\alpha + p\cos^2\alpha = y\sin\alpha - p\sin^2\alpha \Rightarrow x\cos\alpha + y\sin\alpha = p(\sin^2\alpha + \cos^2\alpha)$

$\Rightarrow$ $\mathbf{x\cos\alpha + y\sin\alpha = p}$. This is the required equation of the line.

Remark. In the equation $x\cos\alpha + y\sin\alpha = p$, we observe that :

(*i*) the constant term p on the R.H.S., being the length of perpendicular, is positive.

(*ii*) (coeff. of $x)^2$ + (coeff. of $y)^2 = \cos^2\alpha + \sin^2\alpha = 1$.

Corollary 1. If $\alpha = 0°$, the equation $x\cos\alpha + y\sin\alpha = p$ becomes $x\cos 0° + y\sin 0° = p$ or $x(1) + y(0) = p$ *i.e.*, $x = p$, which is the equation of a line parallel to y-axis.

Corollary 2. If $\alpha = 90°$, the equation $x\cos\alpha + y\sin\alpha = p$ becomes $x\cos 90° + y\sin 90° = p$ or $x(0) + y(1) = p$ *i.e.*, $y = p$, which is the equation of a line parallel to x-axis.

Example 4. *Find the equation of the line for which* $p = 5$ *and* $\alpha = 135°$. *Also sketch the line.* (G.J.U. B.Pharma, 2003)

Sol. We have $p = 5$ and $\alpha = 135°$.

The equation of the line in the **normal form** is

$$x \cos \alpha + y \sin \alpha = p.$$

$$\Rightarrow \quad x \cos 135° + y \sin 135° = 5$$

$$\Rightarrow \quad x\left(-\frac{1}{\sqrt{2}}\right) + y\left(\frac{1}{\sqrt{2}}\right) = 5$$

$$\Rightarrow \quad \mathbf{x - y + 5\sqrt{2} = 0.}$$

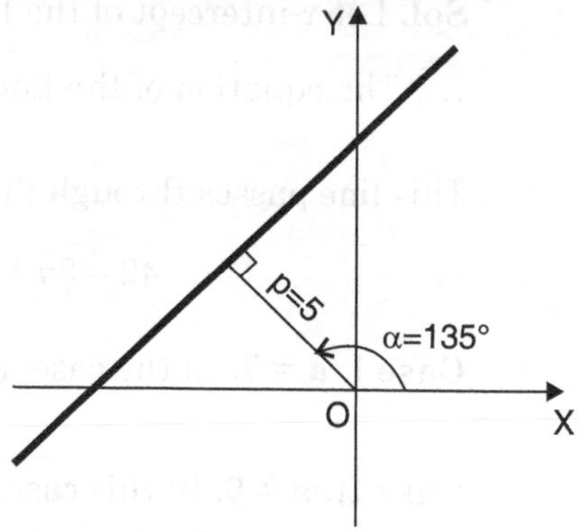

Example 5. *Sketch roughly the lines satisfying the given conditions and write their equations :*

(i) *inclination* $\theta = 150°$ *and distance from origin* $= 3$.

(ii) *x-intercept* $= 7$ *and distance from the origin* $= 2$.

(iii) *nearest point to the origin is* $(3, -4)$.

Sol. (i) The Fig. (i) shows that there are two lines (l_1 and l_2) satisfying the given conditions.

For l_1, $\alpha = 60°$ and $p = 3$.

$\therefore$ The equation of l_1 is $x \cos 60° + y \sin 60° = 3$

or $$x\left(\frac{1}{2}\right) + y\left(\frac{\sqrt{3}}{2}\right) = 3$$

or $$\mathbf{x + \sqrt{3}\,y - 6 = 0.}$$

For l_2, $\alpha = 180° + 60° = 240°$ and $p = 3$.

$\therefore$ The equation of l_2 is

$$x \cos 240° + y \sin 240° = 3$$

or $$x\left(-\frac{1}{2}\right) + y\left(-\frac{\sqrt{3}}{2}\right) = 3$$

or $$\mathbf{x + \sqrt{3}\,y + 6 = 0.}$$

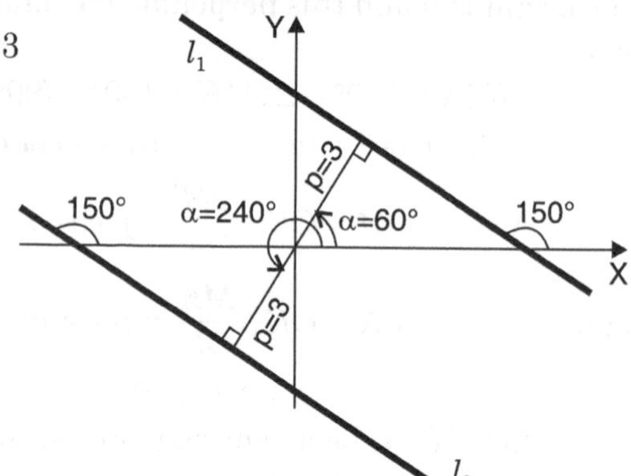

Fig. (i)

(ii) The Fig. (ii) shows that there are two lines (l_1 and l_2) satisfying the given conditions.

For l_1, $\cos \alpha_1 = \frac{2}{7}$ and $p = 2$.

$$\therefore \quad \sin \alpha_1 = \sqrt{1 - \cos^2 \alpha_1} = \sqrt{1 - \frac{4}{49}} = \sqrt{\frac{45}{49}} = \frac{3\sqrt{5}}{7}$$

($\because$ α_1 is in first quadrant)

$\therefore$ The equation of l_1 is $x \cos \alpha_1 + y \sin \alpha_1 = 2$

or $$x\left(\frac{2}{7}\right) + y\left(\frac{3\sqrt{5}}{7}\right) = 2 \quad \text{or} \quad \mathbf{2x + 3\sqrt{5}\,y - 14 = 0.}$$

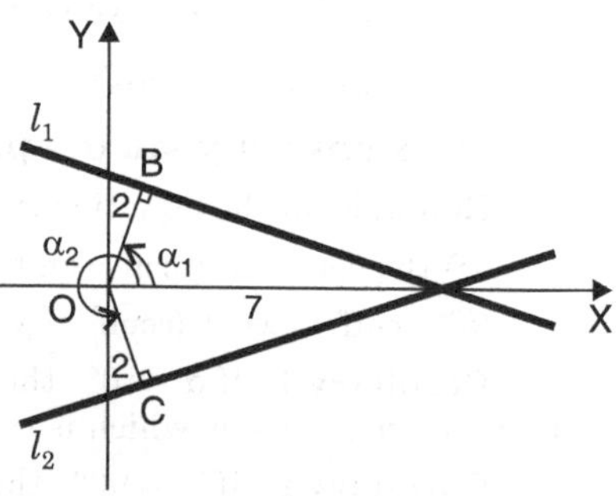

Fig. (ii)

For l_2, $\cos\alpha_2 = \cos(360° - \alpha_1) = \cos\alpha_1 = \frac{2}{7}$ and $p = 2$.

$$\therefore \quad \sin\alpha_2 = -\sqrt{1-\cos^2\alpha_2} = -\sqrt{1-\frac{4}{49}} = -\sqrt{\frac{45}{49}} = -\frac{3\sqrt{5}}{7}$$

($\because$ α_2 is in fourth quadrant)

$\therefore$ The equation of l_2 is $x\cos\alpha_2 + y\sin\alpha_2 = 2$

or $$x\left(\frac{2}{7}\right) + y\left(-\frac{3\sqrt{5}}{7}\right) = 2 \quad \text{or} \quad \mathbf{2x - 3\sqrt{5}\,y - 14 = 0.}$$

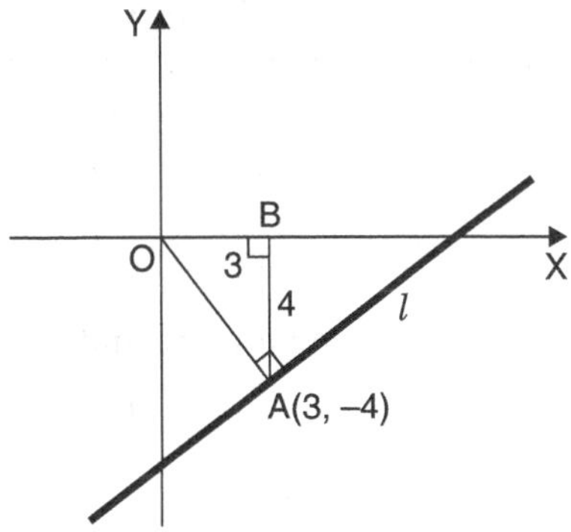

(*iii*) The figure shows that there is only one line (l) satisfying given condition.

Slope of $$OA = \frac{-4-0}{3-0} = -\frac{4}{3}.$$

$$\therefore \quad \text{Slope of } l = -\frac{1}{-4/3} = \frac{3}{4}.$$

$\therefore$ Equation of l is $y - (-4) = \frac{3}{4}(x-3)$

i.e., $$\mathbf{3x - 4y - 25 = 0.}$$

10.24. SYMMETRIC FORM (OR DISTANCE FORM)

To find the equation of a straight line having given one point on the line and its inclination.

Let a non-vertical line passes through the point $A(x_1, y_1)$ and having inclination θ *i.e.,* making an angle θ with the positive direction of x-axis.

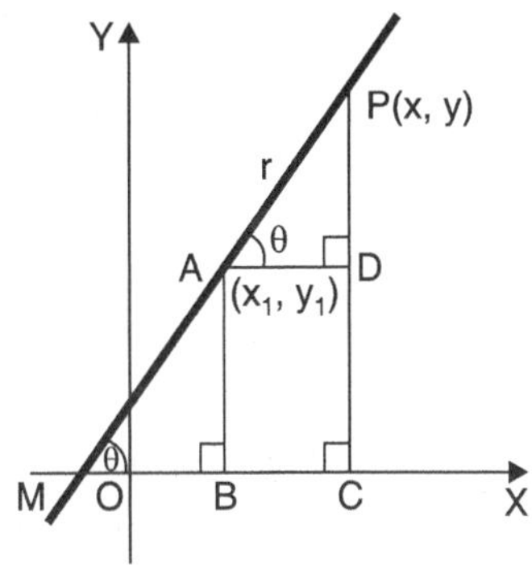

Let $P(x, y)$ be a general point on the line. Let $AP = r$.

$\therefore \quad \angle PAD = \angle PMB = \theta.$

From ΔAPD, we have

$$\cos\theta = \frac{AD}{AP} = \frac{BC}{AP} = \frac{OC - OB}{AP} = \frac{x - x_1}{r}$$

and $$\sin\theta = \frac{PD}{AP} = \frac{PC - CD}{AP} = \frac{PC - AB}{AP} = \frac{y - y_1}{r}.$$

$$\therefore \quad \frac{x - x_1}{\cos\theta} = r \quad \text{and} \quad \frac{y - y_1}{\sin\theta} = r.$$

$\Rightarrow$ $\dfrac{\mathbf{x - x_1}}{\mathbf{\cos\theta}} = \dfrac{\mathbf{y - y_1}}{\mathbf{\sin\theta}} = \mathbf{r}$. This is the required equation of the line.

Remark 1. The distance 'r' is positive for all points lying on one side of the given point A and negative for all points lying on the other side of the given point A.

Remark 2. From the equation $\frac{x - x_1}{\cos\theta} = \frac{y - y_1}{\sin\theta} = r$, we have $x = x_1 + r\cos\theta$ and $y = y_1 + r\sin\theta$.

$\therefore$ The coordinates of any point on this line are $(x_1 + r\cos\theta, y_1 + r\sin\theta)$.

Here 'r' is called the *parameter* and represent the distance between the points

$$(x_1 + r\cos\theta, y_1 + r\sin\theta) \text{ and } (x_1, y_1).$$

Example 6. *Find the equation of a line which passes through the point (– 2, 3) and makes angle 60° with the positive direction of x-axis.*

Sol. Here given point, $(x_1, y_1) = (-2, 3)$ and $\theta = 60°$.

Using $\dfrac{x - x_1}{\cos\theta} = \dfrac{y - y_1}{\sin\theta} = r$, the equation of the line is $\dfrac{y-3}{\cos 60°} = \dfrac{x+2}{\sin 60°} = r$.

or $\quad \mathbf{\dfrac{y-3}{1/2} = \dfrac{x+2}{\sqrt{3/2}} = r}$, **where 'r' is the distance between (x, y) and (– 2, 3).**

Example 7. *Find the equation of the straight line which passes through the point (2, 9) and making an angle of 45° with x-axis. Also find the points on the line which are at the distance of (i) 2 units (ii) 5 units from (2, 9).*

Sol. Here given point, $(x_1, y_1) = (2, 9)$ and $\theta = 45°$.

The equation of the line in the **symmetric form** is $\dfrac{x - x_1}{\cos\theta} = \dfrac{y - y_1}{\sin\theta} = r$, where '*r*' is the distance between (x, y) and (x_1, y_1).

$$\Rightarrow \quad \frac{x-2}{\cos 45°} = \frac{x-9}{\sin 45°} = r \quad \Rightarrow \quad \frac{x-2}{1/\sqrt{2}} = \frac{y-9}{1/\sqrt{2}} = r \qquad \text{...(1)}$$

(1) implies $\quad \sqrt{2}(x-2) = \sqrt{2}(y-9) \quad \Rightarrow \quad \mathbf{x - y + 7 = 0}$. This is the required equation.

$$(1) \Rightarrow \quad x = 2 + \frac{1}{\sqrt{2}}r, \quad y = 9 + \frac{1}{\sqrt{2}}r \qquad \text{...(2)}$$

(*i*) Let $r = 2$. $\therefore$ (2) $\Rightarrow$ $x = 2 + \dfrac{1}{\sqrt{2}} \times 2 = 2 + \sqrt{2}$ and $y = 9 + \dfrac{1}{\sqrt{2}} \times 2 = 9 + \sqrt{2}$.

$\therefore$ The point $(\mathbf{2 + \sqrt{2}, 9 + \sqrt{2}})$ is on the line and is at a distance of 2 units from (2, 9).

(*ii*) Let $r = 5$. $\therefore$ (1) $\Rightarrow$ $x = 2 + \dfrac{1}{\sqrt{2}} \times 5$ and $y = 9 + \dfrac{1}{\sqrt{2}} \times 5$.

$\therefore$ The point $\left(\mathbf{2 + \dfrac{5}{\sqrt{2}}, 9 + \dfrac{5}{\sqrt{2}}}\right)$ is on the line and is at a distance of 5 units from (2, 9).

Example 8. *A straight line is drawn through the point A($\sqrt{3}$, 2) making an angle of π/6 with positive direction of the x-axis. If it meets the straight line $\sqrt{3}x - 4y + 8 = 0$ in B, find the distance between A and B.*

Sol. Here given point, $A = (x_1, y_1) = (\sqrt{3}, 2)$ and $\theta = \pi/6$.

$\therefore$ The equation of the line in the **symmetric form** is

$$\frac{x - x_1}{\cos\theta} = \frac{y - y_1}{\sin\theta} = r,$$

where '*r*' is the distance between (x, y) and (x_1, y_1).

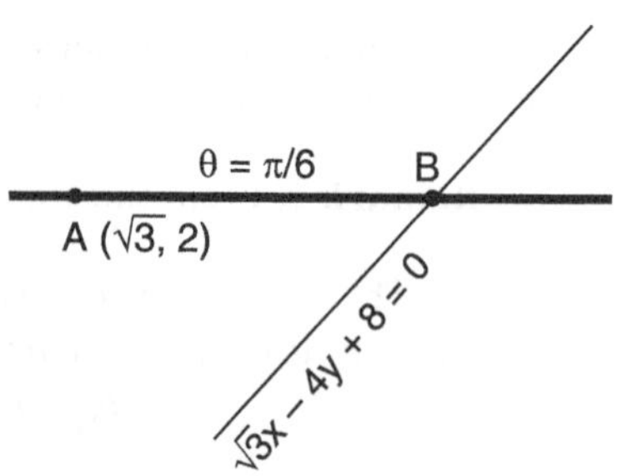

$$\Rightarrow \quad \frac{x - \sqrt{3}}{\cos \pi/6} = \frac{y-2}{\sin \pi/6} = r \quad \Rightarrow \quad \frac{x-\sqrt{3}}{\sqrt{3}/2} = \frac{y-2}{1/2} = r$$

$$\Rightarrow \quad x = \sqrt{3} + \frac{\sqrt{3}}{2}r, \quad y = 2 + \frac{1}{2}r.$$

Let the coordinates of B be $\left(\sqrt{3}+\frac{\sqrt{3}}{2}r,\ 2+\frac{1}{2}r\right)$. This point lies on the line $\sqrt{3}x-4y+8=0$.

$$\Rightarrow \quad \sqrt{3}\left(\sqrt{3}+\frac{\sqrt{3}}{2}r\right)-4\left(2+\frac{1}{2}r\right)+8=0 \quad \Rightarrow \quad 3+\frac{3}{2}r-8-2r+8=0 \quad \Rightarrow \quad r=6.$$

$\therefore$ The point B is at a distance of **6 units** from A.

WORKING RULES FOR SOLVING PROBLEMS

Rule I. *The equation of the line having slope 'm' and y-intercept 'c' is* $\mathbf{y = mx + c}$.

Rule II. *The equation of the line having intercepts 'a' and 'b' is* $\dfrac{\mathbf{x}}{\mathbf{a}}+\dfrac{\mathbf{y}}{\mathbf{b}}=\mathbf{1}$.

Rule III. *The equation of the line for which the length of perpendicular from the origin is 'p' and this perpendicular is inclined at an angle α to the x-axis is*

$$\mathbf{x\cos\alpha + y\sin\alpha = p}.$$

Rule IV. *The equation of the line passing through (x_1, y_1) and having inclination 'θ' is*

$$\frac{\mathbf{x-x_1}}{\mathbf{\cos\theta}}=\frac{\mathbf{y-y_1}}{\mathbf{\sin\theta}}=\mathbf{r}.$$

'r' is distance between the points (x, y) and (x_1, y_1).

EXERCISE 10.8

1. Find the equation of the straight line which makes :

(*i*) an angle 30° with x-axis and cuts off intercept 4 from the positive direction of y-axis.

(*ii*) an angle $\tan^{-1} 2$ with the x-axis and cuts off intercept 5 from the negative side of y-axis.

2. Find the equation of the straight line cutting off intercepts a and b from the axes where :

(*i*) $a = 1,\ b = 4$ (*ii*) $a = 5,\ b = -10$.

3. Find the equation of the straight line for which :

(*i*) $p = 1,\ \alpha = 30°$ (*ii*) $p = 2,\ \alpha = 135°$.

4. Find the equation of the straight line in symmetric form which passes through the point (x_1, y_1) and having inclination θ, where :

(*i*) $(x_1, y_1) = (-2, 1),\ \theta = 45°$ (*ii*) $(x_1, y_1) = (-1, 0),\ \theta = 120°$

(*iii*) $(x_1, y_1) = (0, 5),\ \theta = 60°$.

5. Find the equation of the line which passes through (2, 5) and cuts off equal intercepts on the axes. *(P.T.U. B.Pharma, Dec. 2004)*

6. Find the equation of the line which passes through (3, – 5) and cuts off intercepts on the axes which are equal in magnitude but opposite in sign.

7. A line passes through (1, 3) and its y-intercept is three times its intercept on x-axis. Find the equation of this line.

8. Find the equations of the lines which cut off intercepts on the axes whose sum and product are 1 and – 6 respectively.

9. Find the equation of the line which passes through (4, 1) and is such that the portion of the line intercepted between the axes is divided by this point internally in the ratio 1 : 2.

(P.T.U. B.Pharma, Dec. 2003)

10. (*i*) Find the equation of the line which passes through (2, 2) and the sum of whose intercepts on coordinate axes is 9. (*P.T.U. B.Pharma, June 2003*)
(*ii*) Find the equation of the line passing through the point (2, – 4) and sum of whose interupts on the line is 5. (*P.T.U. B.Pharma, Dec. 2005*)

11. Find the equation of the line which passes through the point (22, – 6) and is such that the intercept on x-axis exceeds the intercept on y-axis by 5.

12. A straight line is such that the segment of the line intercepted between the axes is bisected at the point $P(a, b)$. Prove that its equation is $\frac{x}{a} + \frac{y}{b} = 2$.

13. If the straight line $\frac{x}{a} + \frac{y}{b} = 1$ passes through the points (8, – 9) and (12, – 15), find the values of a and b.

14. Find the equation of the straight line at a distance of 3 units from the origin such that the perpendicular from the origin to the line makes an angle α, given by $\tan \alpha = 5/12$, with the positive direction of x-axis.

15. Find the equation of the line for which $p = 2$, $\sin \alpha = 4/5$.

16. Sketch roughly the lines satisfying the given conditions and write their equations :
(*i*) inclination $\theta = 90°$ and distance from the origin = 5
(*ii*) inclination $\theta = 120°$ and distance from the origin = 4.

17. Sketch roughly and write the equation of the line whose nearest point to the origin is :
(*i*) (2, 4) (*ii*) (1, – 4).

18. Find the equations of two straight lines which are at a distance of 1/2 unit from the origin and passes through the point (0, 1).

19. Slope of a straight line through $A(3, 2)$ is 3/4. Find the coordinates of the points on the line that are 5 units away from A.

20. Find the distance of the point (2, 3) from the line $2x - 3y + 9 = 0$ measured along a line making an angle of 45° with the x-axis.

21. From the point (3, 5), a line is drawn at 45° with OX and terminated by the line $3x + 4y = 50$, find the length of the line.

Answers

1. (*i*) $x - \sqrt{3}\,y + 4\sqrt{3} = 0$ (*ii*) $2x - y - 5 = 0$ 2. (*i*) $4x + y - 4 = 0$ (*ii*) $2x - y - 10 = 0$

3. (*i*) $\sqrt{3}\,x + y - 2 = 0$ (*ii*) $x - y + 2\sqrt{2} = 0$ 4. (*i*) $\frac{x+2}{1/\sqrt{2}} = \frac{y-1}{1/\sqrt{2}} = r$

(*ii*) $\frac{x+1}{-1/2} = \frac{y}{\sqrt{3}/2} = r$ (*iii*) $\sqrt{3}x - y + 5 = 0$ 5. $x + y - 7 = 0$ 6. $x - y - 8 = 0$

7. $3x + y - 6 = 0$ 8. $2x - 3y - 6 = 0$, $3x - 2y + 6 = 0$ 9. $x + 2y - 6 = 0$

10. (*i*) $2x + y = 6$, $x + 2y = 6$ (*ii*) $x - 2y - 10 = 0$, $4x + y - 4 = 0$

11. $x + 2y - 10 = 0$, $6x + 11y - 66 = 0$ 13. 2, 3

14. $12x + 5y - 39 = 0$, $12x + 5y + 39 = 0$ 15. $3x + 4y - 10 = 0$, $3x - 4y + 10 = 0$

16. (*i*) $x \cos 0° + y \sin 0° = 5$ *i.e.,* $x = 5$, $x \cos 180° + y \sin 180° = 5$ *i.e.,* $x = -5$

(*ii*) $x \cos 30° + y \sin 30° = 4$ *i.e.,* $\sqrt{3}\,x + y - 8 = 0$, $x \cos 210° + y \sin 210° = 4$ *i.e.,* $\sqrt{3}\,x + y + 8 = 0$

17. (*i*) $x\left(\frac{2}{2\sqrt{5}}\right) + y\left(\frac{4}{2\sqrt{5}}\right) = 2\sqrt{5}$ *i.e.,* $x + 2y - 10 = 0$

(*ii*) $x\left(\frac{1}{\sqrt{17}}\right) + y\left(-\frac{4}{\sqrt{17}}\right) = \sqrt{17}$ *i.e.,* $x - 4y - 17 = 0$

18. $\sqrt{3}\,x + y - 1 = 0$, $\sqrt{3}\,x - y + 1 = 0$

19. $\dfrac{x-3}{4/5} = \dfrac{y-2}{3/5} = r$ *i.e.,* $3x - 4y - 1 = 0$, (7, 5), (– 1, – 1)

20. $4\sqrt{2}$ units **21.** $3\sqrt{2}$ units.

Hints

15. $\cos \alpha = \pm \sqrt{1 - \sin^2 \alpha} = \pm \dfrac{3}{5}$.

18. If the equation of the line be $x \cos \alpha + y \sin \alpha = p$, then $p = 1/2$ and $(0) \cos \alpha + (1) \sin \alpha = 1/2$ implies $\sin \alpha = 1/2$ and $\cos \alpha = \pm \sqrt{1 - 1/4} = \pm \sqrt{3}/2$.

19. The equation of the line is $\dfrac{x-3}{3/5} = \dfrac{y-2}{4/5} = r$. Take $r = \pm 5$ to find the required points.

10.25. THE GENERAL FORM OF THE EQUATION OF A LINE

We shall prove that the equation of any line can be expressed in the form $Ax + By + C = 0$. This will also hold conversely, *i.e.*, the points whose coordinates satisfies the above equation would all lie on a straight line.

Theorem. Prove that every straight line has equation of the form Ax + By + C = 0 and conversely an equation of the type Ax + By + C = 0 (A, B are not both zero) always represents a straight line.

Proof. Let l be any straight line. The line is either perpendicular to x-axis or not.

Case I. l is perpendicular to x-axis.

Let the line be at a distance k from the y-axis.

∴ The equation of the line $x = k$. This equation can be expressed as $1(x) + 0(y) + (-k) = 0$.

This equation is of the type $Ax + By + C = 0$, where $A = 1, B = 0, C = -k$.

Case II. l is not perpendicular to x-axis.

Let m and c be the slope and y-intercept of the line respectively.

The equation of the line in **slope-intercept** form is $y = mx + c$.

This equation can be expressed as $m.x + (-1)y + c = 0$.

This equation is of the type $Ax + By + C = 0$, where $A = m, B = -1, C = c$.

∴ Every line can be represented by a linear equation of the type $Ax + By + C = 0$.

Converse. Let $Ax + By + C = 0$ be a linear equation in x and y with A, B not both zero.

Case I. B = 0. In this case, the equation reduces to $Ax + 0.y + C = 0$.

$$\Rightarrow \qquad x = -C/A$$

This represents the straight line parallel to y-axis and at a distance $-C/A$ from it.

Case II. B ≠ 0. In this case, the equation reduces to $By = -Ax - C$.

$$\Rightarrow \qquad y = \left(-\frac{A}{B}\right)x + \left(-\frac{C}{B}\right).$$

This represents the straight line whose slope and y-intercept are $-\dfrac{A}{B}$ and $-\dfrac{C}{B}$ respectively.

Hence the result holds.

ILLUSTRATIVE EXAMPLES

Example 1. *Find the equation of the line passing through (3, 5) and (1, – 2).*

Sol. Let the equation of the required line be $Ax + By + C = 0$. ...(1)

(3, 5) and (1, – 2) are on the line.

$\therefore$ $3A + 5B + C = 0$ ...(2)

$A - 2B + C = 0$ ...(3)

$$\therefore \quad \frac{A}{5+2} = \frac{B}{1-3} = \frac{C}{-6-5} = k \text{ (say)}$$

$\therefore$ $A = 7k, B = -2k, C = -11k$

$\therefore$ (1) $\Rightarrow$ $7kx - 2ky - 11k = 0$ or $\mathbf{7x - 2y - 11 = 0}$. This is the required equation.

10.26. REDUCTION OF GENERAL EQUATION TO STANDARD FORM

Let $Ax + By + C = 0$ be the general equation of a straight line.

$\therefore$ A and B are not both zero.

(*i*) **Reduction of 'slope-intercept' form.** Given equation is $Ax + By + C = 0$.

$$\Rightarrow \quad By = -Ax - C \Rightarrow y = \left(-\frac{A}{B}\right)x + \left(-\frac{C}{B}\right). \quad \text{(Assuming } B \neq 0\text{)}$$

Comparing this with $y = mx + c$, we get slope $= -\frac{\mathbf{A}}{\mathbf{B}}$ and y-intercept $= -\frac{\mathbf{C}}{\mathbf{B}}$.

WORKING RULES FOR REDUCING Ax + By + C = 0 TO 'SLOPE-INTERCEPT' FORM

Step I. *Keep only 'By' on the L.H.S. and get* $By = -Ax - C$.

Step II. *Divide both sides by 'B' and get* $y = \left(-\frac{A}{B}\right)x + \left(-\frac{C}{B}\right)$.

Step III. *This expresses the given equation in the 'slope-intercept' form. Here* $m = -A/B$ *and* $c = -C/B$.

(*ii*) **Reduction to 'Intercept' form.** Given equation is $Ax + By + C = 0$.

$$\Rightarrow \quad Ax + By = -C \Rightarrow \frac{A}{-C}x + \frac{B}{-C}y = 1 \quad \text{(Assuming } C \neq 0\text{)}$$

$$\Rightarrow \quad \frac{x}{(-C/A)} + \frac{y}{(-C/B)} = 1 \quad \text{(Assuming } A \neq 0, B \neq 0\text{)}$$

Comparing this with $\frac{x}{a} + \frac{y}{b} = 1$, we get x-intercept $= -\frac{\mathbf{C}}{\mathbf{A}}$ and y-intercept $= -\frac{\mathbf{C}}{\mathbf{B}}$.

WORKING RULES FOR REDUCING Ax + By + C = 0 TO 'INTERCEPT' FORM

Step I. *Shift constant 'C' to the R.H.S. and get* $Ax + By = -C$.

Step II. *Divide both sides by '– C' and get* $\frac{Ax}{-C} + \frac{By}{-C} = 1$.

Step III. *Make coefficients of x and y occur as their denominator and get*

$$\frac{x}{(-C/A)} + \frac{y}{(-C/B)} = 1.$$

Step IV. *This expresses the given equation in the 'intercept' form. Here* $a = -C/A$ *and* $b = -C/B$.

Remark. The x-intercept can be obtained by putting $y = 0$ in the equation and then solving for x. The value of x gives x-intercept.

Similarly, y-intercept can also be obtained by putting $x = 0$ in the equation and then solving for y. The value of y gives y-intercept.

(*iii*) **Reduction to 'Normal' form.** Given equation is $Ax + By + C = 0$.

Let its **normal form** be $x \cos \alpha + y \sin \alpha = p$.

$$\Rightarrow \quad \frac{A}{\cos \alpha} = \frac{B}{\sin \alpha} = \frac{C}{-p} \quad \Rightarrow \quad \cos \alpha = \frac{-Ap}{C} \text{ and } \sin \alpha = \frac{-Bp}{C} \qquad (\text{Assuming } C \neq 0)$$

$$\therefore \quad 1 = \cos^2 \alpha + \sin^2 \alpha = \left(\frac{-Ap}{C}\right)^2 + \left(\frac{-Bp}{C}\right)^2 = \frac{p^2}{C^2}(A^2 + B^2)$$

$$\therefore \quad p = \pm \frac{C}{\sqrt{A^2 + B^2}}.$$

Case I. C is positive. $\therefore \quad p = \dfrac{C}{\sqrt{A^2+B^2}}$. $\qquad (\because \; p$ is to be positive)

$$\therefore \quad \cos \alpha = -\frac{A}{C}\left(\frac{C}{\sqrt{A^2+B^2}}\right) = -\frac{A}{\sqrt{A^2+B^2}} \text{ and } \sin \alpha = -\frac{B}{C}\left(\frac{C}{\sqrt{A^2+B^2}}\right) = -\frac{B}{\sqrt{A^2+B^2}}.$$

$$\therefore \quad x \cos \alpha + y \sin \alpha = p \text{ becomes } \left(-\frac{A}{\sqrt{A^2+B^2}}\right)x + \left(-\frac{B}{\sqrt{A^2+B^2}}\right)y = \frac{C}{\sqrt{A^2+B^2}}.$$

This is the required normal form of the given equation.

Case II. C is negative. $\therefore \quad p = -\dfrac{C}{\sqrt{A^2+B^2}}$ $\qquad (\because \; p$ is to be positive)

$$\therefore \quad \cos \alpha = -\frac{A}{C}\left(-\frac{C}{\sqrt{A^2+B^2}}\right) = \frac{A}{\sqrt{A^2+B^2}} \text{ and } \sin \alpha = -\frac{B}{C}\left(-\frac{C}{\sqrt{A^2+B^2}}\right) = \frac{B}{\sqrt{A^2+B^2}}$$

$$\therefore \quad x \cos \alpha + y \sin \alpha = p \text{ becomes } \left(\frac{A}{\sqrt{A^2+B^2}}\right)x + \left(\frac{B}{\sqrt{A^2+B^2}}\right)y = -\frac{C}{\sqrt{A^2+B^2}}.$$

This is the required normal form of the given equation.

WORKING RULES FOR REDUCING Ax + By + C = 0 TO 'NORMAL' FORM

Step I. *Shift constant 'C' to the R.H.S. and get* $Ax + By = -C$.

Step II. *If the R.H.S. is not positive, then make it so by multiplying the whole equation by '– 1'.*

Step III. *Divide the whole equation by* $\sqrt{A^2 + B^2}$.

Step IV. *This expresses the given equation in the 'normal' form.*

Example 2. *Reduce* $\sqrt{3}x + y + 2 = 0$ *to the 'slope-intercept form' and hence find its slope, inclination and y-intercept. Also sketch the line on the coordinate plane.*

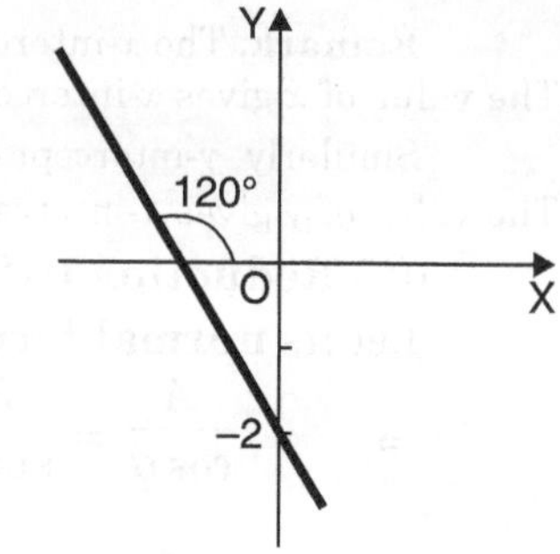

Sol. We have $\sqrt{3}x + y + 2 = 0$. $\therefore$ $y = -\sqrt{3}x - 2$.

Comparing it with $y = mx + c$, we get $m = -\sqrt{3}$ and $c = -2$.

$\therefore$ Slope $= -\sqrt{3}$ and y-intercept $= -2$.

Let inclination of the line be θ.

$\therefore$ $\tan\theta = -\sqrt{3} = -\tan 60° = \tan(180° - 60°) = \tan 120°$

$\therefore$ $\theta = 120°$.

$\therefore$ Inclination = **120°.** The sketch of the line is shown in the figure.

Example 3. *Reduce $4x - 3y - 12 = 0$ to the 'intercept form' and hence find its intercepts on the axes. Also sketch the line on the coordinate plane.*

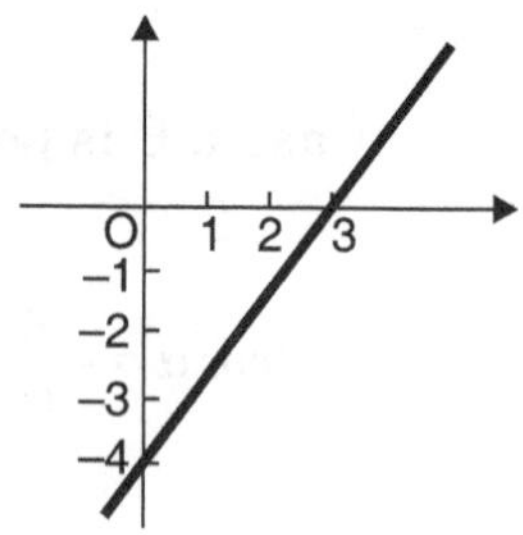

Sol. We have $4x - 3y - 12 = 0$. $\therefore$ $4x - 3y = 12$

Dividing by 12, we get $\frac{4x}{12} - \frac{3y}{12} = 1$ *i.e.,* $\frac{x}{3} + \frac{y}{-4} = 1$.

Comparing it with $\frac{x}{a} + \frac{y}{b} = 1$, we get $a = 3$ and $b = -4$.

$\therefore$ x-intercept = **3** and y-intercept = **– 4**.

The sketch of the line is shown in the figure.

Example 4. *Reduce $\sqrt{3}x + y - 4 = 0$ to the 'normal form' and hence find the values of p and α. Also sketch the line on the coordinate plane.*

Sol. We have $\sqrt{3}x + y - 4 = 0$.

$\Rightarrow$ $\sqrt{3}x + y = 4$ **(Note that R.H.S. is positive)**

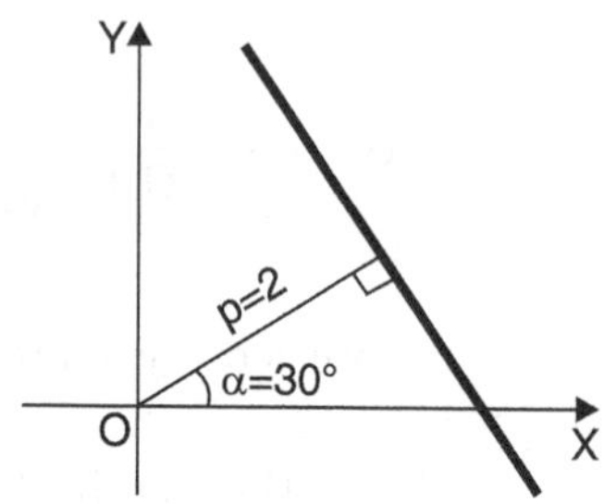

Dividing both sides by $\sqrt{(\sqrt{3})^2 + (1)^2} = \sqrt{4} = 2$, we get

$$\frac{\sqrt{3}}{2}x + \frac{1}{2}y = \frac{4}{2} = 2.$$

Comparing it with $x\cos\alpha + y\sin\alpha = p$, we get

$$\cos\alpha = \frac{\sqrt{3}}{2},\ \sin\alpha = \frac{1}{2},\ p = 2.$$

As $\cos\alpha$ and $\sin\alpha$ are both positive, α lies in I quadrant.

$\alpha = 30°$ satisfies both equations.

$\therefore$ p = **2** and α = **30°.** The sketch of the line is shown in the figure.

Example 5. *Reduce $x + \sqrt{3}y + 4 = 0$ to the ;*

(*i*) *slope-intercept form and find its slope and y-intercept.*

(*ii*) *intercept form and find its intercepts on the axes.*

(*iii*) *normal form and find the values of p and α.*

Sol. (*i*) Given equation is $x + \sqrt{3}y + 4 = 0$.

$$\Rightarrow \quad \sqrt{3}y = -x - 4 \Rightarrow \mathbf{y} = \left(-\frac{1}{\sqrt{3}}\right)\mathbf{x} + \left(-\frac{4}{\sqrt{3}}\right).$$

This is in the form $y = mx + c$ with slope = **– 1/$\sqrt{3}$** and y-intercept = **– 4/$\sqrt{3}$.**

(*ii*) Given equation is $x + \sqrt{3}y + 4 = 0$.

$$\Rightarrow \qquad x + \sqrt{3}y = -4 \Rightarrow \frac{x}{-4} + \frac{\sqrt{3}}{-4}y = 1 \Rightarrow \frac{\mathbf{x}}{-\mathbf{4}} + \frac{\mathbf{y}}{-\mathbf{4}/\sqrt{\mathbf{3}}} = \mathbf{1}.$$

This is in the form $\frac{x}{a} + \frac{y}{b} = 1$ with x-intercept = **– 4** and y-intercept = **$-4\sqrt{3}$.**

(*iii*) Given equation is $x + \sqrt{3}y + 4 = 0$

$\Rightarrow \qquad x + \sqrt{3}y = -4 \Rightarrow -x - \sqrt{3}y = 4$ **(Note that R.H.S. is made positive)**

Dividing both sides by $\sqrt{(-1)^2 + (-\sqrt{3})^2} = \sqrt{4} = 2$, we get $-\frac{1}{2}x - \frac{\sqrt{3}}{2}y = 2$.

Comparing it with $x \cos \alpha + y \sin \alpha = p$, we get $\cos \alpha = -\frac{1}{2}$, $\sin \alpha = -\frac{\sqrt{3}}{2}$ and $p = 2$.

As cos α and sin α are both negative, α lies in III quadrant and $\alpha = 180° + 60° = 240°$.

(Note that cos 240° = – 1/2 and sin 240° = – $\sqrt{3}/2$)

∴ The required normal form is **x cos 240° + y sin 240° = 2.** Here p = **2** and α = **240°**.

Example 6. *Reduce the equation of the straight line 4x + 3y + 10 = 0, to normal form and find its distance from the origin.* (P.T.U. B.Pharma, Dec. 2001)

Sol. Given equation is $4x + 3y + 10 = 0$ ...(1)

$\Rightarrow \qquad -4x - 3y = 10$ (Note that R.H.S. is positive)

Dividing both sides by $\sqrt{(-4)^2 + (-3)^2}$, we get

$$-\frac{4}{5}x - \frac{3}{5}y = 2$$

Comparing it with $x \cos \alpha + y \sin \alpha = p$, we get

$$\cos \alpha = -\frac{4}{5}, \sin \alpha = -\frac{3}{5}$$

As cos α and sin α are both negative therefore α lies in 3rd quadrant.

Also $p = 2$ = distance of the line (1) from the origin.

EXERCISE 10.9

1. Reduce the following equations to the slope-intercept form and find the values of slope and y-intercept :
   (*i*) $x + 3y = 10$ (*ii*) $3x + 3y = 5$ (*iii*) $y = 0$.
2. Reduce the following equations to the intercept form and find its intercepts on the axes :
   (*i*) $2x + y - 6 = 0$ (*ii*) $5x - y - 15 = 0$ (*iii*) $17x - 3y + 45 = 0$.
3. Reduce the following equations to the normal form and find the values of p and α :
   (*i*) $x + y - 2 = 0$ (*ii*) $\sqrt{3}x + y + 6 = 0$ (*iii*) $x - \sqrt{3}y - 12 = 0$.
4. (*i*) Find the slope of the line which cuts off equal intercepts on the axes.
   (*ii*) Find the slope of the line which cuts intercepts, equal in magnitude but opposite in sign, on the axes.

5. Reduce (*i*) $x - 3 = 0$ (*ii*) $y + 4 = 0$ to the normal form. Also sketch these lines on the coordinate plane.

6. Reduce the equation $x \sec \alpha - y \operatorname{cosec} \alpha = a$ to the slope-intercept form and hence find the slope and the coordinates of the point where it meets the *y*-axis.

7. Show that the line joining (9, 3) and (15, – 3) cuts off equal intercepts on the axes.

8. Show that the points (5, 11), (1, – 1) and (11, 29) lie on a line. Find its equation and the intercepts made by it on the axes.

9. Which of the lines $3x + 4y - 9 = 0$ and $2x + 6y + 19 = 0$ is farther from the origin ?

10. Find the intercepts of the line $x \sin \alpha + y \cos \alpha = \sin 2\alpha$ on the axes. Also find the mid-point of the line segment intercepted between the axes.

11. Find the equation of the line passing through (– 1, – 2) and (– 5, – 3), assuming the equation of the line to be $Ax + By + C = 0$. Also reduce this equation to the intercept form.

12. The line $Ax + By + C = 0$ passes through the points (– 1, 8) and (4, – 2). Find the equation of the line. Also reduce this equation to the intercept form.

13. Show that the origin is equidistant from the straight lines $4x + 3y + 10 = 0$, $5x - 12y + 26 = 0$ and $7x + 24y = 50$.

Answers

1. (*i*) $y = -\frac{1}{3}x + \frac{10}{3}$, slope $= -\frac{1}{3}$, *y*-intercept $= \frac{10}{3}$

   (*ii*) $y = -x + \frac{5}{3}$, slope = – 1, *y*-intercept $= \frac{5}{3}$

   (*iii*) $y = 0x + 0$, slope = 0, *y*-intercept = 0

2. (*i*) $\frac{x}{3} + \frac{y}{6} = 1$, *x*-intercept = 3, *y*-intercept = 6

   (*ii*) $\frac{x}{3} + \frac{y}{-15} = 1$, *x*-intercept = 3, *y*-intercept = – 15

   (*iii*) $\frac{x}{-45/17} + \frac{y}{15} = 1$, *x*-intercept $= -\frac{45}{17}$, *y*-intercept = 15

3. (*i*) $x \cos 45° + y \sin 45° = \sqrt{2}$, $p = \sqrt{2}$, $\alpha = 45°$

   (*ii*) $x \cos 210° + y \sin 210° = 3$, $p = 3$, $\alpha = 210°$

   (*iii*) $x \cos 300° + y \sin 300° = 6$, $p = 6$, $\alpha = 300°$

4. (*i*) – 1 (*ii*) 1

5. (*i*) $x \cos 0° + y \sin 0° = 3$ (*ii*) $x \cos 270° + y \sin 270° = 4$.

6. $y = x \tan \alpha - a \sin \alpha$, $\tan \alpha$, $(0, -a \sin \alpha)$

8. $3x - y - 4 = 0$, $a = 4/3$, $b = -4$

9. $2x + 6y + 19 = 0$

10. $a = 2 \cos \alpha$, $b = 2 \sin \alpha$, $(\cos \alpha, \sin \alpha)$

11. $x - 4y - 7 = 0$, $\frac{x}{7} + \frac{y}{-7/4} = 1$

12. $2x + y - 6 = 0$, $\frac{x}{3} + \frac{y}{6} = 1$.

Hint

9. $3x + 4y - 9 = 0 \Rightarrow 3x + 4y = 9 \Rightarrow \frac{3}{5}x + \frac{4}{5}y = \frac{9}{5}$ $\quad (\because \sqrt{(3)^2 + (4)^2} = 5)$

$\therefore \quad p = 9/5 = 1.8$

$$2x + 6y + 19 = 0 \Rightarrow -2x - 6y = 19 \Rightarrow -\frac{2}{2\sqrt{10}}x - \frac{6}{2\sqrt{10}}y = \frac{19}{2\sqrt{10}}$$

$$(\because \sqrt{(-2)^2 + (-6)^2} = 2\sqrt{10})$$

$\therefore \quad p = \dfrac{19}{2\sqrt{10}} = \dfrac{19\sqrt{10}}{20} = 3.004.$ $\therefore$ The second line is farther from the origin.

10.27. ANGLE BETWEEN TWO LINES

Two intersecting lines intersect at two angles, which are supplementary to each other. For example, if angle between two intersecting lines is 60°, then the other angle would be 180° – 60° = 120°. Generally, the angle which is not greater than 90° is taken as the angle between the lines.

Now, we shall find the angle between two lines with given slopes.

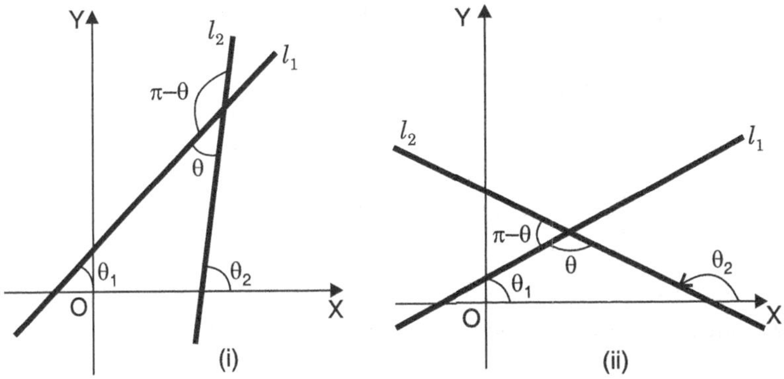

Let l_1 and l_2 be two non-vertical lines with slopes m_1 and m_2 respectively. Let θ_1, θ_2 be the inclinations of these lines.

$$\therefore \quad m_1 = \tan\theta_1 \quad \text{and} \quad m_2 = \tan\theta_2.$$

Let θ and $\pi - \theta$ be the angles between the lines $(\theta \neq \pi/2)$.

$\therefore$ Exterior angle, $\theta_2 = \theta_1 + \theta$ implies $\theta = \theta_2 - \theta_1$

$$\therefore \quad \tan\theta = \tan(\theta_2 - \theta_1) = \frac{\tan\theta_2 - \tan\theta_1}{1 + \tan\theta_2 \tan\theta_1}$$

$$\therefore \quad \mathbf{\tan\theta = \frac{m_2 - m_1}{1 + m_2 m_1}}.$$

Remark. In Fig. (*i*), angle θ is acute, so $\tan\theta > 0$.

$$\therefore \quad \frac{m_2 - m_1}{1 + m_2 m_1} > 0 \text{ and we have } \tan\theta = \left|\frac{m_2 - m_1}{1 - m_2 m_1}\right|.$$

In Fig. (*ii*), angle θ is obtuse, so $\tan\theta < 0$.

$$\therefore \quad \frac{m_2 - m_1}{1 + m_2 m_1} < 0.$$

In this case, the other angle $\pi - \theta$ would be acute.

Also, $$\tan(\pi - \theta) = -\tan\theta = -\frac{m_2 - m_1}{1 + m_2 m_1} = \left|\frac{m_2 - m_1}{1 + m_2 m_1}\right|.$$

∴ **The tangent of the acute angle between the lines with slopes m_1 and m_2 is equal to** $\left|\dfrac{m_2 - m_1}{1 + m_2 m_1}\right|$, **which is also equal to** $\left|\dfrac{m_1 - m_2}{1 + m_1 m_2}\right|$.

WORKING RULES TO FIND THE ANGLE BETWEEN GIVEN LINES

Step I. *Find the slopes of the given lines. Let these slopes be m_1 and m_2.*

Step II. *Assume θ to be acute angle between the given lines.*

Step III. *Put* $\tan\theta = \left|\dfrac{m_1 - m_2}{1 + m_1 m_2}\right|$ *and simplify the R.H.S. to get the value of θ.*

Corollary. If m_1 and m_2 are the slopes of the lines $A_1x + B_1y + C_1 = 0$ and $A_2x + B_2y + C_2 = 0$, then $m_1 = -\dfrac{A_1}{B_1}$ and $m_2 = -\dfrac{A_2}{B_2}$. If θ be the acute angle between these lines, then

$$\tan\theta = \left|\frac{m_1 - m_2}{1 + m_1 m_2}\right| = \left|\frac{-\dfrac{A_1}{B_1} + \dfrac{A_2}{B_2}}{1 + \left(-\dfrac{A_1}{B_1}\right)\left(-\dfrac{A_2}{B_2}\right)}\right| = \left|\frac{\mathbf{A_2B_1 - A_1B_2}}{\mathbf{A_1A_2 + B_1B_2}}\right|.$$

ILLUSTRATIVE EXAMPLES

Example 1. *Find the angle between the lines $3x + y - 7 = 0$ and $x + 2y + 9 = 0$.*

Sol. Let m_1 and m_2 be the slopes of the given lines $3x + y - 7 = 0$ and $x + 2y + 9 = 0$ respectively.

∴ $m_1 = -3$ and $m_2 = -1/2$.

[∵ $3x + y - 7 = 0 \Rightarrow y = (-3)x + 7$ and $x + 2y + 9 = 0 \Rightarrow y = (-1/2)x - (9/2)$.]

Let θ be the acute angle between the given lines.

$$\therefore \quad \tan\theta = \left|\frac{m_1 - m_2}{1 + m_1 m_2}\right| = \left|\frac{-3 + 1/2}{1 + (-3)(-1/2)}\right| = \left|\frac{-5/2}{5/2}\right| = |-1| = 1.$$

∴ $\theta = \mathbf{45°}$.

Example 2. *Find the angle between the lines joining the points (0, 0), (2, 3) and (2, – 2), (3, 5).*

Sol. Let the given points be $A(0, 0)$, $B(2, 3)$, $C(2, -2)$, $D(3, 5)$. Let m_1 and m_2 be the slopes of the lines AB and CD respectively.

$$\therefore \quad m_1 = \frac{3 - 0}{2 - 0} = \frac{3}{2} \quad \text{and} \quad m_2 = \frac{5 + 2}{3 - 2} = \frac{7}{1} = 7.$$

Let θ be the acute angle between the lines.

$$\therefore \quad \tan\theta = \left|\frac{m_1 - m_2}{1 + m_1 m_2}\right| = \left|\frac{\dfrac{3}{2} - 7}{1 + \left(\dfrac{3}{2}\right)(7)}\right| = \left|\frac{-\dfrac{11}{2}}{\dfrac{23}{2}}\right| = \left|-\frac{11}{23}\right| = \frac{11}{23}.$$

∴ $\theta = \mathbf{\tan^{-1}(11/23)}$.

Example 3. *The acute angle between two lines is $\pi/4$ and slope of one of them is 1/2. Find the slope of the other line.*

Sol. If θ be the acute angle between the lines with slopes m_1 and m_2, then

$$\tan\theta = \left|\frac{m_1 - m_2}{1 + m_1 m_2}\right|.$$

Let $$\theta = \frac{\pi}{4} \text{ and } m_1 = \frac{1}{2}.$$

$\therefore$ $$\tan\frac{\pi}{4} = \left|\frac{\frac{1}{2} - m_2}{1 + \frac{1}{2}m_2}\right| \Rightarrow 1 = \left|\frac{1 - 2m_2}{2 + m_2}\right| \Rightarrow \frac{1 - 2m_2}{2 + m_2} = +1 \text{ or } -1$$

Now, $$\frac{1 - 2m_2}{2 + m_2} = 1 \Rightarrow m_2 = -\frac{1}{3} \quad \text{and} \quad \frac{1 - 2m_2}{2 + m_2} = -1 \Rightarrow m_2 = 3.$$

$\therefore$ The slope of the other line is either **– 1/3** or **3**.

Example 4. *Find the equation of the straight line which passes through the origin and making angle 60° with the line $x + \sqrt{3}y + 3\sqrt{3} = 0$.*

Sol. Given line is $x + \sqrt{3}y + 3\sqrt{3} = 0$ ...(1)

$\Rightarrow$ $$y = \left(-\frac{1}{\sqrt{3}}\right)x - 3. \quad \therefore \quad \text{Slope of (1)} = -\frac{1}{\sqrt{3}}.$$

Let slope of the required line be m. Also between these lines is given to be 60°.

$\Rightarrow$ $$\tan 60° = \left|\frac{m - (-1/\sqrt{3})}{1 + m(-1/\sqrt{3})}\right| \Rightarrow \sqrt{3} = \left|\frac{\sqrt{3}m + 1}{\sqrt{3} - m}\right| \Rightarrow \frac{\sqrt{3}m + 1}{\sqrt{3} - m} = \pm\sqrt{3}.$$

$$\frac{\sqrt{3}m + 1}{\sqrt{3} - m} = \sqrt{3} \Rightarrow \sqrt{3}m + 1 = 3 - \sqrt{3}m \Rightarrow m = \frac{1}{\sqrt{3}}.$$

Using $y = mx + c$, the equation of the required line is

$$y = \frac{1}{\sqrt{3}}x + 0 \quad i.e., \quad \mathbf{x - \sqrt{3}y = 0.}$$

($\because$ This passes through origin, so $c = 0$)

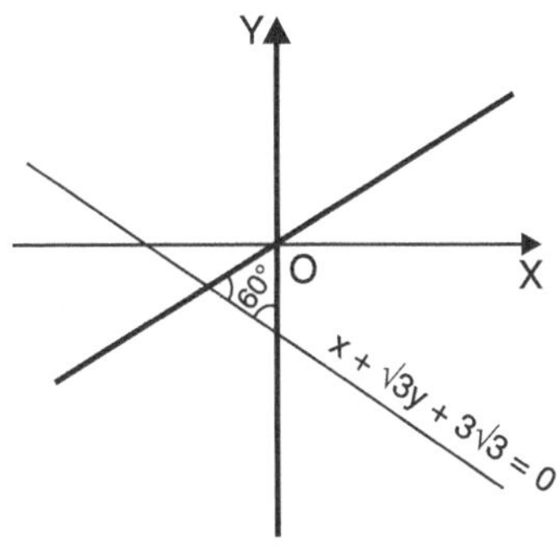

$$\frac{\sqrt{3}m + 1}{\sqrt{3} - m} = -\sqrt{3} \Rightarrow \sqrt{3}m + 1 = -3 + \sqrt{3}m$$

$\Rightarrow$ m is not defined.

$\therefore$ The slope of the required line is not defined. Thus, the required line is a vertical line. This line is to pass through the origin.

$\therefore$ The equation of the required line is **x = 0.**

Example 5. *Find the equations of the lines which pass through (4, 5) and make equal angles with the lines $5x - 12y + 6 = 0$ and $3x = 4y + 7$.*

Sol. Given lines are

$5x - 12y + 6 = 0$ ...(1) and $3x = 4y + 7$ ...(2)

Let m_1 and m_2 be the slopes of (1) and (2) respectively.

$\therefore$ $$m_1 = \frac{5}{12} \quad \text{and} \quad m_2 = \frac{3}{4}$$

Let m be the slope of the required line.

Let θ be the acute angle which the required line make given lines.

$$\therefore \qquad \tan\theta = \left|\frac{m - m_1}{1 + m_2}\right| = \left|\frac{m - \frac{5}{12}}{1 + m\,.\,\frac{5}{12}}\right| = \left|\frac{12m - 5}{12 + 5m}\right|$$

Also,
$$\tan\theta = \left|\frac{m - m_2}{1 + mm_2}\right| = \left|\frac{m - \frac{3}{4}}{1 + m\,.\,\frac{3}{4}}\right| = \left|\frac{4m - 3}{4 + 3m}\right|$$

$$\Rightarrow \quad \left|\frac{12m - 5}{12 + 5m}\right| = \left|\frac{4m - 3}{4 + 3m}\right| \Rightarrow \frac{12m - 5}{12 + 5m} = \pm\frac{4m - 3}{4 + 3m}$$

$$\frac{12m - 5}{12 + 5m} = \frac{4m - 3}{4 + 3m} \Rightarrow m^2 = -1, \text{ which is impossible.}$$

$$\frac{12m - 5}{12 + 5m} = -\frac{4m - 3}{4 + 3m} \Rightarrow 28\,m^2 + 33\,m - 28 = 0 \Rightarrow m = \frac{4}{7}, -\frac{7}{4}$$

If $m = \frac{4}{7}$, the equation of the required line is $y - 5 = \frac{4}{7}(x - 4)$ or $\mathbf{4x - 7y + 19 = 0.}$

If $m = -\frac{7}{4}$, the equation of the required line is $y - 5 = -\frac{7}{4}(x - 4)$ or $\mathbf{7x + 4y - 48 = 0.}$

Example 6. *Find the angles of the triangle whose vertices are (2, 4) (– 4, 2) and (0, 0).*

Sol. Let the vertices of the triangle be A (2, 4), B(– 4, 2) and C (0, 0).

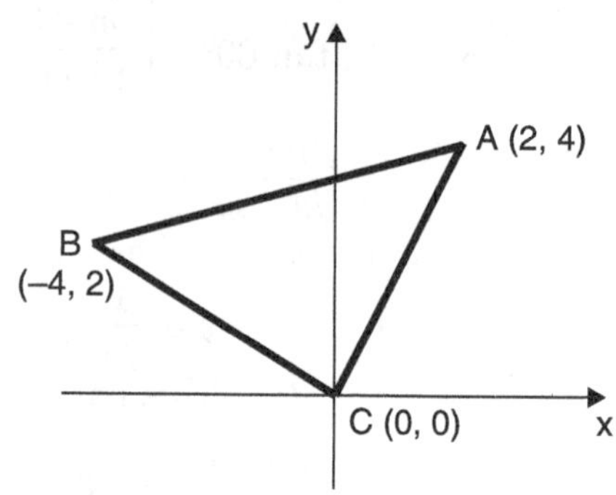

To find A. Slope of $AB = \frac{2 - 4}{-4 - 2} = \frac{1}{3}$

and slope of $AC = \frac{0 - 4}{0 - 2} = 2.$

Inclinations of AB and AC are both acute angles.

∴ A, being difference of acute angles is also acute.

$$\therefore \quad \tan A = \left|\frac{2 - \frac{1}{3}}{1 + 2\left(\frac{1}{3}\right)}\right| = \left|\frac{5/3}{5/3}\right| = 1 \quad \therefore \quad A = \mathbf{45°}.$$

To find B. Slope of BA = slope of $AB = \frac{1}{3}$ and slope of $BC = \frac{0 - 2}{0 + 4} = -\frac{1}{2}$.

Now,
$$\left|\frac{\frac{1}{3} - \left(-\frac{1}{2}\right)}{1 + \frac{1}{3}\left(-\frac{1}{2}\right)}\right| = \left|\frac{\frac{5}{6}}{\frac{5}{6}}\right| = |\,1\,| = 1 = \tan 45°$$

∴ Angle between lines BA and BC is 45°. Now 180° – 45° = 135°.

If B = 135°, then C = 180° – (A + B) = 180° – (45° + 135°) = 0°, which is impossible.

∴ B = **45°**.

To find C. $A + B + C$ = 180° implies C = 180° – (A + B) = 180° – (45° + 45°) = **90°**.

Example 7. *Two opposite vertices of a square are (3, 4) and (1, – 1). Find the coordinates of other vertices.*

Sol. Let the square be with vertices $A(3, 4)$, $B(x_1, y_1)$, $C(1, -1)$ and $D(x_2, y_2)$.

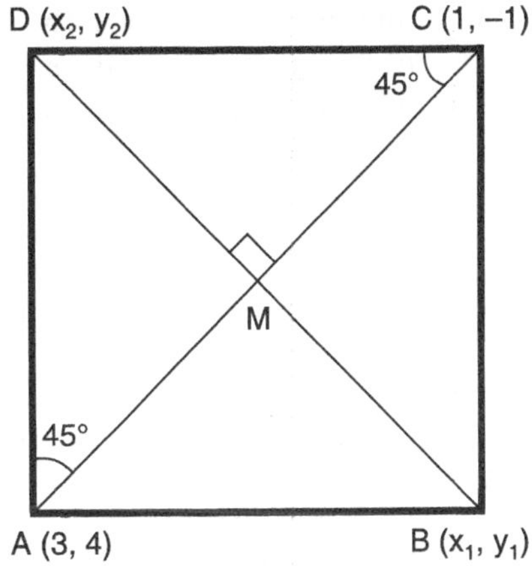

Slope of $AC = \frac{4-(-1)}{3-1} = \frac{5}{2}$.

Let m be the slope of a line making angle of 45° with AC.

$$\therefore \quad \tan 45° = \left| \frac{m - \frac{5}{2}}{1 + m \cdot \frac{5}{2}} \right|$$

$$\Rightarrow \quad \frac{m - \frac{5}{2}}{1 + \frac{5m}{2}} = \pm 1 \Rightarrow 2m - 5 = \pm (2 + 5m)$$

$2m - 5 = 2 + 5m \Rightarrow 3m = 7 \Rightarrow m = -7/3$ and $2m - 5 = -(2 + 5m) \Rightarrow 7m = 3 \Rightarrow m = 3/7$.
Let slope of AD be $-7/3$. $\therefore$ Slope of $CD = 3/7$.

$\therefore$ Equation of AD is $y - 4 = -\frac{7}{3}(x - 3)$ or $7x + 3y - 33 = 0$ ...(1)

Equation of CD is $y + 1 = \frac{3}{7}(x - 1)$ or $3x - 7y - 10 = 0$ ...(2)

Solving (1) and (2), we get $\frac{x}{-30-231} = \frac{y}{-99+70} = \frac{1}{-49-9}$

$$\Rightarrow \quad x = \frac{-261}{-58} = \frac{9}{2} \text{ and } y = \frac{-29}{-58} = \frac{1}{2}$$

$\therefore$ Coordinates of D are **(9/2, 1/2).**
M is mid-point of BD and AC both.

$$\therefore \quad \left(\frac{x_1 + \frac{9}{2}}{2}, \frac{y_1 + \frac{1}{2}}{2} \right) = \left(\frac{3+1}{2}, \frac{4+(-1)}{2} \right)$$

$$\Rightarrow \quad \frac{2x_1 + 9}{4} = 2, \frac{2y_1 + 1}{4} = \frac{3}{2} \Rightarrow x_1 = -\frac{1}{2}, y_1 = \frac{5}{2}$$

$\therefore$ Coordinates of B are **(– 1/2, 5/2).**

EXERCISE 10.10

1. Find the acute angle between the lines whose slopes are :

(*i*) 3 and 1/2 (*ii*) $\sqrt{3}$ and $1/\sqrt{3}$.

2. Find the obtuse angle between the lines whose slopes are :
(i) $\sqrt{3}$ and $1/\sqrt{3}$ (ii) $2 - \sqrt{3}$ and $2 + \sqrt{3}$.
3. Find the angle between the lines :
(i) $3x + y - 8 = 0$ and $x + 2y + 2 = 0$ (ii) $2x - y + 1 = 0$ and $x + y + 8 = 0$.
4. Find the angle between the lines :
(i) $\frac{x}{a} + \frac{y}{b} = 1$ and $\frac{x}{b} - \frac{y}{a} = 1$
(ii) $x \cos \alpha_1 + y \sin \alpha_1 = p_1$ and $x \cos \alpha_2 + y \sin \alpha_2 = p_2$ where $\alpha_1 > \alpha_2$.
5. The angle between two lines is $\pi/4$ and the slope of one of them is 1. Find the inclination of the other line.
6. Two lines passing through the point (2, 3) make an angle of 45°. If the slope of one of the lines is 2, find the slope of the other.
7. Find the equation of the straight line which passes through (4, 5) and making angle 45° with the straight line $2x + y + 1 = 0$.
8. The line through (4, 3) and (– 6, 0) intersects the line $5x + y = 0$. Find the angle of intersection.
9. Find the angles of the triangle whose vertices are (3, 4), (4, 4) and (4, 5). It is given that the triangle is not an obtused angled.
10. Find the angles of the acute angled triangle whose vertices are (1, 2), (3, – 2) and (– 3, 2).
11. Two opposite vertices of a square are (2, 1) and (– 2, 1). Find the other vertices.

Answers

1. (i) 45° (ii) 30°
2. (i) 150° (ii) 120°
3. (i) 45° (ii) $\tan^{-1} 3$
4. (i) 90° (ii) $\alpha_1 - \alpha_2$
5. 0°, 90°
6. $\frac{1}{3}, -3$
7. $3x - y - 7 = 0$, $x + 3y - 19 = 0$
8. $\tan^{-1} \frac{\sqrt{3}}{5}$
9. 45°, 45°, 90°
10. $\tan^{-1} \frac{4}{7}, \tan^{-1} \frac{2}{3}, \tan^{-1} 2$
11. (0, – 1), (0, 3).

10.28. CONDITION FOR PARALLELISM OF LINES

Let two lines with slopes m_1 and m_2 be parallel.

∴ Angle between these lines is 0°.

$$\therefore \quad \tan 0^\circ = \left| \frac{m_1 - m_2}{1 + m_1 m_2} \right|$$

$$\Rightarrow \quad \frac{m_1 - m_2}{1 + m_1 m_2} = 0 \quad i.e., \quad m_1 - m_2 = 0 \quad \text{or} \quad \mathbf{m_1 = m_2}.$$

∴ **If two lines are parallel then their slopes are equal.**

Corollary I. The lines $A_1x + B_1y + C_1 = 0$ and $A_2x + B_2y + C_2 = 0$ are **parallel** if their slopes are equal

$$i.e., \quad -\frac{A_1}{B_1} = -\frac{A_2}{B_2} \quad \text{or} \quad \frac{\mathbf{A_1}}{\mathbf{A_2}} = \frac{\mathbf{B_1}}{\mathbf{B_2}} \quad \text{or} \quad \mathbf{A_1B_2 - A_2B_1 = 0}.$$

Corollary II. The slope of the line $Ax + By + C = 0$ is $-A/B$.

$\therefore$ Any line parallel to this line will have slope $-A/B$.

$\therefore$ The equation of a line parallel to $Ax + By + C = 0$ is of the form $y = -\frac{A}{B}x + C'$ ...(1)

where C' is the y-intercept of the line (1)

$$\Rightarrow \quad Ax + By - BC' = 0 \;\Rightarrow\; Ax + By + k = 0 \text{ where } k = -BC' \text{ is some constant.}$$

$\therefore$ **The equation of a line parallel to Ax + By + C = 0 is of the type Ax + By + k = 0, where k is some constant to be determined by using other given conditions.**

Equivalently, $Ax + By + k = 0$ represents the family of lines parallel to the line

$$Ax + By + C = 0.$$

ILLUSTRATIVE EXAMPLES

Example 1. (*i*) *Find the slope of a line which is parallel to the line* $4x + 3y - 9 = 0$.

(*ii*) *Prove that the lines* $x + 3y + 4 = 0$ *and* $2x + 6y - 7 = 0$, *are parallel.*

(P.T.U. B.Pharma, Dec. 2004)

Sol. (*i*) Given line is $\quad 4x + 3y - 9 = 0.$

$$\Rightarrow \quad 3y = -4x + 9 \;\Rightarrow\; y = -\frac{4}{3}x + 3$$

$\therefore$ Slope of given line $= -4/3$.

The required line is parallel to the given line, so the slope of required line is also **– 4/3.**

(*ii*) Given lines are $x + 3y + 4 = 0$ ...(1)

$$2x + 6y - 7 = 0 \qquad ...(2)$$

Slope of line (1) $= -\frac{1}{3} = m_1$

Slope of line (2) $= -\frac{2}{6} = -\frac{1}{3} = m_2$,

$\Rightarrow$ $m_1 = m_2$. Hence the given lines (1) and (2) are parallel.

Example 2. *Find the equation of the straight line that has y-intercept 4 and is parallel to the straight line* $2x - 3y = 7$.

Sol. Given line is $2x - 3y = 7$ ...(1)

$$(1) \Rightarrow \quad 3y = 2x - 7 \;\Rightarrow\; y = \frac{2}{3}x - \frac{7}{3}$$

$\therefore$ Slope of (1) is 2/3.

The required line is parallel to (1), so its slope is also 2/3. y-intercept of required line = 4

$\therefore$ By using '$y = mx + c$' form, the equation of the required line is

$$y = \frac{2}{3}x + 4 \quad \text{or} \quad \mathbf{2x - 3y + 12 = 0.}$$

10.29. CONDITION FOR PERPENDICULARITY OF LINES

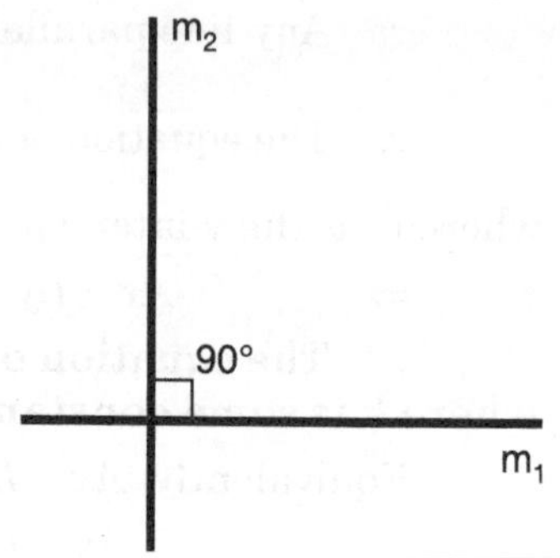

Let two lines with slopes m_1 and m_2 be perpendicular.

∴ Angle between these lines is 90°.

$$\therefore \quad \tan 90° = \left|\frac{m_1 - m_2}{1 + m_1 m_2}\right| \text{ is not defined.}$$

$$\Rightarrow \quad 1 + m_1 m_2 = 0 \quad i.e., \quad \mathbf{m_1 m_2 = -1.}$$

∴ **If two lines are perpendicular then product of their slopes is '– 1'.**

Corollary I. The lines $A_1x + B_1y + C_1 = 0$ and $A_2x + B_2y + C_2 = 0$ are **perpendicular** if product of their slopes is '– 1'

$$i.e., \quad \left(-\frac{A_1}{B_1}\right)\left(-\frac{A_2}{B_2}\right) = -1 \quad \text{or} \quad \mathbf{A_1A_2 + B_1B_2 = 0.}$$

Corollary II. The slope of the line $Ax + By + C = 0$ is $-A/B$.

∴ Any line perpendicular to this line will have slope equal to negative reciprocal of $-A/B$, which is B/A.

∴ The equation of a line perpendicular to $Ax + By + C = 0$ is of the form $y = \frac{B}{A}x + C'$ ...(1)

where C' is the y-intercept of the line.

(1) $\Rightarrow \quad Bx - Ay + AC' = 0 \quad \Rightarrow \quad Bx - Ay + k = 0$, where $k = AC'$ is some constant.

∴ **The equation of a line perpendicular to Ax + By + C = 0 is of the type Bx – Ay + k = 0, where k is some constant to be determined by using other given conditions.**

Equivalently, $Bx - Ay + k = 0$ represents the family of lines perpendicular to the line $Ax + By + C = 0$.

Example 3. *Find the slope of a line which is perpendicular to the line 7x + 4y + 11 = 0.*

Sol. Given line is $7x + 4y + 11 = 0$.

$$\Rightarrow \quad 4y = -7x - 11 \quad \Rightarrow \quad y = \left(-\frac{7}{4}\right)x - \frac{11}{4}$$

∴ Slope of given line $= -\frac{7}{4}$. Let slope of required line be m.

The required line is perpendicular to the given line, so product of their slopes is '– 1'.

$$\therefore \quad m\left(-\frac{7}{4}\right) = -1 \quad i.e., \quad m = \mathbf{\frac{4}{7}}.$$

Alternatively, the slope of required line = negative reciprocal of slope of given line

$$= -\frac{1}{(-7/4)} = \mathbf{\frac{4}{7}}.$$

Example 4. *Find the equation of the straight line passing through (2, 3) and perpendicular to 4x – 3y = 10.*

Sol. Given line is $4x - 3y = 10$ ...(1)

(1) $\Rightarrow$ $3y = 4x - 10 \Rightarrow y = \frac{4}{3}x - \frac{10}{3}$

$\therefore$ Slope of (1) is 4/3.

The required line is perpendicular to (1), so its slope is negative reciprocal of 4/3, which is $-\frac{1}{4/3} = -\frac{3}{4}$. The required line is also to pass through (2, 3).

$\therefore$ By using $y - y_1 = m(x - x_1)$ form, the equation of the required line is

$$y - 3 = -\frac{3}{4}(x - 2) \quad i.e., \quad 4y - 12 = -3x + 6 \quad \text{or} \quad \mathbf{3x + 4y - 18 = 0.}$$

Example 5. *Find the equation of the straight line which is (i) parallel (ii) perpendicular to the line $2x + 3y - 8 = 0$ and passing through the mid-point of the line segment joining (0, 7) and (8, 7).* (P.T.U. B.Pharma, Dec. 2004)

Sol. Let the given points be $A(0, 7)$ and $B(8, 7)$. Let M be its mid-point.

$\therefore$ Coordinates of M are $\left(\frac{0+8}{2}, \frac{7+7}{2}\right)$ *i.e.*, (4, 7).

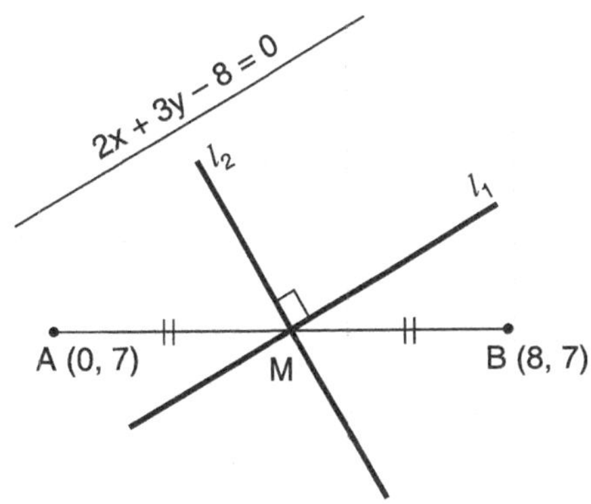

Given line is $2x + 3y - 8 = 0$ ...(1)

$\Rightarrow$ $3y = -2x + 8 \Rightarrow y = \left(-\frac{2}{3}\right)x + \frac{8}{3}$

$\therefore$ Slope of (1) = – 2/3

(*i*) Slope of required line $(l_1) = -2/3$ and this is to pass through M (4, 7).

$\therefore$ By using $y - y_1 = m(x - x_1)$ form, the equation of the required line is

$$y - 7 = -\frac{2}{3}(x - 4) \quad i.e., \quad 3y - 21 = -2x + 8$$

or $\mathbf{2x + 3y - 29 = 0.}$

(*ii*) Slope of required line (l_2) = negative reciprocal of $-2/3 = -\frac{1}{-2/3} = \frac{3}{2}$ and this is to pass through $M(4, 7)$.

$\therefore$ By using '$y - y_1 = m(x - x_1)$' form, the equation of the required line is

$$y - 7 = \frac{3}{2}(x - 4) \quad i.e., \quad 2y - 14 = 3x - 12 \quad \text{or} \quad \mathbf{3x - 2y + 2 = 0.}$$

Example 6. *Find the equation of the right bisector of the line segment joining (1, 1) and (2, 3).*

Sol. Let the given points be $A(1, 1)$ and $B(2, 3)$.
Let M be its mid-point.

$\therefore$ Coordinates of M are $\left(\frac{1+2}{2}, \frac{1+3}{2}\right)$ *i.e.*, $\left(\frac{3}{2}, 2\right)$.

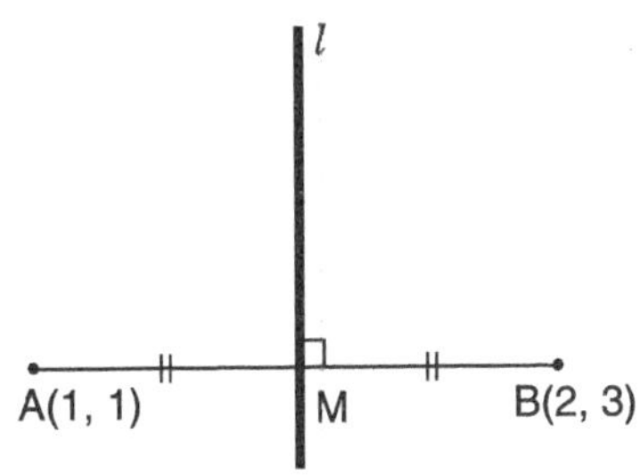

$$\text{Slope of } AB = \frac{3-1}{2-1} = 2$$

$\therefore$ Slope of required line (l) = negative reciprocal of 2 $= -1/2$.

∴ By using '$y - y_1 = m\,(x - x_1)$' form, the equation of the right bisector (l) of AB is

$$y - 2 = -\frac{1}{2}\left(x - \frac{3}{2}\right) \quad i.e., \quad 4y - 8 = -2x + 3 \quad \text{or} \quad \mathbf{2x + 4y - 11 = 0.}$$

Example 7. *Let A(1, 1), B(2, – 1) and C(– 1, 3) be the vertices of ΔABC. Find the equations of the perpendiculars drawn from the vertices to the opposite sides.*

Sol. Given vertices are $A(1, 1)$, $B(2, -1)$ and $C(-1, 3)$. Let AL, BM and CN be the perpendiculars from the vertices.

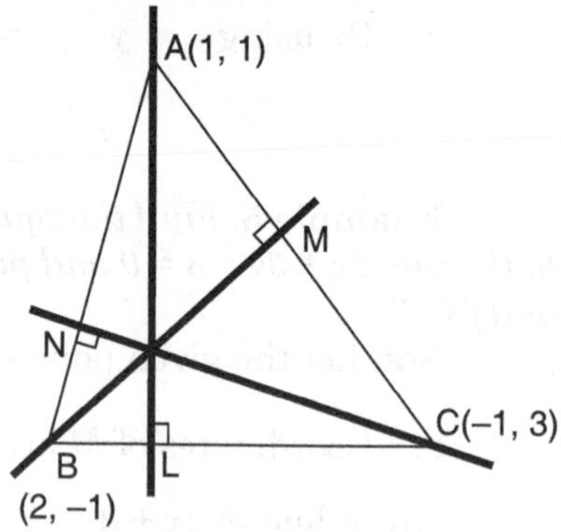

Equation of AL. Slope of $BC = \dfrac{3+1}{-1-2} = -\dfrac{4}{3}$.

∴ Slope of $AL = -\dfrac{1}{-4/3} = \dfrac{3}{4}$

∴ Equation of AL is $y - 1 = \dfrac{3}{4}(x - 1)$

i.e., **3x – 4y + 1 = 0.**

Equation of BM. Slope of $CA = \dfrac{1-3}{1+1} = -1$

∴ Slope of $BM = -\dfrac{1}{-1} = 1$

∴ Equation of BM is $y + 1 = 1.(x - 2)$ *i.e.,* **x – y – 3 = 0.**

Equation of CN. Slope of $AB = \dfrac{-1-1}{2-1} = -2$. ∴ Slope of $CN = -\dfrac{1}{-2} = \dfrac{1}{2}$

∴ Equation of CN is $y - 3 = \dfrac{1}{2}(x + 1)$ *i.e.,* **x – 2y + 7 = 0.**

Example 8. *One side of a rectangle lies along the line 4x + 7y + 5 = 0. Two of its vertices are (– 3, 1) and (1, 1). Find the equation of the other three sides.*

Sol. Let $ABCD$ be the rectangle.

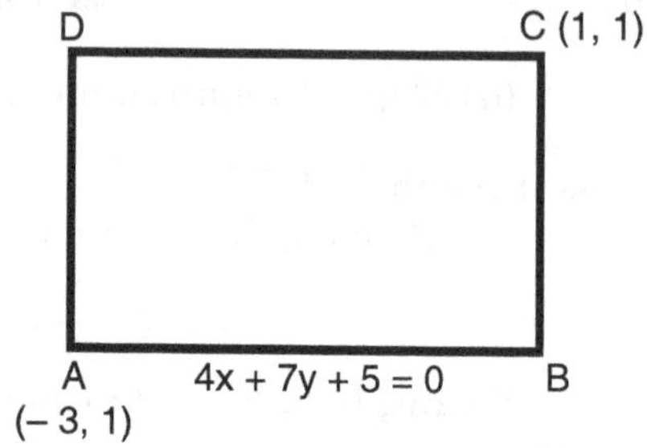

Let equation of AB be $4x + 7y + 5 = 0$.

Since $(-3, 1)$ satisfies the given equation, it is either A or B.

Let $(-3, 1)$ be the coordinates of A.

Slope of $AB = -\dfrac{4}{7}$.

Let coordinates of D be $(1, 1)$.

∴ Slope of $AD = \dfrac{1-1}{1+3} = 0$

But $AB \perp AD$, so $(1, 1)$ cannot be the coordinates of D.

∴ Coordinates of C are $(1, 1)$.

Equation of AD. Let the equation of AD be $7x - 4y + k = 0$. ($\because AD \perp AB$)

$A(-3, 1)$ is on AD.

∴ $7(-3) - 4(1) + k = 0$ or $k = 25$

∴ Equation of AD is **7x – 4y + 25 = 0.**

Equation of CD. Let equation of CD be $4x + 7y + k = 0$. ($\because$ $CD \parallel AB$)

$C(1, 1)$ is on CD.

$\therefore$ $4(1) + 7(1) + k = 0$ or $k = -11$

$\therefore$ Equation of CD is $\mathbf{4x + 7y - 11 = 0}$

Equation of BC. Let equation of BC be $7x - 4y + k = 0$. ($\because$ $BC \perp AB$)

$C(1, 1)$ is on BC.

$\therefore$ $7(1) - 4(1) + k = 0$ or $k = -3$

$\therefore$ Equation of BC is $\mathbf{7x - 4y - 3 = 0}$.

WORKING RULES FOR SOLVING PROBLEMS

Rule I. *Lines with slopes m_1 and m_2 are :*

(i) parallel if and only if $m_1 = m_2$.

(ii) perpendicular if and only if $m_1 m_2 = -1$.

Rule II. *The equation of a straight line parallel to $Ax + By + C = 0$ is of the form*

$$Ax + By + k = 0.$$

Rule III. *The equation of a straight line perpendicular to $Ax + By + C = 0$ is of the form $Bx - Ay + k = 0$.*

EXERCISE 10.11

1. Find the slope of the line which is parallel to the line whose slope is :

(i) 1/2 (ii) – 3/7.

2. Find the slope of the line which is perpendicular to the line whose slope is :

(i) 4 (ii) – 5/7.

3. Find the slope of a line which is parallel to the line :

(i) $2x - y + 8 = 0$ (ii) $\dfrac{x}{a} + \dfrac{y}{b} = 1$.

4. Find the slope of a line which is perpendicular to the line :

(i) $x + y - 9 = 0$ (ii) $x \cos \alpha + y \sin \alpha = p$.

5. (i) Find the equation of the straight line parallel to $2x + 5y = 7$ and passing through (– 1, 4).

(ii) Find the equation of the straight line parallel to $2x - y + 8 = 0$ and having y-intercept 4.

6. (i) Find the equation of the straight line perpendicular to $2x + 3y - 8 = 0$ and passing through (9, 2).

(ii) Find the equation of the straight line perpendicular to $2x + 4y - 7 = 0$ and having y-intercept 5.

(iii) Find the equation of the line passing through the point (2, – 1) and has x-intercept as 4.

(P.T.U. B.Pharma, Dec. 2005)

7. Find the equation of the straight line which is (i) parallel (ii) perpendicular to the line $4x - y + 8 = 0$ and passing through the mid-point of the line segment joining (1, 5) and (3, 11) (iii) find the equation of the line passing through the mid-point of the line segment joining the points (1, 3) and (2, – 1) and parallel to the line $3x - y = 7$ *(P.T.U. B.Pharma, Dec. 2004)*

8. Find the equation of the straight line which is (i) parallel (ii) perpendicular to the line $Ax + By + C = 0$ and passing through the point (x_1, y_1).

9. Find the equation of the right bisector of the line segment joining the points :

(i) (0, 9) and (6, 1) (ii) (α, β) and (β, α).

10. For the triangle ABC whose vertices are $A(-2, 3)$, $B(4, -3)$ and $C(4, 5)$, find the equation of the :
(*i*) right bisector of BC (*ii*) altitude from A.
(*iii*) straight line parallel to BC and passing through A.

11. Find the equation of the straight line joining the points $(3, -1)$ and $(2, 3)$. Also find the equation of another straight line perpendicular to this line and passing through $(5, 2)$.

12. Show that the equation of the straight line passing through $(a \cos^3 \theta, a \sin^3 \theta)$ and perpendicular to the line $x \sec \theta + y \operatorname{cosec} \theta = a$ is $x \cos \theta - y \sin \theta = a \cos 2\theta$.

Answers

1. (*i*) 1/2 (*ii*) – 3/7 **2.** (*i*) – 1/4 (*ii*) 7/5
3. (*i*) 2 (*ii*) $-b/a$ **4.** (*i*) 1 (*ii*) $\tan \alpha$
5. (*i*) $2x + 5y - 18 = 0$ (*ii*) $2x - y + 4 = 0$ **6.** (*i*) $3x - 2y - 23 = 0$ (*ii*) $2x - y + 5 = 0$
(*iii*) $y + 1 = \frac{1}{2}(x - 2)$
7. (*i*) $4x - y = 0$ (*ii*) $x + 4y - 34 = 0$ (*iii*) $6x - 2y - 7 = 0$
8. (*i*) $Ax + By - (Ax_1 + By_1) = 0$ (*ii*) $Bx - Ay - (Bx_1 - Ay_1) = 0$
9. (*i*) $3x - 4y + 11 = 0$ (*ii*) $x - y = 0$ **10.** (*i*) $y - 1 = 0$ (*ii*) $y - 3 = 0$
(*iii*) $x + 2 = 0$ **11.** $4x + y - 11 = 0$, $x - 4y + 3 = 0$.

Hint

12. Slope of required line $= -\dfrac{1}{-\sec \theta/\operatorname{cosec} \theta} = \dfrac{\cos \theta}{\sin \theta}$.

$\therefore$ Required line is $y - a \sin^3 \theta = \dfrac{\cos \theta}{\sin \theta}(x - a \cos^3 \theta)$ or $x \cos \theta - y \sin \theta = a(\cos^4 \theta - \sin^4 \theta)$
$= a \cos 2\theta$.

11

Functions and Limits

11.1. DEFINITIONS

(*i*) **Constant.** A symbol representing exactly one number is called a **constant.** For example, 2, -4, π are constants.

(*ii*) **Variable.** A symbol representing any one of a set of numbers is called a **variable.**

For example :

(*a*) If x represents any number between 4 and 7, then x is a variable.

(*b*) If x represents a day in April, then x is a variable and takes value in the set {1, 2, 3, ..., 30}.

open interval (a, b)

(*iii*) **Intervals.** If a and b are real numbers such that $a < b$, then the set of all numbers between a and b is called the **open interval** from a to b and is written (a, b). The points a and b are called the **end points** of the open interval (a, b). An open interval does not contain its end points *i.e.*, $a \notin (a, b)$ and $b \notin (a, b)$.

For example, (2, 5) denotes the set $\{x \in \mathbf{R} : 2 < x < 5\}$ of all real numbers between 2 and 5.

closed interval [a, b]

If a and b are real numbers such that $a < b$, then the set of all numbers from a to b is called the **closed interval** from a to b and is written $[a, b]$. The points a and b are called the **end points** of the closed interval $[a, b]$. A closed interval contains its end points *i.e.*, $a \in [a, b]$ and $b \in [a, b]$.

For example, [2, 4] denotes the set $[x \in \mathbf{R} : 2 \le x \le 4\}$ of all real numbers from 2 to 4.

Similarly, for $a < b$, we define $[a, b) = \{x : a \le x < b\}$ and $(a, b] = \{x : a < x \le b\}$.

These intervals are called **semi-open** or **semi-closed intervals.**

If $a < b$, then the intervals (a, b), $[a, b]$, $[a, b)$, $(a, b]$ are called **finite intervals.**

If a is any real number, then the sets of the type $\{x : x < a\}$, $\{x : x \le a\}$, $\{x : x > a\}$, and $\{x : x \ge a\}$, are called **infinite intervals** and are respectively written as $(-\infty, a)$, $(-\infty, a]$, (a, ∞) and $[a, \infty)$.

The set **R** of real numbers is also an infinite set, written as $(-\infty, \infty)$.

Caution. Open intervals do not contain their end-points whereas closed intervals do.

For example, $7 \notin (3, 7)$ and $7 \in [3, 7]$.

11.2. FUNCTION

(P.T.U. May, 2007)

We know that the area of a circle depends upon its radius. If the radius of a circle be 4 cm, then its area is uniquely determined and is equal to 16π cm^2. If the variables x and y respectively represent the radius and the area of a circle, then we have $y = \pi x^2$. For any positive value of x, the value of area, y is uniquely determined. The value of the variable y $(= \pi x^2)$ depends upon the value of the variable x.

The square root a positive real number is not unique whereas the positive square root of a positive number is uniquely determined. Presently, we are interested in variables where any particular value of one variable, say x, gives a uniquely determined value of another variable, say y. Such a relation between two variables is called a *function*. The function concept lays the foundation of the study of the most important branch 'Calculus' of mathematics.

Let A and B be two non-empty sets. Let variables x and y represent the elements of A and B respectively. If there is given a rule or device denoted by 'f' so that for each value of x, there is a uniquely determined value of y, then 'f' is said to be a **function** from A to B and written as $f : A \to B$. The variables x and y are respectively called the **independent variable** and the **dependent variable** of the function. This is so, because each y-value depend on the corresponding x-value. It is customary to say that the dependent variable 'y' is a function of the independent variable 'x' and written as $y = f(x)$. This is read as "y is equal to f of x". The letter f in the notation $y = f(x)$ indicates that some kind of operations must be performed on the value of x to obtain the value of y.

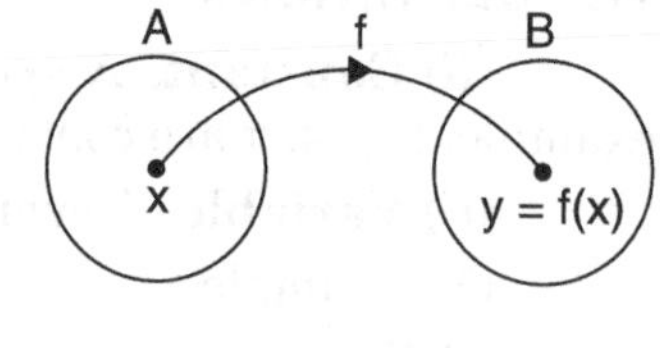

For example, let x and y be variables and $y = f(x)$, where $f(x) = \frac{x^2 + x}{x - 7}$. Here, for each admissible value of x, we get a unique value of y. Thus, y is a function of x. If $x = 2$, then $y = \frac{(2)^2 + 2}{2 - 7} = -\frac{6}{5}$. We say that the value of the function $f(x)$ at $x =$ '2' is $-\frac{6}{5}$. Notationally, this is written as $f(2) = -\frac{6}{5}$. For a value 'a' of x, the value $f(a)$ is also called the **image** of 'a' under f.

Thus, we can say that the image of '2' is '$-\frac{6}{5}$'. Similarly, we have $f(0) = 0$, $f(8) = 72$, etc.

Remark 1. If at a point a, the function $f(x)$ takes any of the indeterminate forms : $\frac{0}{0}, \frac{\infty}{\infty}, \infty - \infty$, 0°, 1^∞, ∞°, etc., then we say that $f(x)$ is not defined at $x = a$. For example, the functions $\frac{x^2 - 4}{x - 2}$, x^x takes indeterminate forms $\frac{0}{0}$, 0^0 at $x = 2$, $x = 0$ respectively.

Remark 2. The symbol $f(x)$ is read as "f of x" but never "f times x". It is used to denote a given function of x. If in the same problem, another function of x occurs, then that function may be denoted by any of the following ways : $f_1(x)$, $g(x)$, $h(x)$, $\phi(x)$, etc.

Remark 3. Imaginary numbers have no role in calculus and the term will be used only for pointing out their exclusion.

Illustration :

Let A = {1, 2, 3} and B = {1, 2, 3, ..., 10}.

Let $f : A \to B$ be the correspondence which assigns to each elements in A, its square.

Hence, we have $f(1) = 1^2 = 1$, $f(2) = 2^2 = 4$, $f(3) = 3^2 = 9$.

Since to each element (1 or 2 or 3) of A, there correspond exactly one element of B, so f is a function.

This association can also be shown diagrammatically. In this case, every element of B is not the image of some element of A.

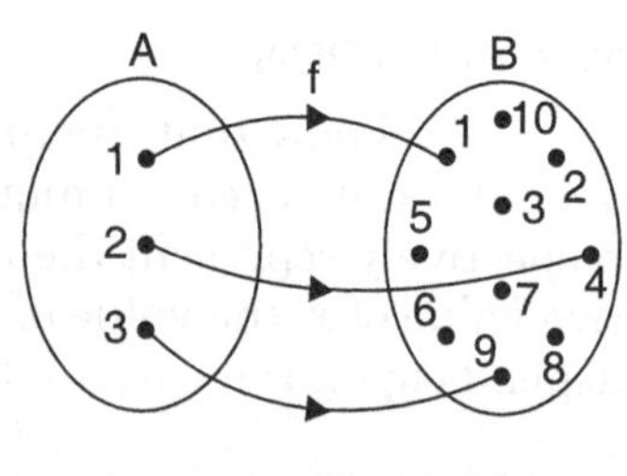

For this function, we have

$$f(A) = \{1^2, 2^2, 3^2\} = \{1, 4, 9\} \subseteq B.$$

Sometimes independent and dependent variables of a function are connected by algebraic equations.

For example, $y = f(x) = x^2 + 6x + 7$ represents a function, because for every value of x, there exists exactly one value of y.

If $y = \pm\ 2x$, then for every non-zero value of x, there will correspond two values of y. Therefore, y is not a function of x.

Some more examples :

1. If y be the volume of the sphere whose radius is x, then y is related with x as $y = \frac{4}{3}\pi x^3$. Here, to each **positive** value of x, there is a unique value of y.

$\therefore$ y is a function of x.

2. $y = \begin{cases} 1+x & \text{if} \quad x \geq 2 \\ 2x+7 & \text{if} \quad x \leq 2 \end{cases}$,

is not a function of x, for if $x = 2$, then $y = 3, 11$. If we modify the equations as

$$y = \begin{cases} 1+x & \text{if} \quad x > 2 \\ 2x+7 & \text{if} \quad x \leq 2 \end{cases}, \text{ then } y \text{ becomes a function of } x.$$

11.3. REAL FUNCTION

Let f be a function from set A to set B. If A and B are subsets of the real number system **R**, then the function $f : A \to B$ is called a **real valued function of a real variable** and in short, we call such a function as a **real function**. For example, the function $\phi : \mathbf{R} \to \mathbf{R}$ defined by $\phi(x) = x^2 + 3x + 7$, $x \in \mathbf{R}$ is a real function.

11.4. EQUAL FUNCTIONS

Let f and g be two functions from set A to set B. The functions f and g are said to be **equal functions** if $f(x) = g(x)$ for all $x \in A$.

For example, the functions $f(x) = \dfrac{x^2 - 1}{x - 1}$, $x \in \mathbf{R} - \{1\}$ and $g(x) = x + 1$, $x \in \mathbf{R}$ are not equal functions because $f(1)$ is not defined whereas $g(1) = 2$ and for all real numbers other than '1' we have $f(x) = g(x)$.

In calculus, we consider only real functions. From now onwards, a 'real functions' would be referred just as a 'function'.

ILLUSTRATIVE EXAMPLES

Example 1. *If* $f(x) = \dfrac{4x - 3}{1 + 7x}$, *find* $f(0)$, $f(5)$, $f(1/7)$, $f(f(2))$.

Sol. We have $f(x) = \dfrac{4x - 3}{1 + 7x}$

$$f(0) = \frac{4(0) - 3}{1 + 7(0)} = \frac{0 - 3}{1 + 0} = -\mathbf{3} \qquad f(5) = \frac{4(5) - 3}{1 + 7(5)} = \frac{20 - 3}{1 + 35} = \mathbf{\frac{17}{36}}$$

$$f\left(\frac{1}{7}\right) = \frac{4(1/7) - 3}{1 + 7(1/7)} = \frac{4/7 - 3}{1 + 1} = -\mathbf{\frac{17}{14}}$$

$$f(f(2)) = f\left[\frac{4(2)-3}{1+7(2)}\right] = f\left(\frac{8-3}{1+14}\right) = f\left(\frac{5}{15}\right) = f\left(\frac{1}{3}\right) = \frac{4\left(\frac{1}{3}\right)-3}{1+7\left(\frac{1}{3}\right)} = \frac{\frac{4-9}{3}}{\frac{3+7}{3}} = \frac{-5}{10} = -\,\mathbf{\frac{1}{2}}.$$

Example 2. *If* $f(x) = \begin{cases} 4x-1 & \text{if } -3 < x < 0 \\ 1+x & \text{if } 0 \le x < 2 \\ x^2+9 & \text{if } 4 \le x < 5 \end{cases}$*, find f(4), f(– 2), f(0), f(1), f(4.9).*

Sol. We have

$$f(x) = \begin{cases} 4x-1 & \text{if } -3 < x < 0 \\ 1+x & \text{if } 0 \le x < 2 \\ x^2+9 & \text{if } 4 \le x < 5 \end{cases}.$$

Now $f(4) = (4)^2 + 9 = \mathbf{25}$ $(\because\ 4 \le 4 < 5)$

$f(-2) = 4(-2) - 1 = \mathbf{-9}$ $(\because\ -3 < -2 < 0)$

$f(0) = 1 + 0 = \mathbf{1}$ $(\because\ 0 \le 0 < 2)$

$f(1) = 1 + 1 = \mathbf{2}$ $(\because\ 0 \le 1 < 2)$

$f(4.9) = (4.9)^2 + 9 = \mathbf{33.01}.$ $(\because\ 4 \le 4.9 < 5)$

Example 3. *If* $f(x) = \log \frac{1+x}{1-x}$*, show that* $f\left(\frac{2x}{1+x^2}\right) = 2f(x)$.

Sol. We have $f(x) = \log \frac{1+x}{1-x}$.

$$f\left(\frac{2x}{1+x^2}\right) = \text{value of '}f\text{' at } \frac{2x}{1+x^2} = \log \frac{1+\left(\frac{2x}{1+x^2}\right)}{1-\left(\frac{2x}{1+x^2}\right)} = \log \frac{1+x^2+2x}{1+x^2-2x}$$

$$= \log \frac{(1+x)^2}{(1-x)^2} = \log \left(\frac{1+x}{1-x}\right)^2 = 2 \log \frac{1+x}{1-x} = 2f(x).$$

EXERCISE 11.1

1. If $f : \mathbf{R} \to \mathbf{R}$ is a function defined by $f(x) = x^2 + 5x + 6$, $x \in \mathbf{R}$, find $f(0)$, $f(2)$, $f(-3)$, $f(4)$.
2. If $f(x) = \begin{cases} 4x+1 & x < 2 \\ 1-x & x \ge 2 \end{cases}$, find $f(0)$, $f(4)$, $f(1)$, $f(5)$.
3. If $f(x) = \sqrt{x^2+5x+7}$, find $f(2)$, $f(-3)$, $f(7)$.
4. If $f(x) = |\,x+2\,| + |\,x-3\,|$, find $f(-3)$, $f(0)$, $f(2)$, $f(5)$.
5. If $f(x) = x^3$, find $\frac{f(x+h)-f(x)}{h}$.

6. Let f be defined by $f(x) = x - 4$ and g be defined by $g(x) = \begin{cases} \dfrac{x^2 - 16}{x + 4}, & x \neq -4 \\ k, & x = -4 \end{cases}$. Find k such that $f(x) = g(x)$, for all x.

7. If $f(x) = \dfrac{1 + e^x}{1 - e^x}$, then show that $f(-x) = -f(x)$.

8. If $f(x) = \dfrac{5x + 3}{4x - 5}$, $x \neq \dfrac{5}{4}$, show that $f(f(x))$ is an identity function.

9. If $f(x) = x^3 - \dfrac{1}{x^3}$, then show that $f(x) + f\left(\dfrac{1}{x}\right) = 0$.

10. A function $f(x)$ is said to be an **even** function if $f(-x) = f(x)$. Show that the following functions are even functions :

(*i*) $f(x) = x^2 + \cos x$ (*ii*) $g(x) = |x| + 4$

(*iii*) $h(x) = x^4 + 2x^2 + 9$.

11. A function $f(x)$ is said to be an **odd** function if $f(-x) = -f(x)$. Show that the following functions are odd functions :

(*i*) $f(x) = x^3 + x$ (*ii*) $g(x) = \sin x$

(*iii*) $\phi(x) = x - x^5 + x^7$ (*iv*) $f(x) = \log(x + \sqrt{x^2 + 1})$.

Answers

1. 6, 20, 0, 42 **2.** 1, – 3, 5, – 4 **3.** $\sqrt{21}$, 1, $\sqrt{91}$ **4.** 7, 5, 5, 9

5. $3x^2 + 3xh + h^2$ **6.** – 8.

Hints

6. $x \neq -4 \Rightarrow g(x) = \dfrac{x^2 - 16}{x + 4} = x - 4 = f(x)$. $\therefore$ $f(x) = g(x)$ for $x = -4$

Also $f(-4) = -4 - 4 = -8$. $\therefore$ $f = g$ if $g(-4) = -8$.

11. (*iv*) $f(-x) = \log(-x + \sqrt{x^2 + 1}) = \log \dfrac{(\sqrt{x^2 + 1} - x)(\sqrt{x^2 + 1} + x)}{\sqrt{x^2 + 1} + x} = -\log(x + \sqrt{x^2 + 1})$.

11.5. DOMAIN AND RANGE OF A FUNCTION

Let $y = f(x)$ be a real function. The **domain** of this function is defined as the set of all values of the independent variable x for which the function is defined. The domain of the function $y = f(x)$ is written as $D(f)$.

$\therefore$ **D(f) = {x ∈ R : f(x) is defined i.e., it is a real number}.**

For example, if $y = f(x) = \sqrt{x + 2}$, then $f(x)$ is defined only when $x + 2$ is not negative *i.e.*, when $x \geq -2$. Thus, we have $D(f) = [-2, \infty)$.

The **range** of the real function $y = f(x)$ is defined as the set of all values assumed by the dependent variable y as its independent variable x runs over the domain of the function. The range of the function is written as $R(f)$.

$\therefore$ **R(f) = {f(x) : x ∈ D (f)}.**

For example, if $y = f(x) = |x|$, then the dependent variable y takes only non-negative real number values. Thus, $R(f) = [0, \infty)$.

WORKING RULES FOR SOLVING PROBLEMS

Step I. *Find the values of x for which the given function takes the form :*

$$\frac{1}{zero}, \sqrt{a \text{ negative number}}, \frac{1}{\sqrt{a \text{ non-positive number}}}, \text{ etc.}$$

Step II. *The set of real numbers, excluding those obtained in* **step I**, *is the domain of the given function.*

Step III. *For the range, find the value of x in terms of the dependent variable y and identify the values of y for which the variable x is not in the domain of the function.*

Step IV. *The set of real numbers, excluding those obtained in* **step III**, *is the range of the given function.*

ILLUSTRATIVE EXAMPLES

Example 1. *Find the domain and range of the following functions :*

(*i*) $\frac{1}{x-2}$ (*ii*) $\frac{x}{x+5}$ (*iii*) $\frac{x}{2+x^2}$ (*iv*) $\frac{x^2}{4+x^2}$.

Sol. (*i*) Let $y = f(x) = \frac{1}{x-2}$ ...(1)

Domain. $f(x)$ is defined only when $x - 2 \neq 0$ *i.e.*, when $x \neq 2$. $\therefore$ $D(f) = \mathbf{R} - \{2\}$

Range. (1) $\Rightarrow$ $xy - 2y = 1$ $\Rightarrow$ $xy = 1 + 2y$ $\Rightarrow$ $x = \frac{1+2y}{y}$

$\therefore$ $y \neq 0$, otherwise x is not defined. $\therefore$ $R(f) = \mathbf{R} - \{0\}$.

(*ii*) Let $y = f(x) = \frac{x}{x+5}$ ...(1)

Domain. $f(x)$ is defined only when $x + 5 \neq 0$ *i.e.*, when $x \neq -5$. $\therefore$ $D(f) = \mathbf{R} - \{-5\}$.

Range. (1) $\Rightarrow$ $xy + 5y = x$ $\Rightarrow$ $x - xy = 5y$ $\Rightarrow$ $x = \frac{5y}{1-y}$

$\therefore$ x is not defined when $y = 1$. $\therefore$ $R(f) = \mathbf{R} - \{1\}$.

(*iii*) Let $y = f(x) = \frac{x}{2+x^2}$ ...(1)

Domain. $2 + x^2 \neq 0$ for all $x \in \mathbf{R}$.

$\therefore$ $f(x)$ is defined for all real numbers. $\therefore$ $D(f) = \mathbf{R}$.

Range. (1) $\Rightarrow$ $2y + x^2y = x$ $\Rightarrow$ $yx^2 - x + 2y = 0$

$\therefore$ Disc. $= b^2 - 4ac = (-1)^2 - 4y \,.\, 2y \geq 0$ ($\because$ x is real)

$$\Rightarrow \quad 1 - 8y^2 \geq 0 \quad \Rightarrow \quad y^2 \leq \frac{1}{8} \quad \Rightarrow \quad -\frac{1}{\sqrt{8}} \leq y \leq \frac{1^*}{\sqrt{8}} \text{ i.e., } -\frac{1}{2\sqrt{2}} \leq y \leq \frac{1}{2\sqrt{y}}$$

$$\therefore \quad R(f) = \left[-\frac{1}{2\sqrt{2}}, \frac{1}{2\sqrt{2}}\right].$$

* **Why this step.** If a is any positive number, then $y^2 \leq a$ implies $-\sqrt{a} \leq y \leq \sqrt{a}$ and $y^2 \geq a$ implies $y \leq -\sqrt{a}$ or $y \geq \sqrt{a}$.

(*iv*) Let $y = f(x) = \dfrac{x^2}{4 + x^2}$ ...(1)

Domain. $4 + x^2 \neq 0$ for all $x \in \mathbf{R}$. $\therefore$ $D(f) = \mathbf{R}$

Range. (1) $\Rightarrow$ $4y + yx^2 = x^2$ $\Rightarrow$ $x^2 - yx^2 = 4y$ $\Rightarrow$ $x^2 = \dfrac{4y}{1-y}$.

$x \in \mathbf{R}$ $\Rightarrow$ $x^2 \geq 0 \Rightarrow$ $\dfrac{4y}{1-y} \geq 0$ $\Rightarrow$ either $4y \geq 0,\ 1 - y > 0$ or $4y \leq 0,\ 1 - y < 0$*

$\Rightarrow$ $4y \geq 0,\ 1 - y > 0$ $\Rightarrow$ $y \geq 0,\ y < 1$ $\Rightarrow$ $y \in [0, 1)$

$4y \leq 0,\ 1 - y < 0$ $\Rightarrow$ $y \leq 0,\ y > 1$, which is impossible.

$\therefore$ $y \in [0, 1)$ $\therefore$ $R(f) = \mathbf{[0, 1)}$.

Alternative method

(1) $\Rightarrow$ $y = \dfrac{4 + x^2 - 4}{4 + x^2} = 1 - \dfrac{4}{4 + x^2} < 1$. $\therefore$ $y < 1$ $\left(\because \dfrac{4}{4 + x^2} > 0\right)$

Also $y = \dfrac{x^2}{4 + x^2} \geq 0$. $\therefore$ $y \geq 0$

$\therefore$ $R(f) = \mathbf{[0, 1)}$.

Example 2. *Find the domain and range of the following functions :*

(*i*) $\sqrt{x + 2}$ (*ii*) $\dfrac{1}{\sqrt{x - 4}}$ (*iii*) $\dfrac{1}{(2x - 3)(x + 1)}$ (*iv*) $\sqrt{(x - 1)(3 - x)}$.

Sol. (*i*) Let $y = f(x) = \sqrt{x + 2}$ ...(1)

Domain. $f(x)$ is defined only when $x + 2 \geq 0$ *i.e.*, when $x \geq -2$. $\therefore$ $D(f) = \mathbf{[-2, \infty)}$.

Range. (1) $\Rightarrow$ $x + 2 = y^2$ $\Rightarrow$ $x = y^2 - 2$ $\Rightarrow$ $y^2 - 2 \geq -2$ $\Rightarrow$ $y^2 \geq 0$ $\Rightarrow$ $y \geq 0$

because $y = \sqrt{x + 2}$ is non-negative.

$\therefore$ $R(f) = \mathbf{[0, \infty)}$.

(*ii*) Let $y = f(x) = \dfrac{1}{\sqrt{x - 4}}$ ...(1)

Domain. $f(x)$ is defined only when $x - 4 > 0$ *i.e.*, when $x > 4$. $\therefore$ $D(f) = \mathbf{(4, \infty)}$.

Range. (1) $\Rightarrow$ $y^2 = \dfrac{1}{x - 4}$ $\Rightarrow$ $x - 4 = \dfrac{1}{y^2}$ $\Rightarrow$ $x = \dfrac{1}{y^2} + 4$.

$\therefore$ $\dfrac{1}{y^2} + 4 > 4$ $\Rightarrow$ $\dfrac{1}{y^2} > 0$ $\Rightarrow$ $y^2 > 0$ $\Rightarrow$ $y > 0$ because $y = \dfrac{1}{\sqrt{x - 4}}$ is positive.

$\therefore$ $R(f) = \mathbf{(0, \infty)}$.

(*iii*) Let $y = f(x) = \dfrac{1}{(2x - 3)(x + 1)}$. ...(1)

Domain. $f(x)$ is not defined when $2x - 3 = 0$ or $x + 1 = 0$ *i.e.*, when $x = 3/2$ or $x = -1$.

$\therefore$ $D(f) = \mathbf{R - \{-1, 3/2\}}$.

Range. (1) $\Rightarrow$ $(2x - 3)(x + 1)\,y = 1$ *i.e.*, $2yx^2 - yx - (3y + 1) = 0$

$\therefore$ Disc. $= (b^2 - 4ac) = (-y)^2 + 4(2y)(3y + 1) \geq 0$. ($\because$ x is real)

$\Rightarrow$ $25y^2 + 8y \geq 0$ $\Rightarrow$ $y(25y + 8) \geq 0$

***Why this step.** If $\dfrac{a}{b} \geq 0$, then either $a \geq 0,\ b > 0$ or $a \leq 0,\ b < 0$.

Either $y \geq 0$, $25y + 8 \geq 0$	or $y \leq 0$, $25y + 8 \leq 0$
$\Rightarrow$ $y \geq 0$, $y \geq -8/25$	$\Rightarrow$ $y \leq 0$, $y \leq -8/25$
$\Rightarrow$ $y \geq 0$	$\Rightarrow$ $y \leq -8/25$

$\therefore$ $\quad$ **R(f) = $(-\infty, -8/25] \cup [0, \infty)$.**

(*iv*) Let $y = f(x) = \sqrt{(x-1)(3-x)}$ ...(1)

Domain. $f(x)$ is defined when $(x - 1)(3 - x) \geq 0$.

$\therefore$ $x - 1 \geq 0$, $3 - x \geq 0$	or $x - 1 \leq 0$, $3 - x \leq 0$
$\Rightarrow$ $x \geq 1$, $x \leq 3$	$\Rightarrow$ $x \leq 1$, $x \geq 3$
$\Rightarrow$ $x \in [1, 3]$	This is impossible.

$\therefore$ $\quad$ **D(f) = [1, 3].**

Range. (1) $\Rightarrow$ $y \geq 0$ and $y^2 = (x - 1)(3 - x) = -x^2 + 4x - 3$

$\Rightarrow$ $x^2 - 4x + (3 + y^2) = 0$

$\therefore$ Disc. $= (-4)^2 - 4(1)(3 + y^2) \geq 0$ $\quad$ ($\because$ x is real)

$\Rightarrow$ $16 - 12 - 4y^2 \geq 0$

$\Rightarrow$ $4y^2 \leq 4$ $\Rightarrow$ $y^2 \leq 1$

$\Rightarrow$ $-1 \leq y \leq 1$ $\Rightarrow$ $0 \leq y \leq 1$ because $y \geq 0$.

$\therefore$ $\quad$ **R(f) = [0, 1].**

Example 3. *Find the domain and range of the following functions :*

(*i*) $-|x|$ $\quad$ (*ii*) $|x - 3|$ $\quad$ (*iii*) $1 - |x - 2|$ $\quad$ (*iv*) $\dfrac{x+4}{|x+4|}$.

Sol. (*i*) Let $y = f(x) = -|x|$. ...(1)

Domain. $f(x)$ is defined for all real numbers. $\quad$ $\therefore$ D(f) = **R**.

Range. (1) $\Rightarrow$ $|x| = -y$ $\therefore$ $-y \geq 0$ *i.e.*, $y \leq 0$. $\quad$ $\therefore$ R(f) = **$(-\infty, 0]$.**

(*ii*) Let $y = f(x) = |x - 3|$. ...(1)

Domain. $f(x)$ is defined for all real numbers. $\quad$ $\therefore$ D(f) = **R**.

Range. (1) $\Rightarrow$ $y \geq 0$ because $|x - 3| \geq 0$. $\quad$ $\therefore$ R(f) = **$[0, \infty)$.**

(*iii*) Let $y = f(x) = 1 - |x - 2|$. ...(1)

Domain. $f(x)$ is defined for all real numbers. $\quad$ $\therefore$ D(f) = **R**.

Range. (1) $\Rightarrow$ $|x - 2| = 1 - y$ $\Rightarrow$ $1 - y \geq 0$ $\Rightarrow$ $y \leq 1$ $\quad$ $\therefore$ R(f) = **$(-\infty, 1]$.**

(*iv*) Let $y = f(x) = \dfrac{x+4}{|x+4|}$.

Domain. $f(x)$ is not defined when $x + 4 = 0$ *i.e.*, when $x = -4$. $\quad$ $\therefore$ D(f) = **R – {– 4}.**

Range. $x \neq -4$ $\Rightarrow$ $x < -4$ or $x > -4$.

Now, $\quad x < -4 \Rightarrow x + 4 < 0 \Rightarrow |x + 4| = -(x + 4) \Rightarrow f(x) = \dfrac{x+4}{-(x+4)} = -1$

and $\quad x > -4 \Rightarrow x + 4 > 0 \Rightarrow |x + 4| = x + 4 \Rightarrow f(x) = \dfrac{x+4}{x+4} = 1$

This function can take only two values – 1 and 1. $\quad$ $\therefore$ R(f) = **{– 1, 1}.**

Example 4. *Find the domain and range of the following functions :*

(*i*) $f(x) = \begin{cases} 2 - x & \text{if} \quad 0 < x < 2 \\ x - 1 & \text{if} \quad 3 \le x < 4 \end{cases}$ (*ii*) $f(x) = \begin{cases} x + 7 & \text{if} \quad -3 \le x < 5 \\ x^2 & \text{if} \quad 5 \le x < 7 \\ 6 - 2x & \text{if} \quad x \ge 7 \end{cases}$

Sol. (*i*) We have $f(x) = \begin{cases} 2 - x & \text{if} \quad 0 < x < 2 \\ x - 1 & \text{if} \quad 3 \le x < 4 \end{cases}$.

Domain. $f(x)$ is defined if $0 < x < 2$ or $3 \le x < 4$.

$\therefore$ $D(f) = \mathbf{(0, 2) \cup [3, 4)}$.

Range. $0 < x < 2 \Rightarrow 0 > -x > -2 \Rightarrow 2 + 0 > 2 - x > 2 - 2$

$\Rightarrow 0 < f(x) < 2 \Rightarrow f(x) \in (0, 2)$.

$3 \le x < 4 \Rightarrow 3 - 1 \le x - 1 < 4 - 1$

$\Rightarrow 2 \le f(x) < 3 \Rightarrow f(x) \in [2, 3)$

$\therefore$ $R(f) = (0, 2) \cup [2, 3) = \mathbf{(0, 3)}$.

(*ii*) We have $f(x) = \begin{cases} x + 7 & \text{if} \quad -3 \le x < 5 \\ x^2 & \text{if} \quad 5 \le x < 7 \\ 6 - 2x & \text{if} \quad x \ge 7 \end{cases}$

Domain. $f(x)$ is defined if $-3 \le x < 5$ or $5 \le x < 7$ or $x \ge 7$.

$\therefore$ $D(f) = [-3, 5) \cup [5, 7) \cup [7, \infty) = \mathbf{[-3, \infty)}$.

Range. $-3 \le x < 5 \Rightarrow -3 + 7 \le x + 7 \le 5 + 7$

$\Rightarrow 4 \le f(x) \le 12 \Rightarrow f(x) \in [4, 12]$.

$5 \le x < 7 \Rightarrow 25 \le x^2 < 49$

$\Rightarrow 25 \le f(x) < 49 \Rightarrow f(x) \in [25, 49)$

$x \ge 7 \Rightarrow -2x \le -2(7) \Rightarrow 6 - 2x \le 6 - 14$

$\Rightarrow f(x) \le -8 \Rightarrow f(x) \in (-\infty, -8]$

$\therefore$ $R(f) = [4, 12] \cup [25, 49) \cup (-\infty, -8] = \mathbf{(-\infty, -8] \cup [4, 12] \cup [25, 49)}$.

EXERCISE 11.2

Find the domain of the following functions :

1. $f(x) = 2x + 1$
2. $f(x) = x^2 + 1$
3. $f(x) = \dfrac{x}{x^2 + 3x + 2}$
4. $f(x) = \dfrac{x + 7}{x^2 - 8x + 4}$
5. $f(x) = \dfrac{x^3 - 5x + 3}{x^2 - 1}$
6. $f(x) = \sqrt{x + 4}$
7. $f(x) = \sqrt{x^2 + 4}$
8. $f(x) = \dfrac{1}{\sqrt{x - 5}}$
9. $f(x) = -|x - 1|$
10. $f(x) = \dfrac{x + 3}{|x + 4|}$
11. $f(x) = x|x|$
12. $f(x) = \dfrac{1}{\sqrt{x + |x|}}$.

Find the domain and range of the following functions :

13. $f(x) = \dfrac{5}{2x + 1}$
14. $f(x) = \dfrac{3x}{2x - 8}$
15. $f(x) = \dfrac{3}{2 - x^2}$
16. $f(x) = \dfrac{x}{1 + x^2}$.
17. $f(x) = \dfrac{x^2 - 9}{x - 3}$
18. $\dfrac{|x - 4|}{x - 4}$
19. $f(x) = \sqrt{16 - x^2}$
20. $f(x) = \dfrac{1}{\sqrt{x - 5}}$.

21. $f(x) = 1 + 8x - 2x^2$ **22.** $f(x) = \frac{1}{x^2 - 5x + 6}$ **23.** $f(x) = \sqrt{x^2 - 6x + 8}$

24. $f(x) = \frac{1}{\sqrt{x^2 + 4x + 3}}$ **25.** $f(x) = \begin{cases} 3x + 4 & -1 \le x < 2 \\ 1 + x & 2 \le x \le 5 \end{cases}$ **26.** $f(x) = \begin{cases} x + 4 & x < 3 \\ 2x + 9 & 3 \le x \le 9 \\ 5 & x > 9 \end{cases}$

27. $f(x) = x\,!$.

Answers

1. **R** **2.** **R** **3.** $\mathbf{R} - \{-1, -2\}$

4. $\mathbf{R} - \{4 - 2\sqrt{3}, 4 + 2\sqrt{3}\}$ **5.** $\mathbf{R} - \{-1, 1\}$ **6.** $[-4, \infty)$ **7.** **R**

8. $(5, \infty)$ **9.** **R** **10.** $\mathbf{R} - \{-4\}$ **11.** **R**

12. $(0, \infty)$ **13.** $\mathbf{R} - \{-1/2\}$, $\mathrm{R} - \{0\}$ **14.** $\mathbf{R} - \{4\}$, $\mathbf{R} - \{3/2\}$

15. $\mathrm{R} - \{-\sqrt{2}, \sqrt{2}\}$, $(-\infty, 0) \cup [3/2, \infty)$ **16.** **R,** $\left[-\frac{1}{2}, \frac{1}{2}\right]$

17. $\mathbf{R} - \{3\}$, $\mathbf{R} - \{6\}$ **18.** $\mathbf{R} - \{4\}$, $\{-1, 1\}$ **19.** $[-4, 4]$, $[0, 4]$ **20.** $(5, \infty)$, $(0, \infty)$

21. **R**, $[0, 9]$ **22.** $\mathbf{R} - \{2, 3\}$, $(-\infty, -4\} \cup (0, \infty)$

23. $(-\infty, 2\} \cup \{4, \infty)$, $\{0, \infty)$ **24.** $(-\infty, -3) \cup (-1, \infty)$, $(0, \infty)$

25. $[-1, 5]$, $[1, 10)$ **26.** $(-\infty, 9)$, $(-\infty, 7) \cup (15, 27)$

27. $\mathbf{N} \cup \{0\}$, $\{n\,! : n = 0, 1, 2, ...\}$.

11.6. OPERATIONS ON REAL FUNCTIONS

In the present section, we shall learn the method of defining new functions from given functions :

Let f and g be real functions.

(*i*) If $D(f) \cap D(g) \neq \phi$, then the **sum function** '$f + g$' of f and g is defined by

$$(f + g)(x) = f(x) + g(x) \text{ for all } x \in D(f) \cap D(g).$$

(*ii*) If $D(f) \cap D(g) \neq \phi$, then the **difference function** '$f - g$' of f and g is defined by

$$(f - g)(x) = f(x) - g(x) \text{ for all } x \in D(f) \cap D(g).$$

(*iii*) If $D(f) \cap D(g) \neq \phi$, then the **product function** 'fg' of f and g is defined by

$$(fg)(x) = f(x) \,.\, g(x) \text{ for all } x \in D(f) \cap D(g).$$

(*iv*) If $D(f) \cap D(g) - \{x : g(x) = 0\} \neq \phi$, then the **quotient function** 'f/g' of f and g is defined by

$$\left(\frac{f}{g}\right)(x) = \frac{f(x)}{g(x)} \text{ for all } x \in D(f) \cap D(g) - \{x : g(x) = 0\}.$$

(*v*) For any scalar k, the **scalar multiple function** 'kf' of f is defined by

$$(kf)(x) = k \,.\, f(x) \text{ for all } x \in D(f).$$

(*vi*) If $R(f) \subseteq D(g)$, then the **composite function** gof of f and g is defined by

$$(gof)(x) = g(f(x)) \text{ for all } x \in D(f).$$

(*vii*) If $R(g) \subseteq D(f)$, then the **composite function** fog of g and f is defined by

$$(fog)(x) = f(g(x)) \text{ for all } x \in D(g).$$

Remarks 1. Sometimes, composite functions *gof* and *fog* are defined by restricting the domains of *f* and *g* respectively. The functions *gof* and *fog* need not be equal.

2. If $f(x)$ and $g(x)$ are polynomials, then their quotient function $\frac{f(x)}{g(x)}$, $x \in \mathbf{R} - \{x ; g(x) = 0\}$ is called a **rational function.**

ILLUSTRATIVE EXAMPLES

Example 1. *Find the domain of the function* $\frac{1}{x+2} + \log(x+4) + \sqrt{x^2+4} - 4^{\sin^{-1}x}$.

Sol. Let $f(x) = \frac{1}{x+2} + \log(x+4) + \sqrt{x^2+4} - 4^{\sin^{-1}x}$.

Let $f_1(x) = \frac{1}{x+2}$, $f_2(x) = \log(x+4)$, $f_3(x) = \sqrt{x^2+4}$, $f_4(x) = 4^{\sin^{-1}x}$.

$f_1(x)$ is defined if $x + 2 \neq 0$ *i.e.*, if $x \neq -2$. $\therefore$ $D(f_1) = (-\infty, -2) \cup (-2, \infty)$

$f_2(x)$ is defined if $x + 4 > 0$ *i.e.*, if $x > -4$. $\therefore$ $D(f_2) = (-4, \infty)$

$f_3(x)$ is defined if $x \in \mathbf{R}$. $\therefore$ $D(f_3) = (-\infty, \infty)$

$f_4(x)$ is defined if $x \in [-1, 1]$. $\therefore$ $D(f_4) = [-1, 1]$

$\therefore$ $D(f) = D(f_1) \cap D(f_2) \cap D(f_3) \cap D(f_4) = \mathbf{[-1, 1]}$.

Example 2. *If* $f(x) = x^2 + x - 1$ *and* $g(x) = 4x - 7$, *then find :*

(*i*) $(f+g)(2)$ (*ii*) $(f-g)(7)$

(*iii*) $(fg)(-5)$ (*iv*) $\left(\frac{f}{g}\right)(4)$.

Sol. We have $f(x) = x^2 + x - 1$, $g(x) = 4x - 7$.

(*i*) $(f+g)(x) = f(x) + g(x) = (x^2 + x - 1) + (4x - 7) = x^2 + 5x - 8$

$\therefore$ $(f+g)(2) = (2)^2 + 5(2) - 8 = \mathbf{6}$.

Alternative method

$(f+g)(2) = f(2) + g(2) = (2^2 + 2 - 1) + (4(2) - 7) = 5 + 1 = \mathbf{6}$.

(*ii*) $(f-g)(7) = f(7) - g(7) = (7^2 + 7 - 1) - (4(7) - 7) = 55 - 21 = \mathbf{34}$.

(*iii*) $(fg)(-5) = f(-5)\, g(-5) = ((-5)^2 + (-5) - 1\ (4(-5) - 7) = (25 - 5 - 1)(-20 - 7) = \mathbf{-513}$.

(*iv*) $\left(\frac{f}{g}\right)(4) = \frac{f(4)}{g(4)} = \frac{4^2 + 4 - 1}{4(4) - 7} = \mathbf{\frac{19}{9}}$.

Example 3. *Let f be the exponential function and g be the logarithmic function.*

(*i*) *What is* $(f+g)(1)$? (*ii*) *What is* $(fg)(1)$?

(*iii*) *What is* $(fog)(1)$? (*iv*) *What is* $(2f)(1)$?

Sol. We have $f(x) = e^x$ and $g(x) = \log x$.

(*i*) $(f+g)(1) = f(1) + g(1) = e^1 + \log 1 = e + 0 = \mathbf{e}$.

(*ii*) $(fg)(1) = f(1) \cdot g(1) = e^1 \cdot \log 1 = e \cdot (0) = \mathbf{0}$.

(*iii*) $(fog)(1) = f(g(1)) = f(\log 1) = f(0) = e^0 = \mathbf{1}$.

(*iv*) $(2f)(1) = 2(f(1)) = 2(e^1) = \mathbf{2e}$.

Example 4. *If f(x) = x and g(x) = | x |. Find the functions (i) f + g (ii) f – g (iii) fg (iv) f/g.*

Sol. We have $f(x) = x$ and $g(x) = |x|$.

(*i*) $(f + g)(x) = f(x) + g(x) = x + |x|$

$$= \begin{cases} x + x & \text{if } x \geq 0 \\ x + (-x) & \text{if } x < 0 \end{cases} = \begin{cases} \mathbf{2x} & \textbf{if } \mathbf{x \geq 0} \\ \mathbf{0} & \textbf{if } \mathbf{x < 0} \end{cases}$$

(*ii*) $(f - g)(x) = f(x) - g(x) = x - |x|$

$$= \begin{cases} x - x & \text{if } x \geq 0 \\ x - (-x) & \text{if } x < 0 \end{cases} = \begin{cases} \mathbf{0} & \textbf{if } \mathbf{x \geq 0} \\ \mathbf{2x} & \textbf{if } \mathbf{x < 0} \end{cases}$$

(*iii*) $(fg/(x)) = f(x)\, g(x) = x|x|$

$$= \begin{cases} x(x) & \text{if } x \geq 0 \\ x(-x) & \text{if } x < 0 \end{cases} = \begin{cases} \mathbf{x^2} & \textbf{if } \mathbf{x \geq 0} \\ \mathbf{-x^2} & \textbf{if } \mathbf{x < 0} \end{cases}$$

(*iv*) $\left(\dfrac{f}{g}\right)(x) = \dfrac{f(x)}{g(x)} = \dfrac{x}{|x|}, \; x \neq 0$

$$= \begin{cases} x/x & \text{if } x > 0 \\ x/(-x) & \text{if } x < 0 \end{cases} = \begin{cases} \mathbf{1} & \textbf{if } \mathbf{x > 0} \\ \mathbf{-1} & \textbf{if } \mathbf{x < 0.} \end{cases}$$

Example 5. *Let f be the sine function and let g be the function 2x. Find the functions fog and gof.*

Sol. We have $f(x) = \sin x$ and $g(x) = 2x$.

$\therefore$ D(f) = **R**, R(f) = [– 1, 1], D(g) = **R,** R(g) = **R.**

$\Rightarrow$ R(f) $\subseteq$ D(g). $\therefore$ *gof* is defined

$\therefore$ $(gof)(x) = g(f(x)) = g(\sin x) = \mathbf{2 \sin x}$, for $x \in$ D(f) *i.e.*, $x \in$ **R.**

Also R(g) $\subseteq$ D(f). $\therefore$ *fog* is defined.

$\therefore$ $(fog)(x) = f(g(x)) = f(2x) = \mathbf{\sin 2x}$, for $x \in$ D(g) *i.e.*, $x \in$ **R.**

Example 6. *Describe fog and gof giving their domains, where f(x) = eˣ, g(x) = log x.*

Sol. $f(x) = e^x$, $g(x) = \log x$

$\therefore$ D(f) = (– ∞, ∞), R(f) = (0, ∞) and D(g) = (0, ∞), R(g) = (– ∞, ∞)

fog. Here, R(g) $\subseteq$ D(f). $\therefore$ *fog* is defined.

For $x \in (0, \infty)$, $(fog)(x) = f(g(x)) = f(\log x) = e^{\log x} = x$

$\therefore$ **(fog) (x) = x** $\forall$ **x ∈ (0, ∞).**

gof. Here, R(f) $\subseteq$ D(g). $\therefore$ *gof* is defined.

For $x \in (-\infty, \infty)$, $(gof)(x) = g(f(x)) = g(e^x) = \log(e^x) = x \log e = x(1) = x$.

$\therefore$ **(gof)(x) = x** $\forall$ **x ∈ (– ∞, ∞).**

Example 7. *For the following functions f and g, find the functions, f + g, f – g, fg, f/g, fog and gof :*

(*i*) $f(x) = \sin x, x \in$ **R** (*ii*) $f(x) = \sqrt{x+4}$ $\quad x \geq -4$

$g(x) = x^2 + 1, x \in$ **R** $\quad$ $g(x) = \sqrt{x-4}$ $\quad x \geq 4$.

Sol. (*i*) We have $f(x) = \sin x$, $x \in$ **R** and $g(x) = x^2 + 1$, $x \in$ **R.**

(*a*) For $x \in$ **R,** $(f + g)(x) = f(x) + g(x) = \mathbf{\sin x + x^2 + 1}$.

(*b*) For $x \in \mathbf{R}$, $(f-g)(x) = f(x) - g(x) = \sin x - (x^2+1) = \mathbf{\sin x - x^2 - 1}$.

(*c*) For $x \in \mathbf{R}$, $(fg)(x) = f(x) \, . \, g(x) = \mathbf{(\sin x)(x^2+1)}$.

(*d*) For $x \in \mathbf{R}$, $\left(\dfrac{f}{g}\right)(x) = \dfrac{f(x)}{g(x)} = \dfrac{\mathbf{\sin x}}{\mathbf{x^2+1}}$. $\qquad (x^2 + 1 \neq 0, \text{ for } x \in \mathbf{R})$

(*e*) For $x \in \mathbf{R}$, $(fog)(x) = f(g(x)) = f(x^2+1) = \mathbf{\sin (x^2+1)}$. $\qquad$ (Here, $R(g) \subseteq D(f)$)

(*f*) For $x \in \mathbf{R}$, $(gof)(x) = g(f(x)) = g(\sin x) = (\sin x)^2 + 1 = \mathbf{1 + \sin^2 x}$. $\qquad$ (Here, $R(f) \subseteq D(g)$)

(*ii*) We have $f(x) = \sqrt{x+4}$, $x \geq -4$, $\quad g(x) = \sqrt{x-4}$, $x \geq 4$.

Now $D(f) = [-4, \infty)$ and $D(g) = [4, \infty)$

$\therefore \quad D(f+g) = D(f-g) = D(fg) = D(f) \cap D(g) = [-4, \infty) \cap [4, \infty) = [4, \infty)$

(*a*) For $x \in [4, \infty)$, $(f+g)(x) = f(x) + g(x) = \mathbf{\sqrt{x+4} + \sqrt{x-4}}$.

(*b*) For $x \in [4, \infty)$, $(f-g)(x) = f(x) - g(x) = \mathbf{\sqrt{x+4} - \sqrt{x-4}}$.

(*c*) For $x \in [4, \infty)$, $(fg)(x) = f(x) \, . \, g(x) = \sqrt{x+4} \, . \sqrt{x-4} = \mathbf{\sqrt{x^2-16}}$.

(*d*) $D\left(\dfrac{f}{g}\right) = D(f) \cap D(g) — \{x : g(x) = 0\} = [4, \infty) — \{4\} = (4, \infty)$.

For $x \in (4, \infty)$, $\left(\dfrac{f}{g}\right)(x) = \dfrac{f(x)}{g(x)} = \dfrac{\mathbf{\sqrt{x+4}}}{\mathbf{\sqrt{x-4}}}$.

(*e*) $D(f) = [-4, \infty)$, $R(f) = [0, \infty)$ and $D(g) = [4, \infty)$, $R(g) = [0, \infty)$.

Here, $R(g) \subseteq D(f)$. $\qquad \therefore$ *fog* is defined,

For $x \in [4, \infty)$, $(fog)(x) = f(g(x)) = f(\sqrt{x-4}) = \mathbf{\sqrt{\sqrt{x-4}+4}}$.

(*f*) Here, $R(f) \not\subseteq D(g)$. $\quad \therefore \quad$ *gof* is not defined for all x in $D(f)$.

$(gof)(x) = g(f(x)) = g\left(\sqrt{x+4}\right) = \sqrt{\sqrt{x+4}-4}$.

$\therefore \quad (gof)(x)$ is defined when $\sqrt{x+4} \geq 4$ *i.e.*, $x \geq 12$.

$\therefore \quad$ For $x \in [(12, \infty)$, $(gof)(x) = \mathbf{\sqrt{\sqrt{x+4}-4}}$.

11.7. INVERSE OF A FUNCTION

Let $f : X \to Y$ be a one-one onto function.

$\therefore \quad$ For each $y \in Y$, there exists exactly one $x \in X$ such that $f(x) = y$.

$\therefore \quad$ We get a function, denoted as f^{-1} defined as :

$$f^{-1} : Y \to X \text{ and } f^{-1}(y) = x \text{ iff } f(x) = y.$$

The function f^{-1} is called the **inverse function** of f.

WORKING RULES FOR SOLVING PROBLEMS

Step I. *Take $f(x_1) = f(x_2)$ and show that $x_1 = x_2$. $\therefore$ f is one-one.*

Step II. *Take $k \in R(f)$ and solve $f(x) = k$, to get the value of x $\therefore$ f is onto.*

Step III. *To define f^{-1} let $f^{-1}(x) = k$, where $x \in R(f)$.*

Step IV. *$f^{-1}(x) = k$ implies $f(k) = x$. Solve this equation to find the value of k in terms of x.*

Example 8. *Find the inverse of the function $f(x) = 4x - 7$, $x \in$ **R**.*

Sol. We have $f(x) = 4x - 7$, $x \in \mathbf{R}$.

f(x) is one-one. For $x_1, x_2 \in \mathbf{R}$,

$f(x_1) = f(x_2) \Rightarrow 4x_1 - 7 = 4x_2 - 7 \Rightarrow 4x_1 = 4x_2 \Rightarrow x_1 = x_2$

$\therefore$ $f(x)$ is one-one.

f(x) is onto. Let $k \in \mathbf{R}$.

$$f(x) = k \Rightarrow 4x - 7 = k \Rightarrow x = \frac{k+7}{4} \in \mathbf{R}$$

$\therefore$ $f(x)$ is onto.

$\therefore$ $f(x) = 4x - 7$ is invertible *i.e.*, f^{-1} exists.

For $x \in \mathbf{R}$, let $f^{-1}(x) = k$.

$$\Rightarrow f(k) = x \Rightarrow 4k - 7 = x \Rightarrow k = \frac{x+7}{4}.$$

$\therefore$ For $x \in \mathbf{R}$, $f^{-1}(x) = \mathbf{\frac{x+7}{4}}$.

Example 9. *Given a function defined by $y = f(x) = \sqrt{4 - x^2}$, $0 \le x \le 2$, $0 \le y \le 2$. Find the inverse of f.*

Sol. We have $y = f(x) = \sqrt{4 - x^2}$, $0 \le x \le 2$, $0 \le y \le 2$.

f(x) is one-one. For $0 \le x_1, x_2 \le 2$,

$$f(x_1) = f(x_2) \Rightarrow \sqrt{4 - x_1^2} = \sqrt{4 - x_2^2} \Rightarrow 4 - x_1^2 = 4 - x_2^2 \Rightarrow x_1^2 = x_2^2$$

$$\Rightarrow x_1 = \pm x_2 \Rightarrow x_1 = x_2 \qquad (\because \; x_1 \text{ and } x_2 \text{ are non-negative})$$

$\therefore$ $f(x)$ is one-one.

f(x) is onto. Let $k \in [0, 2]$.

$$f(x) = k \Rightarrow \sqrt{4 - x^2} = k \Rightarrow 4 - x^2 = k^2 \Rightarrow x^2 = 4 - k^2$$

$$\Rightarrow x = \sqrt{4 - k^2} \Rightarrow 0 \le x \le 2 \qquad (\because \; 0 \le \sqrt{4 - k^2} \le 2)$$

$\therefore$ $f(x)$ is onto. $\therefore$ $f(x) = \sqrt{4 - x^2}$ is invertible *i.e.*, f^{-1} exists.

For $0 \le x \le 2$, let $f^{-1}(x) = k$ and $0 \le k \le 2$

$\Rightarrow f(k) = x \Rightarrow \sqrt{4 - k^2} = x \Rightarrow 4 - k^2 = x^2 \Rightarrow k^2 = 4 - x^2 \Rightarrow k = \sqrt{4 - x^2}$ because $0 \le k \le 2$.

$\therefore$ For $0 \le x \le 2$, $f^{-1}(x) = \mathbf{\sqrt{4 - x^2}}$.

Example 10. *If $f(x) = \frac{x-1}{x+1}$ $(x \ne 1, -1)$, show that fof^{-1} is an identity function.*

Sol. We have $f(x) = \frac{x-1}{x+1}$ $(x \ne 1, -1)$

f(x) is one-one. For $x_1, x_2 \in \mathbf{R} - \{1, -1\}$,

$$f(x_1) = f(x_2) \Rightarrow \frac{x_1 - 1}{x_1 + 1} = \frac{x_2 - 1}{x_2 + 1} \Rightarrow (x_1 - 1)(x_2 + 1) = (x_2 - 1)(x_1 + 1)$$

$$\Rightarrow x_1x_2 + x_1 - x_2 - 1 = x_2x_1 + x_2 - x_1 - 1 \Rightarrow x_1 = x_2.$$

$\therefore$ $f(x)$ is one-one.

Range of f(x). Let $f(x) = k$.

$$\Rightarrow \frac{x-1}{x+1} = k \Rightarrow x - 1 = k(x+1) \Rightarrow x = \frac{1+k}{1-k}. \quad \therefore \; x \text{ is not defined when } k = 1$$

$$\text{Also } f(1) = \frac{1-1}{1+1} = 0$$

$\therefore$ Range of $f(x) = \mathbf{R} - \{0, 1\}$ $\qquad$ ($\because$ $D(f) = \mathbf{R} - \{1, -1\}$)

$\therefore$ $f(x) : \mathbf{R} - \{1, -1\} \to \mathbf{R} - \{0, 1\}$ is one-one onto.

$\therefore$ $f(x)$ is invertible *i.e.*, f^{-1} exists.

For $x \in \mathbf{R} - \{0, 1\}$, let $f^{-1}(x) = k$.

$$\Rightarrow f(k) = x \Rightarrow \frac{k-1}{k+1} = x \Rightarrow k = \frac{1+x}{1-x}. \quad \therefore \text{ For } x \in \mathbf{R} - \{0, 1\}, f^{-1}(x) = \frac{1+x}{1-x}.$$

fof^{-1}. For $x \in \mathbf{R} - \{0, 1\}$,

$$(fof^{-1})(x) = f(f^{-1}(x)) = f\left(\frac{1+x}{1-x}\right) = \frac{\frac{1+x}{1-x} - 1}{\frac{1+x}{1-x} + 1} = \frac{1+x-1+x}{1+x+1-x} = \frac{2x}{2} = x.$$

$\therefore$ fof^{-1} is an identity function.

EXERCISE 11.3

1. If $f(x) = x + 1$, $g(x) = 2x + 8$, find $(f + g)(5)$ and $(f - g)(8)$.
2. If $f(x) = |x|$ and $g(x) = |x + 1|$, find $(f + g)(1)$, $(f - g)(7)$, $(fg)(-4)$ and $(f/g)(2)$.
3. If $f(x) = 4x$, $g(x) = |1 - x|$, find $(fog)(3)$ and $(gof)(3)$.
4. If $f(x) = 3 \sin x$ and $g(x) = \cos^2 x$, find $(f + g)(\pi/3)$ and $(fg)(\pi/6)$.
5. If $f(x) = 2x - 1$ and $g(x) = x^2$, find $(f + g)(0)$, $(f - g)(2)$, $(fg)(1)$, $(f/g)(4)$.
6. If $f(x) = \dfrac{1}{1-x}$, show that $(fof)\left(\dfrac{1}{2}\right) = -1$.
7. If $f(x) = \dfrac{x-1}{x+1}$, show that $(fof)(4) = -\dfrac{1}{4}$.
8. If $f(x) = e^x$, $g(x) = \log x$ $(x > 0)$, show that $fog = gof$ on $(0, \infty)$.
9. If $f(x) = \sqrt{x}$ $(x > 0)$ and $g(x) = x^2 - 1$, find if $fog = gof$.
10. If $f(x) = x$, $x \le 10$ and $g(x) = 4$, $x \ge 0$, then find $f + g$, $f - g$, fg, $\dfrac{f}{g}$.
11. If $f(x) = \dfrac{4}{x}$ and $g(x) = 0$, show that fog is not defined.
12. If $f(x) = x^2 - 1$ and $g(x) = 3x + 1$, find the functions fog, gof, fof and gog.
13. If $f(x) = \dfrac{x}{\sqrt{1+x^2}}$, then show that $(fofof)(x) = \dfrac{x}{\sqrt{1+3x^2}}$.
14. Find the inverse of the function $f(x) = x - 4$.

15. Find the inverse of the function $f(x) = 1 - 6x$.

16. Find the inverse of the function $f(x) = \sqrt{1 - x^2}$, $x \in [0, 1]$ $y \in [0, 1]$.

Answers

1. 24, – 15 **2.** 3, – 1, 12, 2/3 **3.** 8, 11 **4.** $\frac{6\sqrt{3}+1}{4}, \frac{9}{8}$

5. – 1, – 1, 1, 7/16

9. fog is not defined because $R(g) \not\subseteq D(f)$.

10. $(f + g)(x) = x + 4, x \in [0, 10]$, $(f - g)(x) = x - 4, x \in [0, 10]$,

$(fg)(x) = 4x, x \in [0, 10]$, $\left(\frac{f}{g}\right)(x) = \frac{x}{4}, x \in [0, 10]$

12. $(fog)(x) = 9x^2 + 6x, x \in \mathbf{R}$, $(gof)(x) = 3x^2 - 2, x \in \mathbf{R}$

$(fof)(x) = x^4 - 2x^2, x \in \mathbf{R}$, $(gog)(x) = 9x + 4, x \in \mathbf{R}$

14. $f^{-1}(x) = x + 4, x \in \mathbf{R}$ **15.** $f^{-1}(x) = \frac{1-x}{6}, x \in \mathbf{R}$ **16.** $f^{-1}(x) = \sqrt{1 - x^2}, x \in [0, 1]$

Hints

11. $R(g) = \{0\}$, $D(f) = \mathbf{R} - \{0\}$. $\therefore$ $R(g) \not\subseteq D(f)$.

12. $D(f) = \mathbf{R}$ and $D(g) = \mathbf{R}$. $\therefore$ $R(g) \subseteq D(f)$, $R(f) \subseteq D(g)$, $R(f) \subseteq D(f)$, $R(g) \subseteq D(g)$.

11.8. LIMIT

The limit of a variable is a fundamental concept in the study of both Differential calculus and Integral calculus. In the present chapter, we shall study the methods of evaluating limits of function (*i.e.*, of dependent variables) of various kinds.

11.9. MEANING OF $x \to a$

Let x be a variable and we take the values of x as 1.9, 1.99, 1.999, 1.99999, 1.99999999, ... From these values it is clear that as x takes these values, from left to right, the numerical difference between x and 2 gets closer and closer to 0. Similarly, if x take values 2.01, 2.0001, 2,000001, 2.0000001, ... even then the numerical difference between x and 2 gets closer to 0. In such a case, we say that x approaches 2 and write $x \to 2$.

In general, when we say x approaches a, then we mean that the variable x takes those values which are either less than a or greater than a and the numerical difference between x and a can be made as small as we please.

In other words, the statement $x \to a$ implies that $|x - a|$ can be made as small as we like. When we say $x \to a$, then it is always understood that x will not take the value a, but rather would assume values which are very close to a.

11.10. LIMIT OF A FUNCTION AT A POINT

Let $f(x)$ be a function of x. Let a and l be constants such that as $x \to a$, we have $f(x) \to l$, *i.e.*, the numerical difference between $f(x)$ and l can be made as small as we please by taking x sufficiently close to a. In such a case, we say that the **limit of function f(x)** as x approaches a is l. We write this as $\lim_{x \to a} f(x) = l$ or as $\operatorname{Lt}_{x \to a} f(x) = l$.

In case, no such number l exist, then we say that $\lim_{x \to a} f(x)$ does not exist finitely.

11.11. SOME STANDARD RESULTS

In this section, we list some of the results involving limits of functions. The proof of these are beyond the scope of this book.

1. If f(x) = k, a constant function, then $\lim_{x \to a} f(x) = k$.

2. $\lim_{x \to a} k(f(x)) = k \lim_{x \to a} f(x)$

3. $\lim_{x \to a} (f(x) + g(x)) = \lim_{x \to a} f(x) + \lim_{x \to a} g(x)$

4. $\lim_{x \to a} (f(x) - g(x)) = \lim_{x \to a} f(x) - \lim_{x \to a} g(x)$

5. $\lim_{x \to a} (f(x) \,.\, g(x)) = \lim_{x \to a} f(x) \,.\, \lim_{x \to a} g(x)$

6. $\lim_{x \to a} \dfrac{f(x)}{g(x)} = \dfrac{\lim_{x \to a} f(x)}{\lim_{x \to a} g(x)}$, **provided** $\lim_{x \to a} g(x) \neq 0$.

7. $\lim_{x \to a} \sqrt[n]{f(x)} = \sqrt[n]{\lim_{x \to a} f(x)}$, **provided** $\sqrt[n]{\lim_{x \to a} f(x)}$ **is a real number.**

ILLUSTRATIVE EXAMPLES

Example 1. *Evaluate :*

(i) $\lim_{x \to 1} (x^2 + 6x + 4)$ (ii) $\lim_{x \to 4} \dfrac{x^3 + 4}{1 - x}$ (iii) $\lim_{x \to 2} \sqrt{36 - x^2}$.

Sol. (i) $\lim_{x \to 1} (x^2 + 6x + 4) = \lim_{x \to 1} x^2 + \lim_{x \to 1} 6x \lim_{x \to 1} 4$

$= \lim_{x \to 1} x \,.\, \lim_{x \to 1} x + 6 \lim_{x \to 1} x + 4$

$= 1 \times 1 + 6 \times 1 + 4 = \mathbf{11.}$

(ii) $\lim_{x \to 4} \dfrac{x^3 + 4}{1 - x} = \dfrac{\lim_{x \to 4} (x^3 + 4)}{\lim_{x \to 4} (1 - x)}$ $(\because \lim_{x \to 4} (1 - x) = 1 - 4 = -3 \neq 0)$

$= \dfrac{4(4)(4) + 4}{1 - 4} = -\dfrac{\mathbf{68}}{\mathbf{3}}$.

(iii) $\lim_{x \to 2} \sqrt{36 - x^2} = \sqrt{\lim_{x \to 2} (36 - x^2)} = \sqrt{36 - \lim_{x \to 2} x \,.\, \lim_{x \to 2} x}$

$= \sqrt{36 - (2)(2)} = \sqrt{32} = \mathbf{4\sqrt{2}}$.

Example 2. *Evaluate :*

(i) $\lim_{x \to 1} \dfrac{x^3 - 1}{x - 1}$ (ii) $\lim_{x \to 1} \dfrac{\sqrt{x^2 - 1} + \sqrt{x - 1}}{\sqrt{x^2 - 1}}$.

Sol. (i) $\lim_{x \to 1} \dfrac{x^3 - 1}{x - 1} = \lim_{x \to 1} \dfrac{(x - 1)(x^2 + x + 1)}{x - 1} = \lim_{x \to 1} (x^2 + x + 1) = (1)^2 + 1 + 1 = \mathbf{3}$.

(ii) $\lim_{x \to 1} \dfrac{\sqrt{x^2 - 1} + \sqrt{x - 1}}{\sqrt{x^2 - 1}} = \lim_{x \to 1} \dfrac{\sqrt{x + 1}\,\sqrt{x - 1} + \sqrt{x - 1}}{\sqrt{x + 1}\,\sqrt{x - 1}}$

$$= \lim_{x \to 1} \frac{\sqrt{x-1}\,(\sqrt{x+1}+1)}{\sqrt{x+1}\sqrt{x-1}} = \lim_{x \to 1} \frac{\sqrt{x+1}+1}{\sqrt{x+1}} \quad \text{(Cancelling the non-zero factor } \sqrt{x-1}\text{)}$$

$$= \frac{\lim\limits_{x \to 1} (\sqrt{x+1}+1)}{\lim\limits_{x \to 1} \sqrt{x+1}} = \frac{\sqrt{\lim\limits_{x \to 1} (x+1)} + \lim\limits_{x \to 1} 1}{\sqrt{\lim\limits_{x \to 1} (x+1)}} = \frac{\sqrt{1+1}+1}{\sqrt{1+1}} = \frac{\sqrt{2}+1}{\sqrt{2}}.$$

EXERCISE 11.4

Evaluate the following limits (Q. No. 1–3) :

1. (*i*) $\lim\limits_{x \to 2} (4x-5)$ (*ii*) $\lim\limits_{x \to 10} x^3(x^3+x^2+x-7)$

2. (*i*) $\lim\limits_{x \to -2} \dfrac{x^2-4}{x-2}$ (*ii*) $\lim\limits_{x \to 2} \left(\dfrac{x^8-16}{x^4-4} + \dfrac{x^2-9}{x-3} \right)$

3. (*i*) $\lim\limits_{x \to 0} \dfrac{x}{\sqrt{2+x}+\sqrt{2-x}}$ (*ii*) $\lim\limits_{x \to 4} \dfrac{x^2-16}{\sqrt{x^2+9}-4}$

(*iii*) $\operatorname{Lt}\limits_{x \to 4} \dfrac{x-4}{3-\sqrt{13-x}}$. (*P.T.U. B.Pharma Dec. 2005*)

4. Show that the functions '$1-x$' and '$1+x^2$' have the same limit at $x = 0$ and $x = -1$ and at no other point.

Answers

1. (*i*) 3 (*ii*) 11,03,000 2. (*i*) 0 (*ii*) 25
3. (*i*) 0 (*ii*) 0 (*iii*) 6.

Hint

4. Let $\lim\limits_{x \to \lambda} (1-x) = \lim\limits_{x \to \lambda} (1+x^2)$.

$\therefore \quad 1-\lambda = 1+\lambda^2$ *i.e.*, $\lambda^2+\lambda = 0$ or $\lambda = 0, -1$.

11.12. EVALUATION OF $\lim\limits_{x \to a} \frac{f(x)}{g(x)}$ WHERE $\lim\limits_{x \to a} f(x) = 0$ AND $\lim\limits_{x \to a} g(x) = 0$

We know that $\lim\limits_{x \to a} \dfrac{f(x)}{g(x)} = \dfrac{\lim\limits_{x \to a} f(x)}{\lim\limits_{x \to a} g(x)}$ provided $\lim\limits_{x \to a} g(x) \neq 0$.

In case, $\lim\limits_{x \to a} f(x) = 0$ and $\lim\limits_{x \to a} g(x) = 0$, the following three methods are used to evaluate $\lim\limits_{x \to a} \dfrac{f(x)}{g(x)}$.

(*i*) **Method of factors.** In this method, $f(x)$ and $g(x)$ are factorised and the quotient $\dfrac{f(x)}{g(x)}$ is simplified by cancelling the common factors. If the limit of the resulting denominator as $x \to a$ is non-zero, then the *quotient rule of limits* is applied to get the required limit.

WORKING RULES FOR SOLVING PROBLEMS

Step I. *Factorise f(x) and g(x).*

Step II. *The common factors in* $\frac{f(x)}{g(x)}$ *are cancelled. Be sure that the limit of the resulting denominator as* $x \to a$ *is non-zero.*

Step III. *Apply 'quotient rule of limits' at this stage.*

ILLUSTRATIVE EXAMPLES

Example 1. *Evaluate the following limits :*

(*i*) $\lim_{x \to 2} \frac{x^2 - 3x + 2}{x^2 - x - 2}$ (*ii*) $\lim_{x \to -3} \frac{x^3 + 4x^2 + 4x + 3}{x^2 + 2x - 3}$

Sol. (*i*) At $x = 2$, $\frac{x^2 - 3x + 2}{x^2 - x - 2} = \frac{4 - 6 + 2}{4 - 2 - 2} = \text{‘}\frac{0}{0}\text{’}$ form.

$x^2 - 3x + 2 = (x - 2)(x - 1)$ and $x^2 - x - 2 = (x - 2)(x + 1)$.

$$\therefore \quad \lim_{x \to 2} \frac{x^2 - 3x + 2}{x^2 - x - 2} = \lim_{x \to 2} \frac{(x - 2)(x - 1)}{(x - 2)(x + 1)} = \lim_{x \to 2} \frac{x - 1}{x + 1}^{*} = \frac{\lim_{x \to 2} (x - 1)}{\lim_{x \to 2} (x + 1)} = \frac{2 - 1}{2 + 1} = \mathbf{\frac{1}{3}}.$$

(*ii*) At $x = -3$, $\frac{x^3 + 4x^2 + 4x + 3}{x^2 + 2x - 3} = \frac{(-3)^3 + 4(-3)^2 + 4(-3) + 3}{(-3)^2 + 2(-3) - 3}$

$$= \frac{-27 + 36 - 12 + 3}{9 - 6 - 3} = \text{‘}\frac{0}{0}\text{’ form.}$$

Dividing $x^3 + 4x^2 + 4x + 3$ by $x - (-3)$, we get

$$x^3 + 4x^2 + 4x + 3 = (x + 3)(x^2 + x + 1).$$

Also, $x^2 + 2x - 3 = (x + 3)(x - 1)$.

$$\therefore \quad \lim_{x \to -3} \frac{x^3 + 4x^2 + 4x + 3}{x^2 + 2x - 3} = \lim_{x \to -3} \frac{(x + 3)(x^2 + x + 1)}{(x + 3)(x - 1)} = \lim_{x \to -3} \frac{x^2 + x + 1}{x - 1}$$

$$= \frac{(-3)^2 + (-3) + 1}{(-3) - 1} = \frac{9 - 3 + 1}{-4} = \mathbf{-\frac{7}{4}}.$$

Example 2. *Evaluate the following limits :*

(*i*) $\lim_{x \to 2} \left(\frac{x}{x - 2} - \frac{4}{x^2 - 2x} \right)$ (*ii*) $\lim_{x \to 3} (x^2 - 9) \left(\frac{1}{x + 3} + \frac{1}{x - 3} \right)$.

Sol. (*i*) $\frac{x}{x - 2} - \frac{4}{x^2 - 2x} = \frac{x}{x - 2} - \frac{4}{x(x - 2)} = \frac{1}{x - 2}\left(x - \frac{4}{x} \right)$

$$= \frac{1}{x - 2}\left(\frac{x^2 - 4}{x} \right) = \frac{(x - 2)(x + 2)}{(x - 2)x} = \frac{x + 2}{x}$$

$$\therefore \quad \lim_{x \to 2} \left(\frac{x}{x - 2} - \frac{4}{x^2 - 2x} \right) = \lim_{x \to 2} \frac{(x - 2)(x + 2)}{(x - 2)x} = \lim_{x \to 2} \frac{x + 2}{x} = \frac{2 + 2}{2} = \frac{4}{2} = \mathbf{2}.$$

***Why this step.** $x \to 2 \Rightarrow x \neq 2$.

(*ii*) $\lim\limits_{x \to 3} (x^2 - 9)\left(\dfrac{1}{x+3} + \dfrac{1}{x-3}\right) = \lim\limits_{x \to 3} (x^2 - 9)\left(\dfrac{x-3+x+3}{x^2-9}\right) = \lim\limits_{x \to 3} 2x^* = 2(3) = \mathbf{6}.$

Example 3. *If* $\lim\limits_{x \to a} \dfrac{x^9 + a^9}{x + a} = 9$, *find all possible values of a.*

Sol. For $a = 0 \quad \Rightarrow \quad 9 = \lim\limits_{x \to a} \dfrac{x^9 + a^9}{x + a} = \lim\limits_{x \to 0} \dfrac{x^9 + 0}{x + 0} = \lim\limits_{x \to 0} x^8 = 0$, which is impossible.

$\therefore \quad a \neq 0$. The given equation implies,

$$9 = \lim_{x \to a} \frac{x^9 + a^9}{x + a} = \frac{\lim\limits_{x \to a} (x^9 + a^9)}{\lim\limits_{x \to a} (x + a)} = \frac{a^9 + a^9}{a + a} = a^8 \qquad (\because \lim_{x \to a} (x + a) = a + a = 2a \neq 0)$$

$\Rightarrow \qquad a^8 = 9 \quad \Rightarrow \quad a = \mathbf{9^{1/8}}.$

Method of substitution. In this method, we put $x = a + h$.

$\therefore$ As x approaches a, the difference $x - a$ approaches 0.

Since $x = a + h$ implies $h = x - a$, we have h approaches 0 as x approaches a.

$\therefore \quad h \to 0$ as $x \to a$.

$$\therefore \qquad \lim_{x \to a} \frac{f(x)}{g(x)} = \lim_{h \to 0} \frac{f(a+h)}{g(a+h)}.$$

The quotient $\dfrac{f(a+h)}{g(a+h)}$ is simplified by cancelling powers of h. If the limit of the resulting denominator as $h \to 0$ is non-zero, then the *quotient rule of limits* is applied to get the required limit.

WORKING RULES FOR SOLVING PROBLEMS

Step I. *Put* $x = a + h$ *and replace* x *by* $a + h$ *in* $\dfrac{f(x)}{g(x)}$ *to get* $\dfrac{f(a+h)}{g(a+h)}$.

Step II. *Simplify* $\dfrac{f(a+h)}{g(a+h)}$. *Be sure that the limit of the resulting denominator as* $h \to 0$ *is non-zero.*

Step III. *Apply 'quotient rule of limits' at this stage.*

Example 4. *Evaluate the following limits :*

(*i*) $\lim\limits_{x \to 2} \dfrac{x^2 - 3x + 2}{x^2 - x - 2}$ (*ii*) $\lim\limits_{x \to -3} \dfrac{x^3 + 4x^2 + 4x + 3}{x^2 + 2x - 3}$

Sol. (*i*) At $x = 2$, $\dfrac{x^2 - 3x + 2}{x^2 - x - 2} = \dfrac{4 - 6 + 2}{4 - 2 - 2} = \text{‘}\dfrac{0}{0}\text{’}$ form.

Let $x = 2 + h$. $\therefore \quad h \to 0$ as $x \to 2$

$$\therefore \quad \lim_{x \to 2} \frac{x^2 - 3x + 2}{x^2 - x - 2} = \lim_{h \to 0} \frac{(2+h)^2 - 3(2+h) + 2}{(2+h)^2 - (2+h) - 2} = \lim_{h \to 0} \frac{4 + h^2 + 4h - 6 - 3h + 2}{4 + h^2 + 4h - 2 - h - 2}$$

***Why this step,** $x \to 3 \Rightarrow x^2 - 9 \neq 0$.

$$= \lim_{h \to 0} \frac{h^2 + h}{h^2 + 3h} = \lim_{h \to 0} \frac{h(h+1)}{h(h+3)} = \lim_{h \to 0} \frac{h+1}{h+3} \quad (\because \quad h \to 0 \quad \Rightarrow \quad h \neq 0)$$

$$= \frac{\lim_{h \to 0} (h+1)}{\lim_{h \to 0} (h+3)} = \frac{0+1}{0+3} = \mathbf{\frac{1}{3}}.$$

(*ii*) At $x = -3$, $\quad \dfrac{x^3 + 4x^2 + 4x + 3}{x^2 + 2x - 3} = \dfrac{-27 + 36 - 12 + 3}{9 - 6 - 3} = \text{'}\dfrac{0}{0}\text{'}$ form.

Let $x = -3 + h$. $\quad \therefore \quad h \to 0$ as $x \to -3$

$$\therefore \quad \lim_{x \to -3} \frac{x^3 + 4x^2 + 4x + 3}{x^2 + 2x - 3} = \lim_{h \to 0} \frac{(-3+h)^3 + 4(-3+h)^2 + 4(-3+h) + 3}{(-3+h)^2 + 2(-3+h) - 3}$$

$$= \lim_{h \to 0} \frac{(h^3 - 9h^2 + 27h - 27) + 4(h^2 - 6h + 9) - 12 + 4h + 3}{h^2 - 6h + 9 - 6 + 2h - 3}$$

$$= \lim_{h \to 0} \frac{h^3 - 5h^2 + 7h}{h^2 - 4h} = \lim_{h \to 0} \frac{h(h^2 - 5h + 7)}{h(h-4)} = \lim_{h \to 0} \frac{h^2 - 5h + 7}{h - 4}$$

$$= \frac{0 - 0 + 7}{0 - 4} = -\mathbf{\frac{7}{4}}.$$

Important note. While evaluating $\lim_{x \to a} \dfrac{f(x)}{g(x)}$, if $f(x)$ and $g(x)$ appear to be factorised easily, then the **method of factorisation** should be preferred over the **method of substitution.**

(*iii*) **Method of rationalisation.** This method is used when either $f(x)$ or $g(x)$ or both involve expressions consisting of square roots. In this method, the expressions involving square roots are rationalised. After the process of rationalisation, if the limit of the resulting denominator as $x \to a$ is non-zero, then the *quotient rule of limits* is applied to get the required limit.

WORKING RULES FOR SOLVING PROBLEMS

Step I. *Rationalise the expression in* $\dfrac{f(x)}{g(x)}$ *which involve square roots.*

Step II. *Simplify the numerator and the denominator by cancelling factors, by sure that the limit of the resulting denominator as* $x \to a$ *is non-zero.*

Step III. *Apply 'quotient rule of limits' at this stage.*

Example 5. *Evaluate the following limits :*

(*i*) $\displaystyle \lim_{x \to 0} \frac{\sqrt{2-x} - \sqrt{2+x}}{x}$ $\qquad$ (*ii*) $\displaystyle \lim_{x \to 0} \frac{\sqrt{1 + x + x^2} - 1}{x}$

(*iii*) $\displaystyle \lim_{h \to 0} \frac{1}{h}\left(\frac{1}{\sqrt{x+h}} - \frac{1}{\sqrt{x}}\right)$ $\qquad$ (*iv*) $\displaystyle \lim_{x \to a} \frac{\sqrt{a + 2x} - \sqrt{3x}}{\sqrt{3a + x} - 2\sqrt{x}}$, $a \neq 0$.

(G.N.D.U. B.Pharma, April 2004, 2005)

Sol. (*i*) At $x = 0$, $\quad \dfrac{\sqrt{2-x} - \sqrt{2+x}}{x} = \dfrac{\sqrt{2} - \sqrt{2}}{0} = \text{'}\dfrac{0}{0}\text{'}$ form.

$$\therefore \lim_{x\to 0}\frac{\sqrt{2-x}-\sqrt{2+x}}{x} = \lim_{x\to 0}\left(\frac{\sqrt{2-x}-\sqrt{2+x}}{x}\times\frac{\sqrt{2-x}+\sqrt{2+x}}{\sqrt{2-x}+\sqrt{2+x}}\right)$$

$$= \lim_{x\to 0}\frac{(2-x)-(2+x)}{x(\sqrt{2-x}+\sqrt{2+x})} = \lim_{x\to 0}\frac{-2x}{x(\sqrt{2-x}+\sqrt{2+x})} \quad | \ (a-b)(a+b)=a^2-b^2$$

$$= \lim_{x\to 0}\frac{-2}{\sqrt{2-x}+\sqrt{2+x}} = \frac{-2}{\sqrt{2-0}+\sqrt{2+0}} = \frac{-2}{2\sqrt{2}} = -\frac{\mathbf{1}}{\mathbf{\sqrt{2}}}.$$

(*ii*) At $x = 0$, $\dfrac{\sqrt{1+x+x^2}-1}{x} = \dfrac{\sqrt{1+0+0}-1}{0} = \text{‘}\dfrac{0}{0}\text{’}$ form.

$$\therefore \lim_{x\to 0}\frac{\sqrt{1+x+x^2}-1}{x} = \lim_{x\to 0}\left(\frac{\sqrt{1+x+x^2}-1}{x}\times\frac{\sqrt{1+x+x^2}+1}{\sqrt{1+x+x^2}+1}\right)$$

$$= \lim_{x\to 0}\frac{1+x+x^2-1}{x\sqrt{1+x+x^2}+1} = \lim_{x\to 0}\frac{x+x^2}{x(\sqrt{1+x+x^2}+1)}$$

$$= \lim_{x\to 0}\frac{1+x}{\sqrt{1+x+x^2}+1} = \frac{1+0}{\sqrt{1+0+0}+1} = \frac{1}{1+1} = \frac{\mathbf{1}}{\mathbf{2}}.$$

(*iii*) $\displaystyle\lim_{h\to 0}\frac{1}{h}\left(\frac{1}{\sqrt{x+h}}-\frac{1}{\sqrt{x}}\right) = \lim_{h\to 0}\frac{\sqrt{x}-\sqrt{x+h}}{h\sqrt{x+h}\,\sqrt{x}}$.

At $h = 0$, $\dfrac{\sqrt{x}-\sqrt{x+h}}{h(\sqrt{x+h}\,.\sqrt{x})} = \dfrac{\sqrt{x}-\sqrt{x}}{0(\sqrt{x}\,.\sqrt{x})} = \text{‘}\dfrac{0}{0}\text{’}$ form.

$$\therefore \lim_{h\to 0}\frac{1}{h}\left(\frac{1}{\sqrt{x+h}}-\frac{1}{\sqrt{x}}\right) = \lim_{h\to 0}\frac{\sqrt{x}-\sqrt{x+h}}{h\sqrt{x+h}\,\sqrt{x}} = \lim_{h\to 0}\left(\frac{\sqrt{x}-\sqrt{x+h}}{h\sqrt{x+h}\,\sqrt{x}}\cdot\frac{\sqrt{x}+\sqrt{x+h}}{\sqrt{x}+\sqrt{x+h}}\right)$$

$$= \lim_{h\to 0}\frac{-h}{h\sqrt{x+h}\sqrt{x}(\sqrt{x}+\sqrt{x+h})} = \lim_{h\to 0}\frac{-1}{\sqrt{x+h}\sqrt{x}(\sqrt{x}+\sqrt{x+h})}$$

$(\because\ h\to 0 \ \Rightarrow\ h\neq 0)$

$$= \frac{-1}{\sqrt{x}\,.\sqrt{x}(\sqrt{x}+\sqrt{x})} = \frac{-1}{x(2\sqrt{x})} = -\frac{\mathbf{1}}{\mathbf{2x^{3/2}}}.$$

(*iv*) At $x = a$, $\dfrac{\sqrt{a+2x}-\sqrt{3x}}{\sqrt{3a+x}-2\sqrt{x}} = \dfrac{\sqrt{a+2a}-\sqrt{3a}}{\sqrt{3a+a}-2\sqrt{a}} = \dfrac{\sqrt{3a}-\sqrt{3a}}{2\sqrt{a}-2\sqrt{a}} = \text{‘}\dfrac{0}{0}\text{’}$ form.

$$\therefore \lim_{x\to a}\frac{\sqrt{a+2x}-\sqrt{3x}}{\sqrt{3a+x}-2\sqrt{x}} = \lim_{x\to a}\left(\frac{\sqrt{a+2x}-\sqrt{3x}}{\sqrt{3a+x}-2\sqrt{x}}\times\frac{\sqrt{a+2x}+\sqrt{3x}}{\sqrt{a+2x}+\sqrt{3x}}\times\frac{\sqrt{3a+x}+2\sqrt{x}}{\sqrt{3a+x}+2\sqrt{x}}\right)$$

$$= \lim_{x\to 0}\left(\frac{(a+2x)-3x}{(3a+x)-4x}\times\frac{\sqrt{3a+x}+2\sqrt{x}}{\sqrt{a+2x}+\sqrt{3x}}\right) = \lim_{x\to a}\left(\frac{a-x}{3(a-x)}\times\frac{\sqrt{3a+x}+2\sqrt{x}}{\sqrt{a+2x}+3\sqrt{x}}\right)$$

$$= \frac{1}{3} \lim_{x \to a} \frac{\sqrt{3a+x} + 2\sqrt{x}}{\sqrt{a+2x} + \sqrt{3x}} = \frac{1}{3} \cdot \frac{\sqrt{3a+a} + 2\sqrt{a}}{\sqrt{a+2a} + \sqrt{3a}} = \frac{1}{3} \cdot \frac{2\sqrt{a} + 2\sqrt{a}}{\sqrt{3a} + \sqrt{3a}}$$

$$= \frac{1}{3} \cdot \frac{4\sqrt{a}}{2\sqrt{3}\sqrt{a}} = \mathbf{\frac{2\sqrt{3}}{9}}. \qquad (\because a \neq 0)$$

Example 6. *Evaluate the following limits :*

(*i*) $\lim\limits_{x \to \sqrt{2}} \dfrac{\sqrt{3+2x} - (\sqrt{2}+1)}{x^2 - 2}$ (*ii*) $\lim\limits_{x \to 4} \dfrac{3 - \sqrt{5+x}}{1 - \sqrt{5-x}}$.

Sol. (*i*) At $x = \sqrt{2}$, $\dfrac{\sqrt{3+2x} - (\sqrt{2}+1)}{x^2-2} = \dfrac{\sqrt{3+2\sqrt{2}} - (\sqrt{2}+1)}{2-2}$

$$= \frac{\sqrt{(\sqrt{2}+1)^2} - (\sqrt{2}+1)}{0} = \text{`}\frac{0}{0}\text{' form.}$$

$$\therefore \lim_{x \to \sqrt{2}} \frac{\sqrt{3+2x} - (\sqrt{2}+1)}{x^2-2} = \lim_{x \to \sqrt{2}} \left(\frac{\sqrt{3+2x} - (\sqrt{2}+1)}{x^2-2} \times \frac{\sqrt{3+2x} + (\sqrt{2}+1)}{\sqrt{3+2x} + (\sqrt{2}+1)} \right)$$

$$= \lim_{x \to \sqrt{2}} \frac{(3+2x) - (\sqrt{2}+1)^2}{(x^2-2)(\sqrt{3+2x} + \sqrt{2}+1)} = \lim_{x \to \sqrt{2}} \frac{(3+2x) - (2+1+2\sqrt{2})}{(x^2-2)(\sqrt{3+2x} + \sqrt{2}+1)}$$

$$= \lim_{x \to \sqrt{2}} \frac{2(x-\sqrt{2})}{(x-\sqrt{2})(x+\sqrt{2})(\sqrt{3+2x} + \sqrt{2}+1)} = \lim_{x \to \sqrt{2}} \frac{2}{(x+\sqrt{2})(\sqrt{3+2x} + \sqrt{2}+1)}$$

$$= \frac{2}{(\sqrt{2}+\sqrt{2})(\sqrt{3+2\sqrt{2}} + \sqrt{2}+1)} = \frac{2}{2\sqrt{2}\,(\sqrt{(\sqrt{2}+1)^2} + \sqrt{2}+1)}$$

$$= \frac{1}{\sqrt{2}\,(\sqrt{2}+1+\sqrt{2}+1)} = \frac{1}{\sqrt{2}\,(2(\sqrt{2}+1))} = \frac{\sqrt{2}-1}{2\sqrt{2}(2-1)} = \mathbf{\frac{\sqrt{2}-1}{2\sqrt{2}}}.$$

(*ii*) At $x = 4$, $\dfrac{3-\sqrt{5+x}}{1-\sqrt{5-x}} = \dfrac{3-\sqrt{5+4}}{1-\sqrt{5-4}} = \dfrac{3-3}{1-1} = \text{`}\dfrac{0}{0}\text{' form.}$

$$\therefore \lim_{x \to 4} \frac{3-\sqrt{5+x}}{1-\sqrt{5-x}} = \lim_{x \to 4} \left(\frac{3-\sqrt{5+x}}{1-\sqrt{5-x}} \times \frac{3+\sqrt{5+x}}{3+\sqrt{5+x}} \times \frac{1+\sqrt{5-x}}{1+\sqrt{5-x}} \right)$$

$$= \lim_{x \to 4} \left(\frac{9-(5+x)}{1-(5-x)} \times \frac{1+\sqrt{5-x}}{3+\sqrt{5+x}} \right) = \lim_{x \to 4} \left(\frac{4-x}{x-4} \times \frac{1+\sqrt{5-x}}{3+\sqrt{5+x}} \right)$$

$$= \lim_{x \to 4} \left(-\frac{1+\sqrt{5-x}}{3+\sqrt{5+x}} \right) = -\frac{1+\sqrt{5-4}}{3+\sqrt{5+4}} = -\frac{1+1}{3+3} = -\frac{2}{6} = -\mathbf{\frac{1}{3}}.$$

EXERCISE 11.5

1. Evaluate the following limits by using the factorisation method :

(i) $\lim\limits_{x \to 2} \dfrac{x^3 - 8}{x^2 - 4}$ (ii) $\lim\limits_{x \to 0} \dfrac{5x - 7x^2}{7x + 5x^9}$ (iii) $\lim\limits_{x \to 1} \dfrac{x - 1}{2x^2 - 7x + 5}$

(iv) $\lim\limits_{x \to 3} \dfrac{x^2 - 4x + 3}{2x^2 - 3x - 9}$ (v) $\lim\limits_{x \to 1} \dfrac{(2x - 3)(\sqrt{x} - 1)}{3x^2 + 3x - 6}$ (vi) $\lim\limits_{x \to 1} \dfrac{1 - x^{-1/3}}{1 - x^{-2/3}}$.

(P.T.U. B.Pharma, Dec. 2004)

2. Evaluate the following limits by using the factorisation method :

(i) $\lim\limits_{h \to 0} \dfrac{1}{h}\left(\dfrac{1}{x + h} - \dfrac{1}{x}\right)$ (ii) $\lim\limits_{x \to 2} \left(\dfrac{1}{x - 2} - \dfrac{2}{x^2 - 2x}\right)$

(iii) $\lim\limits_{x \to 1} \left(\dfrac{1}{x^2 + x - 2} - \dfrac{x}{x^3 - 1}\right)$ (iv) $\lim\limits_{x \to 2} \left(\dfrac{1}{x - 2} - \dfrac{2(2x - 3)}{x^3 - 3x^2 + 2x}\right)$.

3. Evaluate the following limits by using the substitution method :

(i) $\lim\limits_{x \to 2} \dfrac{x^2 - 5x + 6}{x^2 - 4}$ (ii) $\lim\limits_{x \to -5} \dfrac{2x^2 + 9x - 5}{x + 5}$ (iii) $\lim\limits_{x \to 1} \dfrac{x - 1}{2x^2 - 7x + 5}$

(iv) $\lim\limits_{x \to 2} \dfrac{x^3 + 3x^2 - 9x - 2}{x^3 - x - 6}$ (v) $\lim\limits_{x \to 2} \dfrac{x^2 + 3x - 10}{x^2 - 9x + 14}$ (vi) $\lim\limits_{x \to 1} \dfrac{x^4 - 3x^2 + 2}{x^3 - 5x^2 + 3x + 1}$.

(P.T.U. B.Pharma, Dec. 2004)

(vii) $\underset{x \to 2}{\text{Lt}} \dfrac{x^2 + 6x - 16}{x^4 - 16}$ (P.T.U. B.Pharma, May 2005)

4. Evaluate the following limits by using the rationalisation method :

(i) $\lim\limits_{x \to 0} \dfrac{\sqrt{1 + x} - 1}{x}$ (ii) $\lim\limits_{x \to 2} \dfrac{2 - x}{\sqrt{2 + x} - 2}$ (iii) $\lim\limits_{x \to 0} \dfrac{\sqrt{1 + 3x} - \sqrt{1 - 3x}}{x}$

(iv) $\lim\limits_{h \to 0} \dfrac{\sqrt{x + h} - \sqrt{x}}{h}, x \neq 0$ (v) $\lim\limits_{x \to 0} \dfrac{\sqrt{a + x} - \sqrt{a}}{x\sqrt{a(a + x)}}, a \neq 0$ (vi) $\lim\limits_{x \to 0} \dfrac{\sqrt{1 + x^2} - \sqrt{1 + x}}{\sqrt{1 + x^3} - \sqrt{1 + x}}$.

5. Evaluate the following limits by using the rationalisation method :

(i) $\lim\limits_{x \to 0} \dfrac{\sqrt{1 - x^2} - \sqrt{1 + x^2}}{2x^2}$ (ii) $\lim\limits_{x \to 3} \dfrac{\sqrt{2x + 3} - \sqrt{4x - 3}}{x^2 - 9}$ (iii) $\lim\limits_{x \to 2} \dfrac{x^2 - 4}{\sqrt{x + 2} - \sqrt{3x - 2}}$

(iv) $\lim\limits_{x \to 4} \dfrac{x^2 - 16}{\sqrt{x^2 + 9} - 5}$ (v) $\lim\limits_{x \to \sqrt{10}} \dfrac{\sqrt{7 + 2x} - (\sqrt{5} + \sqrt{2})}{x^2 - 10}$

(vi) $\lim\limits_{x \to \sqrt{6}} \dfrac{\sqrt{5 + 2x} - (\sqrt{3} + \sqrt{2})}{x^2 - 6}$.

6. Is the function $\dfrac{\sqrt{1+x}-\sqrt{1-x}}{x}$ defined for all values of x? Indicate the values of x for which it is defined and real. Find the limit of this function as $x \to 0$.

7. (*i*) Given $f(x) = x^2 - 3x$, find $\lim\limits_{h\to 0} \dfrac{f(x+h)-f(x)}{h}$.

(*ii*) Given $f(x) = \sqrt{x+1}$, find $\lim\limits_{h\to 0} \dfrac{f(x+h)-f(x)}{h}$, where $x > -1$.

8. If $\lim\limits_{x\to\infty}\left(\dfrac{x^2+1}{x+1} - (ax+b)\right) = 2$, find the values of a and b.

Answers

1. (*i*) 3 (*ii*) $\frac{5}{7}$ (*iii*) $-\frac{1}{3}$ (*iv*) $\frac{2}{9}$ (*v*) $-\frac{1}{18}$ (*vi*) $\frac{1}{2}$

2. (*i*) $-\frac{1}{x^2}$ (*ii*) $\frac{1}{2}$ (*iii*) $-\frac{1}{9}$ (*iv*) $-\frac{1}{2}$

3. (*i*) $-\frac{1}{4}$ (*ii*) -11 (*iii*) $-\frac{1}{3}$ (*iv*) $\frac{15}{11}$ (*v*) -1 (*vi*) $\frac{1}{2}$ (*vii*) $\frac{5}{16}$

4. (*i*) $\frac{1}{2}$ (*ii*) -4 (*iii*) 3 (*iv*) $\frac{1}{2\sqrt{x}}$ (*v*) $\frac{1}{2}a^{-3/2}$ (*vi*) 1

5. (*i*) $-\frac{1}{2}$ (*ii*) $-\frac{1}{18}$ (*iii*) -8 (*iv*) 10 (*v*) $\frac{\sqrt{5}-\sqrt{2}}{6\sqrt{10}}$ (*vi*) $\frac{\sqrt{3}-\sqrt{2}}{2\sqrt{6}}$

6. No, Domain = $[-1, 1] - \{0\}$, Limit = 1 **7.** (*i*) $2x - 3$ (*ii*) $\frac{1}{2\sqrt{x+1}}$

8. $a = 1$, $b = -3$.

Hints

1. (*v*) Use $3x^2 + 3x - 6 = 3(x-1)(x+2) = 3(\sqrt{x}-1)(\sqrt{x}+1)(x+2)$.

2. (*iv*) Use $x^3 - 3x^2 + x = x(x-1)(x-2)$.

4. (*vi*) Rationalise num. and den.

5. (*v*) Rationalise num.

8. $$\lim_{x\to\infty}\left[\frac{x^2+1}{x+1} - (ax+b)\right] = \lim_{x\to\infty}\frac{(1-a)x^2-(a+b)x+1-b}{x+1}.$$

Since this limit is a non-zero finite number, we should have deg. numer. = deg. denom.

$\therefore\ 1 - a = 0$ *i.e.*, $a = 1$.

$$\therefore \lim_{x\to\infty}\frac{0-(1+b)x+1-b}{x+1} = 2 \quad\text{or}\quad \lim_{x\to\infty}\frac{-(1+b)+\dfrac{1-b}{x}}{1+\dfrac{1}{x}} = 2.$$

11.13. EVALUATION OF $\lim\limits_{x \to a} \dfrac{x^n - a^n}{x - a}$; n, a ∈ R, a > 0

Theorem I. *If n is any integer and a is any non-zero real number, then prove that*

$$\lim_{x \to a} \frac{x^n - a^n}{x - a} = na^{n-1}.$$

Proof. Case I. n is positive i.e., n ∈ N

Let $x = a + h$. ∴ $h \to 0$ as $x \to a$.

$$\therefore \quad \lim_{x \to a} \frac{x^n - a^n}{x - a} = \lim_{h \to 0} \frac{(a+h)^n - a^n}{(a+h) - a}$$

$$= \lim_{h \to 0} \frac{1}{h} [(a^n + {}^nC_1\, a^{n-1}h + {}^nC_2\, a^{n-2}h^2 + \ldots\ldots + h^n) - a^n]$$

$$= \lim_{h \to 0} \frac{1}{h} [{}^nC_1 a^{n-1}h + {}^nC_2 a^{n-2}h^2 + \ldots\ldots + h^n] = \lim_{h \to 0} [{}^nC_1 a^{n-1} + {}^nC_2 a^{n-2}h + \ldots\ldots + h^{n-1}]$$

$$= {}^nC_1 a^{n-1} + 0 + \ldots\ldots + 0 = na^{n-1}.$$

Case II. n = 0

$$\therefore \quad \lim_{x \to a} \frac{x^n - a^n}{x - a} = \lim_{x \to a} \frac{x^0 - a^0}{x - a} = \lim_{x \to a} \frac{1 - 1}{x - a} = \lim_{x \to a} 0 = 0.$$

Also, $na^{n-1} = 0.a^{0-1} = 0$.

Case III. n is negative integer

Let $n = -m$. ∴ $m \in N$.

$$\therefore \quad \lim_{x \to a} = \frac{x^n - a^n}{x - a} = \lim_{x \to a} \frac{x^{-m} - a^{-m}}{x - a} = \lim_{x \to a} \frac{a^m - x^m}{x^m a^m (x - a)} = -\frac{1}{a^m} \lim_{x \to a} \frac{x^m - a^m}{x^m (x - a)}$$

$$= -\frac{1}{a^m} \left(\lim_{x \to a} \frac{1}{x^m} \right) \left(\lim_{x \to a} \frac{x^m - a^m}{x - a} \right) = -\frac{1}{a^m} \cdot \frac{1}{a^m} \cdot ma^{m-1}$$

(Using Case I, because $m \in$ **N**)

$$= -ma^{-m-1} = na^{n-1}.$$

∴ Combining all cases, we get $\mathbf{\lim\limits_{x \to a} \dfrac{x^n - a^n}{x - a} = na^{n-1}}$, **n ∈ Z.**

Remark. If $a > 0$, then the result $\lim\limits_{x \to a} \dfrac{x^n - a^n}{x - a} = na^{n-1}$ is also true when $n \in$ **R**. We have restricted a to be positive, because if for example, $a = -2$, then numbers like $a^{1/4} = (-2)^{1/4}$ are not real numbers.

Thus, we have $\mathbf{\lim\limits_{x \to a} \dfrac{x^n - a^n}{x - a} = na^{n-1}}$, **n, a ∈ R, a > 0.**

ILLUSTRATIVE EXAMPLES

Example 1. *Evaluate the following limits :*

(i) $\lim_{x \to 2} \dfrac{x^{10} - 1024}{x^4 - 16}$ (ii) $\lim_{x \to a} \dfrac{(x+2)^{5/3} - (a+2)^{5/3}}{x - a}$

Sol. (i) $\lim_{x \to 2} \dfrac{x^{10} - 1024}{x^4 - 16} = \lim_{x \to 2} \dfrac{\dfrac{x^{10} - 2^{10}}{x - 2}}{\dfrac{x^4 - 2^4}{x - 2}} = \dfrac{\lim_{x \to 2} \dfrac{x^{10} - 2^{10}}{x - 2}}{\lim_{x \to 2} \dfrac{x^4 - 2^4}{x - 2}} = \dfrac{10 \,.\, 2^{10-1}}{4 \,.\, 2^{4-1}} = \mathbf{160.}$

$\left| \lim_{x \to a} \dfrac{x^n - a^n}{x - a} = na^{n-1}, a > 0 \right.$

(ii) $\lim_{x \to a} \dfrac{(x+2)^{5/3} - (a+2)^{5/3}}{x - a} = \lim_{x+2 \to a+2} \dfrac{(x+2)^{5/3} - (a+2)^{5/3}}{(x+2) - (a+2)} = \dfrac{5}{3}(a+2)^{\frac{5}{3} - 1}$

$(\because \; x \to a \;\Rightarrow\; x + 2 \to a + 2 \;\Rightarrow\; a \to 0)$

$= \dfrac{\mathbf{5}}{\mathbf{3}} \mathbf{(a + 2)^{2/3}}.$

Example 2. *If* $\lim_{x \to 1} \dfrac{x^4 - 1}{x - 1} = \lim_{x \to k} \dfrac{x^3 - k^3}{x^2 - k^2}$ *, find the value of k.*

Sol. We have $\lim_{x \to 1} \dfrac{x^4 - 1}{x - 1} = \lim_{x \to k} \dfrac{x^3 - k^3}{x^2 - k^2}$ ...(1)

L.H.S. $= \lim_{x \to 1} \dfrac{x^4 - 1}{x - 1} = \lim_{x \to 1} \dfrac{x^4 - 1^4}{x - 1} = 4(1)^{4-1} = 4$

R.H.S. $= \lim_{x \to k} \dfrac{x^3 - k^3}{x^2 - k^2} = \lim_{x \to k} \left(\dfrac{x^3 - k^3}{x^2 - k^2} \times \dfrac{x - k}{x - k} \right) = \lim_{x \to k} \left(\dfrac{x^3 - k^3}{x - k} \div \dfrac{x^2 - k^2}{x - k} \right)$

$= \lim_{x \to k} \dfrac{x^3 - k^3}{x - k} \div \lim_{x \to k} \dfrac{x^2 - k^2}{x - k} = 3k^{3-1} \div 2k^{2-1} = \dfrac{3k^2}{2k} = \dfrac{3}{2}k$

$\therefore$ (1) implies $4 = \dfrac{3}{2}k. \quad \therefore \quad k = \dfrac{\mathbf{8}}{\mathbf{3}}.$

EXERCISE 11.6

Evaluate the following limits (Q. No. 1–3) :

1. (i) $\lim_{x \to 4} \dfrac{x^{\sqrt{2}} - (4)^{\sqrt{2}}}{x - 4}$ (ii) $\lim_{x \to 5} \dfrac{x^7 - 5^7}{x^3 - 5^3}$ (iii) $\lim_{x \to 2} \dfrac{x^8 - 2^8}{x^7 - 2^7}$.

2. (i) $\lim_{x \to \sqrt{2}} \dfrac{x^8 - 16}{x - \sqrt{2}}$ (ii) $\lim_{x \to -1} \dfrac{x^7 + 1}{x^5 + 1}$.

3. (i) $\lim_{x \to a} \dfrac{x^m - a^m}{x^n - a^n}$, $m, n \in \mathbf{R}$ (ii) $\lim_{x \to 0} \dfrac{(1 + x)^n - 1}{x}$.

4. (*i*) If $\lim_{x \to -a} \dfrac{x^9 + a^9}{x + a} = 9$, find all possible values of a.

(*ii*) If $\lim_{x \to a} \dfrac{x^9 - a^9}{x - a} = \lim_{x \to 5} (4 + x)$, find all possible values of a.

5. $\lim_{x \to 2} \dfrac{x^n - 2^n}{x - 2} = 80$, then show that $n = 5$.

Answers

1. (*i*) $\sqrt{2}\,(4)^{\sqrt{2}-1}$ (*ii*) $\dfrac{4375}{3}$ (*iii*) $\dfrac{16}{7}$ **2.** (*i*) $64\sqrt{2}$ (*ii*) $\dfrac{7}{5}$

3. (*i*) $\dfrac{m}{n} a^{m-n}$ (*ii*) n **4.** (*i*) ± 1 (*ii*) ± 1.

Hints

3. (*ii*) $\dfrac{(1+x)^n - 1}{x} = \dfrac{(1+x)^n - 1^n}{(1+x) - 1}$ **5.** $80 = 5(2)^{5-1}$.

11.14. EVALUATION OF TRIGONOMETRICAL LIMITS

In this section, we shall consider limits of functions, involving trigonometric functions.

Theorem I. *If angle θ is measured in radians, then prove that* $\lim_{\theta \to 0} \dfrac{\sin \theta}{\theta} = 1$.

Proof. Let O be the centre of a unit circle. Let us first assume that the angle θ is *positive*. Let ∠AOB = θ radians. From A draw AD ⊥ OA, meeting OB produced to D.

From Fig. 11.1, we have

Area of ΔOAB ≤ area of sector OAB ≤ area of ΔOAD.

$$\Rightarrow \quad \tfrac{1}{2}\,\text{OA.BC} \le \tfrac{1}{2}\,\text{OA}^2.\theta \le \tfrac{1}{2}\,\text{OA.DA}$$

[∵ Area of a sector = $\frac{1}{2}$ (radius)^2 × angle in radians]

$$\Rightarrow \quad \text{BC} \le \theta \le \text{DA} \qquad \Rightarrow \quad \sin\theta \le \theta \le \tan\theta$$

$$\Rightarrow \quad 1 \le \frac{\theta}{\sin\theta} \le \frac{1}{\cos\theta} \qquad \Rightarrow \quad \cos\theta \le \frac{\sin\theta}{\theta} \le 1$$

$$\Rightarrow \quad \lim_{\theta \to 0} \cos\theta \le \lim_{\theta \to 0} \frac{\sin\theta}{\theta} \le \lim_{\theta \to 0} 1$$

$$\Rightarrow \quad 1 \le \lim_{\theta \to 0} \frac{\sin\theta}{\theta} \le 1 \quad \Rightarrow \quad \lim_{\theta \to 0} \frac{\sin\theta}{\theta} = 1.$$

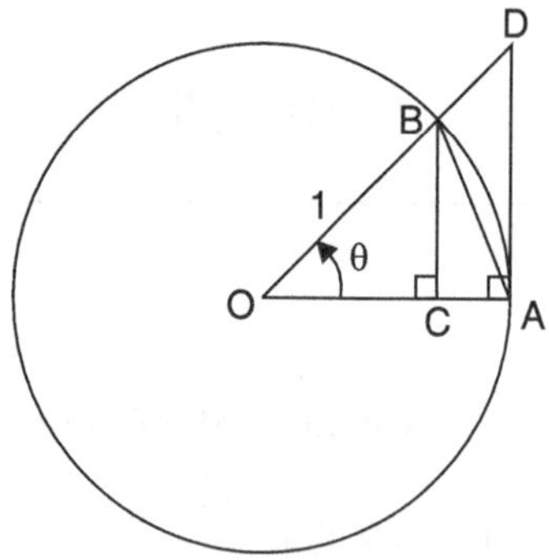

Fig. 11.1

∴ If θ is *positive*, then $\lim_{\theta \to 0} \dfrac{\sin\theta}{\theta} = 1$. ...(1)

If angle θ is *negative*, then – θ is positive and

$$\lim_{\theta \to 0} \frac{\sin\theta}{\theta} = \lim_{\theta \to 0} \frac{-\sin\theta}{-\theta} = \lim_{-\theta \to 0} \frac{\sin(-\theta)}{(-\theta)} = 1.$$

[∵ – θ is + ve and so (1) is applicable.]

∴ For any angle θ, $\mathbf{\lim_{\theta \to 0} \dfrac{\sin\theta}{\theta} = 1.}$

Corollary 1. *Prove that (i)* $\lim_{\theta \to 0} \frac{\theta}{\sin \theta} = 1$ *and (ii)* $\lim_{\theta \to 0} \frac{\tan \theta}{\theta} = 1.$

Proof. *(i)* $\lim_{\theta \to 0} \frac{\theta}{\sin \theta} = \lim_{\theta \to 0} \frac{1}{\frac{\sin \theta}{\theta}} = \frac{\lim_{\theta \to 0} 1}{\lim_{\theta \to 0} \frac{\sin \theta}{\theta}} = \frac{1}{1} = \mathbf{1}.$

(ii) $\lim_{\theta \to 0} \frac{\tan \theta}{\theta} = \lim_{\theta \to 0} \left(\frac{\sin \theta}{\theta} \times \frac{1}{\cos \theta} \right) = \left(\lim_{\theta \to 0} \frac{\sin \theta}{\theta} \right) \left(\lim_{\theta \to 0} \frac{1}{\cos \theta} \right) = (1)\left(\frac{1}{1}\right) = \mathbf{1}.$

Corollary 2. *Prove that (i)* $\lim_{x \to 0} \frac{\sin^{-1} x}{x} = 1$ *and (ii)* $\lim_{x \to 0} \frac{\tan^{-1} x}{x} = 1.$

Proof. *(i)* Let $x = \sin \theta$.

$\therefore \quad x \to 0 \Rightarrow \sin \theta \to 0 \Rightarrow \theta \to 0.$

$$\therefore \quad \lim_{x \to 0} \frac{\sin^{-1} x}{x} = \lim_{\theta \to 0} \frac{\theta}{\sin \theta} = \lim_{\theta \to 0} \frac{1}{\frac{\sin \theta}{\theta}} = \frac{\lim_{\theta \to 0} 1}{\lim_{\theta \to 0} \frac{\sin \theta}{\theta}} = \frac{1}{1} = \mathbf{1}.$$

(ii) Let $x = \tan \theta$.

$\therefore \quad x \to 0 \Rightarrow \tan \theta \to 0 \Rightarrow \theta \to 0.$

$$\therefore \quad \lim_{x \to 0} \frac{\tan^{-1} x}{x} = \lim_{\theta \to 0} \frac{\theta}{\tan \theta} = \lim_{\theta \to 0} \frac{\cos \theta}{\left(\frac{\sin \theta}{\theta}\right)} = \frac{\lim_{\theta \to 0} \cos \theta}{\lim_{\theta \to 0} \frac{\sin \theta}{\theta}} = \frac{1}{1} = \mathbf{1}.$$

ILLUSTRATIVE EXAMPLES

Example 1. *Evaluate :*

(i) $\lim_{x \to 0} \frac{\sin x^\circ}{x}$ *(ii)* $\lim_{x \to 0} \frac{\sin ax}{\sin bx}$ *(iii)* $\lim_{x \to 0} \frac{\cos^2 x}{1 - \sin x}$

(iv) $\lim_{\theta \to 0} \frac{1 - \cos m\theta}{1 - \cos n\theta}$ *(v)* $\lim_{x \to 0} \frac{\sin 5x}{\tan 3x}$ *(vi)* $\lim_{x \to 0} \frac{\tan 8x}{\sin 2x}$.

Sol. *(i)* $180° = \pi$ radians implies $x° = \frac{\pi x}{180}$ radians.

$$\therefore \quad \lim_{x \to 0} \frac{\sin x^\circ}{x} = \lim_{x \to 0} \frac{\sin \frac{x\pi}{180}}{x} = \lim_{x \to 0} \frac{\sin \frac{x\pi}{180}}{\frac{x\pi}{180} \cdot \frac{180}{\pi}} = \frac{\pi}{180} \lim_{\frac{x\pi}{180} \to 0} \frac{\sin \frac{\pi x}{180}}{\frac{x\pi}{180}} = \frac{\pi}{180} . 1 = \frac{\pi}{\mathbf{180}}.$$

(ii) $$\lim_{x \to 0} \frac{\sin ax}{\sin bx} = \lim_{x \to 0} \frac{\frac{\sin ax}{ax} . ax}{\frac{\sin bx}{bx} . bx} = \frac{a}{b} \frac{\lim_{ax \to 0} \frac{\sin ax}{ax}}{\lim_{bx \to 0} \frac{\sin bx}{bx}} = \frac{a}{b} . \frac{1}{1} = \frac{\mathbf{a}}{\mathbf{b}}.$$

(iii) $$\lim_{x \to 0} \frac{\cos^2 x}{1 - \sin x} = \lim_{x \to 0} \frac{1 - \sin^2 x}{1 - \sin x} = \lim_{x \to 0} (1 + \sin x) = 1 + 0 = \mathbf{1}.$$

$(\because \; x \to 0 \Rightarrow 1 - \sin x \neq 0)$

(*iv*) $\lim\limits_{\theta \to 0} \dfrac{1-\cos m\theta}{1-\cos n\theta} = \lim\limits_{\theta \to 0} \dfrac{2\sin^2 \frac{m\theta}{2}}{2\sin^2 \frac{n\theta}{2}} = \lim\limits_{\theta \to 0} \left(\dfrac{\sin \frac{m\theta}{2}}{\sin \frac{n\theta}{2}} \right)^2 = \left[\lim\limits_{\theta \to 0} \dfrac{\sin \frac{m\theta}{2}}{\sin \frac{n\theta}{2}} \right]^2.$

$$= \left[\lim_{\theta \to 0} \left(\frac{\sin \frac{m\theta}{2}}{\frac{m\theta}{2}} \times \frac{\frac{n\theta}{2}}{\sin \frac{n\theta}{2}} \times \frac{m}{n} \right) \right]^2 = \frac{m^2}{n^2} \left[\lim_{\frac{m\theta}{2} \to 0} \frac{\sin \frac{m\theta}{2}}{\frac{m\theta}{2}} \times \lim_{\frac{n\theta}{2} \to 0} \frac{\frac{n\theta}{2}}{\sin \frac{n\theta}{2}} \right]^2$$

$$= \frac{m^2}{n^2}(1 \times 1)^2 = \mathbf{\frac{m^2}{n^2}}.$$

(*v*) $\lim\limits_{x \to 0} \dfrac{\sin 5x}{\tan 3x} = \lim\limits_{x \to 0} \left[\dfrac{\sin 5x}{5x} \times \dfrac{3x}{\tan 3x} \times \dfrac{5}{3} \right] = \dfrac{5}{3} \lim\limits_{5x \to 0} \dfrac{\sin 5x}{5x} \lim\limits_{3x \to 0} \dfrac{3x}{\tan 3x}$

$$= \frac{5}{3} \times 1 \times 1 = \mathbf{\frac{5}{3}}.$$

(*vi*) $\lim\limits_{x \to 0} \dfrac{\tan 8x}{\sin 2x} = \lim\limits_{x \to 0} \left[\dfrac{\tan 8x}{8x} \times \dfrac{2x}{\sin 2x} \times \dfrac{8}{2} \right] = 4 \lim\limits_{8x \to 0} \dfrac{\tan 8x}{8x} \lim\limits_{2x \to 0} \dfrac{2x}{\sin 2x}$

$$= 4 \times 1 \times 1 = \mathbf{4}.$$

Example 2. *Evaluate the following limits :*

(*i*) $\lim\limits_{\theta \to \pi/2} \dfrac{\cot \theta}{\frac{\pi}{2} - \theta}$ (*ii*) $\lim\limits_{x \to 0} \dfrac{\sin (2+x) - \sin (2-x)}{x}$

(*iii*) $\lim\limits_{x \to 0} \dfrac{\tan x - \sin x}{\sin^3 x}$ (*iv*) $\lim\limits_{x \to 0} \dfrac{\cos mx - \cos nx}{x^2}$.

Sol. (*i*) Let $\theta = \dfrac{\pi}{2} + h$. $\therefore$ $h \to 0$ as $\theta \to \dfrac{\pi}{2}$.

$$\therefore \lim_{\theta \to \pi/2} \frac{\cot \theta}{\frac{\pi}{2} - \theta} = \lim_{h \to 0} \frac{\cot \left(\frac{\pi}{2} + h \right)}{\frac{\pi}{2} - \left(\frac{\pi}{2} + h \right)} = \lim_{h \to 0} \frac{-\tan h}{-h} = \lim_{h \to 0} \frac{\tan h}{h} = \mathbf{1}.$$

(*ii*) $\lim\limits_{x \to 0} \dfrac{\sin (2+x) - \sin (2-x)}{x} = \lim\limits_{x \to 0} \dfrac{2\cos \frac{2+x+2-x}{2} \sin \frac{2+x-2+x}{2}}{x}$

$$= \lim_{x \to 0} \frac{2\cos 2 \sin x}{x} = 2\cos 2 \lim_{x \to 0} \frac{\sin x}{x} = 2\cos 2 \,.\, 1 = \mathbf{2\cos 2}.$$

(*iii*) $\lim\limits_{x \to 0} \dfrac{\tan x - \sin x}{\sin^3 x} = \lim\limits_{x \to 0} \dfrac{\frac{\sin x}{\cos x} - \sin x}{\sin^3 x} = \lim\limits_{x \to 0} \dfrac{\sin x - \sin x \cos x}{\cos x \sin^3 x} = \lim\limits_{x \to 0} \dfrac{1 - \cos x}{\cos x \sin^2 x}$

$$= \lim_{x \to 0} \frac{1}{\cos x (1 + \cos x)} = \frac{1}{1(1+1)} = \mathbf{\frac{1}{2}}.$$

(iv) $$\lim_{x\to 0}\frac{\cos mx - \cos nx}{x^2} = \lim_{x\to 0}\frac{2\sin\frac{m+n}{2}x\,.\,\sin\frac{n-m}{2}x}{x^2}$$

$$= 2\lim_{x\to 0}\left(\frac{\sin\frac{m+n}{2}x}{\frac{m+n}{2}x\,.\,\frac{2}{m+n}}\cdot\frac{\sin\frac{n-m}{2}x}{\frac{n-m}{2}x\,.\,\frac{2}{n-m}}\right)$$

$$= \frac{n^2-m^2}{2}\left(\lim_{\frac{m+n}{2}x\to 0}\frac{\sin\frac{m+n}{2}x}{\frac{m+n}{2}x}\right)\times\left(\lim_{\frac{n-m}{2}x\to 0}\frac{\sin\frac{n-m}{2}x}{\frac{n-m}{2}x}\right)$$

$$= \frac{n^2-m^2}{2}\,.\,1\,.\,1 = \mathbf{\frac{n^2-m^2}{2}}.$$

Example 3. *Evaluate :*

(i) $\displaystyle lim_{x\to a}\frac{sin\ x - sin\ a}{\sqrt{x}-\sqrt{a}}$ (ii) $\displaystyle lim_{x\to\frac{\pi}{4}}\frac{sin\ x - cos\ x}{\frac{\pi}{4}-x}$

(iii) $\displaystyle lim_{x\to 0}\frac{sec\ 5x - sec\ 3x}{sec\ 3x - sec\ x}$ (iv) $\displaystyle lim_{x\to 0}\frac{\sqrt{2}-\sqrt{1+cos\ x}}{sin^2\ x}$.

Sol. (i) Let $x = a + h$. $\therefore\ h \to 0$ as $x \to a$

$$\therefore\ \lim_{x\to 0}\frac{\sin x - \sin a}{\sqrt{x}-\sqrt{a}} = \lim_{h\to 0}\frac{\sin(a+h)-\sin a}{\sqrt{a+h}-\sqrt{a}}$$

$$= \lim_{h\to 0}\left(\frac{2\cos\left(a+\frac{h}{2}\right)\sin\frac{h}{2}}{\sqrt{a+h}-\sqrt{a}}\times\frac{\sqrt{a+h}+\sqrt{a}}{\sqrt{a+h}+\sqrt{a}}\right)$$

$$= \lim_{h\to 0}\frac{\left[2\cos\left(a+\frac{h}{2}\right)\sin\frac{h}{2}\right][\sqrt{a+h}+\sqrt{a}]}{(a+h)-a}$$

$$= \lim_{h\to 0}\left(\cos a\cos\frac{h}{2} - \sin a\sin\frac{h}{2}\right)\times\lim_{\frac{h}{2}\to 0}\frac{\sin\frac{h}{2}}{\frac{h}{2}}\times\lim_{h\to 0}[\sqrt{a+h}+\sqrt{a}]$$

$$= ((\cos a)\,.1 - (\sin a)\,.0)\times 1\times[\sqrt{a}+\sqrt{a}] = \mathbf{2\sqrt{a}\ \cos a.}$$

(ii) Let $x = \frac{\pi}{4} + h$ $\therefore\ h \to 0$ as $x \to \frac{\pi}{4}$.

$$\therefore\ \lim_{x\to\pi/4}\frac{\sin x - \cos x}{\frac{\pi}{4}-x} = \lim_{h\to 0}\frac{\sin\left(\frac{\pi}{4}+h\right)-\cos\left(\frac{\pi}{4}+h\right)}{\frac{\pi}{4}-\left(\frac{\pi}{4}+h\right)}$$

$$= \lim_{h \to 0} \frac{\left(\frac{1}{\sqrt{2}} \cos h + \frac{1}{\sqrt{2}} \sin h\right) - \left(\frac{1}{\sqrt{2}} \cos h - \frac{1}{\sqrt{2}} \sin h\right)}{-h}$$

$$= - \lim_{h \to 0} \frac{\sqrt{2} \sin h}{h} = - \sqrt{2} \times 1 = - \sqrt{2}\,.$$

(*iii*) $$\lim_{x \to 0} \frac{\sec 5x - \sec 3x}{\sec 3x - \sec x} = \lim_{x \to 0} \frac{\frac{1}{\cos 5x} - \frac{1}{\cos 3x}}{\frac{1}{\cos 3x} - \frac{1}{\cos x}} = \lim_{x \to 0} \left(\frac{\cos 3x - \cos 5x}{\cos x - \cos 3x} \times \frac{\cos 3x \cos x}{\cos 5x \cos 3x}\right)$$

$$= \lim_{x \to 0} \left(\frac{2 \sin 4x \sin x}{2 \sin 2x \sin x} \times \frac{\cos x}{\cos 5x}\right) = \lim_{x \to 0} \left(\frac{\sin 4x}{4x} \times \frac{2x}{\sin 2x} \times \frac{4}{2}\right) \lim_{x \to 0} \left(\frac{\cos x}{\cos 5x}\right)$$

$$= 2 \left(\lim_{4x \to 0} \frac{\sin 4x}{4x}\right) \left(\lim_{2x \to 0} \frac{2x}{\sin 2x}\right) \times \frac{1}{1} = 2 \times 1 \times 1 \times 1 = \mathbf{2}.$$

(*iv*) $$\lim_{x \to 0} \frac{\sqrt{2} - \sqrt{1 + \cos x}}{\sin^2 x} = \lim_{x \to 0} \left(\frac{\sqrt{2} - \sqrt{1 + \cos x}}{\sin^2 x} \times \frac{\sqrt{2} + \sqrt{1 + \cos x}}{\sqrt{2} + \sqrt{1 + \cos x}}\right)$$

$$= \lim_{x \to 0} \frac{2 - (1 + \cos x)}{\sin^2 x (\sqrt{2} + \sqrt{1 + \cos x})} = \lim_{x \to 0} \frac{1 - \cos x}{(1 - \cos^2 x)(\sqrt{2} + \sqrt{1 + \cos x})}$$

$$= \lim_{x \to 0} \frac{1}{(1 + \cos x)(\sqrt{2} + \sqrt{1 + \cos x})} = \frac{1}{\lim_{x \to 0} (1 + \cos x) \lim_{x \to 0} (\sqrt{2} + \sqrt{1 + \cos x})}$$

$$= \frac{1}{(1 + 1)(\sqrt{2} + \sqrt{1 + 1})} = \frac{1}{2 \times 2\sqrt{2}} = \frac{\sqrt{2}}{8}.$$

Example 4. *Evaluate the following limits :*

(*i*) $\lim_{x \to \frac{\pi}{6}} \frac{\cot^2 x - 3}{\operatorname{cosec} x - 2}$ (*ii*) $\lim_{x \to \frac{\pi}{2}} \frac{1 - \sin^3 x}{\cos^2 x}$.

(*iii*) $\lim_{x \to \frac{\pi}{2}} \frac{\left(\frac{\pi}{2} - x\right) \sin x - 2 \cos x}{\frac{\pi}{2} - x + \cot x}$.

Sol. (*i*) $$\lim_{x \to \frac{\pi}{6}} \frac{\cot^2 x - 3}{\operatorname{cosec} x - 2} = \lim_{x \to \frac{\pi}{6}} \frac{(\cot^2 x + 1) - 4}{\operatorname{cosec} x - 2} = \lim_{x \to \frac{\pi}{6}} \frac{\operatorname{cosec}^2 x - 4}{\operatorname{cosec} x - 2}$$

$$= \lim_{x \to \frac{\pi}{6}} \frac{(\operatorname{cosec} + 2)(\operatorname{cosec} x - 2)}{\operatorname{cosec} x - 2} = \lim_{x \to \frac{\pi}{6}} (\operatorname{cosec} x + 2)$$

$$= \lim_{x \to \pi/6} \left(\frac{1}{\sin x} + 2\right) = \frac{1}{\sin \frac{\pi}{6}} + 2$$ ($\because$ sine is a continuous function)

$$= \frac{1}{1/2} + 2 = \mathbf{4}.$$

(ii) Let $x = \frac{\pi}{2} + h$. $\therefore$ $h \to 0$ as $x \to \frac{\pi}{2}$ $= \lim_{h \to 0} \frac{1 - \cos^3 h}{1 - \cos^2 h}$

$$\therefore \quad \lim_{x \to \frac{\pi}{2}} \frac{1 - \sin^3 x}{\cos^2 x} = \lim_{h \to 0} \frac{1 - \sin^3\left(\frac{\pi}{2} + h\right)}{\cos^2\left(\frac{\pi}{2} + h\right)} = \lim_{h \to 0} \frac{1 - (\cos h)^3}{(-\sin h)^2} = \lim_{h \to 0} \frac{1 - \cos^3 h}{\sin^2 h}$$

$$= \lim_{h \to 0} \frac{(1 - \cos h)(1 + \cos h + \cos^2 h)}{(1 - \cos h)(1 + \cos h)} = \lim_{h \to 0} \frac{1 + \cos h + \cos^2 h}{1 + \cos h} = \frac{1 + 1 + (1)^2}{1 + 1} = \mathbf{\frac{3}{2}}.$$

(iii) Let $x = \frac{\pi}{2} + h$. $\therefore$ $h \to 0$ as $x \to \frac{\pi}{2}$

$$\therefore \quad \lim_{x \to \frac{\pi}{2}} \frac{\left(\frac{\pi}{2} - x\right)\sin x - 2\cos x}{\frac{\pi}{2} - x + \cot x} = \lim_{h \to 0} \frac{\left(\frac{\pi}{2} - \left(\frac{\pi}{2} + h\right)\right)\sin\left(\frac{\pi}{2} + h\right) - 2\cos\left(\frac{\pi}{2} + h\right)}{\frac{\pi}{2} - \left(\frac{\pi}{2} + h\right) + \cot\left(\frac{\pi}{2} + h\right)}$$

$$= \lim_{h \to 0} \frac{-h\cos h + 2\sin h}{-h - \tan h} = \lim_{h \to 0} \frac{h\cos h - 2\sin h}{h + \tan h} = \lim_{h \to 0} \frac{\cos h - 2\left(\frac{\sin h}{h}\right)}{1 + \left(\frac{\tan h}{h}\right)}$$

$$= \frac{1 - 2(1)}{1 + 1} = -\mathbf{\frac{1}{2}}.$$

Example 5. *Evaluate :*

(i) $\lim_{x \to 0} \frac{\sin^{-1} 3x}{\tan 7x}$ (ii) $\lim_{x \to 0} \frac{2\sin^{-1} x + 5\tan^{-1} x}{3x - 2\tan^{-1} x}$

(iii) $\lim_{x \to 0} \frac{\sqrt{1 + x} - \sqrt{1 - x}}{\sin^{-1} x}$.

Sol. (i) $\lim_{x \to 0} \frac{\sin^{-1} 3x}{\tan 7x} = \lim_{x \to 0}\left(\frac{\sin^{-1} 3x}{3x} \times \frac{7x}{\tan 7x} \times \frac{3x}{7x}\right)$

$$= \left(\lim_{3x \to 0} \frac{\sin^{-1} 3x}{3x}\right)\left(\lim_{7x \to 0} \frac{7x}{\tan 7x}\right)\left(\frac{3}{7}\right) = 1 \times 1 \times \frac{3}{7} = \mathbf{\frac{3}{7}}.$$

(ii) $\lim_{x \to 0} \frac{2\sin^{-1} x + 5\tan^{-1} x}{3x - 2\tan^{-1} x} = \lim_{x \to 0} \frac{\frac{2\sin^{-1} x}{x} + \frac{5\tan^{-1} x}{x}}{\frac{3x}{x} - \frac{2\tan^{-1} x}{x}}$

$$= \frac{2\left(\lim_{x \to 0} \frac{\sin^{-1} x}{x}\right) + 5\left(\lim_{x \to 0} \frac{\tan^{-1} x}{x}\right)}{3 - 2\left(\lim_{x \to 0} \frac{\tan^{-1} x}{x}\right)} = \frac{2(1) + 5(1)}{3 - 2(1)} = \mathbf{7}.$$

(*iii*) Let $x = \sin\theta$. $\therefore\ x \to 0 \Rightarrow \sin\theta \to 0 \Rightarrow \theta \to 0$.

$$\therefore \lim_{x\to 0} \frac{\sqrt{1+x}-\sqrt{1-x}}{\sin^{-1} x} = \lim_{\theta\to 0} \frac{\sqrt{1+\sin\theta}-\sqrt{1-\sin\theta}}{\theta}$$

$$= \lim_{\theta\to 0}\left(\frac{\sqrt{1+\sin\theta}-\sqrt{1-\sin\theta}}{\theta} \times \frac{\sqrt{1+\sin\theta}+\sqrt{1-\sin\theta}}{\sqrt{1+\sin\theta}+\sqrt{1-\sin\theta}}\right)$$

$$= \lim_{\theta\to 0} \frac{(1+\sin\theta)-(1-\sin\theta)}{\theta(\sqrt{1+\sin\theta}+\sqrt{1-\sin\theta})} = \lim_{\theta\to 0} \frac{2\sin\theta}{\theta(\sqrt{1+\sin\theta}+\sqrt{1-\sin\theta})}$$

$$= 2\lim_{\theta\to 0}\frac{\sin\theta}{\theta} \cdot \lim_{\theta\to 0}\frac{1}{\sqrt{1+\sin\theta}+\sqrt{1-\sin\theta}} = 2\,.\,1\,.\,\frac{1}{\sqrt{1+0}+\sqrt{1-0}} = \frac{2}{2} = \mathbf{1.}$$

Example 6. *Evaluate :* $\lim_{x\to\infty} x\left(\tan^{-1}\frac{x+1}{x+4} - \frac{\pi}{4}\right)$.

Sol. $$\tan^{-1}\frac{x+1}{x+4} - \frac{\pi}{4} = -\left(\tan^{-1}1 - \tan^{-1}\frac{x+1}{x+4}\right) = -\tan^{-1}\left(\frac{1-\frac{x+1}{x+4}}{1+(1)\left(\frac{x+1}{x+4}\right)}\right)$$

| $\tan^7 A - \tan^{-1}B = \tan^7 \frac{A-B}{A+B}$

$$= -\tan^{-1}\frac{3}{2x+5}.$$

$$\therefore \lim_{x\to\infty} x\left(\tan^{-1}\frac{x+1}{x+4} - \frac{\pi}{4}\right) = \lim_{x\to\infty}\left(-x\tan^{-1}\frac{3}{2x+5}\right) = -\lim_{x\to\infty}\frac{\tan^{-1}\frac{3}{2x+5}}{\frac{3}{2x+5}} \cdot \frac{3x}{2x+5}$$

| $\tan^{-1} 1 = \frac{\pi}{4}$

$$= -\lim_{\frac{3}{2x+5}\to 0}\frac{\tan^{-1}\frac{3}{2x+5}}{\frac{3}{2x+5}} \cdot \lim_{\frac{1}{x}\to 0}\frac{3}{2+\frac{5}{x}} = -(1)\,.\,\frac{3}{2+0} = -\frac{\mathbf{3}}{\mathbf{2}}.$$

EXERCISE 11.7

Evaluate the following limits (Q. No. 1–2) :

1. (*i*) $\lim_{x\to 0}\frac{\sin\frac{x}{4}}{3x}$ (*ii*) $\lim_{x\to 0}\frac{4x}{\tan 3x}$ (*iii*) $\lim_{x\to 0}\frac{\sin x^2}{x}$

(*iv*) $\lim_{x\to 0}\frac{\sin 4x}{\sin 7x}$ (*v*) $\lim_{\theta\to 0}\frac{\sin 3\theta}{\tan 2\theta}$ (*vi*) $\lim_{x\to 0}\frac{\tan 4x}{\tan 3x}$

2. (*i*) $\lim_{x\to 0}\frac{x\cos x + \sin x}{x^2 + \tan x}$ (*ii*) $\lim_{x\to 0}\frac{\sin 3x + 7x}{4x + \sin 2x}$ (*iii*) $\lim_{x\to 0}\frac{\sin 2x + 3x}{4x - \sin 5x}$

(*iv*) $\lim_{x \to 0} \frac{1 - \cos x}{x^2}$ (*v*) $\lim_{x \to 0} \frac{x \tan 4x}{1 - \cos 4x}$ (*vi*) $\lim_{x \to 0} \frac{2 \sin x - \sin 2x}{x^3}$.

(*G.N.D.U. B.Pharma, April 2003, 2001*)

3. If $\lim_{\theta \to 0} (k\theta \operatorname{cosec} \theta) = \lim_{\theta \to 0} (\theta \operatorname{cosec} k\theta)$, show that $k = \pm 1$. (*G.N.D.U. B.Pharma, April 2002*)

Evaluate the following limits (Q. No. 4–8) :

4. (*i*) $\lim_{x \to 0} \frac{1 - \cos 2x}{\sin^2 2x}$ (*ii*) $\lim_{x \to 0} \frac{\sin 2x (1 - \cos 2x)}{x^3}$ (*iii*) $\lim_{x \to 0} \frac{1 - \cos 2x + \tan^2 x}{x \sin x}$

(*iv*) $\lim_{x \to 0} \frac{\sin x - 2 \sin 3x + \sin 5x}{x}$.

5. (*i*) $\lim_{x \to 0} \frac{\operatorname{cosec} x - \cot x}{x}$ (*P.T.U. B.Pharma, May 2003 ; June 2003*)

(*ii*) $\lim_{x \to 0} \frac{\cot 2x - \operatorname{cosec} 2x}{x}$ (*iii*) $\lim_{x \to 0} \frac{\tan 2x - \sin 2x}{x^3}$ (*G.N.D.U. B.Pharma, 2005*)

(*iv*) $\lim_{h \to 0} \frac{\sin (x + h) - \sin x}{h}$.

6. (*i*) $\lim_{x \to 0} \frac{\sin 2x + \sin 6x}{\sin 5x - \sin 3x}$ (*ii*) $\lim_{x \to 0} \frac{\cos ax - \cos bx}{\cos cx - \cos dx}$ (*iii*) $\lim_{x \to 0} \frac{\sec 4x - \sec 2x}{\sec 3x - \sec x}$

(*iv*) $\lim_{x \to 0} \frac{\sqrt{1 + \sin x} - \sqrt{1 - \sin x}}{x}$ (*v*) $\lim_{x \to 0} \frac{\sqrt{1 + x} - \sqrt{1 - x}}{\sin x}$

7. (*i*) $\lim_{x \to \pi} \frac{\sin 3x - 3 \sin x}{(\pi - x)^3}$ (*ii*) $\lim_{x \to \pi/2} \frac{2x - \pi}{\cos x}$ (*iii*) $\lim_{x \to \pi/2} \left(\frac{\pi}{2} - x\right) \tan x$

(*iv*) $\lim_{x \to \pi/4} \frac{1 - \tan x}{x - \pi/4}$ (*v*) $\lim_{x \to \pi/2} \frac{1 + \cos 2x}{(\pi - 2x)^2}$

8. (*i*) $\lim_{x \to \alpha} \frac{\cos x - \cos \alpha}{x - \alpha}$ (*ii*) $\lim_{x \to \pi/2} (\sec x - \tan x)$ (*iii*) $\lim_{x \to \pi/4} \frac{\sin x - \cos x}{\cos 2x}$

(*iv*) $\lim_{x \to \pi/4} \frac{\sin x - \cos x}{x - \frac{\pi}{4}}$ (*v*) $\lim_{x \to a} \frac{\cos x - \cos a}{\cot x - \cot a}$.

Answers

1. (*i*) $\frac{1}{12}$ (*ii*) $\frac{4}{3}$ (*iii*) 0 (*iv*) $\frac{4}{7}$ (*v*) $\frac{3}{2}$ (*vi*) $\frac{4}{3}$

2. (*i*) 2 (*ii*) $\frac{5}{3}$ (*iii*) -5 (*iv*) $\frac{1}{2}$ (*v*) $\frac{1}{2}$ (*vi*) 1

4. (*i*) $\frac{1}{2}$ (*ii*) 4 (*iii*) 3 (*iv*) 0

5. (*i*) $\frac{1}{2}$ (*ii*) -1 (*iii*) 4 (*iv*) $\cos x$

6. (*i*) 4 (*ii*) $\frac{a^2 - b^2}{c^2 - d^2}$ (*iii*) $\frac{3}{2}$ (*iv*) 1 (*v*) 1

7. (*i*) -4 (*ii*) -2 (*iii*) 1 (*iv*) -2 (*v*) $\frac{1}{2}$

8. (*i*) $-\sin \alpha$ (*ii*) 0 (*iii*) $-\frac{1}{\sqrt{2}}$ (*iv*) $\sqrt{2}$ (*v*) $\sin^3 a$.

Hints

2. (*i*) Divide numerator and denominator by x. (*vi*) Use $2 \sin x - \sin 2x = 2 \sin x (1 - \cos x)$.
3. Use $\operatorname{cosec} \theta = 1/\sin \theta$.
4. (*iii*) $1 - \cos 2x + \tan^2 x = 2 \sin^2 x + \tan^2 x = \sin^2 x (2 + \sec^2 x)$.
   (*iv*) $\sin x - 2 \sin 3x + \sin 5x = (\sin 5x + \sin x) - 2 \sin 3x$.
5. (*iv*) Use $\sin (x + h) - \sin x = 2 \cos \left(x + \frac{h}{2}\right) \sin \frac{h}{2}$.
6. (*iv*) Rationalise the numerator.
7. (*i*) Use $\sin 3x = 3 \sin x - 4 \sin^3 x$
8. (*iii*) $\sin x - \cos x = \sqrt{2} \sin \left(x - \frac{\pi}{4}\right)$. (*v*) $\frac{\cos x - \cos a}{\cot x - \cot a} = \frac{(\cos x - \cos a) \sin x \sin a}{\sin (a - x)}$.

11.15. EVALUATION OF $\lim\limits_{x \to 0} \frac{a^x - 1}{x}$, $a > 0$

Theorem I. *If $a > 0$, prove that $\lim\limits_{x \to 0} \frac{a^x - 1}{x} = \log a$.*

Proof. We have

$$a^x = 1 + x \log a + \frac{x^2}{2!} (\log a)^2 + \ldots$$

$$\therefore \quad \lim_{x \to 0} \frac{a^x - 1}{x} = \lim_{x \to 0} \frac{\left(1 + x \log a + \frac{x^2}{2!} (\log a)^2 + \ldots\right) - 1}{x}$$

$$= \lim_{x \to 0} x \left[\frac{\log a + \frac{x}{2!} (\log a)^2 + \ldots}{x}\right]$$

$$= \lim_{x \to 0} \left(\log a + \frac{x}{2!} (\log a)^2 + \ldots\right) = \log a + 0 + \ldots = \log a$$

$$\therefore \quad \lim_{x \to 0} \frac{a^x - 1}{x} = \log a.$$

Corollary 1. *Prove that $\lim\limits_{x \to 0} \frac{e^x - 1}{x} = 1$.*

Proof. We have

$$\lim_{x \to 0} \frac{a^x - 1}{x} = \log a, \text{ where } a > 0.$$

In particular, let $a = e$.

$$\therefore \quad \lim_{x \to 0} \frac{e^x - 1}{x} = \log e = 1$$

$$\therefore \quad \lim_{x \to 0} \frac{e^x - 1}{x} = 1.$$

Theorem II. *Prove that* $\lim\limits_{x \to 0} \dfrac{\log(1+x)}{x} = 1.$

Proof. We have

$$\log(1+x) = x - \frac{x^2}{2} + \frac{x^3}{3} - \frac{x^4}{4} + \ldots$$

$$\therefore \quad \lim_{x \to 0} \frac{\log(1+x)}{x} = \lim_{x \to 0} \frac{1}{x}\left[x - \frac{x^2}{2} + \frac{x^3}{3} - \frac{x^4}{4} + \ldots\right]$$

$$= \lim_{x \to 0}\left[1 - \frac{x}{2} + \frac{x^2}{3} - \frac{x^3}{4} + \ldots\right] = 1 - 0 + 0 - 0 + \ \ldots = 1$$

$$\therefore \quad \lim_{x \to 0} \frac{\log(1+x)}{x} = 1.$$

Corollary 1. *Prove that* $\lim\limits_{x \to 0} (1+x)^{1/x} = e.$

Proof. We have $\lim\limits_{x \to 0} \dfrac{\log(1+x)}{x} = 1.$

$$\Rightarrow \quad \lim_{x \to 0} \log(1+x)^{1/x} = 1 \quad \Rightarrow \quad \log\left(\lim_{x \to 0}(1+x)^{1/x}\right) = 1$$

($\because$ Logarithm is a continuous function)

$$\Rightarrow \quad \lim_{x \to 0} (1+x)^{1/x} = e^1 = e$$

$$\therefore \quad \lim_{x \to 0} (1+x)^{1/x} = e.$$

Corollary 2. *Prove that* $\lim\limits_{n \to \infty}\left(1 + \dfrac{1}{n}\right)^n = e.$

Proof. Let $x = \dfrac{1}{n}$. $\therefore \quad x \to 0$ as $n \to \infty$

$$\Rightarrow \quad \lim_{n \to \infty}\left(1 + \frac{1}{n}\right)^n = \lim_{x \to 0}(1+x)^{1/x} = e$$

$$\therefore \quad \lim_{n \to \infty}\left(1 + \frac{1}{n}\right)^n = e.$$

Remark. In this section, we have proved the following formulae :

(i) $\lim\limits_{x \to 0} \dfrac{a^x - 1}{x} = \log a, a > 0$ *(ii)* $\lim\limits_{x \to 0} \dfrac{e^x - 1}{x} = 1$ *(iii)* $\lim\limits_{x \to 0} \dfrac{\log(1+x)}{x} = 1$

(iv) $\lim\limits_{x \to 0} (1+x)^{1/x} = e$ *(v)* $\lim\limits_{n \to \infty}\left(1 + \dfrac{1}{n}\right)^n = e.$

ILLUSTRATIVE EXAMPLES

Example 1. *Evaluate the following limits :*

(i) $\lim\limits_{x \to 0} \dfrac{e^{2x} - 1}{x}$ (ii) $\lim\limits_{x \to 0} \dfrac{e^{x} - e^{-x}}{x}$ (iii) $\lim\limits_{x \to 0} \dfrac{e^{3+x} - \sin x - e^3}{x}$

(iv) $\lim\limits_{x \to 0} \dfrac{x(e^{x} - 1)}{1 - \cos 2x}$ (v) $\lim\limits_{x \to 0} \dfrac{x(e^{2+x} - e^2)}{1 - \cos x}$ (vi) $\lim\limits_{x \to 0} \dfrac{8^x - 4^x - 2^x + 1}{x \tan x}$.

Sol. (i) $\lim\limits_{x \to 0} \dfrac{e^{2x} - 1}{x} = \lim\limits_{x \to 0} \left(\dfrac{e^{2x} - 1}{2x} \times 2 \right) = 2 \,.\, \lim\limits_{2x \to 0} \dfrac{e^{2x} - 1}{2x} = 2 \times 1 = \mathbf{2.}$

(ii) $\lim\limits_{x \to 0} \dfrac{e^{x} - e^{-x}}{x} = \lim\limits_{x \to 0} \dfrac{e^{x} - 1 + 1 - e^{-x}}{x} = \lim\limits_{x \to 0} \left(\dfrac{e^x - 1}{x} + \dfrac{e^{-x} - 1}{-x} \right)$

$$= \lim_{x \to 0} \frac{e^x - 1}{x} + \lim_{x \to 0} \frac{e^{-x} - 1}{-x} = 1 + 1 = \mathbf{2.}$$

(iii) $\lim\limits_{x \to 0} \dfrac{e^{3+x} - \sin x - e^3}{x} = e^3 \lim\limits_{x \to 0} \dfrac{e^x - 1}{x} - \lim\limits_{x \to 0} \dfrac{\sin x}{x} = e^3 \times 1 - 1 = \mathbf{e^3 - 1.}$

(iv) $\lim\limits_{x \to 0} \dfrac{x(e^x - 1)}{1 - \cos 2x} = \lim\limits_{x \to 0} \dfrac{x(e^x - 1)}{2 \sin^2 x} = \lim\limits_{x \to 0} \left[\dfrac{e^x - 1}{x} \cdot \dfrac{x}{\sin x} \cdot \dfrac{x}{\sin x} \cdot \dfrac{1}{2} \right]$

$$= \frac{1}{2} \left(\lim_{x \to 0} \frac{e^x - 1}{x} \right) \left(\lim_{x \to 0} \frac{x}{\sin x} \right) \cdot \left(\lim_{x \to 0} \frac{x}{\sin x} \right) = \frac{1}{2} \times 1 \times 1 \times 1 = \mathbf{\frac{1}{2}}\,.$$

(v) $\lim\limits_{x \to 0} \dfrac{x(e^{2+x} - e^2)}{1 - \cos x} = \lim\limits_{x \to 0} \dfrac{xe^2(e^x - 1)(1 + \cos x)}{1 - \cos^2 x} = e^2 \lim\limits_{x \to 0} \left[\dfrac{x}{\sin x} \cdot \dfrac{x}{\sin x} \cdot \dfrac{e^x - 1}{x} . (1 + \cos x) \right]$

$$= e^2 \left(\lim_{x \to 0} \frac{x}{\sin x} \right) \left(\lim_{x \to 0} \frac{x}{\sin x} \right) \left(\lim_{x \to 0} \frac{e^x - 1}{x} \right) \left(\lim_{x \to 0} (1 + \cos x) \right)$$

$$= e^2 \times 1 \times 1 \times 1 \times (1 + 1) = \mathbf{2e^2.}$$

(vi) $\lim\limits_{x \to 0} \dfrac{8^x - 4^x - 2^x + 1}{x \tan x} = \lim\limits_{x \to 0} \dfrac{(4^x - 1)(2^x - 1)}{x \tan x} = \lim\limits_{x \to 0} \left(\dfrac{4^x - 1}{x} \cdot \dfrac{2^x - 1}{x} \cdot \dfrac{x}{\tan x} \right)$

| $8^x - 4^x - 2^x + 1 = (2^3)^x - (2^2)^x - 2^x + 1 = 2^{2x}(2^x - 1) - (2^x - 1) = (2^x - 1)(4^x - 1)$

$$= \lim_{x \to 0} \frac{4^x - 1}{x} \,.\, \lim_{x \to 0} \frac{2^x - 1}{x} \,.\, \lim_{x \to 0} \frac{x}{\tan x} = (\log 4)(\log 2).1 = \mathbf{(\log 4)(\log 2).}$$

Example 2. *Evaluate the following limits :*

(i) $\lim\limits_{x \to 2} \dfrac{e^x - e^2}{x - 2}$ (ii) $\lim\limits_{x \to \pi/2} \dfrac{e^{\cos x} - 1}{\cos x}$ (iii) $\lim\limits_{x \to 1} \dfrac{a^{x-1} - 1}{\sin \pi x}, a > 0$

Sol. (*i*) Let $x = 2 + h$. $\therefore$ $h \to 0$ as $x \to 2$

$$\therefore \lim_{x \to 2} \frac{e^x - e^2}{x - 2} = \lim_{h \to 0} \frac{e^{2+h} - e^2}{(2+h) - 2} = \lim_{h \to 0} \frac{e^2(e^h - 1)}{h} = e^2 \lim_{h \to 0} \frac{e^h - 1}{h} = e^2 \times 1 = \mathbf{e^2}.$$

(*ii*) Let $x = \frac{\pi}{2} + h$. $\therefore$ $h \to 0$ as $x \to \frac{\pi}{2}$

$$\therefore \lim_{x \to \frac{\pi}{2}} \frac{e^{\cos x} - 1}{\cos x} = \lim_{h \to 0} \frac{e^{\cos\left(\frac{\pi}{2} + h\right)} - 1}{\cos\left(\frac{\pi}{2} + h\right)} = \lim_{h \to 0} \frac{e^{-\sin h} - 1}{-\sin h} = \lim_{-\sin h \to 0} \frac{e^{-\sin h} - 1}{-\sin h} = \mathbf{1}.$$

(*iii*) Let $x = 1 + h$. $\therefore$ $h \to 0$ as $x \to 1$

$$\therefore \lim_{x \to 1} \frac{a^{x-1} - 1}{\sin \pi x} = \lim_{h \to 0} \frac{a^{(1+h)-1} - 1}{\sin \pi(1+h)} = \lim_{h \to 0} \frac{a^h - 1}{\sin(\pi + \pi h)} = \lim_{h \to 0} \frac{a^h - 1}{-\sin \pi h}$$

$$= -\lim \left(\frac{a^h - 1}{h} \cdot \frac{\pi h}{\sin \pi h} \cdot \frac{1}{\pi}\right) = -\frac{1}{\pi}\left(\lim_{h \to 0} \frac{a^h - 1}{h}\right)\left(\lim_{\pi h \to 0} \frac{\pi h}{\sin \pi h}\right)$$

$$= -\frac{1}{\pi}(\log a) \cdot 1 = -\frac{\mathbf{1}}{\pi} \mathbf{\log a}.$$

Example 3. *Evaluate :*

(*i*) $\lim_{x \to 0} (1 + 5x)^{\frac{1}{2x}}$ (*ii*) $\lim_{x \to \infty} \left(\frac{x+6}{x+1}\right)^{x+4}$

Sol. (*i*) $\lim_{x \to 0} (1 + 5x)^{1/2x} = \lim_{x \to 0} (1 + 5x)^{(1/5x)(5/2)} = \lim_{x \to 0} [(1 + 5x)^{1/5x}]^{5/2}$

$$= \left[\lim_{5x \to \infty} (1 + 5x)^{1/5x}\right]^{5/2} = \mathbf{e^{5/2}} \qquad \Big| \because \lim_{x \to \infty} (1 + x)^{1/x} = e$$

(*ii*) $$\lim_{x \to \infty} \left(\frac{x+6}{x+1}\right)^{x+4} = \lim_{x \to \infty} \left(\frac{x+1+5}{x+1}\right)^{x+4} = \lim_{x \to \infty} \left(1 + \frac{5}{x+1}\right)^{x+4}$$

$$= \lim_{x \to \infty} \left[\left(1 + \frac{5}{x+1}\right)^{x+1} \left(1 + \frac{5}{x+1}\right)^3\right]$$

$$= \lim_{x \to \infty} \left[\left(1 + \frac{5}{x+1}\right)^{\frac{x+1}{5} \cdot 5} \left(1 + \frac{5}{x+1}\right)^3\right]$$

$$= \left[\lim_{\frac{x+1}{5} \to \infty} \left(1 + \frac{5}{x+1}\right)^{\frac{x+1}{5}}\right]^5 \left[\lim_{x \to \infty} \left(1 + \frac{5}{x+1}\right)\right]^3 = e^5 \cdot (1)^3 = \mathbf{e^5}.$$

EXERCISE 11.8

Evaluate the following limits (Q. No. 1–8) :

1. (*i*) $\lim_{x\to 0} \frac{e^{5x}-1}{x}$ (*ii*) $\lim_{x\to 0} \frac{e^{\sin x}-1}{x}$ (*iii*) $\lim_{x\to 0} \frac{2^x-1}{8x}$ (*iv*) $\lim_{x\to 0} \frac{3^{2+x}-9}{x}$

2. (*i*) $\lim_{x\to 0} \frac{e^x-x-1}{x}$ (*ii*) $\lim_{x\to 0} \frac{e^{2x}-e^{-3x}}{x}$ (*iii*) $\lim_{x\to 0} \frac{e^x+5\sin x-1}{2x}$ (*iv*) $\lim_{x\to 0} \frac{e^x+e^{-x}-2}{x^2}$

3. (*i*) $\lim_{x\to 0} \frac{a^x-1}{b^x-1}$ (*ii*) $\lim_{x\to 0} \frac{a^{3x}-b^{4x}}{x}$ (*iii*) $\lim_{x\to 0} \frac{3^x-4^x}{5^x-6^x}$

(*iv*) $\lim_{x\to 0} \frac{a^x+b^x+c^x-3}{x}$ (*v*) $\lim_{x\to 0} \frac{e^x+e^x-2}{x^2}$.

4. (*i*) $\lim_{x\to 5} \frac{e^x-e^5}{x-5}$ (*ii*) $\lim_{x\to 1} \frac{\log x}{x-1}$ (*iii*) $\lim_{x\to 1} \frac{e^{-x}-e^{-1}}{x-1}$ (*iv*) $\lim_{x\to 0} \frac{e^x-e^{\sin x}}{x-\sin x}$

5. (*i*) $\lim_{x\to 0} \frac{\log(1+x)}{e^x-1}$ (*ii*) $\lim_{x\to 0} \frac{2x(e^x-1)}{1-\cos 3x}$ (*iii*) $\lim_{x\to 1} \frac{e^x-e^{-x}}{e^x+e^{-x}}$

6. (*i*) $\lim_{x\to \pi/2} \frac{e^{\cos x}-1}{\frac{\pi}{2}-x}$ (*ii*) $\lim_{x\to \pi/2} \frac{e^{\sin x}-1}{\sin x}$ (*iii*) $\lim_{x\to 5} \frac{\log x-\log 5}{x-5}$

7. (*i*) $\lim_{x\to 0} \frac{7^x+7^{-x}-2}{x^2}$ (*ii*) $\lim_{x\to e} \frac{\log x-1}{x-e}$

8. (*i*) $\lim_{x\to 0} (1+3x)^{1/4x}$ (*ii*) $\lim_{n\to\infty} \left(1+\frac{5}{n}\right)^{2n}$ (*iii*) $\lim_{x\to\infty} \left(\frac{x}{1+x}\right)^x$.

Answers

1. (*i*) 5 (*ii*) 1 (*iii*) $\frac{1}{8}\log 2$ (*iv*) 9 log 3

2. (*i*) 0 (*ii*) 5 (*iii*) 3 (*iv*) 1

3. (*i*) $\frac{\log a}{\log b}$ (*ii*) $\log \frac{a^3}{b^4}$ (*iii*) $\frac{\log(3/4)}{\log(5/6)}$

(*iv*) log *abc* (*v*) 1

4. (*i*) e^5 (*ii*) 1 (*iii*) $-1/e$ (*iv*) 1

5. (*i*) 1 (*ii*) $\frac{4}{9}$ (*iii*) $\frac{e^2-1}{e^2+1}$

6. (*i*) 1 (*ii*) $e-1$ (*iii*) 1/5

7. (*i*) $(\log 7)^2$ (*ii*) $\frac{1}{e}$

8. (*i*) $e^{3/4}$ (*ii*) e^{10} (*iii*) $1/e$.

Hints

3. (*v*) $a^x + b^x + c^x - 3 = (a^x - 1) + (b^x - 1) + (c^x - 1)$

(*vi*) $\dfrac{e^x + e^{-x} - 2}{x^2} = \left(\dfrac{e^{x/2} - e^{-x/2}}{x}\right)^2 = \left(\dfrac{e^x - 1}{x}\right)^2 \cdot \dfrac{1}{e^x}$

4. (*iv*) $\dfrac{e^x - e^{\sin x}}{x - \sin x} = \dfrac{e^{\sin x}(e^{x - \sin x} - 1)}{x - \sin x} = \left[\left(\dfrac{e^{\sin x} - 1}{\sin x}\right)\sin x + 1\right] \cdot \dfrac{e^{x - \sin x} - 1}{x - \sin x}$

5. (*iii*) $x = 1 + h \Rightarrow \dfrac{e^x - e^{-x}}{e^x + e^{-x}} = \dfrac{e^{2x} - 1}{e^{2x} + 1} = \dfrac{e^{2+2h} - 1}{e^{2+2h} + 1} = \dfrac{e^2 . e^{2h} - 1}{e^2 . e^{2h} + 1}$

6. (*ii*) $x = \dfrac{\pi}{2} + h$ implies $\dfrac{e^{\sin x} - 1}{\sin x} = \dfrac{e^{\sin\left(\frac{\pi}{2} + h\right)} - 1}{\sin\left(\dfrac{\pi}{2} + h\right)} = \dfrac{e^{\cos h} - 1}{\cos h}$

7. (*i*) $\dfrac{7^x + 7^{-x} - 2}{x^2} = \left(\dfrac{7^{x/2} - 7^{-x/2}}{x}\right)^2 = \left(\dfrac{1}{2} \cdot \dfrac{7^{x/2} - 1}{x/2} + \dfrac{1}{2} \cdot \dfrac{7^{-x/2} - 1}{-x/2}\right)^2$

(*ii*) $x = e + h \Rightarrow \dfrac{\log x - 1}{x - e} = \dfrac{\log(e + h) - \log e}{h} = \dfrac{1}{e} \cdot \dfrac{\log\left(1 + \frac{h}{e}\right)}{h/e}$

8. (*iii*) $\left(\dfrac{x}{1 + x}\right)^x = \left(1 + \dfrac{1}{x}\right)^{-x}$.

11.16. LIMITS AT INFINITY AND INFINITE LIMITS

(*i*) **Meaning of x → + ∞.** If x is a variable such that it can take any real value howsoever large, then we say that the variable x approaches $+\infty$ and write this as $x \to +\infty$.

In other words, $x \to +\infty$ implies that x can take value greater than any pre-assigned value (howsoever large). Generally the symbol $+\infty$ is written just as ∞.

(*ii*) **Meaning of x → – ∞.** If x is a variable such that it can take any real value howsoever small, then we say that the variable x approaches $-\infty$ and write this as $x \to -\infty$.

In other words, $x \to -\infty$ implies that x can take value less than any pre-assigned value (howsoever small).

For example, $x \to -\infty$, then the variable x would be capable of assuming values even less than – 10000000 or even less than any number which is less than – 10000000.

(*iii*) **Infinite limit of a function.** Let $f(x)$ be a function of x. If the value of $f(x)$ can be made greater than any pre-assigned number by taking x close to a, then we say that the function $f(x)$ becomes positively infinite as x approaches to a. We write this as $\lim\limits_{x \to a} f(x) = +\infty$.

Similarly, if the value of $f(x)$ can be made less than any pre-assigned number by taking x close to a, then we say that the function $f(x)$ becomes negatively infinite as x approaches a. We write this as $\lim\limits_{x \to a} f(x) = -\infty$.

ILLUSTRATIVE EXAMPLES

Example 1. *Evaluate the following limits :*

(*i*) $\lim_{x \to \infty} \frac{1}{x}$ (*ii*) $\lim_{x \to -\infty} \frac{1}{x}$.

Sol. (*i*) We have $x \to \infty$. $\therefore$ The variable x can be made as large as we like.

$\therefore$ $\frac{1}{x}$ is positive and grows smaller and smaller as x becomes larger and larger.

$$\therefore \quad \frac{1}{x} \to 0 \quad i.e., \quad \lim_{x \to \infty} \frac{1}{x} = \mathbf{0}.$$

(*ii*) We have $x \to -\infty$. $\therefore$ The variable x can be made as small as we like.

$\therefore$ $\frac{1}{x}$ is negative and numerically grows smaller and smaller as x becomes smaller and smaller.

$$\therefore \quad \frac{1}{x} \to 0 \quad i.e., \quad \lim_{x \to -\infty} \frac{1}{x} = \mathbf{0}.$$

Example 2. *Evaluate the following limits :*

(*i*) $\lim_{x \to \infty} \frac{4x^2 + 5x + 6}{3x^2 + 4x + 5}$ (*ii*) $\lim_{x \to \infty} \frac{x^7 + x^3 + 1}{x^8 - 3x^6 + 5}$ (*iii*) $\lim_{x \to \infty} \frac{(2x-1)^{20}(3x-1)^{30}}{(2x+1)^{50}}$

(*iv*) $\lim_{x \to \infty} \frac{3x^2 + x + 5}{7x - 9}$ (*v*) $\lim_{x \to \infty} \frac{x^8 - x^3 + 2}{-3x^4 + x + 7}$ (*vi*) $\lim_{x \to -\infty} \frac{x^3 + 2x + 8}{2x^2 + 11}$.

Sol. (*i*) Here, degree of num.(2) = degree of den.(2).

$\therefore$ We divide the numerator and the denominator by x^2.

$$\therefore \quad \lim_{x \to \infty} \frac{4x^2 + 5x + 6}{3x^2 + 4x + 5} = \lim_{x \to \infty} \frac{\frac{4x^2}{x^2} + \frac{5x}{x^2} + \frac{6}{x^2}}{\frac{3x^2}{x^2} + \frac{4x}{x^2} + \frac{5}{x^2}} = \lim_{x \to \infty} \frac{4 + \frac{5}{x} + \frac{6}{x^2}}{3 + \frac{4}{x} + \frac{5}{x^2}} = \frac{4+0+0}{3+0+0} = \mathbf{\frac{4}{3}}.$$

(*ii*) Here, degree of num. (7) < degree of den. (8)

$\therefore$ We divide the numerator and the denominator by x^8.

$$\therefore \quad \lim_{x \to \infty} \frac{x^7 + x^3 + 1}{x^8 - 3x^6 + 5} = \lim_{x \to \infty} \frac{\frac{x^7}{x^8} + \frac{x^3}{x^8} + \frac{1}{x^8}}{\frac{x^8}{x^8} - 3\frac{x^6}{x^8} + \frac{5}{x^8}} = \lim_{x \to \infty} \frac{\frac{1}{x} + \frac{1}{x^5} + \frac{1}{x^8}}{1 - \frac{3}{x^2} + \frac{5}{x^8}} = \frac{0+0+0}{1-0+0} = \mathbf{0}.$$

(*iii*) Here, degree of num. (50) = degree of den. (50)

$\therefore$ We divide the numerator and the denominator by x^{50}.

$$\therefore \quad \lim_{x \to \infty} \frac{(2x-1)^{20}(3x-1)^{30}}{(2x+1)^{50}} = \lim_{x \to \infty} \frac{\frac{(2x-1)^{20}}{x^{20}} \cdot \frac{(3x-1)^{30}}{x^{30}}}{\frac{(2x+1)^{50}}{x^{50}}} = \lim_{x \to \infty} \frac{\left(\frac{2x-1}{x}\right)^{20}\left(\frac{3x-1}{x}\right)^{30}}{\left(\frac{2x+1}{x}\right)^{50}}$$

$$= \lim_{x \to \infty} \frac{\left(2-\frac{1}{x}\right)^{20}\left(3-\frac{1}{x}\right)^{30}}{\left(2+\frac{1}{x}\right)^{50}} = \frac{(2-0)^{20}(3-0)^{30}}{(2+0)^{50}} = \frac{2^{20} \times 3^{30}}{2^{50}} = \frac{3^{30}}{2^{30}} = \left(\frac{\mathbf{3}}{\mathbf{2}}\right)^{\mathbf{30}}.$$

(*iv*) Here, degree of num. (2) > degree of den. (1)

$$\therefore \quad \lim_{x \to \infty} \frac{3x^2+x+5}{7x-9} = \lim_{x \to \infty} \frac{3x^2\left(1+\frac{1}{3x}+\frac{5}{3x^2}\right)}{7x\left(1-\frac{9}{x}\right)} = \lim_{x \to \infty} \left[\frac{3}{7}x \cdot \frac{1+\frac{1}{3x}+\frac{5}{3x^2}}{1-\frac{9}{x}}\right]$$

$$= \infty \times \frac{1+0+0}{1-0} = \infty.$$

(*v*) Here degree of num. (8) > degree of den. (4)

$$\therefore \quad \lim_{x \to \infty} \frac{x^8-x^3+2}{-3x^4+x+7} = \lim_{x \to \infty} \frac{x^8\left(1-\frac{1}{x^5}+\frac{2}{x^8}\right)}{-3x^4\left(1-\frac{1}{3x^3}-\frac{7}{3x^4}\right)} = \lim_{x \to \infty} \left[\left(-\frac{1}{3}x^4\right) \cdot \frac{1-\frac{1}{x^5}+\frac{2}{x^8}}{1-\frac{1}{3x^3}-\frac{7}{3x^4}}\right]$$

$$= -\infty \times \frac{1-0+0}{1-0-0} = -\infty.$$

(*vi*) Here degree of num. (3) > degree of den. (2)

$$\therefore \quad \lim_{x \to -\infty} \frac{x^3+2x+8}{2x^2+11} = \lim_{x \to -\infty} \frac{x^3\left(1+\frac{2}{x^2}+\frac{8}{x^3}\right)}{2x^2\left(1+\frac{11}{2x^2}\right)} = \lim_{x \to -\infty} \left[\frac{x}{2} \cdot \frac{1+\frac{2}{x^2}+\frac{8}{x^3}}{1+\frac{11}{2x^2}}\right]$$

$$= -\infty \times \frac{1+0+0}{1+0} = -\infty.$$

Example 3. *Evaluate the following limits :*

(*i*) $\lim_{n \to \infty} \left(\frac{1}{n^2}+\frac{2}{n^2}+\frac{3}{n^2}+\ldots\ldots+\frac{n-1}{n^2}\right)$ (*ii*) $\lim_{n \to \infty} \frac{\Sigma n^3}{n^4}$.

Sol. (*i*) $\lim_{n \to \infty} \frac{1}{n^2}+\frac{2}{n^2}+\frac{3}{n^2}+\ldots\ldots+\frac{n-1}{n^2} = \lim_{n \to \infty} \frac{1+2+3+\ldots\ldots+(n-1)}{n^2}$

$$= \lim_{n \to \infty} \frac{\frac{n-1}{2}\{2(1)+((n-1)-1).1\}}{n^2} = \lim_{n \to \infty} \frac{\frac{n-1}{2}\{2+n-2\}}{n^2}$$

$$\left| \; S_n = \frac{n}{2}[2a+(n-1)d] \right.$$

$$= \lim_{n\to\infty} \frac{(n-1)n}{2n^2} = \frac{1}{2}\lim_{n\to\infty}\left(1-\frac{1}{n}\right) = \frac{1}{2}(1-0) = \mathbf{\frac{1}{2}}.$$

$$(ii)\ \lim_{n\to\infty} \frac{\Sigma n^3}{n^4} = \lim_{n\to\infty} \frac{\left(\frac{n(n+1)}{2}\right)^2}{n^4} = \lim_{n\to\infty} \frac{n^2(n+1)^2}{4n^4} \qquad \left[\because\ \Sigma n^3 = \left(\frac{n(n+1)}{2}\right)^2\right]$$

$$= \lim_{n\to\infty} \frac{(n+1)^2}{4n^2} = \frac{1}{4}\lim_{n\to\infty} \frac{n^2+2n+1}{n^2}$$

$$= \frac{1}{4}\lim_{n\to\infty}\left(\frac{n^2}{n^2}+\frac{2n}{n^2}+\frac{1}{n^2}\right) = \frac{1}{4}\lim_{n\to\infty}\left(1+\frac{2}{n}+\frac{1}{n^2}\right)$$

$$= \frac{1}{4}\left(\lim_{n\to\infty} 1 + 2\lim_{n\to\infty}\frac{1}{n} + \lim_{n\to\infty}\frac{1}{n^2}\right) = \frac{1}{4}[1+2(0)+0] = \mathbf{\frac{1}{4}}.$$

Example 4. *Evaluate the following limits :*

$$(i)\ \lim_{x\to\infty} \frac{x^2-5}{\sqrt{3x^6+4x^2+2}} \qquad (ii)\ \lim_{x\to\infty} \frac{\sqrt{x^2+1}-\sqrt[3]{x^3+1}}{\sqrt[4]{x^4+1}-\sqrt[5]{x^4+1}}.$$

Sol. (*i*) The highest power of x occurring in the fraction is '3'.

∴ We divide the numerator and the denominator by x^3.

$$\therefore \quad \lim_{x\to\infty} \frac{x^2-5}{\sqrt{3x^6+4x^2+2}} = \lim_{x\to\infty} \frac{\frac{x^2}{x^3}-\frac{5}{x^3}}{\frac{\sqrt{3x^6+4x^2+2}}{x^3}} = \lim_{x\to\infty} \frac{\frac{1}{x}-\frac{5}{x^3}}{\sqrt{3+\frac{4}{x^4}+\frac{2}{x^6}}} = \frac{0}{\sqrt{3}} = \mathbf{0.}$$

(*ii*) The highest degree of x occurring in the fraction is '1'. We divide the numerator and denominator by x.

$$\therefore \quad \lim_{x\to\infty} \frac{\sqrt{x^2+1}-\sqrt[3]{x^3+1}}{\sqrt[4]{x^4+1}-\sqrt[5]{x^4+1}} = \lim_{x\to\infty} \frac{\sqrt{\frac{x^2+1}{x^2}}-\sqrt[3]{\frac{x^3+1}{x^3}}}{\sqrt[4]{\frac{x^4+1}{x^4}}-\sqrt[5]{\frac{x^4+1}{x^5}}} = \lim_{x\to\infty} \frac{\sqrt{1+\frac{1}{x^2}}-\sqrt{1+\frac{1}{x^3}}}{\sqrt{1+\frac{1}{x^4}}-\sqrt{\frac{1}{x}+\frac{1}{x^5}}}$$

$$= \frac{\sqrt{1+0}-\sqrt{1+0}}{\sqrt{1+0}-\sqrt{0+0}} = \frac{1-1}{1-0} = \mathbf{0.}$$

Example 5. *Evaluate the following limits :*

$$(i)\ \lim_{x\to\infty} (\sqrt{x+2}-\sqrt{x}) \qquad (ii)\ \lim_{x\to-\infty} (\sqrt{4x^2+7x}+2x).$$

Sol. (*i*) $\lim_{x\to\infty} (\sqrt{x+2}-\sqrt{x}) = \lim_{x\to\infty}\left(\sqrt{x+2}-\sqrt{x}\times\frac{\sqrt{x+2}+\sqrt{x}}{\sqrt{x+2}+\sqrt{x}}\right) = \lim_{x\to\infty}\frac{(x+2)-x}{\sqrt{x+2}+\sqrt{x}}$

$$= \lim_{x\to\infty} \frac{2}{\sqrt{x+2}+\sqrt{x}} = \lim_{x\to\infty} \frac{\frac{2}{\sqrt{x}}}{\sqrt{\frac{x+2}{x}}+\sqrt{\frac{x}{x}}} = \lim_{x\to\infty} \frac{\frac{2}{\sqrt{x}}}{\sqrt{1+\frac{2}{x}}+1}$$

$$= \frac{0}{\sqrt{1+0}+1} = \frac{0}{2} = \mathbf{0.}$$

(*ii*) $\lim_{x \to -\infty} (\sqrt{4x^2 + 7x} + 2x) = \lim_{x \to -\infty} \left((\sqrt{4x^2 + 7x} + 2x) \times \frac{\sqrt{4x^2 + 7x} - 2x}{\sqrt{4x^2 + 7x} - 2x} \right)$

$$= \lim_{x \to -\infty} \frac{7x}{\sqrt{4x^2 + 7x} - 2x} = \lim_{x \to -\infty} \frac{7x}{(-x)\sqrt{4 + \frac{7}{x}} - 2x} \qquad (\because \; x < 0 \Rightarrow -x > 0)$$

$$= \lim_{x \to -\infty} \frac{7}{-\sqrt{4 + \frac{7}{x}} - 2} = \frac{7}{-\sqrt{4 + 0} - 2} = -\frac{\mathbf{7}}{\mathbf{4}}.$$

Example 6. *Show that* $\lim_{x \to \infty} \left(\sqrt{x^2 + x + 1} - x \right) \neq \lim_{x \to \infty} \left(\sqrt{x^2 + 1} - x \right)$.

Sol. L.H.S. $= \lim_{x \to \infty} \left(\sqrt{x^2 + x + 1} - x \right) = \lim_{x \to \infty} \left[\frac{\sqrt{x^2 + x + 1} - x}{1} \cdot \frac{\sqrt{x^2 + x + 1} + x}{\sqrt{x^2 + x + 1} + x} \right]$

$$= \lim_{x \to \infty} \frac{(x^2 + x + 1) - x^2}{\sqrt{x^2 + x + 1} + x} = \lim_{x \to \infty} \frac{x + 1}{\sqrt{x^2 + x + 1} + x} = \lim_{x \to \infty} \frac{\frac{x + 1}{x}}{\frac{\sqrt{x^2 + x + 1}}{x} + \frac{x}{x}}$$

$$= \lim_{x \to \infty} \frac{1 + \frac{1}{x}}{\sqrt{1 + \frac{1}{x} + \frac{1}{x^2}} + 1} = \frac{1 + 0}{\sqrt{1 + 0 + 0} + 1} = \frac{1}{1 + 1} = \frac{1}{2}.$$

R.H.S. $= \lim_{x \to \infty} (\sqrt{x^2 + 1} - x) = \lim_{x \to \infty} \left(\frac{\sqrt{x^2 + 1} - x}{1} \times \frac{\sqrt{x^2 + 1} + x}{\sqrt{x^2 + 1} + x} \right) = \lim_{x \to \infty} \frac{(x^2 + 1) - x^2}{\sqrt{x^2 + 1} + x}$

$$= \lim_{x \to \infty} \frac{1}{\sqrt{x^2 + 1} + x} = \lim_{x \to \infty} \frac{\frac{1}{x}}{\frac{\sqrt{x^2 + 1}}{x} + \frac{x}{x}} = \lim_{x \to \infty} \frac{\frac{1}{x}}{\sqrt{1 + \frac{1}{x^2}} + 1} = \frac{1}{\sqrt{1 + 0} + 1} = 0.$$

$\therefore$ L.H.S. $\neq$ R.H.S. Heace, the result holds.

EXERCISE 11.9

Evaluate the following limits :

1. (*i*) $\lim_{x \to 0} \frac{1}{x^2}$ (*ii*) $\lim_{x \to 0} \frac{1}{x^3}$ (*iii*) $\lim_{x \to 0} \frac{1}{x^4}$

(*iv*) $\lim_{x \to \infty} \frac{1}{(x - 1)^2}$ (*v*) $\lim_{x \to \infty} \frac{1}{x^3}$ (*vi*) $\lim_{x \to 1} \frac{1}{|1 - x|}$.

2. (*i*) $\lim_{x \to \infty} \frac{x - 1}{x + 1}$ (*ii*) $\lim_{x \to \infty} \frac{x^2 + 1}{2x^2 + x - 1}$ (*iii*) $\lim_{x \to \infty} \frac{x^3 + x^2 - 6x + 8}{4x^3 + 5x - 8}$

(*iv*) $\lim_{x \to \infty} \frac{x^4 + 7x^3 + 46x + a}{x^4 + 6}, a \neq 0$ (*v*) $\lim_{x \to \infty} \frac{(x - 5)^7}{(x + 5)^8}$ (*vi*) $\lim_{x \to \infty} \frac{(3x - 1)(x + 2)(1 - x)}{(3 + x)^2 (x + 1)}$

3. (i) $\lim\limits_{x \to \infty} \dfrac{x^3 + x + 9}{x^2 + 1}$ (ii) $\lim\limits_{x \to \infty} \dfrac{2x^2 + 5x + 9}{1 - 3x}$ (iii) $\lim\limits_{x \to \infty} \dfrac{-x^7 + 6x^3 + 2x + 9}{6 + 5x - 3x^3}$

(iv) $\lim\limits_{x \to -\infty} \dfrac{4x^5 + x + 9}{x^2 + 6}$ (v) $\lim\limits_{x \to -\infty} \dfrac{x^3 + x}{1 - 4x^2}$ (vi) $\lim\limits_{x \to -\infty} \dfrac{-2x^9 + x + 5}{-x^3 + 6x + 8}$

4. (i) $\lim\limits_{n \to \infty} \left(\dfrac{1}{n^2} + \dfrac{2}{n^2} + \dfrac{3}{n^2} + \ldots\ldots + \dfrac{n}{n^2}\right)$ (ii) $\lim\limits_{n \to \infty} \left(\dfrac{1^2 + 2^2 + 3^2 + \ldots\ldots + n^2}{n^3}\right)$

(iii) $\lim\limits_{n \to \infty} \left(\dfrac{1^3}{n^4 + 1} + \dfrac{2^3}{n^4 + 1} + \ldots\ldots + \dfrac{n^3}{n^4 + 1}\right)$ (iv) $\lim\limits_{n \to \infty} \dfrac{\Sigma n^3}{n^5 + 2}$

Evaluate the following limits (Q. No. 5–6) :

5. (i) $\lim\limits_{x \to \infty} (\sqrt{x + 4} - \sqrt{x - 1})$ (ii) $\lim\limits_{x \to \infty} (2x - \sqrt{4x^2 + x})$

(iii) $\lim\limits_{x \to \infty} x[\sqrt{x^2 + 2} - \sqrt{x^2 - 2}]$ (iv) $\lim\limits_{x \to \infty} (\sqrt{x + 2}\, [\sqrt{x + 1} - \sqrt{x}])$

(v) $\lim\limits_{x \to -\infty} (\sqrt{x^2 - 8x} + x)$ (vi) $\lim\limits_{x \to \infty} (\sqrt{x^2 + x - 1} - x)$.

6. (i) $\lim\limits_{x \to \infty} \dfrac{4x^{-3} - 2x^{-2}}{3x^{-3} - 2x^{-2}}$ (ii) $\lim\limits_{x \to \infty} \dfrac{\sqrt{x^2 + \alpha^2} - \sqrt{x^2 + \beta^2}}{\sqrt{x^2 + \gamma^2} - \sqrt{x^2 + \delta^2}}$

(iii) $\lim\limits_{x \to \infty} \dfrac{\sqrt{a + x} - \sqrt{a}}{x\sqrt{a^2 + ax}}$ (iv) $\lim\limits_{n \to \infty} \left(1 + \dfrac{1}{2} + \dfrac{1}{2^2} + \ldots\ldots \dfrac{1}{2^n}\right)$.

7. It is given that $f(x) = \dfrac{ax + b}{x + 1}$, $\lim\limits_{x \to 0} f(x) = 2$ and $\lim\limits_{x \to \infty} f(x) = 1$. Show that $f(-2) = 0$.

Answers

1. (i) ∞ (ii) Limit does not exist (iii) ∞ (iv) 0 (v) 0 (vi) ∞

2. (i) 1 (ii) $\dfrac{1}{2}$ (iii) $\dfrac{1}{4}$ (iv) 1 (v) 0 (vi) -3

3. (i) ∞ (ii) $-\infty$ (iii) ∞ (iv) $-\infty$ (v) ∞ (vi) ∞

4. (i) $\dfrac{1}{2}$ (ii) $\dfrac{1}{3}$ (iii) $\dfrac{1}{4}$ (iv) 0

5. (i) 0 (ii) $-\dfrac{1}{4}$ (iii) 2 (iv) $\dfrac{1}{2}$ (v) 4 (vi) $\dfrac{1}{2}$

6. (i) 1 (ii) $\dfrac{\alpha^2 - \beta^2}{\gamma^2 - \delta^2}$ (iii) 0 (iv) 2.

Hints

1. (vi) $x \to 1 \Rightarrow 1 - x \to 0 \Rightarrow |1 - x| \to 0^+$.

2. (v) $$\frac{(x - 5)^7}{(x + 5)^8} = \frac{\dfrac{1}{x} \cdot \dfrac{(x - 5)^7}{x^7}}{\dfrac{(x + 5)^8}{x^8}} = \frac{\dfrac{1}{x} \cdot \left(1 - \dfrac{5}{x}\right)^7}{\left(1 + \dfrac{5}{x}\right)^8}.$$

4. (iii) $$\frac{1^3}{n^4 + 1} + \frac{2^3}{n^4 + 1} + \ldots + \frac{n^3}{n^4 + 1} = \frac{\Sigma n^3}{n^4 + 1} = \left(\frac{n(n + 1)}{2}\right)^2 \cdot \frac{1}{n^4 + 1}.$$

5. (*iv*) $\sqrt{x+2}\,(\sqrt{x+1}-\sqrt{x}) = \dfrac{\sqrt{x+2}\,(x+1-x)}{\sqrt{x+1}+\sqrt{x}} = \dfrac{\dfrac{\sqrt{x+2}}{\sqrt{x}}}{\dfrac{\sqrt{x+1}}{\sqrt{x}}+\dfrac{\sqrt{x}}{\sqrt{x}}} = \dfrac{\sqrt{1+\dfrac{2}{x}}}{\sqrt{1+\dfrac{1}{x}}+1}$.

6. (*i*) $\dfrac{4x^{-3}-2x^{-2}}{3x^{-3}-2x^{-2}} = \dfrac{\dfrac{4}{x}-2}{\dfrac{3}{x}-2}$. (*ii*) Rationalise numerator and denominator.

11.17. LEFT AND RIGHT LIMITS

We know that $x \to a$ means that x takes values which are very close to a and its value may be greater than or less than a.

For example, when we say $x \to 4$, then x will take values like 4.0001, 3.99997, 3.9900008, 4.00000025, etc. In other words, $x - 4$ may be positive or may be negative but its value will be very close to zero. In case, we restrict x to take only those values which are less than 4 and very near to 4, then we say that the variable x is approaching 4 from left. We write this as $x \to 4^-$. Similarly, if x is restricted to assume values which are close to 4, but greater than 4, then we say that x is approaching 4 from right. We write this as $x \to 4^+$. Now, let us define one-sided limits in a formal way.

If a variable x takes values which are close to a constant a and always remains on the left of a *i.e.,* the value of x is less than a, then we say that x approaches a from left and we write this as $x \to a^-$.

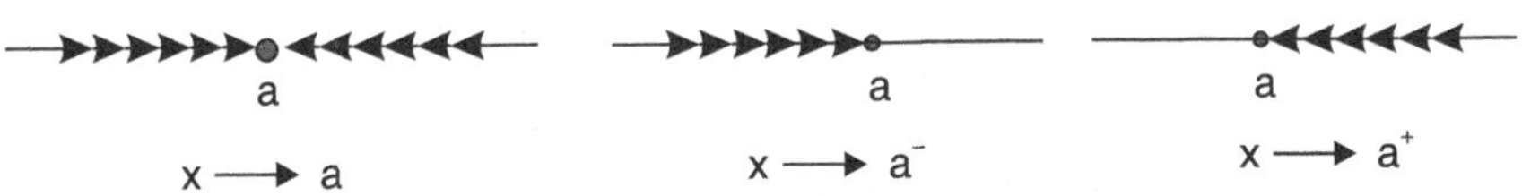

Similarly, if a variable x takes values which are close to a constant a and always remains on the right of a *i.e.,* the value of x is greater than a, then we say that x approaches a from right and we write this as $x \to a^+$.

Let $f(x)$ be a function of x. If $f(x) \to l$ as $x \to a^-$, then l is called the **left limit** of function f at a and we write this as $\lim\limits_{x \to a^-} f(x) = l$.

Similarly, if $f(x) \to l$ as $x \to a^+$, then l is called the **right limit** of f at a and we write this as $\lim\limits_{x \to a^+} f(x) = l$.

If $\lim\limits_{x \to a^-} f(x) \neq \lim\limits_{x \to a^+} f(x)$, then by definition of limit of a function at a point, $\lim\limits_{x \to a} f(x)$ does not exist.

If $\lim\limits_{x \to a^-} f(x) = \lim\limits_{x \to a^+} f(x)$, then by definition of limit of a function at a point, $\lim\limits_{x \to a} f(x)$ exists and $\lim\limits_{x \to a} f(x) = \lim\limits_{x \to a^-} f(x)$ (or $\lim\limits_{x \to a^+} f(x)$).

Remark 1. If $x \to a^-$ and $x = a - h$, then we have $h = a - x$ and as $x \to a^-$, the difference $a - x\ (= h)$ is positive and is close to zero.

$\therefore \quad x \to a^- \quad \Rightarrow \quad h \to 0^+$, where $x = a - h$.

Remark 2. If $x \to a^+$ and $x = a + h$, then we have $h = x - a$ and as $x \to a^+$, the difference $x - a\ (= h)$ is positive and is close to zero.

$\therefore \quad x \to a^+ \quad \Rightarrow \quad h \to 0^+$, where $x = a + h$.

11.18. CALCULATION OF LEFT AND RIGHT LIMITS

Till now, we have been discussing limit of a function as its variable approaching a certain fixed number.

When we say x approaches a, it means that x may x take values which are very close to a and its value is either less than a or greater than a. Thus, $x \to a$ means $x \to a^-$ and $x \to a^+$ both.

In the present section, we shall study the method of finding limits of functions for which direct calculation of $\lim_{x \to a} f(x)$ is not possible and this limit is calculated in two phases.

We calculate $\lim_{x \to a^-} f(x)$ and $\lim_{x \to a^+} f(x)$ separately. If both these limits are equal, then by definition of limit of a function, $\lim_{x \to a} f(x)$ exists and is equal to the value of $\lim_{x \to a^-} f(x)$ (*i.e.,* of $\lim_{x \to a^+} f(x)$). In case, both limits are not equal or any one of these does not exist then by definition, $\lim_{x \to a} f(x)$ does not exist.

Let $f(x)$ be a function of x. We have already seen that

$$x \to a^- \;\Rightarrow\; h \to 0^+, \text{ where } x = a - h$$

and

$$x \to a^+ \;\Rightarrow\; h \to 0^+, \text{ where } x = a + h.$$

$$\therefore \quad \mathbf{\lim_{x \to a^-} f(x) = \lim_{h \to 0^+} f(a - h) \text{ and } \lim_{x \to a^+} f(x) = \lim_{h \to 0^+} f(a + h).}$$

If both the above limits exist and are equal, then $\lim_{x \to a} f(x)$ exist and equals the common limit.

Remark. Left limits and right limits are required to be used when the function under consideration is involving modulus function, bracket function and/or is defined by more than one rule.

WORKING RULES FOR EVALUATING $\lim_{x \to a} f(x)$

Step I. *To find* $\lim_{x \to a^-} f(x)$, *put* $x = a - h$ *in* $f(x)$ *and find* $\lim_{h \to 0^+} f(a - h)$.

Step II. *To find* $\lim_{x \to a^+} f(x)$, *put* $x = a + h$ *in* $f(x)$ *and find* $\lim_{h \to 0^+} f(a + h)$.

Step III. *If* $\lim_{x \to a^-} f(x) = \lim_{x \to a^+} f(x)$, *then* $\lim_{x \to a} f(x)$ *exist and is equal to the common limit.*

Step IV. *If* $\lim_{x \to a^-} f(x) \neq \lim_{x \to a^+} f(x)$, *then* $\lim_{x \to a} f(x)$ *does not exist.*

ILLUSTRATIVE EXAMPLES

Example 1. *Evaluate :*

(*i*) $\lim_{x \to 0^+} \frac{1}{x}$ (*ii*) $\lim_{x \to 0^-} \frac{1}{x}$ (*iii*) $\lim_{x \to 0} \frac{1}{x}$ (*iv*) $\lim_{x \to 0} \frac{1}{|x|}$

(*v*) $\lim_{x \to 0^+} \frac{x}{|x|}$ (*vi*) $\lim_{x \to 0^-} \frac{x}{|x|}$ (*vii*) $\lim_{x \to 0} \frac{x}{|x|}$.

Sol. (*i*) We have $x \to 0^+$.

$\therefore$ x takes values very near to 0 and on right of it.

$\therefore$ x is +ve, so $1/x$ is also +ve.

Now as x approaches 0 from right, $1/x$ remains positive and increases very rapidly.

$$\therefore \quad x \to 0^+ \quad \Rightarrow \quad \frac{1}{x} \to +\infty \quad i.e., \quad \lim_{x \to 0^+} \frac{1}{x} = +\infty.$$

(*ii*) We have $x \to 0^-$.

$\therefore$ x takes values very near to 0 and on left of it.

$\therefore$ x is –ve and consequently $1/x$ is also –ve.

Now as x approaches 0 from left, $1/x$ remains negative and decreases very rapidly.

$$\therefore \quad x \to 0^- \quad \Rightarrow \quad \frac{1}{x} \to -\infty \quad i.e., \quad \lim_{x \to 0^-} \frac{1}{x} = -\infty.$$

(*iii*) We have $x \to 0$. $\therefore$ $x \to 0^-$ and $x \to 0^+$.

$$\text{Now } x \to 0^- \quad \Rightarrow \quad \lim_{x \to 0^-} \frac{1}{x} = -\infty \quad \text{and} \quad x \to 0^+ \quad \Rightarrow \quad \lim_{x \to 0^+} \frac{1}{x} = +\infty.$$

$$\therefore \quad \mathbf{\lim_{x \to 0} \frac{1}{x} \textbf{ does not exist.}} \qquad \left(\lim_{x \to 0^-} \frac{1}{x} \neq \lim_{x \to 0^+} \frac{1}{x}\right)$$

(*iv*) We have $x \to 0$.

$\therefore$ x takes values close to 0 and can take positive as well as negative values.

$\therefore$ $|x|$ takes values close to 0 and is always positive.

$$\therefore \quad \frac{1}{|x|} \text{ increases very rapidly as } x \to 0. \text{ Hence, } \quad \lim_{x \to 0} \frac{1}{|x|} = +\infty.$$

(*v*) We have $x \to 0^+$. $\therefore$ x takes values close to 0 and is positive.

$$\therefore \quad \frac{x}{|x|} = \frac{x}{x} = 1 \qquad \text{Hence, } \quad \lim_{x \to 0^+} \frac{x}{|x|} = \lim_{x \to 0^+} 1 = \mathbf{1}.$$

(*vi*) We have $x \to 0^-$. $\therefore$ x takes values close to 0 and is negative.

$$\therefore \quad \frac{x}{|x|} = \frac{x}{-x} = -1$$

$$\text{Hence, } \lim_{x \to 0^-} \frac{x}{|x|} = \lim_{x \to 0^-} (-1) = \mathbf{-1}.$$

(*vii*) We have $x \to 0$. $\therefore$ $x \to 0^-$ and $x \to 0^+$.

$$\text{Now } x \to 0^- \Rightarrow \lim_{x \to 0^-} \frac{x}{|x|} = -1 \quad \text{and} \quad x \to 0^+ \Rightarrow \lim_{x \to 0^+} \frac{x}{|x|} = 1.$$

$$\therefore \quad \mathbf{\lim_{x \to 0} \frac{x}{|x|} \textbf{ does not exist.}}$$

Example 2. *Let* $f(x) = \begin{cases} \dfrac{x+2}{x^2+3}, & x < 2 \\ \dfrac{2}{7}x, & x > 2 \end{cases}$.

Find $\lim_{x \to 1} f(x)$ *and* $\lim_{x \to 2} f(x)$.

Sol. limit as x → 1. We have $f(x) = \dfrac{x+2}{x^2+3}$, whenever $x < 2$.

$\therefore$ For values of x for which $x \to 1^-$ and for which $x \to 1^+$, the value of $f(x)$ is given by $\dfrac{x+2}{x^2+3}$.

$\therefore$ We need not find $\lim\limits_{x \to 1^-} f(x)$ and $\lim\limits_{x \to 1^+} f(x)$ separately.

$$\text{We have } \lim_{x \to 1} f(x) = \lim_{x \to 1} \frac{x+2}{x^2+3} = \frac{\lim\limits_{x \to 1}(x+2)}{\lim\limits_{x \to 1}(x^2+3)} = \frac{1+2}{(1)^2+3} = \frac{3}{4}.$$

limit as x → 2. We have $f(x) = \dfrac{x+2}{x^2+3}$, whenever $x < 2$ and $f(x) = \dfrac{2}{7}x$, whenever $x > 2$.

$\therefore$ For values of x for which $x \to 2^-$, the value of $f(x)$ is given by $\dfrac{x+2}{x^2+3}$ and for values of x for which $x \to 2^+$, the value of $f(x)$ is given by $\dfrac{2}{7}x$.

$\therefore$ We need to find $\lim\limits_{x \to 2^-} f(x)$ and $\lim\limits_{x \to 2^+} f(x)$ separately.

$$\text{We have } \lim_{x \to 2^-} f(x) = \lim_{x \to 2^-} \frac{x+2}{x^2+3} = \frac{\lim\limits_{x \to 2^-}(x+2)}{\lim\limits_{x \to 2^-}(x^2+3)} = \frac{2+2}{(2)^2+3} = \frac{4}{7}$$

and $$\lim_{x \to 2^+} f(x) = \lim_{x \to 2^+} \left(\frac{2}{7}x\right) = \frac{2}{7}(2) = \frac{4}{7}.$$

$\therefore$ $$\lim_{x \to 2^-} f(x) = \lim_{x \to 2^+} f(x) \quad \left(\text{each} = \frac{4}{7}\right) \text{ Hence, } \lim_{x \to 2} f(x) \text{ exists and equals } \frac{4}{7}.$$

Example 3. (*i*) *Let a function f be defined as* $f(x) = \begin{cases} x & \text{if } 0 \le x < \frac{1}{2} \\ x - 1 & \text{if } \frac{1}{2} < x \le 1 \end{cases}$.

Does $\lim\limits_{x \to \frac{1}{2}} f(x)$ *exist ?*

(*ii*) *Evaluate the left-hand and right-hand limits of the following function at x = 1 :*

$$f(x) = \begin{cases} 1 + x^2 & \text{if } 0 \le x \le 1 \\ 2 - x & \text{if } x > 1 \end{cases}.$$

Does $\lim\limits_{x \to 1} f(x)$ *exist ?*

Sol. (*i*) $\lim\limits_{x \to \frac{1}{2}^-} f(x) = \lim\limits_{x \to \frac{1}{2}^-} x = \dfrac{1}{2}$ and $\lim\limits_{x \to \frac{1}{2}^+} f(x) = \lim\limits_{x \to \frac{1}{2}^+} (x-1) = \dfrac{1}{2} - 1 = -\dfrac{1}{2}$.

$\therefore$ $\lim\limits_{x \to \frac{1}{2}^-} f(x) \ne \lim\limits_{x \to \frac{1}{2}^+} f(x)$. Hence, $\lim\limits_{x \to \frac{1}{2}} f(x)$ **does not exist.**

(*ii*) $\lim\limits_{x \to 1^-} f(x) = \lim\limits_{x \to 1^-} (1 + x^2) = 1 + (1)^2 = 2$

and $\lim_{x \to 1^+} f(x) = \lim_{x \to 1^+} (2 - x) = 2 - 1 = 1.$

$\therefore \lim_{x \to 1^-} f(x) \neq \lim_{x \to 1^+} f(x)$ Hence, $\mathbf{\lim_{x \to 1} f(x)}$ **does not exist.**

Example 4. *Find k so that* $\lim_{x \to 2} f(x)$ *exists, where*

$$f(x) = \begin{cases} 2x + 3 & \text{if } x \le 2 \\ x + k & \text{if } x > 2 \end{cases}.$$

Sol. $\lim_{x \to 2^-} f(x) = \lim_{x \to 2^-} (2x + 3) = 2(2) + 3 = 7$

and $\lim_{x \to 2^+} f(x) = \lim_{x \to 2^+} (x + k) = 2 + k.$

Since $\lim_{x \to 2} f(x)$ exists, we have $\lim_{x \to 2^-} f(x) = \lim_{x \to 2^+} f(x).$

$\therefore$ $7 = 2 + k$ or $k = \mathbf{5}$.

Example 5. *Let* $f(x) = \begin{cases} ax + b & \text{if } x > 1 \\ 3bx - 2a + 1 & \text{if } x < 1 \end{cases}.$

Find a relation between a and b so that $\lim_{x \to 1} f(x)$ *exists.*

Sol. $\lim_{x \to 1^-} f(x) = \lim_{x \to 1^-} (3bx - 2a + 1) = 3b(1) - 2a + 1 = 3b - 2a + 1$

and $\lim_{x \to 1^+} f(x) = \lim_{x \to 1^+} (ax + b) = a(1) + b = a + b.$

Since $\lim_{x \to 1} f(x)$ exists, we have $\lim_{x \to 1^-} f(x) = \lim_{x \to 1^+} f(x).$

$\Rightarrow$ $3b - 2a + 1 = a + b$ *i.e.*, $\mathbf{3a - 2b - 1 = 0.}$

Example 6. *Show that* $\lim_{x \to 2} \frac{|x - 2|}{x - 2}$ *does not exist.*

Sol. Let $f(x) = \frac{|x - 2|}{x - 2}.$

$$\therefore \lim_{x \to 2^-} f(x) = \lim_{h \to 0^+} f(2 - h) = \lim_{h \to 0^+} \frac{|(2 - h) - 2|}{(2 - h) - 2} = \lim_{h \to 0^+} \frac{|-h|}{-h}$$

$$= \lim_{h \to 0^+} \frac{h}{-h} = \lim_{h \to 0^+} \frac{1}{-1} = -1. \qquad \text{(Putting } x = 2 - h\text{)}$$

$$\text{Also } \lim_{x \to 2^+} f(x) = \lim_{h \to 0^+} f(2 + h) = \lim_{h \to 0^+} \frac{|(2 + h) - 2|}{(2 + h) - 2} = \lim_{h \to 0^+} \frac{|h|}{h} = \lim_{h \to 0^+} \frac{h}{h} = \lim_{h \to 0^+} 1 = 1.$$

(Putting $x = 2 + h$)

$\therefore \lim_{x \to 2^-} f(x) \neq \lim_{x \to 2^+} f(x).$ Hence, $\lim_{x \to 2} f(x)$ *i.e.*, $\mathbf{\lim_{x \to 2} \frac{|x - 2|}{x - 2}}$ **does not exist.**

EXERCISE 11.10

1. If $f(x) = \begin{cases} 4 & \text{if } x \geq 3 \\ x+1 & \text{if } x < 3 \end{cases}$, find $\lim_{x \to 3} f(x)$, if exists.

2. If $f(x) = \begin{cases} x+1 & \text{if } x \geq 0 \\ x-1 & \text{if } x < 0 \end{cases}$, find $\lim_{x \to 0} f(x)$, if exists.

3. If $f(x) = \begin{cases} 5x-4 & \text{if } 0 < x \leq 1 \\ 4x^2 - 3x & \text{if } 1 < x < 2 \end{cases}$, find $\lim_{x \to 1} f(x)$, if exists.

4. Evaluate :

   (*i*) $\lim_{x \to 2^+} \dfrac{x-3}{x^2-4}$ (*ii*) $\lim_{x \to (-2)^+} \dfrac{x^2-1}{2x+4}$ (*iii*) $\lim_{x \to 0^-} \dfrac{x^2-3x+2}{x^3-2x^2}$ (*iv*) $\lim_{x \to 0^-} (2 - \cot x)$.

5. (*i*) If f is an even function, then show that $\lim_{x \to 0^+} f(x) = \lim_{x \to 0^-} f(x)$.

   (*ii*) If f is an odd function and if $\lim_{x \to 0} f(x)$ exists, then show that $\lim_{x \to 0} f(x)$ must be zero.

6. If $f(x) = \begin{cases} 5x-4 & \text{if } x \leq 1, \\ 4x^3 - 3x & \text{if } x > 1 \end{cases}$, find $\lim_{x \to -4} f(x)$, $\lim_{x \to 0} f(x)$ and $\lim_{x \to 1} f(x)$, if exists.

7. (*i*) If $f(x) = \begin{cases} \cos x & \text{if } x \geq 0 \\ x+k & \text{if } x < 0 \end{cases}$ and $\lim_{x \to 0} f(x)$ exists, find the value of k.

   (*ii*) Find k so that $\lim_{x \to 2} f(x)$ may exists, where $f(x) = \begin{cases} 2x+3 & \text{if } x \leq 2 \\ x+2k & \text{if } x > 2 \end{cases}$.

8. (*i*) If $f(x) = \sqrt{8-2x}$, find $\lim_{x \to 4} f(x)$, if exists.

   (*ii*) Find $\lim_{x \to 2} \sqrt{6-3x}$, if exists.

Evaluate the following limits, if exists (Q. No. 9–10) :

9. (*i*) $\lim_{x \to 0} |x|$ (*ii*) $\lim_{x \to -5} |x+5|$ (*iii*) $\lim_{x \to 0} \dfrac{|x|}{x}$ (*iv*) $\lim_{x \to 0} \dfrac{x}{|x|}$.

10. (*i*) $\lim_{x \to 0^-} \dfrac{x}{|x| + x^2}$ (*ii*) $\lim_{x \to 0} \dfrac{x}{|x| + x^2}$

11. Show that the following limits do not exist :

    (*i*) $\lim_{x \to 3} \dfrac{|x-3|}{x^2-9}$ (*ii*) $\lim_{x \to 4} \dfrac{[x] + |x-4|}{x^2+5x}$ (*iii*) $\lim_{x \to 2} \dfrac{|x-2|}{x-2}$

Answers

1. 4 **2.** Limit does not exist **3.** 1

4. (*i*) $-\infty$ (*ii*) ∞ (*iii*) $-\infty$ (*iv*) ∞

6. -24, -4, 1 **7.** (*i*) 1 (*ii*) 5/2

8. (*i*) Limit does not exist (*ii*) Limit does not exist

9. (*i*) 0 (*ii*) 0 (*iii*) Limit does not exist (*iv*) Limit does not exist

10. (*i*) -1 (*ii*) Limit does not exist

Hints

4. (*i*) $\lim_{x \to 2^+} \dfrac{x-3}{x^2-4} = \lim_{h \to 0^+} \dfrac{(2+h)-3}{(2+h)^2-4} = \lim_{h \to 0^+} \dfrac{h-1}{h^2+4h} = -\infty$ because $h-1 \to -1$, $h^2+4h \to 0$ and $h^2+4h > 0$.

(*ii*) $\lim_{x \to (-2)^+} \dfrac{x^2-1}{2x+4} = \lim_{h \to 0^+} \dfrac{(-2+h)^2-1}{2(-2+h)+4} = \lim_{h \to 0^+} \dfrac{3+h^2-4h}{2h} = \infty$ because $3+h^2-4h \to 3$, $2h \to 0$ and $2h > 0$.

(*iii*) $\lim_{x \to 0^-} \dfrac{x^2-3x+2}{x^3-2x^2} = \lim_{h \to 0^+} \dfrac{(-h)^2-3(-h)+2}{(-h)^3-2(-h)^2} = \lim_{h \to 0^+} \dfrac{h^2+3h+2}{-h^3-2h^2} = -\infty$ because

$h^2+3h+2 \to 2$, $-h^3-2h^2 \to 0$ and $-h^3-2h^3 < 0$.

5. (*i*) $\lim_{x \to 0^+} f(x) = \lim_{x \to 0^+} f(-x) = \lim_{t \to 0^-} f(t) = \lim_{x \to 0^-} f(x)$. (Putting $t = -x$ and then replacing t by x)

(*ii*) $\lim_{x \to 0^+} f(x) = \lim_{x \to 0^+} (-f(-x)) = -\lim_{t \to 0^-} f(t) = -\lim_{x \to 0^-} f(x)$

(Putting $t = -x$ and then replacing t by x)

Also, $\lim_{x \to 0^+} f(x) = \lim_{x \to 0^-} f(x) = \lim_{x \to 0} f(x)$.

8. (*i*) $\lim_{x \to 4^+} f(x)$ does not exist, because $f(x)$ is defined only when $x \le 4$.

12

Differentiation

12.1. INTRODUCTION

The concept of '*derivative*' is a basic tool in the science of calculus. It also helps to measure the rate of change of variables. The maximum and minimum values of functions can also be evaluated by using derivatives.

In the present chapter, we shall study the following topics :

★ Derivative at a point
★ Derivative of a function
★ Geometrical and physical significance of derivative
★ Derivative of functions from first principle
★ Derivative of sum, difference, product and quotient of functions
★ Derivative of polynomial functions
★ Derivative of function of a function (chain rule).

I. DERIVATIVE AT A POINT

12.2. DERIVATIVE AT A POINT

Let $y = f(x)$ be a function and a be any arbitrary but fixed point in the domain of the function. Let δx represents a numerically small increment (it is immaterial whether it is positive or negative) of x from $x = a$ and let δy be the corresponding increment of y from $y = f(a)$.

$\therefore \quad \delta y = f(a + \delta x) - f(a)$.

The quotient $\frac{\delta y}{\delta x}$ is called the **average rate of change** of $f(x)$ on the interval between $x = a$ and $x = a + \delta x$.

As $\delta x \to 0$, the limit of $\frac{\delta y}{\delta x}$ may or may not exist. In case, $\lim\limits_{\delta x \to 0} \frac{\delta y}{\delta x}$ exists finitely, then this limit is called the **derivative** of the function $y = f(x)$ with respect to x at $x = a$ and is denoted by $f'(a)$.

$$\therefore \qquad f'(a) = \lim_{\delta x \to 0} \frac{\delta y}{\delta x}$$

$$\Rightarrow \qquad f'(a) = \lim_{\delta x \to 0} \frac{f(a + \delta x) - f(a)}{\delta x} \qquad (\because \ \delta y = f(a + \delta x) - f(a))$$

If small increment δx of x is represented by h, then we have

$$\mathbf{f'(a) = \lim_{h \to 0} \frac{f(a + h) - f(a)}{h}}.$$

12.3. METHOD OF FINDING DERIVATIVE OF A FUNCTION

Let $y = f(x)$ be any derivable function of x. Let x be an arbitrary point in its domain. Let δx represent an increment of x and let δy, the corresponding increment of the value of y.

$$\therefore \quad y + \delta y = f(x + \delta x)$$

$$\Rightarrow \quad \delta y = f(x + \delta x) - f(x) \quad \Rightarrow \quad \frac{\delta y}{\delta x} = \frac{f(x+\delta x) - f(x)}{\delta x}$$

By definition, $$\mathbf{f'(x) = \lim_{\delta x \to 0} \frac{\delta y}{\delta x} = \lim_{\delta x \to 0} \frac{f(x+\delta x) - f(x)}{\delta x}}.$$

The operation of finding the derivative of a function is called **differentiation** of the function.

It is worthwhile to note that $\frac{\delta y}{\delta x}$ is the quotient of δy and δx but $\frac{dy}{dx}$ is not the quotient of dy and dx. Instead, $\frac{dy}{dx}$ is to be interpreted as $\frac{d}{dx}(y)$ and read as *dee y by dee x*.

Remark 1. The differentiation of a function by applying above definition is called the differentiation by using the **first principle** or **from definition** or **ab-initio.**

Remark 2. In practice, the functions are differentiated by using certain standard formulae which are established by using principles of differentiation.

WORKING RULES TO FIND DERIVATIVES USING FIRST PRINCIPLE

Step I. *Write $y = f(x)$ and $y + \delta y = f(x + \delta x)$.*

Step II. *Subtract these to get $\delta y = f(x + \delta x) - f(x)$.*

Step III. *Divide δy by δx to get $\frac{\delta y}{\delta x} = \frac{f(x+\delta x) - f(x)}{\delta x}$.*

Step IV. *Take $\lim_{\delta x \to 0} \frac{\delta y}{\delta x}$ with x fixed by simplifying $\frac{f(x+\delta x) - f(x)}{\delta x}$ and cancelling the 'δx' occurring in the denominator. This gives $\frac{dy}{dx}$ i.e., $f'(x)$.*

ILLUSTRATIVE EXAMPLES

Example 1. *Find the differential coefficient of $6x + 11$ w.r.t. x by using the first principle.*

Sol. Let $$y = 6x + 11 \quad \ldots(1)$$

Let δx be an increment of x and δy, the corresponding increment of y.

$$\therefore \quad y + \delta y = 6(x + \delta x) + 11 \quad \ldots(2)$$

$$(2) - (1) \Rightarrow \quad \delta y = (6x + 6\delta x + 11) - (6x + 11) = 6\delta x$$

$$\therefore \quad \frac{\delta y}{\delta x} = \frac{6\delta x}{\delta x} = 6 \qquad \therefore \quad \lim_{\delta x \to 0} \frac{\delta y}{\delta x} = \lim_{\delta x \to 0} 6 = 6$$

$$\therefore \quad \frac{dy}{dx} = \mathbf{6.}$$

Example 2. *Find the differential coefficient of $x^2 + 3x$ w.r.t. x by using the first principle.*

Sol. Let $$y = x^2 + 3x \quad \ldots(1)$$

Let δx be an increment of x and δy, the corresponding increment of y.

$$\therefore \quad y + \delta y = (x + \delta x)^2 + 3(x + \delta x) \quad \ldots(2)$$

$$(2) - (1) \Rightarrow \quad \delta y = (x^2 + (\delta x)^2 + 2x.\delta x + 3x + 3.\delta x) - (x^2 + 3x) = (\delta x)^2 + 2x.\delta x + 3.\delta x$$

$$\therefore \quad \frac{\delta y}{\delta x} = \frac{(\delta x)^2 + 2x.\delta x + 3.\delta x}{\delta x} = \delta x + 2x + 3.$$

$$\therefore \quad \lim_{\delta x \to 0} \frac{\delta y}{\delta x} = \lim_{\delta x \to 0} (\delta x + 2x + 3) = 2x + 3.$$

$$\therefore \quad \frac{dy}{dx} = \mathbf{2x + 3.}$$

Example 3. *Find the derivative of* $\frac{x^2 - 6}{3x}$ *by using the definition.*

Sol. Let $$y = \frac{x^2 - 6}{3x} = \frac{x}{3} - \frac{2^*}{x} \qquad \ldots(1)$$

Let δx be an increment of x and δy, the corresponding increment of y.

$$\therefore \quad y + \delta y = \frac{x + \delta x}{3} - \frac{2}{x + \delta x} \qquad \ldots(2)$$

$$(2) - (1) \Rightarrow \quad \delta y = \left(\frac{x + \delta x}{3} - \frac{2}{x + \delta x}\right) - \left(\frac{x}{3} - \frac{2}{x}\right) = \left(\frac{x + \delta x}{3} - \frac{x}{3}\right) - \left(\frac{2}{x + \delta x} - \frac{2}{x}\right)$$

$$= \frac{x + \delta x - x}{3} - \frac{2x - 2x - 2\delta x}{(x + \delta x)x} = \frac{\delta x}{3} + \frac{2\delta x}{(x + \delta x)x} = \delta x\left[\frac{1}{3} + \frac{2}{(x + \delta x)x}\right]$$

$$\therefore \quad \frac{\delta y}{\delta x} = \frac{1}{3} + \frac{2}{(x + \delta x)x}$$

$$\therefore \quad \lim_{\delta x \to 0} \frac{\delta y}{\delta x} = \lim_{\delta x \to 0} \left[\frac{1}{3} + \frac{2}{(x + \delta x)x}\right] = \frac{1}{3} + \frac{2}{x^2}$$

$$\therefore \quad \frac{dy}{dx} = \mathbf{\frac{1}{3} + \frac{2}{x^2}}.$$

Example 4. *Find the derivative of* $\sqrt{3x - 4}$ *w.r.t. x using the first principle.*

Sol. Let $$y = \sqrt{3x - 4} \qquad \ldots(1)$$

Let δx be an increment of x and δy, the corresponding increment of y.

$$\therefore \quad y + \delta y = \sqrt{3(x + \delta x) - 4} \qquad \ldots(2)$$

$$(2) - (1) \Rightarrow \quad \delta y = \sqrt{3x + 3\delta x - 4} - \sqrt{3x - 4}$$

$$\therefore \quad \frac{\delta y}{\delta x} = \frac{\sqrt{3x + 3\delta x - 4} - \sqrt{3x - 4}}{\delta x}$$

$$= \frac{\sqrt{3x + 3\delta x - 4} - \sqrt{3x - 4}}{\delta x} \cdot \frac{\sqrt{3x + 3\delta x - 4} + \sqrt{3x - 4}}{\sqrt{3x + 3\delta x - 4} + \sqrt{3x - 4}}$$

$$= \frac{(3x + 3\delta x - 4) - (3x - 4)}{\delta x\,(\sqrt{3x + 3\delta x - 4} + \sqrt{3x - 4})} = \frac{3}{\sqrt{3x + 3\delta x - 4} + \sqrt{3x - 4}}$$

$$\therefore \quad \lim_{\delta x \to 0} \frac{\delta y}{\delta x} = \lim_{\delta x \to 0} \frac{3}{\sqrt{3x + 3\delta x - 4} + \sqrt{3x - 4}} = \frac{3}{\sqrt{3x + 0 - 4} + \sqrt{3x - 4}} = \frac{3}{2\sqrt{3x - 4}}$$

$$\therefore \quad \frac{dy}{dx} = \mathbf{\frac{3}{2\sqrt{3x - 4}}}.$$

*Note this step carefully.

Example 5. *Find the derivative of* $\frac{1}{\sqrt{x}}$ *w.r.t. x by using the definition. Also find the derivative at x = 4.*

Sol. Let $$y = \frac{1}{\sqrt{x}} = x^{-1/2} \quad \ldots(1)$$

Let δx be a small increment of x and δy, the corresponding increment of y.

$$\therefore \quad y + \delta y = (x + \delta x)^{-1/2} \quad \ldots(2)$$

$$(2) - (1) \Rightarrow \quad \delta y = (x + \delta x)^{-1/2} - x^{-1/2}$$

$$\therefore \quad \lim_{\delta x \to 0} \frac{\delta y}{\delta x} = \lim_{\delta x \to 0} \frac{(x+\delta x)^{-1/2} - x^{-1/2}}{\delta x} = \lim_{x + \delta x \to x} \frac{(x+\delta x)^{-1/2} - x^{-1/2}}{(x+\delta x) - x}$$

$$= -\frac{1}{2} x^{-\frac{1}{2} - 1} = -\frac{1}{2x^{3/2}}. \quad \left(\text{Using } \lim_{x \to a} \frac{x^n - a^n}{x - a} = na^{n-1}\right)$$

$$\therefore \quad \frac{dy}{dx} = -\frac{\mathbf{1}}{\mathbf{2x^{3/2}}}.$$

At $x = 4$, $$\frac{dy}{dx} = -\frac{1}{2(4)^{3/2}} = -\frac{\mathbf{1}}{\mathbf{16}}.$$

Example 6. *Find the derivative of* $f(x) = x^{1/3}$, *by using the first principle. Examine* $f'(0)$, $f'(1)$.

Sol. Let $$y = f(x) = x^{1/3} \quad \ldots(1)$$

Let δx be an increment of x and δy, the corresponding increment of y.

$$\therefore \quad y + \delta y = (x + \delta x)^{1/3} \quad \ldots(2)$$

$$(2) - (1) \Rightarrow \quad \delta y = (x + \delta x)^{1/3} - x^{1/3}$$

$$\therefore \quad \lim_{\delta x \to 0} \frac{\delta y}{\delta x} = \lim_{\delta x \to 0} \frac{(x+\delta x)^{1/3} - x^{1/3}}{\delta x} = \lim_{x + \delta x \to x} \frac{(x+\delta x)^{1/3} - x^{1/3}}{(x+\delta x) - x}$$

$$= \frac{1}{3} x^{\frac{1}{3} - 1} = \frac{1}{3x^{2/3}}. \quad \left(\text{Using } \lim_{x \to a} \frac{x^n - a^n}{x - a} = na^{n-1}\right)$$

$$\therefore \quad \frac{dy}{dx} \quad i.e., \quad f'(x) = \frac{\mathbf{1}}{\mathbf{3x^{2/3}}}.$$

$$\therefore \quad f'(0) = \frac{1}{3(0)^{2/3}}, \text{ which does not exist and } f'(1) = \frac{1}{3(1)^{2/3}} = \frac{\mathbf{1}}{\mathbf{3}}.$$

Example 7. *Find the derivative of* $x^{-3/4}$ *w.r.t. x by using the definition.*

Sol. Let $$y = x^{-3/4} \quad \ldots(1)$$

Let δx be an increment of x and δy, the corresponding increment of y.

$$\therefore \quad y + \delta y = (x + \delta x)^{-3/4} \quad \ldots(2)$$

$$(2) - (1) \Rightarrow \quad \delta y = (x + \delta x)^{-3/4} - x^{-3/4}$$

$$\therefore \quad \lim_{\delta x \to 0} \frac{\delta y}{\delta x} = \lim_{\delta x \to 0} \frac{(x+\delta x)^{-3/4} - x^{-3/4}}{\delta x} = \lim_{x + \delta x \to x} \frac{(x+\delta x)^{-3/4} - x^{-3/4}}{(x+\delta x) - x}$$

$$= -\frac{3}{4}x^{-\frac{3}{4}-1} = -\frac{3}{4}x^{-7/4}. \qquad \left(\text{Using } \lim_{x \to a} \frac{x^n - a^n}{x - a} = na^{n-1}\right)$$

$$\therefore \quad \frac{dy}{dx} = -\frac{3}{4}\mathbf{x}^{-7/4}.$$

Example 8. *Differentiate* $\frac{ax+b}{cx+d}$ *w.r.t. x by using definition.*

Sol. Let $\qquad y = \frac{ax+b}{cx+d} \qquad \ldots(1)$

Let δx be an increment of x and δy, the corresponding increment of y.

$$\therefore \quad y + \delta y = \frac{a(x+\delta x)+b}{c(x+\delta x)+d} \qquad \ldots(2)$$

$$(2) - (1) \Rightarrow \quad \delta y = \frac{(ax+b)+a\delta x}{(cx+d)+c\delta x} - \frac{ax+b}{cx+d}$$

$$= \frac{(ax+b)(cx+d)+a\delta x(cx+d)-(ax+b)(cx+d)-c\delta x(ax+b)}{(cx+d+c\,\delta x)(cx+d)}$$

$$= \frac{\delta x(acx+ad-acx-bc)}{(cx+d+c\delta x)(cx+d)} = \frac{\delta x(ad-bc)}{(cx+d+c\delta x)(cx+d)}.$$

$$\therefore \quad \frac{\delta y}{\delta x} = \frac{ad-bc}{(cx+d+c\delta x)(cx+d)}.$$

$$\therefore \quad \lim_{\delta x \to 0} \frac{\delta y}{\delta x} = \lim_{\delta x \to 0} \frac{ad-bc}{(cx+d+c\delta x)(cx+d)} = \frac{ad-bc}{(cx+d+0)(cx+d)} = \frac{ad-bc}{(cx+d)^2}$$

$$\therefore \quad \frac{dy}{dx} = \frac{\mathbf{ad-bc}}{(\mathbf{cx+d})^2}.$$

EXERCISE 12.1

Find the derivative of the following functions w.r.t. x by using the first principle (Q. No. 1—4) :

1. (*i*) $4x + 2$ (*ii*) $x^2 + x + 3$ (*iii*) x^3 (*iv*) x^5
2. (*i*) $(x^2 + 1)(x - 5)$ (*ii*) $(x + 1)(x + 2)(x + 3)$ (*iii*) $\frac{x-2}{x+3}$ (*iv*) $\frac{x^2+2}{x+1}$.
3. (*i*) $\frac{1}{x}$ (*ii*) $\frac{1}{x^2}$ (*iii*) $\frac{x^2+1}{x}$ (*iv*) $\frac{x^2-1}{x}$
4. (*i*) $\sqrt{x}$ (*ii*) $\frac{1}{\sqrt{4x-3}}$ (*iii*) $\sqrt{x} + \frac{1}{\sqrt{x}}$ (*iv*) $x^2 - \frac{1}{\sqrt{x}}$.
5. Differentiate x^{28} from the first principle. *(P.T.U. B. Pharma, May 2004)*

Answers

1. (*i*) 4 (*ii*) $2x + 1$ (*iii*) $3x^2$ (*iv*) $5x^4$

2. (*i*) $3x^2 - 10x + 1$ (*ii*) $3x^2 + 12x + 11$ (*iii*) $\frac{5}{(x+3)^2}$ (*iv*) $1 - \frac{3}{(x+1)^2}$

3. (*i*) $-\frac{1}{x^2}$ (*ii*) $-\frac{2}{x^3}$ (*iii*) $1 - \frac{1}{x^2}$ (*iv*) $1 + \frac{1}{x^2}$

4. (*i*) $\frac{1}{2\sqrt{x}}$ (*ii*) $-\frac{2}{(4x-3)^{3/2}}$ (*iii*) $\frac{1}{2\sqrt{x}} - \frac{1}{2x\sqrt{x}}$ (*iv*) $2x + \frac{1}{2x^{3/2}}$.

5. $28\,x^{27}$.

Hints

2. (*ii*) Let $y = f(x) = (x+1)(x+2)(x+3)$. $\therefore\ y = x^3 + 6x^2 + 11x + 6$.

(*iv*) Let $y = \frac{x^2+2}{x+1}$. $\therefore\ y = x - \frac{x-2}{x+1}$.

II. DERIVATIVE OF ALGEBRAIC FUNCTIONS

12.4. DERIVATIVE OF x^n w.r.t. x

Let $y = x^n$, ...(1)

where n is any *real number*.

Let δx be an increment of x and δy, the corresponding increment of y.

$\therefore\ y + \delta y = (x + \delta x)^n$. ...(2)

(2) – (1) $\Rightarrow\ \delta y = (x+\delta x)^n - x^n$

$$\therefore\ \frac{\delta y}{\delta x} = \frac{(x+\delta x)^n - x^n}{\delta x}$$

$$\therefore\ \lim_{\delta x \to 0} \frac{\delta y}{\delta x} = \lim_{\delta x \to 0} \frac{(x+\delta x)^n - x^n}{\delta x} = \lim_{x+\delta x \to x} \frac{(x+\delta x)^n - x^n}{(x+\delta x) - x} = nx^{n-1}$$

$$\left(\text{Using } \lim_{x \to a} \frac{x^n - a^n}{x - a} = na^{n-1}\right)$$

$$\therefore\ \frac{dy}{dx} = nx^{n-1} \quad i.e., \quad \frac{\mathbf{d}}{\mathbf{dx}}(\mathbf{x^n}) = \mathbf{nx^{n-1}}.$$

Thus, the derivative of x^n w.r.t. x is find out by reducing the power of x by unity and multiplying it by the original power.

ILLUSTRATIVE EXAMPLES

Example 1. *Differentiate the following functions :*

(*i*) x^{-11} *w.r.t. x* (*ii*) $s^{3/2}$ *w.r.t. s.*

Sol. (*i*) Let $y = x^{-11}$. $\therefore\ \frac{dy}{dx} = -11x^{-11-1} = -11x^{-12} = -\frac{\mathbf{11}}{\mathbf{x^{12}}}$.

(*ii*) Let $y = s^{3/2}$. $\therefore\ \frac{dy}{ds} = \frac{3}{2}s^{\frac{3}{2}-1} = \frac{\mathbf{3}}{\mathbf{2}}\mathbf{s^{1/2}}$.

12.5. DERIVATIVE OF $(ax + b)^n$ w.r.t. x

Let $y = (ax + b)^n$, ...(1)

where n is any *real number.*

Let δx be an increment of x and δy, the corresponding increment of y.

$\therefore \quad y + \delta y = (a(x + \delta x) + b)^n$ ...(2)

$(2) - (1) \Rightarrow \quad \delta y = (ax + b + a\delta x)^n - (ax + b)^n$

$$\therefore \quad \frac{\delta y}{\delta x} = \frac{(ax+b+a\delta x)^n - (ax+b)^n}{\delta x}$$

$$\therefore \quad \lim_{\delta x \to 0} \frac{\delta y}{\delta x} = \lim_{\delta x \to 0} \frac{(ax+b+a\delta x)^n - (ax+b)^n}{\delta x} = a \lim_{\delta x \to 0} \frac{(ax+b+a\delta x)^n - (ax+b)^n}{a\delta x}$$

$$= a \lim_{(ax+b+a\delta x) \to (ax+b)} \frac{(ax+b+a\delta x)^n - (ax+b)^n}{(ax+b+a\delta x) - (ax+b)} = a \,.\, n(ax+b)^{n-1}.$$

$$\left(\text{Using } \lim_{x \to 0} \frac{x^n - a^n}{x - a} = na^{n-1}\right)$$

$$\therefore \quad \frac{dy}{dx} = n(ax+b)^{n-1}a \quad i.e., \quad \frac{d}{dx}(ax+b)^n = n(ax+b)^{n-1}a.$$

Thus, the derivative of $(ax + b)^n$ w.r.t.x is found out by reducing the power of ax + b by unity and multiplying it by the original power and the coefficient of x.

Example 2. *Differentiate the following functions :*

(*i*) $(3x + 7)^{-2}$ *w.r.t. x* (*ii*) $(2t - 3)^4$ *w.r.t. t*

Sol. (*i*) Let $y = (3x + 7)^{-2}$.

$$\therefore \quad \frac{dy}{dx} = -2(3x+7)^{-2-1} \,.\, 3 = -\frac{6}{(3x+7)^3}.$$

(*ii*) Let $y = (2t - 3)^4$.

$$\therefore \quad \frac{dy}{dt} = 4(2t-3)^{4-1} \,.\, 2 = 8(2t-3)^3.$$

EXERCISE 12.2

1. Differentiate the following functions w.r.t. x :

   (*i*) x^{99} (*ii*) x^{-7} (*iii*) $\sqrt{x^3}$ (*iv*) $\frac{1}{\sqrt{x}}$.

2. If $f(x) = x^n$ and if $f'(1) = 10$, find the value of n.
3. Find a if $f'(a) = 0$, where $f(x) = x^3 - 3x^2 + 3x - 1$.
4. Differentiate the following functions w.r.t. x :

   (*i*) $(2 - 5x)^3$ (*ii*) $(5 + 9x)^{-7/9}$ (*iii*) $\sqrt{7 - 9x}$ (*iv*) $\frac{1}{\sqrt{7x - 9}}$.

Answers

1. (*i*) $99x^{98}$ (*ii*) $-\frac{7}{x^8}$ (*iii*) $\frac{3}{2}\sqrt{x}$ (*iv*) $-\frac{1}{2x^{3/2}}$ **2.** 10

3. 1 **4.** (*i*) $-15(2-5x)^2$ (*ii*) $-\frac{7}{(5+9x)^{16/9}}$ (*iii*) $-\frac{9}{2\sqrt{7-9x}}$ (*iv*) $-\frac{7}{2(7x-9)^{3/2}}$.

III. FUNDAMENTAL THEOREMS

12.6. DERIVATIVE OF A CONSTANT FUNCTION

Let $y = k$ be a constant function. Let δx be an increment of x and δy, the corresponding increment of y.

$$\therefore \quad y + \delta y = k$$

$$\therefore \quad (y + \delta y) - y = k - k$$

$$\Rightarrow \quad \delta y = 0 \quad \Rightarrow \quad \frac{\delta y}{\delta x} = \frac{0}{\delta x} = 0 \quad \Rightarrow \quad \lim_{\delta x \to 0} \frac{\delta y}{\delta x} = 0 \quad i.e., \quad \frac{dy}{dx} = 0.$$

Hence, $$\mathbf{\frac{d}{dx}(k) = 0.}$$

Thus, the derivative of a constant function is always zero.

12.7. DERIVATIVE OF SCALAR MULTIPLE OF A FUNCTION

If u is a differentiable function of x, then for any scalar k, $\frac{d}{dx}(ku) = k\,\frac{du}{dx}$.

Proof. Let $u = f(x)$ and $y = ku$.

Let δx be an increment of x and δu, δy the corresponding increments of u, y respectively.

$$\therefore \quad y = ku \qquad \ldots(1)$$

and $$y + \delta y = k(u + \delta u). \qquad \ldots(2)$$

$$(2) - (1) \quad \Rightarrow \quad \delta y = k\delta u \quad \Rightarrow \quad \frac{\delta y}{\delta x} = k\,\frac{\delta u}{\delta x}$$

$$\therefore \quad \lim_{\delta x \to 0} \frac{\delta y}{\delta x} = \lim_{\delta x \to 0} k\,\frac{\delta u}{\delta x} = k \lim_{\delta x \to 0} \frac{\delta u}{\delta x} = k\,\frac{du}{dx}$$

$$\therefore \quad \frac{dy}{dx} = k\,\frac{du}{dx}.$$

Hence, $$\mathbf{\frac{d}{dx}(ku) = k\,\frac{du}{dx}}.$$

Thus, the derivative of a constant times a differentiable function is the product of the constant and the derivative of the function.

12.8. DERIVATIVE OF SUM OF FUNCTIONS

If u and v are differentiable functions of x, then $\mathbf{\frac{d}{dx}(u + v) = \frac{du}{dx} + \frac{dv}{dx}}$.

Proof. Let $u = f(x)$, $v = g(x)$ and $y = u + v$.

Let δx be an increment of x and δu, δv, δy, the corresponding increment of u, v, y respectively.

$\therefore \quad y = u + v$ ...(1)

and $\quad y + \delta y = (u + \delta u) + (v + \delta v).$ ...(2)

$$(2) - (1) \Rightarrow \quad \delta y = \delta u + \delta v \Rightarrow \frac{\delta y}{\delta x} = \frac{\delta u}{\delta x} + \frac{\delta v}{\delta x}$$

$$\Rightarrow \quad \lim_{\delta x \to 0} \frac{\delta y}{\delta x} = \lim_{\delta x \to 0} \frac{\delta u}{\delta x} + \lim_{\delta x \to 0} \frac{\delta v}{\delta x} = \frac{du}{dx} + \frac{dv}{dx}$$

$$\therefore \quad \frac{dy}{dx} = \frac{du}{dx} + \frac{dv}{dx}$$

Hence, $\quad \mathbf{\frac{d}{dx}(u + v) = \frac{du}{dx} + \frac{dv}{dx}}.$

Thus, the derivative of the sum of two differentiable functions is the sum of the derivatives of the separate functions.

12.9. DERIVATIVE OF DIFFERENCE OF FUNCTIONS

If u and v are differentiable functions of x, then $\frac{d}{dx}(u - v) = \frac{du}{dx} - \frac{dv}{dx}$.

Proof. Let $u = f(x)$, $v = g(x)$ and $y = u - v$.

Let δx be an increment of x and δu, δv, δy the corresponding increments of u, v, y respectively.

$\therefore \quad y = u - v$ ...(1)

and $\quad y + \delta y = (u + \delta u) - (v + \delta v).$ ...(2)

$$(2) - (1) \Rightarrow \quad \delta y = \delta u - \delta v \Rightarrow \frac{\delta y}{\delta x} = \frac{\delta u}{\delta x} - \frac{\delta v}{\delta x}$$

$$\Rightarrow \quad \lim_{\delta x \to 0} \frac{\delta y}{\delta x} = \lim_{\delta x \to 0} \frac{\delta u}{\delta x} - \lim_{\delta x \to 0} \frac{\delta v}{\delta x} = \frac{du}{dx} - \frac{dv}{dx}$$

$$\therefore \quad \frac{dy}{dx} = \frac{du}{dx} - \frac{dv}{dx}.$$

Hence, $\quad \mathbf{\frac{d}{dx}(u - v) = \frac{du}{dx} - \frac{dv}{dx}}.$

Thus, the derivative of the difference of two differentiable functions is the difference of the derivatives of separate functions, in order.

ILLUSTRATIVE EXAMPLES

Example 1. *Differentiate the following functions w.r.t. x :*

(*i*) $x^3 + 2x^2 + 7$ (*ii*) $(x^2 - 4x + 5)(x^3 - 2)$ (*iii*) $\left(x - \frac{1}{x}\right)\left(x^2 - \frac{1}{x^2}\right)$

(*iv*) $\left(\sqrt{x} + \frac{1}{\sqrt{x}}\right)^2$ (*v*) $\frac{2x^2 + 3x + 4}{\sqrt{x}}$.

Sol. (*i*) Let $y = x^3 + 2x^2 + 7$.

$$\therefore \quad \frac{dy}{dx} = \frac{d}{dx}(x^3) + \frac{d}{dx}(2x^2) + \frac{d}{dx}(7) = 3x^2 + 4x + 0 = \mathbf{3x^2 + 4x}.$$

(*ii*) Let $\quad y = (x^2 - 4x + 5)(x^3 - 2)$.

$\therefore \quad y = x^5 - 2x^2 - 4x^4 + 8x + 5x^3 - 10 = x^5 - 4x^4 + 5x^3 - 2x^2 + 8x - 10$

$\therefore \quad \frac{dy}{dx} = \frac{d}{dx}(x^5) - \frac{d}{dx}(4x^4) + \frac{d}{dx}(5x^3) - \frac{d}{dx}(2x^2) + \frac{d}{dx}(8x) - \frac{d}{dx}(10)$

$= 5x^4 - 16x^3 + 15x^2 - 4x + 8 \,.\, 1 - 0 = \mathbf{5x^4 - 16x^3 + 15x^2 - 4x + 8.}$

(*iii*) Let $\quad y = \left(x - \frac{1}{x}\right)\left(x^2 - \frac{1}{x^2}\right).$

$\therefore \quad y = x^3 - \frac{1}{x} - x + \frac{1}{x^3} = x^3 - x^{-1} - x + x^{-3}$

$\therefore \quad \frac{dy}{dx} = \frac{d}{dx}(x^3) - \frac{d}{dx}(x^{-1}) - \frac{d}{dx}(x) + \frac{d}{dx}(x^{-3})$

$= 3x^2 - (-1)x^{-2} - 1 + (-3)\,x^{-4} = \mathbf{3x^2 + \frac{1}{x^2} - 1 - \frac{3}{x^4}}.$

(*iv*) Let $\quad y = \left(\sqrt{x} + \frac{1}{\sqrt{x}}\right)^2. \quad \therefore \quad y = x + \frac{1}{x} + 2 = x + x^{-1} + 2$

$\therefore \quad \frac{dy}{dx} = \frac{d}{dx}(x) + \frac{d}{dx}(x^{-1}) + \frac{d}{dx}(2) = 1 + (-1)\,x^{-2} + 0 = \mathbf{1 - \frac{1}{x^2}}.$

(*v*) Let $\quad y = \frac{2x^2 + 3x + 4}{\sqrt{x}}. \quad \therefore \quad y = \frac{2x^2}{\sqrt{x}} + \frac{3x}{\sqrt{x}} + \frac{4}{\sqrt{x}} = 2x^{3/2} + 3x^{1/2} + 4x^{-1/2}$

$\therefore \quad \frac{dy}{dx} = \frac{d}{dx}(2x^{3/2}) + \frac{d}{dx}(3x^{1/2}) + \frac{d}{dx}(4x^{-1/2})$

$= 3x^{1/2} + \frac{3}{2}\,x^{-1/2} - 2x^{-3/2} = \mathbf{3\sqrt{x} + \frac{3}{2\sqrt{x}} - \frac{2}{x^{3/2}}}.$

Example 2. (*i*) *If $z = t^5 - 3t^4 + 2t^3 + 8$, then find $\frac{dz}{dt}$. Also, find the value of the derivative at $t = 0, 1, 5$.*

(*ii*) *If $\phi(y) = y^3 + y^{3/2} + 2y^{1/2}$, find $\phi'(y)$, $\phi'(1)$, $\phi'(4)$ and $\phi'(16)$.*

Sol. (*i*) We have $\quad z = t^5 - 3t^4 + 2t^3 + 8.$

$\therefore \quad \frac{dz}{dt} = \frac{d}{dt}(t^5) - \frac{d}{dt}(3t^4) + \frac{d}{dt}(2t^3) + \frac{d}{dt}(8)$

$= 5t^4 - 12t^3 + 6t^2 + 0 = \mathbf{5t^4 - 12t^3 + 6t^2}.$

Also, $\quad \left.\frac{dz}{dt}\right|_{t=0} = 5(0)^4 - 12(0)^3 + 6(0)^2 = 0 - 0 + 0 = \mathbf{0}$

$\left.\frac{dz}{dt}\right|_{t=1} = 5(1)^4 - 12(1)^3 + 6(1)^2 = 5 - 12 + 6 = \mathbf{-1}$

and $\quad \left.\frac{dz}{dt}\right|_{t=5} = 5(5)^4 - 12(5)^3 + 6(5)^2 = 3125 - 1500 + 150 = \mathbf{1775.}$

(*ii*) We have $\quad \phi(y) = y^3 + y^{3/2} + 2y^{1/2}.$

$\therefore \quad \phi'(y) = \frac{d}{dy}(y^3) + \frac{d}{dy}(y^{3/2}) + \frac{d}{dy}(2y^{1/2})$

$= 3y^2 + \frac{3}{2}y^{1/2} + 2\left(\frac{1}{2}\right)y^{-1/2} = \mathbf{3y^2 + \frac{3}{2}\sqrt{y} + \frac{1}{\sqrt{y}}}.$

Also $\phi'(1) = 3(1)^2 + \frac{3}{2}\sqrt{1} + \frac{1}{\sqrt{1}} = 3 + \frac{3}{2} + 1 = \mathbf{5\frac{1}{2}}$

$$\phi'(4) = 3(4)^2 + \frac{3}{2}\sqrt{4} + \frac{1}{\sqrt{4}} = 48 + 3 + \frac{1}{2} = \mathbf{51\frac{1}{2}}$$

and $\phi'(16) = 3(16)^2 + \frac{3}{2}\sqrt{16} + \frac{1}{\sqrt{16}} = 768 + 6 + \frac{1}{4} = \mathbf{774\frac{1}{4}}$.

EXERCISE 12.3

1. If $y = x^4 + 7x^3 + 8x^2 + 3x + 2$ find $\frac{dy}{dx}$.

2. If $z = t^3 + 4t + 7$, find $\frac{dz}{dt}$.

3. If $f(x) = x^2 - 9x + 20$, then find $f'(x)$ and hence find $f'(100)$ and $f'(22)$.

4. If for $f(x) = \lambda x^2 + \mu x + 12$, $f'(4) = 15$ and $f'(2) = 11$, then find λ and μ.

5. If $H(y) = 8y^9 + 6y^3 + 8y^2 + 7$, find $H'(y)$ and $H'(0)$.

6. If $u = 7t^4 - 2t^3 - 8t - 5$, find $\frac{du}{dt}$ at $t = 2$.

7. Differentiate the following functions w.r.t. x :

(*i*) $\frac{2x^2 - 3x + 1}{\sqrt{x}}$ (*ii*) $\frac{(x+5)(2x^2-1)}{x}$ (*iii*) $\left(x + \frac{1}{x}\right)\left(x^2 - \frac{1}{x^2}\right)$ (*iv*) $\left(x + \frac{1}{x}\right)\left(\sqrt{x} + \frac{1}{\sqrt{x}}\right)$.

8. If $y = \sqrt{x} + \frac{1}{\sqrt{x}}$, show that $2x\frac{dy}{dx} + y = 2\sqrt{x}$.

9. If $y = x + \frac{1}{x}$, show that $x^2\frac{dy}{dx} - xy + 2 = 0$.

Answers

1. $4x^3 + 21x^2 + 16x + 3$ **2.** $3t^2 + 4$ **3.** $2x - 9$, 191, 35 **4.** 1, 7

5. $72y^8 + 18y^2 + 16y$, 0 **6.** 192 **7.** (*i*) $3\sqrt{x} - \frac{3}{2\sqrt{x}} - \frac{1}{2x^{3/2}}$

(*ii*) $4x + 10 + \frac{5}{x^2}$ (*iii*) $3x^2 + \frac{1}{x^2} + 1 + \frac{3}{x^4}$ (*iv*) $\frac{3}{2}\sqrt{x} + \frac{1}{2\sqrt{x}} - \frac{1}{2x^{3/2}} - \frac{3}{2x^{5/2}}$.

12.10. DERIVATIVE OF PRODUCT OF FUNCTIONS

If u and v are differentiable functions of x, then $\frac{d}{dx}(uv) = u\frac{dv}{dx} + v\frac{du}{dx}$.

Proof. Let $u = f(x)$, $v = g(x)$ and $y = uv$.

Let δx be an increment of x and δu, δv, δy the corresponding increments of u, v, y respectively.

$\therefore \qquad y = uv \qquad \ldots(1)$

and $\qquad y + \delta y = (u + \delta u)(v + \delta v) \qquad \ldots(2)$

(2) – (1) $\Rightarrow \qquad \delta y = uv + u(\delta v) + v(\delta u) + (\delta u)(\delta v) - uv = u(\delta v) + v(\delta u) + (\delta u)(\delta v)$

$$\therefore \qquad \frac{\delta y}{\delta x} = u\frac{\delta u}{\delta x} + v\frac{\delta u}{\delta x} + \frac{\delta u}{\delta x}\,.\,\delta v$$

$$\therefore \qquad \lim_{\delta x \to 0}\frac{\delta y}{\delta x} = u\lim_{\delta x \to 0}\frac{\delta v}{\delta x} + v\lim_{\delta x \to 0}\frac{\delta u}{\delta x} + \left(\lim_{\delta x \to 0}\frac{\delta u}{\delta x}\right)\lim_{\delta x \to 0}\delta v$$

$$= u\frac{dv}{dx} + v\frac{du}{dx} + \frac{du}{dx}\,.\,0 = u\frac{dv}{dx} + v\frac{du}{dx}$$

$$\therefore \qquad \frac{dy}{dx} = u\frac{dv}{dx} + v\frac{du}{dx}$$

Hence, $\qquad \mathbf{\frac{d}{dx}(uv) = u\frac{dv}{dx} + v\frac{du}{dx}}.$

Thus, the derivative of the product of two differentiable functions is equal to the first function times the derivative of the second function plus the second function times the derivative of the first function.

Aid to memory. *The derivative of the product of two functions is equal to*

1st function × derivative of 2nd + 2nd function × derivative of 1st.

Remark. The above formula is generally called the **product rule of differentiation.**

Caution. *The derivative of a product is not equal to the product of the derivatives.*

ILLUSTRATIVE EXAMPLES

Example 1. *Differentiate* $(x^2 + 1)(x^2 + x + 4)$ *w.r.t. x by using product rule and without using product rule. Verify that the derivatives are same.*

Sol. Let $\qquad y = (x^2 + 1)(x^2 + x + 4) \qquad \ldots(1)$

By using 'product rule'

$$(1) \Rightarrow \qquad \frac{dy}{dx} = (x^2 + 1)\frac{d}{dx}(x^2 + x + 4) + (x^2 + x + 4)\frac{d}{dx}(x^2 + 1)$$

$$= (x^2 + 1)(2x + 1 + 0) + (x^2 + x + 4)(2x + 0)$$

$$= (2x^3 + x^2 + 2x + 1) + (2x^3 + 2x^2 + 8x) = \mathbf{4x^3 + 3x^2 + 10x + 1.}$$

Without using 'product rule'

$$(1) \Rightarrow \qquad y = x^4 + x^3 + 4x^2 + x^2 + x + 4 = x^4 + x^3 + 5x^2 + x + 4$$

$$\therefore \qquad \frac{dy}{dx} = 4x^3 + 3x^2 + 10x + 1 + 0 = \mathbf{4x^3 + 3x^2 + 10x + 1.}$$

$\therefore$ The derivatives are same.

Example 2. *Differentiate the following functions w.r.t. x :*

(*i*) $(3x + 4)^2\,(2x - 3)^3$ $\qquad$ (*ii*) $(x + a)^m(x + b)^n$ $\qquad$ (*iii*) $(3x + 8)^{7/3}(x + 7)^{-3}$.

Sol. (*i*) Let $\;y = (3x + 4)^2\,(2x - 3)^3$.

$$\therefore \qquad \frac{dy}{dx} = (3x + 4)^2\frac{d}{dx}(2x - 3)^3 + (2x - 3)^3\frac{d}{dx}(3x + 4)^2 \qquad \text{(Using 'product rule')}$$

$$= (3x + 4)^2\,.\,3(2x - 3)^2\,.\,2 + (2x - 3)^3\,.\,2(3x + 4)^1\,.\,3$$

$$= 6(3x + 4)(2x - 3)^2[(3x + 4) + (2x - 3)] = \mathbf{6(3x + 4)(2x - 3)^2(5x + 1).}$$

(*ii*) Let $y = (x + a)^m (x + b)^n$.

$$\therefore \quad \frac{dy}{dx} = (x+a)^m \frac{d}{dx}(x+b)^n + (x+b)^n \frac{d}{dx}(x+a)^m \quad \text{(Using 'product rule')}$$

$$= (x+a)^m \,.\, n(x+b)^{n-1} \,.\, 1 + (x+b)^n \,.\, m(x+a)^{m-1} \,.\, 1$$

$$= (x+a)^{m-1}(x+b)^{n-1}[n(x+a) + m(x+b)]$$

$$= \mathbf{(x+a)^{m-1}(x+b)^{n-1}((m+n)x + an + bm)}.$$

(*iii*) Let $y = (3x + 8)^{7/3}(x + 7)^{-3}$.

$$\therefore \quad \frac{dy}{dx} = (3x+8)^{7/3} \frac{d}{dx}(x+7)^{-3} + (x+7)^{-3} \frac{d}{dx}(3x+8)^{7/3} \quad \text{(Using 'product rule')}$$

$$= (3x+8)^{7/3} \,.\, (-3)(x+7)^{-4} \,.\, 1 + (x+7)^{-3} \,.\, \frac{7}{3}(3x+8)^{4/3} \,.\, 3$$

$$= (3x+8)^{4/3}(x+7)^{-4}[-3(3x+8) + 7(x+7)]$$

$$= \mathbf{(3x+8)^{4/3}(x+7)^{-4}(-2x+25)}.$$

Example 3. *If u, v, w are three differentiable functions of x, prove that*

$$\frac{d}{dx}(uvw) = \frac{du}{dx} vw + u \frac{dv}{dx} w + uv \frac{dw}{dx}.$$

Use this result to differentiate $x^3(x^2 + 1)(x^4 + 1)$ *w.r.t. x.*

Sol. $\frac{d}{dx}(uvw) = \frac{d}{dx}[(uv)w] = uv \frac{dw}{dx} + w \frac{d}{dx}(uv)$ (Using 'product rule')

$$= uv \frac{dw}{dx} + w\left[u \frac{dv}{dx} + v \frac{du}{dx}\right] = uv \frac{dw}{dx} + uw \frac{dv}{dx} + vw \frac{du}{dx}.$$

$$\frac{d}{dx}[x^3(x^2+1)(x^4+1)]$$

$$= \frac{d}{dx}(x^3) \,.\, (x^2+1)(x^4+1) + x^3 \frac{d}{dx}(x^2+1) \,.\, (x^4+1) + x^3(x^2+1) \,.\, \frac{d}{dx}(x^4+1)$$

$$= 3x^2 \,.\, (x^6 + x^4 + x^2 + 1) + x^3 \,.\, 2x \,.\, (x^4+1) + x^3(x^2+1) \,.\, 4x^3$$

$$= 3x^8 + 3x^6 + 3x^4 + 3x^2 + 2x^8 + 2x^4 + 4x^8 + 4x^6$$

$$= \mathbf{9x^8 + 7x^6 + 5x^4 + 3x^2}.$$

EXERCISE 12.4

1. Differentiate $(x - 2)(x + 3)$ w.r.t. x by using product rule and without using product rule. Verify that the derivatives are same.
2. Differentiate the following functions w.r.t. x :
   (*i*) $(x^2 - 5x + 6)(x^3 + 2)$ (*ii*) $(x^2 - x + 3)(x^2 + 2x - 4)$.
   (*iii*) $(2x + 3)^5 (3 - x)^4$ (*iv*) $(2x + 1)^{1/3}(x + 1)$.
3. Find the derivative of $(x + 1)(2x + 1)^2(3x + 1)^3$ w.r.t. x.
4. Differentiate $2s^2\sqrt{2 - s}$ w.r.t. s.

Answers

1. $2x + 1$ 2. (i) $5x^4 - 20x^3 + 18x^2 + 4x - 10$

(ii) $4x^3 + 3x^2 - 6x + 10$ (iii) $18(1 - x)(2x + 3)^4(3 - x)^3$

(iv) $\dfrac{8x + 5}{3(2x + 1)^{2/3}}$ 3. $(2x + 1)(3x + 1)^2 (36x^2 + 48x + 14)$ 4. $\dfrac{s(8 - 5s)}{\sqrt{2 - s}}$.

12.11. DERIVATIVE OF QUOTIENT OF FUNCTIONS

If u and v are differentiable functions of x, then $\dfrac{d}{dx}\left(\dfrac{u}{v}\right) = \dfrac{v\dfrac{du}{dx} - u\dfrac{dv}{dx}}{v^2}$.

Proof. Let $u = f(x)$ and $v = g(x)$ and $y = u/v$.

Let δx be an increment of x and δu, δv, δy the corresponding increments of u, v, y respectively.

$$\therefore \quad y = \frac{u}{v} \quad \text{...(1)}$$

and

$$y + \delta y = \frac{u + \delta u}{v + \delta v}. \quad \text{...(2)}$$

$$(2) - (1) \Rightarrow \quad \delta y = \frac{u + \delta u}{v + \delta v} - \frac{u}{v} = \frac{uv + v(\delta u) - uv - u(\delta v)}{(v + \delta v)\,v} = \frac{v(\delta u) - u(\delta v)}{(v + \delta v)\,v}.$$

$$\therefore \quad \frac{\delta y}{\delta x} = \frac{1}{\delta x} \cdot \frac{v(\delta u) - u(\delta v)}{(v + \delta v)\,v} = \frac{v\dfrac{\delta u}{\delta x} - u\dfrac{\delta v}{\delta x}}{(v + \delta v)\,v}$$

$$\therefore \quad \lim_{\delta x \to 0} \frac{\delta y}{\delta x} = \frac{v \lim\limits_{\delta x \to 0} \dfrac{\delta u}{\delta x} - u \lim\limits_{\delta x \to 0} \dfrac{\delta v}{\delta x}}{(v + \lim\limits_{\delta x \to 0} \delta v)\,v} = \frac{v\dfrac{du}{dx} - u\dfrac{dv}{dx}}{(v + 0)\,v} = \frac{v\dfrac{du}{dx} - u\dfrac{dv}{dx}}{v^2}$$

$$\therefore \quad \frac{dy}{dx} = \frac{v\dfrac{du}{dx} - u\dfrac{dv}{dx}}{v^2}.$$

Hence, $$\mathbf{\frac{d}{dx}\left(\frac{u}{v}\right) = \frac{v\dfrac{du}{dx} - u\dfrac{dv}{dx}}{v^2}}.$$

Thus, the derivative of the quotient of two differentiable functions is equal to the denominator times the derivative of the numerator minus the numerator times the derivative of the denominator, all divided by the square of the denominator.

Aid to memory. *The differential coefficient of the quotient of two functions is equal to*

$$\frac{\textit{Denom.} \times [\textit{diff. coeff. of Num.}] - \textit{Num.} \times [\textit{diff. coeff. of Denom.}]}{(\textit{Denom.})^2}$$

Remark . The above formula is generally called the '**quotient rule of differentiation**'.

Caution. *The derivative of a quotient is not equal to the quotient of the derivatives.*

ILLUSTRATIVE EXAMPLES

Example. *Differentiate the following functions w.r.t. x :*

(*i*) $\dfrac{4-x^2}{4+x^2}$ (*ii*) $\sqrt{\dfrac{1+x}{1-x}}$ (*iii*) $\dfrac{3x+2}{(x+5)(2x+1)+3}$.

Sol. (*i*) Let $y = \dfrac{4-x^2}{4+x^2}$. $\left(\dfrac{\text{'}u\text{'}}{v}\text{ Form}\right)$

$$\therefore \quad \frac{dy}{dx} = \frac{(4+x^2)\frac{d}{dx}(4-x^2)-(4-x^2)\frac{d}{dx}(4+x^2)}{(4+x^2)^2} \quad \text{(Using 'quotient rule')}$$

$$= \frac{(4+x^2)(0-2x)-(4-x^2)(0+2x)}{(4+x^2)^2} = \frac{-8x-2x^3-8x+2x^3}{(4+x^2)^2} = -\frac{\mathbf{16x}}{\mathbf{(4+x^2)^2}}.$$

(*ii*) Let $y = \sqrt{\dfrac{1+x}{1-x}}$. $\therefore \quad y = \dfrac{\sqrt{1+x}}{\sqrt{1-x}}$ $\left(\dfrac{\text{'}u\text{'}}{v}\text{ Form}\right)$

$$\therefore \quad \frac{dy}{dx} = \frac{\sqrt{1-x}\,\frac{d}{dx}(\sqrt{1+x})-\sqrt{1+x}\,\frac{d}{dx}(\sqrt{1-x})}{\left(\sqrt{1-x}\right)^2} \quad \text{(Using 'quotient rule')}$$

$$= \frac{\sqrt{1-x}\,.\,\frac{1}{2}(1+x)^{-1/2}\,.\,1-\sqrt{1+x}\,.\,\frac{1}{2}(1-x)^{-1/2}\,.\,(-1)}{1-x}$$

$$= \frac{\frac{\sqrt{1-x}}{\sqrt{1+x}}+\frac{\sqrt{1+x}}{\sqrt{1-x}}}{2(1-x)} = \frac{(1-x)+(1+x)}{2\sqrt{1+x}\sqrt{1-x}\,(1-x)}$$

$$= \frac{2}{2\sqrt{1+x}\sqrt{1-x}\,(1-x)} = \frac{\mathbf{1}}{\sqrt{\mathbf{1+x}}\,\mathbf{(1-x)^{3/2}}}.$$

(*iii*) Let $y = \dfrac{3x+2}{(x+5)(2x+1)+3}$. $\therefore \quad y = \dfrac{3x+2}{2x^2+11x+8}$ $\left(\dfrac{\text{'}u\text{'}}{v}\text{ Form}\right)$

$$\therefore \quad \frac{dy}{dx} = \frac{(2x^2+11x+8)\frac{d}{dx}(3x+2)-(3x+2)\frac{d}{dx}(2x^2+11x+8)}{(2x^2+11x+8)^2}$$

(Using 'quotient rule')

$$= \frac{(2x^2+11x+8)(3(1)+0)-(3x+2)(4x+11(1)+0)}{(2x^2+11x+8)^2}$$

$$= \frac{6x^2+33x+24-12x^2-8x-33x-22}{(2x^2+11x+8)^2} = \frac{\mathbf{-6x^2-8x+2}}{\mathbf{(2x^2+11x+8)^2}}.$$

EXERCISE 12.5

1. Differentiate the following functions w.r.t. x :

(*i*) $\dfrac{7x+4}{4x-7}$ (*ii*) $\dfrac{1}{px^2+qx+r}$.

2. We know that $\dfrac{d}{dx}\left(\dfrac{5}{x^2}\right) = 5\dfrac{d}{dx}(x^{-2}) = -10x^{-3}$.

Differentiate $5/x^2$ by quotient rule and verify that the answer is the same.

3. Differentiate the following functions w.r.t. x :

(*i*) $\dfrac{x^n-a^n}{x-a}$ (*ii*) $\dfrac{x-1}{\sqrt{x+1}}$ (*iii*) $\dfrac{\sqrt{a}+\sqrt{x}}{\sqrt{a}-\sqrt{x}}$ (*iv*) $\dfrac{ax^m+bx^n}{ax^m-bx^n}$.

4. If $y = \dfrac{x}{x+4}$, show that $x\dfrac{dy}{dx} = y(1-y)$.

5. Differentiate $(2x+7)(3x+1)^{-1}$ by using (*i*) product rule and (*ii*) quotient rule separately. Verify that the answers obtained are same.

6. Differentiate $\dfrac{1}{(x-a)(x-b)(x-c)}$ w.r.t. x.

Answers

1. (*i*) $-\dfrac{65}{(4x-7)^2}$ (*ii*) $-\dfrac{2px+q}{(px^2+qx+r)^2}$

3. (*i*) $\dfrac{(n-1)\,x^n - a(nx^{n-1}-a^{n-1})}{(x-a)^2}$ (*ii*) $\dfrac{x+3}{2(x+1)^{3/2}}$

(*iii*) $\dfrac{\sqrt{a}}{\sqrt{x}(\sqrt{a}-\sqrt{x})^2}$ (*iv*) $\dfrac{(ax^m-bx^n)(amx^{m-1}+bnx^{n-1})-(ax^m+bx^n)(amx^{m-1}-bnx^{n-1})}{(ax^m-bx^n)^2}$

6. $-\dfrac{[(x-a)(x-b)+(x-a)(x-c)+(x-b)(x-c)]}{(x-a)^2(x-b)^2(x-c)^2}$.

12.12. DERIVATIVE OF FUNCTION OF A FUNCTION (CHAIN RULE)

Let $y = (ax^2+bx+c)^n$.

$u = ax^2+bx+c$ implies $y = u^n$.

$\therefore$ y has been expressed as a function of variable u, which is itself a function of x.

Thus, y is a function of a function of x.

In general, if y is a function of variable u, and u in turn is a function of variable x, then we say that y is a **function of a function** of x.

Let δx be an increment of x and δu, δy, the corresponding increments of u, y respectively. We assume that $\delta u \neq 0$.

$$\therefore \qquad \frac{\delta y}{\delta x} = \frac{\delta y}{\delta u}\cdot\frac{\delta u}{\delta x}$$

$$\Rightarrow \quad \lim_{\delta x \to 0} \frac{\delta y}{\delta x} = \lim_{\delta x \to 0} \left(\frac{\delta y}{\delta u} \cdot \frac{\delta u}{\delta x} \right) = \lim_{\delta u \to 0} \frac{\delta y}{\delta u} \cdot \lim_{\delta x \to 0} \frac{\delta u}{\delta x}$$

$$\therefore \quad \frac{dy}{dx} = \frac{dy}{du} \cdot \frac{du}{dx}.$$

This result is called the **Chain rule.**

Thus, if y is a differentiable function of u and u is a differentiable function of x, then the derivative of y with respect to x is equal to the derivative of y with respect to u times the derivative of u with respect to x.

Remark 1. If u is a differentiable function of x, then by **chain rule**

$$\frac{d}{dx}(u^n) = \frac{d}{du}(u^n) \cdot \frac{du}{dx} = nu^{n-1} \frac{du}{dx}.$$

Remark 2. The derivative of u^n w.r.t. u is nu^{n-1} but w.r.t. x, it is $nu^{n-1} \frac{du}{dx}$.

Remark 3. If $y = f(u)$, $u = g(w)$, $w = h(x)$ are differentiable functions, then by **chain rule,** we have

$$\frac{dy}{dx} = \frac{dy}{du} \cdot \frac{du}{dx} = \frac{dy}{du}\left[\frac{du}{dw} \cdot \frac{dw}{dx}\right] = \frac{dy}{du} \cdot \frac{du}{dw} \cdot \frac{dw}{dx}.$$

ILLUSTRATIVE EXAMPLES

Example 1. *If $y = u^2 + 1$ and $u = x^3 + x + 4$, find the derivative of y w.r.t. x (i) using chain rule and (ii) without using chain rule. Verify that the derivatives are same.*

Sol. We have $y = u^2 + 1$ ...(1) and $u = x^3 + x + 4$. ...(2)

(*i*) **Using chain rule**

(1) $\Rightarrow \quad \frac{dy}{du} = 2u$. Also (2) $\Rightarrow \quad \frac{du}{dx} = 3x^2 + 1$

$$\therefore \quad \frac{dy}{dx} = \frac{dy}{du} \cdot \frac{du}{dx} = 2u.(3x^2 + 1) = 2(x^3 + x + 4)(3x^2 + 1)$$

$$= \mathbf{6x^5 + 8x^3 + 24x^2 + 2x + 8.}$$

(*ii*) **Without using chain rule**

(1) and (2) implies $y = (x^3 + x + 4)^2 + 1$

$$= x^6 + x^2 + 16 + 2x^4 + 8x + 8x^3 + 1 = x^6 + 2x^4 + 8x^3 + x^2 + 8x + 17$$

$$\therefore \quad \frac{dy}{dx} = \mathbf{6x^5 + 8x^3 + 24x^2 + 2x + 8.}$$

We get the same derivatives.

Example 2. *Find $\frac{dy}{dx}$ by using **chain rule** for the following functions :*

(*i*) $y = u^2 + u + 2$, $u = v^2 + 5$, $v = 6x + 7$

(*ii*) $y = 2u^2 + 3$, $u = 12v^2 + v$, $v = 3w + 8$, $w = 2x$.

Sol. (*i*) We have $y = u^2 + u + 2$, $u = v^2 + 5$ and $v = 6x + 7$.

$$\therefore \quad \frac{dy}{du} = 2u + 1, \quad \frac{du}{dv} = 2v, \quad \frac{dv}{dx} = 6.$$

By **chain rule,** $\frac{dy}{dx} = \frac{dy}{du} \cdot \frac{du}{dv} \cdot \frac{dv}{dx} = (2u + 1)(2v)(6) = (2(v^2 + 5) + 1) \cdot 12v$

$$= 12v(2v^2 + 11) = 12(6x + 7)(2(6x + 7)^2 + 11)$$

$$= \mathbf{12(6x + 7)(72x^2 + 168x + 109).}$$

(*ii*) We have $y = 2u^2 + 3,\ u = 12v^2 + v,\ v = 3w + 8,\ w = 2x.$

$$\therefore \quad \frac{dy}{du} = 4u,\ \frac{du}{dv} = 24v + 1,\ \frac{dv}{dw} = 3,\ \frac{dw}{dx} = 2.$$

By **chain rule,** $\dfrac{dy}{dx} = \dfrac{dy}{du}\cdot\dfrac{du}{dv}\cdot\dfrac{dv}{dw}\cdot\dfrac{dw}{dx} = (4u)(24v + 1)(3)(2)$

$$= 24u(24v + 1) = 24(12v^2 + v)(24v + 1)$$
$$= 24(12(3w + 8)^2 + (3w + 8))(24(3w + 8) + 1)$$
$$= 24\,(108\,w^2 + 576\,w + 768)\,(72\,w + 193)$$
$$= 24\,(7776\,w^3 + 41472\,w^2 + 55296\,w + 20844\,w^2 + 111168\,w + 148224)$$
$$= 24\,(7776\,w^3 + 62316\,w^2 + 166464\,w + 148224)$$
$$= 24\,(7776\,(2x)^3 + 62316\,(2x)^2 + 166464\,(2x) + 148224) \qquad | \ w = 2x$$
$$= \mathbf{24\,(62208\,x^3 + 249264\,x^2 + 332928\,x + 148224).}$$

Example 3. *Find* $\dfrac{dy}{dx}$ *if* $y = \dfrac{3-u}{2+u},\ u = \dfrac{4x}{1-x^2}$.

Sol. We have $y = \dfrac{3-u}{2+u},\ u = \dfrac{4x}{1-x^2}$. $\left[\dfrac{u}{v} \text{ Form}\right]$

$$\therefore \quad \frac{dy}{du} = \frac{(2+u)\dfrac{d}{dx}(3-u) - (3-u)\dfrac{d}{du}(2+u)}{(2+u)^2} = \frac{(2+u)(-1)-(3-u)(1)}{(2+u)^2} = \frac{-5}{(2+u)^2}$$

Also, $$\frac{du}{dx} = \frac{(1-x^2)\dfrac{d}{dx}(4x) - 4x\dfrac{d}{dx}(1-x^2)}{(1-x^2)^2} = \frac{(1-x^2)\,4 - 4x\,(-2x)}{(1-x^2)^2} = \frac{4x^2+4}{(1-x^2)^2}.$$

$\therefore$ By **chain rule,**

$$\frac{dy}{dx} = \frac{dy}{du}\cdot\frac{du}{dx} = \frac{-5}{(2+u)^2}\cdot\frac{4x^2+4}{(1-x^2)^2} = \frac{-5}{\left(2+\dfrac{4x}{1-x^2}\right)^2}\cdot\frac{4(x^2+1)}{(1-x^2)^2}$$

$$= -\frac{5(1-x^2)^2}{(2+4x-2x^2)^2}\cdot\frac{4(x^2+1)}{(1-x^2)^2} = -\frac{20(x^2+1)}{4(1+2x-x^2)^2} = -\frac{\mathbf{5(x^2+1)}}{\mathbf{(1+2x-x^2)^2}}.$$

Example 4. *Differentiate the following functions w.r.t. x :*

(*i*) $(2x^3 + 3x^2 - x + 5)^4$ (*ii*) $\dfrac{1}{(ax^2+bx+c)^n}$ (*iii*) $\sqrt{\dfrac{x^2-2ax}{a^2-2ab}}$.

Sol. (*i*) Let $y = (2x^3 + 3x^2 - x + 5)^4$

Let $u = 2x^3 + 3x^2 - x + 5$ ...(1) $\therefore\ y = u^4$ ...(2)

(1) $\Rightarrow \dfrac{du}{dx} = 6x^2 + 6x - 1$ and (2) $\Rightarrow \dfrac{dy}{du} = 4u^3$.

$\therefore$ By **chain rule,**

$$\frac{dy}{dx} = \frac{dy}{du}\cdot\frac{du}{dx} = 4u^3(6x^2 + 6x - 1) = \mathbf{4(2x^3 + 3x^2 - x + 5)^3\,(6x^2 + 6x - 1).}$$

Alternative method

Let $y = (2x^3 + 3x^2 - x + 5)^4$. ('$u^n$' Form)

$$\therefore \quad \frac{dy}{dx} = 4(2x^3 + 3x^2 - x + 5)^3 \frac{d}{dx}(2x^3 + 3x^2 - x + 5) \qquad \left(\text{Using } \frac{d}{dx}(u^n) = nu^{n-1}\frac{du}{dx}\right)$$

$$= \mathbf{4(2x^3 + 3x^2 - x + 5)^3 (6x^2 + 6x - 1).}$$

(*ii*) Let $y = \frac{1}{(ax^2 + bx + c)^n}$. $\therefore \ y = (ax^2 + bx + c)^{-n}$.

$$\therefore \quad \frac{dy}{dx} = -n\,(ax^2 + bx + c)^{-n-1}\frac{d}{dx}(ax^2 + bx + c) \qquad \left(\because \ \frac{d}{dx}(u^n) = nu^{n-1}\frac{du}{dx}\right)$$

$$= \frac{-n}{(ax^2 + bx + c)^{n+1}}\,.\,(2ax + b) = -\frac{\mathbf{n(2ax + b)}}{\mathbf{(ax^2 + bx + c)^{n+1}}}.$$

(*iii*) Let $y = \sqrt{\frac{x^2 - 2ax}{a^2 - 2ab}}$. $\therefore \ y = \frac{1}{\sqrt{a^2 - 2ab}}(x^2 - 2ax)^{1/2}$.

$$\therefore \quad \frac{dy}{dx} = \frac{1}{\sqrt{a^2 - 2ab}}\,.\,\frac{d}{dx}(x^2 - 2ax)^{1/2}$$

$$= \frac{1}{\sqrt{a^2 - 2ab}}\,.\,\frac{1}{2}(x^2 - 2ax)^{-1/2}\frac{d}{dx}(x^2 - 2ax) \qquad \left(\because \ \frac{d}{dx}(u^n) = nu^{n-1}\frac{du}{dx}\right)$$

$$= \frac{1}{2\sqrt{a^2 - 2ab}\,\sqrt{x^2 - 2ax}}\,.\,(2x - 2a) = \frac{\mathbf{x - a}}{\mathbf{\sqrt{a^2 - 2ab}\,\sqrt{x^2 - 2ax}}}.$$

Example 5. *Differentiate the following functions w.r.t. x :*

(*i*) $\left(x + \sqrt{x^2 + a^2}\right)^n$ (*ii*) $\sqrt{\frac{a^2 + x^2}{a^2 - x^2}}\,.\,\sqrt[3]{\frac{a^2 - x^2}{a^2 + x^2}}$.

Sol. (*i*) Let $y = (x + \sqrt{x^2 + a^2})^n$. ('$u^n$' Form)

$$\therefore \quad \frac{dy}{dx} = n(x + \sqrt{x^2 + a^2})^{n-1}\,.\,\frac{d}{dx}[x + \sqrt{x^2 + a^2}] = n(x + \sqrt{x^2 + a^2})^{n-1}\left[1 + \frac{d}{dx}(x^2 + a^2)^{1/2}\right]$$

$$= n\left(x + \sqrt{x^2 + a^2}\right)^{n-1}\left[1 + \frac{1}{2}(x^2 + a^2)^{-1/2}\frac{d}{dx}(x^2 + a^2)\right]$$

$$= n(x + \sqrt{x^2 + a^2})^{n-1}\left[1 + \frac{1}{2\sqrt{x^2 + a^2}}\,.\,(2x + 0)\right]$$

$$= n(x + \sqrt{x^2 + a^2})^{n-1}\left(\frac{\sqrt{x^2 + a^2} + x}{\sqrt{x^2 + a^2}}\right) = \frac{\mathbf{n(x + \sqrt{x^2 + a^2})^n}}{\mathbf{\sqrt{x^2 + a^2}}}.$$

(*ii*) Let $y = \sqrt{\dfrac{a^2+x^2}{a^2-x^2}} \cdot \sqrt[3]{\dfrac{a^2-x^2}{a^2+x^2}}$.

$$\therefore \quad y = \left(\frac{a^2+x^2}{a^2-x^2}\right)^{1/2}\left(\frac{a^2+x^2}{a^2-x^2}\right)^{-1/3} = \left(\frac{a^2+x^2}{a^2-x^2}\right)^{1/6} \qquad (\text{'}u^n\text{' Form})$$

$$\therefore \quad \frac{dy}{dx} = \frac{1}{6}\left(\frac{a^2+x^2}{a^2-x^2}\right)^{1/6-1}\frac{d}{dx}\left(\frac{a^2+x^2}{a^2-x^2}\right) = \frac{1}{6}\left(\frac{a^2-x^2}{a^2+x^2}\right)^{5/6}\cdot\frac{(a^2-x^2)(2x)-(a^2+x^2)(-2x)}{(a^2-x^2)^2}$$

$$= \frac{2x[a^2-x^2+a^2+x^2]}{6(a^2+x^2)^{5/6}(a^2-x^2)^{2-(5/6)}} = \mathbf{\frac{2a^2x}{3(a^2+x^2)^{5/6}(a^2-x^2)^{7/6}}}.$$

EXERCISE 12.6

1. Find $\dfrac{dy}{dx}$ by using chain rule for the following functions :

(*i*) $y = u^2,\ u = x^2 + 4$ (*ii*) $y = \sqrt{z},\ z = 5x^4 + 3$

(*iii*) $y = \sqrt{u},\ u = \sqrt{x}$ (*iv*) $y = u^2 + u + 5,\ u = x^2 - x + 1$.

2. If $y = u^2$ and $u = 1 + x^3$, find the derivative of y w.r.t. x :

(*i*) using chain rule (*ii*) without using chain rule.

Verify that the derivatives are same.

3. Find $\dfrac{dy}{dx}$ by using **chain rule** for the following functions :

(*i*) $y = v^2 + v + 1,\ v = u^2 + 1,\ u = 6x$ (*ii*) $y = v^3 + 2v^2 + 4,\ v = 3u + 7,\ u = 9x + 3$.

4. If $y = z + 1,\ z = w + 2,\ w = u^2,\ u = 4x + 5$, find $\dfrac{dy}{dw}, \dfrac{dw}{dx}, \dfrac{dz}{dx}$ and $\dfrac{dy}{dx}$.

Find the derivatives of the following functions w.r.t. x (Q. No. 5–7) :

5. (*i*) $(4x^3 - 5x^2 + 1)^4$ (*ii*) $(3x^3 - x + 7)^{1/3}$

(*iii*) $\left(\dfrac{3x-1}{2x+1}\right)^3$ (*iv*) $\sqrt{\dfrac{1-x^2}{1+x^2}}$

6. (*i*) $(x^2 + 3)^4\,(x + 1)$ (*ii*) $(3x^2 + 2)^3\,(5x + 3)$

(*iii*) $(2x^2 + 3)^{5/3}\,(x + 5)^{-1/2}$ (*iv*) $(1 + x)\sqrt{1+x^2}$

7. (*i*) $\dfrac{\sqrt{a+x}-\sqrt{a-x}}{\sqrt{a+x}+\sqrt{a-x}}$ (*ii*) $\dfrac{x}{\sqrt{ax^2+bx+c}}$

(*iii*) $\dfrac{x}{(a^2-x^2)^{3/2}}$ (*iv*) $\dfrac{x}{a-\sqrt{a^2-x^2}}$.

8. (*i*) If $y = \sqrt{1+x^2}$, show that $y\dfrac{dy}{dx} - x = 0$.

(*ii*) If $y = x + \sqrt{x^2 - 1}$, show that $(y - x)\dfrac{dy}{dx} - y = 0$.

(*iii*) If $y = \sqrt{\dfrac{1-x}{1+x}}$, show that $(1 - x^2)\dfrac{dy}{dx} + y = 0$.

Answers

1. (*i*) $4x(x^2 + 4)$ (*ii*) $\dfrac{10x^3}{\sqrt{5x^4 + 3}}$ (*iii*) $\dfrac{1}{4x^{3/4}}$

(*iv*) $(2x^2 - 2x + 3)(2x - 1)$ **2.** $6x^2(1 + x^3)$ **3.** (*i*) $72x(72x^2 + 3)$

(*ii*) $27(2187x^2 + 2700x + 832)$ **4.** $1, 8(4x + 5), 8(4x + 5), 8(4x + 5)$

5. (*i*) $8(4x^3 - 5x^2 + 1)^3 (6x^2 - 5x)$ (*ii*) $\dfrac{9x^2 - 1}{3(3x^3 - x + 7)^{2/3}}$

(*iii*) $\dfrac{15(3x - 1)^2}{(2x + 1)^4}$ (*iv*) $-\dfrac{2x}{\sqrt{1 - x^2}\,(1 + x^2)^{3/2}}$

6. (*i*) $(9x^2 + 8x + 3)(x^2 + 3)^3$ (*ii*) $(105x^2 + 54x + 10)(3x^2 + 2)^2$

(*iii*) $\dfrac{(2x^2 + 3)^{2/3}(34x^2 + 200x - 9)}{6(x + 5)^{3/2}}$ (*iv*) $\dfrac{2x^2 + x + 1}{\sqrt{1 + x^2}}$

7. (*i*) $\dfrac{a^2 - a\sqrt{a^2 - x^2}}{x^2\sqrt{a^2 - x^2}}$ (*ii*) $\dfrac{bx + 2c}{2(ax^2 + bx + c)^{3/2}}$ (*iii*) $\dfrac{a^2 + 2x^2}{(a^2 - x^2)^{5/2}}$

(*iv*) $-\dfrac{a}{x^2}\left[1 + \dfrac{a}{\sqrt{a^2 - x^2}}\right]$.

Hint

7. (*i*) Rationalise.

IV. DIFFERENTIATION OF TRIGONOMETRIC FUNCTIONS

12.13. DERIVATIVE OF SINE FUNCTION

Let $y = \sin x.$ ...(1)

Let δx be an increment of x and δy, the corresponding increment of y.

$\therefore \quad y + \delta y = \sin (x + \delta x)$ ...(2)

$(2) - (1) \Rightarrow \delta y = \sin (x + \delta x) - \sin x = 2 \cos (x + \tfrac{1}{2}\delta x) \sin \tfrac{1}{2}\delta x$

$$\therefore \quad \frac{\delta y}{\delta x} = \frac{2\cos(x + \frac{1}{2}\delta x)\sin\frac{1}{2}\delta x}{\delta x} = \cos\left(x + \frac{1}{2}\delta x\right).\frac{\sin\frac{1}{2}\delta x}{\frac{1}{2}\delta x}$$

$| \sin C - \sin D = 2 \cos \dfrac{C + D}{2} \sin \dfrac{C - D}{2}$

$$\therefore \quad \lim_{\delta x \to 0} \frac{\delta y}{\delta x} = \lim_{\delta x \to 0} \cos\left(x + \frac{1}{2}\delta x\right) . \lim_{\delta x \to 0} \frac{\sin \frac{1}{2}\delta x}{\frac{1}{2}\delta x} = \cos(x + 0) . 1 = \cos x$$

$$\therefore \quad \frac{dy}{dx} = \cos x \quad i.e., \quad \frac{\mathbf{d}}{\mathbf{dx}}(\mathbf{sin\ x}) = \mathbf{cos\ x.}$$

Thus, the derivative of sine of an angle with respect to the angle is the cosine of the angle.

12.14. DERIVATIVE OF COSINE FUNCTION

Let $y = \cos x.$ ...(1)

Let δx be an increment of x and δy, the corresponding increment of y.

$\therefore \quad y + \delta y = \cos(x + \delta x)$ ...(2)

(2) – (1) $\Rightarrow \quad \delta y = \cos(x + \delta x) - \cos x = -2 \sin(x + \frac{1}{2}\delta x) \sin \frac{1}{2}\delta x$

| $\cos C - \cos D = -2 \sin \frac{C + D}{2} \sin \frac{C - D}{2}$

$$\therefore \quad \frac{\delta y}{\delta x} = -\frac{2 \sin(x + \frac{1}{2}\delta x) \sin \frac{1}{2}\delta x}{\delta x} = -\sin(x + \frac{1}{2}\delta x)\frac{\sin \frac{1}{2}\delta x}{\frac{1}{2}\delta x}$$

$$\therefore \quad \lim_{\delta x \to 0} \frac{\delta y}{dx} = -\lim_{\delta x \to 0} \sin(x + \frac{1}{2}\delta x) \lim_{\delta x \to 0} \frac{\sin \frac{1}{2}\delta x}{\frac{1}{2}\delta x} = -\sin(x + 0) . 1 = -\sin x$$

$$\therefore \quad \frac{dy}{dx} = -\sin x \quad i.e., \quad \frac{\mathbf{d}}{\mathbf{dx}}(\mathbf{cos\ x}) = -\mathbf{sin\ x.}$$

Thus, the derivative of cosine of an angle with respect to the angle is the negative of sine of the angle.

12.15. DERIVATIVE OF TANGENT FUNCTION

Let $y = \tan x.$ ...(1)

Let δx be an increment of x and δy, the corresponding increment of y.

$\therefore \quad y + \delta y = \tan(x + \delta x)$ ...(2)

$$(2) - (1) \Rightarrow \delta y = \tan(x + \delta x) - \tan x = \frac{\sin(x + \delta x)}{\cos(x + \delta x)} - \frac{\sin x}{\cos x}$$

$$= \frac{\sin(x + \delta x)\cos x - \cos(x + \delta x)\sin x}{\cos(x + \delta x)\cos x} = \frac{\sin(x + \delta x - x)}{\cos(x + \delta x)\cos x} = \frac{\sin \delta x}{\cos(x + \delta x)\cos x}$$

| $\sin(A - B) = \sin A \cos B - \cos A \sin B$

$$\therefore \quad \frac{\delta y}{\delta x} = \frac{\sin \delta x}{\delta x . \cos(x + \delta x)\cos x} = \frac{1}{\cos(x + \delta x)\cos x} . \frac{\sin \delta x}{\delta x}.$$

$$\Rightarrow \quad \lim_{\delta x \to 0} \frac{\delta y}{\delta x} = \lim_{\delta x \to 0} \frac{1}{\cos(x + \delta x)\cos x} . \lim_{\delta x \to 0} \frac{\sin \delta x}{\delta x} = \frac{1}{\cos(x + 0)\cos x} . 1 = \sec^2 x$$

$\therefore \quad \dfrac{dy}{dx} = \sec^2 x$ *i.e.,* $\dfrac{\mathbf{d}}{\mathbf{dx}}(\tan \mathbf{x}) = \sec^2 \mathbf{x}.$

Thus, the derivative of tangent of an angle with respect to the angle is the square of the secant of the angle.

12.16. DERIVATIVE OF COTANGENT FUNCTION

Let $\quad y = \cot x.$ ...(1)

Let δx be an increment of x and δy, the corresponding increment of y.

$\therefore \quad y + \delta y = \cot (x + \delta x)$ ...(2)

$$(2) - (1) \Rightarrow \delta y = \cot (x + \delta x) - \cot x = \frac{\cos (x + \delta x)}{\sin (x + \delta x)} - \frac{\cos x}{\sin x}$$

$$= \frac{\sin x \cos (x + \delta x) - \cos x \sin (x + \delta x)}{\sin (x + \delta x) \sin x} = \frac{\sin [x - (x + \delta x)]}{\sin (x + \delta x) \sin x} = \frac{-\sin \delta x}{\sin (x + \delta x) \sin x}$$

$$\Rightarrow \quad \frac{\delta y}{\delta x} = \frac{-\sin \delta x}{\delta x \,.\, \sin (x + \delta x) \sin x} = \frac{-1}{\sin (x + \delta x) \sin x} \cdot \frac{\sin \delta x}{\delta x} \qquad | \sin (-\theta) = -\sin \theta$$

$$\lim_{\delta x \to 0} \frac{\delta y}{\delta x} = -\lim_{\delta x \to 0} \frac{1}{\sin (x + \delta x) \sin x} \cdot \lim_{\delta x \to 0} \frac{\sin \delta x}{\delta x} = -\frac{1}{\sin (x + 0) \sin x} \cdot 1 = -\operatorname{cosec}^2 x$$

$\therefore \quad \dfrac{dy}{dx} = -\operatorname{cosec}^2 x$ *i.e.,* $\dfrac{\mathbf{d}}{\mathbf{dx}}(\cot \mathbf{x}) = -\operatorname{\mathbf{cosec}}^2 \mathbf{x}.$

Thus, the derivative of cotangent of an angle with respect to the angle is the negative of the square of the cosecant of the angle.

12.17. DERIVATIVE OF SECANT FUNCTION

Let $\quad y = \sec x.$ ...(1)

Let δx be an increment of x and δy, the corresponding increment of y.

$\therefore \quad y + \delta y = \sec (x + \delta x)$ ...(2)

$$(2) - (1) \Rightarrow \delta y = \sec (x + \delta x) - \sec x = \frac{1}{\cos (x + \delta x)} - \frac{1}{\cos x}$$

$$= \frac{\cos x - \cos (x + \delta x)}{\cos (x + \delta x) \cos x} = \frac{2 \sin (x + \frac{1}{2}\delta x) \sin \frac{1}{2}\delta x}{\cos (x + \delta x) \cos x}$$

$$\left| \cos C - \cos D = 2 \sin \frac{C + D}{2} \sin \frac{D - C}{2} \right.$$

$$\therefore \quad \frac{\delta y}{\delta x} = \frac{2 \sin (x + \frac{1}{2}\delta x) \sin \frac{1}{2}\delta x}{\delta x \,.\, \cos (x + \delta x) \cos x} = \frac{\sin (x + \frac{1}{2}\delta x)}{\cos (x + \delta x) \cos x} \cdot \frac{\sin \frac{1}{2}\delta x}{\frac{1}{2}\delta x}$$

$$\Rightarrow \quad \lim_{\delta x \to 0} \frac{\delta y}{\delta x} = \lim_{\delta x \to 0} \frac{\sin (x + \frac{1}{2}\delta x)}{\cos (x + \delta x) \cos x} \cdot \lim_{\delta x \to 0} \frac{\sin \frac{1}{2}\delta x}{\frac{1}{2}\delta x} = \frac{\sin (x + 0)}{\cos (x + 0) \cos x} \cdot 1 = \sec x \tan x$$

$\therefore \quad \dfrac{dy}{dx} = \sec x \tan x$ *i.e.,* $\dfrac{\mathbf{d}}{\mathbf{dx}}(\sec \mathbf{x}) = \sec \mathbf{x} \tan \mathbf{x}.$

Thus, the derivative of secant of an angle with respect to the angle is the product of secant and tangent of the angle.

12.18. DERIVATIVE OF COSECANT FUNCTION

Let $y = \text{cosec } x.$ ...(1)

Let δx be an increment of x and δy, the corresponding increment of y.

$\therefore \quad y + \delta y = \text{cosec } (x + \delta x)$ ...(2)

$$(2) - (1) \Rightarrow \delta y = \text{cosec } (x + \delta x) - \text{cosec } x = \frac{1}{\sin (x + \delta x)} - \frac{1}{\sin x}$$

$$= \frac{\sin x - \sin (x + \delta x)}{\sin (x + \delta x) \sin x} = \frac{2 \cos (x + \frac{1}{2} \delta x) \sin (- \frac{1}{2} \delta x)}{\sin (x + \delta x) \sin x}$$

$$\left| \sin \text{C} - \sin \text{D} = 2 \cos \frac{\text{C} + \text{D}}{2} \sin \frac{\text{C} - \text{D}}{2} \right.$$

$$\therefore \quad \frac{\delta y}{\delta x} = \frac{-2 \cos (x + \frac{1}{2} \delta x) \sin \frac{1}{2} \delta x}{\delta x \,.\, \sin (x + \delta x) \sin x} = \frac{-\cos (x + \frac{1}{2} \delta x)}{\sin (x + \delta x) \sin x} \cdot \frac{\sin \frac{1}{2} \delta x}{\frac{1}{2} \delta x}$$

$$\Rightarrow \quad \lim_{\delta x \to 0} \frac{\delta y}{\delta x} = \lim_{\delta x \to 0} \frac{-\cos (x + \frac{1}{2} \delta x)}{\sin (x + \delta x) \sin x} \cdot \lim_{\delta x \to 0} \frac{\sin \frac{1}{2} \delta x}{\frac{1}{2} \delta x} = \frac{-\cos (x + 0)}{\sin (x + 0) \sin x} \,.\, 1 = -\text{cosec } x \cot x$$

$$\therefore \quad \frac{dy}{dx} = -\text{cosec } x \cot x \quad i.e., \quad \mathbf{\frac{d}{dx} (cosec\ x) = - cosec\ x\ cot\ x.}$$

Thus, the derivative of cosecant of an angle with respect to the angle is the negative of the product of cosecant and cotangent of the angle.

Summary. The following is the summary of the derivatives of trigonometric functions :

Let u be a differentiable function of x.

$\frac{d}{dx} (\sin x) = \cos x$	$\frac{d}{dx} (\sin u) = \cos u \frac{du}{dx}$
$\frac{d}{dx} (\cos x) = -\sin x$	$\frac{d}{dx} (\cos u) = -\sin u \frac{du}{dx}$
$\frac{d}{dx} (\tan x) = \sec^2 x$	$\frac{d}{dx} (\tan u) = \sec^2 u \frac{du}{dx}$
$\frac{d}{dx} (\cot x) = -\text{cosec}^2 x$	$\frac{d}{dx} (\cot u) = -\text{cosec}^2 u \frac{du}{dx}$
$\frac{d}{dx} (\sec x) = \sec x \tan x$	$\frac{d}{dx} (\sec u) = \sec u \tan u \frac{du}{dx}$
$\frac{d}{dx} (\text{cosec } x) = -\text{cosec } x \cot x$	$\frac{d}{dx} (\text{cosec } u) = -\text{cosec } u \cot u \frac{du}{dx}$

In problems involving derivatives of trigonometric functions, the following trigonometrical identities are frequently used :

(*i*) $\sin (A \pm B) = \sin A \cos B \pm \cos A \sin B$ (*ii*) $\cos (A \pm B) = \cos A \cos B \mp \sin A \sin B$

(*iii*) $\tan (A \pm B) = \dfrac{\tan A \pm \tan B}{1 \mp \tan A \tan B}$ (*iv*) $\sin 2A = 2 \sin A \cos A = \dfrac{2 \tan A}{1 + \tan^2 A}$

(*v*) $\cos 2A = \cos^2 A - \sin^2 A = 1 - 2 \sin^2 A = 2 \cos^2 A - 1 = \dfrac{1 - \tan^2 A}{1 + \tan^2 A}$

(*vi*) $\tan 2A = \dfrac{2 \tan A}{1 - \tan^2 A}$ (*vii*) $1 - \cos 2A = 2 \sin^2 A,\ 1 + \cos 2A = 2 \cos^2 A$

(*viii*) $\sin 3A = 3 \sin A - 4 \sin^3 A$ (*ix*) $\cos 3A = 4 \cos^3 A - 3 \cos A$

(*x*) $\tan 3A = \dfrac{3 \tan A - \tan^3 A}{1 - 3 \tan^2 A}$.

ILLUSTRATIVE EXAMPLES

Example 1. *Differentiate the following functions w.r.t. x :*

(*i*) $\sin x + 4 \cos x - 6 \operatorname{cosec} x$ (*ii*) $\tan x \sec x$ (*iii*) $\dfrac{1 - \cos x}{1 + \cos x}$

(*iv*) $\dfrac{\sin x}{1 - \cos x}$ (*v*) $\dfrac{1 + 2 \sec x}{2 + 5 \sec x}$ (*vi*) $\sqrt{\dfrac{1 - \cos 2x}{1 + \cos 2x}}$.

Sol. (*i*) Let $y = \sin x + 4 \cos x - 6 \operatorname{cosec} x$.

$$\therefore \quad \frac{dy}{dx} = \frac{d}{dx}(\sin x) + 4\frac{d}{dx}(\cos x) - 6\frac{d}{dx}(\operatorname{cosec} x)$$

$$= \cos x + 4(-\sin x) - 6(-\operatorname{cosec} x \cot x)$$

$$= \mathbf{\cos x - 4 \sin x + 6 \operatorname{cosec} x \cot x}.$$

(*ii*) Let $y = \tan x \sec x$. ('*uv*' Form)

$$\therefore \quad \frac{dy}{dx} = \tan x \frac{d}{dx}(\sec x) + \sec x \frac{d}{dx}(\tan x)$$

$$= \tan x \,.\, \sec x \tan x + \sec x \,.\, \sec^2 x = \sec x \,.\, \tan^2 + \sec^3 x$$

$$= \sec x (\sec^2 x - 1) + \sec^3 x = \mathbf{2 \sec^3 x - \sec x}.$$

(*iii*) Let $y = \dfrac{1 - \cos x}{1 + \cos x}$ $\left(\dfrac{'u'}{v} \text{ Form}\right)$

$$\therefore \quad \frac{dy}{dx} = \frac{(1 + \cos x)(0 + \sin x) - (1 - \cos x)(0 - \sin x)}{(1 + \cos x)^2}$$

$$= \frac{\sin x (1 + \cos x + 1 - \cos x)}{(1 + \cos x)^2} = \mathbf{\frac{2 \sin x}{(1 + \cos x)^2}}.$$

(*iv*) Let $y = \dfrac{\sin x}{1 - \cos x}$. $\left(\dfrac{'u'}{v} \text{ Form}\right)$

$$\therefore \quad \frac{dy}{dx} = \frac{(1 - \cos x)\dfrac{d}{dx}(\sin x) - \sin x \dfrac{d}{dx}(1 - \cos x)}{(1 - \cos x)^2}$$

$$= \frac{(1-\cos x)\cos x - \sin x\,(0+\sin x)}{(1-\cos x)^2} = \frac{\cos x - \cos^2 x - \sin^2 x}{(1-\cos x)^2} = \frac{\cos x - (1)}{(1-\cos x)^2}$$

$$= \frac{-(1-\cos x)}{(1-\cos x)^2} = \frac{-1}{1-\cos x} = \frac{-1}{2\sin^2 \frac{x}{2}} = -\frac{1}{2}\,\mathbf{cosec^2\,\frac{x}{2}}.$$

(*v*) Let $y = \dfrac{1+2\sec x}{2+5\sec x}$. $\left(\dfrac{'u'}{v}\text{ Form}\right)$

$$\therefore \quad \frac{dy}{dx} = \frac{(2+5\sec x)(0+2\sec x\tan x) - (1+2\sec x)(0+5\sec x\tan x)}{(2+5\sec x)^2}$$

$$= \frac{\sec x\tan x\,(4+10\sec x - 5 - 10\sec x)}{(2+5\sec x)^2} = -\frac{\mathbf{sec\,x\,tan\,x}}{\mathbf{(2+5\,sec\,x)^2}}.$$

(*vi*) Let $y = \sqrt{\dfrac{1-\cos 2x}{1+\cos 2x}}$. $\quad \therefore \quad y = \sqrt{\dfrac{2\sin^2 x}{2\cos^2 x}} = \tan x$

$$\therefore \quad \frac{dy}{dx} = \mathbf{sec^2\,x}.$$

Example 2. *Differentiate the following functions w.r.t. x :*

(*i*) $\sin x^n$ (*ii*) $\cos^n x$ (*iii*) $\tan^n x^n$

(*iv*) $\sqrt{1+\cot x}$ (*v*) $\sec^3 (x^2+1)$ (*vi*) $\sqrt{\mathrm{cosec}\,(x^3+1)}$.

Sol. (*i*) Let $y = \sin x^n$. ('sin *u*' Form)

By *chain rule,*

$$\frac{dy}{dx} = \cos x^n \frac{d}{dx}(x^n) = \cos x^n \,.\, nx^{n-1} = \mathbf{nx^{n-1}\cos x^n}.$$

(*ii*) Let $y = \cos^n x$.

$\therefore \quad y = (\cos x)^n$ ('u^n' Form)

By *chain rule,*

$$\frac{dy}{dx} = n(\cos x)^{n-1}\,.\,\frac{d}{dx}(\cos x) = n\cos^{n-1} x.\,(-\sin x) = \mathbf{-\,n\cos^{n-1} x\sin x}.$$

(*iii*) Let $y = \tan^n x^n$.

$\therefore \quad y = (\tan x^n)^n$ ('u^n' Form)

By *chain rule,*

$$\frac{dy}{dx} = n(\tan x^n)^{n-1}\,.\,\frac{d}{dx}(\tan x^n) \qquad \text{('tan } u\text{' Form)}$$

$$= n\tan^{n-1} x^n\,.\,\sec^2 x^n\,.\,\frac{d}{dx}(x^n)$$

$$= n\tan^{n-1} x^n \sec^2 x^n\,.\,nx^{n-1} = \mathbf{n^2 x^{n-1}\tan^{n-1} x^n \sec^2 x^n}.$$

(*iv*) Let $y = \sqrt{1+\cot x}$. ('$\sqrt{u}$' Form)

By *chain rule,*

$$\frac{dy}{dx} = \frac{1}{2}(1+\cot x)^{-1/2}\,.\,\frac{d}{dx}(1+\cot x)$$

$$= \frac{1}{2\sqrt{1+\cot x}}\,.\,-\mathrm{cosec}^2 x = -\frac{\mathbf{cosec^2\,x}}{\mathbf{2\sqrt{1+cot\,x}}}.$$

(*v*) Let $y = \sec^3 (x^2 + 1)$.

$\therefore \quad y = [\sec (x^2 + 1)]^3$ ('u^3' Form)

By *chain rule,*

$$\frac{dy}{dx} = 3(\sec (x^2 + 1))^2 \frac{d}{dx} (\sec (x^2 + 1)) \quad \text{('sec } u\text{' Form)}$$

$$= 3 \sec^2 (x^2 + 1) . \sec (x^2 + 1) \tan (x^2 + 1) \frac{d}{dx} (x^2 + 1)$$

$$= 3 \sec^3 (x^2 + 1) \tan (x^2 + 1) . 2x = \mathbf{(x^2 + 1) \tan (x^2 + 1).}$$

(*vi*) Let $y = \sqrt{\text{cosec } (x^3 + 1)}$. ('$\sqrt{u}$' Form)

By *chain rule,*

$$\frac{dy}{dx} = \frac{1}{2}(\text{cosec } (x^3 + 1))^{-1/2} . \frac{d}{dx} (\text{cosec } (x^3 + 1)) \quad \text{('cosec } u\text{' Form)}$$

$$= \frac{1}{2\sqrt{\text{cosec } (x^3 + 1)}} . - \text{cosec } (x^3 + 1) \cot (x^3 + 1) . \frac{d}{dx} (x^3 + 1)$$

$$= -\frac{1}{2}\sqrt{\text{cosec } (x^3 + 1)} \cot (x^3 + 1) . 3x^2 = -\frac{\mathbf{3}}{\mathbf{2}} \mathbf{x^2 \sqrt{\textbf{cosec } (x^3 + 1)} \textbf{ cot } (x^3 + 1).}$$

Example 3. *Using the fact : sin (A + B) = sin A cos B + cos A sin B and the technique of differentiation, obtain the sum formula for cosines i.e., prove : cos (A + B) = cos A cos B – sin A sin B.*

Sol. We have sin (A + B) = sin A cos B + cos A sin B.

Considering A as a constant and B as a variable, we have

$$\frac{d}{dB} (\sin (A + B)) = \frac{d}{dB} (\sin A \cos B + \cos A \sin B)$$

$$\Rightarrow \quad \cos (A + B).(0 + 1) = \sin A.(- \sin B) + \cos A . \cos B$$

$$\Rightarrow \quad \mathbf{\cos (A + B) = \cos A \cos B - \sin A \sin B.}$$

Example 4. *We know that* $\frac{\sec x}{\text{cosec } x} = \tan x$ *and* $\frac{d}{dx} (\tan x) = \sec^2 x$.

Differentiate $\frac{\sec x}{\text{cosec } x}$. *by quotient rule and verify that the answer is the same.*

Sol. Let $y = \frac{\sec x}{\text{cosec } x}$. $\left(\frac{\text{'}u\text{'}}{v} \text{ Form}\right)$

$$\therefore \quad \frac{dy}{dx} = \frac{\text{cosec } x \frac{d}{dx} (\sec x) - \sec x \frac{d}{dx} (\text{cosec } x)}{(\text{cosec } x)^2}$$

$$= \frac{\text{cosec } x . \sec x \tan x - \sec x (- \text{cosec } x \cot x)}{\text{cosec}^2 x} = \frac{\sec x \text{ cosec } x (\tan x + \cot x)}{\text{cosec}^2 x}$$

$$= \frac{\sin x}{\cos x}\left[\frac{\sin x}{\cos x} + \frac{\cos x}{\sin x}\right] = \frac{\sin x}{\cos x}\left[\frac{\sin^2 x + \cos^2 x}{\cos x \sin x}\right] = \frac{1}{\cos^2 x} = \mathbf{\sec^2 x.}$$

Example 5. *If* $y = \sqrt{\dfrac{1-\sin 2x}{1+\sin 2x}}$, *show that* $\dfrac{dy}{dx} + \sec^2\left(\dfrac{\pi}{4} - x\right) = 0.$

(G.J.U. B.Pharma, 2005)

Sol. We have $y = \sqrt{\dfrac{1-\sin 2x}{1+\sin 2x}}$.

$$\therefore \quad y = \sqrt{\frac{1-\cos\left(\frac{\pi}{2} - 2x\right)}{1+\cos\left(\frac{\pi}{2} - 2x\right)}} = \sqrt{\frac{2\sin^2\left(\frac{\pi}{4} - x\right)}{2\cos^2\left(\frac{\pi}{4} - x\right)}} = \tan\left(\frac{\pi}{4} - x\right)$$

| $\sin(90 - \theta) = \cos\theta$ and $\cos(90 - \theta) = \sin\theta$

$$\therefore \quad \frac{dy}{dx} = \sec^2\left(\frac{\pi}{4} - x\right) \cdot \frac{d}{dx}\left(\frac{\pi}{4} - x\right) = \sec^2\left(\frac{\pi}{4} - x\right) \cdot (-1) \qquad \text{('}\tan u\text{' Form)}$$

$$\therefore \quad \mathbf{\frac{dy}{dx} + \sec^2\left(\frac{\pi}{4} - x\right) = 0.}$$

Example 6. *Differentiate the following functions w.r.t. x :*

(*i*) $(x^2 + \cos x)^4$ (*ii*) $\cos(\sin x^2)$ (*iii*) $\cot^3 x - \tan\sqrt{x}$

(*iv*) $x^2 \sin\dfrac{1}{x}$ (*v*) $\cos\dfrac{x}{x-1}$ (*vi*) $\sqrt{\sin\sqrt{1+x^2}}$.

Sol. (*i*) Let $y = (x^2 + \cos x)^4$. ('u^n' Form)

$$\therefore \quad \frac{dy}{dx} = 4(x^2 + \cos x)^3 \frac{d}{dx}(x^2 + \cos x) = \mathbf{4(x^2 + \cos x)^3 (2x - \sin x).}$$

(*ii*) Let $y = \cos(\sin x^2)$ ('$\cos u$' Form)

$$\therefore \quad \frac{dy}{dx} = -\sin(\sin x^2) \frac{d}{dx}(\sin x^2) = -\sin(\sin x^2) \cdot \cos x^2 \cdot \frac{d}{dx}(x^2) \quad \text{('}\sin u\text{' Form)}$$

$$= -\sin(\sin x^2) \cdot \cos x^2 \cdot 2x = \mathbf{-2x \sin(\sin x^2) \cos x^2.}$$

(*iii*) Let $y = \cot^3 x - \tan\sqrt{x}$. $\therefore \quad y = (\cot x)^3 - \tan\sqrt{x}$.

$$\therefore \quad \frac{dy}{dx} = 3(\cot x)^2 \frac{d}{dx}(\cot x) - \sec^2\sqrt{x}\frac{d}{dx}(\sqrt{x}) = 3\cot^2 x \cdot (-\text{cosec}^2 x) - \sec^2\sqrt{x} \cdot \frac{1}{2}x^{-1/2}$$

$$= \mathbf{-3\cot^2 x \,\text{cosec}^2 x - \frac{1}{2\sqrt{x}}\sec^2\sqrt{x}.}$$

(*iv*) Let $y = x^2 \sin\dfrac{1}{x}$. ('uv' Form)

$$\therefore \quad \frac{dy}{dx} = x^2 \frac{d}{dx}\left(\sin\frac{1}{x}\right) + \sin\frac{1}{x}\frac{d}{dx}(x^2) = x^2 \cdot \cos\frac{1}{x}\frac{d}{dx}(x^{-1}) + \sin\frac{1}{x} \cdot 2x$$

$$= x^2 \cos\frac{1}{x} \cdot (-1 \cdot x^{-2}) + 2x\sin\frac{1}{x} = \mathbf{2x\sin\frac{1}{x} - \cos\frac{1}{x}.}$$

(*v*) Let $y = \cos\dfrac{x}{x-1}$. ('$\cos u$' Form)

$$\therefore \quad \frac{dy}{dx} = -\sin\frac{x}{x-1}\frac{d}{dx}\left(\frac{x}{x-1}\right) = -\sin\frac{x}{x-1} \cdot \frac{(x-1)\frac{d}{dx}(x) - x\frac{d}{dx}(x-1)}{(x-1)^2}$$

$$= -\sin\frac{x}{x-1}\cdot\frac{(x-1).1-x.(1-0)}{(x-1)^2} = -\sin\frac{x}{x-1}\cdot\frac{-1}{(x-1)^2} = \frac{\mathbf{1}}{\mathbf{(x-1)^2}}\mathbf{\sin\frac{x}{x-1}}.$$

(*vi*) Let $y=\sqrt{\sin\sqrt{1+x^2}}$ $\therefore$ $y=(\sin\sqrt{1+x^2})^{1/2}$ ('u^n' Form)

$$\therefore \quad \frac{dy}{dx}=\frac{1}{2}(\sin\sqrt{1+x^2})^{-1/2}\frac{d}{dx}(\sin\sqrt{1+x^2}) \quad \text{('sin } u\text{' Form)}$$

$$=\frac{1}{2\sqrt{\sin\sqrt{1+x^2}}}\cos\sqrt{1+x^2}\,\frac{d}{dx}(\sqrt{1+x^2})^{1/2} \quad \text{('}\sqrt{u}\text{' Form)}$$

$$=\frac{\cos\sqrt{1+x^2}}{2\sqrt{\sin\sqrt{1+x^2}}}\,\frac{1}{2}(1+x^2)^{-1/2}\frac{d}{dx}(1+x^2)=\frac{\cos\sqrt{1+x^2}}{4\sqrt{1+x^2}\sqrt{\sin\sqrt{1+x^2}}}.(0+2x)$$

$$=\frac{\mathbf{x\cos\sqrt{1+x^2}}}{\mathbf{2\sqrt{1+x^2}\sqrt{\sin\sqrt{1+x^2}}}}.$$

Example 7. *Find the derivatives of the following functions w.r.t. x at the indicated points :*

(*i*) $x+\sin x\cos x$ *at* $x=0$ (*ii*) $\dfrac{1-\sin x}{1+\cos x}$ *at* $x=\dfrac{\pi}{2}$.

Sol. (*i*) Let $y=x+\sin x\cos x$.

$$\therefore \quad \frac{dy}{dx}=1+[\sin x.(-\sin x)+\cos x.\cos x]=1-\sin^2 x+\cos^2 x$$

$$=\cos^2 x+\cos^2 x=2\cos^2 x.$$

$$\therefore \quad \left.\frac{dy}{dx}\right|_{x=0}=2\cos^2 0=2(1)^2=\mathbf{2.}$$

(*ii*) Let $y=\dfrac{1-\sin x}{1+\cos x}$

$$\therefore \quad \frac{dy}{dx}=\frac{(1+\cos x).(-\cos x)-(1-\sin x)(-\sin x)}{(1+\cos x)^2}$$

$$=\frac{-\cos x-\cos^2 x+\sin x-\sin^2 x}{(1+\cos x)^2}$$

$$=\frac{\sin x-\cos x-(\cos^2 x+\sin^2 x)}{(1+\cos x)^2}=\frac{\sin x-\cos x-1}{(1+\cos x)^2}.$$

$$\therefore \quad \left.\frac{dy}{dx}\right|_{x=\pi/2}=\frac{\sin\frac{\pi}{2}-\cos\frac{\pi}{2}-1}{\left(1+\cos\frac{\pi}{2}\right)^2}=\frac{1-0-1}{(1+0)^2}=\frac{0}{1}=\mathbf{0.}$$

Example 8. *Given* $y=\dfrac{5x}{(1-x)^{2/3}}+\cos^2(2x+1)$, *show that*

$$\frac{dy}{dx}=\frac{5}{3}(1-x)^{-5/3}(3-x)-2\sin(4x+2). \quad \text{(P.T.U. B. Pharma, May 2004)}$$

Sol. Given $y=\dfrac{5x}{(1-x)^{2/3}}+\cos^2(2x+1)$ ...(1)

Differentiating (1) w.r.t. x, we get

$$\frac{dy}{dx} = \frac{(1-x)^{2/3}\frac{d}{dx}(5x) - 5x \cdot \frac{d}{dx}(1-x)^{2/3}}{[(1-x)^{2/3}]^2} + 2\cos(2x+1) \cdot \frac{d}{dx}\cos(2x+1)$$

$$= \frac{(1-x)^{2/3} \cdot 5 - 5x \cdot \frac{2}{3}(1-x)^{\frac{2}{3}-1}(-1)}{(1-x)^{4/3}} - 2\cos(2x+1)\sin(2x+1) \cdot \frac{d}{dx}(2x+1)$$

$$= \frac{5(1-x)^{\frac{2}{3}} + \frac{10x}{3(1-x)^{1/3}}}{(1-x)^{4/3}} - 2 \cdot 2\cos(2x+1)\sin(2x+1)$$

$$= \frac{15(1-x) + 10x}{3(1-x)^{1/3}(1-x)^{4/3}} - 2\sin 2(2x+1) \qquad | \sin 2A = 2\sin A \cos A$$

$$= \frac{15 - 5x}{3(1-x)^{5/3}} - 2\sin(4x+2)$$

$$= \frac{5}{3}(3-x)(1-x)^{-5/3} - 2\sin(4x+2).$$

EXERCISE 12.7

Differentiate the following functions w.r.t. x (Q. No. 1—3) :

1. (*i*) $\sin 4x$ (*ii*) $\cos 5x$ (*iii*) $\tan 2x$
   (*iv*) $\cot 3x$ (*v*) $\sec 10x$ (*vi*) $\text{cosec } 2x$
2. (*i*) $\sin^3 x$ (*ii*) $\cos^2 x$ (*iii*) $\tan^4 x$
   (*iv*) $\cot^4 x$ (*v*) $\sec^5 x$ (*vi*) $\text{cosec}^3 x$
3. (*i*) $x^3 + \sin x$ (*ii*) $3\sin x + 2\sin\alpha$ (*iii*) $\text{cosec } x \cot x$
   (*iv*) $x^5 \cot x$ (*v*) $(x^2 - 5x + 6)\sec x$ (*vi*) $(5 - 4\cos x)(1 - 2\tan x)$.
4. If $y = a\sin x + b\cos x$, show that $y^2 + \left(\frac{dy}{dx}\right)^2 = a^2 + b^2$.
5. (*i*) We know that $\sin 2x = 2\sin x\cos x$. Differentiate $\sin 2x$ by chain rule, differentiate $2\sin x\cos x$ by product rule, and verify that the answers are same.

   (*ii*) We know that $\cot x = \frac{\cos x}{\sin x}$ and that $\frac{d}{dx}(\cot x) = -\text{cosec}^2 x$. Differentiate $\frac{\cos x}{\sin x}$ by quotient rule and verify that the answer is same.

Differentiate the following functions w.r.t. x (Q. No. 6—11) :

6. (*i*) $\frac{\sec x - 1}{\sec x + 1}$ (*ii*) $\frac{\sin x + \cos x}{\sin x - \cos x}$ (*iii*) $\frac{x + \sin x}{x + \cos x}$

(*iv*) $\frac{x}{1 + \tan x}$ (*v*) $\frac{\sec x + \tan x}{\sec x - \tan x}$ (*vi*) $\frac{\tan x - \cot x}{\tan x + \cot x}$

7. (*i*) $\sin 2x \cos 3x$ (*ii*) $\sin 2x + (2x - 5)^2$ (*iii*) $\frac{(\sin x^2)^2}{\sqrt{1 + x^2}}$

(*iv*) $\sin^m x \cos^n x$ (*v*) $\sin x \sin 2x$ (*vi*) $\frac{\sin x + x^2}{\cot 2x}$

8. (*i*) $\cos^2 x$ at $x = \frac{\pi}{2}$ (*ii*) $\frac{1 + \sin x}{\cos x}$ at $x = \frac{\pi}{4}$

(*iii*) $\cos\left(2x + \frac{\pi}{2}\right)$ at $x = \frac{\pi}{3}$ (*iv*) $\cot x + \sec^2 x + 5$ at $x = \frac{\pi}{6}$

9. (*i*) $\cos^2 x^3$ (*ii*) $\sec^2 x^2$ (*iii*) $\text{cosec}^3 \frac{1}{x^2}$ (*iv*) $\cot^3 x^2$

10. (*i*) $\sqrt{\frac{1 + \cos x}{1 - \cos x}}$ (*ii*) $\sqrt{\frac{\sec x - 1}{\sec x + 1}}$ (*iii*) $\sqrt{\frac{1 - \sin x}{1 + \sin x}}$

(*iv*) $\sqrt{\frac{1 + \tan x}{1 - \tan x}}$ (*v*) $\sqrt{\frac{\cos x - \sin x}{\cos x + \sin x}}$ (*vi*) $\sqrt{\frac{\sec x + \tan x}{\sec x - \tan x}}$

11. (*i*) $x \cot^3 x$ (*ii*) $\sin \frac{1 + x^2}{1 - x^2}$ (*iii*) $\sin \sqrt{\sin x + \cos x}$

(*iv*) $\sin (\sin (\sin x))$ (*v*) $\frac{\sin^2 x}{1 + \cos^2 x}$ (*vi*) $\frac{\sin \sqrt{x}}{\sqrt{x}}$.

12. (*i*) If $\sin x = y \sin (x + a)$, show that $\frac{dy}{dx} = \frac{\sin a}{\sin^2 (x + a)}$.

(*ii*) If $\cos y = x \cos (a + y)$, with $\cos a \neq \pm 1$, show that $\frac{dy}{dx} = \frac{\cos^2 (a + y)}{\sin a}$.

Answers

1. (*i*) $4 \cos 4x$ (*ii*) $-5 \sin 5x$ (*iii*) $2 \sec^2 2x$ (*iv*) $-3\,\text{cosec}^2 3x$

(*v*) $10 \sec 10x \tan 10x$ (*vi*) $-2\,\text{cosec}\, 2x \cot 2x$ **2.** (*i*) $3 \sin^2 x \cos x$ (*ii*) $-2 \cos x \sin x$

(*iii*) $4 \tan^3 x \sec^2 x$ (*iv*) $-4 \cot^3 x\, \text{cosec}^2 x$ (*v*) $5 \sec^5 x \tan x$ (*vi*) $-3\,\text{cosec}^3 x \cot x$

3. (*i*) $3x^2 + \cos x$ (*ii*) $3 \cos x$

(*iii*) $-\text{cosec}\, x (\cot^2 x + \text{cosec}^2 x)$ (*iv*) $5x^4 \cot x - x^5 \text{cosec}^2 x$

(*v*) $\sec x (\tan x(x^2 - 5x + 6) + 2x - 5)$ (*vi*) $4 \sin x (1 - 2 \tan x) - 2 \sec^2 x (5 - 4 \cos x)$.

6. (*i*) $\tan \frac{x}{2} \sec^2 \frac{x}{2}$ (*ii*) $-\frac{2}{(\sin x - \cos x)^2}$ or $-\sec^2\left(\frac{\pi}{4} + x\right)$

(*iii*) $\frac{x(\sin x + \cos x) + \cos x - \sin x + 1}{(x + \cos x)^2}$ (*iv*) $\frac{1 + \tan x - x \sec^2 x}{(1 + \tan x)^2}$

(*v*) $2 \sec x (\sec x + \tan x)^2$ (*vi*) $2 \sin 2x$

7. (*i*) $2 \cos 2x \cos 3x - 3 \sin 2x \sin 3x$ (*ii*) $2 \cos 2x + 8x - 20$

(*iii*) $\dfrac{2x(1+x^2) \sin 2x^2 - x \sin^2 x^2}{(1+x^2)^{3/2}}$ (*iv*) $\sin^{m-1} x \cos^{n-1} x\, [m \cos^2 x - n \sin^2 x]$

(*v*) $2 \sin x \cos 2x + \sin 2x \cos x$ (*vi*) $\dfrac{(\cot 2x)(\cos x + 2x) + 2\, (\sin x + x^2)\, \text{cosec}^2\, 2x}{\cot^2 2x}$

8. (*i*) 0 (*ii*) $\sqrt{2} + 2$ (*iii*) 1 (*iv*) $\dfrac{8\sqrt{3} - 36}{9}$

9. (*i*) $-6x^2 \cos x^3 \sin x^3$ (*ii*) $4x \sec^2 x^2 \tan x^2$ (*iii*) $\dfrac{6}{x^3} \text{cosec}^3 \dfrac{1}{x^2} \cot \dfrac{1}{x^2}$

(*iv*) $-6x \cot^2 x^2 \text{cosec}^2 x^2$

10. (*i*) $-\text{cosec}\, x\, (\text{cosec}\, x + \cot x)$ (*ii*) $\dfrac{1}{2} \sec^2 \dfrac{x}{2}$

(*iii*) $\sec x\, (\tan x - \sec x)$ or $-\dfrac{1}{2} \sec^2 \left(\dfrac{\pi}{4} - \dfrac{x}{2}\right)$

(*iv*) $\dfrac{\sec^2\left(\dfrac{\pi}{4} + x\right)}{2\sqrt{\tan\left(\dfrac{\pi}{4} + x\right)}}$ (*v*) $\dfrac{-\sec^2\left(\dfrac{\pi}{4} - x\right)}{2\sqrt{\tan\left(\dfrac{\pi}{4} - x\right)}}$ (*vi*) $\sec x\, (\sec x + \tan x)$

11. (*i*) $\cot^2 x\, (-3x\, \text{cosec}^2 x + \cot x)$ (*ii*) $\dfrac{4x}{(1-x^2)^2} \cos \dfrac{1+x^2}{1-x^2}$

(*iii*) $\dfrac{\cos \sqrt{\sin x + \cos x}\, (\cos x - \sin x)}{2\sqrt{\sin x + \cos x}}$ (*iv*) $\cos (\sin (\sin x)).\cos (\sin x).\cos x$

(*v*) $\dfrac{4 \sin x \cos x}{(1 + \cos^2 x)^2}$ (*vi*) $\dfrac{\sqrt{x} \cos \sqrt{x} - \sin \sqrt{x}}{2x^{3/2}}$

Hint

12. (*i*) We have $x = \dfrac{\cos y}{\cos (a + y)}$. $\therefore$ $\dfrac{dx}{dy} = \dfrac{-\cos (a+y) \sin y + \cos y \sin (a+y)}{\cos^2 (a+y)} = \dfrac{\sin a}{\cos^2 (a+y)}$

$\therefore$ $\dfrac{dy}{dx} = 1 \Big/ \dfrac{dx}{dy} = \dfrac{\cos^2 (a+y)}{\sin a}$.

V. DIFFERENTIATION OF INVERSE TRIGONOMETRIC FUNCTIONS

12.19. DERIVATIVE OF INVERSE SINE FUNCTION

Let $y = \sin^{-1} x$. ...(1)

We know that $x \in [-1, 1]$ and $y \in [-\pi/2, \pi/2]$.

Let δx be an increment in x and δy, the corresponding increment of y.

$\therefore$ $y + \delta y = \sin^{-1} (x + \delta x)$ ...(2)

(1) $\Rightarrow$ $x = \sin y$ ...(3)

and (2) $\Rightarrow$ $x + \delta x = \sin (y + \delta y)$ ...(4)

(4) – (3) $\Rightarrow \delta x = \sin (y + \delta y) - \sin y = 2 \cos \left(y + \frac{1}{2}\delta y\right) \sin \frac{1}{2}\delta y$

$$\therefore \quad \frac{\delta y}{\delta x} = \frac{\delta y}{2 \cos \left(y + \frac{1}{2}\delta y\right) \sin \frac{1}{2}\delta y} = \frac{1}{\cos \left(y + \frac{1}{2}\delta y\right)} \cdot \frac{\frac{1}{2}\delta y}{\sin \frac{1}{2}\delta y} \qquad (\because \delta x \neq 0)$$

$$\therefore \quad \lim_{\delta x \to 0} \frac{\delta y}{\delta x} = \lim_{\delta y \to 0} \frac{1}{\cos \left(y + \frac{1}{2}\delta y\right)} \cdot \lim_{\delta y \to 0} \frac{\frac{1}{2}\delta y}{\sin \frac{1}{2}\delta y} \qquad (\delta x \to 0 \Rightarrow \delta y \to 0)$$

$$= \frac{1}{\cos (y + 0)} . 1 = \frac{1}{\cos y}, \text{ provided } y \in (-\pi/2, \pi/2).$$

$$\therefore \quad \frac{dy}{dx} = \frac{1}{\pm\sqrt{1 - \sin^2 y}} = \frac{1}{\sqrt{1 - x^2}}, \text{ because } \cos y \text{ is +ve on } (-\pi/2, \pi/2).$$

$$\therefore \quad \mathbf{\frac{d}{dx} (\sin^{-1} x) = \frac{1}{\sqrt{1 - x^2}}, \ x \in (-1, 1).}$$

12.20. DERIVATIVE OF INVERSE COSINE FUNCTION

Let $y = \cos^{-1} x$. ...(1)

We know that $x \in [-1, 1]$ and $y \in [0, \pi]$.

Let δx be an increment of x and δy, the corresponding increment of y.

$\therefore \quad y + \delta y = \cos^{-1} (x + \delta x)$ ...(2)

(1) $\Rightarrow \quad x = \cos y$ ...(3)

and (2) $\Rightarrow \quad x + \delta x = \cos (y + \delta y)$ ...(4)

(4) – (3) $\Rightarrow \quad \delta x = \cos (y + \delta y) - \cos y = 2 \sin \left(y + \frac{1}{2}\delta y\right) \sin \left(-\frac{1}{2}\delta y\right)$

$$\therefore \quad \frac{\delta y}{\delta x} = \frac{\delta y}{-2 \sin \left(y + \frac{1}{2}\delta y\right) \sin \frac{1}{2}\delta y} = \frac{-1}{\sin \left(y + \frac{1}{2}\delta y\right)} \cdot \frac{\frac{1}{2}\delta y}{\sin \frac{1}{2}\delta y} \qquad (\because \delta x \neq 0)$$

$$\therefore \quad \lim_{\delta x \to 0} \frac{\delta y}{\delta x} = \lim_{\delta y \to 0} \frac{-1}{\sin \left(y + \frac{1}{2}\delta y\right)} \lim_{\delta y \to 0} \frac{\frac{1}{2}\delta y}{\sin \frac{1}{2}\delta y} \qquad (\because \delta x \to 0 \Rightarrow \delta y \to 0)$$

$$= \frac{-1}{\sin (y + 0)} . 1 = \frac{-1}{\sin y}, \text{ provided } y \in (0, \pi).$$

$$\therefore \quad \frac{dy}{dx} = \frac{-1}{\pm\sqrt{1 - \cos^2 y}} = \frac{-1}{\sqrt{1 - x^2}}, \text{ because } \sin y \text{ is + ve on } (0, \pi).$$

$$\therefore \quad \mathbf{\frac{d}{dx} (\cos^{-1} x) = \frac{-1}{\sqrt{1 - x^2}}, \ x \in (-1, 1).}$$

12.21. DERIVATIVE OF INVERSE TANGENT FUNCTION

Let $y = \tan^{-1} x$. ...(1)

We know that $x \in \mathbf{R}$ and $y \in (-\pi/2, \pi/2)$.

Let δx be an increment of x and δy, the corresponding increment of y.

$\therefore \quad y + \delta y = \tan^{-1}(x + \delta x)$ ...(2)

(1) $\Rightarrow \quad x = \tan y$ ...(3)

and (2) $\Rightarrow \quad x + \delta x = \tan(y + \delta y)$ ...(4)

$$(4) - (3) \Rightarrow \quad \delta x = \tan(y+\delta y) - \tan y = \frac{\sin(y+\delta y)}{\cos(y+\delta y)} - \frac{\sin y}{\cos y}$$

$$= \frac{\sin(y+\delta y)\cos y - \cos(y+\delta y)\sin y}{\cos(y+\delta y)\cos y}$$

$$= \frac{\sin(y+\delta y - y)}{\cos(y+\delta y)\cos y} = \frac{\sin \delta y}{\cos(y+\delta y)\cos y}.$$

$$\therefore \quad \frac{\delta y}{\delta x} = \frac{\delta y \,.\, \cos(y+\delta y)\cos y}{\sin \delta y} = \cos(y+\delta y)\cos y \,.\, \frac{\delta y}{\sin \delta y}$$

$$\Rightarrow \quad \lim_{\delta x \to 0} \frac{\delta y}{\delta x} = \lim_{\delta y \to 0} [\cos(y+\delta y)\cos y] \lim_{\delta y \to 0} \frac{\delta y}{\sin \delta y} \qquad (\because \delta x \to 0 \Rightarrow \delta y \to 0)$$

$$= \cos(y+0)\cos y \,.\, 1 = \cos^2 y = \frac{1}{\sec^2 y} = \frac{1}{1+\tan^2 y} = \frac{1}{1+x^2}.$$

$$\therefore \quad \frac{dy}{dx} = \frac{1}{1+x^2} \quad i.e., \quad \mathbf{\frac{d}{dx}(\tan^{-1} x) = \frac{1}{1+x^2}, \ x \in R.}$$

12.22. DERIVATIVE OF INVERSE COTANGENT FUNCTION

Let $\quad y = \cot^{-1} x.$ ...(1)

We know that $x \in \mathbf{R}$ and $y \in (0, \pi)$.

Let δx be an increment of x and δy, the corresponding increment of y.

$\therefore \quad y + \delta y = \cot^{-1}(x + \delta y)$ ...(2)

(1) $\Rightarrow \quad x = \cot y$ ...(3)

and (2) $\Rightarrow x + \delta x = \cot(y + \delta y)$ ...(4)

$$(4) - (3) \Rightarrow \quad \delta x = \cot(y+\delta y) - \cot y = \frac{\cos(y+\delta y)}{\sin(y+\delta y)} - \frac{\cos y}{\sin y}$$

$$= \frac{\sin y \cos(y+\delta y) - \cos y \sin(y+\delta y)}{\sin(y+\delta y)\sin y}$$

$$= \frac{\sin(y-(y+\delta y))}{\sin(y+\delta y)\sin y} = \frac{-\sin \delta y}{\sin(y+\delta y)\sin y}$$

$$\therefore \quad \frac{\delta y}{\delta x} = -\frac{\delta y \,.\, \sin(y+\delta y)\sin y}{\sin \delta y} = -\sin(y+\delta y)\sin y \,.\, \frac{\delta y}{\sin \delta y}.$$

$$\Rightarrow \quad \lim_{\delta x \to 0} \frac{\delta y}{\delta x} = -\lim_{\delta y \to 0} [\sin(y+\delta y)\sin y] \lim_{\delta y \to 0} \frac{\delta y}{\sin \delta y} \qquad (\because \delta x \to 0 \Rightarrow \delta y \to 0)$$

$$= -\sin(y+0)\sin y \,.\, 1 = -\sin^2 y = \frac{-1}{\operatorname{cosec}^2 y} = \frac{-1}{1+\cot^2 y} = \frac{-1}{1+x^2}.$$

$$\therefore \quad \frac{dy}{dx} = \frac{-1}{1+x^2} \quad i.e., \quad \mathbf{\frac{d}{dx}(\cot^{-1} x) = \frac{-1}{1+x^2}, \ x \in R.}$$

12.23. DERIVATIVE OF INVERSE SECANT FUNCTION

Let $y = \sec^{-1} x.$ ...(1)

We know that $x \in \mathbf{R} - (-1, 1)$ and $y \in [0, \pi] - \left\{\frac{1}{2}\pi\right\}$.

Let δx be an increment of x and δy, the corresponding increment of y.

$\therefore \quad y + \delta y = \sec^{-1}(x + \delta x)$ ...(2)

(1) $\Rightarrow \quad x = \sec y$ ...(3)

and (2) $\Rightarrow \quad x + \delta x = \sec(y + \delta y).$ ...(4)

(4) – (3) $\Rightarrow \quad \delta x = \sec(y + \delta y) - \sec y = \dfrac{1}{\cos(y + \delta y)} - \dfrac{1}{\cos y}$

$$= \frac{\cos y - \cos(y + \delta y)}{\cos(y + \delta y)\cos y} = \frac{2\sin(y + \frac{1}{2}\delta y)\sin\frac{1}{2}\delta y}{\cos(y + \delta y)\cos y}$$

$$\therefore \quad \frac{\delta y}{\delta x} = \frac{\delta y \,.\, \cos(y + \delta y)\cos y}{2\sin(y + \frac{1}{2}\delta y)\sin\frac{1}{2}\delta y} = \frac{\cos(y + \delta y)\cos y}{\sin(y + \frac{1}{2}\delta y)} \,.\, \frac{\frac{1}{2}\delta y}{\sin\frac{1}{2}\delta y}$$

$$\Rightarrow \quad \lim_{\delta x \to 0}\frac{\delta y}{\delta x} = \lim_{\delta y \to 0}\frac{\cos(y + \delta y)\cos y}{\sin(y + \frac{1}{2}\delta y)} \,.\, \lim_{\delta y \to 0}\frac{\frac{1}{2}\delta y}{\sin\frac{1}{2}\delta y} \qquad (\delta x \to 0 \Rightarrow \delta y \to 0)$$

$$= \frac{\cos(y + 0)\cos y}{\sin(y + 0)} \,.\, 1 = \frac{1}{\sec y \tan y}, \quad \text{provided } y \in (0, \pi) - \left\{\tfrac{1}{2}\pi\right\}.$$

$$\therefore \quad \frac{dy}{dx} = \frac{1}{\sec y \,.\, \pm\sqrt{\sec^2 y - 1}} = \frac{1}{\pm x\sqrt{x^2 - 1}} = \frac{1}{|x|\sqrt{x^2 - 1}},$$ because $\sec y \tan y$ is +ve on $(0, \pi) - \left\{\frac{1}{2}\pi\right\}$.

$$\therefore \quad \mathbf{\frac{d}{dx}(\sec^{-1} x) = \frac{1}{|x|\sqrt{x^2 - 1}}, \quad x \in R - [-1, 1].}$$

12.24. DERIVATIVE OF INVERSE COSECANT FUNCTION

Let $y = \text{cosec}^{-1} x.$ ...(1)

We know that $x \in \mathbf{R} - (-1, 1)$ and $y \in \left[-\frac{1}{2}\pi, \frac{1}{2}\pi\right] - \{0\}$.

Let δx be an increment of x and δy, the corresponding increment of y.

$\therefore \quad y + \delta y = \text{cosec}^{-1}(x + \delta x)$ ...(2)

(1) $\Rightarrow \quad x = \text{cosec}\, y$ ...(3)

and (2) $\Rightarrow \quad x + \delta x = \text{cosec}(y + \delta y)$ ...(4)

(4) – (3) $\Rightarrow \quad \delta x = \text{cosec}(y + \delta y) - \text{cosec}\, y = \dfrac{1}{\sin(y + \delta y)} - \dfrac{1}{\sin y}$

$$= \frac{\sin y - \sin(y + \delta y)}{\sin(y + \delta y)\sin y} = \frac{2\cos\left(y + \frac{1}{2}\delta y\right) - \sin\left(-\frac{1}{2}\delta y\right)}{\sin(y + \delta y)\sin y}$$

$$\therefore \quad \frac{\delta y}{\delta x} = -\frac{\delta y \,.\, \sin(y+\delta y)\sin y}{2\cos(y+\frac{1}{2}\delta y)\sin\frac{1}{2}\delta y} = -\frac{\sin(y+\delta y)\sin y}{\cos(y+\frac{1}{2}\delta y)} \cdot \frac{\frac{1}{2}\delta y}{\sin\frac{1}{2}\delta y}$$

$$\Rightarrow \quad \lim_{\delta x \to 0}\frac{\delta y}{\delta x} = -\lim_{\delta y \to 0}\frac{\sin(y+\delta y)\sin y}{\cos(y+\frac{1}{2}\delta y)} \lim_{\delta y \to 0}\frac{\frac{1}{2}\delta y}{\sin\frac{1}{2}\delta y} \qquad (\delta x \to 0 \;\Rightarrow\; \delta y \to 0)$$

$$= -\frac{\sin(y+0)\sin y}{\cos(y+0)} \,.\, 1 = \frac{-1}{\text{cosec } y \cot y}, \text{ provided } y \in \left(-\frac{1}{2}\pi, \frac{1}{2}\pi\right) - \{0\}.$$

$$\therefore \quad \frac{dy}{dx} = \frac{-1}{\text{cosec } y \,.\, \pm\sqrt{\text{cosec}^2 y - 1}} = \frac{-1}{\pm x\sqrt{x^2-1}} = \frac{-1}{|x|\sqrt{x^2-1}},$$

because cosec y cot y is +ve on $\left(-\frac{1}{2}\pi, \frac{1}{2}\pi\right) - \{0\}$.

$$\therefore \quad \mathbf{\frac{d}{dx}(\text{cosec}^{-1} x) = \frac{-1}{|x|\sqrt{x^2-1}}, \quad x \in R - [-1, 1].}$$

Let u be a differentiable function of x :

$\frac{d}{dx}(\sin^{-1} x) = \frac{1}{\sqrt{1-x^2}}$	$\frac{d}{dx}(\sin^{-1} u) = \frac{1}{\sqrt{1-u^2}} \frac{du}{dx}$
$\frac{d}{dx}(\cos^{-1} x) = \frac{-1}{\sqrt{1-x^2}}$	$\frac{d}{dx}(\cos^{-1} u) = \frac{-1}{\sqrt{1-u^2}} \frac{du}{dx}$
$\frac{d}{dx}(\tan^{-1} x) = \frac{1}{1+x^2}$	$\frac{d}{dx}(\tan^{-1} u) = \frac{1}{1+u^2} \frac{du}{dx}$
$\frac{d}{dx}(\cot^{-1} x) = \frac{-1}{1+x^2}$	$\frac{d}{dx}(\cot^{-1} u) = \frac{-1}{1+u^2} \frac{du}{dx}$
$\frac{d}{dx}(\sec^{-1} x) = \frac{1}{\|x\|\sqrt{x^2-1}}$	$\frac{d}{dx}(\sec^{-1} u) = \frac{1}{\|u\|\sqrt{u^2-1}} \frac{du}{dx}$
$\frac{d}{dx}(\text{cosec}^{-1} x) = \frac{-1}{\|x\|\sqrt{x^2-1}}$	$\frac{d}{dx}(\text{cosec}^{-1} u) = \frac{-1}{\|u\|\sqrt{u^2-1}} \frac{du}{dx}$

Remark. If $x \in \mathbf{R} - [-1, 1]$ is assumed to be **positive,** then $|x| = x$ and we have

(*i*) $\frac{d}{dx}(\sec^{-1} x) = \frac{1}{x\sqrt{x^2-1}}$ (*ii*) $\frac{d}{dx}(\text{cosec}^{-1} x) = \frac{-1}{x\sqrt{x^2-1}}$.

WORKING RULES FOR SOLVING PROBLEMS

In case of differentiating a function 'f(x)' involving inverse trigonometric functions of algebraic functions, the following rules help a lot in simplifying the function 'f(x)', to be differentiated.

Rule I. *Put* $x = a \sin\theta$ *(or* $a\cos\theta$*) if* $f(x)$ *involve* $\sqrt{a^2 - x^2}$.

Rule II. *Put* $x = a \tan\theta$ *(or* $a\cot\theta$*) if* $f(x)$ *involve* $\sqrt{a^2 + x^2}$.

Rule III. *Put* $x = a \sec\theta$ *(or* $a\,\text{cosec}\,\theta$*) if* $f(x)$ *involve* $\sqrt{x^2 - a^2}$.

Rule IV. *Put* $x = a\cos 2\theta$ *if* $f(x)$ *involve both* $\sqrt{a-x}$ *and* $\sqrt{a+x}$.

Rule V. *Put* $x^2 = a^2\cos 2\theta$ *if* $f(x)$ *involve both* $\sqrt{a^2-x^2}$ *and* $\sqrt{a^2+x^2}$.

ILLUSTRATIVE EXAMPLES

Example 1. *Differentiate the following functions w.r.t. x :*

(*i*) $\sin^{-1}\frac{x}{a}$ (*ii*) $\sqrt{1-x^2}\,\sin^{-1}x - x$ (*iii*) $\cos^{-1}\frac{x-x^{-1}}{x+x^{-1}}$

(*iv*) $\sec(\tan^{-1}x)$ (*v*) $\cos^{-1}\sqrt{\cos x}$ (*vi*) $\tan^{-1}(\cot x)$.

Sol. (*i*) Let $y = \sin^{-1}\frac{x}{a}$. ('$\sin^{-1} u$' Form)

$$\therefore \quad \frac{dy}{dx} = \frac{1}{\sqrt{1-\left(\frac{x}{a}\right)^2}}\,\frac{d}{dx}\left(\frac{x}{a}\right) = \frac{a}{\sqrt{a^2-x^2}}\cdot\frac{1}{a} = \frac{1}{\sqrt{\mathbf{a^2-x^2}}}.$$

(*ii*) Let $y = \sqrt{1-x^2}\,\sin^{-1}x - x.$

$$\therefore \quad \frac{dy}{dx} = \left[\sqrt{1-x^2}\,\frac{d}{dx}(\sin^{-1}x) + \sin^{-1}x\,.\,\frac{d}{dx}(1-x^2)^{1/2}\right] - 1$$

$$= \sqrt{1-x^2}\,.\,\frac{1}{\sqrt{1-x^2}} + \sin^{-1}x \times \frac{1}{2}(1-x^2)^{-1/2}\,\frac{d}{dx}(1-x^2) - 1$$

$$= 1 + \frac{\sin^{-1}x}{2\sqrt{1-x^2}}(-2x) - 1 = -\frac{\mathbf{x \sin^{-1} x}}{\sqrt{\mathbf{1-x^2}}}.$$

(*iii*) Let $y = \cos^{-1}\frac{x-x^{-1}}{x+x^{-1}} = \cos^{-1}\frac{x-\frac{1}{x}}{x+\frac{1}{x}} = \cos^{-1}\frac{x^2-1}{x^2+1}$. Let $x = \cot\theta$.

$$\therefore \quad y = \cos^{-1}\frac{\cot^2\theta - 1}{\cot^2\theta + 1} = \cos^{-1}\frac{1-\tan^2\theta}{1+\tan^2\theta} = \cos^{-1}(\cos 2\theta) = 2\theta = 2\cot^{-1}x.$$

$$\therefore \quad \frac{dy}{dx} = 2\left(\frac{-1}{1+x^2}\right) = -\frac{\mathbf{2}}{\mathbf{1+x^2}}.$$

(*iv*) Let $y = \sec(\tan^{-1}x)$. Let $x = \tan\theta$.

$$\therefore \quad y = \sec\theta = \sqrt{1+\tan^2\theta} = \sqrt{1+x^2}.$$

$$\therefore \quad \frac{dy}{dx} = \frac{1}{2}(1+x^2)^{-1/2}\,\frac{d}{dx}(1+x^2) = \frac{1}{2\sqrt{1+x^2}}(0+2x) = \frac{\mathbf{x}}{\sqrt{\mathbf{1+x^2}}}.$$

(*v*) Let $y = \cos^{-1}\sqrt{\cos x}$. ('$\cos^{-1} u$' Form)

$$\therefore \quad \frac{dy}{dx} = \frac{-1}{\sqrt{1-(\sqrt{\cos x})^2}}\,\frac{d}{dx}(\sqrt{\cos x})$$ ('$\sqrt{u}$' Form)

$$= \frac{-1}{\sqrt{1-\cos x}}\cdot\frac{1}{2}(\cos x)^{-1/2}\,\frac{d}{dx}(\cos x) = \frac{-1}{2\sqrt{1-\cos x}}\cdot\frac{1}{\sqrt{\cos x}}\cdot -\sin x$$

$$= \frac{\sin x}{2\sqrt{\cos x}} \cdot \frac{1}{\sqrt{1-\cos x}} \cdot \frac{\sqrt{1+\cos x}}{\sqrt{1+\cos x}} = \frac{\sin x}{2\sqrt{\cos x}} \cdot \frac{\sqrt{1+\cos x}}{\sqrt{1-\cos^2 x}}$$

$$= \frac{\sin x}{2\sqrt{\cos x}} \cdot \frac{\sqrt{1+\cos x}}{\sin x} = \frac{1}{2}\sqrt{\frac{1+\cos x}{\cos x}} = \frac{\mathbf{1}}{\mathbf{2}}\sqrt{\mathbf{sec\ x+1}}.$$

(*vi*) Let $y = \tan^{-1}(\cot x)$. ('$\tan^{-1} u$' Form)

$$\therefore \quad \frac{dy}{dx} = \frac{1}{1+\cot^2 x} \frac{d}{dx}(\cot x) = \frac{1}{\text{cosec}^2 x} \cdot -\text{cosec}^2 x = \mathbf{-1}.$$

Alternative method

Let $y = \tan^{-1}(\cot x)$. $\therefore \quad y = \tan^{-1}\left(\tan\left(\frac{\pi}{2} - x\right)\right) = \frac{\pi}{2} - x$

$$\therefore \quad \frac{dy}{dx} = 0 - 1 = \mathbf{-1}.$$

Example 2. *Find the derivative of* $\tan^{-1}\frac{1-x}{1+x} + \tan^{-1}\frac{x+2}{1-2x}$ *w.r.t. x.*

Sol. Let $y = \tan^{-1}\frac{1-x}{1+x} + \tan^{-1}\frac{x+2}{1-2x}$.

Let $x = \tan\theta$ and $2 = \tan k$.

$$\therefore \quad y = \tan^{-1}\frac{1-\tan\theta}{1+\tan\theta} + \tan^{-1}\frac{\tan\theta + \tan k}{1 - \tan k \tan\theta}$$

$$= \tan^{-1}\left(\tan\left(\frac{\pi}{4} - \theta\right)\right) + \tan^{-1}(\tan(\theta + k)) = \left(\frac{\pi}{4} - \theta\right) + (\theta + k) = \frac{\pi}{4} + k.$$

$$\therefore \quad \frac{dy}{dx} = \mathbf{0}.$$

Example 3. *Differentiate the following functions w.r.t. x :*

(*i*) $\tan^{-1}\sqrt{\frac{1+\sin x}{1-\sin x}}$ (*ii*) $\tan^{-1}\frac{1+x^2}{1-x^2}$

(*iii*) $\tan^{-1}\frac{\sqrt{1+x}-\sqrt{1-x}}{\sqrt{1+x}+\sqrt{1-x}}$ (*iv*) $\tan^{-1}\frac{\sqrt{1+x^2}-1}{x}$.

Sol. (*i*) Let $y = \tan^{-1}\sqrt{\frac{1+\sin x}{1-\sin x}}$.

$$\therefore \quad y = \tan^{-1}\sqrt{\frac{1-\cos\left(\frac{1}{2}\pi + x\right)}{1+\cos\left(\frac{1}{2}\pi + x\right)}} = \tan^{-1}\sqrt{\frac{2\sin^2\left(\frac{1}{4}\pi + \frac{1}{2}x\right)}{2\cos^2\left(\frac{1}{4}\pi + \frac{1}{2}x\right)}}$$

| $\cos(90° + \theta) = -\sin\theta$; $\sin(90° + \theta) = \cos\theta$

$$= \tan^{-1}\left[\tan\left(\frac{1}{4}\pi + \frac{1}{2}x\right)\right] = \frac{1}{4}\pi + \frac{1}{2}x.$$

$\therefore \quad \dfrac{dy}{dx} = 0 + \dfrac{1}{2} . 1 = \mathbf{\dfrac{1}{2}}.$

(*ii*) Let $\quad y = \tan^{-1} \dfrac{1+x^2}{1-x^2}$

$\therefore \quad y = \tan^{-1} \dfrac{1+\tan\theta}{1-\tan\theta}$, where $x^2 = \tan\theta$

$= \tan^{-1} [\tan (\tfrac{1}{4}\pi + \theta)] = \tfrac{1}{4}\pi + \theta = \tfrac{1}{4}\pi + \tan^{-1} x^2$

$(\because \; x^2 = \tan\theta \Rightarrow \theta = \tan^{-1} x^2)$

$\therefore \quad \dfrac{dy}{dx} = 0 + \dfrac{1}{1+(x^2)^2} \dfrac{d}{dx}(x^2) = \mathbf{\dfrac{2x}{1+x^4}}.$

(*iii*) Let $\quad y = \tan^{-1} \dfrac{\sqrt{1+x} - \sqrt{1-x}}{\sqrt{1+x} + \sqrt{1+x}}.$

$\therefore \quad y = \tan^{-1} \dfrac{\sqrt{1+\cos 2\theta} - \sqrt{1-\cos 2\theta}}{\sqrt{1+\cos 2\theta} + \sqrt{1-\cos 2\theta}}$, where $x = \cos 2\theta$*.

$= \tan^{-1} \dfrac{\sqrt{2\cos^2\theta} - \sqrt{2\sin^2\theta}}{\sqrt{2\cos^2\theta} + \sqrt{2\sin^2\theta}} = \tan^{-1} \dfrac{\cos\theta - \sin\theta}{\cos\theta + \sin\theta} = \tan^{-1} \dfrac{1-\tan\theta}{1+\tan\theta}$

$= \tan^{-1}\left[\tan\left(\dfrac{1}{4}\pi - \theta\right)\right] = \dfrac{1}{4}\pi - \theta = \dfrac{1}{4}\pi - \dfrac{1}{2}\cos^{-1} x.$

$\left(x = \cos 2\theta \Rightarrow 2\theta = \cos^{-1} x \Rightarrow \theta = \dfrac{1}{2}\cos^{-1} x\right)$

$\therefore \quad \dfrac{dy}{dx} = 0 - \dfrac{1}{2} . \dfrac{-1}{\sqrt{1-x^2}} = \mathbf{\dfrac{1}{2\sqrt{1-x^2}}}.$

(*iv*) Let $\quad y = \tan^{-1} \dfrac{\sqrt{1+x^2} - 1}{x}.$

$\therefore \quad y = \tan^{-1} \dfrac{\sqrt{1+\tan^2\theta} - 1}{\tan\theta}$, where $x = \tan\theta$

$= \tan^{-1} \dfrac{\sec\theta - 1}{\tan\theta} = \tan^{-1} \dfrac{1-\cos\theta}{\sin\theta} = \tan^{-1} \dfrac{2\sin^2 \frac{1}{2}\theta}{2\sin\frac{1}{2}\theta \cos\frac{1}{2}\theta}$

$= \tan^{-1}\left(\tan\dfrac{1}{2}\theta\right) = \dfrac{1}{2}\theta = \dfrac{1}{2}\tan^{-1} x.$

$\therefore \quad \dfrac{dy}{dx} = \dfrac{1}{2} . \dfrac{1}{1+x^2} = \mathbf{\dfrac{1}{2(1+x^2)}}.$

---

* **Why this step.** We have taken $x = \cos 2\theta$ instead of $\cos\theta$, just to avoid *t*-functions of $\dfrac{\theta}{2}$.

Example 4. *Differentiate the following functions w.r.t. x :*

(*i*) $sin\left[2\ tan^{-1}\sqrt{\frac{1-x}{1+x}}\right]$ (*ii*) $cot^{-1}\left[\frac{\sqrt{1+\sin x}+\sqrt{1-\sin x}}{\sqrt{1+\sin x}-\sqrt{1-\sin x}}\right]$

(*iii*) $sin^{-1}\sqrt{\frac{1+x^2}{2}}$ (*iv*) $cos^{-1}\left(\frac{3\cos x-4\sin x}{5}\right)$.

Sol. (*i*) Let $y=\sin\left[2\tan^{-1}\sqrt{\frac{1-x}{1+x}}\right]$.

$$\therefore \quad y=\sin\left[2\tan^{-1}\sqrt{\frac{1-\cos 2\theta}{1+\cos 2\theta}}\right], \quad \text{where } x=\cos 2\theta$$

$$=\sin\left[2\tan^{-1}\sqrt{\frac{2\sin^2\theta}{2\cos^2\theta}}\right]=\sin\,[2\tan^{-1}(\tan\theta)]=\sin 2\theta$$

$$=\sqrt{1-\cos^2 2\theta}=\sqrt{1-x^2}. \qquad (\because \ x=\cos 2\theta)$$

$$\therefore \quad \frac{dy}{dx}=\frac{1}{2}(1-x^2)^{-1/2}\cdot\frac{d}{dx}(1-x^2)=\frac{1}{2\sqrt{1-x^2}}\cdot(-2x)=-\frac{\mathbf{x}}{\sqrt{\mathbf{1-x^2}}}. \qquad (\text{`}\sqrt{u}\text{' Form})$$

(*ii*) Let $y=\cot^{-1}\left[\frac{\sqrt{1+\sin x}+\sqrt{1-\sin x}}{\sqrt{1+\sin x}-\sqrt{1-\sin x}}\right]$.

$$\therefore \quad y=\cot^{-1}\left[\frac{\sqrt{1+\sin x}+\sqrt{1-\sin x}}{\sqrt{1+\sin x}-\sqrt{1-\sin x}}\times\frac{\sqrt{1+\sin x}+\sqrt{1-\sin x}}{\sqrt{1+\sin x}+\sqrt{1-\sin x}}\right]$$

$$=\cot^{-1}\left[\frac{(1+\sin x)+(1-\sin x)+2\sqrt{1+\sin x}\,\sqrt{1-\sin x}}{(1+\sin x)-(1-\sin x)}\right]$$

$$=\cot^{-1}\left[\frac{2+2\cos x}{2\sin x}\right]=\cot^{-1}\left[\frac{1+\cos x}{\sin x}\right] \qquad \mid \sqrt{1-\sin^2 x}=\cos x$$

$$=\cot^{-1}\left[\frac{2\cos^2\frac{1}{2}x}{2\sin\frac{1}{2}x\cos\frac{1}{2}x}\right]=\cot^{-1}\left[\cot\frac{x}{2}\right]=\frac{x}{2}.$$

$$\therefore \quad \frac{dy}{dx}=\frac{\mathbf{1}}{\mathbf{2}}.$$

(*iii*) Let $y=\sin^{-1}\sqrt{\frac{1+x^2}{2}}$. Putting $x^2=\cos 2\theta$, we get

$$y=\sin^{-1}\sqrt{\frac{1+\cos 2\theta}{2}}=\sin^{-1}\sqrt{\frac{2\cos^2\theta}{2}}=\sin^{-1}(\cos\theta)=\sin^{-1}\left(\sin\left(\frac{\pi}{2}-\theta\right)\right)$$

$$= \frac{\pi}{2} - \theta = \frac{\pi}{2} - \frac{1}{2}\cos^{-1} x^2.$$

$$\therefore \quad \frac{dy}{dx} = 0 - \frac{1}{2}\left[\frac{-1}{\sqrt{1-(x^2)^2}}\right].2x = \frac{\mathbf{x}}{\sqrt{\mathbf{1-x^4}}}.$$

(*iv*) Let $y = \cos^{-1}\left(\frac{3\cos x - 4\sin x}{5}\right)$. $\therefore \quad y = \cos^{-1}\left(\frac{3}{5}\cos x - \frac{4}{5}\sin x\right)$

We observe that $\left(\frac{3}{5}\right)^2 + \left(\frac{4}{5}\right)^2 = \frac{25}{25} = 1.$

Let $\cos\theta = \frac{3}{5}$ and θ lies in the first quadrant.

$$\therefore \quad \sin^2\theta = 1 - \cos^2\theta = 1 - \left(\frac{3}{5}\right)^2 = \frac{16}{25}. \quad \therefore \quad \sin\theta = \frac{4}{5}$$

$$\therefore \quad y = \cos^{-1}(\cos\theta\cos x - \sin\theta\sin x) = \cos^{-1}(\cos(\theta + x)) = \theta + x = \cos^{-1}\frac{3}{5} + x.$$

$$\therefore \quad \frac{dy}{dx} = 0 + 1 = \mathbf{1}.$$

Example 5. *Show that* $\sec^{-1} x = \tan^{-1}\sqrt{x^2-1}$ *and thus find* $\frac{d}{dx}(\sec^{-1} x)$.

Sol. Let $\sec^{-1} x = \theta$. $\quad \therefore \quad x = \sec\theta$.

$$\Rightarrow \quad \tan\theta = \sqrt{\sec^2\theta - 1} = \sqrt{x^2-1} \quad \Rightarrow \quad \theta = \tan^{-1}\sqrt{x^2-1}.$$

$$\therefore \quad \sec^{-1} x = \tan^{-1}\sqrt{x^2-1}.$$

$$\therefore \quad \frac{d}{dx}(\sec^{-1} x) = \frac{d}{dx}(\tan^{-1}\sqrt{x^2-1}).$$ ('$\tan^{-1} u$' Form)

$$= \frac{1}{1+\left(\sqrt{x^2-1}\right)^2}\frac{d}{dx}\sqrt{x^2-1} = \frac{1}{1+x^2-1}\cdot\frac{1}{2}(x^2-1)^{-1/2}\cdot\frac{d}{dx}(x^2-1)$$ ('$\sqrt{u}$' Form)

$$= \frac{1}{x^2}\cdot\frac{1}{2\sqrt{x^2-1}}\cdot(2x-0) = \frac{\mathbf{1}}{\mathbf{x\sqrt{x^2-1}}}.$$

Example 6. *If* $y = \tan^{-1}\frac{5ax}{a^2-6x^2}$, *then show that* $\frac{dy}{dx} = \frac{3a}{a^2+9x^2} + \frac{2a}{a^2+4x^2}$.

Sol. We have $y = \tan^{-1}\frac{5ax}{a^2-6x^2}$. ('$\tan^{-1} u$' Form)

$$\therefore \quad y = \tan^{-1}\frac{\frac{5x}{a}}{1-\frac{6x^2}{a^2}} = \tan^{-1}\frac{\frac{3x}{a}+\frac{2x}{a}}{1-\left(\frac{3x}{a}\right)\left(\frac{2x}{a}\right)} = \tan^{-1}\frac{3x}{a} + \tan^{-1}\frac{2x}{a}$$

$$\left|\ \tan^{-1}A + \tan^{-1}B = \tan^{-1}\left(\frac{A+B}{1-AB}\right)\right.$$

$$\therefore \quad \frac{dy}{dx} = \frac{1}{1+\left(\frac{3x}{a}\right)^2} \cdot \frac{d}{dx}\left(\frac{3x}{a}\right) + \frac{1}{1+\left(\frac{2x}{a}\right)^2} \cdot \frac{d}{dx}\left(\frac{2x}{a}\right)$$

$$= \frac{a^2}{a^2+9x^2} \cdot \frac{3}{a} + \frac{a^2}{a^2+4x^2} \cdot \frac{2}{a} = \mathbf{\frac{3a}{a^2+9x^2} + \frac{2a}{a^2+4x^2}}.$$

Example 7. *Differentiate* $\cos^{-1} \frac{a+b\cos x}{b+a\cos x}$ *w.r.t. x.*

Sol. Let $\quad y = \cos^{-1} \frac{a+b\cos x}{b+a\cos x}$. $\quad \therefore \quad y = \cos^{-1} u$, where $u = \frac{a+b\cos x}{b+a\cos x}$.

$$\therefore \quad \frac{dy}{dx} = \frac{-1}{\sqrt{1-u^2}} \cdot \frac{du}{dx}. \qquad \ldots(1)$$

Consider $\quad 1-u^2 = 1 - \left(\frac{a+b\cos x}{b+a\cos x}\right)^2 = \frac{(b+a\cos x)^2-(a+b\cos x)^2}{(b+a\cos x)^2}$

$$= \frac{b^2+a^2\cos^2 x - a^2 - b^2\cos^2 x}{(b+a\cos x)^2} = \frac{(b^2-a^2)-(b^2-a^2)\cos^2 x}{(b+a\cos x)^2}$$

$$= \frac{(b^2-a^2)\sin^2 x}{(b+a\cos x)^2}$$

$$\therefore \quad -\frac{1}{\sqrt{1-u^2}} = -\frac{b+a\cos x}{\sqrt{b^2-a^2}\sin x}. \qquad \text{(Assuming } b^2-a^2>0\text{)}$$

Also, $\quad \frac{du}{dx} = \frac{(b+a\cos x)(-b\sin x)-(a+b\cos x)(-a\sin x)}{(b+a\cos x)^2} = \frac{-(b^2-a^2)\sin x}{(b+a\cos x)^2}$.

$$\therefore \quad (1) \Rightarrow \quad \frac{dy}{dx} = \left(\frac{-b+a\cos x}{\sqrt{b^2-a^2}\sin x} \; \frac{-(b^2-a^2)\sin x}{(b+a\cos x)^2}\right) = \mathbf{\frac{\sqrt{b^2-a^2}}{b+a\cos x}}.$$

EXERCISE 12.8

Differentiate the following functions w.r.t. x (Q. No. 1–8) :

1. (*i*) $\sin^{-1} x^2$ (*ii*) $\cos^{-1} x^4$ (*iii*) $\tan^{-1}(4x^3)$
(*iv*) $\cot^{-1}(x^2+1)$ (*v*) $\sec^{-1}(3x^5)$ (*vi*) $\operatorname{cosec}^{-1}(4x^2)$

2. (*i*) $\sin^{-1}(\cos x)$ (*ii*) $\cos^{-1}\frac{x}{x+1}$ (*iii*) $\cos^{-1}(\tan x^2)$
(*iv*) $(\tan^{-1} x)^2$ (*v*) $\cot^{-1}\sqrt{x}$ (*vi*) $x\sec^{-1} x$.

3. (*i*) $\tan^{-1}\frac{a-x}{1+ax}$ (*ii*) $\tan^{-1}\frac{x^{1/3}+a^{1/3}}{1-x^{1/3}a^{1/3}}$ (*iii*) $\tan^{-1}\frac{\cos x-\sin x}{\cos x+\sin x}$
(*iv*) $\sin^{-1}\frac{2x}{1+x^2}$ (*v*) $\cos^{-1}\frac{1-x^2}{1+x^2}$ (*vi*) $\operatorname{cosec}^{-1}\frac{1+x^2}{2x}$

4. (*i*) $\sin^{-1}(3x-4x^3)$ (*ii*) $\cos^{-1}(4x^3-3x)$ (*iii*) $\tan^{-1}\dfrac{3x-x^3}{1-3x^2}$

(*iv*) $\sin^{-1}\sqrt{\dfrac{1-\cos 2x}{2}}$ (*v*) $\cos^{-1}\sqrt{\dfrac{1+\cos 2x}{2}}$ (*vi*) $\cot^{-1}\dfrac{1-x}{1+x}$

5. (*i*) $\sin^{-1}(1-2x^2)$ (*ii*) $\sin^{-1}[2ax\sqrt{1-a^2x^2}]$ (*iii*) $\sin^{-1}\dfrac{x^2}{\sqrt{a^4+x^4}}$

(*iv*) $\sin^{-1}\dfrac{\sqrt{1+x}+\sqrt{1-x}}{2}$ (*v*) $\sin^{-1}\sqrt{1-x^2}$ (*vi*) $\sin^{-1}(\cos\sqrt{2x+3})$

6. (*i*) $\tan^{-1}\dfrac{x}{\sqrt{a^2-x^2}}$ (*ii*) $\tan^{-1}\dfrac{1}{\sqrt{x^2-1}}$ (*iii*) $\tan^{-1}\dfrac{x}{1+\sqrt{1-x^2}}$

(*iv*) $\tan^{-1}\dfrac{\sqrt{1+a^2x^2}-1}{ax}$ (*v*) $\cot^{-1}\dfrac{1+\cos x}{\sin x}$ (*vi*) $\cot^{-1}(\sqrt{1+x^2}-x)$

7. (*i*) $\sin^{-1}\sqrt{\dfrac{1+\cos 2x}{2}}$ (*ii*) $\cos^{-1}\sqrt{\dfrac{1+x}{2}}$ (*iii*) $\tan^{-1}\dfrac{\cos x}{1+\sin x}$

(*iv*) $\tan^{-1}(\sec x+\tan x)$ (*v*) $\tan^{-1}\dfrac{1-\cos x}{\sin x}$ (*vi*) $\tan^{-1}\dfrac{\sqrt{1+x^2}-\sqrt{1-x^2}}{\sqrt{1+x^2}+\sqrt{1-x^2}}$

8. (*i*) $\tan^{-1}(\sqrt{1+x^2}-x)$ (*ii*) $\tan^{-1}\left(\dfrac{1+\cos x}{1-\cos x}\right)^{1/2}$ (*iii*) $\cos^{-1}\dfrac{2x}{1+x^2}$

(*iv*) $\cos^{-1}\left(2x\sqrt{1-x^2}\right)$

9. (*a*) Differentiate $y=\sin^{-1}(\tan x)$ in two ways : (*i*) by chain rule and (*ii*) by considering $\sin y=\tan x$.

(*b*) If $y=\sin^{-1}\dfrac{1}{\sqrt{1+x^2}}+\tan^{-1}\dfrac{\sqrt{1+x^2}-1}{x}$, find $\dfrac{dy}{dx}$.

10. Show that $\dfrac{d}{dx}\left[\dfrac{x\sqrt{a^2-x^2}}{2}+\dfrac{a^2}{2}\sin^{-1}\dfrac{x}{a}\right]=\sqrt{a^2-x^2}$.

Answers

1. (*i*) $\dfrac{2x}{\sqrt{1-x^4}}$ (*ii*) $-\dfrac{4x^3}{\sqrt{1-x^8}}$ (*iii*) $\dfrac{12x^2}{1+16x^6}$ (*iv*) $-\dfrac{2x}{x^4+2x^2+2}$ (*v*) $\dfrac{5}{x\sqrt{9x^{10}-1}}$

(*vi*) $-\dfrac{2}{x\sqrt{16x^4-1}}$ 2. (*i*) -1 (*ii*) $-\dfrac{1}{(x+1)\sqrt{2x+1}}$ (*iii*) $-\dfrac{2x\sec^2 x^2}{\sqrt{1-\tan^2 x^2}}$

(*iv*) $\dfrac{2\tan^{-1}x}{1+x^2}$ (*v*) $-\dfrac{1}{2\sqrt{x}\,(x+1)}$ (*vi*) $\sec^{-1}x+\dfrac{1}{\sqrt{x^2-1}}$ 3. (*i*) $-\dfrac{1}{1+x^2}$ (*ii*) $\dfrac{1}{3x^{2/3}(1+x^{2/3})}$

(*iii*) -1 (*iv*) $\dfrac{2}{1+x^2}$ (*v*) $\dfrac{2}{1+x^2}$ (*vi*) $\dfrac{2}{1+x^2}$

4. (*i*) $\dfrac{3}{\sqrt{1-x^2}}$ (*ii*) $-\dfrac{3}{\sqrt{1-x^2}}$ (*iii*) $\dfrac{3}{1+x^2}$ (*iv*) 1 (*v*) 1

(*vi*) $\dfrac{1}{1+x^2}$

5. (*i*) $-\dfrac{2}{\sqrt{1-x^2}}$ (*ii*) $\dfrac{2a}{\sqrt{1-a^2x^2}}$ (*iii*) $\dfrac{2a^2x}{a^4+x^4}$ (*iv*) $-\dfrac{1}{2\sqrt{1-x^2}}$ (*v*) $-\dfrac{1}{\sqrt{1-x^2}}$

(*vi*) $-\dfrac{1}{\sqrt{2x+3}}$

6. (*i*) $\dfrac{1}{\sqrt{a^2-x^2}}$ (*ii*) $-\dfrac{1}{x\sqrt{x^2-1}}$ (*iii*) $\dfrac{1}{2\sqrt{1-x^2}}$ (*iv*) $\dfrac{a}{2(1+a^2x^2)}$ (*v*) $\dfrac{1}{2}$

(*vi*) $\dfrac{1}{2(1+x^2)}$ **7.** (*i*) -1 (*ii*) $-\dfrac{1}{2\sqrt{1-x^2}}$ (*iii*) $-\dfrac{1}{2}$ (*iv*) $\dfrac{1}{2}$

(*v*) $\dfrac{1}{2}$ (*vi*) $\dfrac{x}{\sqrt{1-x^4}}$ **8.** (*i*) $\dfrac{-1}{2(1+x^2)}$ (*ii*) $-\dfrac{1}{2}$ (*iii*) $-\dfrac{2}{1+x^2}$

(*iv*) $\dfrac{2}{\sqrt{1-x^2}}$

9. (*a*) $\dfrac{\sec^2 x}{\sqrt{1-\tan^2 x}}$ (*b*) $-\dfrac{1}{2(1+x^2)}$.

Hints

3. (*iv*) $x=\tan\theta \Rightarrow \sin^{-1}\dfrac{2\tan\theta}{1+\tan^2\theta}=\sin^{-1}(\sin 2\theta)=2\theta=2\tan^{-1}x.$

4. (*i*) $x=\sin\theta \Rightarrow \sin^{-1}(3x-4x^3)=\sin^{-1}(3\sin\theta-4\sin^3\theta)=\sin^{-1}(\sin 3\theta)=3\theta=3\sin^{-1}x.$

5. (*ii*) Use $ax=\sin\theta$ (*iii*) Use $x^2=a^2\tan\theta$ (*iv*) Use $x=\cos 2\theta$.

8. (*iii*) Recall $\sin^{-1}x+\cos^{-1}x=\dfrac{\pi}{2}$.

VI. DIFFERENTIATION OF EXPONENTIAL FUNCTIONS

12.25. DERIVATIVE OF EXPONENTIAL FUNCTION

Let $y=a^x,\ a>0.$ ...(1)

Let δx be an increment of x and δy, the corresponding increment of y.

$\therefore\quad y+\delta y=a^{x+\delta x}$ ...(2)

(2) – (1) $\Rightarrow\ \delta y=a^{x+\delta x}-a^x=a^x(a^{\delta x}-1)$

$$\Rightarrow\quad \frac{\delta y}{\delta x}=a^x\,.\,\frac{a^{\delta x}-1}{\delta x}$$

$$\therefore\quad \lim_{\delta x\to 0}\frac{\delta y}{\delta x}=a^x\lim_{\delta x\to 0}\frac{a^{\delta x}-1}{\delta x}=a^x\log a \qquad \left(\text{Using } \lim_{\delta x\to 0}\frac{a^x-1}{x}=\log a\right)$$

$$\therefore\quad \frac{dy}{dx}=a^x\log a \quad i.e.,\quad \mathbf{\frac{d}{dx}(a^x)=a^x\log a.}$$ (*P.T.U. B.Pharma, Dec. 2006*)

Remark 1. In particular, if $a=e$, then $\dfrac{d}{dx}(e^x)=e^x\log e=e^x\,.\,1=e^x$ *i.e.,* $\mathbf{\dfrac{d}{dx}(e^x)=e^x}$.

ILLUSTRATIVE EXAMPLES

Example 1. *Differentiate the following functions w.r.t. x :*

(i) $e^{\sin x}$ (ii) $e^{ax}\cos(bx+c)$ (iii) $e^{3\log x}$ (iv) $e^{3x}\sin 4x.\ 2^x$

(v) $e^{m\cos^{-1}x}\sin nx$ (vi) $\dfrac{e^{ax}-e^{-ax}}{e^{ax}+e^{-ax}}$ (vii) $\dfrac{e^x}{1+\sin x}$ (viii) $e^{\sin\sqrt{x}}$.

(G.N.D.U. B.Pharma, 2005)

Sol. (i) Let $y = e^{\sin x}$. ('e^{u}' Form)

$\therefore \quad \dfrac{dy}{dx} = e^{\sin x}\dfrac{d}{dx}(\sin x) = \mathbf{e^{\sin x}\cos x}.$

(ii) Let $y = e^{ax}\cos(bx+c)$. ('uv' Form)

$$\therefore \quad \frac{dy}{dx} = e^{ax}\frac{d}{dx}(\cos(bx+c)) + \cos(bx+c)\frac{d}{dx}(e^{ax})$$

$$= e^{ax}[-\sin(bx+c)]\frac{d}{dx}(bx+c) + \cos(bx+c)\,.\,e^{ax}\frac{d}{dx}(ax)$$

$$= -e^{ax}\sin(bx+c)\,.\,b + \cos(bx+c)\,.e^{ax}\,.\,a$$

$$= \mathbf{e^{ax}[a\cos(bx+c) - b\sin(bx+c)]}.$$

(iii) Let $y = e^{3\log x}$. $\therefore$ $y = e^{\log x^3} = x^3$ $(\because\ e^{\log k} = k)$

$\therefore \quad \dfrac{dy}{dx} = \mathbf{3x^2}.$

(iv) Let $y = e^{3x}\sin 4x.\ 2^x$. *i.e.*, y is product of three functions.

$$\therefore \quad \frac{dy}{dx} = \left(\frac{d}{dx}(e^{3x})\right).\sin 4x\,.\,2^x + e^{3x}\frac{d}{dx}(\sin 4x)\,.\,2^x + e^{3x}\sin 4x\,.\,\frac{d}{dx}(2^x)$$

$$= (e^{3x}.3)\sin 4x\,.\,2^x + e^{3x}(\cos 4x\,.\,4)\,2^x + e^{3x}\sin 4x\,.\,2^x\log 2$$

$$= \mathbf{e^{3x}\,2^x\,[3\sin 4x + 4\cos 4x + \log 2\,.\,\sin 4x]}.$$

(v) Let $y = e^{m\cos^{-1}x}\sin nx$. *i.e.*, y is product of two functions.

$$\therefore \quad \frac{dy}{dx} = e^{m\cos^{-1}x}\frac{d}{dx}(\sin nx) + \sin nx\frac{d}{dx}\left(e^{m\cos^{-1}x}\right)$$

$$= e^{m\cos^{-1}x}\cos nx\frac{d}{dx}(nx) + \sin nx\,.\,e^{m\cos^{-1}x}\frac{d}{dx}(m\cos^{-1}x)$$

$$= e^{m\cos^{-1}x}\left[\cos nx.n + \sin nx.m\left(\frac{-1}{\sqrt{1-x^2}}\right)\right] = \mathbf{e^{m\cos^{-1}x}\left[n\cos nx - \frac{m\sin nx}{\sqrt{1-x^2}}\right]}.$$

(vi) Let $y = \dfrac{e^{ax}-e^{-ax}}{e^{ax}+e^{-ax}}$.

$$\therefore \quad \frac{dy}{dx} = \frac{(e^{ax}+e^{-ax})\frac{d}{dx}(e^{ax}-e^{-ax}) - (e^{ax}-e^{-ax})\frac{d}{dx}(e^{ax}+e^{-ax})}{(e^{ax}+e^{-ax})^2}$$

$$= \frac{(e^{ax}+e^{-ax})(e^{ax}.a - e^{-ax}.-a) - (e^{ax}-e^{-ax})(e^{ax}.a + e^{-ax}.-a)}{(e^{ax}+e^{-ax})^2}$$

$$= \frac{a(e^{ax}+e^{-ax})^2 - a(e^{ax}-e^{-ax})^2}{(e^{ax}+e^{-ax})^2} = \frac{a[(e^{2ax}+e^{-2ax}+2)-(e^{2ax}+e^{-2ax}-2)]}{(e^{ax}+e^{-ax})^2}$$

$$= \frac{\mathbf{4a}}{(\mathbf{e^{ax}}+\mathbf{e^{-ax}})^2}.$$

(*vii*) Let $y = \dfrac{e^x}{1+\sin x}$.

$$\therefore \quad \frac{dy}{dx} = \frac{(1+\sin x)e^x - e^x(0+\cos x)}{(1+\sin x)^2} = \frac{\mathbf{e^x(1+\sin x - \cos x)}}{\mathbf{(1+\sin x)^2}}.$$

(*viii*) Let $y = e^{\sin\sqrt{x}}$. ('e^u Form)

$$\therefore \quad \frac{dy}{dx} = e^{\sin\sqrt{x}} \frac{d}{dx}(\sin\sqrt{x}) \qquad \text{('}\sin u\text{' Form)}$$

$$= e^{\sin\sqrt{x}} . \cos\sqrt{x} . \frac{d}{dx}(\sqrt{x}) = \frac{\mathbf{e^{\sin\sqrt{x}}\cos\sqrt{x}}}{\mathbf{2\sqrt{x}}}.$$

Example 2. *Differentiate the following functions w.r.t. x :*

(*i*) $e^{\tan^{-1}(\cos\sqrt{x})}$ (*ii*) $e^{ax} \sin(b \tan^{-1} x)$.

Sol. (*i*) Let $y = e^{\tan^{-1}(\cos\sqrt{x})}$. ('$e^u$' Form)

$$\therefore \quad \frac{dy}{dx} = e^{\tan^{-1}(\cos\sqrt{x})} \frac{d}{dx}(\tan^{-1}(\cos\sqrt{x})) \qquad \text{('}\tan^{-1} u\text{' Form)}$$

$$= e^{\tan^{-1}(\cos\sqrt{x})} . \frac{1}{1+(\cos\sqrt{x})^2} \frac{d}{dx}(\cos\sqrt{x}) \qquad \text{('}\cos u\text{' Form)}$$

$$= \frac{e^{\tan^{-1}(\cos\sqrt{x})}}{1+\cos^2\sqrt{x}} . -\sin\sqrt{x} \frac{d}{dx}(\sqrt{x}) = -\frac{\sin\sqrt{x} . e^{\tan^{-1}(\cos\sqrt{x})}}{1+\cos^2\sqrt{x}} . \frac{1}{2} x^{-1/2}$$

$$= -\frac{\mathbf{\sin\sqrt{x} . e^{\tan^{-1}\cos\sqrt{x}}}}{\mathbf{2\sqrt{x}(1+\cos^2\sqrt{x})}}.$$

(*ii*) Let $y = e^{ax} \sin(b \tan^{-1} x)$.

$$\therefore \quad \frac{dy}{dx} = e^{ax} \frac{d}{dx}(\sin(b\tan^{-1} x)) + \sin(b\tan^{-1} x) \frac{d}{dx}(e^{ax})$$

$$= e^{ax} . \cos(b\tan^{-1} x) \frac{d}{dx}(b\tan^{-1} x) + \sin(b\tan^{-1} x). e^{ax} . a$$

$$= \mathbf{e^{ax}} \left[\frac{\mathbf{b\cos(b\tan^{-1} x)}}{\mathbf{1+x^2}} + \mathbf{a\sin(b\tan^{-1} x)}\right].$$

Example 3. *Differentiate $e^{2x} \sin 3x$ w.r.t. x.*

Sol. Let $y = e^{2x} \sin 3x$

Differentiating w.r.t. x, we get

$$\frac{dy}{dx} = \frac{d}{dx}(e^{2x}\sin 3x) = e^{2x}\frac{d}{dx}(\sin 3x) + \sin 3x \frac{d}{dx}(e^{2x})$$

$$= e^{2x}\cos 3x \frac{d}{dx}(3x) + \sin 3x . e^{2x} . \frac{d}{dx}(2x) = 3e^{2x}\cos 3x + 2\sin 3x\, e^{2x}.$$

EXERCISE 12.9

Find the derivatives of the following functions w.r.t. x (Q. No. 1–4) :

1. (*i*) e^{x^2} (*ii*) 4^{x+9} (*iii*) $e^{\cos x}$

   (*iv*) $a^{\tan x}$ (*v*) a^{7x+4} (*vi*) $3^{\sin^2 x}$

2. (*i*) $\frac{1}{3}e^x - 5e$ (*ii*) $e^x \sin x$ (*iii*) $\frac{1}{x}e^x$

   (*iv*) $e^{\sin^{-1}(x+1)}$ (*v*) $e^{x \cos x}$ (*vi*) 10^{10x}

3. (*i*) $x^3 e^x \cos x$ (*ii*) $\frac{e^x}{1+\cos x}$ (*iii*) $\frac{e^x(x-1)}{x^2+1}$

   (*iv*) $\frac{e^{ax}}{\sin(bx+c)}$ (*v*) $\frac{e^x+e^{-x}}{e^x-e^{-x}}$ (*vi*) $e^{\cos^{-1}(x+1)}$

4. (*i*) $2^x e^{\sin^{-1} x}$ (*ii*) $e^{\sqrt{1-x^2}} \tan x$ (*iii*) $xe^{\sqrt{\sin x}}$

   (*iv*) $x^{-1/3} e^x$ (*v*) $e^{\sin x} \sin e^x$ (*vi*) $e^{\cot^{-1} x^2}$.

Answers

1. (*i*) $2xe^{x^2}$ (*ii*) $4^{x+9} \log 4$ (*iii*) $-(\sin x)\, e^{\cos x}$ (*iv*) $a^{\tan x} \sec^2 x \log a$

   (*v*) $7a^{7x+4} \log a$ (*vi*) $3^{\sin^2 x} \sin 2x \log 3$

2. (*i*) $\frac{1}{3}e^x$ (*ii*) $e^x(\sin x + \cos x)$ (*iii*) $\frac{e^x(x-1)}{x^2}$ (*iv*) $\frac{e^{\sin^{-1}(x+1)}}{\sqrt{1-(x+1)^2}}$

   (*v*) $e^{x\cos x}(\cos x - x \sin x)$ (*vi*) $10^{10x}(10 \log 10)$

3. (*i*) $x^2e^x(-x \sin x + (x+3)\cos x)$ (*ii*) $\frac{e^x(1+\cos x+\sin x)}{(1+\cos x)^2}$

   (*iii*) $\frac{xe^x(x^2-2x+3)}{(x^2+1)^2}$ (*iv*) $\frac{e^{ax}[a \sin(bx+c) - b\cos(bx+c)]}{\sin^2(bx+c)}$ (*v*) $-\frac{4}{(e^x-e^{-x})^2}$

   (*vi*) $-\frac{1}{\sqrt{1-(x+1)^2}} e^{\cos^{-1}(x+1)}$

4. (*i*) $2^x e^{\sin^{-1} x}\left(\frac{1}{\sqrt{1-x^2}} + \log 2\right)$ (*ii*) $e^{\sqrt{1-x^2}}\left(\sec^2 x - \frac{x \tan x}{\sqrt{1-x^2}}\right)$

   (*iii*) $e^{\sqrt{\sin x}}\left(\frac{x \cos x}{2\sqrt{\sin x}} + 1\right)$ (*iv*) $e^x\left(x^{-1/3} - \frac{1}{3}x^{-4/3}\right)$

   (*v*) $e^{\sin x}(e^x \cos e^x + \cos x \sin e^x)$ (*vi*) $-\frac{2x}{1+x^4} e^{\cot^{-1} x^2}$.

VII. DIFFERENTIATION OF LOGARITHMIC FUNCTIONS

12.26. DERIVATIVE OF LOGARITHMIC FUNCTION

Let $y = \log_a x, \ (a, x > 0, a \neq 1).$ ...(1)

Let δx be an increment of x and δy, the corresponding increment of y.

$\therefore \quad y + \delta y = \log_a (x + \delta x)$ ...(2)

$$(2) - (1) \Rightarrow \delta y = \log_a (x + \delta x) - \log_a x = \log_a \left(\frac{x + \delta x}{x}\right) = \log_a \left(1 + \frac{\delta x}{x}\right)$$

$$\Rightarrow \quad \frac{\delta y}{\delta x} = \frac{1}{\delta x} \log_a \left(1 + \frac{\delta x}{x}\right) = \frac{1}{x} \cdot \frac{x}{\delta x} \log_a \left(1 + \frac{\delta x}{x}\right) = \frac{1}{x} \log_a \left(1 + \frac{\delta x}{x}\right)^{x/\delta x}$$

$$\therefore \quad \lim_{\delta x \to 0} \frac{\delta y}{\delta x} = \frac{1}{x} \lim_{\delta x \to 0} \left[\log_a \left(1 + \frac{\delta x}{x}\right)^{x/\delta x}\right] = \frac{1}{x} \log_a \left[\lim_{\delta x/x \to 0} \left(1 + \frac{\delta x}{x}\right)^{x/\delta x}\right]$$

($\because$ Logarithm is a continuous function)

$$= \frac{1}{x} \log_a e. \qquad \left(\text{Using } \lim_{x \to 0} (1 + x)^{1/x} = e\right)$$

$$\therefore \quad \frac{dy}{dx} = \frac{1}{x} \log_a e \quad i.e., \quad \mathbf{\frac{d}{dx} (\log_a x) = \frac{1}{x} \log_a e.}$$

Remark 1. In particular, if $a = e$, then $\frac{d}{dx} (\log_e x) = \frac{1}{x} \log_e e$ *i.e.,* $\mathbf{\frac{d}{dx} (\log x) = \frac{1}{x}}$.

Remark 2. Logarithms to the base 'e' are called **natural logarithms.** The natural logarithm of x is denoted by '$\log x$' or by '$\ln x$'.

Remark 3. The following formulae will be found very useful in differentiation of functions :

(*i*) $\log_a mn = \log_a m + \log_a n$ (*ii*) $\log_a \frac{m}{n} = \log_a m - \log_a n$

(*iii*) $\log_a m^n = n \log_a m$ (*iv*) $\log_a m = \log_b m \times \log_a b$.

ILLUSTRATIVE EXAMPLES

Example 1. *If* $y = \log \tan \left(\frac{\pi}{4} + \frac{x}{2}\right)$, *show that* $\frac{dy}{dx} - \sec x = 0$.

Sol. We have $y = \log \tan \left(\frac{\pi}{4} + \frac{x}{2}\right)$. ('$\log u$' Form)

$$\therefore \quad \frac{dy}{dx} = \frac{1}{\tan\left(\frac{\pi}{4} + \frac{x}{2}\right)} \cdot \frac{d}{dx} \left(\tan\left(\frac{\pi}{4} + \frac{x}{2}\right)\right) \qquad (\text{'}\tan u\text{' Form})$$

$$= \frac{1}{\tan\left(\frac{\pi}{4}+\frac{x}{2}\right)} . \sec^2\left(\frac{\pi}{4}+\frac{x}{2}\right) . \frac{d}{dx}\left(\frac{\pi}{4}+\frac{x}{2}\right)$$

$$= \frac{\cos\left(\frac{\pi}{4}+\frac{x}{2}\right)}{\sin\left(\frac{\pi}{4}+\frac{x}{2}\right)} . \frac{1}{\cos^2\left(\frac{\pi}{4}+\frac{x}{2}\right)} . \left(0+\frac{1}{2}\right)$$

$$= \frac{1}{2\sin\left(\frac{\pi}{4}+\frac{x}{2}\right)\cos\left(\frac{\pi}{4}+\frac{x}{2}\right)} = \frac{1}{\sin\left(\frac{\pi}{2}+x\right)} = \frac{1}{\cos x} = \sec x.$$

$\therefore \quad \mathbf{\frac{dy}{dx} - \sec x = 0.}$ | $\sin 2A = 2 \sin A \cos A$

Example 2. *Differentiate the following functions w.r.t. x :*

(*i*) $\log \sin 3x$ (*ii*) $\log (x + 5)^6$ (*iii*) $\log (x + \sqrt{1 + x^2})$

(*iv*) $e^x \log \sin 2x$ (*v*) $\dfrac{\log x + e^{-x}}{\sin 2x}$ (*vi*) $\sin [\sin (\log 3x)]$.

Sol. (*i*) Let $y = \log \sin 3x$. ('log *u*' Form)

$$\therefore \quad \frac{dy}{dx} = \frac{1}{\sin 3x} . \frac{d}{dx} (\sin 3x) \qquad \text{('sin } u\text{' Form)}$$

$$= \frac{1}{\sin 3x} . \cos 3x \frac{d}{dx}(3x) = \mathbf{3 \cot 3x.}$$

(*ii*) Let $y = \log (x + 5)^6$. $\quad \therefore \quad y = 6 \log (x + 5)$ $\quad (\because \log m^n = n \log m)$

$$\therefore \quad \frac{dy}{dx} = 6 . \frac{1}{x+5} \frac{d}{dx} (x + 5) = \frac{6}{x+5} . (1 + 0) = \mathbf{\frac{6}{x+5}}.$$

(*iii*) Let $y = \log \left(x + \sqrt{1+x^2}\right)$. ('log *u*' Form)

$$\therefore \quad \frac{dy}{dx} = \frac{1}{x+\sqrt{1+x^2}} . \frac{d}{dx}\left(x+\sqrt{1+x^2}\right) = \frac{1}{x+\sqrt{1+x^2}}\left[1+\frac{d}{dx}(1+x^2)^{1/2}\right]$$

$$= \frac{1}{x+\sqrt{1+x^2}}\left[1+\frac{1}{2\sqrt{1+x^2}}(0+2x)\right] = \frac{1}{x+\sqrt{1+x^2}}\left[1+\frac{x}{\sqrt{1+x^2}}\right]$$

$$= \frac{1}{x+\sqrt{1+x^2}}\left[\frac{\sqrt{1+x^2}+x}{\sqrt{1+x^2}}\right] = \mathbf{\frac{1}{\sqrt{1+x^2}}}.$$

(*iv*) Let $y = e^x \log \sin 2x$. ('*uv*' Form)

$$\therefore \quad \frac{dy}{dx} = e^x \frac{d}{dx} (\log \sin 2x) + \log \sin 2x \frac{d}{dx}(e^x)$$

$$= e^x \frac{1}{\sin 2x} \frac{d}{dx} (\sin 2x) + \log \sin 2x . e^x$$

$$= \frac{e^x}{\sin 2x} \cos 2x.2 + \log \sin 2x \,.\, e^x = \mathbf{e^x\,[\,2\cot 2x + \log \sin 2x]}.$$

(*v*) Let $\quad y = \dfrac{\log x + e^{-x}}{\sin 2x}$.

$$\therefore \quad \frac{dy}{dx} = \frac{\sin 2x\left[\dfrac{1}{x} + e^{-x}(-1)\right] - [\log x + e^{-x}]\,(2\cos 2x)}{\sin^2 2x}$$

$$= \frac{\dfrac{1}{x} - e^{-x}}{\sin 2x} - \frac{2\cot 2x\,[\log x + e^{-x}]}{\sin 2x} = \frac{1 - xe^{-x}}{x \sin 2x} - \frac{2\cot 2x\,[\log x + e^{-x}]}{\sin 2x}$$

$$= \mathbf{\frac{1 - xe^{-x} - 2x\cot 2x\,[\log x + e^{-x}]}{x \sin 2x}}.$$

(*vi*) Let $\quad y = \sin[\sin(\log 3x)]$.

$$\therefore \quad \frac{dy}{dx} = \cos[\sin(\log 3x)]\,\frac{d}{dx}[\sin(\log 3x)]$$

$$= \cos[\sin(\log 3x)]\cos(\log 3x)\,\frac{d}{dx}(\log 3x)$$

$$= \cos[\sin(\log 3x)]\cos(\log 3x)\,.\,\frac{1}{3x}\,.\,3$$

$$= \mathbf{\frac{1}{x}\cos[\sin(\log 3x)]\cos(\log 3x)}.$$

Example 3. *Differentiate the following functions w.r.t. x :*

(*i*) $\log_{\sin x} \cos x$ $\qquad$ (*ii*) $\log \sin \sqrt{x^2+1}$.

Sol. (*i*) Let $y = \log_{\sin x} \cos x$.

$$\therefore \quad y = \frac{\log \cos x}{\log \sin x} \qquad \left(\because \; \log_n m = \frac{\log m}{\log n}\right)$$

$$\therefore \quad \frac{dy}{dx} = \frac{(\log \sin x)\dfrac{d}{dx}(\log \cos x) - \log \cos x\,.\,\dfrac{d}{dx}(\log \sin x)}{(\log \sin x)^2}$$

$$= \frac{(\log \sin x)\dfrac{1}{\cos x}.(-\sin x) - (\log \cos x)\dfrac{1}{\sin x}.\cos x}{(\log \sin x)^2}$$

$$= -\mathbf{\frac{(\log \sin x)\tan x + (\log \cos x)\cot x}{(\log \sin x)^2}}.$$

(*ii*) Let $\quad y = \log \sin \sqrt{x^2+1}$. $\qquad$ ('log *u*' Form)

$$\therefore \quad \frac{dy}{dx} = \frac{1}{\sin\sqrt{x^2+1}}\,\frac{d}{dx}(\sin\sqrt{x^2+1}) \qquad \text{('sin } u\text{' Form)}$$

$$= \frac{1}{\sin\sqrt{x^2+1}}\,.\,\cos\sqrt{x^2+1}\,.\,\frac{d}{dx}(\sqrt{x^2+1}) \qquad \text{('}\sqrt{u}\text{' Form)}$$

$$= \cot\sqrt{x^2+1}\,.\,\frac{1}{2}(x^2+1)^{-1/2}(2x+0) = \mathbf{\frac{x\cot\sqrt{x^2+1}}{\sqrt{x^2+1}}}.$$

Example 4. *Differentiate the following functions w.r.t. x :*

(*i*) $\log \left[e^x \left(\frac{x-2}{x+2} \right)^{3/4} \right]$ (*ii*) $\log \sqrt{\frac{1+\cos^2 x}{1-e^{2x}}}$.

Sol. (*i*) Let $y = \log \left[e^x \left(\frac{x-2}{x+2} \right)^{3/4} \right]$

$$\therefore \quad y = \log e^x + \log \left(\frac{x-2}{x+2} \right)^{3/4} = x \log e + \frac{3}{4} \log \frac{x-2}{x+2}$$

$$= x + \frac{3}{4} [\log (x-2) - \log (x+2)]$$

$$\therefore \quad \frac{dy}{dx} = 1 + \frac{3}{4} \left[\frac{1}{x-2}.1 - \frac{1}{x+2}.1 \right] = 1 + \frac{3}{4} \left[\frac{(x+2)-(x-2)}{x^2-4} \right] = 1 + \frac{3}{x^2-4} = \mathbf{\frac{x^2-1}{x^2-4}}.$$

(*ii*) Let $y = \log \sqrt{\frac{1+\cos^2 x}{1-e^{2x}}}$

$$\therefore \quad y = \frac{1}{2} \log \frac{1+\cos^2 x}{1-e^{2x}} = \frac{1}{2} [\log (1+\cos^2 x) - \log (1-e^{2x})]$$

$$\therefore \quad \frac{dy}{dx} = \frac{1}{2} \left[\frac{1}{1+\cos^2 x} \cdot \frac{d}{dx}(1+\cos^2 x) - \frac{1}{1-e^{2x}} \frac{d}{dx}(1-e^{2x}) \right]$$

$$= \frac{1}{2} \left[\frac{0 + 2\cos x \cdot (-\sin x)}{1+\cos^2 x} - \frac{0 - e^{2x}.2}{1-e^{2x}} \right]$$

$$= \mathbf{-\frac{\cos x \sin x}{1+\cos^2 x} + \frac{e^{2x}}{1-e^{2x}}}.$$

EXERCISE 12.10

Find the derivatives of the following functions w.r.t. x :

1. (*i*) $\log x^2$ (*ii*) $\log \sqrt{x}$ (*iii*) $\log \cos x$
   (*iv*) $\log_a \sec x$ (*v*) $\cos (\log x)$ (*vi*) $\log (x^2 + a^2)$
2. (*i*) $\cos (\log x)^2$ (*ii*) $\log \tan \left(\frac{\pi}{4} + \frac{x}{2} \right)$ (*iii*) $\frac{1}{\log \cos x}$
   (*iv*) $5^{\log \sin x}$ (*v*) $\frac{e^{2x}}{\log x}$ (*vi*) $\frac{1}{x} \log x$.
3. (*i*) $x^5 + \log x + 7$ (*ii*) $3 \tan x + 5 \log_a x + \sqrt{x} - 3e^x + \frac{1}{x}$
   (*iii*) $e^x + 3 \sec x + \log x^2$ (*iv*) $\cot x + 2 \cos x + 3 \sin x - \frac{1}{2} e^x - \log_a x$.

4. (i) $\dfrac{x^2 + e^x}{\log x + 20}$ (ii) $\dfrac{\log x - 4x^3}{\log x + 4x^3}$ (iii) $\dfrac{\sin x + \log x}{x^7 + e^x}$.

Find the derivatives of the following functions w.r.t. x (Q. No. 5–6) :

5. (i) $e^{-ax^2} \sin \log x$ (ii) $\sin (e^x \log x)$ (iii) $\dfrac{\log x}{1 + x \log x}$

(iv) $\dfrac{e^x + \log x}{\sin 3x}$ (v) $e^{\log (x + \sqrt{x^2 + a^2})}$ (vi) $\log \sqrt{\dfrac{1 + \sin^2 x}{1 - \tan x}}$

6. (i) $\log \dfrac{a + b \tan x}{a - b \tan x}$ (ii) $\log \dfrac{1 + \sqrt{x}}{1 - \sqrt{x}}$ (iii) $\log \dfrac{1 - x \sin x}{1 + x \sin x}$

(iv) $\log \sqrt{\dfrac{x + ab}{x - ab}}$ (v) $\log \sqrt{\dfrac{1 + \sin x}{1 - \sin x}}$ (vi) $\log \sqrt{\dfrac{1 - \cos x}{1 + \cos x}}$

7. Find $\dfrac{ds}{dt}$ for the function $s = \tan^{-1} t + \log \sqrt{\dfrac{t - 1}{t + 1}}$.

Answers

1. (i) $\dfrac{2}{x}$ (ii) $\dfrac{1}{2x}$ (iii) $- \tan x$ (iv) $\tan x \,.\, \log_a e$

(v) $-\dfrac{\sin (\log x)}{x}$ (vi) $\dfrac{2x}{x^2 + a^2}$ **2.** (i) $- 2 \sin (\log x)^2 \,.\, \dfrac{\log x}{x}$

(ii) $\sec x$ (iii) $\dfrac{\tan x}{(\log \cos x)^2}$ (iv) $5^{\log \sin x} \cot x \,.\, \log 5$ (v) $\dfrac{e^{2x}(2x \log x - 1)}{x(\log x)^2}$

(vi) $\dfrac{1 - \log x}{x^2}$ **3.** (i) $5x^4 + \dfrac{1}{x}$ (ii) $3 \sec^2 x + \dfrac{5}{x} \log_a e + \dfrac{1}{2\sqrt{x}} - 3e^x - \dfrac{1}{x^2}$

(iii) $e^x + 3 \sec x \tan x + \dfrac{2}{x}$ (iv) $- \operatorname{cosec}^2 x - 2 \sin x + 3 \cos x - \dfrac{1}{2} e^x - \dfrac{1}{x} \log_a e$

4. (i) $\dfrac{x(\log x + 20)(2x + e^x) - x^2 - e^x}{x(\log x + 20)^2}$ (ii) $\dfrac{8x^2 (1 - 3 \log x)}{(\log x + 4x^3)^2}$

(iii) $\dfrac{(x^7 + e^x)(x \cos x + 1) - x(\sin x + \log x)(7x^6 + e^x)}{x(x^7 + e^x)^2}$

5. (i) $e^{-ax^2} \left(\dfrac{\cos (\log x)}{x} - 2ax \sin (\log x) \right)$ (ii) $\cos (e^x \log x).\, e^x \left(\dfrac{1 + x \log x}{x} \right)$

(iii) $\dfrac{1 - x (\log x)^2}{x (1 + x \log x)^2}$ (iv) $\dfrac{\sin 3x \,.\, (xe^x + 1) - 3x \cos 3x \,.\, (e^x + \log x)}{x \sin^2 3x}$

(v) $1 + \dfrac{x}{\sqrt{x^2 + a^2}}$ (vi) $\dfrac{1}{2} \left[\dfrac{\sin 2x}{1 + \sin^2 x} + \dfrac{\sec^2 x}{1 - \tan x} \right]$

6. (i) $\dfrac{2ab \sec^2 x}{a^2 - b^2 \tan^2 x}$ (ii) $\dfrac{1}{\sqrt{x}(1 - x)}$ (iii) $\dfrac{2 (\sin x + x \cos x)}{x^2 \sin^2 x - 1}$ (iv) $\dfrac{ab}{a^2 b^2 - x^2}$

(v) $\sec x$ (vi) $\operatorname{cosec} x$ **7.** $\dfrac{2t^2}{t^4 - 1}$.

VIII. DIFFERENTIATION OF IMPLICIT FUNCTIONS

12.27. IMPLICIT FUNCTIONS

An equation of the form $f(x, y) = 0$, not expressing y explicitly in terms of x is called an **implicit equation.**

An implicit equation may determine one or more functions. A function determined by an implicit equation is called an **implicit function.**

Examples. (*i*) The implicit equation $2x - 3y + 7 = 0$ determine one implicit function :

$$y = \frac{2x + 7}{3}.$$

(*ii*) The implicit equation $x^2 + y^2 - 4 = 0$ determine two implicit functions :

$$y = \sqrt{4 - x^2},\ -2 \le x \le 2$$

and $$y = -\sqrt{4 - x^2},\ -2 \le x \le 2.$$

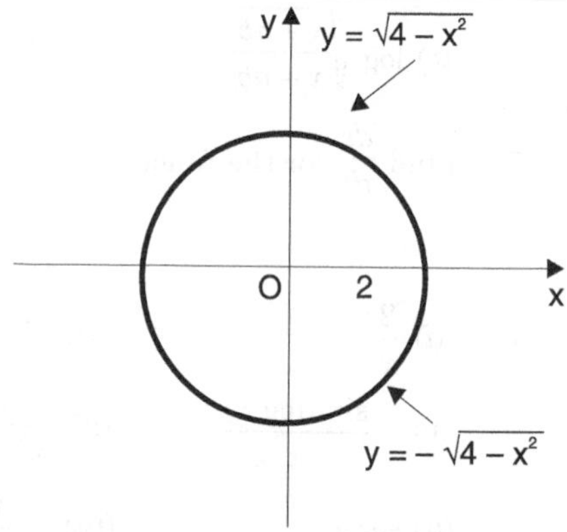

The graph of the first function is the upper semi-circle and the graph of the second function is lower semi-circle as shown in the figure.

It is worth while to note that the implicit equation $x^2 + y^2 - 4 = 0$ represent the whole circle with centre at (0, 0) and radius 2.

(*iii*) The implicit equation $x^2 + y^2 + 4 = 0$ does not represent any implicit function, because $y = \pm\sqrt{-4 - x^2}$ is not defined for any real number value of x.

(*iv*) The implicit equation $x^6 + 5y^6 + 2y^5x^3 - xy + 9 = 0$ cannot be solved for y. The process of showing whether implicit equations of this type represent implicit functions or not is beyond the scope of this book. Our intention is to find derivatives of implicit functions. The derivative $\frac{dy}{dx}$ of y w.r.t. x, in case of implicit functions, is carried by using *implicit differentiation*. In implicit differentiation, we need not express y explicitly in terms of x.

12.28. IMPLICIT DIFFERENTIATION

The process of finding derivatives of implicit functions is called **implicit differentiation.** For this purpose, any of the following methods is followed :

(*i*) In this method, the implicit equation is solved for y and it is expressed explicitly in terms of x and differentiation of y is carried. This method is found useful only when y is easily expressible in terms of x.

(*ii*) In this method, y is thought of as a function of x, and both members of the implicit equation are differentiated w.r.t. x. The resulting equation is solved to find the value of $\frac{dy}{dx}$.

Remark 1. The following derivatives will be found very useful in implicit differentiation :

(*i*) $\frac{d}{dx}(y^n) = ny^{n-1}\frac{dy}{dx}$

(*ii*) $\frac{d}{dx}(x^m y^n) = x^m \frac{d}{dx}(y^n) + y^n \frac{d}{dx}(x^m) = x^m . ny^{n-1}\frac{dy}{dx} + y^n . mx^{m-1} = x^{m-1}\ y^{n-1}\left[nx\frac{dy}{dx} + my\right]$

(*iii*) $\frac{d}{dx}(\sin y^n) = \cos y^n \,.\, ny^{n-1}\frac{dy}{dx}$. (*iv*) $\frac{d}{dx}(e^y) = e^y\frac{dy}{dx}$, etc.

Caution. It should be observed that $\frac{d}{dy}(y^2 + 2y) = 2y + 2$ but $\frac{d}{dx}(y^2 + 2y) = (2y + 2)\frac{dy}{dx}$.

Similarly, $$\frac{d}{dy}(\tan y) = \sec^2 y \quad \text{and} \quad \frac{d}{dx}(\tan y) = \sec^2 y\,\frac{dy}{dx}.$$

Remark 2. If $\frac{dx}{dy}$ exists and is non-zero, then we have $\frac{dy}{dx} = \frac{1}{dx/dy}$. This is so because

$$\frac{dy}{dx} = \lim_{\delta x \to 0}\frac{\delta y}{\delta x} = \lim_{\delta x \to 0}\frac{\delta y/\delta y}{\delta x/\delta y} = \frac{1}{\lim\limits_{\delta y \to 0}\frac{\delta x}{\delta y}} = \frac{1}{\frac{dx}{dy}}.$$

WORKING RULES FOR IMPLICIT DIFFERENTIATION

Step I. *Write the implicit equation for which* $\frac{dy}{dx}$ *is to be found out.*

Step II. *Differentiate both members of the implicit equation w.r.t. x.*

Step III. *Take all terms containing* $\frac{dy}{dx}$ *to one side and all other terms to the other side.*

Step IV. *Find* $\frac{dy}{dx}$ *using this equation.*

ILLUSTRATIVE EXAMPLES

Example 1. *If* $x^2 + y^2 + 6x - 16 = 0$, *find* $\frac{dy}{dx}$ *when* $x = 0$.

Sol. We have $x^2 + y^2 + 6x - 16 = 0$. ...(1)

Differentiating w.r.t. x, we get

$$\frac{d}{dx}(x^2) + \frac{d}{dx}(y^2) + \frac{d}{dx}(6x) - \frac{d}{dx}(16) = \frac{d}{dx}(0).$$

$$\Rightarrow \quad 2x + 2y\frac{dy}{dx} + 6(1) - 0 = 0 \quad \Rightarrow \quad \frac{dy}{dx} = \frac{-6-2x}{2y} = \frac{-x-3}{y}.$$

When $x = 0$, (1) $\Rightarrow 0 + y^2 + 6(0) - 16 = 0 \Rightarrow y = \pm 4$.

At $(0, 4)$, $\frac{dy}{dx} = \frac{-0-3}{4} = -\mathbf{\frac{3}{4}}$. At $(0, -4)$, $\frac{dy}{dx} = \frac{-0-3}{-4} = \mathbf{\frac{3}{4}}$.

Example 2. *Find* $\frac{dy}{dx}$ *for the following implicit equations :*

(*i*) $x^2 + y^2 = 3xy + 5$ (*ii*) $x^n + y^n = a^n$

(*iii*) $ax^2 + 2hxy + by^2 + 2gx + 2fy + c = 0$ (*iv*) $\log xy = x^2 + y^2$.

Sol. (*i*) We have $x^2 + y^2 = 3xy + 5$.

Differentiating w.r.t. x, we get $2x + 2y\frac{dy}{dx} = 3\left[x\frac{dy}{dx} + y.1\right] + 0$

$$\Rightarrow \quad 2y\frac{dy}{dx} - 3x\frac{dy}{dx} = 3y - 2x \quad \Rightarrow \quad \frac{dy}{dx} = \mathbf{\frac{3y - 2x}{2y - 3x}}.$$

(*ii*) We have $x^n + y^n = a^n$.

Differentiating w.r.t. x, we get $\quad nx^{n-1} + ny^{n-1}\dfrac{dy}{dx} = 0.$

$$\Rightarrow \qquad ny^{n-1}\frac{dy}{dx} = -nx^{n-1} \;\Rightarrow\; \frac{dy}{dx} = -\frac{nx^{n-1}}{ny^{n-1}} = -\frac{\mathbf{x^{n-1}}}{\mathbf{y^{n-1}}}.$$

(*iii*) We have $ax^2 + 2hxy + by^2 + 2gx + 2fy + c = 0.$

Differentiating w.r.t. x, we get

$$a.2x + 2h\left[x\frac{dy}{dx} + y.1\right] + b.2y\frac{dy}{dx} + 2g.1 + 2f\frac{dy}{dx} + 0 = 0.$$

$$\Rightarrow \qquad \frac{dy}{dx}(2hx + 2by + 2f) = -(2ax + 2hy + 2g)$$

$$\Rightarrow \qquad \frac{dy}{dx} = -\frac{2ax + 2hy + 2g}{2hx + 2by + 2f} = -\frac{\mathbf{ax + hy + g}}{\mathbf{hx + by + f}}.$$

(*iv*) We have $\quad \log xy = x^2 + y^2. \quad \therefore \quad \log x + \log y = x^2 + y^2.$

Differentiating w.r.t. x, we get $\quad \dfrac{1}{x} + \dfrac{1}{y}.\dfrac{dy}{dx} = 2x + 2y\dfrac{dy}{dx}.$

$$\Rightarrow \qquad \frac{dy}{dx}\left(\frac{1}{y} - 2y\right) = 2x - \frac{1}{x} \qquad \Rightarrow \qquad \frac{dy}{dx}\left(\frac{1-2y^2}{y}\right) = \frac{2x^2-1}{x}$$

$$\therefore \qquad \frac{dy}{dx} = \frac{\mathbf{y(2x^2 - 1)}}{\mathbf{x(1 - 2y^2)}}.$$

Example 3. *Find* $\dfrac{dy}{dx}$ *for the following implicit equations :*

(*i*) *cos* $(x + y) = y$ *sin* $2x$ (*ii*) *sin* $(xy) + \dfrac{x}{y} = x^2 - y$

(*iii*) y *sec* x + *tan* $y + xy = 0$ (*iv*) $xy + xe^{-y} + ye^x = x^2$.

Sol. (*i*) We have $\cos(x + y) = y \sin 2x$. Differentiating w.r.t. x, we get

$$-\sin(x + y)\frac{d}{dx}(x + y) = y.2\cos 2x + \sin 2x\frac{dy}{dx}.$$

$$\Rightarrow \qquad -\sin(x + y).\left(1 + \frac{dy}{dx}\right) = 2y\cos 2x + \sin 2x\frac{dy}{dx}$$

$$\Rightarrow \qquad \frac{dy}{dx}(-\sin(x + y) - \sin 2x) = 2y\cos 2x + \sin(x + y)$$

$$\Rightarrow \qquad \frac{dy}{dx} = -\frac{\mathbf{2y\cos 2x + \sin(x + y)}}{\mathbf{\sin(x + y) + \sin 2x}}.$$

(*ii*) We have $\sin(xy) + \dfrac{x}{y} = x^2 - y$. Differentiating w.r.t. x, we get

$$\cos(xy)\left[x\frac{dy}{dx} + y.1\right] + \frac{y.1 - x\dfrac{dy}{dx}}{y^2} = 2x - \frac{dy}{dx}.$$

$$\Rightarrow \quad x\cos(xy)\frac{dy}{dx} + y\cos(xy) + \frac{1}{y} - \frac{x}{y^2}\frac{dy}{dx} = 2x - \frac{dy}{dx}$$

$$\Rightarrow \quad \frac{dy}{dx}\left[x\cos(xy) - \frac{x}{y^2} + 1\right] = 2x - \frac{1}{y} - y\cos(xy)$$

$$\Rightarrow \quad \frac{dy}{dx}\left[\frac{xy^2\cos(xy) - x + y^2}{y^2}\right] = \frac{2xy - 1 - y^2\cos(xy)}{y}$$

$$\Rightarrow \quad \frac{dy}{dx} = \frac{\mathbf{y\,[2xy - 1 - y^2\cos(xy)]}}{\mathbf{xy^2\cos(xy) - x + y^2}}.$$

(*iii*) We have $y \sec x + \tan y + xy = 0$. Differentiating w.r.t. x, we get

$$\left(y\sec x\tan x + \sec x\frac{dy}{dx}\right) + \sec^2 y\frac{dy}{dx} + \left(x\frac{dy}{dx} + y\,.\,1\right) = 0.$$

$$\Rightarrow \quad \frac{dy}{dx}(\sec x + \sec^2 y + x) = -(y\sec x\tan x + y)$$

$$\therefore \quad \frac{dy}{dx} = -\frac{\mathbf{y\sec x\tan x + y}}{\mathbf{\sec x + \sec^2 y + x}}.$$

(*iv*) We have $xy + xe^{-y} + ye^x = x^2$. Differentiating w.r.t. x, we get

$$\left(x\frac{dy}{dx} + y.1\right) + \left(xe^{-y}(-1)\frac{dy}{dx} + e^{-y}.1\right) + \left(ye^x + e^x\frac{dy}{dx}\right) = 2x.$$

$$\Rightarrow \quad \frac{dy}{dx}(x - xe^{-y} + e^x) = 2x - ye^x - e^{-y} - y$$

$$\therefore \quad \frac{dy}{dx} = \frac{\mathbf{2x - ye^x - e^{-y} - y}}{\mathbf{x - xe^{-y} + e^x}}.$$

Example 4. *If* $x\sqrt{1+y} + y\sqrt{1+x} = 0$, *show that* $\frac{dy}{dx} = -\frac{1}{(x+1)^2}$.

Sol. We have $\quad x\sqrt{1+y} + y\sqrt{1+x} = 0.$

$\Rightarrow \quad x\sqrt{1+y} = -y\sqrt{1+x} \quad \Rightarrow \quad x^2(1+y) = y^2(1+x)$

$\Rightarrow \quad x^2 + x^2y = y^2 + xy^2 \quad \Rightarrow \quad x^2 - y^2 = xy^2 - x^2y \quad \Rightarrow \quad (x-y)(x+y) = -xy(x-y)$

$\Rightarrow \quad (x-y)(x+y+xy) = 0 \quad \Rightarrow \quad x + y = -xy$

(Since $x = y$ does not satisfy the given implicit equation, we have $x - y \neq 0$)

$$\Rightarrow \quad y(1+x) = -x \quad \Rightarrow \quad y = -\frac{x}{1+x}$$

$$\therefore \quad \frac{dy}{dx} = -\frac{(1+x).1 - x.1}{(1+x)^2} = -\frac{\mathbf{1}}{\mathbf{(1+x)^2}}.$$

Example 5. *If* $e^x + e^y = e^{x+y}$, *show that* $\frac{dy}{dx} + e^{y-x} = 0$.

Sol. We have $\quad e^x + e^y = e^{x+y}.$

$$\Rightarrow \quad e^x + e^y = e^xe^y \quad \Rightarrow \quad e^x = e^y(e^x - 1) \quad \Rightarrow \quad e^y = \frac{e^x}{e^x - 1} \qquad \ldots(1)$$

Differentiating w.r.t. x, we get

$$e^y\frac{dy}{dx} = \frac{(e^x-1)e^x - e^x(e^x - 0)}{(e^x-1)^2} = \frac{e^{2x} - e^x - e^{2x}}{(e^x-1)^2} = \frac{-e^x}{(e^x-1)^2}$$

$$\therefore \quad \frac{dy}{dx} = \frac{1}{e^y}\left(\frac{-e^x}{(e^x-1)^2}\right) = -\frac{e^{x-y}}{(e^x-1)^2} = -\frac{e^{x-y}}{(e^x/e^y)^2} \quad \text{(Using (1))}$$

$$= -\frac{e^{x-y}}{(e^{x-y})^2} = -\frac{1}{e^{x-y}} = -e^{-(x-y)} = -e^{y-x}$$

$$\mathbf{\frac{dy}{dx} + e^{y-x} = 0.}$$

Example 6. *If $x \sin (a+y) = \sin y$, show that $\frac{dy}{dx} = \frac{\sin^2 (a+y)}{\sin a}$.*

Sol. Given

$$\sin y = x \sin (a+y) \quad \text{...(1)}$$

$$= x(\sin a \cos y + \cos a \sin y)$$

$$\sin y = x \sin a \cos y + x \cos a \sin y$$

$$\Rightarrow \quad (1 - x \cos a) \sin y = x \sin a \cos y \quad \text{...(2)}$$

Differentiating (2) w.r.t. x, we get

$$(1 - x \cos a) \cos y \frac{dy}{dx} + \sin y (-\cos a) = \sin a \left(-x \sin y \frac{dy}{dx} + \cos y\right)$$

$$\Rightarrow \quad [(1 - x \cos a) \cos y + x \sin a \sin y]\frac{dy}{dx} = \sin a \cos y + \cos a \sin y$$

$$\Rightarrow \quad (\cos y - x \cos a \cos y + x \sin a \sin y) \frac{dy}{dx} = \sin (a+y)$$

$$\Rightarrow \quad [\cos y - x (\cos a \cos y - \sin a \sin y)] \frac{dy}{dx} = \sin (a+y)$$

$$\Rightarrow \quad (\cos y - x \cos (a+y)) \frac{dy}{dx} = \sin (a+y)$$

$$\Rightarrow \quad \frac{dy}{dx} = \frac{\sin (a+y)}{\cos y - x \cos (a+y)} = \frac{\sin (a+y)}{\cos y - \dfrac{\sin y \cos (a+y)}{\sin (a+y)}} \quad |\ \text{Using (1)}$$

$$= \frac{\sin^2 (a+y)}{\sin (a+y) \cos y - \sin y \cos (a+y)}$$

$$= \frac{\sin^2 (a+y)}{\sin (a+y-y)} = \frac{\sin^2 (a+y)}{\sin a}$$

Alternative. Given

$$x = \frac{\sin y}{\sin (a+y)} \quad \text{...(1)}$$

Differentiating (1) w.r.t. y, we get

$$\frac{dx}{dy} = \frac{\sin (a+y) \dfrac{d}{dy} (\sin y) - \sin y \cos(a+y)}{\sin^2 (a+y)}$$

$$= \frac{\sin (a+y) \cos y - \sin y \cos (a+y)}{\sin^2 (a+y)}$$

$$= \frac{\sin(a+y-y)}{\sin^2(a+y)} = \frac{\sin a}{\sin^2(a+y)}$$

$$\therefore \quad \frac{dy}{dx} = \frac{1}{\frac{dx}{dy}} = \frac{\sin^2(a+y)}{\sin a}.$$

EXERCISE 12.11

1. Find $\frac{dy}{dx}$ for the following implicit equations :

(i) $4x - 7y + 15 = 0$ (ii) $x^2 - y^2 + 9 = 0$ (iii) $xy = c^2$

(iv) $y^2 - 4ax = 0$ (v) $\frac{x^2}{a^2} + \frac{y^2}{b^2} = 2003$ (vi) $x^{2/3} + y^{2/3} = a^{2/3}$.

2. If $\sqrt{x} + \sqrt{y} = 5$, find the value of $\frac{dy}{dx}$ at (4, 9).

3. Find $\frac{dy}{dx}$ for the following implicit equations :

(i) $2x^2 - 3xy + 4y^2 = 8$ (ii) $xy - x^3 + y^2 = 0$ (iii) $x^3 + y^3 = 3axy$ (*P.T.U. B.Pharma, May 2006*)

(iv) $xy^2 - x^2y - 4 = 0$ (v) $(x^2 + y^2)^2 = xy$ (vi) $4x + 3y = \log(4x - 3y)$

(vii) $y(y + 1) = x(x + 1)(x + 2)$ (viii) $xy \log(x + y) = 1$ (ix) $\log\sqrt{x^2 + y^2} = \tan^{-1}\frac{y}{x}$

(x) $xy = 100(x + y)$ (xi) $x^3 + y^2 = 30xy$ (xii) $\sin y + \log y = x^2 + 18x + 3$.

4. (i) If $y^2 \sin x + y \tan x + (1 + x^2)\cos x = 0$, find $\frac{dy}{dx}$.

(ii) If $xy = \tan(xy)$, show that $\frac{dy}{dx} = -\frac{y}{x}$.

(iii) If $y = x \sin y$, show that $x\frac{dy}{dx} = \frac{y}{1 - x\cos y}$.

5. (i) If $\sec y = x^3$, show that $x\frac{dy}{dx} = \frac{3}{\sqrt{x^6 - 1}}$.

(ii) If $y \tan x - y^2 \cos x + 2x = 0$, find $\frac{dy}{dx}$.

(iii) If $\tan(x + y) + \tan(x - y) = 1$, find $\frac{dy}{dx}$.

6. If $\sin y = x \sin(a + y)$, show that $\frac{dy}{dx} = \frac{\sin^2(a+y)}{\sin a}$.

7. (i) If $y = \sqrt{x + \sqrt{x + \sqrt{x + \ldots\ldots \infty}}}$, show that $(2y - 1)\frac{dy}{dx} = 1$.

(ii) If $y = \sqrt{\cos x + \sqrt{\cos x + \sqrt{\cos x + \ldots\ldots \infty}}}$, show that $\frac{dy}{dx} = \frac{\sin x}{1 - 2y}$.

(*iii*) If $y = \sqrt{\sin x + \sqrt{\sin x + \sqrt{\sin x + \ldots\ldots \infty}}}$, show that $(2y - 1)\dfrac{dy}{dx} = \cos x$.

(*iv*) If $y = \sqrt{\log x + \sqrt{\log x + \sqrt{\log x + \ldots\ldots \infty}}}$, show that $(2y - 1)\dfrac{dy}{dx} = \dfrac{1}{x}$.

(*P.T.U. B.Pharma, June 2003*)

(*v*) $y = x + \cfrac{1}{x + \cfrac{1}{x + \ldots\ldots \text{to } \infty}}$, show that $\dfrac{dy}{dx} = \dfrac{y}{2y - x}$.

Answers

1. (*i*) $\frac{4}{7}$ (*ii*) $\frac{x}{y}$ (*iii*) $-\frac{c^2}{x^2}$ (*iv*) $\frac{2a}{y}$ (*v*) $-\frac{b^2x}{a^2y}$

(*vi*) $-\left(\frac{y}{x}\right)^{1/3}$ **2.** $-\frac{3}{2}$ **3.** (*i*) $\frac{4x - 3y}{3x - 8y}$ (*ii*) $\frac{3x^2 - y}{x + 2y}$ (*iii*) $\frac{ay - x^2}{y^2 - ax}$

(*iv*) $\frac{y(2x - y)}{x(2y - x)}$ (*v*) $\frac{y - 4x(x^2 + y^2)}{4y(x^2 + y^2) - x}$ (*vi*) $\frac{4(1 - 4x + 3y)}{3(1 + 4x - 3y)}$

(*vii*) $\frac{3x^2 + 6x + 2}{2y + 1}$ (*viii*) $-\frac{y(x + (x + y)\log(x + y))}{x(y + (x + y)\log(x + y))}$ (*ix*) $\frac{x + y}{x - y}$

(*x*) $\frac{100 - y}{x - 100}$ (*xi*) $\frac{10y - x^2}{y^2 - 10x}$ (*xii*) $\frac{2y(x + 9)}{y\cos y + 1}$

4. (*i*) $\frac{(1 + x^2)\sin x - (2x + y^2)\cos x - y\sec^2 x}{2y\sin x + \tan x}$ **5.** (*ii*) $\frac{y\sec^2 x + y^2 \sin x + 2}{2y\cos x - \tan x}$

(*iii*) $-\frac{\sec^2(x + y) + \sec^2(x - y)}{\sec^2(x + y) - \sec^2(x - y)}$.

Hints

6. $$x = \frac{\sin y}{\sin(a + y)} \Rightarrow \frac{dx}{dy} = \frac{\sin(a + y)\cos y - \sin y\cos(a + y)}{\sin^2(a + y)} = \frac{\sin a}{\sin^2(a + y)}$$

$$\therefore \quad \frac{dy}{dx} = \frac{1}{dx/dy} = \frac{\sin^2(a + y)}{\sin a}.$$

Alternative method

$$\sin y = x\sin(a + y) \Rightarrow \cos y\frac{dy}{dx} = x\cos(a + y)\frac{dy}{dx} + 1\,.\,\sin(a + y)$$

$$\Rightarrow \frac{dy}{dx} = \frac{\sin(a + y)}{\cos y - x\cos(a + y)} = \frac{\sin(a + y)}{\cos y - \dfrac{\sin y}{\sin(a + y)}\,.\cos(a + y)}$$

$$= \frac{\sin^2(a + y)}{\sin(a + y)\cos y - \sin y\cos(a + y)}.$$

7. (*i*) We have $y = \sqrt{x + y}$. $\therefore$ $y^2 = x + y$.

$$\Rightarrow \quad 2y\frac{dy}{dx} = 1 + \frac{dy}{dx}.$$

IX. LOGARITHMIC DIFFERENTIATION

12.29. DERIVATIVE OF u^v, WHERE u AND v ARE DIFFERENTIABLE FUNCTIONS

Let $y = u^v$.

Taking natural logarithms on both sides, we get

$$\log y = \log u^v = v \log u^*.$$

Differentiating w.r.t. x, we get $\frac{d}{dx}(\log y) = \frac{d}{dx}(v \log u)$.

$$\Rightarrow \quad \frac{1}{y}\frac{dy}{dx} = v\frac{d}{dx}(\log u) + \log u \frac{dv}{dx} = v.\frac{1}{u}.\frac{du}{dx} + \log u \frac{dv}{dx}$$

$$\Rightarrow \quad \frac{dy}{dx} = y\left[\frac{v}{u}\frac{du}{dx} + \log u \frac{dv}{dx}\right]$$

$$\therefore \quad \mathbf{\frac{d}{dx}(u^v) = u^v\left[\frac{v}{u}\frac{du}{dx} + \log u \frac{dv}{dx}\right]}.$$

The process of differentiation after taking logarithms is called **logarithmic differentiation.**

Logarithmic differentiation is also used when the function to be differentiated is a product of several factors.

WORKING RULES FOR SOLVING PROBLEMS

Step I. *Take natural logarithms (i.e., with base 'e') on both sides.*

Step II. *Simplify using the formulae* $\log mn = \log m + \log n$, $\log \frac{m}{n} = \log m - \log n$ *and* $\log m^n = n \log m$.

Step III. *Differentiate both sides w.r.t. x.*

Step IV. *Collect the terms containing* $\frac{dy}{dx}$ *and find* $\frac{dy}{dx}$.

ILLUSTRATIVE EXAMPLES

Example 1. *Differentiate the following functions w.r.t. x :*

(*i*) x^x (*ii*) $(\sin x)^{\cos x}$ (*iii*) $(x^x)\sqrt{x}$

(*G.N.D.U. B.Pharma, April 2003 ; P.T.U. May 2006*)

(*iv*) $(10)^{(10x)}$ (*v*) $x^{(x^x)}$ (*vi*) $(\sin x)^{\cos^{-1} x}$.

(*P.T.U. B.Pharma, May 2005*)

Sol. (*i*) Let $y = x^x$. $\therefore \log y = \log x^x = x \log x$.

Differentiating w.r.t. x, we get $\frac{1}{y}\frac{dy}{dx} = x.\frac{1}{x} + \log x . 1$

$$\Rightarrow \quad \frac{dy}{dx} = y(1 + \log x) = \mathbf{x^x (1 + \log x)}.$$

*We know that $\log x$ is defined only when $x > 0$. So whenever we take logarithm of a 'quantity' then it is always assumed that the 'quantity' would now be allowed to take only positive values.

(*ii*) Let $y = (\sin x)^{\cos x}$ $\therefore$ $\log y = \log (\sin x)^{\cos x} = \cos x \,.\, \log \sin x.$

Differentiating w.r.t. x, we get

$$\frac{1}{y}\frac{dy}{dx} = \cos x \frac{d}{dx}(\log \sin x) + \log \sin x \frac{d}{dx}(\cos x)$$

$$= \cos x \left[\frac{1}{\sin x} . \cos x\right] + \log \sin x \,.\, - \sin x$$

$$\therefore \quad \frac{dy}{dx} = y(\cot x \cos x - \sin x \log \sin x) = \mathbf{(\sin x)^{\cos x} [\cot x \cos x - \sin x \log \sin x].}$$

(*iii*) Let $y = (x^x)\sqrt{x}$. $\therefore$ $y = x^x \,.\, x^{1/2} = x^{x+1/2}$

$$\log y = \log x^{\left(x+\frac{1}{2}\right)} = \left(x + \frac{1}{2}\right) \log x.$$

Differentiating w.r.t. x, we get $\frac{1}{y}\frac{dy}{dx} = \left(x + \frac{1}{2}\right) . \frac{1}{x} + \log x \,.\, (1 + 0).$

$$\Rightarrow \quad \frac{dy}{dx} = y\left[1 + \frac{1}{2x} + \log x\right] = \mathbf{x^{x+1/2}\left[1 + \frac{1}{2x} + \log x\right].}$$

(*iv*) Let $y = 10^{(10^x)}$. $\therefore$ $\log y = \log 10^{(10^x)} = 10^x \log 10.$

Differentiating w.r.t. x, we get $\frac{1}{y}\frac{dy}{dx} = \log 10\, [10^x \log 10].$

$$\therefore \quad \frac{dy}{dx} = y\, 10^x (\log 10)^2 = \mathbf{10^{(10^x)}\, 10^x (\log 10)^2.}$$

Alternative method

Let $y = 10^{(10^x)}$. ('10^u' Form)

$$\therefore \quad \frac{dy}{dx} = 10^{(10^x)} \log 10 \frac{d}{dx}(10^x) = 10^{(10^x)} \log 10 . 10^x \log 10$$

$$= \mathbf{10^{(10^x)}\, 10^x (\log 10)^2.}$$

(*v*) Let $y = x^{(x^x)}$. $\therefore$ $\log y = \log x^{(x^x)} = x^x \log x.$

Differentiating w.r.t. x, we get $\frac{1}{y}\frac{dy}{dx} = x^x . \frac{1}{x} + \log x \,.\, \frac{d}{dx}(x^x)$ ...(1)

Let $u = x^x$.

$\therefore$ $\log u = x \log x.$

$$\Rightarrow \quad \frac{1}{u}\frac{du}{dx} = x . \frac{1}{x} + \log x \,.\, 1 \Rightarrow \frac{du}{dx} = u(1 + \log x) \text{ i.e., } \frac{d}{dx}(x^x) = x^x(1 + \log x).$$

$$\therefore \; (1) \Rightarrow \frac{1}{y}\frac{dy}{dx} = \frac{x^x}{x} + \log x \,.\, x^x (1 + \log x)$$

$$\therefore \quad \frac{dy}{dx} = yx^x\left[\frac{1}{x} + \log x \,.\, (1 + \log x)\right] = \mathbf{x^{(x^x)}\, x^x \left[\frac{1}{x} + \log x \,.\, (1 + \log x)\right].}$$

(*vi*) Let $y = (\sin x)^{\cos^{-1} x}$. $\therefore$ $\log y = \log (\sin x)^{\cos^{-1} x} = \cos^{-1} x \log \sin x$

Differentiating w.r.t. x, we get

$$\frac{1}{y}\frac{dy}{dx} = \cos^{-1} x \,.\, \frac{1}{\sin x} . \cos x + \log \sin x \left(-\frac{1}{\sqrt{1-x^2}}\right) = \cos^{-1} x \cot x - \frac{\log \sin x}{\sqrt{1-x^2}}.$$

$$\therefore \quad \frac{dy}{dx} = y\left[\cos^{-1} x \cot x - \frac{\log \sin x}{\sqrt{1-x^2}}\right] = (\sin x)^{\cos^{-1} x}\left[\cos^{-1} x \cot x - \frac{\log \sin x}{\sqrt{1-x^2}}\right].$$

Example 2. *Differentiate the following functions w.r.t. x :*

(*i*) $(\log x)^x$ (*ii*) $\cos (x^x)$

(*iii*) $(\sec^2 x)^{1/x}$ (*iv*) $x^{\sin 2x + \cos 2x}$.

Sol. (*i*) Let $y = (\log x)^x$. $\therefore \log y = \log (\log x)^x = x \log (\log x)$.

Differentiating w.r.t. x, we get $\frac{1}{y}\frac{dy}{dx} = x\left(\frac{1}{\log x}\cdot\frac{1}{x}\right) + 1 \,.\, \log (\log x)$.

$$\frac{dy}{dx} = y\left[\frac{1}{\log x} + \log \log x\right] = (\log x)^x\left[\frac{1}{\log x} + \log \log x\right].$$

(*ii*) Let $y = \cos (x^x)$. $\therefore \frac{dy}{dx} = -\sin (x^x)\frac{d}{dx}(x^x)$ ...(1)

Let $u = x^x$. $\therefore \log u = \log (x^x) = x \log x$.

Differentiating w.r.t. x, we get $\frac{1}{u}\frac{du}{dx} = x \,.\, \frac{1}{x} + (\log x) \,.\, 1$.

$$\therefore \quad \frac{du}{dx} = u\,[1 + \log x] \text{ i.e., } \frac{d}{dx}(x^x) = x^x\,[1 + \log x].$$

Putting this value in (1), we get $\frac{dy}{dx} = -\sin (x^x) \,.\, x^x\,[1 + \log x]$.

(*iii*) Let $y = (\sec^2 x)^{1/x}$.

$$\therefore \quad \log y = \log (\sec^2 x)^{1/x} = \frac{1}{x}\log (\sec^2 x) = \frac{1}{x} \,.\, 2 \log \sec x = \frac{2}{x}\log \sec x.$$

Differentiating w.r.t. x, we get

$$\frac{1}{y}\frac{dy}{dx} = \frac{2}{x}\left(\frac{1}{\sec x}\,.\,\sec x \tan x\right) + \left(-\frac{2}{x^2}\right)\log \sec x = \frac{2}{x^2}(x \tan x - \log \sec x)$$

$$\therefore \quad \frac{dy}{dx} = y \,.\, \frac{2}{x^2}(x \tan x - \log \sec x) = \frac{2(\sec^2 x)^{1/x}}{x^2}(x \tan x - \log \sec x).$$

(*iv*) Let $y = x^{\sin 2x + \cos 2x}$.

$$\therefore \quad \log y = \log x^{\sin 2x + \cos 2x} = (\sin 2x + \cos 2x) \log x.$$

Differentiating w.r.t. x, we get

$$\frac{1}{y}\frac{dy}{dx} = (\sin 2x + \cos 2x)\frac{1}{x} + (2 \cos 2x - 2 \sin 2x) \log x$$

$$\therefore \quad \frac{dy}{dx} = y\left[\frac{\sin 2x + \cos 2x}{x} + 2(\cos 2x - \sin 2x) \log x\right]$$

$$= x^{\sin 2x + \cos 2x}\left[\frac{\sin 2x + \cos 2x}{x} + 2(\cos 2x - \sin 2x) \log x\right].$$

Example 3. *Differentiate sin (x^x) + $(\sin x)^x$ w.r.t. x.*

Sol. Let $y = \sin(x^x) + (\sin x)^x$. $\therefore$ $\dfrac{dy}{dx} = \dfrac{d}{dx}(\sin(x^x)) + \dfrac{d}{dx}((\sin x)^x)$. ...(1)

Let $u = \sin(x^x)$.

$$\therefore \quad \frac{du}{dx} = \cos(x^x)\frac{d}{dx}(x^x). \quad ...(2)$$

Let $v = x^x$.

$$\Rightarrow \quad \log v = x \log x \Rightarrow \frac{1}{v}\frac{dv}{dx} = x \cdot \frac{1}{x} + \log x \cdot 1 \Rightarrow \frac{d}{dx}(x^x) = x^x(1 + \log x).$$

$$\therefore \ (2) \Rightarrow \frac{du}{dx} = \cos(x^x) \cdot x^x(1 + \log x)$$

or $$\frac{d}{dx}(\sin(x^x)) = x^x \cos(x^x) \cdot (1 + \log x). \quad ...(3)$$

Let $w = (\sin x)^x$. $\therefore$ $\log w = x \log \sin x$.

$$\Rightarrow \quad \frac{1}{w}\frac{dw}{dx} = x \cdot \frac{1}{\sin x}\cos x + \log \sin x \cdot 1 \Rightarrow \frac{dw}{dx} = w\,[x \cot x + \log \sin x]$$

or $$\frac{d}{dx}((\sin x)^x) = (\sin x)^x [x \cot x + \log \sin x]. \quad ...(4)$$

Using (3) and (4) in (1), we get

$$\frac{dy}{dx} = \mathbf{x^x \cos(x^x)(1 + \log x) + (\sin x)^x [x \cot x + \log \sin x].}$$

Example 4. *Differentiate $(\sin x)^{\cos x} + (\cos x)^{\sin x}$ w.r.t. x.*

Sol. Let $y = (\sin x)^{\cos x} + (\cos x)^{\sin x}$.

$$\therefore \quad \frac{dy}{dx} = \frac{d}{dx}((\sin x)^{\cos x}) + \frac{d}{dx}((\cos x)^{\sin x}) \quad ...(1)$$

Let $u = (\sin x)^{\cos x}$. $\therefore$ $\log u = \cos x \log \sin x$

$$\Rightarrow \quad \frac{1}{u}\frac{du}{dx} = \cos x \cdot \frac{1}{\sin x} \cdot \cos x + (\log \sin x)(-\sin x) = \cos x \cot x - \sin x \log \sin x$$

$$\therefore \quad \frac{du}{dx} = u\,[\cos x \cot x - \sin x \log \sin x]$$

$$\Rightarrow \quad \frac{d}{dx}((\sin x)^{\cos x}) = (\sin x)^{\cos x}[\cos x \cot x - \sin x \log \sin x] \quad ...(2)$$

Let $v = (\cos x)^{\sin x}$. $\therefore$ $\log v = \sin x \log \cos x$

$$\Rightarrow \quad \frac{1}{v}\frac{dv}{dx} = \sin x \cdot \frac{1}{\cos x} \cdot (-\sin x) + (\log \cos x)\cos x$$

$$= -\sin x \tan x + \cos x \log \cos x$$

$$\therefore \quad \frac{dv}{dx} = v\,[\cos x \log \cos x - \sin x \tan x]$$

$$\Rightarrow \quad \frac{d}{dx}((\cos x)^{\sin x}) = (\cos x)^{\sin x}[\cos x \log \cos x - \sin x \tan x] \quad ...(3)$$

Using (2) and (3) in (1), we get

$$\frac{dy}{dx} = \mathbf{(\sin x)^{\cos x}[\cos x \cot x - \sin x \log \sin x]}$$
$$\mathbf{+ (\cos x)^{\sin x}[\cos x \log \cos x - \sin x \tan x].}$$

Example 5. *Find* $\frac{dy}{dx}$ *if* $x^y + y^x = 1.$

Sol. We have $x^y + y^x = 1$.

Differentiating w.r.t. x, we get $\frac{d}{dx}(x^y) + \frac{d}{dx}(y^x) = 0.$ ...(1)

Let $u = x^y$. $\therefore$ $\log u = y \log x$.

$\Rightarrow \quad \frac{1}{u}\frac{du}{dx} = y \cdot \frac{1}{x} + \log x \cdot \frac{dy}{dx} \quad \Rightarrow \quad \frac{du}{dx} = u\left(\frac{y}{x} + \log x \frac{dy}{dx}\right)$

or $\quad \frac{d}{dx}(x^y) = x^y\left(\frac{y}{x} + \log x \frac{dy}{dx}\right)$ ...(2)

Let $v = y^x$. $\therefore$ $\log v = x \log y$.

$\Rightarrow \quad \frac{1}{v}\frac{dv}{dx} = x \cdot \frac{1}{y}\frac{dy}{dx} + \log y \cdot 1 \quad \Rightarrow \quad \frac{dv}{dx} = v\left(\frac{x}{y}\frac{dy}{dx} + \log y\right)$

or $\quad \frac{d}{dx}(y^x) = y^x\left(\frac{x}{y}\frac{dy}{dx} + \log y\right).$ ...(3)

Using (2) and (3) in (1), we get $x^y\left(\frac{y}{x} + \log x \frac{dy}{dx}\right) + y^x\left(\frac{x}{y}\frac{dy}{dx} + \log y\right) = 0.$

$\Rightarrow \quad (x^y \log x + xy^{x-1})\frac{dy}{dx} = -(y^x \log y + yx^{y-1})$ or $\frac{dy}{dx} = -\frac{\mathbf{y^x \log y + yx^{y-1}}}{\mathbf{x^y \log x + xy^{x-1}}}.$

Example 6. *If* $y = e^{x + e^{x + e^{x + \ldots\ldots \infty}}}$, *show that* $\frac{dy}{dx} = \frac{y}{1-y}.$

Sol. We have $\quad y = e^{x + e^{x + e^{x + \ldots\ldots \infty}}} \quad \therefore \quad y = e^{x+y}$

$\Rightarrow \quad \log y = \log e^{x+y} = (x + y)\log e = x + y$

$\therefore \quad \log y = x + y.$

Differentiating w.r.t. x, we get $\frac{1}{y}\frac{dy}{dx} = 1 + \frac{dy}{dx}.$

$\Rightarrow \quad \left(\frac{1}{y} - 1\right)\frac{dy}{dx} = 1 \quad \Rightarrow \quad \frac{dy}{dx} = \frac{\mathbf{y}}{\mathbf{1-y}}.$

EXERCISE 12.12

Differentiate the following functions w.r.t. x (Q. No. 1–6) :

1. (*i*) $x^{(x^2)}$ (*ii*) $x^{1/x}$ (*iii*) $x^{\sqrt{x}}$

(*iv*) $x^{\log x}$ (*v*) $x^{\sin x}$ (*vi*) $x^{\sin^{-1} x}$.

Note. $(x^x)^x$ and $x^{(x^2)}$ are same functions.

2. (*i*) $(\sin x)^x$ (*ii*) $(\tan x)^{1/x}$ (*iii*) $(\tan x)^{\cot x}$

(*iv*) $(\tan x)^{\log x}$ (*v*) $(\sin x)^{\log x}$ (*vi*) $x^{\sin x - \cos x}$.

3. (*i*) $(x^2 + 3)^4 (x^2 + 5)^2$ (*ii*) $\sqrt{1-x^2}\, e^{5x^2}$ (*iii*) $e^{2x} . \sin 4x . 2^x$

(*iv*) $3^x x^{x+5} \cos^{-1} x$ (*v*) $(x + 1)^2 (x - 2)^3 (x + 4) \log x$ (*vi*) $\cos x \cos 2x \cos 3x$

(*vii*) $\sqrt{(x-1)(x-2)(x-3)(x-4)}$ (*viii*) $\sqrt{\dfrac{(x-3)(x^2+4)}{3x^2+4x+5}}$ (*ix*) $\sqrt{\dfrac{(x-1)(x-2)}{(x-3)(x-4)(x-5)}}$

4. (*i*) $x^x e^{2(x+3)}$ (*ii*) $\dfrac{e^x \log x}{x^2}$ (*iii*) $\dfrac{e^{x^2} \tan^{-1} x}{\sqrt{1+x^2}}$

(*iv*) $\dfrac{x^4 \sqrt{x+6}}{(3x+5)^2}$ (*v*) $\dfrac{x \cos^{-1} x}{\sqrt{1-x^2}}$

5. (*i*) $x^{\log x} + (\log x)^x$ (*ii*) $(\sin x)^x + \sin^{-1}\sqrt{x}$ (*iii*) $(\sin x)^{\tan x} + (\cos x)^{\sec x}$

(P.T.U. B.Pharma, Dec. 2006, Dec. 2002)

(*iv*) $x^{\cot x} + (\cos x)^{\sin x}$ (*v*) $x^{1/x} + (1/x)^x$ (*vi*) $x^{\cos x} + (\cos x)^x$

6. (*i*) $y = x^{\cot x} + \dfrac{2x^2 - 3}{x^2 + x + 2}$ (*ii*) $y = x^{\sin x - \cos x} + \dfrac{x^2 - 1}{x^2 + 1}$ (*iii*) $y = x^{\tan x} + \sqrt{x + \dfrac{1}{x}}$

(*iv*) $x^{x \cos x} + \dfrac{x^2+1}{x^2-1}$ (*v*) $y = x^{\sin x} + (\sin x)^x$ (*vi*) $e^{\sin x} + (\tan x)^x$

(*vii*) $x^{\cos x} + (\sin x)^{\tan x}$ (*viii*) $(\sin x)^x + (\cos x)^{\tan x}$ (*ix*) $(x \cos x)^x + (x \sin x)^{1/x}$.

7. Find $\dfrac{dy}{dx}$ for the following implicit equations :

(*i*) $x^y = y^x$, (*ii*) $(\cos x)^y = (\sin y)^x$ (*iii*) $(\tan x)^y = (\tan y)^x$

8. If $x^y = e^{x-y}$, then show that $\dfrac{dy}{dx} = \dfrac{\log x}{(1+\log x)^2}$.

9. (*i*) If $x^p y^q = (x + y)^{p+q}$, then show that $\dfrac{dy}{dx} = \dfrac{y}{x}$.

(*ii*) If $x^{7/3} y^{2/3} = (x + y)^3$, then show that $\dfrac{dy}{dx} = \dfrac{y}{x}$.

10. (*i*) If $y = x^{x^{x^{\cdots\infty}}}$, show that $x\dfrac{dy}{dx} = \dfrac{y^2}{1 - y \log x}$.

(*ii*) If $y = (\cos x)^{(\cos x)^{(\cos x)\cdots\infty}}$, show that $\dfrac{dy}{dx} = \dfrac{y^2 \tan x}{y \log \cos x - 1}$.

11. If $y^x + x^y + x^x = a^b$, find $\dfrac{dy}{dx}$.

12. If u, v, w are differentiable functions of x, then show that $\dfrac{d}{dx}(u.v.w) = \dfrac{du}{dx}.v.w + v.\dfrac{dv}{dx}.w + u.v.\dfrac{dv}{dx}$ in two ways first by repeated application of product rule, second by logarithmic differentiation.

Answers

1. (*i*) $x^{x^2+1}(1 + 2 \log x)$ (*ii*) $x^{1/x}\left(\dfrac{1 - \log x}{x^2}\right)$ (*iii*) $x^{\sqrt{x}}\left(\dfrac{2 + \log x}{2\sqrt{x}}\right)$

(*iv*) $x^{\log x}\left(\dfrac{2 \log x}{x}\right)$ (*v*) $x^{\sin x}\left(\dfrac{\sin x}{x} + \cos x . \log x\right)$ (*vi*) $x^{\sin^{-1} x}\left(\dfrac{\sin^{-1} x}{x} + \dfrac{\log x}{\sqrt{1-x^2}}\right)$

2. (*i*) $(\sin x)^x (x \cot x + \log \sin x)$

(*ii*) $(\tan x)^{1/x} \left(\frac{2x \operatorname{cosec} 2x - \log \tan x}{x^2} \right)$

(*iii*) $(\tan x)^{\cot x} \operatorname{cosec}^2 x (1 - \log \tan x)$

(*iv*) $(\tan x)^{\log x} \left(\frac{\log x}{\sin x \cos x} + \frac{\log \tan x}{x} \right)$

(*v*) $(\sin x)^{\log x} \left(\frac{\log \sin x}{x} + \log x \,.\, \cot x \right)$

(*vi*) $x^{\sin x - \cos x} \left(\frac{\sin x - \cos x}{x} + (\cos x + \sin x) \log x \right)$

3. (*i*) $(x^2 + 3)^4 (x^2 + 5)^2 \left[\frac{8x}{x^2 + 3} + \frac{4x}{x^2 + 5} \right]$

(*ii*) $\sqrt{1 - x^2}\; e^{5x^2} \left(10x - \frac{x}{1 - x^2} \right)$

(*iii*) $e^{3x} \sin 4x \,.\, 2^x (3 + 4 \cot 4x + \log 2)$

(*iv*) $3^x x^{x+5} \cos^{-1} x \left(\log 3x + \frac{x + 5}{x} - \frac{1}{\sqrt{1 - x^2} \cos^{-1} x} \right)$

(*v*) $(x + 1)^2 (x - 2)^3 (x + 4) \log x \left(\frac{2}{x + 1} + \frac{3}{x - 2} + \frac{1}{x + 4} + \frac{1}{x \log x} \right)$

(*vi*) $-\cos x \cos 2x \cos 3x (\tan x + 2 \tan 2x + 3 \tan 3x)$

(*vii*) $\frac{1}{2} \sqrt{(x - 1)(x - 2)(x - 3)(x - 4)} \left(\frac{1}{x - 1} + \frac{1}{x - 2} + \frac{1}{x - 3} + \frac{1}{x - 4} \right)$

(*viii*) $\frac{1}{2} \sqrt{\frac{(x - 3)(x^2 + 4)}{3x^2 + 4x + 5}} \left[\frac{1}{x - 3} + \frac{2x}{x^2 + 4} - \frac{6x + 4}{3x^2 + 4x + 5} \right]$

(*ix*) $\frac{1}{2} \sqrt{\frac{(x - 1)(x - 2)}{(x - 3)(x - 4)(x - 5)}} \left[\frac{1}{x - 1} + \frac{1}{x - 2} - \frac{1}{x - 3} - \frac{1}{x - 4} - \frac{1}{x - 5} \right]$

4. (*i*) $x^x e^{2(x+2)} (3 + \log x)$

(*ii*) $\frac{e^x \log x}{x^2} \left(1 + \frac{1}{x \log x} - \frac{2}{x} \right)$

(*iii*) $\frac{e^{x^2} \tan^{-1} x}{\sqrt{1 + x^2}} \left(2x + \frac{1}{(1 + x^2) \tan^{-1} x} - \frac{x}{1 + x^2} \right)$

(*iv*) $\frac{x^4 \sqrt{x + 6}}{(3x + 5)^2} \left(\frac{4}{x} + \frac{1}{2(x + 6)} - \frac{6}{3x + 5} \right)$

(*v*) $\frac{x \cos^{-1} x}{\sqrt{1 - x^2}} \left(\frac{1}{x} - \frac{1}{\sqrt{1 - x^2} \cos^{-1} x} + \frac{x}{1 - x^2} \right)$

5. (*i*) $x^{\log x} \left(\frac{2 \log x}{x} \right) + (\log x)^x \left(\frac{1}{\log x} + \log (\log x) \right)$

(*ii*) $(\sin x)^x (x \cot x + \log \sin x) + \frac{1}{2\sqrt{x}\, \sqrt{1 - x}}$

(*iii*) $(\sin x)^{\tan x} (1 + \sec^2 x \,.\, \log \sin x) + (\cos x)^{\sec x} \sec x \tan x (\log \cos x - 1)$

(*iv*) $x^{\cot x} \left(\frac{\cot x}{x} - \operatorname{cosec}^2 x \log x \right) + (\cos x)^{\sin x} (-\sin x \tan x + \cos x \log \cos x)$

(*v*) $x^{\frac{1}{x} - 2} (1 - \log x) - (1/x)^x (1 + \log x)$

(*vi*) $x^{\cos x} \left(\frac{\cos x}{x} - \sin x \,.\, \log x \right) + (\cos x)^x (\log \cos x - x \tan x)$

6. (i) $x^{\cot x}\left[\frac{\cot x}{x} - \text{cosec}^2 x \log x\right] + \frac{2x^2+14x+3}{(x^2+x+2)^2}$

(ii) $x^{\sin x - \cos x}\left[\frac{\sin x - \cos x}{x} + (\cos x + \sin x)\log x\right] + \frac{4x}{(x^2+1)^2}$

(iii) $x^{\tan x}\left[\sec^2 x \log x + \frac{\tan x}{x}\right] + \frac{x^2-1}{2x\sqrt{x}\sqrt{x^2+1}}$

(iv) $x^{x\cos x}[\cos x.(1+\log x) + \sin x - x\sin x \log x] - \frac{4x}{(x^2-1)^2}$

(v) $x^{\sin x}\left[\frac{\sin x}{x} + \log x.\cos x\right] + (\sin x)^x[x\cot x + \log \sin x]$

(vi) $e^{\sin x}\cos x + (\tan x)^x[\log \tan x + 2x\,\text{cosec}\,2x]$.

(vii) $x^{\cos x}\left[\frac{\cos x}{x} - \sin x \log x\right] + (\sin x)^{\tan x}[1 + \sec^2 x \log \sin x]$

(viii) $(\sin x)^x[\log \sin x + x\cot x] + (\cos x)^{\tan x}[\sec^2 x \log \cos x - \tan^2 x]$

(ix) $(x\cos x)^x[1 - x\tan x + \log(x\cos x)] + \frac{(x\sin x)^{1/x}}{x^2}[x\cot x + 1 - \log(x\sin x)]$

7. (i) $\frac{y(x\log y - y)}{x(y\log x - x)}$ (ii) $\frac{\log \sin y + y\tan x}{\log \cos x - x\cot y}$ (iii) $\frac{\log \tan y - 2y\,\text{cosec}\,2x}{\log \tan x - 2x\,\text{cosec}\,2y}$

11. $-\frac{y^x \log y + yx^{y-1} + x^x(1+\log x)}{x(y^{x-1} + x^{y-1}\log x)}$.

Hints

8. $x^y = e^{x-y} \Rightarrow y\log x = x - y \Rightarrow y = \frac{x}{1+\log x}$.

9. $x^p y^q = (x+y)^{p+q} \Rightarrow p\log x + q\log y = (p+q)\log(x+y) \Rightarrow \frac{p}{x} + \frac{q}{y}\frac{dy}{dx} = \frac{p+q}{x+y}\left(1+\frac{dy}{dx}\right)$.

10. (i) We have $y = x^y$.

11. Let $u = y^x$, $v = x^y$, $w = x^x$ $\therefore$ $\frac{du}{dx} + \frac{dv}{dx} + \frac{dw}{dx} = 0$

Now calculate each term using logarithmic differentiation.

12. (i) $\frac{d}{dx}(u(vw)) = u\frac{d}{dx}(vw) + (vw)\frac{du}{dx} = u\left[v\frac{dw}{dx} + w\frac{dv}{dx}\right] + vw\frac{dv}{dx}$

(ii) $y = uvw \Rightarrow \log y = \log u + \log v + \log w$

$$\Rightarrow \frac{1}{y}\frac{dy}{dx} = \frac{1}{u}\frac{du}{dx} + \frac{1}{v}\frac{dv}{dx} + \frac{1}{w}\frac{dw}{dx} \Rightarrow \frac{dy}{dx} = \frac{y}{u}\frac{du}{dx} + \frac{y}{v}\frac{dv}{dx} + \frac{y}{w}\frac{dw}{dx}.$$

X. DIFFERENTIATION OF FUNCTIONS EXPRESSED IN PARAMETRIC FORM

12.30. PARAMETRIC FORM OF A FUNCTION

Let y be a function of x. If $x = f(t)$ and $y = g(t)$ are functions of a variable t, then the functions $x = f(t)$ and $y = g(t)$ are said to represent the function "y of x" in the **parametric form.** The variable 't' is called the **parameter** of the function "y of x".

For example, the functions $x = t^2 + t - 3$ and $y = t - 1$ represent y as a function of x, in the parametric form.

12.31. DERIVATIVE OF FUNCTION EXPRESSED IN PARAMETRIC FORM

Let $x = f(t)$, $y = g(t)$ represent y as a function of x in the parametric form. Let x and y be differentiable functions of t.

Let δt be an increment of t and δx, δy, the corresponding increments of x, y respectively. We assume, $\delta x \neq 0$.

$$\therefore \quad \frac{\delta y}{\delta x} = \frac{\delta y / \delta t}{\delta x / \delta t} \Rightarrow \lim_{\delta t \to 0} \frac{\delta y}{\delta x} = \lim_{\delta t \to 0} \frac{\delta y / \delta t}{\delta x / \delta t}$$

$$\Rightarrow \quad \lim_{\delta x \to 0} \frac{\delta y}{\delta x} = \frac{\lim_{\delta t \to 0} \delta y / \delta t}{\lim_{\delta t \to 0} \delta x / \delta t}, \text{ provide } \lim_{\delta t \to 0} \frac{\delta x}{\delta t} \neq 0$$

$$\Rightarrow \quad \mathbf{\frac{dy}{dx} = \frac{dy/dt}{dx/dt}}, \textbf{ provided } \mathbf{\frac{dx}{dt} \neq 0.}$$

WORKING RULES FOR SOLVING PROBLEMS

Step I. *Write the functions x and y in terms of the parameter, t (say).*

Step II. *Find* $\frac{dx}{dt}$ *and* $\frac{dy}{dt}$.

Step III. *Use the formula :* $\frac{dy}{dx} = \frac{dy/dt}{dx/dt}$.

Step IV. *Substitute the values of dy/dt and dx/dt and simplify.*

ILLUSTRATIVE EXAMPLES

Example 1. *Find* $\frac{dy}{dx}$ *for the following functions expressed in parametric form :*

(*i*) $x = t^3$, $y = t^2$

(*ii*) $x = \frac{3at}{1+t^3}$, $y = \frac{3at^2}{1+t^3}$

(*iii*) $x = a(1 - \cos\theta)$, $y = a(\theta + \sin\theta)$. (P.T.U. B.Pharma May 2005)

Sol. (*i*) We have $x = t^3$, $y = t^2$.

$$\therefore \quad \frac{dx}{dt} = 3t^2 \text{ and } \frac{dy}{dx} = 2t.$$

$$\therefore \quad \frac{dy}{dx} = \frac{dy/dt}{dx/dt} = \frac{2t}{3t^2} = \mathbf{\frac{2}{3t}}.$$

(*ii*) We have $x = \frac{3at}{1+t^3}$, $y = \frac{3at^2}{1+t^3}$

$$\therefore \quad \frac{dx}{dt} = \frac{(1+t^3).3a - 3at.3t^2}{(1+t^3)^2} = \frac{3a + 3at^3 - 9at^3}{(1+t^3)^2} = \frac{3a(1-2t^3)}{(1+t^3)^2}$$

and

$$\frac{dy}{dt} = \frac{(1+t^3).6at - 3at^2.3t^2}{(1+t^3)^2} = \frac{6at + 6at^4 - 9at^4}{(1+t^3)^2} = \frac{3at(2-t^3)}{(1+t^3)^2}.$$

$$\therefore \quad \frac{dy}{dx} = \frac{dy/dt}{dx/dt} = \frac{3at(2-t^3)}{(1+t^3)^2} \div \frac{3a(1-2t^3)}{(1+t^3)^2} = \mathbf{\frac{t(2-t^3)}{1-2t^3}}.$$

(*iii*) We have $x = a(1 - \cos\theta)$, $y = a(\theta + \sin\theta)$.

$$\therefore \quad \frac{dx}{d\theta} = a(0 + \sin\theta) = a \sin\theta \quad \text{and} \quad \frac{dy}{d\theta} = a(1 + \cos\theta)$$

$$\therefore \quad \frac{dy}{dx} = \frac{dy/d\theta}{dx/d\theta} = \frac{a(1+\cos\theta)}{a\sin\theta} = \frac{1+\cos\theta}{\sin\theta} = \frac{2\cos^2\frac{1}{2}\theta}{2\sin\frac{1}{2}\theta\cos\frac{1}{2}\theta} = \frac{\cos\frac{1}{2}\theta}{\sin\frac{1}{2}\theta} = \mathbf{\cot\frac{1}{2}\theta}.$$

Example 2. *Find* $\frac{dy}{dx}$ *for the following functions given in parametric form :*

(*i*) $x = a\left(\cos\theta + \log\tan\frac{1}{2}\theta\right)$, $y = a\sin\theta$ at $\theta = \pi/4$

(*ii*) $x = 2\cos\theta - \cos 2\theta$, $y = 2\sin\theta - \sin 2\theta$ at $\theta = \pi/2$.

Sol. (*i*) We have $x = a\left(\cos\theta + \log\tan\frac{1}{2}\theta\right)$.

$$\therefore \quad \frac{dx}{d\theta} = a\left[-\sin\theta + \frac{1}{\tan\frac{1}{2}\theta}.\sec^2\frac{1}{2}\theta.\frac{1}{2}\right] = a\left[-\sin\theta + \frac{1}{2\sin\frac{1}{2}\theta\cos\frac{1}{2}\theta}\right]$$

$$= a\left[-\sin\theta + \frac{1}{\sin\theta}\right] = a\left[\frac{1-\sin^2\theta}{\sin\theta}\right] = \frac{a\cos^2\theta}{\sin\theta}. \qquad | \sin 2\theta = 2\sin\theta\cos\theta$$

Also, $y = a\sin\theta$. $\therefore \frac{dy}{d\theta} = a\cos\theta$.

$$\frac{dy}{dx} = \frac{dy/d\theta}{dx/d\theta} = a\cos\theta \times \frac{\sin\theta}{a\cos^2\theta} = \tan\theta.$$

$$\therefore \text{ At } \theta = \frac{\pi}{4}, \frac{dy}{dx} = \tan\frac{\pi}{4} = \mathbf{1}.$$

(*ii*) We have $x = 2\cos\theta - \cos 2\theta$.

$$\therefore \quad \frac{dx}{d\theta} = 2 . -\sin\theta - (-\sin 2\theta) . 2 = 2(\sin 2\theta - \sin\theta).$$

Also, $y = 2\sin\theta - \sin 2\theta$.

$$\therefore \quad \frac{dy}{d\theta} = 2\cos\theta - \cos 2\theta . 2 = 2(\cos\theta - \cos 2\theta).$$

$$\therefore \quad \frac{dy}{dx} = \frac{dy/d\theta}{dx/d\theta} = \frac{2(\cos\theta - \cos 2\theta)}{2(\sin 2\theta - \sin\theta)} = \frac{\cos\theta - \cos 2\theta}{\sin 2\theta - \sin\theta}.$$

$$\therefore \quad \text{At } \theta = \frac{\pi}{2}, \frac{dy}{dx} = \frac{\cos\frac{1}{2}\pi - \cos\pi}{\sin\pi - \sin\frac{1}{2}\pi} = \frac{0-(-1)}{0-1} = \mathbf{-1}.$$

Example 3. *Find* $\frac{dy}{dx}$ *for the following functions given in parametric form :*

(*i*) $x = 3 \cos \theta - 2 \cos^3 \theta, \quad y = 3 \sin \theta - 2 \sin^3 \theta$

(*ii*) $x = \cos^{-1} \frac{1}{\sqrt{1+t^2}}, \qquad y = \sin^{-1} \frac{t}{\sqrt{1+t^2}}$.

Sol. (*i*) We have $x = 3 \cos \theta - 2 \cos^3 \theta$ and $y = 3 \sin \theta - 2 \sin^3 \theta$.

$$\therefore \quad \frac{dx}{d\theta} = -3 \sin \theta - 6 \cos^2 \theta \,.\, (-\sin \theta) = 3 \sin \theta \,(2 \cos^2 \theta - 1) = 3 \sin \theta \cos 2\theta$$

and
$$\frac{dy}{d\theta} = 3 \cos \theta - 6 \sin^2 \theta \,.\, (\cos \theta) = 3 \cos \theta \,(1 - 2 \sin^2 \theta) = 3 \cos \theta \cos 2\theta.$$

$$\therefore \quad \frac{dy}{dx} = \frac{dy/d\theta}{dx/d\theta} = \frac{3 \sin \theta \cos 2\theta}{3 \cos \theta \cos 2\theta} = \mathbf{\tan \theta.}$$

(*ii*) We have $x = \cos^{-1} \frac{1}{\sqrt{1+t^2}}, \quad y = \sin^{-1} \frac{t}{\sqrt{1+t^2}}$. Let $t = \tan \theta$.

$$\therefore \quad x = \cos^{-1} \frac{1}{\sqrt{1+\tan^2 \theta}} = \cos^{-1} \frac{1}{\sec \theta} = \cos^{-1} (\cos \theta) = \theta = \tan^{-1} t.$$

$$\therefore \quad \frac{dx}{dt} = \frac{1}{1+t^2}.$$

Also
$$y = \sin^{-1} \frac{\tan \theta}{\sqrt{1+\tan^2 \theta}} = \sin^{-1} \frac{\tan \theta}{\sec \theta} = \sin^{-1} (\sin \theta) = \theta = \tan^{-1} t.$$

$$\therefore \quad \frac{dy}{dt} = \frac{1}{1+t^2}.$$

$$\therefore \quad \frac{dy}{dx} = \frac{dy/dt}{dx/dt} = \frac{1}{1+t^2} \Big/ \frac{1}{1+t^2} = \mathbf{1.}$$

EXERCISE 12.13

Find $\frac{dy}{dx}$ for the following functions given in parametric form (Q. No. 1–2) :

1. (*i*) $x = ct, \ y = c/t$ (*ii*) $x = t^2 + 2t - 1, \ y = 2t^2 - t + 5$
   (*iii*) $x = \log t, \ y = \sin t$ (*iv*) $x = a \sec \theta, \ y = b \tan \theta$.
2. (*i*) $x = at^2, \ y = 2at$ (*P.T.U. B.Pharma, May 2006*) (*ii*) $x = \sin 2t, \ y = 2 \cos t$
   (*iii*) $x = a \cos^2 \theta, \ y = b \sin^2 \theta$ (*iv*) $x = 4 \sin \theta, \ y = 7 \cos \theta$
   (*v*) $x = a \cos 2\theta, \ y = a \sin 2\theta$.
3. Find the value of $\frac{dy}{dx}$ at $t = 4$, where $x = \sqrt{t}$ and $y = t - \frac{1}{\sqrt{t}}$.

Find $\frac{dy}{dx}$ for the following functions given in parametric form (Q. No. 4–7) :

4. (*i*) $x = a \tan^2 \theta, \ y = b \sec^2 \theta$ (*ii*) $x = a \cot^2 \theta, \ y = b \operatorname{cosec}^2 \theta$
   (*iii*) $x = a \cos^3 \theta, \ y = a \sin^3 \theta$ (*iv*) $x = a \sec^3 \theta, \ y = b \tan^3 \theta$

5. (*i*) $x = a(\theta + \sin\theta), y = a(1 + \cos\theta)$ (*ii*) $x = a(\cos\theta + \theta\sin\theta), y = a(\sin\theta - \theta\cos\theta)$
(*iii*) $x = \cos\theta + \cos 2\theta, y = \sin\theta + \sin 2\theta$ (*iv*) $x = a\sin 2\theta(1 + \cos 2\theta), y = a\cos 2\theta(1 - \cos 2\theta)$

6. (*i*) $x = e^t\cos t, y = e^t\sin t$ (*ii*) $x = e^t(\sin t + \cos t), y = e^t(\sin t - \cos t)$
(*iii*) $x = a\left(\cos t + \log\tan\frac{t}{2}\right), y = a\sin t$ (*iv*) $x = \sqrt{\sin 2\theta}, y = \sqrt{\cos 2\theta}$
(*P.T.U. B.Pharma, 2002*)

7. (*i*) $x = \frac{2a}{1+t^2}, y = \frac{2a^2}{1+t^2}$ (*ii*) $x = \frac{a(1-t^2)}{1+t^2}, y = \frac{2bt}{1+t^2}$ (*P.T.U. B.Pharma, Dec. 2005*)
(*iii*) $y = \frac{3t^2}{1+t^3}, x = \frac{3}{1+t^3}$ (*P.T.U. B.Pharma, May 2007*)

8. Find $\frac{dy}{dx}$ when $x = a\sqrt{\frac{t^2-1}{t^2+1}}$ and $y = at\sqrt{\frac{t^2-1}{t^2+1}}$.

9. Find $\frac{dy}{dx}$ when $x = e^{\theta}\left(\theta + \frac{1}{\theta}\right)$ and $y = e^{-\theta}\left(\theta - \frac{1}{\theta}\right)$.

10. If $x = \frac{1+\log t}{t^2}, y = \frac{3+2\log t}{t}, t > 0$, then show that $y\frac{dy}{dx} = 2x\left(\frac{dy}{dx}\right)^2 + 1$.

11. If $x = \frac{1-t^2}{1+t^2}, y = \frac{2t}{1+t^2}$, show that $\frac{dy}{dx} + \frac{x}{y} = 0$. (*P.T.U. B.Pharma, Dec. 2002, 2004*)

Answers

1. (*i*) $-\frac{1}{t^2}$ (*ii*) $\frac{4t-1}{2t+2}$ (*iii*) $t\cos t$ (*iv*) $\frac{b}{a}\operatorname{cosec}\theta$

2. (*i*) $\frac{1}{t}$ (*ii*) $-\sin t\sec 2t$ (*iii*) $-\frac{b}{a}$ (*iv*) $-\frac{7}{4}\tan\theta$ (*v*) $-\tan 2\theta$

3. $\frac{17}{4}$ 4. (*i*) $\frac{b}{a}$ (*ii*) $\frac{b}{a}$ (*iii*) $-\tan\theta$ (*iv*) $\frac{b}{a}\sin\theta$

5. (*i*) $-\tan\frac{1}{2}\theta$ (*ii*) $\tan\theta$ (*iii*) $-\frac{\cos\theta + 2\cos 2\theta}{\sin\theta + 2\sin 2\theta}$ (*iv*) $\tan\theta$

6. (*i*) $\frac{\cos t + \sin t}{\cos t - \sin t}$ (*ii*) $\tan t$ (*iii*) $\tan t$ (*iv*) $-(\tan 2\theta)^{3/2}$

7. (*i*) a (*ii*) $\frac{b(t^2-1)}{2at}$ (*iii*) $\frac{t^2}{3} - \frac{2}{3t}$ 8. $\frac{t^4 + 2t^2 - 1}{2t}$

9. $\frac{\theta + \theta^2 - \theta^3 + 1}{(\theta + \theta^2 + \theta^3 - 1)e^{2\theta}}$.

XI. DIFFERENTIATION OF A FUNCTION w.r.t. ANOTHER FUNCTION

12.32. DERIVATIVE OF A FUNCTION w.r.t. ANOTHER FUNCTION

Let $u = f(x)$ and $v = g(x)$ be differentiable functions of x.

Let δx be an increment of x and δu, δv, the corresponding increments of u, v respectively. We assume that $\delta v \neq 0$.

$$\therefore \quad \frac{\delta u}{\delta v} = \frac{\delta u/\delta x}{\delta v/\delta x} \quad \Rightarrow \quad \lim_{\delta x \to 0} \frac{\delta u}{\delta v} = \lim_{\delta x \to 0} \frac{\delta u/\delta x}{\delta v/\delta x}$$

$$\Rightarrow \quad \lim_{\delta v \to 0} \frac{\delta u}{\delta v} = \frac{\lim_{\delta x \to 0} \delta u/\delta x}{\lim_{\delta x \to 0} \delta v/\delta x}, \text{ provided } \lim_{\delta x \to 0} \frac{\delta v}{\delta x} \neq 0.$$

$$\Rightarrow \quad \mathbf{\frac{du}{dv} = \frac{du/dx}{dv/dx}}, \textbf{ provided } \mathbf{\frac{dv}{dx} \neq 0.}$$

WORKING RULES FOR SOLVING PROBLEMS

Step I. *Of the given functions, if the first function is to be differentiated w.r.t. the second function, then put these functions as u and v respectively.*

$\therefore$ *We require* $\frac{du}{dv}$.

Step II. *Find* $\frac{du}{dx}$ *and* $\frac{dv}{dx}$.

Step III. *Use the formula :* $\frac{du}{dv} = \frac{du/dx}{dv/dx}$.

Step IV. *Substitute the values of du/dx and dv/dx and simplify.*

ILLUSTRATIVE EXAMPLES

Example 1. *Differentiate :*

(*i*) $4x^3 + 4x$ *w.r.t.* $x^2 + 2$ (*ii*) $\sin^3 x$ *w.r.t.* $\cos^3 x$ (*iii*) $\sec x°$ *w.r.t.* $\tan x°$.

Sol. (*i*) Let $u = 4x^3 + 4x$ and $v = x^2 + 2$.

$$\therefore \quad \frac{du}{dx} = 12x^2 + 4 \text{ and } \frac{dv}{dx} = 2x$$

$$\therefore \quad \frac{du}{dv} = \frac{du/dx}{dv/dx} = \frac{12x^2 + 4}{2x} = \mathbf{\frac{6x^2 + 2}{x}}.$$

(*ii*) Let $u = \sin^3 x$ and $v = \cos^3 x$.

$\therefore$ $u = (\sin x)^3$ and $v = (\cos x)^3$

$$\therefore \quad \frac{du}{dx} = 3(\sin x)^2 \frac{d}{dx}(\sin x) = 3 \sin^2 x \cos x$$

and $$\frac{dv}{dx} = 3(\cos x)^2 \frac{d}{dx}(\cos x) = -3 \cos^2 x \sin x$$

$$\therefore \quad \frac{du}{dv} = \frac{du/dx}{dv/dx} = \frac{3 \sin^2 \cos x}{-3 \cos^2 x \sin x} = \mathbf{-\tan x}.$$

(*iii*) Let $u = \sec x°$ and $v = \tan x°$.

$$\therefore \quad u = \sec \frac{\pi x}{180} \text{ and } v = \tan \frac{\pi x}{180} \qquad \left[\because \quad x° = \frac{\pi x}{180} \text{ radians}\right]$$

$$\therefore \quad \frac{du}{dx} = \sec \frac{\pi x}{180} \tan \frac{\pi x}{180} \cdot \frac{\pi}{180} = \frac{\pi}{180} \sec x° \tan x°$$

and $$\frac{dv}{dx} = \sec^2 \frac{\pi x}{180} \cdot \frac{\pi}{180} = \frac{\pi}{180} \sec^2 x°.$$

$$\therefore \quad \frac{du}{dv} = \frac{du/dx}{dv/dx} = \frac{\frac{\pi}{180} \sec x° \tan x°}{\frac{\pi}{180} \sec^2 x°} = \mathbf{\sin x°}.$$

Example 2. *Differentiate $x^{\sin x}$ w.r.t. $(\sin x)^x$.*

Sol. Let $u = x^{\sin x}$ and $v = (\sin x)^x$.

$\therefore$ $\log u = \sin x \,.\, \log x$. Differentiating w.r.t. x, we get

$$\frac{1}{u}\frac{du}{dx} = \sin x \,.\, \frac{1}{x} + \log x \,.\, \cos x.$$

$$\therefore \quad \frac{du}{dx} = u\left(\frac{\sin x}{x} + \log x \,.\, \cos x\right) = x^{\sin x}\left(\frac{\sin x}{x} + \log x \,.\, \cos x\right).$$

Also $v = (\sin x)^x$.

$\therefore$ $\log v = x \log \sin x$. Differentiating w.r.t. x, we get

$$\frac{1}{v}\frac{dv}{dx} = x\left(\frac{1}{\sin x} \,.\, \cos x\right) + \log \sin x \,.\, 1$$

$$\therefore \quad \frac{dv}{dx} = v\,(x \cot x + \log \sin x) = (\sin x)^x\,(x \cot x + \log \sin x).$$

$$\therefore \quad \frac{du}{dv} = \frac{du/dx}{dv/dx} = \frac{\mathbf{x^{\sin x}\left(\frac{\sin x}{x} + \log x \,.\, \cos x\right)}}{\mathbf{(\sin x)^x\,(x \cot x + \log \sin x)}}.$$

Example 3. *Show that the derivative of $\tan^{-1} \frac{\sqrt{1+x^2} - 1}{x}$ w.r.t. $\tan^{-1} x$ is independent of x.*

Sol. Let $u = \tan^{-1} \frac{\sqrt{1+x^2} - 1}{x}$ and $v = \tan^{-1} x$.

Let $x = \tan \theta$.

$$\therefore \quad u = \tan^{-1} \frac{\sqrt{1+\tan^2\theta} - 1}{\tan\theta} = \tan^{-1} \frac{\sec\theta - 1}{\tan\theta} = \tan^{-1} \frac{1 - \cos\theta}{\sin\theta}$$

$$= \tan^{-1} \frac{2\sin^2 \frac{1}{2}\theta}{2 \sin \frac{1}{2}\theta \cos \frac{1}{2}\theta} = \tan^{-1} (\tan \tfrac{1}{2}\theta) = \tfrac{1}{2}\theta = \tfrac{1}{2}\tan^{-1} x.$$

$$\therefore \quad \frac{du}{dx} = \frac{1}{2} \cdot \frac{1}{1+x^2}.$$

$$v = \tan^{-1} x \text{ implies } \frac{dv}{dx} = \frac{1}{1+x^2}.$$

$$\therefore \quad \frac{du}{dv} = \frac{du/dx}{dv/dx} = \frac{1}{2(1+x^2)} \Big/ \frac{1}{1+x^2} = \frac{1}{2}, \text{ which is independent of } x.$$

Example 4. *Differentiate $\tan^{-1}\sqrt{\frac{1-x^2}{1+x^2}}$ w.r.t. $\cos^{-1} x^2$.*

Sol. Let $u = \tan^{-1}\sqrt{\frac{1-x^2}{1+x^2}}$ and $v = \cos^{-1} x^2$.

Let $x^2 = \cos\theta$.

$$\therefore \quad u = \tan^{-1}\sqrt{\frac{1-\cos\theta}{1+\cos\theta}} = \tan^{-1}\sqrt{\frac{2\sin^2\frac{1}{2}\theta}{2\cos^2\frac{1}{2}\theta}} = \tan^{-1}(\tan\tfrac{1}{2}\theta) = \tfrac{1}{2}\theta = \tfrac{1}{2}\cos^{-1}x^2.$$

$$\frac{du}{dx} = \frac{1}{2}\cdot\frac{-1}{\sqrt{1-(x^2)^2}}\cdot 2x = \frac{-x}{\sqrt{1-x^4}}.$$

Also $v = \cos^{-1}x^2$ implies $\dfrac{dv}{dx} = \dfrac{-1}{\sqrt{1-(x^2)^2}}\cdot 2x = \dfrac{-2x}{\sqrt{1-x^4}}$

$$\therefore \quad \frac{du}{dv} = \frac{du/dx}{dv/dx} = \frac{-x}{\sqrt{1-x^4}}\Big/\frac{-2x}{\sqrt{1-x^4}} = \mathbf{\frac{1}{2}}.$$

Remark. In the above question, we obtained $u = \dfrac{1}{2}\cos^{-1}x^2$.

We could have alternatively proceeded as follows :

$$u = \frac{1}{2}\cos^{-1}x^2 = \frac{1}{2}v. \quad \therefore \quad \frac{du}{dv} = \frac{1}{2}.$$

Example 5. *Differentiate* $\tan^{-1}\dfrac{x}{1+\sqrt{1+x^2}}$ *w.r.t.* $\sin\left(2\cot^{-1}\sqrt{\dfrac{1+x}{1-x}}\right)$.

Sol. Let $u = \tan^{-1}\dfrac{x}{1+\sqrt{1+x^2}}$ and $v = \sin\left(2\cot^{-1}\sqrt{\dfrac{1+x}{1-x}}\right)$.

Let $x = \tan\theta$.

$$\therefore \quad u = \tan^{-1}\frac{\tan\theta}{1+\sqrt{1+\tan^2\theta}} = \tan^{-1}\frac{\tan\theta}{1+\sec\theta} = \tan^{-1}\frac{\sin\theta}{\cos\theta+1}$$

$$= \tan^{-1}\frac{2\sin\frac{\theta}{2}\cos\frac{\theta}{2}}{2\cos^2\frac{\theta}{2}} = \tan^{-1}\left(\tan\frac{\theta}{2}\right) = \frac{\theta}{2} = \frac{1}{2}\tan^{-1}x.$$

$$\therefore \quad \frac{du}{dx} = \frac{1}{2}\cdot\frac{1}{1+x^2} = \frac{1}{2(1+x^2)}.$$

We have $v = \sin\left(2\cot^{-1}\sqrt{\dfrac{1+x}{1-x}}\right)$.

Let $x = \cos\theta$.

$$\therefore \quad v = \sin\left(2\cot^{-1}\sqrt{\frac{1+\cos\theta}{1-\cos\theta}}\right) = \sin\left(2\cot^{-1}\sqrt{\frac{2\cos^2\theta/2}{2\sin^2\theta/2}}\right)$$

$$= \sin\left(2\cot^{-1}\left(\cot\frac{\theta}{2}\right)\right) = \sin\left(2\left(\frac{\theta}{2}\right)\right) = \sin\theta = \sqrt{1-\cos^2\theta} = \sqrt{1-x^2}.$$

$$\therefore \quad \frac{dv}{dx} = \frac{1}{2}(1-x^2)^{-1/2}\frac{d}{dx}(1-x^2) = \frac{-2x}{2\sqrt{1-x^2}} = -\frac{x}{\sqrt{1-x^2}}$$

$$\therefore \quad \frac{du}{dv} = \frac{du/dx}{dv/dx} = \frac{\frac{1}{2(1+x^2)}}{\frac{-x}{\sqrt{1-x^2}}} = -\mathbf{\frac{\sqrt{1-x^2}}{2x(1+x^2)}}.$$

EXERCISE 12.14

Differentiate :

1. (*i*) x^2 w.r.t. x^3 (*ii*) $\sin x$ w.r.t. $\cos x$
(*iii*) $\theta^2 + \theta + 7$ w.r.t. $\theta + 1$ (*iv*) $\log x$ w.r.t. e^x

2. (*i*) e^{2x} w.r.t. e^x (*ii*) $\sin x^2$ w.r.t. x^2
(*iii*) x^{-2} w.r.t. x^2 (*iv*) $\sec x$ w.r.t. $\operatorname{cosec} x$

3. (*i*) $\sin x^3$ w.r.t. $\cos x^3$ (*ii*) $\tan^2 x$ w.r.t. $\cos^2 x$
(*iii*) $\log \sec x$ w.r.t. $\tan x$ (*iv*) $\log x^6$ w.r.t. $\log x^5$

4. (*i*) $\sin^2 x$ w.r.t. $(\log x)^2$ (*ii*) $e^{\tan x}$ w.r.t. $\tan^{-1} x$
(*iii*) $\log_{10} x$ w.r.t. x^2 (*iv*) $\sin x^2$ w.r.t. $\sin^2 x$.

5. (*i*) $\tan^{-1} \dfrac{\sqrt{1+x^2}-1}{x}$ w.r.t. $\sin^{-1} \dfrac{2x}{1+x^2}$ (*ii*) $\tan^{-1} \dfrac{\sqrt{1+x^2}-1}{x}$ w.r.t. $\cos^{-1} \dfrac{1-x^2}{1+x^2}$
(*iii*) $\tan^{-1} \dfrac{\sqrt{1+x^2}-1}{x}$ w.r.t. $\tan^{-1} \dfrac{2x}{1-x^2}$ (*iv*) $\tan^{-1} \dfrac{2x}{1-x^2}$ w.r.t. $\tan^{-1} x$

6. (*i*) $\cos^{-1} \dfrac{1-x^2}{1+x^2}$ w.r.t. $\tan^{-1} x$ (*ii*) $\sin^{-1} \dfrac{2x}{1+x^2}$ w.r.t. $\tan^{-1} x$
(P.T.U. B.Pharma, Dec. 2002)
(*iii*) $\tan^{-1} \dfrac{2x}{1+x^2}$ w.r.t. $\sin^{-1} \dfrac{2x}{1+x^2}$ (*iv*) $\tan^{-1} \dfrac{3x-x^3}{1-3x^2}$ w.r.t. $\tan^{-1} \dfrac{2x}{1-x^2}$

7. (*i*) $\tan^{-1} \sqrt{\dfrac{1-\cos x}{1+\cos x}}$ w.r.t. $\tan^{-1} x$ (*ii*) $\sec^{-1} \dfrac{1}{2x^2-1}$ w.r.t. $\sqrt{1-x^2}$ at $x = \dfrac{1}{2}$
(*iii*) $\tan^{-1} \dfrac{\sqrt{1+x^2}-\sqrt{1-x^2}}{\sqrt{1+x^2}+\sqrt{1-x^2}}$ w.r.t. $\sin^{-1} \dfrac{2x}{1+x^2}$
(*iv*) $\sec^{-1} \dfrac{1}{2x^2-1}$ w.r.t. $\sin^{-1} (3x - 4x^3)$

8. (*i*) $x^{\sin^{-1} x}$ w.r.t. $\sin^{-1} x$ (*ii*) $x^{\cos x}$ w.r.t. $(\cos x)^x$
(*iii*) $\sin x^x$ w.r.t. $\tan^{-1} \sqrt{\dfrac{1-x^2}{1+x^2}}$.

Answers

1. (*i*) $\dfrac{2}{3x}$ (*ii*) $-\cot x$ (*iii*) $2\theta + 1$ (*iv*) $\dfrac{1}{xe^x}$

2. (*i*) $2e^x$ (*ii*) $\cos x^2$ (*iii*) $-\dfrac{1}{x^4}$ (*iv*) $-\tan^3 x$

3. (*i*) $-\cot x^3$ (*ii*) $-\sec^4 x$ (*iii*) $\dfrac{1}{2} \sin 2x$ (*iv*) $\dfrac{6}{5}$

4. (*i*) $\dfrac{x \sin 2x}{2 \log x}$ (*ii*) $e^{\tan x} (\sec^2 x)(1 + x^2)$ (*iii*) $\dfrac{\log_{10} e}{2x^2}$ (*iv*) $\dfrac{2x \cos x^2}{\sin 2x}$

5. (i) $\frac{1}{4}$ (ii) $\frac{1}{4}$ (iii) $\frac{1}{4}$ (iv) 2

6. (i) 2 (ii) 2 (iii) 1 (iv) 3/2

7. (i) $\frac{1+x^2}{2}$ (ii) 4 (iii) $\frac{x(1+x^2)}{2\sqrt{1-x^4}}$ (iv) $-\frac{2}{3}$

8. (i) $x^{\sin^{-1}x}\left[\frac{\sqrt{1-x^2}\sin^{-1}x}{x}+\log x\right]$ (ii) $\frac{x^{\cos x}\left[\frac{\cos x}{x}-\log x \,.\, \sin x\right]}{(\cos x)^x\,[\log\cos x - x\tan x]}$

(iii) $-x^{x-1}(1+\log x)\sqrt{1-x^4}\cos x^x$.

XII. DIFFERENTIATION OF HIGHER ORDER

12.33. DEFINITIONS

If $y = f(x)$ is a differentiable function of x then its derivative $\frac{dy}{dx}$ *i.e.*, $f'(x)$ is also a function of x.

If the function $f'(x)$ of x is also differentiable, then its derivative is denoted by $\frac{d^2y}{dx^2}$ or by $f''(x)$. The symbol $\frac{d^2y}{dx^2}$ is called the **second order derivative** of $y = f(x)$ w.r.t. x and read as *dee two y over dee x squared.* $\frac{d^2y}{dx^2}$ is also denoted by y'' or by y_2.

In general, if $y = f(x)$ is a differentiable function of x, then for $n\ (> 1) \in \mathbf{N}$, the **n^{th} order derivative** of y w.r.t. x exists if the $(n-1)^{th}$ order derivative of y is differentiable w.r.t. x.

For example, the 3rd order derivative of $y = f(x)$ exists if $\frac{d^2y}{dx^2}$ is differentiable. If 3rd order derivative of $y = f(x)$ is differentiable, then we can talk of 4th order derivative of y.

The nth order derivative of $y\ (= f(x))$ is denoted by any one of the following symbols :

$$\frac{d^ny}{dx^n},\quad f^{(n)}(x),\quad y_n,\quad \frac{d^n}{dx^n}(f(x)) \quad\text{or}\quad D^n y.$$

ILLUSTRATIVE EXAMPLES

Example 1. *Find the second derivative of the following functions :*

(i) $x \sin x$ (ii) $\sin^{-1} x$

(iii) $\log \log x$ (iv) $\cos^2 x + 3^x$.

Sol. (i) Let $y = x \sin x$.

$$\therefore \quad \frac{dy}{dx} = x\frac{d}{dx}(\sin x) + \sin x \frac{d}{dx}(x) = x\cos x + \sin x \,.\, 1.$$

$$\therefore \quad \frac{d^2y}{dx^2} = [x\,.\,(-\sin x) + \cos x\,.\,1] + \cos x = \mathbf{2\cos x - x\sin x.}$$

(*ii*) Let $y = \sin^{-1} x$.

$$\therefore \quad \frac{dy}{dx} = \frac{1}{\sqrt{1-x^2}} = (1-x^2)^{-1/2} \qquad \text{('}u^n\text{' Form)}$$

$$\therefore \quad \frac{d^2y}{dx^2} = -\frac{1}{2}(1-x^2)^{-3/2}\frac{d}{dx}(1-x^2) = -\frac{1}{2(1-x^2)^{3/2}}(0-2x) = \mathbf{\frac{x}{(1-x^2)^{3/2}}}.$$

(*iii*) Let $y = \log \log x$. ('log *u*' Form)

$$\therefore \quad \frac{dy}{dx} = \frac{1}{\log x}\cdot\frac{d}{dx}(\log x) = \frac{1}{\log x}\cdot\frac{1}{x} = \frac{1}{x\log x}.$$

$$\therefore \quad \frac{d^2y}{dx^2} = \frac{(x\log x)\frac{d}{dx}(1) - 1\frac{d}{dx}(x\log x)}{(x\log x)^2} = \frac{0-\left[x\cdot\frac{1}{x}+\log x\cdot 1\right]}{(x\log x)^2} = -\mathbf{\frac{1+\log x}{(x\log x)^2}}.$$

(*iv*) Let $y = \cos^2 x + 3^x$.

$$\therefore \quad \frac{dy}{dx} = 2(\cos x)^1(-\sin x) + 3^x\log 3 = -2\sin x\cos x + 3^x\log x = -\sin 2x + 3^x\log 3$$

$$\therefore \quad \frac{d^2y}{dx^2} = -\cos 2x\,.\,(2) + (3^x\log 3)\log 3 = \mathbf{-2\cos 2x + 3^x(\log 3)^2}.$$

Example 2. (*i*) *If* $y = A\cos nx + B\sin nx$, *show that* $\frac{d^2y}{dx^2} + n^2y = 0$.

(G.N.D.U. 2002 ; P.T.U. May 2006)

(*ii*) *If* $y = x + \tan x$, *then show that* $\cos^2 x\,\frac{d^2y}{dx^2} - 2y + 2x = 0$.

(*iii*) *If* $y = \operatorname{cosec} x + \cot x$, *then show that* $\sin x\,\frac{d^2y}{dx^2} = y^2$. (P.T.U. B.Pharma, May 2005)

Sol. (*i*) We have $y = A\cos nx + B\sin nx$.

$$\therefore \quad \frac{dy}{dx} = A(-\sin nx)\,.\,n + B\cos nx\,.\,n = -An\sin nx + Bn\cos nx$$

$$\frac{d^2y}{dx^2} = -An\cos nx\,.\,n + Bn(-\sin nx)\,.\,n = -An^2\cos nx - Bn^2\sin nx$$

$$= -n^2(A\cos nx + B\sin nx) = -n^2y$$

$$\therefore \quad \mathbf{\frac{d^2y}{dx^2} + n^2y = 0.}$$

(*ii*) We have $y = x + \tan x$.

$$\therefore \quad \frac{dy}{dx} = 1 + \sec^2 x \quad \text{and} \quad \frac{d^2y}{dx^2} = 0 + 2\sec x.\sec x\tan x = 2\sec^2 x\tan x$$

$$\therefore \quad \cos^2 x\frac{d^2y}{dx^2} - 2y + 2x = \cos^2 x(2\sec^2 x\tan x) - 2(x+\tan x) + 2x$$

$$= 2\tan x - 2x - 2\tan x + 2x = 0.$$

$$\therefore \quad \mathbf{\cos^2 x\frac{d^2y}{dx^2} - 2y + 2x = 0.}$$

(*iii*) We have $y = \operatorname{cosec} x + \cot x$.

$$\therefore \quad \frac{dy}{dx} = -\operatorname{cosec} x\cot x - \operatorname{cosec}^2 x$$

$$\Rightarrow \quad \frac{d^2y}{dx^2} = -(\operatorname{cosec} x\,.\,-\operatorname{cosec}^2 x + -\operatorname{cosec} x\cot x\,.\,\cot x) - 2\operatorname{cosec} x(-\operatorname{cosec} x\cot x)$$

$$= \text{cosec}^3 x + \text{cosec } x \cot^2 x + 2 \text{ cosec}^2 x \cot x$$

$\therefore$ $$\sin x \frac{d^2y}{dx^2} = \sin x \text{ cosec } x (\text{cosec}^2 x + \cot^2 x + 2 \text{ cosec } x \cot x) = (\text{cosec } x + \cot x)^2 = y^2$$

$\therefore$ $$\mathbf{\sin x \frac{d^2y}{dx^2} = y^2.}$$

Example 3. *(i) If* $y = x \log \dfrac{x}{a + bx}$, *show that* $x^3 \dfrac{d^2y}{dx^2} = \left(x \dfrac{dy}{dx} - y\right)^2$.

(ii) If $y = x^x$, *show that* $\dfrac{d^2y}{dx^2} - \dfrac{1}{y}\left(\dfrac{dy}{dx}\right)^2 - \dfrac{y}{x} = 0.$

(iii) If $x = a \cos \theta + b \sin \theta$ *and* $y = a \sin \theta - b \cos \theta$, *show that* $y^2 \dfrac{d^2y}{dx^2} - x \dfrac{dy}{dx} + y = 0.$

Sol. *(i)* We have $\quad y = x \log \dfrac{x}{a + bx}$.

$\Rightarrow$ $$\frac{y}{x} = \log x - \log (a + bx)$$

Differentiating w.r.t. x, we get $\dfrac{x \dfrac{dy}{dx} - y . 1}{x^2} = \dfrac{1}{x} - \dfrac{1}{a + bx} . (0 + b.1)$.

$\Rightarrow$ $$x\frac{dy}{dx} - y = x^2 \left(\frac{1}{x} - \frac{b}{a + bx}\right) = \frac{ax^2}{x(a + bx)} = \frac{ax}{a + bx}$$

$\therefore$ $$x \frac{dy}{dx} - y = \frac{ax}{a + bx} \qquad ...(1)$$

Differentiating w.r.t. x, we get $\left(x \dfrac{d^2y}{dx^2} + \dfrac{dy}{dx} . 1\right) - \dfrac{dy}{dx} = \dfrac{(a + bx) . a - ax . (0 + b)}{(a + bx)^2}$

$\therefore$ $$x \frac{d^2y}{dx^2} = \frac{a^2}{(a + bx)^2}$$

Multiplying both sides by x^2, we get

$$x^3 \frac{d^2y}{dx^2} = \frac{a^2x^2}{(a + bx)^2} = \left(\frac{ax}{a + bx}\right)^2 = \left(x \frac{dy}{dx} - y\right)^2 \qquad \text{(Using (1))}$$

$\therefore$ $$\mathbf{x^3 \frac{d^2y}{dx^2} = \left(x \frac{dy}{dx} - y\right)^2.}$$

(ii) We have $y = x^x$. Taking logarithms on both sides, we get $\log y = x \log x$.
Differentiating w.r.t. x, we get

$$\frac{1}{y} \frac{dy}{dx} = x . \frac{1}{x} + \log x . 1 = 1 + \log x \; i.e., \; \frac{dy}{dx} = y (1 + \log x) \qquad ...(1)$$

Differentiating w.r.t. x, we get

$$\frac{d^2y}{dx^2} = y \left(0 + \frac{1}{x}\right) + (1 + \log x) \frac{dy}{dx} = \frac{y}{x} + \left(\frac{1}{y} \frac{dy}{dx}\right) \frac{dy}{dx} = \frac{y}{x} + \frac{1}{y} \left(\frac{dy}{dx}\right)^2 \qquad \text{(Using (1))}$$

$\therefore$ $$\mathbf{\frac{d^2y}{dx^2} - \frac{1}{y} \left(\frac{dy}{dx}\right)^2 - \frac{y}{x} = 0.}$$

(*iii*) We have $x = a\cos\theta + b\sin\theta$ and $y = a\sin\theta - b\cos\theta$.

$$\therefore \quad \frac{dy}{dx} = \frac{dy/d\theta}{dx/d\theta} = \frac{a\cos\theta + b\sin\theta}{-a\sin\theta + b\cos\theta} = -\frac{a\cos\theta + b\sin\theta}{a\sin\theta - b\cos\theta} = -\frac{x}{y}$$

$$\therefore \quad y\frac{dy}{dx} = -x$$

Differentiating w.r.t. x, we get $y\dfrac{d^2y}{dx^2} + \dfrac{dy}{dx}\cdot\dfrac{dy}{dx} = -1$

Multiplying by y, we get $y^2\dfrac{d^2y}{dx^2} + y\dfrac{dy}{dx}\cdot\dfrac{dy}{dx} = -y$

$$\Rightarrow \quad y^2\frac{d^2y}{dx^2} + y\left(-\frac{x}{y}\right)\left(\frac{dy}{dx}\right) + y = 0 \qquad \left(\because \frac{dy}{dx} = -\frac{x}{y}\right)$$

$$\Rightarrow \quad \mathbf{y^2\frac{d^2y}{dx^2} - x\frac{dy}{dx} + y = 0.}$$

Example 4. (*i*) *If* $y = e^{ax}\sin bx$, *show that* $\dfrac{d^2y}{dx^2} - 2a\dfrac{dy}{dx} + (a^2 + b^2)y = 0.$

(*ii*) *If* $y = \dfrac{\sin^{-1}x}{\sqrt{1-x^2}}$, *then show that* $(1 - x^2)\dfrac{d^2y}{dx^2} - 3x\dfrac{dy}{dx} - y = 0.$

Sol. (*i*) We have $y = e^{ax}\sin bx$.

$$\therefore \quad \frac{dy}{dx} = e^{ax}\frac{d}{dx}(\sin bx) + \sin bx\frac{d}{dx}(e^{ax}) = e^{ax}\cdot\cos bx\cdot b + \sin bx\cdot e^{ax}\cdot a$$

$$\therefore \quad \frac{dy}{dx} = e^{ax}(b\cos bx + a\sin bx).$$

Differentiating again, we get

$$\frac{d^2y}{dx^2} = e^{ax}[b\,.\,-\sin bx\,.\,b + a\cos bx\,.\,b] + (b\cos bx + a\sin bx)[e^{ax}\,.\,a]$$

$$= e^{ax}[-b^2\sin bx + ab\cos bx + ab\cos bx + a^2\sin bx]$$

$$\therefore \quad \frac{d^2y}{dx^2} = e^{ax}[(a^2 - b^2)\sin bx + 2ab\cos bx].$$

$$\therefore \quad \frac{d^2y}{dx^2} - 2a\frac{dy}{dx} + (a^2 + b^2)y$$

$$= e^{ax}[(a^2 - b^2)\sin bx + 2ab\cos bx] - 2ae^{ax}[b\cos bx + a\sin bx] + (a^2 + b^2)e^{ax}\sin bx$$

$$= e^{ax}[(a^2 - b^2)\sin bx + 2ab\cos bx - 2ab\cos bx - 2a^2\sin bx + (a^2 + b^2)\sin bx]$$

$$= e^{ax}(a^2 - b^2 - 2a^2 + a^2 + b^2)\sin bx = 0.$$

$$\therefore \quad \mathbf{\frac{d^2y}{dx^2} - 2a\frac{dy}{dx} + (a^2 + b^2)y = 0.}$$

(*ii*) We have $y = \dfrac{\sin^{-1} x}{\sqrt{1-x^2}}$. $\therefore$ $(1 - x^2)\, y^2 = (\sin^{-1} x)^2$

Differentiating w.r.t. x, we get

$$(1 - x^2) \,.\, 2yy_1 + (-2x)y^2 = 2 \sin^{-1} x \,.\, \frac{1}{\sqrt{1-x^2}}.$$

$$\Rightarrow \quad 2(1 - x^2)yy_1 - 2xy^2 = 2y$$

$$\Rightarrow \quad (1 - x^2)y_1 - xy = 1$$

Differentiating again w.r.t. x, [Dividing by $2y$]

$$(1 - x^2)y_2 + (-2x)y_1 - (xy_1 + 1.y) = 0 \quad \Rightarrow \quad (1 - x^2)y_2 - 3xy_1 - y = 0$$

$$\therefore \quad \mathbf{(1 - x^2)\frac{d^2y}{dx^2} - 3x\frac{dy}{dx} - y = 0.}$$

Example 5. *If* $x = a\,(\theta + \sin\theta)$, $y = a\,(1 + \cos\theta)$, *find* $\dfrac{d^2y}{dx^2}$ *at* $\theta = \dfrac{\pi}{2}$.

(P.T.U. B.Pharma, Dec. 2006)

Sol. We have $x = a\,(\theta + \sin\theta)$ and $y = a(1 + \cos\theta)$.

$$\Rightarrow \quad \frac{dx}{d\theta} = a\,(1 + \cos\theta) \quad \text{and} \quad \frac{dy}{d\theta} = -a\sin\theta$$

$$\therefore \quad \frac{dy}{dx} = \frac{dy/d\theta}{dx/d\theta} = \frac{-a\sin\theta}{a\,(1+\cos\theta)} = \frac{-2\sin\frac{\theta}{2}\cos\frac{\theta}{2}}{2\cos^2\frac{\theta}{2}} = -\tan\frac{\theta}{2}$$

Differentiating dy/dx w.r.t. x, we get

$$\frac{d^2y}{dx^2} = \frac{d}{dx}\left(-\tan\frac{\theta}{2}\right) = \frac{d}{d\theta}\left(-\tan\frac{\theta}{2}\right).\frac{d\theta}{dx} = \frac{-\sec^2\frac{\theta}{2}\cdot\frac{1}{2}}{dx/d\theta} = \frac{-\frac{1}{2}\sec^2\frac{1}{2}\theta}{a\,.\,2\cos^2\frac{1}{2}\theta} = -\frac{\sec^4\frac{1}{2}\theta}{4a}.$$

At $\theta = \dfrac{\pi}{2}$, $\quad \dfrac{d^2y}{dx^2} = -\dfrac{\sec^4(\pi/4)}{4a} = -\dfrac{(\sqrt{2})^4}{4a} = \mathbf{-\dfrac{1}{a}}.$

Example 6. *If* $y^3 + x^3 - 3axy = 0$, *show that* $\dfrac{d^2y}{dx^2} = -\dfrac{2a^3xy}{(y^2 - ax)^3}$.

Sol. We have $y^3 + x^3 - 3axy = 0$.

Differentiating w.r.t. x, we get $3y^2\dfrac{dy}{dx} + 3x^2 - 3a\left[x\dfrac{dy}{dx} + y\,.\,1\right] = 0.$

$$\Rightarrow \quad \frac{dy}{dx}[3y^2 - 3ax] = 3ay - 3x^2 \quad \Rightarrow \quad \frac{dy}{dx} = \frac{ay - x^2}{y^2 - ax}. \qquad \ldots(1)$$

Now
$$\frac{d^2y}{dx^2} = \frac{d}{dx}\left(\frac{dy}{dx}\right) = \frac{(y^2 - ax)\left(a\frac{dy}{dx} - 2x\right) - (ay - x^2)\left(2y\frac{dy}{dx} - a\right)}{(y^2 - ax)^2}$$

$$= \frac{\frac{dy}{dx}[ay^2 - a^2x - 2ay^2 + 2x^2y] - 2xy^2 + 2ax^2 + a^2y - ax^2}{(y^2 - ax)^2}$$

$$= \frac{\left(\frac{ay - x^2}{y^2 - ax}\right)[-ay^2 - a^2x + 2x^2y] + [a^2y + ax^2 - 2xy^2]}{(y^2 - ax)^2} \qquad | \text{ Using (1)}$$

$$= \frac{(ay - x^2)(-ay^2 - a^2x + 2x^2y) + (y^2 - ax)(a^2y + ax^2 - 2xy^2)}{(y^2 - ax)^3}$$

$$= \frac{(-a^2y^3 + ax^2y^2 - a^3xy + a^2x^3 + 2ax^2y^2 - 2x^4y) + (a^2y^3 - a^3xy + ax^2y^2 - a^2x^3 - 2xy^4 + 2ax^2y^2)}{(y^2 - ax)^3}$$

$$= \frac{6ax^2y^2 - 2a^3xy - 2x^4y - 2xy^4}{(y^2 - ax)^3} = \frac{-2xy(x^3 + y^3 - 3axy) - 2a^3xy}{(y^2 - ax)^3}$$

$$= \frac{-2xy(0) - 2a^3xy}{(y^2 - ax)^3} = -\frac{\mathbf{2a^3\,xy}}{\mathbf{(y^2 - ax)^3}}.$$

Example 7. *If* $y = e^{m\sin^{-1} x}$, *show that* $(1 - x^2)\dfrac{d^2y}{dx^2} - x\dfrac{dy}{dx} - m^2y = 0$.

Sol. Given $y = e^{m \sin^{-1} x}$ ...(1)

Differentiating (1) w.r.t. x, we get

$$\frac{dy}{dx} = e^{m \sin^{-1} x} \cdot \frac{d}{dx}(m \sin^{-1} x) = e^{m \sin^{-1} x} \cdot \frac{m}{\sqrt{1 - x^2}}$$

$\Rightarrow$ $(\sqrt{1 - x^2})\dfrac{dy}{dx} = my$ ...(2)

Squaring, $(1 - x^2)\left(\dfrac{dy}{dx}\right)^2 = m^2y^2$ ...(3)

Differentiating (3) w.r.t. x, we get

$$(1 - x^2)\,2\left(\frac{dy}{dx}\right)\frac{d^2y}{dx^2} + \left(\frac{dy}{dx}\right)^2(-2x) = 2m^2y\frac{dy}{dx}$$

Dividing by $2\dfrac{dy}{dx}$, we get

$$(1 - x^2)\frac{d^2y}{dx^2} - x\frac{dy}{dx} = m^2y$$

or $$(1 - x^2)\frac{d^2y}{dx^2} - x\frac{dy}{dx} - m^2y = 0.$$

EXERCISE 12.15

Find the second order derivative w.r.t. x of the following functions (Q. No. 1–2) :

1. (*i*) $4x^3 + x + 7$ (*ii*) $\log x$ (*iii*) $x^3 + \tan x$
2. (*i*) $y = \cos^2 x + 3x$ (*ii*) $y = 2 \log x + x^3$ (*iii*) $y = \cos^{-1} x$
   (*iv*) $y = \log \sin x$ (*v*) $y = \log x + e^{2x}$ (*vi*) $y = \log (x^2/e^x)$.
3. (*i*) If $y = \tan x$, show that $y_2 = 2yy_1$.

(*ii*) If $y = \cot x$, show that $\dfrac{d^2y}{dx^2} + 2y\dfrac{dy}{dx} = 0$.

(*iii*) If $e^y(x + 1) = 1$, show that $\dfrac{d^2y}{dx^2} = \left(\dfrac{dy}{dx}\right)^2$.

(*iv*) If $y = \dfrac{\log x}{x}$, show that $\dfrac{d^2y}{dx^2} = \dfrac{2\log x - 3}{x^3}$.

4. (*i*) If $y = e^x(\sin x + \cos x)$, show that $\dfrac{d^2y}{dx^2} - 2\dfrac{dy}{dx} + 2y = 0$.

(*ii*) If $y = x + \cot x$, show that $\sin^2 x\dfrac{d^2y}{dx^2} - 2y + 2x = 0$.

(*iii*) If $y = A\,e^{mx} + B\,e^{nx}$, show that $\dfrac{d^2y}{dx^2} - (m + n)\dfrac{dy}{dx} + mny = 0$.

(*iv*) If $y = a\sin 2x + b\cos 2x$, show that $\dfrac{d^2y}{dx^2} + 4y = 0$.

5. Find the third derivative of the function $x^7 + \sin x$ w.r.t. x.

6. (*i*) Find the second derivative of the function $e^{-x}\cos x$ w.r.t. x.

(*ii*) Find the second derivative of the function $e^{x\sin x}$ w.r.t. x.

7. (*i*) If $y = \log(\cos x)$, find y_3. (*ii*) If $y = x^3\log x$, find y_3.

(*iii*) If $y = \dfrac{x}{\sqrt{1 - x^2}}$, find y_3. (*iv*) If $y = \cos 2x$, find y_3.

8. (*i*) If $y = \sin^{-1} x$, show that $(1 - x^2)\dfrac{d^2y}{dx^2} - x\dfrac{dy}{dx} = 0$.

(*ii*) If $y = \tan^{-1} x$, show that $(1 + x^2)\dfrac{d^2y}{dx^2} + 2x\dfrac{dy}{dx} = 0$.

(*P.T.U. B.Pharma, Dec. 2002 ; G.N.D.U. B.Pharma, April 2003, 2005*)

(*iii*) If $y = (\tan^{-1} x)^2$, show that $(x^2 + 1)^2 y_2 + 2x(x^2 + 1)y_1 - 2 = 0$.

(*iv*) If $y = \log(x + \sqrt{x^2 + 1})$, show that $(x^2 + 1)y_2 + xy_1 = 0$.

(*v*) If $y = [\log(x + \sqrt{1 + x^2})]^2$, show that $(1 + x^2)\dfrac{d^2y}{dx^2} + x\dfrac{dy}{dx} = 2$.

9. (*i*) If $\log y = \tan^{-1} x$, show that $(1 + x^2)y_2 + (2x - 1)y_1 = 0$.

(*ii*) If $y = e^{m\cos^{-1} x}$, show that $(1 - x^2)y_2 - xy_1 - m^2y = 0$. (P.T.U. B.Pharma, June 2003)

(*iii*) If $y = \sin(n\sin^{-1} x)$, show that $(1 - x^2)y_2 - xy_1 + n^2y = 0$.

(*iv*) If $y = \sec x - \tan x$, show that $\cos x\dfrac{d^2y}{dx^2} = y^2$.

(*v*) If $y = 3\cos\log x + 4\sin\log x$, show that $x^2y_2 + xy_1 + y = 0$.

10. Find $\dfrac{d^2y}{dx^2}$, if :

(*i*) $x = at^2$, $y = 2at$ (*ii*) $x = a\cos^3\theta$, $y = a\sin^3\theta$ (*iii*) $x = \dfrac{2at^2}{1+t}$, $y = \dfrac{3at}{1+t}$.

11. Find $\dfrac{d^2y}{dx^2}$ at $\theta = \dfrac{\pi}{2}$ for the parametric functions :

(*i*) $x = a(\theta - \sin\theta),\ y = a(1 + \cos\theta)$ (*ii*) $x = a(\theta + \sin\theta),\ y = a(1 - \cos\theta)$.

12. (*i*) If $x = a(\cos\theta + \theta\sin\theta)$ and $y = a(\sin\theta - \theta\cos\theta)$, find the value of $\dfrac{d^2y}{dx^2}$. Also find its value at $\theta = \dfrac{\pi}{4}$.

(*ii*) If $x = 2\cos t - \cos 2t$ and $y = 2\sin t - \sin 2t$, find the value of $\dfrac{d^2y}{dx^2}$ at $t = \dfrac{\pi}{2}$.

Answers

1. (*i*) $24x$ (*ii*) $-\dfrac{1}{x^2}$ (*iii*) $6x + 2\sec^2 x\tan x$

2. (*i*) $-2\cos 2x$ (*ii*) $6x - \dfrac{2}{x^2}$ (*iii*) $-\dfrac{x}{(1-x^2)^{3/2}}$ (*iv*) $-\text{cosec}^2 x$ (*v*) $-\dfrac{1}{x^2} + 4e^{2x}$

(*vi*) $-\dfrac{2}{x^2}$ **5.** $210x^4 - \cos x$ **6.** (*i*) $2e^{-x}\sin x$

(*ii*) $e^{x\sin x}[(x\cos x + \sin x)^2 + 2\cos x - x\sin x]$

7. (*i*) $-2\sec^2 x\tan x$ (*ii*) $11 + 6\log x$ (*iii*) $\dfrac{3(4x^2+1)}{(1-x^2)^{7/2}}$ (*iv*) $8\sin 2x$

10. (*i*) $-\dfrac{1}{2at^3}$ (*ii*) $\dfrac{\sec^4\theta}{3a\sin\theta}$ (*iii*) $-\dfrac{3}{2a}\left(\dfrac{1+t}{t^2+2t}\right)^3$

11. (*i*) $\dfrac{1}{a}$ (*ii*) $\dfrac{1}{a}$ **12.** (*i*) $\dfrac{\sec^3\theta}{a\theta}$, $\dfrac{8\sqrt{2}}{\pi a}$ (*ii*) $-\dfrac{3}{2}$.

Hints

3. (*iii*) $e^y(x+1) = 1 \Rightarrow e^y = \dfrac{1}{x+1} \Rightarrow y\log e = -\log(x+1)$.

8. (*i*) Differentiate y w.r.t. x ; cross multiply and square ; again differentiate.

(*v*) $y = [\log(x + \sqrt{1+x^2})]^2$ implies

$$y_1 = 2\log(x + \sqrt{1+x^2}).\ \frac{1}{x+\sqrt{1+x^2}} \cdot \left(1 + \frac{2x}{2\sqrt{1+x^2}}\right) = \frac{2\log(x+\sqrt{1+x^2})}{\sqrt{1+x^2}}$$

$$\Rightarrow (1+x^2)\,y_1^{\,2} = 4y \Rightarrow (1+x^2)\,2y_1y_2 + (0+2x)\,y_1^{\,2} = 4y_1 \Rightarrow (1+x^2)y_2 + xy_1 = 2.$$

13

Integration as Inverse of Differentiation

13.1. INTRODUCTION

In Differential Calculus, we studied the methods of differentiating various types of functions. We also saw that the derivatives are used in finding the slope of tangents to curves and for finding the rate of change of variables w.r.t. other variables. There are also so many other applications of derivatives, some of those we have already studied in the preceding chapters.

We turn our attention now to another aspect of calculus. Our new problem will involve the finding of a function whose derivative is known. This operation is the inverse of differentiation and gives rise to the branch of calculus called **Integral Calculus.** We know that the derivative of x^4 w.r.t. x is $4x^3$. But now we shall be interested in finding that function, whose derivative is $4x^3$. This process of finding functions with given derivatives is called **integration**.

The techniques of integration are found very useful in finding plane areas, lengths of arcs, volumes of solid of revolution, etc. In this chapter, we shall confine ourselves to fundamental formulae of integration.

13.2. DEFINITIONS

If the function $f(x)$ be the derivative of the function $g(x)$ w.r.t. x on a certain interval of the x-axis, then we say that an **integral** of $f(x)$ w.r.t. x is $g(x)$ and write it symbolically as

$$\int \mathbf{f(x)\ dx = g(x).}$$

The variable x is called the **variable of integration.** It is customary to use the differential 'dx' of the variable 'x' in the process of integration. The differential 'dx' indicate that the integration is to be carried w.r.t. x. The symbol '$\int$' (elongated S) denote the process of integration. The function whose integral is to be found out is called the **integrand.** If the integral of $f(x)$ w.r.t. x is $g(x)$, then we also say that $g(x)$ is an **antiderivative** of $f(x)$.

13.3. CONSTANT OF INTEGRATION

We know that $\frac{d}{dx}(x^3) = 3x^2$ and so by definition, x^3 is an integral of $3x^2$ w.r.t. x. Also for any constant C, we have $\frac{d}{dx}(x^3 + C) = 3x^2 + 0 = 3x^2$.

$$\therefore \qquad \int 3x^2\, dx = x^3 + C.$$

$\therefore$ $x^3 + C$ is an integral of $3x^2$ for any constant 'C'.

Thus, there will be infinitely many integrals of the function $3x^2$, because we shall get an integral for each value of C.

In general, if $f(x)$ be the derivative of the function $g(x)$ w.r.t. x on a certain interval of the x-axis, then for any constant C, we have

$$\frac{d}{dx}(g(x) + C) = \frac{d}{dx}(g(x)) + \frac{d}{dx}(C) = \frac{d}{dx}(g(x)) = f(x).$$

$$\therefore \quad \int \mathbf{f(x)\ dx = g(x) + C.}$$

The function $f(x)$ is the *integrand*. The function $g(x)$ is called a **particular integral** of $f(x)$ w.r.t. x, the function $g(x) + C$ is called the **indefinite integral** of $f(x)$ w.r.t. x and the constant C is called the **constant of integration.**

In the next theorem, we shall prove that $g(x) + C$ represents all possible integrals of the function $f(x)$.

Theorem. Let $\int f(x)\,dx = g(x)$. If G(x) be any other integral of f(x), then

G(x) = g(x) + C, for some constant.

Proof. We have $\int f(x)\,dx = g(x)$ and $\int f(x)\,dx = G(x)$.

$\Rightarrow \quad g'(x) = f(x)$ and $G'(x) = f(x)$.

Define $\quad h(x) = G(x) - g(x)$.

$\therefore \quad h'(x) = G'(x) - g'(x) = f(x) - f(x) = 0$ ...(1)

Let a and b $(a < b)$ be any two real numbers.

$\therefore$ (*i*) $h(x)$ continuous on $[a, b]$ and (*ii*) $h(x)$ is derivable on (a, b).

$\therefore$ By **L.M.V. theorem,** there exists at least one $c \in (a, b)$ such that

$$\frac{h(b) - h(a)}{b - a} = h'(c).$$

$\Rightarrow \quad h(b) - h(a) = (b - a) \,.\, h'(c) = (b - a)(0) = 0$ [using (1)]

$\Rightarrow \quad h(b) = h(a)$

$\therefore$ For any points a, b, we have $h(a) = h(b)$. Hence, $h(x)$ is a constant function.

$\Rightarrow$ $G(x) - g(x)$ is a constant function.

$\therefore$ $G(x) - g(x) = C$ *i.e.*, $G(x) = g(x) + C$, for some constant C.

13.4. FUNDAMENTAL INTEGRATION FORMULAE

1. $\frac{d}{dx}(C) = 0 \qquad \therefore \int \mathbf{0dx = C.}$

2. $\frac{d}{dx}(x) = 1 \qquad \therefore \int \mathbf{1dx = x + C.}$

3. $\frac{d}{dx}(kx) = k. \qquad \therefore \int \mathbf{kdx = kx + C.}$

4. $\frac{d}{dx}\left(\frac{x^{n+1}}{n+1}\right) = \frac{(n+1)x^n}{n+1} = x^n$, provided $n \neq -1$.

$\therefore$ **For n ≠ – 1,** $\int \mathbf{x^n\ dx = \frac{x^{n+1}}{n+1} + C.}$

5. $\dfrac{d}{dx}\left(\dfrac{(ax+b)^{n+1}}{(n+1)a}\right) = \dfrac{(n+1)(ax+b)^n \cdot a}{(n+1)a} = (ax+b)^n$, provided $n \neq -1$.

$\therefore$ **For $\mathbf{n \neq -1}$, $\displaystyle\int (\mathbf{ax+b})^{\mathbf{n}}\, \mathbf{dx} = \dfrac{(\mathbf{ax+b})^{\mathbf{n+1}}}{(\mathbf{n+1})\,\mathbf{a}} + \mathbf{C}$.**

Remark. Result (4) can be also obtained from result (5) by taking $a = 1$ and $b = 0$.

6. $x > 0 \Rightarrow \dfrac{d}{dx}(\log|x|) = \dfrac{d}{dx}(\log x) = \dfrac{1}{x}$.

Also, $x < 0 \Rightarrow \dfrac{d}{dx}(\log|x|) = \dfrac{d}{dx}(\log(-x)) = \dfrac{1}{-x}.(-1) = \dfrac{1}{x}$.

$\therefore \quad x \neq 0 \Rightarrow \dfrac{d}{dx}(\log|x|) = \dfrac{1}{x}$.

$\therefore$ **For $\mathbf{x \neq 0}$, $\displaystyle\int \frac{1}{\mathbf{x}}\, \mathbf{dx} = \mathbf{\log|x| + C}$.**

7. $ax + b > 0 \Rightarrow \dfrac{d}{dx}\left(\dfrac{\log|ax+b|}{a}\right) = \dfrac{d}{dx}\left(\dfrac{\log(ax+b)}{a}\right) = \dfrac{1}{ax+b}\cdot\dfrac{a}{a} = \dfrac{1}{ax+b}$

Also, $ax + b < 0 \Rightarrow \dfrac{d}{dx}\left(\dfrac{\log|ax+b|}{a}\right) = \dfrac{d}{dx}\left(\dfrac{\log[-(ax+b)]}{a}\right) = \dfrac{1}{-(ax+b)}\cdot\dfrac{-a}{a} = \dfrac{1}{ax+b}$.

$\therefore \quad ax + b \neq 0 \Rightarrow \dfrac{d}{dx}\left(\dfrac{\log|ax+b|}{a}\right) = \dfrac{1}{ax+b}$.

$\therefore$ **For $\mathbf{ax+b \neq 0}$, $\displaystyle\int \frac{1}{\mathbf{ax+b}}\, \mathbf{dx} = \frac{\mathbf{\log|ax+b|}}{\mathbf{a}} + \mathbf{C}$.**

Remark. Result (6) can also be obtained from result (7) by taking $a = 1$ and $b = 0$.

8. $\dfrac{d}{dx}\left(\dfrac{a^x}{\log a}\right) = \dfrac{a^x \log a}{\log a} = a^x$, provided $a > 0$, $a \neq 1$.

$\therefore$ **For $\mathbf{a > 0, a \neq 1}$, $\displaystyle\int \mathbf{a^x\, dx} = \frac{\mathbf{a^x}}{\mathbf{\log a}} + \mathbf{C}$.**

In particular, $\displaystyle\int e^x\, dx = \frac{e^x}{\log e} + C = \frac{e^x}{1} + C$. $\quad [\because \log e = 1]$

$\therefore \quad$ **$\displaystyle\int \mathbf{e^x\, dx = e^x + C}$.**

9. $\dfrac{d}{dx}(\sin x) = \cos x$. $\quad \therefore$ **$\displaystyle\int \mathbf{\cos x\, dx = \sin x + C}$.**

10. $\dfrac{d}{dx}(-\cos x) = \sin x$. $\quad \therefore$ **$\displaystyle\int \mathbf{\sin x\, dx = -\cos x + C}$.**

11. $\dfrac{d}{dx}(\tan x) = \sec^2 x$. $\quad \therefore$ **$\displaystyle\int \mathbf{\sec^2 x\, dx = \tan x + C}$.**

12. $\dfrac{d}{dx}(-\cot x) = \operatorname{cosec}^2 x$. $\quad \therefore$ **$\displaystyle\int \mathbf{\operatorname{cosec}^2 x\, dx = -\cot x + C}$.**

13. $\frac{d}{dx}$ (sec x) = sec x tan x. $\therefore \int \mathbf{sec\,x\,tan\,x\,dx = sec\,x + C.}$

14. $\frac{d}{dx}$ (– cosec x) = cosec x cot x. $\therefore \int \mathbf{cosec\,x\,cot\,x\,dx = -\,cosec\,x + C.}$

15. $\frac{d}{dx}(\sin^{-1} x) = \frac{1}{\sqrt{1-x^2}}$. $\therefore \int \frac{1}{\sqrt{1-x^2}}\,dx = \sin^{-1} x + C.$

16. $\frac{d}{dx}(\tan^{-1} x) = \frac{1}{1+x^2}$. $\therefore \int \frac{1}{1+x^2}\,dx = \tan^{-1} x + C.$

17. $\frac{d}{dx}(\sec^{-1} x) = \frac{1}{x\sqrt{x^2-1}}$. $\therefore \int \frac{1}{x\sqrt{x^2-1}}\,dx = \sec^{-1} x + C.$

Remark. Results (5) and (7) are extensions of results (4) and (6) respectively. Similar extensions of results (8) to (17) are also valid.

ILLUSTRATIVE EXAMPLES

Example 1. *Integrate the following functions w.r.t. x :*

(*i*) x^{-6} (*ii*) $\frac{1}{\sqrt{x}}$ (*iii*) $\frac{1}{\sqrt[n]{x}}$

(*iv*) $e^{3\log x}$ (*v*) $\sqrt{2x+5}$ (*vi*) $\frac{1}{(2x+9)^{5/7}}$.

Sol. (*i*) $\int x^{-6}\,dx = \frac{x^{-6+1}}{-6+1} + C = -\frac{\mathbf{x^{-5}}}{\mathbf{5}} + \mathbf{C}.$

(*ii*) $\int \frac{1}{\sqrt{x}}\,dx = \int x^{-1/2}\,dx = \frac{x^{-\frac{1}{2}+1}}{-\frac{1}{2}+1} + C = \frac{x^{1/2}}{1/2} + C = \mathbf{2\sqrt{x} + C.}$

(*iii*) $\int \frac{1}{\sqrt[n]{x}}\,dx = \int x^{-1/n}\,dx = \frac{x^{-\frac{1}{n}+1}}{-\frac{1}{n}+1} + C = \frac{\mathbf{n}}{\mathbf{n-1}}\,\mathbf{x}^{1-\frac{1}{n}} + \mathbf{C}.$

(*iv*) $\int e^{3\log x}\,dx = \int e^{\log x^3}\,dx = \int x^3\,dx = \frac{\mathbf{x^4}}{\mathbf{4}} + \mathbf{C}.$ $[\because\ e^{\log f(x)} = f(x)]$

(*v*) $\int \sqrt{2x+5}\,dx = \int (2x+5)^{1/2}\,dx = \frac{(2x+5)^{\frac{1}{2}+1}}{(\frac{1}{2}+1)\,2} + C = \frac{\mathbf{1}}{\mathbf{3}}\,\mathbf{(2x+5)^{3/2} + C.}$

(*vi*) $\int \frac{dx}{(2x+9)^{5/7}} = \int (2x+9)^{-5/7}\,dx = \frac{(2x+9)^{-(5/7)+1}}{\left(-\frac{5}{7}+1\right)2} + C = \frac{\mathbf{7}}{\mathbf{4}}\,\mathbf{(2x+9)^{2/7} + C.}$

Example 2. *Evaluate :*

(*i*) $\int \sqrt{z^3}\,dz$ (*ii*) $\int \sqrt[3]{s^2}\,ds$ (*iii*) $\int [(1-4x)^{-3}]^4\,dx$ (*iv*) $\int \frac{dy}{y^3}$.

Sol. (*i*) $\int \sqrt{z^3}\, dz = \int (z^3)^{1/2}\, dz = \int z^{3/2}\, dz = \dfrac{z^{\frac{3}{2}+1}}{\frac{3}{2}+1} + C = \mathbf{\dfrac{2z^{5/2}}{5} + C.}$

(*ii*) $\int \sqrt[3]{s^2}\, ds = \int (s^2)^{1/3}\, ds = \int s^{2/3}\, ds = \dfrac{s^{\frac{2}{3}+1}}{\frac{2}{3}+1} + C = \mathbf{\dfrac{3s^{5/3}}{5} + C.}$

(*iii*) $\int [(1-4x)^{-3}]^4\, dx = \int (1-4x)^{-12}\, dx = \dfrac{(1-4x)^{-12+1}}{(-12+1)(-4)} + C = \mathbf{\dfrac{(1-4x)^{-11}}{44} + C.}$

(*iv*) $\int \dfrac{dy}{y^3} = \int y^{-3}\, dy = \dfrac{y^{-3+1}}{-3+1} + C = \dfrac{y^{-2}}{-2} + C = \mathbf{-\dfrac{1}{2y^2} + C.}$

Theorem I. *If f(x) be an integrable function of x, then* $\dfrac{d}{dx}\left(\int f(x)\, dx\right) = f(x)$.

Proof. Let $\int f(x)\, dx = g(x) + C$ ...(1) $\therefore$ By definition, $\dfrac{d}{dx}(g(x)) = f(x)$ ...(2)

Now $\dfrac{d}{dx}\left(\int f(x)\, dx\right) = \dfrac{d}{dx}(g(x) + C)$ [Using (1)]

$= \dfrac{d}{dx}(g(x)) + \dfrac{d}{dx}(C) = f(x) + 0 = f(x).$ [Using (2)]

$\therefore$ $\mathbf{\dfrac{d}{dx}\left(\int f(x)\, dx\right) = f(x).}$

Remark. The above result shows that *integration* and *differentiation* are inverse processes and they neutralise the effect of each other.

Theorem II. *If* $\int f(x)\, dx = g(x) + C$, *then* $\int f(ax+b)\, dx = \dfrac{g(ax+b)}{a} + C.$

Proof. $\int f(x)\, dx = g(x) + C \Rightarrow \dfrac{d}{dx}(g(x)) = f(x).$

$\therefore$ $\dfrac{d}{dz}(g(z)) = f(z)$, for any variable z. ...(1)

Let $z = ax + b.$

$\therefore$ $\dfrac{d}{dx}\left(\dfrac{g(ax+b)}{a} + C\right) = \dfrac{1}{a}\dfrac{d}{dx}(g(z)) = \dfrac{1}{a}\dfrac{d}{dz}(g(z)) \cdot \dfrac{dz}{dx}$

$= \dfrac{1}{a} f(z).(a.1 + 0) = f(z) = f(ax+b).$ [Using (1)]

$\therefore$ $\mathbf{\int f(ax+b)\, dx = \dfrac{g(ax+b)}{a} + C.}$

Corollary. By using this theorem we have following extensions of formulae :

1. $\int x^n\, dx = \dfrac{x^{n+1}}{n+1} + C,\ n \neq -1 \Rightarrow \mathbf{\int (ax+b)^n\, dx = \dfrac{(ax+b)^{n+1}}{(n+1)\,a} + C,\ n \neq -1}$

2. $\int \dfrac{1}{x}\, dx = \log|x| + C \Rightarrow \mathbf{\int \dfrac{1}{ax+b}\, dx = \dfrac{\log|ax+b|}{a} + C}$

3. $\int a^x\,dx = \frac{a^x}{\log a} + C$, where $a > 0, a \neq 1$

$$\Rightarrow \int a^{mx+k}\,dx = \frac{a^{mx+k}}{m \log a} + C, \text{ where } a > 0, a \neq 1$$

4. $\int e^x\,dx = e^x + C \quad \Rightarrow \int e^{mx+k}\,dx = \frac{e^{mx+k}}{m} + C$

5. $\int \cos x\,dx = \sin x + C \quad \Rightarrow \int \cos(ax+b)\,dx = \frac{\sin(ax+b)}{a} + C$

6. $\int \sin x\,dx = -\cos x + C \quad \Rightarrow \int \sin(ax+b)\,dx = -\frac{\cos(ax+b)}{a} + C$

7. $\int \sec^2 x\,dx = \tan x + C \quad \Rightarrow \int \sec^2(ax+b)\,dx = \frac{\tan(ax+b)}{a} + C$

8. $\int \text{cosec}^2 x\,dx = -\cot x + C \quad \Rightarrow \int \text{cosec}^2(ax+b)\,dx = -\frac{\cot(ax+b)}{a} + C$

9. $\int \sec x \tan x\,dx = \sec x + C \quad \Rightarrow \int \sec(ax+b)\tan(ax+b)\,dx = \frac{\sec(ax+b)}{a} + C$

10. $\int \text{cosec}\,x \cot x\,dx = -\text{cosec}\,x + C$

$$\Rightarrow \int \text{cosec}(ax+b)\cot(ax+b)\,dx = -\frac{\text{cosec}(ax+b)}{a} + C.$$

For example,

(*a*) $\int \cos(4x+1)\,dx = \frac{\sin(4x+1)}{4} + C$, because $\int \cos x\,dx = \sin x + C$

(*b*) $\int \sec^2(1-7z)\,dz = \frac{\tan(1-7z)}{-7} + C$, because $\int \sec^2 z\,dz = \tan z + C$.

Example 3. *Evaluate the following integrals :*

(*i*) $\int \sin(4x+5)\,dx$ (*ii*) $\int \tan(3-4z)\sec(3-4z)\,dz$

(*iii*) $\int \text{cosec}^2(10s-3)\,ds$ (*iv*) $\int 7^{8-9x}\,dx$.

Sol. (*i*) We have $\int \sin x\,dx = -\cos x + C$. $\therefore \int \sin(4x+5)\,dx = -\frac{\cos(4x+5)}{4} + C$.

(*ii*) We have $\int \sec z \tan z\,dz = \sec z + C$. $\therefore \int \tan(3-4z)\sec(3-4z)\,dz = \frac{\sec(3-4z)}{-4} + C$.

(*iii*) We have $\int \text{cosec}^2 s\,ds = -\cot s + C$. $\therefore \int \text{cosec}^2(10s-3)\,ds = -\frac{\cot(10s-3)}{10} + C$.

(*iv*) We have $\int 7^x\,dx = \frac{7^x}{\log 7} + C$. $\therefore \int 7^{8-9x}\,dx = \frac{7^{8-9x}}{(-9)\log 7} + C$.

IMPORTANT FORMULAE

1. (i) $\int 0\,dx = C$ (ii) $\int k\,dx = kx + C$

2. (i) $\int x^n\,dx = \dfrac{x^{n+1}}{n+1} + C,\ n \neq -1$ (ii) $\int (ax+b)^n\,dx = \dfrac{(ax+b)^{n+1}}{(n+1)a} + C,\ n \neq -1$

3. (i) $\int \dfrac{1}{x}\,dx = \log|\,x\,| + C$ (ii) $\int \dfrac{1}{ax+b}\,dx = \dfrac{\log|ax+b|}{a} + C$

4. (i) $\int a^x\,dx = \dfrac{a^x}{\log a} + C,\ a > 0\ a \neq 1$ (ii) $\int a^{mx+k}\,dx = \dfrac{a^{mx+k}}{m\log a} + C,\ a > 0,\ a \neq 1$

5. (i) $\int e^x\,dx = e^x + C$ (ii) $\int e^{mx+k}\,dx = \dfrac{e^{mx+k}}{m} + C$

6. (i) $\int \cos\,dx = \sin x + C$ (ii) $\int \cos(ax+b)\,dx = \dfrac{\sin(ax+b)}{a} + C$

7. (i) $\int \sin x\,dx = -\cos x + C$ (ii) $\int \sin(ax+b)\,dx = -\dfrac{\cos(ax+b)}{a} + C$

8. (i) $\int \sec^2 x\,dx = \tan x + C$ (ii) $\int \sec^2(ax+b)\,dx = \dfrac{\tan(ax+b)}{a} + C$

9. (i) $\int \operatorname{cosec}^2 x\,dx = -\cot x + C$ (ii) $\int \operatorname{cosec}^2(ax+b)\,dx = -\dfrac{\cot(ax+b)}{a} + C$

10. (i) $\int \sec x \tan x\,dx = \sec x\,dx + C$

(ii) $\int \sec(ax+b)\tan(ax+b)\,dx = \dfrac{\sec(ax+b)}{a} + C$

11. (i) $\int \operatorname{cosec} x \cot x\,dx = -\operatorname{cosec} x + C$

(ii) $\int \operatorname{cosec}(ax+b)\cot(ax+b)\,dx = -\dfrac{\operatorname{cosec}(ax+b)}{a} + C.$

EXERCISE 13.1

Integrate the following functions w.r.t. x (Q. No. 1–6) :

1. (i) x^4 (ii) $\sqrt{x}$ (iii) $x^{1/7}$ (iv) $\dfrac{\sqrt[5]{x^2}}{\sqrt{x^{14/5}}}$

2. (i) $(1-8x)^{-3}$ (ii) $\sqrt[4]{(1-x)^3}$ (iii) $\dfrac{(2+7x)^{1/4}}{(2+7x)^{1/3}}$ (iv) $\dfrac{(25x+6)^{-7}}{(25x+6)^{-6}}$.

3. (i) $\int z^{1/4}\,dz$ (ii) $\int \dfrac{s^{3/2}}{s^{2/3}}\,ds$ (iii) $\int z|\,z\,|\,dz$ (iv) $\int e^{-\log x}\,dx$

4. (i) $\int \sin(7x-5)\,dx$ (ii) $\int \cos(4t-3)\,dt$ (iii) $\int \sec^2(7-4x)\,dx$

(iv) $\int \operatorname{cosec}^2(2x+3)\,dx$ (v) $\int \sec 7x \tan 7x\,dx$ (vi) $\int \operatorname{cosec} 9x \cot 9x\,dx$

5. (*i*) $\int \frac{dx}{\sqrt{1-(4x+5)^2}}$ (*ii*) $\int \frac{dx}{1+(15x+9)^2}$ (*iii*) $\int \frac{dx}{4x\sqrt{16x^2-1}}$

6. (*i*) $\int e^{6s+9}\, ds$ (*ii*) $\int \frac{e^{4x+3}}{e^{3x+7}}\, dx$ (*iii*) $\int 7^{5x+9}\, dx.$

Answers

1. (*i*) $\frac{x^5}{5} + C$ (*ii*) $\frac{2}{3} x^{3/2} + C$ (*iii*) $\frac{7}{8} x^{8/7} + C$ (*iv*) $\log |x| + C$

2. (*i*) $\frac{(1-8x)^{-2}}{16} + C$ (*ii*) $-\frac{4}{7}(1-x)^{7/4} + C$ (*iii*) $\frac{12}{77}(2+7x)^{11/12} + C$ (*iv*) $\frac{1}{25}\log|25x+6| + C$

3. (*i*) $\frac{4}{5} z^{5/4} + C$ (*ii*) $\frac{6}{11} s^{11/6} + C$ (*iii*) $\frac{|z|^3}{3} + C$ (*iv*) $\log x + C$

4. (*i*) $-\frac{\cos(7x-5)}{7} + C$ (*ii*) $\frac{\sin(4t-3)}{4} + C$ (*iii*) $-\frac{\tan(7-4x)}{4} + C$

(*iv*) $-\frac{\cot(2x+3)}{2} + C$ (*v*) $\frac{\sec 7x}{7} + C$ (*vi*) $-\frac{\text{cosec } 9x}{9} + C$

5. (*i*) $\frac{\sin^{-1}(4x+5)}{4} + C$ (*ii*) $\frac{\tan^{-1}(15x+9)}{15} + C$ (*iii*) $\frac{\sec^{-1} 4x}{4} + C$

6. (*i*) $\frac{e^{6s+9}}{6} + C$ (*ii*) $e^{x-4} + C$ (*iii*) $\frac{7^{5x+9}}{5 \log 7} + C.$

Hints

3. (*iii*) **Case I. z > 0**

$$\int z\,|z|\,dz = \int z \,.\, z\, dz = \frac{z^3}{3} + C = \frac{|z|^3}{3} + C$$

Case II. z < 0

$$\int z\,|z|\,dz = \int z\,(-z)\, dz = -\frac{z^3}{3} + C = \frac{(-z)^3}{3} + C = \frac{|z|^3}{3} + C$$

(*iv*) $e^{-\log x} = e^{\log(1/x)} = \frac{1}{x}$. Also, $x > 0$ because $\log x$ is defined.

13.5. INTEGRAL OF SUM OR DIFFERENCE OF TWO FUNCTIONS

Theorem. *If f(x) and g(x) be any two integrable functions of x, then*

$$\int (f(x) \pm g(x))\, dx = \int f(x)\, dx \pm \int g(x)\, dx.$$

Proof. $\frac{d}{dx}\left(\int f(x)\, dx \pm \int g(x)\, dx\right) = \frac{d}{dx}\left(\int f(x)\, dx\right) \pm \frac{d}{dx}\left(\int g(x)\, dx\right) = f(x) \pm g(x).$

$\therefore$ By definition, $\int \mathbf{(f(x) \pm g(x))\, dx} = \int \mathbf{f(x)\, dx} \pm \int \mathbf{g(x)\, dx.}$

Remark. The above result can also be extended as follows :

$$\int (f_1(x) \pm f_2(x) \pm \ldots\ldots \pm f_n(x))\, dx = \int f_1(x)\, dx \pm \int f_2(x)\, dx \pm \ldots\ldots \pm \int f_n(x)\, dx.$$

For example,

$$\int (x^4 + \sin x + e^{2x})\, dx = \int x^4\, dx + \int \sin x\, dx + \int e^{2x}\, dx$$

$$= \frac{x^5}{5} + (-\cos x) + \frac{e^{2x}}{2} + C = \frac{x^5}{5} - \cos x + \frac{e^{2x}}{2} + C.$$

13.6. INTEGRAL OF CONSTANT MULTIPLE OF A FUNCTION

Theorem. *If f(x) be any integrable function of x, then for any constant a*

$$\int a.f(x)\, dx = a \int f(x)\, dx.$$

Proof. $\frac{d}{dx}\left(a.\int f(x)\, dx\right) = a\, \frac{d}{dx}\left(\int f(x)\, dx\right) = a\, .\, f(x).$

$\therefore$ By definition, $\int \mathbf{a\, .\, f(x)\, dx = a \int f(x)\, dx.}$

For example, $\int 6x^{11}\, dx = 6\int x^{11}\, dx = 6\left(\frac{x^{12}}{12}\right) + C = \frac{x^{12}}{2} + C.$

In practice, we directly write $\int 6x^{11}\, dx = \frac{6x^{12}}{12} + C = \frac{x^{12}}{2} + C.$

ILLUSTRATIVE EXAMPLES

Example 1. *Integrate the following functions w.r.t. x :*

(*i*) $x^2 + 7x + 5$ (*ii*) $(x + 2)(x - 1)(x + 5)(x - 7)$ (*iii*) $\left(x + \frac{1}{x}\right)^2 (x - 7)$

(*iv*) $\frac{x^2 + 3x + 4}{\sqrt{x}}$ (*v*) $\left(\sqrt{x} - \frac{1}{\sqrt{x}}\right)^3$ (*vi*) $e^{x \log a} + e^{a \log x} + e^{a \log a}$.

Sol. (*i*) $\int (x^2 + 7x + 5)\, dx = \int x^2\, dx + 7\int x\, dx + \int 5\, dx = \mathbf{\frac{x^3}{3} + 7\frac{x^2}{2} + 5x + C.}$

(*ii*) $(x + 2)(x - 1)(x + 5)(x - 7) = [(x + 2)(x - 1)][(x + 5)(x - 7)]$

$= (x^2 + x - 2)(x^2 - 2x - 35) = x^4 + x^3 - 2x^2 - 2x^3 - 2x^2 + 4x - 35x^2 - 35x + 70$

$= x^4 - x^3 - 39x^2 - 31x + 70.$

$$\therefore \quad \int (x + 2)(x - 1)(x + 5)(x - 7)\, dx = \int (x^4 - x^3 - 39x^2 - 31x + 70)\, dx$$

$$= \frac{x^5}{5} - \frac{x^4}{4} - \frac{39}{3}x^3 - \frac{31}{2}x^2 + 70x + C = \mathbf{\frac{x^5}{5} - \frac{x^4}{4} - 13x^3 - \frac{31}{2}x^2 + 70x + C.}$$

(*iii*) $\left(x + \frac{1}{x}\right)^2 (x - 7) = \left(x^2 + \frac{1}{x^2} + 2\right)(x - 7) = x^3 - 7x^2 + \frac{1}{x} - \frac{7}{x^2} + 2x - 14$

$$= x^3 - 7x^2 + 2x + \frac{1}{x} - 7x^{-2} - 14.$$

$$\therefore \quad \int \left[\left(x + \frac{1}{x}\right)^2 (x - 7)\right] dx = \int \left(x^3 - 7x^2 + 2x + \frac{1}{x} - 7x^{-2} - 14\right) dx$$

$$= \frac{x^4}{4} - 7\left(\frac{x^3}{3}\right) + 2\left(\frac{x^2}{2}\right) + \log|x| - 7\left(\frac{x^{-1}}{-1}\right) - 14x + C$$

$$= \mathbf{\frac{x^4}{4} - \frac{7}{3}x^3 + x^2 + \log|x| + \frac{7}{x} - 14x + C.}$$

(iv) $$\frac{x^2+3x+4}{\sqrt{x}} = \frac{x^2}{\sqrt{x}} + \frac{3x}{\sqrt{x}} + \frac{4}{\sqrt{x}} = x^{3/2} + 3x^{1/2} + 4x^{-1/2}.$$

$$\therefore \int \frac{x^2+3x+4}{\sqrt{x}}\,dx = \int (x^{3/2} + 3x^{1/2} + 4x^{-1/2})\,dx = \frac{x^{5/2}}{5/2} + 3.\frac{x^{3/2}}{3/2} + 4.\frac{x^{1/2}}{1/2} + C$$

$$= \mathbf{\frac{2}{5}x^{5/2} + 2x^{3/2} + 8x^{1/2} + C.}$$

(v) $$\left(\sqrt{x} - \frac{1}{\sqrt{x}}\right)^3 = (\sqrt{x})^3 - \left(\frac{1}{\sqrt{x}}\right)^3 - 3\sqrt{x}\,.\frac{1}{\sqrt{x}}\left(\sqrt{x} - \frac{1}{\sqrt{x}}\right)$$

$$= x^{3/2} - x^{-3/2} - 3\sqrt{x} + \frac{3}{\sqrt{x}} = x^{3/2} - x^{-3/2} - 3x^{1/2} + 3x^{-1/2}.$$

$$\therefore \int \left(\sqrt{x} - \frac{1}{\sqrt{x}}\right)^3 dx = \int (x^{3/2} - x^{-3/2} - 3x^{1/2} + 3x^{-1/2})\,dx$$

$$= \frac{x^{5/2}}{5/2} - \frac{x^{-1/2}}{-1/2} - 3.\frac{x^{3/2}}{3/2} + 3.\frac{x^{1/2}}{1/2} + C$$

$$= \mathbf{\frac{2}{5}x^{5/2} + 2x^{-1/2} - 2x^{3/2} + 6x^{1/2} + C.}$$

(vi) $$\int (e^{x\log a} + e^{a\log x} + e^{a\log a})\,dx = \int (e^{\log a^x} + e^{\log x^a} + e^{\log a^a})\,dx$$

$$= \int (a^x + x^a + a^a)\,dx \qquad (\because\ e^{\log k} = k)$$

$$= \mathbf{\frac{a^x}{\log a} + \frac{x^{a+1}}{a+1} + a^a x + C.}$$

Example 2. *If $f'(x) = \sqrt{x}$ and $f(1) = 2$, find $f(x)$.*

Sol. We have $f'(x) = \sqrt{x}$.

$$\therefore \quad f(x) = \int f'(x)\,dx = \int \sqrt{x}\,dx = \frac{x^{3/2}}{3/2} + C = \frac{2}{3}x^{3/2} + C$$

Also, $f(1) = 2. \quad \therefore \quad 2 = \frac{2}{3}(1)^{3/2} + C$ or $C = 2 - \frac{2}{3} = \frac{4}{3}$

$$\therefore \quad f(x) = \mathbf{\frac{2}{3}x^{3/2} + \frac{4}{3}}.$$

Example 3. *Evaluate the following integrals :*

(i) $\int (e^x + 3\cos x - 4x^3 + 2)\,dx$ (ii) $\int (x^3 + a^x + e^x a^x + \sin a)\,dx$

(iii) $\int \frac{1}{\sin^2 x \cos^2 x}\,dx$ (iv) $\int \frac{\sin^3 x + \cos^3 x}{\sin^2 x \cos^2 x}\,dx.$

Sol. (*i*) $\int (e^x + 3\cos x - 4x^3 + 2)\, dx = \int e^x\, dx + 3\int \cos x\, dx - 4\int x^3\, dx + 2\int 1 \,.\, dx$

$$= e^x + 3\sin x - 4 \,.\, \frac{x^4}{4} + 2x + C = \mathbf{e^x + 3\sin x - x^4 + 2x + C.}$$

(*ii*) $\int (x^3 + a^x + e^x a^x + \sin a)\, dx = \int x^3\, dx + \int a^x\, dx + \int (ea)^x\, dx + \sin a \int 1 \,.\, dx$

$$= \frac{x^4}{4} + \frac{a^x}{\log a} + \frac{(ea)^x}{\log ea} + (\sin a)\, x + C = \mathbf{\frac{x^4}{4} + \frac{a^x}{\log a} + \frac{e^x a^x}{1 + \log a} + x \sin a + C.}$$

$\mid \log ea = \log e + \log a = 1 + \log a$

(*iii*) $\int \frac{1}{\sin^2 x \cos^2 x}\, dx = 4\int \frac{dx}{(2\sin x \cos x)^2} = 4\int \frac{dx}{\sin^2 2x} = 4\int \text{cosec}^2\, 2x\, dx$

$$= \frac{4(-\cot 2x)}{2} + C = \mathbf{-2\cot 2x + C.}$$

(*iv*) $\int \frac{\sin^3 x + \cos^3 x}{\sin^2 x \cos^2 x}\, dx = \int \left(\frac{\sin^3 x}{\sin^2 x \cos^2 x} + \frac{\cos^3 x}{\sin^2 x \cos^2 x} \right) dx$

$$= \int \left(\frac{\sin x}{\cos x \cos x} + \frac{\cos x}{\sin x \sin x} \right) dx = \int (\sec x \tan x + \text{cosec}\, x \cot x)\, dx$$

$$= \sec x + (-\text{cosec}\, x) + C = \mathbf{\sec x - \text{cosec}\, x + C.}$$

Example 4. *Integrate the following functions w.r.t. x :*

(*i*) $\frac{x^4}{x+1}$ (*ii*) $\frac{x^4 + x^2 - 1}{2(x^2 + 1)}$.

Sol.

(*i*) $\int \frac{x^4}{x+1}\, dx = \int \left(x^3 - x^2 + x - 1 + \frac{1}{x+1} \right) dx$

$$= \mathbf{\frac{x^4}{4} - \frac{x^3}{3} + \frac{x^2}{2} - x + \log | x + 1 | + C.}$$

$$x + 1 \overline{)\, x^4 \,(}\, x^3 - x^2 + x - 1$$
$$\underline{x^4 + x^3}$$
$$-x^3$$
$$\underline{-x^3 - x^2}$$
$$x^2$$
$$\underline{x^2 + x}$$
$$-x$$
$$\underline{-x - 1}$$
$$1$$

(*ii*) $\int \frac{x^4 + x^2 - 1}{2(x^2 + 1)}\, dx = \frac{1}{2}\int \frac{x^2(x^2 + 1) - 1}{x^2 + 1}\, dx$

$$= \frac{1}{2}\int \left(x^2 - \frac{1}{x^2 + 1} \right) dx = \mathbf{\frac{1}{2}\left[\frac{x^3}{3} - \tan^{-1} x \right] + C.}$$

Remark. In case of rational functions, if the degree of the numerator of the integrand is equal to or greater than that of the denominator, then divide the numerator by the denominator until the degree of the remainder is less than that of the denominator and then use the formula :

$$\mathbf{\frac{Numerator}{Denominator} = Quotient + \frac{Remainder}{Denominator}.}$$

Example 5. *Integrate the following functions w.r.t. x :*

(*i*) $\frac{1}{\sqrt{ax+b} - \sqrt{ax+c}}$ (*ii*) $\frac{x}{\sqrt{1+x}}$.

Sol. (*i*) $\int \frac{1}{\sqrt{ax+b} - \sqrt{ax+c}}\, dx = \int \left[\frac{1}{\sqrt{ax+b} - \sqrt{ax+c}} \right] \left[\frac{\sqrt{ax+b} + \sqrt{ax+c}}{\sqrt{ax+b} + \sqrt{ax+c}} \right] dx$

$$= \int \frac{\sqrt{ax+b}+\sqrt{ax+c}}{(ax+b)-(ax+c)}\,dx = \frac{1}{b-c}\int (\sqrt{ax+b}+\sqrt{ax+c})\,dx$$

$$= \frac{1}{b-c}\left[\frac{(ax+b)^{3/2}}{(3/2)\,.\,a}+\frac{(ax+c)^{3/2}}{(3/2)\,.\,a}\right] + C = \frac{\mathbf{2}}{\mathbf{3a\,(b-c)}}\,\mathbf{[(ax+b)^{3/2}+(ax+c)^{3/2}]+C.}$$

$$(ii)\ \int \frac{x}{\sqrt{1+x}}\,dx = \int \frac{(1+x)-1}{\sqrt{1+x}}\,dx = \int \left[\frac{1+x}{\sqrt{1+x}}-\frac{1}{\sqrt{1+x}}\right]dx = \int [(1+x)^{1/2}-(1+x)^{-1/2}]\,dx$$

$$= \frac{(1+x)^{3/2}}{(3/2)\,.\,1}-\frac{(1+x)^{1/2}}{(1/2)\,.\,1}+C = \frac{\mathbf{2}}{\mathbf{3}}\,\mathbf{(1+x)^{3/2}-2(1+x)^{1/2}+C.}$$

Example 6. *Evaluate the following integrals :*

(i) $\int \frac{1}{1-\sin x}\,dx$ (ii) $\int \sqrt{1+\sin x}\,dx$

(iii) $\int \tan^{-1}\sqrt{\frac{1-\sin x}{1+\sin x}}\,dx$ (iv) $\int \sin x\,\sqrt{1+\cos 2x}\,dx.$

Sol. (i) $\int \frac{1}{1-\sin x}\,dx = \int \left(\frac{1}{1-\sin x}\cdot\frac{1+\sin x}{1+\sin x}\right)dx = \int \frac{1+\sin x}{\cos^2 x}\,dx$

$$= \int (\sec^2 x + \sec x \tan x)\,dx = \mathbf{\tan x + \sec x + C.}$$

$$(ii)\ \int \sqrt{1+\sin x}\,dx = \int \sqrt{\sin^2\frac{x}{2}+\cos^2\frac{x}{2}+2\sin\frac{x}{2}\cos\frac{x}{2}}\,dx = \int \sqrt{\left(\sin\frac{x}{2}+\cos\frac{x}{2}\right)^2}\,dx$$

$$= \int \left(\sin\frac{x}{2}+\cos\frac{x}{2}\right)dx = \frac{-\cos x/2^{*}}{1/2}+\frac{\sin x/2}{1/2}+C = \mathbf{2\left(\sin\frac{x}{2}-\cos\frac{x}{2}\right)+C.}$$

Alternative method

$$\int \sqrt{1+\sin x}\,dx = \int \sqrt{1+\cos\left(\frac{\pi}{2}-x\right)}\,dx = \int \sqrt{2\cos^2\left(\frac{\pi}{4}-\frac{\pi}{2}\right)}\,dx$$

$$= \sqrt{2}\int \cos\left(\frac{\pi}{4}-\frac{\pi}{2}\right)dx = \sqrt{2}\,\frac{\sin\left(\frac{\pi}{4}-\frac{x}{2}\right)}{-1/2}+C = -\,2\sqrt{2}\left(\sin\frac{\pi}{4}\cos\frac{x}{2}-\cos\frac{\pi}{4}\sin\frac{x}{2}\right)+C$$

$$= -\,2\sqrt{2}\left(\frac{1}{\sqrt{2}}\cos\frac{x}{2}-\frac{1}{\sqrt{2}}\sin\frac{x}{2}\right)+C = \mathbf{2\left(\sin\frac{x}{2}-\cos\frac{x}{2}\right)+C.}$$

$$(iii)\ \tan^{-1}\sqrt{\frac{1-\sin x}{1+\sin x}} = \tan^{-1}\sqrt{\frac{1-\cos\left(\frac{\pi}{2}-x\right)}{1+\cos\left(\frac{\pi}{2}-x\right)}} = \tan^{-1}\sqrt{\frac{2\sin^2\left(\frac{\pi}{4}-\frac{x}{2}\right)}{2\cos^2\left(\frac{\pi}{4}-\frac{x}{2}\right)}}$$

---

* We have $\int \sin(ax+b)\,dx = -\frac{\cos(ax+b)}{a}+C$ and $\int \cos(ax+b)\,dx = \frac{\sin(ax+b)}{a}+C.$

In such integrals, the students are advised not to forget to divide by 'a', the coefficient of x.

$$= \tan^{-1}\left(\tan\left(\frac{\pi}{4}-\frac{x}{2}\right)\right) = \frac{\pi}{4}-\frac{x}{2}.$$

$$\therefore \quad \int \tan^{-1}\sqrt{\frac{1-\sin x}{1+\sin x}}\,dx = \int\left(\frac{\pi}{4}-\frac{x}{2}\right)dx = \frac{\pi}{4}x - \frac{1}{2}\left(\frac{x^2}{2}\right) + C = \frac{\pi}{4}\mathbf{x} - \frac{\mathbf{x^2}}{4} + \mathbf{C}.$$

(*iv*) $$\int \sin x\sqrt{1+\cos 2x}\,dx = \int \sin x \,.\, \sqrt{2\cos^2 x\,dx} = \sqrt{2}\int \sin x\cos x\,dx$$

$$= \frac{1}{\sqrt{2}}\int 2\sin x\cos x\,dx = \frac{1}{\sqrt{2}}\int \sin 2x\,dx$$

$$= \frac{1}{\sqrt{2}}\left(-\frac{\cos 2x}{2}\right) + C = -\frac{1}{2\sqrt{2}}\cos 2x + C.$$

13.7. EVALUATION OF $\int \sin^n x\,dx$ AND $\int \cos^n x\,dx$

In order to evaluate $\int \sin^n x\,dx$ and $\int \cos^n x\,dx$, for $n = 2, 3, 4$, the integrands $\sin^n x$ and $\cos^n x$ are expressed in terms of sines and cosines of multiples of x by using the following formulae :

(*i*) $\cos 2x = 1 - 2\sin^2 x$ (*ii*) $\cos 2x = 2\cos^2 x - 1$

(*iii*) $\sin 3x = 3\sin x - 4\sin^3 x$ (*iv*) $\cos 3x = 4\cos^3 x - 3\cos x$.

Example 7. *Evaluate the following integrals :*

(*i*) $\int \sin^2 4x\,dx$ (*ii*) $\int \cos^2 \frac{x}{2}\,dx$ (*iii*) $\int \sin^3 x\,dx$

(*iv*) $\int \cos^4 x\,dx$ (*v*) $\int \sin^4 x\cos^4 x\,dx$ (*vi*) $\int \frac{\sin^8 x - \cos^8 x}{1-2\sin^2 x\cos^2 x}\,dx$.

Sol. (*i*) $$\int \sin^2 4x\,dx = \int \frac{1-\cos 8x}{2}\,dx \qquad \left(\because \cos 2\theta = 1-2\sin^2\theta \Rightarrow \sin^2\theta = \frac{1-\cos 2\theta}{2}\right)$$

$$= \frac{1}{2}\left(x - \frac{\sin 8x}{8}\right) + C.$$

(*ii*) $$\int \cos^2\frac{x}{2}\,dx = \int \frac{1+\cos x}{2}\,dx \qquad \left(\because \cos 2\theta = 2\cos^2\theta - 1 \Rightarrow \cos^2\theta = \frac{1+\cos 2\theta}{2}\right)$$

$$= \frac{1}{2}(x + \sin x) + C.$$

(*iii*) We have $\sin 3x = 3\sin x - 4\sin^3 x$.

$$\Rightarrow \quad 4\sin^3 x = 3\sin x - \sin 3x \Rightarrow \sin^3 x = \frac{3}{4}\sin x - \frac{1}{4}\sin 3x.$$

$$\therefore \quad \int \sin^3 x\,dx = \int\left(\frac{3}{4}\sin x - \frac{1}{4}\sin 3x\right)dx = \frac{3}{4}(-\cos x) - \frac{1}{4}\left(-\frac{\cos 3x}{3}\right) + C$$

$$= -\frac{3}{4}\cos x + \frac{1}{12}\cos 3x + C.$$

(*iv*) $\cos^4 x = (\cos^2 x)^2 = \left(\frac{1+\cos 2x}{2}\right)^2 = \frac{1}{4}(1+\cos^2 2x + 2\cos 2x)$

$$= \frac{1}{4}\left(1+\frac{1+\cos 4x}{2} + 2\cos 2x\right) = \frac{1}{4}\left(\frac{3}{2}+\frac{1}{2}\cos 4x + 2\cos 2x\right)$$

$$= \frac{3}{8}+\frac{1}{8}\cos 4x + \frac{1}{2}\cos 2x.$$

$$\therefore \quad \int \cos^4 x\, dx = \int\left(\frac{3}{8}+\frac{1}{8}\cos 4x + \frac{1}{2}\cos 2x\right)dx = \frac{3}{8}x + \frac{1}{8}\cdot\frac{\sin 4x}{4} + \frac{1}{2}\cdot\frac{\sin 2x}{2} + C$$

$$= \mathbf{\frac{3}{8}x + \frac{1}{32}\sin 4x + \frac{1}{4}\sin 2x + C.}$$

(*v*) $\int \sin^4 x\cos^4 x\, dx = \int \frac{1}{16}(2\sin x\cos x)^4\, dx = \frac{1}{16}\int \sin^4 2x\, dx = \frac{1}{16}\int (\sin^2 2x)^2\, dx$

$$= \frac{1}{16}\int\left(\frac{1-\cos 4x}{2}\right)^2 dx = \frac{1}{64}\int (1+\cos^2 4x - 2\cos 4x)\, dx$$

$$= \frac{1}{64}\int\left(1+\frac{1+\cos 8x}{2} - 2\cos 4x\right)dx = \frac{1}{64}\int\left(\frac{3}{2}+\frac{1}{2}\cos 8x - 2\cos 4x\right)dx$$

$$= \mathbf{\frac{1}{64}\left[\frac{3}{2}x + \frac{1}{16}\sin 8x - \frac{1}{2}\sin 4x\right] + C.}$$

(*vi*) Let $\quad I = \int \frac{\sin^8 x - \cos^8 x}{1-2\sin^2 x\cos^2 x}$

$$\frac{\sin^8 x - \cos^8 x}{1-2\sin^2 x\cos^2 x} = \frac{(\sin^4 x - \cos^4 x)(\sin^4 x + \cos^4 x)}{1-2\sin^2 x\cos^2 x}$$

$$= \frac{(\sin^2 x + \cos^2 x)(\sin^2 x - \cos^2 x)\,[(\sin^2 x + \cos^2 x)^2 - 2\sin^2 x\,.\cos^2 x)]}{1-2\sin^2 x\cos^2 x}$$

$$= \frac{1.(-\cos 2x)(1-2\sin^2 x\cos^2 x)}{1-2\sin^2\cos^2 x} = -\cos 2x \qquad | \ (a^2+b^2)^2 = a^4+b^4+2a^2b^2$$

$$\therefore \quad I = \int -\cos 2x\, dx = -\mathbf{\frac{\sin 2x}{2} + C.}$$

13.8. EVALUATION OF ∫ sin mx cos nx dx, ∫ cos mx cos nx dx AND ∫ sin mx sin nx dx

In order to evaluate ∫ sin mx cos nx dx, ∫ cos mx cos nx dx, ∫ sin mx sin nx dx, the integrands sin mx cos nx, cos mx cos nx, sin mx sin nx are expressed in terms of the sum (or difference) of sines and cosines of multiplies of x by using the formulae :

(*i*) **2 sin A cos B = sin (A + B) + sin (A – B)**

(*ii*) **2 cos A sin B = sin (A + B) – sin (A – B)**

(*iii*) **2 cos A cos B = cos (A + B) + cos (A – B)**

(*iv*) **2 sin A sin B = cos (A – B) – cos (A + B)**

In applying these formulae, it is advisable to have A > B, because A > B would imply A – B > 0.

Example 8. *Evaluate the following integrals :*

(*i*) ∫ *sin 4x cos 7x dx* (*ii*) ∫ *sin 2x sin 3x dx.*

(*iii*) ∫ *sin mx sin nx dx, where m and n are distinct positive numbers*

(*iv*) ∫ *cos 2x cos 4x cos 6x dx.*

Sol. (*i*) $\int \sin 4x \cos 7x \, dx = \frac{1}{2} \int 2 \cos 7x \sin 4x^* \, dx$

$$= \frac{1}{2} \int (\sin (7x + 4x) \, dx - \sin (7x - 4x)) \, dx$$

$$= \frac{1}{2} \int (\sin 11x - \sin 3x) \, dx = \frac{1}{2}\left(-\frac{\cos 11x}{11} + \frac{\cos 3x}{3}\right) + C$$

$$= \mathbf{\frac{\cos 3x}{6} - \frac{\cos 11x}{22} + C.}$$

(*ii*) $\int \sin 2x \sin 3x \, dx = \frac{1}{2} \int 2 \sin 3x \sin 2x \, dx = \frac{1}{2} \int (\cos x - \cos 5x) \, dx$

$= \mathbf{\frac{1}{2}\left(\sin x - \frac{\sin 5x}{5}\right) + C.}$ [Using 2 sin A sin B = cos (A – B) – cos (A + B)]

(*iii*) $\int \sin mx \sin nx \, dx = \frac{1}{2} \int 2 \sin mx \sin nx \, dx = \frac{1}{2} \int (\cos (mx - nx) - \cos (mx + nx)) \, dx$

$$= \frac{1}{2} \int (\cos (m - n)x - \cos (m + n)x) \, dx$$

$$= \mathbf{\frac{1}{2}\left(\frac{\sin (m-n)x}{m-n} - \frac{\sin (m+n)x}{m+n}\right) + C.} \quad (m \neq n \Rightarrow m - n \neq 0)$$

(*iv*) $\cos 2x \cos 4x \cos 6x = \frac{1}{2} (2 \cos 4x \cos 2x) \cos 6x = \frac{1}{2} (\cos 6x + \cos 2x) \cos 6x$

$$= \frac{1}{2} (\cos^2 6x + \cos 2x \cos 6x) = \frac{1}{4} (2 \cos^2 6x + 2 \cos 6x \cos 2x)$$

$$= \frac{1}{4} ((1 + \cos 12x) + (\cos 8x + \cos 4x)) = \frac{1}{4} (1 + \cos 4x + \cos 8x + \cos 12x).$$

$\therefore \quad \int \cos 2x \cos 4x \cos 6x \, dx = \int \frac{1}{4} (1 + \cos 4x + \cos 8x + \cos 12x) \, dx$

$$= \mathbf{\frac{1}{4}\left(x + \frac{\sin 4x}{4} + \frac{\sin 8x}{8} + \frac{\sin 12x}{12}\right) + C.}$$

Example 9. *If* $f'(x) = e^x + \frac{1}{1+x^2}$ *and* $f(0) = 1$, *find* $f(x)$.

Sol. We have $f'(x) = e^x + \frac{1}{1+x^2}$.

$\therefore \quad f(x) = \int f'(x) \, dx = \int \left(e^x + \frac{1}{1+x^2}\right) dx = e^x + \tan^{-1} x + C$

Now $f(0) = 1$ implies $e^0 + \tan^{-1} 0 + C = 1$.

$\Rightarrow \quad 1 + 0 + C = 1 \quad \Rightarrow \quad C = 0$

$\therefore \quad f(x) = \mathbf{e^x + \tan^{-1} x.}$

***Why this step**. We have multiplied and divided the integrand by 2 and interchanged sine and cosine functions because we are interested in using the formula 2 cos A sin B = sin (A + B) – sin (A – B) with angle A greater than angle B. If A > B, then A – B is a + ve angle.

EXERCISE 13.2

Integrate the following functions w.r.t. x (Q. No. 1–15) :

1. (i) $x^4 + 3x$ (ii) $\sqrt{x} - \dfrac{1}{\sqrt{x}}$ (iii) $\sqrt{x}\,(ax^2 + bx + c)$

(iv) $\sqrt{x+1} - \sqrt{x+2}$ (v) $x^3 + x + \dfrac{1}{x} + \dfrac{1}{x^3}$ (vi) $(1+\sqrt{x})(1-\sqrt{x})$

2. (i) $\dfrac{x^3 + 3x^2 + 4}{\sqrt{x}}$ (ii) $\left(x + \dfrac{1}{x^2}\right)\left(x^2 + \dfrac{1}{x^2}\right)$ (iii) $\dfrac{(x+1)^2}{\sqrt{x}}$

(iv) $\left(x + \dfrac{1}{x}\right)^3$ (v) $\dfrac{(x^2+1)^2}{x^3}$ (vi) $\dfrac{(x+1)^3}{x^2}$

3. (i) $\dfrac{x^2}{1+x}$ (ii) $\dfrac{x^3+1}{x-1}$ (iii) $\dfrac{x^4+1}{x^2+1}$

(iv) $\dfrac{x^5+2}{x+1}$ (v) $\dfrac{x^2+2x}{(x+1)^2}$ (vi) $\dfrac{x^4+x^2+1}{x^2-x+1}$

4. (i) $\dfrac{1}{\sqrt{x+1}+\sqrt{x+2}}$ (ii) $\dfrac{1}{\sqrt{5x+3}+\sqrt{5x-2}}$ (iii) $\dfrac{1}{\sqrt{2x+1}+\sqrt{2x+2}}$

(iv) $\dfrac{1}{\sqrt{3x+2}+\sqrt{3x-2}}$ (v) $\dfrac{1}{\sqrt{x+3}-\sqrt{x+2}}$ (vi) $\dfrac{7}{\sqrt{x+5}+\sqrt{x-3}}$

(P.T.U. B.Pharma, May 2005)

5. (i) $\dfrac{x-1}{\sqrt{x+4}}$ (ii) $\dfrac{\sec^2 x}{\text{cosec}^2 x}$ (iii) $\dfrac{2 - 3\sin x}{\cos^2 x}$

(iv) $\sin x + e^x + 4^x + x^4$ (v) $\sec^2 x + \text{cosec}^2 x$

(vi) $\sqrt{1 + 2\tan x\,(\sec x + \tan x)}$

(vii) $\text{cosec}\, x\,(\text{cosec}\, x + \cot x)$ (viii) $a^{2x} - \text{cosec}\, x \cot x$ (ix) $\dfrac{1+\sin x}{1-\sin x}\,dx.$

6. (i) $\displaystyle\int \left(\dfrac{2\cos x}{3\sin^2 x} + 5x^2 - 1\right) dx$ (ii) $\displaystyle\int (\sin x - \cos x)^2\, dx$ (iii) $\displaystyle\int \dfrac{\sin^2 x}{(1+\cos x)^2}\, dx$

(iv) $\displaystyle\int \dfrac{\tan x}{\sec x + \tan x}\, dx$ (v) $\displaystyle\int \dfrac{\cos 2x + 2\sin^2 x}{\cos^2 x}\, dx$ (vi) $\displaystyle\int \dfrac{\cos^2 x - \sin^2 x}{\sqrt{1+\cos 4x}}\, dx$

7. (i) $\displaystyle\int (2\tan x - 3\cot x)^2\, dx$ (ii) $\displaystyle\int \sec^2 x\, \text{cosec}^2 x\, dx$ *(P.T.U. B.Pharma, May 2007)*

(iii) $\displaystyle\int \dfrac{\sin^2 x - \cos^2 x}{\sin^2 x \cos^2 x}\, dx$ (iv) $\displaystyle\int \dfrac{\cos 2x - \cos 2\alpha}{\cos x - \cos \alpha}\, dx$

8. (i) $\displaystyle\int \sin^2 \theta\, d\theta$ (ii) $\displaystyle\int \cos^3 x\, dx$

(P.T.U. B.Pharma, May 2007)

(iii) $\displaystyle\int \sin^4 x\, dx$ (iv) $\displaystyle\int \cos^4 \dfrac{x}{2}\, dx$

9. (*i*) $\int \frac{1}{1+\cos x}\,dx$ (*ii*) $\int \frac{\cos 2x}{1-\cos 2x}\,dx$

(*iii*) $\int \frac{1}{1+\sin x}\,dx$ (*iv*) $\int \frac{\sin x}{1+\sin x}\,dx$

(*G.N.D.U. B.Pharma, 2000*)

10. (*i*) $\int \frac{1+\cos x}{1-\cos x}\,dx$ (*ii*) $\int \sqrt{1-\cos x}\,dx$

(*iii*) $\int \sqrt{1+\sin 2x}\,dx$ (*iv*) $\int \cos x\sqrt{1+\cos 2x}\,dx$

11. (*i*) $\int \sin 4x \cos 7x\,dx$ (*ii*) $\int \cos 5x \cos 2x\,dx$

(*iii*) $\int \cos 7x \cos x\,dx$ (*iv*) $\int \sin 5x \sin 3x\,dx$

(*v*) $\int \sin 3x \cos 4x\,dx$ (*vi*) $\int \sin x \sin 2x \sin 3x\,dx$

12. (*i*) $\int \frac{\sin 4x}{\sin x}\,dx$ (*ii*) $\int \sin^2 x \cos^2 x\,dx$

(*iii*) $\int \frac{\cos x - \cos 2x}{1-\cos x}\,dx$ (*iv*) $\int \frac{\sin^6 x + \cos^6 x}{\sin^2 x \cos^2 x}\,dx$

13. (*i*) $\int \frac{(a^x+b^x)^2}{a^x b^x}\,dx$ (*ii*) $\int (e^{2a \log x} + e^{2x \log a})\,dx$

(*iii*) $\int ((e^{2x})^2 + (a^{2x})^2)\,dx$ (*iv*) $\int \frac{e^{5\log x} - e^{4\log x}}{e^{3\log x} - e^{2\log x}}\,dx$

14. (*i*) $\int \sin^{-1}(\cos x)\,dx$ (*ii*) $\int \cot^{-1}(\operatorname{cosec} x + \cot x)\,dx$

(*iii*) $\int \tan^{-1}\frac{\sin 2x}{1+\cos 2x}\,dx$ (*iv*) $\int \tan^{-1}\sqrt{\frac{1-\cos 2x}{1+\cos 2x}}\,dx$

15. (*i*) $\int \sin x\sqrt{1-\cos 2x}\,dx$ (*ii*) $\int \sin^3 x \cos^3 x\,dx$

(*iii*) $\int \sin^4 x \cos^4 x\,dx$ (*iv*) $\int \sin^3 (2x+1)\,dx.$

Answers

1. (*i*) $\frac{x^5}{5} + \frac{3x^2}{2} + C$ (*ii*) $\frac{2x^{3/2}}{3} - 2x^{1/2} + C$

(*iii*) $\frac{2a}{7}x^{7/2} + \frac{2bx^{5/2}}{5} + \frac{2c}{3}x^{3/2}$ + K, where K is any constant

(*iv*) $\frac{2(x+1)^{3/2}}{3} - 2\frac{(x+2)^{3/2}}{3} + C$ (*v*) $\frac{x^4}{4} + \frac{x^2}{2} + \log|x| - \frac{1}{2x^2} + C$ (*vi*) $x - \frac{x^2}{2} + C$

2. (*i*) $\frac{2}{7}x^{7/2} + \frac{6}{5}x^{5/2} + 8x^{1/2} + C$ (*ii*) $\frac{1}{4}x^4 + \log|x| + x - \frac{1}{3x^3} + C$

(*iii*) $\frac{2}{5}x^{5/2} + \frac{4}{3}x^{3/2} + 2x^{1/2} + C$ (*iv*) $\frac{1}{4}x^4 + \frac{3}{2}x^2 + 3\log|x| - \frac{1}{2x^2} + C$

(*v*) $\frac{x^2}{2} + 2\log|x| - \frac{1}{2x^2} + C$ (*vi*) $\frac{x^2}{2} + 3x + 3\log|x| - \frac{1}{x} + C$

3. (*i*) $\frac{x^2}{2} - x + \log|1+x| + C$ (*ii*) $\frac{x^3}{3} + \frac{x^2}{2} + x + 2\log|x-1| + C$

(*iii*) $\frac{x^3}{3} - x + 2\tan^{-1}x + C$ (*iv*) $\frac{x^5}{5} - \frac{x^4}{4} + \frac{x^3}{3} - \frac{x^2}{2} + x + \log|x+1| + C$

(*v*) $x + \frac{1}{x+1} + C$ (*vi*) $\frac{x^3}{3} + \frac{x^2}{2} + x + C$

4. (*i*) $-\frac{2}{3}[(x+1)^{3/2} - (x+2)^{3/2}] + C$ (*ii*) $\frac{2}{75}[(5x+3)^{3/2} - (5x-2)^{3/2}] + C$

(*iii*) $\frac{1}{3}[(2x+2)^{3/2} - (2x+1)^{3/2}] + C$ (*iv*) $\frac{1}{18}[(3x+2)^{3/2} - (3x-2)^{3/2}] + C$

(*v*) $\frac{2}{3}[(x+3)^{3/2} + (x+2)^{3/2}] + C$ (*vi*) $\frac{7}{12}[(x+5)^{3/2} - (x-3)^{3/2}] + C$

5. (*i*) $\frac{2}{3}(x+4)^{3/2} - 10\sqrt{x+4} + C$ (*ii*) $\tan x - x + C$ (*iii*) $2\tan x - 3\sec x + C$

(*iv*) $-\cos x + e^x + \frac{4^x}{\log 4} + \frac{x^5}{5} + C$ (*v*) $\tan x - \cot x + C$ (*vi*) $\log|\sec x(\sec x + \tan x)| + C$

(*vii*) $-\cot x - \operatorname{cosec} x + C$ (*viii*) $\frac{a^{2x}}{2\log a} + \operatorname{cosec} x + C$ (*ix*) $2\tan x + 2\sec x - x + C$

6. (*i*) $-\frac{2}{3}\operatorname{cosec} x + \frac{5}{3}x^3 - x + C$ (*ii*) $x + \frac{1}{2}\cos 2x + C$ (*iii*) $2\tan\frac{x}{2} - x + C$

(*iv*) $\sec x - \tan x + x + C$ (*v*) $\tan x + C$ (*vi*) $\frac{x}{\sqrt{2}} + C$

7. (*i*) $4\tan x - 9\cot x - 25x + C$ (*ii*) $\tan x - \cot x + C$

(*iii*) $\tan x + \cot x + C$ (*iv*) $2(\sin x + x\cos\alpha) + C$

8. (*i*) $\frac{\theta}{2} - \frac{\sin 2\theta}{4} + C$ (*ii*) $\frac{3}{4}\sin x + \frac{1}{12}\sin 3x + C$

(*iii*) $\frac{3}{8}x - \frac{\sin 2x}{4} + \frac{\sin 4x}{32} + C$ (*iv*) $\frac{3}{8}x + \frac{\sin x}{2} + \frac{\sin 2x}{16} + C$

9. (*i*) $\operatorname{cosec} x - \cot x + C$ or $\tan\frac{x}{2} + C$ (*ii*) $-x - \frac{1}{2}\cot x + C$

(*iii*) $\tan x - \sec x + C$ (*iv*) $x - \tan x + \sec x + C$

10. (*i*) $-x - 2\cot\frac{x}{2} + C$ (*ii*) $-2\sqrt{2}\cos\frac{x}{2} + C$

(*iii*) $\sin x - \cos x + C$ (*iv*) $\frac{1}{\sqrt{2}}(x + \sin x\cos x) + C$

11. (*i*) $\frac{1}{6}\cos 3x - \frac{1}{22}\cos 11x + C$ (*ii*) $\frac{1}{14}\sin 7x + \frac{1}{6}\sin 3x + C$

(*iii*) $\frac{1}{12}\sin 6x + \frac{1}{16}\sin 8x + C$ (*iv*) $\frac{1}{4}\sin 2x - \frac{1}{16}\sin 8x + C$

(*v*) $\frac{1}{2}\cos x - \frac{1}{14}\cos 7x + C$ (*vi*) $\frac{1}{24}\cos 6x - \frac{1}{16}\cos 4x - \frac{1}{8}\cos 2x + C$

12. (*i*) $\frac{2}{3}\sin 3x + 2\sin x + C$ (*ii*) $\frac{1}{8}\left(x - \frac{\sin 4x}{4}\right) + C$

(*iii*) $x + 2\sin x + C$ (*iv*) $\tan x - \cot x - 3x + C$

13. (*i*) $\frac{(a/b)^x}{\log \frac{a}{b}} + \frac{(b/a)^x}{\log \frac{b}{a}} + 2x + C$ (*ii*) $\frac{x^{2a+1}}{2a+1} + \frac{a^{2x}}{2\log a} + C$

(*iii*) $\frac{e^{4x}}{4} + \frac{a^{4x}}{4\log a} + C$ (*iv*) $\frac{x^3}{3} + C$

14. (*i*) $\frac{\pi}{2}x - \frac{x^2}{2} + C$ (*ii*) $\frac{x^2}{4} + C$

(*iii*) $\frac{x^2}{2} + C$ (*iv*) $\frac{x^2}{2} + C$

15. (*i*) $\frac{1}{\sqrt{2}}\left(x - \frac{\sin 2x}{2}\right) + C$ (*ii*) $\frac{1}{64}\left(\frac{1}{3}\cos 6x - 3\cos 2x\right) + C$

(*iii*) $\frac{1}{128}\left(3x - \sin 4x + \frac{1}{8}\sin 8x\right) + C$ (*iv*) $-\frac{3}{8}\cos(2x+1) + \frac{1}{24}\cos(6x+3) + C$

Hints

5. (*i*) $\frac{x-1}{\sqrt{x+4}} = \frac{(x+4)-5}{\sqrt{x+4}} = (x+4)^{1/2} - 5(x+4)^{-1/2}$

(*ii*) $\frac{\sec^2 x}{\mathrm{cosec}^2 x} = \tan^2 x = \sec^2 x - 1.$

(*vi*) $1 + 2\tan x(\sec x + \tan x) = 1 + 2\tan x \sec x + 2\tan^2 x$

$= (1 + \tan^2 x) + \tan^2 x + 2\tan x \sec x = (\sec x + \tan x)^2$

(*ix*) Multiply and divide by $1 + \sin x$.

6. (*ii*) Integrand $= 1 - \sin 2x$ (*iii*) Integrand $= \frac{1-\cos x}{1+\cos x} = \sec^2 \frac{x}{2} - 1$

(*iv*) Multiply and divide by $\sec x - \tan x$ (*v*) Use : $\cos 2x = 1 - 2\sin^2 x$

(*vi*) Use : $\cos 4x = 2\cos^2 2x - 1$.

7. (*ii*) $\sec^2 x\,\mathrm{cosec}^2 x = (1 + \tan^2 x)\,\mathrm{cosec}^2 x = \mathrm{cosec}^2 x + \sec^2 x$

(*iv*) $\cos 2x - \cos 2\alpha = (2\cos^2 x - 1) - (2\cos^2 \alpha - 1)$.

9. (*ii*) $\frac{\cos 2x}{1-\cos 2x} = \frac{\cos^2 x - \sin^2 x}{2\sin^2 x}$ (*iv*) $\frac{\sin x}{1+\sin x} = 1 - \frac{1}{1+\sin x} = 1 - \frac{1-\sin x}{\cos^2 x}$.

10. (*iii*) $1 + \sin 2x = (\sin x + \cos x)^2$ (*iv*) Use $1 + \cos 2x = 2\cos^2 x$.

12. (*i*) $\sin 4x = 2(2\sin x\cos x)\cos 2x$ (*ii*) Use : $\sin x\cos x = \frac{1}{2}\sin 2x$

(*iii*) $\cos x - \cos 2x = \cos x - (2\cos^2 x - 1)$

(*iv*) Use : $a^3 + b^3 = (a+b)^3 - 3ab(a+b)$ and $\sec^2 x\,\mathrm{cosec}^2 x = (1+\tan^2 x)\,\mathrm{cosec}^2 x = \mathrm{cosec}^2 x + \sec^2 x$.

13. (*iv*) Integrand $= \frac{e^{\log x^5} - e^{\log x^4}}{e^{\log x^3} - e^{\log x^2}} = \frac{x^5 - x^4}{x^3 - x^2} = x^2.$

14. (*i*) $\sin^{-1}(\cos x) = \sin^{-1}\left[\sin\left(\frac{\pi}{2} - x\right)\right]$

(*ii*) $\cot^{-1}(\text{cosec } x + \cot x) = \cot^{-1}\left(\frac{1+\cos x}{\sin x}\right) = \cot^{-1}\left(\frac{2\cos^2\frac{x}{2}}{2\sin\frac{x}{2}\cos\frac{x}{2}}\right) = \cot^{-1}\left(\cot\frac{x}{2}\right) = \frac{x}{2}.$

15. (*iii*) $\sin^4 x \cos^4 x = \left(\frac{1}{2}\sin 2x\right)^4 = \frac{1}{16}(\sin^2 2x)^2.$

INTEGRATION BY SUBSTITUTIONS

13.9. INTRODUCTION

In the preceding topic, we studied the method of finding the integral of a given function, provided the function to be integrated *i.e.*, the integrand is given in the standard form. Generally the integrand in integrals are not given in standard forms and these are reduced to standard form by using the following methods :

I. Integration by substitutions

II. Integration by parts

III. Integration by partial fractions

IV. Integration using miscellaneous substitutions.

In the present chapter, we shall study 'integration by substitutions' in detail.

13.10. FIRST SUBSTITUTION METHOD

Let the integral to be evaluated be $\int f(x)\,dx$.

If the integrand $f(x)$ of the integral is not in an integrable form, then the variable of integration (x) is changed suitably into a new variable (z) by substitution and on differentiation and simplification, the new integrand is found integrable.

Let $y = \int f(x)\,dx$ and we choose to substitute $x = \phi(z)$.

$\therefore \quad \frac{dy}{dx} = f(x)$

Also $\quad \frac{dy}{dz} = \frac{dy}{dx}\cdot\frac{dx}{dz} = f(x)\frac{dx}{dz} = f[\phi(z)]\,\phi'(z)$

$\therefore \quad y = \int [f[\phi(z)]\,\phi'(z)\,dz$

or $$\int \mathbf{f(x)\,dx} = \int \mathbf{f[\phi(z)]\,\phi'(z)\,dz}, \textbf{ where } \mathbf{x = \phi(z)}.$$

13.11. SECOND SUBSTITUTION METHOD

Let the integral to be evaluated be $\int f(x)\,dx$.

If the integrand $f(x)$ of the integral is not in an integrable form, then some function of the variable of integration (x) is taken as a new variable.

This method is suitable when the given integrand is completely expressible in terms of the new variable.

Let $\quad y = \int f(x)\,dx$ and we choose to substitute $g(x) = z$.

$$\therefore \quad \frac{dz}{dx} = g'(x) \text{ i.e., } dz = g'(x)\,dx$$

$$\therefore \quad y = \int \frac{f(x)}{g'(x)}\,dz$$

or $$\int \mathbf{f(x)\,dx} = \int \frac{\mathbf{f(x)}}{\mathbf{g'(x)}}\,\mathbf{dz}, \textbf{ where } \mathbf{g(x) = z.}$$

The substitution $g(x) = z$ is used only when the integrand $\frac{f(x)}{g'(x)}$ is expressible in terms of z only.

Corollary 1. If $f(x) = \frac{g'(x)}{g(x)}$, then

$$\int f(x)\,dx = \int \frac{g'(x)}{g(x)}\,dx = \int \frac{1}{z}\,dz, \text{ where } z = g(x)$$

$$(\because \quad g(x) = z \Rightarrow dz = g'(x)\,dx)$$

$$= \log|z| + C = \log|g(x)| + C$$

$$\therefore \quad \int \frac{\mathbf{g'(x)}}{\mathbf{g(x)}}\,\mathbf{dx} = \mathbf{\log|g(x)| + C.}$$

Corollary 2. If $f(x) = (g(x))^n g'(x)$, where $n \neq -1$, then

$$\int f(x)\,dx = \int (g(x))^n g'(x)\,dx = \int z^n\,dz, \text{ where } z = g(x)$$

$$= \frac{z^{n+1}}{n+1} + C = \frac{(g(x))^{n+1}}{n+1} + C.$$

$$\therefore \quad \int (\mathbf{g(x)})^{\mathbf{n}}\,\mathbf{g'(x)\,dx} = \frac{(\mathbf{g(x)})^{\mathbf{n+1}}}{\mathbf{n+1}} + \mathbf{C}, \textbf{ provided } \mathbf{n \neq -1.}$$

Remark. In practical problems, that function of variable of integration is taken as new variable of integration whose derivative is available in the numerator of the integrand.

For example, $\int \sin^5 x \cos x\,dx$ can be evaluated by putting $z = \sin x$ ($= g(x)$), because $(\sin x)' = \cos x$ is in the numerator of the integrand *i.e.*, in $\sin^5 x \cos x$.

ILLUSTRATIVE EXAMPLES

Example 1. *Evaluate :*

(*i*) $\int \frac{x^3}{1+x^4}\,dx$ (*ii*) $\int \frac{1-\tan x}{1+\tan x}\,dx$ (*iii*) $\int \frac{\tan x}{\log \sec x}\,dx$ (*iv*) $\int \frac{dx}{\sqrt{x}\,(1+\sqrt{x})}$.

(*v*) $\int \frac{1}{x^2}\cos\frac{1}{x}\,dx$ (P.T.U. B.Pharma, Dec. 2004)

Sol. (*i*) Let $I = \int \frac{x^3}{1+x^4}\, dx.$

$\therefore \quad I = \frac{1}{4}\int \frac{4x^3}{1+x^4}\, dx.$ We observe that $\frac{d}{dx}(1 + x^4) = 4x^3.$

Let $\quad z = 1 + x^4. \quad \therefore \quad dz = 4x^3\, dx. \qquad \left(\because \quad dz = \left(\frac{dz}{dx}\right) dx\right)$

$\therefore \quad I = \frac{1}{4}\int \frac{dz}{z} = \frac{1}{4} \log |z| + C = \frac{1}{4} \log |1 + x^4| + C = \mathbf{\frac{1}{4} \log (1 + x^4) + C.}$

$(\because \quad 1 + x^4 > 0)$

(*ii*) Let $I = \int \frac{1 - \tan x}{1 + \tan x}\, dx. \quad \therefore \quad I = \int \frac{1 - \frac{\sin x}{\cos x}}{1 + \frac{\sin x}{\cos x}}\, dx = \int \frac{\cos x - \sin x}{\cos x + \sin x}\, dx$

We observe that $\frac{d}{dx}(\cos x + \sin x) = - \sin x + \cos x = \cos x - \sin x.$

Let $z = \cos x + \sin x. \quad \therefore \quad dz = (- \sin x + \cos x)\, dx$

$\therefore \quad I = \int \frac{dz}{z} = \log |z| + C = \mathbf{\log | \cos x + \sin x | + C.}$

(*iii*) Let $I = \int \frac{\tan x}{\log \sec x}\, dx.$

We observe that $(\log \sec x)' = \frac{1}{\sec x} . \sec x \tan x = \tan x.$

Let $z = \log \sec x. \quad \therefore \quad dz = \frac{1}{\sec x} . \sec x \tan x\, dx = \tan x\, dx$

$\therefore \quad I = \int \frac{dz}{z} = \log |z| + C = \mathbf{\log | \log \sec x | + C.}$

(*iv*) Let $I = \int \frac{dx}{\sqrt{x}\,(1 + \sqrt{x})}$. We observe that $(1 + \sqrt{x})' = 0 + \frac{1}{2} x^{-1/2} = \frac{1}{2\sqrt{x}}.$

We write, $\quad I = 2\int \frac{1}{(1+\sqrt{x})} . \frac{1}{2\sqrt{x}}\, dx.$

Let $z = 1 + \sqrt{x} \quad \therefore \quad dz = \frac{1}{2\sqrt{x}}\, dx$

$\therefore \quad I = 2\int \frac{1}{z} . dz = 2 \log |z| + C = 2 \log |1 + \sqrt{x}| + C = \mathbf{2 \log (1 + \sqrt{x}) + C.}$

Remark. In the above example, the integrand in each integral is of the type '$\left(\frac{g'(x)}{g(x)}\right)$'.

(*v*) Put $\frac{1}{x} = t \quad \Rightarrow \quad -\frac{1}{x^2}\, dx = dt$

$\Rightarrow \quad \frac{1}{x^2}\, dx = - dt$

$\therefore \quad \int \frac{1}{x^2} \cos \frac{1}{x}\, dx = \int - \cos t\, dt = - \sin t + c = - \sin \frac{1}{x} + c$

Example 2. *Evaluate :*

(*i*) $\int (2x+4)\sqrt{x^2+4x+3}\, dx$ (*ii*) $\int \frac{(1+\sqrt{x})^n}{\sqrt{x}} dx$

(*iii*) $\int sec^4 x \tan x\, dx$ (*iv*) $\int \frac{(tan^{-1} x)^2}{1+x^2} dx.$

Sol. (*i*) Let I = $\int (2x + 4) \sqrt{x^2 + 4x + 3}\, dx$. We observe that $\frac{d}{dx}(x^2 + 4x + 3) = 2x + 4$.

Let $z = x^2 + 4x + 3$. $\therefore$ $dz = (2x + 4)\, dx$

$$\therefore \quad I = \int \sqrt{z}\, dz = \frac{z^{3/2}}{3/2} + C = \frac{\mathbf{2}}{\mathbf{3}} \mathbf{(x^2 + 4x + 3)^{3/2} + C.}$$

(*ii*) Let I = $\int \frac{(1+\sqrt{x})^n}{\sqrt{x}} dx$. We observe that $(1 + \sqrt{x})' = 0 + \frac{1}{2} x^{-1/2} = \frac{1}{2\sqrt{x}}$.

We write, $$I = 2 \int (1+\sqrt{x})^n \cdot \frac{1}{2\sqrt{x}} dx$$

Let $z = 1 + \sqrt{x}$. $\therefore$ $dz = \frac{1}{2\sqrt{x}} dx$

$$\therefore \quad I = 2\int z^n\, dz = \frac{2z^{n+1}}{n+1} + C = \frac{\mathbf{2}}{\mathbf{n+1}} \mathbf{(1 + \sqrt{x})^{n+1} + C.}$$

(*iii*) Let I = $\int \sec^4 x \tan x\, dx$. $\therefore$ I = $\int (\sec x)^3 \cdot \sec x \tan x^* dx$.

We observe that $(\sec x)' = \sec x \tan x$.
Let $z = \sec x$. $\therefore$ $dz = \sec x \tan x\, dx$

$$\therefore \quad I = \int z^3\, dz = \frac{z^4}{4} + C = \frac{\mathbf{sec^4 x}}{\mathbf{4}} \mathbf{+ C.}$$

(*iv*) Let I = $\int \frac{(\tan^{-1} x)^2}{1+x^2} dx$. We observe that $(\tan^{-1} x)' = \frac{1}{1+x^2}$.

Let $z = \tan^{-1} x$. $\therefore$ $dz = \frac{1}{1+x^2} dx$.

$$\therefore \quad I = \int z^2\, dz = \frac{z^3}{3} + C = \frac{\mathbf{(tan^{-1} x)^3}}{\mathbf{3}} \mathbf{+ C.}$$

Remark. In the above example, the integrand in each integral is of the type '$(g(x))^n g'(x)$'.

13.12. STANDARD INTEGRALS ∫ tan x dx, ∫ cot x dx, ∫ sec x dx AND ∫ cosec x dx

We shall prove that :

1. $\int \mathbf{tan\ x\ dx = log\ |\ sec\ x\ | + C}$

***Why this step**. We have written $\sec^4 x \tan x$ as $(\sec x)^3 \sec x \tan x$ because, $(\sec x)' = \sec x \tan x$.
$\therefore$ If we take $g(x) = \sec x$, then the integrand takes the form $(g(x))^3 g'(x)$.

2. $\int \cot x \, dx = \log |\sin x| + C$

3. $\int \sec x \, dx = \log |\sec x + \tan x| + C = \log \left| \tan\left(\frac{\pi}{4} + \frac{x}{2}\right) \right| + C$

4. $\int \operatorname{cosec} x \, dx = \log |\operatorname{cosec} x - \cot x| + C = \log \left| \tan \frac{x}{2} \right| + C.$

Proof. 1. Let $I = \int \tan x \, dx$. $\therefore \quad I = \int \frac{\tan x \sec x}{\sec x} dx$

Let $z = \sec x$. $\therefore \quad dz = \sec x \tan x \, dx$

$$\therefore \qquad I = \int \frac{dz}{z} = \log |z| + C = \log |\sec x| + C.$$

2. Let $I = \int \cot x \, dx$. $\therefore \quad I = \int \frac{\cos x}{\sin x} dx$

Let $z = \sin x$. $\therefore \quad dz = \cos x \, dx$

$$\therefore \qquad I = \int \frac{dz}{z} = \log |z| + C = \log |\sin x| + C.$$

3. Let $I = \int \sec x$. $\therefore \quad I = \int \frac{\sec x (\sec x + \tan x)}{\sec x + \tan x} dx$

Let $z = \sec x + \tan x$. $\therefore \quad dz = (\sec x \tan x + \sec^2 x) \, dx = \sec x (\sec x + \tan x) \, dx$

$$\therefore \qquad I = \int \frac{dz}{z} = \log |z| + C = \log |\sec x + \tan x| + C.$$

We have

$$\sec x + \tan x = \frac{1 + \sin x}{\cos x} = \frac{1 - \cos\left(\frac{\pi}{2} + x\right)}{\sin\left(\frac{\pi}{2} + x\right)} = \frac{2 \sin^2\left(\frac{\pi}{4} + \frac{x}{2}\right)}{2 \sin\left(\frac{\pi}{4} + \frac{x}{2}\right) \cos\left(\frac{\pi}{4} + \frac{x}{2}\right)} = \tan\left(\frac{\pi}{4} + \frac{x}{2}\right)$$

$$\therefore \quad \int \sec x \, dx = \log |\sec x + \tan x| + C = \log \left| \tan\left(\frac{\pi}{4} + \frac{x}{2}\right) \right| + C.$$

4. Let $I = \int \operatorname{cosec} x \, dx$. $\therefore \quad I = \int \frac{\operatorname{cosec} x (\operatorname{cosec} x - \cot x)}{\operatorname{cosec} x - \cot x} dx$

Let $z = \operatorname{cosec} x - \cot x$. $\therefore \quad dz = (-\operatorname{cosec} x \cot x + \operatorname{cosec}^2 x) \, dx = \operatorname{cosec} x (\operatorname{cosec} x - \cot x) dx$

$$\therefore \qquad I = \int \frac{dz}{z} = \log |z| + C = \log |\operatorname{cosec} x - \cot x| + C.$$

We have $\operatorname{cosec} x - \cot x = \dfrac{1 - \cos x}{\sin x} = \dfrac{2 \sin^2 \frac{1}{2} x}{2 \sin \frac{1}{2} x \cos \frac{1}{2} x} = \tan \dfrac{1}{2} x.$

$$\therefore \quad \int \operatorname{cosec} x \, dx = \log |\operatorname{cosec} x - \cot x| + C = \log \left| \tan \frac{x}{2} \right| + C.$$

Remark 1. $\int \tan x \, dx$ can also be evaluated by writing $\tan x$ as $\frac{\sin x}{\cos x}$ and letting $z = \cos x$.

Remark 2. $\int \cot x \, dx$ can also be evaluated by writing $\cot x$ as $\frac{\operatorname{cosec} x \cot x}{\operatorname{cosec} x}$ and letting $z = \operatorname{cosec} x$.

Important Note. The students are advised to remember the following results :

(*i*) $\int \mathbf{\sin x\, dx = -\cos x + C}$ (*ii*) $\int \mathbf{\cos x\, dx = \sin x + C}$

(*iii*) $\int \mathbf{\tan x\, dx = \log | \sec x | + C}$ (*iv*) $\int \mathbf{\cot x\, dx = \log | \sin x | + C}$

(*v*) $\int \mathbf{\sec x\, dx = \log | \sec x + \tan x | + C = \log \left| \tan\left(\frac{\pi}{4} + \frac{x}{2}\right) \right| + C}$

(*vi*) $\int \mathbf{\operatorname{cosec} x\, dx = \log | \operatorname{cosec} x - \cot x | + C = \log \left| \tan \frac{x}{2} \right| + C.}$

Example 3. *Evaluate :*

(*i*) $\int \operatorname{cosec}^4 x\, dx$ (*ii*) $\int x^3 \sin x^4\, dx$ (*iii*) $\int \frac{1 + \tan x}{x + \log \sec x} dx$

(*iv*) $\int \frac{e^x (1 + x)}{\cos (xe^x)} dx$ (*v*) $\int \frac{\tan x}{a + b \tan^2 x} dx$ (*vi*) $\int \frac{\tan^5 \sqrt{x} \sec^2 \sqrt{x}}{\sqrt{x}} dx$

(*vii*) $\int \frac{1}{x + x \log x} dx$ (*viii*) $\int \frac{2^x}{1 + 4^x} dx$ (*ix*) $\int \frac{\sin (2 \tan^{-1} x)}{1 + x^2} dx$

(*x*) $\int \frac{(x + 1)}{x (1 + xe^x)^2} dx$ (P.T.U. B. Pharma, June 2003)

Sol. (*i*) Let $I = \int \operatorname{cosec}^4 x\, dx.$

$\therefore \quad I = \int \operatorname{cosec}^2 x \operatorname{cosec}^2 x\, dx = \int (1 + \cot^2 x) \operatorname{cosec}^2 x\, dx$

Let $z = \cot x. \quad \therefore \quad dz = -\operatorname{cosec}^2 x\, dx.$

$\therefore \quad I = \int (1 + z^2)(-dz) = -\int (1 + z^2)\, dz = -\left(z + \frac{z^3}{3}\right) + C = -\left(\mathbf{\cot x + \frac{\cot^3 x}{3}}\right) \mathbf{+ C.}$

(*ii*) Let $I = \int x^3 \sin x^4\, dx. \quad \therefore \quad I = \int (\sin x^4)\, x^3\, dx$

Let $z = x^4. \quad \therefore \quad dz = 4x^3\, dx$ *i.e.,* $x^3\, dx = \frac{dz}{4}$

$\therefore \quad I = \int \sin z \,.\, \frac{dz}{4} = \frac{1}{4}(-\cos z) + C = \mathbf{-\frac{1}{4} \cos x^4 + C.}$

(*iii*) Let $I = \int \frac{1 + \tan x}{x + \log \sec x} dx.$

Let $z = x + \log \sec x \quad \therefore \quad dz = \left(1 + \frac{1}{\sec x} . \sec x \tan x\right) dx = (1 + \tan x)\, dx$

$\therefore \quad I = \int \frac{dz}{z} = \log | z | + C = \mathbf{\log | x + \log \sec x | + C.}$

(*iv*) Let $I = \int \frac{e^x (1 + x)}{\cos (xe^x)} dx.$

Let $z = xe^x. \quad \therefore \quad dz = (1 \,.\, e^x + xe^x)\, dx = e^x (1 + x)\, dx$

$\therefore \quad I = \int \frac{1}{\cos z} dz = \int \sec z\, dz = \log | \sec z + \tan z | + C$

$= \mathbf{\log | \sec (xe^x) + \tan (xe^x) | + C.}$

(*v*) Let $I = \int \frac{\tan x}{a + b\tan^2 x} dx.$ $\therefore$ $I = \int \frac{\sin x/\cos x}{a + b(\sin^2 x/\cos^2 x)} dx = \int \frac{\sin x \cos x}{a\cos^2 x + b\sin^2 x} dx$

Let $z = a\cos^2 x + b\sin^2 x.$

$\therefore$ $dz = (2a\cos x(-\sin x) + 2b\sin x\cos x)\, dx = 2(b - a)\sin x\cos x\, dx$

$\therefore$ $\sin x\cos x\, dx = \frac{dz}{2(b-a)}$

$\therefore$ $I = \int \frac{1}{z} \cdot \frac{dz}{2(b-a)} = \frac{1}{2(b-a)} \int \frac{1}{z} dz = \frac{1}{2(b-a)} \log|z| + C$

$= \mathbf{\frac{1}{2(b-a)} \log|a\cos^2 x + b\sin^2 x| + C.}$

(*vi*) Let $I = \int \frac{\tan^5 \sqrt{x}\sec^2 \sqrt{x}}{\sqrt{x}} dx.$

Let $z = \sqrt{x}.$ $\therefore$ $dz = \frac{1}{2\sqrt{x}} dx$ or $\frac{dx}{\sqrt{x}} = 2dz.$

$\therefore$ $I = \int \tan^5 z\sec^2 z\,(2dz) = 2\int \tan^5 z\sec^2 z\, dz$

Let $t = \tan z.$ $\therefore$ $dt = \sec^2 z\, dz$

$\therefore$ $I = 2\int t^5\, dt = 2 \cdot \frac{t^6}{6} + C = \frac{1}{3}\tan^6 z + C = \mathbf{\frac{1}{3}\tan^6\sqrt{x} + C.}$

(*vii*) Let $I = \int \frac{1}{x + x\log x} dx.$ $\therefore$ $I = \int \frac{1}{x(1 + \log x)} dx.$

Let $z = 1 + \log x.$ $\therefore$ $dz = \frac{1}{x} dx.$

$\therefore$ $I = \int \frac{1}{z} dz = \log|z| + C = \mathbf{\log|1 + \log x| + C.}$

(*viii*) Let $I = \int \frac{2^x}{1 + 4^x} dx.$

Let $z = 2^x.$ $\therefore$ $dz = (2^x \log 2)\, dx.$

$\therefore$ $I = \int \frac{1}{1 + z^2} \cdot \frac{dz}{\log 2} = \frac{1}{\log 2}\tan^{-1} z + C = \mathbf{\frac{1}{\log 2}\tan^{-1}(2^x) + C.}$

(*ix*) Let $I = \int \frac{\sin(2\tan^{-1} x)}{1 + x^2} dx.$

Let $z = \tan^{-1} x.$ $\therefore$ $dz = \frac{1}{1 + x^2} dx$

$\therefore$ $I = \int \sin 2z\, dz = -\frac{\cos 2z}{2} + C = -\frac{1}{2} \cdot \frac{1 - \tan^2 z}{1 + \tan^2 z} + C = \mathbf{-\frac{1}{2}\left(\frac{1 - x^2}{1 + x^2}\right) + C.}$

(*x*) $I = \int \frac{(x+1)}{x(1 + xe^x)^2} dx = \int \frac{e^x(x+1)}{xe^x(1 + xe^x)^2} dx$

Put $xe^x = t \Rightarrow (xe^x + e^x)\, dx = dt$

$\Rightarrow \quad e^x (x + 1)\, dx = dt$

$$\therefore \quad I = \int \frac{dt}{t(1+t)^2} \quad ...(1)$$

Consider $\dfrac{1}{t(t+1)^2} = \dfrac{A}{t} + \dfrac{B}{t+1} + \dfrac{C}{(t+1)^2}$

Multiplying by $t(t + 1)^2$, we get

$$1 = A(t+1)^2 + B\,t(t+1) + Ct$$

$$1 = A(t^2 + 1 + 2t) + B(t^2 + t) + Ct$$

Equating the co-efficient of t^2; $\quad 0 = A + B \quad ...(i)$

Equating the co-efficient of t; $\quad 0 = 2A + B + C \quad ...(ii)$

Equating the constant term, $\quad 1 = A \quad ...(iii)$

$\therefore$ from (*i*) $B = -1$

from (*ii*) $C = -2A - B = -2 + 1 = -1$

$\therefore$ from (1)

$$I = \int \frac{dt}{t(t+1)^2} = \int \left(\frac{1}{t} - \frac{1}{t+1} - \frac{1}{(t+1)^2} \right) dt$$

$$= \log t - \log(t+1) + \frac{1}{t+1} + c$$

$$= \log \frac{t}{t+1} + \frac{1}{t+1} + c = \log \frac{xe^x}{xe^x + 1} + \frac{1}{xe^x + 1} + c.$$

Example 4. *Evaluate :*

(*i*) $\int \cos^3 x \sin^4 x\, dx$ (*ii*) $\displaystyle\int \frac{\sqrt{\tan x}}{\sin x \cos x}\, dx$ (*iii*) $\displaystyle\int \frac{1}{1 - \tan x}\, dx.$

Sol. (*i*) Let $I = \int \cos^3 x \sin^4 x\, dx.$

$\therefore \quad I = \int (1 - \sin^2 x) \cos x \sin^4 x\, dx = \int (1 - \sin^2 x) \sin^4 x \cos x\, dx$

Let $\quad z = \sin x. \quad \therefore \quad dz = \cos x\, dx.$

$$\therefore \quad I = \int (1 - z^2) z^4\, dz = \int (z^4 - z^6)\, dx = \frac{z^5}{5} - \frac{z^7}{7} + C = \mathbf{\frac{\sin^5 x}{7} - \frac{\sin^7 x}{7} + C.}$$

Remark. It may be noted that the substitution $z = \cos x$ will not be found useful. The students are advised to check it.

(*ii*) Let $\quad I = \displaystyle\int \frac{\sqrt{\tan x}}{\sin x \cos x}\, dx.$

$$\therefore \quad I = \int \frac{\sqrt{\tan x}}{\left(\frac{\sin x}{\cos x} . \cos x \right) \cos x}\, dx = \int \frac{\sqrt{\tan x}}{\tan x \cos^2 x}\, dx = \int \frac{\sec^2 x}{\sqrt{\tan x}}\, dx$$

Let $z = \tan x. \quad \therefore \quad dz = \sec^2 x\, dx$

$$\therefore \quad I = \int \frac{dz}{\sqrt{z}} = \int z^{-1/2}\, dz = \frac{z^{1/2}}{1/2} + C = \mathbf{2\sqrt{\tan x} + C.}$$

(*iii*) Let $\quad I = \int \frac{1}{1-\tan x}\, dx.$

$$\therefore \quad I = \int \frac{1}{1-\frac{\sin x}{\cos x}}\, dx = \int \frac{\cos x}{\cos x - \sin x}\, dx = \frac{1}{\sqrt{2}} \int \frac{\cos x}{\left(\frac{1}{\sqrt{2}}\cos x - \frac{1}{\sqrt{2}}\sin x\right)}\, dx$$

$$= \frac{1}{\sqrt{2}} \int \frac{\cos x}{\cos\left(x + \frac{\pi}{4}\right)}\, dx$$

Let $\quad z = x + \frac{\pi}{4}. \quad \therefore \quad dz = dx.$

$$\therefore \quad I = \frac{1}{\sqrt{2}} \int \frac{\cos\left(z - \frac{\pi}{4}\right) dx}{\cos z} = \frac{1}{\sqrt{2}} \int \frac{\cos z \cos \frac{\pi}{4} + \sin z \sin \frac{\pi}{4}}{\cos z}\, dz$$

$$= \frac{1}{\sqrt{2}} \int \left(\frac{1}{\sqrt{2}} + \frac{1}{\sqrt{2}} \tan z\right) dz = \frac{1}{2} \int (1 + \tan z)\, dz = \frac{1}{2}\, [z + \log |\sec z|] + C$$

$$= \mathbf{\frac{1}{2}\left[x + \frac{\pi}{4} + \log\left|\sec\left(x + \frac{\pi}{4}\right)\right|\right] + C.}$$

13.13. EVALUATION OF ∫ sin^m x cos^n x dx

The integral $\int \sin^m x \cos^n x\, dx$ is evaluated by using following rules :

(*i*) If $m, n \in \mathbf{N}$ and m is odd, then the substitution $z = \cos x$ is used.

(*ii*) If $m, n \in \mathbf{N}$ and n is odd, then substitution $z = \sin x$ is used.

(*iii*) If $m, n \in \mathbf{N}$ and both are odd, then either $z = \sin x$ or $z = \cos x$ is used. It is advisable to use $z = \sin x$ if $m \geq n$ and $z = \cos x$ if $n \geq m$.

(*iv*) If $m, n \in \mathbf{N}$ and both are even, then $\sin^m x$ and $\cos^n x$ are expressed in terms of sines and cosines of multiples of x.

(*v*) If $m, n \in \mathbf{Q}$ and $m + n$ is a negative even integer, then the substitution $z = \tan x$ is used.

Example 5. *Evaluate :*

(*i*) $\int \sin^3 x \cos^{10} x\, dx$ (*ii*) $\int \sin^4 x \cos^3 x\, dx$ (*iii*) $\int \sin^2 x \cos^2 x\, dx$

(*iv*) $\int \sin^2 x \cos^4 x\, dx.$

Sol. (*i*) Let $\quad I = \int \sin^3 x \cos^{10} x\, dx.$ (Here $m = 3$ is odd)

Let $\quad z = \cos x. \quad \therefore \quad dz = -\sin x\, dx.$

$$\therefore \quad I = \int \sin^2 x \cos^{10} x \sin x\, dx = \int (1 - \cos^2 x) \cos^{10} x \sin x\, dx$$

$$= -\int (1 - z^2)\, z^{10}.dz$$

$$= \int (z^{12} - z^{10})\, dz = \frac{z^{13}}{13} - \frac{z^{11}}{11} + C = \mathbf{\frac{\cos^{13} x}{13} - \frac{\cos^{11} x}{11} + C.}$$

(*ii*) Let $\quad I = \int \sin^4 x \cos^3 x\, dx.$ (Here $n = 3$ is odd)

Let $\quad z = \sin x. \quad \therefore \quad dz = \cos x\, dx.$

$\therefore$ $$I = \int \sin^4 x \cos^2 x \cos x \, dx = \int \sin^4 x (1 - \sin^2 x) \cos x \, dx = \int z^4 (1 - z^2) \, dz$$

$$= \int (z^4 - z^6) \, dz = \frac{z^5}{5} - \frac{z^7}{7} + C = \frac{\sin^5 x}{5} - \frac{\sin^7 x}{7} + C.$$

(*iii*) Let $$I = \int \sin^2 x \cos^2 x \, dx.$$

$\therefore$ $$I = \int \frac{1 - \cos 2x}{2} \cdot \frac{1 + \cos 2x}{2} \, dx = \frac{1}{4} \int (1 - \cos^2 2x) \, dx = \frac{1}{4} \int \sin^2 2x \, dx$$

$$= \frac{1}{4} \int \frac{1 - \cos 4x}{2} \, dx = \frac{1}{8} \int (1 - \cos 4x) \, dx = \frac{1}{8} \left[x - \frac{\sin 4x}{4} \right] + C.$$

Alternative method

Let $$I = \int \sin^2 x \cos^2 x \, dx$$

$\therefore$ $$I = \frac{1}{4} \int (2 \sin x \cos x)^2 \, dx = \frac{1}{4} \int (\sin 2x)^2 \, dx = \frac{1}{4} \int \frac{1 - \cos 4x}{2} \, dx$$

$$= \frac{1}{8} \int (1 - \cos 4x) \, dx = \frac{1}{8} \left[x - \frac{\sin 4x}{4} \right] + C.$$

(*iv*) Let $$I = \int \sin^2 x \cos^4 x \, dx.$$

$\therefore$ $$I = \int \left(\frac{1 - \cos 2x}{2} \right) \left(\frac{1 + \cos 2x}{2} \right)^2 dx$$

$$= \frac{1}{8} \int (1 - \cos 2x) (1 + \cos^2 2x + 2 \cos 2x) \, dx$$

$$= \frac{1}{8} \int 1 + \cos^2 2x + 2 \cos 2x - \cos 2x - \cos^3 2x - 2 \cos^2 2x) \, dx$$

$$= \frac{1}{8} \int (1 + \cos 2x - \cos^2 2x - \cos^3 2x) \, dx$$

$$= \frac{1}{8} \int 1 \, dx + \frac{1}{8} \int \cos 2x - \frac{1}{8} \int \frac{1 + \cos 4x}{2} \, dx - \frac{1}{8} \int (1 - \sin^2 2x) \cos 2x \, dx$$

$$= \frac{1}{8} x + \frac{1}{8} \cdot \frac{\sin 2x}{2} - \frac{1}{16} \left(x + \frac{\sin 4x}{4} \right) - \frac{1}{8} \int (1 - z^2) \frac{dz}{2} \text{ where } z = \sin 2x$$

$$= \frac{x}{8} + \frac{\sin 2x}{16} - \frac{x}{16} - \frac{\sin 4x}{64} - \frac{1}{16} \left(z - \frac{z^3}{3} \right) + C$$

$$= \frac{x}{16} + \frac{\sin 2x}{16} - \frac{\sin 4x}{64} - \frac{\sin 2x}{16} + \frac{\sin^3 2x}{48} + C$$

$$= \frac{x}{16} - \frac{\sin 4x}{64} + \frac{\sin^3 2x}{48} + C.$$

EXERCISE 13.3

Evaluate the following integrals (Q. No. 1–8) :

1. (i) $\int \frac{x+1}{x^2+2x-3} dx$ (ii) $\int f'(ax+b)[f(ax+b)]^n dx$ (iii) $\int \frac{x^2}{\sqrt{1-x^6}} dx$

(iv) $\int (2x+1)\sqrt{x^2+x+1}\, dx$ (v) $\int (x^3-1)^{1/3} x^5 dx$ (vi) $\int \frac{4x+1}{\sqrt{2x^2+x-3}} dx$

(vii) $\int x \log x^2 dx$

2. (i) $\int \frac{1}{x(\log x)^m} dx$ (ii) $\int \frac{\log x^2}{x} dx$ (iii) $\int \frac{(\log x)^2}{x} dx$

(iv) $\int \frac{\sqrt{3+\log x}}{x} dx$ (v) $\int \frac{\cot(\log x)}{x} dx$ (vi) $\int \frac{(x+1)(x+\log x)^2}{x} dx$

3. (i) $\int \frac{\sin x}{1+\cos x} dx$ (ii) $\int \frac{\tan x \sec^2 x}{1-\tan^2 x} dx$ (iii) $\int \frac{\sec^2 x}{\tan x + 2} dx$

(iv) $\int \frac{\left(\log \tan \frac{x}{2}\right)^2}{\sin x} dx$ (v) $\int \frac{1-\cot x}{1+\cot x} dx$ (vi) $\int \frac{(x+1)e^x}{\sin^2(xe^x)} dx$

(P.T.U. B.Pharma, Dec. 2005)

4. (i) $\int \frac{\cos x}{1+\sin^2 x} dx$ (ii) $\int \frac{\tan x}{\sec x + \cos x} dx$ (iii) $\int \frac{\sin 2x}{(a+b\cos 2x)^2} dx$

(iv) $\int \sec x \log(\sec x + \tan x)\, dx$ (v) $\int \frac{\sin x \cos x\, dx}{a^2 \sin^2 x + b^2 \cos^2 x}$ (vi) $\int \cot x \log \sin x\, dx.$

5. (i) $\int e^{\cos^2 x} \sin 2x\, dx$ (ii) $\int e^x \operatorname{cosec}^2(e^x)\, dx$ (iii) $\int \sin x \sin(\cos x)\, dx$

(iv) $\int \frac{1}{x^3} \cos\left(\frac{1}{x^2}\right) dx$ (v) $\int \frac{\sin x}{(a+b\cos x)^2} dx$ (vi) $\int \sqrt{\tan x}\,(1+\tan^2 x)\, dx$

6. (i) $\int \frac{x^{e-1}+e^{x-1}}{x^e+e^x} dx$ (ii) $\int (a+be^x)^6 e^x dx$ (iii) $\int \frac{dx}{e^x+e^{-x}}$

(iv) $\int \frac{1-e^x}{1+e^x} dx$ (v) $\int \frac{e^{2x}-1}{e^{2x}+1} dx$ (vi) $\int \sqrt{e^x-1}\, dx$

7. (i) $\int \frac{(\tan^{-1} x)^3}{1+x^2} dx$ (ii) $\int \frac{\sec^2(2\tan^{-1} x)}{1+x^2} dx$ (iii) $\int \frac{x^3 \sin(\tan^{-1} x^4)}{1+x^8} dx$

(P.T.U. B. Pharma, Dec. 2001)

(iv) $\int \frac{x \sin^{-1} x^2}{\sqrt{1-x^4}} dx$ (v) $\int \frac{e^{\sin^{-1} x}}{\sqrt{1-x^2}} dx$ (vi) $\int \frac{e^{m \tan^{-1} x}}{1+x^2} dx$

(K.U.K. B. Pharma, 2002)

8. (i) $\int \frac{1}{\sqrt{1+\cos 2x}} dx$ (ii) $\int \frac{1}{\sqrt{1-\cos 2x}} dx$ (iii) $\int \frac{1}{\sqrt{1+\sin 2x}} dx$

(*iv*) $\int \frac{1}{\sqrt{x}} \cos \sqrt{x}\, dx$ (*v*) $\int \left(\frac{\tan (1/x)}{x}\right)^2 dx$ (*vi*) $\int x^2 e^{x^3} \cos (e^{x^3}) dx$

(*vii*) $\int \frac{1+\cot x}{x+\log \sin x} dx$ (*viii*) $\int \frac{1-\sin 2x}{x+\cos^2 x} dx$ (*ix*) $\int \frac{1+\sin 2x}{x+\sin^2 x} dx$

Evaluate the following integrals (Q. No. 9–16) :

9. (*i*) $\int \frac{x^2}{\sqrt{1-x}} dx$ (*ii*) $\int \frac{4x^2}{\sqrt{2+x}} dx$ (*iii*) $\int \frac{x^2-x}{\sqrt{x+1}} dx$

(*iv*) $\int \frac{x^3}{\sqrt{1-x^8}} dx$ (*v*) $\int \cos^3 x\, e^{\log \sin x} dx$ (*vi*) $\int e^{3 \log x} (x^4+1)^{-1} dx$.

(*vii*) $\int \frac{1}{x+\sqrt{x}} dx$ (*P.T.U. B.Pharma, Dec. 2005*)

10. (*i*) $\int \cos^3 (ax+b) \sin (ax+b)\, dx$ (*ii*) $\int x \sin^3 x^2 \cos x^2\, dx$ (*iii*) $\int \frac{\cos^9 x}{\sin x} dx$

(*iv*) $\int \tan^3 x\, dx$ (*v*) $\int \tan^4 x\, dx$ (*vi*) $\int \cot^5 x\, dx$

11. (*i*) $\int \frac{\sin x}{\sin (x+a)} dx$ (*ii*) $\int \frac{\cos x}{\cos (x-a)} dx$ (*iii*) $\int \frac{dx}{1+\cot x}$

(*iv*) $\int \frac{\sin (x+a)}{\sin (x-a)} dx$ (*v*) $\int \frac{\cos (x+a)}{\cos (x+b)} dx$ (*vi*) $\int \frac{\sin (x+a)}{\cos (x+b)} dx$

12. (*i*) $\int \frac{\cos 2x}{\sin x} dx$ (*ii*) $\int \frac{\cos 2x}{\cos x} dx$ (*iii*) $\int \frac{\sin 8x - \sin 3x}{\cos 8x - \cos 3x} dx$

(*iv*) $\int \frac{\cos ax - \cos bx}{\sin ax + \sin bx} dx$ (*v*) $\int \frac{\cos x + x \sin x}{x (x+\cos x)} dx$

13. (*i*) $\int \sin^3 x \cos^2 x\, dx$ (*ii*) $\int \sin^3 x \cos^4 x\, dx$ (*iii*) $\int \sin^5 x \cos^8 x\, dx$

(*iv*) $\int \sin^5 x \cos^3 x\, dx$ (*v*) $\int \frac{\sin^2 x}{\cos^8 x} dx$

14. (*i*) $\int \frac{dx}{\sqrt{x}+\sqrt[3]{x}}$ (*ii*) $\int \frac{\sin x \cos^3 x}{1+\cos^2 x} dx$ (*iii*) $\int \sqrt{\frac{1-x}{1+x}} dx$

15. (*i*) $\int \sin^5 x\, dx$ (*ii*) $\int \cos^7 x\, dx$

16. (*i*) $\int \operatorname{cosec}^6 x\, dx$ (*ii*) $\int \sin^4 x \cos^2 x\, dx.$

Answers

1. (*i*) $\frac{1}{2} \log | x^2+2x-3 | + C$ (*ii*) $\frac{[f(ax+b)]^{n+1}}{(n+1)\, a} + C$ (*iii*) $\frac{1}{3} \sin^{-1} x^3 + C$

(*iv*) $\frac{2}{3} (x^2+x+1)^{3/2} + C$ (*v*) $\frac{(x^3-1)^{7/3}}{7} + \frac{(x^3-1)^{4/3}}{4} + C$ (*vi*) $2\sqrt{2x^2+x-3} + C$

(*vii*) $-\frac{1}{2} x^2 (\log x^2 - 1) + C$

2. (*i*) $\dfrac{(\log x)^{1-m}}{1-m} + C$ (*ii*) $(\log x)^2 + C$ (*iii*) $\dfrac{1}{3}(\log x)^3 + C$

(*iv*) $\dfrac{2}{3}(3 + \log x)^{3/2} + C$ (*v*) $\log(\sin(\log x)) + C$ (*vi*) $\dfrac{1}{3}(x + \log x)^3 + C$

3. (*i*) $-\log(1 + \cos x) + C$ (*ii*) $-\dfrac{1}{2}\log|1 - \tan^2 x| + C$ (*iii*) $\log|\tan x + 2| + C$

(*iv*) $\dfrac{\left(\log \tan \dfrac{x}{2}\right)^3}{3} + C$ (*v*) $-\log|\cos x + \sin x| + C$ (*vi*) $-\cot(xe^x) + C$

4. (*i*) $\tan^{-1}(\sin x) + C$ (*ii*) $-\tan^{-1}(\cos x) + C$ (*iii*) $\dfrac{1}{2b(a + b\cos 2x)} + C$

(*iv*) $\dfrac{1}{2}(\log(\sec x + \tan x))^2 + C$ (*v*) $\dfrac{1}{2(a^2 - b^2)}\log(a^2 \sin x^2 + b^2\cos^2 x) + C$

(*vi*) $\dfrac{1}{2}(\log \sin x)^2 + C$ **5.** (*i*) $-e^{\cos^2 x} + C$ (*ii*) $-\cot e^x + C$

(*iii*) $\cos(\cos x) + C$ (*iv*) $-\dfrac{1}{2}\sin\left(\dfrac{1}{x^2}\right) + C$ (*v*) $\dfrac{1}{b(a + b\cos x)} + C$

(*vi*) $\dfrac{2}{3}\tan^{3/2} x + C$ **6.** (*i*) $\dfrac{1}{e}\log|x^e + e^x| + C$ (*ii*) $\dfrac{1}{7b}(a + be^x)^7 + C$

(*iii*) $\tan^{-1} e^x + C$ (*iv*) $-2\log|e^{-x/2} + e^{x/2}| + C$ (*v*) $\log(e^x + e^{-x}) + C$

(*vi*) $2[\sqrt{e^x - 1} - \tan^{-1}\sqrt{e^x - 1}] + C$

7. (*i*) $\dfrac{1}{4}(\tan^{-1} x)^4 + C$ (*ii*) $\dfrac{x}{1 - x^2} + C$ (*iii*) $-\dfrac{1}{4}\cos(\tan^{-1} x^4) + C$

(*iv*) $\dfrac{1}{4}(\sin^{-1} x^2)^2 + C$ (*v*) $e^{\sin^{-1} x} + C$ (*vi*) $\dfrac{1}{m}e^{m\tan^{-1} x} + C$

8. (*i*) $\dfrac{1}{\sqrt{2}}\log|\sec x + \tan x| + C$ (*ii*) $\dfrac{1}{\sqrt{2}}\log|\operatorname{cosec} x - \cot x| + C$

(*iii*) $-\dfrac{1}{\sqrt{2}}\log\left|\sec\left(\dfrac{\pi}{4} - x\right) + \tan\left(\dfrac{\pi}{4} - x\right)\right| + C$ (*iv*) $2\sin\sqrt{x} + C$

(*v*) $\dfrac{1}{x} - \tan\dfrac{1}{x} + C$ (*vi*) $\dfrac{1}{3}\sin(e^{x^3}) + C$ (*vii*) $\log|x + \log \sin x| + C$

(*viii*) $\log|x + \cos^2 x| + C$ (*ix*) $\log|x + \sin^2 x| + C$

9. (*i*) $2\sqrt{1 - x}\left[\dfrac{2}{3}(1 - x) - \dfrac{1}{5}(1 - x)^2 - 1\right] + C$

(*ii*) $8\sqrt{2 + x}\left[\dfrac{1}{5}(2 + x)^2 - \dfrac{4}{3}(2 + x) + 4\right] + C$

(*iii*) $2\sqrt{x + 1}\left[\dfrac{1}{5}(x + 1)^2 - x + 1\right] + C$ (*iv*) $\dfrac{1}{4}\sin^{-1} x^4 + C$

(*v*) $-\dfrac{1}{4}\cos^4 x + C$ (*vi*) $\dfrac{1}{4}\log(x^4 + 1) + C$ (*vii*) $\dfrac{1}{2}\log\left(\dfrac{1 + x^2}{3 + x^2}\right) + C.$

10. (*i*) $-\dfrac{1}{4a}\cos^4(ax + b) + C$ (*ii*) $\dfrac{1}{8}\sin^4 x^2 + C$

(*iii*) $\log|\sin x| - 2\sin^2 x + \dfrac{3}{2}\sin^4 x - \dfrac{2}{3}\sin^6 x + \dfrac{1}{8}\sin^8 x + C$

(*iv*) $\frac{1}{2}\tan^2 x - \log|\sec x| + C$ (*v*) $\frac{1}{3}\tan^3 x - \tan x + x + C$

(*vi*) $-\frac{1}{4}\cot^4 x + \frac{1}{2}\cot^2 x + \log|\sin x| + C$

11. (*i*) $(x + a)\cos a - \sin a \log|\sin(x + a)| + C$ (*ii*) $(x - a)\cos a - \sin a \log|\sec(x - a)| + C$

(*iii*) $\frac{1}{2}\left[x + \frac{\pi}{4} - \log\left|\sin\left(x + \frac{\pi}{4}\right)\right|\right] + C$ (*iv*) $(x - a)\cos 2a + \sin 2a \,.\, \log|\sin(x - a)| + C$

(*v*) $(x + b)\cos(a - b) - \sin(a - b)\log|\sec(x + b)| + C$

(*vi*) $(x + b)\sin(a - b) + \cos(a - b)\log|\sec(x + b)| + C$

12. (*i*) $\log|\operatorname{cosec} x - \cot x| + 2\cos x + C$ (*ii*) $-\log|\sec x + \tan x| + 2\sin x + C$

(*iii*) $-\frac{2}{11}\log\left|\sin\frac{11}{2}x\right| + C$ (*iv*) $\frac{2}{b-a}\log\left|\sec\left(\frac{b-a}{2}\right)x\right| + C$

(*v*) $\log\left|\frac{x}{x + \cos x}\right| + C$

13. (*i*) $\frac{1}{5}\cos^5 x - \frac{1}{3}\cos^3 x + C$ (*ii*) $\frac{\cos^7 x}{7} - \frac{\cos^5 x}{5} + C$

(*iii*) $-\frac{1}{9}\cos^9 x + \frac{2}{11}\cos^{11} x - \frac{1}{13}\cos^{13} x + C$ (*iv*) $\frac{\sin^6 x}{6} - \frac{\sin^8 x}{8} + C$

(*v*) $\frac{1}{7}\tan^7 x + \frac{2}{5}\tan^5 x + \frac{1}{3}\tan^3 x + C$

14. (*i*) $2\sqrt{x} - 3\sqrt[3]{x} + 6\sqrt[6]{x} - 6\log|\sqrt[6]{x+1}| + C$ (*ii*) $-\frac{1}{2}\cos^2 x + \frac{1}{2}\log(1 + \cos^2 x) + C$

(*iii*) $\sin^{-1} x + \sqrt{1 - x^2} + C$

15. (*i*) $-\frac{\cos^5 x}{5} + \frac{2\cos^3 x}{3} - \cos x + C$ (*ii*) $\sin x - \sin^3 x + \frac{3\sin^5 x}{5} - \frac{\sin^7 x}{7} + C$

16. (*i*) $-\cot x - \frac{2}{3}\cot^3 x - \frac{1}{5}\cot^5 x + C$ (*ii*) $\frac{x}{16} - \frac{\sin 4x}{64} - \frac{\sin^3 2x}{48} + C.$

Hints

1. (*ii*) Take $z = f(ax + b)$

(*v*) $z = x^3 - 1 \Rightarrow dz = 3x^2\,dx$

$$\therefore \int (x^3 - 1)^{1/3} x^5\,dx = \int z^{1/3} x^5 \frac{dz}{3x^2} = \frac{1}{3}\int z^{1/3}(z + 1)\,dz.$$

2. (*i*) Take $z = \log x$ (*ii*) Use $\log x^2 = 2\log x$

(*v*) Take $z = \log x$ (*vi*) Take $z = x + \log x$.

3. (*ii*) Take $z = 1 - \tan^2 x$

(*iv*) $z = \log\tan\frac{x}{2} \Rightarrow dz = \frac{1}{\tan\frac{x}{2}} \cdot \frac{1}{2}\sec^2\frac{x}{2}\,dx = \frac{1}{2\sin\frac{x}{2}\cos\frac{x}{2}}\,dx = \frac{dx}{\sin x}$

(*vi*) Take $z = xe^x$.

4. (*ii*) Change the integrand into $\sin x$ and $\cos x$

(*iv*) Take $z = \log(\sec x + \tan x)$. (*v*) Take $z = a^2\sin^2 x + b^2\cos^2 x$.

5. (*i*) Take $z = \cos^2 x$ (*iii*) Take $z = \cos x$ (*vi*) Take $z = \tan x$.

6. (*ii*) Take $z = a + be^x$ (*iii*) $\dfrac{1}{e^x + e^{-x}} = \dfrac{e^x}{(e^x)^2 + 1}$

(*iv*) $\dfrac{1-e^x}{1+e^x} = \dfrac{e^{-x/2}(1-e^x)}{e^{-x/2}(1+e^x)} = \dfrac{e^{-x/2} - e^{x/2}}{e^{-x/2} + e^{x/2}}$

$z = e^{-x/2} + e^{x/2} \Rightarrow dz = \dfrac{1}{2}[-e^{-x/2} + e^{x/2}]\,dx = -\dfrac{1}{2}[e^{-x/2} - e^{x/2}]\,dx$

(*vi*) $z = \sqrt{e^x - 1} \Rightarrow e^x = 1 + z^2 \Rightarrow x = \log(1 + z^2)$

$\therefore \quad dx = \dfrac{2z}{1+z^2}\,dz$ and $\displaystyle\int \sqrt{e^x - 1}\,dx = \int z \cdot \frac{2z}{1+z^2}\,dz = 2\int\left(1 - \frac{1}{1+z^2}\right)dz.$

7. (*i*) Take $z = \tan^{-1} x$ (*ii*) Take $z = \tan^{-1} x$
(*iii*) Take $z = \tan^{-1} x^4$ (*iv*) Take $z = \sin^{-1} x^2$.

8. (*i*) $1 + \cos 2x = 2\cos^2 x$.

(*iii*) $1 + \sin 2x = 1 + \cos\left(\dfrac{\pi}{2} - 2x\right) = 2\cos^2\left(\dfrac{\pi}{4} - x\right)$

(*iv*) Take $z = \sqrt{x}$ (*v*) Take $z = \dfrac{1}{x}$

(*vi*) Take $z = e^{x^3}$ (*vii*) Take $z = x + \log \sin x$.

9. (*i*) $z = \sqrt{1-x} \Rightarrow x = 1 - z^2 \Rightarrow dx = -2z\,dz$

$\therefore \quad \displaystyle\int \frac{x^2}{\sqrt{1-x}}\,dx = \int \frac{(1-z^2)^2}{z}\cdot(-2z\,dz) = -2\int (1-z^2)^2\,dz$

(*v*) Use $e^{\log \sin x} = \sin x$ (*vi*) $e^{3\log x}(x^4+1)^{-1} = \dfrac{x^3}{x^4+1}$.

10. (*i*) Take $z = \cos(ax + b)$ (*ii*) Take $z = \sin x^2$

(*iii*) Write $\dfrac{\cos^9 x}{\sin x} = \dfrac{(1-\sin^2 x)^4}{\sin x} \cdot \cos x$ and put $z = \sin x$

(*iv*) Write $\tan^3 x = \tan x(\sec^2 x - 1) = \tan \sec^2 x - \tan x$

(*vi*) $\cot^5 x = \cot^3 x(\text{cosec}^2 x - 1) = \cot^3 x \,\text{cosec}^2 x - \cot^3 x = \cot^3 x\,\text{cosec}^2 x - \cot x(\text{cosec}^2 x - 1)$.

11. (*i*) Take $z = x + a$

(*iii*) Write $\dfrac{1}{1+\cot x} = \dfrac{\sin x}{\cos x + \sin x} = \dfrac{\sin x}{\sqrt{2}\sin\left(x + \dfrac{\pi}{4}\right)}$ and put $z = x + \dfrac{\pi}{4}$

(*iv*) Take $z = x - a$.

12. (*i*) $\dfrac{\cos 2x}{\sin x} = \dfrac{1 - 2\sin^2 x}{\sin x} = \text{cosec}\, x - 2\sin x$

(*iii*) $\dfrac{\sin 8x - \sin 3x}{\cos 8x - \cos 3x} = \dfrac{2\cos\dfrac{11}{2}x \sin\dfrac{5}{2}x}{2\sin\dfrac{11}{2}x \sin\left(-\dfrac{5}{2}x\right)} = -\cot\dfrac{11}{2}x$

(*v*) $\dfrac{\cos x + x\sin x}{x(x + \cos x)} = \dfrac{(\cos x + x) - (x - x\sin x)}{x(x + \cos x)} = \dfrac{1}{x} - \dfrac{1 - \sin x}{x + \cos x}$

13. (*i*) Take $z = \cos x$ (*iv*) Take $z = \sin x$

14. (*i*) Take $z = x^{1/6}$, 6 being the L.C.M. of 2 and 3

(*ii*) Take $z = \cos x$

(iii) $\sqrt{\dfrac{1-x}{1+x}} = \sqrt{\dfrac{1-x}{1+x}\cdot\dfrac{1-x}{1-x}} = \dfrac{1-x}{\sqrt{1-x^2}} = \dfrac{1}{\sqrt{1-x^2}} + \dfrac{1}{2}\left(\dfrac{-2x}{\sqrt{1-x^2}}\right)$.

15. (i) Write $\sin^5 x = (1-\cos^2 x)^2 \sin x$ and take $z = \cos x$.

16. (i) Write $\operatorname{cosec}^6 x = (1+\cot^2 x)^2 \operatorname{cosec}^2 x$ and take $z = \cot x$

(ii) $\sin^4 x \cos^2 x = \sin^4 x\,(1-\sin^2 x) = \sin^4 x - \sin^6 x = \left(\dfrac{1-\cos 2x}{2}\right)^2 - \left(\dfrac{1-\cos 2x}{2}\right)^3$ etc.

13.14. STANDARD INTEGRALS $\int \frac{1}{a^2-x^2}\,dx$, $\int \frac{1}{x^2-a^2}\,dx$ AND $\int \frac{1}{a^2+x^2}\,dx$

We shall prove that :

1. $\displaystyle\int \frac{1}{a^2-x^2}\,dx = \frac{1}{2a}\log\left|\frac{a+x}{a-x}\right| + C$ **2.** $\displaystyle\int \frac{1}{x^2-a^2}\,dx = \frac{1}{2a}\log\left|\frac{x-a}{x+a}\right| + C$

3. $\displaystyle\int \frac{1}{a^2+x^2}\,dx = \frac{1}{a}\tan^{-1}\frac{x}{a} + C.$

Proof 1. $\displaystyle\frac{1}{a^2-x^2} = \frac{1}{(a-x)(a+x)} = \frac{1}{2a}\left[\frac{(a-x)+(a+x)}{(a-x)(a+x)}\right] = \frac{1}{2a}\left[\frac{1}{a+x}+\frac{1}{a-x}\right]$

$$\therefore \quad \int \frac{1}{a^2-x^2}\,dx = \frac{1}{2a}\int\left[\frac{1}{a+x}+\frac{1}{a-x}\right]dx = \frac{1}{2a}\left[\frac{\log|a+x|}{1}+\frac{\log|a-x|}{-1}\right]+C$$

$$= \frac{1}{2a}\log\left|\frac{a+x}{a-x}\right| + C.$$

$$\therefore \quad \int \frac{1}{\mathbf{a}^2-\mathbf{x}^2}\,\mathbf{dx} = \frac{1}{2\mathbf{a}}\log\left|\frac{\mathbf{a}+\mathbf{x}}{\mathbf{a}-\mathbf{x}}\right| + \mathbf{C}.$$

2. $\displaystyle\frac{1}{x^2-a^2} = \frac{1}{2a}\left[\frac{(x+a)-(x-a)}{(x-a)(x+a)}\right] = \frac{1}{2a}\left[\frac{1}{x-a}-\frac{1}{x+a}\right]$

$$\therefore \quad \int \frac{1}{x^2-a^2}\,dx = \frac{1}{2a}\int\left[\frac{1}{x-a}-\frac{1}{x+a}\right]dx = \frac{1}{2a}\left[\frac{\log|x-a|}{1}-\frac{\log|x+a|}{1}\right]+C$$

$$= \frac{1}{2a}\log\left|\frac{x-a}{x+a}\right| + C.$$

$$\therefore \quad \int \frac{1}{\mathbf{x}^2-\mathbf{a}^2}\,\mathbf{dx} = \frac{1}{2\mathbf{a}}\log\left|\frac{\mathbf{x}-\mathbf{a}}{\mathbf{x}+\mathbf{a}}\right| + \mathbf{C}.$$

Remark. The above formula can also be deduced from formula 1 as follows :

$$\int \frac{1}{x^2-a^2}\,dx = -\int \frac{1}{a^2-x^2}\,dx = -\frac{1}{2a}\log\left|\frac{a+x}{a-x}\right| + C = \frac{1}{2a}\log\left|\frac{a+x}{a-x}\right|^{-1} + C$$

$$= \frac{1}{2a}\log\left|\frac{a-x}{a+x}\right| + C = \frac{1}{2a}\log\left|\frac{x-a}{x+a}\right| + C. \qquad (\because\ |a-x| = |x-a|)$$

3. Let $\quad I = \int \frac{1}{a^2 + x^2}\, dx.$ Let $\quad x = a \tan\theta \quad \therefore \quad dx = a \sec^2\theta\, d\theta.$

$$\therefore \quad I = \int \frac{1}{a^2 + a^2 \tan^2\theta} \cdot a \sec^2\theta\, d\theta = \int \frac{a \sec^2\theta}{a^2 (1 + \tan^2\theta)}\, d\theta = \frac{1}{a} \int 1 \cdot d\theta = \frac{1}{a} \cdot \theta + C$$

$$= \frac{1}{a} \tan^{-1} \frac{x}{a} + C.$$

$$\therefore \quad \int \frac{1}{\mathbf{a^2 + x^2}}\, \mathbf{dx} = \frac{\mathbf{1}}{\mathbf{a}} \tan^{-1} \frac{\mathbf{x}}{\mathbf{a}} + \mathbf{C}.$$

ILLUSTRATIVE EXAMPLES

Example 1. *Evaluate :*

(*i*) $\int \frac{dx}{1 - 9x^2}$ (*ii*) $\int \frac{dx}{4 + 25x^2}$ (*iii*) $\int \frac{dx}{9x^2 - 7}$

(*iv*) $\int \frac{dx}{50 + 2x^2}$ (*v*) $\int \frac{dx}{32 - 2x^2}$ (*vi*) $\int \frac{dx}{1 + 5 \sin^2 x}$.

Sol. (*i*) Let $\quad I = \int \frac{dx}{1 - 9x^2}. \quad \therefore \quad I = \frac{1}{9} \int \frac{dx}{1/9 - x^2} = \frac{1}{9} \int \frac{dx}{(1/3)^2 - x^2}$

$$\therefore \quad I = -\frac{1}{9} \cdot \frac{1}{2(1/3)} \log \left| \frac{(1/3) + x}{(1/3) - x} \right| + C = \frac{1}{6} \log \left| \frac{1 + 3x}{1 - 3x} \right| + C.$$

$$\left(\text{Using} \int \frac{dx}{a^2 - x^2} = \frac{1}{2a} \log \left| \frac{a + x}{a - x} \right| + C \right)$$

(*ii*) Let $\quad I = \int \frac{dx}{4 + 25x^2}. \quad \therefore \quad I = \frac{1}{25} \int \frac{dx}{4/5 + x^2} = \frac{1}{25} \int \frac{dx}{(2/5)^2 + x^2}$

$$\therefore \quad I = \frac{1}{25} \cdot \frac{1}{2/5} \tan^{-1} \frac{x}{2/5} + C = \frac{\mathbf{1}}{\mathbf{10}} \tan^{-1} \frac{\mathbf{5x}}{\mathbf{2}} + \mathbf{C}.$$

$$\left(\text{Using} \int \frac{dx}{a^2 + x^2} = \frac{1}{a} \tan^{-1} \frac{x}{a} + C \right)$$

(*iii*) Let $\quad I = \int \frac{dx}{9x^2 - 7}. \quad \therefore \quad I = \frac{1}{9} \int \frac{dx}{x^2 - (7/9)} = \frac{1}{9} \int \frac{dx}{x^2 - (\sqrt{7}/3)^2}$

$$\therefore \quad I = \frac{1}{9} \cdot \frac{1}{2(\sqrt{7}/3)} \log \left| \frac{x - \sqrt{7}/3}{x + \sqrt{7}/3} \right| + C = \frac{\mathbf{1}}{\mathbf{6\sqrt{7}}} \log \left| \frac{\mathbf{3x - \sqrt{7}}}{\mathbf{3x + \sqrt{7}}} \right| + \mathbf{C}.$$

$$\left(\text{Using} \int \frac{dx}{x^2 - a^2} = \frac{1}{2a} \log \left| \frac{x - a}{x - a} \right| + C \right)$$

(*iv*) Let $\quad I = \int \frac{dx}{50 + 2x^2}. \quad \therefore \quad I = \frac{1}{2} \int \frac{dx}{(5)^2 + x^2}$

$$\therefore \quad I = \frac{1}{2} \cdot \frac{1}{5} \tan^{-1} \frac{x}{5} + C = \frac{\mathbf{1}}{\mathbf{10}} \tan^{-1} \frac{\mathbf{x}}{\mathbf{5}} + \mathbf{C}.$$

(*v*) Let $\quad I = \int \frac{dx}{32 - 2x^2}. \quad \therefore \quad I = \frac{1}{2} \int \frac{dx}{(4)^2 - x^2}$

$\therefore \quad I = \frac{1}{2} \cdot \frac{1}{2(4)} \log \left| \frac{4+x}{4-x} \right| + C = \frac{1}{16} \log \left| \frac{4+x}{4-x} \right| + C.$

(*vi*) Let $\quad I = \int \frac{dx}{1 + 5 \sin^2 x}$. Dividing num. and denom. by $\cos^2 x$, we get

$$I = \int \frac{\frac{1}{\cos^2 x}}{\frac{1}{\cos^2 x} + 5 \frac{\sin^2 x}{\cos^2 x}} dx = \int \frac{\sec^2 x}{\sec^2 x + 5 \tan^2 x} dx = \int \frac{\sec^2 x}{1 + 6 \tan^2 x} dx$$

Let $z = \tan x$. $\therefore \quad dz = \sec^2 x \, dx$ $\quad | \sec^2 x - \tan^2 x = 1$

$$\therefore \quad I = \int \frac{dz}{1 + 6z^2} = \frac{1}{6} \int \frac{dz}{(1/6) + z^2} = \frac{1}{6} \int \frac{dz}{(1/\sqrt{6})^2 + z^2}$$

$$= \frac{1}{6} \cdot \frac{1}{1/\sqrt{6}} \tan^{-1} \frac{z}{1/\sqrt{6}} + C = \frac{1}{\sqrt{6}} \tan^{-1} (\sqrt{6} \tan x) + C.$$

13.15. EVALUATION OF $\int \frac{1}{ax^2 + bx + c} dx$, $\int \frac{px + q}{ax^2 + bx + c} dx$ AND $\int \frac{f(x)}{ax^2 + bx + c} dx$, WHERE f(x) IS A POLYNOMIAL OF DEGREE ≥ 2

The quadratic expression $ax^2 + bx + c$ may or may not be expressible as the product of linear factors. In case, $ax^2 + bx + c$ is easily reducible to two linear factors, then these integrals can be easily evaluated by using the method "**integration by partial fractions**", which we shall study in our due course. Thus ;

(*a*) If $ax^2 + bx + c$ is easily reducible to linear factors, then either the method of this section or the method of integration by partial fractions can be used. Out of these two methods, the method of integration by partial fractions would be found more convenient and easy.

(*b*) If $ax^2 + bx + c$ is not easily reducible to linear factors, then the method of integration by partial fractions is not applied and the method of this section is the only choice at our disposal.

(*i*) Evaluation of $\int \frac{1}{ax^2 + bx + c} dx$.

The integral $\int \frac{1}{ax^2 + bx + c} dx$ is evaluated by expressing its integrand either in the form

$\frac{1}{A^2 - z^2}$ or $\frac{1}{z^2 - A^2}$ or $\frac{1}{A^2 + z^2}$ by using suitable substitution.

WORKING RULES FOR EVALUATING $\int \frac{1}{ax^2 + bx + c} dx$

Step I. *Write* $ax^2 + bx + c$ *as* $a\left(x^2 + \frac{b}{a} x + \frac{c}{a}\right)$ *and take a out of the integral sign.*

Step II. $x^2 + \frac{b}{a} x$ *is expressed in terms of a perfect square by adding and subtracting the square of half of the coefficient of x.*

Step III. *Linear expression in x is taken as z and the integral is expressed in terms of z. This resultant integral will be in one of the standard forms.*

Step IV. *Use the standard formula and express the result in terms of x.*

Example 2. *Evaluate :*

(*i*) $\int \frac{dx}{x^2 - 4x + 8}$ (*ii*) $\int \frac{dx}{5 - 8x - x^2}$ (*iii*) $\int \frac{dx}{1 - 6x - 9x^2}$ (*iv*) $\int \frac{dx}{4x^2 - 4x + 3}$.

Sol. (*i*) Let $I = \int \frac{dx}{x^2 - 4x + 8}$ $\therefore$ $I = \int \frac{dx}{(x^2 - 4x + 4) + 4} = \int \frac{dx}{(x-2)^2 + 4}$

Let $z = x - 2$. $\therefore$ $dz = dx$

$$\therefore \quad I = \int \frac{dz}{z^2 + 4} = \int \frac{dz}{z^2 + 2^2} = \frac{1}{2} \tan^{-1} \frac{z}{2} + C = \mathbf{\frac{1}{2} \tan^{-1} \frac{x - 2}{2} + C.}$$

(*ii*) Let $I = \int \frac{dx}{5 - 8x - x^2}$. $\therefore$ $I = -\int \frac{dx}{x^2 + 8x - 5} = -\int \frac{dx}{(x^2 + 8x + 16) - 5 - 16}$

Add and sub. $\left(\frac{1}{2} \text{ coeff. of } x\right)^2$

$$= -\int \frac{dx}{(x + 4)^2 - 21} = \int \frac{dx}{21 - (x + 4)^2}$$

Let $z = x + 4$. $\therefore$ $dz = dx$

$$\therefore \quad I = \int \frac{dz}{21 - z^2} = \int \frac{dz}{(\sqrt{21})^2 - z^2} = \frac{1}{2\sqrt{21}} \log \left| \frac{\sqrt{21} + z}{\sqrt{21} - z} \right| + C$$

$$\left(\text{Using} \int \frac{dx}{a^2 - x^2} = \frac{1}{2a} \log \left| \frac{a + x}{a - x} \right| + C \right)$$

$$= \frac{1}{2\sqrt{21}} \log \left| \frac{\sqrt{21} + x + 4}{\sqrt{21} - (x + 4)} \right| + C = \mathbf{\frac{1}{2\sqrt{21}} \log \left| \frac{\sqrt{21} + 4 + x}{\sqrt{21} - 4 - x} \right| + C.}$$

Alternative method

Let $I = \int \frac{dx}{5 - 8x - x^2}$. $\therefore$ $I = -\int \frac{dx}{x^2 + 8x - 5} = -\int \frac{dx}{(x^2 + 8x + 16) - 21} = -\int \frac{dx}{(x + 4)^2 - 21}$

Let $z = x + 4$. $\therefore$ $dz = dx$

$$\therefore \quad I = -\int \frac{dz}{z^2 - 21} = -\int \frac{dz}{z^2 - (\sqrt{21})^2} = -\frac{1}{2\sqrt{21}} \log \left| \frac{z - \sqrt{21}}{z + \sqrt{21}} \right| + C$$

$$\left(\text{Using} \int \frac{dx}{x^2 - a^2} = \frac{1}{2a} \log \left| \frac{x - a}{x + a} \right| + C \right)$$

$$= \mathbf{-\frac{1}{2\sqrt{21}} \log \left| \frac{x + 4 - \sqrt{21}}{x + 4 + \sqrt{21}} \right| + C.}$$

Remark. This answer is same as obtained by using the first method because

$$-\frac{1}{2\sqrt{21}} \log \left| \frac{x + 4 - \sqrt{21}}{x + 4 + \sqrt{21}} \right| = \frac{1}{2\sqrt{21}} \log \left| \frac{x + 4 - \sqrt{21}}{x + 4 + \sqrt{21}} \right|^{-1} = \frac{1}{2\sqrt{21}} \log \left| \frac{x + 4 + \sqrt{21}}{x + 4 - \sqrt{21}} \right|$$

$$= \frac{1}{2\sqrt{2}} \log \left| \frac{x+4+\sqrt{21}}{\sqrt{21}-(x+4)} \right| = \frac{1}{2\sqrt{21}} \log \left| \frac{\sqrt{21}+4+x}{\sqrt{21}-4-x} \right|.$$

(*iii*) Let $\quad I = \int \frac{dx}{1-6x-9x^2}.$

$$\therefore \quad I = -\frac{1}{9} \int \frac{dx}{x^2 + \frac{2}{3}x - \frac{1}{9}} = -\frac{1}{9} \int \frac{dx}{\left(x^2 + \frac{2}{3}x + \frac{1}{9}\right) - \frac{2}{9}} = -\frac{1}{9} \int \frac{dx}{\left(x+\frac{1}{3}\right)^2 - \frac{2}{9}}$$

Let $z = x + \frac{1}{3}$. $\quad \therefore \quad dz = dx.$ $\qquad \left(\because \left(\frac{1}{2}(\text{coeff. of } x)\right)^2 = \frac{1}{9}\right)$

$$\therefore \quad I = -\frac{1}{9} \int \frac{dz}{z^2 - \left(\frac{\sqrt{2}}{3}\right)^2} = -\frac{1}{9} \cdot \frac{1}{2\left(\frac{\sqrt{2}}{3}\right)} \log \left| \frac{z - \left(\frac{\sqrt{2}}{3}\right)}{z + \left(\frac{\sqrt{2}}{3}\right)} \right| + C$$

$$= -\frac{1}{6\sqrt{2}} \log \left| \frac{x + \frac{1}{3} - \frac{\sqrt{2}}{3}}{x + \frac{1}{3} + \frac{\sqrt{2}}{3}} \right| + C$$

$$= -\frac{\mathbf{1}}{\mathbf{6\sqrt{2}}} \mathbf{\log} \left| \frac{\mathbf{3x+1-\sqrt{2}}}{\mathbf{3x+1+\sqrt{2}}} \right| + \mathbf{C}.$$

(*iv*) Let $\quad I = \int \frac{dx}{4x^2 - 4x + 3}.$

$$\therefore \quad I = \frac{1}{4} \int \frac{dx}{x^2 - x + \frac{3}{4}} = \frac{1}{4} \int \frac{dx}{x^2 - x + \frac{1}{4} + \frac{2}{4}} = \frac{1}{4} \int \frac{dx}{\left(x - \frac{1}{2}\right)^2 + \frac{1}{2}}.$$

$$\left(\because \left(\frac{1}{2}(\text{coeff. of } x)\right)^2 = \frac{1}{4}\right)$$

Let $z = x - \frac{1}{2}$. $\quad \therefore \quad dz = dx.$

$$\therefore \quad I = \frac{1}{4} \int \frac{dz}{z^2 + (1/\sqrt{2})^2} = \frac{1}{4} \cdot \frac{1}{(1/\sqrt{2})} \tan^{-1} \frac{z}{1/\sqrt{2}} + C = \frac{1}{2\sqrt{2}} \tan^{-1} \frac{x - \frac{1}{2}}{1/\sqrt{2}} + C$$

$$= \frac{\mathbf{1}}{\mathbf{2\sqrt{2}}} \mathbf{\tan^{-1}} \frac{\mathbf{2x-1}}{\mathbf{\sqrt{2}}} + \mathbf{C}.$$

EXERCISE 13.4

Evaluate the following integrals (Q. No. 1–2) :

1. (i) $\int \frac{dx}{1-4x^2}$ (ii) $\int \frac{dx}{1+2(x+2)^2}$ (iii) $\int \frac{dx}{1+\frac{x^2}{4}}$

(iv) $\int \frac{x\,dx}{1-x^4}$ (v) $\int \frac{x^2}{1-x^6}dx$ (vi) $\int \frac{x^3+x}{x^4-9}dx$

2. (i) $\int \frac{\sin\theta}{1-4\cos^2\theta}d\theta$ (ii) $\int \frac{dx}{4\sin^2 x+5\cos^2 x}$ (iii) $\int \frac{dx}{\cos^2 x+2\sin^2 x}$

(iv) $\int \frac{dx}{1+3\sin^2 x}$ (v) $\int \frac{dx}{a^2\sin^2 x+b^2\cos^2 x}$ (vi) $\int \frac{\text{cosec}^2 x}{1-\cot^2 x}dx.$

Evaluate the following integrals (Q. No. 3–6) :

3. (i) $\int \frac{dx}{x^2+2x-4}$ (ii) $\int \frac{dx}{x^2+4x+7}$ (iii) $\int \frac{dx}{x^2+8x+20}$

(iv) $\int \frac{dx}{10+2x+x^2}$ (v) $\int \frac{dx}{9x^2-12x+8}$ (vi) $\int \frac{dx}{9x^2+6x+10}$

4. (i) $\int \frac{x\,dx}{x^4+2x^2+3}$ (ii) $\int \frac{x\,dx}{x^4-x^2+1}$ (iii) $\int \frac{\sec^2 x\,dx}{5\tan^2 x-12\tan x+14}$

(iv) $\int \frac{dx}{x\{6(\log x)^2+7\log x+2\}}$ (v) $\int \frac{\cos x\,dx}{\sin^2 x-3\sin x+2}$ (vi) $\int \frac{5^x\,dx}{1+5^x-5^{2x}}$

5. (i) $\int \frac{2x}{x^2+2x+2}dx$ (ii) $\int \frac{x}{x^2+4x+5}dx$ (iii) $\int \frac{5x-2}{1+2x+3x^2}dx$

(iv) $\int \frac{1-3x}{3x^2+4x+2}dx$ (v) $\int \frac{2x+1}{4-3x-x^2}dx$ (vi) $\int \frac{2x+1}{18-4x-x^2}dx$

6. (i) $\int \frac{(3\sin x-2)\cos x}{5-\cos^2 x-4\sin x}dx$ (ii) $\int \frac{2e^{2x}+5e^x}{e^{2x}-e^x-2}dx$ (iii) $\int \frac{7^{2x}}{7^{2x}+7^x+1}dx$

Answers

1. (i) $\frac{1}{4}\log\left|\frac{1+2x}{1-2x}\right|+C$ (ii) $\frac{1}{\sqrt{2}}\tan^{-1}\sqrt{2}\,(x+2)+C$ (iii) $2\tan^{-1}\frac{x}{2}+C$

(iv) $\frac{1}{4}\log\left|\frac{1+x^2}{1-x^2}\right|+C$ (v) $\frac{1}{6}\log\left|\frac{1+x^3}{1-x^3}\right|+C$

(vi) $\frac{1}{4}\log|x^4-9|+\frac{1}{12}\log\left|\frac{x^2-3}{x^2+3}\right|+C$

2. (i) $-\frac{1}{4}\log\left|\frac{1+2\cos\theta}{1-2\cos\theta}\right|+C$ (ii) $\frac{1}{2\sqrt{5}}\tan^{-1}\left(\frac{2}{\sqrt{5}}\tan x\right)+C$ (iii) $\frac{1}{\sqrt{2}}\tan^{-1}(\sqrt{2}\tan x)+C$

(iv) $\frac{1}{2}\tan^{-1}(2\tan x)+C$ (v) $\frac{1}{ab}\tan^{-1}\left(\frac{a}{b}\tan x\right)+C$ (vi) $-\frac{1}{2}\log\left|\frac{1+\cot x}{1-\cot x}\right|+C$

3. (*i*) $\frac{1}{2\sqrt{5}} \log \left| \frac{x+1-\sqrt{5}}{x+1+\sqrt{5}} \right| + C$ (*ii*) $\frac{1}{\sqrt{3}} \tan^{-1} \frac{x+2}{\sqrt{3}} + C$ (*iii*) $\frac{1}{2} \tan^{-1} \frac{x+4}{2} + C$

(*iv*) $\frac{1}{3} \tan^{-1} \frac{x+1}{3} + C$ (*v*) $\frac{1}{6} \tan^{-1} \frac{3x-2}{2} + C$ (*vi*) $\frac{1}{9} \tan^{-1} \frac{3x+1}{3} + C$

4. (*i*) $\frac{1}{2\sqrt{2}} \tan^{-1} \left(\frac{x^2+1}{\sqrt{2}} \right) + C$ (*ii*) $\frac{1}{\sqrt{3}} \tan^{-1} \left(\frac{2x^2-1}{\sqrt{3}} \right) + C$ (*iii*) $\frac{1}{\sqrt{34}} \tan^{-1} \left(\frac{5 \tan x - 6}{\sqrt{34}} \right) + C$

(*iv*) $\log \left| \frac{2 \log x + 1}{3 \log x + 2} \right| + C$ (*v*) $\log \frac{2 - \sin x}{1 - \sin x} + C$

(*vi*) $\frac{1}{\sqrt{5} \log 5} \log \left| \frac{\sqrt{5} - 1 + 2(5^x)}{\sqrt{5} + 1 - 2(5^x)} \right| + C$

5. (*i*) $\log (x^2 + 2x + 2) - 2 \tan^{-1} (x+1) + C$ (*ii*) $\frac{1}{2} \log (x^2 + 4x + 5) - 2 \tan^{-1} (x+2) + C$

(*iii*) $\frac{5}{6} \log (1 + 2x + 3x^2) - \frac{11}{3\sqrt{2}} \tan^{-1} \frac{3x+1}{\sqrt{2}} + C$

(*iv*) $-\frac{1}{2} \log | 3x^2 + 4x + 2 | + \frac{3}{\sqrt{2}} \tan^{-1} \left(\frac{3x+2}{\sqrt{2}} \right) + C$

(*v*) $-\log | 4 - 3x - x^2 | - \frac{2}{5} \log \left| \frac{4+x}{1-x} \right| + C$

(*vi*) $-\log | 18 - 4x - x^2 | - \frac{3}{2\sqrt{22}} \log \left| \frac{\sqrt{22} + x + 2}{\sqrt{22} - x - 2} \right| + C$

6. (*i*) $3 \log (2 - \sin x) + \frac{4}{2 - \sin x} + C$ (*ii*) $\log | e^{2x} - e^x - 2 | + 2 \log \left| \frac{e^x - 2}{e^x + 1} \right| + C$

(*iii*) $\frac{1}{2 \log 7} \log | 7^{2x} + 7^x + 1 | - \frac{1}{\sqrt{3} \log 7} \tan^{-1} \left(\frac{2(7^x) + 1}{\sqrt{3}} \right) + C.$

Hints

1. (*i*) Write $\frac{1}{1-4x^2} = \frac{1}{4} \left(\frac{1}{(1/2)^2 - x^2} \right)$ or take $z = 2x$

(*iv*) Take $z = x^2$ (*vi*) $\frac{x^3 + x}{x^4 - 9} = \frac{x^3}{x^4 - 9} + \frac{x}{x^4 - 9}$.

2. (*i*) Take $z = \cos \theta$

(*ii*) Write $\frac{1}{4 \sin^2 x + 5 \cos^2 x} = \frac{\sec^2 x}{4 \tan^2 x + 5}$ and take $z = \tan x$

(*iv*) Divide num. and den. by $\cos^2 x$ and take $z = \tan x$

(*vi*) Take $z = \cot x$.

4. (*i*) Take $z = x^2$ (*iii*) Take $z = \tan x$

(*iv*) Take $z = \log x$ (*vi*) Take $z = 5^x$.

6. (*ii*) Take $z = e^x$ (*iii*) Take $z = 7^x$.

13.16. EVALUATION OF RATIONAL FUNCTIONS OF sin x AND cos x

In this section, we shall study the evaluation of the integrals of the types :

(i) $\int \frac{dx}{a \sin x + b \cos x + c}$ (ii) $\int \frac{a \sin x + b \cos x}{c \sin x + d \cos x} dx$

and (iii) $\int \frac{a \sin x + b \cos x + c}{d \sin x + e \cos x + f} dx.$

(i) **Evaluation of $\int \frac{dx}{a \sin x + b \cos x + c}$.** This type of integral is evaluated by putting

$\sin x = \frac{2 \tan \frac{1}{2} x}{1 + \tan^2 \frac{1}{2} x}$ and $\cos x = \frac{1 - \tan^2 \frac{1}{2} x}{1 + \tan^2 \frac{1}{2} x}$ and then putting $\tan \frac{1}{2} x$ equal to some other variable.

WORKING RULES FOR EVALUATING $\int \frac{dx}{a \sin x + b \cos x + c}$

Step I. *Put* $\sin x = \frac{2 \tan \frac{1}{2} x}{1 + \tan^2 \frac{1}{2} x}$ *and* $\cos x = \frac{1 - \tan^2 \frac{1}{2} x}{1 + \tan^2 \frac{1}{2} x}$ *in the integrand and simplify it.*

Step II. *Replace* $1 + \tan^2 \frac{1}{2} x$ *in the numerator by* $\sec^2 \frac{1}{2} x$.

Step III. *Put* $\tan \frac{1}{2} x = z$ *and express the integral in terms of z. The resultant integral in terms of z will be of the form* $\int \frac{dz}{Az^2 + Bz + C}$.

Step IV. *Evaluate the integral in z variable and replace z by* $\tan \frac{1}{2} x$.

A particular case. When $c = 0$, the integral $\int \frac{dx}{a \sin x + b \cos x + c}$ takes the form :

$\int \frac{dx}{a \sin x + b \cos x}$. This integral can also be evaluated as follows :

Let $a = r \cos \theta$ and $b = r \sin \theta$, where $r > 0$.

$\therefore \quad r = \sqrt{a^2 + b^2}$ and $\theta = \tan^{-1} (b/a)$.

$$\therefore \quad \int \frac{dx}{a \sin x + b \cos x} = \int \frac{dx}{r \cos \theta \sin x + r \sin \theta \cos x}$$

$$= \frac{1}{r} \int \frac{dx}{\sin (x + \theta)} = \frac{1}{r} \int \operatorname{cosec} (x + \theta)\, dx = \frac{1}{r} \log \left| \tan \frac{x + \theta}{2} \right| + C$$

$$= \frac{1}{\sqrt{a^2 + b^2}} \log \left| \tan \left(\frac{x}{2} + \frac{1}{2} \tan^{-1} \frac{b}{a} \right) \right| + C.$$

ILLUSTRATIVE EXAMPLES

Example 1. *Evaluate :*

(i) $\int \frac{dx}{4+5\sin x}$ (ii) $\int \frac{d\theta}{4\cos\theta - 1}$

(iii) $\int \frac{dx}{3+2\sin x+\cos x}$

Sol. (i) Let $I = \int \frac{dx}{4+5\sin x}$.

$$\therefore \quad I = \int \frac{dx}{4+5\left(\dfrac{2\tan\frac{1}{2}x}{1+\tan^2\frac{1}{2}x}\right)} = \int \frac{\left(1+\tan^2\frac{1}{2}x\right)dx}{4+4\tan^2\frac{1}{2}x+10\tan\frac{1}{2}x}.$$

$$= \frac{1}{2}\int \frac{\sec^2\frac{1}{2}x}{2\tan^2\frac{1}{2}x+5\tan\frac{1}{2}x+2}\,dx.$$

Let $z = \tan\frac{1}{2}x. \quad \therefore \quad dz = \frac{1}{2}\sec^2\frac{1}{2}x\,dx$

$$\therefore \quad I = \frac{1}{2}\int \frac{2\,dz}{2z^2+5z+2} = \frac{1}{2}\int \frac{dz}{z^2+\frac{5}{2}z+1} = \frac{1}{2}\int \frac{dz}{z^2+\frac{5}{2}z+\frac{25}{16}+1-\frac{25}{16}}$$

$$\Big| \because \left(\frac{1}{2}\text{ coeff. of } z\right)^2 = \frac{25}{16}$$

$$= \frac{1}{2}\int \frac{dz}{\left(z+\frac{5}{4}\right)^2-\frac{9}{16}}. \quad \text{Let } t = z+\frac{5}{4}. \quad \therefore \quad dt = dz.$$

$$\therefore \quad I = \frac{1}{2}\int \frac{dt}{t^2-(3/4)^2} = \frac{1}{2(3/2)}\log\left|\frac{t-3/4}{t+3/4}\right| + C = \frac{1}{3}\log\left|\frac{z+\frac{5}{4}-\frac{3}{4}}{z+\frac{5}{4}+\frac{3}{4}}\right| + C$$

$$= \frac{1}{3}\log\left|\frac{\tan\frac{1}{2}x+\frac{1}{2}}{\tan\frac{1}{2}x+2}\right| + C = \boldsymbol{\frac{1}{3}\log\left|\frac{2\tan\frac{1}{2}x+1}{2\tan\frac{1}{2}x+4}\right| + C.}$$

(ii) Let $I = \int \frac{d\theta}{4\cos\theta-1}$.

$$\therefore \quad I = \int \frac{d\theta}{4\left(\dfrac{1-\tan^2\frac{1}{2}\theta}{1+\tan^2\frac{1}{2}\theta}\right)-1} = \int \frac{\left(1+\tan^2\frac{1}{2}\theta\right)d\theta}{4-4\tan^2\frac{1}{2}\theta-1-\tan^2\frac{1}{2}\theta} = \int \frac{\sec^2\frac{1}{2}\theta\,d\theta}{-5\tan^2\frac{1}{2}\theta+3}.$$

Let $z = \tan \frac{1}{2}\theta$. $\therefore$ $dz = \frac{1}{2}\sec^2 \frac{1}{2}\theta\, d\theta$

$$\therefore \quad I = \int \frac{2\,dz}{-5z^2+3} = -\frac{2}{5}\int \frac{dz}{z^2-3/5} = -\frac{2}{5}\cdot\frac{1}{2(\sqrt{3/5})}\log\left|\frac{z-\sqrt{3/5}}{z+\sqrt{3/5}}\right| + C$$

$$= -\frac{1}{\sqrt{15}}\log\left|\frac{\tan\frac{1}{2}\theta - \sqrt{\frac{3}{5}}}{\tan\frac{1}{2}\theta + \sqrt{\frac{3}{5}}}\right| + \mathbf{C}.$$

(*iii*) Let $I = \int \frac{dx}{3+2\sin x+\cos x}$.

$$\therefore \quad I = \int \frac{dx}{3+2\left(\frac{2\tan\frac{1}{2}x}{1+\tan^2\frac{1}{2}x}\right)+\left(\frac{1-\tan^2\frac{1}{2}x}{1+\tan^2\frac{1}{2}x}\right)}$$

$$= \int \frac{(1+\tan^2\frac{1}{2}x)\,dx}{3(1+\tan^2\frac{1}{2}x)+2(2\tan\frac{1}{2}x)+(1-\tan^2\frac{1}{2}x)}$$

$$= \int \frac{\sec^2\frac{1}{2}x\,dx}{3+3\tan^2\frac{1}{2}x+4\tan\frac{1}{2}x+1-\tan^2\frac{1}{2}x} = \int \frac{\sec^2\frac{1}{2}x\,dx}{2\tan^2\frac{1}{2}+4\tan\frac{1}{2}x+4}$$

Let $z = \tan\frac{1}{2}x$. $\therefore$ $dz = \frac{1}{2}\sec^2\frac{1}{2}x\,dx$

$$\therefore \quad I = \int \frac{2\,dz}{2z^2+4z+4} = \int \frac{dz}{z^2+2z+2} = \int \frac{dz}{(z+1)^2+1}$$

Let $z+1 = \tan\theta$ $\therefore$ $dz = \sec^2\theta\, d\theta$.

$$\therefore \quad I = \int \frac{\sec^2\theta\,d\theta}{\tan^2\theta+1} = \int 1\,.\,d\theta = \theta + C = \tan^{-1}(z+1) + C = \mathbf{\tan^{-1}\left(\tan\frac{1}{2}x+1\right)+C.}$$

Example 2. *Evaluate :*

(*i*) $\int \frac{dx}{a+b\sin x}$ (*ii*) $\int \frac{dx}{a+b\cos x}$.

Sol. (*i*) Let $I = \int \frac{dx}{a+b\sin x}$. $\therefore$ $I = \int \frac{dx}{a+b\left[\frac{2\tan\frac{1}{2}x}{1+\tan^2\frac{1}{2}x}\right]} = \int \frac{\sec^2\frac{1}{2}x}{a+a\tan^2\frac{1}{2}x+2b\tan\frac{1}{2}x}\,dx$

Let $z = \tan\frac{1}{2}x$. $\therefore$ $dz = \frac{1}{2}\sec^2\frac{1}{2}x\,dx$.

$$\therefore \quad I = \int \frac{2\,dz}{a+az^2+2bz} = \frac{2}{a}\int \frac{dz}{z^2+\frac{2b}{a}z+1} = \frac{2}{a}\int \frac{dz}{\left(z^2+\frac{2b}{a}z+\frac{b^2}{a^2}\right)+\left(1-\frac{b^2}{a^2}\right)}$$

$$\Bigg|\ \left(\because\ \frac{1}{2}\text{coeff. of } z\right)^2 = \frac{b^2}{a^2}$$

$$= \frac{2}{a}\int \frac{dz}{\left(z+\frac{b}{a}\right)^2 + \left(\frac{a^2-b^2}{a^2}\right)}$$

Let $t = z + \frac{b}{a}$. $\therefore$ $dt = dz$. $\therefore$ $I = \frac{2}{a}\int \frac{dt}{t^2 + \frac{a^2-b^2}{a^2}}$.

Case I. a > b

$$\therefore \quad I = \frac{2}{a}\int \frac{dt}{t^2 + \frac{a^2-b^2}{a^2}} + C = \frac{2}{a}\cdot\frac{1}{\sqrt{a^2-b^2}/a}\tan^{-1}\frac{t}{\sqrt{a^2-b^2}/a} + C \quad \left(\because \frac{a^2-b^2}{a^2} > 0\right)$$

$$= \frac{2}{\sqrt{a^2-b^2}}\tan^{-1}\frac{z+\frac{b}{a}}{\sqrt{a^2-b^2}/a} + C = \frac{2}{\sqrt{a^2-b^2}}\tan^{-1}\frac{az+b}{\sqrt{a^2-b^2}} + C$$

$$= \mathbf{\frac{2}{\sqrt{a^2-b^2}}\tan^{-1}\frac{a\tan\frac{1}{2}x+b}{\sqrt{a^2-b^2}} + C.}$$

Case II. a < b

$$\therefore \quad I = \frac{2}{a}\int \frac{dt}{t^2 + \frac{a^2-b^2}{a^2}} = \frac{2}{a}\int \frac{dt}{t^2 - \frac{b^2-a^2}{a^2}} = \frac{2}{a}\cdot\frac{1}{2\sqrt{b^2-a^2}/a}\log\left|\frac{t-\frac{\sqrt{b^2-a^2}}{a}}{t+\frac{\sqrt{b^2-a^2}}{a}}\right| + C$$

$$= \frac{1}{\sqrt{b^2-a^2}}\log\left|\frac{z+\frac{b}{a}-\frac{\sqrt{b^2-a^2}}{a}}{z+\frac{b}{a}+\frac{\sqrt{b^2-a^2}}{a}}\right| + C$$

$$= \mathbf{\frac{1}{\sqrt{b^2-a^2}}\log\left|\frac{a\tan\frac{1}{2}x+b-\sqrt{b^2-a^2}}{a\tan\frac{1}{2}x+b+\sqrt{b^2-a^2}}\right| + C.}$$

Case III. a = b

$$\therefore \quad I = \int \frac{dx}{a+b\sin x} = \int \frac{dx}{a+a\sin x} = \frac{1}{a}\int \frac{1}{(1+\sin x)}\cdot\frac{1-\sin x}{1-\sin x}\,dx$$

$$= \frac{1}{a}\int \frac{1-\sin x}{\cos^2 x}\,dx = \frac{1}{a}\int (\sec^2 x - \sec x\tan x)\,dx = \mathbf{\frac{1}{a}(\tan x - \sec x) + C.}$$

Corollary. Evaluate $\int \frac{1}{1+\sin x}\,dx$

Ans. $\tan x - \sec x + c$.

(*ii*) Let $\quad I = \int \frac{dx}{a + b\cos x}$.

$$\therefore \quad I = \int \frac{dx}{a + b\left[\dfrac{1-\tan^2 \frac{1}{2}x}{1+\tan^2 \frac{1}{2}x}\right]} = \int \frac{\sec^2 \frac{1}{2}x}{a + a\tan^2 \frac{1}{2}x + b - b\tan^2 \frac{1}{2}x}\,dx$$

$$= \int \frac{\sec^2 \frac{1}{2}x}{(a+b) + (a-b)\tan^2 \frac{1}{2}x}\,dx. \quad \text{Let } z = \tan \frac{1}{2}x. \quad \therefore \quad dz = \frac{1}{2}\sec^2 \frac{1}{2}x\,dx$$

$$\therefore \quad I = \int \frac{2\,dz}{(a+b) + (a-b)z^2} = \frac{2}{a-b}\int \frac{dz}{\frac{a+b}{a-b} + z^2}$$

Case I. a > b

$$\therefore \quad I = \frac{2}{a-b}\int \frac{dz}{\frac{a+b}{a-b} + z^2} \qquad \left(\frac{a+b}{a-b} > 0\right)$$

$$= \frac{2}{a-b}\cdot\frac{1}{\sqrt{(a+b)/(a-b)}}\tan^{-1}\frac{z}{\sqrt{(a+b)/(a-b)}} + C$$

$$= \frac{2}{a-b}\cdot\sqrt{\frac{a-b}{a+b}\cdot\frac{a-b}{a-b}}\cdot\tan^{-1}\left(\sqrt{\frac{a-b}{a+b}}\cdot z\right) + C$$

$$= \frac{\mathbf{2}}{\sqrt{\mathbf{a^2 - b^2}}}\tan^{-1}\left(\sqrt{\frac{\mathbf{a-b}}{\mathbf{a+b}}}\tan\frac{\mathbf{1}}{\mathbf{2}}\mathbf{x}\right) + \mathbf{C}.$$

Case II. a < b

$$\therefore \quad I = \frac{2}{a-b}\int \frac{dz}{\frac{a+b}{a-b} + z^2} = \frac{2}{b-a}\int \frac{dz}{\frac{b+a}{b-a} - z^2}$$

$$= \frac{2}{b-a}\cdot\frac{1}{2\sqrt{(b+a)/(b-a)}}\log\left|\frac{\sqrt{\frac{b+a}{b-a}} + z}{\sqrt{\frac{b+a}{b-a}} - z}\right| + C$$

$$= \frac{2}{b-a}\cdot\sqrt{\frac{b-a}{b+a}\cdot\frac{b-a}{b-a}}\log\left(\sqrt{\frac{b+a}{b-a}}\cdot z\right) + C$$

$$= \frac{\mathbf{1}}{\sqrt{\mathbf{b^2 - a^2}}}\log\left|\frac{\sqrt{\mathbf{b+a}} + \sqrt{\mathbf{b-a}}\tan\frac{1}{2}\mathbf{x}}{\sqrt{\mathbf{b+a}} - \sqrt{\mathbf{b-a}}\tan\frac{1}{2}\mathbf{x}}\right| + \mathbf{C}.$$

Case III. a = b

$$\therefore \qquad I = \int \frac{dx}{a + b\cos x} = \int \frac{dx}{a + a\cos x} = \frac{1}{a}\int \frac{dx}{2\cos^2 \frac{1}{2}x} = \frac{1}{2a}\int \sec^2 \frac{1}{2}x\, dx$$

$$= \frac{1}{2a} \cdot \frac{\tan \frac{1}{2}x}{1/2} = \frac{\mathbf{1}}{\mathbf{a}}\, \mathbf{tan}\, \frac{\mathbf{1}}{\mathbf{2}}\mathbf{x} + \mathbf{C}.$$

EXERCISE 13.5

Evaluate the following integrals :

1. (i) $\int \frac{dx}{1 + 2\cos x}$ (ii) $\int \frac{dx}{2 + \cos x}$ (iii) $\int \frac{dx}{5 + 4\cos x}$

(iv) $\int \frac{dx}{1 - 3\sin x}$ (v) $\int \frac{dx}{5 - 4\sin x}$ (vi) $\int \frac{dx}{2 - 7\sin x}$

2. (i) $\int \frac{dx}{1 + \sin x + \cos x}$ (ii) $\int \frac{dx}{1 - \sin x + \cos x}$ (iii) $\int \frac{dx}{2 + \cos x - \sin x}$

(iv) $\int \frac{dx}{3\cos x - 4\sin x + 5}$ (v) $\int \frac{dx}{a\sin x + b\cos x}$ (vi) $\int \frac{dx}{3\cos x + 4\sin x}$

3. (i) $\int \frac{\sin x\, dx}{\sin x + \cos x}$ (ii) $\int \frac{dx}{1 + \cot x}$ (iii) $\int \frac{d\theta}{1 - \tan \theta}$

Answers

1. (i) $\frac{1}{\sqrt{3}} \log \left| \frac{\sqrt{3} + \tan \frac{1}{2}x}{\sqrt{3} - \tan \frac{1}{2}x} \right| + C$ (ii) $\frac{2}{\sqrt{3}} \tan^{-1}\left(\frac{1}{\sqrt{3}} \tan \frac{x}{2}\right) + C$

(iii) $\frac{2}{3} \tan^{-1}\left(\frac{1}{3} \tan \frac{1}{2}x\right) + C$ (iv) $\frac{1}{2\sqrt{2}} \log \left| \frac{\tan \frac{x}{2} - 3 - 2\sqrt{2}}{\tan \frac{x}{2} - 3 + 2\sqrt{2}} \right| + C$

(v) $\frac{2}{3} \tan^{-1}\left(\frac{5\tan \frac{1}{2}x - 4}{3}\right) + C$ (vi) $\frac{1}{3\sqrt{5}} \log \left| \frac{2\tan \frac{x}{2} - 7 - 3\sqrt{5}}{2\tan \frac{x}{2} - 7 + 3\sqrt{5}} \right| + C$

2. (i) $\log \left| 1 + \tan \frac{1}{2}x \right| + C$ (ii) $-\log \left| 1 - \tan \frac{1}{2}x \right| + C$ (iii) $\sqrt{2} \tan^{-1}\left(\frac{\tan \frac{x}{2} - 1}{\sqrt{2}}\right) + C$

(iv) $\frac{1}{2 - \tan \frac{x}{2}} + C$ (v) $\frac{1}{\sqrt{a^2 + b^2}} \log \left| \frac{\sqrt{a^2 + b^2} - a + b\tan \frac{1}{2}x}{\sqrt{a^2 + b^2} + a - b\tan \frac{1}{2}x} \right| + C$

(vi) $\frac{1}{5}\log\left|\frac{3\tan\frac{1}{2}x+1}{3-\tan\frac{1}{2}x}\right|+C$

3. (i) $\frac{1}{2}x-\frac{1}{2}\log|\sin x+\cos x|+C$ (ii) $\frac{1}{2}x-\frac{1}{2}\log|\sin x+\cos x|+C$

(iii) $\frac{1}{2}\theta-\frac{1}{2}\log|\cos\theta-\sin\theta|+C.$

13.17. STANDARD INTEGRALS $\int\frac{1}{\sqrt{a^2-x^2}}dx$, $\int\frac{1}{\sqrt{x^2-a^2}}dx$ AND $\int\frac{1}{\sqrt{a^2+x^2}}dx$

We shall prove that :

1. $\int\frac{1}{\sqrt{a^2-x^2}}dx=\sin^{-1}\frac{x}{a}+C$

2. $\int\frac{1}{\sqrt{x^2-a^2}}dx=\log|x+\sqrt{x^2-a^2}|+C$

3. $\int\frac{1}{\sqrt{a^2+x^2}}dx=\log|x+\sqrt{a^2+x^2}|+C.$

Proof 1. Let $I=\int\frac{1}{\sqrt{a^2-x^2}}dx$. Let $x=a\sin\theta$. $\therefore$ $dx=a\cos\theta\,d\theta$.

$$\therefore \quad I=\int\frac{a\cos\theta\,d\theta}{\sqrt{a^2-a^2\sin^2\theta}}=\int\frac{a\cos\theta\,d\theta}{a\sqrt{1-\sin^2\theta}}=\int 1.\,d\theta=\theta+C=\sin^{-1}\frac{x}{a}+C.$$

$$\therefore \quad \int\frac{1}{\sqrt{a^2-x^2}}dx=\sin^{-1}\frac{x}{a}+C.$$

2. Let $I=\int\frac{1}{\sqrt{x^2-a^2}}dx$. Let $x=a\sec\theta$. $\therefore$ $dx=a\sec\theta\tan\theta\,d\theta$.

$$\therefore \quad I=\int\frac{a\sec\theta\tan\theta\,d\theta}{\sqrt{a^2\sec^2\theta-a^2}}=\int\frac{a\sec\theta\tan\theta\,d\theta}{a\sqrt{\sec^2\theta-1}}=\int\sec\theta\,d\theta=\log|\sec\theta+\tan\theta|+C_1$$

$$=\log\left|\frac{x}{a}+\sqrt{\frac{x^2}{a^2}-1}\right|+C_1=\log\left|\frac{x+\sqrt{x^2-a^2}}{a}\right|+C_1$$

$$=\log|x+\sqrt{x^2-a^2}|-\log|a|+C_1$$

$$\therefore \quad \int\frac{1}{\sqrt{x^2-a^2}}dx=\log|x+\sqrt{x^2-a^2}|+C,\ \text{where } C=-\log|a|+C_1.$$

3. Let $I=\int\frac{1}{\sqrt{a^2+x^2}}dx$. Let $x=a\tan\theta$. $\therefore$ $dx=a\sec^2\theta\,d\theta$.

$$\therefore \quad I=\int\frac{a\sec^2\theta\,d\theta}{\sqrt{a^2+a^2\tan^2\theta}}=\int\frac{a\sec^2\theta}{a\sec\theta}d\theta=\int\sec\theta\,d\theta=\log|\sec\theta+\tan\theta|+C_1$$

$$= \log \left| \sqrt{1 + \frac{x^2}{a^2}} + \frac{x}{a} \right| + C_1 = \log \mid x + \sqrt{a^2 + x^2} \mid - \log \mid a \mid + C_1$$

$$\left| \sec^2 \theta = 1 + \tan^2 \theta = 1 + \frac{x^2}{a^2} \right.$$

$$\therefore \quad \int \frac{1}{\sqrt{\mathbf{a}^2 + \mathbf{x}^2}} \mathbf{dx} = \mathbf{log} \mid \mathbf{x} + \sqrt{\mathbf{a}^2 + \mathbf{x}^2} \mid + \mathbf{C}, \text{ where } \mathbf{C} = -\mathbf{log} \mid \mathbf{a} \mid + \mathbf{C}_1.$$

ILLUSTRATIVE EXAMPLES

Example 1. *Evaluate :*

(*i*) $\int \frac{dx}{\sqrt{x^2 + 16}}$ (*ii*) $\int \frac{dx}{\sqrt{9x^2 - 7}}$ (*iii*) $\int \frac{dx}{\sqrt{15 - 8x^2}}$

Sol. (*i*) Let $I = \int \frac{dx}{\sqrt{x^2 + 16}}$. $\therefore \quad I = \int \frac{dx}{\sqrt{x^2 + (4)^2}}$.

$$\therefore \quad I = \log \mid x + \sqrt{x^2 + (4)^2} \mid + C = \mathbf{log} \mid \mathbf{x} + \sqrt{\mathbf{x}^2 + \mathbf{16}} \mid + \mathbf{C}.$$

(*ii*) Let $I = \int \frac{dx}{\sqrt{9x^2 - 7}}$. $\therefore \quad I = \frac{1}{3} \int \frac{dx}{\sqrt{x^2 - 7/9}} = \frac{1}{3} \int \frac{dx}{\sqrt{x^2 - (\sqrt{7}/3)^2}}$

$$\therefore \quad I = \frac{1}{3} \log \mid x + \sqrt{x^2 - (\sqrt{7}/3)^2} \mid + C$$

$$\left(\because \int \frac{dx}{\sqrt{x^2 - a^2}} = \log |x + \sqrt{x^2 - a^2}| + C \right)$$

$$= \frac{\mathbf{1}}{\mathbf{3}} \mathbf{log} \mid \mathbf{x} + \sqrt{\mathbf{x}^2 - \mathbf{7/9}} \mid + \mathbf{C}.$$

(*iii*) Let $I = \int \frac{dx}{\sqrt{15 - 8x^2}}$. $\therefore \quad I = \frac{1}{\sqrt{8}} \int \frac{dx}{\sqrt{15/8 - x^2}} = \frac{1}{2\sqrt{2}} \int \frac{dx}{\sqrt{(\sqrt{15/8})^2 - x^2}}$

$$\therefore \quad I = \frac{1}{2\sqrt{2}} \sin^{-1} \frac{x}{\sqrt{15/8}} + C = \frac{\mathbf{1}}{\mathbf{2}\sqrt{\mathbf{2}}} \mathbf{sin}^{-1} \sqrt{\frac{\mathbf{8}}{\mathbf{15}}}\, \mathbf{x} + \mathbf{C}.$$

$$\left(\text{Using} \int \frac{dx}{\sqrt{a^2 - x^2}} = \sin^{-1} \frac{x}{a} + C \right)$$

13.18. EVALUATION OF $\int \frac{1}{\sqrt{ax^2 + bx + c}} dx$ AND $\int \frac{px + q}{\sqrt{ax^2 + bx + c}} dx$

(*i*) **Evaluation of** $\int \frac{1}{\sqrt{\mathbf{ax}^2 + \mathbf{bx} + \mathbf{c}}}$ **dx.** This integral is evaluated by expressing its integrand either in the form $\frac{1}{\sqrt{A^2 - z^2}}$ or $\frac{1}{\sqrt{z^2 - A^2}}$ or $\frac{1}{\sqrt{A^2 + z^2}}$ by using suitable substitution.

WORKING RULES FOR EVALUATING $\int \frac{1}{\sqrt{ax^2 + bx + c}} dx$

Step I. *Take $\frac{1}{\sqrt{|a|}}$ out of the integral sign. For example, if $I = \int \frac{dx}{\sqrt{7 - 3x - 2x^2}}$ then write $I = \frac{1}{\sqrt{2}} \int \frac{dx}{\sqrt{\frac{7}{2} - \frac{3}{2}x - x^2}}$.*

Step II. *Express the terms containing x^2 and x in the form of a perfect square by adding and subtracting the square of half of the coefficient of x.*

Step III. *Take linear expression in x as z and express the integral in terms of z. This resultant integral will be in one of the standard forms.*

Step IV. *Use the standard formula and express the result in terms of x.*

Example 2. *Evaluate :*

(*i*) $\int \frac{dx}{\sqrt{7 - 6x - x^2}}$ (*ii*) $\int \frac{dx}{\sqrt{2x^2 + 3x - 2}}$ (*iii*) $\int \frac{dx}{\sqrt{(x-1)(x-2)}}$.

Sol. (*i*) Let $I = \int \frac{dx}{\sqrt{7 - 6x - x^2}}$.

$$\therefore \quad I = \int \frac{dx}{\sqrt{7 - (x^2 + 6x)}} = \int \frac{dx}{\sqrt{7 - (x^2 + 6x + 9) + 9}} = \int \frac{dx}{\sqrt{16 - (x+3)^2}}$$

Let $z = x + 3$. $\therefore$ $dz = dx$.

$$\therefore \quad I = \int \frac{dz}{\sqrt{16 - z^2}} = \sin^{-1} \frac{z}{4} + C = \mathbf{\sin^{-1} \frac{x+3}{4} + C.}$$

$\left| \left(\frac{1}{2} \text{coeff. of } x \right)^2 = 9 \right.$

(*ii*) Let $I = \frac{dx}{\sqrt{2x^2 + 3x - 2}}$.

$$\therefore \quad I = \frac{1}{\sqrt{2}} \int \frac{dx}{\sqrt{x^2 + \frac{3}{2}x - 1}} = \frac{1}{\sqrt{2}} \int \frac{dx}{\sqrt{x^2 + \frac{3}{2}x + \frac{9}{16} - \frac{9}{16} - 1}} = \frac{1}{\sqrt{2}} \int \frac{dx}{\sqrt{\left(x + \frac{3}{4}\right)^2 - \frac{25}{16}}}$$

Let $z = x + \frac{3}{4}$. $\therefore$ $dz = dx$

$\left| \left(\frac{1}{2} \text{coeff. of } x \right)^2 = \frac{9}{16} \right.$

$$\therefore \quad I = \frac{1}{\sqrt{2}} \int \frac{dz}{\sqrt{z^2 - (5/4)^2}} = \frac{1}{\sqrt{2}} \log |z + \sqrt{z^2 - (5/4)^2}| + C$$

$$= \mathbf{\frac{1}{\sqrt{2}} \log \left| x + \frac{3}{4} + \sqrt{x^2 + \frac{3}{2}x - 1} \right| + C.}$$

(*iii*) Let $I = \int \frac{dx}{\sqrt{(x-1)(x-2)}}$.

$$\therefore \quad I = \int \frac{dx}{\sqrt{x^2 - 3x + 2}} = \frac{dx}{\sqrt{x^2 - 3x + \frac{9}{4} - \frac{9}{4} + 2}} = \int \frac{dx}{\sqrt{\left(x - \frac{3}{2}\right)^2 - \left(\frac{1}{2}\right)^2}}$$

Let $z = x - \frac{3}{2}$. $\therefore$ $dz = dx$ $\quad \left| \left(\frac{1}{2} \text{ coeff. of } x\right)^2 = \frac{9}{4}\right.$

$$\therefore \quad I = \int \frac{dz}{\sqrt{z^2 - (1/2)^2}} = \log \mid z + \sqrt{z^2 - (1/2)^2} \mid + C$$

$$= \mathbf{\log \left| x - \frac{3}{2} + \sqrt{x^2 - 3x + 2} \right| + C.}$$

Example 3. *Evaluate :*

(*i*) $\int \frac{2x\,dx}{\sqrt{1 - x^2 - x^4}}$ (*ii*) $\int \frac{\cos x\,dx}{\sqrt{\sin^2 x - 2\sin x - 3}}$ (*iii*) $\int \frac{e^x\,dx}{\sqrt{5 - 4e^x - e^{2x}}}$.

Sol. (*i*) Let $I = \int \frac{2x\,dx}{\sqrt{1 - x^2 - x^4}}$. Let $z = x^2$. $\therefore$ $dz = 2x\,dx$.

$$\therefore \quad I = \int \frac{dz}{\sqrt{1 - z - z^2}} = \int \frac{dx}{\sqrt{1 - (z^2 + z)}} = \int \frac{dz}{\sqrt{1 - \left(z^2 + z + \frac{1}{4}\right) + \frac{1}{4}}} = \int \frac{dz}{\sqrt{\frac{5}{4} - \left(z + \frac{1}{2}\right)^2}}.$$

$\left| \left(\frac{1}{2} \text{ coeff. of } z\right)^2 = \frac{1}{4}\right.$

Let $t = z + \frac{1}{2}$. $\therefore$ $dt = dz$.

$$\therefore \quad I = \int \frac{dt}{\sqrt{(\sqrt{5}/2)^2 - t^2}} = \sin^{-1} \frac{t}{\sqrt{5}/2} + C = \sin^{-1} \frac{z + \frac{1}{2}}{\sqrt{5}/2} + C = \sin^{-1} \frac{2z + 1}{\sqrt{5}} + C$$

$$= \mathbf{\sin^{-1} \frac{2x^2 + 1}{\sqrt{5}} + C.}$$

(*ii*) Let $I = \int \frac{\cos x}{\sqrt{\sin^2 x - 2\sin x - 3}}\,dx$. Let $z = \sin x$. $\therefore$ $dz = \cos x\,dx$.

$$\therefore \quad I = \int \frac{dz}{\sqrt{z^2 - 2z - 3}} = \int \frac{dz}{\sqrt{(z^2 - 2z + 1) - 4}} = \int \frac{dz}{\sqrt{(z - 1)^2 - 4}}.$$

Let $t = z - 1$. $\therefore$ $dt = dz$.

$$\therefore \quad I = \int \frac{dt}{\sqrt{t^2 - (2)^2}} = \log \mid t + \sqrt{t^2 - (2)^2} \mid + C = \log \mid z - 1 + \sqrt{(z - 1)^2 - 4} \mid + C$$

$$= \log \mid z - 1 + \sqrt{z^2 - 2z - 3} \mid + C$$

$$= \mathbf{\log \mid \sin x - 1 + \sqrt{\sin^2 x - 2\sin x - 3} \mid + C.}$$

(*iii*) Let $I = \int \frac{e^x}{\sqrt{5 - 4e^x - e^{2x}}} dx$. Let $z = e^x$. $\therefore$ $dz = e^x \, dx$.

$$\therefore \quad I = \int \frac{dz}{\sqrt{5 - 4z - z^2}} = \int \frac{dz}{\sqrt{5 - (z^2 + 4z)}} = \int \frac{dz}{\sqrt{5 - (z^2 + 4z + 4) + 4}} = \int \frac{dz}{\sqrt{9 - (z + 2)^2}}.$$

Let $t = z + 2$. $\therefore$ $dt = dz$.

$$\therefore \quad I = \int \frac{dt}{\sqrt{3^2 - t^2}} = \sin^{-1} \frac{t}{3} + C = \sin^{-1} \frac{z + 2}{3} + C = \mathbf{\sin^{-1} \frac{e^x + 2}{3} + C.}$$

(*ii*) **Evaluation of** $\int \frac{\mathbf{px + q}}{\sqrt{\mathbf{ax^2 + bx + c}}} \mathbf{dx}$**.** This integral is evaluated by expressing the integrand in the form $\lambda \left(\frac{\frac{d}{dx}(ax^2 + bx + c)}{\sqrt{ax^2 + bx + c}} \right) + \mu \left(\frac{1}{\sqrt{ax^2 + bx + c}} \right)$ for some constants λ and μ. The first part is integrated by letting $ax^2 + bx + c$ equal to some other variable and the second part can be integrated by expressing it in the form of either $\frac{1}{\sqrt{A^2 - z^2}}$ or $\frac{1}{\sqrt{z^2 - A^2}}$ or $\frac{1}{\sqrt{A^2 + z^2}}$ by using suitable substitution.

WORKING RULES FOR EVALUATING $\int \frac{\mathbf{px + q}}{\sqrt{\mathbf{ax^2 + bx + c}}} \mathbf{dx}$

Step I. *Put* $px + q = \lambda \left(\frac{d}{dx}(ax^2 + bx + c) \right) + \mu$ *and find the values of* λ *and* μ *by comparing the coefficients of x and the constant terms.*

Step II. *Given integral* $= \lambda \int \frac{2ax + b}{\sqrt{ax^2 + bx + c}} dx + \mu \int \frac{dx}{\sqrt{ax^2 + bx + c}}$.

Step III. *Evaluate the first integral by putting* $z = ax^2 + bx + c$.

Step IV. *Evaluate the second integral by using the method discussed in part* (*i*).

Example 4. *Evaluate :*

(*i*) $\int \frac{6x + 7}{\sqrt{(x - 5)(x - 4)}} dx$ (*ii*) $\int \frac{2x + 3}{\sqrt{x^2 + x + 1}} dx$

(*iii*) $\int \frac{5 - 2x}{\sqrt{6 + x - x^2}} dx$ (*iv*) $\int \frac{3 - 2x}{\sqrt{(2 - x)(1 + x)}} dx.$

Sol. (*i*) Let $I = \int \frac{6x + 7}{\sqrt{(x - 5)(x - 4)}} dx$. $\therefore$ $I = \int \frac{6x + 7}{\sqrt{x^2 - 9x + 20}} dx$

Let $6x + 7 = \lambda \dfrac{d}{dx}(x^2 - 9x + 20) + \mu$. $\therefore$ $6x + 7 = \lambda(2x - 9) + \mu$

$\Rightarrow$ $6 = 2\lambda$ and $7 = -9\lambda + \mu$ $\Rightarrow$ $\lambda = 3$ and $\mu = 34$.

$\therefore$ $6x + 7 = 3(2x - 9) + 34$

$$\therefore \quad I = \int \frac{3(2x-9)+34}{\sqrt{x^2 - 9x + 20}}\,dx = 3\int \frac{2x-9}{\sqrt{x^2-9x+20}}\,dx + 34\int \frac{dx}{\sqrt{x^2-9x+20}}$$

$$= 3I_1 + 34I_2 \text{ (say)}$$

$$I_1 = \int \frac{2x-9}{\sqrt{x^2-9x+20}}\,dx. \qquad \text{Let } z = x^2 - 9x + 20. \quad \therefore \quad dz = 2x - 9.$$

$$\therefore \quad I_1 = \int \frac{dz}{\sqrt{z}} = \int z^{-1/2}\,dz = 2z^{1/2} + C_1 = 2\sqrt{x^2-9x+20} + C_1.$$

$$I_2 = \int \frac{dx}{\sqrt{x^2-9x+20}} = \int \frac{dx}{\sqrt{\left(x^2 - 9x + \frac{81}{4}\right) - \frac{81}{4} + 20}} \qquad \left|\ \left(\frac{1}{2}\text{ coeff. of } x\right)^2 = \frac{81}{4}\right.$$

$$= \int \frac{dx}{\sqrt{\left(x - \frac{9}{2}\right)^2 - \frac{1}{4}}} = \int \frac{dz}{\sqrt{z^2 - (\frac{1}{2})^2}}, \text{ where } z = x - \frac{9}{2}.$$

$$= \log\left| z + \sqrt{z^2 - (\tfrac{1}{2})^2}\right| + C_2$$

$$= \log\left| x - \frac{9}{2} + \sqrt{\left(x - \frac{9}{2}\right)^2 - \frac{1}{4}}\right| + C_2 = \log\left| x - \frac{9}{2} + \sqrt{x^2 - 9x + 20}\right| + C_2$$

$$\therefore \quad I = 3I_1 + 34I_2 = 3(2\sqrt{x^2-9x+20} + C_1) + 34\left(\log\left| x - \frac{9}{2} + \sqrt{x^2-9x+20}\right| + C_2\right)$$

$$= \mathbf{6\sqrt{x^2 - 9x + 20} + 34\log\left| x - \frac{9}{2} + \sqrt{x^2 - 9x + 20}\right| + C, \text{ where } C = 3C_1 + 34C_2.}$$

(*ii*) Let $I = \displaystyle\int \frac{2x+3}{\sqrt{x^2+x+1}}\,dx$.

Let $2x + 3 = \lambda \dfrac{d}{dx}(x^2 + x + 1) + \mu$. $\therefore$ $2x + 3 = \lambda(2x + 1) + \mu = 2\lambda x + (\lambda + \mu)$

$\therefore$ $2 = 2\lambda$ and $3 = \lambda + \mu$ $\therefore$ $\lambda = 1$ and $\mu = 3 - 1 = 2$.

$\therefore$ $2x + 3 = 1(2x + 1) + 2$

$$\therefore \quad I = \int \frac{2x+1}{\sqrt{x^2+x+1}}\,dx + 2 \cdot \int \frac{1}{\sqrt{x^2+x+1}}\,dx = I_1 + 2I_2 \text{ (say)}$$

$$I_1 = \int \frac{2x+1}{\sqrt{x^2+x+1}}\,dx. \quad \text{Let } z = x^2 + x + 1. \quad \therefore \quad dz = (2x+1)\,dx$$

$$\therefore \quad I_1 = \int \frac{dz}{\sqrt{z}} = \int z^{-1/2}\, dz = 2z^{1/2} + C_1 = 2\sqrt{x^2+x+1} + C_1.$$

$$I_2 = \int \frac{dx}{\sqrt{x^2+x+1}} = \int \frac{dx}{\sqrt{\left(x^2+x+\frac{1}{4}\right)+1-\frac{1}{4}}} = \int \frac{dx}{\sqrt{\left(x+\frac{1}{2}\right)^2+\frac{3}{4}}}$$

Let $z = x + \frac{1}{2}$. $\therefore$ $dz = dx$

$$\therefore \quad I_2 = \int \frac{dz}{\sqrt{z^2+(\sqrt{3}/2)^2}} = \log|z+\sqrt{z^2+(\sqrt{3}/2)^2}| + C_2 = \log\left|x+\frac{1}{2}+\sqrt{x^2+x+1}\right| + C_2$$

$$\therefore \quad I = I_1 + 2I_2 = (2\sqrt{x^2+x+1}+C_1) + 2\left(\log\left|x+\frac{1}{2}+\sqrt{x^2+x+1}\right| + C_2\right)$$

$$= \mathbf{2\sqrt{x^2+x+1} + 2\log\left|x+\frac{1}{2}+\sqrt{x^2+x+1}\right| + C, \text{ where } C = C_1 + 2C_2.}$$

(*iii*) Let $I = \int \frac{5-2x}{\sqrt{6+x-x^2}}\, dx.$

Let $5 - 2x = \lambda \frac{d}{dx}(6+x-x^2) + \mu.$ $\quad\therefore\quad 5 - 2x = \lambda(1-2x) + \mu$

$\Rightarrow \quad -2 = -2\lambda$ and $5 = \lambda + \mu \quad \Rightarrow \quad \lambda = 1$ and $\mu = 4$

$\therefore \quad 5 - 2x = 1(1-2x) + 4$

$$\therefore \quad I = \int \frac{(1-2x)+4}{\sqrt{6+x-x^2}}\, dx = \int \frac{1-2x}{\sqrt{6+x-x^2}}\, dx + 4\int \frac{dx}{\sqrt{6+x-x^2}} = I_1 + 4I_2 \text{ (say)}$$

$I_1 = \int \frac{1-2x}{\sqrt{6+x-x^2}}\, dx.$ Let $z = 6 + x - x^2$. $\therefore$ $dz = (1-2x)\, dx.$

$$\therefore \quad I_1 = \int \frac{1}{\sqrt{z}}\, dz = 2\sqrt{z} = 2\sqrt{6+x-x^2} + C_1.$$

$$I_2 = \int \frac{dx}{\sqrt{6+x-x^2}} = \int \frac{dx}{\sqrt{6-(x^2-x)}} = \int \frac{dx}{\sqrt{6-\left(x^2-x+\frac{1}{4}\right)+\frac{1}{4}}}$$

$$= \int \frac{dx}{\sqrt{\frac{25}{4}-\left(x-\frac{1}{2}\right)^2}} = \int \frac{dx}{\sqrt{(5/2)^2-z^2}}, \text{ where } z = x - \frac{1}{2}.$$

$$= \sin^{-1}\frac{z}{5/2} + C_2 = \sin^{-1}\frac{2}{5}\left(x-\frac{1}{2}\right) + C_2 = \sin^{-1}\frac{2x-1}{5} + C_2.$$

$\therefore \quad I = I_1 + 4I_2 = (2\sqrt{6+x-x^2} + C_1) + 4\left(\sin^{-1}\frac{2x-1}{5} + C_2\right)$

$= \mathbf{2\sqrt{6+x-x^2} + 4\sin^{-1}\frac{2x-1}{5} + C}$, **where** $\mathbf{C = C_1 + 4C_2}$.

(*iv*) Let $I = \int \frac{3-2x}{\sqrt{(2-x)(1+x)}}\,dx$. $\quad \therefore \quad I = \int \frac{3-2x}{\sqrt{2+x-x^2}}\,dx$

Let $3 - 2x = \lambda \frac{d}{dx}(2 + x - x^2) + \mu$. $\quad \therefore \quad 3 - 2x = \lambda(1 - 2x) + \mu$.

$\Rightarrow \quad -2 = -2\lambda$ and $3 = \lambda + \mu \quad \Rightarrow \quad \lambda = 1$ and $\mu = 2$.

$\therefore \quad 3 - 2x = 1(1 - 2x) + 2$

$\therefore \quad I = \int \frac{(1-2x)+2}{\sqrt{2+x-x^2}}\,dx = \int \frac{1-2x}{\sqrt{2+x-x^2}}\,dx + 2\int \frac{dx}{\sqrt{2+x-x^2}} = I_1 + 2I_2$ (say)

$I_1 = \int \frac{1-2x}{\sqrt{2+x-x^2}}\,dx$. Let $z = 2 + x - x^2$. $\therefore \quad dz = (1 - 2x)\,dx$

$\therefore \quad I_1 = \int \frac{dz}{\sqrt{z}} = 2\sqrt{z} + C_1 = 2\sqrt{2+x-x^2} + C_1$.

$I_2 = \int \frac{dx}{\sqrt{2+x-x^2}} = \int \frac{dx}{\sqrt{2-(x^2-x)}} = \int \frac{dx}{\sqrt{2-\left(x^2-x+\frac{1}{4}\right)+\frac{1}{4}}}$

$= \int \frac{dx}{\sqrt{\frac{9}{4}-\left(x-\frac{1}{2}\right)^2}} = \int \frac{dx}{\sqrt{(3/2)^2 - z^2}}$, where $z = x - \frac{1}{2}$.

$= \sin^{-1}\frac{z}{3/2} + C_2 = \sin^{-1}\frac{2}{3}\left(x - \frac{1}{2}\right) + C_2 = \sin^{-1}\frac{2x-1}{3} = C_2$.

$\therefore \quad I = I_1 + 2I_2 = 2\sqrt{2+x-x^2} + C_1 + 2\left(\sin^{-1}\frac{2x-1}{3} + C_2\right)$

$= \mathbf{2\sqrt{2+x-x^2} + 2\sin^{-1}\frac{2x-1}{3} + C}$, **where** $\mathbf{C = C_1 + 2C_2}$.

EXERCISE 13.6

Evaluate the following integrals (Q. No. 1–6) :

1. (*i*) $\int \frac{dx}{\sqrt{4x^2-9}}$ (*ii*) $\int \frac{dx}{\sqrt{(2-x)^2+1}}$ (*iii*) $\int \frac{dx}{\sqrt{9-25x^2}}$

(iv) $\int \frac{dx}{\sqrt{1+4x^2}}$ (v) $\int \frac{dx}{\sqrt{a^2+b^2x^2}}$; $a, b > 0$ (vi) $\int \frac{dx}{\sqrt{a^2-b^2x^2}}$; $a, b > 0$

2. (i) $\int \frac{\sec^2 x\, dx}{\sqrt{4+\tan^2 x}}$ (ii) $\int \frac{x\, dx}{\sqrt{9-x^4}}$ (iii) $\int \frac{\sqrt{x}}{\sqrt{a^3-x^3}} dx$

3. (i) $\int \frac{dx}{\sqrt{2x-x^2}}$ (ii) $\int \frac{dx}{\sqrt{7-6x-x^2}}$ (iii) $\int \frac{dx}{\sqrt{x^2+12x+11}}$

(iv) $\int \frac{dx}{\sqrt{2-4x+x^2}}$ (v) $\int \frac{dx}{\sqrt{2x^2+4x+6}}$ (vi) $\int \frac{dx}{\sqrt{3x^2+6x+12}}$

4. (i) $\int \frac{dx}{\sqrt{x(1-2x)}}$ (ii) $\int \frac{dx}{\sqrt{8+x-x^2}}$ (iii) $\int \frac{1}{\sqrt{9+8x-x^2}} dx$

(iv) $\int \frac{dx}{\sqrt{8+3x-x^2}}$ (v) $\int \frac{dx}{\sqrt{x^2+x+1}}$ (vi) $\int \frac{dx}{\sqrt{4x-3x^2+5}}$

5. (i) $\int \frac{dx}{\sqrt{3+4x-2x^2}}$ (ii) $\int \frac{dx}{\sqrt{2+2x-x^2}}$ (iii) $\int \frac{dx}{\sqrt{4+2x-x^2}}$

(iv) $\int \frac{dx}{\sqrt{x}\sqrt{5-x}}$ (v) $\int \frac{dx}{\sqrt{16-2x-2x^2}}$ (vi) $\int \frac{dx}{\sqrt{4-2x-2x^2}}$.

6. (i) $\int \frac{6x+5}{\sqrt{6+x-2x^2}} dx$ (ii) $\int \frac{x}{\sqrt{8+x-x^2}} dx$ (iii) $\int \frac{2x+3}{\sqrt{(x+1)(x+3)}} dx$

(iv) $\int \frac{2x+1}{\sqrt{x^2+4x+3}} dx$ (v) $\int \frac{x}{\sqrt{x^2+x+1}} dx$ (vi) $\int \frac{x+2}{\sqrt{x^2+2x-1}} dx$.

Answers

1. (i) $\frac{1}{2}\log\left| x+\sqrt{x^2-\frac{9}{4}} \right| + C$ (ii) $-\log|2-x+\sqrt{x^2-4x+5}| + C$

(iii) $\frac{1}{5}\sin^{-1}\frac{5x}{3} + C$ (iv) $\frac{1}{2}\log|2x+\sqrt{1+4x^2}| + C$

(v) $\frac{1}{b}\log\left| x+\sqrt{\frac{a^2}{b^2}+x^2} \right| + C$ (vi) $\frac{1}{b}\sin^{-1}\frac{bx}{a} + C$

2. (i) $\log|\tan x+\sqrt{4+\tan^2 x}| + C$ (ii) $\frac{1}{2}\sin^{-1}\frac{x^2}{3} + C$ (iii) $\frac{2}{3}\sin^{-1}\left(\frac{x}{a}\right)^{3/2} + C$

3. (i) $\sin^{-1}(x-1) + C$ (ii) $\sin^{-1}\frac{x+3}{4} + C$ (iii) $\log|x+6+\sqrt{x^2+12x+11}| + C$

(iv) $\log|x-2+\sqrt{x^2-4x+2}| + C$ (v) $\frac{1}{\sqrt{2}}\log(x+1+\sqrt{x^2+2x+3}) + C$

(vi) $\frac{1}{\sqrt{3}}\log(x+1+\sqrt{x^2+2x+4}) + C$

4. (*i*) $\frac{1}{\sqrt{2}}\sin^{-1}(4x-1)+C$ (*ii*) $\sin^{-1}\frac{2x-1}{\sqrt{33}}+C$ (*iii*) $\sin^{-1}\frac{x-4}{5}+C$

(*iv*) $\sin^{-1}\frac{2x-5}{\sqrt{41}}+C$ (*v*) $\log\left|x+\frac{1}{2}+\sqrt{x^2+x+1}\right|+C$

(*vi*) $\frac{1}{\sqrt{3}}\sin^{-1}\frac{3x-2}{\sqrt{19}}+C$

5. (*i*) $\frac{1}{\sqrt{2}}\sin^{-1}\frac{\sqrt{2}(x-1)}{\sqrt{5}}+C$ (*ii*) $\sin^{-1}\frac{x-1}{\sqrt{3}}+C$ (*iii*) $\sin^{-1}\frac{x-1}{\sqrt{5}}+C$

(*iv*) $\sin^{-1}\frac{2x-5}{5}+C$ (*v*) $\frac{1}{\sqrt{2}}\sin^{-1}\frac{2x+1}{\sqrt{33}}+C$ (*vi*) $\frac{1}{\sqrt{2}}\sin^{-1}\frac{2x+1}{3}+C$

6. (*i*) $-3\sqrt{6+x-2x^2}+\frac{13}{2\sqrt{2}}\sin^{-1}\frac{4x-1}{7}+C$ (*ii*) $-\sqrt{8+x-x^2}+\frac{1}{2}\sin^{-1}\frac{2x-1}{\sqrt{33}}+C$

(*iii*) $2\sqrt{x^2+4x+3}-\log\left|x+2+\sqrt{x^2+4x+3}\right|+C$

(*iv*) $2\sqrt{x^2+4x+3}-3\log\left|x+2+\sqrt{x^2+4x+3}\right|+C$

(*v*) $\sqrt{x^2+x+1}-\frac{1}{2}\log\left|x+\frac{1}{2}+\sqrt{x^2+x+1}\right|+C$

(*vi*) $\sqrt{x^2+2x-1}+\log\left|x+1+\sqrt{x^2+2x-1}\right|+C.$

Hints

1. (*i*) Take $z=2x$. **2.** (*i*) Take $z=\tan x$ (*ii*) Take $z=x^2$

(*iii*) Take $z=x^{3/2}$. **3.** (*i*) $2x-x^2=-(x^2-2x)=-(x^2-2x+1)+1=1-(x-1)^2$.

5. (*iv*) $\sqrt{x}\sqrt{5-x}=\sqrt{-(x^2-5x)}=\sqrt{-\left(x^2-5x+\frac{25}{4}\right)+\frac{25}{4}}=\sqrt{\frac{25}{4}-\left(x-\frac{5}{2}\right)^2}$.

INTEGRATION BY PARTS

13.19. INTRODUCTION

In this topic, we shall learn the method of integration when the integrand is a product of two functions.

13.20. METHOD OF INTEGRATION BY PARTS

Let $u=f(x)$ and $w=g(x)$ be two differentiable functions of x.

$$\therefore \quad \frac{d}{dx}(uw)=u\frac{dw}{dx}+w\frac{du}{dx} \quad \text{or} \quad u\frac{dw}{dx}=\frac{d}{dx}(uw)-w\frac{du}{dx}.$$

$$\therefore \quad \int\left(u\frac{dw}{dx}\right)dx=\int\left(\frac{d}{dx}(uw)\right)dx-\int\left(w\frac{du}{dx}\right)dx \qquad \text{...(1)}$$

$$\frac{dw}{dx} = v \quad \Rightarrow \quad w = \int v\, dx.$$

$$\therefore \quad (1) \quad \Rightarrow \quad \int uv\, dx = \int \left(\frac{d}{dx}\left(u \int v\, dx \right) \right) dx - \int \left(\frac{du}{dx} \int v\, dx \right) dx$$

$$\therefore \quad \int \mathbf{uv\, dx} = \mathbf{u} \int \mathbf{v\, dx} - \int \left(\frac{\mathbf{du}}{\mathbf{dx}} \int \mathbf{v\, dx} \right) \mathbf{dx}.$$

In words : Integral of the product of two functions
= 1st function × Integral of 2nd – Integral of
[Diff. coeff. of 1st × Integral of 2nd].

From the above formula, it is clear that the success of the method of **integration by parts** depends on the choice of **first function.** For the choice of first function, the following rules are used :

Rule (*i*) If the integrand contains both functions integrable, take that function as the first which can be finished by repeated differentiation and the other as the second.

For example, in $\int x^2 \sin x\, dx$ or in $\int x^2 e^x\, dx$, take x^2 as the first function and the other as the second.

Rule (*ii*) If the integrand contains one unintegrable, *e.g.*, inverse circular function or a logarithmic function, take unintegrable as the first function and other as second. For example, in $\int x^2 \log x\, dx$, $\int x \sin^{-1} x\, dx$, take $\log x$, $\sin^{-1} x$ as the first function and other as the second respectively.

Rule (*iii*) If the integrand contains only one unintegrable function, then take the given function as the first function and unity as the second. For example, in $\int \tan^{-1} x\, dx$, $\int \log (x^2 + a^2)\, dx$, take $\tan^{-1} x$, $\log (x^2 + a^2)$ as the first function and unity as the second function respectively.

Rule (*iv*) If the integrand contains both the functions integrable and none can be finished by repeated differentiation, then take any one as the first function and other as the second one. Repeat the rule of **integration by parts.**

Remark 1. If the integrand is the product of two functions of different types then their order is determined by the word **ILATE** where the letters **I, L, A, T** and **E** stands respectively for inverse circular function, logarithmic function, algebraic function, trigonometric function and exponential function. In the integrand, that function is taken as the first function which comes first in the word **ILATE.**

ILLUSTRATIVE EXAMPLES

Example 1. *Evaluate :*

(*i*) $\int x \sin x\, dx$ (*ii*) $\int \frac{x}{\cos^2 x}\, dx$ (*iii*) $\int x \cos^2 x\, dx$

(*iv*) $\int x \sin^3 x\, dx$ (*v*) $\int (1 - x^2) \sin 2x\, dx$ (*vi*) $\int x^2 \cos^2 x\, dx.$

Sol. (*i*) Let $I = \int \underset{\text{I}}{x} \; \underset{\text{II}}{\sin x}\, dx.$. [Using **Rule** (*i*)]

Integrating by parts, we get

$$I = x \int \sin x\, dx - \int \left[\frac{d}{dx}(x) . \int \sin x\, dx \right] dx = x . (-\cos x) - \int (1 . -\cos x)\, dx$$

$$= -x \cos x + \int \cos x\, dx = \mathbf{-x \cos x + \sin x + C.}$$

(*ii*) Let $I = \int \frac{x}{\cos^2 x}\, dx$. $\therefore$ $I = \int x \sec^2 x\, dx$

Integrating by parts, we get

$$I = x\int \sec^2 x\, dx - \int \left[\frac{d}{dx}(x) . \int \sec^2 x\, dx\right] dx$$

$$= x \tan x - \int 1 . \tan x\, dx = \mathbf{x \tan x - \log | \sec x | + C.}$$

(*iii*) Let $I = \int x \cos^2 x\, dx$.

$$\therefore \quad I = \int x\left(\frac{1+\cos 2x}{2}\right) dx = \frac{1}{2}\int x\, dx + \frac{1}{2}\int \underset{\text{I}}{x} \underset{\text{II}}{\cos 2x}\, dx \qquad \text{[Using **Rule** (i)]}$$

$$= \frac{1}{2} . \frac{x^2}{2} + \frac{1}{2}\left[x\int \cos 2x\, dx - \int\left[\frac{d}{dx}(x)\int \cos 2x\, dx\right] dx\right]$$

$$= \frac{x^2}{4} + \frac{1}{2}\left[x . \frac{\sin 2x}{2} - \int \frac{1 . \sin 2x}{2}\, dx\right] = \frac{x^2}{4} + \frac{1}{4} x \sin 2x - \frac{1}{4} . \frac{-\cos 2x}{2} + C$$

$$= \mathbf{\frac{x^2}{4} + \frac{1}{4} x \sin 2x + \frac{1}{8}\cos 2x + C.}$$

(*iv*) Let $I = \int x \sin^3 x\, dx$.

We have $\sin 3x = 3 \sin x - 4 \sin^3 x$. $\therefore$ $\sin^3 x = \frac{3}{4}\sin x - \frac{1}{4}\sin 3x$.

$$\therefore \quad I = \int x\left[\frac{3}{4}\sin x - \frac{1}{4}\sin 3x\right] dx = \frac{3}{4}\int \underset{\text{I}}{x} \underset{\text{II}}{\sin x}\, dx - \frac{1}{4}\int \underset{\text{I}}{x} \underset{\text{II}}{\sin 3x}\, dx \qquad \text{[Using **Rule** (i)]}$$

$$= \frac{3}{4}\left[x(-\cos x) - \int 1(-\cos x)\, dx\right] - \frac{1}{4}\left[x\left(-\frac{\cos 3x}{3}\right) - \int 1\left(-\frac{\cos 3x}{3}\right) dx\right]$$

$$= \mathbf{\frac{3}{4}[-x \cos x + \sin x] - \frac{1}{4}\left[-\frac{x \cos 3x}{3} + \frac{\sin 3x}{9}\right] + C.}$$

(*v*) Let $I = \int \underset{\text{I}}{(1 - x^2)} \underset{\text{II}}{\sin 2x}\, dx$. Integrating by parts, we get

$$I = (1 - x^2)\left(\frac{-\cos 2x}{2}\right) - \int (-2x)\left(\frac{-\cos 2x}{2}\right) dx = \frac{x^2 - 1}{2}\cos 2x - \int x \cos 2x\, dx$$

$$= \frac{x^2 - 1}{2}\cos 2x - \left[x\frac{\sin 2x}{2} - \int 1 . \frac{\sin 2x}{2}\, dx\right]$$

$$= \frac{x^2 - 1}{2}\cos 2x - \frac{x}{2}\sin 2x + \frac{1}{2}\left(-\frac{\cos 2x}{2}\right) = \frac{x^2 - 1}{2}\cos 2x - \frac{x}{2}\sin 2x - \frac{1}{4}\cos 2x$$

$$= \mathbf{\frac{2x^2 - 3}{4}\cos 2x - \frac{x}{2}\sin 2x + C.}$$

(*vi*) Let $I = \int x^2 \cos^2 x\, dx.$

$$\therefore \quad I = \int x^2 \left(\frac{1+\cos 2x}{2}\right) dx = \frac{1}{2}\int x^2\, dx + \frac{1}{2}\int x^2 \cos 2x\, dx$$

$$= \frac{1}{2}\left(\frac{x^3}{3}\right) + \frac{1}{2}\left[x^2 \cdot \frac{\sin 2x}{2} - \int 2x \cdot \frac{\sin 2x}{2}\, dx\right]$$

$$= \frac{x^3}{6} + \frac{x^2}{4}\sin 2x - \frac{1}{2}\int x \sin 2x\, dx$$

$$= \frac{x^3}{6} + \frac{x^2}{4}\sin 2x - \frac{1}{2}\left[x\left(-\frac{\cos 2x}{2}\right) - \int 1.\left(\frac{-\cos 2x}{2}\right) dx\right]$$

$$= \frac{x^3}{6} + \frac{x^2}{2}\sin 2x + \frac{x}{4}\cos 2x - \frac{1}{4}\int \cos 2x\, dx$$

$$= \frac{x^3}{6} + \frac{x^2}{4}\sin 2x + \frac{x}{4}\cos 2x - \frac{\sin 2x}{8}$$

$$= \mathbf{\frac{x^3}{6} + \frac{2x^2 - 1}{8}\sin 2x + \frac{x}{4}\cos 2x + C.}$$

Example 2. *Evaluate :*

(*i*) $\int \log x\, dx$ (*ii*) $\int \log (1 + x^2)\, dx$

(*iii*) $\int x^n \log x\, dx$ (*iv*) $\int x \log (1 + x)\, dx$.

Sol. (*i*) $\int \log x\, dx = \int \underset{\text{I}}{\log x} \,.\, \underset{\text{II}}{1}\, dx = \log x \,.\, x - \int \frac{1}{x}.x\, dx$ [Using **Rule** (*iii*)]

$= \mathbf{x \log x - x + C.}$

(*ii*) $\int \log (1 + x^2)\, dx = \int \underset{\text{I}}{\log (1 + x^2)} \,.\, \underset{\text{II}}{1}\, dx$ [Using **Rule** (*iii*)]

$$= \log (1 + x^2) \,.\, x - \int \frac{2x}{1+x^2} \,.\, x\, dx = x \log (1 + x^2) - 2\int \frac{1+x^2-1}{1+x^2}\, dx$$

$$= x \log (1 + x^2) - 2\int \left(1 - \frac{1}{1+x^2}\right) dx$$

$= x \log (1 + x^2) - 2\, [x - \tan^{-1} x] + C$

$= \mathbf{x \log (1 + x^2) - 2x + 2 \tan^{-1} x + C.}$

(*iii*) $\int x^n \log x\, dx = \int \underset{\text{I}}{\log x} \,.\, \underset{\text{II}}{x^n}\, dx$ [Using **Rule** (*ii*)]

$$= \log x \,.\, \frac{x^{n+1}}{n+1} - \int \frac{1}{x} \,.\, \frac{x^{n+1}}{n+1}\, dx = \frac{1}{n+1} x^{n+1} \log x - \frac{1}{n+1}\int x^n \,.\, dx$$

$$= \frac{1}{n+1} x^{n+1} \log x - \frac{1}{n+1} \,.\, \frac{x^{n+1}}{n+1} + C = \mathbf{\frac{1}{n+1} x^{n+1} \log x - \frac{x^{n+1}}{(n+1)^2} + C.}$$

(*iv*) Let $\quad I = \int x \log (1+x)\, dx. \quad \therefore \quad I = \int \underset{I}{\log (1+x)} \,.\, \underset{II}{x}\, dx$ [Using **Rule** (*ii*)]

Integrating by parts, we get

$$I = \log (1+x) \,.\, \frac{x^2}{2} - \int \left(\frac{1}{1+x} \,.\, 1\right) \frac{x^2}{2}\, dx$$

$$= \frac{1}{2}x^2 \log (1+x) - \frac{1}{2}\int \frac{x^2}{1+x}\, dx = \frac{1}{2}x^2 \log (1+x) - \frac{1}{2}\int \frac{(x^2-1)+1}{1+x}\, dx$$

$$= \frac{1}{2}x^2 \log (1+x) - \frac{1}{2}\int \left((x-1) + \frac{1}{1+x}\right) dx = \frac{1}{2}x^2 \log (1+x) - \frac{1}{2}\left[\frac{x^2}{2} - x + \log (1+x)\right] + C$$

$$= \mathbf{\frac{1}{2}(x^2 - 1) \log (1+x) - \frac{1}{4}x^2 + \frac{1}{2}x + C.}$$

Example 3. *Evaluate :*

(*i*) $\int xe^x\, dx$ (*ii*) $\int x^2 e^{-3x}\, dx$ (*iii*) $\int x^3 e^x\, dx.$

Sol. (*i*) $\int \underset{I\,II}{xe^x}\, dx = xe^x - \int 1 \,.\, e^x\, dx$ [Using **Rule** (*i*)]

$$= xe^x - e^x + C = \mathbf{(x-1)e^x + C.}$$

(*ii*) $\int \underset{I}{x^2}\, \underset{II}{e^{-3x}}\, dx = x^2 \,.\left(\frac{e^{-3x}}{-3}\right) - \int 2x \,.\left(\frac{e^{-3x}}{-3}\right) dx = -\frac{x^2 e^{-3x}}{3} + \frac{2}{3}\int \underset{I}{x}\, \underset{II}{e^{-3x}}\, dx$

[Using **Rule** (*i*)]

$$= -\frac{x^2 e^{-3x}}{3} + \frac{2}{3}\left[x \,.\left(\frac{e^{-3x}}{-3}\right) - \int 1 \,.\left(\frac{e^{-3x}}{-3}\right) dx\right] = -\frac{x^2 e^{-3x}}{3} - \frac{2}{9}x e^{-3x} + \frac{2}{9}\int e^{-3x}\, dx$$

$$= -\frac{x^2 e^{-3x}}{3} - \frac{2}{9}x e^{-3x} + \frac{2}{9} \,.\left(\frac{e^{-3x}}{-3}\right) + C = \mathbf{-\frac{e^{-3x}}{3}\left[x^2 + \frac{2}{3}x + \frac{2}{9}\right] + C.}$$

(*iii*) $\int \underset{I}{x^3}\, \underset{II}{e^x}\, dx = x^3 e^x - \int 3x^2 \,.\, e^x\, dx$ [Using **Rule** (*i*)]

$$= x^3 e^x - 3\int \underset{I}{x^2}\, \underset{II}{e^x}\, dx = x^3 e^x - 3\left[x^2 e^x - \int 2x . e^x\, dx\right]$$

$$= x^3 e^x - 3x^2 e^x + 6\left[\int \underset{I}{x}\, \underset{II}{e^x}\, dx\right] = x^3 e^x - 3x^2 e^x + 6\left[xe^x - \int 1 \,.\, e^x\, dx\right]$$

$$= x^3 e^x - 3x^2 e^x + 6xe^x - 6e^x + C = \mathbf{e^x (x^3 - 3x^2 + 6x - 6) + C.}$$

Example 4. *Evaluate :*

(*i*) $\int \sin^{-1} x\, dx$ (*ii*) $\int \tan^{-1} x\, dx$ (*iii*) $\int \sec^{-1} x\, dx.$

Sol. (*i*) Let $\quad I = \int \sin^{-1} x\, dx.$ Let $x = \sin \theta. \quad \therefore \quad dx = \cos \theta\, d\theta.$

$\therefore \quad I = \int \underset{I}{\theta} \,.\, \underset{II}{\cos \theta}\, d\theta = \theta \sin \theta - \int 1 \,.\, \sin \theta\, d\theta.$ [Using **Rule** (*i*)]

$$= \theta \sin \theta + \cos \theta = \theta \sin \theta + \sqrt{1 - \sin^2 \theta} = \mathbf{x \sin^{-1} x + \sqrt{1 - x^2} + C.}$$

(*ii*) Let $\quad I = \int \tan^{-1} x\, dx.$ Let $x = \tan \theta. \quad \therefore \quad dx = \sec^2 \theta\, d\theta.$

$\therefore \quad I = \int \underset{I}{\theta}\, \underset{II}{\sec^2 \theta}\, d\theta = \theta \tan \theta - \int 1 \,.\, \tan \theta\, d\theta$

$= \theta \tan\theta - \log|\sec\theta| = \theta\tan\theta - \log\sqrt{1+\tan^2\theta}$

$= \mathbf{x \tan^{-1} x - \tfrac{1}{2}\log(1+x^2) + C.}$

(*iii*) Let $\quad I = \int \sec^{-1} x\, dx. \quad \therefore \quad I = \int \underset{I}{\theta} \,.\, \underset{II}{\sec\theta} \tan\theta\, d\theta$, where $x = \sec\theta$.

$= \theta\sec\theta - \int 1 \,.\, \sec\theta\, d\theta = \theta\sec\theta - \log|\sec\theta + \tan\theta|$

$= \theta\sec\theta - \log|\sec\theta + \sqrt{\sec^2\theta - 1}| = \mathbf{x \sec^{-1} x - \log|x + \sqrt{x^2-1}| + C.}$

Example 5. *Evaluate :*

(*i*) $\int x \cos^{-1} x\, dx$ (*ii*) $\int x \cot^{-1} x\, dx$ (*iii*) $\int x \operatorname{cosec}^{-1} x\, dx.$

Sol. (*i*) Let $\quad I = \int x\cos^{-1} x\, dx. \quad \therefore \quad I = \int \underset{I}{\cos^{-1} x} \,.\, \underset{II}{x}\, dx$

Integrating by parts, we get $\quad I = \cos^{-1} x \,.\, \dfrac{x^2}{2} - \int \dfrac{-1}{\sqrt{1-x^2}} \,.\, \dfrac{x^2}{2} dx$

$\therefore \quad I = \dfrac{x^2}{2}\cos^{-1} x + \dfrac{1}{2}\int \dfrac{x^2}{\sqrt{1-x^2}} dx \quad ...(1)$

Let $\quad I_1 = \int \dfrac{x^2}{\sqrt{1-x^2}} dx. \quad$ Let $x = \cos\theta. \quad \therefore \quad dx = -\sin\theta\, d\theta$

$\therefore \quad I_1 = \int \dfrac{\cos^2\theta}{\sqrt{1-\cos^2\theta}} \,.\, (-\sin\theta)\, d\theta = -\int \cos^2\theta\, d\theta$

$= -\int \dfrac{1+\cos 2\theta}{2} d\theta = -\dfrac{1}{2}\theta - \dfrac{1}{2} \,.\, \dfrac{\sin 2\theta}{2} = -\dfrac{1}{2}\theta - \dfrac{\sin\theta\cos\theta}{2}$

$= -\dfrac{1}{2}\theta - \dfrac{\cos\theta\sqrt{1-\cos^2\theta}}{2} = -\dfrac{1}{2}\cos^{-1} x - \dfrac{x\sqrt{1-x^2}}{2}.$

Putting the value of I_1 in (1), we get

$$I = \frac{x^2}{2}\cos^{-1} x + \frac{1}{2}\left[-\frac{1}{2}\cos^{-1} x - \frac{x\sqrt{1-x^2}}{2}\right]$$

$$= \mathbf{\frac{x^2}{2}\cos^{-1} x - \frac{1}{4}\cos^{-1} x - \frac{1}{4} x\sqrt{1-x^2} + C.}$$

Alternative method

Let $\quad I = \int x\cos^{-1} x\, dx. \quad$ Let $x = \cos\theta. \quad \therefore \quad dx = -\sin\theta\, d\theta$

$\therefore \quad I = \int \cos\theta \,.\, \theta\,(-\sin\theta\, d\theta) = -\dfrac{1}{2}\int \underset{I}{\theta} \underset{II}{\sin 2\theta}\, d\theta$

Integrating by parts, we get, $I = -\dfrac{1}{2}\left[\theta \,.\, \left(\dfrac{-\cos 2\theta}{2}\right) - \int 1 \,.\, \left(-\dfrac{\cos 2\theta}{2}\right) d\theta\right]$

$$= \frac{\theta\cos 2\theta}{4} - \frac{1}{4}\int \cos 2\theta\, d\theta = \frac{\theta\cos 2\theta}{4} - \frac{1}{4} \,.\, \frac{\sin 2\theta}{2} + C$$

$$= \frac{\theta(2\cos^2\theta - 1)}{4} - \frac{1}{8} \cdot 2\sin\theta\cos\theta + C = \frac{\theta\cos^2\theta}{2} - \frac{\theta}{4} - \frac{\cos\theta\sqrt{1-\cos^2\theta}}{4} + C$$

$$= \frac{\mathbf{(\cos^{-1} x)x^2}}{\mathbf{2}} - \frac{\mathbf{\cos^{-1} x}}{\mathbf{4}} - \frac{\mathbf{x\sqrt{1-x^2}}}{\mathbf{4}} + \mathbf{C.}$$

(*ii*) Let $\quad I = \int x \cot^{-1} x \, dx. \quad \therefore \quad I = \int \underset{I}{\cot^{-1} x} \, . \, \underset{II}{x} \, dx$ [Using **Rule** (*iii*)]

Integrating by parts, we get

$$I = \cot^{-1} x \, . \, \frac{x^2}{2} - \int \frac{-1}{1+x^2} \cdot \frac{x^2}{2} \, dx = \frac{x^2 \cot^{-1} x}{2} + \frac{1}{2} \int \frac{x^2}{1+x^2} \, dx$$

$$= \frac{x^2 \cot^{-1} x}{2} + \frac{1}{2} \int \left(1 - \frac{1}{1+x^2}\right) dx = \frac{x^2 \cot^{-1} x}{2} + \frac{1}{2} [x - \tan^{-1} x]$$

$$\left| \because \frac{x^2}{1+x^2} = \frac{1+x^2-1}{1+x^2} = 1 - \frac{1}{1+x^2} \right.$$

$$= \frac{x^2 \cot^{-1} x}{2} + \frac{1}{2} x - \frac{1}{2}\left(\frac{\pi}{2} - \cot^{-1} x\right) = \frac{(x^2+1)\cot^{-1} x}{2} + \frac{1}{2} x - \frac{\pi}{4} + C$$

$$\left| \because \tan^{-1} x + \cot^{-1} x = \frac{\pi}{2} \right.$$

$$= \frac{\mathbf{(x^2+1)\cot^{-1} x}}{\mathbf{2}} + \frac{\mathbf{1}}{\mathbf{2}} \mathbf{x} + \mathbf{C_1}, \textbf{where } \mathbf{C_1 = C - \frac{\pi}{4}}.$$

(*iii*) Let $\quad I = \int x \operatorname{cosec}^{-1} x \, dx. \quad \therefore \quad I = \int \underset{I}{\operatorname{cosec}^{-1} x} \, . \, \underset{II}{x} \, dx$

Integrating by parts, we get

$$I = \operatorname{cosec}^{-1} x \, . \, \frac{x^2}{2} - \int \frac{-1}{x\sqrt{x^2-1}} \cdot \frac{x^2}{2} \, dx$$

$$\therefore \quad I = \frac{x^2 \operatorname{cosec}^{-1} x}{2} + \frac{1}{2} \int \frac{x}{\sqrt{x^2-1}} \, dx \quad \ldots(1)$$

Let $\quad I_1 = \int \frac{x}{\sqrt{x^2-1}} \, dx. \quad$ Let $z = x^2 - 1. \quad \therefore \quad dz = 2x \, dx$

$$\therefore \quad I_1 = \int \frac{1}{\sqrt{z}} \cdot \frac{dz}{2} = \frac{1}{2} \cdot \frac{z^{1/2}}{1/2} = \sqrt{z} = \sqrt{x^2-1}$$

$\therefore$ Putting the value of I_1 in (1), we get

$$\mathbf{I} = \frac{\mathbf{x^2 \operatorname{cosec}^{-1} x}}{\mathbf{2}} + \frac{\mathbf{1}}{\mathbf{2}} \sqrt{\mathbf{x^2-1}} + \mathbf{C.}$$

Example 6. *Evaluate :*

(*i*) $\int e^x \sin x \, dx$ (*ii*) $\int e^{2x} \sin 3x \, dx$

(*iii*) $\int e^{-x} \cos 4x \cos 2x \, dx$ (*iv*) $\int e^{ax} \sin (bx + c) \, dx$.

Sol. (*i*) Let $\quad I = \int \underset{I}{e^x} \underset{II}{\sin x} \, dx.$ [Using **Rule** (*iv*)]

$$\therefore \quad I = e^x . (-\cos x) - \int e^x . (-\cos x) \, dx = -e^x \cos x + \int e^x \cos x \, dx$$

$$= -e^x \cos x + [e^x \sin x - \int e^x \sin x \, dx]$$

$\therefore$ $\quad I = e^x (\sin x - \cos x) - I. \quad \therefore \quad 2I = e^x (\sin x - \cos x)$

$\therefore$ $\quad \mathbf{I = \frac{1}{2} e^x (\sin x - \cos x) + C.}$

(*ii*) Let $\quad I = \int \underset{I}{e^{2x}} \underset{II}{\sin 3x}\, dx.$ [Using **Rule** (*iv*)]

$$\therefore \quad I = e^{2x}\left(-\frac{\cos 3x}{3}\right) - \int 2e^{2x}\left(-\frac{\cos 3x}{3}\right) dx = -\frac{e^{2x}\cos 3x}{3} + \frac{2}{3}\int \underset{I}{e^{2x}} \underset{II}{\cos 3x}\, dx$$

$$= -\frac{e^{2x}\cos 3x}{3} + \frac{2}{3}\left[e^{2x}\frac{\sin 3x}{3} - \int 2e^{2x}\left(\frac{\sin 3x}{3}\right) dx\right]$$

$$= -\frac{e^{2x}\cos 3x}{3} + \frac{2}{9}e^{2x}\sin 3x - \frac{4}{9}\int e^{2x}\sin 3x\, dx$$

$$\therefore \quad I = \frac{e^{2x}}{9}[2\sin 3x - 3\cos 3x] - \frac{4}{9}I \quad \Rightarrow \quad \left(1+\frac{4}{9}\right)I = \frac{e^{2x}}{9}[2\sin 3x - 3\cos 3x]$$

$$\Rightarrow \quad \frac{13}{9}I = \frac{e^{2x}}{9}[2\sin 3x - 3\cos 3x] \qquad \therefore \quad \mathbf{I = \frac{e^{2x}}{13}[2\sin 3x - 3\cos 3x] + C.}$$

(*iii*) Let $I = \int e^{-x} \cos 4x \cos 2x\, dx.$

$$\therefore \quad I = \frac{1}{2}\int e^{-x}(2\cos 4x \cos 2x)\, dx = \frac{1}{2}\int e^{-x}[\cos(4x+2x) + \cos(4x-2x)]\, dx$$

$$= \frac{1}{2}\int e^{-x}(\cos 6x + \cos 2x)\, dx = \frac{1}{2}\int e^{-x}\cos 6x\, dx + \frac{1}{2}\int e^{-x}\cos 2x\, dx = \frac{1}{2}I_1 + \frac{1}{2}I_2$$

(say).

$I_1 = \int \underset{I}{e^{-x}} \underset{II}{\cos 6x}\, dx$ [Using **Rule** (*iv*)]

$$= e^{-x}.\frac{\sin 6x}{6} - \int e^{-x}.-1.\frac{\sin 6x}{6}\, dx = \frac{e^{-x}\sin 6x}{6} + \frac{1}{6}\int e^{-x}\sin 6x\, dx$$

$$= \frac{e^{-x}\sin 6x}{6} + \frac{1}{6}\left[e^{-x}.\left(-\frac{\cos 6x}{6}\right) - \int e^{-x}.-1.\left(-\frac{\cos 6x}{6}\right)dx\right]$$

$$= \frac{e^{-x}\sin 6x}{6} - \frac{e^{-x}\cos 6x}{36} - \frac{1}{36}\int e^{-x}\cos 6x\, dx.$$

$$\therefore \quad I_1 = \frac{e^{-x}}{36}(6\sin 6x - \cos 6x) - \frac{1}{36}I_1$$

$$\therefore \quad \left(1+\frac{1}{36}\right)I_1 = \frac{e^{-x}}{36}(6\sin 6x - \cos 6x) \qquad \therefore \quad I_1 = \frac{e^{-x}}{37}(6\sin 6x - \cos 6x).$$

$I_2 = \int \underset{I}{e^{-x}} \underset{II}{\cos 2x}\, dx$ [Using **Rule** (*iv*)]

$$= e^{-x}.\frac{\sin 2x}{2} - \int e^{-x}.-1.\frac{\sin 2x}{2}\, dx = \frac{e^{-x}\sin 2x}{2} + \frac{1}{2}\int e^{-x}\sin 2x\, dx$$

$$= \frac{e^{-x}\sin 2x}{2} + \frac{1}{2}\left[e^{-x}.\left(-\frac{\cos 2x}{2}\right) - \int e^{-x}.-1.\left(-\frac{\cos 2x}{2}\right)dx\right]$$

$$= \frac{e^{-x}\sin 2x}{2} - \frac{e^{-x}\cos 2x}{4} - \frac{1}{4}\int e^{-x}\cos 2x\, dx.$$

$$\therefore \quad I_2 = \frac{e^{-x}}{4}(2\sin 2x - \cos 2x) - \frac{1}{4} I_2$$

$$\therefore \quad \left(1+\frac{1}{4}\right) I_2 = \frac{e^{-x}}{4}(2\sin 2x - \cos 2x) \qquad \therefore \quad I_2 = \frac{e^{-x}}{5}(2\sin 2x - \cos 2x).$$

$$\therefore \quad I = \frac{1}{2} I_1 + \frac{1}{2} I_2 = \frac{1}{2}\left[\frac{e^{-x}}{37}(6\sin 6x - \cos 6x)\right] + \frac{1}{2}\left[\frac{e^{-x}}{5}(2\sin 2x - \cos 2x)\right] + C$$

$$= \mathbf{e^{-x}\left[\frac{1}{74}(6\sin 6x - \cos 6x) + \frac{1}{10}(2\sin 2x - \cos 2x)\right] + C.}$$

(*iv*) Let $I = \int \underset{I}{e^{ax}} \underset{II}{\sin (bx+c)}\, dx$. Integrating, by parts, we get

$$I = e^{ax}\left(-\frac{\cos (bx+c)}{b}\right) - \int a e^{ax}\left(-\frac{\cos (bx+c)}{b}\right) dx$$

$$= \frac{-e^{ax}\cos (bx+c)}{b} + \frac{a}{b}\int \underset{I}{e^{ax}} \underset{II}{\cos (bx+c)}\, dx$$

$$= \frac{-e^{ax}\cos (bx+c)}{b} + \frac{a}{b}\left[e^{ax}\frac{\sin (bx+c)}{b} - \int a e^{ax} \cdot \frac{\sin (bx+c)}{b}\, dx\right]$$

$$= \frac{-e^{ax}(bx+c)}{b} + \frac{a}{b^2} e^{ax}\sin (bx+c) - \frac{a^2}{b^2}\int e^{ax}\sin (bx+c)\, dx$$

$$\therefore \quad I = \frac{e^{ax}}{b^2}(a\sin (bx+c) - b\cos (bx+c)) - \frac{a^2}{b^2} I$$

$$\therefore \quad \left(1+\frac{a^2}{b^2}\right) I = \frac{e^{ax}}{b^2}(a\sin (bx+c) - b\cos (bx+c))$$

$$\Rightarrow \quad \mathbf{I = \frac{e^{ax}}{a^2+b^2}(a\sin (bx+c) - b\cos (bx+c)) + C.}$$

The value of 'I' can also be expressed alternatively as follows :

[Let $a = r\cos\theta,\ b = r\sin\theta,\ r > 0. \quad \therefore \quad r = \sqrt{a^2+b^2}$ and $\tan\theta = b/a$.

$$\therefore \quad I = \frac{e^{ax}}{r^2}(r\cos\theta\sin (bx+c) - r\sin\theta\cos (bx+c)) = \frac{e^{ax}}{r}\sin (bx+c-\theta)]$$

$$\therefore \quad \mathbf{I = \frac{e^{ax}}{\sqrt{a^2+b^2}}\sin\left(bx + c - \tan^{-1}\frac{b}{a}\right) + C.}$$

Example 7. *Evaluate :*

(*i*) $\int e^x \sec x\,(1+\tan x)\, dx$ (*ii*) $\int \frac{2+\sin 2x}{1+\cos 2x} e^x\, dx$

(*iii*) $\int e^x\left(\frac{x^2+1}{(x+1)^2}\right) dx$ (*iv*) $\int e^{2x}\left(\frac{2x-1}{4x^2}\right) dx$

(*v*) $\int e^x\left(\frac{2-x}{(1-x)^2}\right) dx.$

Sol. (*i*) Let $I = \int e^x \sec x\,(1 + \tan x)\,dx$. $\therefore$ $I = \int e^x (\sec x + \sec x \tan x)\,dx$

Now $(\sec x)' = \sec x \tan x$.

$\therefore$ We write $I = \int \underset{\text{I}}{\sec x} \,.\, \underset{\text{II}}{e^x}\,dx + \int e^x \sec x \tan dx.$

Evaluating first integral of I by parts, we get

$$I = \left[\sec x \,.\, e^x - \int \sec x \tan x \,.\, e^x\,dx\right] + \int e^x \sec x \tan x\,dx = \mathbf{\sec x e^x + C.}$$

(*ii*) Let $I = \int \frac{2 + \sin 2x}{1 + \cos 2x} e^x\,dx.$

$$\therefore \qquad I = \int \frac{2 + 2\sin x \cos x}{2\cos^2 x} e^x\,dx = \int \left(\frac{1}{\cos^2 x} + \frac{\sin x}{\cos x}\right) e^x\,dx = \int (\sec^2 x + \tan x)\,e^x\,dx$$

Now $(\tan x)' = \sec^2 x$.

$\therefore$ We write $I = \int \underset{\text{I}}{\tan x}\, \underset{\text{II}}{e^x}\,dx + \int \sec^2 x\, e^x\,dx.$

Evaluating first integral of I by parts, we get

$$I = \left[\tan x . e^x - \int \sec^2 x \,.\, e^x\,dx\right] + \int \sec^2 x . e^x\,dx = \mathbf{\tan x e^x + C.}$$

(*iii*) Let $I = \int e^x \left(\frac{x^2 + 1}{(x + 1)^2}\right) dx.$

$$\therefore \qquad I = \int e^x \left(\frac{x^2 - 1 + 2}{(x + 1)^2}\right) dx = \int e^x \left[\frac{x^2 - 1}{(x + 1)^2} + \frac{2}{(x + 1)^2}\right] dx = \int e^x \left[\frac{x - 1}{x + 1} + \frac{2}{(x + 1)^2}\right] dx$$

Now $\left(\frac{x - 1}{x + 1}\right)' = \frac{(x + 1)\,.\,1 - (x - 1)\,.\,1}{(x + 1)^2} = \frac{2}{(x + 1)^2}.$ [′means differentiation]

$\therefore$ We write $I = \int \underset{\text{I}}{\frac{x - 1}{x + 1}}\, \underset{\text{II}}{e^x}\,dx + \int e^x \,.\, \frac{2}{(x + 1)^2}\,dx$

$\therefore$ Evaluating first integral of I by parts, we get

$$I = \left[\frac{x - 1}{x + 1} e^x - \int \frac{2}{(x + 1)^2} \,.\, e^x\,dx\right] + \int e^x \,.\, \frac{2}{(x + 1)^2}\,dx = \mathbf{\frac{x - 1}{x + 1} e^x + C.}$$

(*iv*) Let $I = \int e^{2x} \left(\frac{2x - 1}{4x^2}\right) dx.$ Let $z = 2x$. $\therefore$ $dz = 2dx$.

$$\therefore \qquad I = \int e^z \left(\frac{z - 1}{z^2}\right) \frac{dz}{2} = \frac{1}{2} \int e^z \left(\frac{1}{z} - \frac{1}{z^2}\right) dz. \qquad \text{Now } \left(\frac{1}{z}\right)' = -\frac{1}{z^2}.$$

$\therefore$ We write $I = \frac{1}{2} \int \underset{\text{I}}{\frac{1}{z}}\, \underset{\text{II}}{e^z}\,dz - \frac{1}{2} \int e^z \,.\, \frac{1}{z^2}\,dz.$

Evaluating first integral of I by parts, we get

$$I = \frac{1}{2}\left[\frac{1}{z}e^z - \int -\frac{1}{z^2}e^z\,dz\right] - \frac{1}{2}\int e^z \cdot \frac{1}{z^2}\,dz = \frac{1}{2}\left(\frac{1}{z}e^z\right) = \frac{\mathbf{e^{2x}}}{\mathbf{4x}} + \mathbf{C}.$$

(*v*) Let $\quad I = \int e^x \left(\frac{2-x}{(1-x)^2}\right) dx.$

$$\therefore \qquad I = \int e^x \left(\frac{(1-x)+1}{(1-x)^2}\right) dx = \int e^x \left(\frac{1}{1-x} + \frac{1}{(1-x)^2}\right) dx$$

Now $\left(\frac{1}{1-x}\right)' = (-1)(1-x)^{-2}(-1) = \frac{1}{(1-x)^2}$ ['means differentiation]

We write $\quad I = \int \underset{\text{I}}{\frac{1}{1-x}} \cdot \underset{\text{II}}{e^x}\,dx + \int e^x \cdot \frac{1}{(1-x)^2}\,dx$

$\therefore$ Evaluating first integral of I by parts, we get

$$I = \left[\frac{1}{1-x}e^x - \int \frac{1}{(1-x)^2}e^x\,dx\right] + \int e^x \cdot \frac{1}{(1-x)^2}\,dx = \frac{\mathbf{1}}{\mathbf{1-x}}\,\mathbf{e^x} + \mathbf{C}.$$

EXERCISE 13.7

Evaluate the following integrals (Q. No. 1–3) :

1. (*i*) $\int x \cos x\,dx$ (*ii*) $\int x \sin 3x\,dx$ (*iii*) $\int x \sec^2 x\,dx$

   (*iv*) $\int (2-x)\sin x\,dx$ (*v*) $\int x\,\text{cosec}^2 x\,dx$ (*vi*) $\int x \cos 3x\,dx$

2. (*i*) $\int x \log x\,dx$ (*ii*) $\int x^3 \log x\,dx$ (*iii*) $\int x^3 \log 2x\,dx$

   (*iv*) $\int \frac{\log x}{x^n}\,dx$ (*v*) $\int \log(x+1)\,dx$ (*vi*) $\int x \log x^2\,dx$

3. (*i*) $\int xe^{2x}\,dx$ (*ii*) $\int (1+5x)\,e^{2x}\,dx$ (*iii*) $\int 2x\,e^{5x}\,dx$

   (*iv*) $\int (x-7)\,e^{3x}\,dx$ (*v*) $\int x \cdot 5^x\,dx$ (*vi*) $\int (x+1)\,4^x\,dx.$

Evaluate the following integrals (Q. No. 4–23) :

4. (*i*) $\int x \sin^2 x\,dx$ (*ii*) $\int x^2 \cos 2x\,dx$ (*iii*) $\int x^2 \sin x\,dx$

   (*iv*) $\int x \cos^3 x\,dx$ (*v*) $\int x^3 \cos x^2\,dx$ (*vi*) $\int (1+x^2)\cos 2x\,dx$

5. (*i*) $\int (\log x)^2\, dx$ (*G.N.D.U. B.Pharma, 2004*) (*ii*) $\int (1+x) \log x\, dx$ (*iii*) $\int (\log x)^3\, dx$

(*iv*) $\int x^2 \log (1+x)\, dx$ (*v*) $\int \log (2+x^2)\, dx$ (*vi*) $\int \frac{\log (x+1)}{(x+1)^2}\, dx$

6. (*i*) $\int x^2 e^x\, dx$ (*G.N.D.U. B.Pharma, 2004 ; P.T.U. Dec. 2006*) (*ii*) $\int x^2 e^{3x}\, dx$ (*iii*) $\int (x+1)^2 e^x\, dx$

(*iv*) $\int 2x^3 e^{x^2}\, dx$ (*v*) $\int x^3 e^{-x}\, dx$ (*vi*) $\int x^3 e^{4x}\, dx$

(*vii*) $e^{x^3} . x^5 dx$ (*P.T.U. B.Pharma, May 2007*)

7. (*i*) $\int \cos^{-1} x\, dx$ (*ii*) $\int \cot^{-1} x\, dx$ (*iii*) $\int \text{cosec}^{-1} x\, dx$

8. (*i*) $\int \sin^{-1} \frac{2x}{1+x^2}\, dx$ (*ii*) $\int \cos^{-1} \frac{1-x^2}{1+x^2}\, dx$ (*iii*) $\int \tan^{-1} \frac{2x}{1-x^2}\, dx$

9. (*i*) $\int \cos^{-1} (4x^3 - 3x)\, dx$ (*ii*) $\int \tan^{-1} \frac{3x - x^3}{1-3x^2}\, dx$ (*iii*) $\int \tan^{-1} \frac{x-5}{1+5x}\, dx$

(*iv*) $\int \tan^{-1} \sqrt{\frac{1-x}{1+x}}\, dx$ (*v*) $\int \sin^{-1} \sqrt{\frac{x}{a+x}}\, dx$ (*vi*) $\int (\cos^{-1} x)^2\, dx$

10. (*i*) $\int x \sin^{-1} x\, dx$ (*P.T.U. B.Pharma, May 2001*) (*ii*) $\int x \tan^{-1} x\, dx$ (*iii*) $\int x \sec^{-1} x\, dx$

11. (*i*) $\int x^2 \sin^{-1} x\, dx$ (*ii*) $\int x^2 \tan^{-1} x\, dx$ (*iii*) $\int x^2 \sec^{-1} x\, dx$

12. (*i*) $\int e^x \cos x\, dx$ (*ii*) $\int e^{-x} \sin x\, dx$ (*iii*) $\int e^x \cos 3x\, dx$

(*iv*) $\int e^x \sin 2x\, dx$ (*v*) $\int e^{2x} \sin x\, dx$ (*P.T.U. B.Pharma, May 2002*) (*vi*) $\int e^x \cos 2x\, dx$.

13. (*i*) $\int e^x (\sin x + \cos x)\, dx$ (*ii*) $\int e^x \left(\tan^{-1} x + \frac{1}{1+x^2} \right) dx$ (*iii*) $\int e^x (\tan x - \log \cos x)\, dx$

(*iv*) $\int e^x (\sin 2x + 2 \cos 2x)\, dx$ (*v*) $\int e^x (\cot x + \log \sin x)\, dx$

(*vi*) $\int e^x (\sec x + \log (\sec x + \tan x))\, dx$

14. (*i*) $\int e^x \left(\sin^{-1} x + \frac{1}{\sqrt{1-x^2}} \right) dx$ (*ii*) $\int \frac{(x-3)\, e^x}{(x-1)^3}\, dx$ (*iii*) $\int \frac{xe^x}{(x+1)^2}\, dx$

(*iv*) $\int e^x \left(\frac{1+x}{(2+x)^2} \right) dx$ (*v*) $\int e^x \left(\frac{1+\sin x}{1+\cos x} \right) dx$ (*vi*) $\int e^{2x} \left(\frac{1+\sin 2x}{1+\cos 2x} \right) dx$

15. (i) $\int e^{2x}(-\sin x + 2\cos x)\,dx$ (ii) $\int e^{2x}\left(\frac{\sin 4x - 2}{1-\cos 4x}\right)dx$ (iii) $\int e^{\tan^{-1}x}\left(\frac{1+x+x^2}{1+x^2}\right)dx$

(iv) $\int \frac{\log x - 1}{(\log x)^2}\,dx$ (v) $\int \frac{\log x}{(1+\log x)^2}\,dx$ (vi) $\int (\sin \log x + \cos \log x)\,dx$

16. (i) $\int e^{ax}\sin bx\,dx$ (ii) $\int e^{ax}\cos bx\,dx$

(iii) $\int e^{ax}\cos(bx+c)\,dx$ (iv) $\int e^{3x}\cos 3x\cos x\,dx$

17. (i) $\int \sec^3 x\,dx$ (ii) $\int \text{cosec}^3 x\,dx.$

(G.N.D.U. B.Pharma, P.T.U. B.Pharma, May 2005)

18. (i) $\int \frac{x}{1+\cos x}\,dx$ (ii) $\int \frac{x}{1-\cos x}\,dx$

(iii) $\int \frac{x}{1-\sin x}\,dx$ (iv) $\int \frac{x}{1+\sin 4x}\,dx$

19. (i) $\int (\sin^{-1}x)^2\,dx$ (ii) $\int \sin(\log x)\,dx$

(iii) $\int \frac{\sin^{-1}x}{(1-x^2)^{3/2}}\,dx$ (iv) $\int \sin\sqrt{x}\,dx$

20. (i) $\int \frac{x-\sin x}{1-\cos x}\,dx$ (ii) $\int x^2\cos^2 x\,dx$

(iii) $\int \cos 2x \log \sin x\,dx$ (iv) $\int x^3\sin^{-1}\frac{1}{x}\,dx$

21. (i) $\int x\cos 2x\sin 4x\,dx$ (ii) $\int x^{2n-1}\cos x^n\,dx$

(iii) $\int x^2\sin x\cos x\,dx$ (iv) $\int \sin x^{1/3}\,dx$

22. (i) $\int \frac{\log(\log x)}{x}\,dx$ (ii) $\int \log\left(x+\sqrt{x^2+a^2}\right)dx$

(iii) $\int e^x(1+x)\log(xe^x)\,dx$ (iv) $\int \log(1+x)^{(1+x)}\,dx$

23. (i) $\int x\sin x\sin 2x\sin 3x\,dx$ (ii) $\int \frac{\sqrt{1-\sin x}}{1+\cos x}\,e^{-x/2}\,dx.$

(iii) $\int \frac{x^2}{(x\sin x+\cos x)^2}\,dx$ (iv) $\int \frac{\sqrt{x^2+1}\,(\log(x^2+1)-2\log x)}{x^4}\,dx$

24. We know that x^3 is the integral of $3x^2$. Taking this as $(3x)(x)$, verify that the integration by parts gives the same answer.

25. Integrate $\sin x\cos x$ by parts. If the integral is I, you get $I = \sin^2 x - I$, from this find I. Check your answer by finding I by another method.

Answers

1. (*i*) $x \sin x + \cos x + C$ (*ii*) $-\frac{x \cos 3x}{3} + \frac{\sin 3x}{9} + C$ (*iii*) $x \tan x - \log |\sec x| + C$

(*iv*) $(x-2) \cos x - \sin x + C$ (*v*) $-x \cot x + \log |\sin x| + C$

(*vi*) $\frac{x \sin 3x}{3} + \frac{\cos 3x}{9} + C$ **2.** (*i*) $\frac{1}{2}x^2 \log x - \frac{1}{4}x^2 + C$ (*ii*) $\frac{x^4}{4} \log x - \frac{x^4}{16} + C$

(*iii*) $\frac{x^4}{4} \log 2x - \frac{x^4}{16} + C$ (*iv*) $\frac{x^{1-n}}{(1-n)^2}((1-n) \log x - 1) + C$

(*v*) $(x+1) \log (x+1) - x + C$ (*vi*) $x^2 \log x - \frac{x^2}{2} + C$ **3.** (*i*) $\frac{e^{2x}}{4}(2x-1) + C$

(*ii*) $\frac{e^{2x}}{4}(10x-3) + C$ (*iii*) $\frac{e^{5x}}{25}(10x-2) + C$ (*iv*) $\frac{e^{3x}}{9}(3x-22) + C$

(*v*) $\frac{5^x}{(\log 5)^2}(x \log 5 - 1) + C$ (*vi*) $\frac{4^x}{(\log 4)^2}[(x+1) \log 4 - 1] + C$

4. (*i*) $\frac{x^2}{4} - \frac{x \sin 2x}{4} - \frac{\cos 2x}{8} + C$ (*ii*) $\frac{x^2 \sin 2x}{2} + \frac{x \cos 2x}{2} - \frac{\sin 2x}{4} + C$

(*iii*) $-x^2 \cos x + 2x \sin x + 2 \cos x + C$

(*iv*) $\frac{1}{4}\left[3x \sin x + 3 \cos x + \frac{x \sin 3x}{3} + \frac{\cos 3x}{9}\right] + C$ (*v*) $\frac{1}{2}[x^2 \sin x^2 + \cos x^2] + C$

(*vi*) $\frac{2x^2+1}{4} \sin 2x + \frac{x}{2} \cos 2x + C$ **5.** (*i*) $x(\log x)^2 - 2x \log x + 2x + C$

(*ii*) $\left(x + \frac{x^2}{2}\right) \log x - \left(x + \frac{x^2}{4}\right) + C$ (*iii*) $x[(\log x)^3 - 3 (\log x)^2 + 6 \log x - 6] + C$

(*iv*) $\frac{x^3+1}{3} \log (x+1) - \frac{x^3}{9} + \frac{x^2}{6} - \frac{x}{3} + C$ (*v*) $x \log (2 + x^2) - 2x + 2\sqrt{2} \tan^{-1} \frac{x}{\sqrt{2}} + C$

(*vi*) $-\frac{1}{x+1}(\log (x+1) + 1) + C$ **6.** (*i*) $e^x (x^2 - 2x + 2) + C$ (*ii*) $\frac{e^{3x}}{27}(9x^2 - 6x + 2) + C$

(*iii*) $e^x (x^2 + 1) + C$ (*iv*) $(x^2 - 1) e^{x^2} + C$ (*v*) $-e^{-x}(x^3 + 3x^2 + 6x + 6) + C$

(*vi*) $\frac{e^{4x}}{128}[32x^3 - 24x^2 + 12x - 3] + C$ (*vii*) $\frac{1}{3}(x^3 - 1)e^{x^3} + C$

7. (*i*) $x \cos^{-1} x - \sqrt{1-x^2} + C$

(*ii*) $x \cot^{-1} x + \frac{1}{2} \log (1 + x^2) + C$ (*iii*) $x \operatorname{cosec}^{-1} x + \log |x + \sqrt{x^2 - 1}| + C$

8. (*i*) $2x \tan^{-1} x - \log (1 + x^2) + C$ (*ii*) $2x \tan^{-1} x - \log (1 + x^2) + C$

(*iii*) $2x \tan^{-1} x - \log (1 + x^2) + C$ **9.** (*i*) $3 [x \cos^{-1} x - \sqrt{1-x^2}] + C$

(*ii*) $3 [x \tan^{-1} x - \frac{1}{2} \log (1 + x^2)] + C$ (*iii*) $x \tan^{-1} x - \frac{1}{2} \log (1 + x^2) - x \tan^{-1} 5 + C$

(*iv*) $\frac{1}{2}[x\cos^{-1}x-\sqrt{1-x^2}]+C$ (*v*) $(x+a)\tan^{-1}\sqrt{\frac{x}{a}}-\sqrt{ax}+C$

(*vi*) $x(\cos^{-1}x)^2-2\sqrt{1-x^2}\cos^{-1}x-2x+C$

10. (*i*) $\frac{1}{4}[x\sqrt{1-x^2}-(1-2x^2)\sin^{-1}x]+C$ (*ii*) $\frac{x^2+1}{2}\tan^{-1}x-\frac{x}{2}+C$

(*iii*) $\frac{x^2\sec^{-1}x}{2}-\frac{1}{2}\sqrt{x^2-1}+C$

11. (*i*) $\frac{x^3\sin^{-1}x}{3}+\frac{1}{9}(x^2+2)\sqrt{1-x^2}+C$ (*ii*) $\frac{x^3\tan^{-1}x}{3}-\frac{1}{6}x^2+\frac{1}{6}\log(1+x^2)+C$

(*iii*) $\frac{x^3\sec^{-1}x}{3}-\frac{x\sqrt{x^2-1}}{6}-\frac{1}{6}\log|x+\sqrt{x^2-1}|+C$

12. (*i*) $\frac{e^x}{2}(\sin x+\cos x)+C$ (*ii*) $-\frac{e^{-x}}{2}(\sin x+\cos x)+C$ (*iii*) $\frac{e^x}{10}(3\sin 3x+\cos 3x)+C$

(*iv*) $\frac{e^x}{5}(\sin 2x-2\cos 2x)+C$ (*v*) $\frac{e^{2x}}{5}(2\sin x-\cos x)+C$ (*vi*) $\frac{e^x}{5}(2\sin 2x+\cos 2x)+C$

13. (*i*) $e^x\sin x+C$ (*ii*) $e^x\tan^{-1}x+C$ (*iii*) $e^x\log\sec x+C$

(*iv*) $e^x\sin 2x+C$ (*v*) $e^x\log\sin x+C$ (*vi*) $e^x\log(\sec x+\tan x)+C$

14. (*i*) $e^x\sin^{-1}x+C$ (*ii*) $\frac{e^x}{(x-1)^2}+C$ (*iii*) $\frac{e^x}{x+1}+C$

(*iv*) $\frac{e^x}{x+2}+C$ (*v*) $e^x\tan\frac{x}{2}+C$ (*vi*) $\frac{1}{2}e^{2x}\tan x+C$

15. (*i*) $e^{2x}\cos x+C$ (*ii*) $\frac{1}{2}e^{2x}\cot 2x+C$ (*iii*) $x\,e^{\tan^{-1}x}+C$

(*iv*) $\frac{x}{\log x}+C$ (*v*) $\frac{x}{\log x+1}+C$ (*vi*) $x\sin\log x+C$

16. (*i*) $\frac{e^{ax}}{a^2+b^2}(a\sin bx-b\cos bx)+C$ (*ii*) $\frac{e^{ax}}{a^2+b^2}(a\cos bx+b\sin bx)+C$

(*iii*) $\frac{e^{ax}}{a^2+b^2}(a\cos(bx+c)+b\sin(bx+c))+C$

(*iv*) $\frac{e^{3x}}{50}(3\cos 4x+4\sin 4x)+\frac{e^{3x}}{26}(3\cos 2x+2\sin 2x)+C$

17. (*i*) $\frac{\sec x\tan x}{2}+\frac{1}{2}\log|\sec x+\tan x|+C$ (*ii*) $-\frac{\operatorname{cosec} x\cot x}{2}+\frac{1}{2}\log|\operatorname{cosec} x-\cot x|+C$

18. (*i*) $x\tan\frac{x}{2}-2\log\left|\sec\frac{x}{2}\right|+C$ (*ii*) $-x\cot\frac{x}{2}+2\log\left|\sin\frac{x}{2}\right|+C$

(*iii*) $x\cot\left(\frac{\pi}{4}-\frac{x}{2}\right)+2\log\left|\sin\left(\frac{\pi}{4}-\frac{x}{2}\right)\right|+C$ (*iv*) $-\frac{x}{4}\tan\left(\frac{\pi}{4}-2x\right)-\frac{1}{8}\log\left|\sec\left(\frac{\pi}{4}-2x\right)\right|+C$

19. (*i*) $x(\sin^{-1}x)^2 + 2\sqrt{1-x^2}\,\sin^{-1}x - 2x + C$ (*ii*) $\frac{x}{2}(\sin(\log x) - \cos(\log x)) + C$

(*iii*) $\frac{x}{\sqrt{1-x^2}}\sin^{-1}x + \log\sqrt{1-x^2} + C$ (*iv*) $2[\sin\sqrt{x} - \sqrt{x}\cos\sqrt{x}] + C$

20. (*i*) $-x\cot\frac{x}{2} + C$ (*ii*) $\frac{x^3}{6} + \frac{x^2\sin 2x}{4} + \frac{x\cos 2x}{4} - \frac{\sin 2x}{8} + C$

(*iii*) $\frac{(2\log\sin x - 1)\sin 2x}{4} - \frac{x}{2} + C$ (*iv*) $\frac{1}{4}\left[x^4\sin^{-1}\frac{1}{x} + \frac{(x^2+2)\sqrt{x^2-1}}{3}\right] + C$

21. (*i*) $-\frac{1}{12}x\cos 6x + \frac{1}{72}\sin 6x - \frac{1}{4}x\cos 2x + \frac{1}{8}\sin 2x + C$

(*ii*) $\frac{1}{n}(\cos x^n + x^n\sin x^n) + C$ (*iii*) $-\frac{x^2\cos 2x}{4} + \frac{x\sin 2x}{4} + \frac{\cos 2x}{8} + C$

(*iv*) $3(2 - x^{2/3})\cos x^{1/3} + 6x^{1/3}\sin x^{1/3} + C$

22. (*i*) $(\log x)(\log\log x - 1) + C$ (*ii*) $x\log[x + \sqrt{x^2+a^2}] - \sqrt{x^2+a^2} + C$

(*iii*) $xe^x(\log xe^x - 1) + C$ (*iv*) $\frac{1}{2}(1+x)^2\log(1+x) - \frac{1}{4}(1+x)^2 + C$

23. (*i*) $\frac{1}{4}\left[\left(\frac{\sin 2x}{4} + \frac{\sin 4x}{16} - \frac{\sin 6x}{36}\right) - x\left(\frac{\cos 2x}{2} + \frac{\cos 4x}{4} - \frac{\cos 6x}{6}\right)\right] + C$

(*ii*) $-e^{-x/2}\sec\frac{x}{2} + C$ (*iii*) $\frac{\sin x - x\cos x}{x\sin x + \cos x} + C$

(*iv*) $-\frac{1}{3}\left(1+\frac{1}{x^2}\right)^{3/2}\left[\log\left(1+\frac{1}{x^2}\right) - \frac{2}{3}\right] + C$ **25.** $\frac{\sin^2 x}{2} + C.$

Hints

4. (*i*) use $\sin^2 x = \frac{1-\cos 2x}{2}$ (*iv*) $\cos 3x = 4\cos^3 x - 3\cos x$ implies $\cos^3 x = \frac{3}{4}\cos x + \frac{1}{4}\cos 3x$

(*v*) Take $z = x^2$.

5. (*iii*) $$\int(\log x)^3\,dx = \int\underset{\text{I}}{(\log x)^3}\,.\,\underset{\text{II}}{1}\,dx = (\log x)^3\,.\,x - \int 3(\log x)^2\,.\,\frac{1}{x}\,.\,x\,dx$$

$$= x(\log x)^3 - 3\int(\log x)^2\,.\,1\,dx$$

(*vi*) $$\int\frac{\log(x+1)}{(x+1)^2}\,dx = \int\underset{\text{I}}{\log(x+1)}\,.\,\underset{\text{II}}{(x+1)^{-2}}\,dx$$

$$= \log(x+1)\,.\left(\frac{(x+1)^{-1}}{-1}\right) - \int\frac{1}{x+1}\cdot\left(\frac{(x+1)^{-1}}{-1}\right)dx$$

$$= -\frac{\log(x+1)}{x+1} + \int(x+1)^{-2}\,dx.$$

6. (*iv*) Take $z = x^2$. **8.** (*i*) Take $x = \tan\theta$.

9. (*i*) Take $x = \cos\theta$ (*ii*) Take $x = \tan\theta$

(*iii*) Take $x = \tan\theta$ and $5 = \tan k$.

$$\therefore \quad \tan^{-1}\frac{x-5}{1+5x} = \tan^{-1}\tan(\theta - k) = \theta - k = \tan^{-1}x - \tan^{-1}5$$

(*iv*) Take $x = \cos 2\theta$.

(*v*) $\sin^{-1}\sqrt{\dfrac{x}{a+x}} = \theta \Rightarrow \dfrac{x}{a+x} = \sin^2\theta \Rightarrow x = a\tan^2\theta \Rightarrow dx = 2a\tan\theta\sec^2\theta\,d\theta.$

$$\therefore \quad \int \sin^{-1}\sqrt{\frac{x}{a+x}}\,dx = \int \underset{\text{I}}{\theta}\,.\,\underset{\text{II}}{2a\tan\theta\sec^2\theta}\,d\theta = 2a\left[\theta\,.\frac{\tan^2\theta}{2} - \int 1\,.\frac{\tan^2\theta}{2}\,d\theta\right]$$

$$= a\theta\tan^2\theta - a\,[\tan\theta - \theta] = a\,[\theta\tan^2\theta - \tan\theta + \theta].$$

(*vi*) $x = \cos\theta \Rightarrow \displaystyle\int (\cos^{-1}x)^2\,dx = \int \theta^2\,.-\sin\theta\,d\theta = -\int \theta^2\sin\theta\,d\theta$

14. (*ii*) $\dfrac{x-3}{(x-1)^3} = \dfrac{(x-1)-2}{(x-1)^3} = \dfrac{1}{(x-1)^2} - \dfrac{2}{(x-1)^3}$ (*iv*) $\dfrac{1+x}{(2+x)^2} = \dfrac{(2+x)-1}{(2+x)^2} = \dfrac{1}{2+x} - \dfrac{1}{(2+x)^2}$

(*v*) $\dfrac{1+\sin x}{1+\cos x} = \dfrac{1+2\sin\frac{x}{2}\cos\frac{x}{2}}{2\cos^2\frac{x}{2}} = \dfrac{1}{2}\sec^2\dfrac{x}{2} + \tan\dfrac{x}{2}$ (*vi*) Take $z = 2x$.

15. (*i*) Take $z = 2x$ (*ii*) Take $z = 2x$

(*iii*) $z = \tan^{-1}x \Rightarrow \displaystyle\int e^{\tan^{-1}x}\left(\frac{1+x+x^2}{1+x^2}\right)dx = \int e^z\left(1 + \frac{\tan z + \tan^2 z}{1+\tan^2 z}\right).\sec^2 z\,dz$

$$= \int e^z(\tan z + \sec^2 z)\,dz$$

(*iv*) $z = \log x \Rightarrow x = e^z \Rightarrow dx = e^z\,dz$

$$\therefore \int \frac{\log x - 1}{(\log x)^2}\,dx = \int \frac{z-1}{z^2}e^z\,dz = \int e^z\left(\frac{1}{z} - \frac{1}{z^2}\right)dz$$

(*vi*) $z = \log x \Rightarrow x = e^z \Rightarrow dx = e^z\,dz$

$$\therefore \int (\sin\log x + \cos\log x)\,dx = \int (\sin z + \cos z)\,e^z\,dz.$$

16. (*iv*) Use $\cos 3x\cos x = \dfrac{1}{2}(\cos 4x + \cos 2x)$.

17. (*i*) Let $\quad \text{I} = \displaystyle\int \sec^3 x\,dx$

$$\therefore \quad \text{I} = \int \underset{\text{I}}{\sec x}\,\underset{\text{II}}{\sec^2 x}\,dx = \sec x\tan x - \int \sec x\tan x\,.\tan x\,dx$$

$$= \sec x\tan x - \int \sec^3 x\,dx + \int \sec x\,dx$$

$$\therefore \quad 2\text{I} = \sec x\tan x + \log|\sec x + \tan x| + \text{C}.$$

18. (*i*) $\dfrac{x}{1+\cos x} = \dfrac{x}{2\cos^2\frac{x}{2}} = \dfrac{1}{2}\left(x.\sec^2\dfrac{x}{2}\right)$

(iii) $\dfrac{x}{1-\sin x}=\dfrac{x}{1-\cos\left(\dfrac{\pi}{2}-x\right)}=\dfrac{x}{2\sin^2\left(\dfrac{\pi}{4}-\dfrac{x}{2}\right)}=\dfrac{1}{2}\left(x\,.\,\text{cosec}^2\left(\dfrac{\pi}{4}-\dfrac{x}{2}\right)\right).$

19. (i) Take $z=\sin^{-1}x$ (ii) Take $z=\log x$

(iii) Take $z=\sin^{-1}x$ (iv) Take $z=\sqrt{x}$.

20. (i) $\dfrac{x-\sin x}{1-\cos x}=\dfrac{x}{1-\cos x}-\dfrac{\sin x}{1-\cos x}$ (ii) Use $\cos^2 x=\dfrac{1+\cos 2x}{2}$

(iii) $\displaystyle\int \underset{\text{I}}{\log\sin x}\,.\,\underset{\text{II}}{\cos 2x}\,dx=\log\sin x\,.\,\frac{\sin 2x}{2}-\int \cot x\,\frac{\sin 2x}{2}\,dx=\frac{\sin 2x\log\sin x}{2}-\int \cos^2 x\,dx$

(iv) $z=\sin^{-1}\dfrac{1}{x}\ \Rightarrow\ x=\text{cosec}\,z\ \Rightarrow\ dx=-\,\text{cosec}\,z\cot z\,dz$

$$\therefore\quad \int x^3\sin^{-1}\frac{1}{x}\,dx=\int \text{cosec}^3 z\,.\,z\,.-\text{cosec}\,z\cot z\,dz$$

$$=\int \underset{\text{I}}{z}\underset{\text{II}}{\left(\frac{\text{cosec}^4 z}{4}\right)'}dz=z\,.\,\frac{\text{cosec}^4 z}{4}-\int 1\,.\,\frac{\text{cosec}^4 z}{4}\,dz$$

$$=\frac{z\,\text{cosec}^4 z}{4}+\frac{1}{4}\int(1+\cot^2 z)(-\text{cosec}^2 z)\,dz$$

$$=\frac{z\,\text{cosec}^4 z}{4}+\frac{1}{4}\left[\cot z+\frac{\cot^3 z}{3}\right]+\text{C}.$$

21. (i) Use $\cos 2x\sin 4x=\dfrac{1}{2}(2\sin 4x\cos 2x)=\dfrac{1}{2}(\sin 6x+\sin 2x)$

(ii) Take $z=x^n$ (iii) $\sin x\cos x=\dfrac{\sin 2x}{2}$ (iv) Take $z=x^{1/3}$.

22. (i) Take $z=\log x$

(ii) $\displaystyle\int \underset{\text{I}}{\log(x+\sqrt{x^2+a^2})}\,.\,\underset{\text{II}}{1}\,dx=\log(x+\sqrt{x^2+a^2})\,.\,x-\int\frac{1}{x+\sqrt{x^2+a^2}}\left(1+\frac{2x}{2\sqrt{x^2+a^2}}\right).\,x\,dx$

$$=x\log(x+\sqrt{x^2+a^2})-\int\frac{x}{\sqrt{x^2+a^2}}\,dx$$

(iii) Take $z=xe^x$

(iv) $\displaystyle\int \log(1+x)^{(1+x)}dx=\int(1+x)\log(1+x)\,dx=\int \underset{\text{I}}{\log(1+x)}\,.\,\underset{\text{II}}{(1+x)}\,dx$

$$=\log(1+x)\,.\,\frac{(1+x)^2}{2}-\int\frac{1}{1+x}\,.\,\frac{(1+x)^2}{2}\,dx$$

23. (i) $\sin x\sin 2x\sin 3x=\dfrac{1}{2}(2\sin 2x\sin x)\sin 3x$

$$=\frac{1}{2}(\cos x-\cos 3x)\sin 3x=\frac{1}{4}(2\sin 3x\cos x-2\sin 3x\cos 3x)$$

$$= \frac{1}{4} (\sin 4x + \sin 2x - \sin 6x).$$

(*ii*) $\text{Integrand} = \dfrac{\cos \frac{1}{2}x - \sin \frac{1}{2}x}{2\cos^2 \frac{1}{2}x} e^{-\frac{x}{2}} = \left(\frac{1}{2}\sec\frac{1}{2}x - \frac{1}{2}\sec\frac{1}{2}x\tan\frac{1}{2}x\right)e^{-\frac{x}{2}}.$

Now put $z = -\dfrac{x}{2}$ and simplify.

(*iii*) $\displaystyle\int \frac{x^2}{(x\sin x + \cos x)^2}dx = \int \underset{\text{I}}{\frac{x}{\cos x}} \cdot \underset{\text{II}}{\frac{x\cos x}{(x\sin x + \cos x)^2}} dx$

$$= \frac{x}{\cos x} \cdot \int \frac{x\cos x}{(x\sin x+\cos x)^2}dx - \int \left(\frac{(\cos x + x\sin x)}{\cos^2 x}\int \frac{x\cos x}{(x\sin x + \cos x)^2}dx\right)dx$$

$$z = x\sin x + \cos x \quad \Rightarrow \quad dz = x\cos x\, dx$$

$$\therefore \quad \int \frac{x\cos x}{(x\sin x+\cos x)^2}dx = \frac{(x\sin x + \cos x)^{-1}}{-1} = -\frac{1}{x\sin x + \cos x}$$

$$\therefore \quad \text{Given integral} = \frac{x}{\cos x} \cdot -\frac{1}{x\sin x+\cos x} + \int \frac{dx}{\cos^2 x} + C$$

(*iv*) $\displaystyle\frac{\sqrt{x^2+1}\,[\log(x^2+1) - 2\log x]}{x^4}dx = \int \frac{\sqrt{x^2+1}}{x^4}\log\left(\frac{x^2+1}{x^2}\right)dx$

$$= \int \sqrt{1+\frac{1}{x^2}}\log\left(1+\frac{1}{x^2}\right)\cdot\frac{1}{x^3}dx = -\frac{1}{2}\int \sqrt{z}\log z\, dz, \text{ where } z = 1 + \frac{1}{x^2}.$$

13.21. STANDARD INTEGRALS $\int \sqrt{a^2 - x^2}\,dx$, $\int \sqrt{x^2 - a^2}\,dx$ AND $\int \sqrt{a^2 + x^2}\,dx$

We shall prove that :

1. $\displaystyle\int \sqrt{a^2 - x^2}\,dx = \frac{x\sqrt{a^2-x^2}}{2} + \frac{a^2}{2}\sin^{-1}\frac{x}{a} + C$ (*G.N.D.U. B.Pharma, 2005*)

2. $\displaystyle\int \sqrt{x^2 - a^2}\,dx = \frac{x\sqrt{x^2-a^2}}{2} - \frac{a^2}{2}\log\left|x + \sqrt{x^2-a^2}\right| + C$

3. $\displaystyle\int \sqrt{a^2 + x^2}\,dx = \frac{x\sqrt{a^2+x^2}}{2} + \frac{a^2}{2}\log\left|x + \sqrt{a^2+x^2}\right| + C.$

Proof. 1. Let $I = \int \sqrt{a^2-x^2}\,dx. \quad \therefore \quad I = \int \underset{\text{I}}{\sqrt{a^2-x^2}} \cdot \underset{\text{II}}{1}\,dx.$

Integrating by parts, we get

$$I = \sqrt{a^2-x^2}\,.\,x - \int \frac{1}{2\sqrt{a^2-x^2}}.(-2x).x\,dx$$

$$= x\sqrt{a^2-x^2} - \int \frac{-x^2}{\sqrt{a^2-x^2}}dx = x\sqrt{a^2-x^2} - \int \frac{(a^2-x^2)-a^2}{\sqrt{a^2-x^2}}dx$$

$$= x\sqrt{a^2-x^2} - \int \sqrt{a^2-x^2}\,dx + a^2\int \frac{1}{\sqrt{a^2-x^2}}dx$$

$$= x\sqrt{a^2 - x^2} - \int \sqrt{a^2 - x^2}\, dx + a^2 \sin^{-1}\frac{x}{a}.$$

$$\therefore \quad 2\int \sqrt{a^2 - x^2}\, dx = x\sqrt{a^2 - x^2} + a^2 \sin^{-1}\frac{x}{a}$$

or
$$\int \sqrt{\mathbf{a^2 - x^2}}\, \mathbf{dx} = \frac{\mathbf{x\sqrt{a^2 - x^2}}}{\mathbf{2}} + \frac{\mathbf{a^2}}{\mathbf{2}} \mathbf{\sin^{-1}\frac{x}{a} + C.}$$

2. Let $I = \int \sqrt{x^2 - a^2}\, dx. \quad \therefore \quad I = \int \underset{\text{I}}{\sqrt{x^2 - a^2}} \,.\, \underset{\text{II}}{1}\, dx$

Integrating by parts, we get

$$I = \sqrt{x^2 - a^2}\,.\,x - \int \frac{1}{2\sqrt{x^2 - a^2}}\,.\,2x\,.\,x\, dx = x\sqrt{x^2 - a^2} - \int \frac{x^2}{\sqrt{x^2 - a^2}}\, dx$$

$$= x\sqrt{x^2 - a^2} - \int \frac{(x^2 - a^2) + a^2}{\sqrt{x^2 - a^2}}\, dx = x\sqrt{x^2 - a^2} - \int \sqrt{x^2 - a^2}\, dx - a^2 \int \frac{dx}{\sqrt{x^2 - a^2}}$$

$$\therefore \quad I = x\sqrt{x^2 - a^2} - I - a^2 \log|x + \sqrt{x^2 - a^2}|$$

or
$$2I = x\sqrt{x^2 - a^2} - a^2 \log|x + \sqrt{x^2 - a^2}|$$

or
$$I = \frac{x\sqrt{x^2 - a^2}}{2} - \frac{a^2}{2} \log|x + \sqrt{x^2 - a^2}| + C$$

$$\therefore \quad \int \sqrt{\mathbf{x^2 - a^2}}\, \mathbf{dx} = \frac{\mathbf{x\sqrt{x^2 - a^2}}}{\mathbf{2}} - \frac{\mathbf{a^2}}{\mathbf{2}} \mathbf{\log|x + \sqrt{x^2 - a^2}| + C.}$$

3. Let $I = \int \sqrt{a^2 + x^2}\, dx. \quad \therefore \quad I = \int \underset{\text{I}}{\sqrt{a^2 + x^2}} \,.\, \underset{\text{II}}{1}\, dx$

Integrating by part, we get $I = \sqrt{a^2 + x^2}\,.\,x - \int \frac{1}{2\sqrt{a^2 + x^2}}\,.\,2x\,.\,x\, dx$

$$= x\sqrt{a^2 + x^2} - \int \frac{(x^2 + a^2) - a^2}{\sqrt{a^2 + x^2}}\, dx = x\sqrt{a^2 + x^2} - \int \sqrt{a^2 + x^2}\, dx + a^2 \int \frac{dx}{\sqrt{a^2 + x^2}}$$

$$\therefore \quad I = x\sqrt{a^2 + x^2} - I + a^2 \log|x + \sqrt{a^2 + x^2}|$$

$$\therefore \quad 2I = x\sqrt{a^2 + x^2} + a^2 \log|x + \sqrt{a^2 + x^2}|$$

$$\therefore \quad I = \frac{x\sqrt{a^2 + x^2}}{2} + \frac{a^2}{2} \log|x + \sqrt{a^2 + x^2}| + C$$

$$\therefore \quad \int \sqrt{\mathbf{a^2 + x^2}}\, \mathbf{dx} = \frac{\mathbf{x\sqrt{a^2 + x^2}}}{\mathbf{2}} + \frac{\mathbf{a^2}}{\mathbf{2}} \mathbf{\log|x + \sqrt{a^2 + x^2}| + C.}$$

Remark. The integrals 1, 2 and 3 can also be evaluated by substituting $x = a \sin \theta$, $x = a \sec \theta$ and $a \tan \theta$ respectively.

ILLUSTRATIVE EXAMPLES

Example 1. *Evaluate :*

(*i*) $\sqrt{9-x^2}\,dx$ (*ii*) $\int \sqrt{8+7x^2}\,dx$ (*iii*) $\int \sqrt{17x^2-11}\,dx$ (*iv*) $\int \sqrt{5-\dfrac{x^2}{12}}\,dx.$

Sol. (*i*) Let $I = \int \sqrt{9-x^2}\,dx. \quad \therefore \quad I = \int \sqrt{(3)^2-x^2}\,dx$

$$= \frac{x\sqrt{(3)^2-x^2}}{2} + \frac{(3)^2}{2}\sin^{-1}\frac{x}{3} + C$$

$$\left| \int \sqrt{a^2-x^2}\,dx = \frac{x}{2}\sqrt{a^2-x^2} + \frac{a^2}{2}\sin^{-1}\frac{x}{a} \right.$$

$$= \mathbf{\frac{x\sqrt{9-x^2}}{2} + \frac{9}{2}\sin^{-1}\frac{x}{3} + C.}$$

(*ii*) Let $I = \int \sqrt{8+7x^2}\,dx. \quad \therefore \quad I = \sqrt{7}\int \sqrt{\frac{8}{7}+x^2}\,dx = \sqrt{7}\int \sqrt{(\sqrt{8/7})^2+x^2}\,dx$

$$= \sqrt{7}\left[\frac{x}{2}\sqrt{\left(\sqrt{\frac{8}{7}}\right)^2+x^2} + \frac{(\sqrt{8/7})^2}{2}\log|x+\sqrt{(\sqrt{8/7})^2+x^2}|\right] + C$$

$$= \sqrt{7}\left[\frac{x\sqrt{8/7+x^2}}{2} + \frac{8}{14}\log\left|x+\sqrt{\frac{8}{7}+x^2}\right|\right] + C$$

$$= \mathbf{\frac{x\sqrt{8+7x^2}}{2} + \frac{4\sqrt{7}}{7}\log\left|x+\sqrt{\frac{8}{7}+x^2}\right| + C.}$$

(*iii*) Let $I = \int \sqrt{17x^2-11}\,dx. \quad \therefore \quad I = \sqrt{17}\int \sqrt{x^2-11/17}\,dx$ $\left(\text{Here } a^2 = \dfrac{11}{17}\right)$

$$= \sqrt{17}\left[\frac{x\sqrt{x^2-11/17}}{2} - \frac{11/17}{2}\log|x+\sqrt{x^2-11/17}|\right] + C$$

$$= \mathbf{\frac{x\sqrt{17x^2-11}}{2} - \frac{11\sqrt{17}}{34}\log|x+\sqrt{x^2-11/17}| + C.}$$

(*iv*) Let $I = \int \sqrt{5-\frac{x^2}{12}}\,dx. \quad \therefore \quad I = \frac{1}{\sqrt{12}}\int \sqrt{60-x^2}\,dx$ (Here $a^2 = 60$)

$$= \frac{1}{\sqrt{12}}\left[\frac{x\sqrt{60-x^2}}{2} + \frac{60}{2}\sin^{-1}\frac{x}{\sqrt{60}}\right] + C \quad \left(\text{By using formula for } \int\sqrt{a^2-x^2}\,dx\right)$$

$$= \mathbf{\frac{x\sqrt{60-x^2}}{4\sqrt{3}} + 5\sqrt{3}\sin^{-1}\frac{x}{\sqrt{60}} + C.}$$

13.22. EVALUATION OF $\int \sqrt{ax^2 + bx + c}\, dx$ AND $\int (px + q)\sqrt{ax^2 + bx + c}\, dx$

(*i*) **Evaluation of $\int \sqrt{ax^2 + bx + c}\, dx$.** This integral is evaluated by expressing its integrand either in the form $\sqrt{A^2 - z^2}$ or $\sqrt{z^2 - A^2}$ or $\sqrt{A^2 + z^2}$ by using suitable substitution.

WORKING RULES FOR EVALUATING $\int \sqrt{ax^2 + bx + c}\, dx$

Step I. *Take $\sqrt{|a|}$ out of the integral sign. For example, if $I = \int \sqrt{7 - 3x - 2x^2}\, dx$, then write $I = \sqrt{2} \int \sqrt{\frac{7}{2} - \frac{3}{2}x - x^2}\, dx$.*

Step II. *Express the terms containing x^2 and x in the form of a perfect square by adding and subtracting the square of half of the coefficient of x.*

Step III. *Take linear expression in x as z and express the integral in terms of z. This resultant integral will be in one of the standard forms.*

Step IV. *Use the standard formula and express the result in terms of x.*

Example 2. *Evaluate :*

(*i*) $\int \sqrt{x^2 - 4x + 2}\, dx$ (*ii*) $\int \sqrt{2x^2 + 3x + 4}\, dx$ (*iii*) $\int \sqrt{3 - 2x - 2x^2}\, dx$.

Sol. (*i*) Let $I = \int \sqrt{x^2 - 4x + 2}\, dx$. $\therefore$ $I = \int \sqrt{(x^2 - 4x + 4) + (2 - 4)}\, dx = \int \sqrt{(x - 2)^2 - 2}\, dx$

Let $z = x - 2$. $\therefore$ $dz = dx$

$$\therefore \quad I = \int \sqrt{z^2 - 2}\, dz = \frac{z\sqrt{z^2 - 2}}{2} - \frac{2}{2} \log |z + \sqrt{z^2 - 2}| + C \qquad \text{(Here } a^2 = 2\text{)}$$

$$= \frac{(x - 2)\sqrt{(x - 2)^2 - 2}}{2} - \log |x - 2 + \sqrt{(x - 2)^2 - 2}| + C$$

$$= \frac{(\mathbf{x - 2})\sqrt{\mathbf{x^2 - 4x + 2}}}{\mathbf{2}} - \mathbf{\log |x - 2 + \sqrt{x^2 - 4x + 2}| + C.}$$

(*ii*) Let $I = \int \sqrt{2x^2 + 3x + 4}\, dx$.

$$\therefore \quad I = \sqrt{2} \int \sqrt{x^2 + \frac{3}{2}x + 2}\, dx = \sqrt{2} \int \sqrt{\left(x^2 + \frac{3}{2}x + \frac{9}{16}\right) + 2 - \frac{9}{16}}\, dx$$

$$= \sqrt{2} \int \sqrt{\left(x + \frac{3}{4}\right)^2 + \frac{23}{16}}\, dx. \quad \text{Let } z = x + \frac{3}{4} \quad \therefore \quad dz = dx.$$

$$\therefore \quad I = \sqrt{2} \int \sqrt{z^2 + \frac{23}{16}}\, dz \qquad \left(\text{Integrand is of the type } \sqrt{x^2 + a^2} \text{ with } a^2 = \frac{23}{16}\right)$$

$$= \sqrt{2} \left[\frac{z\sqrt{\frac{23}{16} + z^2}}{2} + \frac{\frac{23}{16}}{2} \log \left| z + \sqrt{\frac{23}{16} + z^2} \right| \right] + C$$

$$= \frac{\left(x+\frac{3}{4}\right)\sqrt{\frac{23}{16}+\left(x+\frac{3}{4}\right)^2}}{\sqrt{2}} + \frac{23\sqrt{2}}{32}\log\left|x+\frac{3}{4}+\sqrt{\frac{23}{16}+\left(x+\frac{3}{4}\right)^2}\right| + C$$

$$= \frac{(4x+3)\sqrt{x^2+\frac{3}{2}x+2}}{4\sqrt{2}} + \frac{23\sqrt{2}}{32}\log\left|x+\frac{3}{4}+\sqrt{x^2+\frac{3}{2}x+2}\right| + C$$

$$= \mathbf{\frac{(4x+3)\sqrt{2x^2+3x+4}}{8} + \frac{23\sqrt{2}}{32}\log\left|\frac{4x+3}{4}+\sqrt{\frac{2x^2+3x+4}{2}}\right| + C.}$$

(*iii*) Let $I = \int \sqrt{3-2x-2x^2}\,dx$.

$\therefore \quad I = \sqrt{2}\sqrt{\frac{3}{2}-x-x^2}\,dx = \sqrt{2}\int\sqrt{\frac{3}{2}-(x^2+x)}\,dx = \sqrt{2}\int\sqrt{\frac{3}{2}-\left(x^2+x+\frac{1}{4}\right)+\frac{1}{4}}\,dx$

$\therefore \quad I = \sqrt{2}\int\sqrt{\frac{7}{4}-\left(x+\frac{1}{2}\right)^2}\,dx.$ Let $z = x+\frac{1}{2}$. $\therefore \quad dz = dx$

$\therefore \quad I = \sqrt{2}\int\sqrt{\frac{7}{4}-z^2}\,dz = \sqrt{2}\left[\frac{z\sqrt{\frac{7}{4}-z^2}}{2} + \frac{7/4}{2}\sin^{-1}\frac{z}{\sqrt{7/4}}\right] + C \qquad \left(\text{Here } a^2 = \frac{7}{4}\right)$

$$= \frac{1}{\sqrt{2}}\left(x+\frac{1}{2}\right)\sqrt{\frac{7}{4}-\left(x+\frac{1}{2}\right)^2} + \frac{7\sqrt{2}}{8}\sin^{-1}\frac{2\left(x+\frac{1}{2}\right)}{\sqrt{7}} + C$$

$$= \mathbf{\frac{(2x+1)\sqrt{3-2x-2x^2}}{4} + \frac{7\sqrt{2}}{8}\sin^{-1}\frac{2x+1}{\sqrt{7}} + C.}$$

EXERCISE 13.8

1. **Evaluate the following integrals :**

(*i*) $\int \sqrt{16-9x^2}\,dx$ (*ii*) $\int \sqrt{2x^2-3}\,dx$ (*iii*) $\int \sqrt{\frac{x^2}{2}+4}\,dx$

(*iv*) $\int \cos x\sqrt{9-\sin^2 x}\,dx$ (*v*) $\int x\sqrt{x^4+a^4}\,dx$ (*vi*) $\int x\sqrt{x^4-1}\,dx.$

2. (*i*) $\int \frac{x^2}{\sqrt{x^2+1}}\,dx$ (*ii*) $\int \frac{x^2+x+3}{\sqrt{x^2+1}}\,dx$ (*iii*) $\int (x+1)\sqrt{\frac{x+2}{x-2}}\,dx$

3. (*i*) $\int \sqrt{x^2+6x-4}\,dx$ (*ii*) $\int \sqrt{x^2+2x+3}\,dx$ (*iii*) $\int \sqrt{x^2+8x+4}\,dx$

(*iv*) $\int \sqrt{x^2+8x+12}\,dx$ (*v*) $\int \sqrt{x^2-4x+8}\,dx$ (*vi*) $\int \sqrt{x^2+4x+1}\,dx$

4. (*i*) $\int \sqrt{2ax - x^2}\, dx$ (*ii*) $\sqrt{8 + 2x - x^2}\, dx$ (*iii*) $\int \sqrt{x^2 + x}\, dx$

(*iv*) $\int \sqrt{1 - 4x - x^2}\, dx$ (*v*) $\int \sqrt{4 - 3x - 2x^2}\, dx$ (*vi*) $\int \sqrt{3 + 8x - 3x^2}\, dx$

Answers

1. (*i*) $\frac{x\sqrt{16 - 9x^2}}{2} + \frac{8}{3}\sin^{-1}\frac{3x}{4} + C$ (*ii*) $\frac{x\sqrt{2x^2 - 3}}{2} - \frac{3\sqrt{2}}{4}\log|x + \sqrt{x^2 - 3/2}| + C$

(*iii*) $\frac{1}{2\sqrt{2}}[x\sqrt{x^2 + 8} + 8\log|x + \sqrt{x^2 + 8}| + C$ (*iv*) $\frac{\sin x\sqrt{9 - \sin^2 x}}{2} + \frac{9}{2}\sin^{-1}\left(\frac{\sin x}{3}\right) + C$

(*v*) $\frac{x^2\sqrt{x^4 + a^4}}{4} + \frac{a^4}{4}\log|x^2 + \sqrt{x^4 + a^4}| + C$ (*vi*) $\frac{x^2\sqrt{x^4 - 1}}{4} - \frac{1}{4}\log|x^2 + \sqrt{x^4 - 1}| + C$

2. (*i*) $\frac{x\sqrt{x^2 + 1}}{2} - \frac{1}{2}\log|x + \sqrt{x^2 + 1}| + C$ (*ii*) $\frac{(x + 2)\sqrt{x^2 + 1}}{2} + \frac{5}{2}\log|x + \sqrt{x^2 + 1}| + C$

(*iii*) $\frac{(x + 6)\sqrt{x^2 - 4}}{2} + 4\log|x + \sqrt{x^2 - 4}| + C$

3. (*i*) $\frac{(x + 3)\sqrt{x^2 + 6x - 4}}{2} - \frac{13}{2}\log|x + 3 + \sqrt{x^2 + 6x - 4}| + C$

(*ii*) $\frac{(x + 1)\sqrt{x^2 + 2x + 3}}{2} + \log|x + 1 + \sqrt{x^2 + 2x + 3}| + C$

(*iii*) $\frac{(x + 4)\sqrt{x^2 + 8x + 4}}{2} - 6\log|x + 4 + \sqrt{x^2 + 8x + 4}| + C$

(*iv*) $\frac{(x + 4)\sqrt{x^2 + 8x + 12}}{2} - 2\log|x + 4 + \sqrt{x^2 + 8x + 12}| + C$

(*v*) $\frac{(x - 2)\sqrt{x^2 - 4x + 8}}{2} + 2\log|x - 2 + \sqrt{x^2 - 4x + 8}| + C$

(*vi*) $\frac{(x + 2)\sqrt{x^2 + 4x + 1}}{2} - \frac{3}{2}\log|x + 2 + \sqrt{x^2 + 4x + 1}| + C$

4. (*i*) $\frac{x - a}{2}\sqrt{2ax - x^2} + \frac{a^2}{2}\sin^{-1}\frac{x - a}{a} + C$ (*ii*) $\frac{x - 1}{2}\sqrt{8 + 2x - x^2} + \frac{9}{2}\sin^{-1}\frac{x - 1}{3} + C$

(*iii*) $\frac{2x + 1}{4}\sqrt{x^2 + x} - \frac{1}{8}\log\left|x + \frac{1}{2} + \sqrt{x^2 + x}\right| + C$

(*iv*) $\frac{(x + 2)\sqrt{1 - 4x - x^2}}{2} + \frac{5}{2}\sin^{-1}\frac{x + 2}{\sqrt{5}} + C$

(*v*) $\frac{(4x + 3)\sqrt{4 - 3x - 2x^2}}{8} + \frac{41\sqrt{2}}{32}\sin^{-1}\frac{4x + 3}{\sqrt{41}} + C$

(*vi*) $\frac{(3x - 4)\sqrt{3 + 8x - 3x^2}}{6} + \frac{25\sqrt{3}}{18}\sin^{-1}\frac{3x - 4}{5} + C$

Hints

1. (*iv*) Take $z = \sin x$ (*v*) Take $z = x^2$.

2. (*i*) $\dfrac{x^2}{\sqrt{x^2+1}} = \dfrac{(x^2+1)-1}{\sqrt{x^2+1}} = \sqrt{x^2+1} - \dfrac{1}{\sqrt{x^2+1}}$

(*ii*) $\dfrac{x^2+x+3}{\sqrt{x^2+1}} = \dfrac{x^2+1}{\sqrt{x^2+1}} + \dfrac{x}{\sqrt{x^2+1}} + \dfrac{2}{\sqrt{x^2+1}} = \sqrt{x^2+1} + \dfrac{1}{2}\left(\dfrac{2x}{\sqrt{x^2+1}}\right) + 2\left(\dfrac{1}{\sqrt{x^2+1}}\right)$

(*iii*) $(x+1)\sqrt{\dfrac{x+2}{x-2}} = \dfrac{(x+1)(x+2)}{\sqrt{x^2-4}} = \dfrac{x^2+3x+2}{\sqrt{x^2-4}} = \dfrac{(x^2-4)+3x+6}{\sqrt{x^2-4}}$

$= \sqrt{x^2-4} + \dfrac{3}{2}\left(\dfrac{2x}{\sqrt{x^2-4}}\right) + 6\left(\dfrac{1}{\sqrt{x^2-4}}\right).$

4. (*i*) $2ax - x^2 = -(x^2 - 2ax) = -(x^2 - 2ax + a^2) + a^2 = a^2 - (x-a)^2$.

INTEGRATION BY PARTIAL FRACTIONS

13.23. INTRODUCTION

We know that the quotient of any two polynomials is called a rational function. In the present topic, we shall learn the method of integration by partial fractions. This method will help us to integrate any given rational function.

13.24. METHOD OF RESOLUTION INTO PARTIAL FRACTIONS

If $f(x)$ and $g(x)$ are polynomials, then $\dfrac{f(x)}{g(x)}$ is a rational function or a **rational fraction.**

If deg. $f(x)$ < dog. $f(x)$, then the rational fraction $\dfrac{f(x)}{g(x)}$ is called **proper,** otherwise $\dfrac{f(x)}{g(x)}$ is called **improper.**

For example, the rational fraction $\dfrac{x^2+x+3}{(x+2)(x^2+7)}$ is *proper* and the rational function $\dfrac{x^3+x+9}{x^2+5x-6}$ is *improper.*

If $\dfrac{f(x)}{g(x)}$ is an improper rational fraction, then we can divide $f(x)$ by $g(x)$ so as to write $\dfrac{f(x)}{g(x)}$ as the sum of a polynomial and a proper rational fraction.

∴ **Any improper fraction can be expressed as the sum of a polynomial and a proper fraction.**

It can be proved mathematically that any proper fraction may be resolved into partial fractions and :

(*i*) if $ax + b$ is any linear non-repeated factor in the denominator, then there corresponds a partial fraction of the form $\dfrac{A}{ax+b}$.

(*ii*) if $ax + b$ is any linear factor repeated r ($\in$ N) times in the denominator, then there corresponds partial fractions of the form $\frac{A}{ax+b}, \frac{B}{(ax+b)^2}, \frac{C}{(ax+b)^3}$, ... r terms.

(*iii*) if $ax^2 + bx + c$ is any irreducible quadratic factor in the denominator, then there corresponds a partial fraction of the form $\frac{Ax+B}{ax^2+bx+c}$, and

(*iv*) if $ax^2 + bx + c$ is any irreducible factor repeated r ($\in$ N) times in the denominator, then there corresponds partial fractions of the form $\frac{Ax+B}{ax^2+bx+c}, \frac{Cx+D}{(ax^2+bx+c)^2}$, ... r terms.

The quantities A, B, C, D, ... are all constants independent of x.

$\therefore$ The given proper fraction can be expressed as the sum of its partial fractions. **The constants A, B, C, D, ... occurring in the numerators of the partial fractions are determined by simplifying the sum of partial fractions and then giving various values to x, to obtain equations involving unknown constants or by comparing the coefficients of like powers of x.**

Illustrations. (*i*) The partial fractions of the proper fraction $\frac{x-1}{(x+1)(x-2)(x+3)^3}$ are of the types $\frac{A}{x+1}, \frac{B}{x-2}, \frac{C}{x+3}, \frac{D}{(x+3)^2}$ and $\frac{E}{(x+3)^3}$.

(*ii*) The partial fractions of the proper fraction $\frac{x^2+4}{(x-7)(x+2)^2(x^2+2x+3)}$ are of the types $\frac{A}{x-7}, \frac{B}{x+2}, \frac{C}{(x+2)^2}$ and $\frac{Dx+E}{x^2+2x+3}$.

(*iii*) The partial fractions of the proper fraction $\frac{x^3+x+7}{(x-1)(x+2)^2(x^2+5)^2}$ are of the types $\frac{A}{x-1}, \frac{B}{x+2}, \frac{C}{(x+2)^2}, \frac{Dx+E}{x^2+5}$ and $\frac{Fx+G}{(x^2+5)^2}$.

ILLUSTRATIVE EXAMPLES

Example 1. *Resolve the following fractions into partial fractions :*

(*i*) $\frac{3x-1}{x^2-1}$ (*ii*) $\frac{6x^3+5x^2-7}{3x^2-2x-1}$ (*iii*) $\frac{2x+1}{(x+2)(x-3)^2}$

(*iv*) $\frac{2x+3}{(x^2+1)(x+1)}$ (*v*) $\frac{x^2+5}{(x^2+x+1)(x^2-x+2)}$ (*vi*) $\frac{x}{(x-1)(x^2+1)^2}$.

Sol. (*i*) $\frac{3x-1}{x^2-1}$ is a proper fraction and the denominator has linear non-repeated factors.

Let $\frac{3x-1}{x^2-1} = \frac{3x-1}{(x-1)(x+1)} = \frac{A}{x-1} + \frac{B}{x+1}$. $\therefore$ $\frac{3x-1}{(x-1)(x+1)} = \frac{A(x+1)+B(x-1)}{(x-1)(x+1)}$

Multiplying by $(x-1)(x+1)$, we get, $3x - 1 = A(x + 1) + B(x - 1)$ ...(1)

Now we find the values of A and B.

Method I. By comparing the coefficients of like powers of x.

(1) $\Rightarrow$ $3x - 1 = (A + B)x + (A - B)$.

$\therefore$ $A + B = 3$ and $A - B = -1$. Solving these equations, we get $A = 1$ and $B = 2$.

Method II. By giving specific values to x.

Let us put $x = 1$ and $x = 2$ in (1).

$x = 1 \Rightarrow 3(1) - 1 = A(1 + 1) + B(1 - 1) \Rightarrow 2A = 2 \Rightarrow A = 1$

$x = 2 \Rightarrow 3(2) - 1 = A(2 + 1) + B(2 - 1) \Rightarrow 3A + B = 5 \Rightarrow B = 5 - 3A = 5 - 3(1) = 2.$

$\therefore$ $A = 1$ and $B = 2$.

In this method, it is always convenient to use those values of x which make linear factors in the denominator zero.

Here $x - 1 = 0 \Rightarrow x = 1$ and $x + 1 = 0 \Rightarrow x = -1$. We put $x = 1$ and $x = -1$ in (1).

$x = 1 \Rightarrow 3(1) - 1 = A(1 + 1) + B(0) \Rightarrow A = 2/2 = 1$

$x = -1 \Rightarrow 3(-1) - 1 = A(0) + B(-1 - 1) \Rightarrow B = -4/-2 = 2.$

$\therefore$ $A = 1$ and $B = 2$.

$$\therefore \quad \frac{3x-1}{(x-1)(x+1)} = \frac{\mathbf{1}}{\mathbf{x-1}} + \frac{\mathbf{2}}{\mathbf{x+1}}.$$

Remark. In the second method, we observe that

$$A = \text{Value of } \frac{3x-1}{x+1} \text{ when } x = 1 \quad \text{and} \quad B = \text{Value of } \frac{3x-1}{x-1} \text{ when } x = -1.$$

$\therefore$ The value of A can be found out by putting $x = 1$ in the given proper fraction, after omitting $x - 1$ from the denominator. Similarly, the value of B can be found out by putting $x = -1$ in the given proper fraction, after omitting $x + 1$ from the denominator.

Thus, when the factors in the denominator are *linear* and *non-repeated,* we can decompose the given proper fraction into partial fractions as follows :

$$\frac{3x-1}{(x-1)(x+1)} = \frac{3(1)-1}{(x-1)(1+1)} + \frac{3(-1)-1}{(-1-1)(x+1)} = \frac{2}{(x-1)2} + \frac{-4}{(-2)(x+1)} = \frac{1}{x-1} + \frac{2}{x+1}.$$

This is called the **short cut method** of finding the values of A, B etc. It is very important to bear in mind that this short cut method is to be used only when the denominator has *only linear and non-repeated factors.*

(*ii*) Given fraction is not a proper fraction. Dividing $6x^3 + 5x^2 - 7$ by $3x^2 - 2x - 1$, we get

$$6x^3 + 5x^2 - 7 = (2x + 3)(3x^2 - 2x - 1) + (8x - 4).$$

$$\therefore \quad \frac{6x^3+5x^2-7}{3x^2-2x-1} = 2x + 3 + \frac{8x-4}{(x-1)(3x+1)} \qquad (\because \; 3x^2 - 2x - 1 = (x-1)(3x+1))$$

$$= 2x + 3 + \left(\frac{8(1)-4}{(x-1)(3(1)+1)} + \frac{8\left(-\frac{1}{3}\right)-4}{\left(-\frac{1}{3}-1\right)(3x+1)} \right)$$

(Using **shortcut method**)

$$= 2x + 3 + \frac{4}{(x-1)4} + \frac{-20/3}{\left(-\frac{4}{3}\right)(3x+1)} = \mathbf{2x + 3 + \frac{1}{x-1} + \frac{5}{3x+1}}.$$

(iii) $\dfrac{2x+1}{(x+2)(x-3)^2}$ is a proper fraction. Let the partial fraction corresponding to the factor $x + 2$ be $\dfrac{A}{x+2}$ and the partial fractions, corresponding to the factor $(x-3)^2$ be $\dfrac{B}{x-3}$ and $\dfrac{C}{(x-3)^2}$.

$$\therefore \quad \frac{2x+1}{(x+2)(x-3)^2} = \frac{A}{x+2} + \frac{B}{x-3} + \frac{C}{(x-3)^2}.$$

Multiplying both sides by $(x+2)(x-3)^2$, we get

$$2x + 1 = A(x-3)^2 + B(x+2)(x-3) + C(x+2) \quad \text{...(1)}$$

Now $x + 2 = 0 \Rightarrow x = -2$ and $x - 3 = 0 \Rightarrow x = 3$.

$$x = -2 \text{ in (1) implies } -3 = A(-5)^2 + B(0) + C(0) \Rightarrow A = \frac{-3}{25}$$

$$x = 3 \text{ in (1) implies } 7 = A(0) + B(0) + C(5) \Rightarrow C = \frac{7}{5}$$

Comparing the coefficients of x^2 in (1), we get $0 = A + B$.

$$\therefore \quad B = -A = -\left(\frac{-3}{25}\right) = \frac{3}{25}.$$

$$\therefore \quad \frac{2x+1}{(x+2)(x-3)^2} = \frac{\frac{-3}{25}}{x+2} + \frac{\frac{3}{25}}{x-3} + \frac{\frac{7}{5}}{(x-3)^2} = -\frac{\mathbf{3}}{\mathbf{25(x+2)}} + \frac{\mathbf{3}}{\mathbf{25(x-3)}} + \frac{\mathbf{7}}{\mathbf{5(x-3)^2}}.$$

(iv) $\dfrac{2x+3}{(x+1)(x^2+1)}$ is a proper fraction. Let the partial fraction corresponding to the factor $x + 1$ be $\dfrac{A}{x+1}$ and the partial fraction corresponding to the factor $x^2 + 1$ be $\dfrac{Bx+C}{x^2+1}$.

$$\therefore \quad \frac{2x+3}{(x+1)(x^2+1)} = \frac{A}{x+1} + \frac{Bx+C}{x^2+1}.$$

Multiplying both sides by $(x+1)(x^2+1)$, we get

$$2x + 3 = A(x^2+1) + (Bx+C)(x+1) \quad \text{...(1)}$$

Put $x + 1 = 0$ or $x = -1$ in (1) $\Rightarrow 1 = A(2) + (-B + C)(0) \Rightarrow A = \dfrac{1}{2}$

Comparing the coefficients of x^2 in (1), we get $0 = A + B$. $\therefore\ B = -A = -\dfrac{1}{2}$.

Comparing the coefficients of x, we get $2 = B + C$. $\therefore\ C = 2 - B = 2 + \dfrac{1}{2} = \dfrac{5}{2}$.

$$\therefore \quad \frac{2x+3}{(x+1)(x^2+1)} = \frac{1/2}{x+1} + \frac{(-1/2)x + 5/2}{x^2+1} = \frac{\mathbf{1}}{\mathbf{2(x+1)}} + \frac{\mathbf{5 - x}}{\mathbf{2(x^2+1)}}.$$

(v) $\dfrac{x^2+5}{(x^2+x+1)(x^2-x+2)}$ is a proper fraction with non-repeated irreducible quadratic factors $x^2 + x + 1$ and $x^2 - x + 2$.

Let $\dfrac{x^2+5}{(x^2+x+1)(x^2-x+2)} = \dfrac{Ax+B}{x^2+x+1} + \dfrac{Cx+D}{x^2-x+2}$.

Multiplying both sides by $(x^2+x+1)(x^2-x+2)$, we get

$$x^2+5 = (Ax+B)(x^2-x+2)+(Cx+D)(x^2+x+1) \qquad \ldots(1)$$

Comparing the coefficients of x^3, x^2, x and the constant terms, we get

$0 = A + C \quad \ldots(1) \qquad 1 = -A+B+C+D \quad \ldots(2)$

$0 = 2A - B + C + D \quad \ldots(3) \qquad 5 = 2B + D \quad \ldots(4)$

(1) $\Rightarrow$ $C = -A$. $\therefore$ (2) $\Rightarrow$ $-A+B-A+D = 1$ *i.e.*, $-2A+B+D=1$ ...(5)

(3) $\Rightarrow$ $2A-B-A+D=0$ *i.e.*, $A-B+D=0$ ...(6)

(5) + 2(6) $\Rightarrow$ $-B+3D=1$ ...(7)

Solving (4) and (7), we get $B = 2$ and $D = 1$.

(5) $\Rightarrow$ $-2A+2+1=1$ or $A=1$. (1) $\Rightarrow$ $C=-A=-1$.

$\therefore$ $A=1$, $B=2$, $C=-1$, $D=1$.

$$\therefore \quad \frac{x^2+5}{(x^2+x+1)(x^2-x+2)} = \frac{1.x+2}{x^2+x+1} + \frac{-1.x+1}{x^2-x+2} = \mathbf{\frac{x+2}{x^2+x+1} - \frac{x-1}{x^2-x+2}}.$$

(*vi*) $\dfrac{x}{(x-1)(x^2+1)^2}$ is a proper fraction. Let $\dfrac{x}{(x-1)(x^2+1)^2} = \dfrac{A}{x-1} + \dfrac{Bx+C}{x^2+1} + \dfrac{Dx+E}{(x^2+1)^2}$.

Multiplying by $(x-1)(x^2+1)^2$, we get

$$x = A(x^2+1)^2 + (Bx+C)(x-1)(x^2+1) + (Dx+E)(x-1) \qquad \ldots(1)$$

$x = 1$ in (1) $\Rightarrow$ $1 = A(4) + (B+C)(0)(2) + (D+E)(0)$ $\Rightarrow$ $A = \dfrac{1}{4}$

Comparing the coefficients x^4, x^3, x^2, x, in (1), we get

$0 = A + B \quad \ldots(2) \qquad 0 = -B + C \quad \ldots(3)$

$0 = 2A + B - C + D \quad \ldots(4) \qquad 1 = -B + C - D + E \quad \ldots(5)$

(2) $\Rightarrow$ $B = -A = -\dfrac{1}{4}$, (3) $\Rightarrow$ $C = B = -\dfrac{1}{4}$

$$\Rightarrow D = -2A - B + C = -\frac{1}{2} + \frac{1}{4} - \frac{1}{4} = -\frac{1}{2}$$

(5) $\Rightarrow$ $E = 1 + B - C + D = 1 - \dfrac{1}{4} + \dfrac{1}{4} - \dfrac{1}{2} = \dfrac{1}{2}$

$\therefore$ $A = \dfrac{1}{4}$, $B = -\dfrac{1}{4}$, $C = -\dfrac{1}{4}$, $D = -\dfrac{1}{2}$, $E = \dfrac{1}{2}$.

$$\therefore \quad \frac{x}{(x-1)(x^2+1)^2} = \frac{1/4}{x-1} + \frac{(-1/4)x + (-1/4)}{x^2+1} + \frac{(-1/2)x + 1/2}{(x^2+1)^2}$$

$$= \mathbf{\frac{1}{4(x-1)} - \frac{x+1}{4(x^2+1)} - \frac{x-1}{2(x^2+1)^2}}.$$

13.25. INTEGRATION BY PARTIAL FRACTIONS

We have seen that if a rational fraction is not a proper fraction, then we can write that rational fraction as the sum of a polynomial and a proper fraction. Further, a proper fraction can be expressed as the sum of its partial fractions. The partial fractions are of the following types :

(a) $\dfrac{\text{Constant}}{\text{Linear polynomial}}$ (b) $\dfrac{\text{Constant}}{(\text{Linear polynomial})^n}$

(c) $\dfrac{\text{Linear polynomial}}{\text{Irreducible quadratic polynomial}}$ (d) $\dfrac{\text{Linear polynomial}}{(\text{Irreducible quadratic polynomial})^n}$

where n (> 1) is any natural number.

WORKING RULES FOR INTEGRATING RATIONAL FRACTIONS

Step I. *Let* $\dfrac{f(x)}{g(x)}$ *be the given rational fraction. If* $\dfrac{f(x)}{g(x)}$ *is not proper, then divide* $f(x)$ *by* $g(x)$ *to obtain the quotient* $q(x)$ *and the remainder* $r(x)$.

Step II. *Write* $\dfrac{f(x)}{g(x)} = q(x) + \dfrac{r(x)}{g(x)}$. *In case* $\dfrac{f(x)}{g(x)}$ *is proper, then we have* $q(x) = 0$ *and* $r(x) = f(x)$.

Step III. *Factorize* $g(x)$ *into linear and irreducible quadratic factors. Express* $\dfrac{r(x)}{g(x)}$ *as the sum of its partial fractions.*

Step IV. *Write* $\dfrac{f(x)}{g(x)}$ *as the sum of* $q(x)$ *and the sum of partial fractions of* $\dfrac{r(x)}{g(x)}$. *Integrate each part on the right hand side and this gives the required integral.*

Example 2. *Evaluate :* $\displaystyle\int \frac{3x+4}{x^2-5x+6}\,dx.$

Sol. Let $\displaystyle I = \int \frac{3x+4}{x^2-5x+6}\,dx. \quad \therefore \quad I = \int \frac{3x+4}{(x-2)(x-3)}\,dx$

The integrand is a *proper fraction*.

Let $\dfrac{3x+4}{(x-2)(x-3)} = \dfrac{A}{x-2} + \dfrac{B}{x-3}$.

Multiplying by $(x-2)(x-3)$, we get $3x + 4 = A(x-3) + B(x-2)$ ...(1)

$x - 2 = 0 \Rightarrow x = 2$ and $x - 3 = 0 \Rightarrow x = 3$.

We put $x = 2$ and $x = 3$ in (1)

$x = 2 \Rightarrow 3(2) + 4 = A(2-3) + B(2-2) \Rightarrow 10 = -A + 0 \Rightarrow A = -10$

$x = 3 \Rightarrow 3(3) + 4 = A(3-3) + B(3-2) \Rightarrow 13 = 0 + B \Rightarrow B = 13$

$$\therefore \quad \frac{3x+4}{(x-2)(x-3)} = \frac{-10}{x-2} + \frac{13}{x-3}$$

$$\therefore \quad I = \int \left(\frac{-10}{x-2} + \frac{13}{x-3}\right) dx = \mathbf{-10 \log|x-2| + 13 \log|x-3| + C.}$$

Alternative method

Let $\displaystyle I = \int \frac{3x+4}{x^2-5x+6}\,dx$. Let $3x + 4 = p\dfrac{d}{dx}(x^2 - 5x + 6) + q. \quad \therefore \quad 3x + 4 = p(2x-5) + q.$

Comparing coefficients, we get $3 = 2p$ and $4 = -5p + q$.

Solving these, we get $p = \frac{3}{2}$ and $q = \frac{23}{2}$. $\therefore$ $3x + 4 = \frac{3}{2}(2x - 5) + \frac{23}{2}$

$$\therefore \quad I = \int \frac{\frac{3}{2}(2x-5)+\frac{23}{2}}{x^2-5x+6}\,dx = \frac{3}{2}\int \frac{2x-5}{x^2-5x+6}\,dx + \frac{23}{2}\int \frac{dx}{x^2-5x+6} = \frac{3}{2}I_1 + \frac{23}{2}I_2 \text{ (say)}$$

$$I_1 = \int \frac{2x-5}{x^2-5x+6}\,dx = \log|x^2-5x+6| + C_1.$$

$$I_2 = \int \frac{dx}{x^2-5x+6} = \int \frac{dx}{\left(x^2-5x+\frac{25}{4}\right)+6-\frac{25}{4}} = \int \frac{dx}{\left(x-\frac{5}{2}\right)^2-\frac{1}{4}}$$

Let $z = x - \frac{5}{2}$. $\therefore$ $dz = dx$.

$$\therefore \quad I_2 = \int \frac{dz}{z^2-(1/2)^2} = \frac{1}{2\left(\frac{1}{2}\right)}\log\left|\frac{z-1/2}{z+1/2}\right| + C_2 = \log\left|\frac{x-\frac{5}{2}-\frac{1}{2}}{x-\frac{5}{2}+\frac{1}{2}}\right| + C_2 = \log\left|\frac{x-3}{x-2}\right| + C_2.$$

$$\therefore \quad I = \frac{3}{2}I_1 + \frac{23}{2}I_2 = \frac{3}{2}(\log|x^2-5x+6| + C_1) + \frac{23}{2}\left(\log\left|\frac{x-3}{x-2}\right| + C_2\right)$$

$$= \frac{3}{2}\log|(x-2)(x-3)| + \frac{23}{2}\log\left|\frac{x-3}{x-2}\right| + C, \text{ where } C = \frac{3}{2}C_1 + \frac{23}{2}C_2$$

$$= \frac{3}{2}\log|x-2| + \frac{3}{2}\log|x-3| + \frac{23}{2}\log|x-3| - \frac{23}{2}\log|x-2| + C$$

$$= \mathbf{-10\log|x-2| + 13\log|x-3| + C.}$$

A special note on $\int \frac{px+q}{ax^2+bx+c}\,dx$**.** The quadratic polynomial $ax^2 + bx + c$ may or may not be irreducible.

Case I. ax² + bx + c is either irreducible or not easily reducible. In this case $\int \frac{px+q}{ax^2+bx+c}\,dx$ is evaluated by expressing it as $\lambda\int \frac{2ax+b}{ax^2+bx+c}\,dx + \mu\int \frac{1}{ax^2+bx+c}\,dx$ for some scalars λ and μ. This method has already been discussed.

Case II. ax² + bx + c is easily reducible. In this case, either the method discussed in **case I** is used or the method of partial fractions is used. The reader would find that the method of integrating by partial fractions is comparatively very easy.

$\therefore$ If $ax^2 + bx + c$ is easily reducible into linear factors, then $\int \frac{px+q}{ax^2+bx+c}\,dx$ must be evaluated through partial fractions.

Example 3. *Evaluate :*

(i) $\int \frac{dx}{(x+1)(2x+1)}$ (ii) $\int \frac{(2x+1)\,dx}{(x+1)(x+2)}$ (iii) $\int \frac{x^2}{(x-1)(x-2)}\,dx$

(*iv*) $\int \frac{2x-1}{(x-1)(x+2)(x-3)}\,dx$ (*v*) $\int \frac{x^2+2}{(x^2+1)(x^2+4)}\,dx$ (*vi*) $\int \frac{e^x}{e^{2x}-4}\,dx.$

Sol. (*i*) Let $I = \int \frac{dx}{(x+1)(2x+1)}$. The integrand is a *proper* fraction.

Let $\frac{1}{(x+1)(2x+1)} = \frac{A}{x+1} + \frac{B}{2x+1}.$

Multiplying by $(x + 1)(2x + 1)$, we get $1 = A(2x + 1) + B(x + 1)$ ...(1)

$x + 1 = 0 \Rightarrow x = -1$ and $2x + 1 = 0 \Rightarrow x = -1/2.$

In (1), $x = -1 \Rightarrow 1 = A(-2 + 1) + B(0) \Rightarrow A = -1$

$x = -1/2 \Rightarrow 1 = A(0) + B\left(-\tfrac{1}{2} + 1\right) \Rightarrow B = 2$

$$\therefore \quad \frac{1}{(x+1)(2x+1)} = \frac{-1}{x+1} + \frac{2}{2x+1}$$

$$\therefore \quad I = \int\left(\frac{-1}{x+1} + \frac{2}{2x+1}\right)dx = -\log|x+1| + 2 \cdot \frac{\log|2x+1|}{2} + C = \mathbf{\log\left|\frac{2x+1}{x+1}\right| + C.}$$

(*ii*) Let $I = \int \frac{2x+1}{(x+1)(x-2)}\,dx$. The integrand is a *proper* fraction.

The factors in the denominator are *linear and non-repeated.* We use the *short cut method* to break up the integrand.

$$\frac{2x+1}{(x+1)(x-2)} = \frac{2(-1)+1}{(x+1)(-1-2)} + \frac{2(2)+1}{(2+1)(x-2)} = \frac{-1}{-3(x+1)} + \frac{5}{3(x-2)} = \frac{1}{3}\left[\frac{1}{x+1} + \frac{5}{x-2}\right]$$

$$\therefore \quad I = \frac{1}{3}\int\left[\frac{1}{x+1} + \frac{5}{x-2}\right]dx = \mathbf{\frac{1}{3}\,[\log|x+1| + 5\log|x-2|] + C.}$$

(*iii*) Let $I = \int \frac{x^2}{(x-1)(x-2)}\,dx.$

Since the integrand is not a *proper* fraction, we have by long division,

$$\frac{x^2}{(x-1)(x-2)} = \frac{x^2}{x^2-3x+2} = \frac{x^2-3x+2+3x-2}{x^2-3x+2}$$

$$= 1 + \frac{3x-2}{x^2-3x+2} = 1 + \frac{3x-2}{(x-1)(x-2)} \qquad ...(1)$$

$$\left[\text{By } \textit{short cut method}: \frac{3x-2}{(x-1)(x-2)} = \frac{3(1)-2}{(x-1)(1-2)} + \frac{3(2)-2}{(2-1)(x-2)} = \frac{1}{(x-1)(-1)} + \frac{4}{1(x-2)} = -\frac{1}{x-1} + \frac{4}{x-2}\right]$$

$$\therefore \quad (1) \Rightarrow \frac{x^2}{(x-1)(x-2)} = 1 - \frac{1}{x-1} + \frac{4}{x-2}$$

$$\therefore \quad I = \int\left(1 - \frac{1}{x-1} + \frac{4}{x-2}\right)dx = \mathbf{x - \log|x-1| + 4\log|x-2| + C.}$$

(*iv*) Let $I = \int \frac{2x-1}{(x-1)(x+2)(x-3)}\,dx$. The integrand is a *proper* fraction.

The factors in the denominator are *linear* and *non-repeated.* By short cut method,

$$\therefore \quad \frac{2x-1}{(x-1)(x+2)(x-3)} = \frac{2(1)-1}{(x-1)(1+2)(1-3)} + \frac{2(-2)-1}{(-2-1)(x+2)(-2-3)} + \frac{2(3)-1}{(3-1)(3+2)(x-3)}$$

$$= \frac{1}{-6(x-1)} + \frac{-5}{15(x+2)} + \frac{5}{10(x-3)}$$

$$\therefore \quad I = \int \left[-\frac{1}{6(x-1)} - \frac{1}{3(x+2)} + \frac{1}{2(x-3)}\right] dx$$

$$= -\frac{1}{6}\log|x-1| - \frac{1}{3}\log|x+2| + \frac{1}{2}\log|x-3| + C.$$

(*v*) Let $I = \int \frac{x^2+2}{(x^2+1)(x^2+4)}\,dx$. Let $z = x^2$. $\therefore \quad \frac{x^2+2}{(x^2+1)(x^2+4)} = \frac{z+2}{(z+1)(z+4)}$

Let $\frac{z+2}{(z+1)(z+4)} = \frac{A}{z+1} + \frac{B}{z+4}$. $\therefore \quad z+2 = A(z+4) + B(z+1)$

$z = -1 \Rightarrow \quad 1 = 3A + B(0) \quad \Rightarrow \quad A = 1/3$

$z = -4 \Rightarrow \quad -2 = A(0) + (-3)B \quad \Rightarrow \quad B = 2/3$

$$\therefore \quad \frac{z+2}{(z+1)(z+4)} = \frac{1/3}{z+1} + \frac{2/3}{z+4}. \quad \therefore \quad \frac{x^2+2}{(x^2+1)(x^2+4)} = \frac{1}{3(x^2+1)} + \frac{2}{3(x^2+4)}.$$

$$\therefore \quad I = \frac{1}{3}\int\left[\frac{1}{x^2+1} + \frac{2}{x^2+4}\right] dx = \frac{1}{3}\left[\tan^{-1}x + \frac{2}{2}\tan^{-1}\frac{x}{2}\right] = \frac{1}{3}\left[\tan^{-1}x + \tan^{-1}\frac{x}{2}\right] + C.$$

(*vi*) Let $I = \int \frac{e^x}{e^{2x}-4}\,dx$. Let $z = e^x$. $\therefore \quad dz = e^x\,dx$.

$$\therefore \quad I = \int \frac{dz}{z^2-4} = \int \frac{dz}{(z-2)(z+2)} = \int\left[\frac{1}{(z-2)(2+2)} + \frac{1}{(-2-2)(z+2)}\right] dz$$

$$= \frac{1}{4}\int\left[\frac{1}{z-2} - \frac{1}{z+2}\right] dz = \frac{1}{4}[\log|z-2| - \log|z+2|] + C$$

$$= \frac{1}{4}\log\left|\frac{z-2}{z+2}\right| + C = \frac{1}{4}\log\left|\frac{e^x-2}{e^x+2}\right| + C.$$

Example 4. *Evaluate :*

(*i*) $\int \frac{12x^2-2x-9}{(4x^2-1)(x+3)}\,dx$ (*ii*) $\int \frac{x^2+8x+4}{x^3-4x}\,dx$

(*iii*) $\int \frac{2x\,dx}{(x^2+2)(x^2+3)}$ (*iv*) $\int \frac{x^2}{(1+x^3)(2+x^3)}\,dx$.

Sol. (*i*) Let $I = \int \frac{12x^2 - 2x - 9}{(4x^2 - 1)(x + 3)} dx$. $\therefore$ $I = \int \frac{12x^2 - 2x - 9}{(2x - 1)(2x + 1)(x + 3)} dx$

The integrand is a *proper* fraction.

The factors of $(4x^2 - 1)(x + 3)$ are *linear* and *non-repeated.*

Let $\frac{12x^2 - 2x - 9}{(2x - 1)(2x + 1)(x + 3)} = \frac{A}{2x - 1} + \frac{B}{2x + 1} + \frac{C}{x + 3}$

$$= \frac{A(2x + 1)(x + 3) + B(2x - 1)(x + 3) + C(2x - 1)(2x + 1)}{(2x - 1)(2x + 1)(x + 3)}$$

$\therefore$ $12x^2 - 2x - 9 = A(2x + 1)(x + 3) + B(2x - 1)(x + 3) + C(2x - 1)(2x + 1)$ ...(1)

Now, $2x - 1 = 0 \Rightarrow x = \frac{1}{2}$, $2x + 1 = 0 \Rightarrow x = -\frac{1}{2}$ and $x + 3 = 0 \Rightarrow x = -3$.

$x = \frac{1}{2}$ in (1) $\Rightarrow 3 - 1 - 9 = A(2)\left(\frac{7}{2}\right) + 0 + 0 \Rightarrow -7 = 7A \Rightarrow \mathbf{A = -1}$

$x = -\frac{1}{2}$ in (1) $\Rightarrow 3 + 1 - 9 = 0 + B(-2)\left(\frac{5}{2}\right) + 0 \Rightarrow -5 = -5B \Rightarrow \mathbf{B = 1}$

and $x = -3$ in (1) $\Rightarrow 108 + 6 - 9 = 0 + 0 + C(-7)(-5) \Rightarrow 105 = 35C \Rightarrow \mathbf{C = 3.}$

$\therefore$ $\frac{12x^2 - 2x - 9}{(2x - 1)(2x + 1)(x + 3)} = \frac{-1}{2x - 1} + \frac{1}{2x + 1} + \frac{3}{x + 3}$

$\therefore$ $I = \int \left[-\frac{1}{2x - 1} + \frac{1}{2x + 1} + \frac{3}{x + 3}\right] dx$

$$= -\frac{\log|2x - 1|}{2} + \frac{\log|2x + 1|}{2} + 3\log|x + 3| + C.$$

Alternative method

Let $I = \int \frac{12x^2 - 2x - 9}{(4x^2 - 1)(x + 3)} dx$. $\therefore$ $I = \int \frac{12x^2 - 2x - 9}{(2x - 1)(2x + 1)(x + 3)} dx.$

Since the factors of $(4x^2 - 1)(x + 3)$ are *linear* and *non-repeated*, we have by *short cut method :*

$$I = \int \left[\frac{12(\frac{1}{2})^2 - 2(\frac{1}{2}) - 9}{(2x - 1)(2(\frac{1}{2}) + 1)(\frac{1}{2} + 3)} + \frac{12(-\frac{1}{2})^2 - 2(-\frac{1}{2}) - 9}{(2(-\frac{1}{2}) - 1)(2x + 1)(-\frac{1}{2} + 3)} + \frac{12(-3)^2 - 2(-3) - 9}{(2(-3) - 1)(2(-3) + 1)(x + 3)}\right] dx$$

$$= \int \left[\frac{3 - 1 - 9}{(2x - 1)(2)\left(\frac{7}{2}\right)} + \frac{3 + 1 - 9}{(-2)(2x + 1)\left(\frac{5}{2}\right)} + \frac{108 + 6 - 9}{(-7)(-5)(x + 3)}\right] dx$$

$$= \int \left[\frac{-7}{(2x - 1)(7)} + \frac{-5}{(2x + 1)(-5)} + \frac{105}{35(x + 3)}\right] dx = \int \left[-\frac{1}{2x - 1} + \frac{1}{2x + 1} + \frac{3}{x + 3}\right] dx$$

$$= -\frac{\log|2x - 1|}{2} + \frac{\log|2x + 1|}{2} + 3\log|x + 3| + C.$$

(*ii*) Let $I = \int \frac{x^2 + 8x + 4}{x^3 - 4x} dx$. The integrand is a *proper* fraction.

To resolve the integrand into *partial fractions,* let

$$\frac{x^2 + 8x + 4}{x^3 - 4x} = \frac{x^2 + 8x + 4}{x(x + 2)(x - 2)} = \frac{A}{x} + \frac{B}{x + 2} + \frac{C}{x - 2} = \frac{A(x + 2)(x - 2) + Bx(x - 2) + Cx(x + 2)}{x(x + 2)(x - 2)}$$

$\therefore \quad x^2 + 8x + 4 = A(x + 2)(x - 2) + Bx(x - 2) + Cx(x + 2)$ ...(1)

$x = 0$ in (1) $\Rightarrow$ $0 + 0 + 4 = -4A$ or $A = -\mathbf{1}$

$x = -2$ in (1) $\Rightarrow$ $4 - 16 + 4 = 8B$ or $B = -\mathbf{1}$

$x = 2$ in (1) $\Rightarrow$ $4 + 16 + 4 = 8C$ or $C = \mathbf{3}$.

Hence (1) can be written as $\frac{x^2 + 8x + 4}{x^3 - 4x} = \frac{-1}{x} - \frac{1}{x + 2} + \frac{3}{x - 2}$.

$$\therefore \quad \int \frac{x^2 + 8x + 4}{x^3 - 4x} dx = -\int \frac{dx}{x} - \int \frac{dx}{x + 2} + 3\int \frac{dx}{x - 2}$$

$$= \mathbf{-\log|x| - \log|x + 2| + 3\log|x - 2| + C.}$$

(*iii*) Let $\quad I = \int \frac{2x\,dx}{(x^2 + 2)(x^2 + 3)}$. Let $z = x^2$. $\therefore \quad dz = 2x\,dx$.

$\therefore \quad I = \int \frac{dz}{(z + 2)(z + 3)}$. The integrand is a *proper* fraction.

Since the factors of $(z + 2)(z + 3)$ are *linear* and *non-repeated,* we have by *short-cut method* :

$$I = \int \left[\frac{1}{(z + 2)(-2 + 3)} + \frac{1}{(-3 + 2)(z + 3)}\right] dz \qquad | \text{ Short cut method}$$

$$= \int \left[\frac{1}{z + 2} - \frac{1}{z + 3}\right] dz = [\log|z + 2| - \log|z + 3|] + C$$

$$= \log\left|\frac{z + 2}{z + 3}\right| + C = \log\left|\frac{x^2 + 2}{x^2 + 3}\right| + C = \mathbf{\log\frac{x^2 + 2}{x^2 + 3} + C.}$$

(*iv*) Let $\quad I = \int \frac{x^2}{(1 + x^3)(2 + x^3)} dx$. Let $z = x^3$. $\therefore \quad dz = 3x^2\,dx$.

$\therefore \quad I = \frac{1}{3}\int \frac{dz}{(1 + z)(2 + z)}$. The integrand is a *proper* fraction.

Since the factors of $(1 + z)(2 + z)$ are *linear* and *non-repeated*, we have by *shortcut method* :

$$I = \frac{1}{3}\int \left[\frac{1}{(1 + z)(2 - 1)} + \frac{1}{(1 - 2)(2 + z)}\right] dz$$

$$= \frac{1}{3}\int \left[\frac{1}{1 + z} - \frac{1}{2 + z}\right] dz = \frac{1}{3}[\log|1 + z| - \log|2 + z|] + C$$

$$= \frac{1}{3}\log\left|\frac{1 + z}{2 + z}\right| + C = \mathbf{\frac{1}{3}\log\left|\frac{1 + x^3}{2 + x^3}\right| + C.}$$

Example 5. *Evaluate :*

(i) $\int \frac{5x^3 + 18x^2 - 10x - 6}{x(x+3)(5x-2)} dx$ (ii) $\int \frac{x^4 + x^3 + 2x^2 + 4x + 1}{x(x+1)} dx$

(iii) $\int \frac{dx}{\sin x\,(3 + 2\cos x)}$ (iv) $\int \frac{x^2\, dx}{x^2 - 4x + 3}$.

Sol. (i) Since the integrand is *not a proper* fraction, we have by long division,

$$\frac{5x^3 + 18x^2 - 10x - 6}{x(x+3)(5x-2)} = \frac{5x^3 + 18x^2 - 10x - 6}{5x^3 + 13x^2 - 6x}$$

$$= \frac{5x^3 + 13x^2 - 6x + 5x^2 - 4x - 6}{5x^3 + 13x^2 - 6x} = 1 + \frac{5x^2 - 4x - 6}{5x^3 + 13x^2 - 6x} \quad ...(1)$$

We resolve $\frac{5x^2 - 4x - 6}{5x^3 + 13x^2 - 6x}$ into partial fractions.

Let $\frac{5x^2 - 4x - 6}{x(x+3)(5x-2)} = \frac{A}{x} + \frac{B}{x+3} + \frac{C}{5x-2}$ ...(2)

$$= \frac{A(x+3)(5x-2) + Bx(5x-2) + Cx(x+3)}{x(x+3)(5x-2)}.$$

$\therefore$ $5x^2 - 4x - 6 = A(x+3)(5x-2) + Bx(5x-2) + Cx(x+3)$ ...(3)

When $x = 0$, (3) $\Rightarrow$ $-6 = -6A$ $\Rightarrow$ $A = 1$

When $x = -3$, (3) $\Rightarrow$ $45 + 12 - 6 = B(-3)(-15-2)$ $\Rightarrow$ $B = 1$

When $x = \frac{2}{5}$, (3) $\Rightarrow$ $5\left(\frac{2}{5}\right)^2 - 4\left(\frac{2}{5}\right) - 6 = C\left(\frac{2}{5}\right)\left(\frac{2}{5} + 3\right)$ $\Rightarrow$ $C = -5$.

$\therefore$ (3) $\Rightarrow$ $\frac{5x^2 - 4x - 6}{x(x+3)(5x-2)} = \frac{1}{x} + \frac{1}{x+3} - \frac{5}{5x-2}$

$\therefore$ (1) $\Rightarrow$ $\int \frac{5x^3 + 18x^2 - 10x - 6}{x(x+3)(5x-2)} dx = \int \left(1 + \frac{1}{x} + \frac{1}{x+3} - \frac{5}{5x-2}\right) dx$

$$= \int dx + \int \frac{dx}{x} + \int \frac{dx}{x+3} - 5\int \frac{dx}{5x-2}$$

$$= x + \log|x| + \log|x+3| - 5 \cdot \frac{\log|5x-2|}{5} + C$$

$$= \mathbf{x + \log|x| + \log|x+3| - \log|5x-2| + C.}$$

(ii) Since the integrand is *not a proper* fraction, we use long division to get

$$\frac{x^4 + x^3 + 2x^2 + 4x + 1}{x(x+1)} = \frac{x^2(x^2+x) + 2(x^2+x) + 2x + 1}{x^2 + x} = x^2 + 2 + \frac{2x+1}{x^2+x} \quad ...(1)$$

We resolve $\frac{2x+1}{x^2+x}$ into partial fractions.

Let $\dfrac{2x+1}{x(x+1)} = \dfrac{A}{x} + \dfrac{B}{x+1} = \dfrac{A(x+1)+Bx}{x(x+1)}$ ...(2)

$\therefore \quad 2x + 1 = A(x + 1) + Bx.$ ...(3)

When $x = 0$, (3) $\Rightarrow \; 1 = A$

When $x = -1$, (3) $\Rightarrow \; -2 + 1 = -B$ or $B = 1$.

$\therefore$ (2) $\Rightarrow \dfrac{2x+1}{x(x+1)} = \dfrac{1}{x} + \dfrac{1}{x+1}$

$\therefore$ (1) $\Rightarrow \displaystyle\int \frac{x^4+x^3+2x^2+4x+1}{x(x+1)}\,dx = \int\left(x^2+2+\frac{1}{x}+\frac{1}{x+1}\right)dx$

$$= \mathbf{\frac{x^3}{3} + 2x + \log|x| + \log|x+1| + C.}$$

(*iii*) Let $I = \displaystyle\int \frac{dx}{\sin x\,(3+2\cos x)}$.

$\therefore \quad I = \displaystyle\int \frac{\sin x\,dx}{(1-\cos^2 x)(3+2\cos x)} = \int \frac{\sin x\,dx}{(1-\cos x)(1+\cos x)(3+2\cos x)}$

Let $z = \cos x$. $\therefore \; dz = -\sin x\,dx$

$\therefore \quad I = \displaystyle\int \frac{-dz}{(1-z)(1+z)(3+2z)} = \int\left(\frac{-1}{(1-z)(2)(5)} + \frac{-1}{(2)(1+z)(1)} + \frac{1}{(\frac{5}{2})(-\frac{1}{2})(3+2z)}\right)dz$

(Using shortcut method)

$\therefore \quad I = \displaystyle\int\left[\frac{-1}{10(1-z)} - \frac{1}{2(1+z)} + \frac{4}{5(3+2z)}\right]dz$

$= -\dfrac{1}{10}\cdot\dfrac{\log|1-z|}{-1} - \dfrac{1}{2}\log|1+z| + \dfrac{4}{5}\cdot\dfrac{\log|3+2z|}{2} + C$

$= \dfrac{1}{10}\log|1-\cos x| - \dfrac{1}{2}\log|1+\cos x| + \dfrac{2}{5}\log|3+2\cos x| + C$

$= \mathbf{\dfrac{1}{10}\log(1-\cos x) - \dfrac{1}{2}\log(1+\cos x) + \dfrac{2}{5}\log|3+2\cos x| + C.}$

(*iv*) Let $I = \displaystyle\int \frac{x^2\,dx}{x^2-4x+3}$. The integrand is *not a proper* fraction.

By long division $\dfrac{x^2}{x^2-4x+3} = \dfrac{x^2-4x+3+4x-3}{x^2-4x+3} = 1 + \dfrac{4x-3}{x^2-4x+3}$ ...(1)

By *short cut method* :

$\dfrac{4x-3}{x^2-4x+3} = \dfrac{4x-3}{(x-1)(x-3)} = \dfrac{4(1)-3}{(x-1)(1-3)} + \dfrac{4(3)-3}{(3-1)(x-3)}$

$= \dfrac{1}{(x-1)(-2)} + \dfrac{9}{2(x-3)} = -\dfrac{1}{2(x-1)} + \dfrac{9}{2(x-3)}$

$\therefore$ (1) $\Rightarrow \dfrac{x^2}{x^2-4x+3} = 1 - \dfrac{1}{2(x-1)} + \dfrac{9}{2(x-3)}$.

$\therefore \quad I = \displaystyle\int\left(1 - \frac{1}{2(x-1)} + \frac{9}{2(x-3)}\right)dx = \mathbf{x - \frac{1}{2}\log|x-1| + \frac{9}{2}\log|x-3| + C.}$

EXERCISE 13.9

Evaluate the following integrals (Q. No. 1–7) :

1. (i) $\int \frac{(3x+2)\,dx}{(x-1)(2x+3)}$ (ii) $\int \frac{x\,dx}{(x+2)(3-2x)}$ (iii) $\int \frac{(x-1)\,dx}{(x+1)(x-2)}$

(iv) $\int \frac{x\,dx}{x^2-x-2}$ (v) $\int \frac{(x+7)\,dx}{x^2+2x-8}$ (vi) $\int \frac{dx}{x-x^3}$

2. (i) $\int \frac{dx}{(x-1)(x-2)(x-3)}$ (ii) $\int \frac{x^2\,dx}{(x-1)(x-2)(x-3)}$ (iii) $\int \frac{(2x-3)\,dx}{(x^2-1)(2x+3)}$

(iv) $\int \frac{dx}{(x-1)(x+2)(x-3)}$

3. (i) $\int \frac{\sin x\,dx}{(2+\cos x)(5+\cos x)}$ (ii) $\int \frac{\cos x\,dx}{(1+\sin x)(2+\sin x)}$ (iii) $\int \frac{5x}{(x+1)(x^2-4)}\,dx$

(iv) $\int \frac{dx}{x[6(\log x)^2+7\log x+2]}$ (v) $\int \frac{x}{(x^2+1)(x^2+2)}\,dx$ (vi) $\int \frac{dx}{\sin x(1+2\cos x)}$

4. (i) $\int \frac{x^2}{(x^2+2)(x^2+3)}\,dx$ (ii) $\int \frac{(x^2+1)\,dx}{(x^2+2)(2x^2+1)}$ (iii) $\int \frac{dx}{x^4-a^4}$

(iv) $\int \frac{x^2\,dx}{(x^2+2)(2x^2+1)}$ (v) $\int \frac{(2x^2-1)\,dx}{(3x^2+7)(x^2-8)}$ (vi) $\int \frac{(x^2+1)(x^2+2)}{(x^2+3)(x^2+4)}\,dx$

5. (i) $\int \frac{dx}{x(x^4+1)}$ (ii) $\int \frac{dx}{x(x^4-1)}$ (iii) $\int \frac{dx}{e^x-1}$

(iv) $\int \frac{2x}{(1+x^2)(3+x^2)}\,dx$ (v) $\int \frac{\sin 2x}{(1+\sin x)(2+\sin x)}\,dx$

(P.T.U. B.Pharma Dec. 2005)

6. (i) $\int \frac{x^3\,dx}{x^2-4}$ (ii) $\int \frac{x^2+x+3}{(x-2)(x+1)}\,dx$ (iii) $\int \frac{x^2+5x+3}{x^2+3x+2}\,dx$

(P.T.U. B.Pharma May 2005)

(iv) $\int \frac{x^3}{(x-1)(x-2)}\,dx$ (v) $\int \frac{x^2+5x+2}{x^2+3x+2}\,dx$ (vi) $\int \frac{x^4}{(x+1)(x-2)}\,dx$

7. (i) $\int \frac{dx}{\sin x+\sin 2x}$ (ii) $\int \frac{\cos x}{(1-\sin x)(2-\sin x)}\,dx$ (iii) $\int \frac{\sin x}{1-4\cos^2 x}\,dx$

(iv) $\int \frac{\sin x}{\sin 4x}\,dx$ (v) $\int \frac{dx}{1+3e^x+2e^{2x}}$ (vi) $\int \frac{dx}{x(1+4x^3+3x^6)}$.

Evaluate the following integrals (Q. No. 8–12) :

8. (i) $\int \frac{3x-2}{(x+3)(x+1)^2}\,dx$ (ii) $\int \frac{\cos x}{(1-\sin x)^3(2+\sin x)}\,dx$ (iii) $\int \frac{5x^2+18x+17}{(x+1)^2(2x+3)}\,dx$

(iv) $\int \frac{x^2}{(x-1)^3(x+1)}\,dx$ (v) $\int \frac{3-2x^2}{(x^2-3x+2)^2}\,dx$ (vi) $\int \frac{3x+5}{x^3-x^2-x+1}\,dx$

9. (*i*) $\int \frac{dx}{(e^x - 1)^2}$ (*ii*) $\int \frac{dx}{e^{2x} - 3e^x}$ (*iii*) $\int \frac{x^2 + x - 1}{(x+1)^2 (x+2)} dx$
(*P.T.U. B.Pharma Dec. 2005*)

10. (*i*) $\int \frac{x^2 - 2}{x^5 - x} dx$ (*ii*) $\int \frac{\sin x}{\cos x (1 + \cos^2 x)} dx$ (*iii*) $\int \frac{(1+x)^3}{(1-x)^3} dx$

11. (*i*) $\int \frac{3\, dx}{(1-x)(1+x^2)}$ (*ii*) $\int \frac{dx}{(2-x)(x^2+3)}$ (*iii*) $\int \frac{dx}{x^3 + 1}$

(*iv*) $\int \frac{dx}{1 + x + x^2 + x^3}$ (*v*) $\int \frac{5x\, dx}{(x+1)(x^2+9)}$ (*vi*) $\int \frac{5}{(x^2+1)(x+2)} dx$

12. (*i*) $\int \frac{8}{(x^2+4)(x+2)} dx$ (*ii*) $\int \frac{2x^2 + x + 7}{(x^2+1)(x+5)} dx$ (*iii*) $\int \frac{x^3 - 1}{x^3 + x} dx$

(*iv*) $\int \frac{18}{(x^2+2)(x+4)} dx$ (*v*) $\int \frac{10}{(x-1)(x^2+9)} dx$ (*vi*) $\int \frac{(2 + \tan^2 \theta) \sec^2 \theta}{1 + \tan^3 \theta} d\theta.$

Answers

1. (*i*) $\frac{1}{2} \log [(x-1)^2 \mid 2x+3 \mid] + C$ (*ii*) $-\frac{2}{7} \log \mid x+2 \mid - \frac{3}{14} \log \mid 3 - 2x \mid + C$

(*iii*) $\frac{1}{3} \log (x+1)^2 \mid x-2 \mid + C$ (*iv*) $\frac{1}{3} \log [\mid x+1 \mid (x-2)^2] + C$ (*v*) $\frac{1}{2} \log \frac{|x-2|^3}{|x+4|} + C$

(*vi*) $\frac{1}{2} \log \frac{x^2}{|1-x^2|} + C$ **2.** (*i*) $\frac{1}{2} \log \mid x-1 \mid - \log \mid x-2 \mid + \frac{1}{2} \log \mid x-3 \mid + C$

(*ii*) $\frac{1}{2} \log \mid x-1 \mid - 4 \log \mid x-2 \mid + \frac{9}{2} \log \mid x-3 \mid + C$

(*iii*) $-\frac{1}{10} \log \mid x-1 \mid + \frac{5}{2} \log \mid x+1 \mid - \frac{12}{5} \log \mid 2x+3 \mid + C$

(*iv*) $-\frac{1}{6} \log \mid x-1 \mid + \frac{1}{15} \log \mid x+2 \mid + \frac{1}{10} \log \mid x-3 \mid + C$

3. (*i*) $\frac{1}{3} \log \left| \frac{5 + \cos x}{2 + \cos x} \right| + C$ (*ii*) $\log (1 + \sin x) - \log (2 + \sin x) + C$

(*iii*) $\frac{5}{6} \log \frac{(x+1)^2 |x-2|}{|(x+2)|^3} + C$ (*iv*) $\log \left| \frac{2 \log x + 1}{3 \log x + 2} \right| + C$ (*v*) $\frac{1}{2} \log \frac{x^2 + 1}{x^2 + 2} + C$

(*vi*) $\frac{1}{6} \log (1 - \cos x) + \frac{1}{2} \log (1 + \cos x) - \frac{2}{3} \log \mid 1 + 2 \cos x \mid + C$

4. (*i*) $-\sqrt{2} \tan^{-1} \frac{x}{\sqrt{2}} + \sqrt{3} \tan^{-1} \frac{x}{\sqrt{3}} + C$ (*ii*) $\frac{1}{3\sqrt{2}} \tan^{-1} \frac{x}{\sqrt{2}} + \frac{1}{3\sqrt{2}} \tan^{-1} (\sqrt{2}\, x) + C$

(*iii*) $\frac{1}{4a^3} \left[\log \left| \frac{x-a}{x+a} \right| - 2 \tan^{-1} \frac{x}{a} \right] + C$ (*iv*) $\frac{2}{3\sqrt{2}} \tan^{-1} \frac{x}{\sqrt{2}} - \frac{1}{3\sqrt{2}} \tan^{-1} \sqrt{2}\, x + C$

(*v*) $\frac{17}{31\sqrt{21}} \tan^{-1} (\sqrt{3/7}\, x) + \frac{15}{124\sqrt{2}} \log \left| \frac{x - 2\sqrt{2}}{x + 2\sqrt{2}} \right| + C$

(*vi*) $x + \frac{2}{\sqrt{3}} \tan^{-1} \frac{x}{\sqrt{3}} - 3 \tan^{-1} \frac{x}{2} + C$

5. (*i*) $\frac{1}{4}\log\frac{x^4}{x^4+1}+C$ (*ii*) $\frac{1}{4}\log\frac{|x^4-1|}{x^4}+C$ (*iii*) $\log\frac{|e^x-1|}{e^x}+C$

(*iv*) $\frac{1}{2}\log\frac{1+x^2}{3+x^2}+C$ (*v*) $-2\log(1+\sin x)+4\log(2+\sin x)+C.$

6. (*i*) $\frac{x^2}{2}+2\log|x^2-4|+C$ (*ii*) $x+3\log|x-2|-\log|x+1|+C$

(*iii*) $x-\log|x+1|+3\log|x+2|+C$ (*iv*) $\frac{x^2}{2}+3x-\log|x-1|+8\log|x-2|+C$

(*v*) $x-2\log|x+1|+4\log|x+2|+C$

(*vi*) $\frac{x^3}{3}+\frac{x^2}{2}+3x+\frac{16}{3}\log|x-2|-\frac{1}{3}\log|x+1|+C$

7. (*i*) $\frac{1}{6}\log(1-\cos x)+\frac{1}{2}\log(1+\cos x)-\frac{2}{3}\log|1+2\cos x|+C$

(*ii*) $\log\frac{2-\sin x}{1-\sin x}+C$ (*iii*) $\frac{1}{4}\log\left|\frac{1-2\cos x}{1+2\cos x}\right|+C$

(*iv*) $-\frac{1}{8}\log\frac{1+\sin x}{1-\sin x}+\frac{1}{4\sqrt{2}}\log\left|\frac{1+\sqrt{2}\sin x}{1-\sqrt{2}\sin x}\right|+C$

(*v*) $x+\log(1+e^x)-2\log(1+2e^x)+C$

(*vi*) $\frac{1}{3}\log|x^3|+\frac{1}{6}\log|1+x^3|-\frac{1}{2}\log|1+3x^3|+C$

8. (*i*) $\frac{11}{4}\log\left|\frac{x+1}{x+3}\right|+\frac{5}{2(x+1)}+C$

(*ii*) $\frac{1}{27}\log\frac{(\sin x+2)}{|1-\sin x|}-\frac{1}{9(\sin x-1)}+\frac{1}{6(\sin x-1)^2}+C$

(*iii*) $\frac{-4}{x+1}+\frac{5}{2}\log|2x+3|+C$ (*iv*) $\frac{1}{8}\log\left|\frac{x-1}{x+1}\right|-\frac{3}{4(x-1)}-\frac{1}{4(x-1)^2}+C$

(*v*) $2\log\left|\frac{x-2}{x-1}\right|-\frac{1}{x-1}+\frac{5}{x-2}+C$ (*vi*) $\frac{1}{2}\log\left|\frac{x+1}{x-1}\right|-\frac{4}{x-1}+C$

9. (*i*) $\log\left|\frac{e^x}{e^x-1}\right|-\frac{1}{e^x-1}+C$ (*ii*) $\frac{1}{3e^x}+\frac{1}{9}\log\left|\frac{e^x-3}{e^x}\right|+C$

(*iii*) $\frac{1}{x+1}+\log|x+2|+C$

10. (*i*) $2\log|x|-\frac{1}{4}\log|x^2-1|-\frac{3}{4}\log(x^2+1)+C$

(*ii*) $\frac{1}{2}\log(1+\cos^2 x)-\log|\cos x|+C$ (*iii*) $-x-6\log|1-x|-\frac{12}{1-x}+\frac{4}{(1-x)^2}+C$

11. (*i*) $-\frac{3}{2}\log|1-x|+\frac{3}{4}\log(1+x^2)+\frac{3}{2}\tan^{-1}x+C$

(*ii*) $-\frac{1}{7}\log|2-x|+\frac{1}{14}\log(x^2+3)+\frac{2}{7\sqrt{3}}\tan^{-1}\frac{x}{\sqrt{3}}+C$

(*iii*) $\frac{1}{3}\log|x+1| - \frac{1}{6}\log|x^2-x+1| + \frac{1}{\sqrt{3}}\tan^{-1}\frac{2x-1}{\sqrt{3}} + C$

(*iv*) $\frac{1}{2}\log|1+x| - \frac{1}{4}\log(1+x^2) + \frac{1}{2}\tan^{-1}x + C$

(*v*) $-\frac{1}{2}\log|x+1| + \frac{1}{4}\log(x^2+9) + \frac{3}{2}\tan^{-1}\frac{x}{3} + C$

(*vi*) $\log|x+2| - \frac{1}{2}\log(x^2+1) + 2\tan^{-1}x + C$

12. (*i*) $\log|x+2| - \frac{1}{2}\log(x^2+4) + \tan^{-1}\frac{x}{2} + C$ (*ii*) $2\log|x+5| + \tan^{-1}x + C$

(*iii*) $x - \log|x| - \tan^{-1}x + \frac{1}{2}\log(x^2+1) + C$

(*iv*) $-\frac{1}{2}\log(x^2+2) + 2\sqrt{2}\tan^{-1}\frac{x}{\sqrt{2}} + \log|x+4| + C$

(*v*) $-\frac{1}{2}\log(x^2+9) - \frac{1}{3}\tan^{-1}\frac{x}{3} + \log|x-1| + C$

(*vi*) $\log|1+\tan\theta| + \frac{2}{\sqrt{3}}\tan^{-1}\left(\frac{2\tan\theta-1}{\sqrt{3}}\right) + C$

Hints

1. (*vi*) $\frac{1}{x-x^3} = \frac{1}{x(1-x)(1+x)}$

3. (*ii*) Take $z = \sin x$ (*iv*) Take $z = \log x$ (*v*) Take $z = x^2$

(*vi*) write $\frac{1}{\sin x\,(1+2\cos x)} = \frac{\sin x}{(1-\cos^2 x)(1+2\cos x)}$

$= \frac{\sin x}{(1-\cos x)(1+\cos x)(1+2\cos x)}$ and take $z = \cos x$.

4. (*i*) $\frac{x^2}{(x^2+2)(x^2+3)} = \frac{y}{(y+2)(y+3)}$, $y = x^2$

$= \frac{-2}{(y+2)\,.\,1} + \frac{-3}{(-1)\,.\,(y+3)} = \frac{-2}{x^2+2} + \frac{3}{x^2+3}$.

(*vi*) $y = x^2$ implies $\frac{(x^2+1)(x^2+2)}{(x^2+3)(x^2+4)} = \frac{(y+1)(y+2)}{(y+3)(y+4)} = 1 + \frac{A}{y+3} + \frac{B}{y+4}$

$\therefore$ $(y+1)(y+2) = (y+3)(y+4) + A(y+4) + B(y+3)$

$y = -3 \Rightarrow (-2)(-1) = 0 + A(1) + 0 \Rightarrow A = 2$

$y = -4 \Rightarrow (-3)(-2) = 0 + 0 + B(-1) \Rightarrow B = -6$

$\therefore$ Integrand $= 1 + \frac{2}{y+3} - \frac{6}{y+4} = 1 + \frac{2}{x^2+3} - \frac{6}{x^2+4}$.

5. (*ii*) $z = x^4 \Rightarrow dz = 4x^3\,dx$

$$\therefore \int \frac{dx}{x(x^4-1)} = \int \frac{x^3\,dx}{x^4(x^4-1)} = \frac{1}{4}\int \frac{dz}{z(z-1)}$$

(*iii*) $z = e^x - 1 \Rightarrow e^x = z + 1 \Rightarrow x = \log(z+1) \Rightarrow dx = \dfrac{1}{z+1}\, dz$

$$\therefore \quad \int \frac{dx}{e^x - 1} = \int \frac{1}{z} \cdot \frac{1}{z+1}\, dz.$$

7. (*i*) $\dfrac{1}{\sin x + \sin 2x} = \dfrac{1}{\sin x + 2 \sin x \cos x} = \dfrac{\sin x}{\sin^2 x\,(1 + 2\cos x)}$

(*iv*) $\displaystyle\int \frac{\sin x}{\sin 4x}\, dx = \int \frac{\sin x}{2 \sin 2x \cos 2x}\, dx = \int \frac{\sin x\, dx}{2\,.\,2 \sin x \cos x \cos 2x}$

$$= \int \frac{dx}{4 \cos x\,(1 - 2\sin^2 x)} = \int \frac{\cos x\, dx}{4\,(1 - \sin^2 x)(1 - 2 \sin^2 x)}$$

$$= \frac{1}{4} \int \frac{dz}{(1 - z^2)(1 - 2z^2)} \text{ where } z = \sin x.$$

Also, $\dfrac{1}{(1 - z^2)(1 - 2z^2)} = \dfrac{1}{(1 - y)(1 - 2y)}$, $y = z^2$

$$= \frac{1}{(1-y)(-1)} + \frac{1}{(1/2)(1 - 2y)} = -\frac{1}{1 - z^2} + \frac{2}{1 - 2z^2} = -\frac{1}{1-z^2} + \frac{1}{(1/2) - z^2}.$$

(*v*) $z = e^x \Rightarrow \displaystyle\int \frac{dx}{1 + 3e^x + 2e^{2x}} = \int \frac{1}{1 + 3z + 2z^2} \cdot \frac{dz}{z} = \int \frac{dz}{z\,(z+1)(2z+1)}$

(*vi*) Write $\dfrac{1}{x\,(1 + 4x^3 + 3x^6)} = \dfrac{x^2}{x^3\,(1 + 4x^3 + 3x^6)}$ and take $z = x^3$.

8. (*ii*) Take $z = \sin x$

9. (*i*) $z = e^x - 1 \Rightarrow x = \log(z+1) \Rightarrow dx = \dfrac{dz}{z+1}$

$$\therefore \quad \int \frac{dx}{(e^x - 1)^2} = \int \frac{1}{z^2} \cdot \frac{dz}{z+1}.$$

(*ii*) Write $\dfrac{1}{e^{2x} - 3e^x} = \dfrac{1}{e^x\,(e^x - 3)} = \dfrac{e^x}{(e^x)^2\,(e^x - 3)}$ and take $z = e^x$.

10. (*i*) $\displaystyle\int \frac{x^2 - 2}{x^5 - x}\, dx = \int \frac{x\,(x^2 - 2)\, dx}{x^2\,(x^2 - 1)(x^2 + 1)} = \frac{1}{2} \int \frac{(z - 2)\, dz}{z\,(z - 1)(z + 2)}$, where $z = x^2$

(*ii*) $\displaystyle\int \frac{\sin x\, dx}{\cos x\,(1 + \cos^2 x)} = \int \frac{\sin x \cos x\, dx}{\cos^2 x\,(1 + \cos^2 x)} = -\frac{1}{2} \int \frac{dz}{z\,(1+z)}$, where $z = \cos^2 x$

(*iii*) Let $y = 1 - x$.

$$\therefore \quad \frac{(1+x)^3}{(1-x)^3} = \frac{(1 + 1 - y)^3}{y^3} = \frac{8 - 12y + 6y^2 - y^3}{y^3} = \frac{8}{y^3} - \frac{12}{y^2} + \frac{6}{y} - 1$$

$$= \frac{8}{(1-x)^3} - \frac{12}{(1-x)^2} + \frac{6}{1-x} - 1.$$

12. (*vi*) Take $z = \tan \theta$.

14
Definite Integral

14.1. INTRODUCTION

In the preceding chapter, we learnt the method evaluating definite integral as the limit of a sum. This method is very cumbersome and can help evaluating definite integral only for very simple types of integrands. Now we shall study another method of evaluating definite integrals. This method would use the indefinite integral of the concerned definite integral.

14.2. FUNDAMENTAL THEOREM OF CALCULUS

If f(x) is a function continuous on [a, b] and if $\int f(x)\,dx = F(x) + C$, then

$$\int_a^b f(x)\,dx = F(b) - F(a).$$

The proof of this theorem is beyond the scope of this book.

14.3. DEFINITE INTEGRAL BY USING THE CORRESPONDING INDEFINITE INTEGRAL

We know that if $f(x)$ is a function continuous on the closed interval $[a, b]$, then the definite integral of $f(x)$ over the interval $[a, b]$ *i.e.*, $\int_a^b f(x)\,dx$ is equal to the limit of the sum :

$$\int_a^b f(x)\,dx = \lim_{h \to 0} \{h[f(a) + f(a+h) + f(a+2h) + \ldots\ldots + f(a+(n-1)h)]\}$$

where $n \in \mathbf{N}$ and $h = \dfrac{b-a}{n}$.

In practice, the evaluation of the limit of this sum as $h \to 0$ is very typical. This is why, we can calculate the definite integrals of very simple functions by using this method.

The **fundamental theorem of calculus** has given us a very beautiful alternative to find the definite integrals of continuous functions.

Let $\quad \int f(x)\,dx = F(x) + C.$

$\therefore$ By the **fundamental theorem of calculus,** we have

$$\int_a^b f(x)\,dx = F(b) - F(a).$$

Remark. The difference $F(b) - F(a)$ is written as $F(x)\Big|_a^b$

WORKING RULES FOR EVALUATING $\int_a^b f(x)\,dx$

Step I. *Evaluate the indefinite integral* $\int f(x)\,dx$. *Let* $\int f(x)\,dx = F(x) + C$.

Step II. *Write :* $\int_a^b f(x)\,dx = F(x)\Big|_a^b = F(b) - F(a)$.

Step III. *Find the value of* $F(b) - F(a)$. *This gives the value of* $\int_a^b f(x)\,dx$.

ILLUSTRATIVE EXAMPLES

Example 1. *Evaluate the following definite integrals :*

(*i*) $\int_0^1 (2x+3)\,dx$ (P.T.U. B.Pharma, Dec. 2003) (*ii*) $\int_2^4 \frac{1}{x}\,dx$ (*iii*) $\int_2^5 (x^3 + x)\,dx$

(*iv*) $\int_0^{\pi} \sin\frac{x}{2}\,dx$ (*v*) $\int_0^{\pi/4} \tan^2\theta\,d\theta$ (*vi*) $\int_{\pi/6}^{\pi/2} \frac{\cos x}{\sin^2 x}\,dx.$

Sol. (*i*) $\int (2x+3)\,dx = x^2 + 3x + c$

$\therefore$ By **fundamental theorem of calculus,** we have

$$\int_0^1 (2x+3)\,dx = \Big| x^2 + 3x \Big|_0^1 = (1+3) - 0 = \mathbf{4}.$$

(*ii*) $\int \frac{1}{x}\,dx = \log|x| + C.$

$\therefore$ By **fundamental theorem of calculus,** we have

$$\int_2^4 \frac{1}{x}\,dx = \log|x|\Big|_2^4 = \log\ |\ 4\ |\ - \log\ |\ 2\ |\ = \log 4 - \log 2 = \log\frac{4}{2} = \mathbf{\log 2}.$$

(*iii*) $\int (x^3 + x)\,dx = \frac{x^4}{4} + \frac{x^2}{2} + C.$

$\therefore$ By **fundamental theorem of calculus,** we have

$$\int_2^5 (x^3 + x)\,dx = \left(\frac{x^4}{4} + \frac{x^2}{2}\right)\Bigg|_2^5 = \left(\frac{(5)^4}{4} + \frac{(5)^2}{2}\right) - \left(\frac{(2)^4}{4} + \frac{(2)^2}{2}\right) = \frac{625}{4} + \frac{25}{2} - 4 - 2 = \mathbf{\frac{651}{4}}.$$

(*iv*) $\int \sin\frac{x}{2}\,dx = -2\cos\frac{x}{2} + C.$

$\therefore$ By **fundamental theorem of calculus,** we have

$$\int_0^{\pi} \sin\frac{x}{2}\,dx = -2\cos\frac{x}{2}\Big|_0^{\pi} = -2\left(\cos\frac{\pi}{2} - \cos 0\right) = -2(0-1) = \mathbf{2}.$$

(*v*) $\int \tan^2\theta\,d\theta = \int (\sec^2\theta - 1)\,d\theta = \tan\theta - \theta + C.$

$\therefore$ By **fundamental theorem of calculus,** we have

$$\int_0^{\pi/4} \tan^2 \theta \, d\theta = (\tan \theta - \theta)\Big|_0^{\pi/4} = \left(\tan \frac{\pi}{4} - \frac{\pi}{4}\right) - (\tan 0 - 0) = 1 - \frac{\pi}{4} - 0 = \mathbf{1 - \frac{\pi}{4}}.$$

(*vi*) $\int \frac{\cos x}{\sin^2 x} dx = \int \frac{1}{\sin x} \cdot \frac{\cos x}{\sin x} dx = \int \operatorname{cosec} x \cot x \, dx = -\operatorname{cosec} x + C.$

$\therefore$ By **fundamental theorem of calculus,** we have

$$\int_{\pi/6}^{\pi/2} \frac{\cos x}{\sin^2 x} dx = -\operatorname{cosec} x \Big|_{\pi/6}^{\pi/2} = -\left(\operatorname{cosec} \frac{\pi}{2} - \operatorname{cosec} \frac{\pi}{6}\right) = -(1 - 2) = \mathbf{1}.$$

Example 2. *Evaluate the following definite integrals :*

(*i*) $\int_0^{\pi/2} \cos^2 x \, dx$ (*ii*) $\int_0^1 x e^{x^2} dx$ (*iii*) $\int_1^3 \frac{1}{x^2(x+1)} dx$

(*iv*) $\int_0^{\pi/2} \frac{d\theta}{5 + 3\cos\theta}$ (*v*) $\int_0^{\pi/4} 2\tan^3 x \, dx$ (*vi*) $\int_0^{\pi/4} \sqrt{1 - \sin 2x} \, dx.$

Sol. (*i*) $\int \cos^2 x \, dx = \int \frac{1 + \cos 2x}{2} dx = \frac{1}{2}\left[x + \frac{\sin 2x}{2}\right] + C.$

$\therefore$ By **fundamental theorem of calculus,** we have

$$\int_0^{\pi/2} \cos^2 x \, dx = \frac{1}{2}\left[x + \frac{\sin 2x}{2}\right]\Bigg|_0^{\pi/2} = \frac{1}{2}\left[\frac{\pi}{2} + \frac{\sin \pi}{2}\right] - \frac{1}{2}\left[0 + \frac{\sin 0}{2}\right]$$

$$= \frac{1}{2}\left[\frac{\pi}{2} + 0\right] - \frac{1}{2}(0) = \frac{\pi}{4}. \qquad | \because \sin \pi = 0$$

(*ii*) Let $I = \int x e^{x^2} dx.$ Let $z = x^2$. $\therefore$ $dz = 2x \, dx.$

$$\therefore \quad I = \int e^z \frac{dz}{2} = \frac{1}{2}\int e^z \, dz = \frac{1}{2} e^z + C = \frac{1}{2} e^{x^2} + C.$$

$\therefore$ By **Fundamental Theorem of Calculus,** $\int_0^1 x e^{x^2} dx = \frac{1}{2} e^{x^2}\Big|_0^1 = \frac{1}{2}(e^1 - e^0) = \mathbf{\frac{e - 1}{2}}.$

(*iii*) Let $I = \int \frac{1}{x^2(x+1)} dx.$ Let $\frac{1}{x^2(x+1)} = \frac{A}{x} + \frac{B}{x^2} + \frac{C}{x+1}.$

$\therefore$ $1 = Ax(x + 1) + B(x + 1) + Cx^2$

$x = 0 \Rightarrow 1 = 0 + B + 0 \Rightarrow B = 1$

$x = -1 \Rightarrow 1 = 0 + 0 + C(-1)^2 \Rightarrow C = 1$

$x = 2 \Rightarrow 1 = 6A + 3B + 4C \Rightarrow 6A + 3(1) + 4(1) = 1 \Rightarrow A = -1.$

$$\therefore \quad \frac{1}{x^2(x+1)} = \frac{-1}{x} + \frac{1}{x^2} + \frac{1}{x+1}$$

$$\therefore \quad I = \int \left(-\frac{1}{x} + \frac{1}{x^2} + \frac{1}{x+1}\right) dx = -\log|x| + \frac{x^{-1}}{-1} + \log|x+1| + C = \log\left|\frac{x+1}{x}\right| - \frac{1}{x} + C.$$

$\therefore$ By **Fundamental Theorem of Calculus,** we have

$$\int_1^3 \frac{1}{x^2(x+1)}\,dx = \left(\log\left|\frac{x+1}{x}\right| - \frac{1}{x}\right)\Bigg|_1^3 = \left(\log\frac{4}{3} - \frac{1}{3}\right) - \left(\log\frac{2}{1} - \frac{1}{1}\right)$$

$$= \left(\log\frac{4}{3} - \log 2\right) - \frac{1}{3} + 1 = \mathbf{\frac{2}{3} + \log\frac{2}{3}}.$$

(*iv*) Let $I = \int \frac{d\theta}{5 + 3\cos\theta}$.

$$\therefore \qquad I = \int \frac{d\theta}{5 + 3\left(\frac{1-\tan^2\frac{1}{2}\theta}{1+\tan^2\frac{1}{2}\theta}\right)} = \int \frac{(1+\tan^2\frac{1}{2}\theta)\,d\theta}{5 + 5\tan^2\frac{1}{2}\theta + 3 - 3\tan^2\frac{1}{2}\theta} = \int \frac{\sec^2\frac{1}{2}\theta\,d\theta}{8 + 2\tan^2\frac{1}{2}\theta}$$

$$= \frac{1}{2}\int \frac{\sec^2\frac{1}{2}\theta}{4 + \tan^2\frac{1}{2}\theta}\,d\theta$$

Let $\quad z = \tan\frac{1}{2}\theta. \quad \therefore \quad dz = \frac{1}{2}\sec^2\frac{1}{2}\theta\,d\theta$

$$\therefore \qquad I = \frac{1}{2}\int \frac{2\,dz}{4+z^2} = \int \frac{dz}{4+z^2} = \frac{1}{2}\tan^{-1}\frac{z}{2} + C = \frac{1}{2}\tan^{-1}\left(\frac{1}{2}\tan\frac{\theta}{2}\right) + C.$$

$\therefore$ By **Fundamental Theorem of Calculus,** we have

$$\int_0^{\pi/2} \frac{d\theta}{5+3\cos\theta} = \frac{1}{2}\tan^{-1}\left(\frac{1}{2}\tan\frac{\theta}{2}\right)\Bigg|_0^{\pi/2} = \frac{1}{2}\tan^{-1}\left(\frac{1}{2}\tan\frac{\pi}{4}\right) - \frac{1}{2}\tan^{-1}\left(\frac{1}{2}\tan 0\right)$$

$$= \frac{1}{2}\tan^{-1}\left(\frac{1}{2}\times 1\right) - \frac{1}{2}\tan^{-1}(0) = \mathbf{\frac{1}{2}\tan^{-1}\frac{1}{2}}. \qquad \mid \tan^{-1} 0 = 0$$

(*v*) $\int_0^{\pi/4} 2\tan^3 x\,dx = 2\int_0^{\pi/4} \tan x \,.\, \tan^2 x\,dx = 2\int_0^{\pi/4} \tan x \,.\,(\sec^2 x - 1)\,dx$

$$= 2\int_0^{\pi/4} \tan x\sec^2 x\,dx - 2\int_0^{\pi/4} \tan x\,dx = 2\,.\,\frac{(\tan x)^2}{2}\Bigg|_0^{\pi/4} - 2\Big[\log|\sec x|\Big]_0^{\pi/4}$$

$$= \left(\tan^2\frac{\pi}{4} - \tan^2 0\right) - 2\left(\log\left|\sec\frac{\pi}{4}\right| - \log|\sec 0|\right)$$

$$= (1-0) - 2(\log\sqrt{2} - \log 1) = 1 - 2\log\sqrt{2} + 0 = \mathbf{1 - \log 2.}$$

(*vi*) $\int_0^{\pi/4} \sqrt{1 - \sin 2x}\,dx = \int_0^{\pi/4} \sqrt{\sin^2 x + \cos^2 x - 2\sin x\cos x}\,dx$

$$= \int_0^{\pi/4} \sqrt{(\sin x - \cos x)^2}\,dx = \int_0^{\pi/4} |\sin x - \cos x|\,dx$$

$$= \int_0^{\pi/4} -(\sin x - \cos x)\,dx$$

For $0 \le x \le \frac{\pi}{4}$, $\sin x \le \cos x$
$\Rightarrow \quad \sin x - \cos x \le 0$
$\Rightarrow \quad |\sin x - \cos x| = -(\sin x - \cos x)$

$$= \int_0^{\pi/4} (\cos x - \sin x)\, dx$$

$$= \Big[(\sin x + \cos x)\Big]_0^{\pi/4} = \left(\sin \frac{\pi}{4} + \cos \frac{\pi}{4}\right) - (\sin 0 + \cos 0)$$

$$= \left(\frac{1}{\sqrt{2}} + \frac{1}{\sqrt{2}}\right) - (0 + 1) = \mathbf{\sqrt{2} - 1.}$$

Example 3. *Evaluate the following definite integrals :*

(*i*) $\int_0^{\pi} \frac{dx}{1 + \sin x}$ (*ii*) $\int_1^{2} \frac{5x^2}{x^2 + 4x + 3}\, dx$ (*iii*) $\int_{\pi/4}^{\pi/2} \cos 2x \log \sin x\, dx.$

Sol. (*i*) Let $I = \int_0^{\pi} \frac{dx}{1 + \sin x}.$

$$\therefore \quad I = \int_0^{\pi} \frac{1}{1 + \sin x} \cdot \frac{1 - \sin x}{1 - \sin x}\, dx = \int_0^{\pi} \left(\frac{1 - \sin x}{1 - \sin^2 x}\right) dx = \int_0^{\pi} \left(\frac{1}{\cos^2 x} - \frac{\sin x}{\cos^2 x}\right) dx$$

$$= \int_0^{\pi} (\sec^2 x - \tan x \sec x)\, dx = \Big[(\tan x - \sec x)\Big]_0^{\pi}$$

$$= (\tan \pi - \sec \pi) - (\tan 0 - \sec 0) = (0 - (-1)) - (0 - 1) = \mathbf{2.}$$

(*ii*) Let $I = \int_1^{2} \frac{5x^2}{x^2 + 4x + 3}\, dx$. The integrand is not a proper fraction.

$\therefore$ By *long division,* we have

$$\frac{5x^2}{x^2 + 4x + 3} = \frac{5(x^2 + 4x + 3) - 20x - 15}{x^2 + 4x + 3} = 5 - \frac{20x + 15}{x^2 + 4x + 3} \qquad \ldots(1)$$

Also, $\frac{20x + 15}{x^2 + 4x + 3} = \frac{20x + 15}{(x + 1)(x + 3)} = \frac{(-5)}{(x + 1)(2)} + \frac{-45}{(-2)(x + 3)} = \frac{-5}{2(x + 1)} + \frac{45}{2(x + 3)}$

(Using short cut method)

$$\therefore \quad (1) \Rightarrow \quad \frac{5x^2}{x^2 + 4x + 3} = 5 - \left(\frac{-5}{2(x + 1)} + \frac{45}{2(x + 3)}\right)$$

$$\therefore \quad \int_1^{2} \frac{5x^2}{x^2 + 4x + 3}\, dx = \int_1^{2} \left(5 + \frac{5}{2(x + 1)} - \frac{45}{2(x + 3)}\right) dx$$

$$= \left[5x + \frac{5}{2} \log |x + 1| - \frac{45}{2} \log |x + 3|\right]_1^2$$

$$= \left(10 + \frac{5}{2} \log 3 - \frac{45}{2} \log 5\right) - \left(5 + \frac{5}{2} \log 2 - \frac{45}{2} \log 4\right)$$

$$= 5 + \frac{5}{2} (\log 3 - \log 2) - \frac{45}{2} (\log 5 - \log 4)$$

$$= \mathbf{5 + \frac{5}{2} \log \frac{3}{2} - \frac{45}{2} \log \frac{5}{4}.}$$

(*iii*) Let $I = \int_{\pi/4}^{\pi/2} \cos 2x \log \sin x \, dx$.

$$\therefore \quad I = \int_{\pi/4}^{\pi/2} \underset{\text{I}}{\log \sin x} \, . \underset{\text{II}}{\cos 2x} \, dx = \log \sin x \, . \frac{\sin 2x}{2}\Bigg|_{\pi/4}^{\pi/2} - \int_{\pi/4}^{\pi/2} \frac{1}{\sin x} \, . \cos x \, . \frac{\sin 2x}{2} \, dx$$

$$= \log \sin \frac{\pi}{2} \, . \frac{\sin \pi}{2} - \log \sin \frac{\pi}{4} \, . \frac{\sin \frac{1}{2}\pi}{2} - \int_{\pi/4}^{\pi/2} \frac{\cos x}{\sin x} \, . \frac{2 \sin x \cos x}{2} \, dx$$

$$= 0 - \left(\log \frac{1}{\sqrt{2}}\right)\frac{1}{2} - \int_{\pi/4}^{\pi/2} \cos^2 x \, dx = \frac{1}{4}\log 2 - \int_{\pi/4}^{\pi/2} \frac{1 + \cos 2x}{2} \, dx$$

$$= \frac{1}{4}\log 2 - \frac{1}{2}\left[x + \frac{\sin 2x}{2}\right]\Bigg|_{\pi/4}^{\pi/2} = \frac{1}{4}\log 2 - \frac{1}{2}\left[\frac{\pi}{2} + 0\right] + \frac{1}{2}\left[\frac{\pi}{4} + \frac{1}{2}\right]$$

$$= \mathbf{\frac{1}{4}\log 2 - \frac{\pi}{8} + \frac{1}{4}}.$$

Example 4. *Show that* $\int_0^{\pi/2} \sin^6 \theta \, d\theta = \frac{5}{6}\int_0^{\pi/2} \sin^4 \theta \, d\theta$.

Sol. Let $I = \int_0^{\pi/2} \sin^6 \theta \, d\theta$.

$$\therefore \quad I = \int_0^{\pi/2} \sin^5 \theta \, . \sin \theta \, d\theta, \text{ Integrating by parts, we get}$$

$$I = \left[\sin^5 \theta \, (-\cos \theta)\right]_0^{\pi/2} - \int_0^{\pi/2} (5 \sin^4 \theta \cos \theta)(-\cos \theta) \, d\theta$$

$$= 0 - 0 + 5\int_0^{\pi/2} \sin^4 \theta \cos^2 \theta \, d\theta = 5\int_0^{\pi/2} \sin^4 \theta \, (1 - \sin^2 \theta) \, d\theta$$

$$= 5\int_0^{\pi/2} \sin^4 \theta \, d\theta - 5\int_0^{\pi/2} \sin^6 \theta \, d\theta$$

$$\therefore \quad I = 5\int_0^{\pi/2} \sin^4 \theta \, d\theta - 5I$$

$$\Rightarrow \quad 6I = 5\int_0^{\pi/2} \sin^4 \theta \, d\theta, \; i.e., I = \frac{5}{6}\int_0^{\pi/2} \sin^4 \theta \, d\theta.$$

Example 5. *Evaluate the following definite integrals :*

(*i*) $\int_0^{\pi/2} x^2 \cos 2x \, dx$ (*ii*) $\int_0^{\pi/2} x^2 \sin 3x \, dx$

(*iii*) $\int_0^{\pi/2} \frac{\cos x}{1 + \sin^2 x} \, dx.$ (*iv*) $\int_0^{\pi/2} \sin^3 x \, dx$

Sol. (*i*) Let $I = \int_0^{\pi/2} \underset{\text{I}}{x^2} \underset{\text{II}}{\cos 2x} \, dx$. Integrating by parts, we get

$$I = \left[x^2\left(\frac{\sin 2x}{2}\right)\right]_0^{\pi/2} - \int_0^{\pi/2} 2x\left(\frac{\sin 2x}{2}\right) dx = \frac{\pi^2}{4}\left(\frac{\sin \pi}{2}\right) - 0 - \int_0^{\pi/2} x \sin 2x \, dx$$

$$= 0 - \left[x\left(-\frac{\cos 2x}{2}\right)\right]\Bigg|_0^{\pi/2} - \int_0^{\pi/2} 1.\left(-\frac{\cos 2x}{2}\right) dx \qquad | \because \sin \pi = 0$$

$$= \frac{\pi}{2}\left(\frac{\cos \pi}{2}\right) - 0 + \frac{1}{2}\int_0^{\pi/2} \cos 2x\, dx = \frac{\pi}{2}\left(-\frac{1}{2}\right) + \frac{1}{2}\left[\frac{\sin 2x}{2}\right]_0^{\pi/2}$$

$$= -\frac{\pi}{4} + \frac{1}{4}[\sin \pi - \sin 0] = -\frac{\pi}{4} + 0 = -\frac{\pi}{\mathbf{4}}.$$

(*ii*) Let $I = \int_0^{\pi/2} \underset{I}{x^2} \underset{II}{\sin 3x}\, dx$. Integrating by parts, we get

$$I = x^2\left[-\frac{\cos 3x}{3}\right]_0^{\pi/2} - \int_0^{\pi/2} 2x\left(-\frac{\cos 3x}{3}\right)dx$$

$$= \frac{\pi^2}{4}\left(-\frac{\cos(3\pi/2)}{3}\right) - 0 + \frac{2}{3}\int_0^{\pi/2} x \cos 3x\, dx$$

$$= -\frac{\pi^2}{12}(0) + \frac{2}{3}\left[x\left(\frac{\sin 3x}{3}\right)\Bigg|_0^{\pi/2} - \int_0^{\pi/2} 1\left(\frac{\sin 3x}{3}\right)dx\right]$$

$$= 0 + \frac{2}{9}\Big[(x \sin 3x)\Big]_0^{\pi/2} - \frac{2}{9}\int_0^{\pi/2} \sin 3x\, dx = \frac{2}{9}\left(\frac{\pi}{2}\sin\frac{3\pi}{2} - 0\right) - \frac{2}{9}\left[\left(-\frac{\cos 3x}{3}\right)\right]_0^{\pi/2}$$

$$= \frac{2}{9}\left[\left(\frac{\pi}{2}(-1)\right) + \frac{1}{3}\cos 3x\right]_0^{\pi/2} = -\frac{\pi}{9} + \frac{2}{27}\left(\cos\frac{3\pi}{2} - \cos 0\right)$$

$$= -\frac{\pi}{9} + \frac{2}{27}(0 - 1) = -\frac{\pi}{9} - \frac{2}{27} = -\frac{\mathbf{3\pi + 2}}{\mathbf{27}}.$$

(*iii*) Let $I = \int \frac{\cos x}{1 + \sin^2 x}\, dx$. Let $z = \sin x$. $\therefore$ $dz = \cos x\, dx$

$$\therefore \quad I = \int \frac{dz}{1 + z^2} = \tan^{-1} z + C = \tan^{-1}(\sin x) + C.$$

$$\therefore \quad \int_0^{\pi/2} \frac{\cos x}{1 + \sin^2 x}\, dx = \Big[\tan^{-1}(\sin x)\Big]_0^{\pi/2} = \tan^{-1}\left(\sin\frac{\pi}{2}\right) - \tan^{-1}(\sin 0)$$

$$= \tan^{-1}(1) - \tan^{-1}(0) = \frac{\pi}{4} - 0 = \frac{\pi}{\mathbf{4}}.$$

(*iv*) Consider $\sin 3x = 3 \sin x - 4 \sin^3 x$

$\Rightarrow$ $4 \sin^3 x = 3 \sin x - \sin 3x$

$\Rightarrow$ $\sin^3 x = \frac{1}{4}(3 \sin x - \sin 3x)$

$$\therefore \quad \int_0^{\pi/2} \sin^3 x\, dx = \frac{1}{4}\int_0^{\pi/2}(3 \sin x - \sin 3x)\, dx = \frac{1}{4}\left| -3\cos x + \frac{\cos 3x}{3}\right|_0^{\pi/2}$$

$$= \frac{1}{4}\left[\left(-3\cos\frac{\pi}{2} + \frac{1}{3}\cos\frac{3\pi}{2}\right) - \left(-3\cos 0 + \frac{\cos 0}{3}\right)\right]$$

$$= \frac{1}{4}\left[0 + 3 - \frac{1}{3}\right] = \frac{1}{4}\cdot\frac{8}{3} = \frac{2}{3}$$

$\cos\frac{\pi}{2} = 0$
$\cos\frac{3\pi}{2} = 0$
$\cos 0 = 1$

EXERCISE 14.1

Evaluate the following definite integrals (Q. No. 1–9) :

1. (*i*) $\int_3^4 \frac{1}{x}\,dx$ (*ii*) $\int_{-4}^{-1} \frac{1}{x}\,dx$ (*iii*) $\int_a^b x^n\,dx$

(*iv*) $\int_0^1 \frac{dx}{2x-3}$ (*v*) $\int_0^2 \sqrt{6x+4}\,dx$ (*vi*) $\int_0^1 \frac{dx}{3x-4}$

2. (*i*) $\int_{-1}^1 e^x\,dx$ (*ii*) $\int_0^1 e^{2x}(e^{2x}+3)\,dx$ (*iii*) $\int_{-\pi/2}^{\pi/2} \cos x\,dx$

(*iv*) $\int_a^b (x+\cos x)\,dx$ (*v*) $\int_{\pi/6}^{\pi/4} \text{cosec}^2 x\,dx$ (*vi*) $\int_{\pi/6}^{\pi/3} \frac{\sin x}{\cos^2 x}\,dx.$

3. (*i*) $\int_0^{\pi/2} \frac{d\theta}{1+\cos\theta}$ (*ii*) $\int_{\pi/3}^{\pi/2} \frac{d\theta}{1-\cos\theta}$ (*iii*) $\int_{\pi/6}^{\pi/4} \text{cosec}\, x\,dx$

(*iv*) $\int_0^{\sqrt{3}} \frac{dx}{1+x^2}$ (*v*) $\int_0^{\pi/2} \sqrt{1-\cos 2x}\,dx$ (*vi*) $\int_0^{\pi/4} \sqrt{1+\sin 2x}\,dx$

4. (*i*) $\int_0^4 (\sqrt{x}-2x+x^2)\,dx$ (*ii*) $\int_1^2 \left(\sqrt{2x}-\frac{1}{\sqrt{2x}}\right)dx$ (*iii*) $\int_1^4 \frac{x^4+2x^2-1}{\sqrt{x}}\,dx$

(*iv*) $\int_0^1 \frac{1-x}{1+x}\,dx$ (*v*) $\int_0^{\pi/2} (x+6\cos 2x)\,dx$ (*vi*) $\int_0^{\pi/2} \sin^2 \frac{x}{2}\,dx$

5. (*i*) $\int_0^{\pi/2} \sin x \sin 2x\,dx$ (*ii*) $\int_0^{\pi/4} (2\tan^2 x + x^3 + 2)\,dx$ (*iii*) $\int_0^{\pi/4} \sin^4 x\,dx$

(*iv*) $\int_0^{\pi/2} \frac{\sin^2\theta}{(1+\cos\theta)^2}\,d\theta$ (*v*) $\int_0^{\pi/4} \sin 2x \sin 3x\,dx$ (*vi*) $\int_0^{\pi/6} \cos x \cos 2x\,dx$

6. (*i*) $\int_3^4 \frac{dx}{\sqrt{x^2+4}}$ (*ii*) $\int_0^3 \sqrt{9-x^2}\,dx$ (*iii*) $\int_0^{\sqrt{2}} \sqrt{2-x^2}\,dx$

(*iv*) $\int_{\sqrt{2}}^2 \frac{dx}{x\sqrt{x^2-1}}$ (*v*) $\int_0^{\pi/2} (a\cos^2 x + b\sin^2 x)\,dx$ (*vi*) $\int_0^{\pi/2} \cos^4 x\,dx$

7. (*i*) $\int_1^2 \log x\,dx$ (*ii*) $\int_1^2 \frac{\log x}{x^2}\,dx$ (*iii*) $\int_1^3 \frac{\log x}{(1+x)^2}\,dx$

(*iv*) $\int_0^1 xe^x\,dx$ (*v*) $\int_0^{\pi/2} x\cos x\,dx$ (*vi*) $\int_0^{\pi/4} x^2 \sin x\,dx$

8. (*i*) $\int_1^e \frac{e^x}{x}(1+x\log x)\,dx$ (*ii*) $\int_0^1 \frac{xe^x}{(1+x)^2}\,dx$ (*iii*) $\int_1^2 \frac{x-1}{x^2}e^x\,dx$

9. (*i*) $\int_0^{\pi/2} x^2\cos^2 x\,dx$ (*ii*) $\int_1^2 \frac{x+3}{x(x+2)}\,dx$

(*iii*) $\int_1^5 \frac{dx}{x^2(1+x)}$ (*iv*) $\int_0^4 \frac{x}{(1+x)(1+x^2)}\,dx.$

10. If $\int_0^k \frac{dx}{2+8x^2} = \frac{\pi}{16}$, find the value of k.

11. If $\int_0^a \sqrt{x}\, dx = 2a \int_0^{\pi/2} \sin^3 x\, dx$, find the value of $\int_a^{a+1} x\, dx$. *(P.T.U. B.Pharma, Dec. 2002)*

12. Evaluate $\int_{\pi/6}^{\pi/4} \tan\theta(1+\tan\theta)\, d\theta$ *(P.T.U. B.Pharma, May 2007)*

Answers

1. (*i*) $\log\frac{4}{3}$ (*ii*) $-\log 4$ (*iii*) $\frac{1}{n+1}(b^{n+1}-a^{n+1})$

(*iv*) $-\frac{1}{2}\log 3$ (*v*) $\frac{56}{9}$ (*vi*) $-\frac{1}{3}\log 4$

2. (*i*) $e-\frac{1}{e}$ (*ii*) $\frac{e^4+6e^2-7}{4}$ (*iii*) 2

(*iv*) $\frac{b^2-a^2}{2}+\sin b-\sin a$ (*v*) $\sqrt{3}-1$ (*vi*) $2-\frac{2}{\sqrt{3}}$

3. (*i*) 1 (*ii*) $\sqrt{3}-1$ (*iii*) $\log((\sqrt{2}-1)(\sqrt{3}+2))$

(*iv*) $\pi/3$ (*v*) $\sqrt{2}$ (*vi*) 1

4. (*i*) $\frac{32}{3}$ (*ii*) $\frac{2+\sqrt{2}}{3}$ (*iii*) $\frac{6136}{45}$

(*iv*) $2\log 2-1$ (*v*) $\frac{\pi^2}{8}$ (*vi*) $\frac{\pi-2}{4}$

5. (*i*) $\frac{2}{3}$ (*ii*) $\frac{\pi^4}{1024}+2$ (*iii*) $\frac{3\pi}{32}-\frac{1}{4}$

(*iv*) $2-\frac{\pi}{2}$ (*v*) $\frac{3\sqrt{2}}{10}$ (*vi*) $\frac{5}{12}$

6. (*i*) $\log\frac{4+2\sqrt{5}}{3+\sqrt{13}}$ (*ii*) $\frac{9\pi}{4}$ (*iii*) $\frac{\pi}{2}$

(*iv*) $\frac{\pi}{12}$ (*v*) $\frac{(a+b)\pi}{4}$ (*vi*) $\frac{3\pi}{16}$

7. (*i*) $2\log 2-1$ (*ii*) $\frac{1-\log 2}{2}$ (*iii*) $\frac{3}{4}\log 3-\log 2$

(*iv*) 1 (*v*) $\frac{\pi}{2}-1$ (*vi*) $\sqrt{2}-2+\frac{\pi}{2\sqrt{2}}-\frac{\pi^2}{16\sqrt{2}}$

8. (*i*) e^e (*ii*) $\frac{e-2}{2}$ (*iii*) $\frac{e(e-2)}{2}$

9. (*i*) $\frac{\pi^3}{48}-\frac{\pi}{8}$ (*ii*) $\frac{1}{2}\log 6$ (*iii*) $\frac{4}{5}+\log\frac{3}{5}$

(*iv*) $\frac{1}{4}\log\frac{17}{25}+\frac{1}{2}\tan^{-1}4$ **10.** $\frac{1}{2}$ **11.** $\frac{1}{2}, \frac{9}{2}$

12. $\frac{1}{2}\log\frac{3}{2}+1-\frac{1}{\sqrt{3}}-\frac{\pi}{2}$.

Hints

2. (*ii*) $e^{2x}(e^{2x} + 3) = e^{4x} + 3e^{2x}$ (*vi*) $\dfrac{\sin x}{\cos^2 x} = \sec x \tan x$.

3. (*vi*) $1 + \sin 2x = (\cos x + \sin x)^2$

5. (*i*) $\sin x \sin 2x = \dfrac{1}{2}(\cos x - \cos 3x)$

(*iii*) $\sin^4 x = \left(\dfrac{1-\cos 2x}{2}\right)^2 = \dfrac{1}{4}\left(1 + \dfrac{1+\cos 4x}{2} - 2\cos 2x\right)$

(*iv*) $\dfrac{\sin^2\theta}{(1+\cos\theta)^2} = \dfrac{1-\cos^2\theta}{(1+\cos\theta)^2} = \dfrac{1-\cos\theta}{1+\cos\theta} = \dfrac{2\sin^2\frac{1}{2}\theta}{2\cos^2\frac{1}{2}\theta} = \sec^2\dfrac{1}{2}\theta - 1.$

6. (*v*) $a\cos^2 x + b\sin^2 x = a\left(\dfrac{1+\cos 2x}{2}\right) + b\left(\dfrac{1-\cos 2x}{2}\right) = \dfrac{a+b}{2} + \dfrac{a-b}{2}\cos 2x$.

8. (*i*) $\dfrac{e^x}{x}(1 + x\log x) = e^x\left(\log x + \dfrac{1}{x}\right)$ and $(\log x)' = \dfrac{1}{x}$

(*ii*) $\dfrac{xe^x}{(1+x)^2} = e^x\left(\dfrac{x+1-1}{(1+x)^2}\right) = e^x\left(\dfrac{1}{1+x} + \dfrac{-1}{(1+x)^2}\right)$

9. (*i*) Write $\cos^2 x = \dfrac{1+\cos 2x}{2}$

10. $\displaystyle\int_0^k \frac{dx}{2+8x^2} = \frac{1}{8}\int_0^k \frac{dx}{(1/2)^2 + x^2} = \frac{1}{8}\cdot\frac{1}{1/2}\tan^{-1}\frac{x}{1/2}\Bigg|_0^k = \frac{1}{4}\tan^{-1} 2k - 0$.

14.4. EVALUATION OF DEFINITE INTEGRAL BY CHANGING LIMITS AFTER SUITABLE SUBSTITUTION

To compute $\int_a^b f(x)\,dx$, we evaluate the corresponding indefinite integral $\int f(x)\,dx$. Sometimes, $\int f(x)\,dx$ is evaluated by using some substitution. The definite integral $\int_a^b f(x)\,dx$ can also be evaluated by retaining the new variable and changing the limits accordingly.

WORKING RULES FOR EVALUATING $\int_a^b f(x)\,dx$

Step I. *Let $z = g(x)$ be the desired substitution, for evaluating $\int f(x)\,dx$.*

$$z = g(x) \Rightarrow dz = g'(x)\,dx.$$

Step II. *Find the values $g(a)$ and $g(b)$ of the variable z, when x is equal to a and b respectively.*

Step III. *Write :* $\displaystyle\int_a^b f(x)\,dx = \int_{g(a)}^{g(b)} \frac{f(x)}{g'(x)}\,dz$ *and express* $\dfrac{f(x)}{g'(x)}$ *in terms of z only.*

Step IV. *Integrate* $\dfrac{f(x)}{g'(x)}$ *(expressed in terms of z) w.r.t. z and find the value of the integral between new limits $g(a)$ and $g(b)$. This gives the value of* $\int_a^b f(x)\,dx$.

ILLUSTRATIVE EXAMPLES

Example 1. *Evaluate the following definite integrals :*

(i) $\int_0^{1/\sqrt{2}} \frac{\sin^{-1} x}{(1-x^2)^{3/2}}\,dx$ (ii) $\int_0^1 x(1-x)^5\,dx$

(iii) $\int_0^{\pi/2} \frac{\sin x \cos x}{1+\sin^4 x}\,dx$ (iv) $\int_0^{\pi/2} \sin 2x \tan^{-1}(\sin x)\,dx.$

Sol. (i) Let $I = \int_0^{1/\sqrt{2}} \frac{\sin^{-1} x}{(1-x^2)^{3/2}}\,dx$. Let $x = \sin\theta$. $\therefore$ $dx = \cos\theta\,d\theta$

When $x = \frac{1}{\sqrt{2}} \Rightarrow \sin\theta = \frac{1}{\sqrt{2}} \Rightarrow \theta = \frac{\pi}{4}$ and when $x = 0 \Rightarrow \sin\theta = 0 \Rightarrow \theta = 0.$

$$\therefore \quad I = \int_0^{\pi/4} \frac{\theta}{(1-\sin^2\theta)^{3/2}}.\cos\theta\,d\theta = \int_0^{\pi/4} \frac{\theta}{\cos^3\theta}.\cos\theta\,d\theta = \int_0^{\pi/4} \underset{\text{I}}{\theta}.\underset{\text{II}}{\sec^2\theta}\,d\theta$$

Integrating by parts, we get

$$I = \theta\tan\theta\Big|_0^{\pi/4} - \int_0^{\pi/4} 1.\tan\theta\,d\theta = \frac{\pi}{4}\tan\frac{\pi}{4} - 0 - \Big[\log|\sec\theta|\Big]_0^{\pi/4}$$

$$= \frac{\pi}{4}(1) - \left[\log\left|\sec\frac{\pi}{4}\right| - \log|\sec 0|\right] = \frac{\pi}{4} - [\log\sqrt{2} - \log 1]$$

$$= \frac{\pi}{4} - \frac{1}{2}\log 2 + 0 = \mathbf{\frac{\pi - 2\log 2}{4}}.$$

(ii) Let $I = \int_0^1 x(1-x)^5\,dx$. Let $z = 1-x$. $\therefore$ $dz = -dx$

When $x = 1 \Rightarrow z = 1-1 = 0$ and when $x = 0 \Rightarrow z = 1-0 = 1.$

$$\therefore \quad I = \int_1^0 (1-z)\,z^5(-dz) = \int_1^0 (z-1)z^5\,dz = \int_1^0 (z^6 - z^5)\,dz$$

$$= \left(\frac{z^7}{7} - \frac{z^6}{6}\right)\Bigg|_1^0 = (0-0) - \left(\frac{1}{7} - \frac{1}{6}\right) = \mathbf{\frac{1}{42}}.$$

(iii) Let $I = \int_0^{\pi/2} \frac{\sin x\cos x}{1+\sin^4 x}\,dx$. Let $z = \sin^2 x$. $\therefore$ $dz = 2\sin x\cos x\,dx$

When $x = \frac{\pi}{2} \Rightarrow z = \sin^2\frac{\pi}{2} = 1$ and when $x = 0 \Rightarrow z = \sin^2 0 = 0.$

$$\therefore \quad I = \int_0^1 \frac{1}{1+z^2}.\frac{dz}{2} = \left[\frac{1}{2}\tan^{-1} z\right]_0^1 = \frac{1}{2}[\tan^{-1}1 - \tan^{-1}0] = \frac{1}{2}\left[\frac{\pi}{4} - 0\right] = \mathbf{\frac{\pi}{8}}.$$

(iv) Let $I = \int_0^{\pi/2} \sin 2x\tan^{-1}(\sin x)\,dx$.

$\therefore$ $I = \int_0^{\pi/2} 2\sin x\cos x\tan^{-1}(\sin x)\,dx$. Let $z = \sin x$. $\therefore$ $dz = \cos x\,dx$

When $x = \frac{\pi}{2} \Rightarrow z = \sin\frac{\pi}{2} = 1$ and when $x = 0 \Rightarrow z = \sin 0 = 0.$

$\therefore \quad I = \int_0^1 2z \tan^{-1} z \, dz = 2\int_0^1 \underset{\text{I}}{(\tan^{-1} z)} \underset{\text{II}}{z} \, dz$. Integrating by parts, we get

$$I = 2\left[(\tan^{-1} z)\frac{z^2}{2}\right]_0^1 - 2\int_0^1 \frac{1}{1+z^2} \cdot \frac{z^2}{2} dz = \left[z^2 \tan^{-1} z\right]_0^1 - \int_0^1 \frac{1+z^2-1}{1+z^2} dz$$

$$= 1 . \tan^{-1} 1 - 0 - \int_0^1 \left(1 - \frac{1}{1+z^2}\right) dz = \frac{\pi}{4} - [z - \tan^{-1} z]\Big|_0^1$$

$$= \frac{\pi}{4} - [(1 - \tan^{-1} 1) - (0 - 0)] = \frac{\pi}{4} - 1 + \frac{\pi}{4} = \boldsymbol{\frac{\pi}{2} - 1}.$$

Example 2. *Evaluate the following definite integrals :*

(*i*) $\int_0^{\pi/2} \frac{dx}{6 - \cos x}$ (*ii*) $\int_0^{\pi/2} \frac{dx}{7 + 2 \sin x}$

(*iii*) $\int_0^{\pi/2} \frac{dx}{5 \cos x + 3 \sin x}$.

Sol. (*i*) Let $I = \int_0^{\pi/2} \frac{dx}{6 - \cos x}$. $\therefore \; I = \int_0^{\pi/2} \frac{dx}{6 - \frac{1 - \tan^2 \frac{x}{2}}{1 + \tan^2 \frac{x}{2}}} = \int_0^{\pi/2} \frac{\sec^2 \frac{x}{2}}{5 + 7\tan^2 \frac{x}{2}} dx$

Let $z = \tan \frac{x}{2}$ $\therefore$ $dz = \frac{1}{2} \sec^2 \frac{x}{2} dx$ When $x = \frac{\pi}{2} \Rightarrow z = \tan \frac{\pi}{4} = 1$
and when $x = 0 \Rightarrow z = \tan 0 = 0$.

$$\therefore \quad I = \int_0^1 \frac{2dz}{5 + 7z^2} = \frac{2}{7}\int_0^1 \frac{dz}{\frac{5}{7} + z^2} = \frac{2}{7} \cdot \frac{1}{\sqrt{5/7}}\left[\tan^{-1} \frac{z}{\sqrt{5/7}}\right]_0^1 = \frac{2}{\sqrt{35}}\left[\tan^{-1} \sqrt{\frac{7}{5}} z\right]_0^1$$

$$= \frac{2}{\sqrt{35}}\left(\tan^{-1}\left(\sqrt{\frac{7}{5}} \times 1\right) - \tan^{-1} 0\right) = \boldsymbol{\frac{2}{\sqrt{35}} \tan^{-1} \sqrt{\frac{7}{5}}}.$$

(*ii*) Let $I = \int_0^{\pi/2} \frac{dx}{7 + 2 \sin x}$. $\therefore \; I = \int_0^{\pi/2} \frac{dx}{7 + 2\left(\frac{2 \tan \frac{x}{2}}{1 + \tan^2 \frac{x}{2}}\right)} = \int_0^{\pi/2} \frac{\sec^2 \frac{x}{2}}{7 + 7\tan^2 \frac{x}{2} + 4 \tan \frac{x}{2}} dx$

Let $z = \tan \frac{x}{2}$. $\therefore$ $dz = \frac{1}{2} \sec^2 \frac{x}{2} dx$.

When $x = \frac{\pi}{2} \Rightarrow z = \tan \frac{\pi}{4} = 1$ and when $x = 0 \Rightarrow z = \tan 0 = 0$.

$$\therefore \quad I = \int_0^1 \frac{2\,dz}{7 + 7z^2 + 4z} = \frac{2}{7}\int_0^1 \frac{dz}{z^2 + \frac{4}{7}z + 1} = \frac{2}{7}\int_0^1 \frac{dz}{\left(z^2 + \frac{4}{7}z + \frac{4}{49}\right) + 1 - \frac{4}{49}}$$

$$= \frac{2}{7}\int_0^1 \frac{dz}{\left(z + \frac{2}{7}\right)^2 + \frac{45}{49}}$$

Let $t = z + \frac{2}{7}$. $\therefore\ dt = dz.$

when $z = 1 \Rightarrow t = 1 + \frac{2}{7} = \frac{9}{7}$ and when $z = 0 \Rightarrow t = 0 + \frac{2}{7} = \frac{2}{7}.$

$$\therefore \quad I = \frac{2}{7}\int_{2/7}^{9/7} \frac{dt}{t^2 + \frac{45}{49}} = \frac{2}{7}\cdot\frac{1}{\sqrt{45/7}}\left[\tan^{-1}\frac{t}{\sqrt{45/7}}\right]_{2/7}^{9/7}$$

$$= \frac{2}{3\sqrt{5}}\left[\tan^{-1}\frac{7}{3\sqrt{5}}\left(\frac{9}{7}\right) - \tan^{-1}\frac{7}{3\sqrt{5}}\left(\frac{2}{7}\right)\right] = \mathbf{\frac{2}{3\sqrt{5}}\left[\tan^{-1}\frac{3}{\sqrt{5}} - \tan^{-1}\frac{2}{3\sqrt{5}}\right]}.$$

(*iii*) Let $I = \int_0^{\pi/2} \frac{dx}{5\cos x + 3\sin x}$.

$$\therefore \quad I = \int_0^{\pi/2} \frac{dx}{5\,\frac{1-\tan^2\frac{x}{2}}{1+\tan^2\frac{x}{2}} + 3\,\frac{2\tan\frac{x}{2}}{1+\tan^2\frac{x}{2}}} = \int_0^{\pi/2} \frac{\sec^2\frac{x}{2}\,dx}{5 - 5\tan^2\frac{x}{2} + 6\tan\frac{x}{2}}$$

Let $\quad z = \tan\frac{x}{2}$. $\therefore\ dz = \frac{1}{2}\sec^2\frac{x}{2}\,dx$

When $x = \frac{\pi}{2} \Rightarrow z = \tan\frac{\pi}{4} = 1$ and when $x = 0 \Rightarrow z = \tan 0 = 0.$

$$\therefore \quad I = \int_0^1 \frac{2\,dz}{5 - 5z^2 + 6z} = -\frac{2}{5}\int_0^1 \frac{dz}{z^2 - \frac{6}{5}z - 1} = -\frac{2}{5}\int_0^1 \frac{dz}{z^2 - \frac{6}{5}z + \frac{9}{25} - 1 - \frac{9}{25}}$$

$$= -\frac{2}{5}\int_0^1 \frac{dz}{\left(z - \frac{3}{5}\right)^2 - \frac{34}{25}}.$$

$t = z - \frac{3}{5} \quad \Rightarrow \quad dt = dz$

When $z = 1 \Rightarrow t = 1 - \frac{3}{5} = \frac{2}{5}$ and when $z = 0 \Rightarrow t = 0 - \frac{3}{5} = -\frac{3}{5}.$

$$\therefore \quad I = -\frac{2}{5}\int_{-3/5}^{2/5} \frac{dt}{t^2 - \frac{34}{25}} = -\frac{2}{5}\cdot\frac{1}{2\cdot\frac{\sqrt{34}}{5}}\log\left[\frac{t - \sqrt{34}/5}{t + \sqrt{34}/5}\right]_{-3/5}^{2/5}$$

$$= -\frac{1}{\sqrt{34}}\log\left[\left|\frac{5t - \sqrt{34}}{5t + \sqrt{34}}\right|\right]_{-3/5}^{2/5} = -\frac{1}{\sqrt{34}}\left[\log\left|\frac{2 - \sqrt{34}}{2 + \sqrt{34}}\right| - \log\left|\frac{-3 - \sqrt{34}}{-3 + \sqrt{34}}\right|\right]$$

$$= -\frac{1}{\sqrt{34}}\left[\log\frac{\sqrt{34} - 2}{\sqrt{34} + 2} - \log\frac{\sqrt{34} + 3}{\sqrt{34} - 3}\right] = -\frac{1}{\sqrt{34}}\log\left(\frac{\sqrt{34} - 2}{\sqrt{34} + 2}\times\frac{\sqrt{34} - 3}{\sqrt{34} + 3}\right)$$

$$= -\frac{1}{\sqrt{34}}\log\left(\frac{40 - 5\sqrt{34}}{40 + 5\sqrt{34}}\right) = -\frac{1}{\sqrt{34}}\log\left(\frac{8 - \sqrt{34}}{8 + \sqrt{34}}\times\frac{8 - \sqrt{34}}{8 - \sqrt{34}}\right) = \mathbf{-\frac{1}{\sqrt{34}}\log\frac{49 - 8\sqrt{34}}{15}}.$$

Example 3 *Evaluate the following definite integrals :*

(*i*) $\int_0^a \sin^{-1}\sqrt{\frac{x}{a+x}}\,dx$ (*ii*) $\int_0^{\pi/2} (\sqrt{\tan x}+\sqrt{\cot x})\,dx$

(*iii*) $\int_0^{\pi/4} \frac{\tan^3 x}{1+\cos 2x}\,dx$ (*iv*) $\int_0^1 x(\tan^{-1} x)^2\,dx.$

Sol. (*i*) Let $I = \int_0^a \sin^{-1}\sqrt{\frac{x}{a+x}}\,dx.$

Let $x = a\tan^2\theta \quad \therefore \quad dx = 2a\tan\theta\sec^2\theta\,d\theta$

when $x = a \Rightarrow a\tan^2\theta = a \Rightarrow \tan^2\theta = 1 \Rightarrow \tan\theta = 1 \Rightarrow \theta = \pi/4$

when $x = 0 \Rightarrow a\tan^2\theta = 0 \Rightarrow \tan^2\theta = 0 \Rightarrow \tan\theta = 0 \Rightarrow \theta = 0$

$$\therefore \quad I = \int_0^{\pi/4} \sin^{-1}\left(\sqrt{\frac{a\tan^2\theta}{a+a\tan^2\theta}}\right).\,2a\tan\theta\sec^2\theta\,d\theta$$

$$= 2a\int_0^{\pi/4} \sin^{-1}\left(\frac{\tan\theta}{\sec\theta}\right).\tan\theta\sec^2\theta\,d\theta = 2a\int_0^{\pi/4} \sin^{-1}(\sin\theta).\tan\theta\sec^2\theta\,d\theta$$

$$= 2a\int_0^{\pi/4} \underset{\text{I}}{\theta}\,.\,\underset{\text{II}}{\tan\theta\sec^2\theta}\,d\theta = 2a\left[\theta.\frac{\tan^2\theta}{2}\Bigg|_0^{\pi/4} - \int_0^{\pi/4} 1.\frac{\tan^2\theta}{2}\,d\theta\right]$$

$$= a\theta\tan^2\theta\Big|_0^{\pi/4} - a\int_0^{\pi/4}(\sec^2\theta - 1)\,d\theta = a\left[\frac{\pi}{4}\tan^2\frac{\pi}{4} - 0\right] - a\,[\tan\theta - \theta]\Big|_0^{\pi/4}$$

$$= a.\frac{\pi}{4} - a\left[\left(\tan\frac{\pi}{4} - \frac{\pi}{4}\right) - (\tan 0 - 0)\right] = \frac{\pi a}{4} - a\left(1 - \frac{\pi}{4}\right) = \frac{\pi a}{4} - a + \frac{\pi a}{4} = \mathbf{a}\left(\frac{\pi - 2}{2}\right).$$

(*ii*) Let $I = \int_0^{\pi/2} (\sqrt{\tan x} + \sqrt{\cot x})\,dx.$

$$\therefore \quad I = \int_0^{\pi/2}\left(\frac{\sqrt{\sin x}}{\sqrt{\cos x}} + \frac{\sqrt{\cos x}}{\sqrt{\sin x}}\right)dx = \int_0^{\pi/2}\frac{\sin x + \cos x}{\sqrt{\sin x\cos x}}\,dx$$

$$= \sqrt{2}\int_0^{\pi/2}\frac{\sin x + \cos x}{\sqrt{2\sin x\cos x}}\,dx = \sqrt{2}\int_0^{\pi/2}\frac{\sin x + \cos x}{\sqrt{1-(1-2\sin x\cos x)}}\,dx \qquad \textbf{(Note this step)}$$

$$= \sqrt{2}\int_0^{\pi/2}\frac{\sin x + \cos x}{\sqrt{1-(\sin^2 x + \cos^2 x - 2\sin x\cos x}}\,dx = \sqrt{2}\int_0^{\pi/2}\frac{\sin x + \cos x}{\sqrt{1-(\sin x - \cos x)^2}}\,dx.$$

Let $z = \sin x - \cos x. \quad \therefore \quad dz = (\cos x + \sin x)\,dx$

When $x = \frac{\pi}{2} \Rightarrow z = \sin\frac{\pi}{2} - \cos\frac{\pi}{2} = 1 - 0 = 1$ and when $x = 0 \Rightarrow z = \sin 0 - \cos 0 = 0 - 1 = -1.$

$\therefore \quad I = \sqrt{2}\int_{-1}^{1}\frac{dz}{\sqrt{1-z^2}} = \sqrt{2}\left[\sin^{-1} z\right]_{-1}^{1} = \sqrt{2}\,[\sin^{-1}(1) - \sin^{-1}(-1)] = \sqrt{2}\left[\frac{\pi}{2} - \left(-\frac{\pi}{2}\right)\right] = \pi\sqrt{\mathbf{2}}.$

(*iii*) Let $I = \int_0^{\pi/4}\frac{\tan^3 x}{1+\cos 2x}\,dx.$

$\therefore \quad I = \int_0^{\pi/4}\frac{\tan^3 x}{2\cos^2 x}\,dx = \frac{1}{2}\int_0^{\pi/4}\tan^3 x\sec^2 x\,dx$

Let $z = \tan x \Rightarrow dz = \sec^2 x\,dx$

When $x = \frac{\pi}{4} \Rightarrow z = \tan\frac{\pi}{4} = 1$ and when $x = 0 \Rightarrow z = \tan 0 = 0.$

$\therefore \quad I = \frac{1}{2}\int_0^1 z^3\,dz = \frac{1}{2}\cdot\left[\frac{z^4}{4}\right]_0^1 = \frac{1}{8}[(1)^4 - (0)^4] = \frac{\mathbf{1}}{\mathbf{8}}.$

(*iv*) Let $I = \int_0^1 x(\tan^{-1} x)^2\,dx.$ $\therefore$ $I = \int_0^1 \underset{\text{I}}{(\tan^{-1} x)^2}\,.\,\underset{\text{II}}{x}\,dx$

Integrating by parts, we get

$$I = \left[(\tan^{-1} x)^2\frac{x^2}{2}\right]_0^1 - \int_0^1\left(2\tan^{-1} x\,.\,\frac{1}{1+x^2}\right)\frac{x^2}{2}\,dx$$

$$= (\tan^{-1} 1)^2\,.\,\frac{1}{2} - 0 - \int_0^1\frac{x^2}{1+x^2}\tan^{-1} x\,dx = \frac{\pi^2}{16}\cdot\frac{1}{2} - \int_0^1\left(1 - \frac{1}{1+x^2}\right)\tan^{-1} x\,dx$$

$$= \frac{\pi^2}{32} - \int_0^1\tan^{-1} x\,dx + \int_0^1\frac{\tan^{-1} x}{1+x^2}\,dx = \frac{\pi^2}{32} - \int_0^1(\tan^{-1} x)\,.\,1\,dx + \left[\frac{(\tan^{-1} x)^2}{2}\right]_0^1$$

$$= \frac{\pi^2}{32} - \left[(\tan^{-1} x)\,.\,x\Big|_0^1 - \int_0^1\frac{1}{1+x^2}\,.\,x\,dx\right] + \frac{1}{2}[(\tan^{-1} 1)^2 - (\tan^{-1} 0)^2]$$

$$= \frac{\pi^2}{32} - \left[(\tan^{-1} 1)\,.\,1 - 0 - \frac{1}{2}\int_0^1\frac{2x}{1+x^2}\,dx\right] + \frac{1}{2}\left[\frac{\pi^2}{16} - 0\right]$$

$$= \frac{\pi^2}{32} - \frac{\pi}{4} + \frac{1}{2}\log(1+x^2)\Big|_0^1 + \frac{\pi^2}{32} = \frac{\pi^2}{16} - \frac{\pi}{4} + \frac{1}{2}[\log 2 - \log 1] = \frac{\pi}{\mathbf{4}}\left(\frac{\pi}{\mathbf{4}} - \mathbf{1}\right) + \frac{\mathbf{1}}{\mathbf{2}}\,\mathbf{\log 2}.$$

EXERCISE 14.2

Evaluate the following definite integrals (Q. No. 1–12) :

1. (*i*) $\int_0^1\frac{2x}{5x^2+1}\,dx$ (*ii*) $\int_1^2\frac{3x}{9x^2-1}\,dx$ (*iii*) $\int_2^4\frac{x}{x^2+1}\,dx$

(*iv*) $\int_1^3(x^2+e^x)(x^3+3e^x+4)\,dx$ (*v*) $\int_0^2 x\sqrt{2-x}\,dx$ (*vi*) $\int_{-1}^2\frac{x}{(x^2+1)^2}\,dx$

2. (i) $\int_1^2 \frac{dx}{x(1+\log x)^2}$ (ii) $\int_{\pi/6}^{\pi/2} \frac{\operatorname{cosec} x \cot x}{1+\operatorname{cosec}^2 x} dx$ (iii) $\int_0^{\pi/6} \cos^{-3} 2\theta \sin 2\theta \, d\theta$

(iv) $\int_0^1 \frac{e^x}{1+e^{2x}} dx$ (v) $\int_0^{\pi/2} \frac{\sin x}{1+\cos^2 x} dx$ (vi) $\int_0^{\pi/3} \frac{\sec x \tan x}{1+\sec^2 x} dx$

3. (i) $\int_1^2 \frac{dx}{\sqrt{x^2+4x+3}}$ (ii) $\int_0^{\pi/2} \sqrt{\cos\theta} \sin^3 \theta \, d\theta$ (iii) $\int_0^1 \frac{\tan^{-1} x}{1+x^2} dx$

(iv) $\int_{\pi/4}^{\pi/2} \frac{\cos x \, dx}{(\cos (x/2) + \sin (x/2))^2}$ (v) $\int_0^{(\pi/2)^{1/3}} x^2 \sin x^3 \, dx$ (vi) $\int_0^1 \sin^{-1} x \, dx$

4. (i) $\int_0^{\pi/2} \frac{\sin\theta}{\sqrt{1+\cos\theta}} d\theta$ (ii) $\int_0^1 \frac{\sqrt{\tan^{-1} x}}{1+x^2} dx$ (iii) $\int_0^{\pi/4} \frac{dx}{9+16\cos^2 x}$

(iv) $\int_0^{2\pi} e^{x/2} \sin\left(\frac{\pi}{4}+\frac{x}{2}\right) dx$ (v) $\int_0^4 \frac{dx}{\sqrt{x^2+2x+3}}$ (vi) $\int_0^a \frac{x^4}{\sqrt{a^2-x^2}} dx$

5. (i) $\int_0^{\pi/2} \frac{dx}{5+4\cos x}$ (ii) $\int_0^{\pi/2} \frac{dx}{5+4\sin x}$ (iii) $\int_0^{\pi/2} \frac{dx}{1-2\sin x}$

(iv) $\int_0^{\pi/2} \frac{dx}{3+2\cos x}$ (v) $\int_0^{\pi/2} \frac{dx}{5+3\cos x}$ (vi) $\int_0^{\pi/2} \frac{dx}{2+\cos x}$

(P.T.U. B.Pharma, June 2003)

6. (i) $\int_0^{\pi/2} \frac{dx}{5+2\cos x}$ (ii) $\int_0^{\pi/2} \frac{dx}{4\cos x+2\sin x}$ (iii) $\int_0^{\pi/2} \frac{dx}{2\cos x+4\sin x}$

(iv) $\int_0^{\pi/4} \frac{dx}{1+\cos^2 x}$ (v) $\int_0^{\pi/4} \frac{dx}{4+9\cos^2 x}$

7. (i) $\int_0^1 x\sqrt{\frac{1-x^2}{1+x^2}} dx$ (ii) $\int_0^{\pi/2} \frac{\cos x}{(2+\sin x)(1+\sin x)} dx$ (iii) $\int_0^{\pi/2} x^3 \sin 3x \, dx$

(iv) $\int_0^4 \frac{dx}{1+\sqrt{x}}$ (v) $\int_0^1 \sqrt{\frac{1-x}{1+x}} dx$ (vi) $\int_0^1 \frac{1-x^2}{(1+x^2)^2} dx$

8. (i) $\int_1^2 \frac{dx}{(x+1)\sqrt{x^2-1}}$ (ii) $\int_0^{\pi/2} \cos^7 x \, dx$ (iii) $\int_0^1 \sin^{-1}\sqrt{x} \, dx$

(iv) $\int_{-a}^a \sqrt{\frac{a-x}{a+x}} dx$ (v) $\int_0^{\pi/4} \frac{\sin x+\cos x}{9+16\sin 2x} dx$ (vi) $\int_0^{\pi/2} x \sin x \sin 2x \, dx.$

9. (i) $\int_0^{\pi} 5(5-4\cos\theta)^{1/4} \sin\theta \, d\theta$ (ii) $\int_{\pi/6}^{\pi/3} \frac{\sin x+\cos x}{\sqrt{\sin 2x}} dx$ (iii) $\int_0^{\pi/2} \frac{\cos^2 x}{\cos^2 x+4\sin^2 x} dx$

(iv) $\int_{\pi/2}^{\pi} e^x \frac{1-\sin x}{1-\cos x} dx.$

Answers

1. (*i*) $\frac{1}{5}\log 6$ (*ii*) $\frac{1}{6}\log\frac{35}{8}$ (*iii*) $\frac{1}{2}\log\frac{17}{5}$

(*iv*) $\frac{1}{6}[(31+3e^3)^2-(5+3e)^2]$ (*v*) $\frac{16\sqrt{2}}{15}$ (*vi*) $\frac{3}{20}$

2. (*i*) $\frac{\log 2}{1+\log 2}$ (*ii*) $\tan^{-1}2-\frac{\pi}{4}$ (*iii*) $\frac{3}{4}$

(*iv*) $\tan^{-1}e-\frac{\pi}{4}$ (*v*) $\frac{\pi}{4}$ (*vi*) $\tan^{-1}2-\frac{\pi}{4}$

3. (*i*) $\log\frac{4+\sqrt{15}}{3+2\sqrt{2}}$ (*ii*) $\frac{8}{21}$ (*iii*) $\frac{\pi^2}{32}$

(*iv*) $\log(4-2\sqrt{2})$ (*v*) $\frac{1}{3}$ (*vi*) $\frac{\pi}{2}-1$

4. (*i*) $2(\sqrt{2}-1)$ (*ii*) $\frac{\pi^{3/2}}{12}$ (*iii*) $\frac{1}{15}\tan^{-1}\frac{3}{5}$

(*iv*) 0 (*v*) $\log(2+\sqrt{3})$ (*vi*) $\frac{3\pi a^4}{16}$

5. (*i*) $\frac{2}{3}\tan^{-1}\frac{1}{3}$ (*ii*) $\frac{2}{3}\tan^{-1}\frac{1}{3}$ (*iii*) $\frac{1}{\sqrt{3}}\log(2-\sqrt{3})$

(*iv*) $\frac{2}{\sqrt{5}}\tan^{-1}\frac{1}{\sqrt{5}}$ (*v*) $\frac{1}{2}\tan^{-1}\frac{1}{2}$ (*vi*) $\frac{\pi}{3\sqrt{3}}$

6. (*i*) $\frac{2}{\sqrt{21}}\tan^{-1}\sqrt{\frac{3}{7}}$ (*ii*) $\frac{1}{\sqrt{5}}\log\frac{3+\sqrt{5}}{2}$ (*iii*) $\frac{1}{2\sqrt{5}}\log\frac{7+3\sqrt{5}}{2}$

(*iv*) $\frac{1}{\sqrt{2}}\tan^{-1}\frac{1}{\sqrt{2}}$ (*v*) $\frac{1}{2\sqrt{13}}\tan^{-1}\frac{2}{\sqrt{13}}$

7. (*i*) $\frac{\pi-2}{4}$ (*ii*) $\log\frac{4}{3}$ (*iii*) $\frac{2}{27}-\frac{\pi^2}{12}$

(*iv*) $4-2\log 3$ (*v*) $\frac{\pi}{2}-1$ (*vi*) $\frac{1}{2}$

8. (*i*) $\frac{1}{\sqrt{3}}$ (*ii*) $\frac{16}{35}$ (*iii*) $\frac{\pi}{4}$

(*iv*) πa (*v*) $\frac{\log 9}{40}$ (*vi*) $\frac{3\pi-4}{9}$

9. (*i*) $9\sqrt{3}-1$ (*ii*) $2\sin^{-1}\frac{\sqrt{3}-1}{2}$ (*iii*) $\frac{\pi}{6}$

(*iv*) $e^{\pi/2}$.

Hints

1. (*v*) $z = \sqrt{2-x} \Rightarrow x = 2 - z^2 \Rightarrow dx = -2z\,dz$

$$\therefore \int_0^2 x\sqrt{2-x}\,dx = \int_{\sqrt{2}}^0 (2-z^2)\,z\,.\,2z\,dz = 2\int_{\sqrt{2}}^0 (z^4 - 2z^2)\,dz$$

(*vi*) Take $z = x^2 + 1$.

2. (*i*) $z = 1 + \log x \Rightarrow \int_1^2 \frac{dx}{x(1+\log x)^2} = \int_0^{1+\log 2} \frac{dz}{z^2}$

(*ii*) Take $z = \operatorname{cosec} x$ (*iii*) Take $z = \cos 2\theta$.

3. (*i*) Write $x^2 + 4x + 3 = (x+2)^2 - 1$ and take $z = x + 2$

(*iii*) Take $z = \tan^{-1} x$

(*iv*) Write $\left(\cos\frac{x}{2} + \sin\frac{x}{2}\right)^2 = 1 + 2\sin\frac{x}{2}\cos\frac{x}{2} = 1 + \sin x$ and take $z = 1 + \sin x$

(*v*) Take $z = x^3$ (*vi*) Take $z = \sin^{-1} x$.

4. (*i*) Take $z = 1 + \cos\theta$ (*ii*) Take $z = \tan^{-1} x$

(*iii*) Write $\frac{1}{9 + 16\cos^2 x} = \frac{1/\cos^2 x}{9\sec^2 x + 16} = \frac{\sec^2 x}{9 + 9\tan^2 x + 16}$ and take $z = \tan x$

(*iv*) $z = \frac{x}{2} \Rightarrow \int_0^{2\pi} e^{x/2}\sin\left(\frac{\pi}{4} + \frac{x}{2}\right)dx = 2\int_0^{\pi} e^z \sin\left(\frac{\pi}{4} + z\right)dz$

$$= \frac{2}{\sqrt{2}}\int_0^{\pi} e^z(\sin z + \cos z)\,dz = \sqrt{2}\,e^z \sin z\Big|_0^{\pi}$$

(*vi*) $x = a\sin\theta \Rightarrow \int_0^a \frac{x^4}{\sqrt{a^2 - x^2}}dx = \int_0^{\pi/2} \frac{a^4\sin^4\theta}{a\cos\theta}\,.\,a\cos\theta\,d\theta = a^4\int_0^{\pi/2}\left(\frac{1-\cos 2\theta}{2}\right)^2 d\theta$.

5. (*i*) Take $z = \tan\frac{x}{2}$ after expressing the integrand in terms of $\tan\frac{x}{2}$.

7. (*i*) $x^2 = \cos\theta$ implies $\int_0^1 x\sqrt{\frac{1-x^2}{1+x^2}}\,dx = \int_{\pi/2}^0 \sqrt{\frac{1-\cos\theta}{1+\cos\theta}}\,.-\frac{\sin\theta}{2}\,d\theta$

$$= -\int_{\pi/2}^0 \frac{\sin\frac{1}{2}\theta}{\cos\frac{1}{2}\theta}\,.\sin\frac{1}{2}\theta\cos\frac{1}{2}\theta\,d\theta = -\int_{\pi/2}^0 \sin^2\frac{1}{2}\theta\,d\theta = -\int_{\pi/2}^0 \frac{1-\cos\theta}{2}\,d\theta$$

(*iii*) Integrate by parts three times

(*iv*) Put $z = \sqrt{x}$ (*v*) $\sqrt{\frac{1-x}{1+x}} = \frac{1-x}{\sqrt{1-x^2}}$

(*vi*) Let $x = \tan\theta$. $\therefore \int_0^1 \frac{1-x^2}{(1+x^2)^2}dx = \int_0^{\pi/4}\frac{1-\tan^2\theta}{\sec^4\theta}\,.\sec^2\theta\,d\theta = \int_0^{\pi/4}\cos 2\theta\,d\theta$.

8. (*i*) Put $z = \frac{1}{x+1}$ (*iii*) Put $z = \sin^{-1}\sqrt{x}$

(*iv*) $x = a\cos\theta \Rightarrow \int_{-a}^{a}\sqrt{\dfrac{a-x}{a+x}}\,dx = \int_{\pi}^{0}\sqrt{\dfrac{1-\cos\theta}{1+\cos\theta}}\,.-a\sin\theta\,d\theta$

$$= a\int_0^{\pi}\frac{\sin\frac{1}{2}\theta}{\cos\frac{1}{2}\theta}\,.\,2\sin\frac{1}{2}\theta\cos\frac{1}{2}\theta\,d\theta = 2a\int_0^{\pi}\sin^2\frac{1}{2}\theta\,d\theta = a\int_0^{\pi}(1-\cos\theta)\,d\theta$$

(*v*) We have $(\sin x - \cos x)' = \cos x + \sin x$ and $(\sin x - \cos x)^2 = 1 - \sin 2x$.

$\therefore\ \sin 2x = 1 - (\sin x - \cos x)^2$ and $9 + 16\sin 2x = 9 + 16\,(1 - (\sin x - \cos x)^2)$

$$= 25 - 16\,(\sin x - \cos x)^2$$

(*vi*) $x\sin x\sin 2x = \dfrac{1}{2}\,x\,(\cos x - \cos 3x)$

9. (*i*) $z = \cos\theta \Rightarrow \int_0^{\pi} 5\,(5 - 4\cos\theta)^{1/4}\sin\theta\,d\theta$

$$= -\int_1^{-1} 5\,(5-4z)^{1/4}\,dz = -5\,.\,\frac{(5-4z)^{5/4}}{(5/4)(-4)}\Bigg|_1^{-1}$$

(*ii*) Write $\sin 2x = 1 - (1 - \sin 2x) = 1 - (\cos^2 x + \sin^2 x - 2\sin x\cos x) = 1 - (\cos x - \sin x)^2$.

Now take $\quad z = \cos x - \sin x$

(*iii*) $\dfrac{\cos^2 x}{\cos^2 x + 4\sin^2 x} = \dfrac{1-\sin^2 x}{1+3\sin^2 x} = -\dfrac{1}{3} + \dfrac{4}{3\,(1+3\sin^2 x)}$

$$= -\frac{1}{3} + \frac{4\sec^2 x}{3\,(\sec^2 x + 3\tan^2 x)} = -\frac{1}{3} + \frac{4}{3}\left(\frac{\sec^2 x}{1+4\tan^2 x}\right) = -\frac{1}{3} + \frac{1}{3}\left(\frac{\sec^2 x}{(1/2)^2 + \tan^2 x}\right)$$

(*iv*) $\displaystyle\int_{\pi/2}^{\pi} e^x\,\frac{1-\sin x}{1-\cos x}\,dx = \int_{\pi/2}^{\pi} e^x\left(\frac{1}{2\sin^2\frac{x}{2}} - \frac{\cos\frac{x}{2}}{\sin\frac{x}{2}}\right)dx$

$$= \int_{\pi/2}^{\pi} e^x\left(-\cot\frac{x}{2} + \frac{1}{2}\operatorname{cosec}^2\frac{x}{2}\right)dx = -e^x\cot\frac{x}{2}\Bigg|_{\pi/2}^{\pi}.$$

14.5. PROPERTIES OF DEFINITE INTEGRALS

In this section, we shall establish some properties of definite integrals, which are very useful in solving definite integrals :

(*i*) $\displaystyle\int_a^b f(x)\,dx = -\int_b^a f(x)\,dx,\ a > b$ (*ii*) $\displaystyle\int_a^b f(x)\,dx = \int_a^c f(x)\,dx + \int_c^b f(x)\,dx,\ a < c < b$

(*iii*) $\displaystyle\int_a^b f(x)\,dx = \int_a^b f(a+b-x)\,dx$ (*iv*) $\displaystyle\int_0^a f(x)\,dx = \int_0^a f(a-x)\,dx$

(*v*) $\displaystyle\int_0^{2a} f(x)\,dx = \int_0^a f(x)\,dx + \int_0^a f(2a-x)\,dx$

(*vi*) $\displaystyle\int_0^{2a} f(x)\,dx = \begin{cases} 2\displaystyle\int_0^a f(x)\,dx & \text{if } f(2a-x) = f(x) \\ 0 & \text{if } f(2a-x) = -f(x)\end{cases}$

(*vii*) $\int_{-a}^{a} f(x)\,dx = \begin{cases} 2\int_0^a f(x)\,dx & \text{if } f(x) \text{ is even} \\ 0 & \text{if } f(x) \text{ is odd.} \end{cases}$

14.6. EVALUATION OF DEFINITE INTEGRALS USING PROPERTIES

The above properties are very useful in solving definite integrals. Now we shall see how one can make use of these properties in evaluating definite integrals. The reader is strongly advised to master the results of these properties.

WORKING RULES FOR EVALUATING DEFINITE INTEGRALS

Rule I. *If x occurs in the numerator and it is desired to remove it, then the property*

$$\int_0^a f(x)\,dx = \int_0^a f(a-x)\,dx \text{ is generally used.}$$

Rule II. *If f(x) is an odd function, then* $\int_{-a}^{a} f(x)\,dx$ *is always zero.*

Rule III. *If f(x) is an even function then,* $\int_{-a}^{a} f(x)\,dx = 2\int_0^a f(x)\,dx.$

Rule IV. *In case f(x) is expressed by two or more rules or if f(x) involves 'modulus' in its definition, then the formula* $\int_a^b f(x)\,dx = \int_a^c f(x)\,dx + \int_c^b f(x)\,dx$ *is found useful.*

ILLUSTRATIVE EXAMPLES

Example 1. *Evaluate :*

(*i*) $\int_0^{\pi/2} \frac{\sin x}{\sin x + \cos x}\,dx$ (*ii*) $\int_0^{\pi/2} \frac{dx}{1+\tan x}$ (*iii*) $\int_0^{\pi/2} \frac{dx}{1+\cot x}$.

Sol. (*i*) Let $I = \int_0^{\pi/2} \frac{\sin x}{\sin x + \cos x}\,dx.$ ...(1)

Using $\int_0^a f(x)\,dx = \int_0^a f(a-x)\,dx$, we get

$$I = \int_0^{\pi/2} \frac{\sin\left(\frac{\pi}{2}-x\right)}{\sin\left(\frac{\pi}{2}-x\right)+\cos\left(\frac{\pi}{2}-x\right)}\,dx = \int_0^{\pi/2} \frac{\cos x}{\cos x + \sin x}\,dx.$$

$$\therefore \quad I = \int_0^{\pi/2} \frac{\cos x}{\cos x + \sin x}\,dx \quad ...(2)$$

$$(1) + (2) \Rightarrow 2I = \int_0^{\pi/2} \frac{\sin x + \cos x}{\cos x + \sin x}\,dx = \int_0^{\pi/2} 1\,.\,dx = x\Big|_0^{\pi/2} = \frac{\pi}{2}$$

$$\therefore \quad I = \frac{1}{2}\left(\frac{\pi}{2}\right) = \frac{\pi}{4}.$$

(*ii*) $\int_0^{\pi/2} \frac{1}{1+\tan x}\,dx = \int_0^{\pi/2} \frac{1}{1+\frac{\sin x}{\cos x}}\,dx = \int_0^{\pi/2} \frac{\cos x}{\sin x + \cos x}\,dx = \frac{\pi}{4}$. [Proceed as in part (*i*)]

(*iii*) $\int_0^{\pi/2} \frac{1}{1+\cot x}\,dx = \int_0^{\pi/2} \frac{1}{1+\frac{\cos x}{\sin x}}\,dx = \int_0^{\pi/2} \frac{\sin x}{\sin x + \cos x}\,dx = \frac{\pi}{4}$. [Proceed as in part (*i*)]

Example 2. *Evaluate :*

(*i*) $\int_0^{\pi/2} \frac{\sqrt{\sin x}}{\sqrt{\sin x} + \sqrt{\cos x}}\,dx$ (*ii*) $\int_0^{\pi/2} \frac{dx}{1+\sqrt{\tan x}}$ (*iii*) $\int_0^{\pi/2} \frac{\sqrt{\tan x}}{1+\sqrt{\tan x}}\,dx$.

Sol. (*i*) Let $\quad I = \int_0^{\pi/2} \frac{\sqrt{\sin x}}{\sqrt{\sin x} + \sqrt{\cos x}}\,dx. \qquad \ldots(1)$

Using $\int_0^a f(x)\,dx = \int_0^a f(a-x)\,dx$, we get

$$I = \int_0^{\pi/2} \frac{\sqrt{\sin\left(\frac{\pi}{2}-x\right)}}{\sqrt{\sin\left(\frac{\pi}{2}-x\right)} + \sqrt{\cos\left(\frac{\pi}{2}-x\right)}}\,dx = \int_0^{\pi/2} \frac{\sqrt{\cos x}}{\sqrt{\cos x} + \sqrt{\sin x}}\,dx$$

$$\therefore \quad I = \int_0^{\pi/2} \frac{\sqrt{\cos x}}{\sqrt{\cos x} + \sqrt{\sin x}}\,dx \qquad \ldots(2)$$

$$(1) + (2) \Rightarrow \quad 2I = \int_0^{\pi/2} \frac{\sqrt{\sin x} + \sqrt{\cos x}}{\sqrt{\cos x} + \sqrt{\sin x}}\,dx = \int_0^{\pi/2} 1 \,.\, dx = \frac{\pi}{2}.$$

$$\therefore \quad I = \frac{1}{2}\left(\frac{\pi}{2}\right) = \frac{\pi}{4}.$$

(*ii*) $\int_0^{\pi/2} \frac{dx}{1+\sqrt{\tan x}} = \int_0^{\pi/2} \frac{dx}{1+\frac{\sqrt{\sin x}}{\sqrt{\cos x}}} = \int_0^{\pi/2} \frac{\sqrt{\cos x}}{\sqrt{\cos x} + \sqrt{\sin x}}\,dx = \frac{\pi}{4}$. [Proceed as in part (*i*)]

(*iii*) $\int_0^{\pi/2} \frac{\sqrt{\tan x}}{1+\sqrt{\tan x}}\,dx = \int_0^{\pi/2} \frac{\sqrt{\sin x}/\sqrt{\cos x}}{1+(\sqrt{\sin x}/\sqrt{\cos x})}\,dx = \int_0^{\pi/2} \frac{\sqrt{\sin x}}{\sqrt{\cos x} + \sqrt{\sin x}}\,dx = \frac{\pi}{4}$.

[Proceed as in part (*i*)]

Example 3. *Evaluate :*

(*i*) $\int_0^{\pi} \frac{x\,dx}{1+\sin x}$ (*ii*) $\int_0^{\pi} \frac{x \tan x}{\sec x + \tan x}\,dx$ (*iii*) $\int_1^{3} \frac{\sqrt{4-x}}{\sqrt{x} + \sqrt{4-x}}\,dx$.

Sol. (*i*) Let $I = \int_0^{\pi} \frac{x\,dx}{1+\sin x}$. Using $\int_0^a f(x)\,dx = \int_0^a f(a-x)\,dx$, we get

$$I = \int_0^\pi \frac{(\pi - x)}{1 + \sin(\pi - x)}\,dx = \int_0^\pi \frac{\pi\,dx}{1 + \sin x} - \int_0^\pi \frac{x\,dx}{1 + \sin x} = \pi \int_0^\pi \frac{dx}{1 + \sin x} - I$$

$$\Rightarrow \qquad 2I = \pi \int_0^\pi \frac{dx}{1 + \sin x}.$$

$$\therefore \qquad I = \frac{\pi}{2} \int_0^\pi \frac{dx}{1 + \sin x} = \frac{\pi}{2} \int_0^\pi \frac{(1 - \sin x)}{(1 + \sin x)(1 - \sin x)}\,dx$$

$$= \frac{\pi}{2} \int_0^\pi \frac{1 - \sin x}{1 - \sin^2 x}\,dx = \frac{\pi}{2} \int_0^\pi \frac{1 - \sin x}{\cos^2 x}\,dx = \frac{\pi}{2} \int_0^\pi \left(\frac{1}{\cos^2 x} - \frac{\sin x}{\cos x \cos x} \right) dx$$

$$= \frac{\pi}{2} \int_0^\pi (\sec^2 x - \sec x \tan x)\,dx = \frac{\pi}{2} \Big[\tan x - \sec x \Big]_0^\pi$$

$$= \frac{\pi}{2} [\tan \pi - \sec \pi] - \frac{\pi}{2} [\tan 0 - \sec 0] = \frac{\pi}{2} (0 - (-1) - 0 + 1) = 2 . \frac{\pi}{2} = \pi.$$

(*ii*) Let $I = \int_0^\pi \frac{x \tan x}{\sec x + \tan x}\,dx$. Using $\int_0^a f(x)\,dx = \int_0^a f(a - x)\,dx$, we get

$$I = \int_0^\pi \frac{(\pi - x) \tan(\pi - x)}{\sec(\pi - x) + \tan(\pi - x)}\,dx = \int_0^\pi \frac{(\pi - x) . - \tan x}{- \sec x - \tan x}\,dx = \int_0^\pi \frac{(\pi - x) \tan x}{\sec x + \tan x}\,dx$$

$$= \pi \int_0^\pi \frac{\tan x}{\sec x + \tan x}\,dx - \int_0^\pi \frac{x \tan x}{\sec x + \tan x}\,dx = \pi \int_0^\pi \frac{\dfrac{\sin x}{\cos x}}{\dfrac{1}{\cos x} + \dfrac{\sin x}{\cos x}}\,dx - I.$$

$$\therefore \qquad I = \pi \int_0^\pi \frac{\sin x}{1 + \sin x}\,dx - I$$

$$\therefore \qquad 2I = \pi \int_0^\pi \frac{\sin x}{1 + \sin x}\,dx$$

$$\therefore \qquad I = \frac{\pi}{2} \int_0^\pi \frac{\sin x}{1 + \sin x}\,dx = \frac{\pi}{2} \int_0^\pi \frac{(1 + \sin x) - 1}{1 + \sin x}\,dx = \frac{\pi}{2} \int_0^\pi \frac{1 + \sin x}{1 + \sin x}\,dx - \frac{\pi}{2} \int_0^\pi \frac{1}{1 + \sin x}\,dx$$

$$= \frac{\pi}{2} \int_0^\pi 1 . dx - \frac{\pi}{2} \int_0^\pi \frac{1 - \sin x}{1 - \sin^2 x}\,dx = \frac{\pi}{2} [x]_0^\pi - \frac{\pi}{2} \int_0^\pi \frac{1 - \sin x}{\cos^2 x}\,dx$$

$$= \frac{\pi}{2} [\pi - 0] - \frac{\pi}{2} \int_0^\pi (\sec^2 x - \sec x \tan x)\,dx = \frac{\pi^2}{2} - \frac{\pi}{2} [\tan x - \sec x]_0^\pi$$

$$= \frac{\pi^2}{2} - \frac{\pi}{2} [(\tan \pi - \sec \pi) - (\tan 0 - \sec 0)] = \frac{\pi^2}{2} - \frac{\pi}{2} [0 - (-1) - 0 + 1] = \pi \left(\frac{\pi}{2} - 1 \right).$$

(*iii*) Let $\quad I = \int_1^3 \frac{\sqrt{4 - x}}{\sqrt{x} + \sqrt{4 - x}}\,dx \qquad \ldots(1)$

Using $\int_a^b f(x)\,dx = \int_a^b f(a+b-x)\,dx$, we get

$$I = \int_1^3 \frac{\sqrt{4-(1+3-x)}}{\sqrt{1+3-x}+\sqrt{4-(1+3-x)}}\,dx \quad \text{or} \quad I = \int_1^3 \frac{\sqrt{x}}{\sqrt{4-x}+\sqrt{x}}\,dx \qquad \ldots(2)$$

$$(1) + (2) \Rightarrow \quad 2I = \int_1^3 \frac{\sqrt{4-x}+\sqrt{x}}{\sqrt{x}+\sqrt{4-x}}\,dx = \int_1^3 1 \,.\, dx = x\Big|_1^3 = 3 - 1 = 2$$

$$\therefore \quad I = \frac{1}{2}(2) = \mathbf{1.}$$

Example 4. *Evaluate :*

(*i*) $\int_{-\pi}^{\pi} x^{10} \sin^7 x\,dx$ (*ii*) $\int_{-1}^{1} \log \frac{4-x}{4+x}\,dx$ (*iii*) $\int_{-a}^{a} \sqrt{\frac{a-x}{a+x}}\,dx.$

Sol. (*i*) Let $I = \int_{-\pi}^{\pi} x^{10} \sin^7 x\,dx.$ Let $f(x) = x^{10} \sin^7 x.$

$\therefore \quad f(-x) = (-x)^{10} \sin^7 (-x) = x^{10} (-\sin x)^7 = x^{10} (-\sin^7 x) = -x^{10} \sin^7 x = -f(x).$

$\therefore \quad f(x)$ is an *odd function.*

$$\therefore \quad \int_{-\pi}^{\pi} x^{10} \sin^7 x\,dx = \mathbf{0}. \qquad \left(\text{By using } \int_{-a}^{a} f(x) = 0, \text{ where } f(x) \text{ is an odd function.}\right)$$

(*ii*) Let $I = \int_{-1}^{1} \log \frac{4-x}{4+x}\,dx.$ Let $f(x) = \log \frac{4-x}{4+x}$

$$\therefore \quad f(-x) = \log \frac{4-(-x)}{4+(-x)} = \log \frac{4+x}{4-x} = \log \left(\frac{4-x}{4+x}\right)^{-1} = -\log \frac{4-x}{4+x} = -f(x).$$

$\therefore \quad f(x)$ is an *odd function.*

$$\therefore \quad \int_{-1}^{1} \log \frac{4-x}{4+x}\,dx = \mathbf{0}. \qquad \left(\text{By using } \int_{-a}^{a} f(x)\,dx = 0, \text{ where } f(x) \text{ is an odd function.}\right)$$

(*iii*) Let $I = \int_{-a}^{a} \sqrt{\frac{a-x}{a+x}}\,dx.$

$$\therefore \quad I = \int_{-a}^{a} \sqrt{\frac{a-x}{a+x} \times \frac{a-x}{a-x}}\,dx = \int_{-a}^{a} \frac{a-x}{\sqrt{a^2-x^2}}\,dx$$

$$= a\int_{-a}^{a} \frac{dx}{\sqrt{a^2-x^2}} - \int_{-a}^{a} \frac{x\,dx}{\sqrt{a^2-x^2}} = aI_1 - I_2 \qquad \text{(say)}$$

Now $$I_1 = \int_{-a}^{a} \frac{dx}{\sqrt{a^2-x^2}} = \left[\sin^{-1} \frac{x}{a}\right]_{-a}^{a} = \sin^{-1} 1 - \sin^{-1}(-1) = \frac{\pi}{2} - \left(-\frac{\pi}{2}\right) = \pi.$$

$$I_2 = \int_{-a}^{a} \frac{x\,dx}{\sqrt{a^2-x^2}}. \quad \text{Let } x = a \sin \theta. \quad \therefore \quad dx = a \cos \theta\, d\theta$$

When $x = a \Rightarrow a \sin \theta = a \Rightarrow \sin \theta = 1 \Rightarrow \theta = \pi/2$

When $x = -a \Rightarrow a \sin \theta = -a \Rightarrow \sin \theta = -1 \Rightarrow \theta = -\pi/2$

$$\therefore \quad I_2 = \int_{-\pi/2}^{\pi/2} \frac{a \sin \theta}{\sqrt{a^2 - a^2 \sin^2 \theta}} \cdot a \cos \theta \, d\theta = a \int_{-\pi/2}^{\pi/2} \sin \theta \, d\theta$$

$$\left[a(-\cos \theta) \right]_{-\pi/2}^{\pi/2} = -a\left(\cos \frac{\pi}{2} - \cos\left(-\frac{\pi}{2}\right)\right) = -a(0-0) = 0.$$

$$\therefore \quad I = aI_1 - I_2 = a\pi - 0 = \mathbf{a\pi}.$$

Example 5. *Evaluate :*

(*i*) $\int_{-1}^{1} f(x)\,dx$, *where* $f(x) = \begin{cases} 1-2x, & x \le 0 \\ 1+2x, & x \ge 0 \end{cases}$

(*ii*) $\int_{-2}^{2} f(x)\,dx$, *where* $f(x) = \begin{cases} 2x-1, & -2 \le x \le 1 \\ 3x+2, & 1 \le x \le 2 \end{cases}$.

Sol. (*i*) $\int_{-1}^{1} f(x)\,dx = \int_{-1}^{0} f(x)\,dx + \int_{0}^{1} f(x)\,dx\,^* = \int_{-1}^{0} (1-2x)\,dx + \int_{0}^{1} (1+2x)\,dx$

$$= (x - x^2)\Big|_{-1}^{0} + (x + x^2)\Big|_{0}^{1} = (0 - (-1-1)) + ((1+1) - 0) = 2 + 2 = \mathbf{4}.$$

(*ii*) $\int_{-2}^{2} f(x)\,dx = \int_{-2}^{1} f(x)\,dx + \int_{1}^{2} f(x)\,dx = \int_{-2}^{1} (2x-1)\,dx + \int_{1}^{2} (3x+2)\,dx$

$$= \left| x^2 - x \right|_{-2}^{1} + \left| \frac{3x^2}{2} + 2x \right|_{1}^{2} = 0 - (4+2) + (6+4) - \left(\frac{3}{2} + 2\right) = -6 + 10 - \frac{7}{2} = 4 - \frac{7}{2} = \mathbf{\frac{1}{2}}.$$

Example 6. *Verify that :*

(*i*) $\int_0^2 (x^2+2)\,dx = \int_0^1 (x^2+2)\,dx + \int_1^2 (x^2+2)\,dx$

(*ii*) $\int_0^1 (e^x+1)\,dx = \int_0^1 e^x\,dx + \int_0^1 1 \,.\, dx.$

Sol. (*i*) L.H.S $= \int_0^2 (x^2+2)\,dx = \left(\frac{x^3}{3} + 2x\right)\Bigg|_0^2 = \left(\frac{8}{3} + 4\right) - (0+0) = \frac{20}{3}.$

$$\text{R.H.S.} = \int_0^1 (x^2+2)\,dx + \int_1^2 (x^2+2)\,dx = \left(\frac{x^3}{3} + 2x\right)\Bigg|_0^1 + \left(\frac{x^3}{3} + 2x\right)\Bigg|_1^2$$

$$= \left[\left(\frac{1}{3} + 2\right) - (0+0)\right] + \left[\left(\frac{8}{3} + 4\right) - \left(\frac{1}{3} + 2\right)\right] = \left(\frac{7}{3} - 0\right) + \left(\frac{20}{3} - \frac{7}{3}\right) = \frac{20}{3}.$$

$\therefore$ The result holds.

(*ii*) L.H.S. $= \int_0^1 (e^x+1)\,dx = (e^x + x)\Big|_0^1 = (e^1+1) - (e^0+0) = e + 1 - 1 = e.$

***Why this step.** We have used the result :

$\int_a^b f(x)\,dx = \int_a^c f(x)\,dx + \int_c^b f(x)\,dx$ with $c = 0$, because $f(x)$ is defined separately for $x \le 0$ and $x \ge 0$.

$$\text{R.H.S.} = \int_0^1 e^x\,dx + \int_0^1 1\,.\,dx = e^x\Big|_0^1 + x\Big|_0^1 = (e^1 - e^0) + (1-0) = e - 1 + 1 = e.$$

$\therefore$ The result holds.

Example 7. *Evaluate :*

(*i*) $\int_{-4}^{4} |x+1|\,dx$ (*ii*) $\int_0^3 |x-2|\,dx$ (*iii*) $\int_{1/4}^{1} |2x-1|\,dx.$

Sol. (*i*) Let $\text{I} = \int_{-4}^{4} |x+1|\,dx.$ $x + 1 = 0 \Rightarrow x = -1$

$$\therefore \quad \text{I} = \int_{-4}^{-1} |x+1|\,dx + \int_{-1}^{4} |x+1|\,dx^* = \int_{-4}^{-1} -(x+1)\,dx + \int_{-1}^{4} (x+1)\,dx$$

($\because$ $x + 1 \le 0$ on $[-4, -1]$ and $x + 1 \ge 0$ on $[-1, 4]$)

$$= -\left[\frac{(x+1)^2}{2}\right]_{-4}^{-1} + \left[\frac{(x+1)^2}{2}\right]_{-1}^{4} = -\frac{1}{2}(0-9) + \frac{1}{2}(25-0) = \mathbf{17}.$$

(*ii*) Let $\text{I} = \int_0^3 |x-2|\,dx.$ $x - 2 = 0 \Rightarrow x = 2.$

$$\therefore \quad \text{I} = \int_0^2 |x-2|\,dx + \int_2^3 |x-2|\,dx = \int_0^2 (2-x)\,dx + \int_2^3 (x-2)\,dx$$

($\because$ $x - 2 \le 0$ on $[0, 2]$ and $x - 2 \ge 0$ on $[2, 3]$)

$$= \left[\frac{(2-x)^2}{-2}\right]_0^2 + \left[\frac{(x-2)^2}{2}\right]_0^2 = -\frac{1}{2}(0-4) + \frac{1}{2}(1-0) = \frac{\mathbf{5}}{\mathbf{2}}.$$

(*iii*) Let $\text{I} = \int_{1/4}^{1} |2x-1|\,dx.$ $2x - 1 = 0 \Rightarrow x = \frac{1}{2}.$

$$\therefore \quad \text{I} = \int_{1/4}^{1/2} |2x-1|\,dx + \int_{1/2}^{1} |2x-1|\,dx = \int_{1/4}^{1/2} [-(2x-1)]\,dx + \int_{1/2}^{1} (2x-1)\,dx$$

$\left(\because\ 2x - 1 \le 0 \text{ on } \left[\frac{1}{4}, \frac{1}{2}\right] \text{ and } 2x - 1 \ge 0 \text{ on } \left[\frac{1}{2}, 1\right]\right)$

$$= -\left[\frac{(2x-1)^2}{4}\right]_{1/4}^{1/2} + \left[\frac{(2x-1)^2}{4}\right]_{1/2}^{1} = -\frac{1}{4}\left(0 - \frac{1}{4}\right) + \frac{1}{4}(1-0) = \frac{\mathbf{5}}{\mathbf{16}}.$$

***Why this step.** We have chosen the value $c = -1$ for using $\int_a^b f(x)\,dx = \int_a^c f(x)\,dx + \int_c^b f(x)\,dx$ because on $[-4, -1]$, we have $x + 1 \le 0$ *i.e.,* $|x+1| = -(x+1)$ and on $[-1, 4]$, we have $x + 1 \ge 0$ *i.e.,* $|x+1| = x + 1$.

Example 8. *Evaluate (i)* $\int_0^{\pi/2} \frac{\sin^2 x}{\sin x + \cos x} dx$ *(ii)* $\int_0^{\pi/2} \log \sin x \, dx.$

Sol. (*i*) Let $I = \int_0^{\pi/2} \frac{\sin^2 x}{\sin x + \cos x} dx$. ...(1)

Using $\int_0^a f(x)\, dx = \int_0^a f(a - x)\, dx$, we get

$$I = \int_0^{\pi/2} \frac{\sin^2\left(\frac{\pi}{2} - x\right)}{\sin\left(\frac{\pi}{2} - x\right) + \cos\left(\frac{\pi}{2} - x\right)} dx \quad \text{or} \quad I = \int_0^{\pi/2} \frac{\cos^2 x}{\cos x + \sin x} dx \qquad ...(2)$$

$$(1) + (2) \Rightarrow 2I = \int_0^{\pi/2} \frac{\sin^2 x + \cos^2 x}{\sin x + \cos x} dx = \int_0^{\pi/2} \frac{1}{\sin x + \cos x} dx$$

$$\therefore \quad I = \frac{1}{2}\int_0^{\pi/2} \frac{1}{\sin x + \cos x} dx = \frac{1}{2}\int_0^{\pi/2} \frac{dx}{\sqrt{2}\left(\sin x \sin \frac{\pi}{4} + \cos x \cos \frac{\pi}{4}\right)}$$

$$= \frac{1}{2\sqrt{2}}\int_0^{\pi/2} \frac{dx}{\cos\left(x - \frac{\pi}{4}\right)} = \frac{1}{2\sqrt{2}}\int_0^{\pi/2} \sec\left(x - \frac{\pi}{4}\right) dx$$

$$= \frac{1}{2\sqrt{2}}\left[\log\left|\sec\left(x - \frac{\pi}{4}\right) + \tan\left(x - \frac{\pi}{4}\right)\right|\right]_0^{\pi/2}$$

$$= \frac{1}{2\sqrt{2}}\left[\log\left|\sec\frac{\pi}{4} + \tan\frac{\pi}{4}\right| - \log\left|\sec\left(-\frac{\pi}{4}\right) + \tan\left(-\frac{\pi}{4}\right)\right|\right]$$

$$= \frac{1}{2\sqrt{2}}[\log|\sqrt{2} + 1| - \log|\sqrt{2} - 1|] = \frac{1}{2\sqrt{2}}[\log(\sqrt{2} + 1) - \log(\sqrt{2} - 1)]$$

$$= \frac{1}{2\sqrt{2}}\log\frac{\sqrt{2}+1}{\sqrt{2}-1} = \frac{1}{2\sqrt{2}}\log\left(\frac{\sqrt{2}+1}{\sqrt{2}-1}\cdot\frac{\sqrt{2}+1}{\sqrt{2}+1}\right) = \frac{1}{2\sqrt{2}}\log\frac{(\sqrt{2}+1)^2}{1} = \mathbf{\frac{1}{\sqrt{2}}\log(\sqrt{2}+1).}$$

(*ii*) Let $I = \int_0^{\pi/2} \log \sin x \, dx$. ...(1)

$$\therefore \quad I = \int_0^{\pi/2} \log \sin\left(\frac{\pi}{2} - x\right) dx$$

$$\therefore \quad I = \int_0^{\pi/2} \log \cos x \, dx \qquad ...(2)$$

$$(1) + (2) \Rightarrow 2I = \int_0^{\pi/2} (\log \sin x + \log \cos x)\, dx$$

$$\Rightarrow \quad I = \frac{1}{2}\int_0^{\pi/2} \log \frac{\sin 2x}{2} dx$$

Let $z = 2x. \quad \therefore \quad dz = 2dx$

$$x = \frac{\pi}{2} \Rightarrow z = \pi \quad \text{and} \quad x = 0 \Rightarrow z = 0.$$

$$\therefore \quad I = \frac{1}{2}\int_0^{\pi} \log \frac{\sin z}{2} \frac{dz}{2} = \frac{1}{4}\int_0^{\pi} \log \frac{\sin z}{2}\, dz$$

$$= \frac{1}{4} . 2\int_0^{\pi/2} \log \frac{\sin z}{2}\, dz \qquad \left(\because \int_0^{2a} f(x)\, dx = 2\int_0^{a} f(x)\, dx \text{ if } f(2a - x) = f(x)\right)$$

$$= \frac{1}{2}\int_0^{\pi/2} \log \sin z\, dz - \frac{1}{2}\int_0^{\pi/2} \log 2\, dz$$

$$= \frac{1}{2} I - \frac{1}{2} (\log 2)\, z \Big|_0^{\pi/2} = \frac{1}{2} I - \left(\frac{1}{2}\log 2\right)\left(\frac{\pi}{2} - 0\right)$$

$$\therefore \quad I = \frac{1}{2} I - \frac{\pi}{4}\log 2 \Rightarrow \frac{1}{2} I = -\frac{\pi}{4}\log 2 \Rightarrow I = -\frac{\pi}{2}\log 2$$

$$\therefore \quad \int_0^{\pi/2} \mathbf{\log \sin x\, dx} = -\frac{\pi}{2}\mathbf{\log 2}.$$

EXERCISE 14.3

Evaluate the following definite integrals (Q. No. 1–3) :

1. (*i*) $\int_0^{\pi/2} \frac{\sin^3 x\, dx}{\sin^3 x + \cos^3 x}$ (*ii*) $\int_0^{\pi/2} \frac{\sin^5 x\, dx}{\sin^5 x + \cos^5 x}$ (*iii*) $\int_0^{\pi/2} \frac{\sqrt{\cos x}\, dx}{\sqrt{\sin x} + \sqrt{\cos x}}$

(*iv*) $\int_0^{\pi/2} \frac{dx}{1 + \tan^3 x}$ (*v*) $\int_0^{\pi/2} \frac{dx}{1 + \sqrt{\cot x}}$ (*vi*) $\int_0^{\pi/2} \frac{\sin^{5/3} x\, dx}{\sin^{5/3} x + \cos^{5/3} x}$

2. (*i*) $\int_0^{\pi/2} \frac{(\sin x - \cos x)\, dx}{1 + \sin x \cos x}$ (*ii*) $\int_{\pi/6}^{\pi/3} \frac{\sqrt{\sin x}}{\sqrt{\sin x} + \sqrt{\cos x}}\, dx$

(*iii*) $\int_0^{\pi/2} (\sin x - \cos x) \ln (\sin x + \cos x)\, dx$ (*iv*) $\int_0^{\pi} \sin^2 x \cos^3 x\, dx$

(*v*) $\int_0^{\pi/2} \sin 2x \log \tan x\, dx$ (*vi*) $\int_0^{1} \log\left(\frac{1}{x} - 1\right) dx$

3. (*i*) $\int_0^{a} \frac{\sqrt{x}\, dx}{\sqrt{x} + \sqrt{a - x}}$ (*ii*) $\int_0^{\pi} \sin^{2m} x \cos^{2n+1} x\, dx$ (*iii*) $\int_0^{1} \frac{\log (1 + x)}{1 + x^2}\, dx$

(*iv*) $\int_0^{7} \frac{\sqrt[3]{x}\, dx}{\sqrt[3]{x} + \sqrt[3]{7 - x}}$ (*v*) $\int_0^{5} \frac{\sqrt[4]{x + 4}\, dx}{\sqrt[4]{x + 4} + \sqrt[4]{9 - x}}$ (*vi*) $\int_1^{4} \frac{\sqrt{x}\, dx}{\sqrt{5 - x} + \sqrt{x}}$.

4. (*i*) Show that $\int_0^{\pi/2} \sin^n x\, dx = \int_0^{\pi/2} \cos^n x\, dx.$

(*ii*) Show that $\int_0^{2a} f(x)\, dx = \int_0^{2a} f(2a - x)\, dx.$

Evaluate the following definite integrals (Q. No. 5–6) :

5. (*i*) $\int_{-\pi/2}^{\pi/2} \sin^7 x\,dx$ (*ii*) $\int_{-99}^{99} (x^{99} + x^{49} + x^{19})\,dx$ (*iii*) $\int_{-1}^{1} \sin^5 x \cos^4 x\,dx$

6. (*i*) $\int_{-a}^{a} x^3\sqrt{a^2 - x^2}\,dx$ (*ii*) $\int_{-\pi}^{\pi} \frac{\sin 2x}{a - b\cos x}\,dx.$

Evaluate the following definite integrals (Q. No. 7–11) :

7. (*i*) $\int_{1}^{4} f(x)$, where $f(x) = \begin{cases} 7x + 3, & \text{if } 1 \le x \le 3 \\ 8x, & \text{if } 3 \le x \le 4 \end{cases}$

(*ii*) $\int_{-3}^{2} f(x)\,dx$, where $f(x) = \begin{cases} x^2, & \text{if } -3 \le x \le 1 \\ 2 - x, & \text{if } 1 \le x \le 2 \end{cases}$

(*iii*) $\int_{1}^{4} f(x)\,dx$, where $f(x) = \begin{cases} 4x + 3 & \text{if } 1 \le x \le 2 \\ 3x + 5 & \text{if } 2 \le x \le 4 \end{cases}$

8. (*i*) $\int_{0}^{8} |x - 5|\,dx$ (*ii*) $\int_{0}^{7} |x - 4|\,dx$ (*iii*) $\int_{-5}^{5} |x + 2|\,dx$

(*P.T.U. B.Pharma, Dec. 2001*) (*P.T.U. B.Pharma, May 2002*)

9. (*i*) $\int_{-a}^{a} (\sqrt{1 + x + x^2} - \sqrt{1 - x + x^2})\,dx$ (*ii*) $\int_{0}^{a} \frac{dx}{x + \sqrt{a^2 - x^2}}$

(*iii*) $\int_{0}^{\pi} \frac{x \sin x}{1 + \sin x}\,dx$

10. (*i*) $\int_{0}^{3} x^2(3 - x)^{1/2}\,dx$ (*ii*) $\int_{0}^{1} x(1 - x)^{3/2}\,dx$ (*iii*) $\int_{0}^{a} x^2(a - x)^{3/2}\,dx$

11. (*i*) $\int_{0}^{\pi/2} \frac{x}{\sin x + \cos x}\,dx$ (*ii*) $\int_{0}^{\pi} \frac{x}{a^2\cos^2 x + b^2 \sin^2 x}\,dx$.

12. Prove $\int_{0}^{\pi} \frac{dx}{5 + 3\cos x} = \frac{\pi}{4}$ (*P.T.U. B.Pharma, June 2003*)

Answers

1. (*i*) π/4 (*ii*) π/4 (*iii*) π/4 (*iv*) π/4 (*v*) π/4 (*vi*) π/4

2. (*i*) 0 (*ii*) π/12 (*iii*) 0 (*iv*) 0 (*v*) 0 (*vi*) 0

3. (*i*) $\frac{a}{2}$ (*ii*) 0 (*iii*) $\frac{\pi}{8}\log 2$ (*iv*) $\frac{7}{2}$ (*v*) $\frac{5}{2}$ (*vi*) $\frac{3}{2}$

5. (*i*) 0 (*ii*) 0 (*iii*) 0 **6.** (*i*) 0 (*ii*) 0

7. (*i*) 62 (*ii*) 59/6 (*iii*) 37

8. (*i*) 17 (*ii*) 25/2 (*iii*) 29

9. (*i*) 0 (*ii*) $\frac{\pi}{4}$ (*iii*) $\pi\left(\frac{\pi}{2} - 1\right)$ **10.** (*i*) $\frac{144\sqrt{3}}{35}$

(*ii*) $\frac{4}{35}$ (*iii*) $\frac{16}{315}a^{9/2}$

11. (*i*) $\frac{\pi}{2\sqrt{2}}\log(\sqrt{2} + 1)$ (*ii*) $\frac{\pi^2}{2ab}$.

Hints

2. (*i*) Let $\quad I = \int_0^{\pi/2} \frac{(\sin x - \cos x)\, dx}{1 + \sin x \cos x}$. $\quad \therefore \quad I = \int_0^{\pi/2} \frac{\sin\left(\frac{\pi}{2} - x\right) - \cos\left(\frac{\pi}{2} - x\right)}{1 + \sin\left(\frac{\pi}{2} - x\right)\cos\left(\frac{\pi}{2} - x\right)}\, dx$

$$= \int_0^{\pi/2} \frac{\cos x - \sin x}{1 + \cos x \sin x}\, dx = -\int_0^{\pi/2} \frac{\sin x - \cos x}{1 + \sin x \cos x}\, dx = -I$$

(*ii*) Let $\quad I = \int_{\pi/6}^{\pi/3} \frac{\sqrt{\sin x}}{\sqrt{\sin x} + \sqrt{\cos x}}\, dx$

$$\therefore \quad I = \int_{\pi/6}^{\pi/3} \frac{\sqrt{\sin\left(\frac{\pi}{6} + \frac{\pi}{3} - x\right)}}{\sqrt{\sin\left(\frac{\pi}{6} + \frac{\pi}{3} - x\right)} + \sqrt{\cos\left(\frac{\pi}{6} + \frac{\pi}{3} - x\right)}}\, dx = \int_{\pi/6}^{\pi/3} \frac{\sqrt{\cos x}}{\sqrt{\cos x} + \sqrt{\sin x}}\, dx$$

Adding, we get $2I = \int_{\pi/6}^{\pi/3} 1 \,.\, dx$

(*iii*) Use $\quad \int_a^b f(x)\, dx = \int_a^b f(a + b - x)\, dx$

(*iv*) Let $\quad I = \int_0^{\pi} \sin^2 x \cos^3 x\, dx$

$$\therefore \quad I = \int_0^{\pi} \sin^2 (\pi - x) \cos^3 (\pi - x)\, dx = \int_0^{\pi} \sin^2 x\, (-\cos x)^3\, dx = -I$$

(*v*) Let $\quad I = \int_0^{\pi/2} \sin 2x \log \tan x\, dx$

$$\therefore \quad I = \int_0^{\pi/2} \sin 2\left(\frac{\pi}{2} - x\right) \log \tan\left(\frac{\pi}{2} - x\right) dx = \int_0^{\pi/2} \sin 2x \log \cot x\, dx$$

$$= -\int_0^{\pi/2} \sin 2x \log \tan x\, dx = -I$$

(*vi*) Let $\quad I = \int_0^1 \log\left(\frac{1}{x} - 1\right) dx$

$$\therefore \quad I = \int_0^1 \log\left(\frac{1}{1 - x} - 1\right) dx = \int_0^1 \log \frac{x}{1 - x}\, dx = -\int_0^1 \log \frac{1 - x}{x}\, dx$$

$$= -\int_0^1 \log\left(\frac{1}{x} - 1\right) dx = -I.$$

3. (*i*), (*ii*) Use $\quad \int_0^a f(x)\, dx = \int_0^a f(a - x)\, dx$

(*iii*) $x = \tan\theta \quad \Rightarrow \quad \int_0^1 \frac{\log(1 + x)}{1 + x^2}\, dx = \int_0^{\pi/4} \log(1 + \tan\theta)\, d\theta$

Now use $\quad \int_0^a f(x)\, dx = \int_0^a f(a - x)\, dx$.

5. The integrand in each part is an odd function.

6. (*ii*) $\int_{-\pi}^{\pi} \frac{\sin 2x}{a - b\cos x}\, dx = 0$, because the integrand is odd.

7. (*i*) $\int_1^4 f(x)\,dx = \int_1^3 (7x+3)\,dx + \int_3^4 8x\,dx$

8. (*i*) $\int_0^8 |x-5|\,dx = \int_0^5 -(x-5)\,dx + \int_5^8 (x-5)\,dx$

(*ii*) $\int_0^7 |x-4|\,dx = \int_0^4 -(x-4)\,dx + \int_4^7 (x-4)\,dx$

(*iii*) $\int_{-5}^5 |x+2|\,dx = \int_{-5}^{-2} -(x+2)\,dx + \int_{-2}^5 (x+2)\,dx$.

9. (*i*) The integrand is an *odd* function.

(*ii*) $x = a\sin\theta \Rightarrow \int_0^a \frac{dx}{x+\sqrt{a^2-x^2}} = \int_0^{\pi/2} \frac{\cos\theta}{\sin\theta+\cos\theta}\,d\theta$

(*iii*) Use $\int_0^a f(x)\,dx = \int_0^a f(a-x)\,dx$

10. Use $\int_0^a f(x)\,dx = \int_0^a f(a-x)\,dx$ in each part.

11. Use $\int_0^a f(x)\,dx = \int_0^a f(a-x)\,dx$.

15*A

Evaluation of $\Sigma n, \Sigma n^2, \Sigma n^3$

15.1. SEQUENCE OF NATURAL NUMBERS : 1, 2, 3, ...

The sum of first n terms of the sequence 1, 2, 3, ... is denoted by Σn.

$\therefore$ $\Sigma n = 1 + 2 + 3 + ... + n$. This is an arithmetic series of n terms with $a = 1$ and $d = 1$.

$\therefore$ $\Sigma n = \frac{n}{2}[2(1) + (n-1)1]$ (Using $S_n = \frac{n}{2}[2a + (n-1)d]$)

$\therefore$ $\mathbf{\Sigma n = \frac{n(n+1)}{2}, \quad n \in N.}$

Under the notation Σn, we write $\sum_{k=1}^{8} k$ as $(\Sigma n)_{n=8}$.

Remark. Σn *i.e.,* $1 + 2 + 3 + ... + n$ can also be written as $\sum_{k=1}^{n} k$.

15.2. SEQUENCE OF SQUARES OF NATURAL NUMBERS : $1^2, 2^2, 3^2$, ...

The sum of first n terms of the sequence $1^2, 2^2, 3^2$, ... is denoted by Σn^2.

$\therefore$ $\Sigma n^2 = 1^2 + 2^2 + 3^2 + ... + n^2$.

In the next theorem, we establish a formula to find the value of Σn^2.

Theorem. Prove that $\mathbf{\Sigma n^2 = 1^2 + 2^2 + 3^2 + ... + n^2 = \frac{n(n+1)(2n+1)}{6}, n \in N.}$

Proof. For any x, we have $(x+1)^3 = x^3 + 3x^2 + 3x + 1$.

$\Rightarrow$ $(x+1)^3 - x^3 = 3x^2 + 3x + 1$. ...(1)

Putting $x = 1, 2, 3, ..., n$ in (1), we get

$$2^3 - 1^3 = 3.1^2 + 3.1 + 1$$
$$3^3 - 2^3 = 3.2^2 + 3.2 + 1$$
$$4^3 - 3^3 = 3.3^2 + 3.3 + 1$$
$$......................................$$
$$(n+1)^3 - n^3 = 3.n^2 + 3.n + 1$$

---

* This chapter is included in the current syllabus of P.U., Chandigarh ; M.D.U., Rohtak ; G.J.U., Hisar and K.U., Kurukshetra.

Adding these vertically, we get

$$(n+1)^3 - 1^3 = 3(1^2 + 2^2 + 3^2 + \ldots\ldots + n^2) + 3(1 + 2 + 3 + \ldots\ldots + n) + n$$

$$\Rightarrow \quad (n+1)^3 - 1 = 3\Sigma n^2 + 3\Sigma n + n$$

$$\Rightarrow \quad 3\Sigma n^2 = (n+1)^3 - 3\Sigma n - n - 1 = (n+1)^3 - 3\,\frac{n(n+1)}{2} - (n+1)$$

$$= \frac{(n+1)}{2}\,[2(n+1)^2 - 3n - 2] = \frac{n+1}{2}\,[2n^2 + n] = \frac{(n+1)\,n\,(2n+1)}{2}.$$

$$\therefore \quad \mathbf{\Sigma n^2 = \frac{n(n+1)(2n+1)}{6}, \quad n \in N.}$$

Remark. Σn^2 *i.e.,* $1^2 + 2^2 + 3^2 + \ldots\ldots + n^2$ can also be written as $\sum_{k=1}^{n} k^2$.

Illustration. $\sum_{k=1}^{10} k^2 = 1^2 + 2^2 + 3^2 + \ldots\ldots + 10^2 = \frac{10\,(10+1)\,(20+1)}{6} = 385.$

Under the notation Σn^2, we write $\sum_{k=1}^{10} k^2$ as $(\Sigma n^2)_{n=10}$.

15.3. SEQUENCE OF CUBES OF NATURAL NUMBERS : 1^3, 2^3, 3^3, ...

The sum of first n terms of the sequence 1^3, 2^3, 3^3, ... is denoted by Σn^3.

$$\therefore \quad \Sigma n^3 = 1^3 + 2^3 + 3^3 + \ldots\ldots + n^3.$$

Now, we prove a result regarding the value of Σn^3.

Theorem. Prove that : $\mathbf{\Sigma n^3 = 1^3 + 2^3 + 3^3 + \ldots\ldots + n^3 = \left[\frac{n(n+1)}{2}\right]^2, \quad n \in N.}$

Proof. For any x, we have $(x+1)^4 = x^4 + 4x^3 + 6x^2 + 4x + 1$.

$$\Rightarrow \quad (x+1)^4 - x^4 = 4x^3 + 6x^2 + 4x + 1 \quad \ldots(1)$$

Putting $x = 1, 2, 3, \ldots\ldots, n$ in (1), we get

$$2^4 - 1^4 = 4.1^3 + 6.1^2 + 4.1 + 1$$

$$3^4 - 2^4 = 4.2^3 + 6.2^2 + 4.2 + 1$$

$$4^4 - 3^4 = 4.3^3 + 6.3^2 + 4.3 + 1$$

....

$$(n+1)^4 - n^4 = 4.n^3 + 6.n^2 + 4.n + 1$$

Adding these vertically, we get

$$(n+1)^4 - 1^4 = 4(1^3 + 2^3 + 3^3 + \ldots\ldots + n^3) + 6(1^2 + 2^2 + 3^2 + \ldots\ldots + n^2) + 4(1 + 2 + 3 + \ldots\ldots + n) + n$$

$$\Rightarrow \quad (n+1)^4 - 1 = 4\Sigma n^3 + 6\Sigma n^2 + 4\Sigma n + n$$

$$\Rightarrow \quad 4\Sigma n^3 = (n+1)^4 - 6\Sigma n^2 - 4\Sigma n - n - 1$$

$$= (n+1)^4 - 6\,\frac{n(n+1)(2n+1)}{6} - 4\,\frac{n(n+1)}{2} - (n+1)$$

$$= (n+1)\,[(n+1)^3 - n(2n+1) - 2n - 1]$$

$$= (n + 1)\,[n^3 + 3n^2 + 3n + 1 - 2n^2 - n - 2n - 1]$$
$$= (n + 1)\,(n^3 + n^2) = (n + 1)n^2\,(n + 1) = [n(n + 1)]^2.$$

$\therefore$
$$\mathbf{\Sigma n^3 = \left[\frac{n\,(n+1)}{2}\right]^2.}$$

Remark. Σn^3 *i.e.,* $1^3 + 2^3 + 3^3 + \ldots\ldots + n^3$ can also be written as $\sum_{k=1}^{n} k^3$.

Illustration. $\sum_{k=1}^{7} k^3 = 1^3 + 2^3 + 3^3 + \ldots\ldots + 7^3 = \left[\frac{7(7+1)}{2}\right]^2 = 784.$

Under the notation Σn^3, we write $\sum_{k=1}^{7} k^3$ as $(\Sigma n^3)_{n=7}$.

Theorem. If for a sequence, $T_n = an^3 + bn^2 + cn + d$, then prove that $S_n = a\Sigma n^3 + b\Sigma n^2 + c\Sigma n + nd$.

Proof. We have $T_n = an^3 + bn^2 + cn + d.$

$\therefore$
$$T_1 = a.1^3 + b.1^2 + c.1 + d.$$
$$T_2 = a.2^3 + b.2^2 + c.2 + d.$$
$$T_3 = a.3^3 + b.3^2 + c.3 + d.$$
$$\ldots\ldots\ldots\ldots\ldots\ldots\ldots\ldots$$
$$T_n = an^3 + bn^2 + cn + d.$$

Adding vertically, we get $S_n = T_1 + T_2 + T_3 + \ldots\ldots + T_n$

$$= a(1^3 + 2^3 + 3^3 + \ldots\ldots + n^3) + b(1^2 + 2^2 + 3^2 + \ldots\ldots + n^2) + c\,(1 + 2 + 3 + \ldots\ldots + n) + nd$$

$\therefore$ $\mathbf{S_n = a\Sigma n^3 + b\Sigma n^2 + c\Sigma n + nd.}$

$\therefore$ The result holds.

WORKING RULES TO FIND THE SUM OF SERIES

Step I. *Write the nth term T_n of the given series.*

Step II. *Simplify T_n and write it as a polynomial in n. Let $T_n = an^3 + bn^2 + cn + d$.*

Step III. *Write the sum S_n of the first n terms of the series as $a\Sigma n^3 + b\Sigma n^2 + c\Sigma n + nd$.*

Step IV. *Put* $\Sigma n = \frac{n(n+1)}{2}$, $\Sigma n^2 = \frac{n\,(n+1)\,(2n+1)}{6}$ *and* $\Sigma n^3 = \left[\frac{n\,(n+1)}{2}\right]^2$ *in the expression :* $S_n = a\Sigma n^3 + b\Sigma n^2 + c\Sigma n + nd$ *and simplify.*

ILLUSTRATIVE EXAMPLES

Example 1. *Find the following sums :*

(*i*) $12 + 13 + 14 + \ldots\ldots + 37$

(*ii*) $1^2 + 2^2 + 3^2 + \ldots\ldots + 20^2$

(*iii*) $18^2 + 19^2 + 20^2 + \ldots\ldots + 47^2$

(*iv*) $11^3 + 12^3 + 13^3 + \ldots\ldots + 25^3$

(*v*) $2^3 + 4^3 + 6^3 + \ldots\ldots + 18^3$.

Sol. (*i*) $12 + 13 + 14 + \ldots\ldots + 37 = (1 + 2 + \ldots\ldots + 11 + 12 + 13 + 14 + \ldots\ldots + 37) - (1 + 2 + \ldots\ldots + 11)$

$$= (\Sigma n)_{n=37} - (\Sigma n)_{n=11} = \frac{37(37+1)}{2} - \frac{11(11+1)}{2} = 703 - 66 = \mathbf{637.}$$

(*ii*) $1^2 + 2^2 + 3^2 + \ldots\ldots + 20^2 = (\Sigma n^2)_{n=20}$

$$= \frac{20(20+1)(2(20)+1)}{6} = \frac{20(21)(41)}{6} = \mathbf{2870.}$$

(*iii*) $18^2 + 19^2 + 20^2 + \ldots\ldots + 47^2 = (1^2 + 2^2 + 3^2 + \ldots\ldots + 47^2) - (1^2 + 2^2 + 3^2 + \ldots\ldots + 17^2)$

$$= (\Sigma n^2)_{n=47} - (\Sigma n^2)_{n=17}$$

$$= \frac{47(47+1)(2(47)+1)}{6} - \frac{17(17+1)(2(17)+1)}{6}$$

$$= \frac{47(48)(95)}{6} - \frac{17(18)(35)}{6} = 35720 - 1785 = \mathbf{33935.}$$

(*iv*) $11^3 + 12^3 + 13^3 + \ldots\ldots + 25^3 = (1^3 + 2^3 + 3^3 + \ldots\ldots + 25^3) - (1^3 + 2^3 + 3^3 + \ldots\ldots + 10^3)$

$$= (\Sigma n^3)_{n=25} - (\Sigma n^3)_{n=10}$$

$$= \left[\frac{25(25+1)}{2}\right]^2 - \left[\frac{10(10+1)}{2}\right]^2 = 105625 - 3025 = \mathbf{102600.}$$

(*v*) $2^3 + 4^3 + 6^3 + \ldots\ldots + 18^3 = (2 \times 1)^3 + (2 \times 2)^3 + (2 \times 3)^3 + \ldots\ldots (2 \times 9)^3$

$$= 2^3[1^3 + 2^3 + 3^3 + \ldots\ldots + 9^3] = 8(\Sigma n^3)_{n=9} = 8\left[\frac{9(9+1)}{2}\right]^2 = \mathbf{16200.}$$

Example 2. *Find the sum of first n terms of the series : $1^2 + 3^2 + 5^2 + \ldots\ldots$*

Sol. The given series is $1^2 + 3^2 + 5^2 + \ldots\ldots$

Let T_n be the nth term.

$\therefore \quad T_n = (n\text{th term of } 1, 3, 5, \ldots\ldots)^2$

$= (1 + (n-1)2)^2 = (2n-1)^2 = 4n^2 - 4n + 1.$

$\therefore$ Sum of n terms $= S_n = \Sigma T_n{}^* = \Sigma(4n^2 - 4n + 1) = 4\Sigma n^2 - 4\Sigma n + n$

$$= 4\left(\frac{n(n+1)(2n+1)}{6}\right) - 4\left(\frac{n(n+1)}{2}\right) + n$$

$$= \frac{n}{3}[2(n+1)(2n+1) - 6(n+1) + 3] = \mathbf{\frac{n(4n^2-1)}{3}}.$$

Alternative method. $1^2 + 3^2 + 5^2 + \ldots\ldots n$ terms.

$= (1^2 + 2^2 + 3^2 + 4^2 + 5^2 + 6^2 + \ldots\ldots 2n \text{ terms}) - (2^2 + 4^2 + 6^2 + \ldots\ldots n \text{ terms})$

$= (1^2 + 2^2 + 3^2 + \ldots\ldots 2n \text{ terms}) - 2^2(1^2 + 2^2 + 3^2 + \ldots\ldots n \text{ terms})$

$$= \frac{2n(2n+1)(2(2n)+1)}{6} - 4\,\frac{n(n+1)(2n+1)}{6}$$

$$= \frac{2n(2n+1)}{6}[4n + 1 - 2(n+1)] = \frac{n(2n+1)}{3}[2n-1] = \mathbf{\frac{n(4n^2-1)}{3}}.$$

* ΣT_n stands for $T_1 + T_2 + T_3 + \ldots\ldots + T_n$.

Example 3. *Find the sum of the series : 1.2.3 + 2.3.4 + 3.4.5 + ... to n terms.*

Sol. The given series is 1.2.3 + 2.3.4 + 3.4.5 + ... Let T_n be the nth term.

$$\therefore \quad T_n = \begin{pmatrix} n\text{th term of} \\ 1, 2, 3, \ldots\ldots \end{pmatrix} \begin{pmatrix} n\text{th term of} \\ 2, 3, 4, \ldots\ldots \end{pmatrix} \begin{pmatrix} n\text{th term of} \\ 3, 4, 5, \ldots\ldots \end{pmatrix}$$

$$= n(n+1)(n+2) = n^3 + 3n^2 + 2n.$$

$$\therefore \quad \text{Sum of } n \text{ terms} = S_n = \Sigma T_n = \Sigma n^3 + 3\Sigma n^2 + 2\Sigma n$$

$$= \left[\frac{n(n+1)}{2}\right]^2 + 3\frac{n(n+1)(2n+1)}{6} + 2\frac{n(n+1)}{2}$$

$$= \frac{n(n+1)}{4}[n(n+1) + 2(2n+1) + 4] = \mathbf{\frac{n(n+1)(n^2+5n+6)}{4}}.$$

Example 4. *Sum the series : $1.3^2 + 2.5^2 + 3.7^2 + \ldots\ldots$ to 20 terms.*

Sol. The given series is $1.3^2 + 2.5^2 + 3.7^2 + \ldots\ldots$

Let T_n be the nth term

$$\therefore \quad T_n = \begin{pmatrix} n\text{th term of} \\ 1, 2, 3, \ldots\ldots \end{pmatrix} \begin{pmatrix} n\text{th term of} \\ 3, 5, 7, \ldots\ldots \end{pmatrix}^2 = n\,(3 + (n-1)\,2)^2 = n(2n+1)^2 = 4n^3 + 4n^2 + n.$$

$$\therefore \quad \text{Sum of 20 terms} = S_{20} = (\Sigma T_n)_{n=20} = 4(\Sigma n^3)_{n=20} + 4(\Sigma n^2)_{n=20} + (\Sigma n)_{n=20}.$$

$$= 4\left[\frac{20\,(20+1)}{2}\right]^2 + 4\left[\frac{20(20+1)\,(2(20)+1)}{6}\right] + \frac{20(20+1)}{2}$$

$$= 176400 + 11480 + 210 = \mathbf{188090.}$$

Example 5. *If S_1, S_2, S_3 are the sums of the first n natural numbers, their squares and their cubes respectively, show that* $S_3\,(1 + 8S_1) = 9S_2^2$.

Sol. We have $\quad S_1 = \dfrac{n(n+1)}{2}, \quad S_2 = \dfrac{n(n+1)(2n+1)}{6}, \quad S_3 = \left(\dfrac{n(n+1)}{2}\right)^2$

$$\therefore \quad S_3(1 + 8S_1) = \left[\frac{n(n+1)}{2}\right]^2 \left[1 + 8\,.\,\frac{n(n+1)}{2}\right] = \frac{n^2\,(n+1)^2}{4}\,[1 + 4n^2 + 4n]$$

$$= \frac{n^2\,(n+1)^2\,(2n+1)^2}{4}$$

and $$9S_2^{\,2} = 9\left[\frac{n(n+1)(2n+1)}{6}\right]^2 = \frac{n^2(n+1)^2\,(2n+1)^2}{4}.$$

$$\therefore \quad \mathbf{S_3(1 + 8S_1) = 9S_2^{\,2}}.$$

Example 6. *Find the sum of all possible products of the first n natural numbers taken two at a time.*

Sol. Let S = Sum of all possible products of the first n natural numbers taken two at a time.

We have $\quad (1 + 2 + 3 + \ldots\ldots + n)^2 = (1^2 + 2^2 + 3^2 + \ldots\ldots + n^2) + 2S$ **(Note this step)**

$$\Rightarrow \quad (\Sigma n)^2 = \Sigma n^2 + 2S \;\Rightarrow\; 2S = (\Sigma n)^2 - \Sigma n^2$$

$$\therefore \quad S = \frac{1}{2}\,[(\Sigma n)^2 - (\Sigma n^2)] = \frac{1}{2}\left[\left(\frac{n(n+1)}{2}\right)^2 - \frac{n(n+1)(2n+1)}{6}\right]$$

$$= \frac{n(n+1)}{4}\left[\frac{n(n+1)}{2} - \frac{2n+1}{3}\right]$$

$$= \frac{n(n+1)}{24}[3n^2 + 3n - 4n - 2] = \mathbf{\frac{n(n+1)(3n^2 - n - 2)}{24}}.$$

Example 7. *Sum of the series :* $\frac{1^3}{1} + \frac{1^3 + 2^3}{1+3} + \frac{1^3 + 2^3 + 3^3}{1+3+5} + \ldots\ldots$ *to n terms.*

Sol. Let T_n be the nth term of the given series.

$$\therefore \quad T_n = \frac{1^3 + 2^3 + 3^3 + \ldots\ldots n \text{ terms}}{1 + 3 + 5 + \ldots\ldots n \text{ terms}} = \frac{\Sigma n^3}{\frac{n}{2}[2(1) + (n-1)2]} = \frac{\left[\frac{n(n+1)}{2}\right]^2}{\frac{n}{2}(2n)} = \frac{(n+1)^2}{4}$$

$$= \frac{1}{4}(n^2 + 2n + 1).$$

$$\therefore \quad \text{Sum of } n \text{ terms} = S_n = \Sigma T_n = \frac{1}{4}\Sigma(n^2 + 2n + 1)$$

$$= \frac{1}{4}(\Sigma n^2 + 2\Sigma n + n] = \frac{1}{4}\left[\frac{n(n+1)(2n+1)}{6} + 2\left(\frac{n(n+1)}{2}\right) + n\right]$$

$$= \frac{n}{24}[(n+1)(2n+1) + 6(n+1) + 6] = \mathbf{\frac{n(2n^2 + 9n + 13)}{24}}.$$

EXERCISE

Find the following sums (Q. No. 1–3) :

1. (*i*) $1 + 2 + 3 + \ldots\ldots + 30$ (*ii*) $1^2 + 2^2 + 3^2 + \ldots\ldots + 45^2$
   (*iii*) $1^3 + 2^3 + 3^3 + \ldots\ldots + 22^3$.
2. (*i*) $5 + 6 + 7 + \ldots\ldots + 100$ (*ii*) $5^2 + 6^2 + 7^2 + \ldots\ldots + 20^2$
   (*iii*) $15^3 + 16^3 + 17^3 + \ldots\ldots + 30^3$.
3. (*i*) $15 + 30 + 45 + \ldots\ldots + 390$ (*ii*) $3^2 + 6^2 + 9^2 + \ldots\ldots + 42^2$
   (*iii*) $4^3 + 8^3 + 12^3 + \ldots\ldots + 40^3$.

Find the sum of the first n terms of the series whose nth term is given by (Q. No. 4—9) :

4. $n(n + 2)$ **5.** $n(n + 3)$ **6.** $n^2 + n + 1$
7. $n(n + 1)(n + 4)$ **8.** $n^2 + 2^n$ **9.** $n^3 - 4^n$
10. Find the sum of the series $1^2 + 4^2 + 7^2 + \ldots\ldots$ to n terms.
11. Find the sum of the series 3.8 + 6.11 + 9.14 + ... to n terms.
12. Find the sum of the first n terms of the series $3.1^2 + 5.2^2 + 7.3^2 + \ldots\ldots$
13. Find the sum of the first 25 terms of the series 5.6 + 6.7 + 7.8 + ...
14. Find the sum of the first n terms of the series $1^2 + (1^2 + 2^2) + (1^2 + 2^2 + 3^2) + \ldots\ldots$
15. Evaluate : $1^3 + 3^3 + 5^3 + \ldots\ldots + (2n - 1)^3$.
16. Show that $\frac{1.2^2 + 2.3^2 + \ldots\ldots + n(n+1)^2}{1^2.2 + 2^2.3 + \ldots\ldots + n^2(n+1)} = \frac{3n+5}{3n+1}$.
17. Show that the sum of the cubes of any number of consecutive positive integers is divisible by the sum of those integers.

Answers

1. (*i*) 465 (*ii*) 31395 (*iii*) 64009

2. (*i*) 5040 (*ii*) 2840 (*iii*) 205200

3. (*i*) 5265 (*ii*) 9135 (*iii*) 193600

4. $\dfrac{n(n+1)(2n+7)}{6}$ **5.** $\dfrac{n(n+1)(n+5)}{3}$ **6.** $\dfrac{n^3+3n^2+5n}{3}$ **7.** $\dfrac{n(n+1)(3n^2+23n+34)}{12}$

8. $\dfrac{n(n+1)(2n+1)}{6}+2(2^n-1)$ **9.** $\dfrac{n^2(n+1)^2}{4}-\dfrac{4(4^n-1)}{3}$

10. $\dfrac{n(6n^2-3n-1)}{2}$ **11.** $3n(n+1)(n+3)$ **12.** $\dfrac{n(n+1)(3n^2+5n+1)}{6}$

13. 8950 **14.** $\dfrac{n(n+1)^2(n+2)}{12}$ **15.** $n^2(2n^2-1)$.

Hints

14. $T_n = 1^2+2^2+3^2+\ldots\ldots+n^2=\Sigma n^2=\dfrac{n(n+1)(2n+1)}{6}$.

16. $$\text{L.H.S.}=\frac{\sum\limits_{k=1}^{n} k(k+1)^2}{\sum\limits_{k=1}^{n} k^2(k+1)}=\frac{\sum\limits_{k=1}^{n} k^3+2\sum\limits_{k=1}^{n} k^2+\sum\limits_{k=1}^{n} k}{\sum\limits_{k=1}^{n} k^3+\sum\limits_{k=1}^{n} k^2}=\frac{\Sigma n^3+2\Sigma n^2+\Sigma n}{\Sigma n^3+\Sigma n^2}.$$

17. Let the positive numbers be $m+1, m+2, \ldots\ldots, m+n$.

∴ Sum of cubes $=(m+1)^3+(m+2)^3+\ldots\ldots+(m+n)^3$

$=[1^3+2^3+\ldots\ldots+(m+n)^3]-[1^3+2^3+\ldots\ldots+m^3]$

$=\Sigma(m+n)^3-\Sigma m^3=\left[\dfrac{(m+n)(m+n+1)}{2}\right]^2-\left[\dfrac{m(m+1)}{2}\right]^2$

Also, sum of these numbers $=(m+1)+(m+2)+\ldots\ldots+(m+n)$

$=[1+2+\ldots\ldots+(m+n)]-[1+2+\ldots\ldots+m]$

$=\Sigma(m+n)-\Sigma m=\dfrac{(m+n)(m+n+1)}{2}-\dfrac{m(m+1)}{2}$

∴ $\dfrac{\text{Sum of cubes of numbers}}{\text{Sum of numbers}}=\dfrac{(m+n)(m+n+1)}{2}+\dfrac{m(m+1)}{2}$ $[\because\ x^2-y^2=(x-y)(x+y)]$

$=[1+2+\ldots\ldots+(m+n)]+[1+2+\ldots\ldots+m]\in \mathbf{N}$.

∴ Sum of cubes of numbers is divisible by sum of numbers.

15*B

Partial Fractions

15.4. INTRODUCTION

We know that the quotient of any two polynomials is called a rational function. In the present chapter, we shall learn the techniques of resolving a fraction into partial fractions. This method will help us to integrate any given rational function.

15.5. RESOLUTION OF A FRACTION INTO PARTIAL FRACTIONS

We know the method of finding the sum of two or more algebraic fractions by reducing the denominators of fractions to a common denominator, which is their L.C.M. For example :

$$\frac{2}{x-3}+\frac{5}{x+2}=\frac{2x+4+5x-15}{(x-3)(x+2)}=\frac{7x-11}{x^2-x-6}.$$

The reverse process of breaking up a single fraction into simpler fractions whose denominators are the factors of the denominator of the given fraction is called the **resolution of a fraction into its partial fractions.**

For example, $\frac{2x}{x^2-1}=\frac{1}{x-1}+\frac{1}{x+1}$.

Here $\frac{1}{x-1}$ and $\frac{1}{x+1}$ are the partial fractions of the fraction $\frac{2x}{x^2-1}$.

15.6. METHOD OF RESOLUTION INTO PARTIAL FRACTIONS

If $f(x)$ and $g(x)$ are polynomials, then $\frac{f(x)}{g(x)}$ is a rational function or a **rational fraction.**

If deg. $f(x)$ < deg. $f(x)$, then the rational fraction $\frac{f(x)}{g(x)}$ is called **proper,** otherwise $\frac{f(x)}{g(x)}$ is called **improper.**

For example, the rational fraction $\frac{x^2+x+3}{(x+2)(x^2+7)}$ is *proper* and the rational fraction $\frac{x^3+x+9}{x^2+5x-6}$ is *improper.*

If $\frac{f(x)}{g(x)}$ is an improper rational fraction, then we can divide $f(x)$ by $g(x)$ so as to write $\frac{f(x)}{g(x)}$ as the sum of a polynomial and a proper rational fraction.

$\therefore$ **Any improper fraction can be expressed as the sum of a polynomial and a proper fraction.**

---

* This chapter is included in the current syllabus of P.U., Chandigarh ; M.D.U., Rohtak ; G.J.U., Hisar ; and K.U., Kurukshetra.

It can be proved mathematically that any proper fraction may be resolved into partial fractions and :

(*i*) if $ax + b$ is any linear non-repeated factor in the denominator, then there corresponds a partial fraction of the form $\frac{A}{ax+b}$.

(*ii*) if $ax + b$ is any linear factor repeated r ($\in$ N) times in the denominator, then there corresponds partial fractions of the form $\frac{A}{ax+b}, \frac{B}{(ax+b)^2}, \frac{C}{(ax+b)^3}, \ldots\ldots$ r terms.

(*iii*) if $ax^2 + bx + c$ is any irreducible quadratic factor in the denominator, then there corresponds a partial fraction of the form $\frac{Ax+B}{ax^2+bx+c}$, and

(*iv*) if $ax^2 + bx + c$ is any irreducible factor repeated r ($\in$ N) times in the denominator, then there corresponds partial fractions of the form $\frac{Ax+B}{ax^2+bx+c}, \frac{Cx+D}{(ax^2+bx+c)^2}, \ldots\ldots$ r terms.

The quantities A, B, C, D, ... are all constants independent of x.

∴ The given proper fraction can be expressed as the sum of its partial fractions. **The constants A, B, C, D, ... occurring in the numerators of the partial fractions are determined by simplifying the sum of partial fractions and then giving various values to x, to obtain equations involving unknown constants or by comparing the coefficients of like powers of x.**

Illustrations. (*i*) The partial fractions of the proper fraction $\frac{x-1}{(x+1)(x-2)(x+3)^3}$ are of the types $\frac{A}{x+1}, \frac{B}{x-2}, \frac{C}{x+3}, \frac{D}{(x+3)^2}$ and $\frac{E}{(x+3)^3}$.

(*ii*) The partial fractions of the proper fraction $\frac{x^2+4}{(x-7)(x+2)^2(x^2+2x+3)}$ are of the types $\frac{A}{x-7}, \frac{B}{x+2}, \frac{C}{(x+2)^2}$ and $\frac{Dx+E}{x^2+2x+3}$.

(*iii*) The partial fractions of the proper fraction $\frac{x^3+x+7}{(x-1)(x+2)^2(x^2+5)^2}$ are of the types $\frac{A}{x-1}, \frac{B}{x+2}, \frac{C}{(x+2)^2}, \frac{Dx+E}{x^2+5}$ and $\frac{Fx+G}{(x^2+5)^2}$.

ILLUSTRATIVE EXAMPLES

Example 1. *Resolve the following fractions into partial fractions :*

(*i*) $\frac{3x-1}{x^2-1}$ (*ii*) $\frac{6x^3+5x^2-7}{3x^2-2x-1}$ (*iii*) $\frac{2x+1}{(x+2)(x-3)^2}$

(*iv*) $\frac{2x+3}{(x^2+1)(x+1)}$ (*v*) $\frac{x^2+5}{(x^2+x+1)(x^2-x+2)}$ (*vi*) $\frac{x}{(x-1)(x^2+1)^2}$.

Sol. (*i*) $\frac{3x-1}{x^2-1}$ is a proper fraction and the denominator has linear non-repeated factors.

Let $\frac{3x-1}{x^2-1} = \frac{3x-1}{(x-1)(x+1)} = \frac{A}{x-1} + \frac{B}{x+1}$. $\therefore \quad \frac{3x-1}{(x-1)(x+1)} = \frac{A(x+1)+B(x-1)}{(x-1)(x+1)}$

Multiplying by $(x-1)(x+1)$, we get, $3x-1 = A(x+1) + B(x-1)$ ...(1)

Now we find the values of A and B.

Method I. By comparing the coefficients of like powers of x.

(1) $\Rightarrow \quad 3x - 1 = (A + B)x + (A - B).$

$\therefore$ A + B = 3 and A – B = – 1. Solving these equations, we get A = 1 and B = 2.

Method II. By giving specific values to x.

Let us put $x = 1$ and $x = 2$ in (1).

$x = 1 \quad \Rightarrow \quad 3(1) - 1 = A(1+1) + B(1-1) \quad \Rightarrow \quad 2A = 2 \quad \Rightarrow \quad A = 1$

$x = 2 \quad \Rightarrow \quad 3(2) - 1 = A(2+1) + B(2-1) \quad \Rightarrow \quad 3A + B = 5 \quad \Rightarrow \quad B = 5 - 3A = 5 - 3(1) = 2.$

$\therefore \quad A = 1$ and $B = 2$.

Method III. In this method, it is always convenient to use those values of x which make linear factors in the denominator zero.

Here $x - 1 = 0 \Rightarrow x = 1$ and $x + 1 = 0 \Rightarrow x = -1$. We put $x = 1$ and $x = -1$ in (1).

$x = 1 \quad \Rightarrow \quad 3(1) - 1 = A(1+1) + B(0) \quad \Rightarrow \quad A = 2/2 = 1$

$x = -1 \quad \Rightarrow \quad 3(-1) - 1 = A(0) + B(-1-1) \quad \Rightarrow \quad B = \frac{-4}{-2} = 2.$

$\therefore \quad A = 1$ and $B = 2$.

$$\therefore \quad \frac{3x-1}{(x-1)(x+1)} = \frac{\mathbf{1}}{\mathbf{x-1}} + \frac{\mathbf{2}}{\mathbf{x+1}}.$$

Remark. In the second method, we observe that

$$A = \text{Value of } \frac{3x-1}{x+1} \text{ when } x = 1 \quad \text{and} \quad B = \text{Value of } \frac{3x-1}{x-1} \text{ when } x = -1.$$

$\therefore$ The value of A can be found out by putting $x = 1$ in the given proper fraction, after omitting $x - 1$ from the denominator. Similarly, the value of B can be found out by putting $x = -1$ in the given proper fraction, after omitting $x + 1$ from the denominator.

Thus, when the factors in the denominator are *linear* and *non-repeated,* we can decompose the given proper fraction into partial fractions as follows :

$$\frac{3x-1}{(x-1)(x+1)} = \frac{3(1)-1}{(x-1)(1+1)} + \frac{3(-1)-1}{(-1-1)(x+1)} = \frac{2}{(x-1)2} + \frac{-4}{(-2)(x+1)} = \frac{1}{x-1} + \frac{2}{x+1}.$$

This is called the **short cut method** of finding the values of A, B etc. It is very important to bear in mind that this short cut method is to be used only when the denominator has *only linear and non-repeated factors.*

(*ii*) Given fraction is not a proper fraction. Dividing $6x^3 + 5x^2 - 7$ by $3x^2 - 2x - 1$, we get

$$6x^3 + 5x^2 - 7 = (2x + 3)(3x^2 - 2x - 1) + (8x - 4).$$

$$\therefore \quad \frac{6x^3 + 5x^2 - 7}{3x^2 - 2x - 1} = 2x + 3 + \frac{8x-4}{(x-1)(3x+1)} \qquad (\because \; 3x^2 - 2x - 1 = (x-1)(3x+1))$$

$$= 2x + 3 + \left(\frac{8(1) - 4}{(x-1)(3(1)+1)} + \frac{8\left(-\frac{1}{3}\right) - 4}{\left(-\frac{1}{3} - 1\right)(3x+1)} \right)$$

(Using **short-cut method**)

$$= 2x + 3 + \frac{4}{(x-1)4} + \frac{-20/3}{\left(-\frac{4}{3}\right)(3x+1)} = \mathbf{2x + 3 + \frac{1}{x-1} + \frac{5}{3x+1}}.$$

(iii) $\frac{2x+1}{(x+2)(x-3)^2}$ is a proper fraction. Let the partial fraction corresponding to the factor $x + 2$ be $\frac{A}{x+2}$ and the partial fractions, corresponding to the factor $(x-3)^2$ be $\frac{B}{x-3}$ and $\frac{C}{(x-3)^2}$.

$$\therefore \qquad \frac{2x+1}{(x+2)(x-3)^2} = \frac{A}{x+2} + \frac{B}{x-3} + \frac{C}{(x-3)^2}.$$

Multiplying both sides by $(x + 2)(x - 3)^2$, we get

$$2x + 1 = A(x-3)^2 + B(x+2)(x-3) + C(x+2) \qquad \ldots(1)$$

Now $x + 2 = 0 \Rightarrow x = -2$ and $x - 3 = 0 \Rightarrow x = 3$.

$$x = -2 \text{ in (1) implies } -3 = A(-5)^2 + B(0) + C(0) \Rightarrow A = \frac{-3}{25}$$

$$x = 3 \text{ in (1) implies } 7 = A(0) + B(0) + C(5) \Rightarrow C = \frac{7}{5}$$

Comparing the coefficients of x^2 in (1), we get $0 = A + B$.

$$\therefore \qquad B = -A = -\left(\frac{-3}{25}\right) = \frac{3}{25}.$$

$$\therefore \qquad \frac{2x+1}{(x+2)(x-3)^2} = \frac{\frac{-3}{25}}{x+2} + \frac{\frac{3}{25}}{x-3} + \frac{\frac{7}{5}}{(x-3)^2} = \mathbf{-\frac{3}{25(x+2)} + \frac{3}{25(x-3)} + \frac{7}{5(x-3)^2}}.$$

(iv) $\frac{2x+3}{(x+1)(x^2+1)}$ is a proper fraction. Let the partial fraction corresponding to the factor $x + 1$ be $\frac{A}{x+1}$ and the partial fraction corresponding to the factor $x^2 + 1$ be $\frac{Bx+C}{x^2+1}$.

$$\therefore \qquad \frac{2x+3}{(x+1)(x^2+1)} = \frac{A}{x+1} + \frac{Bx+C}{x^2+1}.$$

Multiplying both sides by $(x + 1)(x^2 + 1)$, we get

$$2x + 3 = A(x^2+1) + (Bx+C)(x+1) \qquad \ldots(1)$$

$$x = -1 \text{ in (1)} \Rightarrow 1 = A(2) + (-B + C)(0) \Rightarrow A = \frac{1}{2}$$

Comparing the coefficients of x^2 in (1), we get $0 = A + B$. $\therefore$ $B = -A = -\frac{1}{2}$.

Comparing the coefficients of x, we get $2 = B + C$. $\therefore$ $C = 2 - B = 2 + \frac{1}{2} = \frac{5}{2}$.

$$\therefore \quad \frac{2x+3}{(x+1)(x^2+1)} = \frac{1/2}{x+1} + \frac{(-1/2)x + 5/2}{x^2+1} = \mathbf{\frac{1}{2(x+1)} + \frac{5-x}{2(x^2+1)}}.$$

(*v*) $\frac{x^2+5}{(x^2+x+1)(x^2-x+2)}$ is a proper fraction with non-repeated irreducible quadratic factors $x^2 + x + 1$ and $x^2 - x + 2$.

Let $\frac{x^2+5}{(x^2+x+1)(x^2-x+2)} = \frac{Ax+B}{x^2+x+1} + \frac{Cx+D}{x^2-x+2}$.

Multiplying both sides by $(x^2 + x + 1)(x^2 - x + 2)$, we get

$$x^2 + 5 = (Ax + B)(x^2 - x + 2) + (Cx + D)(x^2 + x + 1) \quad ...(1)$$

Comparing the coefficients of x^3, x^2, x and the constant terms, we get

$0 = A + C$ ...(1) $\quad 1 = -A + B + C + D$ ...(2)

$0 = 2A - B + C + D$ ...(3) $\quad 5 = 2B + D$ ...(4)

(1) $\Rightarrow$ $C = -A$. $\therefore$ (2) $\Rightarrow$ $-A + B - A + D = 1$ *i.e.*, $-2A + B + D = 1$ ...(5)

(3) $\Rightarrow$ $2A - B - A + D = 0$ *i.e.*, $A - B + D = 0$ ...(6)

(5) + 2(6) $\Rightarrow$ $-B + 3D = 1$ ...(7)

Solving (4) and (7), we get $B = 2$ and $D = 1$.

(5) $\Rightarrow$ $-2A + 2 + 1 = 1$ or $A = 1$. (1) $\Rightarrow$ $C = -A = -1$.

$\therefore$ $A = 1, B = 2, C = -1, D = 1$.

$$\therefore \quad \frac{x^2+5}{(x^2+x+1)(x^2-x+2)} = \frac{1.x+2}{x^2+x+1} + \frac{-1.x+1}{x^2-x+2} = \mathbf{\frac{x+2}{x^2+x+1} - \frac{x-1}{x^2-x+2}}.$$

(*vi*) $\frac{x}{(x-1)(x^2+1)^2}$ is a proper fraction. Let $\frac{x}{(x-1)(x^2+1)^2} = \frac{A}{x-1} + \frac{Bx+C}{x^2+1} + \frac{Dx+E}{(x^2+1)^2}$.

Multiplying by $(x - 1)(x^2 + 1)^2$, we get

$$x = A(x^2 + 1)^2 + (Bx + C)(x - 1)(x^2 + 1) + (Dx + E)(x - 1) \quad ...(1)$$

$x = 1$ in (1) $\Rightarrow$ $1 = A(4) + (B + C)(0)(2) + (D + E)(0)$ $\Rightarrow$ $A = \frac{1}{4}$

Comparing the coefficients x^4, x^3, x^2, x, in (1), we get

$0 = A + B$ ...(2) $\quad 0 = -B + C$ ...(3)

$0 = 2A + B - C + D$ ...(4) $\quad 1 = -B + C - D + E$ ...(5)

(2) $\Rightarrow$ $B = -A = -\frac{1}{4}$, (3) $\Rightarrow C = B = -\frac{1}{4}$ (4) $\Rightarrow$ $D = -2A - B + C = -\frac{1}{2} + \frac{1}{4} - \frac{1}{4} = -\frac{1}{2}$

(5) $\Rightarrow$ $E = 1 + B - C + D = 1 - \frac{1}{4} + \frac{1}{4} - \frac{1}{2} = \frac{1}{2}$

$\therefore$ $A = \frac{1}{4}, B = -\frac{1}{4}, C = -\frac{1}{4}, D = -\frac{1}{2}, E = \frac{1}{2}$.

$$\therefore \quad \frac{x}{(x-1)(x^2+1)^2} = \frac{1/4}{x-1} + \frac{(-1/4)x + (-1/4)}{x^2+1} + \frac{(-1/2)x + 1/2}{(x^2+1)^2}$$

$$= \mathbf{\frac{1}{4(x-1)} - \frac{x+1}{4(x^2+1)} - \frac{x-1}{2(x^2+1)^2}}.$$

Example 2. *Resolve the following fractions into partial fractions*

(*i*) $\dfrac{1}{(x+1)(2x+1)}$ (*ii*) $\dfrac{2x+1}{(x+1)(x-2)}$ (*iii*) $\dfrac{x^2}{(x-1)(x-2)}$

(*iv*) $\dfrac{2x-1}{(x-1)(x+2)(x-3)}$ (*v*) $\dfrac{x^2+2}{(x^2+1)(x^2+4)}$ (*vi*) $\dfrac{3x+4}{x^2-5x+6}$.

Sol. (*i*) Let $\dfrac{1}{(x+1)(2x+1)} = \dfrac{A}{x+1} + \dfrac{B}{2x+1}$.

Multiplying by $(x + 1)(2x + 1)$, we get $1 = A(2x + 1) + B(x + 1)$ ...(1)

$x + 1 = 0 \quad \Rightarrow \quad x = -1$ and $2x + 1 = 0 \quad \Rightarrow \quad x = -1/2$.

In (1), $x = -1 \quad \Rightarrow \quad 1 = A(-2 + 1) + B(0) \quad \Rightarrow \quad A = -1$

$x = -1/2 \quad \Rightarrow \quad 1 = A(0) + B(-\tfrac{1}{2} + 1) \quad \Rightarrow \quad B = 2$

$$\therefore \quad \frac{1}{(x+1)(2x+1)} = -\frac{1}{x+1} + \frac{2}{2x+1}.$$

(*ii*) The factors in the denominator are *linear* and *non-repeated*. We use the *short cut method* to break up the given fraction

$$\frac{2x+1}{(x+1)(x-2)} = \frac{2(-1)+1}{(x+1)(-1-2)} + \frac{2(2)+1}{(2+1)(x-2)} = \frac{-1}{-3(x+1)} + \frac{5}{3(x-2)} = \frac{1}{3}\left[\frac{1}{x+1} + \frac{5}{x-2}\right].$$

(*iii*) Since the given fraction is not a *proper* fraction, we have by long division,

$$\frac{x^2}{(x-1)(x-2)} = \frac{x^2}{x^2-3x+2} = 1 + \frac{3x-2}{x^2-3x+2} = 1 + \frac{3x-2}{(x-1)(x-2)} \qquad ...(1)$$

By *short cut method* :

$$\frac{3x-2}{(x-1)(x-2)} = \frac{3(1)-2}{(x-1)(1-2)} + \frac{3(2)-2}{(2-1)(x-2)} = \frac{1}{(x-1)(-1)} + \frac{4}{1(x-2)} = -\frac{1}{x-1} + \frac{4}{x-2}$$

$$\therefore \quad (1) \quad \Rightarrow \quad \frac{x^2}{(x-1)(x-2)} = 1 - \frac{1}{x-1} + \frac{4}{x-2}.$$

(*iv*) The factors in the denominator are *linear* and *non-repeated.*

$$\therefore \quad \frac{2x-1}{(x-1)(x+2)(x-3)} = \frac{2(1)-1}{(x-1)(1+2)(1-3)} + \frac{2(-2)-1}{(-2-1)(x+2)(-2-3)} + \frac{2(3)-1}{(3-1)(3+2)(x-3)}$$

$$= \frac{1}{-6(x-1)} + \frac{-5}{15(x+2)} + \frac{5}{10(x-3)}.$$

(*v*) Let $z = x^2$. $\quad \therefore \quad \dfrac{x^2+2}{(x^2+1)(x^2+4)} = \dfrac{z+2}{(z+1)(z+4)}$

Let $\dfrac{z+2}{(z+1)(z+4)} = \dfrac{A}{z+1} + \dfrac{B}{z+4}$. $\quad \therefore \quad z + 2 = A(z + 4) + B(z + 1)$

$z = -1 \quad \Rightarrow \quad 1 = 3A + B(0) \quad \Rightarrow \quad A = 1/3$

$z = -4 \quad \Rightarrow \quad -2 = A(0) + (-3)B \quad \Rightarrow \quad B = 2/3$

$\therefore \quad \dfrac{z+2}{(z+1)(z+4)} = \dfrac{1/3}{z+1} + \dfrac{2/3}{z+4}$. $\quad \therefore \quad \dfrac{x^2+2}{(x^2+1)(x^2+4)} = \dfrac{1}{3(x^2+1)} + \dfrac{2}{3(x^2+4)}$.

(*vi*) Let $\dfrac{3x+4}{x^2-5x+6} = \dfrac{3x+4}{(x-2)(x-3)} = \dfrac{A}{x-2} + \dfrac{B}{x-3}$.

Multiplying by $(x-2)(x-3)$, we get $3x+4 = A(x-3) + B(x-2)$ ...(1)

$x-2=0 \Rightarrow x=2$ and $x-3=0 \Rightarrow x=3$.

We put $x=2$ and $x=3$ in (1)

$x=2 \quad \Rightarrow \quad 3(2)+4 = A(2-3)+B(2-2) \Rightarrow 10 = -A+0 \quad \Rightarrow A = -10$

$x=3 \quad \Rightarrow \quad 3(3)+4 = A(3-3)+B(3-2) \Rightarrow 13 = 0+B \quad \Rightarrow B = 13$

$\therefore \quad \dfrac{3x+4}{(x-2)(x-3)} = \dfrac{-10}{x-2} + \dfrac{13}{x-3}$.

Example 3. *Resolve the following fractions into partial fractions*

(*i*) $\dfrac{12x^2-2x-9}{(4x^2-1)(x+3)}$ (*ii*) $\dfrac{x^2+8x+4}{x^3-4x}$

(*iii*) $\dfrac{2x}{(x^2+2)(x^2+3)}$.

Sol. (*i*) The factors of $(4x^2-1)(x+3)$ are *linear* and *non-repeated*.

Let $\dfrac{12x^2-2x-9}{(2x-1)(2x+1)(x+3)} = \dfrac{A}{2x-1} + \dfrac{B}{2x+1} + \dfrac{C}{x+3}$

$$= \frac{A(2x+1)(x+3) + B(2x-1)(x+3) + C(2x-1)(2x+1)}{(2x-1)(2x+1)(x+3)}$$

$$\therefore \quad 12x^2-2x-9 = A(2x+1)(x+3) + B(2x-1)(x+3) + C(2x-1)(2x+1) \quad ...(1)$$

Now, $2x-1=0 \Rightarrow x=\dfrac{1}{2}$, $2x+1=0 \Rightarrow x=-\dfrac{1}{2}$ and $x+3=0 \Rightarrow x=-3$.

$x=\dfrac{1}{2} \Rightarrow 3-1-9 = A(2)\left(\dfrac{7}{2}\right)+0+0 \Rightarrow -7 = 7A \Rightarrow A = -1$

$x=-\dfrac{1}{2} \Rightarrow 3+1-9 = 0+B(-2)\left(\dfrac{5}{2}\right)+0 \Rightarrow -5 = -5B \Rightarrow B = 1$

and $x=-3 \Rightarrow 108+6-9 = 0+0+C(-7)(-5) \Rightarrow 105 = 35C \Rightarrow C = 3$.

$\therefore \quad \dfrac{12x^2-2x-9}{(2x-1)(2x+1)(x+3)} = \dfrac{-1}{2x-1} + \dfrac{1}{2x+1} + \dfrac{3}{x+3}$.

(*ii*) $\dfrac{x^2+8x+4}{x^3-4x} = \dfrac{x^2+8x+4}{x(x+2)(x-2)} = \dfrac{A}{x} + \dfrac{B}{x+2} + \dfrac{C}{x-2}$ ...(1)

$$= \frac{A(x+2)(x-2) + Bx(x-2) + Cx(x+2)}{x(x+2)(x-2)}$$

$\therefore \quad x^2 + 8x + 4 = A(x+2)(x-2) + Bx(x-2) + Cx(x+2)$

$$x = 0 \quad \Rightarrow \quad 0 + 0 + 4 = -4A \quad \text{or} \quad A = -1$$

$$x = -2 \quad \Rightarrow \quad 4 - 16 + 4 = 8B \quad \text{or} \quad B = -1$$

$$x = 2 \quad \Rightarrow \quad 4 + 16 + 4 = 8C \quad \text{or} \quad C = 3.$$

Hence (1) can be written as $\dfrac{x^2+8x+4}{x^3-4x} = \dfrac{-1}{x} - \dfrac{1}{x+2} + \dfrac{3}{x-2}$.

(*iii*) $$\frac{2x}{(x^2+2)(x^2+3)} = \frac{Ax+B}{x^2+2} + \frac{Cx+D}{x^2+3}$$

$$= \frac{(Ax+B)(x^2+3) + (Cx+D)(x^2+2)}{(x^2+2)(x^2+3)}$$

Multiplying by $(x^2+2)(x^2+3)$, we get

$$2x = (Ax+B)(x^2+3) + (Cx+D)(x^2+2)$$

Equating the coefficient of x^3, both sides, we get

$$0 = A + C \qquad ...(1)$$

Equating the coefficient of x^2, both sides, we get

$$0 = B + D \qquad ...(2)$$

Equating the coefficient of x, both sides, we get

$$2 = 3A + 2C \qquad ...(3)$$

Equating the constant term, we get

$$0 = 3B + 2D \qquad ...(4)$$

From (1), $C = -A$ and from (3), $2 = 3A - 2A \;\Rightarrow\; A = 2$

$\therefore \qquad C = -2$

From (2) $\;B = -D\;$ and $\;$ from (4) $-3D + 2D = 0$

$\Rightarrow \qquad D = 0, B = 0$

$$\therefore \quad \frac{2x}{(x^2+2)(x^2+3)} = \frac{2x}{x^2+2} - \frac{2x}{x^2+3}.$$

Example 4. *Resolve the following fractions into partial fractions :*

(*i*) $\dfrac{5x^3+18x^2-10x-6}{x(x+3)(5x-2)}$ (*ii*) $\dfrac{x^4+x^3+2x^2+4x+1}{x(x+1)}$

Sol. (*i*) Since the given fraction is *not a proper* fraction, we have by long division,

$$\frac{5x^3+18x^2-10x-6}{x(x+3)(5x-2)} = 1 + \frac{5x^2-4x-6}{5x^3+13x^2-6x} \qquad ...(1)$$

We resolve $\dfrac{5x^2-4x-6}{5x^2+13x^2-6x}$ into partial fractions.

Let $$\frac{5x^2-4x-6}{x(x+3)(5x-2)} = \frac{A}{x} + \frac{B}{x+3} + \frac{C}{5x-2} \qquad ...(2)$$

$$= \frac{A(x+3)(5x-2) + Bx(5x-2) + Cx(x+3)}{x(x+3)(5x-2)}.$$

$\therefore \quad 5x^2 - 4x - 6 = A(x+3)(5x-2) + Bx(5x-2) + Cx(x+3)$ ...(3)

When $x = 0$, (3) $\Rightarrow -6 = -6A \Rightarrow A = 1$

When $x = -3$, (3) $\Rightarrow 45 + 12 - 6 = B(-3)(-15-2) \Rightarrow B = 1$

When $x = \frac{2}{5}$, (3) $\Rightarrow 5\left(\frac{2}{5}\right)^2 - 4\left(\frac{2}{5}\right) - 6 = C\left(\frac{2}{5}\right)\left(\frac{2}{5}+3\right) \Rightarrow C = -5.$

$\therefore$ (3) $\Rightarrow \frac{5x^2 - 4x - 6}{x(x+3)(5x-2)} = \frac{1}{x} + \frac{1}{x+3} - \frac{5}{5x-2}$

(*ii*) Since the given fraction is *not a proper* fraction, we use long division to get

$$\frac{x^4 + x^3 + 2x^2 + 4x + 1}{x(x+1)} = x^2 + 2 + \frac{2x+1}{x^2+x} \qquad ...(1)$$

We resolve $\frac{2x+1}{x^2+x}$ into partial fractions.

Let $\frac{2x+1}{x(x+1)} = \frac{A}{x} + \frac{B}{x+1} = \frac{A(x+1)+Bx}{x(x+1)}$ ...(2)

$\therefore \quad 2x + 1 = A(x+1) + Bx.$ ...(3)

When $x = 0$, (3) $\Rightarrow 1 = A$

When $x = -1$, (3) $\Rightarrow -2 + 1 = -B$ or $B = 1.$

$\therefore$ (2) $\Rightarrow \frac{2x+1}{x(x+1)} = \frac{1}{x} + \frac{1}{x+1}.$

Example 5. *Resulve the following fractions into partial fractions :*

(*i*) $\frac{x^2+1}{(x-1)^2(x+3)}$ (*ii*) $\frac{3x+1}{(x-2)^2(x+2)}$.

Sol. (*i*) Let $\frac{x^2+1}{(x-1)^2(x+3)} = \frac{A}{x-1} + \frac{B}{(x-1)^2} + \frac{C}{x+3}$.

$\therefore \quad x^2 + 1 = A(x-1)(x+3) + B(x+3) + C(x-1)^2$ ...(1)

$x = 1 \Rightarrow 1 + 1 = A(0)(4) + B(4) + C(0) \Rightarrow B = 1/2$

$x = -3 \Rightarrow 9 + 1 = A(-4)(0) + B(0) + C(16) \Rightarrow C = 5/8$

Let us put $x = 0$ in (1).

$\therefore \quad 0 + 1 = A(-1)(3) + B(3) + C(1)$

$$\Rightarrow \quad 3A = 3B + C - 1 = 3\left(\frac{1}{2}\right) + \frac{5}{8} - 1 = \frac{9}{8} \quad \Rightarrow \quad A = \frac{3}{8}.$$

$$\therefore \quad \frac{x^2+1}{(x-1)^2(x+3)} = \frac{3/8}{x-1} + \frac{1/2}{(x-1)^2} + \frac{5/8}{x+3}$$

(*ii*) Let $\frac{3x+1}{(x-2)^2(x+2)} = \frac{A}{x-2} + \frac{B}{(x-2)^2} + \frac{C}{x+2}$.

$\therefore \quad 3x + 1 = A(x-2)(x+2) + B(x+2) + C(x-2)^2$ ...(1)

$x = 2 \Rightarrow 7 = A(0)(4) + B(4) + C(0) \Rightarrow B = 7/4$

$x = -2 \Rightarrow -5 = A(-4)(0) + B(0) + C(16) \Rightarrow C = -5/16$

Let us put $x = 0$ in (1).

$\therefore \quad 1 = A(-2)(2) + B(2) + C(4)$

$\Rightarrow \quad 1 = -4A + 2\left(\frac{7}{4}\right) + 4\left(-\frac{5}{16}\right) \quad \Rightarrow \quad 4A = \frac{7}{2} - \frac{5}{4} - 1 = \frac{5}{4} \quad \Rightarrow \quad A = \frac{5}{16}$

$$\therefore \quad \frac{3x+1}{(x-2)^2(x+2)} = \frac{5/16}{x-2} + \frac{7/4}{(x-2)^2} + \frac{-5/16}{x+2}$$

Example 6. *Resolve the following fractions into partial fractions:*

(*i*) $\frac{x^4}{x^4-1}$ (*ii*) $\frac{x^4}{x^4-16}$

(*iii*) $\frac{x}{x^3+x^2+x+1}$

Sol. (*i*) By long division, $\frac{x^4}{x^4-1} = 1 + \frac{1}{x^4-1}$

Let $\frac{1}{x^4-1} = \frac{1}{(x^2-1)(x^2+1)} = \frac{1}{(x-1)(x+1)(x^2+1)} = \frac{A}{x-1} + \frac{B}{x+1} + \frac{Cx+D}{x^2+1}$ ...(1)

Multiplying by $(x-1)(x+1)(x^2+1)$, we get

$\therefore \quad 1 = A(x+1)(x^2+1) + B(x-1)(x^2+1) + (Cx+D)(x^2-1)$

$x = 1 \quad \Rightarrow \quad 1 = A(2)(2) + B(0)(2) + (C+D)(0) \quad \Rightarrow \quad 1 = 4A \quad \Rightarrow \quad A = \frac{1}{4}$

$x = -1 \quad \Rightarrow \quad 1 = A(0)(2) + B(-2)(2) + (-C+D)(0) \quad \Rightarrow \quad 1 = -4B \quad \Rightarrow \quad B = -\frac{1}{4}$

Comparing the coefficient of x^3, we get $0 = A + B + C$.

$\therefore \quad C = -A - B = -\frac{1}{4} + \frac{1}{4} = 0.$

Comparing the coefficient of x^2, we get $0 = A - B + D$.

$\therefore \quad D = B - A = -\frac{1}{4} - \frac{1}{4} = -\frac{1}{2}.$

$\therefore \quad A = \frac{1}{4},\ B = -\frac{1}{4},\ C = 0,\ D = -\frac{1}{2}.$

$$\therefore \quad \frac{1}{x^4-1} = \frac{1/4}{x-1} + \frac{-1/4}{x+1} + \frac{0.x+(-1/2)}{x^2+1} = \frac{1}{4(x-1)} - \frac{1}{4(x+1)} - \frac{1}{2(x^2+1)}$$

$$\therefore \quad (1) \Rightarrow \quad \frac{x^4}{x^4-1} = 1 + \frac{1}{4(x-1)} - \frac{1}{4(x+1)} - \frac{1}{2(x^2+1)}$$

(*ii*) By long division, $\frac{x^4}{x^4-16} = 1 + \frac{16}{x^4-16}$ ...(1)

Consider $\frac{16}{x^4-16} = \frac{16}{(x^2-4)(x^2+4)} = \frac{16}{(y-4)(y+4)}$ where $y = x^2$.

$$= \frac{16}{(y-4)(y+4)} + \frac{16}{(-4+4)(y+4)} = \frac{2}{y-4} - \frac{2}{y+4} = \frac{2}{x^2-4} - \frac{2}{x^2+4}$$

$\therefore$ (1) $\Rightarrow$ $\dfrac{16}{x^4-16} = 1 + \dfrac{2}{x^2-4} - \dfrac{2}{x^2+4}.$

(*iii*) Let $\dfrac{x}{(x^2+1)(x+1)} = \dfrac{Ax+B}{x^2+1} + \dfrac{C}{x+1}.$

$\therefore$ $x = (Ax + B)(x + 1) + C(x^2 + 1)$ ...(1)

$x = -1 \Rightarrow -1 = (-A + B)(0) + C(2) \Rightarrow C = -1/2$

Comparing the coefficients of like powers of x in (1), we get

$0 = A + C$ ...(2) $1 = A + B$ ...(3)

(2) $\Rightarrow$ $A = -C = -(-1/2) = 1/2$

(3) $\Rightarrow$ $B = 1 - A = 1 - (1/2) = 1/2$

$$\therefore \quad \frac{x}{(x^2+1)(x+1)} = \frac{\frac{1}{2}x + \frac{1}{2}}{x^2+1} + \frac{-\frac{1}{2}}{x+1} = \frac{x+1}{2(x^2+1)} - \frac{1}{2(x+1)}.$$

Example 7. *Resolve the following fractions into partial fractions :*

(*i*) $\dfrac{2x^3}{(x^2+1)^2}$ (*ii*) $\dfrac{x^5-x^4+4x^3-4x^2+8x-4}{(x^2+2)^3}.$

Sol. (*i*) Let $\dfrac{2x^3}{(x^2+1)^2} = \dfrac{Ax+B}{x^2+1} + \dfrac{Cx+D}{(x^2+1)^2}$

$\therefore$ $2x^3 = (Ax + B)(x^2 + 1) + (Cx + D)$

$\Rightarrow$ $2x^3 = Ax^3 + Bx^2 + (A + C)x + B + D$

Comparing the coefficients, we get $A = 2, B = 0, A + C = 0, B + D = 0$

Solving, we get $A = 2, B = 0, C = -2, D = 0$.

$$\therefore \quad \frac{2x^3}{(x^2+1)^2} = \frac{2x+0}{x^2+1} + \frac{-2x+0}{(x^2+1)^2} = \frac{2x}{x^2+1} - \frac{2x}{(x^2+1)^2}$$

(*ii*) Let $\dfrac{x^5-x^4+4x^3-4x^2+8x-4}{(x^2+2)^3} = \dfrac{Ax+B}{x^2+2} + \dfrac{Cx+D}{(x^2+2)^2} + \dfrac{Ex+F}{(x^2+2)^3}$

$\therefore$ $x^5 - x^4 + 4x^3 - 4x^2 + 8x - 4 = (Ax + B)(x^2 + 2)^2 + (Cx + D)(x^2 + 2) + (Ex + F)$

$= Ax^5 + Bx^4 + (4A + C)x^3 + (4B + D)x^2 + (4A + 2C + E)x + (4B + 2D + F)$

Comparing the coefficients, we get

$A = 1, B = -1, 4A + C = 4, 4B + D = -4, 4A + 2C + E = 8, 4B + 2D + F = -4.$

$\Rightarrow$ $A = 1, B = -1, C = 0, D = 0, E = 4, F = 0.$

$$\therefore \quad \frac{x^5 - x^4 + 4x^3 - 4x^2 + 8x - 4}{(x^2+2)^3} = \frac{x-1}{x^2+2} + \frac{4x}{(x^2+2)^3}.$$

EXERCISE 15.2

1. Resolve the following fractions into partial fractions :

(i) $\dfrac{3x+2}{(x-1)(2x+3)}$ (ii) $\dfrac{x}{(x+2)(3-2x)}$ (iii) $\dfrac{x-1}{(x+1)(x-2)}$

(iv) $\dfrac{2x-5}{(x-1)^2}$ (v) $\dfrac{x^3-5x^2+6x-2}{x(x-2)^2(x-1)}$.

2. Resolve the following fractions into partial fractions :

(i) $\dfrac{x^2+6x+6}{(x+1)^2(x+3)}$ (ii) $\dfrac{x^3+x^2+1}{x^2(x-1)(x+2)}$ (iii) $\dfrac{2x^2+9x-6}{(x-1)(x+3)(x+2)}$

(iv) $\dfrac{x}{(x^2+4)(x^2+9)}$.

3. Resolve the following fractions into partial fractions :

(i) $\dfrac{3x^2-x^3-4x+52}{(x^2+4)(x+1)(x-2)}$ (ii) $\dfrac{17}{(x^2+4)(x^2+2x+5)}$ (iii) $\dfrac{3x-33}{(x-1)(x+2)(x^2+9)}$

(iv) $\dfrac{1}{x^2(x^2+1)}$.

4. Resolve the following fractions into partial fractions :

(i) $\dfrac{1}{(x-1)(x+1)(x^2+1)}$ (ii) $\dfrac{x^2-3x+1}{(x-1)^3}$ (iii) $\dfrac{5x}{(1-x)(x+1)(x^2+4)}$

(iv) $\dfrac{x^2+1}{x^2(x+1)^2}$ (v) $\dfrac{(x^2+1)(x^2+2)}{(x^2+3)(x^2+4)}$.

Answers

1. (i) $\dfrac{1}{x-1} + \dfrac{1}{2x+3}$ (ii) $\dfrac{2}{7(x+2)} + \dfrac{3}{7(3-2x)}$ (iii) $\dfrac{2}{x-1} - \dfrac{3}{(x-1)^2}$

(iv) $\dfrac{2}{x-1} - \dfrac{3}{(x-1)^2}$ (v) $\dfrac{1}{2x} + \dfrac{1}{2(x-2)} - \dfrac{1}{(x-2)^2}$.

2. (i) $\dfrac{7}{4(x+1)} + \dfrac{1}{2(x+1)^2} - \dfrac{3}{4(x+1)^3}$ (ii) $-\dfrac{1}{4x} - \dfrac{1}{2x^2} + \dfrac{1}{x-1} + \dfrac{1}{4(x+2)}$

(iii) $\dfrac{5}{12(x-1)} - \dfrac{15}{4(x+3)} + \dfrac{16}{3(x+2)}$ (iv) $\dfrac{1}{5(x^2+4)} - \dfrac{1}{5(x^2+9)}$

3. (*i*) $\dfrac{-4}{x+1} + \dfrac{2}{x-1} + \dfrac{x-6}{x^2+4}$ (*ii*) $\dfrac{-2x+1}{x^2+4} + \dfrac{2x+3}{x^2+2x+5}$

(*iii*) $\dfrac{-1}{x-1} + \dfrac{1}{x+2} + \dfrac{3}{x^2+9}$ (*iv*) $\dfrac{1}{x^2} - \dfrac{1}{x^2+1}$

4. (*i*) $\dfrac{1}{4(x-1)} - \dfrac{1}{4(x+1)} - \dfrac{1}{2(x^2+1)}$ (*ii*) $\dfrac{1}{x-1} - \dfrac{1}{(x-1)^2} - \dfrac{1}{(x-1)^3}$

(*iii*) $-\dfrac{1}{2(x-1)} - \dfrac{1}{2(x+1)} + \dfrac{x}{(x^2+4)}$ (*iv*) $-\dfrac{2}{x} + \dfrac{1}{x^2} + \dfrac{2}{x+1} + \dfrac{2}{(x+1)^3}$

(*v*) $1 + \dfrac{2}{x^2+3} - \dfrac{6}{x^2+4}$.

Index

M

N

O

P

Q

R

S

T

U

Z